Industrial Tomography

Woodhead Publishing Series in
Electronic and Optical Materials

Industrial Tomography

Systems and Applications

Second Edition

Edited by

Mi Wang
University of Leeds, Leeds, West Yorkshire,
United Kingdom

Woodhead Publishing is an imprint of Elsevier
50 Hampshire Street, 5th Floor, Cambridge, MA 02139, United States
The Boulevard, Langford Lane, Kidlington, OX5 1GB, United Kingdom

ISBN: 978-0-12-823015-2

For information on all Woodhead Publishing publications visit our website at https://www.elsevier.com/books-and-journals

Publisher: Matthew Deans
Acquisitions Editor: Kayla Dos Santos
Editorial Project Manager: Rachel Pomery
Production Project Manager: Surya Narayanan Jayachandran
Cover Designer: Miles Hitchen

Typeset by TNQ Technologies

Contents

List of contributors

David Alumbaugh Earth & Environmental Sciences Area – Berkeley Lab, Berkeley, CA, United States

Yong Bao University of Edinburgh, Edinburgh, United Kingdom

Zhang Cao School of Instrumentation and Opto-Electronic Engineering, Beihang University, Beijing, China

Shah M. Chowdhury The Ohio State University, Columbus, OH, United States

Sean M. Collins University of Leeds, Leeds, United Kingdom

Kyle Daun Department of Mechanical and Mechatronics Engineering, University of Waterloo, Waterloo, ON, Canada

Feng Dong Tianjin Key Laboratory of Proces Measurement and Control, School of Electrical and Information Engineering, Tianjin University, Tianjin, China

Liang-Shih Fan The Ohio State University, Columbus, OH, United States

Yousef Faraj Department of Chemical Engineering, Faculty of Science and Engineering, University of Chester, Chester, United Kingdom

Samuel J. Grauer Department of Mechanical Engineering, The Pennsylvania State University, University Park, PA, United States

Uwe Hampel Helmholtz-Zentrum Dresden-Rossendorf, Dresden, Germany; Technische Universität Dresden, Dresden, Germany

Daniel J. Holland Department of Chemical and Process Engineering, University of Canterbury, Christchurch, New Zealand

Brian S. Hoyle University of Leeds, Leeds, United Kingdom

Jiabin Jia University of Edinburgh, Edinburgh, United Kingdom

Apostolos Kantzas University of Calgary, Calgary, AB, Canada

Daisuke Kawashima Chiba University, Inage-ku, Chiba-shi, Chiba, Japan

Anil Kumar Khambampati Department of Electronic Engineering, Jeju National University, Jeju, South Korea

Kyung Youn Kim Department of Electronic Engineering, Jeju National University, Jeju, South Korea

Sin Kim School of Energy Systems Engineering, Chung-Ang University, Seoul, South Korea

Chang Liu School of Engineering, The University of Edinburgh, Edinburgh, Scotland, United Kingdom

Thomas D. Machin Stream Sensing, Manchester, United Kingdom

Qussai Marashdeh Tech4Imaging LLC, Columbus, OH, United States

Hugh McCann School of Engineering, The University of Edinburgh, Edinburgh, Scotland, United Kingdom

Junita Mohamad-Saleh Universiti Sains Malaysia, Penang, Malaysia

Volodymyr Mosorov Institute of Applied Computer Science, Lodz University of Technology, Lodz, Poland

Saba Mylvaganam Department of Electrical Engineering, IT and Cybernetics, Faculty of Technology, Natural Sciences and Maritime Sciences, University of South-Eastern Norway, Campus Porsgrunn, Norway

David Parker University of Birmingham, Birmingham, United Kingdom

Anthony J. Peyton Department of Electrical and Electronic Engineering, University of Manchester, Manchester, United Kingdom

Ken Primrose Industrial Tomography Systems Ltd, Manchester, United Kingdom; Stream Sensing Ltd, Manchester, United Kingdom

Andrew J. Sederman Department of Chemical Engineering and Biotechnology, University of Cambridge, Cambridge, United Kingdom

Mohadeseh Sharifi Department of Chemical & Materials Engineering, University of Auckland, Auckland, New Zealand

Benjamin Straiton Tech4Imaging LLC, Columbus, OH, United States

Masahiro Takei Chiba University, Inage-ku, Chiba-shi, Chiba, Japan

Chao Tan Tianjin Key Laboratory of Proces Measurement and Control, School of Electrical and Information Engineering, Tianjin University, Tianjin, China

Fernando L. Teixeira The Ohio State University, Columbus, OH, United States

Steven Wagner Institute for Reactive Flows and Diagnostics, Technical University of Darmstadt, Darmstadt, Hesse, Germany

Aining Wang The Ohio State University, Columbus, OH, United States

Mi Wang University of Leeds, School of Chemical and Process Engineering, United Kingdom

Qiang Wang University of Edinburgh, Edinburgh, United Kingdom

Nicholas James Watson Faculty of Engineering, University of Nottingham, Nottingham, United Kingdom

Michael Wilt Earth & Environmental Sciences Area – Berkeley Lab, Berkeley, CA, United States

Paul Wright School of Electrical and Electronic Engineering, University of Manchester, Manchester, United Kingdom

Cheng-gang Xie Schlumberger, Singapore

Lijun Xu School of Instrumentation and Opto-Electronic Engineering, Beihang University, Beijing, China

Brent Young Department of Chemical & Materials Engineering, University of Auckland, Auckland, New Zealand

Yanlin Zhao China University of Petroleum, Beijing, China

Preface

Tomography is a fast, nondestructive visualization method which provides a detailed picture of the internal structure or workings of an imaged object. Tomography systems can be used to optimize the performance of a wide variety of industrial processes, producing effective imaging of complex internal structures. *Industrial Tomography* is contributed by chapter authors from about 30 universities and industries worldwide and provides an indispensable guide for applied physicists, materials scientists and engineers working in the process imaging and industrial inverse problems fields, or in such applications industries as the chemical, oil and gas, pharmaceutical, mineral, and nuclear industries.

To reflect on the rapid progresses and development in process tomography, most of the chapters are updated in the second edition. A number of new chapters are added to the second edition, including Chapter 10 featuring electron tomography, Chapter 13 elucidating machine learning process information from tomography data, Chapter 14 on the methods of advanced electrical tomography visualization, Chapter 15 on recent electrical resistance tomography applications to chemical engineering, Chapter 16 discussing some examples of applications in industry, Chapter 20 detailing slurry flow characterization with ERT, Chapter 26 on emerging applications of AI and possibilities for process control, and finally Chapter 27 on other diverse tomography applications.

Part One reviews key tomographic techniques and modalities for industrial applications in depth, with chapters focused on electrical capacitance tomography (ECT), electrical impedance tomography (EIT, or sometimes referred to as electrical resistance tomography ERT), electromagnetic induction tomography (EMT), magnetic resonance imaging (MRI), chemical species, X-ray tomography (XRT), radioisotope tracer techniques, ultrasound and spectroscopic tomography, and electron tomography. Part Two goes on to discuss image reconstruction, outlining mathematical concepts, and both hard-field and direct image reconstruction algorithms, machine learning, and advanced visualization process of tomography data. Finally, Part Three explores systems and applications across a variety of industries, providing visualization, characterization, and measurement for chemical and process engineering such as mixing processes, fluidized bed, trickle bed, and microreactors; multiphase flow transportation, typically, in mineral, dredging, and oil—gas applications; process control approached with artificial intelligent methods approach; and ideas and concepts for diverse tomography applications.

I would like to express my sincere gratitude to about 50 chapter authors for their great efforts and invaluable contributions during the COVID-19 pandemic, particularly, to those who and/or whose families were overcoming coronavirus infection during their writing up. It is an unforgettable cooperation, thank you.

Mi Wang

He is a Professor of Process Tomography and Sensing at the University of Leeds, UK, where he leads the On-Line Instrumentation Laboratory (OLIL), promoting and facilitating progress and accessibility of process tomography across engineering science.

Introduction

Richard A. Williams
Heriot-Watt University, Edinburgh, United Kingdom

The second edition of this volume speaks to the extent of industrial application and the global maturing community working in the science, technology, and engineering of tomographic sensor technology. The inherent appeal of making measurements that reveal the *changing properties* of materials in a *hidden space* is mesmerizing. The aspiration that sensing techniques such as those used in medical engineering and chemical sciences might be able to diagnose and enhance the performance of dynamic industrial processes remains a natural driver for exploring tomographic methods that might be deployed in such complex environments. This volume provides a guide to the state of this art and surveys the art of the possible.

In essence, the ability to gain useful information on the properties of any process is defined by the purpose of the information required. In the following chapters, there are examples of tomographic techniques being used to

- understand internal behaviors/phenomena,
- model phenomena (of both known and unknown phenomena),
- assure correct or safe operation of processes,
- undertake static material inspection for quality assurance or porosity mapping,
- enable operation control of a process to maximize efficiency,
- fit a model of known behavior to a process based on measurable characteristics.

Such a diverse range of purposes require careful consideration of the information that is required and hence judicious selection of the sensing methods and methods of data analysis. At its simplest, it could be answering a question such as "is the process behaving correctly?" by reporting "yes" or "no." This can be invaluable tool for industrial operation where deviations from the normal need to quickly identified and remedied. At the other extreme, the question might be "what is going on inside this process and how can understand and model it?." Such questions are evidently the most challenging for tomographic analysis since the length- and temporal scales of inspection may not be known and may vary with location in the process and will be complicated in the case of multiphase and complex composition materials. The process conditions of temperature, pressure, and toxicity may present severe constraints to access, since many tomographic sensors are still invasive to the process.

The examples illustrated here in general fall into two categories, now well established. The first involves extracting information from pixelated data sets from which a visual image representing a given process characteristic (material density, conductivity, atomic number, etc.) is obtained as a result of a selected nonlinear reconstruction process. The image is then interpreted in some way. An example of this may be the analysis of a material (a tablet or drug) using x-ray microtomography from which different components can be identified in 3-dimension (3d) and then this information used to predict subsequent process behavior can be predicted (dissolution, breakage, etc.). I will refer to this as a *Type 1* approach. The second category may use data to fit system behavior to process parameters or models that are well known, or assumed, to represent characteristics of the systems and can be fitted with a statistical confidence. Fitting these parameters yields the visualization. An example of this may be the estimation of concentration gradients of a suspended material in a complex reactor or separator that can be used to develop or inform computational fluid mechanics models or other process behaviors. I will refer to this as a *Type 2* approach. Beyond this, there are further options to use different numbers of sensors arranged around a process volume and also different sensor modalities (or sensor types) and then see if registration of these data can be used to enhance understanding of the phenomena.

Some methods may use a tomographic sensor array coupled with other single points of measurement with a different sensor type providing augmented information or to enable calibration. Information from these hybrid methods can also be treated via *Type 1 or 2 protocols.*

The introduction here is intended to provide a brief overview and context to the structure and contents of the book in its three parts that explore the sensors and measurement principles (Chapters 1–10), more advanced aspects of measurement and interpretation (Chapters 11–14) and some applications (Chapters 15–27). Some key messages and attributes of the chapters are highlighted in this overview.

Understanding the basics of sensor design and reconstruction

Sensor design and adaption into the industrial process is obviously a critical step but needs to be carefully considered along with selection of the measurement regime and analysis. We see in Chapter 1 the importance of measurement hardware and its intimate association with the method of reconstruction, in this case an inherently (3d) volumetric approach applied to ECT. As shown in this chapter, such advanced deployments can be benchmarked against other chemical species methods such as MRI, for appropriate fluid/materials properties. Chapter 2 presents impedance methods, arguably the most researched electrical tomographic modality, since the electrical sensors and industrial deployment at first-sight look simple! The alignment between purpose of the measurement and the sensor design and excitation is, however, extremely critical and recent advances pioneered by the author demonstrate how EIT can be used to abstract pixel-based velocity. This is a major achievement and advance. Chapter 3 surveys magnetic induction methods MIT and flags their wide range of applications well suited to it. In common with all so-called soft-field methods, the challenges of data reconstruction, and interdependencies between the creation of the sensing field and the sensitivity of measurements within it, Chapter 4 brings the well-developed techniques of a ubiquitous clinical imaging method to industrial use. The requirement for very intense magnetic fields generally restricts this method to a laboratory setting and hence its use for material testing, process discovery, and understanding. It is a powerful method and with the sophisticated bespoke methodologies developed and explained in this chapter it is possible to extract detailed spatial and dynamic chemical-specific and phase-specific information. Moving to chemical species that are predominantly in gaseous phase, Chapter 5 presents the development of CST using optical methods that are obviously rapid and accurate provided that optical access can be gained. There are interesting applications in combustion and environmental domains. X-rays are perhaps the best known tomographic modality (XCT) and have been used in a wide range of industrial processes some at considerable scale, for example, in large diameter (0.3−3 m) fluidized bed systems. While great care must be taken due to radiation, significant advances have been made, as Chapter 6 reports, both at industrial scale and in the advent of micro-x-ray tomography. The latter offer benchtop or in-lab use and such systems are in widespread use for nondestructive testing and material modeling of porous materials, granules, and tablets. Chapter 7 presents a sophisticated form of this known as single photo positron emission (SPECT) is described. This has great process scale accuracy in mapping positions of particles and/or fluids and has found great merit in verifying fluid and granular flow models. It is possible to use other high intensity radiation fields such as those derived from synchrotron sources, but these are not so readily available except at national laboratories, but have been used for industrial purposes such as mapping crystal polymorphs during product manufacture. The theory of ultrasound tomography (UST) is stepped-through in Chapter 8 and, but again, translating this from an ambient static clinical setting into a noisy industrial process can be challenging but has been achieved for sold−gas and gas−liquid systems. A further sophistication is featured in the next Chapter 9 in which the multiple frequency approach enables extraction of information in a timely, but computationally demanding, manner using multimodal approaches. This is illustrated here using impedance spectroscopic tomography. The notion is to be able to fully resolve the identity of a material in a process, its location, and its journey. Chapter 10 takes us to the world of using sophisticated techniques based on super-high resolution electron microscopy to reconstruct three dimensional volumes of static materials, both hard (particles, crystal) and soft (biological structures, emulsions). This is a laboratory-based method.

Optimizing data collection and analysis for industrial information

From a primer on image reconstruction methodologies for soft (Chapter 11) to hard fields (Chapter 13 in the 1st edition of the book), approaches to tomography have tended to be driven from *Type I* methods described above. Starting from the instrument and ending up with the data is a classic clinical approach for analysis of the tomographic behavior of the space. For some applications a particular "region of interest" (ROI) in the process space is of concern or changes within a particular ROI. This can be approached either by reconstructing data for the whole volume then selecting the region for subsequent inspection/analysis, or by deploying efforts for data measurement, collection, and analysis solely on that region. In processes that have complex geometries, requiring special consideration in measurement and analysis, an ROI approach is essential. Some principles around such methodologies are described in Chapter 12.

Advances can also be made with a *Type 2* approach that starts by defining the process measurement need and then ways of sensing those parameters. Chapter 13 cites some examples of methods that range from the deployment of neural networks for image reconstruction and interpretation. Early work in industrial process tomography was focused on the thorny challenge of mapping multiphase flow (oil/water/gas flows, hydraulic conveying, process separators) and simpler single phase flows (pneumatic conveying, liquid pipeline flow). Latterly, the design of (very smart) machine learning—based methods has been expanded and used to great effect to abstract detailed process information. A leading example being the use of electrical resistance tomography to extract velocity data that is then used to infer rheological properties through an assumed material flow model, also described in this chapter. Chapter 14 introduces a further important concept of using measurements from different sensors and sensor types observing the same process space. It indicates how these data can be "fused" to provide overall enhanced measurements of a complex (flow) process or to better extract quantitative information on the composition and flow of a particular phase. These techniques have been widely used in 3d industrial process tomography and in multisensor tomography.

A compendium of applications examples

A diverse range of applications have been demonstrated for tomographic methods ranging from those at the benchtop for materials analysis or laboratory testing through to installations in pilot and commercial manufacturing plant. The scope of this section of the book is to provide insight to the variety of ways in which the techniques can be applied.

Chapter 15 provides a helpful guide to the range of specific industrial chemical process engineering applications for ERT, demonstrating the versatility of the method and those systems for which it is best suited. Some of the most advanced and values examples of taking tomography into online use in working industrial process are described in Chapter 16 which presents a fascinating summary of advanced applications of electrical methods for fluid systems oriented toward process control. This contains some contemporary examples of advances installations. Electrical methods applied in geoscience applications in boreholes are described in Chapter 17 based on seismology style deployments under sea. Chapter 18 continues this theme of electrical mapping in porous materials in marine environments and then takes us into the world of upstream flow measure in pipes. These represent some of the most challenging flow measurement problems. Chapter 19 presents and reviews classical flow measurement techniques using a variety of tomographic sensor and multimodal measurements for pipelines, including coupled measurements using dual measurement systems. Slurries of liquids and particulates are ubiquitous in marine, metallurgical, and environmental systems and are often optically opaque and therefore not accessible using optical sensors. Chapter 20 described some applications of mostly ERT methods to such systems. Many specialist and high throughput manufacturing systems involve multiple microreactors requiring analysis at small lengths scales, but often in confined known geometries. Chapter 21 identifies applications for tomographic methods in these important systems ranging from x-ray spectroscopy to electrical methods and Chapter 22 looks at x-ray methods at large diameter process scale and Chapter 25 at large fluidized systems focusing on ECT. Fixed bed reactors are very important in a range of chemical and manufacturing processes and there is a significant advantage in being able to understand the flow conditions and conversion states within them. Chapter 23 provides a thorough analysis of methods suited to such systems and cites the various advantages and limitations. Effectiveness of mixing is notoriously difficult to assess and has therefore attracted significant work examining reactors, pipelines, and swirl inducement technologies, as reviewed in Chapter 24. The application of different knowledge-based intelligence methods are illustrated in Chapter 26, mostly applied to flow systems, these being high complex under varying industrial flow conditions and hence requiring an adaptivity and flexibility of both measurement systems and information handling. Tomographic mapping of temperature, pressure, and material growth in forestry is among examples in a wide-ranging set of illustrative examples in Chapter 27 in both laboratory and outdoor settings. This chapter expresses well the many different ways in which ingenious sensor hardware can be configured and used to obtain practical engineering information in which the focus is on gaining the measurement/feature changes of interest for process understanding not just obtaining "the best possible image."

Part One

Tomographic modalities

Electrical Capacitance Tomography

Shah M. Chowdhury [1], *Qussai Marashdeh* [2], *Fernando L. Teixeira* [1] *and Liang-Shih Fan* [1]
[1]The Ohio State University, Columbus, OH, United States; [2]Tech4Imaging LLC, Columbus, OH, United States

1.1 Introduction

Multiphase flow systems are a critical element of many industrial processes as they constitute the medium through which basic ingredients are processed to yield the final product(s). Examples of their use include energy generating processes, food processing, and drug manufacturing, among others. The ability to image multiphase flow interactions in real time has always been a highly desirable capability to further understand the complex dynamics among interacting phases in any flow system. Such understanding is critical, for example, to effectively model, optimize, and scale-up the reactors that host the process. From early on, electrical sensing techniques have attracted much attention as a noninvasive means for imaging of multiphase flow systems. In addition, the rates in which phase interactions occur often demand fast imaging modalities, again making electric sensing techniques a natural choice.

Electrical Capacitance Tomography (ECT) is an electric sensing modality that easily meets the high-speed demands of multiphase flow real-time imaging. ECT has also a noninvasive characteristic, a feature much desired in industrial applications as noted. Interest in using ECT can be traced back to the early 80s at the US Department of Energy and at University of Manchester in the United Kingdom. Both institutions deployed capacitance plates around a process column and reconstructed images of phase flow distributions from the measured mutual capacitances obtained from different plate pairs (Halow & Nicoletti, 1992; Halow, Fasching, & Nicoletti, 1990; Huang, Xie, Thorn, Snowden, & Beck, 1992; Xie et al., 1992). The basic system components used for ECT then, and still used today, are a set of capacitance plates constituting the ECT sensor, a data acquisition hardware (DAS) for measuring the mutual capacitance between different plate pairs, and a processing device for image reconstruction and visualization. Each of these individual components has been further developed over the years in terms of both hardware and software capabilities. Most notable are the efforts aimed at developing new image reconstruction techniques to extract better images from the limited set of capacitance data (Lionheart, 2001; Marashdeh, Warsito, Fan, & Teixeira, 2006a, 2006b; Sattar et al., 2020; Yang & Peng, 2003).

Industrial Tomography. https://doi.org/10.1016/B978-0-12-823015-2.00002-9

Specific applications of ECT include fluidized beds, circulating fluidized beds, pneumatic conveying systems, bubbling beds, trickle beds, and combustion. Favorable features of ECT sensors for these applications are the low profile and, as mentioned before, the noninvasive nature of the sensing apparatus, the fast imaging speed, safety, ease of use, sensor flexibility for fitting around virtually any vessel geometry (including nonstraight cylindrical vessels, T-junctions, L-junctions, etc.), applicability to scale up units, and relatively low complexity and low cost. A present limitation of ECT is the relatively low spatial resolution of the final images compared to some alternative imaging techniques such as magnetic resonance imaging (MRI) or X-rays. ECT is most successful for imaging of low conductivity media. ECT systems operate with excitation frequencies below 10 MHz. This means that for typical industrial size, the field inside the vessels is quasi static and the system is scale invariant in the sense the field distribution inside a vessel is a solution of Laplace's equation and scales directly with the vessel and sensor size (Paris & Hurd, 1979). At these frequencies or lower, ECT sensors can be employed in actual industrial scales for columns up to several meters in diameter or likewise for miniature device applications (Quek, Mohr, Goddard, Fielden, & York, 2010).

In this chapter, the basic operating principles of ECT systems will be discussed and a brief description of available electronic measuring techniques will be presented. Sensor design and its impact on imaging reconstruction will also be reviewed. After that, some sample results and discussions will be provided to illustrate ECT's value and potential in process tomography applications. Recent progress in this technology will also be highlighted, including electrical capacitance volume tomography (ECVT) −based velocimetry methods, three-phase flow decomposition, imaging for water dominated flows, adaptive ECVT, and adaptive relevance vector machine (RVM)−based image reconstruction for uncertainty determination. The main objective of this chapter is to outline the capabilities and strengths of ECT, and to motivate its use in industrial settings. As such, the technical discussion is kept mostly at an introductory level. Further technical details can be found in the list of references appended to this chapter.

1.2 Principle of operation

The ECT sensing problem corresponds to the reconstruction of the dielectric distribution in an imaging domain from a set of capacitance measurements taken at the domain boundary, that is, from between a set of electrode plates placed around the domain.

Fig. 1.1 shows a schematic view of the basic components of ECT, consisting of capacitance sensors, data acquisition electronics, and image reconstruction/visualization software. The capacitive sensors are used here to blanket the region to be imaged with a static electric field, from which sensitivity maps that indicate the regions in the imaging domain from where the changes on the mutual capacitances are being effected. The data acquisition electronics measure the capacitance variations as changes in dielectric material distribution take place inside the imaging domain.

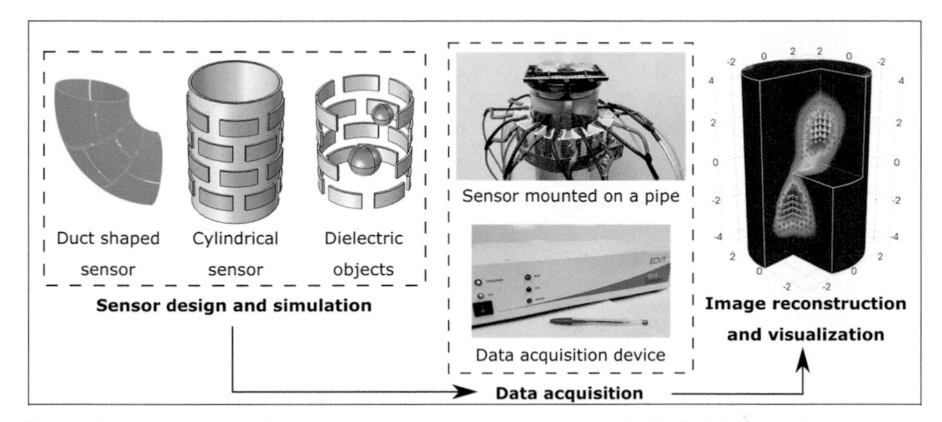

Figure 1.1 Illustration of the basic ECT system components. © 2018, 2019 IEEE. Reprinted, with permission, from Chowdhury, S. M., Marashdeh, Q. M., & Teixeira, F. L. (2018). Inverse normalization method for cross-sectional imaging and velocimetry of two-phase flows based on electrical capacitance tomography. *IEEE Sensors Letter 2*(1), 1–4.; Gunes, C., Chowdhury, S. M., Zuccarelli, C. E., Marashdeh, Q. M., & Teixeira, F. L. (2019). Displacement-current phase tomography for water-dominated two-phase flow velocimetry. *IEEE Sensors Journal 19*(4), 1563–1571.

The set of mutual capacitance data measured by the acquisition system is one dimensional (1D). Nevertheless, imbedded in such data is the spatial information per the sensor design and relative spatial arrangement among the electrode plates. The reconstruction algorithm essentially aims at decoding such two-dimensional (2D) or three-dimensional (3D) spatial information from the 1D capacitance measurements.

Capacitance sensors can be distributed in one plane for 2D imaging. Such 2D arrangement comprises the majority of ECT systems. In this case, the flow is imaged at a particular cross-section inside a vessel. For 3D or volumetric imaging, variations in sensor sensitivity along the axial direction are introduced either by arranging the plates in multiple planes or by varying plate shapes axially to provide 3D sensitivity. Although a time-varying signal is used for measuring the capacitance between sensor plates, a spatial distribution of the electric field can be assumed to follow that of an electrostatic field. This assumption is valid as long as the employed signal is in the order of a few MHz or less and the characteristic dimensions (e.g., diameter) of the cross-section to be imaged are much smaller than the excitation wavelength. These conditions are met in industrial applications. Following the electrostatic approximation, the dielectric constant distribution and electric field distribution inside the sensor domain are governed by the Laplace equation, which relates the dielectric distribution in the imaging domain to the resulting electric field distribution from a given set of boundary conditions.

The electrical capacitance between a pair of electrode plates is defined as the ratio of increase in stored charge relative to an increase in the voltage difference between said plate pair. The measurement acquisition electronics or the data acquisition system (DAS) normally employs a fixed-amplitude voltage signal square or sinusoidal in shape, and measures the mutual capacitance through current integration or by directly measuring the current amplitude. The ECT problem then becomes how to determine the dielectric distribution (or, equivalently, the spatial distribution of the various phases in the flow) within the imaging domain given the measured capacitance between all combinations of plate pairs. Plate pairs are considered independent as long as the sensitivity maps they provide are distinct. As it will be examined below, the sensitivity maps are directly related to the electric field distribution produced by any pair of plates.

In an ECT sensor, the number of independent capacitance measurements available M is a function of the total number of plates n according to the formula

$$M = \frac{n(n-1)}{2}. \tag{1.1}$$

The electric field produced by each plate blankets the imaging domain differently. All capacitance measurements are normalized with respect to the cross-sectional area (or volume) they cover so that they can be used simultaneously for image reconstruction. Different normalization techniques have been developed and described in literature; however, the most accepted are the parallel and series normalization schemes. The parallel scheme assumes that the total capacitance between a pair of plates is the sum of many smaller banks of different dielectric "tubular regions" running in parallel from plate to plate. The normalization in this case is implemented by subtracting the empty cell measurement from the measured capacitance and dividing by the dynamic range between a full and an empty cell. On the other hand, the series normalization scheme assumes a series connection and performs normalization by using the reciprocal of the capacitances.

1.3 Image reconstruction algorithms

In ECT, even though the electric field in the imaging domain obeys Laplace equation, which is a linear partial differential equation, the inverse problem (i.e., image reconstruction) is inherently nonlinear. The nonlinearity here is a result of the how the quantity to be determined (i.e., the spatial dielectric distribution) enters as a local coefficient in the Laplace equation. In other words, a change on such local coefficient produces a change on the underlying field solution (for a given set of boundary conditions) that does not follow a linear relationship with said coefficient. The accuracy of the final solution for the estimated permittivity distribution is determined by the degree of such nonlinear effects (often referred to as the "soft-field effect" inherent in ECT systems, and absent in "hard-field" tomography techniques such as X-ray), the number of independent capacitance measurements available, and the level of ill-posedness of the inverse problem (i.e., the fact that the number of capacitance measurements is typically much smaller than the number of pixels to be determined). The ill-posedness of the problem also affects how noise and small perturbations on the measured data influence the final solution (Hansen, 2010).

Imaging reconstruction algorithms for ECT can be roughly classified into two main types: algebraic and optimization techniques (Lionheart, 2001; Marashdeh, Warsito, Fan, & Teixeira, 2006a, 2006b; Yang & Peng, 2003). Both types start first by characterizing the sensor response under small perturbations of dielectric material. This essentially provides an “impulse response” of the capacitance measurements “sensor output signal” to the dielectric distribution “input signal” under a linear approximation of the problem. For this purpose, the imaging domain is initially divided into small pixels (voxels in the 3D case). Next, the electric field is computed inside the imaging domain when one single plate is excited with a unit potential, while the others are grounded (set to zero potential). This PROCESS is repeated for all combinations of plate pairs. The “impulse response” or sensitivity matrix is then calculated with elements:

$$S_{ij} = V_{0j}\boldsymbol{E}_{si}(x,y,z)\cdot\boldsymbol{E}_{di}(x,y,z) \tag{1.2}$$

where $\boldsymbol{E}_{si}$ is electrical field distribution at the voxel j location when source plate in pair i is activated with unit voltage and other plates are grounded. Similarly, $\boldsymbol{E}_{di}$ is the electrical field distribution at voxel j when the detector plate of pair i is activated with unit voltage and other plates are grounded. V_{0j} is the volume of voxel j.

In summary, Eq. (1.2) provides a linearization of the relationship between the mutual capacitance response to a perturbation on any given pixel permittivity. This is only an approximation because it ignores the nonlinearity caused by the interaction between multiple pixel perturbations at different locations, which also affects the resulting electric field and hence the measured capacitance values. Mathematically, the approximation incurred by the sensitivity matrix in Eq. (1.2) is somewhat akin to that of applying a first-order Born approximation in scattering theory (Marashdeh & Teixeira, 2004). Results produced by applying the sensitivity matrix in Eq. (1.2) can be accurate as long as the contrast between the permittivity of the different phases is not too high and/or if the fractional volume of the perturbed pixel regions is small.

The sensor response to dielectric distribution can also be considered linear with respect to sensitivity matrix as a first-order approximation. This is equivalent to the solution obtained using Linear Back Projection (LPB):

$$\widehat{g} = \boldsymbol{S}^T\boldsymbol{\lambda} \tag{1.3}$$

where $\widehat{g}$ is the image vector (pixel values), $\boldsymbol{S}^T$ is the transpose of the sensitivity matrix, and λ is the measured (mutual) capacitance vector. LBP provides a very fast solution, but because it is a linear approximation, it suffers from visible artifacts. Those artifacts are direct results of the “soft-field” (nonlinear) nature of the ECT problem. To reduce such artifacts, iterative algebraic approaches, for example, can be applied to the ECT problem. In those approaches, the resulting image from the LPB solution is solved for the capacitance values of different plate combinations (i.e., forward solution). The discrepancy between the original measured data and the forward solution is then minimized iteratively. This approach yields significant improvements to the image quality. The Landweber algorithm is perhaps the most common type of iterative algebraic reconstruction utilized in ECT. The iteration can be expressed as

$$\hat{g}_{k+1} = \hat{g}_k - \alpha \boldsymbol{S}^T (\boldsymbol{S}\hat{g}_k - \boldsymbol{\lambda}) \tag{1.4}$$

which converges to a stable solution for the image vector if the parameter α is set according to the following constraint

$$0 < \alpha < \frac{2}{\left\| \boldsymbol{S}^T (\boldsymbol{S}\hat{\boldsymbol{g}}_k - \boldsymbol{\lambda}) \right\|^2}. \tag{1.5}$$

The most straightforward way to solve the forward problem in iterative algebraic techniques is to forward project the resulting image from the last iteration onto the sensitivity matrix. While this method is very fast and convenient (since it utilizes the same sensitivity matrix calculated in advance), it is also limited by the linearization error from representing the soft-field ECT problem using linear superposition of individual perturbation responses (sensitivity matrix). It is possible to increase the accuracy of iterative reconstruction approaches by solving the forward problem using more exact methods such as the Finite Element Methods. However, this strategy incurs in significant computational costs that often render it impractical for use in connection with fast acquisitions and real-time imaging applications.

In addition to LBP and iterative algebraic approaches, other image reconstruction approaches are used in ECT, most prominently optimization approaches that combine, into an objective function, expected image properties such as entropy as a criterion for image reconstruction, in addition to minimizing the iterative error as in iterative algebraic technique. Those techniques assume that there is an inherit limit in reconstructing the ECT image which can be supplemented by some a priori information about the nature of the desired image. For example, the neural network multicriteria optimization image reconstruction technique (NNMOIRT) assumes the desired image as the one that provides the maximum entropy (i.e., image information), least noise (i.e., the objective function includes a smoothness indicator), and also fulfills the iterative minimization of linearized error (Marashdeh, Warsito, Fan, & Teixeira, 2006a, 2006b). The difference between these two approaches—optimization and algebraic—is better illustrated through the following example. Considering an ECT sensor with n electrode plates, the signal-to-noise ratio (SNR) and sensitivity of different plate combinations depend on their geometrical arrangement. Since the sensitivity matrix is ill-posed, noise contamination in certain channels will introduce disproportionate artifacts in the final image. Using algebraic techniques, the reconstruction algorithm will show this artifact as it resolves the final image to best match the measured set of capacitances. On the other hand, optimization algorithms would weight this artifact and attempt to assign a degree of likelihood to it. For example, a single pixel with high dielectric value in a surrounding background of low dielectric value would most likely be interpreted as noise and the algorithm would either ignore it or dilute it. This is implemented in optimization algorithms through the additional set of terms in the objective function that depends on the nature of the final image, in addition to minimizing the error between the measured data and forward solution of the final image. Finally, we note that even for algebraic algorithms it is important to implement regularization techniques to address the ill-posedness of the reconstruction problem. In ECT, this has been achieved with various degrees of success.

1.4 Data acquisition system

As explained above, ECT is concerned with determining dielectric distribution maps in the imaging domain from variations in capacitance signals measured at its boundary. This variation is measured between different pairs of capacitance plates. The DAS is required to record those variations in the presence of parasitic capacitances, which can be many orders of magnitude larger. Parasitic capacitance is the (static) underlying capacitance that exists between the plates used for measurement, or between each plate and its surroundings (ground), and which does not contribute to the varying component being measured. Stray capacitance, on the other hand, is inherent capacitance in the ECT system that is not related to the domain of geometry between the capacitance sensor plates. Stray capacitance is related to coupling between sender/receiver plate paths to the ground. This coupling includes capacitance of cables that connect the DAS to the sensors plates and capacitance between plates and grounded shield. Stray capacitances can also be many orders of magnitude higher than the capacitance variations being measured.

A basic requirement of any good DAS design is the automatic rejection of stray capacitances. Two basic circuit configurations are typically adopted for ECT sensor systems: charge−discharge and AC-based designs (Huang, Xie, Thorn, Snowden, & Beck, 1992; Yang & York, 1999). Both DAS designs rely on amplifying the current through the capacitance sensor plates while automatically rejecting stray capacitance. From the sender plate side, the stray capacitance is in parallel to the capacitance-measuring circuit and, as a result, the current it pulls is diverted from passing through the interplate (desired) capacitive path. Similarly, the stray capacitance on the receiver side is in parallel to the input port of the operational amplifier (Op-Amp) used to amplify the output current coming out of the detector plate. The Op-amp has an input voltage difference of virtual zero. The shunt connection of stray capacitance before and after the targeted interplate capacitance makes the current passing to the Op-Amp a function of interplate capacitance only. For the charge/discharge design, the relation between output of the Op-Amp voltage and interplate capacitance can be written as

$$V_{\text{out}} = 2KfV_cC_xR_f \tag{1.6}$$

where K is a gain factor, f is operating frequency, V_c is source voltage, C_x is capacitance to be measured, and R_f is the value of the feedback resistor. The main disadvantage of the charge−discharge circuit is that it relies on CMOS switches for switching, which have large capacitance values compared to C_x.

As for the AC-based circuit, the interplate capacitance is found from

$$V_{out} = \frac{-C_x}{C_f} V_i \tag{1.7}$$

where C_f is a feedback capacitor for the Op-Amp and V_i is the input excitation voltage. This equation is an approximation that assumes a high-frequency excitation in the 1−10 MHz range, and appropriately selected values for the feedback capacitor and resistor. AC-based circuit designs provide higher SNR compared to charge−discharge at the expense of increased circuit complexity.

The output voltage from both types of circuits varies depending on the flow dynamics between plates and the consequent change in dielectric distribution. The output signal is commonly amplified using programmable gain amplifiers (PGAs) and digitized using analog-to-digital converters (ADCs). Those two steps usually happen in sequence. The PGAs amplify the weak signals to boost their SNR while keeping them within the range of ADCs. PGA values are determined experimentally by a calibration process where the sensor is examined in its full and empty domain states. The full state represents the highest anticipated readings from the sensor interplate combinations and is used to set the PGAs to their maximum gain within the range of ADC. The empty state is used as a floor reference for a later normalization process.

In any ECT sensor, the sensitivity map produced by different pairs of electrode plate depends on the relative distance and angular positioning between the pair. For image reconstruction purposes, the intensity of all plate combinations is "normalized" with respect to overall sensor response so they can be used more uniformly across different sensor deployments. Some normalization strategies have been discussed in Section 1.2 above in relation to the capacitance measurements, and similarly, different normalization approaches are applied to aid in the image reconstruction process. All these normalization strategies all require a measurement of all sensor plate combination when the sensor is full with dielectric material and when it is empty. The simplest and most commonly used capacitance normalization is

$$\widetilde{C}_{ij} = \frac{C_{ij} - C_{ij}^{(e)}}{C_{ij}^{(f)} - C_{ij}^{(e)}} \tag{1.8}$$

where the $\widetilde{C}$, normalized capacitance, value refer to the variable actually employed in the reconstruction algorithm, and the superscripts (*e*) and (*f*) refer to capacitance values with empty and full vessels, respectively.

1.5 Electrical capacitance volume tomography

In conventional ECT applications, the capacitance sensors are designed to surround the 2D imaging domain (vessel cross-section) by means of electrode plates which are symmetrically distributed and equally spaced. The electrode plates are also desired to be of substantial axial length to acquire data with high SNR and reduce the asymmetry of the electric field distribution along the axial direction. The axial asymmetry arises from fringing field effects at the upper and lower edges of activated plates. Since conventional ECT is used for 2D imaging, such fringing effect is viewed as distortion to the desired sensor sensitivity distribution and efforts are made to minimize it. Because of this, elongating the electrode plates along the axial direction is the simplest and most widely used technique to reduce fringing effect.

ECT sensors can also be used to produce quasi-3D images by placing two rows of ECT sensors vertically above each other so as to acquire two simultaneous 2D images. The quasi-3D image is then formed by interpolating the 2D images into one continuous 3D image. Another technique uses time division of a single sensor to achieve the same goal. In such arrangements, successive 2D images acquired by an ECT sensor over a period of time are interpolated to again form a quasi-3D image (Banasiak, Wajman, Betiuk, & Soleimani, 2009). Similar strategies have been successfully used in other hard field imaging modalities as well. However, those techniques pose a challenge in ECT due both to the considerably longer electrodes (and hence the longer 2D "image averaging lengths") and soft-field nature of capacitance measurements.

A technique for capacitance electrode design was devised recently that, instead of attempting to mitigate fringing field effects, instead *exploits* such effects between capacitance plates and volumetric arrangements thereof to construct sensitivity maps that yield inherit 3D features (Li & Holland, 2013; Warsito, Marashdeh, & Fan, 2007). The overall sensor in this case is designed with electrode plates of various shapes ensuring that the aggregate sensitivity map covers a volume and at the same time exhibits sufficient sensitivity variations along both x, y, and z axis. Since the term "3D capacitance tomography" is normally used to refer to 3D or quasi-3D imaging by interpolating successive 2D images as discussed above, the term *electrical capacitance volume tomography* (ECVT) is used to refer to such direct 3D imaging using 3D capacitance sensors.

Fig. 1.2 depicts various sensor designs that exhibit 3D features. Case A provides 3D sensitivity distribution by arranging the plates in a staggered fashion across successive planes. Cases B and C use irregular column shapes to provide sensitivity distributions with exploitable 3D features. Note that these cases also use plates of different shapes and sizes. Case D has a geometrical arrangement similar to Case A, but with an inserted circular hole at one side to accommodate a side flow injection, as opposed to a predominant axial flow. As it can be easily surmised from this figure, ECVT sensors provide substantially more flexibility in plate arrangements, plate shape, and geometric conformality to more general column (vessel) shapes and topologies such as L-junctions and T-junctions.

ECVT sensors also naturally allow for more independent capacitance channels to be used and hence yield higher spatial resolution. For example, in 2D ECT, it is observed that the use of more than 12 capacitance plates around the cylindrical cross-section provides diminishing returns for imaging resolution (Peng, Ye, Lu, & Yang, 2012). This is due in part to the restriction on capacitance plate shapes on the circular boundary and on the decrease on the SNR for each channel when using more plates

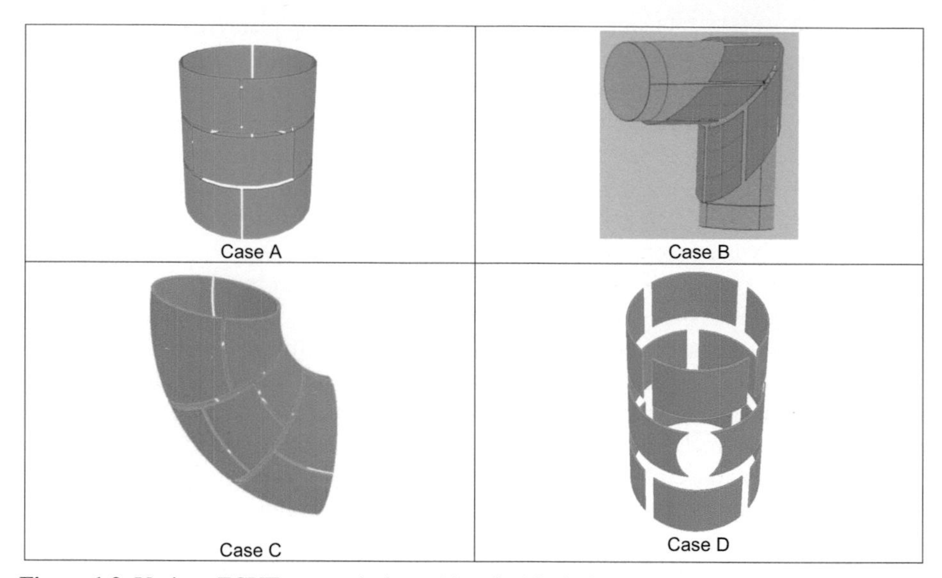

Figure 1.2 Various ECVT sensor designs: (a) cylindrical, (b) sharp bend, (c) duct, and (d) side injection.

(more plates require less area and hence smaller overall capacitances and output signals). In contrast, there is considerably more flexibility in the geometrical distribution of the plates of an ECVT sensor as well as on the ability to choose plates with virtually any 2D shape. It was indeed verified experimentally that this added flexibility provided room for obtaining substantial resolution improvements in ECVT when using more than 12 channels (Wang, Marashdeh, Teixeira, & Fan, 2015). Fig. 1.3 depicts an experimental result illustrating the improvement in resolution that can be obtained in ECVT by increasing the number of channels from 12 to 24. The design depicted in Fig. 1.2, Case A, corresponds to a 12-channels ECVT sensor. A similar design can be adopted for a 24-channels sensor by building the sensor with four planes having six plates in each plane. Both 12-channels and 24-channels sensors were designed to cover the same volume and two spheres filled with sand were used to test the resolution capability of each sensor. Fig. 1.3(a) is a schematic of the objects under imaging, i.e., two adjacent balls of different size in a uniform background. Fig. 1.3(b) shows the imaging result when using the 12-channels sensor, while Fig. 1.3(c) shows the result for the 24-channel sensors. It is clear from these results that using a 24-channels ECVT sensor provides better spatial resolution versus a 12-channels ECVT sensor. Similar experiments were conducted using both sensors and arrived at a similar conclusion (Wang, Marashdeh, Teixeira, & Fan, 2015).

Fig. 1.4 depicts sensitivity map distributions for two pairs of plates in 12-channels and 24-channels ECVT sensors. The sensitivity maps in Fig. 1.4(a) and (b) correspond to pairs of plates in the same plane for the 12-channels and 24-channels cases, respectively. On the other hand, the sensitivity maps in Fig. 1.4(c) and (d) correspond to pairs of plates between the first and third rows of the 12-channels and 24-channels sensors,

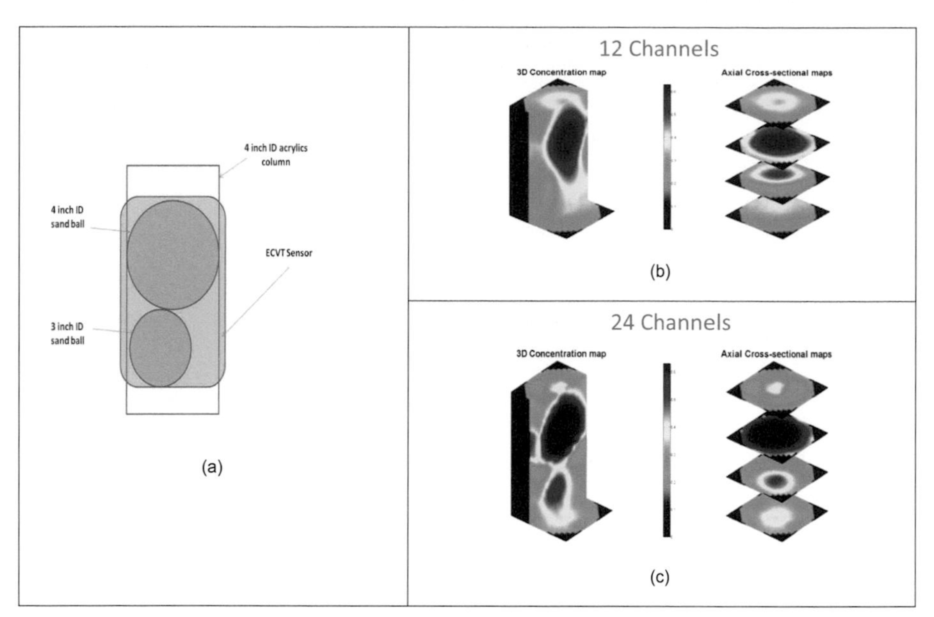

Figure 1.3 Comparison of resolution quality for the imaging of two balls of different sizes in a uniform background, using ECVT sensors with 12- and 24-channels (plate electrodes) (Wang, Marashdeh, Teixeira, & Fan, 2015).

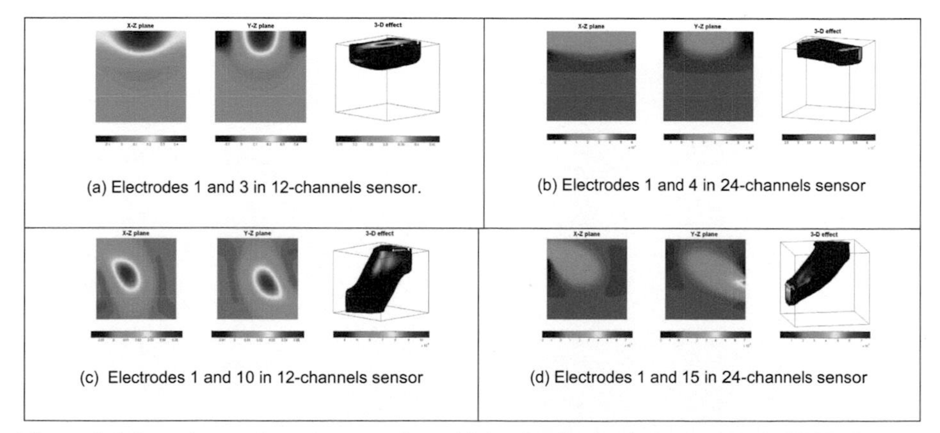

Figure 1.4 Isosurface volume images of sensitivity distribution between selected ECVT sensor plates for the 12-channels sensor (a, c) and the 24-channels (b, d) sensor (Wang, Marashdeh, Teixeira, & Fan, 2015). (a) Electrodes 1 and 3 in 12-channels sensor. (b) Electrodes 1 and 4 in 24-channels sensor. (c) Electrodes 1 and 10 in 12-channels sensor. (d) Electrodes 1 and 15 in 24-channels sensor.

respectively. Both set of figures provide a qualitative assessment of the increase in resolution due to the increase in number of plates in ECVT sensors. In particular, this can be seen from Fig. 1.4 as the volume encapsulated by the iso-surfaces becomes narrower in the case of a 24-channels sensor. This narrowing of the sensitivity maps reflects the ability to provide more independent channels and hence an overall better resolution irrespective of the reconstruction algorithm to be utilized.

1.6 Illustrative examples and discussion

The value of ECT in process tomography is better illustrated through examples that demonstrate feasibility of the technology in real applications. This section will present some results obtained from ECVT sensors as they correspond to state-of-the-art in applied ECT technologies.

The first example is taken from reference Weber, Layfield, Van Essendelft and Mei (2013) and provides a comparison between results obtained from computational fluid dynamics (CFD) simulations versus those from ECVT imaging for a 4″ ID column of glass beads particles. The CPFD software *Barracuda version 15.0.3* was used to carry the CFD simulations in this case (Snider, 2007; O'Rourke & Snider, 2012) and ECVT sensors were used on a column with same dimensions as the simulated case (Weber, Layfield, Van Essendelft, & Mei, 2013). Fig. 1.5 depicts solid fraction results using different simulation models. The last column in this figure presents the ECVT results.

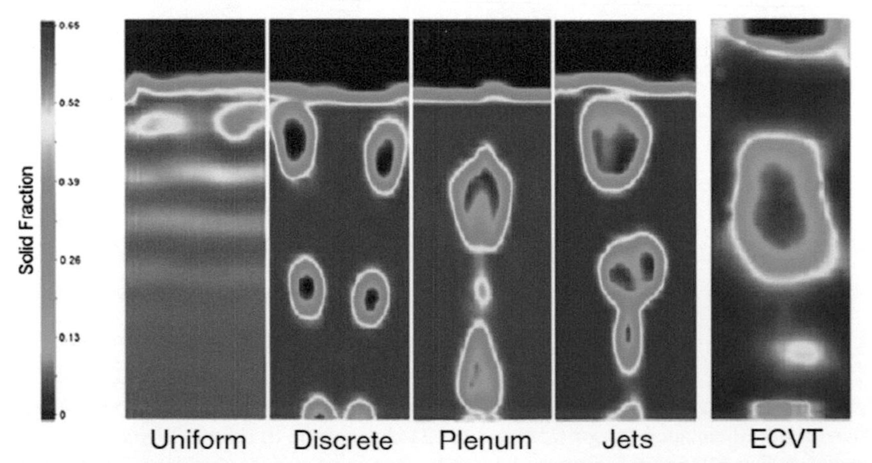

Figure 1.5 Comparison between various CFD simulation models and ECVT for a solid fraction snapshot inside a cylindrical vessel with gas flow. © 2013 Elsevier.
Reprinted, with permission, from Weber, J. M., Layfield, K. J., Van Essendelft, D. T., & Mei, J. S. (2013). Fluid bed characterization using electrical capacitance volume tomography (ECVT), compared to CPFD software's Barracuda. *Powder Technology 250*, 138–146.

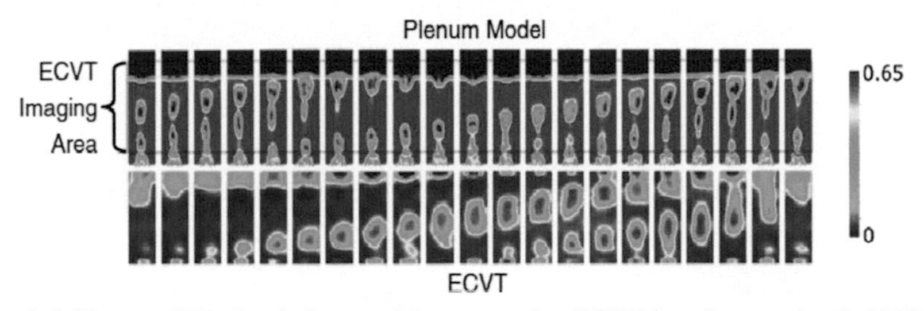

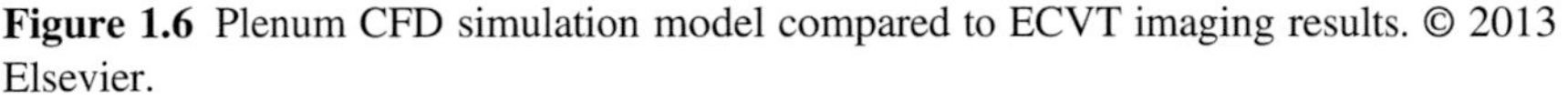

Figure 1.6 Plenum CFD simulation model compared to ECVT imaging results. © 2013 Elsevier.
Reprinted, with permission, from Weber, J. M., Layfield, K. J., Van Essendelft, D. T., & Mei, J. S. (2013). Fluid bed characterization using electrical capacitance volume tomography (ECVT), compared to CPFD software's Barracuda. *Powder Technology 250*, 138−146.

More details on the experiment, as well as further discussions and results, are found in reference (Weber, Layfield, Van Essendelft, & Mei, 2013). This experiment has shown that ECVT can be a valuable guiding tool to selecting/using the proper simulation model for a given set of experimental conditions.

Fig. 1.6 builds on results from the previous figure to determine the appropriate simulation model for comparing real-time results of void fraction and bubble formations using ECVT. This figure shows that ECVT results are able to capture the bubble formation process and the amount of void fraction around the bubbles with a very good agreement against the simulation results.

CFD simulations are carried out based on assumptions related to the process under investigation. They can provide invaluable information if executed properly; however, using them for validation can only be justified with the experimental conditions well matched to the assumptions made in the simulation model. An alternative approach for assessing ECVT results is to use another, established imaging technology that has a quantifiable resolution. MRI in particular has been used extensively in the medical field and has demonstrated an impressive capability to resolve fine details. Recently, MRI has found applications in process tomography in laboratory settings as well (Holland et al., 2009), although important disadvantages of MRI in this case are the relatively high cost and much bulkier equipment, as well as less suitability for high-speed, real-time applications. Nevertheless, the high resolution of MRI makes it an ideal candidate for ECVT validation. A recent study has been conducted to compare fluidized bed results of a 2″ ID column using ECVT and MRI (Holland et al., 2009). The experiments were conducted using each imaging technology independently, but under similar operating conditions. As noted before, ECVT sensors are mostly made of conductive metallic plates. This is a limitation in MRI as metals should be commonly avoided around high intensity magnets. Fig. 1.7 is reproduced from reference Holland et al. (2009) and depicts the ECVT sensor design used in this experiment. Fig. 1.7(b) shows a direct qualitative comparison between ECVT and MRI results through a vertical cross-section of the volume image obtained using each technology. This figure shows clearly that while ECVT can detect voids (bubbles),

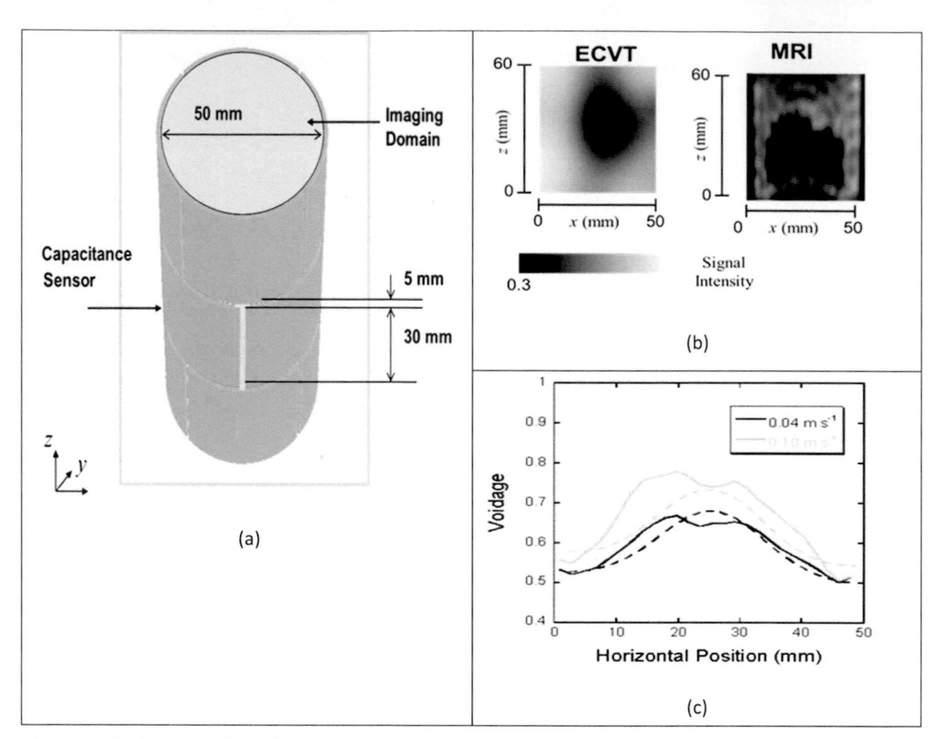

Figure 1.7 A comparison between ECVT and MRI results. (a) ECVT sensor design. (b) Vertical cross-section of imaged bubble. (c) Voidage distribution across imaging domain for ECVT and MRI results, for two flow speeds. The dashed lines refer to ECVT results and the solid lines to MRI results. © 2009 American Chemical Society.
Reprinted (adapted), with permission, from Holland, D. J., Marashdeh, Q. M., Mueller, C. R., Wang, F., Dennis, J. S., Fan, L. -S., et al. (2009). Comparison of ECVT and MR measurements of voidage in a gas-fluidized bed. *Industrial & Engineering Chemistry Research, 48*(1), 172–181.

their location, and size, it has less spatial resolution compared to MRI. Recalling the results presented in Fig. 1.3, one would expect that ECVT can provide better resolution if a 24-channels sensor were to be used, instead of the 12-channels sensor actually utilized in this study. However, it is expected that ECVT still would fall short compared to MRI in capturing the very fine details of the flow structure. For example, the MRI result was able to detect the wake of the bubble rising in the surrounding solid phase, a feature that cannot be well captured by ECVT even when utilizing a 24-channels sensor. A more quantitative comparison is provided in Fig. 1.7(c). Here, the void fraction, as measured by each of the two imaging techniques, is plotted across the diameter of the sensing region. One can see clearly that while both technologies predict accurately the void fraction amplitude, the ECVT results (dashed lines) are a smoothed version of the MRI results (solid lines).

The results illustrated here highlight some of the attractive features of ECT that make it suitable for many industrial applications. To summarize, the main attractive features of ECT sensor technology include the following:

- It operates by noninvasive means.
- It can produce imaging (frame) rates in the hundreds of frames-per-second, thus enabling high-speed real-time measurements of fast reactions and physical flow processes.
- It is safe and nonionizing.
- It is scale invariant, so that the same basic sensor arrangements and imaging reconstruction algorithms can be used equally well in scale-up or scale-down applications.
- It is geometrically flexible and ECVT sensors in particular can be designed to fit around virtually any shape.
- It can be installed and operated at a relatively low cost compared to other imaging technologies.

However, one major disadvantage of ECT/ECVT is its low spatial resolution versus some other modalities such as MRI, as illustrated in Fig. 1.7. Recall that the limitation in the spatial resolution of ECT stems from two basic reasons. First, ECT sensors provide only a limited number of independent measurements, far below the number of unknown variables (i.e., pixels) necessary for high-resolution imaging. This makes the ECT/ECVT problem highly ill-conditioned if higher resolution is required. Secondly, ECT belongs to "soft-field" category of tomography modalities, wherein the field solution is a nonlinear function on dielectric distribution in the imaging domain and the inverse problem (of determining the dielectric distribution from the capacitance measurements) is also inherently nonlinear.

As discussed before, a basic factor that limits the number of independent capacitance measurements in ECT and ECVT is the fact that measured capacitance variations in flow systems are typically very small and there is a need for a minimum electrode plate area to provide sufficient SNR for the output signal. This minimum area requirement sets up a limit on the maximum number of plates for any given vessel size. A recent advancement in ECVT sensor design geared toward mitigating the limited number of independent measurements from a capacitance sensor has been the development of Adaptive Electrical Capacitance Volume Tomography (AECVT), which lifts the limitation on the number of independent measurements by creating synthetic shapes from very small, individual plate segments (Chowdhury, Marashdeh, & Teixeira, 2020; Marashdeh, Teixeira, & Fan, 2014). In other words, in AECVT, each conventional capacitance plate is replaced by an adaptive plate composed of many smaller segments each with a separate signal acquisition circuitry. Such segments can each be activated with different voltage levels. The aggregate voltages on all activated segments form a synthetic plate with a voltage envelope based on activation chosen for each segment. New independent measurements can thus be obtained by changing the synthetic plate activation, i.e., either the voltage envelope or the choice of activated segments, while keeping the minimum area requirement in place for any (synthetic) plate. It is clear that many different choices exist for yielding synthetic plate

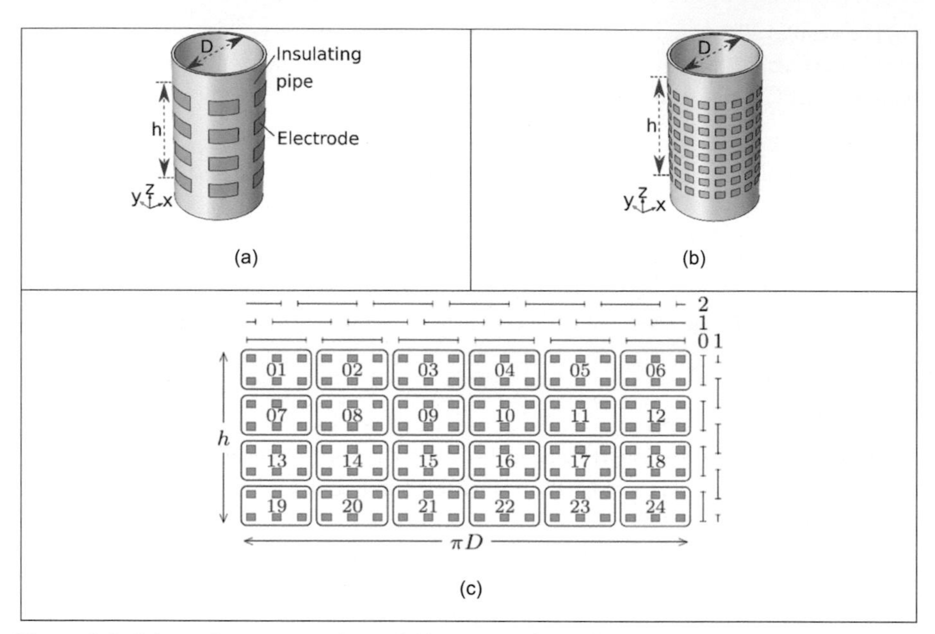

Figure 1.8 Schematic representations of (a) a conventional 6 × 4 ECVT sensor, (b) an adaptive 18 × 8 ECVT sensor, (c) Formation of 3 × 2 synthetic electrodes with electronic scanning. Copyright 2020 IEEE.
Reprinted, with permission, from Chowdhury, S. M., Marashdeh, Q. M., & Teixeira, F. L. (2020). Electronic scanning strategies in adaptive electrical capacitance volume tomography: Tradeoffs and prospects. *IEEE Sensors Journal 20*(16), 9253–9264.

with a multitude of geometries, including noncontiguous shapes. The principle of AECVT is illustrated in Fig. 1.8, where Fig. 1.8(a) depicts a conventional ECVT sensor and Fig. 1.8(b) depicts an adaptive ECVT sensor made of many small segments. In Fig. 1.8(c), the formation of synthetic electrode of size 3 × 2 is illustrated. Here the size indicates how many physical segments are connected horizontally and vertically. The resultant sensor turns out to be 6 × 4 in size, thus conforming to the SNR criterion. The electronic scanning is also illustrated with overhead and side bars.

AECVT requires an acquisition circuit that can accommodate an increased number of segment electrodes and the overall increase in complexity in the sensor design. In AECVT, the aggregate current from the receiver plate is a summation of intercapacitance interaction between each of the selected (activated) segments and the receiver plate. Consequently, the current measured at the receiver plate side using a current-to-voltage amplifier is equivalent to the current produced by the mutual capacitance interaction from a conventional capacitance plate of equivalent shape but with a (spatial) weight provided by the given voltage envelope distribution. Further details on the AECVT technique can be found in Chowdhury, Marashdeh and Teixeira (2020); Marashdeh, Teixeira and Fan (2014).

1.7 Flow velocimetry with ECVT

Determination of the velocity profile in a two-phase flow has been a topic of growing interest due to its importance in understanding the flow dynamics. It also enables the calculation of other flow parameters such as the mass flow rate. Two ECT sensors placed at a certain distance along the flow pipe can be used to calculate the axial velocity (v_z) by cross-correlating the reconstructed images from the sensors (Yang, Stott, Beck, & Xie, 1995). Also, cross-correlation with two successive 3D images from an ECVT sensor can provide a complete velocity profile (v_x, v_y, v_z) (Marashdeh, Wang, Fan, & Warsito, 2007, pp. 1017−1019). Later, a more robust method based on sensitivity gradient is introduced (Chowdhury, Marashdeh, & Teixeira, 2016), which is illustrated in Fig. 1.9 with an ECVT sensor with two dielectric objects. The sensitivity between two electrodes is shown in Fig. 1.9(b), which is required for image reconstruction. The corresponding sensitivity gradient is shown in Fig. 1.9(c), which is required for velocity reconstruction. The reconstructed image is shown in Fig. 1.9(d) and the corresponding velocity profile is shown is Fig. 1.9(e). Note that the two objects

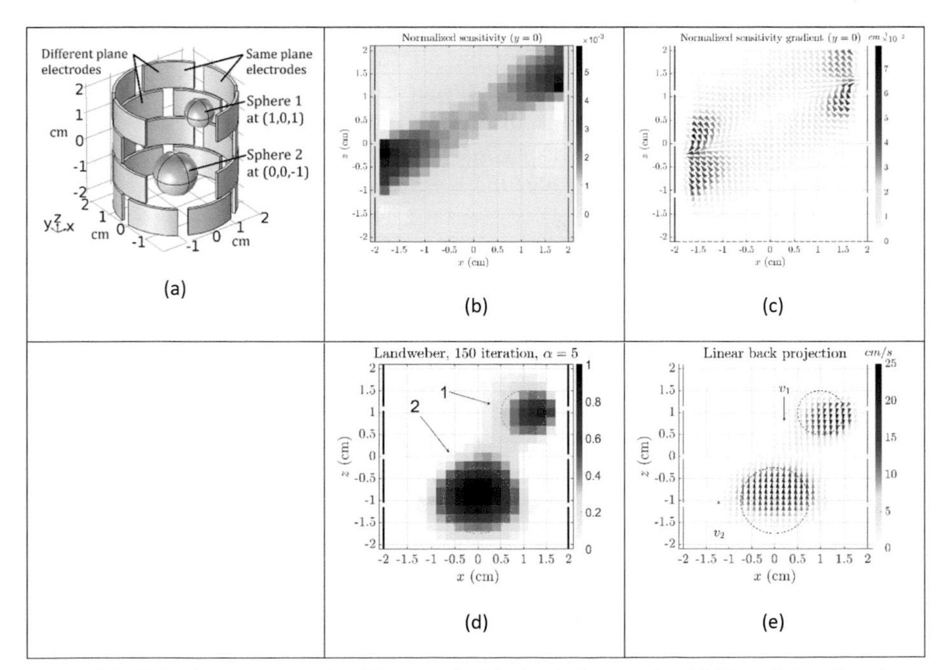

Figure 1.9 Velocity reconstruction illustrated with simulation results. (a) Two dielectric spheres moving in air background, (b) sensitivity distribution, (c) sensitivity gradient, (d) reconstructed image, (e) reconstructed velocity profile. © 2016 IEEE.
Reprinted (adapted), with permission, from Chowdhury, S. M., Marashdeh, Q. M., & Teixeira, F. L. (2016). Velocity profiling of multiphase flows using capacitive sensor sensitivity gradient. *IEEE Sensors Journal. 16*(23), 8365−8373.

are moved toward one another with an average velocity of 20 cm/s in the setup, which can be visualized from the reconstructed velocity profile. A slightly different approach needs to be adapted for flows with higher permittivity continuous phase, e.g., air in oil, which is discussed in Chowdhury, Marashdeh and Teixeira (2018).

1.8 Three-phase flow decomposition by exploiting Maxwell–Wagner–Sillars effect

Due to the single dimensional nature of ECT measurements, which is the capacitance, it is not possible to decompose flows with more than two phases such as air–water–oil. A new dimension can be added by introducing the conductance measurement (Marashdeh , Warsito, Fan, & Teixeira, 2008); however, the overall procedure is not robust. Another new dimension can be the capacitance measurements at a different frequency, which is investigated in Rasel, Zuccarelli, Marashdeh, Fan and Teixeira (2017). This multifrequency method is more robust than the conductance-based method. The multifrequency method requires the presence of a conducting phase such as water, which introduces the frequency-dependent changes in the measured capacitance. Whenever there is a conducting phase, there occurs interfacial polarization which causes a frequency shift in the dispersion curve, known as the Maxwell–Wagner–Sillars (MWS) effect. The dispersion curve for a three phase mixture is shown in Fig. 1.10(a) where the dispersion is occurring at a lower frequency range than that of water alone. Here, the blue (dark gray in print version) curve corresponds to the dispersion curve (permittivity), whereas the red (light gray) curve corresponds to the absorption curve (conductivity). Capacitance data measured at the two static regions of the dispersion curve would facilitate decompose a three-phase flow. A numerical example is presented in b where phase A corresponds to water ($\varepsilon' = 81, \sigma = 5.5$ mS/m), phase B corresponds to air ($\varepsilon' = 1$), and the continuous phase is oil ($\varepsilon' = 5$). The reconstructed images corresponding to phase A and B are shown in d, respectively, hence decomposing the phases individually with respect to the background oil phase. Later, the algorithm proposed in Rasel, Zuccarelli, Marashdeh, Fan and Teixeira (2017) is revised to be applied for flows with water continuous phase (Rasel, Marashdeh, & Teixeira, 2018), an extremely challenging case for ECT.

1.9 Displacement–current phase tomography (DCPT) and water-dominated flow velocimetry

Displacement–current phase tomography (DCPT) has been a spin-off from ECT intended to be applied to medium containing an electrically conductive phase such as water (Gunes, Marashdeh, & Teixeira, 2017; Sines et al., 2020). Application of ECT has been limited for water dominated flows due the presence of conductivity.

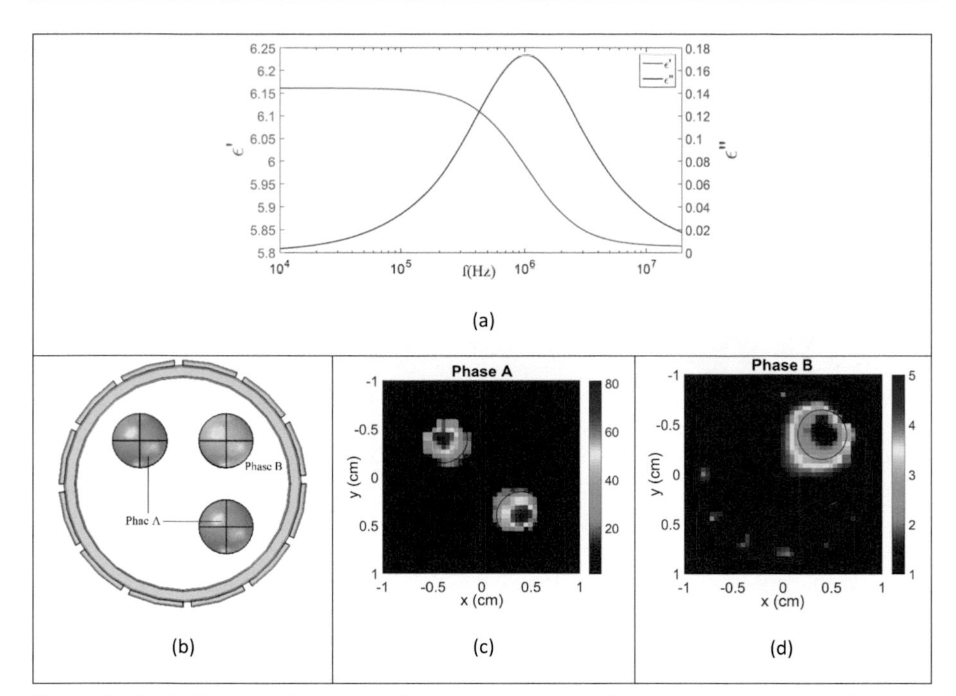

Figure 1.10 MWS-based three-phase flow decomposition. (a) Frequency response of a three-phase mixture consisting of air, water, and oil, (b) simulation setup showing water bubbles as phase A and air bubble as phase B in oil background, (c) reconstructed image showing water bubbles only, (d) reconstructed image showing air bubble only. © 2017 IEEE. Reprinted, with permission, from Rasel, R. K., Zuccarelli, C. E., Marashdeh, Q. M., Fan, L.-S., & Teixeira, F. L. (2017). Toward multiphase flow decomposition based on electrical capacitance tomography sensors. *IEEE Sensors Journal 17*(24), 8027–8036.

In DCPT, the phase angle of the displacement current, i.e., of the admittance, is measured instead of the capacitance. The phase angle is found to be more sensitive to medium with conductivity than the capacitance, which is the reason for DCPT to be suitable for water-dominated flows. The major advantage of DCPT is that it uses the same sensor as ECT, making it nonintrusive, as opposed to electrical impedance tomography which is also applied for imaging conductive medium (Cheney, Isaacson, & Newell, 1999). Two applications of DCPT are shown in Fig. 1.11. First, an ECT sensor with two air bubbles in water is shown in a, with corresponding reconstruction results in Figs. 1.11(b) and (c) for ECT and DCPT, respectively. The image quality for DCPT is much better than that of ECT. Next, a volumetric setup with an ECVT sensor is shown in d containing two moving air bubbles in water. The corresponding reconstructed image and velocity profile are shown in 3D in Fig. 1.11(e). Here DCPT has been applied with an ECVT sensor and combined with the sensitivity gradient based velocimetry method described above (Gunes, Chowdhury, Zuccarelli, Marashdeh, & Teixeira, 2019).

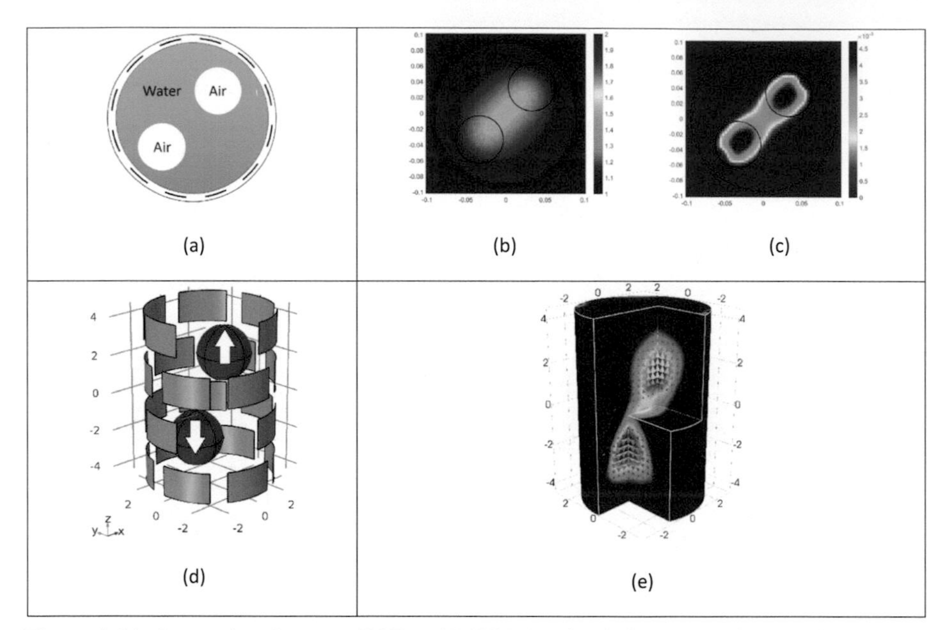

Figure 1.11 Comparison between ECT and DCPT: (a) simulation setup with two air objects in water, (b) ECT reconstruction result, (c) DCPT reconstruction result. DCPT-based volumetric imaging and velocity reconstruction: (d) simulation setup: two air bubbles moving in water background, (e) reconstructed image and velocity profile. © 2017, 2019 IEEE. Reprinted, with permission, from Gunes, C., Chowdhury, S. M., Zuccarelli, C. E., Marashdeh, Q. M., & Teixeira, F. L. (2019). Displacement-current phase tomography for water-dominated two-phase flow velocimetry. *IEEE Sensors Journal 19*(4), 1563−1571; Gunes, C., Marashdeh, Q. M., & Teixeira, F. L. (2017). A comparison between electrical capacitance tomography and displacement-current phase tomography. *IEEE Sensors Journal 17*(24), 8037−8046.

1.10 Recent progress with AECVT

The concept of AECVT is introduced first in Marashdeh, Teixeira and Fan (2014), followed by Zeeshan, Zuccarelli, Acero, Marashdeh and Teixeira (2019) and Chowdhury, Marashdeh and Teixeira (2020) with demonstration of the concept for ECT and ECVT, respectively. Fig. 1.12 shows a demonstration for ECVT. First, a reconstruction case is shown in a for a regular ECVT sensor with 24 electrodes. Next in b, the same scenario is reconstructed for an AECVT sensor showing a clear improvement in resolution. The AECVT sensor is comprised of 288 electrode segments. The acquisition is performed with horizontal- and vertical-shaped synthetic electrodes with electronic scanning, as shown in b. The gain in resolution for AECVT is achieved from the increased number of measurements as compared to ECVT; however, this results in an increase in acquisition time for the former. Therefore, a trade-off has to be established between resolution and acquisition time, which is investigated in Chowdhury, Marashdeh and Teixeira (2020).

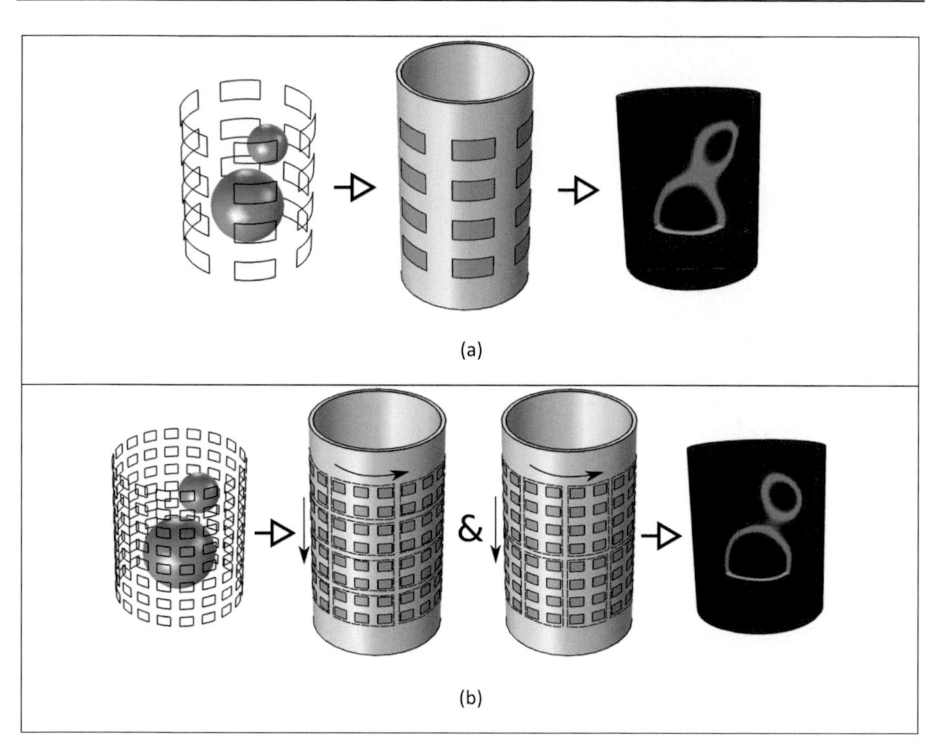

Figure 1.12 Resolution improvement with AECVT. (a) Showing result for regular 6× 4 sensor with electronic scanning, (b) showing result for 24 × 12 sensor with electronic scanning through horizontal and vertical electrode shapes combined. © 2020 IEEE.
Reprinted (adapted), with permission, from Chowdhury, S. M., Marashdeh, Q. M., & Teixeira, F. L. (2020). Electronic scanning strategies in adaptive electrical capacitance volume tomography: Tradeoffs and prospects. *IEEE Sensors Journal 20*(16), 9253–9264.

1.11 RVM for the determination of uncertainty in reconstruction

In ECT/ECVT, due to the spatial behavior of the Laplacian electric fields inside the RoI, the sensitivity map presents higher values near the electrode plates and lower toward the center of the sensing domain (Fang, 2006). This means that the capacitance measurements are not as sensitive to small permittivity variations near the center of the RoI when compared to the ones near the periphery (Ospina-Acero, Marashdeh, & Teixeira, 2020), which in turn translates into lower spatial imaging resolution toward the center as compared to the periphery in ECT/ECVT applications. In other words, the reconstructed image has a higher degree of uncertainty toward the center. An attempt to quantify this uncertainty is undertaken in Ospina-Acero, Marashdeh, and Teixeira (2020) by applying RVM for solving the ECT inverse problem. RVM is a probabilistic approach having its origin in the machine learning world which has been adapted for

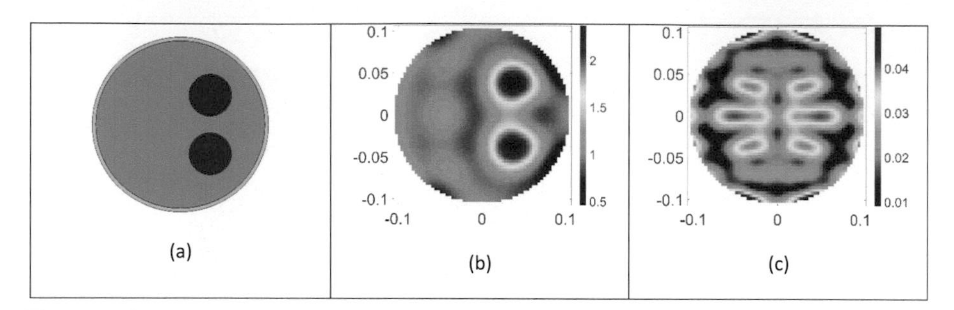

Figure 1.13 Adaptive RVM-based reconstruction results for ECT. (a) Simulation case with two objects, (b) reconstructed image, (c) associated uncertainty. © 2020 IEEE.
Reprinted, with permission, from Ospina-Acero, D., Marashdeh, Q. M., & Teixeira, F. L. (2020). Relevance vector machine image reconstruction algorithm for electrical capacitance tomography with explicit uncertainty estimates. *IEEE Sensors Journal 20*(9), 4925–4939.

the inversion procedure in process tomography applications. One key advantage of RVM in the context of ECT/ECVT is that it provides a pixel-based uncertainty image (as represented by the standard deviation around the estimated value of the pixels) along with the reconstructed image. An example of RVM-based reconstruction is presented in Fig. 1.13. The setup, shown in a, consists of two dielectric objects with a relative permittivity of 3 in a background with a relative permittivity of 1. The reconstructed image is shown in Fig. 1.13(b) and the corresponding standard deviation image in c. In this last image, the uncertainty is higher toward the center and exhibits some symmetry which is due to the symmetry and the displaced position of the objects. More details regarding the procedure to obtain the images in b and c can be found in Ospina-Acero, Marashdeh and Teixeira (2020).

1.12 Conclusion

In this chapter, we have discussed the basic principles of ECT and described the favorable features that make it suitable for process tomography applications, including industrial settings. The basic architecture of an ECT system can be broken down into three main components: the sensing electrodes composed of capacitive plates placed around the domain to be sensed, the data acquisition hardware, and the image reconstruction software. Recent advances in sensor design have enabled direct volume imaging through the suitable distribution of the capacitance plates around a 3D domain, embodied in ECVT systems. We have discussed some practical examples of the application of ECT in industrial settings such as gas–solid systems. We have presented a number of recent developments, including flow velocimetry for two-phase flows, multifrequency ECT for three-phase flow decomposition such as air–water–oil, DCPT for imaging and velocimetry of water-dominated flows, and RVM-based image reconstruction for uncertainty prediction of image reconstruction. All of those have been achieved with the conventional ECT sensor and AC-based data acquisition device. We have also presented the adaptive ECVT, which improves image resolution for ECVT by applying a high electrode number sensor and a more sophisticated data acquisition circuit capable of forming synthetic electrodes.

1.13 Future trends

Image reconstruction techniques represent an active area of research in ECT, and different algorithms have been introduced over the years with varying degrees of success. Landweber's iteration method has been successful for general purposes (Yang & Peng, 2003); however, it cannot provide resolution beyond a certain point due to the physical limits of ECT (Lucas, Margo, Oussar, & Holé, 2015). For this reason, more sophisticated methods are being investigated including multicriterion optimization-based technique (Marashdeh, Warsito, Fan, & Teixeira, 2006a, 2006b), sparse representation−based technique (Ye, Wang, & Yang, 2015), probabilistic methods including Bayesian inferences (Ospina-Acero, Marashdeh, & Teixeira, 2020; Watzenig & Fox, 2009), machine learning based techniques such as the neural network (Lei, Liu, & Wang, 2020; Zheng & Peng, 2020), space-adaptive reconstruction technique (United States Patent No. 9791396, 2017) and so on. With ECVT and AECVT, higher computational resources are required for image and velocity reconstruction. Such computation can be made multiple times faster if the reconstruction is performed in the spatial frequency domain instead of the spatial domain directly, which has been introduced in (Gunes, Ospina-Acero, Marashdeh, & Teixeira, 2018). Therefore, more focus is likely to be given toward reducing computational requirements as algorithms become more sophisticated.

In practice, ECT systems are primarily applied for imaging of permittivity distributions in flow systems involving low conductive materials such as oil−air, but in principle they can also be exploited for imaging conductive phase such as water. Several works have been conducted in this area of research including the dual-modality imaging (Marashdeh, Warsito, Fan, & Teixeira, 2008), MWS effect−based three-phase flow decomposition (Rasel, Zuccarelli, Marashdeh, Fan, & Teixeira, 2017), DCPT (Gunes, Chowdhury, Zuccarelli, Marashdeh, & Teixeira, 2019), simultaneous reconstruction of permittivity and conductivity (Zhang & Soleimani, 2016), and so on. Dual modality imaging methods are essential for water-dominated flow imaging and are likely to receive a greater emphasis in future.

Another active area of research in ECT is the acquisition design for achieving faster sampling rates as well as reducing measurement noise. Emphasis has been given toward digitization of the acquisition device by replacing the analog components with digital counterparts and implementation of parallel data acquisition channels (Cui, Wang, Chen, Xu, & Yang, 2011; Sun, Cao, Huang, Xu, & Yang, 2017). Considerable focus is also being given toward acquisition design for adaptive ECVT (Chowdhury, Marashdeh, & Teixeira, 2020). This progress in acquisition design is likely to continue in future, with greater emphasis toward multifrequency adaptive and reconfigurable acquisition design.

Lastly, considerable focus has been given toward the capacitive sensor design strategies, which includes suitable number of electrodes (Peng, Ye, Lu, & Yang, 2012; Wang, Marashdeh, Teixeira, & Fan, 2015), length of electrodes (Peng, Mou, Yao, Zhang, & Xiao, 2005), required gap between electrodes (Lucas, Margo, Oussar, & Holé, 2015), etc. Specialty sensors like high pressure and high temperature sensor (Yang , 2010) and 3D printing for sensor fabrication are applied (Kowalska, Banasiak, Romanowski, & Sankowski, 2019). Axially elongated sensor with higher number of electrode layers is investigated for covering a larger region of interest (Rasel, Sines, Marashdeh, & Teixeira, 2019). Efforts have been invested for reducing the dynamic range of measured capacitance with differential sensor design (Cui, Wang, & Yin, 2015) and electrode segment combining (Yang, Peng, & Jia, 2017). Adaptive (reconfigurable) sensor is introduced with electronic scanning capabilities for enhancing imaging resolution of ECT and ECVT (Chowdhury, Marashdeh, & Teixeira, 2020; Marashdeh, Teixeira, & Fan, 2014; Yang & Peng, 2013; Zeeshan, Zuccarelli, Acero, Marashdeh, & Teixeira, 2019). Due to the advantages with imaging resolution and the dynamic range of measured capacitance, the adaptive sensor designs is likely to receive a higher emphasis in future.

1.14 Source of further information

An interested reader may pursue some of these topics for further information related to ECT: inverse problems (Hansen, 2010; Tarantola, 2005), electromagnetic fields (Paris & Hurd, 1979), flow measurement (Baker, 2005), multiphase flow (Brennen, 2005), etc.

Acknowledgment

Support from the U.S. Department of Energy and NASA is gratefully acknowledged.

References

Baker, R. C. (2005). *Flow measurement handbook: Industrial designs, operating principles, performance, and applications*. Cambridge University Press. In press.

Banasiak, R., Wajman, R., Betiuk, J., & Soleimani, M. (2009). A new application area for 3D ECT in non-destructive evaluation of dielectric materials. In *3rd international workshop on process tomography. Tokyo, Japan*.

Brennen, C. E. (2005). *Fundamentals of multiphase flow*. Cambridge University Press.

Cheney, M., Isaacson, D., & Newell, J. C. (March 1999). Electrical impedance tomography. *SIAM Review, 41*(1), 85–101.

Chowdhury, S. M., Marashdeh, Q. M., & Teixeira, F. L. (December 2016). Velocity profiling of multiphase flows using capacitive sensor sensitivity gradient. *IEEE Sensors Journal, 16*(23), 8365–8373.

Chowdhury, S. M., Marashdeh, Q. M., & Teixeira, F. L. (March 2018). Inverse normalization method for cross-sectional imaging and velocimetry of two-phase flows based on electrical capacitance tomography. *IEEE Sensors Letter, 2*(1), 1–4.

Chowdhury, S. M., Marashdeh, Q. M., & Teixeira, F. L. (August 2020). Electronic scanning strategies in adaptive electrical capacitance volume tomography: Tradeoffs and prospects. *IEEE Sensors Journal, 20*(16), 9253–9264.

Cui, Z., Wang, H., Chen, Z., Xu, Y., & Yang, W.-Q. (2011). A high-performance digital system for electrical capacitance tomography. *Measurement Science and Technology, 22*(5), 055503.

Cui, Z., Wang, H., & Yin, W. (September 2015). Electrical capacitance tomography with differential sensor. *IEEE Sensors Journal, 15*(9), 5087–5094.

Fang, W. (June 2006). Reconstruction of permittivity profile from boundary capacitance data. *Applied Mathematics and Computation, 177*(1), 178–188.

Gunes, C., Chowdhury, S. M., Zuccarelli, C. E., Marashdeh, Q. M., & Teixeira, F. L. (February 2019). Displacement-current phase tomography for water-dominated two-phase flow velocimetry. *IEEE Sensors Journal, 19*(4), 1563–1571.

Gunes, C., Marashdeh, Q. M., & Teixeira, F. L. (December 2017). A comparison between electrical capacitance tomography and displacement-current phase tomography. *IEEE Sensors Journal, 17*(24), 8037–8046.

Gunes, C., Ospina-Acero, D., Marashdeh, Q. M., & Teixeira, F. L. (December 2018). Acceleration of electrical capacitance volume tomography imaging by fourier-based sparse representations. *IEEE Sensors Journal, 18*(23), 9649–9659.

Halow, J. S., Fasching, G., & Nicoletti, P. (1990). Preliminary capacitance imaging experiments of a fluidized bed. *AlChE Symposium Series, 86*(276), 41–50.

Halow, J. S., & Nicoletti, P. (March 1992). Observations of fluidized bed coalescence using capacitance imaging. *Powder Technology, 69*(3), 255–277.

Hansen, P. C. (2010). In N. J. Higham (Ed.), *Discrete inverse problems: Insight and algorithms*. Philadelphia, PA: SIAM.

Holland, D. J., Marashdeh, Q. M., Mueller, C. R., Wang, F., Dennis, J. S., Fan, L.-S., et al. (January, 2009). Comparison of ECVT and MR measurements of voidage in a gas-fluidized bed. *Industrial & Engineering Chemistry Research, 48*(1), 172–181.

Huang, S. M., Xie, C. G., Thorn, R., Snowden, D., & Beck, M. S. (February 1992). Design of sensor electronics for electrical capacitance tomography. *IEE Proc. G, 139*(1), 83–88.

Kowalska, A., Banasiak, R., Romanowski, A., & Sankowski, D. (August 2019). 3D-printed multilayer sensor structure for electrical capacitance tomography. *Sensors, 19*(15), 3416.

Lei, J., Liu, Q. B., & Wang, X. Y. (June 2020). Ensemble learning-based computational imaging method for electrical capacitance tomography. *Applied Mathematical Modelling, 82*, 521–545.

Li, Y., & Holland, D. J. (October 2013). Fast and robust 3D electrical capacitance tomography. *Measurement Science and Technology, 24*, 105406.

Lionheart, W. R. (2001). Reconstruction algorithms for permittivity and conductivity imaging. *Proc. 2nd World Congr. Ind. Process Tomography*, 4–11 (Hannover, Germany).

Lucas, J., Margo, C., Oussar, Y., & Holé, S. (2015). Physical limitations on spatial resolution in electrical capacitance tomography. *Measurement Science and Technology, 26*(12), 125105.

Marashdeh, Q. M., & Teixeira, F. L. (March 2004). Sensitivity matrix calculation for fast 3-D electrical capacitance tomography (ECT) of flow systems. *IEEE Transactions on Magnetics, 40*(2), 1204–1207.

Marashdeh, Q. M., Teixeira, F. L., & Fan, L.-S. (April 2014). Adaptive electrical capacitance volume tomography. *IEEE Sensors Journal, 14*(4), 1253–1259.

Marashdeh, Q. M., Wang, F., Fan, L.-S., & Warsito, W. (2007). *Velocity measurement of multiphase flows based on electrical capacitance volume tomography*. Atlanta, GA, USA: 2007 IEEE SENSORS.

Marashdeh, Q. M., Warsito, W., Fan, L.-S., & Teixeira, F. L. (April 2006). Nonlinear forward problem solution for electrical capacitance tomography using feed-forward neural network. *IEEE Sensors Journal, 6*(2), 441–449.

Marashdeh, Q. M., Warsito, W., Fan, L.-S., & Teixeira, F. L. (July 2006). A nonlinear image reconstruction technique for ECT using a combined neural network approach. *Measurement Science and Technology, 17*(8), 2097–2103.

Marashdeh, Q. M., Warsito, W., Fan, L.-S., & Teixeira, F. L. (January 2008). Dual imaging modality of granular flow based on ECT sensors. *Granular Matter, 10*(2), 75–80.

O'Rourke, P. J., & Snider, D. M. (October 2012). Inclusion of collisional return-to-isotropy in the MP-PIC method. *Chemical Engineering Science, 80*, 39–54.

Ospina-Acero, D., Marashdeh, Q. M., & Teixeira, F. L. (May 2020). Relevance vector machine image reconstruction algorithm for electrical capacitance tomography with explicit uncertainty estimates. *IEEE Sensors Journal, 20*(9), 4925–4939.

Paris, D. T., & Hurd, F. K. (1979). Quasistatic fields. In *Basic electromagnetic theory* (p. 325368). New York, NY: McGraw-Hill.

Peng, L., Mou, C., Yao, D., Zhang, B., & Xiao, D. (April 2005). Determination of the optimal axial length of the electrode in an electrical capacitance tomography sensor. *Flow Measurement and Instrumentation, 16*(2), 169–175.

Peng, L., Ye, J., Lu, G., & Yang, W. Q. (May 2012). Evaluation of effect of number of electrodes in ECT sensors on image quality. *IEEE Sensors Journal, 12*(5), 1554–1565.

Quek, S., Mohr, S., Goddard, N., Fielden, P., & York, T. (2010). Miniature electrical tomography for microfluidic systems. In *Proc. 6th world congr. Ind. Process tomography. Beijing, China.*

Rasel, R. K., Marashdeh, Q. M., & Teixeira, F. L. (December 2018). Toward electrical capacitance tomography of water-dominated multiphase vertical flows. *IEEE Sensors Journal, 18*(24), 10041–10048.

Rasel, R. K., Sines, J. N., Marashdeh, Q., & Teixeira, F. L. (October 2019). Cross-plane acquisitions in electrical capacitance volume tomography. *IEEE Sensors Journal, 19*(19), 8767–8774.

Rasel, R. K., Zuccarelli, C. E., Marashdeh, Q. M., Fan, L.-S., & Teixeira, F. L. (December 2017). Toward multiphase flow decomposition based on electrical capacitance tomography sensors. *IEEE Sensors Journal, 17*(24), 8027–8036.

Sattar, M. A., Wrasse, A. D., Morales, R. E., Pipa, D. R., Banasiak, R., Da Silva, M. J., et al. (2020). Multichannel capacitive imaging of gas vortex in swirling two-phase flows using parametric reconstruction. *IEEE Access, 8*, 69557–69565.

Sines, J. N., Straiton, B. J., Zuccarelli, C. E., Marashdeh, Q. M., Teixeira, F. L., Fan, L.-S., et al. (July 2020). Study of gas-water flow inside of a horizontal passive cyclonic gas-liquid phase separator system using displacement-current phase tomography. *Gravitational and Space Res, Jul*(2), 28–43.

Snider, D. M. (July 2007). Three fundamental granular flow experiments and CPFD predictions. *Powder Technology, 176*(1), 36–46.

Sun, S., Cao, Z., Huang, A., Xu, L., & Yang, W. (October 2017). A high-speed digital electrical capacitance tomography system combining digital recursive demodulation and parallel capacitance measurement. *IEEE Sensors Journal, 17*(20), 6690–6698.

Tarantola, A. (2005). *Inverse problem theory and methods for model parameter estimation*. Society for Industrial and Applied Mathematics. In press.

Wang, A., Marashdeh, Q. M., Teixeira, F. L., & Fan, L.-S. (2015). Electrical capacitance volume tomography: A comparison between 12- and 24-channels sensor systems. *Progress Electromagn. Res., 41*, 73–84.

Warsito, W., Marashdeh, Q. M., & Fan, L.-S. (March - April 2007). Electrical capacitance volume tomography. *IEEE Sensors Journal, 7*(3–4), 525–535.

Watzenig, D., & Fox, C. (April 2009). A review of statistical modelling and inference for electrical capacitance tomography. *Measurement Science and Technology, 20*(5), 052002.

Weber, J. M., Layfield, K. J., Van Essendelft, D. T., & Mei, J. S. (December 2013). Fluid bed characterization using electrical capacitance volume tomography (ECVT), compared to CPFD software's Barracuda. *Powder Technology, 250*, 138–146.

Xie, C. G., Huang, S. M., Hoyle, B. S., Thorn, R., Lenn, C., Snowden, D., et al. (February, 1992). Electrical capacitance tomography for flow imaging - system model for development of image-reconstruction algorithms and design of primary sensors. *IEE Proceedings G, 139*(1), 89–98.

Yang, W.-Q. (April 2010). Design of electrical capacitance tomography sensors. *Measurement Science and Technology, 21*(4), 042001.

Yang, W. Q., & Peng, L. H. (January 2003). Image reconstruction algorithms for electrical capacitance tomography. *Measurement Science and Technology, 14*(1), R1–R13.

Yang, Y., & Peng, L. (July 2013). A configurable electrical capacitance tomography system using a combining electrode strategy. *Measurement Science and Technology, 24*(7), 1361–6501.

Yang, Y., Peng, L., & Jia, J. (March 2017). A novel multi-electrode sensing strategy for electrical capacitance tomography with ultra-low dynamic range. *Flow Measurement and Instrumentation, 53*, 67–79.

Yang, W. Q., Stott, A. L., Beck, M. S., & Xie, C. G. (August, 1995). Development of capacitance tomographic imaging-systems for oil pipeline measurements. *Review of Scientific Instruments, 66*(8), 4326–4332.

Yang, W. Q., & York, T. A. (January 1999). New AC-based capacitance tomography system. *IEE Proceedings - Science, Measurement and Technology, 146*(1), 47–53.

Ye, J., Wang, H., & Yang, W.-Q. (January 2015). Image reconstruction for electrical capacitance tomography based on sparse representation. *IEEE Transactions on Instrumentation and Measurement, 64*(1), 89–102.

Zeeshan, Z., Zuccarelli, C. E., Acero, D. O., Marashdeh, Q. M., & Teixeira, F. L. (2019 February). Enhancing resolution of electrical capacitive sensors for multiphase flows by fine-stepped electronic scanning of synthetic electrodes. *IEEE Sensors Journal, 68*(2), 462–473.

Zhang, M., & Soleimani, M. (January 2016). Simultaneous reconstruction of permittivity and conductivity using multi-frequency admittance measurement in electrical capacitance tomography. *Measurement Science and Technology, 27*(2), 025405.

Zheng, J., & Peng, L. (2020, May). A deep learning compensated back projection for image reconstruction of electrical capacitance tomography. *IEEE Sensors Journal, 20*(9), 4879–4890.

Electrical impedance tomography

2

Mi Wang
University of Leeds, School of Chemical and Process Engineering, United Kingdom

2.1 Introduction

A variety of types of electrical tomography have evolved since the late 1980s (Bates, McKinnon, & Seagar, 1980; Dines & Lytle, 1981; Lytle & Dines, 1978) to provide alternative, low-cost solutions for clinical, geophysical, and industrial process applications. The first clinical frontal plane electrical impedance imaging technique was proposed by Swanson in 1976 (Swanson, 1976). Henderson, Webster, and Swanson (1976) and Henderson and Webster (1978) designed an "impedance camera" based on the principle of X-ray transmission imaging. In operation, a large electrode was placed on the chest, which resulted in varying currents being collected through a 100-electrode array on the back (Webster, 1990). However, the spatial resolution from the camera was lower than that in X-ray transmission since the current flow in body tissues is not confined to a direct path between the transmitter and the receiver (Brown, Barber, & Freeston, 1983). Brown et al. proposed a tetrapolar-based measurement strategy for electrical impedance tomography (EIT) in 1983 and refined it in 1985 (Brown & Barber, 1985; Brown et al., 1983). This strategy is well known in EIT and is referred to as the adjacent electrode or neighboring pair's strategy. The first commercial EIT for medical research was reported in 1987 and was referred to as the Sheffield APT (applied potential tomography) system (Brown & Seagar, 1987), and then the first commercial EIT for process application was launched in 1997 and was referred to as the Manchester system (Boone, Barber, & Brown, 1997; Wang, Dickin, & Beck, 1993). It was also called electrical resistance tomography (ERT) in process tomography since the early work only measured the resistance components. A number of EIT systems in various forms were developed over the past 3 decades. Some of them are given in Table 2.1, showing the development of EIT at its early stage in biomedical research and for process engineering in the past (Boone et al., 1997). The goal of EIT is to obtain the impedance distribution in the domain of interest. This chapter will address the basic principles and elementary functions of EIT systems, providing inside knowledge of EIT.

Industrial Tomography. https://doi.org/10.1016/B978-0-12-823015-2.00019-4

Table 2.1 Electrical impedance tomography (EIT) at its early stage for biomedical research and development for process engineering.

Authors	Institution	System name	Applied areas	Design frequency	Frame speed	Excitation
Brown and Seagar (1987)	University of Sheffield	Mark I	Biomedical	50 kHz		Floating current
Brown et al. (1994)	University of Sheffield	Mark III	Biomedical	9.6 kHz–1.2 MHz		Dual currents
Wilson et al. (2001)	University of Sheffield	Mark III.5	Biomedical	2 kHz–1.6 MHz	25 fps	Floating current
Lidgey et al. (1992)	Oxford Brookes University	OXPACT-II	Biomedical	9.6 kHz		Constant voltage
Zhu, McLeod, Denyer, Lidgey, and Lionheart (1994)	Oxford Brookes University	OXBACT-III	Biomedical	10–160 kHz		Multiple currents
Newell et al. (1988), Gisser et al. (1991), Cook et al. (1994)	Rensselaer Polytechnic Institute	ACT-III	Biomedical	30 kHz	480 fps	Multiple currents
Wang et al. (1993)	University of Manchester	MARK I-b	Process engineering	1.2–72.8 kHz	20 fps	Dual currents
Ma et al. (2003), Wang et al. (2005), Schlaberg, Jia, Qiu, Wang, and Li (2008)	University of Leeds	FICA	Process engineering	10–320 kHz	1000 dfps	Dual currents
Jia et al. (2010)	University of Leeds	Voltage-applied ERT	Process engineering	10 kHz	300 dfps	Dual voltages
Cilliers et al. (2001), Wilkinson et al. (2005)	University of Manchester, University of Cape Town		Process engineering	12 μs pulse/64 μs circle	1000 fps	Bipolar DC current pulse

2.2 Fundamentals of measurement

2.2.1 Electrode–electrolyte interface modeling

The electrode used in EIT is a transducer that converts the electronic current in a wire to an ionic current in an electrolyte. The behavior at the electrode–electrolyte interface is a predominantly electrochemical reaction. A metal electrode immersed in an electrolyte is polarized when its potential is different from its open-circuit potential (Weinman & Mahler, 1964). The difference between the open-circuit potential V_e (equilibrium potential) and the actual potential V is the overpotential, η, relating the net current density, j, of the electrochemical reaction to the overpotential of the Butler–Volmer equation as described by Eq. (2.1) and illustrated in Fig. 2.1 (Bockris & Drazic, 1972):

$$j = j_0\{\exp[(1-\beta)F\eta / GT] - \exp[(-\beta F\eta / GT)]\}\left(\mathrm{A}/\mathrm{cm}^2\right) \tag{2.1}$$

where j_0 is the exchange current density and a function of the electrode–electrolyte system and, to a limited extent, of the surface structure of the metal. As an example, for a platinum electrode immersed in 75 Ω cm resistivity saline, j_0 is in the order of 5×10^{-5} A/cm^2 (Pollak, 1974a,b), F is the Faraday constant (96,500 C/mol), G is the gas constant (8.317 J/K/mol), T is the temperature in Kelvin (298 K at room temperature), and β is a factor close to 0.5 related to the electrical energy in the passage of the ion across the electrode–electrolyte interface (Bockris & Drazic, 1972).

For small overpotentials (-5 mV $< \eta <$ 5 mV), the relation between overpotential and current density is linear as given by Eq. (2.2) and illustrated in Fig. 2.1. The limits of linearity in Eq. (2.2) are typically reached at current density levels of up to 1 mA/cm^2 when a direct current is applied (Schwan & Ferris, 1968; Weinman & Mahler,

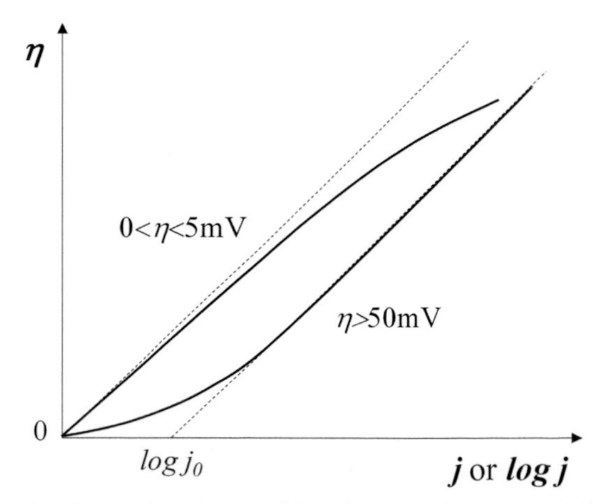

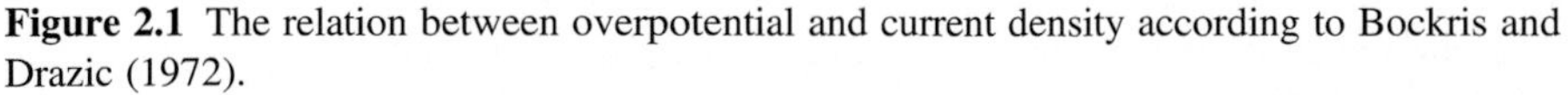
Figure 2.1 The relation between overpotential and current density according to Bockris and Drazic (1972).

1964). When overpotentials are greater than 50 mV or less than −50 mV, the relation to current density is described by Eq. (2.3) as shown in Fig. 2.1:

$$\eta = \frac{GT}{j_0 F} j[\mathrm{V}] \tag{2.2}$$

$$\eta = a + b \ln j(\mathrm{V}) \tag{2.3}$$

where $a = \frac{-GTj_0}{(1-\beta)F}$; $b = \frac{GT}{(1-\beta)F}$.

By analogy with Ohm's law ($V = IR$), the gradient of the $\eta - j$ line in Fig. 2.1 can be interpreted as the equivalent resistance per unit area, R_C, of the charge transfer process on the electrode as given in Eq. (2.4) and which is called the charge transfer resistance for small overpotentials:

$$R_C = \frac{\partial \eta}{\partial j} = \frac{GT}{j_0 F} = \frac{E_T}{j_0} = \frac{0.026}{j_0} \left(\Omega \mathrm{cm}^2\right) \tag{2.4}$$

where E_T is called the temperature voltage. Its value at room temperature is approximately 0.026 V.

Geddes, Costa, and Wise (1971) reported that when current density was increased up to 25 mA/cm^2, a definite nonlinear behavior was observed during his experiments at frequencies from 20 Hz to 10 kHz, which resulted in considerable distortion of the applied sine wave.

Both theory and experiment confirm that the amount of charge stored in the region where the metal electrode is in contact with the electrolyte varies as a function of the potential difference across it (Bockris & Drazic, 1972; Crow, 1974). A simple model of the double layer was suggested by Helmholtz and Perrin at the end of the 19th century (Bockris & Conway, 1969; Bockris & Reddy, 1970). This represents the double layer in terms of the simple plates of a capacitor's point of view. The metal electrode, with its excess charge, forms one side of the double layer, and the electrolyte, with its excess countercharge, forms the other side of the double layer. Of course, there is no real plate on the electrolyte side, but it is imagined that the excess charges in the electrolyte make up a layer of charge just as if they were attached to a plate there (Bockris & Drazic, 1972). The Helmholtz double-layer capacitance, C_H, is given by the following:

$$C_H = \frac{dq_m}{dV} = \frac{\varepsilon_r \varepsilon_0}{\delta^H} \left[\mu \mathrm{Fcm}^{-2}\right] \tag{2.5}$$

In Eq. (2.5), q_m is the excess charge on the metal electrode, V is the potential across the double layer, ε_0 is the permittivity of a vacuum (8.85×10^{-8} μF/cm), ε_r is the relative permittivity of the material between the plates, and δ^H is the effective distance between the plates (e.g., δ^H in water is around 2.8 E $\times 10^{-8}$ cm) (Bockris & Drazic, 1972). In reality, however, the double-layer capacitance is not constant with potential, and, for most electrode systems of practical importance, the capacitance in the absence of larger externally applied potential differences can be estimated as about 10 μF/cm^2 (Pollak, 1974a,b) or 15.9 μF/cm^2 (Bockris & Drazic, 1972) irrespective of the ratio of the ions (Bockris & Reddy, 1970).

The Gouy–Chapman diffuse model (Bockris & Drazic, 1972) may present a better description of the double-layer structure, in which it was assumed that there would be no "sticking" of ions to the electrode and no contact adsorption, and suggested a diffuse charge model as an addition to the Helmholtz theory. The diffuse capacitance is given in Eq. (2.6) by Crow (1974):

$$C_G = \varepsilon_r \varepsilon_0 \kappa = \frac{\varepsilon_r \varepsilon_0}{\delta^G} \left(\mu F / cm^2\right) \tag{2.6}$$

where $1/\kappa$ is the effective radius of the ion atmosphere, here it is identified with δ^G the diffuse length of double layer.

The total capacitance of the double layer, C, is made up of C_H from the inner (adsorption) layer and C_G from the diffuse layer. It is assumed that these capacitances are connected in series:

$$\frac{1}{C} = \frac{1}{C_H} + \frac{1}{C_G} \text{ or } C = \frac{C_H C_G}{C_H + C_G} \tag{2.7}$$

Consequently, if the electrolyte solution is very dilute, $C_G \ll C_H$, then $C \approx C_G$. The double layer is now essentially all diffuse, which is the Gouy–Chapman diffuse model. On the other hand, when the solution is very concentrated, $C_G \gg C_H$, then $C \approx C_H$, which satisfies Helmholtz's earliest model of the double layer.

An equivalent circuit of the electrode–electrolyte given by Pollak (1974a,b) is shown in Fig. 2.2. The total impedance is made up of the interface, the diffusion, and the inner resistance, which are connected in series.

The response of the diffusion layer to an AC current injection was given as follows (Cobbold, 1974):

$$|Z_D| \approx \frac{E_T}{z^2 c_0 F \sqrt{2\pi D}} \frac{1}{\sqrt{f}} \left(\Omega cm^2\right) \tag{2.8}$$

where E_T is the temperature voltage (0.026 V), F the Faraday constant (96,500 C/mol), D the diffusion coefficient (cm^2/s), z the charge transfer number, and c_0 the concentration of the electrolyte under equilibrium conditions (mol/cm^3).

For typical electrolyte in process applications, e.g., NaCl solution, the diffusion coefficient is in the order of 1.5×10^{-5} cm^2/s and a charge transfer number in the order of 0.4 as concentration is below 0.2 mol/L (Conway, 1952). With these parameters, the equation yields

$$|Z_D| \approx \frac{7 \times 10^{-5}}{c_0 \sqrt{f}} \left(\Omega cm^2\right) \tag{2.9}$$

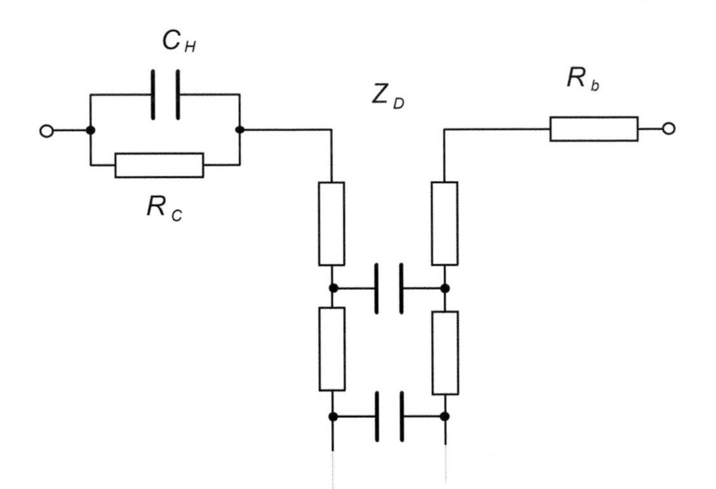

Figure 2.2 Equivalent circuit of the electrode–electrolyte impedance (R_C is the charge transfer resistance, C_H is the Helmholtz double layer capacitance, Z_D is the diffusion layer impedance, and R_b is the bulk resistance of the electrolyte (according to Pollak, 1974a,b)).

The expression given by Eq. (2.9) can be employed down to frequencies where the diffusion length δ^{G} becomes comparable with the electrode diameter d (Pollak, 1974a,b):

$$\delta^{G} = \frac{\sqrt{D}}{2\pi f} < d(\text{cm}) \tag{2.10}$$

$$f > \frac{D}{2\pi d^2} \approx \frac{8 \times 10^{-6}}{d^2}(\text{Hz})(d \text{ in units of cm}) \tag{2.11}$$

For the instance of a concentration 1×10^{-5} mol/cm^3 at a frequency 100 Hz, $|Z_D| = 0.7\ \Omega\ \text{cm}^2$ obtained from Eq. (2.9).

Combining all of the above analyses, the total impedance, Z, of the electrode–electrolyte, at very low frequencies, is approximately equal to the diffusion-layer impedance Z_D (Cobbold, 1974; Pollak, 1974a,b):

$$Z \approx Z_D \propto \frac{1}{\sqrt{f}}\left(\Omega\text{cm}^2\right) \tag{2.12}$$

and at very high frequency:

$$Z \approx R_b + \frac{1}{j\omega C_H}\left(\Omega\text{cm}^2\right) \tag{2.13}$$

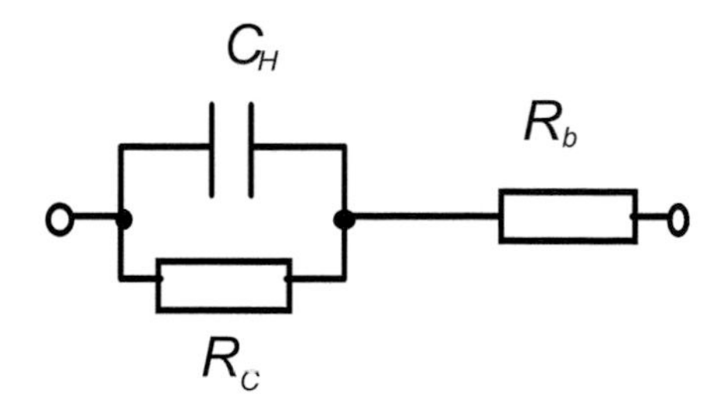

Figure 2.3 Equivalent circuit of the electrode–electrolyte impedance ignoring the diffusion effect.

In the majority of applications, the effect of the diffusion layer can be ignored in order to simplify the solution (Cobbold, 1974; Pollak, 1974a,b). The simplified equivalent circuit is given as in Fig. 2.3.

Only electrode modeling with respect to electrochemistry was discussed in the section. Electrode modeling with respect to electrode size and its relation to the electrical field and boundary voltage as well as to common voltage were also discussed elsewhere (Wang, 2005a,b).

2.2.2 Lead theorem

The lead theorem was derived from the divergence theorem (Eq. 2.14) by Geselowitz (1971) and Lehr (1972) for the impedance plethysmography:

$$\int_{\Omega} \nabla \cdot \boldsymbol{A} d\Omega = \oint_{S} \boldsymbol{A} \cdot \boldsymbol{dS} \tag{2.14}$$

where Ω is a region bounded by a closed surface S. $\boldsymbol{A}$ is a vector function of the position.

Based on the divergence theorem, the mutual impedance change ΔZ due to a change of internal conductivity $\Delta\sigma$ for a four-electrode system was derived by Geselowits and Lehr and later linearized by Murai and Kagawa (1985), given as Eqs. (2.15) and (2.16), respectively. Eq. (2.16) ignores the high-order term and provides a linear relationship with an assumption of $\Delta\sigma << \sigma$:

$$\Delta Z = \frac{\Delta\phi_{\mathrm{AB}}}{I_\phi} = \frac{\Delta\psi_{\mathrm{CD}}}{I_\psi} = -\int \Delta\sigma \frac{\nabla\phi^{\Xi}}{I_\phi} \cdot \frac{\nabla\psi}{I_\psi} d\Omega \tag{2.15}$$

where ψ and ϕ are potential distributions in response to the presence of currents I_ψ and I_ϕ at two ports (A–B and C–D), respectively, as illustrated in Fig. 2.4, and ϕ^{Ξ} is the potential change caused by the conductivity after change $\sigma^{\Xi} = \sigma + \Delta\sigma$.

$$\Delta Z = -\int_{\Omega} \Delta\sigma \frac{\nabla\phi}{I_\phi} \cdot \frac{\nabla\psi}{I_\psi} d\Omega + 0\left((\Delta\sigma)^2\right) \approx -\int_{\Omega} \Delta\sigma \frac{\nabla\phi}{I_\phi} \cdot \frac{\nabla\psi}{I_\psi} d\Omega \; (\Delta\sigma \ll \sigma) \tag{2.16}$$

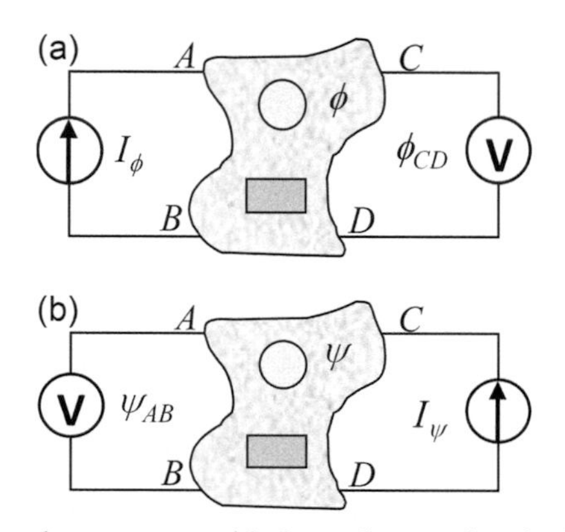

Figure 2.4 Two-port impedance system with the exchange of excitation and measurement positions. (a) original connection, (b) with the exchange.

The mutual impedance Z of a 4-electrode system with a known conductivity distribution can also be described with the similar approach (Wang, Dickin, & Williams, 1995a,b):

$$Z = \frac{\phi_{\mathrm{AB}}}{I_\phi} = \frac{\psi_{\mathrm{CD}}}{I_\psi} = \int \sigma \frac{\nabla\phi}{I_\phi} \cdot \frac{\nabla\psi}{I_\psi} d\Omega \tag{2.17}$$

2.2.3 Reciprocity theorem

According to the lead theorem, the mutual impedance Z for a four-electrode system is the same as expressed by Eq. (2.18):

$$Z = \frac{\phi_{AB}}{I_\phi} = \frac{\psi_{CD}}{I_\psi} \tag{2.18}$$

Based on the reciprocity theorem, for a two-port/four-terminal static impedance system as given in Fig. 2.4, the voltage measurements, V_{CD} and V_{AB}, are the same if the values of the currents remain the same in both Fig. 2.4a and b, which can be expressed by Eq. (2.19) and leads to identify the independent data sets from a measurement system with multiple electrodes:

$$\text{if } I_{AB} = I_{CD} \quad \text{then} \quad V_{CD} \equiv V_{AB} \tag{2.19}$$

2.2.4 4-Electrode method

As illustrated in Fig. 2.5a based on the concepts described in the previous section, the actual voltage presented between two electrodes is function of R_C, C_H, η, and R_b, therefore, function of the charge transfer resistance, double layer capacitance, as well as the ionic concentration of electrolyte, surface area and condition of electrode, and certainly current density over the interface. Therefore, the bulk resistance would

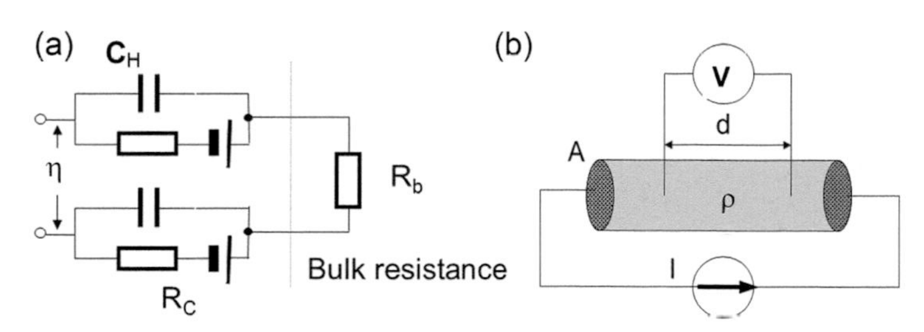

Figure 2.5 (a) The electrode–electrolyte interface. (b) 4-electrode method.

not be simply derived from the voltage measurement. The conventional and effective way to avoid error caused by the interface in the measurement of bulk resistance is to use a 4-electrode method as shown in Fig. 2.5b. In principle, the voltage measurement, *V*, should not be a function of the two interfaces' electrochemical characteristics since the current, *I*, remains constant flowing through any cross-section of the cylindrical object.

2.3 Principle of electrical impedance tomography sensing

2.3.1 Sensing strategies

In EIT, the boundary conditions required to determine the impedance distribution in a domain of interest comprise the injected currents via electrodes and subsequently measured voltages, also via boundary electrodes. A number of excitation and measurement methods or sensing strategies have been reported since the late 1980s. In 1976, Henderson and Webster applied a constant 100 kHz voltage to a large electrode on a patient's chest, which resulted in varying currents collected through a 100-electrode array on the back. All electrodes, except for the measuring electrode, were maintained at 0 V, which directly resulted in an impedance map. This was an early attempt to produce an impedance image of the human chest, however, not the tomography in theory (Webster, 1990). Lytle and Dines (1978) applied stepped voltages to all electrodes that would produce a uniform current distribution in a homogeneous media. They measured the resulting currents at all electrodes and repeated the procedure from several angular directions, which was the first attempt to utilize the tomography technique for electrical impedance imaging. The adjacent electrode pair strategy for EIT was proposed in 1983 (Brown, Leathard, Lu, Wang, & Hampshire, 1983) and refined in 1985 (Brown & Barber, 1985; Brown & Seagar, 1985). With the adjacent pair strategy, current was applied between adjacent electrode pairs and voltages measured between every other pair of adjacent electrodes (Fig. 2.6a). The adjacent electrode pair strategy is now widely used due to the high signal-to-common voltage ratio, high number of independent measurements, and simplicity of hardware and software implementation. Hua, Webster, and Tompkins (1987) used the opposite method of data

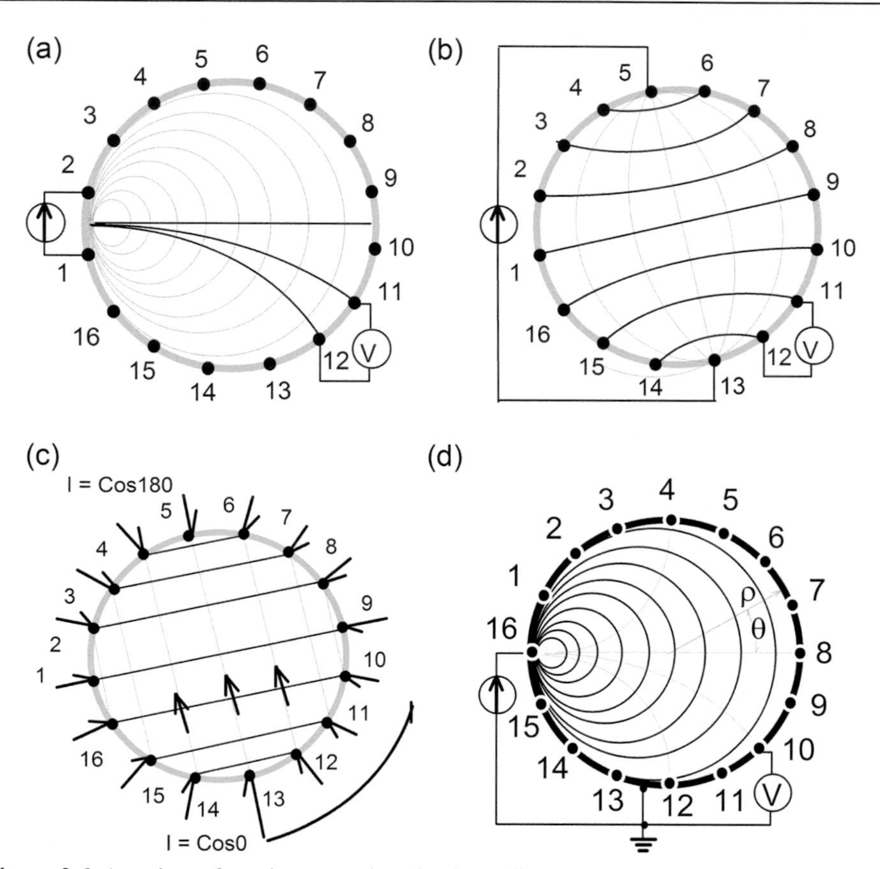

Figure 2.6 A variety of sensing strategies (the dotted line denotes current streamlines and the solid line denotes equal potential lines) (Wang, 1994). (a) adjacent electrode pairs strategy. (b) opposite electrode pairs strategy. (c) adaptive currents strategy and (d) metal wall strategy.

collection as shown in Fig. 2.6b. The main difference between the adjacent electrode pair strategy and the opposite electrode pair strategy is that, in the latter, currents are injected between diametrically opposite electrode pairs rather than between adjacent electrode pairs. They expressed the reason for this simple modification was to enable a relatively higher current density to flow through the center of the vessel and, therefore, expected the imaging sensitivity at the center could be improved. Gisser (1987) proposed the adaptive current strategy where almost any desired current distribution can be obtained by injecting current of appropriate magnitude through all the electrodes simultaneously and measuring voltages between these electrodes and a single grounded electrode (see Fig. 2.6c). It was expressed that this method yields the optimal distinguishability and is the most versatile method of data collection (Webster, 1990). Applying a conventional approach to a conducting walled vessel would result in the excitation currents being of limited use due to both the shunting effect of the conducting wall and the distortion of the electric field distribution. The key step to developing a sensing strategy for a metal-walled vessel was based on the concept of utilizing the shunting effect of the metal wall for the measurement (Wang, Dickin, & Williams, 1994, 1995a,b). In the strategy, the current is injected from one electrode and withdrawn from the whole metal wall. An example of the sensor utilizing this concept is given in Fig. 2.6d.

Applying the reciprocity theorem as given by Eq. (2.19), the number of independent measurements from an EIT sensor with 16 electrodes in the use of the adjacent electrode pair method is given by Eq. (2.20) or (2.21), which are 120 or 104, respectively, depending on those measurements from electrodes involved with or without current excitation. Ignoring the measurements from current drive electrodes, the number of independent measurements is 104, 96, and 120, with respect to Fig. 2.6a, b and d:

$$N_{\text{meas}} = \frac{N_{\text{electrode}}(N_{\text{electrode}} - 1)}{2} \tag{2.20}$$

$$N_{\text{meas}} = \frac{N_{\text{electrode}}(N_{\text{electrode}} - 3)}{2} \tag{2.21}$$

where N_{meas} refers to the number of independent measurements and $N_{\text{electrode}}$ denotes the number of electrodes in the sensor. Eq. (2.21) is employed in most systems to avoid measurements affected by the electrode−electrolyte interface.

A typical measurement data set captured from a miscible liquid mixing process with the P2 + EIT system (Industrial Tomography Systems) is given in Fig. 2.7, which well illustrates the format of the 104 independent measurements, where the horizontal index indicates the pair of electrodes for measurement and the vertical index indicates the pair of electrodes for current injection. A graphic plot of the data set is shown in Fig. 2.8, which reveals the high dynamic range feature of these measurements.

2.3.2 Sensitivity

The change of mutual impedance at the boundary measurement due to the change of conductivity at a pixel inside sensing domain is generally defined as the sensitivity of the conductivity measurement. Supposing that the conductivity distribution is composed of ω small uniform "patches" or pixels, then Eqs. (2.15) and (2.17) can be expressed as Eqs. (2.22) and (2.23), where the sensitivity coefficient s for each discrete pixel is given by Eq. (2.24) (Barber, 1990; Breckon & Pidcock, 1987; Murai & Kagawa, 1985).

$$Z = -\sum_{k=1}^{w} \sigma_k s_{\phi,\psi,k} \tag{2.22}$$

$$\Delta Z = \sum_{k=1}^{w} \Delta\sigma_k s_{\phi,\psi,k} \tag{2.23}$$

Current Injection Pairs	Voltage Measurement Pairs 02 - 03	03 - 04	04 - 05	05 - 06	06 - 07	07 - 08	08 - 09	09 - 10	10 - 11	11 - 12	12 - 13	13 - 14	14 - 15	15 - 16	Measurement Set
16 - 01	4.741e+002	1.606e+002	8.008e+001	4.688e+001	3.359e+001	2.769e+001	2.573e+001	2.822e+001	3.418e+001	4.761e+001	7.617e+001	1.543e+002	4.619e+002		1
01 - 02		4.985e+002	1.689e+002	7.813e+001	4.741e+001	3.438e+001	2.842e+001	2.739e+001	2.920e+001	3.589e+001	4.863e+001	8.057e+001	1.650e+002	5.063e+002	2
02 - 03			4.834e+002	1.538e+002	7.617e+001	4.629e+001	3.340e+001	2.866e+001	2.656e+001	2.837e+001	3.438e+001	4.805e+001	7.910e+001	1.650e+002	3
03 - 04				4.678e+002	1.606e+002	7.764e+001	4.761e+001	3.545e+001	2.896e+001	2.725e+001	2.876e+001	3.501e+001	4.878e+001	8.154e+001	4
04 - 05					4.912e+002	1.655e+002	8.057e+001	5.112e+001	3.564e+001	3.022e+001	2.798e+001	2.949e+001	3.594e+001	5.137e+001	5
05 - 06						4.839e+002	1.621e+002	8.301e+001	4.902e+001	3.545e+001	2.837e+001	2.729e+001	2.886e+001	3.516e+001	6
06 - 07							4.907e+002	1.719e+002	8.154e+001	5.000e+001	3.467e+001	2.974e+001	2.744e+001	2.944e+001	7
07 - 08								5.044e+002	1.626e+002	8.008e+001	4.766e+001	3.477e+001	2.822e+001	2.671e+001	8
08 - 09									4.814e+002	1.621e+002	7.813e+001	4.824e+001	3.462e+001	2.866e+001	9
09 - 10										4.956e+002	1.621e+002	8.154e+001	5.137e+001	3.755e+001	10
10 - 11											4.814e+002	1.655e+002	8.105e+001	5.112e+001	11
11 - 12												4.985e+002	1.646e+002	8.252e+001	12
12 - 13													4.912e+002	1.660e+002	13
13 - 14														4.902e+002	14

Figure 2.7 Format of 104 independent measurements from 16 electrodes sensor with adjacent electrode pair sensing strategy.

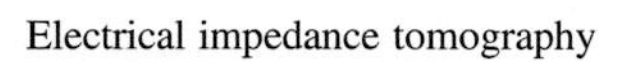

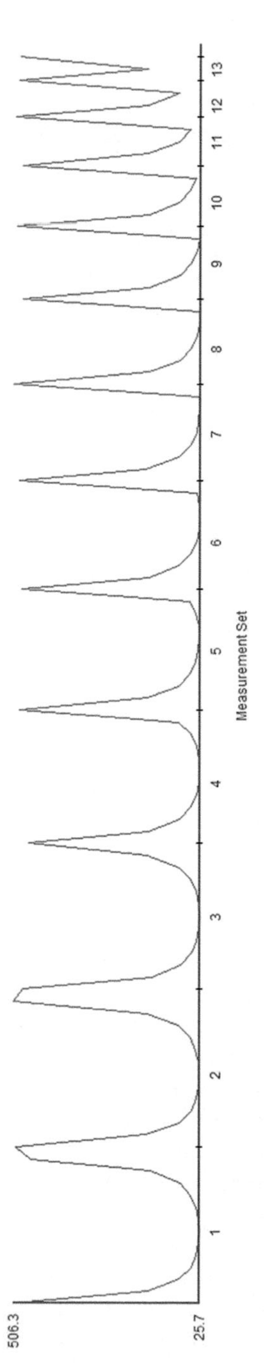

Figure 2.8 A typical measurement profile from a 16-electrodes sensor with adjacent electrode pair sensing strategy.

$$s_{\phi,\psi,k}(\sigma_k) = -\int_{\Omega_k} \frac{\nabla\phi}{I_\phi} \cdot \frac{\nabla\psi}{I_\psi} d\Omega_k \tag{2.24}$$

where Ω_k stands for a discrete 2-dimensional area at location k; σ_k *and* $\Delta\sigma_k$ denote the conductivity and the change of conductivity Ω_k, respectively.

2.3.3 Graphic estimation

A simple graphic method for estimating the sensitivity distribution was reported (Wang, Wang, & Karki, 2016). Eq. (2.24) indicates that the sensitivity at a defined location is a vector scalar production of two electrical fields generated by current excitations on the source and measurement electrode pairs, respectively. In other words, the sensitivity will be maximized at a location where two electric fields are mostly in parallel. With the orthogonal principle between the electrical potential and current density (electrical field intensity) of two electric dipoles, Fig. 2.9 illustrates the sensing "bands" produced by two pairs of electrodes positioned at a nonconductive plate and at an opposite position, respectively. The graphic method (Fig. 2.9a and b) is a simple way to initially estimate the sensitive region from a setting up of electrode allocation for process vessels having different geometry, where the sensing "bands" are highlighted in green (gray in print version). However, no sharp boundary of bands exists in an actual sensitivity distribution due to the use of low frequency electromagnetic excitation. In practice, the sensitivity coefficients are normally computed with either analytical or numerical methods. Fig. 2.9c and d demonstrate the sensitivity distributions simulated with finite element modeling method (FEM) for a 16-electrodes sensing array, where Fig. 2.9c is generated by excitation and measurement electrode pairs 3–4 and 7–6 and Fig. 2.9d is from electrode pair 4–5 and 12–13. For a good sensing strategy, it is necessary the coverage scanning and cross-band scanning can be made over the domain of interests.

2.3.4 Typical sensor

A typical EIT sensor consists of a set of equally spaced electrodes mounted around the periphery of a circular process vessel, in contact with the process fluids inside the vessel. In principle, electrodes could be spread out to any distance and the process vessel could be in any shape depending on the processing domain and imaging sensitivity of interests. The sequence and combination of the measuring electrodes depend on the kind of measurement protocol used, for example, the adjacent electrode pair strategy (Brown & Barber, 1985; Seagar, Barber, & Brown, 1987). In the adjacent electrode pair strategy, the mutual impedance of the measuring electrode pair is obtained by injecting current to a pair of electrodes and taking voltage measurement from other pairs of electrodes. To adapt the specific conditions in process application, the geometric locations of electrodes and the strategy of sensing should be carefully selected. Fig. 2.10 gives an overview of specific sensors for various applications (Wang, 2005a,b). Fig. 2.10a shows an eight-plane sensing array for visualization of

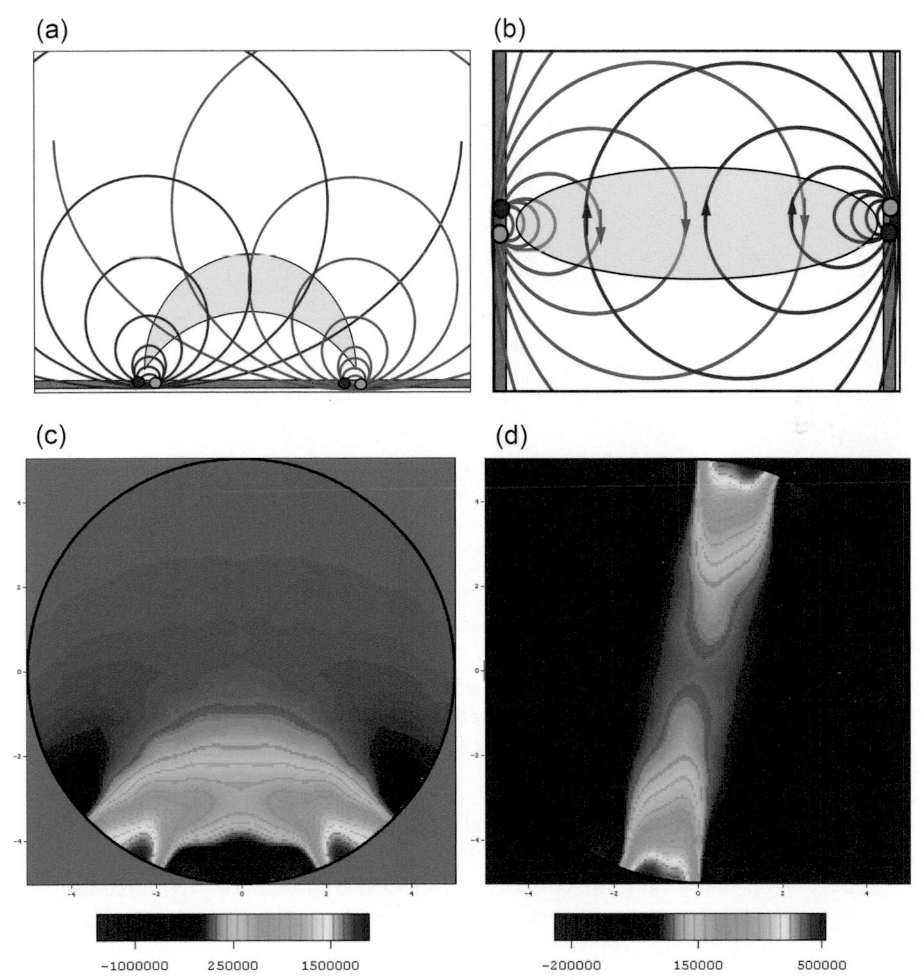

Figure 2.9 Estimation of EIT sensing bands with the graphic analogue method and numerical simulation method. (a) and (b) are from the graphic method where the red (dark gray in print version) and blue (gray in print version) spots denote the current flowing in and out electrodes by applying a current; the red (gray in print version) and blue (dark gray in print version) curves are the equal-current density streams which illustrates the electric field distribution generated by relevant electrode dipoles. (c) and (d) are simulated with FEM for a 2D disk shape domain, where the fields are existed by a current on electrode pair 3–4 and 7–6, 4–5, and 12–13, respectively (Wang et al., 2016).

a large-scale mixing process (Mann et al., 1996a; Wang, Mann, & Dickin, 1996). A total of 128 stainless steel electrodes of 4×10 cm in size were installed on the nonconductive inner wall of a 3^{-m3} mixing tank with a diameter of 1.5 m. Fig. 2.10b shows a sensor for metal-walled containers or pipeline (Wang, Dickin, & Williams, 1994, 1995a,b), which has been proved to be a very fruitful approach

Figure 2.10 EIT sensors in various applications: (a) large-scale mixing (ϕ1500 mm), (b) metal walled vessels and pipeline (ϕ155 mm), (c) conductive ring sensor (diam.ϕ50 mm), (d) miniature slurry pipeline (ϕ6 mm), (e) linear probe for mixing process (provided with courtesy of ITS), (f) solid-water flow pipeline (ϕ50 mm).

(Aw, Rahim, Rahiman, Yunus, & Goh, 2014; Yuen, Mann, York, & Grieveet, 2001). A "single-electrode" sensor such as a conductive ceramic ring shown in Fig. 2.10c was invented to replace the multiplicity of discrete electrodes used in conventional EIT sensors for measurement of stratified flows in horizontal pipelines (Wang & Yin, 2002; Wang, Yin, & Holliday, 2002). The significance of this sensor development is that for the first time a simple electrode can sense complex flows including multiphase mixtures containing stratified gas−water (Wu, Li, Wang, & Williams, 2005) or oil−water flows (Li, Wang, Wu, Ma, & Williams, 2005). Sensor designs suited to the various specific applications are demonstrated in Fig. 2.10d−f, including the miniature sensor (Fig. 2.10d) for the measurement of flow dynamics in an upward slurry flow and the linear electrode array (Fig. 2.10e) for the visualization of mixing process in a pharmaceutical process based on a linear sensor developed for the measurement of subseabed porosity (Qiu, Dickin, James, & Wang, 1993). A similar sensor layout has also been used for the measurement of sediment and sedimentation in a storage tank (Cullivan et al., 2005). Fig. 2.10f shows a dual-plane sensor for measurement of both flow concentration and velocity distributions (Faraj & Wang, 2012; Wang, Jones, & Williams, 2003). Other specific sensors were also reported, e.g., the cross-plane sensor combined with both perpendicular and axial sensing planes for characterization and control of bubble columns (Vijayan, Schlaberg, & Wang, 2007), the U-shaped sensor for the measurement of the transit dynamics in an open channel flow (Schlaberg et al.,

2006), and the parallel-bar linear sensor developed for the visualization of fluid passage in a heap leaching process (Hurry, Wang, Williams, Cross, & Esdaile, 2004). The variety of EIT sensor configurations has demonstrated its excellent adaptability and applicability for process application (Kaipio, Seppänen, & Somersalo, 2005; Ruzinsky & Bennington, 2007; Schlaberg, Wang, Qiu, & Shirhatti, 2005).

The most important issues in EIT sensor design are the materials and the size of electrodes. Other important factors in the selection of materials for electrodes in an EIT sensor are the electrical conductivity, effect of electrochemical polarization, capability of anticorrosion and erosion, and hygiene rate and capabilities for fabrication, depending on the specific requirements in applications. The effect of the overpotential presented at the electrode—electrolyte interface due to different materials is less important in an EIT sensor since alternating current (AC) coupling circuits are used in most process EIT systems. Stainless steel is the most available material for electrode fabrication, but other metallic materials, such as gold, silver, platinum, and silver palladium, can also be used (Dickin & Wang, 1996). In a specific case, conductive ceramic, rubber, or plastic materials can be considered. Electrodes in an EIT sensor have two functions, injecting current and measuring voltage. The size of electrode is more important in the function of current injection than it is in measurement, mainly as the common voltage created in measurement is inversely proportional to the size of electrodes used for current injection. Eq. (2.25) provides a simple relationship between the bulk resistance and the size of electrode, which can be used to estimate the common voltage drop on the electrode simply following Ohm's law. Supposing the surface of a metal electrode in contact with the conductive medium is hemispherically shaped and the radius of the process vessel is far larger than that of the hemisphere, the bulk resistance can be estimated as follows (Weinman & Mahler, 1964):

$$R_b = \frac{1}{2\pi\sigma a p}(\Omega) \tag{2.25}$$

where a denotes the radius of a hemispherically shaped electrode in contact with a conductive medium, and the bulk resistance, R_b, is bulk resistance derived based on Ohm's law. Eq. (2.25) can be extended to encompass other shapes of electrode by employing a shape and surface texture multiplication factor p ($p = 2$ for a metal sphere, 1 for a hemisphere, and 0.6 for a flat-cut disk-shaped electrode) (Weinman & Mahler, 1964). It should be particularly noticed that the change ratio of the bulk resistance is proportional to the change ratio of the dimension (not to the area), and, therefore, a current change in proportion to the dimension will produce the same potential at the electrode after change (not to the current density).

Due to the limited common voltage rejection ratio of sensor electronic circuit (60—80 dB), a large common voltage may greatly affect the measurement sensitivity. Therefore, it is expected that the size of the current injection electrode should be sufficiently large to reduce the voltage drop between the electrode and electrolyte. In contrast, the current absorbed by modern sensor electronics via a measurement electrode is very tiny and thus the voltage drop between the measurement electrode and

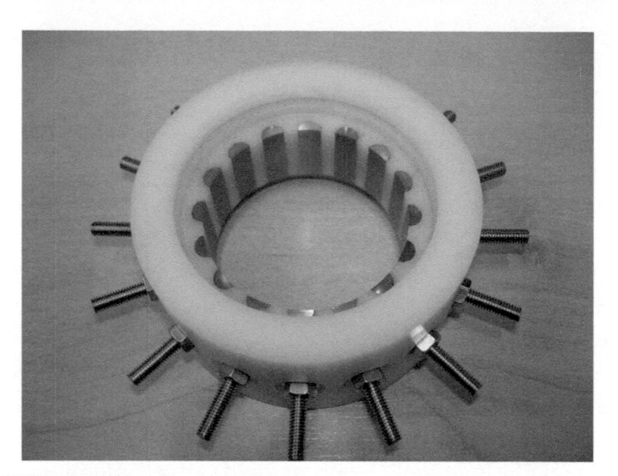

Figure 2.11 An EIT sensor for flow measurement.

electrolyte can be ignored. Therefore, measurement electrodes can be made as small as a needle to enhance their spatial sensitivity. Ideally, the electrodes in an EIT sensor could be in the form of either "compounded"—a measurement electrode with insulation in the center of a current injection electrode—or "interleaved"—both electrodes in an interleave arrangement (Hua, Woo, Webster, & Tompkins, 1993; Paulson, Breckon, & Pidcock, 1992). Dickin and Wang (1996) summarized the variety of electrode designs and commonly acceptable electrode configurations with the same electrode for both functions in most EIT sensors in process engineering applications. As an example, a well-designed and manufactured EIT sensor for pipeline flow measurement is given in Fig. 2.11.

2.4 Data acquisition

2.4.1 Signal sources

Excitation signals in alternative waveforms (AC) have great advantages in reducing the effect of electrode polarization, isolating the direct current (DC) offset potential (e.g., the overpotential), which are used in most current EIT systems. Signals in sinusoidal waveform provide an access to full electrical impedance demodulation and, further, a capability to carry out spectroscopic measurement.

The bidirectional DC pulse excitation was initially used in geophysical tomography (Ronald et al., 1993), where a high voltage (about 100 V) and slow measurement speed (in minutes per frame) are normally used. Recently, a number of bidirectional pulse excitation–based systems for both medical and process applications were developed (Cilliers, Xie, Neethling, Randall, & Wilkinson, 2001) due to the advantage of its simple electronic construction. A recent report indicates it has achieved 1000 frames per second of data acquisition rate (Wilkinson et al., 2005). The system has been demonstrated to work well with a pair of narrow (12 μs) bidirectional current pulses when the medium is assumed to be purely resistive. However, it may not be able to measure, in its current form, both real and imaginary components. With regard to EIT instrumentation simplification, a bipolar DC pulse method would greatly simplify instrumentation as compared to sinusoidal waveform techniques if the targeted fluids of interest can be considered to be a resistive medium, where phase angles do not have to be considered (van Weereld, Collie, & Player, 2001; Wilkinson et al., 2005).

Monolithic function generators were initially used to generate a sinusoidal waveform but were soon replaced by digital forms of "staircase" sinusoidal function in EPROM-based look-up tables, which also provide synchronous sampling signals for demodulation. A quartz-crystal-driver counter is used to clock a stored staircase waveform out of the table repetitively, then convert it to an analogue signal and subsequently filter to remove unwanted harmonics (Brown & Seagar, 1985; Dickin & Wang, 1996; Wang, 1994). More capable monolithic function generators based on direct digital synthesis of a sinusoidal waveform were used (Ma & Wang, 2004) when specific waveform generators were widely commercially available (Analogue Device, 1995).

There is a long-standing use of current sources as the excitation source in EIT (Brown & Seagar, 1985; Dickin & Wang, 1996). The simplest form of current source is shown in Fig. 2.12. The single operational amplifier (op-amp) is arranged in an inverting configuration with the load in the feedback path. Since the gain of an ideal op-amp is considered to be infinite, the inverting input is at virtual earth potential, and, because very little current enters this input, the load current is simply the input voltage divided by the input resistor (Etuke, 1994). The major drawback of this circuit is that it cannot be used in applications with an earthed load or for systems with more than one current source since the output current may not be the same as the input current. In addition, a single current source will generate a high common voltage. Consequently, various designs for float current sources or balanced bipolar current sources were developed to overcome this inherent problem and also to maintain a constant output amplitude over a wide range of load impedances encountered in different process

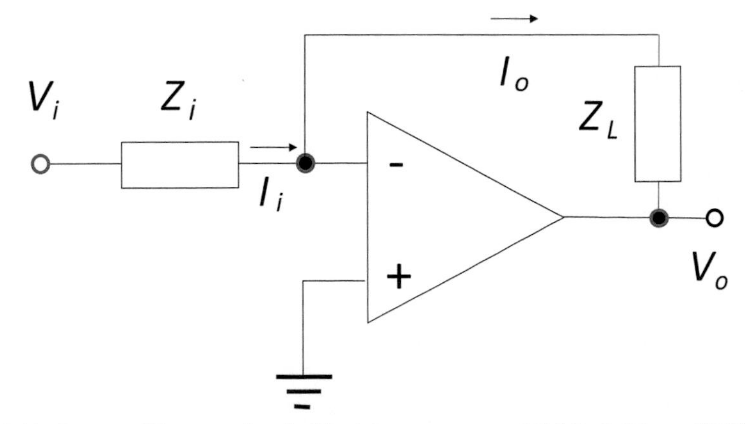

Figure 2.12 One amplifier negative feedback current source (Dickin & Wang, 1996).

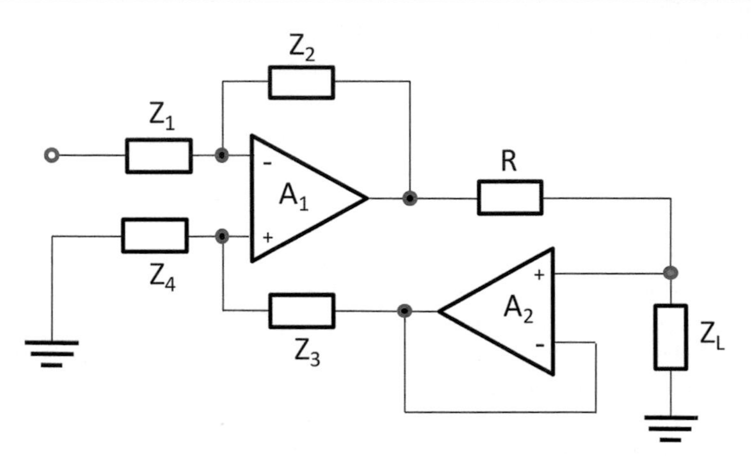

Figure 2.13 The dual operational amplifier positive-feedback VCCS (Dickin & Wang, 1996).

applications. The dual op-amp design developed and shown in Fig. 2.13 (Dickin & Wang, 1996; Wang, Dickin, & Beck, 1993) was based on a positive feedback architecture of Howland's circuit. The bandwidth and current amplitude of the voltage-controlled current source (VCCS) extends linearly from 75 Hz to 153.6 kHz and the adjustable peak-to-peak current amplitude from 0 to 30 mA, respectively. In addition, the output impedance of the VCCS stage is 2.5 MΩ at 76.8 kHz, falling to 1.74 MΩ at 153.6 kHz.

The output admittance and load current of the circuit in Fig. 2.13 are given by Eqs. (2.26) and (2.27):

$$G_{\text{out}} = \frac{1 - [1 + (Z_2/Z_1)]/[1 + (Z_3/Z_4)]}{R} \tag{2.26}$$

$$I_{\text{L}} = \frac{V_1 Z_2}{R Z_1} \tag{2.27}$$

Bragos, Rosell, and Riu (1994) described a wide-band AC-coupled current source that overcomes the problem of saturation at the output DC blocking capacitor. Their solution employs a DC feedback loop in conjunction with a current-sensing VCCS built around a commercially available current feedback amplifier (AD844). The reported output current up to ±10 mA and impedances of their circuit are 6.9 M at 100 kHz, falling to 0.3 M at 1 MHz. To increase the output current, a parallel structure consisting of eight AD844 to constitute four pairs of a bidirection VCCS was developed and is shown in Fig. 2.14 (Wang, 2005a). Two inverting inputs of each pair of AD844s are cascaded together with a current-setting resistor *R*. The positive and negative current outputs of each pair are summed together to form larger current outputs. Two inverse-phased sine wave voltage inputs are buffered and then connected to the noninverting input of each AD844, where two capacitances *C* and two resistances *r* with time constant *RC* can restore the DC components generated in the circuit and

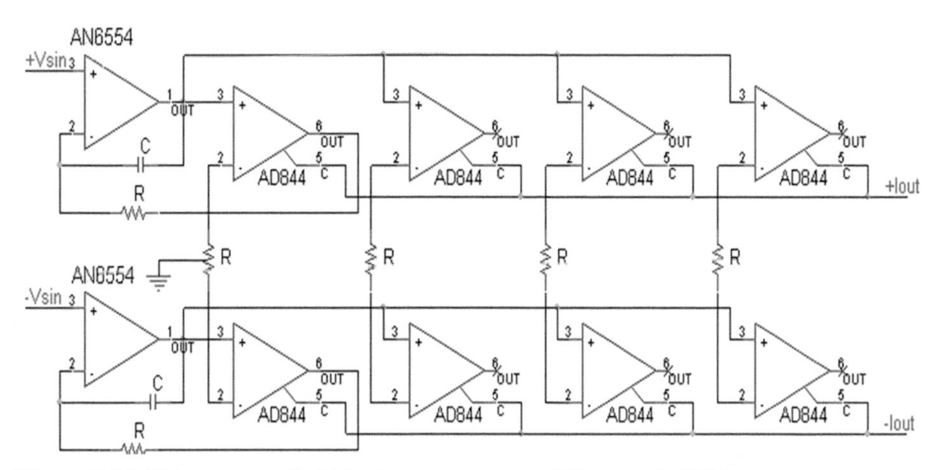

Figure 2.14 Voltage controlled bipolar current source (Wang et al., 2005).

then cancel the DC offset at the current outputs. The total output currents can be estimated using Eq. (2.28). The $\pm I_{out}$ have an ideal 180 degree phase difference with negligible DC biasing, and their amplitudes can be easily balanced with the potentiometer, R_p. The maximum output current amplitude of this circuit is 40 mA. Its output impedance was approximated as 0.75 MΩ//18 pF (Wang, 2005b):

$$\pm I_{out} = \pm 8 \times V_{Sin}/(2 \times R_{in} + R) \tag{2.28}$$

To implement a high specification current source, particularly with high output impedance and large current at high frequency, is still challenging. In contrast, a precise voltage source is generally easier and also less complicated and lower cost in its implementation (Hartov, Demidenko, Soni, Markova, & Paulsen, 2002). The key issue is that the actual current must be known in order to obtain the impedance being measured. Jia, Wang, Schlaberg, and Li (2010) developed a voltage-drive EIT system, in which two power amplifiers (OPA549) are used to deliver a stable bipolar voltage excitation and a current-sensing transform is connected to the voltage output in series for sensing the actual current flow. At the time of applying the voltage source, the current and differential response voltages from electrodes are simultaneously measured to calculate the transimpedance across the whole sensing domain. The voltage excitation can supply bipolar current up to ±320 mA at a frequency of 10 kHz. It has far fewer requirements for a high specification of the amplifiers in comparison with a constant current source since both the voltage and current on the load are measured at the same time and the resultant measurement is in transimpedance.

2.4.2 Sensor electronics and demodulation

Due to the large number of ON−OFF multiplexer channel operations in the DAS, the transient time has to be taken into account and becomes a key issue for maintaining the

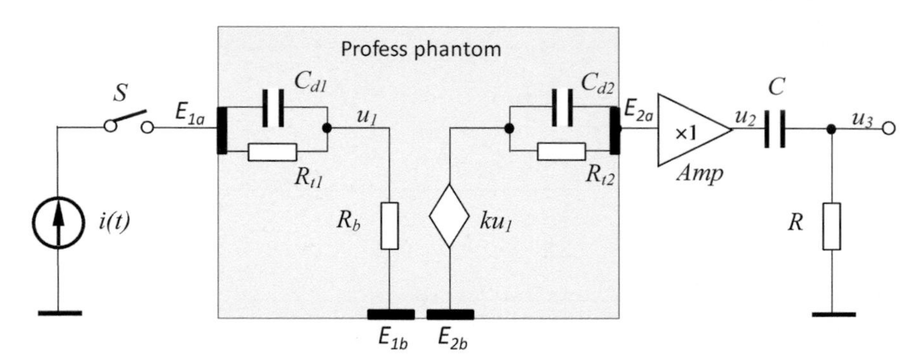

Figure 2.15 The equivalent circuit for one excitation and measurement channel (Wang et al., 2005).

data's accuracy, particularly at a fast data collection rate. To analyze the effects of transient time of the sensing interface, an equivalent circuit for one conductivity measurement channel in EIT (Fig. 2.15) was proposed (Wang et al., 2005; Wang & Ma, 2006). The left side of the figure is for modeling the current excitation and the right side for modeling the voltage measurement. In the figure, $i(t)$ is a sinusoidal current source, S the switch, C_{d1}, R_{t1} the double-layer capacitance and transfer resistance of the electrode–electrolyte interface at the current injection electrode, C_{d2} and R_{t2} the double-layer capacitance and transfer resistance of the electrode–electrolyte interface at the measurement electrode, Amp a voltage buffer with high input impedance and Gain = 1, C and R the AC coupling circuit, R_b the bulk resistance of the process, and ku_1 the transformed voltage at the measured electrode. To simplify the discussion, perfect sinusoidal current source and voltage buffer with infinite output and input impedances, respectively, are assumed in the model. All stray capacitances in circuits and distribution capacitances in aqueous-based processes are also ignored.

Wang and Ma (2006) described the details of the transient process and reported that the response of a sensing system using switched sinusoidal current excitation is approximately the sum of an exponential term and a sinusoidal term (Eq. 2.29). The exponential term is due to the discharging process of the residual potential across C_{d2}, which was charged during the previous switch operation. The value of the residual potential will be uncertain if the switch is operated randomly. The attenuation speed of the charged electrode–electrolyte interface is relatively slow (for example, $\tau_2 = 5$ ms) compared with the fast frame collection rate ($\Delta t_f < 1$ ms) and sampling rate of measurement ($\Delta t_s < 12.5$ μs) (Wang et al., 2005). Therefore, the residual potential (the first term in Eq. 2.29) could produce significant error from a fast data acquisition rate:

$$u_3(t) \approx V_{C_{d2}} e^{-t/\tau_2} + kI_m R_b \cos(\omega t + \theta_1) \tag{2.29}$$

where $\theta_1 = \arctan(\omega\tau)$, $\tau_1 = RC$, $\tau_2 = R_{t2}C_{d2}$, V_{Cd2} is the charged residual voltage due to the previous switch operation, and I_m is the amplitude of $i(t)$.

To eliminate the charged potential, an overzero switching (OZS) scheme was developed by controlling the switching phase angle of the sinusoidal current injection (Jia, Wang, Schlaberg, & Li, 2012; Ma, Lai, Wang, & Jia, 2009; Wang et al., 2005; Wang & Ma, 2006). Both analyses and simulations revealed that the coupling time could be dramatically reduced by the employment of the OZS scheme used now in the Leeds FICA electrical impedance camera to achieve a data acquisition speed of 1164 dual-frames (2.383 million data points) per second with a root-mean-square error of less than 0.6% at 80 kHz in static test application (Ma & Wang, 2004).

One of the most problematic measurement errors arises from common mode voltage (CMV) of the differential measurement because of the limited common mode voltage rejection ratio (CMRR) of the op-amps. The voltage on each measurement electrode is determined from a combination of the injecting current value, the electrode impedance, and the conductivity of the conductive region. The differential voltage between electrode pairs, e.g., in the adjacent electrode pair strategy, is the difference between two voltages on the electrode pair, which is normally much smaller than the common voltage (to the ground) on the electrodes. The error caused by the CMV is mainly due to the limited CMRR of the op-amps, which results in partial transferring of CMV to the output of the op-amps. Besides selecting op-amps with high CMRR and reducing the mismatched impedance of circuits to get a "good" measurement, the better approach is to reduce the CMV. A method used by Brown and Seagar (1985) is the common mode voltage feedback (CMFB). However, because of the phase delay of the feedback circuit, it may cause a phase error in the demodulation and possible feedback oscillation (Murphy & Rolfe, 1988) at high frequency. In most process engineering applications, the potential reference can be obtained by directly grounding an electrical contact to the conductive medium in the process since electrical isolation of the instrument's ground is no longer necessary in most industrial applications, which is referred to as the grounded floating measurement (GFM) (Wang et al., 1993). The feedback path of the CMFB and GFM methods is generally through any unused electrode in contact with the conductive region.

The AC signals must be "readable" by a specific measurement device. The use of the conventional rectification method to convert a weak AC signal to DC is not suitable since its minimum threshold value, e.g., 0.7 V in the use of a diode, is much higher than the signals and also it cannot provide the phase information. Phase-sensitive demodulators (PSDs) in both analogue and digital forms, such as switching demodulators, multipliers, and digital matched filters, are widely used in EIT DASs (Brown & Seagar, 1987; Cook, Saulnier, & Goble, 1991; Goovaerts et al., 1988; Smith, Freeston, Brown, & Sinton, 1992; Webster, 1978). The simplest PSD is illustrated in Fig. 2.16, which includes a positive buffer, a negative buffer, a switch, and a low-pass filter. In operation, it just effectively flips the input signal based on the polarity unity level of a function, e.g., a sequence of square pulses, which equals $V_{mid} = V_{in} \times V_{ref}$. The signal conditions are illustrated in Fig. 2.17, where the input is a sinusoidal signal, $V_{in} = V_m \sin(\omega_0 t + \phi)$, $\omega_0 = 2\pi f_0$, and the reference is a square wave with a unit amplitude and the same angular frequency, ω_0.

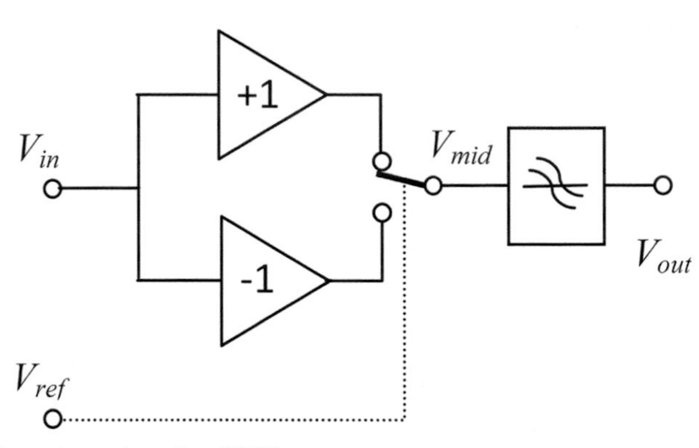

Figure 2.16 Operation principle of PSD.

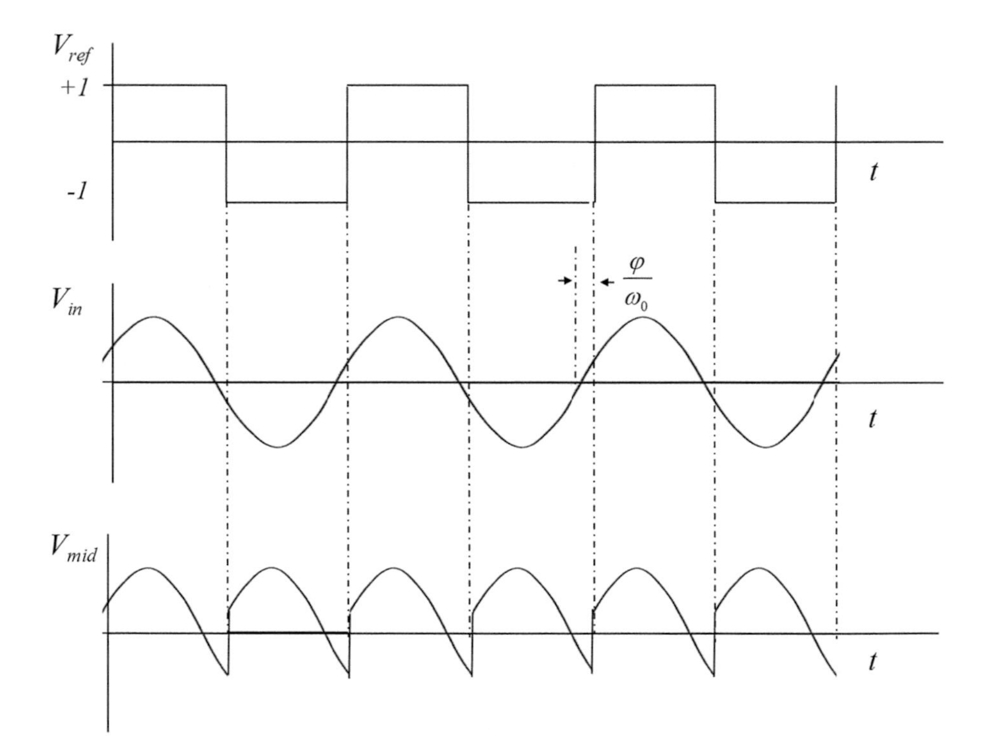

Figure 2.17 Illustration of input and output signals of a PSD (according to Smith et al., 1992).

With Fourier series expansion, V_{mid} can be presented as follows (Wang et al., 1993):

$$V_{\text{mid}}(\omega_0 t, \phi) = \frac{2}{\pi} V_m \cos\phi + \frac{4}{\pi} V_m \frac{\sum_{n=1}^{n} 1}{4n^2 - 1} \{2n\sin(2n\omega_0 t)\sin\phi - \cos(2n\omega_0 t)\cos\phi\} \tag{2.30}$$

Then, in-phase output of the switch-based PSD after the low-pass filter can be expressed as follows:

$$V_{\text{out}}(\phi) = \frac{2}{\pi} V_m \cos\phi \tag{2.31}$$

The quadrature output can be obtained with the same principle when the reference has a phase shift of $\pi/2$.

Using a similar principle but replacing the switch by two multipliers with two sinusoidal references having in-phase and $\pi/2$ phase shift angles, respectively, outputs of the multiplication-based PSD after the low-pass filter are expressed as Eqs. (2.32) and (2.33):

$$V_{\text{out}}(\phi)|_{\text{in-phase}} = \frac{V_r V_m}{2} \cos\phi \tag{2.32}$$

$$V_{\text{out}}(\phi)|_{\text{quadrature}} = \frac{V_r V_m}{2} \sin\phi \tag{2.33}$$

Smith, Freeston, Brown and Sinton (1992) proposed a phase-sensitive detector based on matched filter theory to extract the amplitude and phase of a periodic measurement, which showed that the SNR improvement is equal to its analogue counterpart.

The sampled input, $V_{\text{in}}[n]$ at a frequency f_s, can be presented as follows:

$$V_{\text{in}}[n] = A\sin(2\pi R_f n + \phi) \tag{2.34}$$

where $R_f = f_0/f_s$, $n = 0, 1, 2, \ldots, N-1$, and $N = T_m f_s$ is the number of samples.

The reference sequences, $h[n]$ and $g[h]$, are two sine waveforms with the same frequency as V_{in} but with unit amplitude zero (in-phase) and $\pi/2$ phase shifts, respectively:

$$h[n] = A\sin(2\pi R_f n) \tag{2.35}$$

$$g[n] = A\cos(2\pi R_f n) \tag{2.36}$$

Taking the sums of products from sampled measurements and reference sequences as expressed by Eqs. (2.37) and (2.38),

$$V_{\text{out}}^h = \sum_{n=0}^{N-1} V_{\text{in}}[n] \cdot h[n] \tag{2.37}$$

$$V_{\text{out}}^{g} = \sum_{n=0}^{N-1} V_{\text{in}}[n] \cdot g[n] \tag{2.38}$$

Provided that T_{m} is a multiple of $1/f_0$ and that $f_{\text{s}} > 2f_0$, outputs in Eqs. (2.37) and (2.38) can be expressed as follows:

$$V_{\text{out}}^{h} = \frac{AT_{\text{m}}f_{\text{s}}\cos\phi}{2} \tag{2.39}$$

$$V_{\text{out}}^{g} = \frac{AT_{\text{m}}f_{\text{s}}\sin\phi}{2} \tag{2.40}$$

The phase shift of the input signal is derived from the following:

$$\phi = \text{arctg}\frac{V_{\text{out}}^{g}}{V_{\text{out}}^{h}} \tag{2.41}$$

In the above equations, ϕ is the phase shift between the demodulated signal and the reference signal, T_m is the measurement time of a multiple of the demodulated signal period ($1/f_0$), and f_s is the sample frequency ($f_s > 2f_0$).

The major drawback of analogue PSD is its time cost due to the low-pass filter. It would also cause the complex circuitry and associated cost for demodulation of the signal from a multifrequency excitation. Using digital demodulation, its SNR can be much closer to the theoretical value (Smith et al., 1992) and it has a potential to be further implemented for demodulation of the multifrequency signal in various waveforms.

2.4.3 Data acquisition systems

Various EIT DASs have been developed with specific features and application areas (Brown & Seagar, 1987; Cook et al., 1994; Dong, Xu, Zhang, & Ren, 2012; Gisser et al., 1991; Jia, Wang, Schlaberg, & Li, 2010; Lidgey, Zhu, McLeod, & Breckon, 1992; Ma et al., 2003; Newell, Gisser, & Isaacson, 1988; Wang et al., 1993; Wilkinson et al., 2005; Wilson, Milnes, Waterworth, Smallwood, & Brown, 2001). A typical EIT DAS is made up of six functional parts: voltage generator, current or voltage generator, electrode controller, voltmeter, operation controller, and power supply. Fig. 2.18 shows the block diagram of a DAS (Dickin & Wang, 1996; Wang, 1994; Wang et al., 1993). A single-chip microprocessor, the Intel i8052AH, was employed to manage the operation of the system, which controlled the sine wave generator, the electrode (channel) operation, and analogue-to-digital conversion. Current or voltage was applied to electrodes in the EIT sensor via a multiplexer in a specific scheme. Signals from the EIT sensor were firstly isolated with voltage buffers and then provided to an instrument amplifier. Then, the output of the instrument amplifier was demodulated by either an analogue or digital PSD. These functional parts were built on five EURO

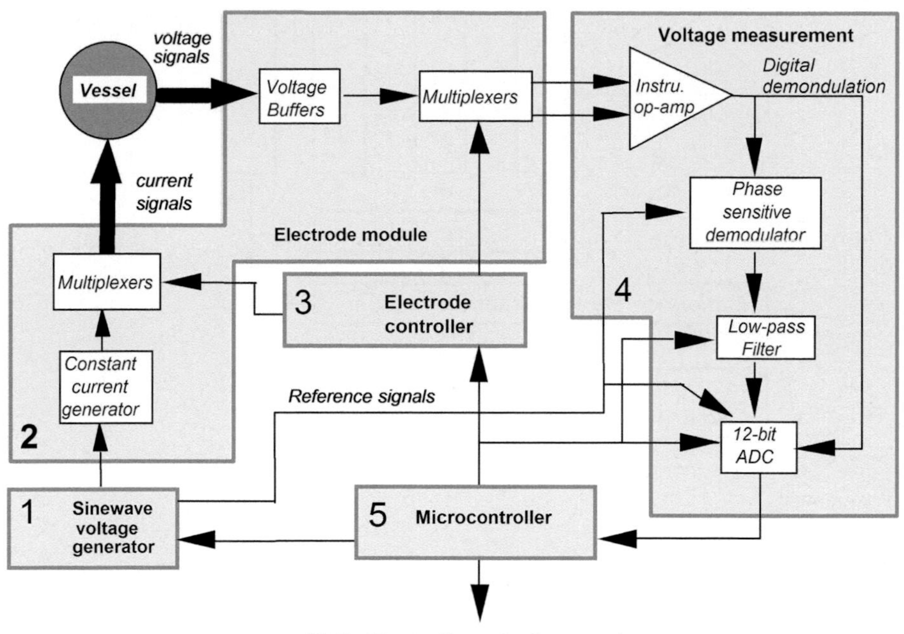

Figure 2.18 Block diagram of the UMIST MK.1b EIT data acquisition system (Wang, 1994).

cards plugged into a 64-way backplane utilized for digital signal transfer and control between each board and the microcontroller. The system can accommodate up to 128 electrodes to form different sensing strategies using a programmable combination.

The diagram of a more sophisticated EIT system (Wang et al., 2007), known as the Leeds FICA system, is shown in Fig. 2.19, which was developed for online measurement of two-phase flows with axial velocities up to 10 m/s (Ma et al., 2003; Wang et al., 2005). The system was designed in a modular fashion and can consist of several data acquisition modules and computing modules. The data acquisition module includes a VCCS, a waveform generator and a parallel measurement module with arrays of programmable gain amplifiers (PGAs), analogue-to-digital convertors (A/D), and first-in-first-out (FIFO) memories. The processes (from sampling to digital data output) in parallel measurement are fully logically operated. Both the excitation and measurement are synchronized by the synchronization unit. Two identical data acquisition subsystems are connected to two electrode arrays. A data acquisition speed of 1164 dual frames (2.383 million data points) per second was achieved with a root-mean-square error of less than 0.6% at 80 kHz in a static set-up (Ma & Wang, 2004). The measurement system, as shown in Fig. 2.19, can consist of up to four digital signal processor (DSP) modules. The third DAS collects data from conductivity, pressure, temperature, and other auxiliary sensors. All three DASs are controlled by one DSP module through a common interface. Each DSP module can be configured to communicate with a remote PC via a fast interface of IEEE1394 but only one at a time (Dai, Ma, & Wang, 2003; Holliday, Ma, Wang, & Williams, 2003; Ma et al., 2003).

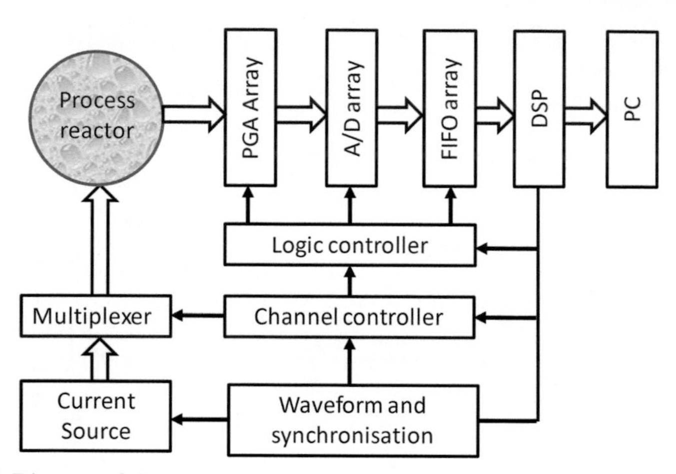

Figure 2.19 Diagram of the Leeds FICA EIT data acquisition system with parallel measurement.

To overcome the limits of the current source for use in process applications involved with high conductive solution, e.g., seawater in 5 S/m conductivity or the 3.5 M nitric acid in nuclear fuel process (Bolton et al., 2007), an EIT system with a bipolar voltage source and current sensing was developed (Jia, Wang, Schlaberg, & Li, 2010; Wang & Jia, 2009). The amplitude of the current output can reach more than 300 mA. Fig. 2.20 illustrates a brief block diagram of the system: a voltage excitation was applied to a pair of electrodes via the first multiplexer and its corresponding

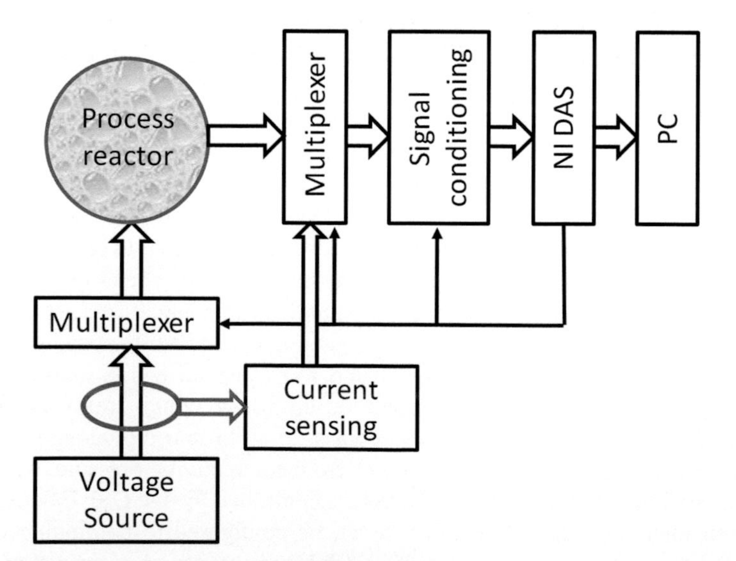

Figure 2.20 Diagram of a voltage-drive EIT system.

current though the target was sensed by a current sensor. Voltage drops between electrode pairs were sampled via the second multiplexer and demodulated by a National Instrument (NI) data acquisition board, which transmitted the acquired data to a PC via a USB 2.0 communication interface for image reconstruction and data fusion.

2.5 Image reconstruction

2.5.1 Inverse problem

For a linear Eq. (2.42), the minimization function (Eq. 2.44) can be obtained from minimizing Eq. (2.43):

$$\mathrm{A} \cdot \mathrm{x} = \mathrm{b} \tag{2.42}$$

$$f(x) = \frac{1}{2}\|\mathrm{Ax} - b\|^2 \tag{2.43}$$

$$\nabla f(\mathbf{x}) = \mathbf{A}^{\mathrm{T}}(\mathbf{Ax} - \mathbf{b}) \tag{2.44}$$

Let $\nabla f = 0$, then,

$$A^T \mathrm{Ax} = A^T b \tag{2.45}$$

If an inverse matrix of $\boldsymbol{A}^T\boldsymbol{A}$ exists, the solution can be made as Eq. (2.46):

$$\mathbf{x} = \left(\mathbf{A}^{\mathrm{T}}\mathbf{A}\right)^{-1}\mathbf{A}^{\mathrm{T}}\mathbf{b} \tag{2.46}$$

It is known the inverse matrix of $\boldsymbol{A}^T\boldsymbol{A}$ is not derivable due to the ill-condition of sensitivity matrix $\boldsymbol{A}$ in EIT. Many indirect methods were developed to solve the linear equation in the past. Typically, they were reported as the back-project method (Kotre, 1994) and the Newton one-step reconstruction (NOSER) method single step method (Cheney, Isaacson, Newell,Simake, & Goble, 1990). However, it should be pointed out the solution for Eq. (2.46), if any exists, only satisfies the change of conductivity $\Delta\sigma << \sigma$. The closest solution may only be obtained from multistep approach (Yorkey, Webster, & Tompkins, 1987).

2.5.2 Sensitivity coefficient back-projection method

Wang et al. (2016) indicated that the sensitivity coefficient back-projection method (SBP) is actually based on an assumption of that the diagonal elements of the transformed matrix, $\boldsymbol{A}^T\boldsymbol{A}$, have the most significant values and the matrix can be approximated to a diagonal matrix. For a particular example of adjacent sensing strategy, the solution to Eq. (2.45) with $\nabla f = 0$ is firstly normalized to Eq. (2.47) then to Eq. (2.48) if the $\tau A^T A$ can be approximated as an identity matrix:

$$\tau \mathbf{A}^T \mathbf{A}\mathbf{x} = \tau \mathbf{A}^T \mathbf{b} \tag{2.47}$$

$$\mathbf{x} = \tau \mathbf{A}^T \mathbf{b} \tag{2.48}$$

where τ is the inversed matrix for the approximated diagonal matrix of $\mathbf{A}^T\mathbf{A}$ ($\mathbf{A} = [a_{ij}]$), in which its diagonal parameters satisfy:

$$\tau_{jj} = \frac{1}{\sum_{i}^{N} (a_{ij})^2} \tag{2.49}$$

The transformed matrix in Eq. (2.47) is also called the resolution matrix (Lévêque, Rivera, & Wittlinger, 1993), $\mathbf{R} = \tau \mathbf{A}^T \mathbf{A}$, and presents the pixels' correlation in EIT, which is the same as that presented by the Hessian matrix in Newton Raphson (NR) method (Yorkey et al., 1987). If none of the pixels is correlated, the Hessian matrix would be a diagonal matrix (Woo, Hua, Webster, & Tompkins 1993). This feature of diagonal domination is commonly interpreted as evidence for high-quality inversion (Lévêque et al. 1993). However, cross-correlations are among all pixels in EIT, which produce a skew form of an optimized matrix, in which the diagonal elements may have most significant values. Fig. 2.21 demonstrates the significance of diagonal components in the resolution matrix from the adjacent electrode pair sensing strategy, which are modeled using FEM with a mesh having 224 pixels for a 16-electrodes sensing array. Eq. (2.48) with the existence of the diagonal matrix approximation provides the mathematical principle for the SBP algorithm.

2.5.3 Multistep methods

The linear approximation, based on ignoring the high-order terms with respect to $\Delta\sigma << \sigma$ (see Eq. 2.17), enables the use of iterative techniques to solve the linear equation (Eq. 2.45) with a precalculated sensitivity matrix based on an estimated initial conductivity distribution, e.g., a homogeneous distribution. However, in most of cases, multisteps of the linear solution process have to be applied in order to achieve better image quality, where the sensitivity matrix should be updated with the change of conductivity distribution obtained from the previous step. Its performance is strongly associated with the methods and the convergent factors used in iterative minimization process as given by Eq. (2.50). The typical examples are the use of NR optimization (Yorkey et al., 1987), Tikhonov regularization method (Lionheart, 2001; Vauhkonen, Vadasz, Karjalainen, Somersalo, & Kaipio, 1998) with the techniques of selection of regularization factors such as the singular value decomposition method and Akaike's information criterion (Akaike, 1974), and the Marquardt method (Marquardt, 1963).

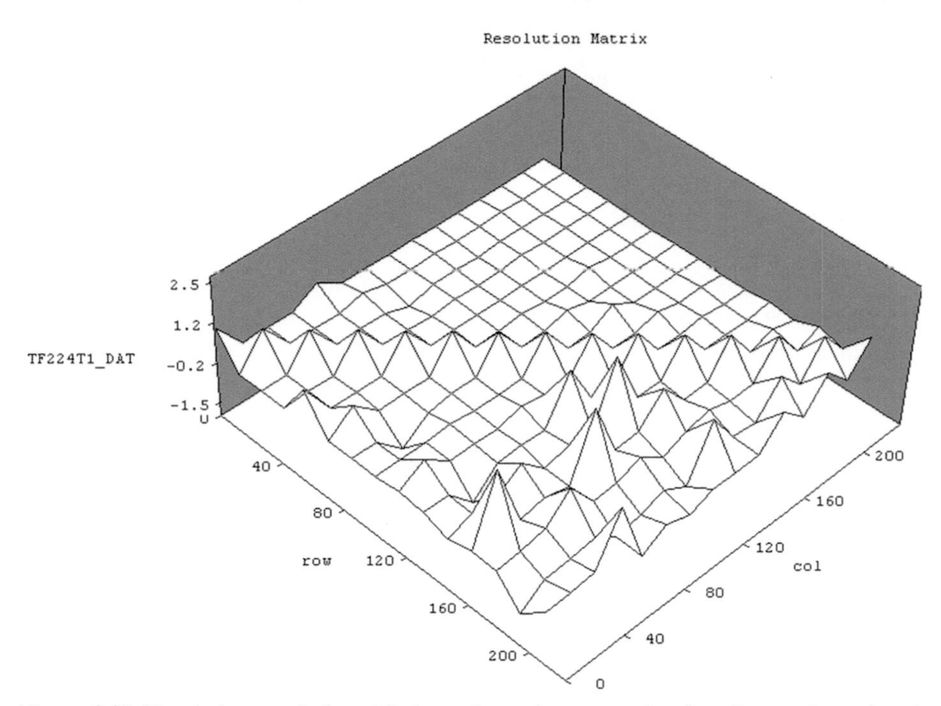

Figure 2.21 Resolution matrix for a 16-electrode sensing array using the adjacent electrode pair sensing strategy with a 2-dimensional mesh with 224 pixels.

The error function is

$$r = \tau A^T Ax - \tau A^T b \tag{2.50}$$

2.5.4 One-step methods

One-step method refers the solution from the iterations at the end of the first step of a multistep method. It is the fact that the result from the first step solution normally provides the most contributive convergence if the regularization factors are properly selected. The quality of resultant images is generally much better than those came from back-projection approximation as results from NOSER method (Cheney et al., 1990) and the SCG method (Wang, 2002). It also runs at a much fast speed than that of multistep method due to the use of the precalculated sensitivity matrix. An iterative solution can be obtained in a kind of the Landweber iteration method as expressed by Eq. (2.51) (Landweber, 1951):

$$x_n \approx x_{n-1} - r_n = x_{n-1} - \tau A^T (Ax_{n-1} - b) \tag{2.51}$$

The selection of τ for the SBP has been suggested by Eq. (2.49). An example for the use of Landweber's method with selection of τ for capacitance tomography was reported by Liu, Fu and Yang (1999) and also many other recent publications. For a general application, the selection of τ can be referred from the Landweber method (1951).

2.6 Imaging capability

Seagar, Barber, and Brown (1987) have proposed definitions of spatial resolution, conductivity contrast, conductivity resolution, and visibility of the reconstructed image for assessment of accuracy in EIT. These definitions cover the major performance in practice. Four major specifications for assessing the accuracy of imaging are defined below:

Spatial resolution: the smallest region of a medium into which the conductivity can be independently determined:

$$\text{spatial resolution} = \frac{r_a}{r_b} \tag{2.52}$$

Conductivity contrast: the ratio between the conductivity of an anomaly to that of its surrounding region:

$$\text{conductivity contrast} = \alpha = \frac{\sigma_a}{\sigma_b} \tag{2.53}$$

Conductivity resolution: the fractional change in conductivity contrast:

$$\text{conductivitycontrast} = \frac{\delta\alpha}{\alpha} \tag{2.54}$$

Distinguishability: the minimum distance between two objects with the same conductivity contrast to their surrounding region:

$$\text{distinguishability}(\text{at}\alpha) = \frac{t_{\text{distinguishabledistance}}}{D_{\text{vesseldiameter}}} \tag{2.55}$$

To have a close view of EIT performance, assessments based on the above definition were carried out (Wang, Mann, & Dickin, 1999). The assessments on imaging capability of EIT were limited to the data simulated with a two-dimensional finite element model in the use of the adjacent electrode pair strategy with 16 electrodes and the images reconstructed using a conjugate gradient method–based algorithm (Wang, 2002). Random errors were added to the simulated measurements. The test set-up is shown in Fig. 2.22, where an object with conductivity σ_a and radius r_a is positioned at the center of the two-dimensional domain with conductivity σ_b and radius r_b.

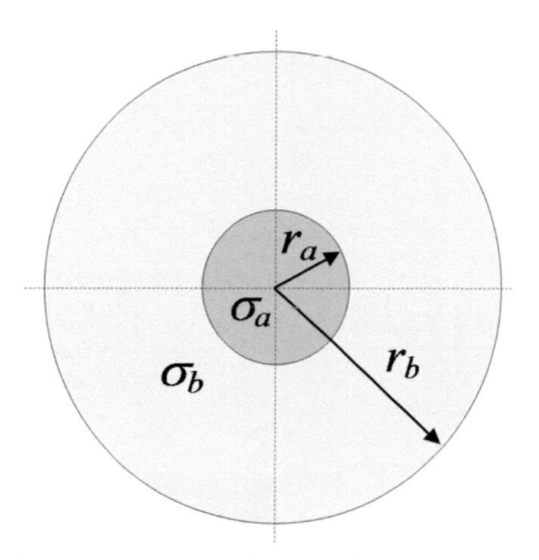

Figure 2.22 The set-up for the assessment of EIT performance.

The preerror (or measurement error) presented at precondition is defined as the error between the boundary voltages obtained from measurement and that simulated from forward solution. The posterror (or conductivity error) presented at the postcondition is defined as the error between the conductivity set-up and that reconstructed from inverse solution (Wang, Mann, & Dickin, 1999). In most applications, only the preerror can be obtained. Fig. 2.23 shows how the error in measurement influences the image (a series of preerrors related to posterrors) by set-ups of different sizes of an object with the same conductivity ratio 1.36:1 (Fig. 2.23a) and different conductivity values of an object with a fixed radius ratio 0.24:1 (Fig. 2.23b), which reveals that for a system with a preerror of 1%, the spatial resolution and the conductivity resolution could achieve 5% and 10% (2.4% of the mean change), respectively. The distinguishability was tested by setting two objects both with area ratio (1/253):1 and conductivity ratio 1.36:1 near the center of the disk domain, which revealed from the simulated results shown in Fig. 2.23c that the imaging distinguishability of the given image in a 16-division of the view scale is around 20%.

2.7 EIT data for process application

Since EIT can detect local changes in electrical conductivity, the technique is used to study unsteady mixing or fluid dynamics of flow mixtures, such as gas–liquid and solid–liquid mixtures, where the dispersed phase of fluids or solids has different conductivities from the aqueous continuous phase. EIT may, therefore, be suitable for numerous aqueous-based processes. However, EIT only produces electrical conductivity distribution or maps. For most process applications, relevant engineering data have to be implemented from conductivity-based information. A few of the typical relationships are introduced here.

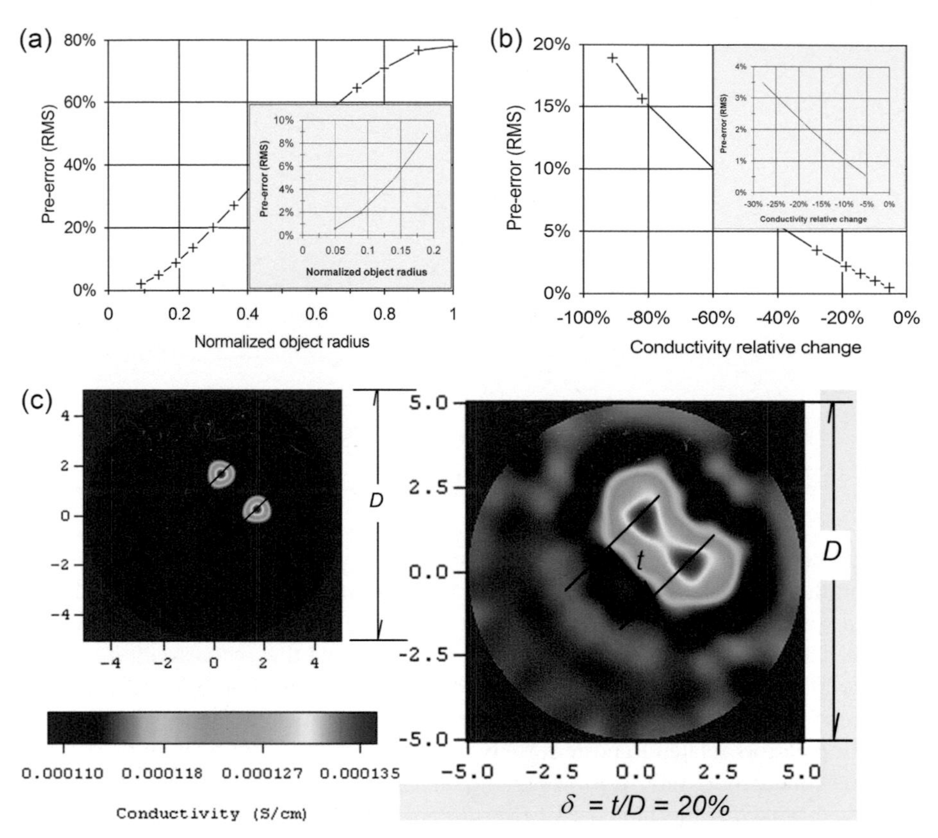

Figure 2.23 Performance of an EIT: (a) spatial resolution, (b) conductivity resolution, (c) distinguish ability (the left: set-up, the right: reconstructed) (Wang et al., 1999). OPA (Overseas Publishers Association)).

2.7.1 Concentration distribution

A number of correlations have been proposed to convert electrical conductivity of a particulate system to volumetric fraction of a dispersed phase (Table 2.2, Shirhatti, 2007). Among them, the Maxwell relationship is the most widely used in EIT application in process engineering applications (Sharifi & Young, 2013).

In research on miscible liquid mixing, the liquids under investigation can be labeled with sufficient conductivity contrast to enhance the imaging performance (Wang et al., 1999). High-concentration saline solution can also be used as a tracer to track the dynamic trajectory of the mixing process (Holden, Wang, Mann, Dickin, & Edwards, 1999). In an instance of single-phase miscible liquid mixing, the conductivity of mixture, σ_m, provides a linear relation to the weight concentration C_w of salt (Holden, Wang, Mann, Dickin, & Edwards, 1998; Mann et al., 1996; Wang et al., 1999):

$$\sigma_m = 1.728636 \times C_w + \sigma_f \tag{2.56}$$

Table 2.2 Correlation functions of conductivity and dispersed phase volumetric fraction (Shirhatti, 2007).

Reference	Correlation
Maxwell (1873)	$C_v = \frac{2\sigma_f+\sigma_s-2\sigma_m-\frac{\sigma_m\sigma_s}{\sigma_f}}{2\sigma_f+\sigma_m-2\sigma_s-\frac{\sigma_m\sigma_s}{\sigma_f}}$ for conductive solid $C_v = \frac{2\sigma_f-2\sigma_m}{2\sigma_f+\sigma_m}$ for Nonconductive solid or liquid
Bruggeman (1935) for particles of different sizes (Mactaggart, Nasr-El-Din, & Masliyah, 1993)	$\frac{\sigma_m}{\sigma_f} = (1-C_v)^{1.5}$(for insulating solid) $\frac{\sigma_m}{\sigma_f} = (1-C_v)^3$ (for conductive solid)
Meredith and Tobias (1962)	$\frac{\sigma_m}{\sigma_f} = 8\frac{(2-C_v)(1-C_v)}{(4+C_v)(4-C_v)}$ (for insulating solid) $\frac{\sigma_m}{\sigma_f} = \frac{(1-C_v)(2+C_v)}{(1-C_v)(2-C_v)}$ (for conductive solid)
Maxwell−Wagner equation or Maxwell−Garnett equation or Wiener's rule (Fricke, 1953)	$\frac{\sigma_m-\sigma_f}{\sigma_m+2\sigma_f} = \left(\frac{\sigma_s-\sigma_f}{\sigma_s+2\sigma_f}\right)C_v$
Rayleigh (Mactaggart et al., 1993)	$\sigma_m = \sigma_f\left[1+\frac{3C_v}{\left(0.394C_v^{10/3}-C_v-2\right)}\right]$
De la Rue and Tobias (Mactaggart et al., 1993)	$\sigma_m = \sigma_f(1-C_v)^m$; $m = 1.5$ for $0.45 \le C_v \le 0.75$
Prager (1963)	$\sigma_m = \frac{\sigma_f(1-C_v)(3-C_v)}{3}$
Weissberg (1963)	$\sigma_m = \sigma_f\left(2\frac{(1-C_v)}{2-\ln(1-C_v)}\right)$

Notes: σ_f is the conductivity of the aqueous continuous phase, σ_s is the conductivity of the dispersed phase, and σ_m is the mixture conductivity. If the dispersed phase is nonconductive, σ_s equals zero, C_v is the volumetric fraction.
Shirhatti, V. S. (2007). *Characterisation and visualisation of particulate solid-liquid mixing using electrical resistance tomography* (Ph.D. thesis). University of Leeds (Provided courtesy of Dr V.S. Shirhatti).

$$C_w = 0.5785 \times \left(\sigma_{mc} - \sigma_f\right) \tag{2.57}$$

where σ_m and σ_f are the mixture's and aqueous conductivity in units of mS/cm, respectively; the weight concentration C_w is in g/L, the range of C_w: $0 \le C_w \le 10$.

Qiu et al. (1994) and Qui, Dickin, James, and Wang (1997) also used Archie's law to convert the conductivity value of a laboratory subseabed set-up to porosity. The conversion is given as follows (Archie, 1942):

$$\phi = \left(\frac{\sigma_m}{\sigma_f}\right)^k \tag{2.58}$$

where ϕ is the porosity (%), σ_m the conductivity of the formation (S/m), σ_f the water conductivity (S/m), and k the cementation index dependent on the shape and packing of the particles ($k = 1.5$ for spherical glass beads).

2.7.2 Velocity distribution

The principle of cross-correlation method has been widely used to derive the velocity component of moving objects from two sets of measurements or images taken with a known time interval, e.g., the velocity distribution derived in Particle image velocimetry (PIV). Unlike PIV to statistically derive velocity component from two photo images of many seeds' over an interrogation region with a known time interval, EIT derives disperse phase velocity (or the structural velocity) based on two series of tomograms obtained from a dual-plane EIT sensor with known frame rate and distance between the two sensing planes. The concept of cross-correlation method applied in PIV and tomography is the same. Therefore, it is possible to derive a velocity map of the dispersed phase-flow component (Lucas, Cory, Waterfall, Loh, & Dickin, 1999; Li, Wang, Wu, & Lucas, 2009, pp. 81–93).

The fundamental principle underpinning cross-correlation flow measurement is the "tagging" of signal's similarity. These concepts are best illustrated in Fig. 2.24.

The basic method is to find τ that can make the difference, ε, minimum. This can be achieved using the least squares criterion as given by Eq. (2.59):

$$\begin{aligned}\varepsilon^2(\tau) &= \int_{-\propto}^{\propto} (f_1(t) - f_2(t-\tau))^2 dt \\ &= \int_{-\propto}^{\propto} (f_1(t))^2 dt + \int_{-\propto}^{\propto} (f_2(t-\tau))^2 dt - 2\int_{-\propto}^{\propto} (f_1(t) f_2(t-\tau))^2 dt \end{aligned} \tag{2.59}$$

The revised error function is given by Eq. (2.60), which only remains the product of two functions since other terms to be constant from the integration over a sufficient time. Hence, the error function is minimum when the last term is maximum. This expression is commonly known as the cross-correlation function denoted by $R_{12}(\tau)$:

$$R_{12}(\tau) = \int_{-\propto}^{\propto} f_1(t) f_2(t-\tau) dt \tag{2.60}$$

$$R_{12}(n) = \sum_{m=1}^{l} f_1(m) f_2(m+n) \tag{2.61}$$

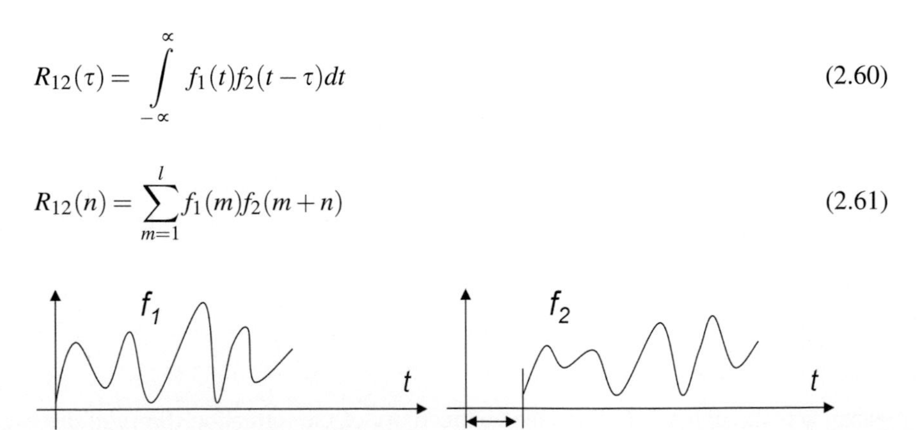

Figure 2.24 Cross-correlated two functions.

where l is the sample length, n is the offset number, and f_1 and f_2 are the values at pixels positioned at the same location in two sequences of the up-flow and down-flow images, respectively.

The disperse phase velocity can be simply derived as follows:

$$v = \frac{d_s}{\tau} = \frac{d_s}{N_p \cdot T_{\text{sampling}}} = \frac{d_s \times f_s}{N_p} \tag{2.62}$$

where v is the velocity, d_s is the distance between two sensors, N_p is the number of offset frames to get the peak value of $R_{12}(N_p)$, and f_s is the sampling frequency of ERT.

The fractional velocity discriminant, κ, can be estimated with Eq. (2.63), or the necessary data collection speed (sec./dual-frame), δ, for a certain κ can be expressed as Eq. (2.64):

$$\kappa = \delta / 2\tau \tag{2.63}$$

$$\delta = 2\tau\kappa \tag{2.64}$$

The abovementioned point-by-point cross-correlation technique for flow velocity measurement is based on an assumption that flow trajectories are parallel to each other and perpendicular to the sensor plane. However, in most cases, this ignores the fact that the trajectories of particles of the dispersed phase exhibit a complex three-dimensional behavior. The "best-correlated pixels" method overcomes this problem by proposing that a signal from one pixel on plane X is somehow better correlated with a signal from a nonaxially corresponding pixel on the plane Y (Mosorov, Sankowski, Mazurkiewicz, & Dyakowski, 2002; Wang, Lucas, Dai, Panayotopoulos, & Williams, 2006). The pixels from the second plane are chosen from the axially corresponding pixel and its neighbors, as described by Eq. (2.65):

$$R_{x[n,m]y[n+i,m+j]}[p] = \sum_{k=0}^{T-1} x_{[n,m]}[k] y_{[n+i,m+j]}[k+p] \quad (i,j) \in B \tag{2.65}$$

where T is the number of the images, for which the cross-correlation is calculated; $P = -(T-1), \ldots 0, 1, 2, 3 \ldots T-1$; $k = 0, 1, 2, 3 \ldots T-1$; n, m are the coordinates of the pixel; x, y are the values of the pixels on plane X and plane Y and B is the group of neighboring pixels on plane Y.

2.8 Future trends

EIT is based on measurements of electrical properties of materials by applying a low-frequency (from DC up to a few MHz) electric field (e.g., current or voltage). They are inexpensive and relatively straightforward to implement with the potential for submillisecond temporal resolution. Due to the limited number of measurements and the propagation nature of low-frequency electromagnetic waves, they normally provide images with a spatial resolution around 5% (the diameter of the object to the diameter of the vessel) and a homogeneity resolution better than 1% (e.g., the mean concentration of gas in water) for an EIT system having 1% measurement error. The use of the high-rank DSP and field-programmable gate array technique has brought remarkable performance improvement to EIT with high speed and/or high flexibility for onboard control and management of various sensing strategies and measurement protocols (Dong, Xu, Zhang, & Ren, 2012; Wang et al., 2005). However, in the consideration of financial, time, and maintenance costs for new development, EIT systems with a commercially available DAS with well-supported software, e.g., Labview, have become more popular (Jia et al., 2010). To break the limitation of EIT in process application due to the request of electrodes to be invasive, an EIT system based on capacitively coupled contactless conductivity detection (C4D) was developed (Wang, Hu, Ji, Huang, & Li, 2012). The system uses metallic electrodes installed on the outside wall of a nonconductive pipe, which forms coupling capacitances between the metal electrodes and electrolyte inside the pipe and therefore allows excitation signal penetration into the process. Most of the current EIT techniques in process engineering only use the resistance part or magnitude of electrical impedance to implement engineering data, e.g., the concentration of a dispersed phase in a mixing process. Electrical impedance tomography spectroscopy (EITS) was proposed for imaging the human thorax (Brown, Leathard, Lu, Wang, & Hampshire, 1995) and received serious attention from many other researchers (Gary et al., 2007; McEwan, Tapson, Schaik, & Holder, 2009; Oh et al., 2008; Saulnier et al., 2007). The impedance spectrum can provide not only the electric but also the dielectric property distributions of the object under investigation, which will be particularly important in colloid-based process engineering since the surface charges of particles present a significant contribution to the imaginary part of impedance (Zhao, Wang, & Hammond, 2011, 2013). EITS is extending the boundary of information from time domain to frequency domain. It is expected that significant new knowledge will be revealed with EITS in a completely different direction.

2.9 Sources of further information

An early book edited by Webster (1990) provided a systematic view of electrical impedance tomography in medical applications, and has been a significant influence for the development of EIT for process engineering applications. The book edited by Williams and Beck (1995), which included EIT as part of its contents, discussed the principle and application of process tomography at its early stage. Boone, Barber and Brown (1997) presented an excellent review and summary on EIT in medical research and diagnoses. It may be a good starting point for beginners to read several of these articles. Brown (2001) addressed a brief review of development and specific features of EIT in both medicine and process engineering. Articles by Dickin and

Wang (1996), York (2001), and Wang (2005a,b) provided specific focus on EIT techniques for process engineering. Dyakowski, Jeanmeure, and Jaworski (2000) presented a good review on electrical tomography for gas–solids and liquid–solids flows, and the most recent review paper by Sharifi and Young (2013) extensively reviewed and summarized the applications of EIT in chemical engineering. Readers can also review specific techniques and featured applications from books edited by Holder (2005), Sikora and Wojtowicz (2010) and Kak and Slaney (2001).

References

Akaike, H. (1974). A new look at statistical model identification. *IEEE Transactions on Automatic Control, 19*, 716–730.

Analog devices datasheet. AD7008/PCB DDS evaluation board.

Archie, G. E. (1942). The electrical resistivity log as an aid in determining some reservoir characteristics. *Transactions of American Institute of Minning Metallurgical Engineers, 146*, 54–62.

Aw, S. R., Rahim, R. A., Rahiman, M. H. F., Yunus, F. R. M., & Goh, C. L. (2014). Electrical resistance tomography: A review of the application of conducting vessel walls. *Powder Technology, 254*, 256–264.

Barber, D. C. (1990). Quantification in impedance imaging. *Clinical Physics & Physiological Measurement, 11*(Suppl. A), 45–56.

Bates, R. H. T., McKinnon, G. C., & Seagar, A. D. (1980). A limitation on systems for imaging electrical conductivity distributions. *IEEE Transactions on Biomedical Engineering BME, 27*, 418–420.

Bockris, J. O.'M., & Conway, B. E. (1963). *Modern aspects of electrochemistry*. New York, NY: Plenum Press.

Bockris, J. O.'M., & Drazic, D. M. (1972). *Electro-chemical science*. Taylor and Francis Ltd.

Bockris, J. O.'M., & Reddy, A. K. N. (1970). *Modern electrochemistry*. London: Macdonald.

Bolton, G., Bennett, M. A., Wang, M., Qiu, C., Wright, M., Primrose, K., et al. (2007). Development of an electrical tomographic system for operation in a remote, acidic and radioactive environment. *Chemical Engineering Journal, 130*, 165–169.

Boone, R., Barber, D., & Brown, B. R. (1997). Review: Imaging with electricity: Report of the European concerted action on impedance tomography. *Journal of Medical Engineering & Technology, 21*(6), 201–232.

Bragos, R., Rosell, I., & Riu, P. (1994). A wide-band a.c.-coupled current source for electrical impedance tomography. *Physiological Measurement*, (Suppl. 2a), A91–A99.

Breckon, W. R., & Pidcock, M. K. (1987). Mathematical aspects of impedance imaging. *Clinical Physics & Physiological Measurement, 8*(Suppl. A), 77–84.

Brown, B. H. (2001). Medical impedance tomography and process impedance tomography: A brief review. *Measurement Science and Technology, 12*, 991–996.

Brown, B. H., Barbar, D. C., Wang, W., Lu, L., Leathard, A. D., Smallwood, R. H., et al. (1994). Multi-frequency imaging and modelling of respiratory related electrical impedance changes. *Physiological Measurement, 15*(Suppl. 2A), A1–A12.

Brown, B. H., & Barber, D. C. (1985). *Tomography, UK patent No. GB2160323A*.

Brown, B. H., Barber, D. C., & Freeston, I. L. (1983). *Tomography, UK patent No. GB2119520A, 1983*.

Brown, B. H., Leathard, A. D., Lu, L., Wang, W., & Hampshire, A. (1995). Measured and expected cole parameters from electrical impedance tomographic spectroscopy images of the human thorax. *Physiological Measurement, 16*, A57–A67.

Brown, B. H., & Seagar, A. D. (1985). Applied potential tomography: Data collection problems. In *IEE international conference on electric and magnetic fields in medicine and biology. London, Dec 79–82.*

Brown, B. H., & Seagar, A. D. (1987). The Sheffield data collection system. *Clinical Physics & Physiological Measurement, 8*(Suppl. A), 91–97.

Bruggeman, D. A. G. (1935). Calculation of various physical constants of heterogeneous substances, part 1, constant and conductivity of mixtures of isotropic substances. *Annual Physics, Leipzig, 24*, 636.

Cheney, M., Isaacson, D., Newell, J. G., Simake, S., & Goble, J. (1990). Noser: An algorithm for solving the inverse conductivity problem. *International Journal of Imaging Systems and Technology, 2*, 66–75.

Cilliers, J. J., Xie, W., Neethling, S. J., Randall, E. W., & Wilkinson, A. J. (2001). Electrical resistance tomography using a bi-directional current pulse technique. *Measurement Science and Technology, 12*(8), 997–1001.

Cobbold, R. S. C. (1974). *Transducers for biomedical measurements: Principles and applications*. John Wiley & Sons.

Conway, B. E. (1952). *Electrochemical data*. Elsevier Publishing Company.

Cook, R. D., Saulnier, G. J., Gisser, D. G., Goble, J. C., Newell, J. C., & Isaacson, D. (1994). Act3-a high-speed, high-precision electrical-impedance tomograph. *IEEE Transactions on Biomedical Engineering, 41*(8), 713–722.

Cook, R. D., Saulnier, G. J., & Goble, J. C. (1991). A phase sensitive voltmeter for a high-speed, high-precision electrical impedance tomography. *Annual International Conference of the IEEE Engineering in Medicine and Biology Society, 13*(1), 22–23.

Crow, D. R. (1974). *Principles and applications of electrochemistry*. London: Chapman and Hall.

Cullivan, J. C., Wang, M., Bolton, G., Baker, G., Clarke, W., & Williams, R. A. (2005). Linear EIT for sedimentation and sediment bed characterization. In *Proceedings of 4th World congress on industrial process tomography* (pp. 910–915). Aizu: VCIPT.

Dai, Y., Ma, Y., Wang, M. A specific signal conditioner for electrical impedance tomography. In: *Proceedings of the 3rd World congress on industrial process tomography*, Banff, VCIPT, pp. 45–49.

Dickin, F. J., & Wang, M. (1996). Electrical resistance tomography for process applications. *Measurement Science and Technology IOP, 7*, 247–260.

Dines, K. A., & Lytle, R. J. (1981). Analysis of electrical conductivity imaging. *Geophysics, 46*, 1025–1036.

Dong, F., Xu, C., Zhang, Z., & Ren, S. (2012). Design of parallel electrical resistance tomography system for measuring multiphase flow. *Chinese Journal of Chemical Engineering, 20*(2), 368–379.

Dyakowski, T., Jeanmeure, L. F. C., & Jaworski, A. J. (2000). Applications of electrical tomography for gas–solids and liquid–solids flows — a review. *Powder Technology, 112*.

Etuke, E. O. (1994). *Impedance spectroscopy form component specificity in tomography imaging* (Ph.D. thesis). UMIST (now University of Manchester).

Faraj, Y., & Wang, M. (2012). ERT investigation on horizontal and vertical counter-gravity slurry flow in pipelines. In *20th international congress of chemical and process engineering CHISA 2012. Procedia engineering*. Elsevier.

Fricke, H. (1953). The Maxwell–Wagner dispersion in a suspension of ellipsoids. *The Journal of Physical Chemistry A, 57*(9), 934–937. https://doi.org/10.1021/j150510a018. Publication Date: September 1953.

Gary, J., Liu, S. N., Tamma, C., Xia, H., Kao, T. J., Newell, J. C., et al. (2007). An electrical impedance spectroscopy system for breast cancer detection. In *Proceedings of the 29th annual international conference of the IEEE EMBS* (pp. 4154–4157). Lyon: France: CitéInternationale.

Geddes, L. A., Costa, C. P. D., & Wise, G. (1971). The impedance of stainless-steel electrodes. *Medical & Biological Engineering, 9*, 511–521.

Geselowitz, D. B. (1971). An application of electrocardiographic lead theory to impedance plethysmography. *IEEE Transactions on Biomedical Engineering, 18*, 38–41.

Gisser, D. G. (1987). Current topics in impedance imaging. *Clinical Physics & Physiological Measurement, 8*(Suppl. A), 39–46.

Gisser, D. G., Newell, J. C., Salunier, G., Hochgraf, C., Cook, R. D., & Goble, J. C. (1991). Analog electronics for a high-speed high-precision electrical impedance tomography. In *Proceedings of the annual international conference of the IEEE EMBS (Engineering in Medicine and Biology Society)* (Vol. 13, pp. 23–24).

Goovaerts, H. G., de Vries, F. R., Meijer, J. H., de Vries, P. M. J. M., Donker, A. J. M., & Schneider, H. (1988). Microprocessor-based system for measurement of electrical impedances during haemodialysis and in postoperative care. *Medical, & Biological Engineering & Computing, 26*, 75–80.

Hartov, A., Demidenko, E., Soni, N., Markova, M., & Paulsen, K. (2002). Using voltage sources as current drivers for electrical impedance tomography. *Measurement Science and Technology, 13*, 1425–1430.

Henderson, P. R., & Webster, J. G. (1978). An impedance camera for spatial specific measurements of the thorax. *IEEE Transactions on Biomedical Engineering BME-, 25*, 250–254.

Henderson, P. R., Webster, J. G., & Swanson, D. K. (1976). A thoracic electrical impedance camera. In *Proceedings of the 29th annual conference on engineering in medicine and biology, Nov. 6–10, Boston* (p. 18 322).

Holden, P. J., Wang, M., Mann, R., Dickin, F. J., & Edwards, R. B. (1998). Imaging stirred-vessel macromixing using electrical resistance tomography. *AIChE Journal*, 780–790.

Holden, P. J., Wang, M., Mann, R., Dickin, F. J., & Edwards, R. B. (1999). On detecting mixing pathologies inside a stirred vessel using electrical resistance tomography. *Transactions of IChemE, 77*(A), 709–712.

Holder, D. S. (Ed.). (2005). *Electrical impedance tomography, methods, history and applications*. Bristol: IOP Publishing.

Holliday, N. T., Ma, Y., Wang, M., & Williams, R. A. (2003). A powerful computational DSP module for multiphase flow measurement. In *Proceedings of the 3rd World congress on industrial process tomography* (pp. 62–67). Banff: VCIPT.

Hua, P., Webster, J. G., & Tompkins, W. J. (1987). Effect of the measurement method on noise handling and image quality of EIT imaging. In *Proceedings of the annual international conference of the IEEE engineering in medicine and biology society* (Vol. 9, pp. 1429–1430).

Hua, P., Woo, E. J., Webster, J. G., & Tompkins, W. J. (1993). Using compound electrodes in electrical impedance tomography. *IEEE Transactions on Biomedical Engineering, 40*, 29–34.

Hurry R.; Wang M.; Williams R.A.; Cross C.; Esdaile L. Development and application of industrial process tomography to monitoring liquid distribution in heap and column leaching. In: JKMRC, international student conference, Brisbane, Australia. Industrial tomography systems. Sunlight House, Quay Street, Manchester, M3 3JZ, UK(2004)

Jia, J., Wang, M., Schlaberg, H. I., & Li, H. (2010). A novel tomographic sensing system for high electrically conductive multiphase flow measurement. *Flow Measurement and Instrumentation, 21*, 184–190.

Jia, J., Wang, M., Schlaberg, H. I., & Li, H. (2012). An optimisation method for the overzero switching scheme. *Flow Measurement and Instrumentation, 27*, 47–52.

Kaipio, J. P., Seppänen, A., & Somersalo, E. (2005). Inference of velocity fields with tomographic measurements. In *Proceedings of 4th World congress on industrial process tomography. Aizu: Japan* (pp. 625–630).

Kak, A. C., & Slaney, M. (2001). *Principles of computerized tomographic imaging*. Philadelphia, PA: The Society for Industrial and Applied Mathematics (SIAM).

Kotre, C. J. (1994). EIT image reconstruction using sensitivity weighted filtered back-projection. *Physiological Measurement, 15*, A125–A136.

Landweber, L. (1951). An iterative formula for Fredholm integral equations of the first kind. *American Journal of Mathematics, 73*, 615–624.

Lehr, J. (1972). A vector derivation useful in impedance plethysmographic field calculations. *IEEE Transactions on Biomedical Engineering, 19*, 156–157.

Lévêque, J., Rivera, L., & Wittlinger, G. (1993). On the use of the checher-board test to assess the resolution of tomographic inversions. *Geophysical Journal International, 115*, 313–318.

Lidgey, F. J., Zhu, Q. S., McLeod, C. N., & Breckon, W. R. (1992). Electrical impedance tomography. Electrode current determination from programmable voltage sources. *Clinical Physics & Physiological Measurement, 13*(Suppl. A), 43–46.

Lionheart, W. R. B. (2001). Regularization, constraints and iterative reconstruction algorithms for EIT. In *Scientific abstracts of 3rd EPSRC engineering network meeting*. London: UCL.

Liu, S., Fu, L., & Yang, W. Q. (1999). Optimization of an iterative image reconstruction algorithm for electrical capacitance tomography. *Measurement Science and Technology, 10*, L37–L39.

Li, H., Wang, M., Wu, Y., & Lucas, G. P. (2009). *Volume flow rate measurement in vertical oil-in-water pipe flow using electrical impedance tomography and a local probe*. https://doi.org/10.1615/MultScienTechn.v21.i1–2.70

Li, H., Wang, M., Wu, Y., Ma, Y., & Williams, R. (2005). Measurement of oil volume fraction and velocity distributions in vertical oil-in-water flows using electrical resistance tomography and a local probe. *Journal of Zhejiang University - Science, 6A*(12), 1412–1415.

Lucas, G. P., Cory, J., Waterfall, R., Loh, W. W., & Dickin, F. J. (1999). Measurement of the solids volume fraction and velocity distributions in solid-liquid flows using dual-plane electrical resistance tomography. *Journal of Flow Measurement and Instrumentation, 10*, 249–258.

Lytle, R. J., & Dines, K. A. (1978). *An impedance camera: A system for determining the spatial variation of electrical conductivity*. Livermore, CA: UCPL-52413 Lawrence Livermore Lab.

Mactaggart, Nasr-El-Din, & Masliyah. (1993). A conductivity probe for measuring local solids concentration in a slurry mixing vessel. *Separations Technology, 3*, 151–160.

Ma, Y., Holliday, N., Dai, Y., Wang, M., Williams, R. A., & Lucas, G. (2003). A high performance online data processing EIT system. In *Proceedings of the 3rd World congress on industrial process tomography* (pp. 27–32). Banff: VCIPT.

Ma, Y., Lai, Y., Wang, M., & Jia, J. (2009). Experimental validation of over-zero switching method. In *What where when multi-dimensional advances for industrial process monitoring, Leeds, UK, June 23–24* (pp. 408–416).

Mann, R., Wang, M., Dickin, F. J., Dyakowski, T., Holden, P. J., Forrest, A. E., et al. (1996). Resistance tomography imaging of stirred vessel mixing at plant scale. In *Fluid mixing 5. IChemE symp., series* (Vol. 140, pp. 155–166).

Marquardt, B. W. (1963). An algorithm for least-squares estimation of non-linear parameters. *SIAM Journal on Applied Mathematics, 11*, 431–441.

Ma, Y., & Wang, M. (2004). Performance of a high-speed impedance camera for flow informatics. In *Proceedings of EDSA04*. Manchester: ASME.

Maxwell, J. C. A. (1873). *Treatise on electricity and magnetism* (Vol. 1). New York, NY: Dover Publications Inc.

McEwan, A., Tapson, J., van Schaik, A., & Holder, D. S. (2009). Code-division-multiplexed electrical impedance tomography spectroscopy. *IEEE Transactions on Biomedical Circuits and Systems, 3*(5), 332–338.

Meredith, R. E., & Tobias, C. W. (1986). *Advances in electrochemistry and electrochemical engineering* (2nd ed.). New York, NY: Interscience.

Mosorov, V., Sankowski, D., Mazurkiewicz, L., & Dyakowski, T. (2002). The 'best-correlated pixels' method for solid mass flow measurements using electrical capacitance tomography. *Measurement Science and Technology, 13*, 1810–1814.

Murai, T., & Kagawa, Y. (1985). Electrical impedance computed tomography based on a finite element model. *IEEE Transactions on Biomedical Engineering, 32*, 177–184.

Murphy, D., & Rolfe, P. (1988). Aspects of instrumentation design for impedance imaging. *Clinical Physics & Physiological Measurement, 9A*, 5–14.

Newell, J. C., Gisser, D. G., & Isaacson, D. (1988). An electric current tomography. *IEEE Transactions on Biomedical Engineering, 35*, 828–833.

Oh, T. I., Koo, H., Lee, K. H., Kim, S. M., Lee, J., Kim, S. W., et al. (2008). Validation of a multi-frequency electrical impedance tomography (mfEIT) system KHU Mark1: Impedance spectroscopy and time-difference imaging. *Physiological Measurement, 29*, 295–307.

Paulson, K. S., Breckon, W. R., & Pidcock, M. K. (1992). A hybrid phantom for electrical impedance tomography. *Clinical Physics & Physiological Measurement, A13*, 155–161.

Pollak, V. (1974a). An equivalent diagram for the interface impedance of metal needle electrodes. *Medical & Biological Engineering*, 454–459.

Pollak, V. (1974b). Computation of the impedance characteristics of metal electrodes for biological investigations. *Medical & Biological Engineering*, 460–464.

Prager, S. (1963). Diffusion and viscous flow in concentrated suspensions. *Physica, 29*, 129–139.

Qiu, C. H., Dickin, F. J., James, A. E., & Wang, M. (1994). Electrical resistance tomography for imaging sub-seabed sediment porosity: Initial findings from laboratory-scale experiments using spherical glass beads. In M. S. Beck, E. Campogrande, M. Morris, R. A. Williams, & R. C. Waterfall (Eds.), *Process tomography–a strategy for industrial exploitation* (pp. 33–41). Oporto: UMIST.

Qui, C., Dickin, F., James, A., & Wang, M. (1997). Electrical resistance tomography for imaging sub-seabed sediment porosity. In *Proceedings of ICEMI '97, Beijing* (pp. 666–669).

Ronald, S. B., et al. (1993). Geopulse – a modular resistivity meter for electrical imaging (tomography). In *Proceedings of the symposium on the application of geophysics to engineering and environmental problems, enviromental and engineering geophysical society, 1, San Diego, CA, USA*.

Ruzinsky, F., & Bennington, C. P. J. (2007). Aspects of liquor flow in a model digester measured using electrical resistance tomography. *Chemical Engineering Journal, 130*(2–3), 67–74.

Saulnier, G. J., Liu, N., Tamma, C., Xia, H., Kao, T., Newell, J. C., & Isaacson, D. (2007). An electrical impedance spectroscopy system for breast cancer detection. In *Proceedings of the 29th annual international conference of the IEEE EMBS CitéInternationale, Lyon, France, August 23–26.*

Schlaberg, H. I., Baas, J. H., Wang, M., Best, J. L., Williams, R. A., & Peakall, J. (2006). Electrical resistance tomography for suspended sediment measurement in open channel flows using a novel sensor design. *Particle & Particle Systems Characterization, 23*(3/4), 78.

Schlaberg, H. I., Jia, J., Qiu, C., Wang, M., & Li, H. (2008). Development and application of the fast impedance camera - a high performance dual-plane electrical impedance tomography system. In *Proceedings of 5th international symposium on process tomography, Zakopane, Poland August.*

Schlaberg, H. I., Wang, M., Qiu, C., & Shirhatti, V. (2005). Parallel array sensors for process visualisation. In *Proceedings of 4th World congress on industrial process tomography. Aizu* (pp. 177–182).

Schwan, H. P., & Ferris, C. D. (1968). Four-electrode null techniques for impedance measurement with high resolution. *The Review of Scientific Instruments, 39*(4), 481–485.

Seagar, A., Barber, D. C., & Brown, B. H. (1987). Theoretical limits to sensitivity and resolution in impedance imaging. *Clinical Physics & Physiological Measurement, 8*(Suppl. A), 13–31.

Sharifi, M., & Young, B. (2013). Electrical resistance tomography (ERT) applications to chemical engineering. *Chemical Engineering Research and Design, 91*(9), 1625–1645.

Shirhatti, V. S. (2007). *Characterisation and visualisation of particulate solid-liquid mixing using electrical resistance tomography* (Ph.D. thesis). University of Leeds.

Sikora, J., & Wojtowicz, S. (Eds.). (2010). *Industrial and biological tomography –theoretical basis and applications.* Warszawa: Electrotechnical Institute.

Smith, R. W. M., Freeston, I. L., Brown, B. H., & Sinton, A. M. (1992). Design of a phase-sensitive detector to maximize signal-to-noise ratio in the presence of Gaussian wide-band noise. *Measurement Science and Technology, 3*, 1054–1062.

Swanson, D. K. (1976). *Measurement errors and the origin of electrical impedance changes in the limb* (Ph.D. thesis). Madison, WI: Department Electrical and Computer Engineering University of Wisconsin.

vanWeereld, J. J. A., Collie, D. A. L., & Player, M. A. (2001). A fast resistance measurement system for impedance tomography using a bipolar DC pulse method. *Measurement Science and Technology, 12*, 1002–1011.

Vauhkonen, M., Vadasz, D., Karjalainen, P. A., Somersalo, E., & Kaipio, J. P. (1998). Tokhonov regularization and prior information in electrical impedance tomography. *IEEE Transactions on Medical Imaging, 17*(2), 285–293.

Vijayan, M., Schlaberg, H. I., & Wang, M. (2007). Effects of sparger geometry on the mechanism of flow pattern transition in a bubble column. *Chemical Engineering Journal, 130*, 171–178.

Wang, M. (1994). *Electrical impedance tomography on conducting walled vessels* (Ph.D. thesis). UK: UMIST (now University of Manchester).

Wang, M. (2002). Inverse solutions for electrical impedance tomography based on conjugate gradients methods. *Measurement Science and Technology IOP, 13*, 101–117.

Wang, M. (2005a). Seeing a new dimension – the past decade's developments on electrical impedance tomography. *Progress in Natural Science, 15*(13), 1–13.

Wang, M. (2005b). Electrode models in electrical impedance tomography. *Journal of Zhejiang University – Science, 6A*(12), 1386–1389.

Wang, M., Dickin, F. J., & Beck, M. S. (1993). Improved electrical impedance tomography data collection system and measurement protocols. In M. S. Beck, E. Campogrande, M. Morris, R. A. Williams, & R. C. Waterfall (Eds.), *Tomography technique and process design and operation* (pp. 75–88). Manchester: Computational Mechanics Publications.

Wang, M., Dickin, F. J., & Williams, R. A. (1994). Electrical resistance tomography of metal walled vessels and pipelines. *Electronics Letters, 30*(10), 771–773.

Wang, M., Dickin, F., & Williams, R. (1995a). *Electrical impedance tomography*. PCT/GB95/00520, GB2300927, AU18570/95, EP749285, US5807251, Publication number WO 95/24155.

Wang, M., Dickin, F.J., & Williams, R.A. (1995b). Modelling and analysis of electrically conducting vessels and pipelines in electrical resistance process tomography. *IEE Proceedings – Science, Measurement and Technology, 142*(4), 313–322.

Wang, B., Hu, Y., Ji, H., Huang, Z., & Li, H. (2012). A novel electrical resistance tomography system based on C4DTechnique. *Proceedings of 17th IEEE Instrumentation and Measurement Technology. Conf. 1929–32, 62*(5), 1017–1024.

Wang, M., & Jia, J. B. (2009). *EIT system for highly conductive medium. PCT/GB2010/051035, GB0910704.6, Publication number CA2803667 (A1).*

Wang, M., Jones, T. F., & Williams, R. A. (2003). Visualisation of asymmetric solids distribution in horizontal swirling flows using electrical resistance tomography. *Transactions on IChemE, 81*(Part A), 854–861.

Wang, M., Lucas, G., Dai, Y., Panayotopoulos, N., & Williams, A. (2006). Visualisation of bubbly velocity distribution in a swirling flow using electrical resistance tomography. *Particle & Particle Systems Characterization, 23*, 321–329.

Wang, M., & Ma, Y. (2006). Over-zero switching scheme for fast data collection operation in electrical impedance tomography. *Measurement Science and Technology IOP, 17*(8), 2078–2082.

Wang, M., Ma, Y., Dai, Y., & Holliday, N. (2007). *EIT data processing system & methods. PCT/GB04/003670, GB2422676, US 10/570,054.*

Wang, M., Ma, Y., Holliday, N., Dai, Y., Williams, R. A., & Lucas, G. (2005). A high performance EIT system. *IEEE Sensors Journal, 5*(2), 289–299.

Wang, M., Mann, R., & Dickin, F. J. (1996). A large scale tomographic sensing system for mixing processes. In *Proceedings for IWISP'96 (the 3rd int. Workshop on image and signal processing advances in computational intelligence in manchester)* (pp. 647–650). Manchester: Elsevier Science B.V.

Wang, M., Mann, R., & Dickin, F. J. (1999). Electrical resistance tomographic sensing systems for industrial applications. *Chemical Engineering Communications, 175*, 49–70. OPA (Overseas Publishers Association).

Wang, M., Wang, Q., & Karki, B. (2016). Arts of electrical impedance tomographic sensing. *Philosophical Transactions of the Royal Society A, 374*, 20150329.

Wang, M., & Yin, W. (2002). *Electrical impedance tomography. PCT/GB01/05636, GB0129772.9, EP1347706, US 6940286, Publication number WO 02/053029.*

Wang, M., Yin, W., & Holliday, N. (2002). A highly adaptive electrical impedance sensing system for flow measurement. *Measurement Science and Technology IOP, 13*, 1884–1889.

Webster, J. G. (1978). Amplifiers and signal processing. In J. G. Webster (Ed.), *Medical instrumentation: Application and design*. Boston: Houghton Mifflin.

Webster, J. G. (Ed.). (1990). *Electrical impedance tomography*. Bristol: Adam Hilger.

Weinman, J., & Mahler, J. (1964). An analysis of electrical properties of metal electrodes. *Medical Electronics & Biological Engineering, 2*, 299–310.

Weissberg, H. L. (1963). Effective diffusion coefficient in porous media. *Journal of Applied Physics, 34*, 2636–2639.

Wilkinson, A. J., Randall, E. W., Cilliers, J. J., Durrett, D. R., Naidoo, T., & Long, T. A. (2005). 1000-measurement frames/second ERT data capture system with real-time visualization. *IEEE Sensors Journal, 5*(2), 300–307.

Williams, R. A., & Beck, M. S. (Eds.). (1995). *Process tomography- principles, techniques and applications*. Amsterdam: Butterworth-Heinemann.

Wilson, A. J., Milnes, P., Waterworth, A. R., Smallwood, R. H., & Brown, B. H. (2001). Mk3.5: A modular, multi-frequency successor to the Mk3a EIS/EIT system. *Physiological Measurement, 22*(1), 49–54.

Woo, E. J., Hua, P., Webster, J. G., & Tompkins, J. (1993). A robust image reconstruction algorithm and its parallel implementation in electrical impedance tomography. *IEEE Transactions on Medical Imaging, 12*(2), 137–146.

Wu, Y., Li, H., Wang, M., & Williams, R. A. (2005). Characterization of air-water two-phase vertical flow by using electrical resistance imaging. *Canadian Journal of Chemical Engineering, 83*(1), 37–41.

York, T. (2001). Status of electrical tomography in industrial applications. *Journal of Electronic Imaging, 10*(3), 608–619.

Yorkey, T. J., Webster, J. G., & Tompkins, W. J. (1987). Comparing reconstruction algorithm for electrical impedance tomography. *IEEE Transactions on Biomedical Engineering, 34*, 843–851.

Yuen, E. L., Mann, R., York, T. A., & Grieveet, B. D. (2001). Electrical resistance tomography imaging of a metal walled solid-liquid filter. In *Proceedings of 2nd World congress on industrial process tomography* (pp. 183–190). Hanover: VCIPT.

Zhao, Y., Wang, M., & Hammond, R. B. (2011). Characterisation of crystallisation processes with electrical impedance spectroscopy. *Nuclear Engineering & Design Journal, 241*, 1938–1944.

Zhao, Y., Wang, M., & Hammond, R. B. (2013). Characterisation of nano-particles in colloids: Relationship between particle size and electrical impedance spectra. *Journal of Nanoscience and Nanotechnology, 13*(2), 808–812.

Zhu, Q. S., McLeod, C. N., Denyer, C. W., Lidgey, F. J., & Lionheart, W. R. (1994). Development of a real-time adaptive current tomograph. *Physiological Measurement, 15*(Suppl. 2a), A37–A43.

Electromagnetic induction tomography

Anthony J. Peyton
Department of Electrical and Electronic Engineering, University of Manchester, Manchester, United Kingdom

3.1 Introduction

Electromagnetic induction tomography is similar in terms of operation to the other electrical tomography methods and complements electrical resistance tomography (ERT), electrical impedance tomography (EIT), and electrical capacitance tomography (ECT) in completing a set of techniques based on the measurement of electrical impedance (be it resistance, capacitance, and inductance). The technique is known loosely as electromagnetic tomography (EMT), but may be better referred to as magnetic induction tomography (MIT) or eddy current tomography by different workers in the field (Griffiths, 2001; Norton & Bowler, 1993). Electromagnetic induction tomography operates by exciting the region of interest with an alternating magnetic field, whereas ECT and ERT both use electric field excitation. MIT exploits the principles of magnetic polarization and Faraday induction and is sensitive to both magnetically permeable material due to the polarization of magnetic dipoles and electrically conductive material due to the flow of magnetically induced eddy currents. A comparison of ECT, ERT, and MIT is given in Table 3.1 showing their principal material sensitivities, nature of their sensing element, and examples of applications.

The operation of an MIT system can be split into four generic functions:

1. The excitation of the region of interest with an energizing magnetic field, using alternating current flowing through a coil (or coils)
2. Modification of this field resulting from the action of the material distribution within the object space.
3. Boundary measurement of the resulting modified field distribution. The transmitters and receivers are usually placed on external boundary based on accessibility to the process and not necessarily in a circular configuration.
4. Image reconstruction, often termed the inverse problem or inversion, which involves converting the measurements back into an image or measurements of the original material distribution.

In many cases, the system output will not be in the form of an image, but may be a parameter or profile of interest. This output may be either calculated directly by the inversion process without an image, or may be obtained indirectly via an intermediary step involving image reconstruction.

Industrial Tomography. https://doi.org/10.1016/B978-0-12-823015-2.00005-4

Table 3.1 A comparison of ECT, ERT and MIT.

Method	Sensor elements	Typical arrangement	Measure values	Typical material properties	Typical material
ECT	Capacitive plates		Capacitance **C**	ε_r 1–100 $\sigma < 0.1$ S/m Low	Oil, water, nonmetallic powders, polymers, burning gasses
ERT (EIT)	Electrodes		Resistance (Impedance) **R/Z**	σ 0.001–10^7 S/m Wide – determined by drive and detection Circuitry ε_r 1–100	Water/saline, biological tissue, rock/geological materials, semiconductors e.g., silicon
MIT	Coils		Self/mutual Inductance **L/M**	σ 0.01–107 S/m depends on size and frequency range μ_r 1–10,000	Metals, some minerals, magnetic materials, biological tissue, graphite.

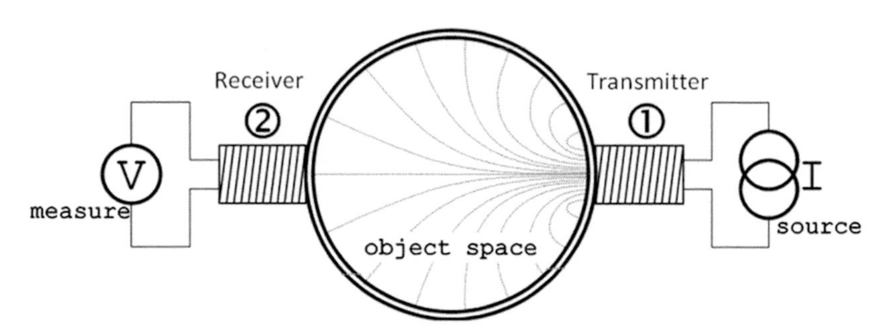

Figure 3.1 Simple concept of an electromagnetic induction tomography system.

Referring to the twin coil example shown in Fig. 3.1, a sinusoidal current is applied to the transmitter (the energizing coil) to generate the magnetic field within the object space. We can assume a background condition with a relative permeability of unity when the object space is empty (filled with air). In this condition, a measurable voltage will be induced in the receiver coil, corresponding to a background or empty space measurement. The relationship between the received voltage v and the transmitter current i is given by

$$v = j\omega Mi \tag{3.1}$$

where M is the mutual inductance between the transmitter and receiver coils.

The spatial distribution of the magnetic field and hence the mutual coupling between the coils is altered by the introduction of magnetic (relative permeability >1) and/or metallic (high conductivity) items inside the object space. This gives rise to a change in the signal Δv in the receiver coil caused by the change in mutual inductance, ΔM, as given by

$$\Delta v = j\omega \, \Delta M \, i \tag{3.2}$$

Magnetic material generally increases the mutual coupling between the coils, resulting in an increased measurement, whereas conducting objects generally attenuate the measurement due to eddy current losses in the material itself. Note ΔM can be complex representing both reactive and resistive coupling. These effects will be described in more detail shortly in Section 3.2. In order to acquire sufficient data for image reconstruction, several successive excitation profiles, loosely termed projections, must be applied to the object space and for each projection a number of receiver measurement should be made. This can be achieved using a single transmitter–receiver pair and indeed one of the earliest reports of MIT (Al-Zeibak & Saunders, 1993) described an experimental system in which the object was physically moved through a transmitter–receiver coil pair to obtain one projection; then the object was rotated and the process was repeated for another projection; and this step was repeated several times to obtain the multiple projections for image reconstruction. Clearly, this mechanical procedure is time consuming and only suited for a limited number of laboratory

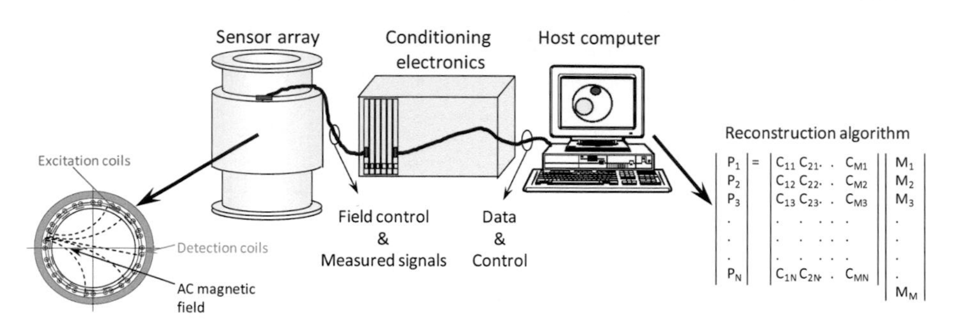

Figure 3.2 Block diagram of a typical electromagnetic induction tomography system.

applications. Virtually, all practical system use multiple transmitters and receivers and multiplex electronically to acquire the measure the mutual coupling between receiver and transmitter elements in real time.

Fig. 3.2 shows the main elements of an MIT. The system is split into three main subsystems (sensor array, conditioning electronics, and host computer). The sensor array is energized with an AC magnetic field created by one or more of the excitation coils. Electrically conductive or ferromagnetic objects within the space modify the distribution of the applied magnetic field and the resultant field changes are measured on the periphery of the space with an array of detection coils. These measurements are fed into a host computer. A set of such measurements is taken for a variety of flux injection patterns, and from the acquired data, the host computer is able to produce an image of the material distribution by using suitable reconstruction software, which is shown in a much simplified form in the diagram as a matrix multiplication relating measured data value to pixel values. In practice, more sophisticated algorithms are used as described later in Section 3.4.

In the context of this chapter, MIT generally refers to systems with excitation coils as the transmitters and detection coils as the receivers; so in effect, the MIT system exploits the changes in the complex mutual induction between the transmitter and receiver coils. In some systems, the coils may perform both transmit and receive functions, for example (Ma, Peyton, Higson, Lyons, & Dickinson, 2006a,b), although in general separate coils are preferred as this avoids the need for circuitry to perform the switching function. Coils are also relatively easy to wind and coils can often be wound on top of each other. Virtually, all applications use coils as opposed to permanent magnets for excitation simply because coils can be electronically controlled and operated at high frequencies (MHz's) if necessary. Coils are also widely used as receivers and are generally preferred over other forms of magnetometers such as flux gates, magneto-resistors, and hall sensors because most MIT systems operate in the range of tens of kHz to tens of MHz, where induction coils have benefits in terms of signal to noise performance (Tumanski, 2007).

MIT, ERT, and ECT are often seen as distinct techniques, with ERT and ECT using plates or electrodes and MIT employing coils or magnetometers. However, there are reports of hybrid systems, where the excitation is achieved by induction and the detection is accomplished using contact electrodes. One such example is induced current ERT where AC magnetic fields are used to induce eddy currents in the object and

ERT is performed using contact electrodes as detectors (Tanguay, Gagnon, & Guardo, 2007). An alternative example (Resagk, Men, Ziolkowski, Kuilekov, & Brauer, 2006) employed direct current injection for excitation and magnetometers to detect the resulting magnetic field caused by the current flow. In this study, two conducting fluids were contained in a cylindrical cell and the tomographic reconstruction problem was to determine the shape of the 2D interface between the two fluids.

In addition to AC induction, the motion of the object material through the sensor array can also create a measureable secondary magnetic field, as a result of Faraday's law of induction. This approach is particularly appropriate for measuring velocity flow in conducting fluids. The excitation field is typically applied with a DC or very low frequency electromagnet and the resulting secondary field can be measured using either (i) contract electrodes or (ii) very sensitive magnetometers, such as, for instance, flux gates. An example of (i) is the measurement of flow rate through pipes. In particular, conventional electromagnetic flow meters consisting of a perpendicular arrangement of flow, magnetic field, and measurement electrodes have been widely used. Unfortunately, conventional flow meters are sensitive to the flow profile within the liquid and consequently must be installed away from bends. The tomographic approach has been reported (Horner, Mesch, & Trächtler, 1996) to improve the response of these measurement devices. Recently, this approach has been extended to allow single and dual phase flow profiles to be imaged (Zhao, Lucas, Leeungculsatien, & Zhang, 2012). An example of (ii) is contactless inductive flow tomography (CIFT) and in one report (Stefani, Gundrum, & Gerbeth, 2004) the 3D velocity profile of a room temperature liquid metal, InGaSn, has been demonstrated when the liquid InGaSn is driven by an impeller. The CIFT technique has also been applied to models of continuous casting of the type commonly used in the steel industry (Timmel, Eckert, Gerbeth, Stefani, & Wondrak, 2010).

The noncontacting and noninvasive nature of inductive tomography systems has made these systems attractive to research in a number of applications where conductive or magnetic materials are involved. The ability to vary the excitation frequency, which in turn determines the depth of penetration of the field into a conducting object, allows frequency to be used for depth profiling and hence to provide additional depth information. Examples of applications include nondestructive testing (NDT), such as imaging cracks in metallic components and depth profiling of conducting coatings; imaging the flow of conducting fluids and in particular the flow of liquid metals; security applications such as characterizing metal objects in walk through metal detectors (WRMDs); biomedical applications; and inspection of nuclear graphite. These are described in more detail later in Section 3.6.

3.2 Principle of operation and governing equations

A formal description of this measuring system can be derived from Maxwell's equations, as follows:

$$\nabla \times \boldsymbol{E} = -\frac{\partial \boldsymbol{B}}{dt} \tag{3.3a}$$

$$\nabla \times \boldsymbol{H} = \boldsymbol{J} - \frac{\partial \boldsymbol{D}}{dt} \tag{3.3b}$$

$$\nabla \cdot \boldsymbol{D} = \rho \tag{3.3c}$$

$$\nabla \cdot \boldsymbol{B} = 0 \tag{3.3d}$$

where $\boldsymbol{B}$ is the magnetic flux density vector, $\boldsymbol{H}$ is the magnetic field strength vector, $\boldsymbol{D}$ is the electrical displacement, $\boldsymbol{E}$ the electric field strength, $\boldsymbol{J}$ is the current density, and ρ the charge density at position $\boldsymbol{r}$ in space. If we assume the materials have linear and isotropic electrical and magnetic properties then:

$$\boldsymbol{B} = \mu_0 \mu_r \boldsymbol{H} \tag{3.4a}$$

$$\boldsymbol{D} = \varepsilon_0 \varepsilon_r \boldsymbol{E} \tag{3.4b}$$

$$\boldsymbol{J} = \sigma \ \boldsymbol{E} \tag{3.4c}$$

where μ and ε represent permeability and permittivity, respectively, with subscripts 0 and r in turn describing free space and relative values and σ is the electrical conductivity. For many practical applications, several fundamental simplifications can be made, such as (i) ignoring the effects of the displacement current which may be negligible for the materials and signal frequencies of interest. i.e.,

$\sigma \gg j\omega \ \varepsilon_0 \varepsilon_r$ and (ii) ignoring the velocity terms, which are mainly of interest to CIFT. So Maxwell's equations for the induction problem can be rewritten in complex phasor notation as

$$\nabla \times \left[\left[\nabla \times \frac{\boldsymbol{B}}{\mu} \right] \cdot \sigma^{-1} \right] = -j\omega \boldsymbol{B} \tag{3.5a}$$

$$\nabla . \boldsymbol{B} = 0 \tag{3.5b}$$

$$\nabla . \boldsymbol{D} = \rho \tag{3.5c}$$

These set of simultaneous differential equations describe the general forward MIT problem and should be solved with the appropriate boundary conditions, which express the particular MIT application and sensor geometry, and an outer boundary condition that ensures that the magnetic field tends to zero as $|\boldsymbol{r}| \rightarrow \infty$. The Maxwell expression $\nabla . \boldsymbol{B} = 0$ is a statement of the continuity of the magnetic field. The electrostatic conditions represented Gauss's Law, i.e., $\nabla . \boldsymbol{D} = \rho$, must also be solved together with the induction effect described by Eq. (3.5a). The free charge, ρ, builds up on the surface of between conductors and insulators to ensure that current flow is parallel to the surface. In some special cases, which have two-dimensional symmetry, there may be no accumulation of surface charge; the magnetically induced electric field is tangential to the surface and then electrostatic effects can be ignored. However, in general, a full 3D solution to the field problem is essential.

The vector field expressions are often expressed in terms of vector magnetic potential, $\boldsymbol{A}$, and scalar potential, V, defined as

$$\boldsymbol{B} = \nabla \times \boldsymbol{A} \tag{3.6a}$$

$$\boldsymbol{E} = -\nabla V - j\omega \boldsymbol{A} \tag{3.6b}$$

Considering individual regions in the object space with constant electrical conductivity and magnetic permeability, then Eqs. (3.5a–c) can be expressed as follows:

$$\nabla^2 \boldsymbol{A} + j\omega\mu\sigma \boldsymbol{A} = \mu \boldsymbol{J} \tag{3.7a}$$

$$\nabla^2 V = -\frac{\rho}{\varepsilon} \tag{3.7b}$$

Consideration of Eqs. (3.7a and b) reveals the scaling relations for the ECT, ERT, and MIT modalities. For ERT and ECT, which obey the Laplace/Poisson Eq. (3.7b), a simple scaling relation exists and the geometry of the sensor can be scaled to suit the application as long as all the dimensions are scaled in proportion. For MIT, the scaling relationship is different. Considering Eq. (3.7a), and assuming as dimensional scaling such that

$$\boldsymbol{r} = k\boldsymbol{r}' \tag{3.8}$$

and letting $\boldsymbol{A}'(\boldsymbol{r}')$ be a solution to Eqs. (3.7a and b) and defining $\boldsymbol{A}'(\boldsymbol{r}') = \boldsymbol{A}(k\boldsymbol{r}')$. Since $\nabla^2 \boldsymbol{A} = \frac{1}{k^2} \nabla^2 \boldsymbol{A}'$, Eq. (3.7a) becomes

$$\nabla^2 \boldsymbol{A}' + j\omega\mu\sigma\, k^2 \boldsymbol{A}' = k^2 \mu \boldsymbol{J} \tag{3.9}$$

which is a diffusion equation. Eq. (3.9) indicates that if the current density $\boldsymbol{J}$ and the frequency increase by k^2 times, then the magnetic potential in the new system is a just a geometrically scaled version of the original system. For an inflated system ($k < 1$), if the current density reduces by $\frac{1}{k^2}$ times, then the current remains the same considering the area also increases by $\frac{1}{k^2}$ times. From $\boldsymbol{B}' = \nabla \times \boldsymbol{A}' = k\nabla \times \boldsymbol{A} = k\boldsymbol{B}$, it can be seen that the magnetic flux density reduces by $\frac{1}{k}$ times. However, the area the flux flows through increases by $\frac{1}{k^2}$ times, hence the inductance increases. The induced voltage and the inductance can also be expressed in terms of a line integration of $\boldsymbol{A}$, where an inflated system increases the integral line length and results in an increase in inductance. These conclusions are true for 3D cases. In a 2D situation, the induced voltage and the inductance are expressed as the difference of at the two end of the coil and therefore the inductance remains the same irrespective of any geometrical scaling.

In general eddy current problems, the depth of penetration, also termed the skin depth, is expressed as

$$\delta = \sqrt{\frac{1}{f\mu\sigma\pi}} \tag{3.10}$$

therefore, $\frac{1}{k^2}$ times increase in frequency corresponds to k times decrease in the skin depth, which essentially is a geometrical quantity. Therefore, by scaling the frequency with a square term of the geometrical scale, a simple relation in terms its primary quantities (magnetic field intensity, flux density, etc.) and secondary quantities (inductance, coupling coefficient) can also be found. These relations provide useful guidelines in process scaling-up and scaling down (Yin & Peyton, 2010a).

The depth of penetration, δ, is an important parameter in MIT because it describes the distance the applied electromagnetic field can penetrate into electrically conducting material. Eq. (3.10) shows that δ is inversely proportional to the square root of frequency. For the plane geometry case, the field will fall away exponentially from the surface of the material such that by a distance of δ from the surface, the field will be $^1/_e$ (36.79%) times the surface value. Clearly, this effect dictates the required range of frequencies for an MIT system. Normally, the depth of penetration is chosen either to be comparable with the dimensions of the largest conducting objects in the object space, or if the space is completely filled with conductive material, to be comparable with the largest dimension of the object space. In the limiting case when δ is very much smaller than the object dimensions, the field penetration is limited to just the very surfaces of the objects and the received signals contain only information on the shape and surface impedance of the conductive material in the object space. On the other hand, when δ is very much larger than the object dimensions, then the field may completely penetrate the object, but the field interactions will be relatively small and therefore the signals caused by the object will also be very small. This is the case for lower conductivity material, for instance, biological tissues. Fig. 3.3 shows a plot of penetration depth versus frequency for a range of nonmagnetic materials of different conductivities. As can be seen, metals (with conductivities in the MS/m range) require low excitation frequencies, kHz or less, whereas biological tissues and aqueous objects (with conductivities of the order of 1 S/m) require frequencies in the MHz range. It is important to note that Fig. 3.3 represents the situation for a nonmagnetic material. For magnetic materials with $\mu_r \gg 1$, such as ferritic steel, then δ will be much smaller than the values in the graph due to the permeability term in Eq. (3.10).

The basis interactions of magnetic and conducting materials with an AC magnetic field, which are relevant to MIT, are summarized in Figs. 3.4–3.6. There are two fundamental physical processes in play, namely polarization and induced eddy currents.

Magnetic materials ($\mu_r > 1$) can polarize due to the interaction of the applied field with the magnetic dipoles at the electronic and atomic scales within the material. Depending on the nature of the dipole interactions, the response of the material may

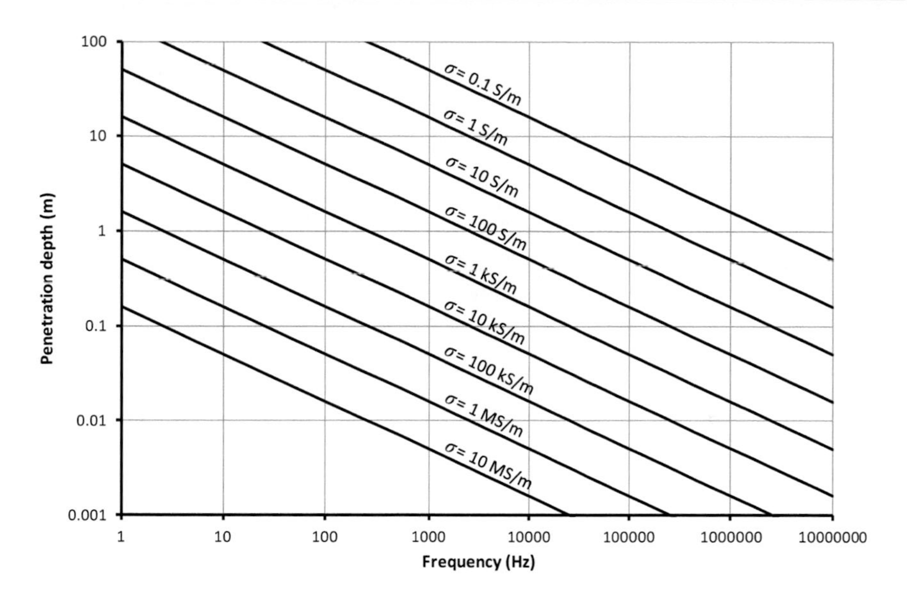

Figure 3.3 Depth of penetration versus frequency for non-magnetic materials of different electrical conductivity.

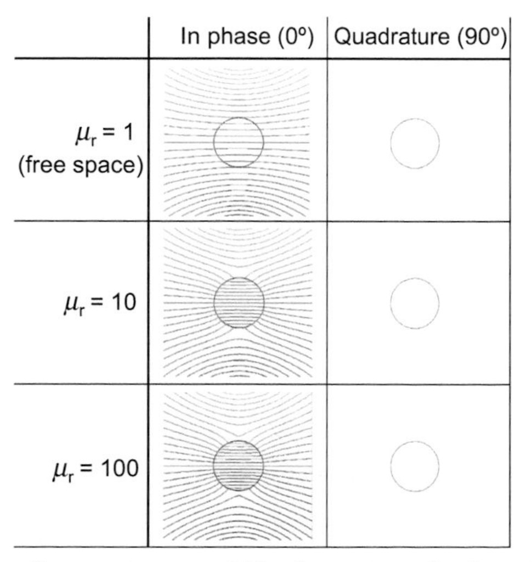

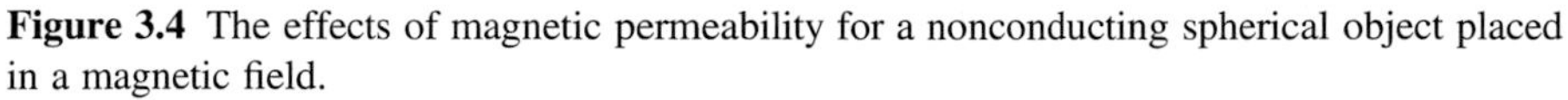

Figure 3.4 The effects of magnetic permeability for a nonconducting spherical object placed in a magnetic field.

be generally classed as either ferrimagnetic or ferromagnetic. In both cases, the material tends to concentrate the magnetic flux and a simple illustration is given in Fig. 3.4. Here a nonconducting sphere is shown in an applied AC magnetic field and the effects of three different values of μ_r are shown on the distribution of the lines of magnetic

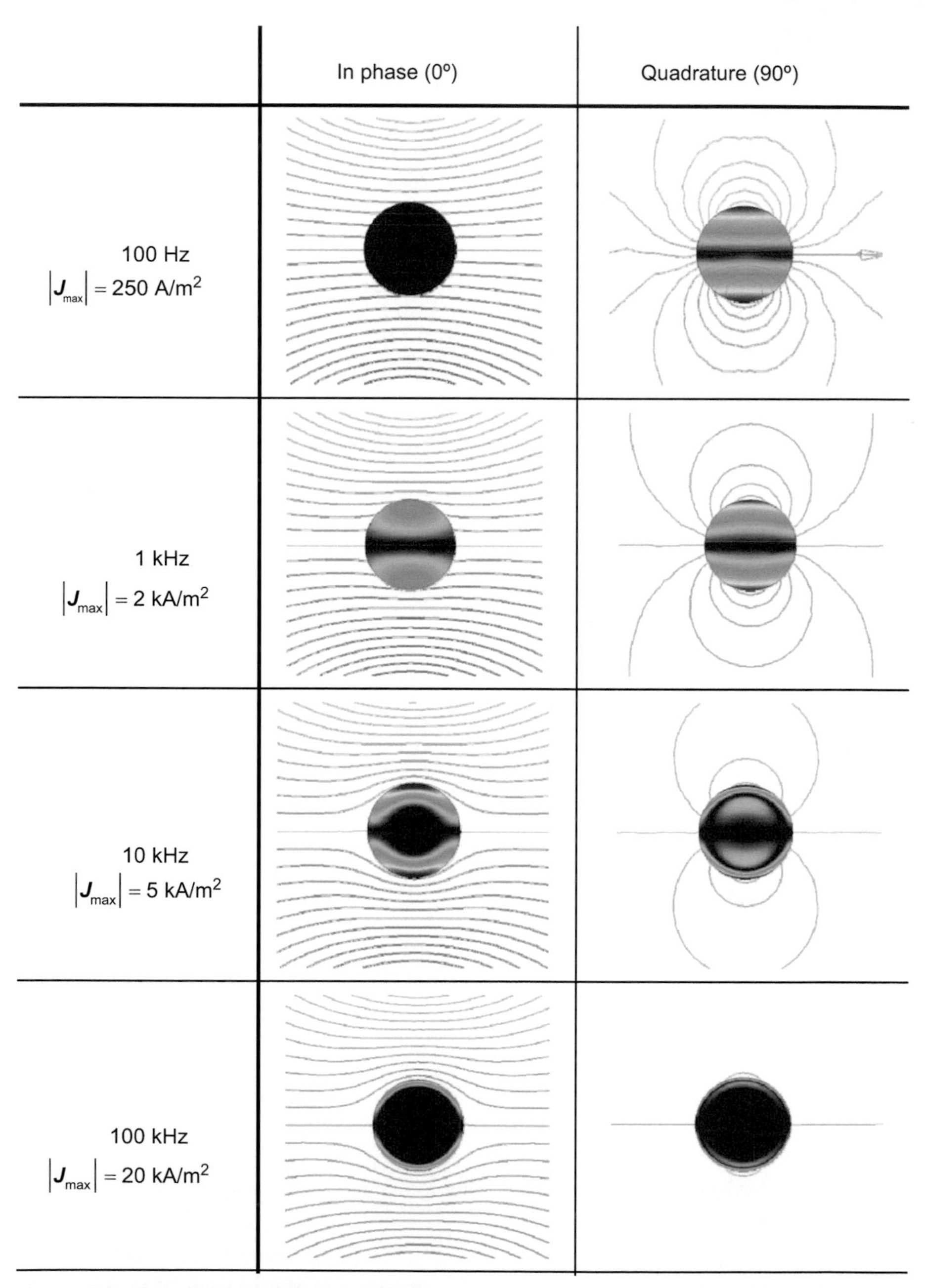

Figure 3.5 The effects of diffusion with frequency for a conducting, non magnetic spherical object placed in an AC magnetic field.

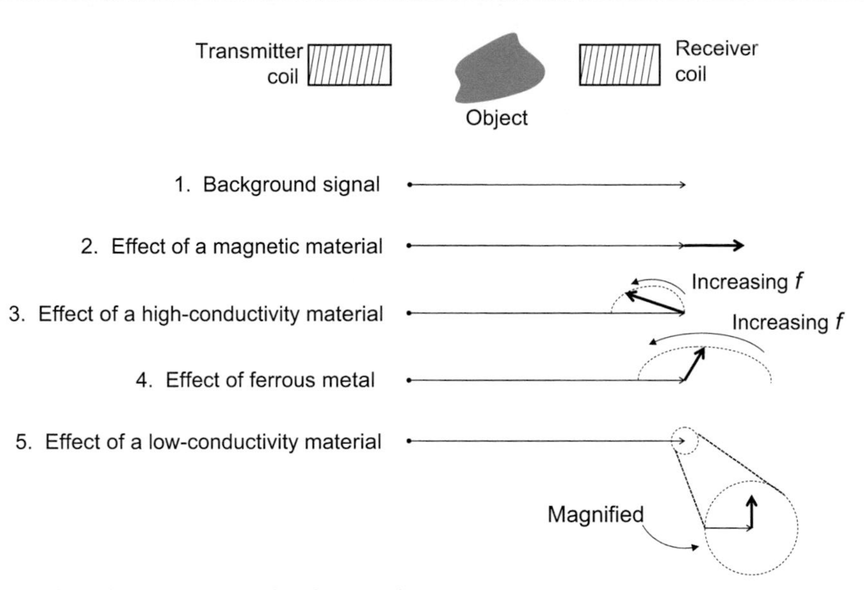

Figure 3.6 Phasor response of different objects in MIT.

flux for both the in-phase and quadrature components of the field. The free space condition, i.e., when $\mu_r = 1$ for the sphere, shows the situation if the sphere was not present. As μ_r increases, the magnetic flux becomes more concentrated within the sphere as shown for the cases with $\mu_r = 10$ and $\mu_r = 100$ where the sphere offers an easier path for the flux to circulate through. Note that the majority of the flux concentration occurs for $1 > \mu_r > 10$. Note also the quadrature component of the field remains zero.

Eddy currents are induced in electrically conducting materials when an AC magnetic field is applied. The eddy currents flow as to oppose the field that has produced them resulting in a net energy loss in system. A simple illusion of the effects of the induced eddy currents in a conducting, nonmagnetic sphere is given in Fig. 3.5. Here the sphere is aluminum and has a diameter of 2 cm. The magnitude of the induced eddy currents within the sphere is shown as a color scale. The color scale is light spectra coded, where blue is zero, green is middle values, and red is maximum. The maximum eddy current magnitudes, $|\boldsymbol{J}_{\max}|$, are also given for each example. At low frequencies (100 Hz), the in-phase component of the field is virtually the same as for free space case, but there is a small quadrature component to the field caused by the induced eddy currents. The induced eddy currents are also the smallest of the four cases shown. As frequency increases, the eddy currents also increase in magnitude, but they are distributed closer to the upper and lower surfaces of the sphere as shown in this cross-section, which illustrates the diffusion effect associated with induced fields. Also as frequency increases, the eddy currents become stronger for the in-phase compared to the quadrature component. At the highest frequency, 100 kHz, the eddy currents flow on the outer skin of the sphere and there is virtually no penetration of the magnetic flux inside the sphere; all the magnetic flux circulates around the object.

Fig. 3.6 shows the effect of a magnetic and/or conducting object on the measured signal as represented by a complex phasor. The diagram shows an idealized situation with the coupling between a transmitter and a receiver with the object positioned in between. Five situations are shown:

1. When there is no object, there is only the background signal resulting from the magnetic flux which directly couples the two coils.
2. For a magnetic, nonconducting object, there is an additional phasor, known as the object signal, which is in-phase with the background signal. Here the object signal is caused by the increased flux passing thought the object. This was also shown earlier in Fig. 3.4.
3. For a conducting, nonmagnetic material, the phasor caused by the object has both in-phase and quadrature components. This situation was also shown earlier in Fig. 3.5. Note the phase angle between the object signal and background signal is approximately 90 degrees at zero frequency and 180 degrees as frequency approaches infinity.
4. The effect of a magnetic and conducting material such as a ferrous metal is the combination of the two cases above. Note that the phase angle between the object signal and background signal is approximately 0 degrees at zero frequency and 180 degrees as frequency approaches infinity.
5. This is a special case for low conducting material, typical of biological tissues. Here there is only a weak interaction between the applied field and the object. The resulting object signal is approximately 90 degrees out of phase with the background. Note that this illustration is idealized and ignores the effects of any diamagnetic or paramagnetic properties and also the flow of capacitive (displacement) current, which will produce a small (and potentially measureable) in-phase signal.

The discussion so far in this section has only covered the two coil case with a single transmitter and receiver. The tomographic problem for MIT is therefore to excite the object space with a number of spatially distinct magnetic field patterns which can be measured at accessible positions on the boundary of the object space. A sufficient number of independent field patterns and field measurements must be used to provide an adequate measurement set to enable images of an acceptable quality to be produced. To achieve this objective, the object space is surrounded by an array of coils. It is usual in MIT to use dedicated excitation and detection coils/channels; however, it is possible to operate coils in both modes using suitable switching circuitry. To obtain the maximum amount of information from the object space measurements, the in-phase (or real) and quadrature (or imaginary) components of both the radial and two tangential field components should be acquired. However, most practical systems ignore the tangential field components and only use coils facing in a radial direction.

3.3 Solution to the forward problem

MIT involves the solution of both the forward and inverse problems. The forward problem is to calculate the received signals for all relevant transmitter/receiver coil combinations if the distribution and electrical properties of the object material are known. This is usually a well-defined problem with a unique solution based on the description of physical laws, as may be expressed by Maxwell's equations. The inverse problem is the converse and involves estimating the object distribution from the measured mutual inductances between the coils, which is both an ill-posed and ill-conditioned problem.

Often the object space is divided into pixel (2D) or voxel (3D) elements and the purpose of the image reconstruction or inversion algorithm is to determine the electrical properties, μ_r or σ of each element to produce a 2D image or 3D representation of the object space. However, the use of pixels or voxels is just one way of representing the object, and in many cases, the end user is interested in particular parameters such as, for example, a conductivity gradient or the position of a boundary between two phases or the position and size of a crack in metallic component. In these cases, it may be more efficient and accurate to represent the object material in a parameterized form and develop a forward solution specifically for that particular description of the problem.

The key requirements of the forward solution are to be accurate and computationally efficient. Unfortunately these are usually conflicting requirements, and over the past 2 decades, there have been reports of several different methods, including the following:

i. Linear approximations
ii. Analytical approach
iii. Finite element method (FEM)
iv. Boundary element method (BEM)
v. Full wave solutions
vi. Impedance method
vii. Dipole approximation

These will be described briefly in the following sections.

A crucial task in the formulation and implementation of the forward problem is to validate and calibrate the model against measurements from the real system. One example of this process for an inductive sensor is described by (Johnstone & Peyton, 2001). For MIT, a calibration process is often needed to fine tune parameters such as the exact position of the coils between measurements and model (Gürsoy & Sharfetter, 2011) and also to account for simplifications assumed in the analysis.

3.3.1 Linear approximation

This approximation may be applied only to very small perturbations in voxel values and a linear response is assumed between changes in the values of the pixels (or voxels) and the consequential changes in the measured mutual inductances between the coils. If there are k pixels, each of value p_i, which represents a required electromagnetic property of the pixel, then the pixels values can be arranged into a vector **P**. Similarly, if there are n measured coil combinations, each with a measured mutual inductance of m_j, then the measured values can be arranged into a vector, **M**. The linear approximation simply relates small changes in **M**, i.e., $\delta\mathbf{M}$, to the change in **P**, i.e., $\delta\mathbf{P}$ via a sensitivity matrix, **S**, as follows:

$$\delta\mathbf{M} = \mathbf{S}.\delta\mathbf{P} \tag{3.11}$$

S is also the Jacobian matrix of the MIT system, where $\delta\mathbf{M}$, **S**, and $\delta\mathbf{P}$ are given by

$$\delta\mathbf{P} = \begin{vmatrix} \delta p_1 \\ \vdots \\ \delta p_i \\ \vdots \\ \delta p_k \end{vmatrix} \quad \delta\mathbf{P} = \begin{vmatrix} \delta m_1 \\ \vdots \\ \delta m_j \\ \vdots \\ \delta m_n \end{vmatrix} \quad \delta\mathbf{P} = \begin{vmatrix} \frac{\partial m_1}{\partial p_1} & \cdots & \frac{\partial m_1}{\partial p_k} \\ \vdots & \ddots & \vdots \\ \frac{\partial m_n}{\partial p_1} & \cdots & \frac{\partial m_n}{\partial p_k} \end{vmatrix} \tag{3.12}$$

Often the δ notation is omitted as the signal changes with respect to a background are implicit and

$\delta\mathbf{P}$ and $\delta\mathbf{M}$ are simply referred to as **P** and **M**, respectively,. The coefficients in **S** can be obtained experimentally by sweeping a small test object, which represents a perturbation, over the object space and measuring the resulting response of each relevant coil combination. This is a laborious process unless automated and so the usual approach is to compute the $\frac{\partial m_j}{\partial p_i}$ coefficients either by perturbing each pixel in turn using, for instance, a FEM model, which is very computationally intensive, or by using the vector fields from a full field solution as described in Section 3.4.

Each row in **S** can be rearranged into its original 2D or 3D pixel structure to represent a sensitivity map of the response of a particular transmitter−receiver coil combination to the spatial position of material in the object space. Fig. 3.7 shows an example of a sensitivity map for a circular 16 pole MIT array in which each pole contained both a transmitter and a receiver coil. In this example, the empty space sensitivity map between poles 1 (12 o'clock position) and 3 (2 o'clock position) is shown. As can be seen, the response if very high for the pixel positions directly in front of the poles.

The advantages of the linear approximation method are ease of implementation and computational simplicity; it is one of the simplest forward models for MIT. Although the linear approximately may be satisfactory for other forms of electrical tomography, such as ECT, where the changes of material properties are relatively small, it does not describe the physics of eddy current induction particular well. Consequently, the disadvantages are manifold. The method is inaccurate, especially for material with a high conductivity or permeability because the object material significantly changes the applied field distribution. Overall, though, the linear forward model does tend to

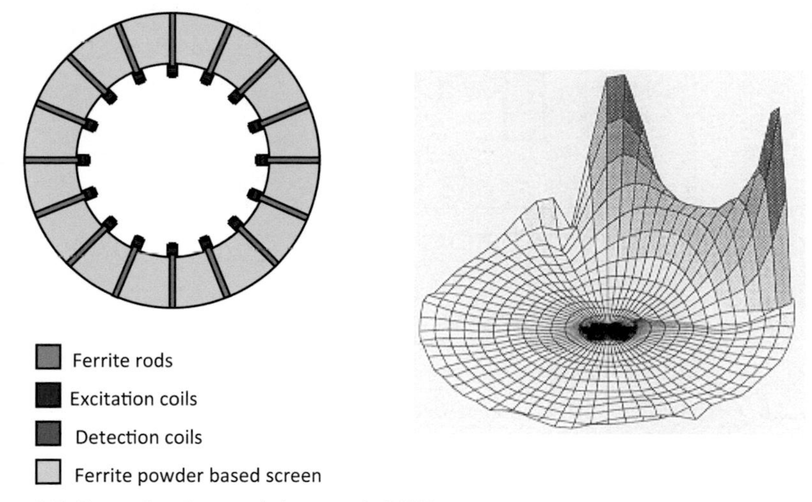

Figure 3.7 Example of a senstivity map in MIT.

produce qualitative images showing roughly where the material is within the object space. It performs better for identifying isolated regions of high conductibility material against an empty space background. For imaging inside conducting regions, the sensitivity matrix should be calculated for a nominal baseline object distribution, which assumes that the external shape of the conducting object is known.

3.3.2 Analytical approach

In cases where the application has a high degree of symmetry or a simple geometry, such as a circular coil positioned in parallel above a large conducting plate or coaxial coil placed on inside a circular pipe, analytical solutions may exist. These can be used either to solve the forward problem directly or to validate the solutions from other computational models by providing examples of known solutions. One popular example is the solution by Dodd and Deeds (1968). This has been extended (Cheng, Dodd, & Deeds, 1971) for circular coil above a layered metal, which is illustrated in schematic form in Fig. 3.8 and can form a basis for tomographic inversion for depth profiling (Yin, Dickinson, & Peyton, 2005).

An advantage of the analytical approach can be computation speed. Unfortunately, the approach can only be used for simple geometries and the method has the disadvantages of having to deal with complicated algebraic solutions which typically involve integrating sums of Bessel functions.

3.3.3 Finite element method

FEM has been widely reported in MIT research, both as a tool to assist sensor design and also as forward solver as part of an iterative reconstruction algorithm (Hollaus, Magele, Merwa, & Scharfetter, 2004a,b; Soleimani, Lionheart, Peyton,

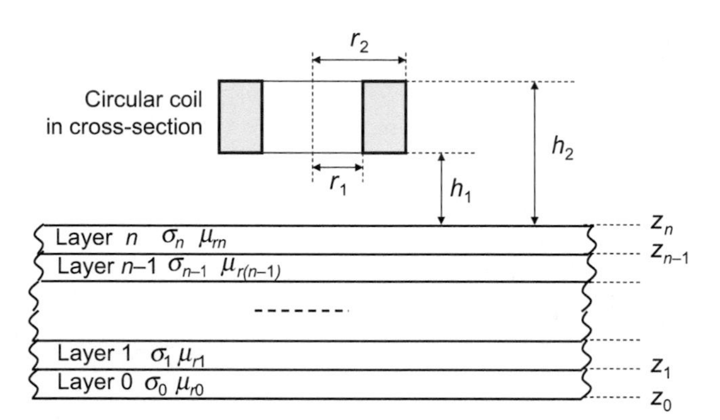

Figure 3.8 A schematic of the solution of Deeds and Dodd for a circular coil above a infinite flat plate.

Ma, & Higson, 2006). There is a considerable amount of literature on the FEM; one good text is Silvester and Ferrari (1996). There are also many commercial suppliers of FEM software, so it is unusual for workers to construct FEM code from first principles. Commercial software packages vary in terms of their features; however, of particular importance for tomography is the ability to access solution data at the mesh level, which is vital for the image reconstruction or inversion process.

The finite element method involves discretizing the problem space in discrete elements, usually triangles for 2D applications and tetrahedra for 3D geometries. The FEM describes the field solution for each element with simple linear or quadratic equations, which are solved simultaneously for the complete system. In MIT, the complete sensor geometry, the material in the object space, and, if the sensor array is not shielded, sufficient external space to account for the field distribution must be included in the model. Eddy current phenomena also present particular difficulties for FEM modeling as the field penetration should well represented. Typically, the tetrahedra should be smaller than the penetration depth described earlier in Eq. (3.10). Regions where the skin depth is small, for example, metal components, and especially ferritic steel, may require a large number of tetrahedra to accurately describe the diffusion of the field, which can result in a problem with a very large number of tetrahedra. In addition, the tetrahedral elements in the skin region may be very small (e.g., < 0.1 mm), whereas the elements elsewhere in free space regions may be large >100 mm. This large differential in element size can also lead to numerical instabilities in the FEM computational processes.

A good example of an FEM mesh applied to a biomedical problem is shown in Fig. 3.9. In the case, the FEM mesh was used in the study of MIT for imaging cerebral hemorrhage. The image on the left shows an enlarged view of the tetrahedral elements on the surface of the head and internal tissues, whereas the right hand shows the complete mesh (Zolgharni, Ledger, Armitage, Holder, & Griffiths, 2009).

The advantages of FEM are the widespread availability of commercial and research codes and the method can give a physically accurate description of most problem

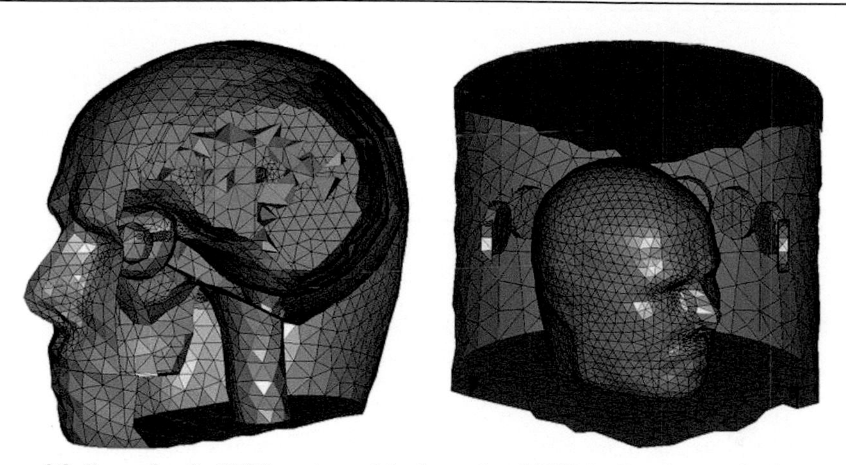

Figure 3.9 Example of a FEM mesh used in the study of MIT for imaging cerebral haemorrhage. Left: enlarged view of the boundary elements on the surface of the head and internal tissues. Right: the complete mesh. Reproduced courtesy Zolgharni et al. (2009).

geometries. Unfortunately, the computation demands from FEM can be significant and there is usually a trade-off between accuracy and computational speed. On the one hand, the forward problem should be represented with a large number of elements in the finite element mesh to ensure adequate accuracy. Normally, this involves a study of mesh convergence, whereby meshes of different sizes and distributions are tested to determine the point at which the solution converges to a common value within error limits. On the other hand, computation time increases markedly with mesh size. At present, FEM cannot offer real-time operation (e.g., video rate) on mid-performance workstations; models with >100 k tetrahedra are often required with computational times per model in the minutes to tens of minutes range. Specialist hardware, especially the use of graphic processing units, and optimization of the algorithms and coding can bring these timescales down significantly.

3.3.4 Boundary element method

Boundary methods can offer a more efficient approach for applications where there are a small number of distinct regions and the tomographic problem is to determine either the boundary or the electromagnetic properties of the material. The approach now is to use surface integrals and Green functions for the particular geometry and solve the fields on the boundaries between the different regions. Once the fields on the boundary are known, the fields everywhere can be determined from the Green functions.

A good example of this approach was reported (Pham, Hua, & Gray, 2000a,b) for imaging solidification of molten metal inside a metal pipe. This is of particular interest to the metal production industry of furnace tapping and molten metal flow control. Fig. 3.10 shows a schematic of this measurement system which contains solenoid carrying AC current placed around the pipe. The impedances of the solenoids are sensitive to the electrical conductivity of the metal pipe, the solidified metal, and the liquid

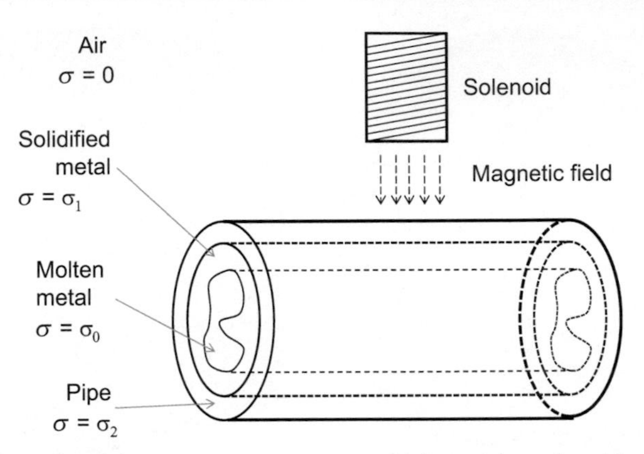

Figure 3.10 Schematic of the measurement system which contains solenoid carrying AC current placed around the pipe as used by Pham et al. (2000).

metal. The tomographic problem is to determine the position of the boundary between the liquid and solid metals inside the pipe. To achieve this, the sensing technique exploits the difference in electrical conductivity between the solid and liquid metals.

The boundary approach can offer an efficient, bespoke solution for applications where the geometry can be split into well-defined regions and the tomographic inversion problem can be specified in these terms. Of course, the solution is tailored to the needs of the application and therefore requires a high investment of effort to formulate and prove the electromagnetic formulations and the detailed implementation of the model. As a final comment, it is possible to combine BEM/FEM approaches, so that boundary methods are used for large regions of uniform properties and FEM is used to describe regions with internal detail.

3.3.5 Full wave solutions

There are a variety of techniques for solving the full set of Maxwell's equations, which include, for example, both frequency and time domain methods. FEM is popular for frequency domain solutions where the fields are calculated for a defined frequency. Time domain techniques directly model the propagation of electromagnetic waves. Here, the problem space is typically divided into a Cartesian grid of cubic elements and the solution involved calculating the propagation of the electric and magnetic fields between neighboring elements in discrete time steps. Popular methods include the finite difference time domain (FDTD) technique and the transmission line matrix method (TLM).

Generally, full wave solvers are very inefficient for MIT because the eddy current and magnetic polarization effects occur at much lower frequencies than wave propagation phenomena and consequently the solution process must continue until all transients have dissipated and the model settled to quasi-DC conditions. Nevertheless, full wave solvers do offer the potential to account for parasitic coupling due to electric fields, which can be important for electrostatic screening and also the effects of displacement current, which can be important as MIT is used for increasingly higher frequencies. Full wave solvers also provide an independent means of validating other forward models as reported by Zolgharni et al. (2009).

3.3.6 Impedance method

For applications where the skin depth, δ, is very much larger than the object dimensions, such as low conductivity applications in biology and processing aqueous solutions, the field changes caused by the object may be very small compared to the applied field. This is also known as the "weakly coupled" case. Here, the excitation field can be assumed to be unaffected by the object and therefore can be computed a priori. The object field resulting from this excitation can be computed independently and typically acts with a phase angle in quadrature to the applied field, as illustrated earlier in Fig. 3.5 (top row) and Fig. 3.6, example 5.

The weakly coupled approximation is used by a number of researchers working on medial MIT applications (Dekdouk, Wuliang Yin, Ktistis, Armitage, & Peyton, 2010; Ktistis, Armitage, & Peyton, 2008; Morris, Griffiths, & Gough, 2001) in order to reduce the computation load of the forward problem. The impedance method is an alternative to edge FEM for solving the induced field inside the object. The method involves dividing the problem space into cubic voxels, with the nodes of the impedance mesh at the voxel centers. The voxels are represented as an electrical network consisting of six admittances and six induced voltages as shown in Fig. 3.11. Thus, the forward problem is reduced to an electrical network equivalent which is solved accordingly. More details on the impedance method are given in Armitage, Le Veen, and Pethig (1983), Gandhi, DeFord, and Kanai (1984).

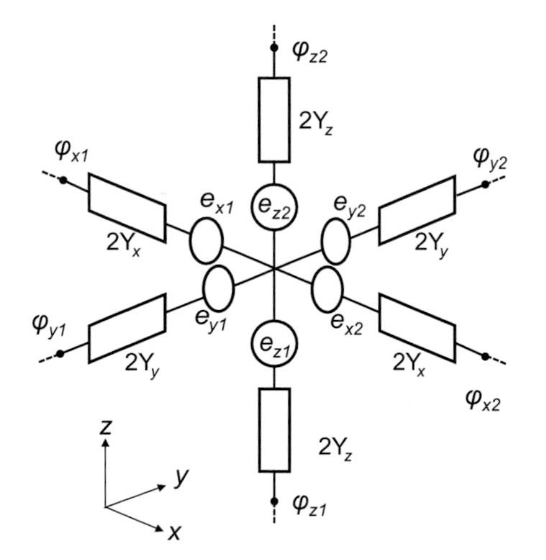

Figure 3.11 Method for dividing the problem space into cubic voxels, with the nodes of the impedance mesh at the voxel centres. The voxels are represented as an electrical network consisting of six admittances and six induced voltages.

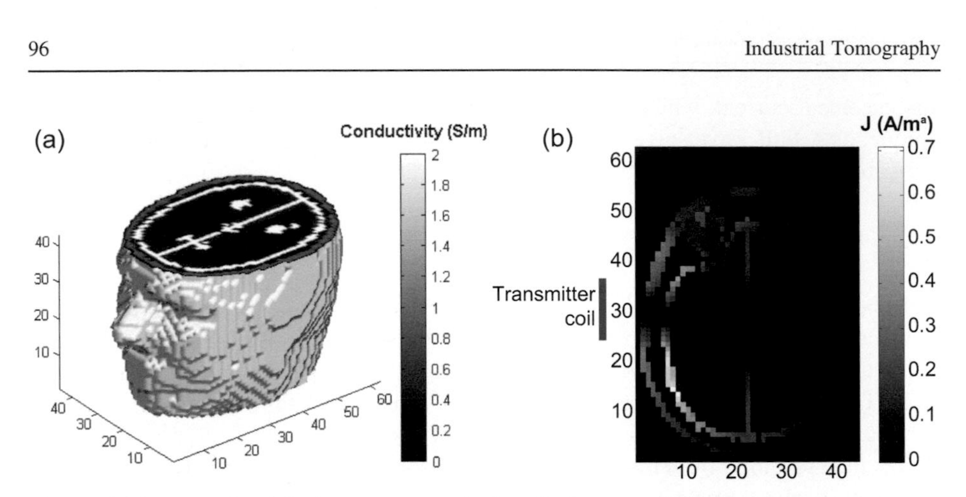

Figure 3.12 Example of the impedance method applied to a study of head imaging courtesy (Dekdouk et al., 2010). (a) Head model with coronal cutaway plane showing conductivity distribution for a normal head (b) contains a plot of the current density on the sagittal sectional.

Fig. 3.12 shows an example of the impedance method applied to a study of head imaging (Dekdouk et al., 2010). Fig. 3.12a shows the head model with coronal cutaway plane showing the conductivity distribution for a normal head with the electrical conductivity of the tissues labeled by the gray scale. Fig. 312b contains a plot of the current density on the sagittal sectional plane of the head. As expected, the current flow is highest in regions with relatively high conductivities notably in the cerebrospinal fluid as well as near the excitation coil where the flux density is strong. The eddy currents also tend to circulate in loops parallel to the current flow in the excitation coil and tend to zero on the axis of the coil, which is also indicated on the plot.

3.3.7 Dipole approximation

The dipole approximation offers a very efficient solution to the forward problem in special situations where the location and characterization of a metallic or magnetic object is required, and the object is relatively small (i.e., $<$ factor of 10) compared to the dimensions of the sensor array. These situations are typical of metal detection, for instance, in WRMDs and systems for unexploded remnants of war (ARW) such as landmines.

In the time harmonic regime, the projection of a conducting target with an alternating primary magnetic field by an MIT sensor generates, according to Faraday's induction law, a flow of eddy currents inside the volume of the target. These currents, in turn, produce a perturbation secondary field which is sensed by the inductive sensor. In theory, the eddy currents can be expanded into a series of multipole moments of currents, and similar representation also applies to the associated secondary field. For a magnetic multipole expansion series of order n, the field terms will have strengths falling off as $\frac{1}{r^{n+2}}$ at distances r away from the currents. The lowest order term in this series is called the dipole moment. If the target is placed far enough from the sensor, the secondary field can be approximated by an induced dipole moment. Fig. 3.13 shows an

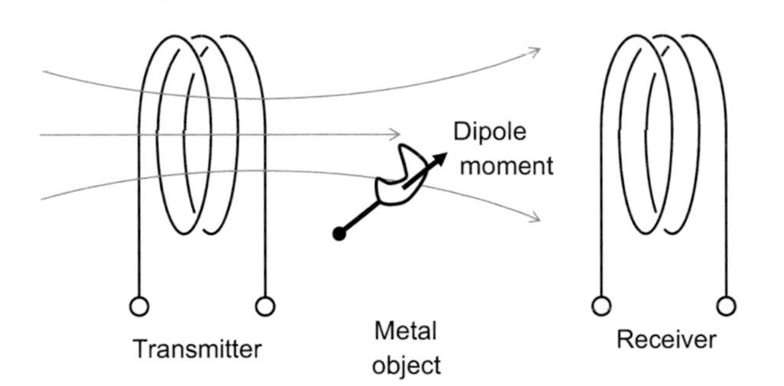

Figure 3.13 Schematic illustration of the dipole approximation with the field from the target being generated by an equivalent magnetic dipole moment.

illustration of this approximation with the field from the target being generated by a dipole moment.

For a sensor with colocated excitation and receiver coils, the well-known Lorentz reciprocity principle, which states that the sensor response is invariant when the excitation and receiver coils are interchanged (Dyck, Lowther, & Freeman, 1994), can be used to express the induced impedance due to the presence of the target as follows:

$$Z(\boldsymbol{r}, \omega) = \frac{j\omega\mu_0}{I_R\, I_T}\, \boldsymbol{H}_R^T(\boldsymbol{r})\, \boldsymbol{M}(j\omega)\, \boldsymbol{H}_T(\boldsymbol{r}) \tag{3.13}$$

where $Z(\boldsymbol{r}_1, \omega)$ is the complex mutual impedance between the transmitter and receiver coils, $\boldsymbol{H}_T(\boldsymbol{r})$ and $\boldsymbol{H}_R(\boldsymbol{r})$ are the background magnetic field strengths generated by the transmitter and the receiver coils, respectively, at the center of the target location, $\boldsymbol{r}$, relative to the sensor, and I_R is the current through the receiver coil. Note that the background terminology refers to the fields being evaluated in space in the absence of the target. In Eq. (3.13), $\boldsymbol{M}(j\omega)$ is a 3×3 symmetric matrix which denotes the magnetic polarizability tensor characterizing the target. It should be noted that the engineering adoption of this tensor structure as described by nine elements is a scope of current investigation in applied mathematics. Recent progress in the area (Ledger & Lionheart, 2015) examined the derivation of an asymptotic formula for the perturbed secondary field due to the presence of a target and suggested that an earlier derivation of tensor represented as a rank 4 tensor (Ammari, Chen, Chen, Garnier, & Volkov, 2014)] with 81 coefficients can reduce to a rank 2 tensor with only 9 coefficients, which provides a solid foundation for the current configuration. These coefficients forming the matrix M are complex valued and frequency dependent, which, as a result, introduces a frequency dependent phase shift between the incident field and the induced impedance. The fields $\boldsymbol{H}_T(\boldsymbol{r})$ and $\boldsymbol{H}_R(\boldsymbol{r})$, however, are virtually frequency independent given the electric size in the medium in which the target buried is usually long compared to the target's depth, which is usually valid for typical MIT frequencies ($\leq$1 MHz). Therefore, using the dipole model, the frequency and target positional dependences in the MIT response are separated between the polarizability tensor and the fields, respectively, where $\boldsymbol{M}(j\omega)$ is invariant with target position but contains its spectral behavior and the fields are only a function of target position. This has the advantage to help in the estimation of the target's tensor and location.

Examples of the use of the dipole approximation for WTMDs have been reported by Marsh, Ktistis, Järvi, Armitage, and Peyton (2013), Marsh, Ktistis, Järvi, Armitage, and Peyton (2014) for single and multiple objects, respectively, and by Makkonen et al. (2014) using classification techniques. Several workers have investigated the dipole model for identification of buried ARW, notably Norton and Won (2001).

3.4 Solution to the inverse problem

The previous section considered several key techniques used to describe the physics of the MIT sensing process. Of course tomography requires this process to be inverted. In general, the MIT inverse problem shares similarities with the inversion problems from other so called *soft field* tomographic techniques, especially the electrical methods; it is usually ill-posed, ill-conditioned, and nonlinear. The aim of this section is to give a summary of the main approaches used for inversion in MIT.

3.4.1 Linear case

The linear case has limited application for many MIT problems as described earlier in Section 3.3.1 and this should be used with caution; however, the method can give crude qualitative images showing rough approximations of the material distribution. The method works best for small perturbations in material, the properties against a known background for which the sensitivity matrix can be calculated. Assuming the linear case, for small changes in the pixel values, the system can be described as $\delta\mathbf{M} = \mathbf{S}.\delta\mathbf{P}$. Normally, the δ notation is omitted as it is implicit that the measurement and pixel values are with respect to a fixed background value, so $\mathbf{M} = \mathbf{S}.\mathbf{P}$. For inversion, it is necessary to determine $\mathbf{P}$ from the measurements, $\mathbf{M}$. Unfortunately, the sensitivity matrix $\mathbf{S}$ cannot be directly inverted. A natural solution, $\mathbf{S}^{\dagger}$, would be to choose the Moore–Penrose generalized inverse, i.e.,

$$\mathbf{S}^{*} = \left(\mathbf{S}^{T}.\mathbf{S}\right)^{-1}.\mathbf{S}^{T} \tag{3.14}$$

where $\mathbf{P} = \mathbf{S}^{*}.\mathbf{M}$ is the least squares solution to $\mathbf{M} = \mathbf{S}.\mathbf{P}$, i.e., min $\mathbf{M} - \mathbf{S}.\mathbf{P}^{2}$. Many MIT applicationsare underdetermined $k > n$ so the problem is usually extremely ill-posed and the calculation of.

$\left(\mathbf{S}^{T}.\mathbf{S}\right)^{-1}$ or $\left(\mathbf{S}.\mathbf{S}^{T}\right)^{-1}$ is normally swamped by numerical error. Consequently regularization is required, which not only seeks to minimize the least squares error, but also penalize large pixel/parameter values in $\mathbf{M} = \mathbf{S}.\mathbf{P}$, stated as

$$\min\left\{\|\mathbf{M} - \mathbf{S}.\mathbf{P}\|^{2} + \propto^{2}\|\mathbf{P}\|^{2}\right\} \tag{3.15}$$

The coefficient $\propto^2$ controls a compromise between fitting the data and controlling the magnitude of the solution. The choice of $\propto^2$ represents the introduction of a degree of a priori knowledge relating to the magnitude of the solution. The result is the commonly used Tikhonov regularization

$$\mathbf{P} = \left(\mathbf{S}^T.\mathbf{S} + \alpha^2\mathbf{I}\right)^{-1}\mathbf{S}^T\mathbf{M} \tag{3.16}$$

Other forms of regularization are also popular in MIT and also electrical tomography such as truncated singular value decomposition, which are well covered in the literature.

3.4.2 Iterative solutions

For many MIT applications, the simple linear solutions described in the previous section do not accurately express the problem. In these cases, an iterative solution is sought, which seeks to minimize an objective function. The objective function, also known as a cost function, describes the fit between the real measured data and the simulated data produced the forward problem. This procedure is summarized by the block diagram in Fig. 3.14.

The algorithm is initiated with a first guess of the pixel or parameter values in **P**, which may either be the background (baseline) values or, for a temporally continuous application, the previously converged solution from the last set of measured data. Typically, some form of constraining is applied to limit or cap extreme values in **P**, such as excluding nonphysical values, for instance, negative electrical conductivities. The previous estimate is then updated to produce the $K+1$ estimate. This estimate of **P** is subsequently input to the forward model. The calculated measurements are then subtracted from the actual measurements to produce an error vector which is applied

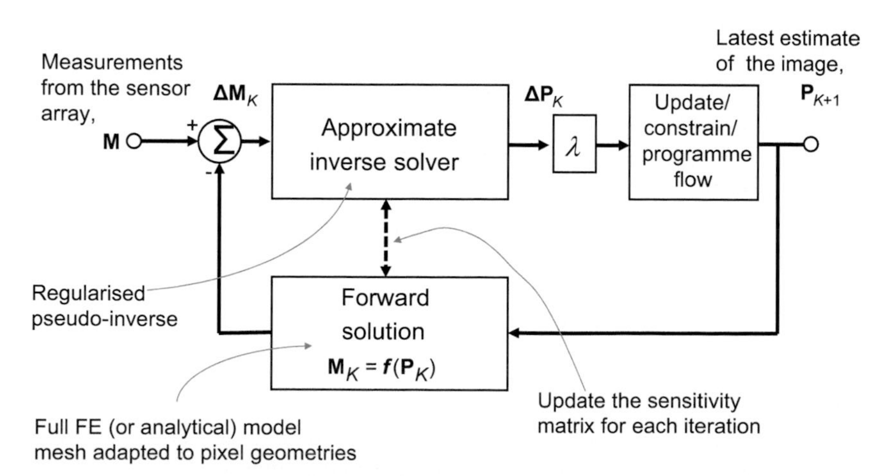

Figure 3.14 Block diagram of a typical MIT inversion algorithm.

to the approximate inverse solution, which is usually based on a piecewise linear approximation, similar to the methods described in the previous Section 3.4.1. The next estimate of **P** is then made and the iterative process continues. A regularization parameter λ is introduced to stabilize the iterative scheme to convergence and usually the value of λ is also adaptively controlled. The inverse problem requires sensitivity values, such as, for example, the sensitivity of a particular measurement to a change is the conductivity of a particular voxel. The sensitivity values are affected by the distribution of the material in the object space and therefore may need to be recalculated regularly and possibly during each iteration, because the material distribution changes as the algorithm converges to a solution.

The block diagram essentially describes a minimization problem, which can be stated as follows:

$$\mathbf{P} = \left(\mathbf{S}^T.\mathbf{S} + \alpha^2\mathbf{I}\right)^{-1}\mathbf{S}^T\mathbf{M} \tag{3.17}$$

$$\mathbf{P}^* = \underset{\mathbf{P}}{\arg\min}\left\{\|f(\mathbf{P}) - \mathbf{M}\|^2\right\} \tag{3.18}$$

The inverse problem is solved by generating a sequence of **P** vectors, i.e., $\mathbf{P}_0$., $\mathbf{P}_1$, $\mathbf{P}_2$, $\mathbf{P}_3$, ... $\mathbf{P}^*$, which converge toward the values which minimize the least squares error between $f(\mathbf{P})$ and **M**. Each iteration of this process requires a new evaluation of the forward problem, which for MIT is generally where the main computational cost of this method occurs. Depending on the algorithm used, **S** may also be updated during each iteration, which has an additional computational cost. The choice as to whether the Jacobian **S** is updated is based on how linear the problem is expected to be, with more highly nonlinear problems requiring a method which incorporates updates to the Jacobian.

All iterative algorithms of this type require an initial estimate to the solution $\mathbf{P}_0$ and then generate a series of vectors, which estimate the solution. These estimates are constructed so as to iteratively approach an overall minimum ($\mathbf{P}^*$) of the objective function, $\{f(\mathbf{P}) - \mathbf{M}^2\}$. Ideally, a global minimum is required; however, very few algorithms are capable of guaranteeing global convergence and so a (sufficiently small) local minimum must often suffice. Which local minimum, or possibly the global minimum, is reached may depend upon the choice of the initial solution.

A new output solution vector is determined by calculating an update to the current output solution:

$$\mathbf{P}_{K+1} = \mathbf{P}_K + \boldsymbol{\lambda}.\Delta\mathbf{P}_K \tag{3.19}$$

where $\Delta\mathbf{P}_K$ is the output solution update at iteration K and $\boldsymbol{\lambda}$ is a relaxation parameter used to control the step size and hence the convergence of the iterative procedure. Most MIT algorithms use methods in which the objective function must decrease after each iteration. These methods are known as descent methods, that is

$$\|f(\mathbf{P}_{K+1}) - \mathbf{M}\|^2 < \|f(\mathbf{P}_K) - \mathbf{M}\|^2 \tag{3.20}$$

The value of $\Delta \mathbf{P}_K$ can be obtained by making a linear approximation for each iterations using techniques, for instance

$$\Delta \mathbf{P}_K = \left(\mathbf{S}^T.\mathbf{S} + \alpha^2 \mathbf{I}\right)^{-1} \mathbf{S}^T \, \Delta \mathbf{M} \tag{3.21}$$

There are a number of well-established algorithms for achieving stable inversion of nonlinear problems such as MIT. One common approach is the regularized Gauss–Newton method in which

$$\Delta \mathbf{P}_K = \left(\mathbf{S}_K^T.\mathbf{S}_K + \alpha^2 \mathbf{R}^T \mathbf{R}\right)^{-1} \left\{\mathbf{S}_K^T\left(\Delta \mathbf{M}_K - \alpha^2 \mathbf{R}^T \mathbf{R} \mathbf{P}_K\right)\right\} \tag{3.22}$$

with α being a regularization parameter and $\mathbf{R}$ is a regularization operator chosen to provide particular types of smoothing.

3.4.3 Sensitivity formulations

The calculation of the sensitivity matrix is fundamental to inversion. As discussed earlier in Section 3.3.1, the coefficients in the sensitivity matrix represent the change caused in a measured parameter, which for MIT is typically the transimpedance between a defined transmitter–receiver coil pair and the change in the a material parameter or pixel value in the object space. Consequently, the sensitivity of a measurement to changes in permeability, conductivity, or velocity, at a particular point in space, is essential to the inversion process. The sensitivity values are dependent on the material distribution in the object space and therefore may need to be recalculated during each iteration of the inversion algorithm. The sensitivity values can be calculated directly from the field values as follows:

Considering a typical setup, a number of coils (for example, 8 or 16) are used as transmitters and receivers in MIT. When a particular coil is used as a transmitter, it is fed by currents of certain values to produce alternating magnetic fields; other coils are used to measure the secondary field due to the velocity-induced magnetic field and the magnetic field produced by the eddy currents in the objects. Several authors (Dyck et al., 1994; Hollaus et al., 2004a,b; Norton & Bowler, 1993; Soleimani & Lionheart, 2006; Yin & Peyton, 2010b) have reported derivations for the sensitivity matrix for a range of electromagnetic problems. These typically follow the approach of defining an adjoint problem to which the reciprocity theorem is then applied. This allows the change in electromagnetic fields to be calculated for a given change in material properties. Following the approach by Dyck et al. (1994), normally the measured variable can be expressed in a form

$$F = \int_V f(\boldsymbol{E}, \boldsymbol{H}) \, dV \tag{3.23}$$

where F is a function of the electric and magnetic fields at a point. In the MIT case, the receiver voltage, V_R, measured when a particular excitation coil is activated can be expressed as

$$V_R = \int_V \boldsymbol{E}_T \cdot \frac{\boldsymbol{J}_R}{I_R}\, dV \tag{3.24}$$

where $\boldsymbol{E}_T$ is the electric field produced by the current flows when one transmitter is activated and $\boldsymbol{J}_R$ is the current flowing in the receiver coil with a current value of I_R. Note that $\frac{\boldsymbol{J}_R}{I_R}$ is a unit vector following the strands of the receiver coil. Parameter changes such as conductivity, permittivity, permeability, and velocity cause a change in $\boldsymbol{E}$ and $\boldsymbol{H}$ which in turn results in a change of F. A variation of F can be expressed as

$$\delta F = \int_V (\nabla_{\boldsymbol{E}} f(\boldsymbol{E}, \boldsymbol{H}).\delta\boldsymbol{E} + \nabla_{\boldsymbol{H}} f(\boldsymbol{E}, \boldsymbol{H}).\delta\boldsymbol{H}).dV \tag{3.25}$$

which is expressed in terms of the functions of the parameters

$$\delta F = \int_V \left(S_\sigma.\delta\sigma + S_\mu.\delta\mu + S_\varepsilon.\delta\varepsilon + S_v.\delta\boldsymbol{v}\right).dV \tag{3.26}$$

where S is the sensitivity of F in response of the concerned parameters, respectively. We may now define a quasi-Poynting vector, $\boldsymbol{S}_{12} = \boldsymbol{E}_1 \times \boldsymbol{H}_2$, where the two fields $\boldsymbol{E}_1$ and $\boldsymbol{H}_2$ are taken from two separate systems (1 and 2) with possibly different material parameters, different source currents and/or different boundary conditions. Taking the volume integral of the divergence of this quasi-Poynting vector and applying the divergence theorem yields

$$\int_V \nabla\cdot\boldsymbol{S}_{12}\, dV = \int_V \nabla\cdot(\boldsymbol{E}_1 \times \boldsymbol{H}_2)\, dV = \oint_S \boldsymbol{E}_1 \times \boldsymbol{H}_2 \cdot d\boldsymbol{S} \tag{3.27}$$

where S is the surface bounding V. Applying the vector identity $\nabla(\boldsymbol{a} \times \boldsymbol{b}) = (\boldsymbol{b} \cdot \nabla \times \boldsymbol{a}) - (\boldsymbol{a} \cdot \nabla \times \boldsymbol{b})$, we have

$$\int_V (\boldsymbol{H}_2 \cdot \nabla \times \boldsymbol{E}_1 - \boldsymbol{E}_1 \cdot \nabla \times \boldsymbol{H}_2)dV = \oint_S \boldsymbol{E}_1 \times \boldsymbol{H}_2 \cdot d\boldsymbol{S} \tag{3.28}$$

Substituting the linear time harmonic Maxwell's Eqs. (3.3a–d) into Eq. (3.28) gives the following Eq. (3.27) and the same is true is when we simply exchange systems 1 and 2 in Eq. (3.27).

$$\int_V \{\boldsymbol{E}_1 \cdot ((\sigma_2 + j\omega_2\varepsilon_2)\boldsymbol{E}_2 + \boldsymbol{J}_{EX2} + \sigma_2(\boldsymbol{v}_2 \times \boldsymbol{B}_2)) + j\omega_1\mu_1\boldsymbol{H}_1 \cdot \boldsymbol{H}_2\}dV = -\oint_S \boldsymbol{E}_1 \times \boldsymbol{H}_2 \cdot d\boldsymbol{S} \tag{3.29a}$$

$$\int_V \{\boldsymbol{E}_2 \cdot ((\sigma_1 + j\omega_1\varepsilon_1)\boldsymbol{E}_1 + \boldsymbol{J}_{EX1} + \sigma_1(\boldsymbol{v}_1 \times \boldsymbol{B}_1)) + j\omega_2\mu_2\boldsymbol{H}_1 \cdot \boldsymbol{H}_2\}dV = -\oint_S \boldsymbol{E}_2 \times \boldsymbol{H}_1 \cdot d\boldsymbol{S} \tag{3.29b}$$

The derivation of the sensitivity formulas now follows from a series of substitutions. First we replace system 1, with a system containing a small variation of the parameters below:

$$\begin{aligned} &\mu_1 \rightarrow \mu + \delta\mu \quad \sigma_1 \rightarrow \sigma + \delta\sigma \\ &\varepsilon_1 \rightarrow \varepsilon + \delta\varepsilon \quad \boldsymbol{v}_1 = \boldsymbol{v} + \delta\boldsymbol{v} \end{aligned} \tag{3.30a}$$

which causes a small change in the field vectors,

$$\boldsymbol{E}_1 \rightarrow \boldsymbol{E}_A + \delta\boldsymbol{E}_A \; \boldsymbol{H}_1 \rightarrow \boldsymbol{H}_A + \delta\boldsymbol{H}_A \tag{3.30b}$$

then setting the material properties for system 2 as

$$\begin{aligned} &\mu_2 \rightarrow \mu \quad \sigma_2 \rightarrow \sigma \\ &\varepsilon_2 \rightarrow \varepsilon \quad \boldsymbol{v}_2 = 0 \end{aligned} \tag{3.30c}$$

with both systems having a common frequency, ω. Making this substitution into Eqs. (3.29a and b), we have, respectively,

$$\int_V \{(\boldsymbol{E}_A + \delta\boldsymbol{E}_A) \cdot ((\sigma + j\omega\varepsilon)\boldsymbol{E}_B + \boldsymbol{J}_{EXB}) + j\omega(\mu + \delta\mu)(\boldsymbol{H}_A + \delta\boldsymbol{H}_A) \cdot \boldsymbol{H}_B\}dV = -\oint_S (\boldsymbol{E}_A + \delta\boldsymbol{E}_A) \times \boldsymbol{H}_B \cdot d\boldsymbol{S} \tag{3.31a}$$

$$\int_V \left\{ \boldsymbol{E}_B \cdot \begin{pmatrix} (\sigma + \delta\sigma + j\omega\varepsilon + j\omega\,\delta\varepsilon)(\boldsymbol{E}_A + \delta\boldsymbol{E}_A) + \boldsymbol{J}_{EXA} \\ +(\sigma + \delta\sigma)((\boldsymbol{v} + \delta\boldsymbol{v}) \times (\boldsymbol{B}_A + \delta\boldsymbol{B}_A)) \end{pmatrix} + j\omega\mu(\boldsymbol{H}_A + \delta\boldsymbol{H}_A) \cdot \boldsymbol{H}_B \right\} dV = -\oint_S \boldsymbol{E}_B \times (\boldsymbol{H}_A + \delta\boldsymbol{H}_A) \cdot d\boldsymbol{S} \tag{3.31b}$$

If we now set the perturbations to zero and subtract this result from Eqs. (3.31a and b), then two expressions for the perturbations follow. Furthermore in the limit as of small perturbations, terms containing the product of more than one δ quantity will vanish. So neglecting these higher order terms results in

$$\int_V \{\delta \boldsymbol{E}_A \cdot ((\sigma + j\omega\varepsilon)\boldsymbol{E}_B + \boldsymbol{J}_{EXB}) + j\omega(\mu\,\delta\boldsymbol{H}_A + \boldsymbol{H}_A\,\delta\mu)\cdot\boldsymbol{H}_B\}dV = -\oint_S \delta\boldsymbol{E}_A \times \boldsymbol{H}_B \cdot d\boldsymbol{S} \tag{3.32a}$$

$$\int_V \left\{ \boldsymbol{E}_B \cdot \left(\begin{array}{c} (\delta\sigma + j\omega\,\delta\varepsilon)\boldsymbol{E}_A \\ +\,\sigma(\delta\boldsymbol{v} \times \boldsymbol{B}_A) + \sigma(\boldsymbol{v} \times \delta\boldsymbol{B}_A) + \delta\sigma(\boldsymbol{v} \times \boldsymbol{B}_A) \end{array} \right) + j\omega\mu(\boldsymbol{H}_A + \delta\boldsymbol{H}_A)\cdot\boldsymbol{H}_B \right\} dV = -\oint_S \boldsymbol{E}_B \times \delta\boldsymbol{H}_A \cdot d\boldsymbol{S} \tag{3.32b}$$

Finally, subtracting Eq. (3.32b) from Eq. (3.32a) in the above equations

$$\begin{aligned} \int_V \boldsymbol{J}_{EXB} \cdot \delta\boldsymbol{E}_A dV = {} & \int_V (\boldsymbol{E}_A \cdot \boldsymbol{E}_B\,\delta\sigma + j\omega\boldsymbol{E}_A\cdot\boldsymbol{E}_B\,\delta\varepsilon - j\omega\boldsymbol{H}_A\cdot\boldsymbol{H}_B\,\delta\mu)dV \\ & + \int_V (\boldsymbol{E}_B \cdot (\sigma(\delta\boldsymbol{v} \times \boldsymbol{B}_A) + \sigma(\boldsymbol{v} \times \delta\boldsymbol{B}_A) + \delta\sigma(\boldsymbol{v} \times \boldsymbol{B}_A)))dV \\ & + \oint_S \delta\boldsymbol{E}_A \times \boldsymbol{H}_B \cdot d\boldsymbol{S} \end{aligned} \tag{3.33}$$

In the case of an inductively coupled system, then system A is considered to be the transmitter coil and system B the receiver coils; then $\boldsymbol{J}_{EXB}$ equals $\frac{\boldsymbol{J}_R}{I_R}$, then $\int_V \boldsymbol{J}_{EXB} \cdot \delta\boldsymbol{E}_A dV$ is the change in receiver voltage cause by the perturbations; see Eq. (3.24) earlier. In addition, if we take the boundary to far enough away such that the fields are completed encompassed, then the surface integral on the right hand of Eq. (3.33) can be neglected.

By identifying the correspondence with Eq. (3.26), the following equalities can be seen to hold, using the vector identity $\boldsymbol{A}\cdot(\boldsymbol{B} \times \boldsymbol{C}) = (\boldsymbol{C} \times \boldsymbol{A})\cdot\boldsymbol{B}$ for S_v

$$
\begin{aligned}
S_\sigma &= \boldsymbol{E}_A \cdot \boldsymbol{E}_B + (\boldsymbol{v} \times \boldsymbol{B}_A) \cdot \boldsymbol{E}_B \\
S_\varepsilon &= j\omega \boldsymbol{E}_A \cdot \boldsymbol{E}_B \\
S_\mu &= -j\omega \boldsymbol{H}_A \cdot \boldsymbol{H}_B \\
\boldsymbol{S}_v &= \sigma\, \boldsymbol{B}_A \times \boldsymbol{E}_B
\end{aligned}
\tag{3.34}
$$

Note that the first three terms are widely used in electrical tomography; the S_σ sensitivity now includes the effects of Faraday induction caused by the movement of conducting material in the magnetic field. Note that the sensitivity to magnetically permeable material is of the form $\boldsymbol{H}$ dot $\boldsymbol{H}$, which is determined by the magnetic fields. In contrast, the sensitivity to electrically conductive material is of the form $\boldsymbol{E}$ dot $\boldsymbol{E}$, which is determined by the induced field and the flow eddy currents within the conductive object.

3.4.4 Neural inversion

Over recent years, there has been growing interest in artificial intelligence (AI) and especially machine learning (ML) and applied to wide range of engineering problems, in particular application where it is infeasible or difficult to formulate conventional algorithms, such as those described in the previous section. Machine learning involves devising algorithms to learn from data provided so that they perform specific tasks, particularly those which can be easily solved using conventional approaches. In the context of electrical tomography, supervised learning, particularly using neural networks, has received attention since the 1990s. A good example of this pioneering work is Alder and Guardo (1994) for EIT showing promising results with computation data in 2D. Interest has been growing again since the first edition of this book with recent publications including Martin and Choi (2016), Rymarczyk, Kłosowski, Kozłowski, and PTchórzewski (2019), showing promising quality of reconstruction when compared with conventional approaches.

A well-trained ML algorithm is capable of mapping nonlinear relationships between the measured responses and pixel or parameter values of interest while improving the solution speed by a significant amount. In effect, the computation heavy work is done during the training phases, with many thousands of training data sets typically required, otherwise, machine learning is also known to over-fit given training dataset that usually results in an undesired solution when tested using unseen test datasets.

3.5 System hardware

MIT systems can be configured in a variety of ways depending of the number of transmitter and receiver coils, size and geometry of the equipment being instrumented, and the frequency range. There is considerable scope of design customization and implementation details can vary significantly from application to application. This section reviews the main considerations and options in the design of an MIT system.

3.5.1 System configurations

MIT systems can use either dual-function transmitter and receivers coils or have dedicated transmit only and receive only coils. Of course system configurations with dual function coils require additional electronic circuitry in order to switch between transmit and receive modes, and the switching circuitry must be designed to minimize any unwanted coupling from the transmitter to the receiver, when the coil is in receiver mode. Coils are usually relatively easy and inexpensive to fabricate and also can be readily overlaid on top of each other. Consequently, dedicated transmitter and receiver coils are preferred over dual function configurations eliminating the need for the switching circuitry. For circular coil arrays, transmitter and receiver coils can be interleaved or wound on the same formers as shown in Fig. 3.15.

A block diagram of a typical transmitter channel is shown in Fig. 3.16 and consists of the following main blocks:

- Digital to analogue converter (DAC)
- Buffer amplifier
- Excitation on/off switch
- Current sense circuitry

A DAC or dedicated waveform generator, such as a direct digital synthesizer, generates the waveform required by the transmitter coil. This signal is amplified/buffered to the required voltage and current levels for the transmitter coil. For dual function coils, an on/off switch provides the necessary isolation when the coil is used in receive mode, which requires a very low off capacitance. The on/off switch must be able to

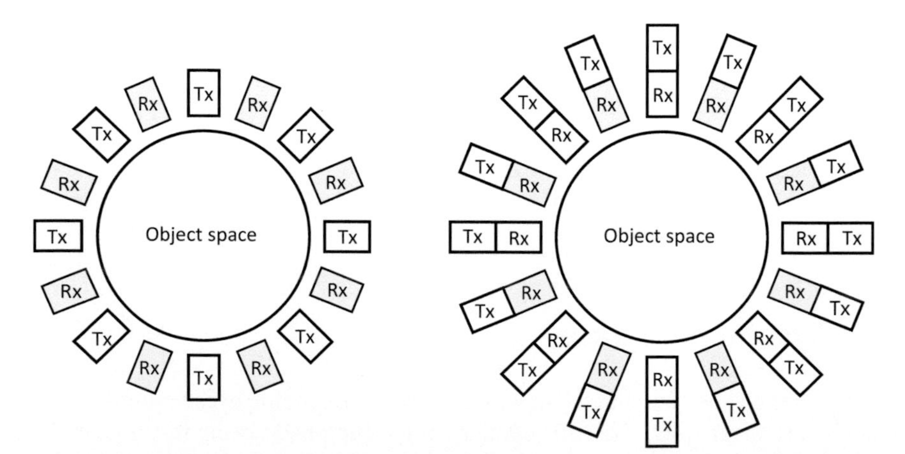

Figure 3.15 Circular MIT arrays with separate transmitter, Tx, and receiver, Rx, coils. (a) Interleaved coils, (b) transmitter and receiver coils sharing the same former.

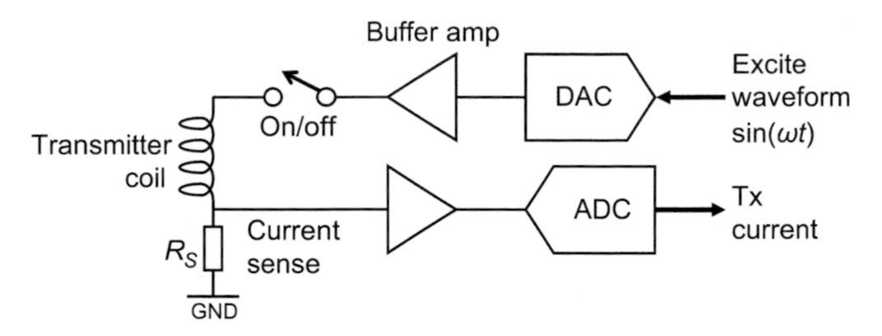

Figure 3.16 A block diagram of a typical MIT transmitter channel which contains a current sense facility.

operate with the full transmitter current, which in some cases can be several amps. A current sense facility is also shown, which can be achieved with a current sense resistor and a current sense amplifier. The output of the current sense amplifier can be digitized and de-odulated in the same way as the signal from the receiver coil, as explained in the following paragraphs. The current sense facility is required to implement a proper transimpedance, $Z_T(j\omega)$, measurement between each receiver and transmitter coil, as given by the complex division

$$Z_T(j\omega) = \frac{v_r(j\omega)}{i_t(j\omega)} \tag{3.35}$$

where, $v_r(j\omega)$ is the voltage detected by the receiver coil and $i_t(j\omega)$ is the current following the transmitter coil. Naturally, the resistance of the sensor resistor, R_S, in Fig. 3.16 should be much less than the reactance of the transmitter coil. A current transformer or current transducer can be used as an alternative to the current sense resistor. MIT systems can operate without the current sense facility, but will then be sensitive to any temperature variations and aging affects in the transmitter coils and transmitter circuitry.

A block diagram of a typical receiver channel is shown in Fig. 3.17 and consists of the following main blocks:

- Front-end amplifier
- Programmable gain amplifier

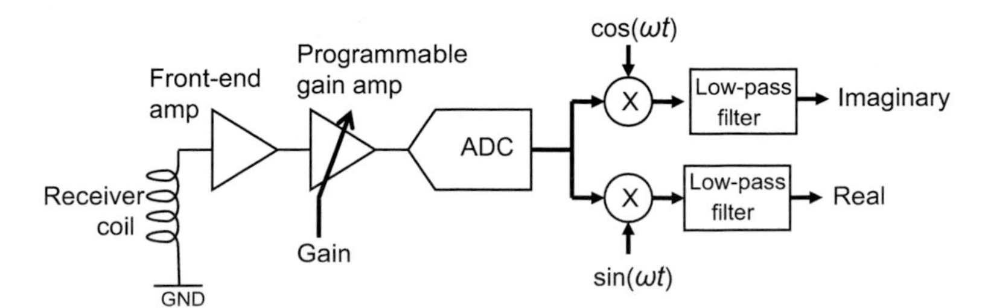

Figure 3.17 A block diagram of a typical MIT receiver channel.

- Analogue to digital converter (ADC)
- In-phase (real) and quadrature (imaginary) demodulation

The front end amplifier is usually positioned relatively close to the receiver in order to minimize the effects of either EM interference or parasitic capacitance in the connecting leads. The capacitance of the connecting leads can introduce an unwanted resonance in the frequency response of the coil, if not properly managed. A programmable gain amplifier (PGA) may be needed to accommodate the difference in the dynamic range of the signal between different coil positions. For instance, the adjacent transmitter/receiver coils in Fig. 3.15a, or the transmitter/receiver coils, sharing the same former in Fig. 3.15b will have a large signal, whereas diametric transmitter/receiver coils will have a much smaller signal. Typically, this can require a gain range of 0–40 dB. Over the past decade, there has been a trend toward digital signal processing systems, in which case the PGA will be followed by a high speed ADC. The resolution of the ADC is usually 14 or 16 bit. The minimum number of samples per cycle of transmitter waveform is four (Smith, Freeston, Brown & Sinton, 1992), but generally, MIT systems employ more than this, exploiting oversampling to increase signal-to-noise ratio (SNR). Typically, 8 or 16 samples per cycle are employed and integer values are preferred to avoid spectral leakage. In-phase and quadrature demodulation consists of a multiplier and low pass filter for both the real and imaginary phases, which are implemented on a digital signal processor (DSP) or field programmable gate array (FPGA). Earlier MIT systems used analogue demodulation, with the analogue to digital conversion being performed after the demodulation. In this analogue case, the outputs of the demodulators can often be multiplexed into a smaller number of ADC's. The analogue scheme is still preferred in some cases as the expense of high speed ADC's and the cost and time associated with implementing the signal processing on DSP's or FPGA's can be decisive factors.

An important consideration for optimizing the SNR performance of an MIT system is to share a common reference voltage between all the channels so that any variations in the primary reference will affect the channels equally in proportion and therefore can be canceled in the impedance division of Eq. (3.33). Similarly, it is also important with the digital architecture, shown in Figs. 3.16 and 3.17, that all sampling processes are synchronized to a common high quality clock to minimize noise related to unwanted phase jitter or spectral leakage introduced during demodulation.

A block diagram of a 16 transmitter and 16 receiver channel MIT system is shown in Fig. 3.18 (Peyton et al., 1999). This system uses an analogue demodulation scheme and indicates some of the complexity required to configure a complete MIT system. Here the circuit functions are split onto different printed circuit boards (PCBs) and the signal to the ADC's are multiplexed to make best use of the sampling rate of the ADCs. A controller, in this case an embedded PC, is used to control the measurement procedure with complete flexibility over the sequencing of excite and receive channels. Communication of the data and top-level control of the acquisition process is achieved over an Ethernet link, which allows the host computer to be positioned anywhere with suitable internet access.

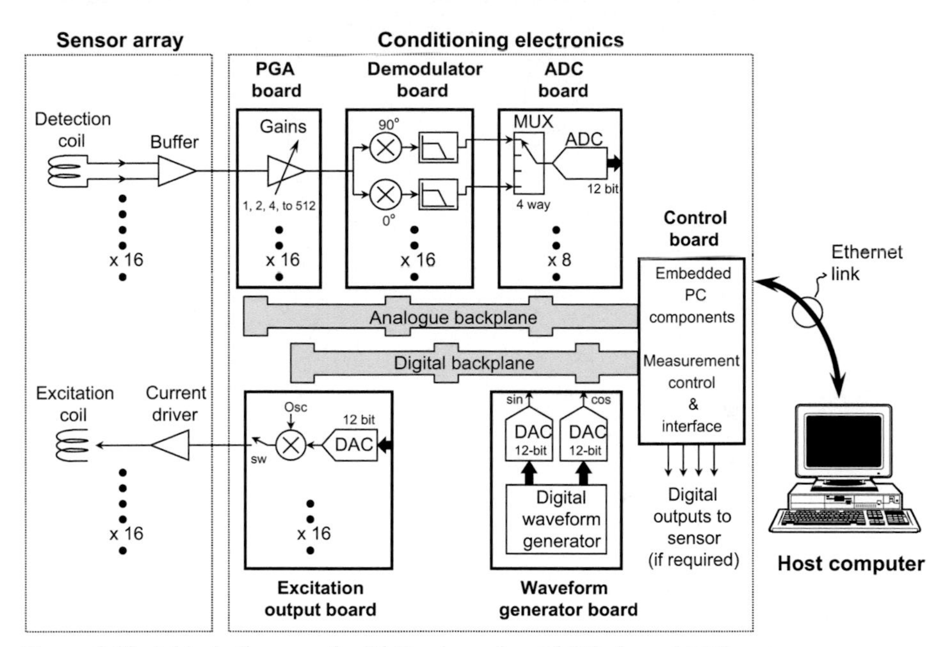

Figure 3.18 A block diagram of a 16 Tx channel – 16 RX channel MIT system.

There are several examples of MIT systems reported in the literature, for example, an 8 channel system used for imaging molten steel (Ma et al., 2006a,b), a 16 channel system for biomedical research (Watson, Williams, Gough, & Griffiths, 2008), which was among the first capable of imaging conductivities below 1 S/m and was built for research on brain edema. A system was reported based on a similar coil array (Vauhkonen, Hamsch, & Igney, 2008). A single channel system based on gradiometers as receivers and capable of multifrequency operation was reported (Scharfetter, Lackner, & Rosell, 2001) and later further advanced to an array (Scharfetter, Köstinger, & Issa, 2008).

There is considerable variety in the geometry of MIT arrays depending on the opportunity to position the coils next to the object space as afforded to by the physical access allowed by the application. In general, the MIT array geometry may be split into three broad categories, namely

i. circular/cylindrical
ii. planar
iii. dual-sided

as shown schematically in Fig. 3.19. Circular arrays were among the first to be researched and are favored for pipeline and many of the biomedical studies. The majority of publications address the 2D circular case and the 3D cylindrical case has received little attention to date. Alternatively, access may be restricted to one surface, which is typical for some NDT applications, geophysical/geological surveying and detection/characterization of buried land mines and other explosive remnants of war

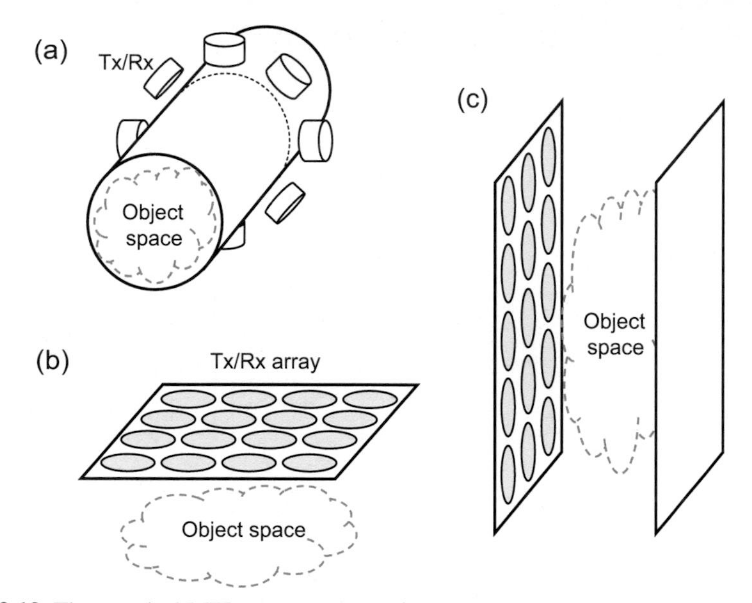

Figure 3.19 Three typical MIT sensor array configurations.

(ERW). In some cases, access is available from two surfaces, a very common example of this type of coil array can be found in WTMDs used for security screening of personnel. Other possible examples include plate and conveyor belt applications.

The majority of publication in MIT research on the arrays described above relate to systems that operate at a single frequency, with very few studies on multifrequency, multicoil arrays. This is partly due to the complexity of the electronic hardware required full spectral and spatial interrogation of the object space, and partly due to the added computational complexity of solving the forward and inverse problems for multiple excitation frequencies. The use of multiple frequencies is also referred to as magnetic induction spectroscopy. One example of a single channel multifrequency problem is depth profiling, where the material properties or thicknesses of subsurface layers are required as depicted in Fig. 3.20, for applications monitoring the

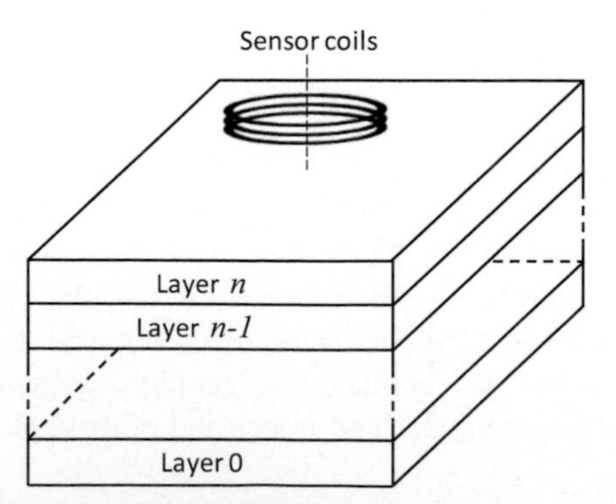

Figure 3.20 Depth profiling using multiple frequency excitation.

thickness of protective metal coatings or noncontact thickness measurement of metal sheet. Here, different excitation frequencies are used to provide different depths of penetration of the EM field and hence probe the depth profile of the conducting layers underneath the coil.

3.5.2 Practical considerations

This section considers a number of practical considerations associated with designing and implementing MIT systems, including coil design, electrostatic screening, resonance, and input stage considerations.

3.5.2.1 Coil design

Coils are typically either wound on a former or realized using PCB techniques. PCBs are attractive because the geometry of the coils can be precisely controlled and the coils can be fabricated in a very repeatable manner. There are a number of PCB substrates available offering different degrees of mechanical stability; however if used, standard FR4 is most common. Thin and flexible substrates are also available, which can be attached directly to standard rigid PCBs in a flexi-rigid process. However, PCBs have their disadvantages: they are restricted in size (typically less than 30 cm); the number of turns is limited by track width and separation specifications from the manufacturer; and the resistance of the tracks is high. Track resistance is a particular limitation for larger coils and at lower frequencies (below say 1 kHz) where higher excitation currents are needed (typically $>$100 mA). Wound coils offer flexibility in the choice of conductor, insulation, and geometry. If appropriate wire and insulators are used, coils can be designed to survive extreme environments and very high temperatures (e.g., up to 1000 °C) and very high radiation fluxes and can extremely rugged. Custom designed and machined or 3D-printed formers can be used to increase the repeatability and precision to which the coils can be built. For very high mechanical stability, coils can be encapsulated in variety of potting compounds such as polyurethane, epoxy, or ceramic and even concrete-based grouts. Hybrid constructions can be used, typically where the transmitter coils are wire wound for high current capacity and receiver coils are printed.

A key design parameter for the coils is the number of turns and there are different considerations for the transmitter and receiver coils. In general, the receiver coils are designed to have as many turns as possible, because the detected voltage, v_r, is approximately directly proportional to the number of turns, N, i.e.,

$$v_r(j\omega) = -\frac{d\lambda}{dt} \approx -\frac{d(N\phi)}{dt} = -N\frac{d}{dt}\int_{S_A} \boldsymbol{B}\cdot d\boldsymbol{S} \tag{3.36}$$

where λ is the flux linkage to the receiver coil, ϕ is the magnetic flux, and S_A is the surface bounded by each turn. The upper limit on the number of turns is dictated by electromagnetic, space, and cost constraints, as winding coils with many thousands of turns can be expensive unless automated. The self-resonant frequency of the coil is the main electromagnetic constraint; increasing the number of turns increases both the self-inductance and self-capacitance of the coil, which in turn reduces the self-resonant frequency. Resonance issues in MIT will be described in more detail shortly. The number of turns on the transmitters is generally chosen to maximize the applied magneto-motive force or ampere turns applied to the object space. This is a case of matching the impedance of each transmitter to the voltage and current drive available from the drive circuitry. High frequency MIT systems (above a few MHz) may be designed to have an impedance close to 50 Ω for maximum power transfer from radio frequency power amplifiers.

There is considerable scope for variety in the geometry of the coils, with dozens of patents claiming inventive coil configurations for inductive applications dating back over many decades. Often the coils are simple circular or rectangular arrangements, sized to maximize the magnetic flux coupling. Ferrite flux concentrators can also be used as well as an external magnetic screen (Yu, Worthington, Stone, & Peyton, 1995); however, the extra weight, cost, and concerns over temperature drift often do not justify this choice. An external electrostatic shield is important to reduce external electromagnetic interference and help to minimize the capacitive coupling between the coils in the array (Peyton, Watson, Williams, Griffiths, & Gough 2003, pp. 352–357).

Gradiometers offer a means of reducing the background coupling between the transmitter and receiver coils to near zero. The subject of background and object signals was discussed earlier in Section 3.2 and illustrated in Fig. 3.6. Reducing the background signal to effectively zero allows the full dynamic range of the system to be used for the object signal and hence increases the SNR of the system. This benefit is not without penalty because gradiometer configurations offer no net DC spatial response, which means that material must be constantly passing through the MIT array or known to occupy a specific volume within the object space, for instance, positioned on one side of the array. Planar and axial gradiometer configurations are shown in Fig. 3.21. The receiver is wound in two halves, Rx1.1 and Rx1.2, which are connected in series opposition so that the induced voltages from the transmitter in each half subtract. Of course all transmitter and receiver coil combinations should benefit from the

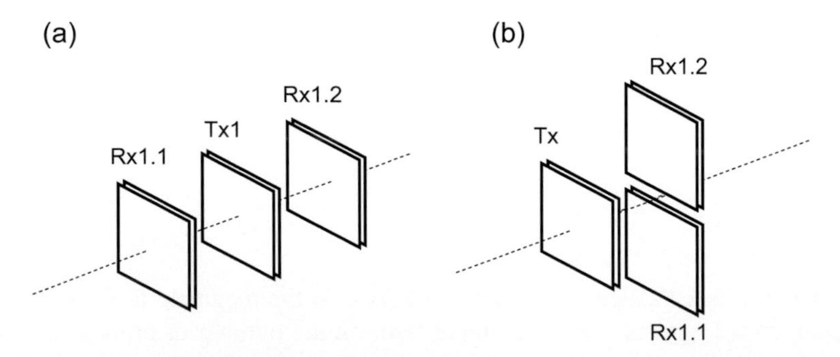

Figure 3.21 Schematic diagram of an axial gradiometer (a) and planar gradiometer (b).

background cancellation property of the gradiometer. Consequently, axial gradiometers are suitable for planar arrays and planar gradiometers for circular arrays.

Coils can also be positioned to reduce the background signal. For example, with phased arrays in nuclear magnetic resonance (NMR), the coils can be overlapped to give zero mutual coupling and hence avoid problematic interactions between nearby coils (Roemer, Edelstein, Hayes, Souza, & Mueller, 1990). Overlapping coils for zero mutual inductance can also be used in EMT in order to reduce the large background measurements that can occur when excitation and detection coils are positioned close together. Fig. 3.22 illustrates this idea using the planar coil array described earlier. In part (a), the measured output of a detection coil is plotted against angular position, as the coil is moved with respect to the excitation coil which is shown in the cross-sectional view. At 0 degrees, the coils are coincident and the output is a maximum. At the opposite extreme, when the coils are furthest apart with an angle of 180 degree, the output is approximately 50 times smaller. At two points in between these extremes, when the coils are partially overlapping, the output is virtually zero. These are null points where there is no net flux passing through the detection coil,

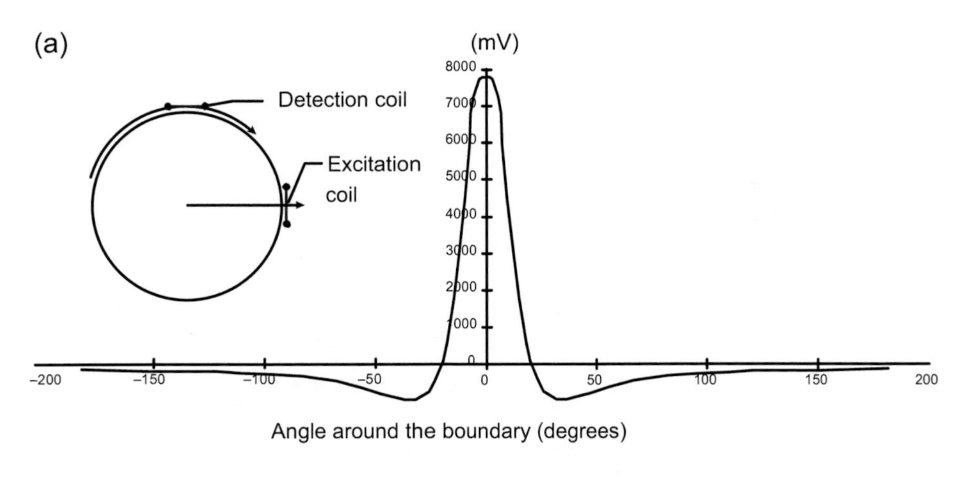

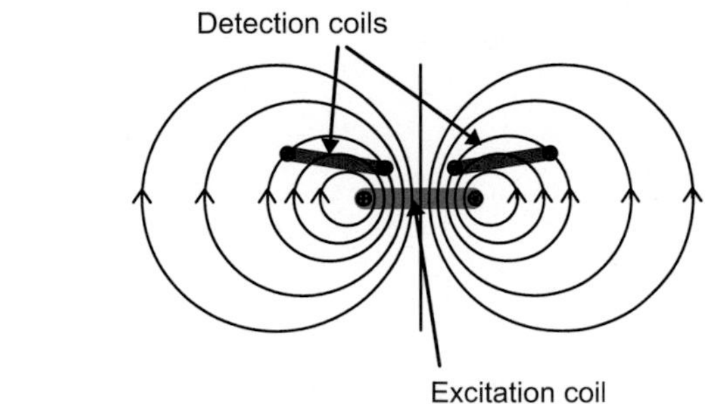

Figure 3.22 Overlapping transmitter and receiver coils to minimise background coupling. (a) Signal response versus angular position showing two zero crossover positions. (b) Coils positioned on lines of magnetic flux.

as shown in an idealized form and with no magnetic screen in Fig. 3.22b. With careful sensor array design, it is possible to ensure that all the detector coils are positioned in the null points of their two neighboring excitation coils. This helps to reduce the variations in signal size, which must be processed by the conditioning electronics.

3.5.2.2 *Electrostatic screening*

For higher frequency (typically over 100 kHz) MIT systems, the effects of parasitic capacitive coupling between the coils become increasing significant. Electrostatic screening can minimize these effects. Three examples of different types of screen are shown in Fig. 3.23. Foil envelopes may be used to screen solenoids; see Fig. 3.23a. Coils are screened both underneath and on top of the winding, to provide a complete Faraday cage around the coil. The screen offers a larger conductive area to its surrounding and as a result may increase the parasitic coupling, particularly for source coils and coils with few turns. The screen also acts as a single turn inductor; therefore, the induced EMF between one end of the screen and the other will be $\frac{1}{N}$ times the source voltage, where again N is the number of turns. Consideration should also be given to which end of the coil is connected to ground, as this first turn can also act as a screen. Combs of various configurations, see Fig. 3.23b, have been used to screen the electric fields produced by the source coil. The thickness and pitch of the tracks must be sufficient to ensure effective screening of the electric field, but must have minimal effect on the magnetic coupling. Ground planes have two roles, as an electromagnetic shield (Yu et al., 1995) and as a defined capacitive route to ground; see Fig. 3.23c. It will however also reduce the magnetic coupling, especially between opposite coils.

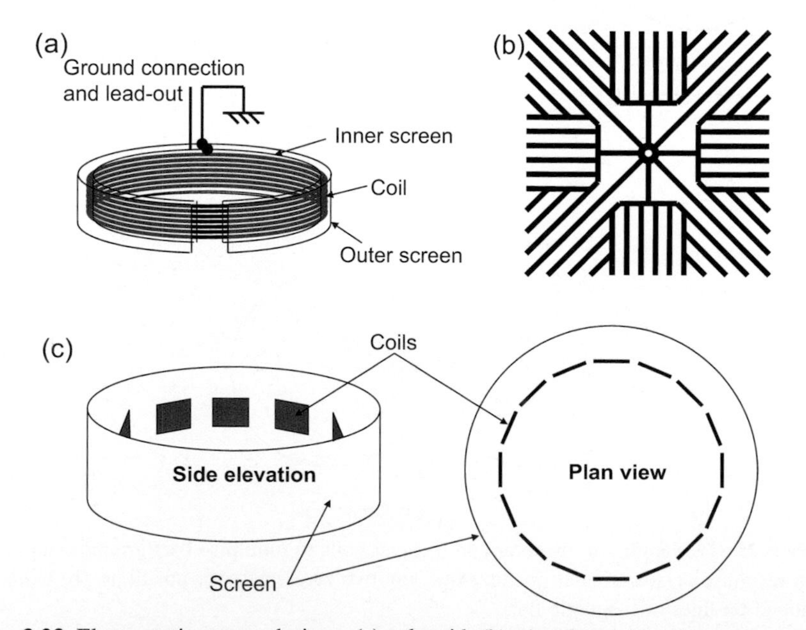

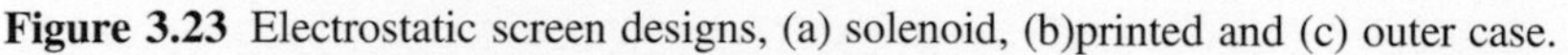
Figure 3.23 Electrostatic screen designs, (a) solenoid, (b)printed and (c) outer case.

The screen—coils separation is therefore a compromise between minimizing the capacitive coupling between coils and maximizing the magnetic field within the sensor. Ground planes have been used to control capacitive coupling in medical EMT systems and its influence on the spatial sensitivity has been simulated (Peyton et al., 2003, pp. 352—357).

Electrostatic shields constructed from resistive materials are popular for some inductive sensor systems, especially conveyor-based metal detectors. Materials such as conducting graphite paint, conducting polymer sheet, and fabrics with woven conducting wires are used in these instruments. It is important to ensure the resistive shield as a sufficiently high surface resistivity (Ω/square) so that it does not affect with applied AC magnetic field, but the surface resistivity is low enough to attenuate the electrostatic field.

3.5.2.3 Resonance

As with many other inductive sensors, coil resonance is also a fundamental concern for MIT. Resonance can be either purposely employed or alternatively completely avoided. The main issues regarding the use of resonance for MIT are summarized in Table 3.2. As can be seen, resonance offers higher signal levels with greater SNR

Table 3.2 Resonant versus non-resonant detection schemes.

	Nonresonant	**Resonant**
Signal quality	Smaller signal	Larger signals, multiplied by the coil Q-factor
	Wider bandwidth presented to noise and interference	Narrow noise window
	Susceptibility to interference increases linearly with frequency due to the differential action of the detector coils	Susceptible to interference is low, governed by the narrow band pass resonant response.
	Lower signal-to-noise ratio	Higher SNR's
Physical principle	Sensor array outputs (e.g., 0 and 90 degrees) are a direct function of the object material and the detected signals can be derived directly from Maxwell's equations using, e.g., FE analysis.	Outputs are dependent on both Maxwell's equations **and** on the transfer function(s) of the detector coil(s). For one resonant coil alone, the transfer function can be readily established. However, if several coupled coils are resonant simultaneously at the same frequency, then the outputs are very difficult to predict theoretically.
		Significant phase changes in the signals can be introduced by the coil transfer functions.

Table 3.2 Resonant versus non-resonant detection schemes.—*Cont'd*

	Nonresonant	Resonant
Hardware issues	Simple parallel data acquisition schemes can be employed (i.e., with measurements taken from all coils simultaneously) because interactions between the individual measurement coils are negligible	Parallel acquisition schemes can be employed, but the interactions between the coils would present a significant problem for image reconstruction, especially for high Q factors.

due to the band pass filtering action that resonance provides. However, the resonant interactions can be difficult to manage. The effects of multiple resonances in MIT interactions are illustrated in Fig. 3.24. The result is that the detected signals are averaged spatially which makes the reconstruction problem more difficult to solve. Only the magnitudes of the signals are sketched in the diagram; however, resonance also introduces significant phase variations.

The detection coils must be connected to a high impedance input stage. Low input capacitance is essential in order to maximize the resonant frequencies. For connecting leads greater than 10 cm and operating frequencies above 100 kHz, the lead capacitance can have a significant parasitic effect, see Fig. 3.25a, which are typically between 50 and 100 pF/m depending on the choice of cable. The resonant frequency, f_r, is given by

$$f_r = \frac{1}{2\pi\sqrt{L.C_{TOT}}} \tag{3.37}$$

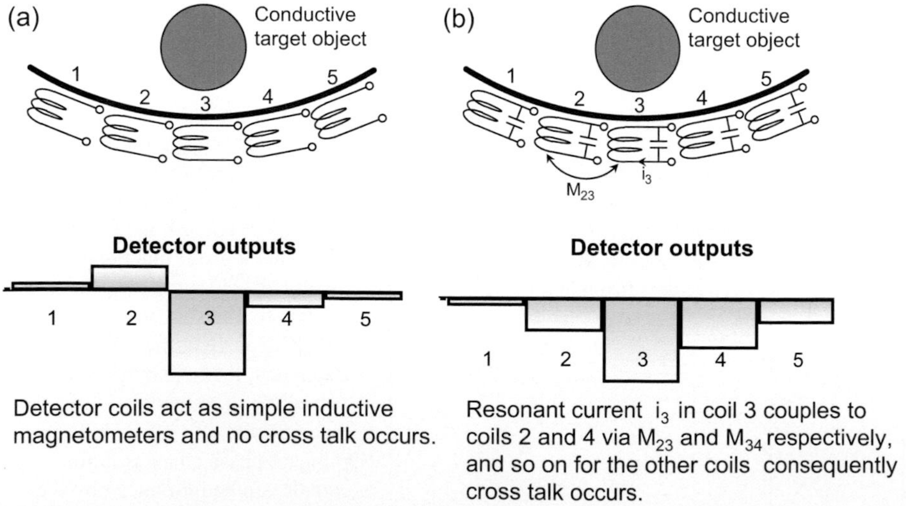

Figure 3.24 The effects of multiple interacting resonances on an MIT sensor, (a) no resonance, (b) resonant case.

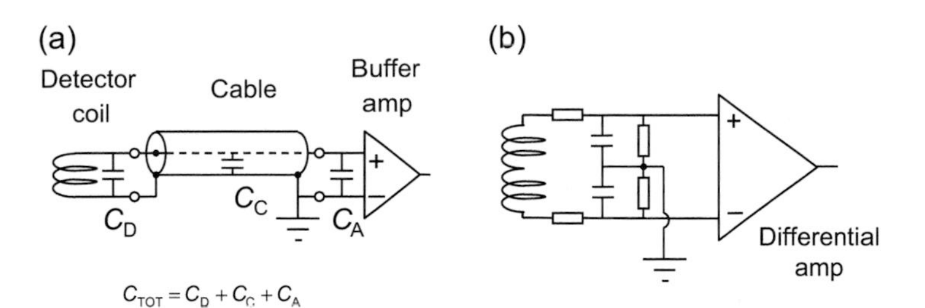

Figure 3.25 Receiver coil input connections, (a) input capacitances (b) differential input connections.

Consequently, the buffer amplifier must often be connected as close as possible to the detector coils. For similar reasons, the driver circuits should also be connected directly to the excitation coils; however, the impedances are often lower so the cabling demands may be less demanding. Differential input configurations such as shown in Fig. 3.25b are often used to minimize interference. Note that radio frequency filtering and a DC bias path are important practical additions as is the use of screened twisted pair cables for the interconnections. Stray coupling with the interconnections between transmitter and receiver circuits should also be carefully controlled either with screens or physical separation.

3.6 Applications

Electromagnetic sensors are used on a wide variety of materials including metals, aqueous fluids, and biological tissues, and consequently MIT applications are reflected across this base. Examples of these include

i. liquid metal processing
ii. nondestructive testing (crack imaging and coating thickness)
iii. conductivity profiling
iv. biomedical
v. oil industry processes
vi. security
vii. geo-physical surveying

A full review of all these application areas is beyond the scope of this chapter; however, selected areas are covered in the following sections to give an indication of the diversity of the use of induction techniques.

3.6.1 Hot metal processing

There are a number of reports of electromagnetic imaging and inversion techniques being applied to high temperature metal process applications due to the robust and simple nature of the sensor arrays. Examples include continuous casting, tapping, and controlled cooling.

Continuous casting is widely used in the steel industry in the production of billets, blooms, and slabs, which are subsequently processed into a variety of products. Increasing demand for tighter manufacturing tolerances, improved steel quality (i.e., cleanliness and surface quality), and more complex compositions has required constant improvements to the entire steel production process. Continuous casting is a critical process as any defects introduced during solidification are difficult to remove in later processes. MIT has been used to study two-phase molten steel—argon flows that occur in the submerged entry nozzle, which is used to transport liquid steel into the mold. Laboratory (Ma, Peyton, Binns, & Higson, 2005) and plant (Higson, Drake, Lyons, Peyton, & Lionheart, 2006) studies have been reported showing that cross-section images of the phase fraction can be obtained under the harsh operating conditions with very high temperatures. CIFT has been demonstrated in a laboratory environment on a 10th scale laboratory model of the continuous casting process, using the room temperature liquid metal, InGaSn (Stefani et al., 2004). Recently, experiments on these laboratory models have shown the potential of combining the MIT technique for monitoring the liquid gas fractions within the SEN with CIFT for imaging the flow within the mold (Wondrak et al., 2012), revealing oscillatory or pulsating behavior of the flow under certain conditions.

The imaging of the solidification of liquid metal in a pipe has been studied (Pham et al., 2000a,b), with the aim of realizing sensor to form part of an electronic flow control system for liquid metal delivery by thermally controlling the degree of solidification inside the refractory pipe. The sensing technique exploits the difference in electrical conductivity of molten metal to solid metal. A later study (Coveney, Gray, & Kyllo, 2005) considered the application of this method to tapping of copper smelters.

Following casting in the production route is hot rolling to form the metal into strip, plate, or a desired section. In the case of steel, hot rolling is also associated with controlled cooling. For some steel grades, such as high strength strip steel used in the automobile industry, the correct cooling process is vital to obtain the required microstructure, and hence, the required mechanical properties in the final steel product. Magnetic induction spectroscopy has been considered for this application and applied on-line (Peyton et al., 2010). The use of an array of electromagnetic sensor monitor the changes of the steel properties as it cools has been reported by imaging the paramagnetic to ferromagnetic phase transition as strip steel cools below the Curie point (Sharif, Bell, Morris, & Peyton, 2001).

3.6.2 Conductivity profiling

Conductivity profiling applications are typically one-dimension inversion problems in which the inductance or eddy current spectrum of a single coil or coil pair is used as the source data for the reconstruction algorithm. Typical applications are conducting coatings or depth profiling. Conducting coatings are used as a protective barrier against corrosion or thermal degradation in a number of engineering applications. One for the first reports (Norton, Kahn, & Mester, 1989) of the use of multifrequency or spectroscopic eddy current methods studied the feasibility of coating measurements on cylindrical geometries such as rods, the so-called pin in sleeve. Later Yin et al. (2005) studied the feasibility of profiling an arbitrary conductivity profile using induction spectroscopy. There are important practical applications for this technique in the power generation industry, in particular, the inspection of the thermal barrier coating, for instance, MCrAlZ, on gas turbine blades (Randazzo & Pignone, 2007) and (Antonelli, Ruzzier, & Necci, 1998) and the inspection of electrical conductivity profile within graphite bricks inside advanced gas cooled nuclear reactors to monitor their density changes (Dekdouk, Chapman, Brown, & Peyton 2012; Tesfalem, Peyton, Fletcher, Brown & Chapman, 2020).

3.6.3 Biomedical

The electrical conductivity and permittivity of biological tissues can vary significantly depending on the tissue type and is typical in the range 0.01−1 S/m. The electrical bioimpedance of tissues has been extensively reported, with (Gabriel, Gabriel, & Corhout, 1996) a relatively comprehensive and well-cited review. Tissues have finite ionic conductivities commensurate with the nature and extent of their ionic content and ionic mobility and their dielectric properties also vary significantly with frequency showing distinct band of dispersion. The low frequency α-dispersion is associated with ionic diffusion processes at the site of the cellular membrane. The β-dispersion, in the tens to hundreds of kilohertz region, is due mainly to the polarization of cellular membranes which act as barriers to the flow of ions between the intra- and extracellular media. Other contributions to the β-dispersion come from the polarization of protein and other organic macromolecules. Finally, the γ-dispersion, in the gigahertz region, is due to the polarization of water molecules.

A number of researchers have tried to exploit the variations in electrical conductivity for biomedical monitoring or tomographic imaging. Applications include life signs monitoring (Tarjan & McFee, 1968), body composition, and brain edema. A good review of MIT within the context of medical EIT is given by Griffiths (2005).

3.6.4 Security

WTMD has proven to be a fundamental part of the screening of large numbers of personnel for potential threat objects and WTMD use is widespread in locations such as airports, prisons, embassies, ports, and government buildings. Rather than imaging the metal in the plane of the detector, Marsh et al. (2013) constructed a forward problem based on the dipole approximation, representing the metal target as an equivalent electromagnetic polarisability tensor. The inversion problem was to determine the tensor components and the 3D coordinate sequence of the metal object as it was carried through the portal. Subsequent report (Makkonen et al., 2014; Marsh et al., 2014) showed the multiple metal objects could be characterized and positioned simultaneously in this manner.

In addition to detection of individual metal items carried on the person, with an array of coils, recently there has been interest in using MIT techniques to image and characterize metal objects within shielded enclosures. Naturally the effect of a metal layer in between the coils and the target or surrounding the target will have a serious confounding effect on the ability to interrogate the target. Despite the difficulties, recent research (Darrer, Watson, Bartlett, & Renzoni, 2015; Wood, Ward, Lloyd, Tatum, & Shenton-Taylor, 2017) has shown that the effects of different metallic shields can be countered with appropriate choice of coil array and frequency range.

3.7 Conclusions and outlook

This chapter has attempted to summarize the current state of the art in MIT technology, covering the key issues including the underlying theory, solutions to the forward and inverse problems, system implementations, and applications. The first reports of MIT date back to the early 1990s with relatively simple systems and crude image reconstruction techniques. There has been considerable progress in the subject over the past decades, and unfortunately, the available space in this chapter does not permit a completely exhaustive review, citing all the relevant MIT literature.

Forward and inverse solutions for MIT are now relatively well understood; however, accurate quantitative forward models are currently very computer intensive. It is likely that progress in computer technology will shorten inversion times, and neural inversion using machine learning offers promise, but nevertheless more progress is needed to formulate algorithms that are more efficient.

As regards hardware platforms, the past decade has seen a slow migration from systems with a high degree of analogue processing, especially demodulation, to digital systems where the ADC is placed as soon in the signal processing chain as possible. Over the time since the first edition of this book, this trend toward more digital systems was continued and they are now the norm. The majority of MIT systems are either multiple channel, single frequency, or single channel, multiple frequency. Some of the single channel, multiple frequency systems also employ simultaneous multifrequency operation, with spectrally rich excitation waveforms. A future direction for MIT hardware innovations is likely to be toward simultaneous multifrequency and multichannel systems, with all the additional complexity, one recent publication (Muttakin & Soleimani, 2020) demonstrating this approach. The majority of MIT research is based on 2D arrays; clearly 3D arrays offer more potential as the forward problem is inherently three dimensional.

This chapter started with a description of MIT in the context of ERT and ECT and the measurement of the three electrical parameters as a family of complementary techniques. With this in mind, multimodal combinations of these techniques may also offer a fruitful direction for future research and development, exploiting commonalities in hardware and a holistic approach to image reconstruction. However, the number of publications reporting dual or multimode systems is very small, possibly limited by the number of applications that could take advantage from this approach. Nevertheless, some demanding applications such as measurement multiphase flow could take benefit and this particular application has attracted considerable interest in past from dual and multimode research including ultrasonic and gamma ray tomography as described later in this book. One particularly noteworthy recent example of a dual mode approach involving MIT (Ma, McCann, & Hunt, 2017) describes a multiphase flow meter prototype, capable of nonintrusively measuring the water volumetric flow rate in a multiphase flow, combining MIT with electromagnetic velocity tomography for simultaneously measuring both the cross-sectional volumetric fraction and the local axial velocity of the water phase, and showing results from a laboratory flow loop.

The number of applications has increased over the past 2 decades with the early work being speculative laboratory based and later publication being more application focused. No MIT system is not likely to offer a global solution and each application likely involves a bespoke solution with a dedicated coil array, data acquisitions scheme, and inversion. Many applications do not require images, but particular parameters. Often a similar inversion process can be used for these cases. This requires a broader interpretation of the term "tomography" in the case of MIT as imagery is not necessarily the required output.

Finally, the number of researchers that have gained PhD's from MIT research is increasing. Some of these have found employment in industry and scientific institutes, which will help the growth of the subject. However, it is crucial that the research becomes application led and focussed on solving problems of high value to industry.

References

Adler, A., & Guardo, R. (1994). A neural network image reconstruction technique for electrical impedance tomography. *IEEE Transactions on Medical Imaging, 13*, 594.

Al-Zeibak, S., & Saunders, N. H. (1993). A feasibility study of in vivo electromagnetic imaging. *Physics in Medicine and Biology, 38*, 151–160.

Ammari, H., Chen, J., Chen, Z., Garnier, J., & Volkov, D. (2014). Target detection and characterization from electromagnetic induction data. *Journal de Mathematiques Pures et Appliquees, 101*, 54–75.

Antonelli, G., Ruzzier, M., & Necci, F. (1998). Thickness measurement of MCrAlY high-temperature coatings by frequency scanning eddy current technique. *Journal of Engineering for Gas Turbines & Power, 120*, 537–542.

Armitage, D. W., Le Veen, H. H., & Pethig, R. (1983). Radiofrequency-induced hyperthermia: Computer simulation of specific absorption rate distribution using realistic anatomic models. *Physics in Medicine and Biology, 28*, 31–42.

Cheng, C. C., Dodd, C. V., & Deeds, W. E. (1971). General analysis of probe coils near stratified conductors. *International Journal of Nondestructive Testing, 3*, 109–130.

Coveney, J. A., Gray, N. B., & Kyllo, A. K. (2005). Applications in the metals production industry. In D. M. Scott, & H. McCann (Eds.), *Book series: Electrical and computer engineeringProcess imaging for automatic control* (pp. 401–433). A Series of Reference Books and Textbooks CRC press-Taylor & Francis.

Darrer, B. J., Watson, J. C., Bartlett, P., & Renzoni, F. (2015). Magnetic imaging: A new tool for UK national nuclear security. *Scientific Reports, 5*, 7944.

Dekdouk, B., Chapman, R., Brown, M., & Peyton, A. J. (2012). Evaluating the conductivity distribution in isotropic polycrystalline graphite using spectroscopic eddy current technique for monitoring weight loss in advanced gas cooled reactors. *NDT&E International, 51*, 150–159.

Dekdouk, B., Wuliang Yin, W., Ktistis, C., Armitage, D. W., & Peyton, A. J. (2010). A method to solve the forward problem in magnetic induction tomography based on the weakly coupled field approximation. *IEEE Transactions on Biomedical Engineering, 57*, 914–921.

Dodd, C. V., & Deeds, W. E. (1968). Analytical solutions to eddy-current probe-coil problem. *Journal of Applied Physics, 39*, 2829–2839.

Dyck, D. N., Lowther, D. A., & Freeman, E. M. (1994). A method of computing the sensitivity of electromagnetic quantities to changes in materials and sources. *IEEE Transactions on Magnetics, 30*, 3415–3418.

Gabriel, C., Gabriel, S., & Corhout, E. (1996). The dielectric properties of biological tissues. *Physics in Medicine and Biology, 41*, 2231–2249.

Gandhi, O. P., DeFord, J. F., & Kanai, H. (1984). Impedance method for calculation of power deposition patterns in magnetically induced hyperthermia. *IEEE Transactions on Biomedical Engineering, 31*, 644–651.

Griffiths, H. (2001). Magnetic induction tomography. *Measurement Science and Technology, 12*, 1126–1131.

Griffiths, H. (2005). Magnetic induction tomography. In D. S. Holder (Ed.), *Electrical impedance tomography: Methods, history and applictions*. IOP Publishing.

Gürsoy, D., & Scharfetter, H. (2011). Imaging artefacts in magnetic induction tomography caused by the structural incorrectness of the sensor model. *Measurement Science and Technology, 22*, 015502 (10pp).

Higson, S. R., Drake, P., Lyons, A., Peyton, A., & Lionheart, B. (2006). Electromagnetic visualisation of steel flow in continuous casting nozzles. *Ironmaking and Steelmaking, 33*, 357–361.

Hollaus, K., Magele, C., Merwa, R., & Scharfetter, H. (2004a). Numerical simulation of the eddy current problem in magnetic induction tomography for biomedical applications by edge elements. *IEEE Transactions on Magnetics, 40*, 623–626.

Hollaus, K., Magele, C., Merwa, R., & Scharfetter, H. (2004b). Fast calculation of the sensitivity matrix in magnetic induction tomography by tetrahedral edge finite elements and the reciprocity theorem. *Physiological Measurement, 25*, 159–165.

Horner, B., Mesch, F., & Trächtler, A. (1996). A multi-sensor induction flowmeter reducing errors due to non-axisymmetric flow profiles. *Measurement Science and Technology, 7*, 354–360.

Johnstone, S., & Peyton, A. J. (2001). The application of parametric 3D finite element modelling techniques to evaluate the performance of a magnetic sensor system. *Sensors and Actuators, 93*, 109–116.

Ktistis, C., Armitage, D. W., & Peyton, A. J. (2008). Calculation of the forward problem for absolute image reconstruction in MIT. *Physiological Measurement, 29*, S455–S464.

Makkonen, J., Marsh, L. A., Vihonen, J., Järvi, A., Armitage, D. W., Visa, A., et al. (2014). KNN classification of metallic targets using the magnetic polarizability tensor. *Measurement Science and Technology, 25*, 055105 (9pp).

Ledger, P. D., & Lionheart, W. R. B. (2015). Characterising the shape and material properties of hidden targets from magnetic induction data. *IMA Journal of Applied Mathematics, 80*(6), 1776–1798. https://doi.org/10.1093/imamat/hxv015

Ma, L., McCann, D., & Hunt, A. (2017). Combining magnetic induction tomography and electromagnetic velocity tomography for water continuous multiphase flows. *IEEE Sensors Journal, 17*, 8271–8281.

Ma, X., Peyton, A. J., Binns, R., & Higson, S. R. (2005). Electromagnetic techniques for imaging the cross-section distribution of molten steel flow in the continuous casting nozzle. *IEEE Sensors Journal, 5*, 224–232.

Ma, X., Peyton, A. J., Higson, S. R., Lyons, A., & Dickinson, S. J. (2006a). Hardware and software design for an electromagnetic induction tomography (EMT) system applied to high contrast metal process applications. *Measurement Science and Technology, 17*, 111–118.

Ma, X., Peyton, A. J., Higson, S. R., Lyons, A., & Dickinson, S. J. (2006b). Hardware and software design for an electromagnetic induction tomography (MIT) system for high contrast metal process applications. *Measurement Science and Technology, 17*, 111–118.

Marsh, L. A., Ktistis, C., Järvi, A., Armitage, D. W., & Peyton, A. J. (2013). Three-dimensional object location and inversion of the magnetic polarizability tensor at a single frequency using a walk-through metal detector. *Measurement Science and Technology, 24*, 045102 (13pp).

Marsh, L. A., Ktistis, C., Järvi, A., Armitage, D. W., & Peyton, A. J. (2014). Determination of the magnetic polarizability tensor and three dimensional object location for multiple objects using a walk-through metal detector. *Measurement Science and Technology, 25*, 055107 (12pp).

Martin, S., & Choi, C. T. M. (2016). Nonlinear electrical impedance tomography reconstruction using artificial neural networks and particle swarm optimization. *IEEE Transactions on Magnetics, 52*, 7203904 (4pp).

Morris, A., Griffiths, H., & Gough, W. (2001). A numerical model for magnetic induction tomographic measurements in biological tissues. *Physiological Measurement, 22*, 113–119.

Muttakin, I., & Soleimani, M. (2020). Magnetic induction tomography spectroscopy for structural and functional characterization in metallic materials. *Materials, 13*, 2639 (21 pp).

Norton, S. J., & Bowler, J. R. (1993). Theory of eddy current inversion. *Journal of Applied Physics, 73*, 501–510.

Norton, S. J., Kahn, A. H., & Mester, M. L. (1989). Reconstructing electrical conductivity profiles from variable-frequency eddy current measurements. *Research in Nondestructive Evaluation, 1*, 167–179.

Norton, S. J., & Won, I. J. (2001). Identification of buried unexploded ordnance from broadband electromagnetic induction data. *IEEE Transactions on Geoscience and Remote Sensing, 39*, 2253–2261.

Peyton, A. J., Beck, M. S., Borges, A. R., de Oliveira, J. D., Lyon, G. M., Yu, Z. Z., et al. (1999). Development of electromagnetic tomography (EMT) for industrial applications. Part 1: Sensor design and instrumentation. In *1st world congress on industrial process tomography, Buxton, Greater Manchester, April 14–17* (pp. 306–312).

Peyton, A. J., Watson, S., Williams, R. J., Griffiths, H., & Gough, W. (2003). Characterising the effects of the external electromagnetic shield on a magnetic induction tomography sensor. In *3rd world congress on industrial process tomography, Banff, Canada*.

Peyton, A. J., Yin, W., Dickinson, S. J., Davis, C. L., Strangwood, M., Hao, X., et al. (2010). Monitoring microstructure changes in rod online by using induction spectroscopy. *Ironmaking and Steelmaking, 37*, 135–139.

Pham, M. H., Hua, Y., & Gray, N. B. (2000a). Imaging the solidification of molten metal by eddy currents: I. *Inverse Problems, 16*, 469–482.

Pham, M. H., Hua, Y., & Gray, N. B. (2000b). Imaging the solidification of molten metal by eddy currents: II. *Inverse Problems, 16*, 483–494.

Randazzo, A., & Pignone, E. (2007). A blade coating inspection method based on an electromagnetic inverse scattering approach. In *IEEE international workshop on imaging systems and techniques - IST 2007, Krakow, Poland.*

Resagk, C., Men, S., Ziolkowski, M., Kuilekov, M., & Brauer, H. (2006). Magnetic field tomography on two electrically conducting fluids. *Measurement Science and Technology, 17*, 2136–2140.

Roemer, P. B., Edelstein, W. A., Hayes, C. E., Souza, S. P., & Mueller, O. M. (1990). The NMR phased array. *Magnetic Resonance in Medicine, 16*, 192–225.

Rymarczyk, T., Kłosowski, G., Kozłowski, E., & PTchórzewski, P. (2019). Comparison of selected machine learning algorithms for industrial electrical tomography. *Sensors, 19*, 1521.

Scharfetter, H., Köstinger, A., & Issa, S. (2008). Hardware for quasi-single-shot multifrequency magnetic induction tomography (MIT): The Graz Mk2 system. *Physiological Measurement, 29*, 431–443.

Scharfetter, H., Lackner, H. K., & Rosell, J. (2001). Magnetic induction tomography: Hardware for multi-frequency measurements in biological tissues. *Physiological Measurement, 22*, 131–146.

Sharif, E., Bell, C., Morris, P. F., Peyton, A. J., et al. (2001). Imaging the transformation of hot strip steel using magnetic techniques. *Journal of Electronic Imaging, 10*(3), 669–678. https://doi.org/10.1117/12.417166

Silvester, P. P., & Ferrari, R. L. (1996). *Finite elements for electrical engineers*. Cambridge: Cambridge University Press.

Smith, R. W. M., Freeston, I., Brown, B. H., & Sinton, A. M. (1992). Design of a phase-sensitive detector to maximise signal-to-noise ratio in the presence of Gaussian wideband noise. *Measurement Science and Technology, 3*, 1054–1062.

Soleimani, M., & Lionheart, W. R. B. (2006). Absolute conductivity reconstruction in magnetic induction tomography using a nonlinear method. *IEEE Transactions on Medical Imaging, 25*, 1521–1530.

Soleimani, M., Lionheart, W. R. B., Peyton, A. J., Ma, X., & Higson, S. R. (2006). A three-dimensional inverse finite-element method applied to experimental eddy-current imaging data. *IEEE Transactions on Magnetics, 42*, 1560–1567.

Stefani, F., Gundrum, T., & Gerbeth, G. (2004). Contactless inductive flow tomography. *Physical Review, 70*, 056306.

Tanguay, L.-F., Gagnon, H., & Guardo, R. (2007). Comparison of applied and induced current electrical impedance tomography. *IEEE Transactions on Biomedical Engineering, 54*, 1643–1649.

Tarjan, P. R., & McFee, R. (1968). Electrodeless measurement of the effective resistivity of the human torso and head by magnetic induction. *IEEE Transactions on Biomedical Engineering, 15*, 266–278.

Tesfalem, H., Peyton, A. J., Fletcher, A. D., Brown, M., & Chapman, B. (2020). Conductivity profiling of graphite moderator bricks from multifrequency eddy current measurements. *IEEE Sensors Journal, 20*, 4840–4849.

Timmel, K., Eckert, S., Gerbeth, G., Stefani, F., & Wondrak, T. (2010). Experimental modeling of the continuous casting process of steel using low melting point metal alloys—the LIMMCAST program. *ISIJ International, 50*, 1134–1141.

Tumanski, S. (2007). Induction coil sensors—a review. *Measurement Science and Technology, 18*, R31–R46.

Vauhkonen, M., Hamsch, M., & Igney, C. H. (2008). A measurement system and image reconstruction in magnetic induction tomography. *Physiological Measurement, 29*, S445–S454.

Watson, S., Williams, R. J., Gough, W., & Griffiths, H. (2008). A magnetic induction tomography system for samples with conductivities below 10 S m^{-1}. *Measurement Science and Technology, 19*, 045501 (11pp).

Wondrak, T., Eckert, S., Galindo, V., Gerbeth, G., Stefani, F., Timmel, K., et al. (2012). Liquid metal experiments with swirling flow submerged entry nozzle. *Ironmaking and Steelmaking, 39*, 1–9.

Wood, J., Ward, R., Lloyd, C., Tatum, P., & Shenton-Taylor, C. (2017). Effect of shielding conductivity on magnetic induction tomographic security imagery. *IEEE Transactions on Magnetics, 53*, 4000406 (6 pp).

Yin, W., Dickinson, S. J., & Peyton, A. J. (2005). Imaging the continuous conductivity profile within layered metal structures using inductance spectroscopy. *IEEE Sensors, 5*, 161–166.

Yin, W., & Peyton, A. (2010a). Dimensional scaling relationships in electrical tomography. In *6th world congress on industrial process tomography, Beijing, China.*

Yin, W., & Peyton, A. J. (2010b). Sensitivity formulation including velocity effects for electromagnetic induction systems. *IEEE Transactions on Magnetics, 46*, 1172–1176.

Yu, Z. Z., Worthington, P. F., Stone, S., & Peyton, A. J. (1995). Electromagnetic screening of inductive tomography sensors. In *European concerted action on process tomography conference, Bergen, Norway* (pp. 300–310).

Zhao, Y. Y., Lucas, G., Leeungculsatien, T., & Zhang, T. (2012). Measurement system design of an imaging electromagnetic flow meter. *AIP Conference Proceedings, 1428*, 258.

Zolgharni, M., Ledger, P. D., Armitage, D. W., Holder, D. S., & Griffiths, H. (2009). Imaging cerebral haemorrhage with magnetic induction tomography: Numerical modelling. *Physiological Measurement, 30*, S187–S200.

Magnetic resonance imaging

4

Andrew J. Sederman
Department of Chemical Engineering and Biotechnology, University of Cambridge, Cambridge, United Kingdom

4.1 Introduction to MRI and NMR

Magnetic resonance imaging (MRI) is now ubiquitous as a diagnostic tool in clinical medicine to the extent that it is recognized by the general public; however, its use in engineering and the physical sciences is much less well known. Despite this, it has many attributes that make it an exceptionally powerful tomography technique due to the different types of contrast and information that it can be used to measure. These include but are not limited to chemical concentration, molecular diffusion, and velocity, all of which can be measured locally in an image of the sample. MRI is also an inherently quantitative measurement tool with well-defined and well-understood physical principles underpinning it. This chapter is aimed principally as a pedagogical introduction to MRI for engineers and will include the basic principles of data acquisition and image reconstruction. It will also consider some of the practicalities and limitations along with its benefits and a few examples of applications to engineering systems.

MRI is, in fact, just a subset of the nuclear magnetic resonance (NMR) technique. NMR encompasses all magnetic resonance (MR) methods, both spatially unresolved and spatially resolved and many readers will be familiar with NMR as a useful analytical tool for spectroscopy in chemistry or for determining the structure of proteins. In general, MR measurements can be divided into two classes—those which do not require any spatial resolution within the data, and those that do. Where spatial encoding information is required, then, in addition to a basic MR equipment, which is typically performed in a large magnetic field, additional smaller constant gradients in the magnetic field will also be used at some point during the experiment. Here we generally consider spatially resolved MR measurements and how to acquire them, but it is useful to note that any information that can be extracted from an NMR experiment can, in theory, be obtained in a spatially resolved form of an image, where signal-to-noise (SNR) and acquisition time are sufficient.

This chapter first explains the basic principles of MRI before looking at specific imaging techniques and how these techniques can be modified to get different image contrast and how to speed up acquisition to enable real-time image collection. This chapter will finish off with some examples of applications of MRI in engineering and some final conclusions.

Industrial Tomography. https://doi.org/10.1016/B978-0-12-823015-2.00003-0

4.2 MRI: basic imaging principles

In this section, there is a brief exposition on the fundamentals of NMR, MRI, and the measurement of diffusion and flow. Where suitable, practical advice on sequence selection and implementation and other aspects of MRI will also be included. There are a number of excellent texts treating the subject in greater detail and the interested reader in the fundamentals of NMR and MRI may find the following texts of interest: Levitt (2002) for an introduction to NMR, and Callaghan (Callaghan, 1991, 2011) for imaging and in particular translational motion. Sodickson and Cory (1998) approach the theory of imaging from a slightly different perspective, and for an alternative medical perspective on MRI physics and pulse sequence selection and design, the reader is referred to the text by Brown, Cheng, Haacke, Thompson, and Venkatesan (2014).

4.2.1 The precessing magnetization vector

Before considering how we get spatial information from the MRI experiment, it is necessary to introduce the basics of the spatially unresolved NMR experiments. This is often taken to suggest a spectroscopy experiment, although other bulk properties such as diffusion and relaxation times may also be measured. The spatially unresolved spectroscopy experiment is the simplest type of MR experiment and the principles upon which this measurement is based underpin all types of MR measurement and so are of relevance here.

All NMR experiments must be performed on nuclei of nonzero nuclear spin, most commonly ^1H, and in the presence of a magnetic field. At equilibrium, a greater proportion of the nuclear spins align with the magnetic field (with the overpopulation defined by the Boltzmann distribution), as a result of which there exists a net magnetization vector aligned parallel to the direction of the magnetic field. This magnetization vector can be manipulated by the application of electromagnetic energy of appropriate frequency. The frequency at which this manipulation occurs is called the resonance frequency, ω_0, and is given by the following:

$$\omega_0 = \gamma B_0 \tag{4.1}$$

where γ is the (nuclei specific) gyromagnetic ratio and B_0 is the magnetic field strength. All NMR sequences have an "excitation pulse" which is used to perturb the magnetization away from parallel to the magnetic field. After the magnetization vector has been perturbed, it will precess about the direction of the magnetic field, $\boldsymbol{B}_0$, at the resonance frequency ω_0 as it returns to its equilibrium magnetization. This precession emits electromagnetic energy which can be detected and forms the basis of the MR measurement. In a standard NMR spectroscopy measurement, the magnetic field observed is slightly modified by the electronic environment of the nucleus under study; thus, ω_0 is also modified and the distribution of frequencies, or spectrum, becomes a fingerprint of the molecular species under investigation.

4.2.2 Magnetic field gradients and imaging

Performing spatially resolved experiments (imaging) requires the additional effect of spatial variations in the magnetic field, or magnetic field gradients. So far we have assumed that B_0 was homogeneous over all experimental space or at least that inhomogeneities were small. Magnetic field gradients are designed such that they add a relatively small magnetic field which varies linearly with position, to the main field such that the resultant field, $B(\boldsymbol{r})$, at position $\boldsymbol{r}$ varies according to

$$B(\boldsymbol{r}) = B_0 + \boldsymbol{G}.\boldsymbol{r} \tag{4.2}$$

where $\boldsymbol{G}$ is the strength of the applied gradient. Practically, the gradient can be applied in any direction, with a known strength and duration. Substituting this into Eq. (4.1), we now see that the frequency is a function of the sample position:

$$\omega(\boldsymbol{r}) = \gamma(B_0 + \boldsymbol{G}.\boldsymbol{r}) \tag{4.3}$$

If we measure the frequency relative to the resonance frequency, ω_0, also known as the rotating frame, we observe that the frequency is directly proportional to the position and this relationship underpins all MRI experiments. If we consider a local volume element dV within our sample, the local spin density, $\rho(\boldsymbol{r})$, for spins within the volume element d_V is then $\rho(\boldsymbol{r})dV$. The NMR signal as a function of time, $dS(t)$, from this element may then be written as follows:

$$dS(t) \propto \rho(\boldsymbol{r})\exp[\mathrm{i}\omega(\boldsymbol{r})t]dV \tag{4.4}$$

Inserting Eq. (4.3) into Eq. (4.4), we obtain

$$dS(t) \propto \rho(\boldsymbol{r})\exp[\mathrm{i}\gamma(B_0 + \boldsymbol{G}.\boldsymbol{r})t]dV \tag{4.5}$$

A transformation into the rotating frame of reference followed by integration over all the samples allows us to write Eq. (4.5) in terms of the overall observed signal $S(t)$ and becomes the following:

$$S(t) = \iiint \rho(\boldsymbol{r})\exp[\mathrm{i}\gamma\boldsymbol{G}.\boldsymbol{r}t]d\boldsymbol{r} \tag{4.6}$$

where $d\boldsymbol{r}$ represents integration over all space. Mansfield & Grannell, 1973 realized that this equation could be simplified by the introduction of the concept of $\boldsymbol{k}$-space defined as follows:

$$\boldsymbol{k} = \frac{\gamma\boldsymbol{G}t}{2\pi} \tag{4.7}$$

where $\boldsymbol{k}$ is known as the reciprocal space vector and has units of m^{-1}. More generally it can be written as the first moment of the gradient with time:

$$\boldsymbol{k}(t) = \frac{\gamma}{2\pi} \int_0^t \boldsymbol{G}(t)dt \tag{4.8}$$

Substituting Eq. (4.7) into Eq. (4.6) gives an expression that describes how the measured signal, S, varies as a function of $\boldsymbol{k}$

$$S(\boldsymbol{k}) = \iiint \rho(\boldsymbol{r})\exp[i2\pi\boldsymbol{k}.\boldsymbol{r}]d\boldsymbol{r} \tag{4.9}$$

And the Fourier transformation of Eq. (4.9) gives

$$\rho(\boldsymbol{r}) = \iiint S(\boldsymbol{k})\exp[-i2\pi\boldsymbol{k}.\boldsymbol{r}]d\boldsymbol{k} \tag{4.10}$$

Taken together, Eqs. (4.9) and (4.10) tell us that $S(\boldsymbol{k})$ and $\rho(\boldsymbol{r})$ form a mutually conjugate Fourier transform pair. It therefore follows that if we acquire our signal over a suitable range of $\boldsymbol{k}$ and then perform a Fourier transform, we will recover, $\boldsymbol{\rho}$, the spin density or nuclei concentration as a function of the position, $\boldsymbol{r}$, that is our concentration image.

This relationship shows that we acquire our image data in the frequency space of the image itself—a fact that can be useful when trying to identify specific repeating features in an image that may show up more clearly in the directly acquired data (Holland, Blake, Tayler, Sederman, & Gladden, 2011). To produce an image of the required resolution and field-of-view (FOV), data must be acquired over a suitable range and increment of $\boldsymbol{k}$. From Fourier theory, it follows that the range of $\boldsymbol{k}$ defines the image resolution and the increment in $\boldsymbol{k}$ defines the FOV such that

$$\text{FOV} = \frac{1}{\Delta \boldsymbol{k}}, \ \Delta l = \frac{1}{\boldsymbol{k}_{\mathbf{range}}} \tag{4.11}$$

where $\Delta\boldsymbol{k}$ is the increment in $\boldsymbol{k}$, Δl is the image resolution and $\boldsymbol{k}_{\mathbf{range}}$ is the total range of $\boldsymbol{k}$ sampled from $-\boldsymbol{k}$ to $\boldsymbol{k}$. Although the majority of MRI measurement is done by fully sampling $\boldsymbol{k}$-space followed by a Fourier transform as described, there has recently been great interest in utilizing other image reconstruction methods such as compressed sensing (Candes, Romberg, & Tao, 2006; Donoho, 2006) where data sampling, and therefore imaging time, is reduced. These will be discussed in more detail in Section 4.2.1.

4.2.3 Frequency and phase encoding

From the previous section, we see that it is sufficient to acquire signal over the correct range of $\boldsymbol{k}$ to produce an image of the desired FOV and resolution. The range of $\boldsymbol{k}$ values that we wish to acquire is termed $\boldsymbol{k}$-space and experimentally there are a number of different ways to traverse through $\boldsymbol{k}$-space. In a typical experiment, an excitation pulse will be followed by the application of magnetic field gradients, of known magnitude and duration, and data acquisition. In conventional imaging, all points in $\boldsymbol{k}$-space must be acquired to produce a fully resolved image and these points would be acquired in the "read" direction and the "phase" direction. In the read (or readout) direction, the gradient is kept constant during the acquisition and $\boldsymbol{k}$-space is traversed, with successive points being acquired at constant time increments. From Eq. (4.7), we see that different values of $\boldsymbol{k}$ are achieved by keeping $\boldsymbol{G}$ constant as the value of t increases in this method. This is also known as the *frequency encode* direction; the spins will have a position-dependent frequency during the readout, as defined by Eq. (4.3). This method assumes that the time evolution of the signal is only due to the position of the nuclei and the applied gradients, which may not always be valid, but it provides a very "cheap" way to acquire points quickly as the time increment between points is typically of the order of a few μs and a whole line of $\boldsymbol{k}$-space of several hundred points can be acquired in less than 1 ms. In conventional imaging, the other spatial dimension would be the *phase encode* direction in which a gradient is applied for a fixed duration at some time between excitation and acquisition, thus imparting a position-dependent phase (angle) as can be seen from a time integral of Eq. (4.3) where $\omega(\boldsymbol{r})$ is the rate of change of phase angle with time. In the phase direction, the gradient is off during the acquisition and usually subsequent $\boldsymbol{k}$-space points in the phase direction are acquired in successive excitations by changing the strength of the applied gradient. The delay between successive excitations is determined by the relaxation of the magnetization back to equilibrium (T_1 relaxation) and is typically of the order of seconds.

4.2.4 Simple 1D and 2D imaging sequences

A conceptually simple imaging sequence is a one-dimensional (1D) image or profile with frequency encoding for position in one direction as shown in Fig. 4.1. In a 1D profile in the x-direction, the intensity at each x position corresponds to integration over y and z at that x-position. In $\boldsymbol{k}$-space, we just need to acquire a single line of points

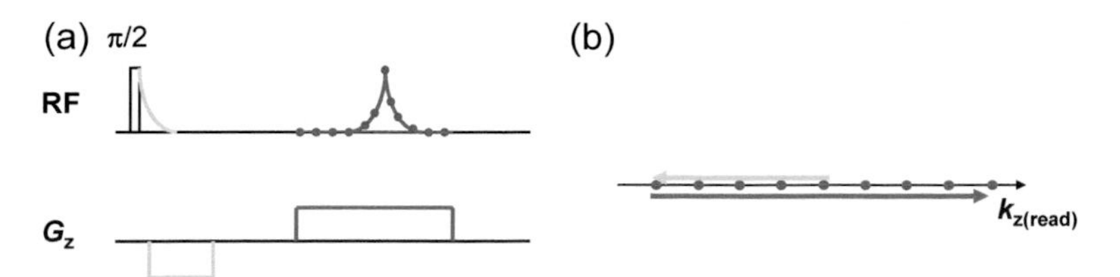

Figure 4.1 (a) Schematic of a simple 1D profile "image" and (b) the corresponding trajectory through $\boldsymbol{k}$-space.

in the K_x direction as shown in Fig. 4.1a where $\Delta \boldsymbol{k}$ and $\boldsymbol{k}_{\mathbf{range}}$ are determined from Eq. (4.11). Initially the magnetization lies along the direction of the magnetic field (z direction), but after perturbation by a 90° excitation pulse, it is in the transverse (x−y) plane and will precess as described by Eq. (4.1). After excitation, the value of $\boldsymbol{k}$ is initialized to zero so a negative dephasing gradient is applied to move the system to $-\boldsymbol{k}_{\mathbf{max}}$. At this point, the polarity of the gradient is switched and as the system moves from $-\boldsymbol{k}_{\mathbf{max}}$ to $\boldsymbol{k}_{\mathbf{max}}$ the digitizer is open to acquire data at fixed time increments that correspond to $\Delta \boldsymbol{k}$. Following Fourier transformation of the data, a 1D profile of the sample is obtained. Timescales for this image are short and excitation, dephasing gradient, and data acquisition can take <1 ms.

More often, we are interested in a 2D image, and using convention imaging this would be done with a frequency encoding direction and a phase direction. A simple 2D imaging sequence and associated 2D $\boldsymbol{k}$-space raster are shown in Fig. 4.2. In this sequence, acquisition of $\boldsymbol{k}$-space in one direction follows in exactly the same way as in the 1D sequence; that is, a dephasing gradient is first applied, followed by a frequency encoding gradient during data acquisition. In the other imaging dimension, the phase direction, a gradient will be applied before acquisition to move the system to the desired $\boldsymbol{k}_{\mathbf{phase}}$ position before acquisition. In this method, a single line of $\boldsymbol{k}$-space is acquired after excitation with the position of that line determined by the strength and duration of the phase gradient (Eq. 4.7). To acquire all points on the $\boldsymbol{k}$-space raster, multiple excitations must be done with the delay between excitations characterized by T_1 (typically of the order of seconds), so the total image acquisition time will often be several minutes. The 2D image is obtained by 2D Fourier transformation of all points in the 2D $\boldsymbol{k}$-space raster.

Here a practical difference between the read and phase directions is seen where points in the read direction may be acquired every few μs, but points in the phase direction are acquired every few seconds which would suggest that we want to acquire points in the read direction wherever possible. This is often the case, but the number of read points that can be acquired from a single excitation is determined by adherence to the previous statement that "the time evolution of the signal is only due to the position of the nuclei and the applied gradients." To understand this and other image acquisition time issues, we must consider signal relaxation (Section 4.2.6).

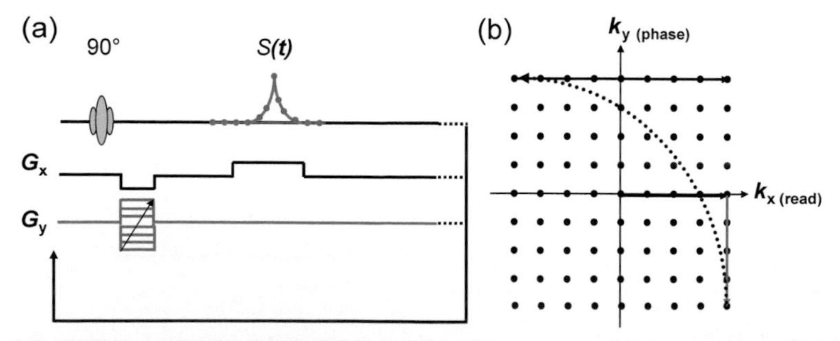

Figure 4.2 (a) Schematic of a simple 2D imaging pulse sequence. (b) the corresponding $\boldsymbol{k}$-space raster used to show how we may interpret the pulse sequence. Following a sufficient T_1 relaxation period the whole sequence is repeated N times to produce the full $\boldsymbol{k}$-space data set.

4.2.5 Slice selection

Although we have seen how to acquire a 2D image in Section 4.2.4, this would correspond to a projection of all material in the direction perpendicular to the imaging plane. To obtain a 2D slice image, we must use a pulse sequence that includes slice selection. Slice selection is obtained by applying an excitation or refocusing pulse that has a narrow excitation band and the application of a gradient field, $\boldsymbol{G}$, simultaneously; the effect of this is shown schematically in Fig. 4.3 where the gradient is in the z-direction. Fig. 4.3a shows the amplitude of the pulse field plotted against the sample frequency. At the same time, the resonance frequencies of the spins along the gradient direction are spread out by the gradient, as shown in Fig. 4.3b. Therefore, the only spins that will be excited are those that resonate within the excitation frequency, $\Delta\omega$, of the pulse and consequently only those spins that lie within the slice defined by Δz are observed. The adjustable parameters $\Delta\omega$ and $\boldsymbol{G}$ control the width of the selected slice, and thin slice selection is obtained by either narrow-band radio frequency (RF) irradiation or a strong gradient $\boldsymbol{G}$. The shape of the slice to be excited is the same as the frequency profile of the pulse, which is related via a Fourier relationship to the modulation of the pulse in time. In most cases, the preferred slice shape would be a perfect rectangle, or Heaviside Hat function. Unfortunately, the Fourier pair of the Hat function is the sinc function $\left(\frac{\sin(x)}{x}\right)$, which has signal intensity extending to infinite time; so it must be truncated to some extent (often severely), as the pulse must be of finite length. This truncation induces artifacts in the frequency domain, which is unsuitable if good slice definition is required. An often used alternative, the Gaussian pulse, gives rise to a Gaussian response. The truncation for this is less severe and the artifacts are almost completely eliminated. It can be seen that soft pulses will be used to form the basis of many experiments. Used alone, they can excite specific regions of the chemical spectrum, and used concurrently with a magnetic gradient, they can be used to excite a localized slice at any place or direction within a sample without the need to reposition the sample.

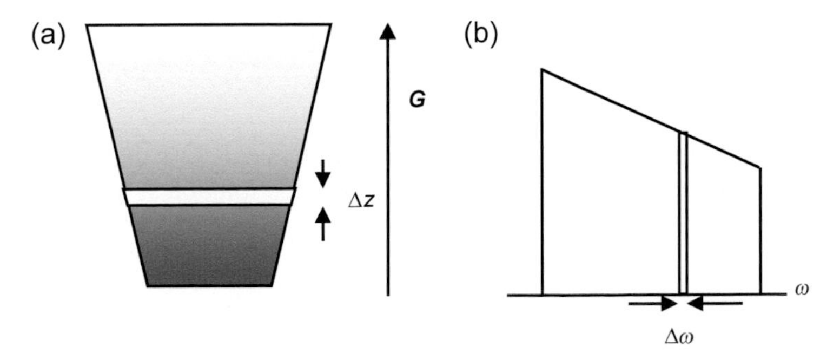

Figure 4.3 Slice selection by selective excitation. When a gradient, $\boldsymbol{G}$, is applied across a sample, the precession frequency is dependent on position. A slice may be selected by a pulse with a narrow excitation frequency, $\Delta\omega$.

4.2.6 Signal relaxation

Following the application of an excitation pulse, the system has excess energy. To return to thermal equilibrium, the system has to redistribute this excess energy—a process known as relaxation. These relaxation times are important in imaging as they will strongly influence how fast images can be acquired and contrast/quantification in those images. Further in-depth discussion on the theories and practicalities of relaxation can be found elsewhere (Abragam, 1961; Keeler, 2011), but a basic introduction is also given here. The most important relaxation times for imaging experiments are the spin–lattice relaxation (T_2) and spin–spin relaxation (T_2) time constants along with the magnetic field inhomogeneity modified spin–spin relaxation (T_2^*). T_1 characterizes the energy exchange between the excited nuclear spin and the surrounding physical environment (that is, the lattice), whereas T_2 and T_2^* characterize the loss of phase angle coherence between nuclear spins. In practical imaging experiments, T_1 affects how fast repeated excitations can be made and T_2 and/or T_2^* affects how long after excitation measurable signal remains. We, therefore, want a short T_1 for fast repetition but a long T_2 so that many points can be acquired after excitation. Because $T_2 \leq T_1$, systems where $T_2 \approx T_1$ offer the best possibilities for fast image acquisition. If a system is characterized by a very small T_2 (e.g., many solids), the coherent signal may decay too fast for it to be possible to perform imaging experiments, whereas a system with very long T_1 will require a long time to come to equilibrium between successive excitations and so imaging may become prohibitively long. Each system, and even species within a system, will have its own T_1 and T_2 relaxation times from which contrast may be produced.

T_1 relaxation: At thermal equilibrium, before the system is perturbed with an excitation pulse, the net magnetization vector is aligned along the direction of the static magnetic field (z). It is the magnitude of this vector that provides the quantitative measurement of the number of nuclear spins excited within the sample. After a 90° excitation pulse, the magnetization vector is rotated through 90° to lie in the x–y plane. As soon as the excitation stops, the magnitude of the magnetization vector increases back along z-direction as it returns to equilibrium. This relaxation process back to its equilibrium magnetization, M_0, can be described by an exponential recovery:

$$\frac{dM_z}{dt} = -\frac{M_z - M_0}{T_1} \tag{4.12}$$

where M_z refers to the z-component of the magnetization and T_1 is the time constant of the recovery. For the 90° excitation described above, the solution at some time, t, after excitation for Eq. (4.12) is as follows:

$$M_z(t) = M_0\left[1 - \exp\left(-\frac{t}{T_1}\right)\right] \tag{4.13}$$

From this, it follows that if we wait a short time between excitations, only a fraction of the magnetization will have been reestablished along z. If we wait times ~5–7 times longer than T_1, the system will have recovered to equilibrium and M_z will have recovered to M_0. In a sequence with repeated excitation pulses such as that shown in Fig. 4.2, the total magnetization available for excitation, and therefore detection, can be calculated from Eq. (4.13) where t is the delay time between excitation pulses. Thus, if we have two species with different T_1 characteristics, by careful selection of the delay time between excitations, signal can be acquired with contrast between the components. This is termed T_1 contrast.

T_2 and T_2^* relaxation: On shorter time scales than T_1, spin−spin relaxation (T_2) occurs. T_2 characterizes the loss of phase angle coherence between spins. During the period following excitation, the individual spins will lose phase coherence with each other as a result of spin−spin interactions and local variations in B_0 causing a decay in the measureable magnetization. This decay of magnetization in the x−y plane due to spin−spin interactions is characterized by the time constant T_2 and is an irreversible decay.

$$M_{xy}(t) = M_0 \exp\left(-\frac{t}{T_2}\right) \tag{4.14}$$

where $M_{xy}(t)$ is the magnetization in the x−y plane at a time t after excitation.

In addition to this decay is the effect of magnetic field heterogeneities—different regions of the sample will have small differences in the magnetic field strength, possibly due to materials with different magnetic susceptibilities, and therefore frequency (Eq. 4.1). This will lead to a loss of phase coherence in addition to T_2—the decay constant used to model this component of the decay is termed T_2' and the combined decay constant, T_2^*, is given by the following:

$$\frac{1}{T_2^*} = \frac{1}{T_2} + \frac{1}{T_2'} \tag{4.15}$$

Thus, the decay of signal after an excitation pulse, in the presence of spin−spin interactions and magnetic field variations, will decay with an envelope of T_2^* time constant. The component due to magnetic field variations (T_2') is recoverable in some imaging sequences, usually using a spin echo. As T_2^* may be much less than T_2 in engineering samples, this is an important consideration when choosing or designing an imaging sequence. As with T_1, it is also possible to obtain T_2 contrast in an image where different components have different T_2. To obtain T_2 or T_2^* contrast, the delay allowed for dephasing between the excitation pulse and data acquisition is adjusted.

4.2.6.1 The spin echo

When a heterogeneous sample is placed in a uniform magnetic field, local magnetic field distortions are created near material boundaries. These field distortions will affect the local frequency, as seen from Eq. (4.1), and so T_2^* may be much shorter than T_2—in such cases, it is beneficial to use the spin echo to refocus dephasing due to these field

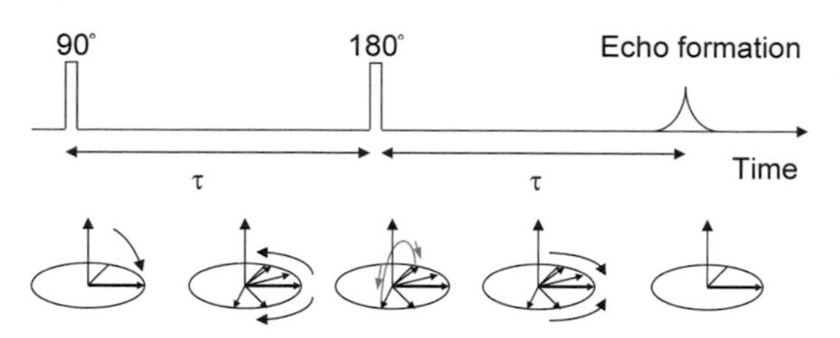

Figure 4.4 The spin echo. The spins are rotated into the transverse plane and dephase due to B_0 inhomogeneities. After time, τ, a 180 degree pulse is applied and the spins begin to refocus. At time 2τ a spin echo is formed.

distortions so the signal decays only with T_2. A schematic of the spin echo is shown in Fig. 4.4. The 180 degree refocusing pulse acts to reverse the dephasing effects due to the local heterogeneities in B_0 such that the final acquired signal (the echo) suffers attenuation resulting from spin–spin interactions only. In short, by using an echo sequence, instead of exciting the system and then allowing the magnetization to decay to zero, the majority of the magnetization can be recovered for use in subsequent measurements. This makes the simple echo sequence shown in Fig. 4.4 a common feature of MRI pulse sequences.

4.2.7 Flow and diffusion

One of the powerful features of MR is the ability to measure incoherent motion (diffusion) and coherent motion (flow). There have been a number of reviews on the use of MRI as a tool to study flow and translational motion in nonmedical research areas (Benders & Blümich, 2019; Callaghan, 2011; Elkins & Alley, 2007; Fukushima, 1999; Mantle & Sederman, 2003; Sederman & Gladden, 2012) and more details and examples can be found in those. Here we introduce time-of-flight (TOF) and phase-shift velocity measurement methods. The diffusion measurement method, also referred to as PFG or PGSE in the literature, is closely associated with the phase shift measurement and is not discussed in detail here, but an in-depth treatise can be found in Chapter 6 of Callaghan's book (Callaghan, 1991).

4.2.7.1 Time-of-flight velocity imaging

TOF or spin tagging methods were first reported in 1959 and their use has been widespread since then. At their simplest, they may just consist of an imaging sequence where the time delay between excitation and acquisition has been increased and only spins that remain within a selected slice will contribute to signal intensity. Therefore, high velocity areas, where fewer spins have remained in the slice, will appear as darker areas in the final image. If the slice thickness and sequence timings are known, then an estimate of the velocity may be made. Although this method is good for identifying flowing regions, accurate quantification can be difficult. Another form of spin tagging, which can aid quantification and provide a very intuitive visualization of the motion, is where the imaging plane is perpendicular to the tagging. An example of this

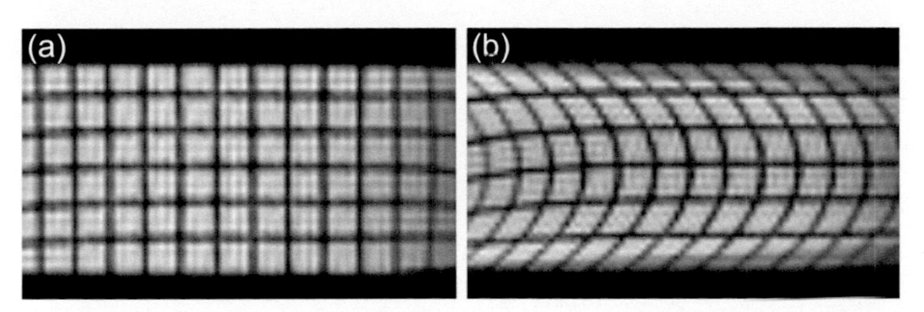

Figure 4.5 Example of a TOF velocity image. Proton image of a longitudinal slice through a pipe (a) stationary and (b) with flow.

is when a grid of spins are tagged by a specially tailored pulse. An example of such a TOF image of pipe flow is shown in Fig. 4.5 where laminar flow in a pipe has been imaged with a time delay between the tagging and the imaging sequence without flow (Fig. 4.5a) and with flow (Fig. 4.5b).

4.2.7.2 Phase-shift velocity imaging

Phase-shift or phase-contrast displacement measurement probably offers the most robust and quantitative way of measuring flow by NMR and was first suggested by Moran (1982). While spin tagging methods are adequate for flow visualizations, their velocity resolution is often limited by the overall spatial imaging resolution. The dynamic range of phase-shift velocity imaging can be adjusted from as little as a few μm/s up to 10–100 m/s (Adair, Mastikhin, & Newling, 2018; van de Meent, Sederman, Gladden, & Goldstein, 2009). It is limited at lower velocities to displacements due to self-diffusion and at higher velocities to residence time in the measurement volume.

The principle of phase-shift velocity imaging can be understood by considering the effects of the applied gradient $\boldsymbol{G}(t)$ on the phase of moving spins. Eq. (4.3) describes the precession frequency at position $\boldsymbol{r}$. Consider this to be time dependent, $\boldsymbol{r}(t)$; then the phase shift $\phi(t)$ is given by the following:

$$\phi(t) = \gamma \int_0^t \boldsymbol{G}(t)\boldsymbol{r}(t)dt \tag{4.16}$$

If the time-dependent position of this spin set is expanded as follows,

$$\boldsymbol{r}(t) = \boldsymbol{r}_0 + \boldsymbol{v}_0 t + \frac{1}{2}\boldsymbol{a}_0 t^2 + \cdots \tag{4.17}$$

then the time-dependent phase accrued by the spins will be the following:

$$\phi(t) = \gamma \left[\boldsymbol{r}_0 \int_0^t \boldsymbol{G}(t)dt + \boldsymbol{v}_0 \int_0^t \boldsymbol{G}(t)tdt + \boldsymbol{a}_0 \int_0^t \boldsymbol{G}(t)t^2 dt + \cdots \right] \tag{4.18}$$

The integrals are the successive moments ($\boldsymbol{M}_0$, $\boldsymbol{M}_1$, $\boldsymbol{M}_2$ ...) of the gradient with time:

$$\text{Zeroth moment, } \boldsymbol{M}_0 = \int_0^t \boldsymbol{G}(t)dt \tag{4.19a}$$

$$\text{First moment, } \boldsymbol{M}_1 = \int_0^t \boldsymbol{G}(t)t\,dt \tag{4.19b}$$

$$\text{Second moment, } \boldsymbol{M}_2 = \int_0^t \boldsymbol{G}(t)t^2dt \tag{4.19c}$$

The zeroth moment, $\boldsymbol{M}_0$, causes a phase shift proportional to position $\boldsymbol{r}_0$ and the first moment, $\boldsymbol{M}_1$, causes a phase shift proportional to velocity $\boldsymbol{v}_0$ etc. Therefore gradient waveforms can be designed that have a nonzero $\boldsymbol{M}_1$ (but zero $\boldsymbol{M}_0$) and the phase, which can be measured, will be proportional to the velocity. Often a bipolar pair as shown in Fig. 4.6 will be used and the velocity can be calculated from the measured phase since $\boldsymbol{M}_1$ is known. Experimentally this is often done by acquiring the image twice with two values of $\boldsymbol{M}_1$ and a velocity is calculated from the difference in phase. The component of the velocity in the direction of the applied gradients is encoded for and a full velocity vector can be measured by repeating the experiment with velocity gradients applied in three orthogonal directions. Fig. 4.7 shows a phase-shift velocity

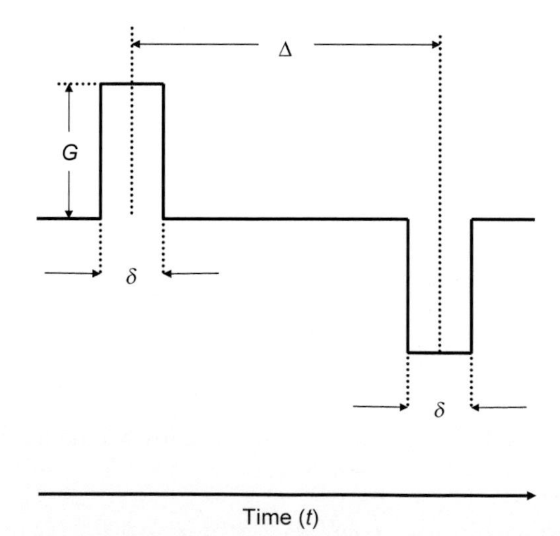

Figure 4.6 A bipolar gradient pulse sequence which has no zero moment with time but nonzero first moment.

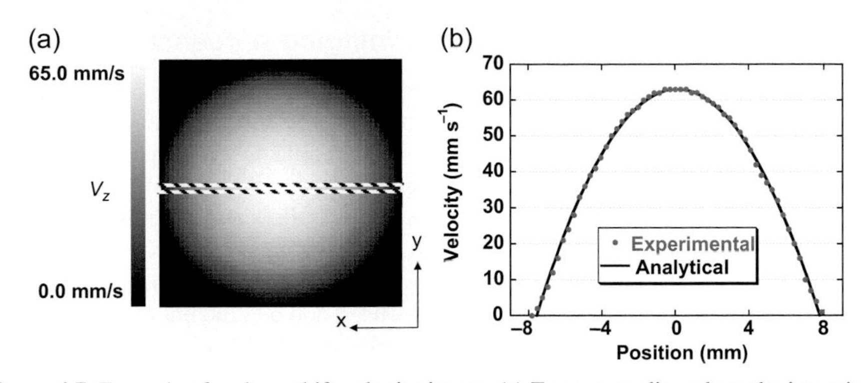

Figure 4.7 Example of a phase shift velocity image. (a) Transverse slice where the intensity is proportional to the signal phase and therefore velocity. (b) Comparison of a slice extracted from (a) with the analytical result for laminar flow in a pipe.

image of parabolic flow in a pipe and a comparison to the analytical velocity distribution; the measurement technique is seen to be highly quantitative.

It should be noted that velocity-induced phase can also cause image artifacts that are undesirable. A detailed discussion of these is beyond the scope of this chapter, but one approach to mitigate against artifacts is to "velocity compensate" the gradient waveforms such that $\boldsymbol{M}_1$ is zero. Several practical methods are well described by Pope and Yao (1993).

4.2.7.3 Diffusion

A very similar experimental approach to phase shift velocity imaging can be used to measure local diffusion coefficients. A bipolar pair of gradients similar to those shown in Fig. 4.6 is again used, but instead of imparting a phase shift, diffusion will reduce, or attenuate, the signal. The diffusion coefficient can be calculated from this attenuation and where coherent flows are also present (Stejskal & Tanner, 1965); the velocity can also be calculated from the phase shift as described above.

4.3 Methods: basic imaging techniques

There are a great variety of possible imaging techniques that are used and that can be designed according to the type of information that is required in the image (concentration, velocity, diffusion, T_1, etc.). To obtain an image, it is sufficient to acquire all points on the $\boldsymbol{k}$-space raster followed by Fourier transformation. The user can then build up the sequence to add contrast, such as including a velocity weighting gradient (nonzero $\boldsymbol{M}_1$) to obtain velocity information. In this section we show some of the basic sequences and how they can be extended to include velocity information.

4.3.1 Gradient echo and spin echo imaging sequences

Most imaging sequences can broadly be described as either gradient echo or spin echo sequences. In gradient echo sequences, the spins are refocused by the gradient alone; an example is shown in Fig. 4.8a. This is very similar to the sequence shown in Fig. 4.2, which is also a gradient echo sequence, but has the addition of a slice selective excitation pulse. In this sequence, the spins are dephased, due to different locations in the applied gradient, when the negative gradient is applied and will then start to rephrase when the gradient polarity is switched. When the total gradient time integral is zero, $\boldsymbol{k} = 0$ (Eq. 4.8), the spins will be refocused and a "gradient echo" is formed. Throughout this time, dephasing of the spins will also occur due to T_2^* such that the acquired points will be T_2^* weighted. As described previously, one direction of $\boldsymbol{k}$-space is acquired in the frequency encode direction with the gradient on during signal detection, while the other is acquired in the phase direction with a gradient applied before acquisition.

An alternative to the gradient echo sequence is the spin echo sequence shown in Fig. 4.8b. In this case a 180 degree refocusing pulse is inserted at time, τ, so that dephasing due to magnetic field inhomogeneities, T_2', is refocused at 2τ and a spin echo is formed; the signal at this point is T_2 weighted. Another difference is the polarity of the dephasing gradient; the 180 degree refocusing pulse inverts the phase of the spins and, therefore, the effective polarity of the gradient. In this case, the dephasing gradient must have the same polarity as the rephasing gradient since they are separated by the 180 degree refocusing pulse. Although this is termed a spin echo experiment, two echoes are formed—the spin echo at 2τ and the gradient echo at $\boldsymbol{k} = 0$—and these are usually chosen to coincide. In this sequence, the data will be T_2 weighted compared to T_2^* for the gradient echo sequence. In engineering samples with several material boundaries, T_2 can be much greater than T_2^*, so a spin echo sequence will be used for quantitative images with minimal contrast in them, whereas a gradient echo sequence may be used to highlight heterogeneities in the sample.

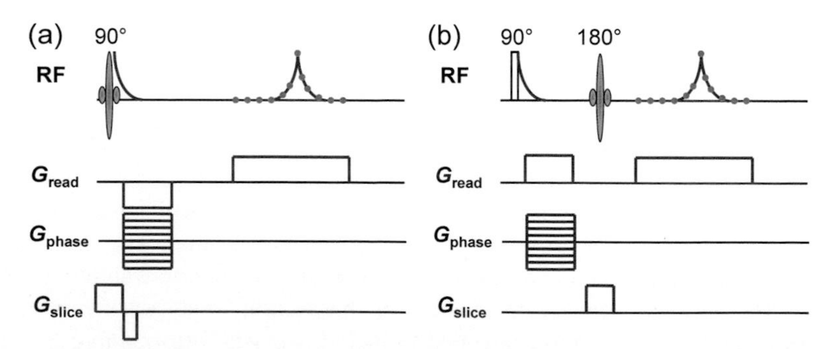

Figure 4.8 (a) Schematic of a simple slice selective two-dimensional (a) gradient echo and (b) spin echo pulse sequence.

4.3.2 Signal relaxation measurement

All MRI images will have some relaxation weighting—signal loss from the effects of relaxation—which will change the image intensity. The most important of these in the spin echo sequence is due to T_2 and the signal will decay according to Eq. (4.14) where t is the time between excitation and data acquisition. Here M_{xy} is the measured signal intensity and M_0 is proportional to the absolute material density. Any signal loss can be minimized by making $\frac{t}{T_2} \ll 1$ such that $M_{xy} \cong M_0$, but this may not be possible in some situations where T_2 is particularly small. In such systems, a series of images can be acquired with increasing values of t and then the pixel measured intensities, M_{xy}, fitted to Eq. (4.14) with M_0 and T_2 as fitted parameters. The final image of material concentration or spin density is proportional to M_0, while the image of T_2 contains information on local molecular mobility (Chen, Hughes, Gladden, & Mantle, 2010).

4.3.3 Velocity imaging

To obtain a quantitative phase-shift velocity image, it is necessary to acquire all points in $\boldsymbol{k}$-space as described in the previous section but also to add a gradient waveform that has a nonzero $\boldsymbol{M}_1$ such as that shown in Fig. 4.6. An example of a simple velocity imaging sequence is shown in Fig. 4.9. This is a spin echo imaging sequence with the addition of a velocity encoding pair of gradients, similar to the bipolar pair shown in Fig. 4.6. But now, both have the same polarity because they are separated by the 180 degree refocusing pulse. The phase of the signal from this sequence will be proportional to the local velocity and the experimental paramcter $\boldsymbol{M}_1$. Although the

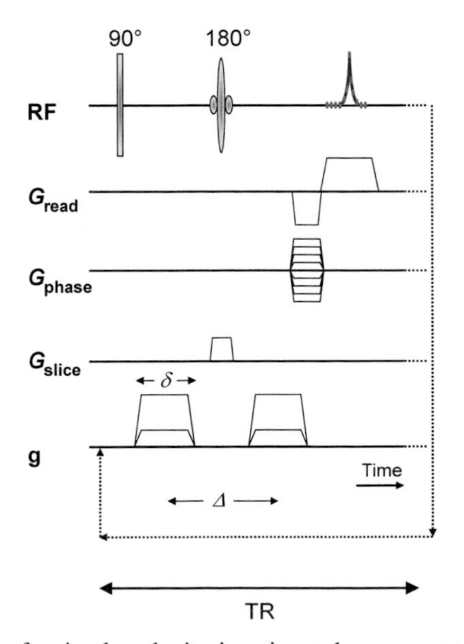

Figure 4.9 A schematic of a simple velocity imaging pulse sequence.

velocity can, in theory, be calculated directly from the phase in the image, it is much more usual to acquire two images with an increment in $\boldsymbol{M}_1$ and use the increment in measured phase to calculate the velocity. The example in Fig. 4.9 illustrates how a velocity encoding pair of gradients can be added to a standard imaging sequence and this method can be used more generally in much more complicated imaging sequences where adding velocity encoding gradients enables velocity measurements to be made.

4.4 Advanced data acquisition: fast imaging approaches

Acquiring images using the simple methodologies described previously will typically take several minutes or longer to acquire, but we are often interested in imaging systems that change over timescales shorter than that. There are a number of families of fast imaging techniques and this section will introduce the most important of these. All of these methods aim to acquire the same points in $\boldsymbol{k}$-space that were being acquired in Section 4.2.4 but to utilize the magnetization in a more efficient way to maximize the temporal resolution of the images. One means to fast or "ultrafast" imaging to acquire the entire 2D $\boldsymbol{k}$-space raster from a single excitation. Echo planar imaging (EPI) and rapid acquisition with relaxation enhancement (RARE) are two such methods and will be discussed further, as will fast low angle shot imaging (FLASH), which still uses multiple excitations but with a very short time between excitations. This chapter will also briefly discuss the use of sparse sampling methods in which the $\boldsymbol{k}$-space raster is undersampled and innovative image reconstruction techniques are used.

4.4.1 Full k-space fast sampling

4.4.1.1 Echo planar techniques

The fastest possible imaging techniques are echo planar or EPI-based schemes that were proposed by Mansfield (Mansfield, 1977), who realized that the complete 2D $\boldsymbol{k}$-space raster could be acquired following a single excitation pulse. After an initial line in $\boldsymbol{k}$-space has been acquired in the presence of a read gradient, the gradient polarity is switched at the same time as a blip in the phase direction is applied. With successive phase blips, the whole of 2D $\boldsymbol{k}$-space is traversed. There are many variations of the EPI technique and one is shown in Fig. 4.10. In the EPI sequence, the digitizer is kept open for the entire image acquisition and is therefore a very efficient method for fast imaging with images often acquired in 5–50 ms. Since there are no 180 degree refocusing pulses during the acquisition, the signal decays with T_2^* during this period making this method unsuitable when the sample has a short T_2^* as is the case for many engineering systems. Despite this, the spiral variant that has very short acquisition times has been successfully used for rapid velocity imaging in two-phase flows (Tayler, Holland, Sederman, & Gladden, 2012b).

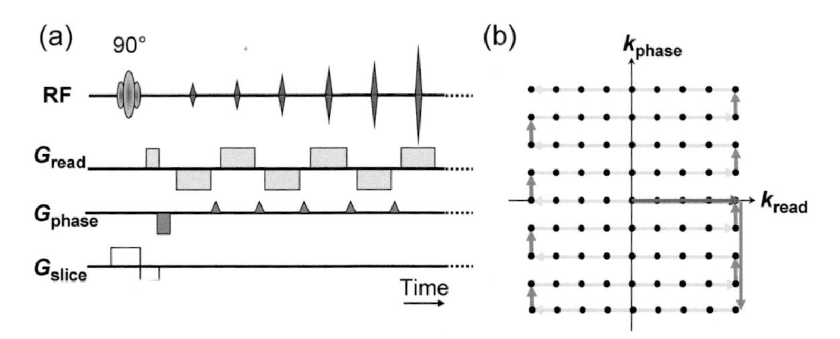

Figure 4.10 EPI (a) slice selective pulse sequence and (b) corresponding $\boldsymbol{k}$-space raster. Although the two dimensions have been designated as read and phase here, they are sometimes both considered as read directions and referred to as the first and second read directions.

4.4.1.2 RARE

Like EPI, the RARE technique (Hennig, Nauerth, & Friedburg, 1986) can acquire a whole 2D image from a single excitation. Unlike EPI, there is a 180 degree refocusing pulse used between each line of $\boldsymbol{k}$-space to form successive spin echoes. The pulse sequence is shown in Fig. 4.11. This pulse sequence is much more robust to local field heterogeneities as the magnetization is continually refocused so that the signal decays with T_2 rather than T_2^*. Although slightly longer than EPI techniques because of the time required for the 180 degree refocusing pulse between lines of $\boldsymbol{k}$-space, this is a more robust technique and well suited to engineering samples where fluid velocities are low. Imaging times for 64×64 acquired points are likely to be 100−200 ms.

4.4.1.3 FLASH

Haase, Frahm, Matthaei, Hanicke and Merboldt (1986) have shown how a fast 2D image may be acquired using the simple gradient echo imaging sequence shown in Fig. 4.8a. In this imaging sequence a single line of $\boldsymbol{k}$-space is acquired after each excitation, but in the FLASH implementation the repetition (recycle) time between excitations is reduced to a few ms and an excitation angle much below 90 degree is used. At

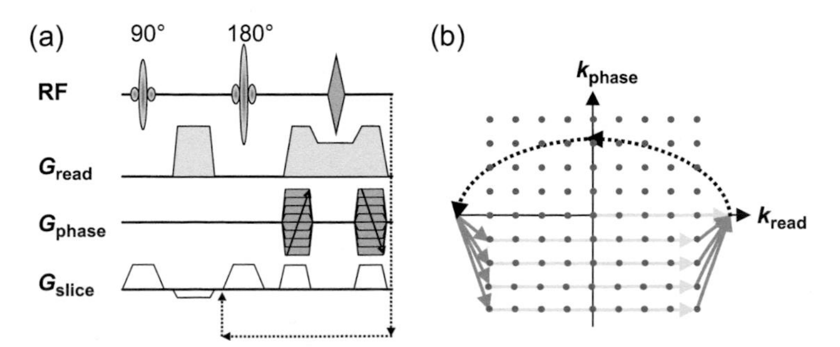

Figure 4.11 Schematic of the RARE pulse sequence with $\boldsymbol{k}$-space raster.

this rate, a 64 $\times$ 64 image would take 192 ms to acquire. Because of the small excitation angle, sometimes only a few degrees, FLASH uses only a small amount of the longitudinal magnetization and SNR is lower than either EPI or RARE. Despite the low SNR of this technique, it is very robust to image artifacts and thus it provides a very useful tool, particularly for low-resolution imaging where SNR is less likely to be a problem.

4.4.2 Undersampled data reconstruction: compressed sensing, bayesian, and deep leaning methods

Acquiring all points on the k-space raster is time consuming and may be unnecessary when there is only a limited amount of data in the reconstructed image. If the $\boldsymbol{k}$-space data can be undersampled, faster imaging techniques can be developed and employed, but, from Nyquist sampling theorem, we then expect that linear recovery methods will lead to image artifacts and a lossy recovery. It is important to note however that not all of the Fourier coefficients are equally important for the reconstruction of the image and several methods have been suggested as ways to reconstruct images from undersampled data. The most successful and promising of these is compressed sensing which is briefly discussed below though more details can be found in Holland and Gladden (2014) and references within. It should be noted that compressed sensing has been found to be useful in many areas of image reconstruction including electrical tomography, radio-astronomy, electron microtomography, and X-ray tomography.

Compressed sensing provides a very widely applicable method to reduce imaging times and can be applied to most MRI imaging strategies with limited knowledge about the sample under investigation, but there are a number of other methods that can provide further reduction in acquisition times. Deep learning and machine learning are increasingly being used to speed up image acquisition or improve image quality in medical MRI. Lundervold and Lundervold (2019), Montalt-Tordera, Muthurangu, Hauptmann, and Steeden (2021), thought their use in MRI applied to engineering applications has been more limited. The large number of datasets required for training are often available for development of clinical methods, but this is more problematic in many engineering research applications which may limit its use. However, this need not restrict its use in industrial monitoring to the same extent, where many measurements can be made over time and it has been used in food applications (Greer, Chen, & Mandal, 2018). The use of Bayesian inference (Holland et al., 2011) has been used to reduce number of data points that need to be acquired by up to two orders of magnitude, though these approaches do not result in a full image and target specific process information, e.g., average bubble size, instead.

4.4.2.1 Compressed sensing

Compressed sensing (Candes et al., 2006; Donoho, 2006; Lustig, Donoho, & Pauly, 2007) has recently attracted much interest in image reconstruction and MR imaging in particular. This reconstruction method utilizes the sparsity of the reconstructed image and can produce good reconstructions from significantly undersampled ***k***-space data. An image is said to be sparse if it can be accurately represented by fewer points in some transform domain of the image, such as a wavelet or Fourier transform. Lustig et al. (2007) showed that the sparsity of MR images in a transform domain could be exploited. Many "natural" images exhibit sparsity in terms of their Fourier or wavelet coefficients, or in terms of discrete gradients. In the same way that photographs can be compressed significantly with minimal loss of information, for example, with JPEG compression, so can many MR images.

Since MR images are compressible, compressed sensing theory (Candes et al., 2006; Donoho, 2006) suggests that taking fewer acquisition samples (undersampling), combined with prior knowledge that the image is sparse in some transform domain, may allow for an accurate image reconstruction. This is achieved using a nonlinear reconstruction method based on optimizing a convex function involving l_1-norms, more details of which can be found elsewhere (Donoho, 2006; Lustig et al., 2007).

When implementing compressed sensing, we must choose which points that we are going to sample. Compressed sensing requires that undersampling causes incoherent artifacts so pseudorandom sampling schemes or radial/spiral sampling schemes are often used.

The principle of the compressed sensing reconstruction is to transform the image into the sparse domain and then, by optimization, maximize the sparsity (minimize the number of nonzero points) of this sparse domain data, while ensuring the image retains fidelity with the acquired data. Thus, if an image that we wish to reconstruct is represented as a stacked vector $\mathbf{x}$, Ψ is the operator that transforms the image to a sparse representation (e.g., the wavelet transform), F is the Fourier transform mapping the image to ***k***-space, and $\mathbf{y}$ is a vector containing all the (undersampled) measured ***k***-space data. The reconstruction is then obtained by solving the following constrained optimization problem:

$$\begin{aligned}&\min \|\Psi\mathbf{x}\|_1\\ &\text{subject to } \|F\mathbf{x}-\mathbf{y}\|_2 \le \varepsilon\end{aligned} \tag{4.20}$$

where ε is a threshold that can be set based on the expected noise level in the data. The l_1-norm

$$\|\mathbf{x}\|_1 = \sum_i |x_i| \tag{4.21}$$

acts as a proxy for sparsity—that is the minimization produces an image that has, in the transform domain, the sparsest representation subject to remaining consistent with the acquired data. There are a variety of methods used to solve the optimization Eq. (4.20). The amount of data sampled is typically reduced by a factor of 2–10, without significant reduction in the quality of the image. A real strength of compressed sensing is that it can be combined with most MRI techniques including the ultrafast single-shot imaging described previously, reducing 2D imaging times to just a few milliseconds with 3D images being acquired in 30–50 ms. An example of such use in engineering applications is shown later in this chapter.

4.4.3 Practicalities and limitations

In principle, MR techniques have many advantages; however, there are a number of practical limitations that must be considered when choosing such a methodology. Advantages of MR include the following:

- The measurement is quantitative. Following calibration we know exactly how much of each chemical species is present.
- The measurement can be made sensitive to a large range of different physical parameters which can then be measured or used to provide image contrast. These include velocity, diffusion, molecular mobility (via relaxation), and chemical shift.
- 2D slice images via slice selection or fully resolved 3D images can be acquired.
- It is a noninvasive measurement with no need for tracers
- It uses low power, nonionizing radiation

Despite these many advantages, it is not a widely used technique in industrial settings for which there may be a number of reasons as detailed below.

Complexity of measurement: The large range of parameters MR can be sensitive to and the countless number of pulse sequences means this is a complex technique to fully understand. Increasingly, vendors are looking to circumvent this by developing application specific instruments with "push-button" operation such that skilled operators are not required. This has been partly driven by the growth in bench-top permanent magnet systems that are relatively cheap and do not need to be kept in a lab environment.

Cost of equipment: Although a number of permanent magnet systems are now available in the price range £30,000–80,000, high-field superconducting magnets that are used for high resolution imaging are typically an order of magnitude more expensive.

Sample environment and material restriction: All samples must be placed in a strong magnetic field which means ferromagnetic materials cannot be used. In addition, the spin manipulation and detection are done with electromagnetic radiation, usually in the RF range and samples must be transparent to these frequency electromagnetic waves. In practice, this means that while stainless steel or other nonferromagnetic metals can be used in the strong magnetic fields, the sample itself usually cannot as it will be invisible to the electromagnetic radiation. Although single-sided magnets, which allow unrestricted access from one direction, are available, most sample environments will have a significant geometric restriction such as a maximum sample diameter that can be passed through the bore of the magnet.

SNR and resolution: There are many things that will affect the resolution and SNR of the image, but typical spatial resolutions may go up to a few tens of microns and 2D images may be acquired in as little as 5 ms. However, there is usually a trade-off between time resolution, spatial resolution, and SNR.

4.5 Applications in engineering

There are a great number of applications of MR in engineering and this chapter does not attempt to review them all here. Instead, three examples are given of the application to gas/liquid bubbly flows and the interested reader may wish to consult more extensive reviews (Benders & Blümich, 2019; Britton, 2017; Fukushima, 1999; Mantle & Sederman, 2003).

Gas–liquid bubbly flows pose a significant problem to many measurement techniques and MRI is no exception. They are highly dynamic both in geometry and velocity with many moving phase boundaries.

4.5.1 Nonspatially resolved measurement

Although this chapter has been about the imaging aspects of MR techniques, useful information can often be acquired from an acquisition that does produce a full image. MR methods using gradients are routinely used in industrial applications such as emulsion droplet sizing and oil well logging. Recently, Holland and coworkers (Holland et al., 2011) have developed a method of directly analyzing the acquired data with Bayesian statistics to extract bubble sizes, rock core grain sizes (Holland, Mitchell, Blake, & Gladden, 2013), and food structure characteristics (Ziovas et al., 2016). The hardware requirements for this measurement are much simpler than for many imaging applications and this technique has been demonstrated both on a low field permanent magnet system and even a system using the earth's magnetic field. Fig. 4.12 shows how this technique has been applied to gas/liquid bubbly flows in a vertical column. Despite this being a highly dynamic system, the Bayesian analysis can be used to extract a probability distribution of the bubble size and the distribution is seen to alter as the gas distributor is changed or the gas flow rate is changed.

4.5.2 MRI and time-of-flight

TOF methods are used less frequently than phase shift methods and it can be more difficult to extract quantitative velocities. Sederman and Gladden (2001) used a variant of the RARE imaging sequence to investigate bubbly flows in ceramic monoliths. In this system, there are three phases making field heterogeneities worse, and the use of a spin echo technique necessary. In the channels, the bubble velocities are reduced to a single direction, parallel to the monolith channels. In this application, the positions of the bubbles in the monolith are imaged, and by repeated image acquisition, all from the same excitation pulse, the bubbles can be tracked. So, bubble velocities are recorded. This can, therefore, be used as a time-of-flight velocity imaging technique. Two images from a single excitation are shown in Fig. 4.13 where each image was acquired in 160 ms (Mantle et al., 2002). TOF measurements in 3D can also be made such as in measurements of oil/water displacement in porous rocks (Ramskill et al., 2017) where compressed sensing has also been utilized to improve the time resolution of the imaging.

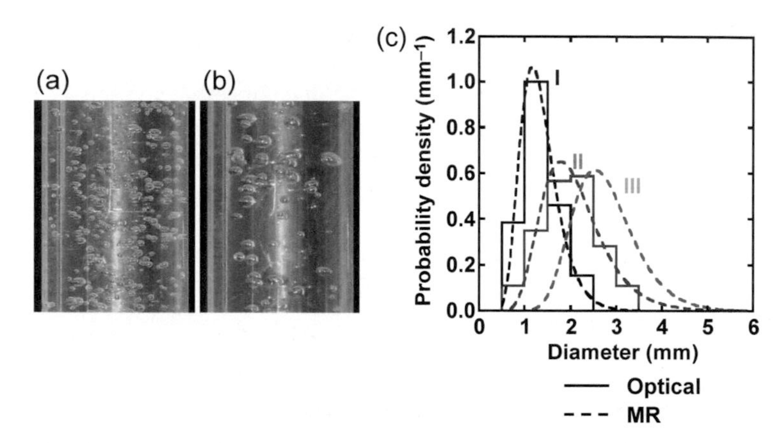

Figure 4.12 Photographs of a swarm of gas bubbles obtained using (a) distributor 1 and (b) distributor 2. The gas flow rate was 100 mL/min, giving a total voidage of 2% in each case. Note the larger bubble size obtained from distributor 2. (c) Comparison of the size distribution obtained in these systems using distributor 1 (black) and distributor 2 (red [light gray in print version]) from both the optical (*solid lines*) and MR (*dashed lines*) techniques. At a gas velocity of 1000 mL/min using distributor 2 (*dotted line*), the gas voidage was 15% and the optical technique could not be used.
Adapted and reproduced with permission from Holland, D. J., Blake, A., Tayler, A. B., Sederman, A. J., & Gladden, L. F. (2011). A Bayesian approach to characterising multi-phase flows using magnetic resonance: Application to bubble flows. *Journal of Magnetic Resonance, 209*, 83–87.

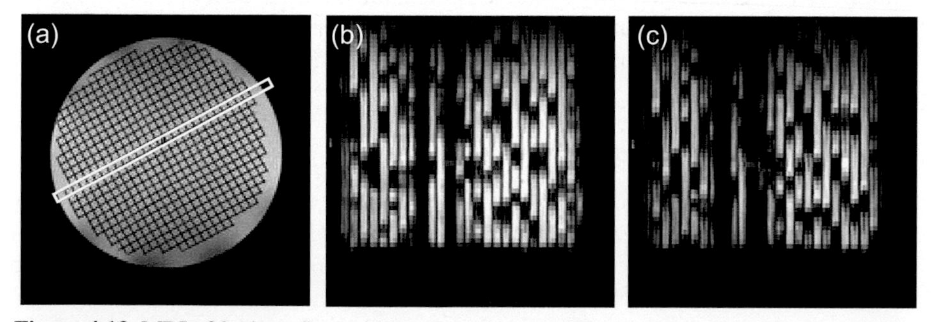

Figure 4.13 MRI of 2-phase flow within a ceramic monolith. A cross-sectional xy image of the fully water-saturated monolith (a) shows the internal structure of the monolith; each channel has a side length 1.2 mm. The highlighted region shows the position of the image slice in the xz-direction for which the visualizations of gas–liquid distribution during two-phase flow are shown in (b) and (c). Data acquisition time for each image was 160 ms; there was a time period of 160 ms between the start of acquisition of each of the images shown in (b) and (c). In all images, the presence of water is indicated by high intensity (white); gas and ceramic are identified by zero intensity (black).
Reproduced with permission from Mantle, M. D., Sederman, A. J., Gladden, L. F., Raymahasay, S., Winterbottom, J. M., & Stitt, E. H. (2002). Dynamic MRI visualization of two-phase flow in a ceramic monolith. *AIChE Journal, 48*, 909–912.

4.5.3 Phase-shift velocity imaging

Phase-shift velocity imaging techniques when allied to an EPI-based sequence are probably the fastest method of MR velocimetry with 2D images being acquired in under 5 ms and a frame rate of 188 frames per second. The spiral EPI-based technique has been applied to turbulent flow in pipes and has also been successfully applied to a single gas bubble (Tayler et al., 2012b) and to gas–liquid bubbly flows (Tayler, Holland, Sederman, & Gladden, 2012a). Quantitative velocity images of gas–liquid distributions during bubbly flow and velocity images of the liquid phase are shown in Fig. 4.14. The high resolution image of liquid velocities around a single bubble shown in Fig. 4.14a shows three orthogonal velocity components and each image was acquired in 5 ms using sparse sampling. The compressed sensing reconstruction technique was then used to reconstruct each velocity component and the three velocity component images were combined to show a full description of the local liquid velocity. More recently this has been applied to bubble burst dynamics (Reci, 2019; Sederman, Gladden, & Reci, 2021) where three component velocity images have been acquired at time intervals of 4 ms. Fig. 4.15 shows 2D images acquired in the transverse and axial direction at 8 ms after a bubble burst event. The ability of MRI to acquire detailed flow information without the need for any tracer, along with the noninvasive acquisition makes it a very powerful tomographic tool for engineering applications.

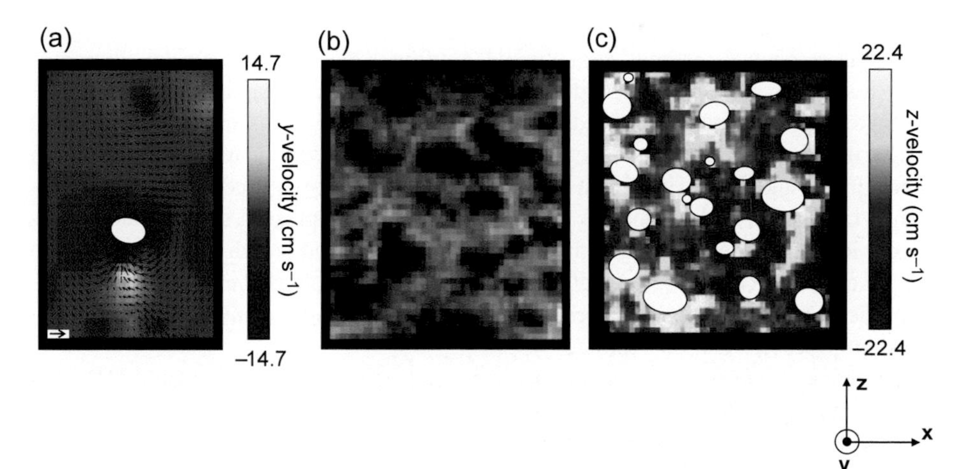

Figure 4.14 (a) Longitudinal image of three orthogonal liquid velocity components for a single bubble rising through water. The vector scale bar on the image corresponds to 15 cm/ s and the three components of the velocity were acquired in a total of 15.9 ms. The approximate location of the bubble is highlighted by the filled white ellipses. Longitudinal (b) magnitude image and (c) velocity image of gas–liquid bubbly flow at a voidage of 28.3%.
Adapted and reproduced with permission from Tayler, A. B., Holland, D. J., Sederman, A. J., & Gladden, L. F. (2012a). Applications of ultra-fast MRI to high voidage bubbly flow: Measurement of bubble size distributions, interfacial area and hydrodynamics. *Chemical Engineering Science*, *71*, 468–483, Tayler, A. B., Holland, D. J., Sederman, A. J., & Gladden, L. F. (2012b). Exploring the origins of turbulence in multiphase flow using compressed sensing MRI. *Physical Review Letters*, *108(26)*, *264505*.

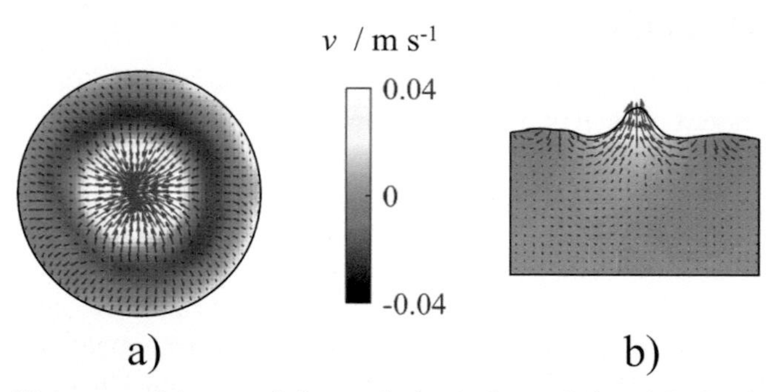

Figure 4.15 2D images of the out-of plane velocity (*color coded*) and the in-plane velocities (*arrow coded*) for (a) a transverse plane cutting through the middle of the bubble and (b) an axial plane cutting through the middle of the bubble at a time of 8 ms after the bubble burst event. The *largest arrow* corresponds to a velocity of 0.15 m/s in the transverse plane image and 0.13 m/s in the axial plane image. The spatial resolution in both dimensions is 265 μm and the pipe diameter is 15.4 mm.
Adapted and reproduced with permission from Sederman, A. J., Gladden, L. F., & Reci, A. (2021). Ultra-fast MR techniques to image multi-phase flows in pipes and reactors: Bubble burst hydrodynamics. *Magnetic Resonance Microscopy: Technology and Applications*, Wiley-VCH.

4.6 Future trends

Magnetic resonance methods are showing a number of very exciting developments most notably in data analysis and hardware in MR and data analysis which are set to make MR methods more widely used in engineering applications. High field MRI with superconducting magnets is likely to remain predominantly in the research environment due to the significant costs and user expertise required to make developments, but with new data analysis techniques such as compressed sensing and probabilistic information extraction methods, new time scales and length scales can now be accessed with MRI. This will expand the range of applications that can be investigated. In particular, with imaging timescales of a few milliseconds, many more multiphase transient processes can be explored.

There are also exciting developments in permanent magnets and low field MR. Permanent magnets offer MR at significantly lower capital costs and are virtually maintenance free, so can be placed where no cryogens or other services are available. They can also be designed with virtually no external magnetic field, so can be placed on site without safety concerns. While these systems will not have the full capability of high field systems, they can be designed to do specific experiments with minimal user input making them useful as monitoring equipment. These are being developed which target a single application and require no user intervention or are "push-button" operated, so can be run in an industrial setting without technical expertise. It is therefore likely that we will see a number of new devices come into the engineering workplace which are all optimized to the specific application that they are used for.

4.7 Conclusions

MRI is an extremely powerful tomography technique which is capable of noninvasively and quantitatively measuring many parameters including flow, chemical composition, and concentration within an image. These benefits have made it ubiquitous in the medical field as a diagnostic tool, but they also make it particularly useful as a research tool in engineering and the physical sciences. Despite this, its use has been restricted due to the costs and complexity required to make full use of its capabilities. It is routinely used in some industries such as oil well characterization where a subset of measurements is made on more restricted hardware.

By understanding the physics underpinning MRI, the user can begin to identify how it can be used and find suitable applications. This chapter aims to provide such a basic understanding of the MRI method and what techniques can be used to speed up acquisition. With this knowledge, the reader can better identify where MRI may provide useful data.

4.8 Sources of further information and advice

There are a number of excellent books on both the theory of MRI and its application in the engineering and physical sciences. Some further texts for the interested reader are Blumich (2000), Brown et al. (2014), Callaghan (1991), (2011), Levitt (2002).

References

Abragam, A. (1961). *The principles of nuclear magnetism*. Oxford: Oxford University Press.

Adair, A., Mastikhin, I. V., & Newling, B. (2018). Motion-sensitized SPRITE measurements of hydrodynamic cavitation in fast pipe flow. *Magnetic Resonance Imaging, 49*, 71–77.

Benders, S., & Blümich, B. (2019). Applications of magnetic resonance imaging in chemical engineering. *Physical Sciences Reviews, 4*.

Blumich, B. (2000). *NMR imaging of materials*. Oxford: Oxford University Press.

Britton, M. M. (2017). MRI of chemical reactions and processes. *Progress in Nuclear Magnetic Resonance Spectroscopy, 101*, 51–70.

Brown, R. W., Cheng, Y.-C. N., Haacke, E. M., Thompson, M. R., & Venkatesan, R. (2014). *Magnetic resonance imaging: Physical principles and sequence design*. New York: John Wiley & Sons, Inc.

Callaghan, P. T. (1991). *Principles of nuclear magnetic resonance microscopy*. Oxford: Oxford University Press.

Callaghan, P. T. (2011). *Translational dynamics and magnetic resonance*. Oxford: Oxford University Press.

Candes, E. J., Romberg, J. K., & Tao, T. (2006). Stable signal recovery from incomplete and inaccurate measurements. *Communications on Pure and Applied Mathematics, 59*, 1207–1223.

Chen, Y. Y., Hughes, L. P., Gladden, L. F., & Mantle, M. D. (2010). Quantitative ultra-fast MRI of HPMC swelling and dissolution. *Journal of Pharmaceutical Sciences, 99*, 3462–3472.

Donoho, D. L. (2006). Compressed sensing. *IEEE Transactions on Information Theory, 52*, 1289–1306.

Elkins, C. J., & Alley, M. T. (2007). Magnetic resonance velocimetry: Applications of magnetic resonance imaging in the measurement of fluid motion. *Experiments in Fluids, 43*, 823–858.

Fukushima, E. (1999). Nuclear magnetic resonance as a tool to study flow. *Annual Review of Fluid Mechanics, 31*, 95–123.

Greer, M., Chen, C., & Mandal, S. (2018). Automated classification of food products using 2D low-field NMR. *Journal of Magnetic Resonance, 294*, 44–58.

Haase, A., Frahm, J., Matthaei, D., Hanicke, W., & Merboldt, K. D. (1986). Flash imaging - rapid NMR imaging using low flip-angle pulses. *Journal of Magnetic Resonance, 67*, 258–266.

Hennig, J., Nauerth, A., & Friedburg, H. (1986). RARE imaging - a fast imaging method for clinical MR. *Magnetic Resonance in Medicine, 3*, 823–833.

Holland, D. J., Blake, A., Tayler, A. B., Sederman, A. J., & Gladden, L. F. (2011). A Bayesian approach to characterising multi-phase flows using magnetic resonance: Application to bubble flows. *Journal of Magnetic Resonance, 209*, 83–87.

Holland, D. J., & Gladden, L. F. (2014). Less is more: How compressed sensing is transforming metrology in chemistry. *Angewandte Chemie International Edition, 53*, 13330–13340.

Holland, D. J., Mitchell, J., Blake, A., & Gladden, L. F. (2013). Grain sizing in porous media using Bayesian magnetic resonance. *Physical Review Letters, 110*(1), Article 018001.

Keeler, J. (2011). *Understanding NMR spectroscopy*. John Wiley & Sons.

Levitt, M. H. (2002). *Spin dynamics*. England: John Wiley and Sons, Ltd.

Lundervold, A. S., & Lundervold, A. (2019). An overview of deep learning in medical imaging focusing on MRI. *Zeitschrift für Medizinische Physik, 29*, 102–127.

Lustig, M., Donoho, D., & Pauly, J. M. (2007). Sparse MRI: The application of compressed sensing for rapid MR imaging. *Magnetic Resonance in Medicine, 58*, 1182–1195.

Mansfield, P. (1977). Multi-planar image formation using NMR spin echoes. *Journal of Physics C: Solid State Physics, 10*, L55–L58.

Mansfield, P., & Grannell, P. K. (1973). NMR "diffraction" in solids? *Journal of Physics C: Solid State Physics, 6*, L422.

Mantle, M. D., & Sederman, A. J. (2003). Dynamic MRI in chemical process and reaction engineering. *Progress in Nuclear Magnetic Resonance Spectroscopy, 43*, 3–60.

Mantle, M. D., Sederman, A. J., Gladden, L. F., Raymahasay, S., Winterbottom, J. M., & Stitt, E. H. (2002). Dynamic MRI visualization of two-phase flow in a ceramic monolith. *AIChE Journal, 48*, 909–912.

Montalt-Tordera, J., Muthurangu, V., Hauptmann, A., & Steeden, J. A. (2021). Machine learning in magnetic resonance imaging: Image reconstruction. *Physica Medica, 83*, 79–87.

Moran, P. R. (1982). A flow velocity zeugmatographic interlace for NMR imaging in humans. *Magnetic Resonance Imaging, 1*, 197–203.

Pope, J. M., & Yao, S. (1993). Quantitative NMR imaging of flow. *Concepts in Magnetic Resonance, 5*, 281–302.

Ramskill, N. P., Sederman, A. J., Mantle, M. D., Appel, M., De Jong, H., & Gladden, L. F. (2017). In situ chemically-selective monitoring of multiphase displacement processes in a carbonate rock using 3D magnetic resonance imaging. *Transport in Porous Media, 121*, 15–35.

Reci, A. (2019). *Signal sampling and processing in magnetic resonance applications* (Ph.D. thesis). University of Cambridge.

Sederman, A. J., & Gladden, L. F. (2001). Magnetic resonance visualisation of single- and two-phase flow in porous media. *Magnetic Resonance Imaging, 19*, 339–343.

Sederman, A. J., & Gladden, L. F. (2012). Velocity imaging of transient flows. eMagRes. Wiley Online Library.

Sederman, A. J., Gladden, L. F., & Reci, A. (2021). Ultra-fast MR techniques to image multi-phase flows in pipes and reactors: Bubble burst hydrodynamics. *Magnetic Resonance Microscopy: Technology and Applications*. Wiley-VCH.

Sodickson, A., & Cory, D. G. (1998). A generalized k-space formalism for treating the spatial aspects of a variety of NMR experiments. *Progress in Nuclear Magnetic Resonance Spectroscopy, 33*, 77–108.

Stejskal, E. O., & Tanner, J. E. (1965). Spin diffusion measurements: Spin echoes in the presence of a time-dependent field gradient. *The Journal of Chemical Physics, 42*, 288–292.

Tayler, A. B., Holland, D. J., Sederman, A. J., & Gladden, L. F. (2012a). Applications of ultra-fast MRI to high voidage bubbly flow: Measurement of bubble size distributions, interfacial area and hydrodynamics. *Chemical Engineering Science, 71*, 468–483.

Tayler, A. B., Holland, D. J., Sederman, A. J., & Gladden, L. F. (2012b). Exploring the origins of turbulence in multiphase flow using compressed sensing MRI. *Physical Review Letters, 108*(26), 264505.

van de Meent, J.-W., Sederman, A. J., Gladden, L. F., & Goldstein, R. E. (2009). Measurement of cytoplasmic streaming in single plant cells by magnetic resonance velocimetry. *Journal of Fluid Mechanics, 642*, 5–14.

Ziovas, K., Sederman, A. J., Gehin-Delval, C., Gunes, D. Z., Hughes, E., & Mantle, M. D. (2016). Rapid sphere sizing using a Bayesian analysis of reciprocal space imaging data. *Journal of Colloid and Interface Science, 462*, 110–122.

Chemical Species Tomography

Hugh McCann[1], *Paul Wright*[2], *Kyle Daun*[3], *Samuel J. Grauer*[4], *Chang Liu*[1] *and Steven Wagner*[5]
[1]School of Engineering, The University of Edinburgh, Edinburgh, Scotland, United Kingdom; [2]School of Electrical and Electronic Engineering, University of Manchester, Manchester, United Kingdom; [3]Department of Mechanical and Mechatronics Engineering, University of Waterloo, Waterloo, ON, Canada; [4]Department of Mechanical Engineering, The Pennsylvania State University, University Park, PA, United States; [5]Institute for Reactive Flows and Diagnostics, Technical University of Darmstadt, Darmstadt, Hesse, Germany

5.1 Introduction

It has been appreciated since at least 1980 (Santoro, Semerjian, Emmerman, & Goulard, 1981) that the principles of X-ray tomographic imaging can be readily adapted to the case where photon absorption depends on the chemical species of the absorbing atom or molecule, opening up the possibility of resolving the distribution of a particular chemical species in gas phase. If many path-integrated measurements can be made through the measurement subject, images may subsequently be reconstructed to reveal the concentration distribution of the target species and, in spectroscopically favorable cases, the temperature distribution. *Chemical species tomography* (CST) has advanced gradually over the intervening period, stimulated not only by the needs of process engineers but also by technological advances in optoelectronics, and has accelerated markedly over the last 20 years.

In this chapter, we aim to describe the state of the art in CST, where photon absorption by vibrational–rotational transitions of the target molecular species are used to achieve chemical specificity, and the contribution of scattered photons to the path-integral measurements can be neglected. Several other tomographic modalities offering some form of chemical specificity are considered to lie beyond the scope of this chapter. The first edition included discussion of chemiluminescence; however, it is "passive" in terms of measurement arrangement, and recent advances in this area merit a separate treatment. Chemically specific imaging using Radio-Frequency transitions of nuclear spin alignment is called magnetic resonance imaging, and is described in Chapter 4. Our focus here is on applications of optical transmission tomography to process engineering, and gaseous flows in particular, typically with negligible photon scattering contribution (see below), placing their study firmly in the *hard-field* category of tomography.

Industrial Tomography. https://doi.org/10.1016/B978-0-12-823015-2.00004-2

Section 5.2 describes the spectroscopic issues involved, ranging from fundamental absorption spectroscopy to the thermodynamic condition of the measured subject, and considers the different spectroscopic measurement techniques that may be used. Since practical implementations of CST systems typically have a very small beam count (a few tens), Section 5.3 discusses in depth the related problems of image reconstruction, where considerable progress has been made since the first edition. Section 5.4 presents methods to optimize the beam arrangement, including recently developed methods to assess spatial resolution as part of the optimization process, and a new case study is included to illustrate the quantitative use of these methods for beam array design. The potential for significant scattering in process engineering subjects means that the minimization of its effect on the measured signals is a primary objective. This is discussed in Section 5.5, which covers several aspects of the hardware technology required for CST, ranging from the crucial issue of optical access to the subject through to the data acquisition (DAQ) methods used to obtain measurements, the latter being expanded in this edition to encompass Wavelength Modulation Spectroscopy (WMS). Section 5.6 is devoted to several case studies on the application of CST, and Section 5.7 presents a brief discussion of future trends.

5.2 Absorption spectroscopy for Chemical Species Tomography

The fundamentals of molecular absorption spectroscopy are discussed at length in many textbooks, e.g., Banwell and McCash (1994). Fig. 5.1 illustrates the absorption

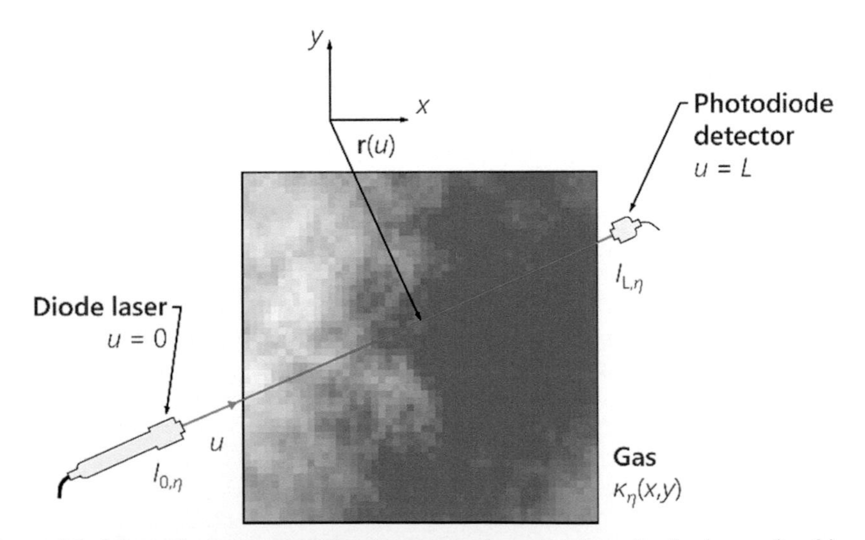

Figure 5.1 Schematic depiction of the absorption of spectral intensity $I_{0,\eta}$ by an absorbing medium containing the target species.

of light incident at intensity $I_{0,\eta}$ and wavenumber η on a gaseous absorbing medium containing the target absorbing species at mole fraction χ_{abs}, where η has been chosen to correspond to a particular spectroscopic transition of the target species, giving transmitted light intensity $I_{L,\eta}$. (Wavenumber η, in vacuo wavelength λ_0, and optical frequency ν, are related by $\eta = 1/\lambda_0 = \nu/c_0$, where c_0 is the speed of light in vacuo. After Section 5.2, symbol λ is not used in text for wavelength, even though the word appears frequently.)

For the case shown in Fig. 5.1, neglecting attenuation by out-scattering or contributions due to blackbody emission or in-scattering,[1] the drop in intensity is governed by the Beer–Lambert law

$$I_{L,\eta} = I_{0,\eta} \exp\left\{ - \int_0^L \kappa_\eta[\mathbf{r}(u)]\, du \right\} = I_{0,\eta} \exp(-\alpha_\eta), \tag{5.1}$$

where L [cm] is the path length through the medium, $\alpha_\eta = -\ln(I_{L,\eta}/I_{0,\eta})$ is the spectral absorbance (or optical density),[2] $\mathbf{r}$ is the position vector corresponding to a parametric location u along the beam, and κ_η [cm^{-1}] is the spectral absorption coefficient (Liu, Zhou, Jeffries, & Hanson, 2007). This coefficient represents the net effect of individual transitions or absorption features, $\kappa_{\eta,k}$ [cm^{-1}], such that

$$\kappa_\eta = \sum_k \kappa_{\eta,k} = P\chi_{abs} \sum_k S_k(T)\phi(T, P, \eta; \eta_k) \tag{5.2}$$

where $S_k(T)$ [cm^{-2}/atm] is the absorption linestrength of the kth spectroscopic transition; $\phi(T, P, \eta, \eta_k)$ [cm] is its lineshape function, positioned at the vacuum linecenter η_k [cm^{-1}]; and P [atm] and T [K] are the pressure and temperature of the absorbing medium, respectively.

Fig. 5.1 also indicates the use of a diode laser as the light source and a photodiode as the detector, which is a common configuration in gas sensing. The ability of diode lasers to be tuned across target absorption features gives rise to the technique's name, tunable diode laser absorption spectroscopy (TDLAS), which is developed and practiced by many groups (Duffin, McGettrick, Johnstone, Stewart, & Moodie, 2007; Goldenstein, Spearrin, Jeffries, & Hanson, 2017; Liu & Xu, 2019; Mihalcea, Baer, & Hanson, 1997). With TDLAS, it is possible to isolate and resolve the lineshape of one absorption feature of the target species. In this context, two further

[1] These assumptions are almost always reasonable for molecular gases, since absorption dominates scattering at the wavelengths used for CST, which are also much shorter than wavelengths important to blackbody emission.

[2] Spectral transmittance/absorptance is often defined as the fractions of intensity transmitted through/absorbed by a gas, i.e., $\tau_\eta = I_{L,\eta}/I_{0,\eta}$ and $\alpha_\eta = 1 - I_{L,\eta}/I_{0,\eta}$, but the definition in Eq. (5.1) is more prevalent in the TDLAS community.

quantities to consider are the local spectrally integrated absorption coefficient of a single transition, K_k,

$$K_k = \int_0^\infty \kappa_{\eta,k} \mathrm{d}\eta = P\chi_{\mathrm{abs}} S_k(T), \tag{5.3}$$

(since the wavenumber integral of the lineshape function is 1) and the corresponding spectrally integrated path absorbance, A_k,

$$A_k = \int_0^\infty \alpha_{\eta,k} \mathrm{d}\eta. \tag{5.4}$$

A_k can be determined from spectrally resolved measurements, $I_{L,\eta}$, using a simultaneous absorbance- and baseline-fitting procedure to infer $I_{L,\eta}$ and α_η. From Eq. (5.1), spectrally integrated absorbance measurements are related to the local spectrally integrated absorption coefficients through an integral equation:

$$A_k = \int_0^\infty \left\{ \int_0^L \kappa_{\eta,k}[\mathbf{r}(u)] \mathrm{d}u \right\} \mathrm{d}\eta = \int_0^L K_k[\mathbf{r}(u)] \mathrm{d}u. \tag{5.5}$$

The path integrals in Eqs. (5.1) and (5.5) constitute the basic measurement model required for tomographic reconstruction, often termed the path concentration integral (PCI) in the literature.

For most molecular species of engineering interest, the practically useful spectroscopies for TDLAS and CST lie in the following regions:

- Mid-infrared (MIR), with wavenumbers in the range $400 \leq \eta \leq 4000$/cm (i.e., $25{,}000 \geq \lambda_0 \geq 2500$ nm), typically corresponding to fundamental vibrational–rotational transitions, or
- Near-infrared (NIR), with wavenumbers in the range $4000 \leq \eta \leq 14{,}000$/cm (i.e., $2500 \geq \lambda_0 \geq 700$ nm), resulting from overtones and combinations of the fundamental transitions.

Rapid development of low-cost, robust light sources, detectors, and fibers for the NIR region, driven by applications in telecommunications and consumer electronics, has resulted in technologies more amenable to engineering application than their MIR counterparts. Hence, notwithstanding recent promising developments (Tancin, Spearrin, & Goldenstein, 2019), the great majority of CST activity to date has been in the NIR, despite the fact that the transition linestrengths are typically 1000 times smaller than those in the MIR, as illustrated in Fig. 5.2 for the case of carbon dioxide (CO_2) at a typical combustion exhaust temperature.

Databases of linestrengths and broadening parameters, such as HITRAN (Gordon et al., 2017), are commonly used to identify candidate absorption lines for a particular

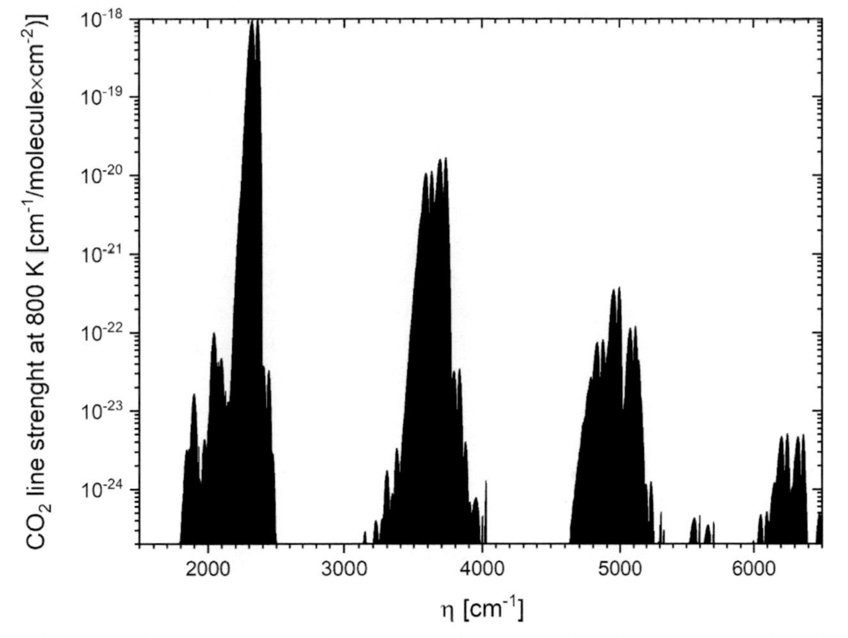

Figure 5.2 Line strength spectrum of CO_2 at 800K from HITRAN 2016 (Gordon et al., 2017).

species measurement, prior to experimental verification at the relevant conditions. As indicated by Eq. (5.2), the detailed shape of the absorption feature is a function of both temperature and pressure, which we illustrate here by continuing to focus on small absorbing molecules. Within a given spectrum, for a chosen target species, the line-strength, S_k, of the kth molecular transition varies with temperature according to (Liu, Zhou, Jeffries, & Hanson, 2007)

$$S_k(T) = S_k(T_0)\frac{Q(T_0)}{Q(T)}\left(\frac{T_0}{T}\right)\frac{1-\exp\left(-\dfrac{h\,c_0\eta_k}{k_B T}\right)}{1-\exp\left(-\dfrac{h\,c_0\eta_k}{k_B T_0}\right)}\exp\left[-\frac{hc_0E_k''}{k_B}\left(\frac{1}{T}-\frac{1}{T_0}\right)\right] \tag{5.6}$$

where h is Planck's constant $[\mathrm{J}\cdot\mathrm{s}]$, k_B is Boltzmann's constant [J/K], $Q(T)$ is the population partition function between the initial states of the absorbing molecule, E_k'' is the lower state energy $[\mathrm{cm}^{-1}]$ of the transition, and T_0 is a reference temperature (typically $T_0 = 296$ K).

Each transition line has a small natural linewidth, typically $\sim 10^{-5}$/cm, associated with Heisenberg's uncertainty principle. This linewidth is increased through the influence of (a) broadening due to molecular collisions, yielding a Lorentzian lineshape and (b) Doppler broadening, due to the random thermal motion of the molecules, yielding a Gaussian lineshape. The convolution of these two effects results in a Voigt profile (Linne, 2002). Temperature dependence is harder to generalize, as increasing temperature may either increase or decrease the population of the lower state, with a corresponding impact on the likelihood of absorption. For the case of water absorption spectroscopy, at around 1385 nm, the combined effects of temperature and pressure variation are illustrated in Fig. 5.3. At the atmospheric or elevated pressures found in the majority of engineering processes, the absorption at any given wavelength is likely to include contributions attributable to several absorption lines. This is evident in Fig. 5.3, where a slight shoulder on the right of the peak absorption of the 1385.12 nm feature at 1 atm reveals the presence of a line that is unresolved in the higher pressure traces. As pressure and temperature increase, the two absorption features at shorter wavelengths begin to merge, while the highest-wavelength feature disappears altogether. Although the 1385.12 nm feature remains distinct at elevated pressures, which is desirable from a measurement perspective (Karagiannopoulos, Cheadle, Wright, & McCann, 2012), it involves more than one molecular state transition, resulting in complex variation of the spectral absorption coefficient with pressure, temperature, and absorber concentration. Temperature sensitivity has been exploited in recent CST systems to enable simultaneous temperature and concentration imaging, as discussed in sections 5.3 and 5.6 below.

For CST imaging of a given target species in a measurement subject, several criteria can be identified for the selection of suitable absorption features for effective isolation

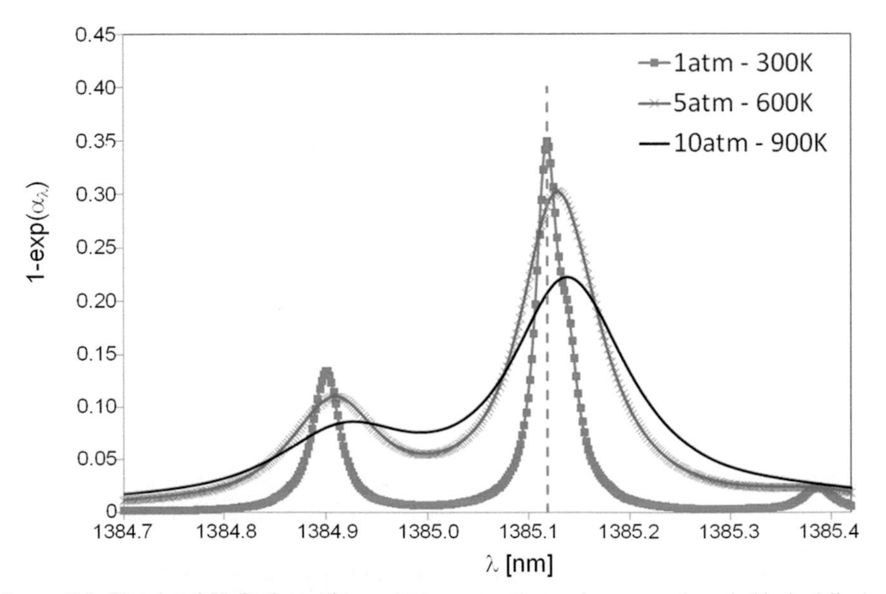

Figure 5.3 Simulated H_2O absorption spectra near a chosen laser wavelength (dashed line) at 1385.12 nm, for 10% mole fraction and 8 cm path length, at temperatures and pressures relevant to automotive engine in-cylinder conditions (Karagiannopoulos, Cheadle, Wright, & McCann, 2012).

of PCI data, but some of them are inevitably case dependent. In general, suitable absorption features should:

(a) give rise to an appropriate range of measured transmittance values, enabling adequate sensitivity, even once the noise properties of the measurement system are considered;
(b) be free from absorption by nontarget species; and
(c) have well-characterized pressure and temperature variation that is consistent with measurement needs.

The line selection process typically requires considerable effort in spectroscopic modeling of both the target and interfering species, combined with experimental validation/calibration. Auxiliary measurements of the subject may be required if a criterion is unable to be met, e.g., in-cylinder pressure measurement during hydrocarbon fuel CST in the cylinder of an operating engine (Wright et al., 2010).

Having identified a suitable absorption feature, an appropriate measurement method must be selected. If the feature is sufficiently broad and stable against varying thermodynamic conditions, it is attractive to carry out a simple direct absorption measurement at a fixed wavelength, typically the peak of the absorption curve. Utilizing an overtone band of the fundamental C−H stretch vibration at 2960/cm, this approach has been applied to CST of propane (Hindle et al., 2001), unleaded gasoline (Wright et al., 2010), and dodecane (Tsekenis et al., 2018). The spectrum of this fundamental absorption, itself relatively smooth due to the different environments of the C−H bonds within the molecule (Klingbeil, Jefferies, & Hanson, 2007), is further smoothed in the overtone band, yielding the broad, relatively unstructured absorption feature shown in Fig. 5.4. Intensity modulation may be used within such fixed-wavelength absorption schemes to aid low-noise detection, but any associated wavelength modulation is incidental to the measurement. Scattering or other nonspecific attenuation effects can be accounted for in such schemes by making a simultaneous measurement at a nonresonant wavelength, using the so-called Dual Wavelength Ratio (DWR) technique (Wright et al., 2010).

Full spectral measurement of an absorption feature by scanning the wavenumber of the incident light (Direct Absorption Spectroscopy, DAS) yields additional information, but its sensitivity can be limited by the need to detect small changes in baseline transmittance. Simultaneous baseline and absorbance fitting can be conducted to extract the absorbance spectrum (i.e., α_η) or integrated absorbances (i.e., A_k) via DAS (Blume, Ebert, Dreizler, & Wagner, 2015; Cole, Makowiecki, Hoghooghi, & Rieker, 2019; Emmert, Blume, Dreizler, & Wagner, 2018; Emmert, Grauer, Wagner, & Daun, 2019). The precision within reconstructed images is severely impacted by measurement noise (Daun, Grauer, & Hadwin, 2016; Hindle et al., 2001) necessitating high signal-to-noise ratio in the PCI measurements. Typically, DAS measurement entails detection of a small change in a large signal, and variation in the baseline occurs in addition to measurement system noise (e.g., ambient, thermal, or relative intensity noise), presenting a difficult noise suppression problem. For these reasons, WMS has become widely used in TDLAS, and it has been deployed extensively during the last decade to

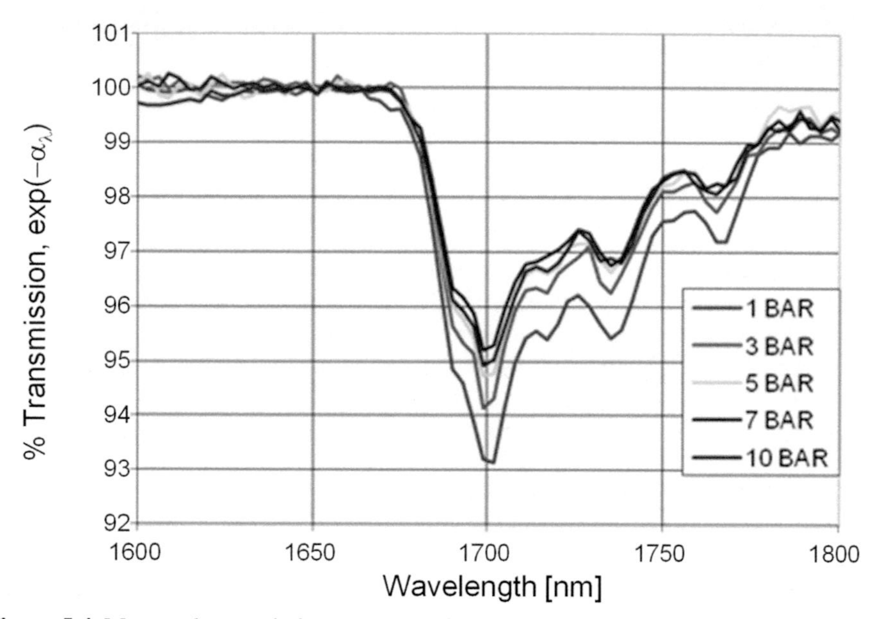

Figure 5.4 Measured transmission spectrum of iso-octane/air mixture at combustion stoichiometry (at 10 bar), for path length 85 mm.

improve detection sensitivity and signal quality in CST. Since CST signals often contain low-frequency components, WMS also helps to eliminate so-called flicker noise, which has a power spectral density proportional to the inverse of the signal frequency.

The principle of WMS is illustrated schematically in Fig. 5.5. A small-amplitude, high-frequency (f) modulation is superimposed on the low-frequency signal that scans the laser emission across an absorption feature, thus modulating rapidly the intensity *and* wavenumber of the laser output. Synchronous signal recovery at f, typically using a lock-in amplifier with narrow bandpass, yields the 1f signal (illustrated in the lower-left panel), which has good noise properties and is proportional to the modulation amplitude and to the first derivative of the absorption spectrum. Based on Rieker, Jeffries and Hanson (2009) and a number of subsequent authors, this approach can be extended using the second harmonic in the detected signal, the so-called 2f signal, which is sensitive to the line center via the second derivative of the absorption spectrum, and has the typical shape shown in the lower-right panel. The (2f/1f) ratio signal is independent of the modulation amplitude and allows for calibration-free quantification of the measured parameters (mole fraction, temperature, and pressure). The benefits of the method are discussed at length in Cai and Kaminski (2014) and include (a) suppression of broadband disturbances of the laser baseline (due to soot or particles); (b) immunity to fluctuations of the detected laser power (vibrations, beam steering); and (c) simplified determination of the line shape function due to the reduced dependence on the feature baseline, which has particular importance for measurements at high pressure. Recent work promises to deliver many of the same

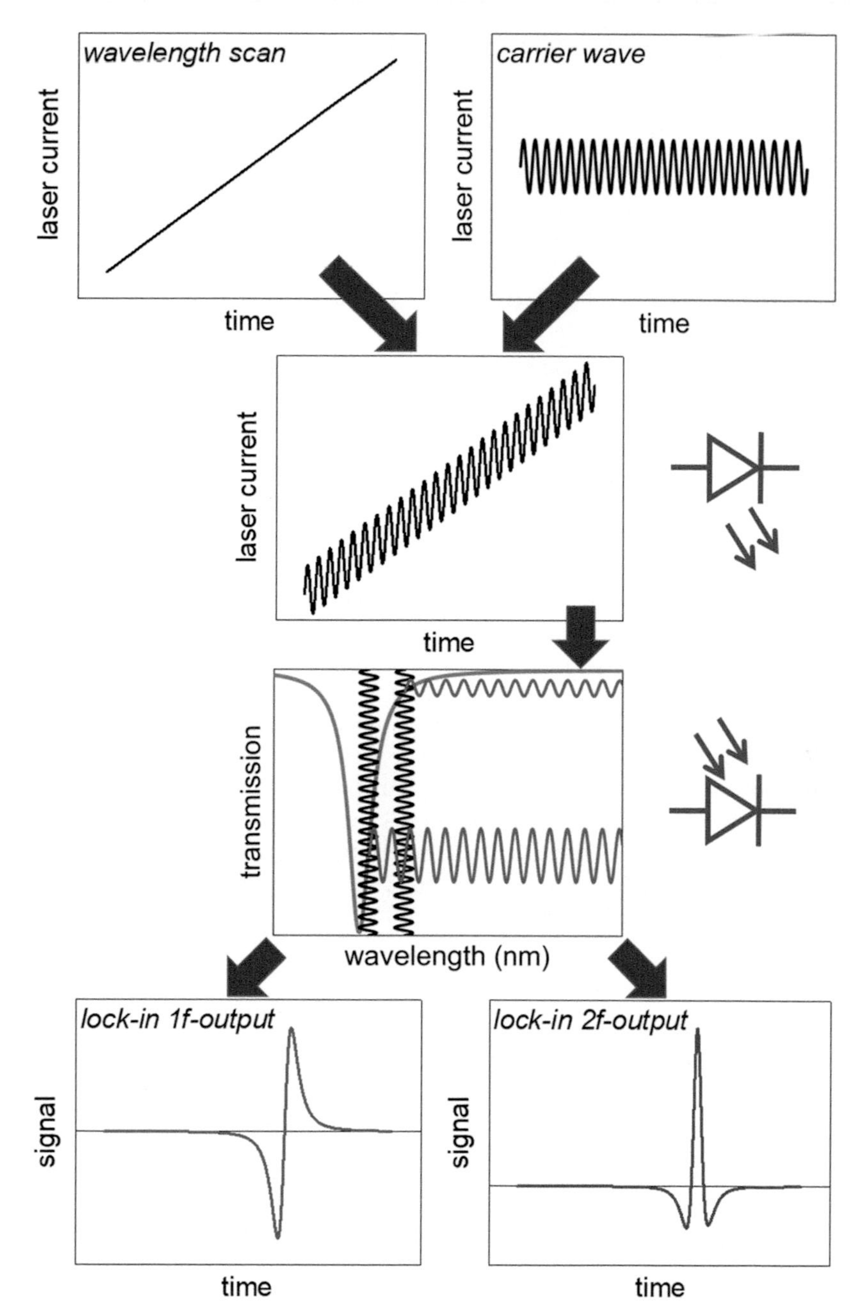

Figure 5.5 Schematic of the WMS measurement method, as described in the text. The effect of modulation at frequency f is illustrated at two points in the slow laser scan through the absorption feature. The f and $2f$ signals are illustrated in the bottom two panels.

benefits, but requires measurement of the $1f$ signal properties only (Upadhyay et al., 2018). WMS has been demonstrated in CST by a number of groups, e.g., Cai and Kaminski (2014), Tsekenis et al. (2016), and Huang, Cao, Zhao, Zhang and Xu (2020). The implications of the particular spectroscopic measurement method adopted for the electronic implementation of a CST system are discussed in Section 5.5.

5.3 Image reconstruction for low beam count systems

Individual line-of-sight (LOS) measurements in CST are governed by Eq. (5.1), which is rearranged to isolate the absorbance,

$$b_i = b(s_i, \theta_i) = \ln\left(\frac{I_{0,i,\eta}}{I_{\mathrm{L},i,\eta}}\right) = \int_0^L \kappa_\eta \left[\mathbf{r}_{(s_i,\theta_i)}(u)\right] \mathrm{d}u, \tag{5.7}$$

where b_i is the absorbance measured along the ith beam and (s_i, θ_i) is a set of *sinogram* coordinates, which specify the trajectory of the beam. Here, s_i is the perpendicular distance between the beam and the origin and θ_i is the angle formed between the beam and the y-axis, as illustrated in Fig. 5.6. An analogous measurement equation can be formed from Eq. (5.4) when the entire lineshape is scanned. A typical measurement scenario features m beams that transect the region of interest (RoI) in the gas, and reconstruction consists of calculating a 2D function, $g(x, y)$, which simultaneously satisfies Eq. (5.4) for each LOS measurement. The function g may be a spectral absorption coefficient κ_η, the integrated coefficient for a given transition K_k, or one or more thermochemical state variables, e.g., χ_{abs}, T, or P such that local values of κ_η in Eq. (5.7) can be calculated via Eq. (5.2).

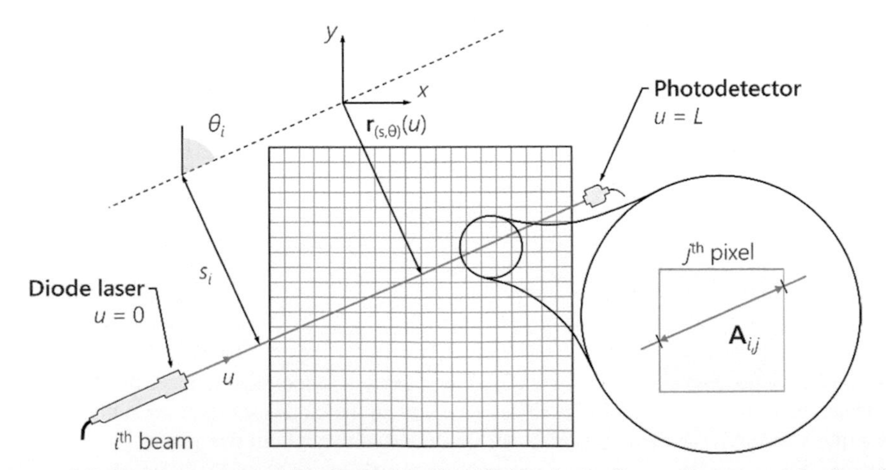

Figure 5.6 Each beam transecting the RoI is identified by its (s, θ) coordinates, and the domain is represented using a set of n basis functions, e.g., pixels.

5.3.1 Reconstruction as an ill-posed inverse problem

Reconstruction is an *ill-posed* problem, since it does not satisfy the conditions of a *well-posed* problem, i.e., one that (i) has a solution, which is (ii) unique, and (iii) stable to small perturbations in the data (Hadamard, 1923). The first criterion is satisfied in CST since there exists a distribution of gas responsible for the observed data. On the other hand, reconstruction clearly violates the uniqueness criterion of well-posed problems since g is unconstrained in regions of the gas not covered by a beam, and there exists an infinite set of field variables that can explain the absorbance contribution along any finite segment of any individual beam.

This ambiguity is reduced by increasing the number of beams in the measurement array. In the limiting (hypothetical) case in which $b(s, \theta)$ is known as a continuous function, the field variable can be reconstructed through an inverse 2D Fourier transform (Bertero & Boccacci, 1998),

$$g(x,y) = \frac{1}{4\pi^2}\int_0^{\pi}\int_{-\infty}^{\infty} b(\omega,\theta)|\omega|e^{\mathrm{i}\omega\,(x\cos\theta + y\sin\theta)}\mathrm{d}\omega\mathrm{d}\theta, \tag{5.8}$$

where $b(\omega, \theta)$ are 1D Fourier transforms of the sinogram,

$$b(\omega,\theta) = \int_{-\infty}^{\infty} b(s,\theta)e^{-\mathrm{i}\omega s}\mathrm{d}s. \tag{5.9}$$

In practice, however, $b(s, \theta)$ is not available as a continuous function. Instead, a related technique called *filtered back projection* works on a discrete set of data, $\{b_i\}$, sampled at uniformly spaced coordinates in the (s, θ) domain. Early lab-scale CST experiments (Beiting, 1992; Santoro, Semerjian, Emmerman, & Goulard, 1981) obtained these data by traversing a single beam through the flow field, or vice versa, but this approach precludes time-resolved estimates of the gas state, which is often the objective of CST experiments. The apparatus needed to carry out a large number of simultaneous measurements along regularly spaced beams can become extremely costly and complex, and may be impossible for CST within combustion devices (e.g., gas turbines (Ma et al., 2013), internal combustion (IC) engine cylinders (Wright et al., 2010)).

For these reasons, an alternative approach called *algebraic reconstruction* has gained favor in CST. Firstly, the RoI is discretized using a finite basis, e.g., a set of n pixels as depicted in Fig. 5.6 or a finite element mesh with n nodes. The field variable is assumed to be uniform within each pixel, or interpolated within each element from the nodal values, and the integral in Eq. (5.7) is approximated as a "ray sum,"

$$b_i = \sum_{j=1}^{n} \mathbf{A}_{i,j}x_j, \tag{5.10}$$

where x_j is the value of g at the jth pixel or node, and $\mathbf{A}_{i,j} = \partial b_i / \partial x_j$ is the sensitivity of the ith measurement to the jth basis function. In the case of a pixel representation, $\mathbf{A}_{i,j}$ amounts to the chord length of the ith beam subtended by the jth pixel, as shown in Fig. 5.6. Eq. (5.10) can be expressed for all m beams and n pixels or nodes in terms of an $m \times n$ matrix equation,

$$\mathbf{Ax} = \mathbf{b}, \tag{5.11}$$

where $\mathbf{x} = \{x_j\}$ is an $n \times 1$ vector of unknowns and $\mathbf{b} = \{b_i\}$ is an $m \times 1$ vector of data. Algebraic reconstruction consists of solving for $\mathbf{x}$ given the observations $\mathbf{b}$. This procedure can accommodate irregular beam arrays, and, especially, supplemental or "prior" information, which is essential when solving ill-posed problems.

The uniqueness and sensitivity of solutions to Eq. (5.11) are governed by the properties of $\mathbf{A}$, which depend upon both the discretization scheme and measurement array. There are two key scenarios (Daun, Grauer, & Hadwin, 2016): When the beam arrangement is sufficiently dense, or the basis sufficiently coarse, the rank of $\mathbf{A}$ equals n and there exists a unique vector $\mathbf{x}$ that solves Eq. (5.11) for each measurement, $\mathbf{b}$. However, even if $\mathbf{A}$ is full rank, it is necessarily ill-conditioned because the path integration in Eq. (5.7) desensitizes b_i to smaller features in g. Consequently, small amounts of measurement noise and model error that overwhelm these small but "true" features of the data are amplified by deconvolution into high frequency artifacts in the reconstruction, thereby violating the stability criterion of ill-posed problems. Such inverse problems are *discrete ill-posed* (Daun, Grauer, & Hadwin, 2016; Hansen, 1999). In almost all CST work, however, the number of basis functions greatly exceeds the number of absorbance measurements ($n \gg m$), so there exists an infinite set of solutions that fully satisfy $\mathbf{b}$, and the uniqueness criterion is violated. For example, in Wright et al. (2010), 27 (or fewer) laser beams transect the RoI, which is discretized into 1844 elements. In such cases, the inverse problem is not discrete ill-posed, but rather *rank deficient* or underdetermined. Additional information is required to obtain a plausible solution to Eq. (5.11) in both the discrete ill-posed and rank-deficient scenarios.

An ill-conditioned matrix can be diagnosed through singular value decomposition (SVD), $\mathbf{A} = \mathbf{U} \sum \mathbf{V}^{\mathrm{T}}$, where $\mathbf{U} \in R^{m \times m}$ and $\mathbf{V} \in R^{n \times n}$ are orthonormal matrices that, respectively, span the row and column spaces of $\mathbf{A}$ and $\sum \in R^{m \times n}$ is a matrix containing min(m,n) singular values along the main diagonal, arranged in decreasing order, $\sigma_1 \geq \sigma_2 \geq \ldots \geq \sigma_{\min(m,n)}$. In the discrete ill-posed case, where $\mathbf{A}$ is full rank, solutions to Eq. (5.11) can be expressed in terms of the SVD of $\mathbf{A}$,

$$\mathbf{x} = \sum_{j=1}^{n} \frac{\mathbf{u}_j^{\mathrm{T}} \mathbf{b}}{\sigma_j} \mathbf{v}_j, \tag{5.12}$$

where $\mathbf{u}_j$ and $\mathbf{v}_j$ are the jth column vectors of $\mathbf{U}$ and $\mathbf{V}$, respectively. The summation terms represent modes of increasing spatial frequency, and, for the summation to converge, the magnitude of the $\mathbf{u}_j^{\mathrm{T}} \mathbf{b}$ terms must approach zero faster than the corresponding singular values, which is the case as long as $\mathbf{x}$ is spatially smooth (Hansen, 1999).

Ill-conditioned matrices feature positive singular values that decay continuously over several orders of magnitude. The unknown vector $\mathbf{x}$ can be found using Eq. (5.12) provided the "exact" projection data are known, but in an experimental setting the data are invariably contaminated with measurement noise as discussed in Section 5.2, and discrepancies between Eqs. (5.7) and (5.10) arise from the discretization of g. The ill-posedness of $\mathbf{A}$ can be illustrated by a conceptual decomposition of the data into an "exact" term, $\mathbf{b}_{\text{exact}}$, and an error term, $\mathbf{e}_{\text{b}}$:

$$\mathbf{b} = \mathbf{b}_{\text{exact}} + \mathbf{e}_{\text{b}}, \tag{5.13}$$

Here, $\mathbf{b}_{\text{exact}} = \mathbf{A}\mathbf{x}_{\text{exact}}$, where $\mathbf{x}_{\text{exact}}$ is an idealized solution defined as the projection of the unknown continuous function g onto the finite basis, and Eq. (5.13) defines the *measurement model*. Accordingly, $\mathbf{x}$ may be expressed as

$$\mathbf{x} = \mathbf{x}_{\text{exact}} + \mathbf{e}_{\text{x}} = \sum_{j=1}^{n} \frac{\mathbf{u}_j^{\text{T}} \mathbf{b}_{\text{exact}}}{\sigma_j} \mathbf{v}_j + \sum_{j=1}^{n} \frac{\mathbf{u}_j^{T} \mathbf{e}_{\text{b}}}{\sigma_j} \mathbf{v}_j, \tag{5.14}$$

where $\mathbf{e}_{\text{x}}$ is a vector of reconstruction errors. While the magnitudes of the $\mathbf{u}_j^{\text{T}} \mathbf{b}_{\text{exact}}$ terms vanish faster than the singular values due to the smoothness of $\mathbf{x}_{\text{exact}}$, the projection of errors, $\mathbf{u}_j^{\text{T}} \mathbf{e}_{\text{b}}$, do not, resulting in large errors in the solution, $\mathbf{e}_{\text{x}}$. Error amplification can be suppressed by truncating the series in Eq. (5.12) or by employing an iterative regularization algorithm, as discussed later.

The SVD can also shed light on rank deficient matrix systems. Formally, the rank of $\mathbf{A}$ cannot exceed min(m,n) and typically rank($\mathbf{A}$) $\leq m \ll n$ (provided each beam path is unique, rank($\mathbf{A}$) $= m$). In this scenario, the set of solutions, $\{\mathbf{x}\}$, which satisfy $\mathbf{A}\mathbf{x} = \mathbf{b}$ is given by

$$\mathbf{x} = \mathbf{x}_{\text{LS}} + \mathbf{x}_{\text{null}} = \underbrace{\sum_{j=1}^{\text{rank}(\mathbf{A})} \frac{\mathbf{u}_j^{\text{T}} \mathbf{b}}{\sigma_j} \mathbf{v}_j}_{\mathbf{x}_{\text{LS}}} + \underbrace{\sum_{j=\text{rank}(\mathbf{A})+1}^{n} C_j \mathbf{v}_j}_{\mathbf{x}_{\text{null}}}, \tag{5.15}$$

where $\{C_j\}$ is any set of real numbers. Eq. (5.14) shows that the solution has two components: $\mathbf{x}_{\text{LS}}$, the unique solution to $\mathbf{A}\mathbf{x} = \mathbf{b}$, having the smallest Euclidean norm, and an infinite set of solutions from the null space of $\mathbf{A}$, $\{\mathbf{x}_{\text{null}} \mid \mathbf{A}\mathbf{x}_{\text{null}} = 0\}$. Fig. 5.7(b) shows an example least-squares solution found through the first summation in Eq. (5.15), and $\mathbf{x}_{\text{null}}$, depicted in Fig. 5.7(d), is the difference between $\mathbf{x}_{\text{LS}}$ and the true solution, $\mathbf{x}$. While $\mathbf{x}_{\text{LS}}$ fully explains the measurements in $\mathbf{b}$, it is obviously nonphysical. Therefore, the data must be supplemented with assumed information about the species distribution in order to arrive at a plausible reconstruction, e.g., Fig. 5.7(c).

Most CST problems are naturally of the rank-deficient type. There are several examples in the CST literature (e.g., (Krämer et al., 1998)) which use a coarse basis/large pixels with the objective of achieving a discrete ill-posed problem. While this approach admits a range of reconstruction techniques intended for discrete ill-posed problems, especially from medical imaging, *we do not recommend this practice* for the following reason: The mesh acts as a form of prior information, but a physically unrealistic one. A better approach is to choose a grid that is segmented to a size that is much smaller than the size of any feature of interest in the subject and then exploit any reliable prior information (Emmert, Wagner, & Daun, 2021). Metrics for quantifying resolution in the tomographic context are discussed later in this chapter.

5.3.2 Reconstruction algorithms

Both rank-deficient and discrete ill-posed CST problems require prior information to generate a unique and stable reconstruction, although the role of added information is significantly different in these scenarios. The least informative priors used in CST ensure that the species concentration distribution is nonnegative, while others promote a degree of spatial smoothness consistent with diffusion-dominated transport. Iterative methods like the algebraic reconstruction technique (ART) (Gordon, Bender, & Herman, 1970; Wang et al., 2015), multiplicative ART (Gordon, Bender, & Herman, 1970; Jeon, Deguchi, Kamimoto, Doh, & Cho, 2017), and simultaneous iterative reconstruction techniques (SIRTs) (Hansen & Jørgensen, 2018; Terzija et al., 2008; Zhao et al., 2020) can be prematurely halted to obtain a spatially smooth solution based on how these algorithms "build up" $\mathbf{x}$, i.e., starting with the low frequency modes and ending with the high frequency components that are more susceptible to noise contamination. Truncating an iterative solver is a method of imposing prior information that is difficult to characterize. Moreover, these approaches are intended for discrete ill-posed problems, although they have also been applied to rank-deficient CST experiments with modifications to "span" the null space of $\mathbf{A}$. For example, Landweber iteration (e.g., (Terzija et al., 2008; Terzija & McCann, 2011)), a form of the SIRT, has been combined with nonnegativity and mean, median, or wavelet filtering, which is applied between successive iterations to "smear" the solution onto pixels that are not transected by beams.

Other CST practitioners, such as Ma et al. (2009), Cai, Ewing and Ma (2008), and Daun (2010), have favored second-order Tikhonov regularization (Tikhonov & Arsenin, 1977), coupled with a nonnegativity constraint. In this technique, the measurement system, $\mathbf{Ax} = \mathbf{b}$, is augmented with a second matrix equation, $\lambda\mathbf{Lx} = \mathbf{0}$, where the smoothing matrix, $\mathbf{L}$, is an approximation to the discrete Laplace operator, e.g., for a pixel representation, elements of the smoothing matrix are given by

$$(\mathbf{L})_{i,j} = \begin{cases} 1, & \text{if } i = j \\ -1/N_i, & \text{if } n \text{ neighbours } m\,, \\ 0, & \text{otherwise} \end{cases} \tag{5.16}$$

where N_i is the number of pixels bordering the ith pixel, and $\mathbf{Lx} = \mathbf{0}$ is satisfied by any uniform vector $\mathbf{x}$. The species concentration distribution is found by solving

$$\mathbf{x}_\lambda = \underset{\mathbf{x}}{\operatorname{argmin}} \left\| \begin{bmatrix} \mathbf{A} \\ \lambda \mathbf{L} \end{bmatrix} \mathbf{x} - \begin{bmatrix} \mathbf{b} \\ 0 \end{bmatrix} \right\|_2^2, \text{ s.t. } \mathbf{x} \geq 0, \tag{5.17}$$

where $\mathbf{0}$ is a vector of zeros and the parameter λ determines the influence of the additional smoothing equations, $\mathbf{Lx} = \mathbf{0}$, relative to the measurement data, $\mathbf{Ax} = \mathbf{b}$. Fig. 5.7(c) shows a reconstruction obtained through this technique; in this example, $\mathbf{x}_\lambda$ accurately captures large-scale structures in $\mathbf{x}$ despite severely limited measurement information, but these results are contingent on proper selection of the parameter λ, which is most often chosen by trial-and-error. Residual-based techniques may be employed when the measurements are spectrally resolved (Emmert, Baroncelli, Kley, Pitsch, & Wagner, 2019).

Many CST implementations solve Eq. (5.17) using an iterative regularization technique, e.g., the SIRT (Grauer et al., 2018). In this case, Tikhonov regularization supplies the prior information, while the iterative solver is solely a vehicle to minimize the residual. Iterative solvers are not prematurely truncated when solving Eq. (5.17) in this context because the augmented matrix system already promotes low frequency solutions via $\mathbf{L}$.

Prior information can also be incorporated through the basis functions used to represent the gas, i.e.,

$$g(\mathbf{r}) \approx \sum_{j=1}^{n} x_j B_j(\mathbf{r}), \tag{5.18}$$

where $\mathbf{r}$ is the position vector, $B_j(\mathbf{r})$ is the jth basis function, x_j is the corresponding coefficient, and the unknown vector $\mathbf{x} = \{x_j\}$ describes a gas state. Typically, the flow field is represented using a pixel basis or triangle element mesh. In the former case, B_j equals unity when $\mathbf{r}$ lies inside the jth pixel, and is zero otherwise. This basis

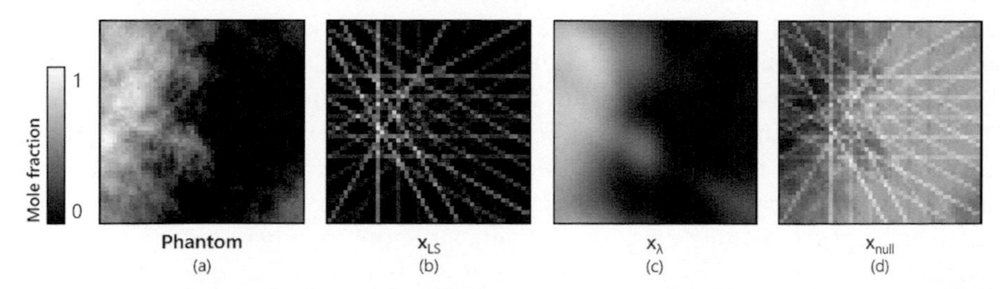

Figure 5.7 Simulation of a limited-data CST scenario with a 50×50 pixel domain ($n = 2500$) and $m = 32$ beams. The nonphysical least-squares solution ($\mathbf{x}_{\mathrm{LS}}$) is confined to pixels transected by a beam, whereas the Tikhonov solution ($\mathbf{x}_\lambda$) is spatially smooth. In this case, there are 2468 spatial modes ($\mathbf{v}_j$) that contribute to $\mathbf{x}_{\mathrm{null}}$.

contributes very little prior information to the reconstruction, and the assumption of a uniform gas distribution within each pixel is unrealistic. However, other, more informative choices are possible, which can lower the number of summation terms needed to obtain a realistic reconstruction. For instance, the triangle element mesh with linear or quadratic interpolation inside each element can improve the approximation in Eq. (5.18) relative to the uniform pixel basis without increasing the number of basis functions (Grauer, Hadwin, Sipkens, & Daun, 2017). Gaussian basis functions may also represent the gas concentration (Drescher, Gadgil, Price, & Nazaroff, 1996; Fischer et al., 2001) A low-dimensional cosine basis has been proposed, to reduce the size of the problem and computational effort needed for reconstruction, along with a technique to enforce bound constraints (Polydorides et al., 2018; Polydorides, Tsekenis, McCann, Prat, & Wright, 2016).

Greater physical fidelity can be obtained by deriving Karhunen–Loève (KL) basis functions from a set of training data (e.g., experimental (Chojnacki, Sarma, Wolga, Torniainen, & Gouldin, 1996; Torniainen & Gouldin, 1998) or Computational Fluid Dynamics (CFD) data (Torniainen, Hinz, & Gouldin, 1998)). This basis is constructed from an SVD on $N \geq n$ column vectors of training data, where column vectors of $\mathbf{U} = [\mathbf{u}_1, \mathbf{u}_2, \ldots, \mathbf{u}_n]$ form the basis functions in Eq. (5.18). The order of the basis functions/columns in $\mathbf{U}$ corresponds to the portion of variation in the gas flow captured by each function. Often the sum in Eq. (5.18) is truncated to obtain a low-dimensional representation of the gas; alternatively, since the variance of the jth coefficient, x_j, is given by the jth singular value, σ_j^2, $\sum$ provides a natural weighting for the KL basis. Note that the accuracy of reconstructions obtained through this approach strongly depends on the veracity of the training set.

Even more physical fidelity may be obtained through artificial neural networks (ANNs) when high-quality training data are available (LeCun, Bengio, & Hinton, 2015). ANNs contain a network of "neurons" arranged and interconnected by "synapses" in a way that mimics a brain. In the case of a feed-forward ANN, neurons are arranged into a sequence of layers. Measurement data are supplied to the first layer of neurons, called the "input layer." Each neuron receives a signal from one or more neurons in the preceding layer; these signals are combined by a nonlinear transfer function and passed on to the next layer. This process is repeated until reaching the output layer, which conveys the field variable, $\mathbf{x}$. ANNs are fitted to a set of training data, which contain sample pairs of inputs and outputs, i.e., $\mathbf{b}$ and $\mathbf{x}$. Prior information is implicitly introduced via the training set and network architecture. Training sets may be assembled from random Gaussian fields (Huang, Liu, & Cai, 2019; Wei, Schwarm, Pineda, & Spearrin, 2020) or experimental data (Huang, Liu, & Dau, 2018; Jin et al., 2019; Ren et al., 2019) . ANN models are inherently interpolative, so reconstruction accuracy hinges on the physical fidelity and comprehensiveness of the training data. The network structure must have sufficient degrees-of-freedom to capture key aspects of the flow, but not so many that the model becomes "overtuned" to experimental artifacts and other errors. In this sense, the ANN complexity (e.g., the number of neurons and layers) is an implicit source of regularization.

Bayesian inference provides a particularly useful viewpoint for CST inversion, since it explicates the role of prior information in the reconstruction and provides means to estimate the reconstruction uncertainty and resolution as well as a framework to optimize the layout of beams. In this approach, the data in **b** and unknown gas state in **x** are envisioned not as fixed parameters but as random variables described by probability density functions (PDFs). These PDFs are related by Bayes' equation,

$$p(\mathbf{x}|\mathbf{b}) = \frac{p(\mathbf{b}|\mathbf{x})p_{\mathrm{pr}}(\mathbf{x})}{p(\mathbf{b})}, \tag{5.19}$$

where $p_{\mathrm{pr}}(\mathbf{x})$ defines what is known about **x** before the measurement (the *prior*), $p(\mathbf{b}|\mathbf{x})$ is the *likelihood* of observing the data in **b** conditional on a hypothetical value of **x** in the context of measurement noise and model error, $p(\mathbf{x}|\mathbf{b})$ is the *posterior* probability of the unknown variables given the observed data and information in the prior, and $p(\mathbf{b})$ is the evidence, which scales the numerator of Eq. (5.19) so that the posterior satisfies the Law of Total Probability.

The likelihood is derived from a measurement model, Eq. (5.13). In many cases $\mathbf{e}_{\mathrm{b}}$ obeys an unbiased multivariate normal distribution, eb $\sim$ (0, $\boldsymbol{\Gamma}_{\mathrm{e}}$), where $\boldsymbol{\Gamma}_{\mathrm{e}}$ is the noise and error covariance matrix. In this case, the likelihood function is given by

$$p(\mathbf{b}|\mathbf{x}) = [\det(2\pi\boldsymbol{\Gamma}_{\mathrm{e}})]^{-1/2} \exp\left\{-\frac{1}{2}[\mathbf{b} - \mathbf{b}_{\mathrm{model}}(\mathbf{x})]^{\mathrm{T}}\boldsymbol{\Gamma}_{\mathrm{e}}^{-1}[\mathbf{b} - \mathbf{b}_{\mathrm{model}}(\mathbf{x})]\right\}, \tag{5.20}$$

and the value of **x** that maximizes $p(\mathbf{b}|\mathbf{x})$ is the maximum likelihood estimate, the solution that maximizes the likelihood of observing the data. In the case of independent and identically distributed measurement noise (i.e., $\boldsymbol{\Gamma}_{\mathrm{e}} = \sigma_{\mathrm{e}}^2\mathbf{I}$), maximizing $p(\mathbf{b}|\mathbf{x})$ is equivalent to minimizing $||\mathbf{Ax} - \mathbf{b}||_2^2$, and for rank-deficient tomography problems, Eq. (5.15) shows that an infinite set of candidate solutions exist that can do this.

Accordingly, the prior is critical for narrowing down this set to solutions that are consistent with the expected attributes of **x**. One common choice is a Gaussian prior (Grauer, Hadwin, & Daun, 2017),

$$p_{\mathrm{pr}}(\mathbf{x}) \propto \exp\left[-\frac{1}{2}(\mathbf{x} - \boldsymbol{\mu}_{\mathrm{pr}})^{\mathrm{T}}\boldsymbol{\Gamma}_{\mathrm{pr}}^{-1}(\mathbf{x} - \boldsymbol{\mu}_{\mathrm{pr}})\right], \tag{5.21}$$

where $\boldsymbol{\mu}_{\mathrm{pr}}$ is a mean distribution vector and $\boldsymbol{\Gamma}_{\mathrm{pr}}$ is a spatial covariance matrix. The spatial covariance matrix serves two purposes: off-diagonal elements encode the spatial structure of **x**, which is crucial for rank-deficient problems, while the magnitude of $\boldsymbol{\Gamma}_{\mathrm{pr}}^{-1}$ compared to $\mathbf{A}^{\mathrm{T}}\boldsymbol{\Gamma}_{\mathrm{e}}^{-1}\mathbf{A}$ indicates the "degree of belief" implied by the prior relative to the measurements. One possibility is to set $\boldsymbol{\mu}_{\mathrm{pr}} = 0$ and define $\boldsymbol{\Gamma}_{\mathrm{pr}}^{-1} = \lambda^2\mathbf{L}^{\mathrm{T}}\mathbf{L}$, which results in the Tikhonov smoothness prior, similar in effect to

Eq. (5.17) only weighted by the measurement error covariance. Another, more informative, choice is an "exponential prior," motivated by the form of autocorrelations commonly observed in turbulent flow fields (Birch, Brown, Dodson, & Thomas, 1978). In this approach, $\boldsymbol{\mu}_{\text{pr}} = 0$ and

$$\boldsymbol{\Gamma}_{\text{pr},i,j} = \sigma_{\text{pr}}^2 \exp\left(-\frac{\|\mathbf{r}_i - \mathbf{r}_j\|_2}{d_{corr}}\right), \tag{5.22}$$

where d_{corr} is a correlation length and σ_{pr}^2 serves to scale the influence of the prior relative to the likelihood. In many scenarios, these parameters can be derived using scale and self-similarity arguments, e.g., expansion of a turbulent jet or plume (Daun, Grauer, & Hadwin, 2016; Grauer, Hadwin, & Daun, 2017). A third choice is a sample-based prior, where $\boldsymbol{\mu}_{\text{pr}}$ and $\boldsymbol{\Gamma}_{\text{pr}}$ are set equal to the sample mean and sample variance of a training set, e.g., CFD data (Daun, Grauer, & Hadwin, 2016), which is equivalent to a weighted KL basis. Gaussian mixture models can be used to relax the reliance on information from the Large Eddy Simulation (Nadir, Brown, Comer, & Bouman, 2015). Often the Gaussian prior in Eq. (5.21) is combined with a nonnegativity prior or a concentration range prior, e.g.,

$$p_{\text{pr}}(\mathbf{x}) = \begin{cases} 1, & x_j \in [x_{\min}, x_{\max}] \quad \forall j \in 1, 2, \ldots, n \\ 0, & \text{otherwise} \end{cases}. \tag{5.23}$$

The prior and likelihood are then incorporated into Bayes' equation, Eq. (5.19).

In the case of tomography, the objective is usually to find the maximum a posteriori (MAP) estimate, $\mathbf{x}_{\text{MAP}}$, which is the most probable value of $\mathbf{x}$ given the data in $\mathbf{b}$ and prior information. In the special case of a Gaussian likelihood and Gaussian prior, this amounts to a constrained least-squares minimization,

$$\mathbf{x}_{\text{MAP}} = \underset{\mathbf{x}}{\operatorname{argmin}} \left\| \begin{bmatrix} \mathbf{L}_{\text{e}}\mathbf{A} \\ \mathbf{L}_{\text{pr}} \end{bmatrix} \mathbf{x} - \begin{bmatrix} \mathbf{L}_{\text{e}}\mathbf{b} \\ \mathbf{L}_{\text{pr}}\boldsymbol{\mu}_{\text{pr}} \end{bmatrix} \right\|_2^2, \text{ s.t. } \mathbf{x} \in [\mathbf{x}_{\min}, \mathbf{x}_{\max}], \tag{5.24}$$

where $\mathbf{L}_{\text{e}}$ and $\mathbf{L}_{\text{pr}}$ are the matrix square roots of $\boldsymbol{\Gamma}_{\text{e}}^{-1}$ and $\boldsymbol{\Gamma}_{\text{pr}}^{-1}$, respectively. The posterior PDF is also Gaussian in this scenario, with a mean of $\mathbf{x}_{\text{MAP}}$ and covariance given by

$$\boldsymbol{\Gamma}_{\text{post}} = \left(\mathbf{A}^{\text{T}}\boldsymbol{\Gamma}_{\text{e}}^{-1}\mathbf{A} + \boldsymbol{\Gamma}_{\text{pr}}^{-1}\right)^{-1}, \tag{5.25}$$

which provides an estimate of uncertainty, as discussed later in the chapter. Tikhonov regularization can be interpreted in a Bayesian way for the purpose of uncertainty quantification by setting $\mathbf{L}_{\text{pr}} = \lambda\mathbf{L}$.

5.3.3 Reconstruction accuracy and resolution

The required spatial resolution of a tomographic sensor depends on the size of the RoI (expressed by its diameter, D) and the smallest size of features to be imaged. For instance, spatial resolution for process control in large-size coal-fired/biomass boilers (Qu et al., 2018) may be several centimeters, but must be at millimeter level for combustion diagnostics in automotive engines (Tsekenis et al., 2018) and swirl combustors (Liu, Cao, Lin, Xu, & McCann, 2018). The term "resolution" is often taken to mean the smallest structure that can be captured by the image, or the minimum distance between two distinguishable objects (Tsekenis, Tait, & McCann, 2015).

In the case of full rank tomography problems, the resolution matrix provides a convenient means to quantify resolution. In this context, resolution can be envisioned using a "pulse-response" thought experiment (Emmert, Wagner, & Daun, 2021), wherein a set of experimental data is generated by $\mathbf{b}_{\text{pulse},j} = \mathbf{A}\mathbf{x}_{\text{pulse},j}$, where $\mathbf{x}_{\text{pulse},j}$ is a vector of zeros except for the jth element, which is unity. The field vector is then reconstructed from the data vector. For linear regularization methods, this step is written explicitly as $\mathbf{x}_{\text{reg}} = \mathbf{A}^{\#}\mathbf{b}_{\text{pulse},j} = \mathbf{A}^{\#}\mathbf{A}\mathbf{x}_{\text{pulse},j} = \mathbf{R}\mathbf{x}_{\text{pulse},j}$, where $\mathbf{A}^{\#}$ is the *regularized pseudoinverse* and $\mathbf{R}$ is the resolution matrix. (For example, in the case of Tikhonov regularization, $\mathbf{A}^{\#} = (\mathbf{A}^{\mathrm{T}}\mathbf{A} + \lambda^2\mathbf{L}^{\mathrm{T}}\mathbf{L})^{-1}\mathbf{A}^{\mathrm{T}}$.) The "blurring" action associated with regularization produces off-diagonal terms that correspond to a finite-width point spread function (PSF), leading to smooth features in the reconstruction; the PSF for the jth basis function can be visualized by plotting the jth column of $\mathbf{R}$ using the basis. Increasing the influence of $\mathbf{L}$ via λ increases the width of the PSFs.

The situation is more complex for rank-deficient tomography problems, since the resolution matrix contains null PSFs that spatially align with "blind spots" in the RoI that fall between the beams. Specifically, if the jth element is not transected by a beam, then $\mathbf{A}\mathbf{x}_{\text{pulse},j} = \mathbf{0}$ and the corresponding row of $\mathbf{R}$ is zero; conceptually, this corresponds to an infinitely wide PSF, or no resolution. This is an overly severe interpretation of resolution, since we have shown that gas concentrations can be resolved between the beams via prior information. Furthermore, $\mathbf{R}$ does not account for the effects of noise on the resolution of a CST sensor.

These problems have been circumvented through an empirical approach (Tsekenis, Tait, & McCann, 2015) whereby reconstructions are carried out on a series of test phantoms, each containing a large rectangular gas plume with straight, sharp edges. The RoI is transected by 31 beams (Fig. 5.8(d)), and is divided into 28 rectangular sectors arranged in 4 annular tracks around the center (Fig. 5.8(e)). The 16 plumes are oriented to ensure that every sector has its long axis perpendicular to a plume edge. Reconstructed images, as illustrated in Fig. 5.8(a–c), reveal blurred edges characterized by an edge spread function in each sector, which is transformed into PSFs. Large plumes are used to ensure that each one is transected by multiple beams. The method can be applied to reconstructions produced by any algorithm. The result is a map of spatial resolution throughout the RoI, but confined to those regions covered by the sectors, as shown in Fig. 5.8(d). The mean value of all the local spatial resolutions is taken to be the global spatial resolution δ_{global} of the CST system, and it

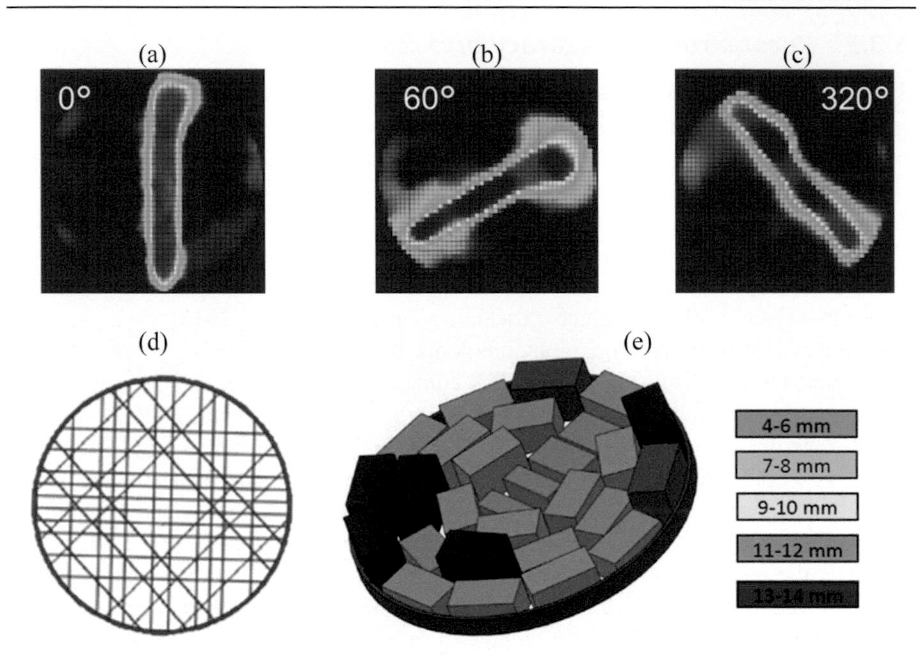

Figure 5.8 Empirical method to determine spatial resolution (Tsekenis, Tait, & McCann, 2015): (a)–(c) examples of the reconstructed phantoms, each comprising of a single rectangular propane plume in a nitrogen coflow; (d) the 31-beam array; (e) for each sector, the spatial resolution result, in a color-coded map for the whole RoI.

includes the effects of noise and choice of reconstruction algorithm. Tsekenis et al. obtained values for spatial resolution that were much better than the previously available theoretical guidance (Bertero & Boccacci, 1998), with δ_{global} about D/9 for the particular 31-beam CST system studied.

The drawbacks of this approach are that (1) it is empirical, so theoretical insight into the factors that govern resolution is limited and (2) being an intrinsic measure of the performance of the CST system, it does not take account of the properties of the gas flow encountered in a given application, e.g., it may be more difficult to resolve closely spaced Gaussian plumes or internal structures in a turbulent flow than an isolated Gaussian structure. Nevertheless, this method provides a route to practical design of CST systems, and a case study is presented in Section 5.4.1 below.

The Bayesian viewpoint presents an entirely different interpretation of uncertainty quantification and resolution. The posterior PDF indicates the likelihood of all possible solutions, including the "true" solution as long as the prior information is consistent with the underlying physics. In principle, $p(\mathbf{x}|\mathbf{b})$ provides unambiguous means to quantify uncertainties, e.g., diagonal elements of $\mathbf{\Gamma}_{post}$ give the width of credibility intervals for each pixel, centered on $\mathbf{x}_{MAP}$, which contain the "true" concentration with a certain probability. Random draws from the posterior PDF can also be used to propagate uncertainties for metrics computed on the reconstructed fields, e.g., the total mass flux of a target species.

In addition, the classical interpretation of resolution can be connected to the Bayesian framework. Consider a camera: photons from multiple lines-of-sight can reach a single pixel due to the finite width of the aperture, which blurs the image and makes it impossible to distinguish features that are sufficiently close. Consequently, the lens and aperture affect the information that can be gleaned from an image; the overall effect is quantified by a PSF, which indicates the correlation of spatial features produced by the imaging system. In tomographic sensors, the laser, optics, DAQ system, and reconstruction algorithm are collectively analogous to a camera's lens and aperture. However, spatial correlations in $\mathbf{x}$ produced by the sensor are affected not only by the regularization technique, as in the resolution matrix derived above, but also by assumptions about the flow built into the prior, which can be assessed using a Bayesian technique.

A Bayesian measure of resolution was proposed in Emmert, Wagner, and Daun (2021), using the covariance of the MAP estimate, $\mathbf{\Gamma}_{\text{MAP}}$ (as opposed to the posterior covariance of $\mathbf{x}$, $\mathbf{\Gamma}_{\text{post}}$). Columns of this matrix are akin to PSFs and can be transformed into a resolution length scale via a discrete Fourier transform. One drawback of this approach is that it conceives of the prior as accurately representing the underlying problem physics, which is not often the case. For example, Tikhonov smoothness priors describe an unrealistically strong correlation structure between all locations in the domain (Daun, Grauer, & Hadwin, 2016), which degrades resolution. In this scenario, the statistical resolution measure represents a conservative estimate of the achievable resolution.

5.3.4 Spectroscopic aspects of reconstruction in Chemical Species Tomography

The technique described thus far, i.e., reconstructing κ_η from α_η or K_k from A_k and then calculating χ_{abs} using Eq. (5.2) or 5.3, respectively, is adequate when (i) there is a single target species, (ii) the temperature and pressure fields are homogeneous, and (iii) the component of gas velocity aligned with the laser beams is minimal. When one or more of these conditions is not met, additional spectral information is needed to reconstruct the variables of interest using a modified algorithm.

There are two techniques by which the spatial distribution of multiple parameters can be determined from LOS attenuation data. The first approach is to construct a sum-of-squares residual between the measurements, $\mathbf{b}$, and a nonlinear measurement model, $\mathbf{b}_{\text{model}}$, which is a function of all the unknown state variables (Ma et al., 2009),

$$F_1(\mathbf{x}) = \|\mathbf{b} - \mathbf{b}_{\text{model}}(\mathbf{x})\|_2^2. \tag{5.27}$$

In this expression, $\mathbf{b}$ contains the absorption measurements from all m beams and w resolved wavenumbers and $\mathbf{x} = [\boldsymbol{\chi}; \mathbf{T}]$, where $\boldsymbol{\chi}$ and $\mathbf{T}$ are $n \times 1$ vectors containing the unknown mole fraction and temperature at each basis function.[3] Measurements are modeled by calculating local spectra via Eq. (5.2) and then conducting LOS integration at each wavelength using the matrix system in Eq. (5.11). When **A** is rank deficient, as is typically the case, Eq. (5.27) has an infinite set of minimizers so it must be supplemented with prior information, usually of the form of additional least-squares functionals derived from applying smoothness priors to $\boldsymbol{\chi}$ and **T**, i.e.,

$$F_2(\boldsymbol{\chi}) = \lambda_\chi^2 \|\mathbf{L}\boldsymbol{\chi}\|_2^2 \text{ and } F_3(\mathbf{T}) = \lambda_\mathrm{T}^2 \|\mathbf{LT}\|_2^2. \tag{5.28}$$

The sum of F_1, F_2, and F_3 is then minimized using a multivariate minimization algorithm. This sum is reportedly nonconvex due to the nonlinear measurement model, so it must be solved using a metaheuristic optimization technique, such as simulated annealing (Ma et al., 2009). Given the strongly nonlinear relationship between the thermodynamic state variables in **x** and absorption spectrum, as well as the high dimension of the problem, with at least $2n$ unknowns, the sum of F_1, F_2, and F_3 is costly to repeatedly compute and thus difficult to minimize (Ma et al., 2009). However, the nonlinear approach enables direct regularization of the gas state, which can be used to encode known relationships between χ_{abs}, T, and P, e.g., stagnation relationships in a high-speed flow (Qu et al., 2019).

In the second approach, linear reconstructions are conducted for either spectrally integrated absorbance coefficients, $\{k_{\eta,1}, k_{\eta,2}, \ldots, k_{\eta,w}\}$, over a set of w wavenumbers, or spectrally integrated absorbances, $\{K_1, K_2, \ldots, K_w\}$, over w absorption features. The unknown variables, i.e., $\{\chi_{\mathrm{abs},i}\}$, T, and P, are computed at each location in a local post processing stage (Grauer, Emmert, Sanders, Wagner, & Daun, 2019; Wood & Ozanyan, 2014). Reconstructions of K_k from two transitions can be leveraged to determine χ_{abs} and T via two-line thermometry (Kasyutich & Martin, 2011), or two *or more* spectrally integrated absorption coefficients can be used to calculate the mole fraction and temperature via local Boltzmann plots (Grauer & Steinberg, 2020). In temperature imaging by two-line thermometry, errors are suppressed through entropy maximization (Bao et al., 2021); the use of additional transitions in a Boltzmann plot analysis can also improve the accuracy of estimates. It is not always convenient to extract spectrally integrated absorbances, e.g., when a broadband hyperspectral source is swept across a large number of overlapping transitions from the target spectrum (Ma et al., 2009), in which case the spectrally resolved data are reconstructed at each wavenumber. This can be quickly accomplished in block form,

$$\mathbf{A}\underbrace{[\mathbf{k}_1, \mathbf{k}_2, \ldots \mathbf{k}_w]}_{\mathbf{K}} = \underbrace{[\mathbf{b}_1, \mathbf{b}_2, \ldots \mathbf{b}_w]}_{\mathbf{B}}, \tag{5.29}$$

[3] In principle, pressure can be included in the nonlinear state vector **x** (Cai & Kaminski, 2014) or in a local spectroscopic regression (Grauer, Emmert, Sanders, Wagner, & Daun, 2019). However, while measurements of inhomogeneous pressure field are possible in principle, they have not yet been demonstrated in practice.

where the $n \times w$ matrix $\mathbf{K}$ contains the unknown spectral field at each of the w measured wavenumbers, with one row per basis function, and the $m \times w$ matrix $\mathbf{B}$ contains the absorbance data, with one row per beam. Eq. (5.29) can be combined with any of the regularization techniques discussed in this chapter. Following reconstruction, a local spectroscopic regression is conducted for each basis function using the rows of $\mathbf{K}$, i.e., values of χ_{abs} and T at the jth basis function are determined by fitting the jth row of $\mathbf{K}$ to the sum of Eq. (5.2) over each transition (Grauer, Emmert, Sanders, Wagner, & Daun, 2019).

5.4 Beam array design and optimization

Reconstruction accuracy depends strongly on the layout of the optical paths that transect the RoI, so CST systems must be designed in a way that optimizes reconstruction accuracy, subject to constraints imposed by the limited optical access offered in many practical scenarios, the high cost of optical components, and the need for these components to be arranged in a noninterfering way. The Fourier-based reconstruction algorithm offers some guiding principles on how beams should be arranged. Eqs. (5.5) and (5.6) imply that tomographic reconstruction amounts to an integration of the projected data over the s and θ coordinates, so beam arrays should fully sample both the axial and angular dimension (Terzija et al., 2008). Thus, while two orthogonal projections of parallel beams, as shown in Fig. 5.9(a), is a convenient array from an experimental perspective, this is done at the cost of reconstruction accuracy. A fan-type arrangement, like that shown in Fig. 5.9(b) (Pal, Ozanyan, & McCann, 2008), provides a much more uniform sampling of (s, θ), as shown in Fig. 5.9(c), and is thus expected to provide better resolution and reconstruction accuracy.

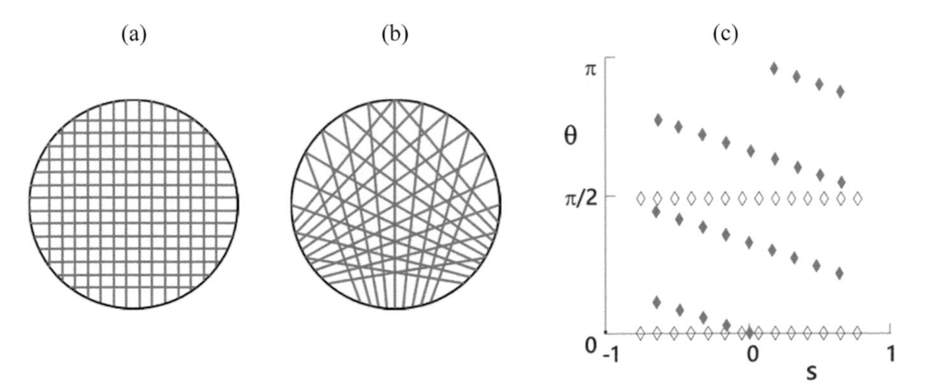

Figure 5.9 Two examples of limited-data beam configurations: (a) a 28-beam array with two orthogonal projections containing 14 parallel beams; (b) a 27-beam array with three fan-shaped projections each containing nine beams; and (c) the sinogram plots for the array in (a) (open symbols) and (b) (solid symbols).

Arrangements that provide the best reconstruction accuracy can be found via a formal design optimization procedure wherein the optical layout is defined by a vector of design parameters, $\mathbf{\Phi}$, and the design performance is quantified by an objective function, $F(\mathbf{\Phi})$, in such a way that the optimal design outcome minimizes the function. The optimal beam layout is then found by solving

$$\mathbf{\Phi}^* = \underset{\mathbf{\Phi}}{\operatorname{argmin}}[F(\mathbf{\Phi})], \text{ s.t. } \mathbf{c}(\mathbf{\Phi}) \leq 0, \tag{5.30}$$

where $\mathbf{c}(\mathbf{\Phi})$ is a vector of bound constraints that ensure that the design can be implemented in a practical setting. The objective function is nonconvex, so the minimization should be carried out using a genetic algorithm or other metaheuristic technique.

The key step in this procedure is to identify an objective function that can predict the performance of a given beam array a priori. It has been argued (Yu, Tian, & Cai, 2017) that beams should be arranged to maximize the orthogonality of rows in the $\mathbf{A}$ matrix, which ensures that the rank of $\mathbf{A}$ is maximized for a given beam count. However, it stands to reason that the beam array optimization should account for prior knowledge incorporated into reconstructions by the chosen algorithm, since we have shown this to be an integral part of the CST system. In the case of second-order Tikhonov regularization, the following functional was proposed (Twynstra & Daun, 2013):

$$F(\mathbf{\Phi}) = \|\mathbf{R}(\mathbf{\Phi}) - \mathbf{I}\|_F^2, \tag{5.31}$$

where $\| \cdot \| F$ denotes the Frobenius norm. The rationale for this choice follows from the concept of resolution and PSFs: minimizing this function minimizes the off-diagonal terms in the resolution matrix which are associated with the smoothing matrix equation, $\mathbf{Lx} = \mathbf{0}$, thereby maximizing the information content of the data relative to the prior.

The Bayesian framework provides an intuitive way to predict reconstruction accuracy via the posterior density width. In the case of a Gaussian error model and a Gaussian prior, the posterior probability density is also a Gaussian distribution with a covariance matrix given by Eq. (5.25). Reconstruction uncertainties at each pixel are indicated by the posterior covariance matrix, and the beam array is selected to minimize the average posterior uncertainty of reconstructions, $F(\mathbf{\Phi}) = \operatorname{tr}\{\mathbf{\Gamma}_{\text{post}}[\mathbf{A}(\mathbf{\Phi})]\}$. It has been shown (Grauer, Hadwin, & Daun, 2016) that this function was a reliable predictor of reconstruction accuracy for imperfect, noisy measurement data, and beam arrays optimized using this approach produced more accurate reconstructions compared to regular beam arrangements. A further advantage of Bayesian optimization is that it explicitly accommodates prior information through the prior covariance; if the prior knowledge includes an approximate plume location, for example, minimizing the posterior covariance will produce a beam arrangement clustered where the plume is thought to be.

5.4.1 Case study: practical beam array design for target spatial resolution

A CST sensor design process to achieve a customized spatial resolution has been presented in Liu, Tsekenis, Polydorides and McCann (2019), using the Frobenius norm beam array optimization method described above (Twynstra & Daun, 2013), followed by system simulations using the quantification method in Tsekenis, Tait and McCann (2015) described in Section 5.3. For a square RoI discretized into 50×50 square cells with side length l_c, they analyze the influence of the number of projections and the total number of beams, aiming to deliver the required spatial resolution using the minimum number of beams. The reconstructed images are divided into 32 sectors on 5 concentric radial tracks. In the case of a 4×30 beam arrangement (4 projections and 30 parallel beams per projection), the Frobenius norm optimization yields an equi-angular separation of 45 degrees between projections, with beam spacing $d = 1.69l_c$ within two orthogonal projections and $2.38l_c$ for the other two. After simulating the system performance, the spatial resolution for each sector ranges from approx. $0.9l_c$ to $1.7l_c$ and the mean spatial resolution, δ_{global}, is $1.33l_c$. Hence, applying this array to a subject with dimension approx. 200 mm, the global spatial resolution is 5 mm. The method has been validated using experimental measurements from fuel injection in a compression ignition engine (Tsekenis et al., 2018), tailoring the spatial resolution to resolve the injection pattern.

For beam arrangements up to 8×40, the resulting values for δ_{global} are shown in Fig. 5.10, revealing the very strong influence of different geometries. For example, the above result for the 4×30 arrangement is much better than the result $\delta_{global} = 2.32l_c$

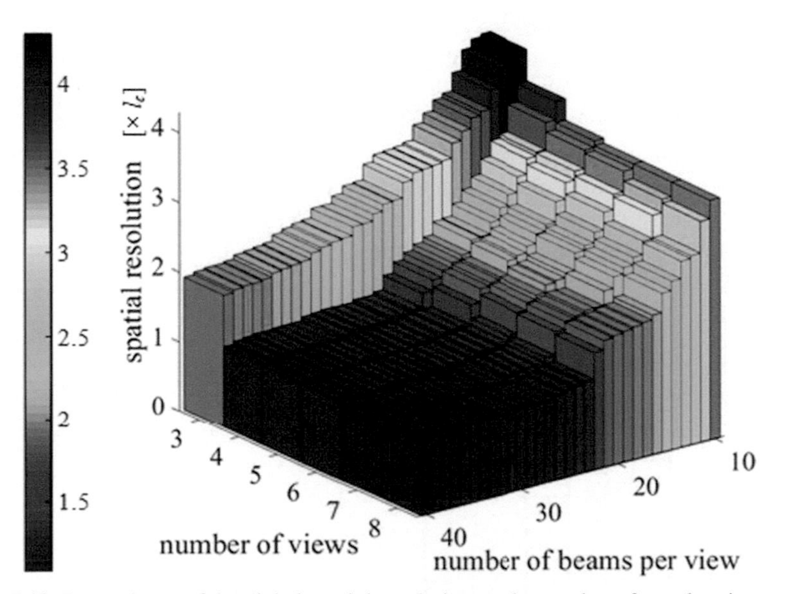

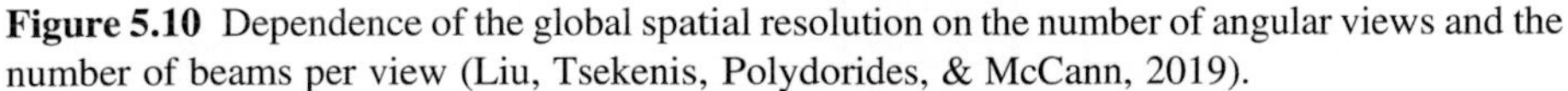
Figure 5.10 Dependence of the global spatial resolution on the number of angular views and the number of beams per view (Liu, Tsekenis, Polydorides, & McCann, 2019).

when the same number of beams are arranged as 8×15, and there is a smooth monotonic variation in between. Fig. 5.10 shows that, generally, for a fixed total beam count, having more beams per projection improves the spatial resolution more quickly than having more angular projections, provided that the number of projections is greater than three. For example, to continue the case of 120 beams, angular under-sampling by a 3×40 arrangement results in worsening of the spatial resolution to $1.91 l_c$. These results further support our remarks in Sections 5.3.3 and 5.4 above concerning the adequacy of *angular* sampling: regardless of the number of beams per projection, δ_{global} is severely degraded by going from four projections to three. In fact, the case of two projections provides, in principle, no spatial resolution at all, and it ought to be avoided.

5.5 Design of Chemical Species Tomography systems

Implementations of CST systems vary widely in form, according to their intended application, but some general comments can be made. Fig. 5.11 shows a hypothetical CST system architecture, intended to illustrate several commonly used design features. One or more optical sources (S1-SM), most often semiconductor diode lasers, are operated under tightly controlled current (I) and temperature (T) conditions, to provide optical radiation of the desired wavelength and intensity (both of which are typically time varying). These sources are multiplexed across a set of launch optics (TX1-TXN), from where the optical radiation transects the measurement space, before arriving at a corresponding set of collection optics (RX1-RXN) and associated photodetectors (PD1-PDN). The output from these photodetectors must then be amplified and the necessary signals, typically representing the transmittance of each beam path at one or more wavelengths, recovered. Many variations of source multiplexing are possible, in terms of both architecture and implementation technology, including the introduction of additional components to enable intensity modulation of individual beams or groups thereof. The corresponding demodulation process is typically the primary function of the signal recovery electronics. The properties of the received optical signals depend upon the propagation characteristics of the medium and the design of the launch and receive optics, so here we shall consider these topics first, before examining optoelectronic issues.

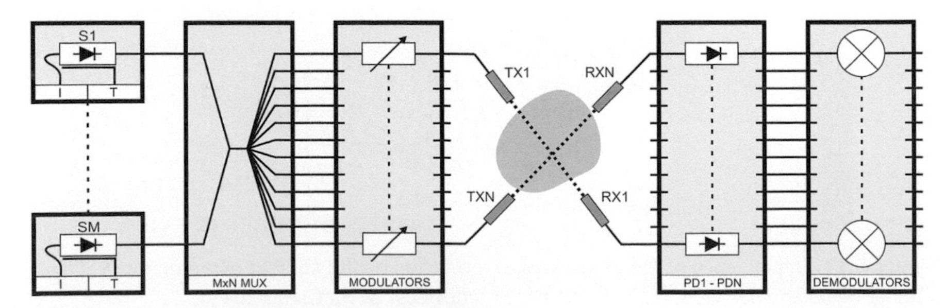

Figure 5.11 One possible CST system architecture.

5.5.1 Optical propagation

As noted in Section 5.2, scattering is largely a nuisance factor in CST, offering little or no useful information. In industrial applications, however, some scattering may occur, due to factors such as particles or droplets in the RoI, spatial variations in the refractive index within the RoI due to species or temperature inhomogeneity, and contamination of the bounding optical surfaces. Scattering may spread or redirect the beam, modifying source-to-detector coupling and increasing the likelihood of interbeam cross-talk. Where scattering is just above negligible levels, a given source will not illuminate receivers other than its LOS counterpart but out-scattering will be nonnegligible so the measured transmittance will include the effects of both absorption and scattering. This is typically addressed using additional wavelengths having a different balance of these two components, ideally including regions of negligible absorption. Depending on the absorption feature under consideration, this may be implemented by chirping a single source, in the manner typical of TDLAS measurements, e.g., Duffin, McGettrick, Johnstone, Stewart and Moodie (2007), or it may require multiple sources, as in the DWR approach used in Wright et al. (2010). Where scattering causes a source to illuminate additional unintended receivers, this cross-talk must be mitigated to maintain the hard-field approximation. In some cases, alternating the direction of source–receive pairs within a projection, effectively doubling the beam pitch for cross-talk purposes, may be adequate. Increasing the directionality of the receivers by the inclusion of lenses or collimating apertures will constrain sensitivity closer to the LOS path, at the cost of increased link losses and more difficult alignment of the launch and receive optics. Orthogonal source modulation schemes can be used to enable rejection of signals from unintended sources during signal recovery, although this does not constrain the region of sensitivity. In some cases, scattering may limit the achievable spatial resolution of a CST system and should be considered when choosing the beam grid and regularization approach.

5.5.2 Optics

The choice of optical components to implement a CST beam array is often influenced by the requirements of the scientific phenomenon or engineering process to which it is being applied. In studies of atmospheric composition (Price, Fischer, Gadgil, & Sextro, 2001), the path integrals may be measured over tens of meters, with their end points located to achieve coverage of some arbitrary RoI, with pressure and temperature at ambient levels. In contrast, industrial applications may entail optical access to processes at nonambient pressure and/or temperature, often across some physical boundary such as a reactor wall, where a component may also be required to act as a pressure window, for example. An ideal optical access arrangement would achieve efficient transmission of light along each measurement path, with negligible cross-talk. Collimation lenses are readily available for such coupling tasks and offer a convenient solution where scattering within the subject can be neglected. Nonnegligible scattering will perturb the (otherwise planar) wavefront reaching the receiver. In such a case, viz. for in-cylinder measurements in combustion engines,

Kranendonk and Sanders (2005) considered explicitly how the properties of the measurement subject must be accounted for in selecting an optical arrangement. Their empirical scattering models led them to adopt a small-diameter collimating launch lens in combination with a larger receive lens, combining optical collection efficiency with the receiver directionality discussed above. Numerous lens options exist, including conventional spherical and ball lenses (as exemplified in Fig. 5.12(a)), aspherics, and gradient-index (GRIN) rod lenses (as in Fig. 5.12(b)). Where multiple wavelengths are used for the separation of scattering and absorption, an achromatic launch condition is clearly desirable. This is relatively straightforward in TDLAS implementations, thanks to the narrow chirp range, but more challenging in the DWR case due to the limited availability of small achromatic lenses and the difficulty of achieving a correspondingly achromatic source distribution. Fan-beam arrangements featuring multiple receivers per source, e.g., as discussed in Case Studies 5.6.2 and 5.6.3 below, are attractive in space-constrained applications. True fan implementations, using line generation optics at each launch, are very wasteful of optical power as much of it falls between receivers, but more efficient embodiments are possible, e.g., using diffractive optical elements to produce a "fan of beams."

In situations where the optical paths must cross a physical boundary, a key decision is whether the relevant elements will function purely as optical access "windows" or also act as lenses. In the latter case, the reduced degrees of freedom within the optical design may make it impossible for the optical elements to have the same form as the internal boundary they replace. In some applications, most obviously food and pharmaceuticals, process-facing crevices may be unacceptable and the use of dedicated windows, matching the internal form of the vessel, may be required. The design freedom gained through these additional optical elements typically increases the cost and complexity of the CST system, and its susceptibility to interferometric noise. An important case arises when the measurement subject must be contained within a transparent annulus, as in Fig. 5.12(b), which implements the 31-beam array that

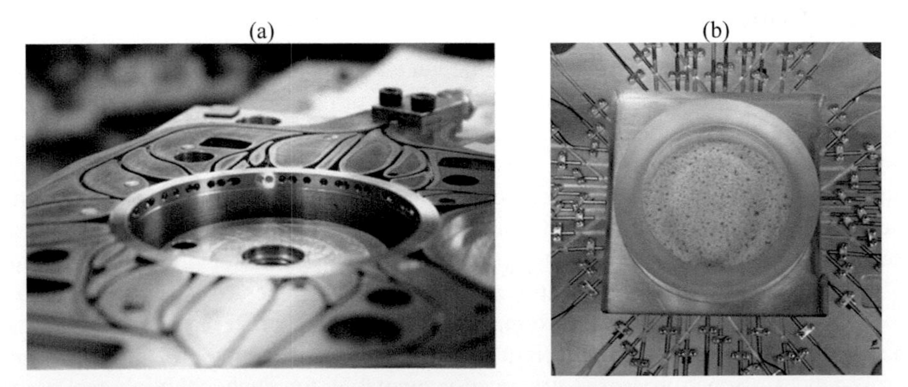

Figure 5.12 Implementations of optical fiber delivery and in-cylinder optical access to (a) a multicylinder gasoline engine, using embedded ball lenses (Wright et al. 2010), and (b) a single-cylinder diesel engine, using a transparent annular section of the cylinder wall and external GRIN lenses (Tsekenis et al., 2018).

was analyzed for spatial resolution performance in Fig. 5.8. Although the annulus serves as a single continuous pressure window, it also acts as a weak meniscal cylindrical lens, complicating source to receiver coupling; additional corrective optics or increased detector area are necessary to avoid increased insertion loss, but these are costly and difficult to miniaturize. Practical demonstrations of this approach, e.g., Terzija et al. (2015) and Tsekenis et al. (2018), have therefore tolerated increased insertion losses, particularly in chord beams, in preference to implementing such corrections. Propagation through cylindrical and tilted optical surfaces also requires greater care in the avoidance of polarization-related errors, especially in systems (e.g., fiber-coupled systems) where polarization drift is expected.

Restricting our concern to the NIR range up to wavelengths of about 2 μm, silica-based fiber optics and components have been used predominantly for beam delivery and/or collection in CST process applications to date. They allow arrangements impossible using free-space optics, particularly in confined or dirty environments, as both cases in Fig. 5.12 illustrate. The ready availability of 1:*N* fiber couplers is particularly useful in systems that share a light source across several launch positions. The use of fiber optics is, however, not without drawbacks: polarization and bend loss effects must be considered; multimode (MM) fibers are susceptible to modal noise; and fiber performance drops off rapidly above about 2 μm. The short fiber lengths and, in receiver applications, poorly controlled launch conditions typical of CST systems, mean that cladding mode propagation may be significant. The weakly guided nature of such propagation gives it the potential to become a significant source of amplitude noise in high vibration environments.

Another key aspect of the optical design of CST systems is alignment. Aiming a source at an omnidirectional receiver is a two-degree of freedom problem, but it increases to four if directional receivers are used. One solution is to carry out one-time alignment during manufacture, as exemplified in the case shown in Fig. 5.12(a), where an Optical Access Layer (OPAL) replaced the equivalent layer along the head of all cylinders of a multicylinder automotive engine (Wright et al., 2010). The 27-beam OPAL was durable for hours of engine operation over a wide range of speed and load conditions, running on retail gasoline (which is a mixture of many hydrocarbon species). The OPAL optical performance was good, but intra-cycle distortions of the cylinder wall required great care in data analysis. Another solution is exemplified in Fig. 5.12(b), which shows the use of a transparent so-called cylinder liner, with external GRIN lenses mounted in fixed launch and receives assemblies laid into accurately machined V-grooves in a flat metal plate. To overcome some of the effects of the optical liner, as discussed above, the groove positions were calculated using ray-tracing software for all 31 beams, and a tolerance of 50 μm on the cylinder liner optical geometry was necessary in order to achieve acceptable alignment. In this case, all of the fibers and lenses are reuseable, with a different grooved plate, for modified beam arrangements and/or deployment on a different engine. A third type of solution, when space is available for human intervention, is to enable alignment adjustment in situ at both the launch and receive ends of the beam, as in the system discussed in Case Study 4.B in Section 5.6.

5.5.3 Optoelectronics and data acquisition

5.5.3.1 Light sources

Semiconductor diode lasers are compact and rugged, and their flexibility in terms of spectral output and frequency response readily enable fixed wavelength, DAS, and WMS measurement. The 1−50 mW output power of a typical NIR diode laser is usually sufficient to ensure that measurement performance is not limited by the available optical power, but rather by interferometric noise or other optical considerations. However, in CST systems, a source may have to be shared among many beam paths and the noise performance of the receiver electronics then assumes much greater importance (Wright, Ozanyan, Carey, & McCann, 2005), unless a means to increase the available optical power is included. Semiconductor optical amplifiers have been used as chirp-free modulators (Karagiannopoulos, Cheadle, Wright, & McCann, 2012), but modest gain and very limited wavelength availability limit their utility. Fiber amplification of modulated diode laser seeds within a TDLAS-based architecture has recently been demonstrated (Wright et al., 2015) and high optical gains are available in several spectral regions, providing sufficient power for CST arrays in excess of 100 beams. Hyperspectral approaches, utilizing more absorption features than would be possible with conventional TDLAS techniques, are being enabled by the emergence of widely tunable sources (An et al., 2011) and frequency combs (Rieker et al., 2014). In the MIR region, source costs remain appreciably higher than in the NIR, despite recent advances.

5.5.3.2 Optical fiber components

Semiconductor diode lasers can be readily coupled into optical fibers, including single mode (SM). Passive distribution of a single source across all paths is a cost-effective architecture, requiring only the source, with associated current and temperature control, and a 1:N coupler to supply all launch lenses. DWR implementation requires the addition of a second source and a WDM coupler to provide the necessary 2:N functionality. Using SM fiber throughout the source and distribution network avoids the modal noise problems associated with MM fibers.

5.5.3.3 Detection

Single-point detection at NIR wavelengths is dominated by Si, Ge, and InGaAs photodiodes. All of these detectors exhibit sensitivity over a wide range of angle of incidence so receiver directionality must be implemented by supplementary optics, which also serve to focus the available light onto a relatively small detector, improving speed and noise performance. Direct illumination of detectors is therefore relatively rare, with the majority of systems including some optics to control the receiver numerical aperture (NA).

Three possible receiver architectures are shown in Fig. 5.13. For the lens detector combination, Fig. 5.13(a), the NA is defined by the diameter and location of the aperture stop (AS), with the optical field-of-view (FOV) being set by either a dedicated

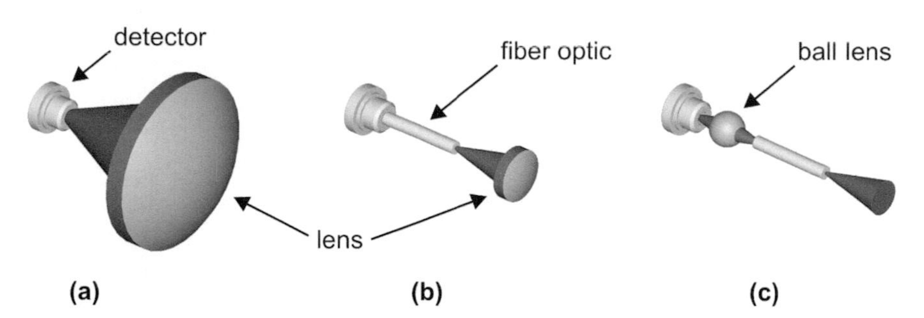

Figure 5.13 (a) lens-detector combination; (b) lensed fiber with butt-coupled detector; (c) unlensed fiber with ball lens coupling to detector.

field stop (FS) or the detector area. For a lensed fiber, Fig. 5.13(b), in the absence of a limiting aperture or vignetting in the lens, the input NA is that of the fiber, as modified by the lens. Where unlensed fibers are used, Fig. 5.13(c), the receiver NA is simply that of the fiber. When designing receivers for CST, the scattering described in Section 5.5.1 should be considered as this may considerably enlarge the focal spot size on the detector. When imaging turbulent media, the detector is typically illuminated by a time-varying speckle pattern. Where optical fibers are used for light collection, the use of MM fiber is typical, due to the difficulties of coupling into SM fiber. This renders the system susceptible to modal noise, as equilibrium mode distributions are unlikely in the relatively short fibers and vibration levels typical of CST systems. This can be partially mitigated by ensuring that the detector captures the entire output of the fiber, either by butt coupling of a large detector, as in Fig. 5.13(b), or using a suitable coupling lens, as in Fig. 5.13(c). Detection bandwidth requirements are dictated by the source modulation used, with up to a few kHz typical for DAS, rising to hundreds of kHz for WMS.

5.5.3.4 Data acquisition

Temporal resolution is often a key factor in CST, e.g., to characterize turbulent flows to facilitate better understanding of reaction mechanisms and the heat transfer process. Fig. 5.14 shows a fully parallel (FP) DAQ system that maximizes the temporal resolution by simultaneous detection of the transmitted intensities for all the laser paths (Fisher et al., 2019). However, its hardware implementation can be complex and expensive, with high power consumption. FP sampling of spectral data from a large number of beams generates a considerable data transfer burden in industrial applications where the in situ signal digitization is generally located remotely from the high-level processors.

These considerations motivated the recent development of a quasiparallel (QP) DAQ system (Enemali, Zhang, McCann, & Liu, 2022), where time-division multiplexing across several beam signals is carried out within each wavelength scan, to enable digitization and demodulation of these beams with a single ADC and a single digital lock-in (DLI) module. The QP scheme maintains the temporal response of the FP

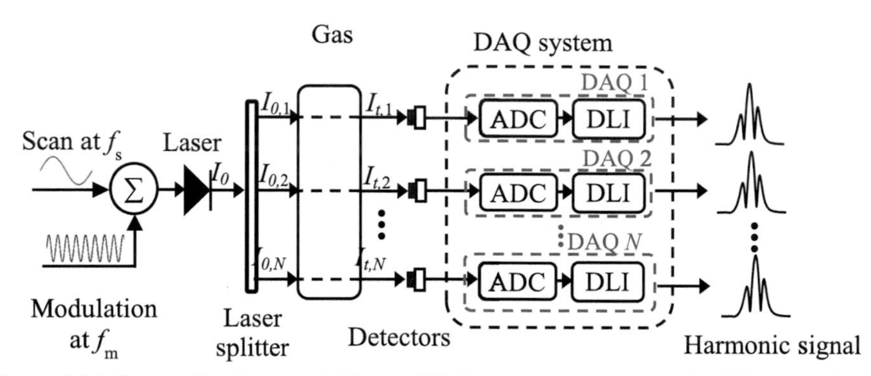

Figure 5.14 Schematic diagram of N-beam WMS measurement system in CST using the fully parallel DAQ system.

scheme, as defined by the wavelength scan rate, but allows the CST system designer to reduce implementation cost and output data rate by reducing the density of spectral sampling. Fig. 5.15 illustrates the QP concept, using four beam paths multiplexed onto a single ADC and DLI. In turn, each signal is sampled and demodulated over a whole number, C, of consecutive fast modulation periods (two in the illustration), providing a measurement at a single wavelength point in the scanned spectrum (the second harmonic signal is given as an example here, i.e., S_{2f}). Beginning with beam

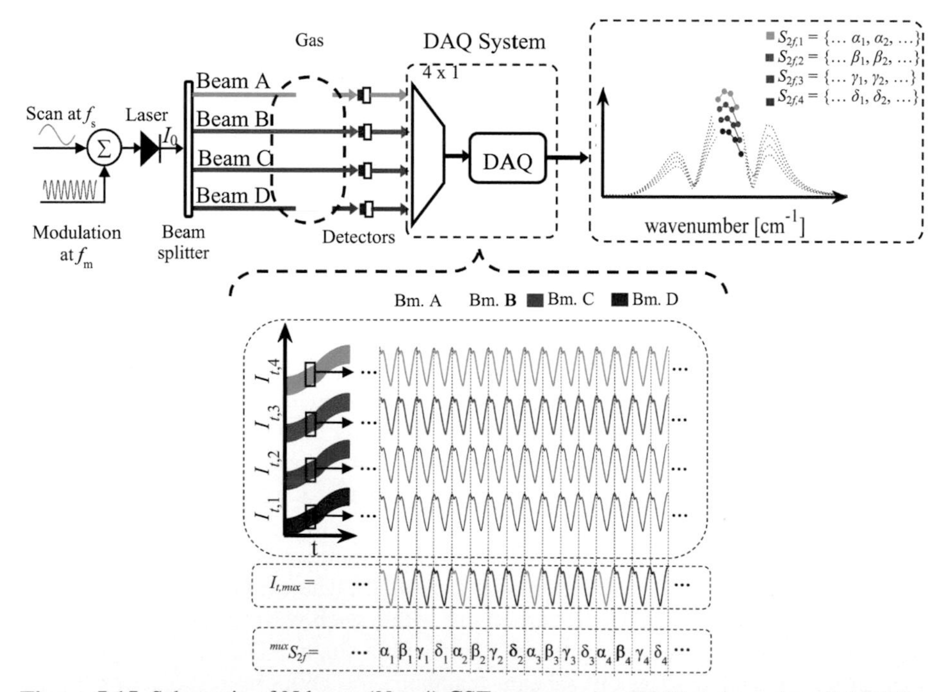

Figure 5.15 Schematic of N-beam ($N = 4$) CST system using WMS and quasiparallel DAQ.

A, the transmitted signal is demodulated over the first two cycles to obtain a wavelength sample α_1; over the next two cycles, the signal from beam B is demodulated to obtain wavelength sample β_1; then wavelength samples γ_1 and δ_1 are obtained from the beam C and beam D signals, respectively, in the same way. Repeating the process, the next wavelength sample α_2 is obtained for beam A, then wavelength sample β_2 for beam B, and so on. The multiplexed WMS-2f signal obtained in Fig. 5.15, denoted ${}^{mux}S_{2f}$, contains the samples from the four multiplexed beams and can be expressed as

$$ {}^{mux}S_{2f} = \{\ldots, \alpha_1, \beta_1, \gamma_1, \delta_1, \alpha_2, \beta_2, \gamma_2, \delta_2, \ldots\}. \quad (5.32) $$

Finally, the ${}^{mux}S_{2f}$ is demultiplexed to recover the S_{2f} signal for each beam as

$$ S_{2f,1} = \{\ldots\alpha_1, \alpha_2, \ldots\}, \quad (5.33) $$

$$ S_{2f,2} = \{\ldots\beta_1, \beta_2, \ldots\}, \quad (5.34) $$

$$ S_{2f,3} = \{\ldots\gamma_1, \gamma_2, \ldots\}, \quad (5.35) $$

$$ S_{2f,4} = \{\ldots\delta_1, \delta_2, \ldots\}, \quad (5.36) $$

5.6 Case studies

The following case studies are intended to illustrate practical application of CST to the imaging and measurement of a range of engineering processes. Given the limitation of space, the reader is encouraged to consult the original literature if further detail is required.

5.6.1 Case study 1: automotive in-cylinder hydrocarbon imaging

IC engines for automotive use must be optimized for a wide range of speed and load conditions, necessitating insight into the underlying in-cylinder gas distributions. For example, air–fuel mixing is critical to efficient combustion, but its quantification in operating engines is problematic, resulting in the use of indirect methods, particularly the addition of tracer species necessary for planar laser-induced fluorescence (PLIF) (Sick, 2013), which is unable to quantify the fuel distribution directly or to provide continuous intracycle imaging. Accordingly, the DWR technique described in Section 5.2 was developed specifically for the study of fuel distribution in-cylinder prior to ignition. For that purpose, the IMAGER system that embodies the DWR method has been applied to:

Case 1.A - retail gasoline in an operating Ford Duratec 2.0L 4-cylinder spark-ignition (SI) engine with port fuel injection (Wright et al., 2010), using the OPAL subsystem shown in Fig. 5.12(a);

Case 1.B - iso-octane (a gasoline surrogate) in a motored (i.e., nonfired) Ricardo Hydra single-cylinder SI research engine with direct injection into the cylinder (Terzija et al., 2015), using the same optical access method as illustrated in Fig. 5.12(b); and

Case 1.C - a 50/50 mixture of iso- and n-dodecane (a diesel surrogate) in a single-cylinder research engine adapted from a Volvo D5 compression-ignition (CI) engine, aspirated with nitrogen (Tsekenis et al., 2018), with the optical access and beam arrangement shown in Fig. 5.12(b) and analyzed in Fig. 5.8.

The IMAGER system exploited hydrocarbon absorption at 1700 nm, with a reference wavelength of 1651 nm, maximizing the differential absorption without introducing unnecessary chromatic aberration. In each of the Cases 1.A−1.C above, different calibration and reference schemes were necessary, e.g., to estimate pressure for absorption calibration, to establish intracycle baseline absorption conditions, and to assess dynamically the signal integrity obtained from each beam. As a result, images were frequently reconstructed in Case 1.A after omitting the data from several beams, typically 6 or more out of the total of 27 installed. Nevertheless, over a wide range of engine operating conditions, robust fuel imaging was achieved at typical rates of 3000 fps (Wright et al., 2010), drawing heavily upon extensive lab benchtop calibration and phantom tests and upon the flexibility of the iterative Landweber image reconstruction scheme (Terzija et al., 2008) (which was used in all three engine cases). The PCI data obtained in Case 1.B were extremely robust, with typically 30 "good" beams out of 32 installed. As discussed in the first edition of this book and in Terzija et al. (2015), the resulting images of fuel vapor dynamics during the engine compression stroke were profoundly revealing of the in-cylinder behavior, and of the underlying relationship with the injection spray dynamics that were also imaged using the reference wavelength data alone.

Turning to Case 1.C, it is first of all important to note that the CI engine principle presents severe challenges for CST: fuel injection directly into the cylinder occurs late in the engine compression stroke, and the fuel vapor evolves from several small jets of fuel spray. In Tsekenis et al. (2018), a 10-cycle operation sequence was developed to provide a robust absorption baseline, and an injection schedule whereby fuel vapor could be imaged reliably for a period in the compression stroke, as shown in Fig. 5.16 when the engine was running at 1200 rpm. The fuel vapor lobe from each of the seven holes in the centrally mounted fuel injector can be identified clearly. The anticlockwise in-cylinder swirl flow is evident from comparison of the two images, and was analyzed in detail by the authors. The fuel contained 1% naphthalene dopant as a tracer for PLIF, which was applied in the same plane when the engine was operated under the same conditions, and provided excellent validation of the CST images. The above Cases 1.A−1.C demonstrate the unique capability of CST to image directly, quantitatively, robustly, and continuously, the dynamic in-cylinder evolution of hydrocarbon fuel distribution in a range of engine modes, including the multicylinder case while running at full load.

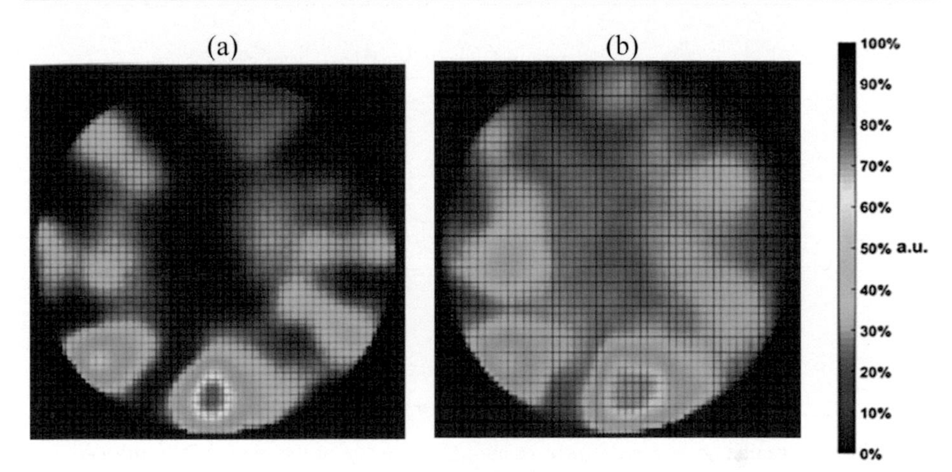

Figure 5.16 Example images of fuel vapor distribution recorded at 3000 fps during the compression stroke of a CI engine, shortly after fuel injection starts at 310 degrees of shaft crank angle (°CA) rotation: (a) at 333 °CA and (b) at 340°CA (Tsekenis et al., 2018).

5.6.2 Lab swirl flame

Swirl injectors are extensively employed to realize lean combustion in practical combustion processes, with high fuel efficiency and reduced NOx emissions. Understanding the reaction mechanism of the swirling flame is important to optimize the structure of the swirl injector. In particular, there is significant risk of lean blowout (LBO), whereby the flame is extinguished if the mixture becomes weaker than the so-called LBO limit, posing a safety hazard for aircraft engines and necessitating an expensive shutdown and restart procedure in land-based engines for power generation (Huang & Yang, 2009). Hence, deeper understanding and in situ real-time active control mechanisms are required for safe and reliable operation of low-emission gas turbine combustors. The dynamic cross-sectional behavior of unconfined swirling flames has been investigated using the fan-beam CST system described in Liu, Cao, Lin, Xu and McCann (2018) to reconstruct the 2D temperature distributions in different cross-sectional planes of the flame. Nonswirling methane (CH_4) fuel is fed through a central nozzle and mixed firstly with air from an inner clockwise swirl generator and then with air from an outer anticlockwise swirl generator, before the mixture is released from the 42 mm diameter injector nozzle. After ignition, an unconfined swirling flame is formed above the nozzle.

The CST sensor utilized coplanar fan-beam illumination in five views simultaneously, as shown in Fig. 5.17, each fan-beam illuminating a span angle of 24 degrees. The RoI diameter was 6 cm and its center was located at the center of the target cross-section of the swirling flame. Transmission of each fan-beam through the RoI was sampled by an array of 12 equally spaced photodiodes, yielding 60 effective beam measurements and spatial resolution of 7.8 mm (Liu et al., 2015). Five height-adjustable supports were used to image the cross-sections of the swirling flame at various desired heights above the nozzle.

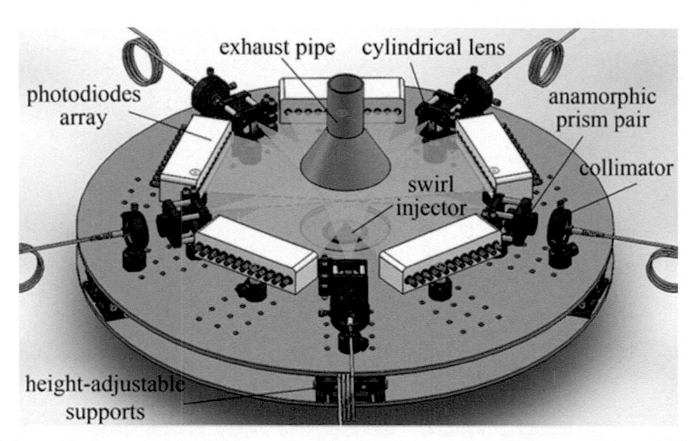

Figure 5.17 Installation of the CST sensor above the nozzle of the model swirl injector (Liu, Cao, Lin, Xu, & McCann, 2018).

Two-line temperature imaging was implemented using DAS of the water transitions at 7444.36 cm^{-1} and 7185.6 cm^{-1}. At an equivalence ratio of 0.205 (for methane and air flows of 8 and 371 L/min, respectively), a stable swirling flame was generated, and its instantaneous (i.e., single frame) temperature distributions were continuously reconstructed using the modified Landweber algorithm. Fig. 5.18 shows sequences of three instantaneous temperature images at 2 cm above the nozzle and at 4 cm above the nozzle, at time intervals of 12 ms. A crescent-shaped high-temperature region is seen to rotate anticlockwise during each sequence. The crescent is formed by the cone-shaped spiral nature of the swirling flow as the methane from the central nozzle is swirled out into the coswirling air. The interaction between the flow field and chemistry becomes significant in reacting flows due to the existence of vortex breakdown that induces flow recirculation. The upstream recirculation zone is close to the swirling

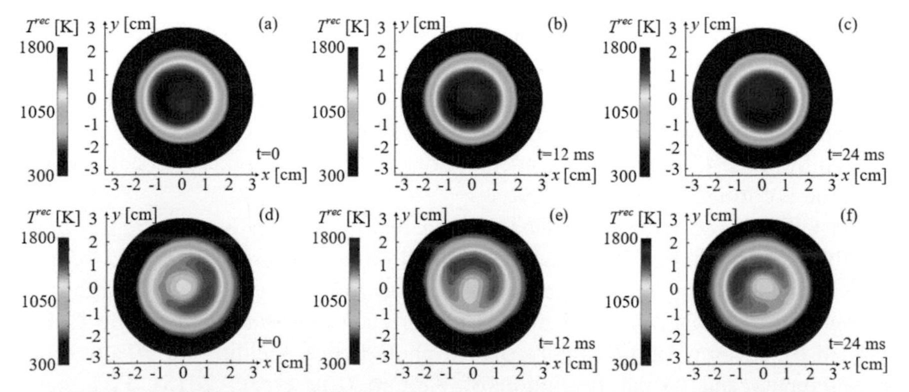

Figure 5.18 Instantaneous temperature images in the swirling flame at heights of 2 cm (a−c) and 4 cm (d−f) above the nozzle. The time interval between successive images is 12 ms (Liu, Cao, Lin, Xu, & McCann, 2018).

inlet, while the downstream recirculation zone is located in the central area. Strong velocity gradients occur in the inner side of the upstream recirculation zone, resulting in vortex breakdown that is generally aligned with the heat release (Syred, 2006). Therefore, a crescent-shaped region of high temperature is observed over the cross-section of the swirling flame. The sense of rotation of the crescent-shaped region is determined by the rotation of the whole spiral, which is anticlockwise in this case. As the heat dissipates downstream of the nozzle (i.e., at greater height above the nozzle in this case), the temperature decreases.

By decreasing the methane flow from 4 L/min while maintaining the above air flow, the dynamics of LBO were revealed, for the first time, in this study; the flame precession is seen to become unstable as the above-described interaction of chemistry and flow reduces the heat generation, and the flame gradually extinguishes. From the onset of the initial instability, the entire LBO event is seen to occur over a period of around 60−70 ms.

5.6.3 Wind tunnel flow

Recently, the fan-beam CST sensor introduced in Case Study 5.6.2 was enhanced by widened fan angle coverage of 50.62 degrees (Zhao et al., 2020), each detected by 24 photodiodes, to yield 120 measurements, as shown in Fig. 5.19. In this case, the five lasers are operated in serial mode to avoid cross-talk between overlapping fans. The laser scan frequency is 5 kHz and the frequency of the modulation signal is 100 kHz, yielding an imaging rate up to 500 fps. This system was experimentally validated by measurement at the exit of a wind tunnel, as shown in Fig. 5.20. Aviation kerosene and air are mixed and burned in the wind tunnel, and the tail gas passes from the 6 cm-diameter exit through the RoI before entering the tail gas recovery

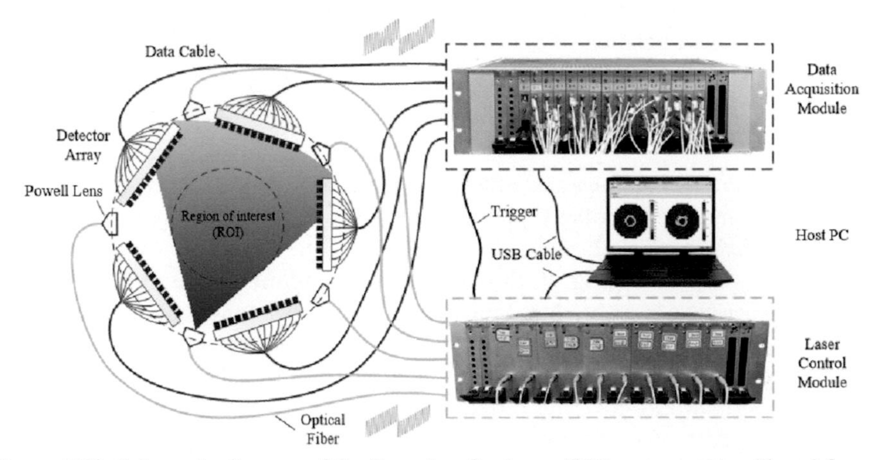

Figure 5.19 Schematic diagram of the five-view fan-beam CST system with widened fan illumination (only one fan is shown) and 24 detectors in each photodiode array (Zhao et al., 2020).

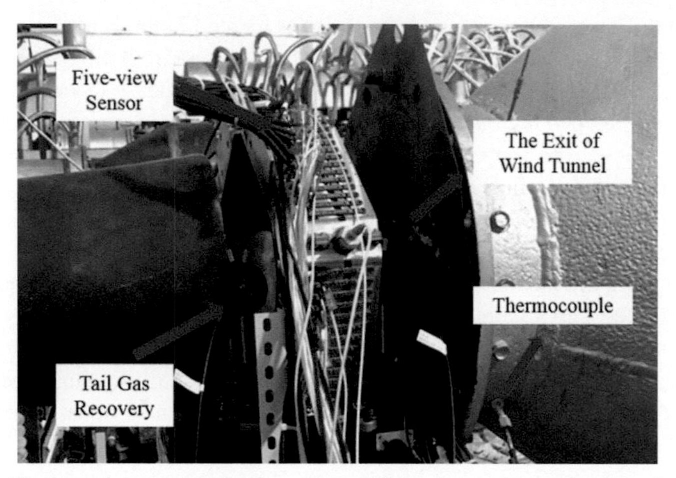

Figure 5.20 The experiment setup for the enhanced five-view sensor at the exit of a wind tunnel (Zhao et al., 2020).

equipment. At the tunnel exit, the air pressure is 1 atm, the gas velocity is around 102 m/s, and the temperature varies from 300 to 1170 K, depending on working conditions.

A thermocouple is placed centrally at the exit of the wind tunnel. Under a certain working condition, the thermocouple reading stabilizes, so the value of the temperature recovered by the CST system at the center of the RoI can be compared with the thermocouple value. In the experiment, 11 working conditions were set, to obtain different temperature distributions at the exit of the wind tunnel. Fig. 5.21 shows the temperature variation with time, ranging from 700K up to 1100K. The central temperature measured by the CST system follows the thermocouple value. Differences between the two values are observed at some conditions, thought to be due to the slightly different installation position of the CST system and the thermocouple.

5.6.4 Turbine exhaust imaging

We describe here two applications of CST: to a turbojet engine that is widely used in small aircraft, and to a turbofan engine that is used in some of the largest civil aircraft in service today.

Case 4.A - H_2O imaging at J85-GE-5 exhaust plane: More than 60,000 J85 turbojet engines have been produced for a variety of applications, including laboratory use for diagnostic equipment development. In the latter setting, Ma et al. (2013) describe a tomographic sensor that uses three synchronized Fourier-domain mode-locked lasers (FDMLs) covering different regions in the NIR spectrum, at 50-kHz repetition rate, with time-division multiplexing between them.

Each FDML sweeps over a unique $\sim 10\ cm^{-1}$ spectral range in the water spectrum. Light is delivered to the measurement location by SM fibers with length of approx. 60 m. The outputs from the lasers were combined and split into 32 beams using a 4×32

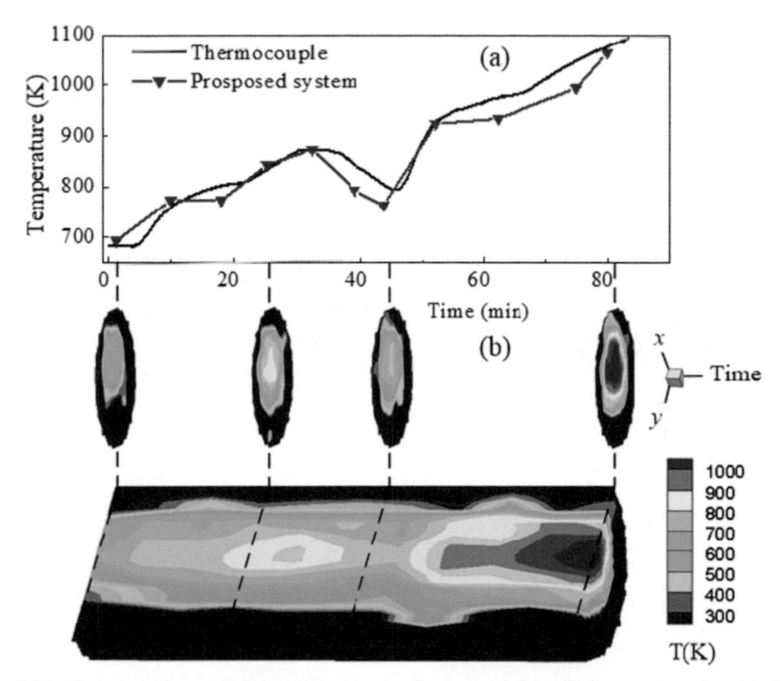

Figure 5.21 Temperature variation with time at the exit of the wind tunnel, under 11 different working conditions (Zhao et al., 2020): (a) measured thermocouple values and the central values obtained from the temperature image reconstructed from the CST system; (b) temperature distributions obtained from the CST system.

optical multiplexer; 15 were installed to probe the RoI horizontally, and 15 vertically, as shown in Fig. 5.22(a), with two beams providing references for laser intensity and wavelength. As shown in Fig. 5.22(b), the electro-optics for the 30 probe beams were mounted on a custom-built aluminum frame that protected them from the high-temperature, high-speed combustion flow. Each of the 30 beams is incident upon a dedicated photodetector, enabling high-speed analogue readout via coaxial cable to a DAQ for digitization and subsequent data storage. In the experiment, DAS was adopted for each LOS measurement.

Fig. 5.22(c) depicts the location of the measurement plane in the exhaust and a sample measurement of the 2D temperature distribution measured at this location, produced using the "simulated annealing" image reconstruction method. While this Case Study demonstrates a number of innovations, it nevertheless suffers from several of the issues discussed in Sections 5.3 and 5.4 above: robust spatial resolution would require rearrangement of the beams into at least three projections (preferably four), and the effective prior information that is imposed by the relatively coarse reconstruction array, with 225 pixels for 30 beams, has a strong but unknown influence on the resultant images. It would also be instructive to carry out a comparison of the simulated annealing reconstruction method with others that have been much more widely used in the field.

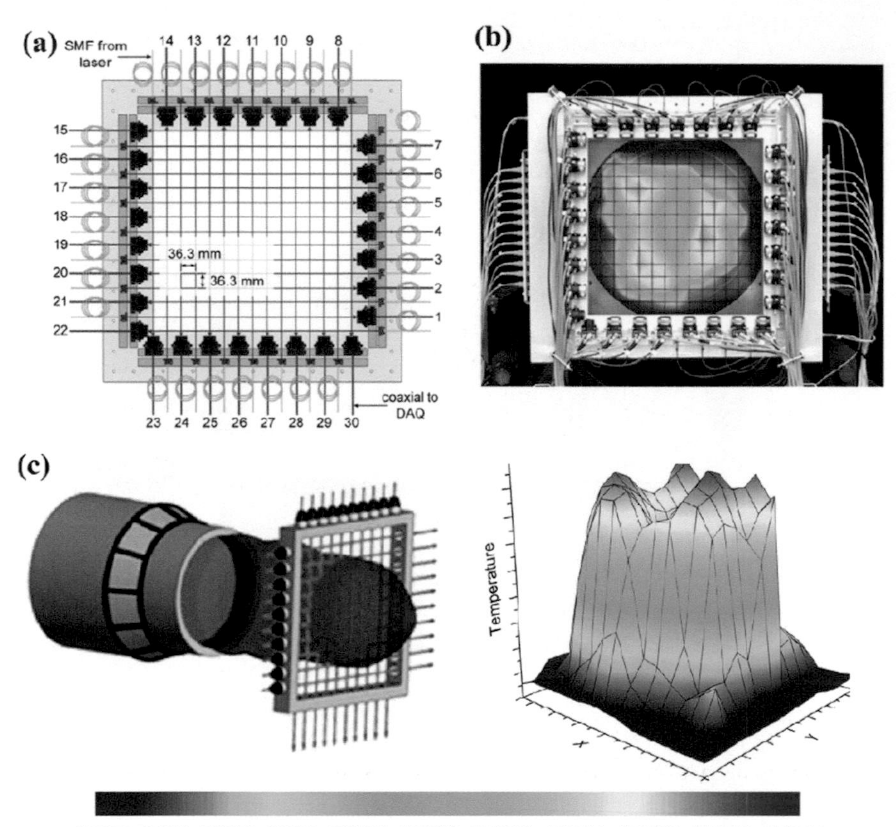

Figure 5.22 Schematic representation of the J85-GE-5 optical test section hardware: (a) beam configuration; (b) photograph of the frame and the optical components overlaid by a sample reconstruction to illustrate the location of the flow field; (c) schematic of the location of the measurements plane in the exhaust and a sample of the recovered 2D temperature (Ma et al., 2013).

Case 4.B - CO_2 Imaging in turbofan engine exhaust plume: Aero engines of this type are deployed in large civil aircraft and use an annular combustor with multiple fuel injectors, an example being the Rolls-Royce Trent XWB, which uses 20 fuel sprays, all of which must operate up to specification, e.g., to maintain fuel efficiency. To image the exhaust-plume distribution of CO_2 from such engines mounted on an industry-standard testbed, a 126-beam CST system has been installed at INTA Madrid (Instituto Nacional de Técnica Aerospacial). The arrangement of six projections, each with 21 parallel beams, is shown in Fig. 5.23 (Fisher et al., 2019). Within each projection, bespoke optical assemblies launch neighboring beams in opposite directions to minimize potential cross-talk. Around the (roughly) 1.2−1.4 m-diameter exhaust plume, the launch and receive optics, light delivery fibers, etc., are housed on a bespoke mounting frame ("the ring") with an internal diameter of 7 m, to withstand

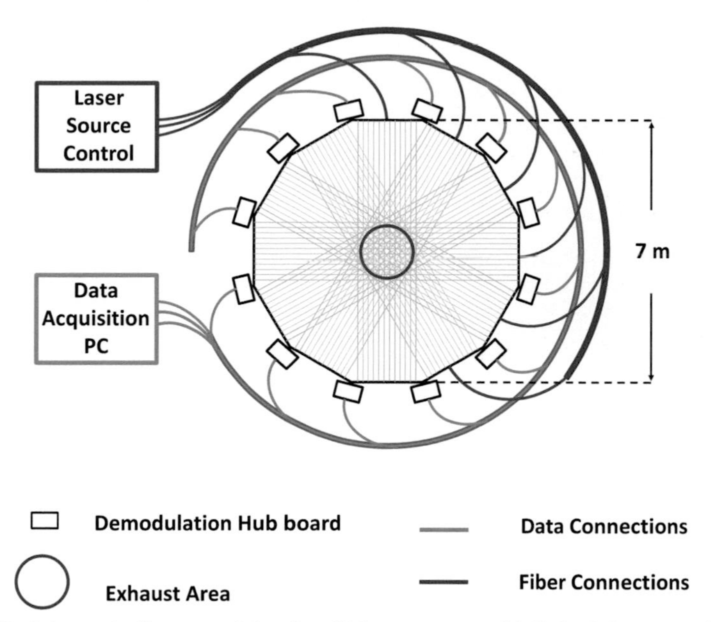

Figure 5.23 Schematic diagram of the 6 × 21 beam array, with light delivery, optics, detectors, and fully parallel DAQ, mounted on the ring around the turbine exhaust plume (Fisher et al. 2019).

flow and vibration effects at all engine conditions. The light source is a Distributed Feedback Interband Cascade Laser, feeding into a Thulium Doped Fiber Amplifier with a 1 × 6 output coupler. The WMS 2*f*/1*f* method was applied to the fully calibrated 1999.4 nm transition of the target species, and the 440-point spectrum measured for each beam was fitted to determine the PCI, taking account of the temperature. The spectroscopic method is well validated for this application by extensive laboratory characterization and by tests on a separate turbine set-up (Benoy et al., 2017). In the preliminary CST tests described below, DAQ consisted of slow multiplexing of the measurements over an 8-channel commercial demodulation unit, yielding a frame rate of 1.25 fps.

Phantom tests of the CST system were carried out prior to its installation in the engine test cell, with the ring lying horizontally, and with circular gas burners (40 cm- and 60 cm-diameter) placed under the measurement plane such that combustion-generated CO_2 plumes flowed buoyantly through the RoI (Tsekenis et al., 2016). These tests enabled the determination of appropriate image reconstruction parameters, using the method described in Polydorides et al. (2018). In all cases, the CO_2 plumes were correctly located in the reconstructed images, as illustrated by the 50 × 50 pixel image in Fig. 5.24(a), where a small burner and a large burner were operated simultaneously and placed with their edges approx. 20 cm apart.

With the ring and equipment installed 3 m downstream of the engine exhaust nozzle in the engine test cell, the exhaust plumes of several aero engines have been measured. Fig. 5.24(b) shows that the typical measured spectrum compares well with simulation

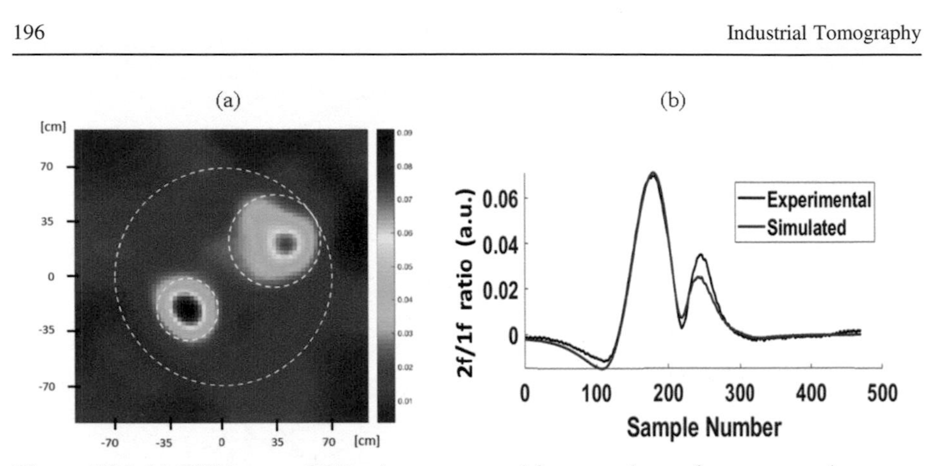

Figure 5.24 (a) CST image of CO_2 phantom comprising two plumes from propane burners (Tsekenis et al. 2016). The white dashed circles show the central 1.4 m-diameter region, and the burner locations; and (b) example of the *2f/1f* absorption spectrum measured on one beam transecting the exhaust plume of a large civil turbofan aero engine (Lengden et al., 2020).

(Lengden et al., 2020). Even at the slow rate of 1.25 fps, very large amounts of data were obtained. For the engine run that yielded the example data shown in Fig. 5.24(b), 24 beams were effectively inoperative, although the remaining 102 beams yielded well-behaved spectra. At the time of writing, detailed analysis is underway and different image reconstruction methods are being explored. Further development of this system at the time of writing aims to establish the facility at Technology Readiness Level 6. This work includes a revised implementation of the multi-FPGA distributed DAQ system that is mounted on the ring and closely coupled with the front-end detection electronics, enabling Fully Parallel measurement of all beams, local demodulation, and much-reduced off-line data load.

5.6.5 Pulverized coal combustion

Coal combustion is a major contributor to global energy supply, and its optimization is crucially important. An extensive program of application-oriented research in CST at Tokushima University in Japan has included several studies of coal combustion. In the particular case discussed here (Wang, Deguchi, Kamimoto, Tainaka, & Tanno, 2020), the distributions of water concentration and temperature in the combustion of pulverized coal with a CH_4 pilot flame are imaged by CST using two diode lasers, each scanning a 0.6 nm wavelength range at 4 kHz, one centered on 1388.3 nm and one centered on 1343.2 nm. In a cylindrical burner of diameter 250 mm, the coal is fed into a 5.7 mm-diameter air jet along the central axis, surrounded by three coaxial air flows, and the CH_4 is introduced at an inclined angle, at the same axial height. Using launch collimators and detectors embedded in the wall of the so-called CT-TDLAS cell, a 32-beam array is placed in a measurement plane 595 mm downstream from the fuel injection plane. The beams are arranged as four projections, equally spaced in angle, each consisting of eight parallel beams, and all measured simultaneously to enable excellent temporal resolution. To eliminate radiation effects, e.g., from soot, the detected light was bandpass filtered (1340−1390 nm), and a nitrogen purge flow prevented any water absorption of the beams in the region outside of the 250 mm burner diameter. The measurement process is temperature calibrated by applying it to a well-controlled laboratory flame with simultaneous thermocouple measurements.

A novel data inversion method is used to produce concentration and temperature images on a 39×39 pixel array. Independent polynomials describe concentration and temperature as a function of position within the RoI, and a joint minimization process includes a spectral absorbance fitting procedure that compares the full absorbance spectrum measured on each beam with a theoretically predicted spectrum. Example images are shown in Fig. 5.25.

The CST system performance was validated by using CFD-based combustion simulations to produce "phantoms". The expected beam absorbance spectra were calculated and passed through the data inversion method described above. Comparison of the recovered distributions with those produced by the simulations enabled the authors to estimate that the CST spatial resolution to be 25 mm, i.e., D/10, consistent with the result from the 31-beam arrangement discussed in Section 5.3.3. The authors concluded that their CST system was capable of application to dusty combustion fields in real time in order to better understand and optimize coal combustion.

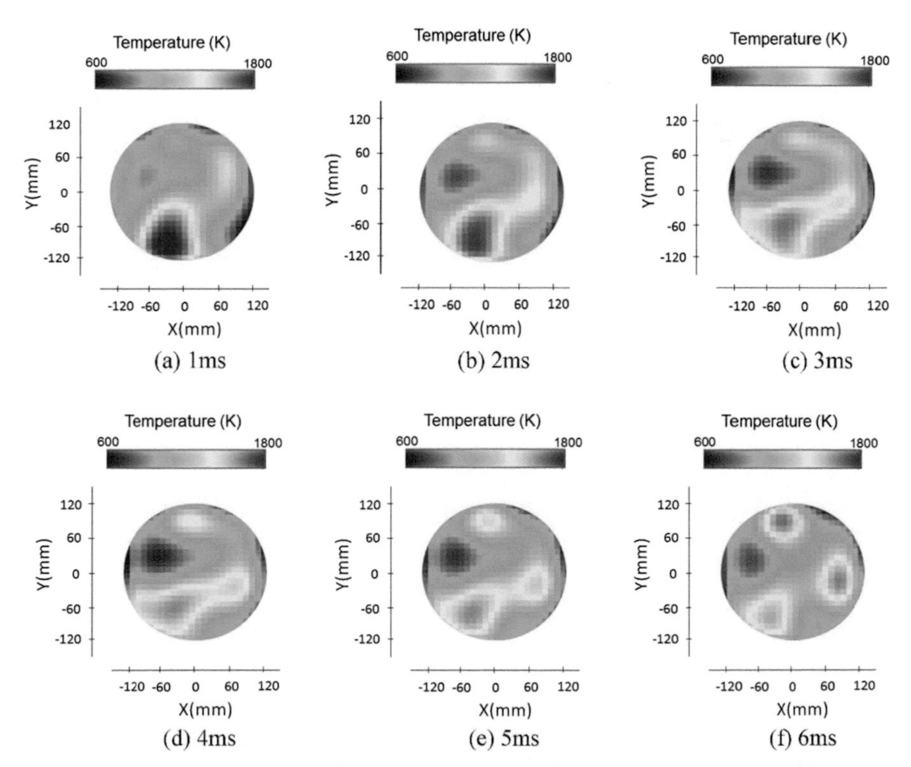

Figure 5.25 Time-resolved images of temperature distribution in a fixed plane during combustion of pulverized coal in a turbulent flow field (Wang, Deguchi, Kamimoto, Tainaka, & Tanno, 2020).

5.7 Future trends

For many decades, health and environmental concerns have resulted in continuous legislative pressure on the owners of chemical processes, and on associated equipment manufacturers, to achieve greater control of chemical emissions and their reduction, particularly in the combustion-based industries. This has been a major motivating factor for the development of many techniques aimed at imaging the spatial distribution of chemical species in a variety of situations, e.g., Sick (2013). Over the last 25 years, great advances in commercially available photonics and electronics hardware, initially facilitated by the manufacturing technology of the optical communication industry, have stimulated the development of Chemical Species Tomography, CST, as described in this chapter. In parallel, the growth and commercialization of several modalities of tomography, described in other Chapters in this book, and applied in situ to a wide range of engineering problems, have driven fundamental research in tomographic reconstruction methods and practical algorithms. These advances in the constituent technologies have been used to synthesize substantial improvements in CST, evidenced by the fact that, in the past 15 years, CST has gone from a specialty diagnostic in mainly experimental and academic settings, to widespread deployment in a diverse range of engineering applications. With many regions of the globe undergoing rapid industrialization, the above legislative pressures will increase and will encompass further industrial sectors, suggesting that CST technology will be exploited further.

Confidence in robust CST system development is enhanced greatly by the establishment of rational methods for beam array optimization, system performance assessment, and fundamental new insights into image reconstruction, as discussed in this chapter. The opportunities facing the CST community, in terms of both "applications pull" and "technology push" are immense. Still a relatively youthful technique, CST clearly stands to gain enormously from exploitation of a huge range of new technologies and methods. Examples of these issues and opportunities are as follows:

- Critical scientific and technological challenges are the driving force for innovation in CST. These challenges range from validation of theoretical models and simulation codes for processes, through the design of process plant, to its optimal control, e.g., for environmental compliance. Potentially pressing examples of the last category are imaging NH_3 and NO distributions for de-NOx optimization in various fossil fuel combustors (Deguchi, Kamimoto, Kiyota, Choi, & Shim, 2013), and the potential for application in the marine propulsion sector (Tsekenis et al., 2017), both of which could yield great societal and commercial benefit.
- Development of robust methods for optical access by many beams (i.e., several hundreds) is also a key goal, perhaps exploiting new technologies such as fiber microprocessing, microelectromechanical systems (MEMS) and electro-optical switching (Tsekenis & Polydorides, 2017). This also requires further development of more cost-effective DAQ systems, as discussed earlier in this chapter.
- Many new parts of the e.m. spectrum are being opened up for CST by continuing developments in lasers and other light sources, e.g., enabling the exploitation of MIR (Goldenstein, Spearrin, Jeffries, & Hanson, 2017) and THz spectroscopy of target species (Ozanyan, Wright, Stringer, & Miles, 2011).

- Progress has been made, as discussed in this chapter, in disentangling the spatial distribution of temperature from that of target species concentration, but extension of that work to high pressures, and to other species, is paramount. This may entail further innovation in spectroscopic measurement methods.
- Challenges in the opto-electronic area include the development of sensitivity to enable CST of (a) species at minor concentrations (<1000 ppm), as discussed for CO measurement in Pal, Ozanyan and McCann (2008) and Pal and McCann (2011), and (b) weak absorbers, an example being O_2, as discussed in Sanders, Wang, Jeffries and Hanson (2001).

We hope that, in this chapter, we have demonstrated the excitement and challenge of CST, and strong progress in its development. We have attempted to set out the concepts and the basic technology that will enable substantial further progress in the near future. Given the growing community in CST, we are confident that a similar review in 10 years' time will present an array of achievements that are unimaginable at the present time.

References

An, X., Kraetschmer, T., Takami, K., Sanders, S. T., Ma, L., Cai, W., et al. (2011). Validation of temperature imaging by H_2O absorption spectroscopy using hyperspectral tomography in controlled experiments. *Applied Optics, 50*, A29–A37.

Banwell, C. N., & McCash, E. M. (1994). *Fundamentals of molecular spectroscopy* (4th ed.). London: McGraw Hill.

Bao, Y., Zhang, R., Enemali, G., Cao, Z., Zhou, B., McCann, H., et al. (2021). Relative entropy regularised TDLAS tomography for robust temperature imaging. *IEEE Transactions on Instrumentation and Measurement, 70*(1), 1–9.

Beiting, E. J. (1992). Fiber-optic fan-beam absorption tomography. *Applied Optics, 31*, 1328–1343.

Benoy, T., Wilson, D., Lengden, M., Armstrong, I., Stewart, G., & Johnstone, W. (2017). Measurement of CO2 concentration and temperature in an aero engine exhaust plume using wavelength modulation spectroscopy. *IEEE Sensors Journal, 17*(19), 6409–6417.

Bertero, M., & Boccacci, P. (1998). *Introduction to inverse problems in imaging*. Boca Raton, FL: CRC Press.

Birch, A., Brown, D., Dodson, M., & Thomas, J. (1978). The turbulent concentration field of a methane jet. *Journal of Fluid Mechanics, 88*(3), 431–449.

Blume, N. G., Ebert, V., Dreizler, A., & Wagner, S. (2015). Broadband fitting approach for the application of supercontinuum broadband laser absorption spectroscopy to combustion environments. *Measurement Science and Technology, 27*(1), 015501.

Cai, W., Ewing, D. J., & Ma, L. (2008). Application of simulated annealing for multispectral tomography. *Computer Physics Communications, 179*, 250–255.

Cai, W., & Kaminski, C. F. (2014). Multiplexed absorption tomography with calibration-free wavelength modulation spectroscopy. *Applied Physics Letters, 104*(15), 154106.

Chojnacki, A. M., Sarma, A., Wolga, G. J., Torniainen, E. D., & Gouldin, F. C. (1996). Infrared tomographic inversion for combustion and incineration. *Combustion Science and Technology, 116–117*, 583–606.

Cole, R. K., Makowiecki, A. S., Hoghooghi, N., & Rieker, G. B. (2019). Baseline-free quantitative absorption spectroscopy based on cepstral analysis. *Optics Express, 27*(26), 37920–37939.

Daun, K. J. (2010). Infrared species limited data tomography through Tikhonov reconstruction. *Journal of Quantitative Spectroscopy & Radiative Transfer, 111*, 105–115.

Daun, K. J., Grauer, S. J., & Hadwin, P. J. (2016). Chemical species tomography of turbulent flows: Discrete ill-posed and rank deficient problems and the use of prior information. *Journal of Quantitative Spectroscopy & Radiative Transfer, 172*, 58–74.

Deguchi, Y., Kamimoto, T., Kiyota, Y., Choi, D., & Shim, J. (2013). Real-time 2D concentration and temperature measurement method using CT tunable diode laser absorption spectroscopy. In *7th world congr. Industrial process tomography, WCIPT7. Krakow, Poland.*

Drescher, A. C., Gadgil, A. J., Price, P. N., & Nazaroff, W. W. (1996). Novel approach for tomographic reconstruction of gas concentration distrivbutions in air: Use of smooth basis functions and simulated annealing. *Atmospheric Environment, 30*, 929–940.

Duffin, K., McGettrick, A., Johnstone, W., Stewart, G., & Moodie, D. G. (2007). Tunable diode laser spectroscopy with wavelength modulation: A calibration-free approach to the recovery of absolute gas absorption line shapes. *Journal of Lightwave Technology, 25*, 3114–3125.

Emmert, J., Baroncelli, M., Kley, S.v., Pitsch, H., & Wagner, S. (2019). Axisymmetric linear hyperspectral absorption spectroscopy and residuum-based parameter selection on a counter flow burner. *Energies, 12*(14), 2786.

Emmert, J., Blume, N. G., Dreizler, A., & Wagner, S. (2018). Data analysis and uncertainty estimation in supercontinuum laser absorption spectroscopy. *Scientific Reports, 8*(1), 1–16.

Emmert, J., Grauer, S. J., Wagner, S., & Daun, K. J. (2019). Efficient Bayesian inference of absorbance spectra from transmitted intensity spectra. *Optics Express, 27*(19), 26893–26909.

Emmert, J., Wagner, S., & Daun, K. J. (2021). Quantifying the spatial resolution of the maximum a posteriori estimate in linear, rank-deficient, Bayesian hard field tomography. *Measurement Science and Technology, 32*(2), Article 025403.

Enemali, G., Zhang, R., McCann, H., & Liu, C. (2022). Cost-effective quasi-parallel sensing instrumentation for industrial chemical species tomography. *IEEE Transactions on Industrial Electronics, 69*, 2107–2116.

Fischer, M. L., Price, P. N., Thatcher, T. L., Schwalbe, C. A., Craig, M. J., Wood, E. E., et al. (2001). Rapid measurements and mapping of tracer gas concentrations in a large indoor space. *Atmospheric Environment, 35*, 2837–2844.

Fisher, E. M., Tsekenis, S., Yang, Y., Chighine, A., Liu, C., Polydorides, N., et al. (2019). A custom, high-channel-count data acquisition system for chemical species tomography of aero-jet engine exhaust plumes. *IEEE Transactions on Instrumentation and Measurement, 2019*(2), 549–558.

Goldenstein, C. S., Spearrin, R. M., Jeffries, J. B., & Hanson, R. K. (2017). Infrared laser-absorption sensing for combustion gases. *Progress in Energy and Combustion Science, 60*, 132–176.

Gordon, R., Bender, R., & Herman, T. G. (1970). Algebraic reconstruction techniques (ART) for three-dimensional electron microscopy and X-ray photography. *Journal of Theoretical Biology, 29*, 471–481.

Gordon, I., Rothman, L., Hill, C., Kochanov, R. V., Tan, Y., Bernath, P., et al. (2017). The HITRAN2016 molecular spectroscopic database. *Journal of Quantitative Spectroscopy & Radiative Transfer, 203*, 3–69.

Grauer, S. J., Emmert, J., Sanders, S. T., Wagner, S., & Daun, K. J. (2019). Multiparameter gas sensing with linear hyperspectral absorption tomography. *Measurement Science and Technology, 30*(10), 105401.

Grauer, S. J., Hadwin, P. J., & Daun, K. J. (2016). Bayesian approach to the design of chemical species tomography experiments. *Applied Optics, 55*(21), 5772–5782.

Grauer, S. J., Hadwin, P. J., & Daun, K. J. (2017). Improving chemical species tomography of turbulent flows using covariance estimation. *Applied Optics, 56*, 3900–3912.

Grauer, S. J., Hadwin, P. J., Sipkens, T. A., & Daun, K. J. (2017). Measurement-based meshing, basis selection, and prior assignment in chemical species tomography. *Optics Express, 25*(21), 25135–25148.

Grauer, S. J., & Steinberg, A. M. (2020). Linear absorption tomography with velocimetry (LATV) for multiparameter measurements in high-speed flows. *Optics Express, 28*(22), 32676–32692.

Grauer, S. J., Unterberger, A., Rittler, A., Daun, K. J., Kempf, A. M., & Mohri, K. (2018). Instantaneous 3D flame imaging by backwards-oriented Schlieren tomography. *Combustion and Flame, 196*, 284–299.

Hadamard, J. S. (1923). *Lectures on cauchy's problem in linear partial differential equations (lectures given at Yale University)*. New Haven, CT: Yale University Press.

Hansen, P. C. (1999). *Rank deficient and discrete ill-posed problems: Numerical aspects of linear inversion*. Philadelphia PA: SIAM.

Hansen, P. C., & Jørgensen, J. S. (2018). AIR tools II: Algebraic iterative reconstruction methods, improved implementation. *Numerical Algorithms, 79*(1), 107–137.

Hindle, F. P., Carey, S. J., Ozanyan, K., Winterbone, D. E., Clough, E., & McCann, H. (2001). Measurement of gaseous hydrocarbon distribution by a near-infrared absorption tomography system. *Journal of Electronic Imaging, 10*, 593–600.

Huang, A., Cao, Z., Zhao, W., Zhang, H., & Xu, L. (2020). Frequency-division multiplexing and main peak scanning WMS method for TDLAS tomography in flame monitoring. *IEEE Transactions on Instrumentation and Measurement, 69*(11), 9087–9096.

Huang, J., Liu, H., & Cai, W. (2019). Online in situ prediction of 3-D flame evolution from its history 2-D projections via deep learning. *Journal of Fluid Mechanics, 875*, R2.

Huang, J., Liu, H., & Dau, J. C. (2018). Reconstruction for limited-data nonlinear tomographic absorption spectroscopy via deep learning. *Journal of Quantitative Spectroscopy & Radiative Transfer, 218*, 187–193.

Huang, Y., & Yang, V. (2009). Dynamics and stability of lean premixed swirl-stabilized combustion. *Progress in Energy and Combustion Science, 35*, 293-264.

Jeon, M.-G., Deguchi, Y., Kamimoto, T., Doh, D.-H., & Cho, G.-R. (2017). Performances of new reconstruction algorithms for CT-TDLAS (computer tomography-tunable diode laser absorption spectroscopy). *Applied Thermal Engineering, 115*, 1148–1160.

Jin, Y., Zhang, W., Song, Y., Qu, X., Li, Z., Ji, Z., et al. (2019). three-dimensional rapid flame chemiluminescence tomography via deep learning. *Optics Express, 27*, 27308–27334.

Karagiannopoulos, S., Cheadle, E. M., Wright, P., & McCann, H. (2012). Multiwavelength diode-laser absorption spectroscopy using external intensity modulation by semiconductor optical amplifiers. *Applied Optics, 51*, 8057–8067.

Kasyutich, V., & Martin, P. (2011). Towards a two-dimensional concentration and temperature laser absorption tomography sensor system. *Applied Physics B, 102*(1), 149–162.

Klingbeil, A. E., Jefferies, J. B., & Hanson, R. K. (2007). Temperature-dependent mid-IR absorption spectra of gaseous hydrocarbons. *Journal of Quantitative Spectroscopy & Radiative Transfer, 107*, 407–420.

Krämer, H., Einecke, S., Schulz, C., Sick, V., Nattrass, S. R., & Kitching, J. S. (1998). Simultaneous mapping of the distribution of different fuel volatility classes using tracer-LIF tomography in an IC engine. *SAE Transactions*, 1049–1060.

Kranendonk, L. A., & Sanders, S. T. (2005). Optical design in beam steering environments with emphasis on laser transmission measurements. *Applied Optics, 44*, 6762–6772.

LeCun, Y., Bengio, Y., & Hinton, G. (2015). Deep learning. *Nature, 521*, 436–444.

Lengden, M., Stewart, G., Johnstone, W., Upadhyay, A., Wilson, D., Polydorides, N., et al. (2020). Recent progress in the development of a chemical species tomographic imaging system to measure carbon dioxide emissions from large-scale commercial aero-engines. *OSA Optical Sensors and Sensing Congress: Applied Industrial Spectroscopy 2020.* Article JM2F.3. https://doi.org/10.1364/AIS.2020.JM2F.3

Linne, M. A. (2002). *Spectroscopic measurement: An introduction to the fundamentals*. London: Academic Press.

Liu, C., Cao, Z., Lin, Y., Xu, L., & McCann, H. (2018). On-line cross-sectional monitoring of a swirling flame using TDLAS tomography. *IEEE Transactions on Instrumentation and Measurement, 67*(6), 1338–1348.

Liu, C., Tsekenis, S. A., Polydorides, N., & McCann, H. (2019). Toward customized spatial resolution in TDLAS tomography. *IEEE Sensors Journal, 19*, 1748–1755.

Liu, C., & Xu, L. (2019). Laser absorption spectroscopy for combustion diagnosis in reactive flows: A review. *Applied Spectroscopy Reviews, 54*, 1–44.

Liu, C., Xu, L., Chen, J., Cao, Z., Lin, Y., & Cai, W. (2015). Development of a fan-beam TDLAS-based tomographic sensor for rapid imaging of temperature and gas concentration. *Optics Express, 23*, 22494–22511.

Liu, X., Zhou, X., Jeffries, J. B., & Hanson, R. K. (2007). Experimental study of H2O spectroscopic parameters in the near-IR (6940-7440 cm̂-1) for gas sensing applications at elevated temperature. *Journal of Quantitative Spectroscopy & Radiative Transfer, 103*, 565–577.

Ma, L., Cai, W., Caswell, A. W., Kraetschmer, T., Sanders, S. T., Roy, S., et al. (2009). Tomographic imaging of temperature and chemical species based on hyperspectral absorption spectroscopy. *Optics Express, 17*(10), 8602–8613.

Ma, L., Li, X., Sanders, S. T., Caswell, A. W., Roy, S., Plemmons, D. H., et al. (2013). 50-kHz-rate 2D imaging of temperature and H_2O concentration at the exhaust plane of a J85 engine using hyperspectral tomography. *Optics Express, 21*(1), 1152–1162.

Mihalcea, R. M., Baer, D. S., & Hanson, R. K. (1997). Diode laser sensor for measurements of CO, CO_2, and CH_4 in combustion flows. *Applied Optics, 36*, 8745–8752.

Nadir, Z., Brown, M. S., Comer, M. L., & Bouman, C. A. (2015). Gaussian mixture prior models for imaging of flow cross sections from sparse hyperspectral measurements. In *2015 IEEE global conference on signal and information processing (GlobalSIP).*

Ozanyan, K. B., Wright, P., Stringer, M. R., & Miles, R. E. (2011). Hard-field THz tomography. *IEEE Sensors Journal, 11*, 2507–2513.

Pal, S., & McCann, H. (2011). Auto-digital gain balancing: A new detection scheme for high-speed chemical species tomography of minor constituents. *Measurement Science and Technology, 22*, 115304.

Pal, S., Ozanyan, K. B., & McCann, H. (2008). A computational study of tomographic measurement of carbon monoxide at minor concentrations. *Measurement Science and Technology, 19*, 094018.

Polydorides, N., Tsekenis, S. A., Fisher, E. F., Chighine, A., McCann, H., Dimiccoli, L., et al. (2018). Constrained models for optical absorption tomography. *Applied Optics, 57*, B1.

Polydorides, N., Tsekenis, S. A., McCann, H., Prat, V.-D. A., & Wright, P. (2016). An efficient approach for limited-data chemical species tomography and its error bounds. *Proc. R. Soc. A., 472*, 20150875.

Price, P. N., Fischer, M. L., Gadgil, A. J., & Sextro, R. G. (2001). An algorithm for real-time tomography of gas concentrations using prior information about spatial derivatives. *Atmospheric Environment, 35*, 2827–2835.

Qu, Q., Cao, Z., Xu, L., Liu, C., Chang, L., & McCann, H. (2019). Reconstruction of two-dimensional velocity distribution in scramjet by laser absorption spectroscopy tomography. *Applied Optics, 58*(1), 205–212.

Qu, Z., Holmgren, p., Skoglund, N., Wagner, D. R., Brostrom, M., & Schmidt, F. M. (2018). Distribution of temperature, H_2O and atomic potassium during entrained flow biomass combustion - coupling in situ TDLAS with modeling approaches and ash chemistry. *Combustion and Flame, 188*, 488–497.

Ren, T., Modest, M. F., Fateev, A., Sutton, G., Zhao, W., & Rusu, F. (2019). Machine learning applied to retrieval of temperature and concentration distributions from infrared emission measurements. *Applied Energy, 252*, 113448.

Rieker, G. B., Giorgetta, F. R., Swann, W. C., Kofler, J., Zolot, A. M., Sinclair, L. C., et al. (2014). Frequency-comb-based remote sensing of greenhouse gases over kilometer air paths. *Optica, 1*(5), 290–298.

Rieker, G. B., Jeffries, J. B., & Hanson, R. K. (2009). Calibration-free wavelength-modulation spectroscopy for measurementsof gas temperature and concentration in harsh environments. *Applied Optics, 48*, 5546–5560.

Sanders, S. T., Wang, J., Jeffries, J. B., & Hanson, R. K. (2001). Diode-laser absorption sensor for line-of-sight gas temperature distributions. *Applied Optics, 40*, 4404–4415.

Santoro, R. J., Semerjian, H. G., Emmerman, P. J., & Goulard, R. (1981). Optical tomography for flow field diagnostics. *International Journal of Heat and Mass Transfer, 24*, 1139–1150.

Sick, V. (2013). High speed imaging in fundamental and applied combustion research. *Proceedings of the Combustion Institute, 34*, 3509–3530.

Syred, N. (2006). A review of oscillation mechanisms and the role of the precessing vortex core (PVC) in swirl combustion systems. *Progress in Energy and Combustion Science, 32*, 93–161.

Tancin, R., Spearrin, R. M., & Goldenstein, C. S. (2019). 2D mid-infrared laser-absorption imaging for tomographic reconstruction of temperature and carbon monoxide in laminar flames. *Optics Express, 27*(10), 14184–14198.

Terzija, N., Davidson, J. L., Garcia-Stewart, C. A., Wright, P., Ozanyan, K. B., Pegrum, S., et al. (2008). Image optimization for chemical species tomography with an irregular and sparse beam array. *Measurement Science and Technology, 19*, 094007 (13 pages).

Terzija, N., Karagiannopoulos, S., Begg, S., Wright, P., Ozanyan, K., & McCann, H. (2015). Tomographic imaging of the liquid and vapour fuel distributions in a single-cylinder direct injection gasoline engine. *International Journal of Engine Research, 16*, 565–579.

Terzija, N., & McCann, H. (2011). Wavelet-based image reconstruction for hard-field tomography with severely limited data. *IEEE Sensors Journal, 11*, 1885–1893.

Tikhonov, A. N., & Arsenin, V. Y. (1977). *Solution of ill-posed problems*. Washington DC: Winston and Sons.

Torniainen, E. D., & Gouldin, F. C. (1998). Tomographic reconstruction of 2-D absorption coefficient distributions from a limited set of infrared absorption data. *Combustion Science and Technology, 131*, 85–105.

Torniainen, E. D., Hinz, A. K., & Gouldin, F. C. (1998). Tomographic analysis of unsteady, reacting flows: Numerical investigation. *AIAA Journal, 36*, 1270–1278.

Tsekenis, S. A., & Polydorides, N. (2017). Optical access schemes for high speed and spatial resolution optical absorption tomography in energy engineering. *IEEE Sensors Journal, 17*(24), 8072–8080.

Tsekenis, S. A., Polydorides, N., Fisher, E. F., Chighine, A., Wilson, D., Humphries, G. S., et al. (2016). Chemical species tomography of carbon dioxide. *8th world congr. Industrial process tomography. WCIPT8, Iguassu falls, Brazil* (p. B17). ISIPT. ISBN 978-0-853-16349-7.

Tsekenis, S. A., Ramaswamy, K. G., Tait, N., Hardalupas, Y., Taylor, A., & McCann, H. (2018). Chemical species tomographic imaging of the vapour fuel distribution in a compression-ignition engine. *International Journal of Engine Research, 19*(7), 718–731.

Tsekenis, S. A., Tait, N., & McCann, H. (2015). Spatially resolved and observer-free experimental quantification of spatial resolution in tomographic images. *Review of Scientific Instruments, 86*, 035104.

Tsekenis, S. A., Wilson, D., Lengden, M., Hyvonen, J., Leinonen, J., Shah, A., et al. (2017). Towards in-cylinder chemical species tomography on large-bore IC engines with pre-chamber. *Flow Measurement and Instrumentation, 53*, 116–125.

Twynstra, M. G., & Daun, K. J. (2013). Laser-absorption tomography beam arrangement optimization using resolution matrices. *Applied Optics, 51*, 7059–7068.

Upadhyay, A., Lengden, M., Wilson, D., Humphries, G. S., Crayford, A. P., Pugh, D. G., et al. (2018). A new RAM normalized 1f-WMS technique for the measurement of gas parameters in harsh environments and a comparison with 2f/1f. *IEEE Photonics J, 10*(6), 6804611.

Wang, Z., Deguchi, Y., Kamimoto, T., Tainaka, K., & Tanno, K. (2020). Pulverized coal combustion application of laser-based temperature sensing system using computed tomography - tunable diode laser absorption spectroscopy (CT-TDLAS). *Fuel, 268*, 117370.

Wang, F., Wu, Q., Huang, Q., Zhang, H., Yan, J., & Cen, K. (2015). Simultaneous measurement of 2-dimensional H2O concentration and temperature distribution in premixed methane/air flame using TDLAS-based tomography technology. *Optics Communications, 346*, 53–63.

Wei, C., Schwarm, K. K., Pineda, D. I., & Spearrin, R. M. (2020). Deap neural network inversion for 3D laser absorption imaging of methane in reacting flows. *Optics Letters, 45*, 2447–2450.

Wood, M. P., & Ozanyan, K. B. (2014). Simultaneous temperature, concentration, and pressure imaging of water vapor in a turbine engine. *IEEE Sensors Journal, 15*(1), 545–551.

Wright, P., McCormick, D., Ozanyan, K. B., Johnson, M., Black, J., Fisher, E., et al. (2015). *Progress towards non-intrusive optical measurement of gas turbine exhaust species distributions*. USA: IEEE Aerospace Conference.

Wright, P., Ozanyan, K. B., Carey, S. J., & McCann, H. (2005). Design of high-performance photodiode receivers. *IEEE Sensors Journal, 5*, 281–288.

Wright, P., Terzija, N., Davidson, J. L., Garcia-Castillo, S., Pegrum, S., Colbourne, S., et al. (2010). High speed chemical species tomography in a multi-cylinder automotive engine. *Chemical Engineering Journal, 158*, 2–10.

Yu, T., Tian, B., & Cai, W. (2017). Development of a beam optimization method for absorption-based tomography. *Optics Express, 25*, 5982–5999.

Zhao, W., Xu, L., Huang, A., Gao, X., Luo, X., Zhang, H., et al. (2020). A WMS based TDLAS tomographic system for distribution retrievals of both gas concentration and temperature in dynamic flames. *IEEE Sensors Journal, 20*(8), 4179–4188.

X-ray computed tomography

6

Uwe Hampel [1,2]
[1]Helmholtz-Zentrum Dresden-Rossendorf, Dresden, Germany; [2]Technische Universität Dresden, Dresden, Germany

6.1 Introduction

The discovery of X-rays by Wilhelm Conrad Röntgen in 1895 was the starting point of an enormously successful development of medical and technical imaging technologies, which has continued through the present day. Röntgen found by chance, in his experimental studies with cathode rays, a new type of electromagnetic radiation. He found that this heretofore unknown radiation penetrates most materials, which are otherwise opaque for visible light, and that its attenuation is rather linear. The new X-rays were found to excite light emission in inorganic crystals and to expose photography plates and films as visible light does. The new radiation appeared particularly attractive for imaging of internal body structures. Röntgen's discovery rapidly led to the development of the first X-ray imaging devices, including X-ray tubes, X-ray films, and, later, X-ray detectors. With the advancements in computer technology in the 1960 and 1970s came another quantum leap in X-ray imaging technology. Computed tomography (CT) for the first time provided physicians with superposition-free cross-sectional images of internal body structures. Again this technique was pioneered by the use of X-rays.

Today X-ray CT is the most widely used imaging technique in medicine and nondestructive testing. In recent years, it has also found its way into process diagnostics. One particular driver is the continuing interest of scientists and engineers to study multiphase flows in energy and process engineering applications (Heindel, 2011). Such flows of gases, liquids, and solids occur in chemical reactors, in coolant circuits of power plants, in heat exchangers and hydrodynamic machines, and also in pipelines and process equipment of the petroleum industry. However, process engineering is a field and market with needs and demands that hardly coincide with those of medicine and nondestructive testing. The main commonality is the need for a nonintrusive spatially resolving measurement technique that gives insight into otherwise nonaccessible and nondetectable structures and processes. While in medicine the key requirement is good contrast for tissue structures in the millimeter and submillimeter range at as low an energy dose as possible, in nondestructive testing the major focus is on highest spatial resolution, which may be compromised by long scanning times. The permissible energy dose here has almost no limitation. Process tomography poses completely different challenges. Industrial process conditions are often hostile and hazardous. Typical are high pressures and temperatures as well as aggressive fluids. Process vessels are commonly large, have thick metallic walls, and often have complex internal

Industrial Tomography. https://doi.org/10.1016/B978-0-12-823015-2.00029-7

structures, which limit the accessibility for instrumentation. Moreover, transient flow conditions in processes make fast imaging capabilities desirable. Radiation protection measures must be provided at the place of installation, which is often difficult and expensive, especially when hard X-rays are being used. Yet the costs of process measurement techniques are in general required to be low. Here, radiation-based measurement techniques have clear disadvantages. X-ray equipment is often bulky and expensive, and radiation protection efforts additionally increase complexity and costs. However, with respect to spatial and temporal resolution as well as capability to penetrate denser construction materials, X-rays are superior to any other type of signal carrier. Therefore, continued progress has been made in the recent past to develop this technology for laboratory and industrial applications.

X-ray attenuation in a material depends on the material's density and atomic number. Hence, X-rays can discriminate structures via their different attenuation, which mostly comes from different densities. Therefore, X-ray imaging is well applicable for the visualization of phase distributions in gas–liquid two-phase flow, for instance, in chemical multiphase reactors, or for particle flows in a fluidized bed. Another example is the measurement of three-phase gas–water–oil flow in a pipeline, for example, in mineral oil processing. In some rare cases, X-rays may disclose properties other than attenuation. Phase-contrast imaging with coherent X-rays enhances boundaries between areas of minute refractive index differences and is of great value in X-ray microscopy and microtomography (Waelchli, von Rohr, Stampanoni, 2005). On the other hand, high-energy X-rays may be used to excite secondary particles or radiation, which can be used to image certain other material properties than density and atomic number. As in medicine, for process diagnostics one therefore needs to critically assess if X-rays are an appropriate means to obtain the required information from the process. In multiphase flows, X-rays are helpful to discriminate phases as long as they possess sufficient radiological contrast. For the discrimination between gas, liquid, and solid content of a flow, X-rays are superiorly suited. Discriminating solid phases and liquid phases is also possible. When density differences between the flow constituents are low, it might be helpful to consider lower X-ray energies and even multienergy techniques. For lower X-ray energies, the photoabsorption becomes the dominant process of radiation–matter interaction; this is unlike Compton scattering for higher energies, strongly dependent on atomic number and therefore on elementary species composition. However, since lower X-ray energies have less overall penetration capability, there are also physical limits to this approach. In cases where contrast of species is too low, addition of contrast agents containing chemical elements of high atomic number, such as barium or iodine, is an option. Moreover, such contrast agents can be deliberately used as flow tracers, for instance, to study dispersion effects in multiphase flows. Limiting factors here are the stability and solubility of the tracer substances under given thermodynamic and chemical conditions. Also, as a matter of course, particles of higher attenuation may be added to particulate flows to gain contrast (Seeger, Kertzscher, Affeld, & Wellnhofer, 2003). Wherever there is no variation in the attenuation properties of the process constituents, X-rays are of no use. This applies, for instance, to pure single-phase flows, gas mixture flows, and very dilute multiphase flows with dispersed particles, droplets, or bubbles. There, as in medicine, complementary methods must be considered. Such methods may be emission tomography, electrical tomography, magnetic tomography, optical tomography, or magnetic resonance imaging, introduced in other chapters of this book.

6.2 Variants of X-ray computed tomography for process applications

Originally, CT imaging was understood as the reconstruction of a parameter distribution within a thin slice of an object from a set of line integrals of this very parameter. Theoretically, all possible line integral values are needed for an accurate reconstruction. But as with any frequency resolution-limited technical sampling process, a subset of line integrals with a homogeneous distribution is sufficient for an approximate reconstruction. In X-ray tomography, the line integrals are obtained by linear attenuation measurements with a set of narrow X-ray beams (Kak & Slaney, 1988). Practically, such beams are formed by the emission area of a focal spot in an X-ray tube and the active area of a radiation detector on the other side of the object. To obtain a "full set" of line integrals, the X-ray tube and detector must be subsequently placed into different positions in space. This can be achieved in multiple ways. Fig. 6.1 shows different scanning principles of increasing complexity. Historically, the configurations were classified as different generations of X-ray tomography scanners (Kalender, 2005). The simplest arrangement is the one with a single beam, for which it is necessary to compile the line integrals sequentially by a combination of rotational and translational movements of the source–detector compound. An improvement in scanning speed can be achieved by using multiple sources and detectors. This gives a large

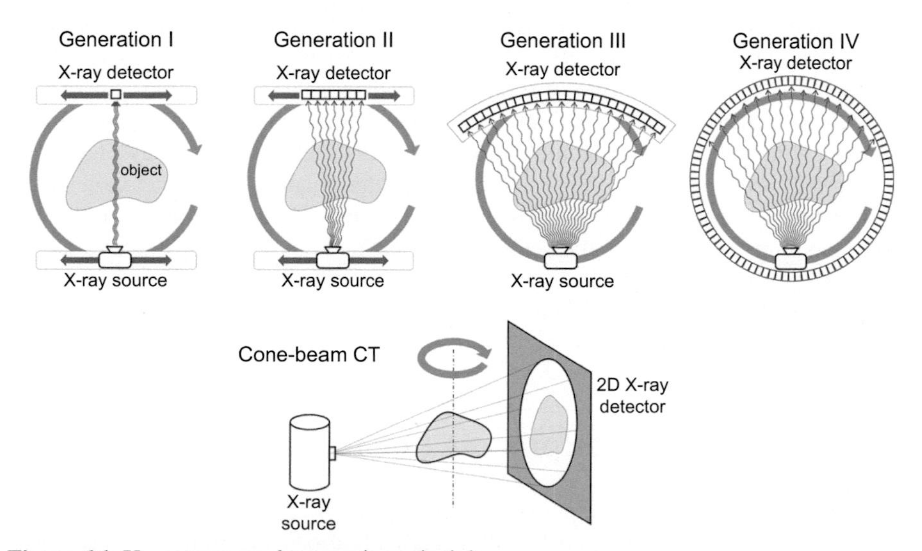

Figure 6.1 X-ray tomography scanning principles.

number of potential spatial–temporal scanning modes. Usually, multiple detectors are more affordable than multiple sources. In generation II scanners, more detectors are used but still the translation–rotation scanning has to be sustained. A rotating single source and multiple detector arrangement of generation III is the most common scanner configuration in X-ray CT today. Here, a single X-ray source and an arc detector, made of many small detector elements, are used. The radiation fan fully encloses the object, such that only rotation of the source–detector compound, or alternatively the object, is required. In principle, this setup can be extended toward incorporation of a second or even more X-ray sources, for instance, sources with different X-ray energy. Generation III scanners have been widely used in multiphase flow studies, for example, to measure gas holdups in gas–liquid contactors (Toye, Marchot, Crine, Pelsser, & L'Homme, 1998), solids distributions in fluidized beds (Deza, Franka, Battaglia, & Heindel, 2009), or void distributions in heated rod bundle assemblies (Inoue et al., 1995). The generation IV arrangement foresees a rotating source and a fixed circular detector arrangement. For mechanically moving X-ray sources, this concept has no particular advantage, so it did not find its way into existing technical solutions. However, it is inherently required for electron beam X-ray tomography, as will be shown further below.

So far, the discussed arrangements allow a two-dimensional (2D) reconstruction of the linear attenuation values in exactly the plane where source and detector are operated. Commonly, this is referred to as transversal scanning. 3D imaging would mathematically require a full set of line integrals in 3D space, which is technically very difficult to achieve (Grangeat, 1991). It can be shown, however, that very good approximation of this ideal sampling scheme can be made by using a rotating compound of a point X-ray source and a 2D X-ray detector, as is also shown in Fig. 6.1. Since the detector records X-ray attenuation along lines within a radiation cone, this type of tomography is frequently referred to as cone-beam tomography. Because of its approximate nature, reconstruction will produce some deviations from the physical parameter distribution, especially in the off-plane regions of the image cube. This has to be considered when using the data for further analysis.

The above classification of tomography arrangements shall serve as a guide to what principle arrangements and scanning schemes are currently used in X-ray process tomography. As we will see in the further course of this chapter, both generation III and IV scanning principles as well as transversal and cone-beam tomography do exist. The salt in the soup, of course, are the details of existing and innovative technical solutions for X-ray generation, X-ray detection, and scanning principles. Hence, the following sections will be devoted to the basic physical and technical principles involved in X-ray process tomography.

6.3 X-ray sources for process tomography

Physically, X-rays emerge whenever charged particles of energy larger than about 1 keV are stopped in a material. In technical X-ray tubes, electrons are accelerated by an electrostatic field over a short distance of a few centimeters and are then shot into the anode material. On a microscopic scale, deceleration of electrons in the Coulomb field of the target nuclei leads to emission of Bremsstrahlung with a continuous-emission energy spectrum between zero and the maximum electron energy $E_{\max} = eU_{\text{acc}}$, which is the product of unit charge and acceleration voltage. In X-ray tubes, the lower part of the spectrum is damped due to absorption of low-energy photons in the anode and in the wall of the tube. Within this energy spectrum, additional emission lines from characteristic radiation appear. This characteristic radiation arises when impinging electrons knock out bound inner shell electrons of target atoms and outer shell electrons refill the inner shell vacancies under emission of fluorescent X-ray photons, whose energy corresponds to the energy gap between the respective electron orbitals. In many process tomography applications, conventional X-ray tubes with such a polyenergetic spectrum are used (Fig. 6.2).

Fig. 6.3 shows schematic drawings of a fixed anode and a rotating anode X-ray tube along with photographs of two technical X-ray tube assemblies. At the cathode side of an X-ray tube, free electrons are generated by thermionic emission from a heated tungsten filament. Filament current control allows adjustment of the electron current and hence X-ray power. The electrical field in front of the cathode is shaped by a Wehnelt cylinder in such a way that the electrons are focused into a small spot on the opposite anode. For tubes with smaller loss power, that is, up to 10 kW, a fixed anode configuration is commonly used. There, the anode is a composite, consisting of a heavy metal conversion target embedded in a bulk copper block for good conductive heat transfer to the environment. In rotating anode tubes, which are made for higher power dissipation, the anode is a round plate-like composite body. Its upper layer is a rhenium–tungsten–molybdenum alloy with a high melting point. Underneath is a pure molybdenum layer, which gives the compound good thermomechanical strength, and further below is a graphite layer, which provides high temporary heat storage capacity. Radial slits in the anode plate provide increased thermomechanical resistance under the high

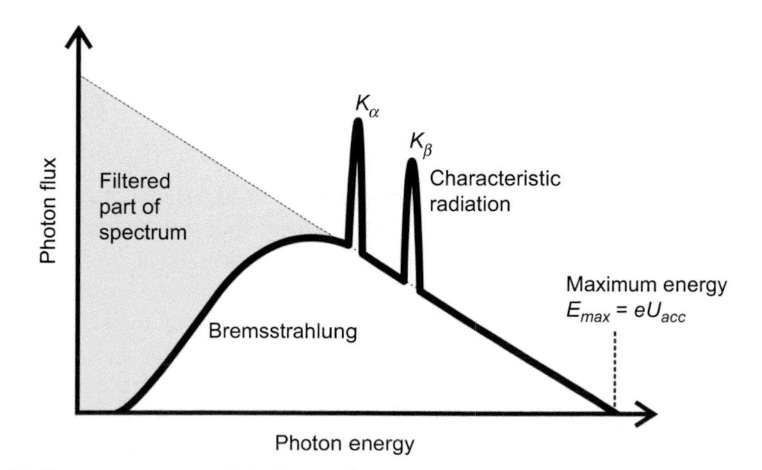

Figure 6.2 Energy spectrum of an X-ray tube.

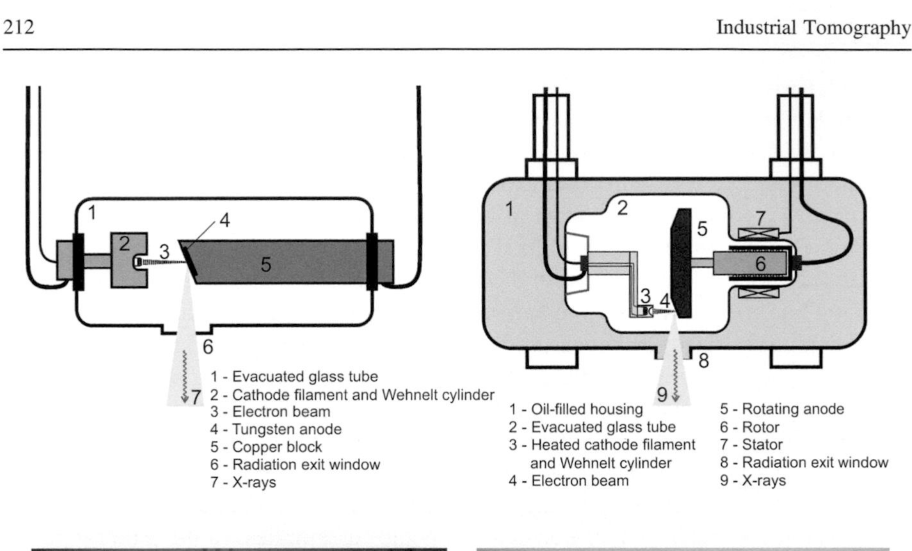

Figure 6.3 X-ray tubes: the top row shows schematics of a fixed anode and a rotating anode X-ray assembly. The bottom row depicts an industrial 450 kV X-ray radiator (left) and a 140 kV/400 W microfocus X-ray tube (right).

centrifugal loads, which occur when the anode is rotating. During operation, the anode plate is rotated at high speed of up to 9000 rpm (in some tubes even higher). It is driven by an electromotor whose stator is outside the tube housing behind the anode. Anode and cathode are supplied with power from a high-voltage generator. To ease electrical isolation, tubes are typically operated in a bipolar mode with $\frac{\pm 1}{2} U_{\text{acc}}$ at both terminals. The focal spot of an X-ray tube has the demagnified bar-like shape of the cathode filament. In the direction of the X-ray window, this focal spot appears reduced in length due to the oblique face of the anode plate. The typical physical size of the spot in industrial X-ray tubes is between $0.5 \times 6\ \text{mm}^2$ and $1.2 \times 12\ \text{mm}^2$. The tube's electrical loss power, which can be up to more than 100 kW, is dissipated mainly as heat in this area. To cope with the high power density, different heat transfer and storage mechanisms cooperate inside the tube. By rotating the anode, the heat produced in the focal spot is distributed across a ring. Heat conduction into the anode metal and subsequently into the graphite support serves to reduce thermal power density on the anode surface. The main heat removal mechanism is radiative heat transfer into a cooling oil bath outside the glass envelope of the tube, which improves with the fourth power of temperature (Stefan–Boltzmann law). A smaller part is conductively removed via the anode shaft. Technical X-ray tubes are moreover enclosed by a protective lead housing with an aluminum window for X-ray extraction. X-ray tube supply electronics allow adjustment and control of tube power, tube voltage, and total exposure time. In addition, it should be noted that modern medical X-ray CT scanners utilize rotating envelope tubes. In such tubes, the anode is firmly connected with the metallic tube envelope such that heat is transferred efficiently by conductance to the outer oil bath. There, not only the anode but the whole envelope is rotated. The cathode in this arrangement is on the tube axis and a magnetic coil arrangement deflects the electrons into the focal spot on the anode plate (for more details, see Schard et al., 2004).

Using industrial X-ray tubes for X-ray inspection and process tomography is quite common. Those types of radiators are well developed and optimized for many applications. The energy range of such tubes is from a few kV up to 450 kV tube voltage. The upper limit is mainly given by increasing technical complexity and effort for high-voltage isolation. High-power X-ray tubes are available with powers above 100 kW and special thermal designs. For high-resolution imaging, tubes have to achieve small focal spot sizes. Such microfocus tubes are available with focal spot sizes down to 1 μm. Because of limited power dissipation capacity in the anode material, the average X-ray power of such tubes is accordingly small, between a few μW and a few hundred watts. Therefore, high-resolution CT imaging with X-ray tubes is always connected with long scanning times, which in turn means that only time-averaged phase fraction distributions of flows can be obtained (Moreno-Atanasio, Williams, & Jia, 2010).

Using a commercial X-ray tube for tomography applications inevitably requires that either the tube or the object of investigation has to be mechanically manipulated to generate the projection data required for reconstruction. Fast tomography, as will be discussed further below, has to abandon this principle, since fast mechanical motion of a heavy X-ray tube or an object has strict physical limits. In principle, there exist three options to avoid mechanical motion (Johansen, Hampel, & Hjertaker, 2010). One way is to use multiple switched X-ray tubes, which requires special nonconventional tube designs. Such solutions were, for instance, proposed by Hori (Hori, Fujimoto, Kawanishi, & Nishikawa, 1998). Another choice is to arrange a limited number of collimated X-ray tubes in such a way that they form disjointed X-ray fans in space (Mudde, 2011). This assures that the detectors placed in the same plane receive radiation from only one source and hence the sources can be operated simultaneously. However, the severe undersampling of projections in this case leads to images with rather low resolution.

Another option to realize a fast-moving X-ray source is to use a steered electron beam (Fischer et al., 2008). This, however, requires a technically different concept of an X-ray source, namely an electron beam generator coupled with a suitable X-ray target. Such a scanner is shown in Fig. 6.4. As in conventional X-ray tubes, the electron current is produced by a heated cathode. A Wehnelt modulator is used to shape the electrostatic field to form the electron beam in front of the anode. Moreover, the Wehnelt electrode is additionally used to adjust the electron current by a

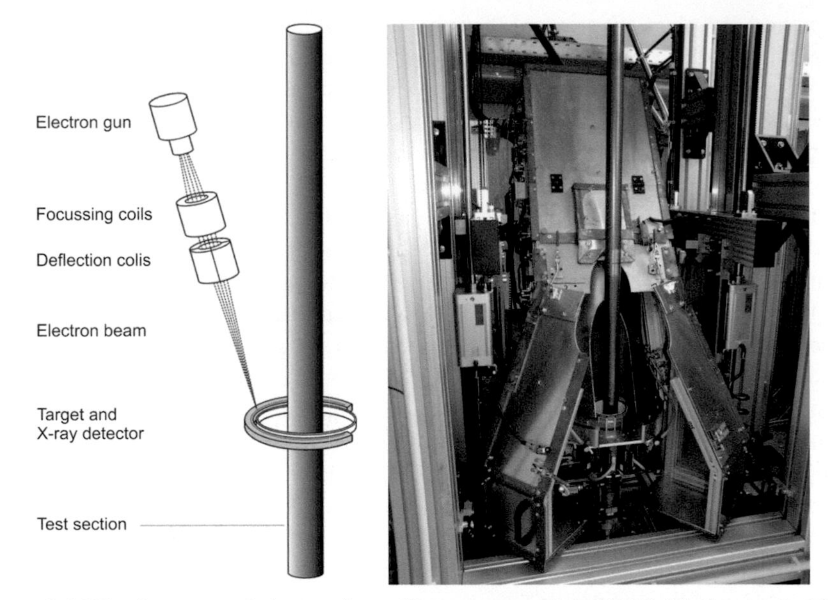

Figure 6.4 Ultrafast scanned electron beam X-ray tomograph ROFEX (Fischer et al., 2008).

negative counter-field that controls the electron penetration to the anode. It can be used either to adjust the beam current or to switch the beam current quickly on and off. The anode, unlike in X-ray tubes, has a penetration hole, which allows the electrons to pass the anode and travel through the field-free space behind it. Here the electrons are no longer accelerated but move toward a target in the tomograph head at constant near-relativistic speed $v_{el} = \sqrt{2eU_{acc}/m_e}$. Behind the anode, an electromagnetic lens system is arranged. This system comprises various electron optics elements, which are required for forming, focusing, and deflecting the beam. Among them is an astigmator to remove beam shape asymmetries, a static focusing coil, lenses for dynamic beam refocusing, and an x−y deflection system for beam steering. The target is a circular tungsten body located at the very bottom of the scanner head. By means of the electron beam optics, the beam is periodically run across the target. Though simple in principle, this concept requires numerous additional technical features, such as beam diagnostics, vacuum system, target cooling, advanced temperature control, safety beam dump, and X-ray power reference measurements, whose descriptions are beyond the scope of this book.

For reasons of completeness, it should be mentioned that higher X-ray powers can be achieved with other types of electron accelerators. For X-ray energies up to 10 MeV electrostatic accelerators of dynamitron type, traveling wave tube accelerators, microtrons, or betatrons may be employed (Hampel et al., 2012). For highest photon flux, coherent synchrotron radiation sources are of special importance (Berg et al., 2013). However, such X-ray sources are today rarely used for process tomography applications, mainly due to their complexity, immobility, and costs.

6.4 X-ray detectors

For X-ray detection in imaging applications, either scintillator-based detectors or room temperature semiconductor detectors are being used (Del Sordo et al., 2009; Shefer et al., 2013; van Eijk, 2002). The physical conversion principles are illustrated in Fig. 6.5. In an inorganic scintillator detector, interacting X-ray photons excite electrons in the valence band of the scintillation crystal lattice toward the conduction band or form excitons (loosely coupled electron–hole pairs). After some time of existence, these electrons or excitons recombine. Activators, which are brought into the lattice in small quantities, deliver intermediate electron state levels at which recombination results in emission of fluorescent light. Examples for technically used scintillator materials are sodium iodine [NaI(Tl)] or cesium iodine [CsI(Tl)] (doped with thallium as activator), bismuth germanium oxide (BGO), cadmium tungstate ($CdWO_4$), or polycrystalline ceramic rare-earth scintillators. In pixel detectors, each scintillator crystal is coupled to a photodiode, which converts the light into an electrical charge.

A direct way of converting X-ray photons into electrical charge is realized in room temperature semiconductor detectors, such as cadmium telluride (CdTe) or cadmium

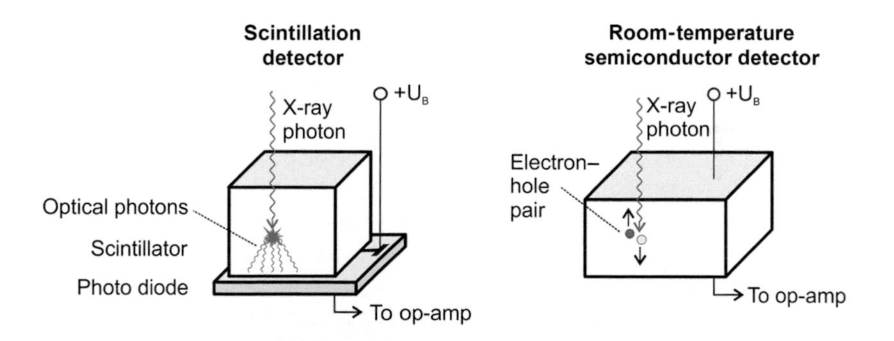

Figure 6.5 X-ray detector principles (top) and photography of a 2D X-ray flat panel imager with active area 20 × 20 cm, 1024 × 1024 pixels, pixel size 200 × 200 μm, and 25 Hz/16 bit read-out.

zinc telluride (CdZnTe) detectors. They are wide (1.44−1.57 eV) band-gap semiconductors with high atomic number and density and thus good quantum efficiency, that is, capability to convert X-ray photons within a given detector volume. In such semiconductor detectors, electron−hole pairs are directly produced from interacting X-rays. While room temperature semiconductor detectors generally have better overall quantum efficiency compared to scintillation detectors, they have some disadvantages with respect to stability, especially in high-flux conditions, due to polarization effects induced by charge-trapping. In X-ray imaging applications, especially for fast imaging techniques, the X-ray photon flux is often rather high. Therefore, X-ray detectors are commonly operated in current mode. That is, the charge delivered by the detector is immediately converted by a transimpedance amplifier into a continuous voltage signal. The bandwidth of this preamplifier is set to a value that fulfills the antialiasing requirements of the subsequent sampling stages. In some particular applications with low photon flux and longer measurement times, pulse processing might be more appropriate. There the detector's preamplifier is operated as a charge-sensitive amplifier, which generates an electrical pulse signal for each converted X-ray photon. X-ray intensities are then measured by counting pulses (Fig. 6.6).

Whereas for transversal 2D tomography applications detector pixel arrangements in the form of custom-made linear, arched, or circular arrays are used, in cone-beam tomography often commercial planar imaging detectors are utilized (Fig. 6.5). Such 2D flat-panel detectors consist of a plate scintillator for X-ray to optical photon conversion, which is coupled to a photodiode matrix with a special read-out circuitry. Such detector arrays were originally developed for digital radiography in medicine. They have a high spatial resolution and size (detector panels with four megapixels with 0.2 mm pitch are available) but are rather slow regarding imaging speed. Other options for 2D detectors are digital X-ray image intensifiers.

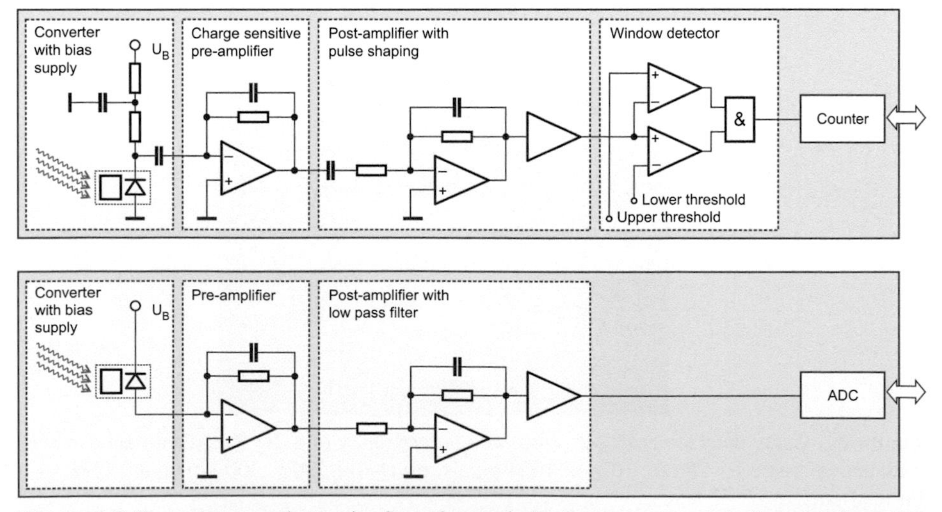

Figure 6.6 X-ray detector electronics for pulse mode (top) and current mode (bottom) detection.

Particularly for fast X-ray imaging and tomography, there are superior requirements for time response, sampling speed, and storage capacity of the X-ray detectors. For example, HZDR's ultrafast ROFEX scanner operates with a dual-plane detector of 576 detector pixels total with 1 MHz sampling speed. The digital interface has to manage data rates of about 1 GB/s. One minute of scanning requires 60 GB memory. With a focal spot speed of up to 3000 m/s at 5000 frame/s, full-scale attenuation signal changes on the detector may be as short as 1 μs

6.5 Attenuation measurement with X-rays

When X-rays penetrate material, the beam intensity decreases along the beam path. As schematically shown in Fig. 6.7, this is due to different physical effects. X-ray photons can be absorbed or elastically and nonelastically scattered at atoms and molecules. At photon energies much higher than twice the rest energy of electrons (1022 keV), there is additional pair creation in the Coulomb field of nuclei. The latter contribution becomes significant only for rather high photon energies, well above 2 MeV, so it is often not considered further in diagnostic X-ray imaging techniques. Also, on the other side of the energy spectrum, the contribution of elastic or Rayleigh scattering can be neglected for typical X-ray transmission measurements. So eventually, X-ray absorption and Compton scattering are the dominating effects of X-ray attenuation in diagnostic imaging techniques (Fig. 6.7).

The most fundamental concept to derive a quantitative description of X-ray attenuation in a material is the concept of the interaction cross-section. Assume that a single

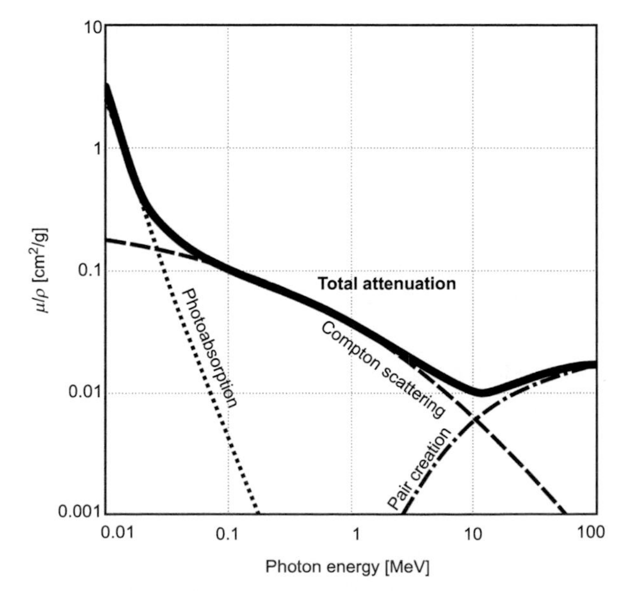

Figure 6.7 Mass attenuation coefficient of water as a function of X-ray energy.

atom of a given chemical element is placed in a unidirectional dense flow of monoenergetic X-ray photons. No matter how complicated the details of the physical interaction between atom and electromagnetic radiation field are, one can simply assume that each atom has an effective cross-sectional target area, called microscopic cross-section, and whenever an X-ray photon propagates through this area, it certainly interacts with the atom. For photoabsorption, we denote the microscopic cross-section as σ_A and for Compton scattering as σ_C. Neglecting Rayleigh scattering and pair creation, the total microscopic cross-section is then $\sigma = \sigma_A + \sigma_C$. Applying more advanced physical analysis, one would find that $\sigma_A \sim Z^4$ and $\sigma_C \sim Z$. Thus, photoabsorption strongly increases with atomic number and hence electron density, while Compton scattering is linearly related to electron density. Further, we assume in the following that Compton scattering decreases the X-ray photon flux. This is not quite correct, since scattered X-rays with lower energy are produced. However, since the radiation is mainly scattered out of the original propagation direction, it can be considered as lost, for example, for a collimated beam. However, the more radiation is scattered, the higher the probability that scattered photons impinge on a detector. This effect is discussed later on.

With the microscopic cross-sections given above, one easily finds that a material slab of a thickness Δx [m], area A [m^2], a density of atoms n [m^{-3}], a molar mass M [kg mol^{-1}], density ρ [kg m^{-3}], and the Avogadro constant N_A [mol^{-1}] has a total cross-section of

$$S = n\sigma A\Delta x = A\sigma\frac{\rho N_A}{M}\Delta x. \tag{6.1}$$

Let us consider an X-ray beam propagating along the direction of x (Fig. 6.8). At the left slab face, it has intensity $I(x)$. Within the thin slab, it loses as much intensity as given by the ratio S/A. Hence, we may write for the small loss ΔI

$$\Delta I = -I(x)\frac{S}{A} = -I(x)\frac{\sigma\rho N_A}{M}\Delta x. \tag{6.2}$$

This holds only for very small Δx; hence, one has to go to the limit $\Delta x \to 0$, giving

$$\Delta I = -I(x)\frac{\sigma\rho N_A}{M}\Delta x = -\mu I(x)\Delta x. \tag{6.3}$$

With $\mu = \sigma\rho N_A/M$, we define the macroscopic cross-section or linear attenuation coefficient, which uniquely defines the ratio dI/dx for a material of certain elementary composition (σ, M) and density (ρ). Since density of a material is not a unique parameter (e.g., depending on temperature, pressure, porosity, etc.), the attenuation is often given as the mass attenuation coefficient, that is, the linear attenuation normalized by density $\mu\rho = \mu/\rho = \sigma N_A/M$.

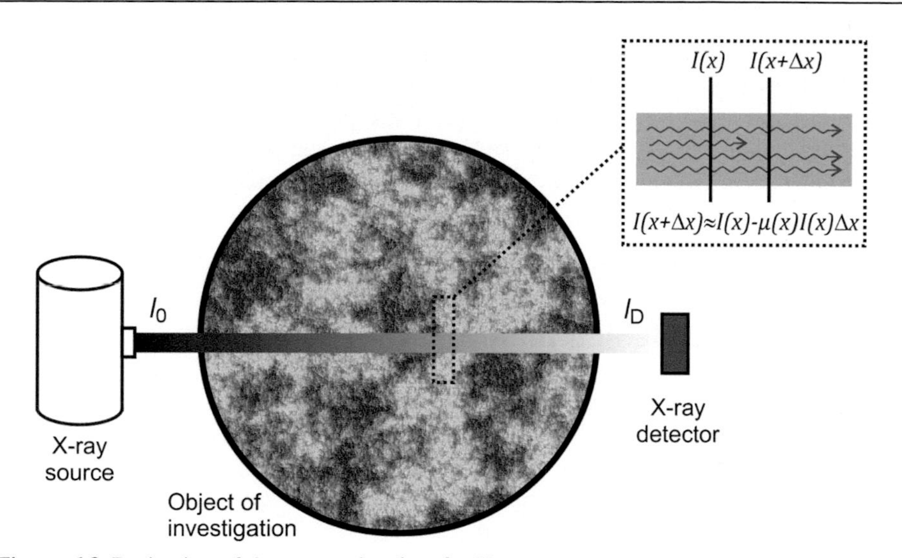

Figure 6.8 Derivation of the attenuation law for X-rays.

Now we consider the case that an X-ray passes an object with a locally changing attenuation coefficient, for instance, a chemical reactor with internals and different gaseous and liquid reactants and products inside. At each position x Eq. (6.3) holds, but now with μ being dependent on x, hence

$$\Delta I = -I(x)\mu(x)\Delta x. \tag{6.4}$$

This is an ordinary first-order linear and inhomogeneous differential equation. If we solve it for the arrangement shown in Fig. 6.8, with $I(0) = I_0$ being the beam intensity at the source and $I(x_D) = I_D$ the beam intensity at the detector, the solution is

$$I_D = I_0 \exp\left(- \int^{D} \mu(x)\Delta x \right). \tag{6.5}$$

For further analysis, it is convenient to introduce a new parameter, the extinction E, which is the negative logarithm of the transmission $T = I_D/I_0$, that is

$$E = -\log(T) = -\log\frac{I_D}{I_0} = \int^{D} \mu(x)\Delta x. \tag{6.6}$$

The indirectly measurable quantity E (which is in fact a function of the directly measurable quantities I_0 and I_D) is simply the integral over the linear attenuation coefficient distribution along the ray. This linearity allows one to elegantly reconstruct

2D or 3D distributions of the linear attenuation coefficient from projections, as will be shown in Chapter 13. However, it should be noted that Eq. (6.6) is a strong simplification of the real physics. At least, there are two phenomena that spoil the linearity, namely beam hardening and radiation scattering. Both phenomena will be discussed briefly in the following section. An example of how to deal with them is given in Section 6.7.

6.6 Beam hardening and radiation scattering

Radiation from a technical X-ray source is, with the exception of synchrotron radiation, polyenergetic. That is, there is a continuous distribution of X-ray energies in the radiation spectrum. The microscopic and macroscopic interaction cross-sections are energy dependent and this complicates the situation. Typically, the X-ray detector provides an energy-integral response, that is, Eq. (6.5) becomes

min

D

$$I_D = I_0 \int \nu_{\max} P(\nu) \exp\left(\int -\mu(x, \nu) \Delta x \right) \Delta \nu, \tag{6.7}$$

with ν being the photon frequency and $P(\nu)$ being the normalized X-ray power spectrum of the source (in other words, the probability distribution of the X-ray energies). Furthermore, the detector itself has an energy-dependent response function, defined among others by the quantum efficiency of radiation conversion. Thus, the detector signal U is in fact given by

min

D

$$U_D = I_0 \int \nu_{\max} h(\nu) P(\nu) \exp\left(\int -\mu(x, \nu) \Delta x \right) \Delta \nu, \tag{6.8}$$

where $h(\nu)$ is the spectral sensitivity of the detector. To circumvent the complications associated with these complex expressions, it is common practice to assume an energy-independent effective attenuation coefficient

min

$$\mu_{\text{eff}}(x) = \int \nu_{\max} h(\nu) P(\nu) \mu(x, \nu) \Delta \nu, \tag{6.9}$$

which then gives

$$I_D \approx I_0 \exp \int^{D} -\mu_{\mathrm{eff}}(x)\Delta x. \tag{6.10}$$

Eqs. (6.9) and (6.10) tell us that this effective attenuation coefficient depends on the overall attenuation coefficient distribution $\mu(x,\nu)$ along the rays. Typically, the attenuation decreases with energy: the more material in the ray path, the lower μ_{eff}. Or, said with other words, the more material is penetrated, the more low-energy radiation is absorbed and the harder is the remaining X-ray spectrum. In X-ray tomography, this nonlinearity results in typical image artifacts, such as streaks along edges of high-attenuating solids (Duerinckx & Macovski, 1978) and overestimated attenuation values in the centroids of continuous phases, the so-called cupping artifacts. A common method for correcting these artifacts is the insertion of a variable c, such that

$$I_D/I_0 + c = \int^{\min} \nu_{\max} h(\nu)P(\nu)\mu(x,\nu)\Delta\nu. \tag{6.11}$$

c is determined by phantom measurements with material distributions close to those of the actual measurements. This method demands preliminary knowledge about the distribution of $\mu(x,\nu)$ (Joseph & Spital, 1978; Nalcioglu & Lou, 1979). The quality of the correction depends on the segmentation of the phases in the initial image. Further, the tomography system property $h(\nu)P(\nu)$ must be known. In Herman, 1979, a method based on polynomial approximation for estimating c is given. The iteration of correcting projection data and image reconstruction converges to a highly artifact-reduced image. More advanced methods (Yan, Whalen, Beaupré, Yen, & Napel, 2000) are based not on the information of the material distribution, but on knowledge of the attenuation values of the present materials. For a two-phase system, it is assumed that every image voxel contains a mixture of phases, so that

$$\mu(x,\nu) = \varepsilon(x)\mu_1(x,\nu) + (1-\varepsilon(x))\mu_2(x,\nu), \varepsilon(x) \in [0,1]. \tag{6.12}$$

Here, the volume fraction function $\varepsilon(x)$ is calculated by an iterative algorithm. Again, for every iteration step, a new image reconstruction has to be performed.

A correction of beam hardening without any knowledge about the present materials can be realized with multienergy tomography. In this case, the iteration process is controlled by the difference values of $\mu(x,\nu_i)$ reconstructed from the difference projection data.

Besides beam hardening, the linearity of integral attenuation and hence quality of reconstructed X-ray CT images depends on scattered radiation. Part of radiation scattered by Compton interactions may accidently fall into the detector and create an intensity offset. For a rough qualitative analysis, we may denote with m_{SD} the total material mass in the spatial channel between the source and a detector and with m_{Exp} the material mass exposed by X-rays. Then, it turns out that scattering increases with the ratio $m_{\mathrm{Exp}}/m_{\mathrm{SD}}$. While the exact quantitative dependency is complex, this simple finding makes clear that limiting the radiation bundle to the available source−detector channels by spatial collimation is useful. Moreover, directional collimation of detectors toward the source is helpful. Energetic collimation, as used in emission or gamma ray tomography, is not applicable for polyenergetic X-ray sources. While transversal scanning allows for effective in-plane collimation, in cone-beam CT, only directional collimation can be applied. But the latter is technically highly challenging. Hence, for high-quality imaging, software-based scattering correction algorithms are frequently being used. They are based on a proper subtraction of the scattered radiation offset, which has to be determined either by separate measurements or by radiation transport simulations. An example for such a correction scheme is given in the next section.

6.7 Cone-beam X-ray computed tomography for gas holdup measurements

One frequently emerging question in process engineering is that of the spatial distribution of the phases within a chemical reaction device. Carefully conducted analysis of the phase distribution enables a critical assessment of the reaction process with respect to process efficiency and discloses the possibilities for further optimization. The aerated continuous stirred-tank reactor is a prominent example for which the knowledge of spatial gas holdup distribution is of essential interest to determine local interfacial area density and, thus, local mass transfer and reaction rate. The following example describes the gas holdup measurement with X-ray tomography in a laboratory-scale cylindrical autoclave reactor vessel with hemispherical bottom (Boden, Bieberle, & Hampel, 2008). Aeration and distribution of the disperse gas phase within the liquid inventory of the reactor is achieved by a multiblade gas-inducing impeller mounted on a hollow shaft, which acts as the inlet for the gaseous phase. The hollow shaft is open to the gas reservoir above the liquid and has holes between the stirrer blades. When the stirrer rotates, centrifugal forces in the liquid result in a pressure drop that sucks gas through the stirrer shaft to the outlets in the impeller region. Due to the high shear stresses in the impeller region, the gas is dispersed and further distributed within the continuous phase by the turbulent liquid flow as well as by buoyancy forces. A "good mixing" means homogeneous distribution of the disperse gas bubbles within the reactor volume with reasonable residence times to enable proper mass transfer from the gas into the liquid phase.

Gas holdup in the reactor can effectively and accurately be measured by X-ray cone-beam CT. The setup consists of a rotating anode X-ray tube and a digital X-ray flat panel detector. The stirred tank reactor vessel is between the two. Since only the gas holdup is of interest, a differential measurement approach is applied. That is, the reactor is scanned with liquid filling at very low stirrer speeds (no gas dispersed in the liquid) as well as in full operation. Then, difference images are computed from both measurements, which give the local gas holdup. Special correction steps were implemented to account for beam hardening and radiation scattering, which is described in the following.

Scatter correction: To resolve small gas fractions reliably, radiation scattering must be corrected. As shown in Fig. 6.9, a movable slit mask is arranged in front of the X-ray tube. This slit mask can be moved in a vertical direction, and depending on its position, only a thin X-ray fan radiographs the object. In this scenario, most of the scattered radiation is blocked off and only in-plane scattering with a much smaller contribution occurs. If scanning time is not a critical factor, one can subsequently scan slices of the object by moving the slit mask and synthesize scatter-free projections from the scans. If scanning time is an issue, this synthetic scanning can be done once for a reference state (e.g., the filled reactor) and repeated without the slit (full cone illumination). With $I_D^{(\mathrm{slit})}$ and $I_D^{(\mathrm{no\ slit})}$ denoting the X-ray intensities at a detector pixel D for both configurations, a scattering offset

$$I_D^{(\mathrm{scat})} = I_D^{(\mathrm{no\ slit})} - I_D^{(\mathrm{slit})} \tag{6.13}$$

can be determined for each detector pixel, which is then subtracted from all other cone illumination measurements later on. Note that this approach is twofold approximate. First, there is some minor remaining scattering fraction within the collimated fan. Secondly, changes in the object distribution, such as different total gas fractions, also change the scattering background. The appropriateness of these approximations must therefore be checked for each application. In this case, with a large source–detector

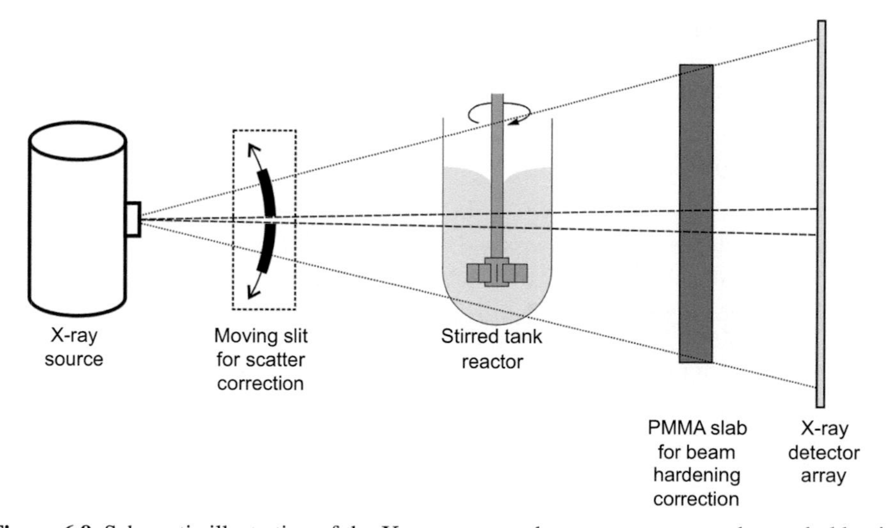

Figure 6.9 Schematic illustration of the X-ray tomography setup to measure the gas holdup in an aerated stirred tank reactor (for further explanation, see text).

distance and only small deviations of the object distribution from the reference state, these approximations are fully sufficient to reduce the overall error from scattered radiation to below 1%.

Beam hardening: An X-ray that propagates straight through the reactor is attenuated along its path by both reactor materials and the liquid. Also, each ray in the X-ray cone penetrates different lengths of these materials. Consequently, due to the beam hardening, for each ray there is a different effective attenuation coefficient $\mu_{\text{eff},D}^{(\text{liq})}$ for the liquid. Now considering that we have made two measurements, one for the gas-free reference state $I_D^{(\varepsilon=0)}$ and one for the aerated reactor $I_D^{(\varepsilon)}$, it follows that

$$\frac{I_D^{(\varepsilon)}}{I_D^{(\varepsilon=0)}} = \exp\left(\mu_{\text{eff},D}^{(\text{liq})} \int \varepsilon(x)\Delta x\right). \tag{6.14}$$

Isolating the line integral of the gas fraction ε, which is input to the image reconstruction, gives

$$\int \varepsilon(x)\Delta x = \frac{1}{\mu_{\text{eff},D}^{(\text{liq})}} \log \frac{I_D^{(\varepsilon)}}{I_D^{(\varepsilon=0)}}. \tag{6.15}$$

As shown in Fig. 6.9, the missing values $\mu_{\text{eff},D}^{(\text{liq})}$ (one for each detector pixel) can be determined by one additional measurement. Therefore, we insert a slab of a material with effective attenuation $\mu_{\text{eff},D}^{(\text{slab})}$ close to the liquid in the reactor between the reactor and the detector. From a calibration measurement $I_D^{(\text{cal})}$ with nonaerated reactor and the slab in the beam, we get

$$\mu_{\text{eff},D}^{(\text{slab})} = \frac{\cos\alpha}{D} \log \frac{I_D^{(\text{cal})}}{I_D^{(\varepsilon=0)}}. \tag{6.16}$$

Here D denotes the slab thickness and α the angle between X-ray and slab normal. For materials with similar atomic number and density, such as liquid and plastics, the ratio $C = \mu_{\text{eff},D}^{(\text{liq})} / \mu_{\text{eff},D}^{(\text{slab})}$ is rather constant over a wide energy range. Hence, having measured $\mu_{\text{eff},D}^{(\text{slab})}$ for each detector pixel with the plate methods and knowing C from either measurement or database values, we can compute the required line integral values of the gas fraction as

$$\int \varepsilon(x)\Delta x = \frac{1}{C\mu_{\text{eff},D}^{(\text{slab})}} \log \frac{I_D^{(\varepsilon)}}{I_D^{(\varepsilon=0)}}. \tag{6.17}$$

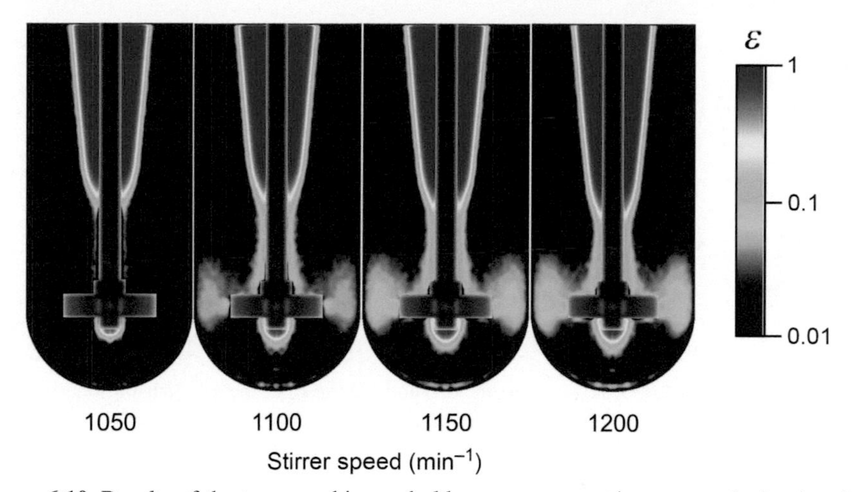

Figure 6.10 Results of the tomographic gas holdup measurement in an aerated stirred tank reactor for different stirrer speeds.

Reconstruction in this experimental work was done with the generalized filtered back-projection algorithm for the 3D cone-beam tomography with circular source trajectory (Feldkamp, Davis, & Kress, 1984). By carefully accounting for beam hardening and radiation scattering effects as described above, gas holdup distribution profiles in the stirred tank reactor were obtained under different operating conditions at and above the critical stirrer speed (Fig. 6.10). As can be seen, cone-beam X-ray tomography resolves well the central vortex around the stirrer shaft in the center of the reactor, as well as the concentration of the gas in the radial region between the stirrer blades and the reactor wall, where the gas holdup is confined by the turbulent flow field. Critical evaluation of the tomographic method shows maximum quantitative errors of the measurements within about 3.5% absolute local gas holdup.

6.8 Static mixer studies with ultrafast electron beam X-ray tomography

Static mixers are an attractive alternative to mechanically stirred vessels (Thakur, Vial, Nigam, Nauman, & Djelveh, 2013). The small space requirement, low equipment cost, good mixing at low shear rates, self-cleaning, sharp residence time distribution, improved selectivity, and byproduct reduction are the main features of static mixers. Over the years, a large number of companies have designed and produced static mixers, including helical mixers, corrugated plate mixers, cross-bar mixers, and others. Static mixers are based on the principles of flow splitting and radial mixing by a series of baffles. Despite its increasing popularity, a detailed quantitative assessment of the mixing mechanism (particularly for gas–liquid systems) and, based on that, an optimal design were so far not possible due to lack of suitable measurement techniques.

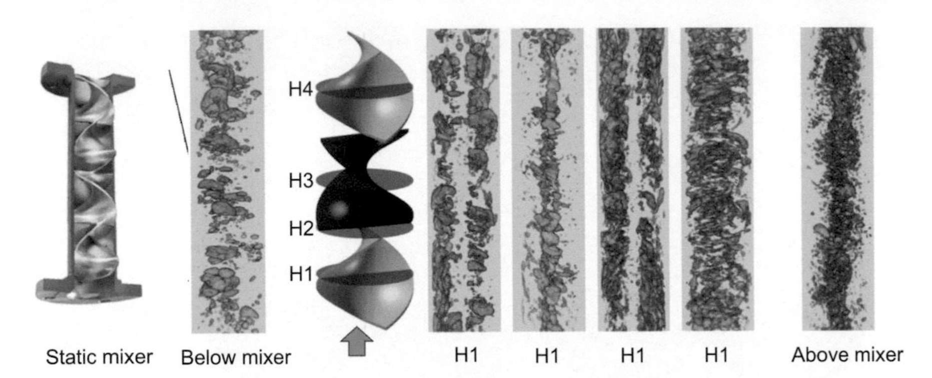

Figure 6.11 View of a helical static mixer (left) and 3D views of the gas distribution in the inlet section, in the outlet section, and within the mixer.

Ultrafast X-ray tomography has been applied to study gas–liquid flow in a helical static mixer. The mixer was built-in in a vertical pipe of 80 mm inner diameter. The objective of the study was to elucidate the flow conditions and develop appropriate scale-up guidelines to ensure proper mixing by increasing the interfacial area and liquid turbulence at economically justifiable energy input. The helical static mixer (Fig. 6.11) consists of segments of twisted blades with alternate clockwise and counterclockwise orientations, which are connected orthogonally with each other. The ultrafast electron beam X-ray tomography scanner ROFEX (Fig. 6.4) was used for this study. It operates at 150 kV acceleration voltage and a maximum beam power of 10 kW. It comprises a dual detector ring with 2×288 room temperature semiconductor pixels, which can be sampled with 1 MSamples per second fully parallel. The scanning speed here was 1000 frames per second in each of the two imaging planes and the plane distance is 11 mm. Fig. 6.11 shows selected results of the static mixer study. Shown are the 3D representations of the interfacial area at different axial locations upstream of, within, and downstream of the static mixer. In this particular case, the mixer was operated at 0.6 m/s liquid superficial velocity and 0.11 m/s gas superficial velocity. The 3D views were created by stacking subsequent cross-sectional images of one scanning plane together and extracting the gas phase, and thus the interfacial area, via a 3D image segmentation algorithm. The interfacial area calculated from the measured gas holdup and bubble diameters was found to increase significantly with the addition of the helical static mixer elements and to decrease with the increase of fluid velocity ratio. The results allowed providing the designer with fluid-dynamic data to optimize the mixer design. Furthermore, such data forms a sound basis for modeling of gas–liquid flows with computational fluid dynamics codes.

6.9 Future trends

X-ray tomography is a very desirable tool for process tomography and multiphase flow analysis. Recent developments have proven that it is a superior tool with respect to achievable spatial and temporal resolution in flow scanning. Nonetheless, expensiveness and efforts for radiation protection make X-ray CT methods less easily applicable, especially as a monitoring instrument in the field. Hence, one of the future trends will be to provide technical solutions for fast X-ray CT that are applicable in real industrial applications, such as in the oil and gas industries. This requires numerous advances, such as high-power, cheap, compact, and robust multisource or moving-source X-ray generators; fully integrated data analysis software; and inherent radiation protection designs. For laboratory applications, where radiation protection and cost issues are somewhat less critical, a future trend is to achieve 3D scanning and increase penetration capability by going to higher X-ray energies. Moreover, interesting developments are to be expected in improving spatial resolution by fast fine-focus sources, multienergy full CT imaging (e.g., for three-phase flow studies), and combination of X-ray CT with other imaging modalities, such as electrical tomography or ultrasound scanning. Lastly, it should be noted that fast high-resolution tomography and its data processing, image reconstruction, and data analysis require ever-increasing computer power as well as intelligent software solutions. Here, parallel computing, for example, on graphics processing unit systems, will certainly evolve its potentials in the future.

6.10 Sources of further information and advice

For the fundamentals of X-ray CT, especially in medicine, there are many good textbooks. Particularly for its application in multiphase flow studies, the comprehensive review (Heindel, 2011) is to be recommended. The author and his group have extensively worked on X-ray CT application in multiphase flow analysis and particularly pioneered fast electron beam X-ray CT for process applications. Hence, further information may be found in the various publications of the group and the internet presentation of Helmholtz-Zentrum Dresden-Rossendorf.

References

Berg, S., Ott, H., Klapp, S. A., Schwing, A., Neiteler, R., Brussee, N., et al. (2013). Real-time 3D imaging of Haines jumps in porous media flow. *PNAS, 110*, 3755–3759.

Boden, S., Bieberle, M., & Hampel, U. (2008). Quantitative measurement of gas holdup distribution in a stirred chemical reactor using X-ray cone-beam computed tomography. *Chemical Engineering Journal, 139*, 351–362.

Del Sordo, S., Abbene, L., Caroli, E., Mancini, A. M., Zappettini, A., & Ubertini, P. (2009). Progress in the development of CdTe and CdZnTe semiconductor radiation detectors for astrophysical and medical applications. *Sensors, 9*, 3491–3526.

Deza, M., Franka, N. P., Battaglia, F., & Heindel, T. J. (2009). CFD modeling and X-ray imaging of biomass in a fluidized bed. *Journal of Fluids Engineering, 131*, 111303.

Duerinckx, A. J., & Macovski, A. (1978). Polychromatic streak artifacts in computed tomography images. *Journal of Computer Assisted Tomography, 2*, 481–487.

van Eijk, C. W. E. (2002). Inorganic scintillators in medical imaging. *Physics in Medical Biology, 47*, R85–R106.

Feldkamp, L. A., Davis, L. C., & Kress, J. W. (1984). Practical cone-beam algorithm. *Journal of Optical Society America A, 1*, 612–691.

Fischer, F., Hoppe, D., Schleicher, E., Mattausch, G., Flaske, H., Bartel, R., et al. (2008). An ultra fast electron beam x-ray tomography scanner. *Measurement of Science and Technology, 19*, 094002.

Grangeat, P. (1991). Mathematical framework of cone beam 3D reconstruction via the first derivative of the Radon transform. *Lecture Notes in Mathematics, 1497*, 66–97.

Hampel, U., Bärtling, Y., Hoppe, D., Kuksanov, N., Fadeev, S., & Salimov, R. (2012). Feasibility study for MeV electron beam tomography. *Review of Scientific Instruments, 83*, 093707.

Heindel, T. J. (2011). A review of X-ray flow visualization with applications to multiphase flows. *Journal of Fluids Engineering, 133*, 074001.

Herman, G. T. (1979). Correction for beam hardening in computed tomography. *Physics in Medicine and Biology, 24*, 81–106.

Hori, K., Fujimoto, T., Kawanishi, K., & Nishikawa, H. (1998). Development of an ultrafast X-ray computed tomography scanner system. *IEEE Transactions on Nuclear Science, 45*, 2089–2094.

Inoue, A., Kurosu, T., Aoki, T., Yagi, M., Mitsutake, T., & Morooka, S. (1995). Void fraction distribution in BWR fuel assembly and evaluation of subchannel code. *Journal of Nuclear Science and Technology, 32*, 629–640.

Johansen, G. A., Hampel, U., & Hjertaker, B. T. (2010). Flow imaging by high speed transmission tomography. *Applied Radiation and Isotopes, 68*, 518–524.

Joseph, P. M., & Spital, R. D. (1978). A method for correcting bone induced artifacts in computed tomography scanners. *Journal of Computer Assisted Tomography, 2*, 100–108.

Kak, A. C., & Slaney, M. (1988) *Principles of computerized tomographic imaging*. New York: IEEE Press.

Kalender, W. A. (2005). Computed tomography. *Fundamentals, system technology, image quality, applications* (2nd ed.). Erlangen: Publicis Corporate Publishing.

Moreno-Atanasio, R., Williams, R. A., & Jia, X. (2010). Combining X-ray microtomography with computer simulations for analysis of granular and porous materials. *Particuology, 8*, 81–99.

Mudde, R. F. (2011). Bubbles in a fluidized bed: A fast X-ray scanner. *AIChE Journal, 57*, 2684–2690.

Nalcioglu, O., & Lou, R. Y. (1979). Post-reconstruction method for beam hardening in computed tomography. *Physics in Medicine and Biology, 24*, 330–340.

Schard, P., Deuringer, J., Freudenberger, J., Hell, E., Knüpfer, W., Mattern, D., et al. (2004). New X-ray tube performance in computed tomography by introducing the rotating envelope tube technology. *Medical Physics, 31*, 2699–2706.

Seeger, A., Kertzscher, U., Affeld, K., & Wellnhofer, E. (2003). Measurement of the local velocity of the solid phase and the local solid holdup in a three-phase flow by X-ray based particle tracking velocimetry (XPTV). *Chemical Engineering Science, 58*, 1721–1729.

Shefer, E., Altman, A., Behling, R., Goshen, R., Gregorian, L., Roterman, Y., et al. (2013). State of the art of CT detectors and sources: A literature review. *Current Radiology Reports, 1*, 76–91.

Thakur, R. K., Vial, C., Nigam, K. D. P., Nauman, E. B., & Djelveh, G. (2013). Static mixers in the process industries - a review. *Chemical Engineering Research and Design, 81*, 787–826.

Toye, D., Marchot, P., Crine, M., Pelsser, A.-M., & L'Homme, G. L. (1998). Local measurement of void fraction and liquid holdup in packed columns using X-ray computed tomography. *Chemical Engineering and Proceedings, 37*, 511–520.

Waelchli, S., von Rohr, R., & Stampanoni, M. (2005). Multiphase flow visualization in microchannels using X-ray tomographic microscopy (XTM). *Journal of Flow Visualization Image Processing, 12*, 1–13.

Yan, C. H., Whalen, R. T., Beaupré, G. S., Yen, S. Y., & Napel, S. (2000). Reconstruction algorithm for polychromatic CT imaging: Application to beam hardening correction. *IEEE Transactions on Medical Imaging, 19*, 1–11.

Radioisotope tracer techniques

7

David Parker
University of Birmingham, Birmingham, United Kingdom

7.1 Nuclear medicine imaging

7.1.1 Single photon emission computed tomography

In medicine, imaging the way in which a biologically interesting molecular species, labeled with a radioactive tracer, distributes through the human body provides one of the most powerful ways of studying normal physiology and identifying pathologies. Since Anger first invented the gamma camera in 1957 (Anger, 1964), the field of diagnostic nuclear medicine has developed into a major international industry.

A gamma camera (Anger, 1964) forms a two-dimensional (2D) image of the radioisotope distribution within the body by detecting emitted gamma rays using a sheet of the scintillator sodium iodide (typically around 10 mm thick and 50 cm in diameter) backed by an array of around 50 photomultiplier tubes (PMTs). Each gamma ray interacting in the scintillator produces a flash of light, which spreads over several PMTs; by comparing the light intensities measured in the different PMTs, the coordinates of the interaction can be determined to within a few mm.

Since it is not possible to focus gamma rays, a collimator is used to define the direction from which each detected gamma has reached the detector. Although the use of a pinhole collimator is possible, generally a parallel-hole collimator is used, consisting of a block of lead between 2.5 and 5 cm thick, through which are drilled a large number of parallel holes designed to transmit only gamma rays close to normal incidence. In practice the collimator transmits a cone of gamma rays, limited by the aspect ratio of the holes, so that the spatial resolution of the resulting image degrades linearly with distance to the source. A typical high-resolution collimator might have holes of 2 mm diameter × 50 mm long, resulting in a spatial resolution of around 12 mm for imaging a source at a distance of 30 cm. The thickness of the lead septa separating the holes is dictated by the need to attenuate nonnormal gamma rays by a sufficient factor, which becomes progressively harder as the gamma ray energy increases. The majority of gamma camera imaging is performed using the radionuclide ^{99m}Tc, which emits 140 keV gamma-rays—this is an ideal energy for medical imaging: high enough to escape the body but low enough to be easily collimated. ^{99m}Tc is also popular because of its convenient half-life (6 h), absence of more damaging emissions, and availability from decay of the radioactive parent ^{99}Mo (half-life 66 h)—this is supplied in the form of a *technetium generator* from which a solution of ^{99m}Tc can be eluted when required.

Industrial Tomography. https://doi.org/10.1016/B978-0-12-823015-2.00030-3

A static gamma camera measures a 2D projected view of the activity distribution within a 3D region. A tomographic image of the 3D volume can be obtained by rotating the camera around the patient, measuring projections at all angles and using a technique such as filtered back-projection or iterative reconstruction to reconstruct the activity within each voxel. This imaging modality is referred to as single photon emission computed tomography (SPECT) to distinguish it from the dual photon imaging used in positron emission tomography (PET) (see below). A complication of SPECT imaging is the need to correct for the attenuation of the gamma rays within the body: the extent of this attenuation depends on the depth of the source and properties of the intervening material, information on which may be provided by complementary imaging techniques. The need to perform a depth-dependent correction for attenuation poses problems for a simple reconstruction technique like filtered back-projection, and with the increased availability of computing power iterative reconstruction techniques are increasingly popular.

7.1.2 Positron emission tomography

PET is a separate branch of nuclear medicine based on the use of positron-emitting radionuclides. These proton-rich nuclides decay by β^+ decay in which a positron (the positively charged antiparticle of the normal electron) is emitted. After slowing down to rest, the positron annihilates with a normal electron producing a pair of 511 keV photons,[1] which are emitted almost exactly back-to-back due to conservation of momentum. If the positron–electron pair had zero momentum prior to annihilation, then the pair of photons would be emitted exactly 180° apart; in practice, the residual momentum of the electron results in an acollinearity of up to about 0.5°.

Simultaneous *coincidence* detection of the two annihilation photons using detectors on either side of the patient defines a *line of response* passing close to the site of the radioactive decay without the need for a physical collimator. For this reason, the coincidence is said to provide *electronic collimation*. Major advantages of PET compared to SPECT include the much higher sensitivity (a typical parallel hole collimator blocks over 99.9% of gamma rays, which would otherwise be detected) and geometric invariance—as noted above, the blurring caused by a parallel hole collimator increases linearly with distance, but the spatial resolution in a PET measurement is approximately independent of depth. Just as in SPECT, it is necessary to correct for attenuation of the 511 keV photons, but, because what matters is the overall probability of either of the two photons being attenuated, for a given line through the body the attenuation correction is independent of the depth of the source and can be accurately determined using an external transmission source.

A PET scanner (Fig. 7.1) generally consists of one or more rings of blocks of thick, high-Z scintillator—20-mm-thick bismuth germanate (BGO) is typical—giving high efficiency for detecting 511 keV photons. Each block is subdivided into smaller

[1] These 511 keV photons are often referred to as gamma rays. Strictly speaking this is incorrect, as the term gamma ray is used to denote photons emitted by an excited nucleus.

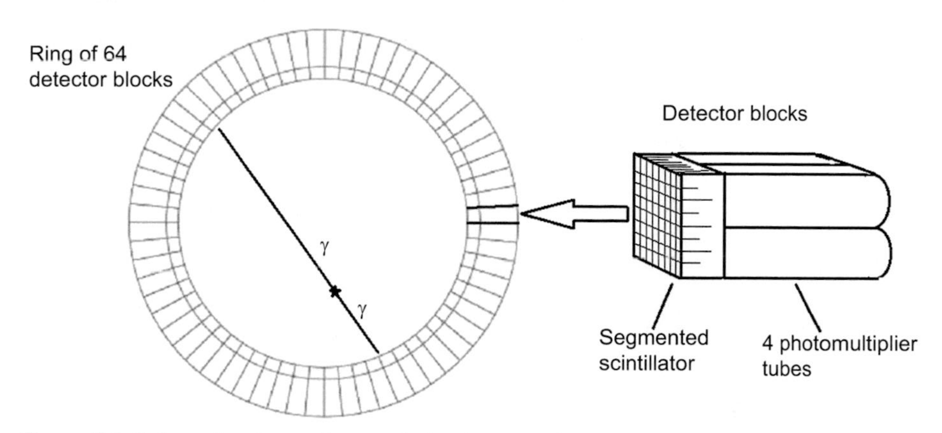

Figure 7.1 Schematic of a positron emission tomography scanner.

elements by cuts that penetrate part-way down, leaving an intact base to which are coupled four PMTs. Comparing the light intensities in the different PMTs allows the individual element in which the photon interacted to be uniquely identified.

Ultimately the spatial resolution of a PET study would be limited by the basic physics: the emitted positron travels some distance before annihilation, and the two photons are not emitted exactly back-to-back. The magnitude of these effects depends on the radioisotope used, the density of the material, and the detector separation, but in a typical study they could be expected to contribute a positional uncertainty of 1–2 mm. In practice, the resolution is usually limited by the size of the detector elements. A typical clinical PET scanner might have blocks around $50 \times 50\ mm^2$ cut into an 8×8 array of elements, each approximately $6.5 \times 6.5\ mm^2$, but high-resolution scanners use $4 \times 4\ mm^2$ elements, and dedicated small animal scanners (intended for drug trials) have elements as small as $1.5 \times 1.5\ mm^2$.

A coincident pair of 511 keV photons is recognized by their near-simultaneous detection in opposing detector blocks. Because of variations in the time taken for a gamma ray interaction to generate a pulse, and to allow for variations in time of arrival of the two photons due to the finite speed of light, some tolerance in timing must be allowed. In practice, two pulses are considered coincident if they occur within a time interval less than a specified value τ, the *resolving time* of the system. For a PET scanner the resolving time is typically around 10 ns. If two unrelated pulses happen to occur within this time interval they will also be recorded— such events are described as *random coincidences*. Random coincidences add background to an image; although for any individual event it is not possible to identify it as *true* or *random*, the contribution of random coincidences to a study can be accurately estimated by repeating the same measurement with a delay ($\gg \tau$) introduced into one side of the coincidence circuit—this eliminates true coincidences but does not alter the rate or distribution of random coincidences. Many PET scanners continuously record both *prompt* and *delayed* coincidences and then reconstruct their images from the difference between the two datasets.

A more intractable problem is the background due to detection of a significant contribution of events where one or both of the photons has scattered prior to detection. For all but very high-Z materials, Compton scattering is the principal interaction mechanism for 511 keV photons. The energy resolution of the detector blocks is generally too poor to discriminate against photons that have scattered through a small angle. In an attempt to reduce the proportion of scattered events in the detected data, some early PET scanners incorporated collimating septa, which essentially subdivided the 3D imaging problem into a stack of separate 2D slices; this design improved the quality of the data, but at the cost of greatly reduced sensitivity, and is no longer popular.

Most clinical PET studies use the nuclide ^{18}F, which has to be produced using a cyclotron. The half-life of ^{18}F is 110 min, long enough to allow transport over extended distances, and a number of cyclotrons supply ^{18}F in the form of fluorodeoxyglucose. Where a local cyclotron is available the shorter lived isotopes, ^{11}C (20 min), ^{13}N (10 min) ,and ^{15}O (2 min) can also be used.

Recently, there has been increased interest in time-of-flight (TOF) PET. Using novel scintillators with faster decay times coupled to silicon photomultipliers, the intrinsic timing resolution of photon detection can be reduced below 1 ns, so that the difference in arrival times of the two annihilation photons provides some localization of the source along the line of response (Surti, 2015).

7.2 Industrial applications

7.2.1 Single photon imaging

Given the widespread availability of gamma cameras in nuclear medicine, it is perhaps surprising that rather few industrial applications of gamma camera imaging have been reported. This may partly be attributed to the relatively low penetration of the low-energy gamma rays for which parallel hole collimators are usually designed (1 cm steel attenuates the 140 keV gamma rays from ^{99m}Tc by a factor of 5). Collimators designed to be used with ^{99m}Tc are generally unsuitable for imaging higher-energy gamma rays.

The low data rates and consequent long imaging times are also a problem, meaning that SPECT has generally been confined to imaging steady-state or slow flows. For example, Perret, Prasher, Kantzas, Hamilton, and Langford (2000) (Calgary) reported the use of SPECT to study breakthrough curves in saturated soil cores. Cores of 77 mm diameter were imaged over a 40 cm length using a rotating gamma camera fitted with a high-resolution collimator. ^{99m}Tc was injected into the flow, and images were acquired. Since each SPECT image took 6 min to acquire, imaging was performed in stop/start mode, with the flow halted during acquisition, and 2 min flow between each image. Images with a spatial resolution of around 4 mm were obtained and clearly showed the flow proceeding through macropore structures.

There have also been some efforts to develop dedicated industrial SPECT systems, suitable for use with higher-energy gamma rays. Legoupil, Pascal, Chambellan, and Bloyet (1997) at Saclay reported an experimental SPECT system for industrial imaging, comprising 36 collimated NaI detectors (6 banks, each of 6 detectors) in a single plane. This system was demonstrated by imaging flow, labeled by injection of ^{99m}Tc, in a 24-cm-diameter vessel. Full-width half-maximum spatial resolutions of approximately 10% of the vessel diameter were reported. More recently, the group from KAERI in Korea have explored designs for industrial SPECT systems based on a 12-gonal geometry (Park et al., 2014) and/or diverging collimators providing optimal coverage of the field of view (Park, Jung, Kim, Moon, & Kim, 2015).

7.2.2 Positron emission tomography

In principle, PET is well suited to studying engineering systems, as the 511 keV photons are reasonably penetrating (50% are transmitted through 11 mm steel). Just as for SPECT, imaging is slow, so PET is most suited for studying steady-state or slow flows.

Positron-emitting nuclides are less widely available (and more expensive) than ^{99m}Tc. The positron emitter ^{68}Ga (half-life 68 min) can be obtained from a generator containing the parent ^{68}Ge (271 days). ^{64}Cu (12.4 h) can be produced by neutron activation of copper in a nuclear reactor. All other positron emitters are cyclotron produced.

Industrial application of PET was pioneered by the Birmingham group in the 1980s. In contrast to the ring scanners now standard in medical PET, the Birmingham work used a "positron camera" consisting of two rectangular detectors (multiwire proportional chambers) operating in coincidence. For 3D tomographic imaging, it was necessary to rotate the camera about the field of view, measuring projected views at all angles, but useful information could also be obtained using 2D projection imaging. For example, Fig. 7.2 shows a sequence of 10 min images of brine flow through a specimen of oil reservoir sandstone (observing the passage of a bolus of ^{18}F fluoride), while Fig. 7.3 shows a sequence of 1 min images of labeled dough being extruded into a metal mold (Hawkesworth et al., 1991).

More recently the Birmingham group has performed some PET imaging using a standard medical PET scanner. For example, Fig. 7.4 shows the dispersion of a bolus of labeled powder within a pharmaceutical blender (Parker et al., 2008). Acquiring the data for each image took 20 min, so this study was performed in stop/start mode with the blender being tumbled for five revolutions between each image acquisition.

A few other groups have reported using PET to explore geotechnical and industrial processes. For over a decade, the Dresden group have applied PET to study flow through soil columns and rock cores, a technique they call GeoPET (Kulenkampff, Grundig, Zakhnini, & Lippmann-Pipke, 2016). Colleagues have also reported the use of a small animal PET scanner to study the dynamics of deposition and resuspension of graphite dust, labeled with ^{18}F, within a working model of a pebble bed reactor (Barth et al., 2014).

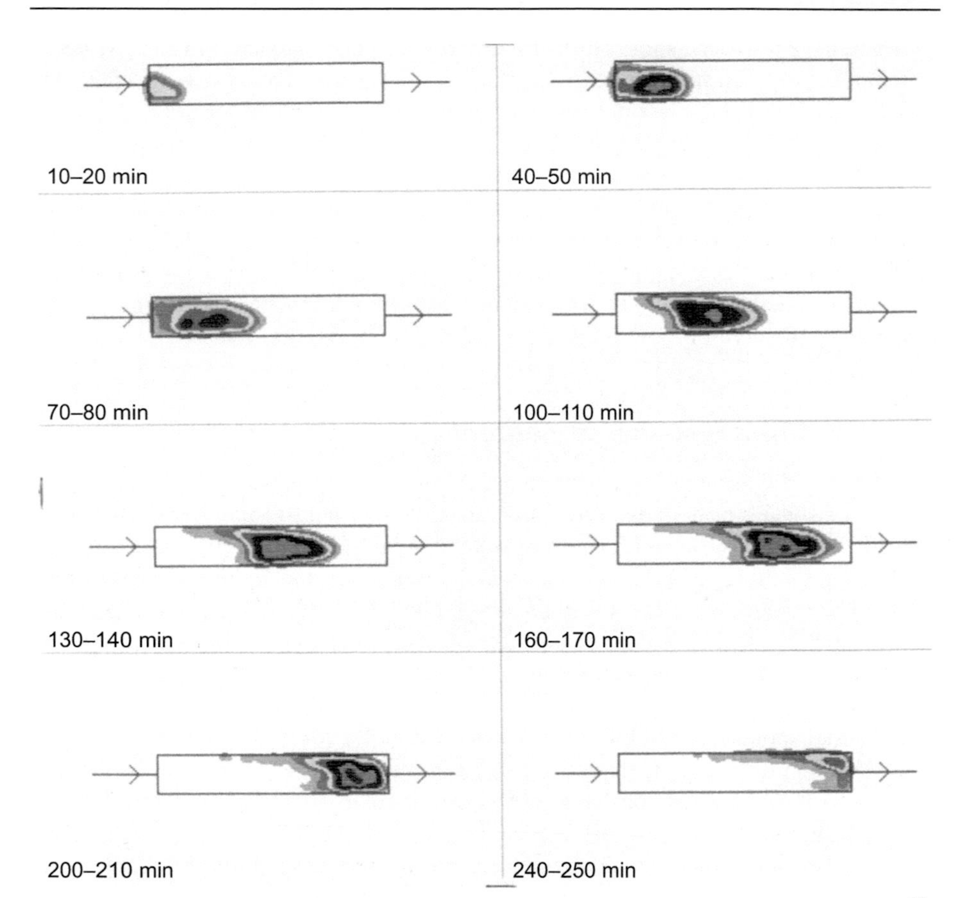

Figure 7.2 Sequence of 10 min positron projection images showing passage of a bolus of ^{18}F brine through a specimen of oil reservoir sandstone ($30 \times 5 \times 5$ cm^3).
From Hawkesworth, M. R., Parker, D. J., Fowles, P., Crilly, J. F., Jefferies, N. L., & Jonkers, G. (1991). Nonmedical applications of a positron camera. *Nuclear Instruments & Methods in Physics Research, Section A, 310*, 423–434.

7.3 Particle tracking

For studying rapid industrial flows, both SPECT and PET are of limited use due to the time required to acquire sufficient counts to enable reconstruction of a 3D tomographic image. Instead, valuable dynamic information can often be obtained by tracking a single radioactively labeled particle. If only a single radioactive source is present, its location can be determined very rapidly, and by repeating this measurement at regular intervals, the track of the particle can be observed. This is a powerful way of studying the behavior of granular systems (if a representative grain is labeled and tracked) or viscous fluids (if a neutrally buoyant flow follower is tracked). In a closed system, where material recirculates, by tracking a single particle over an extended period

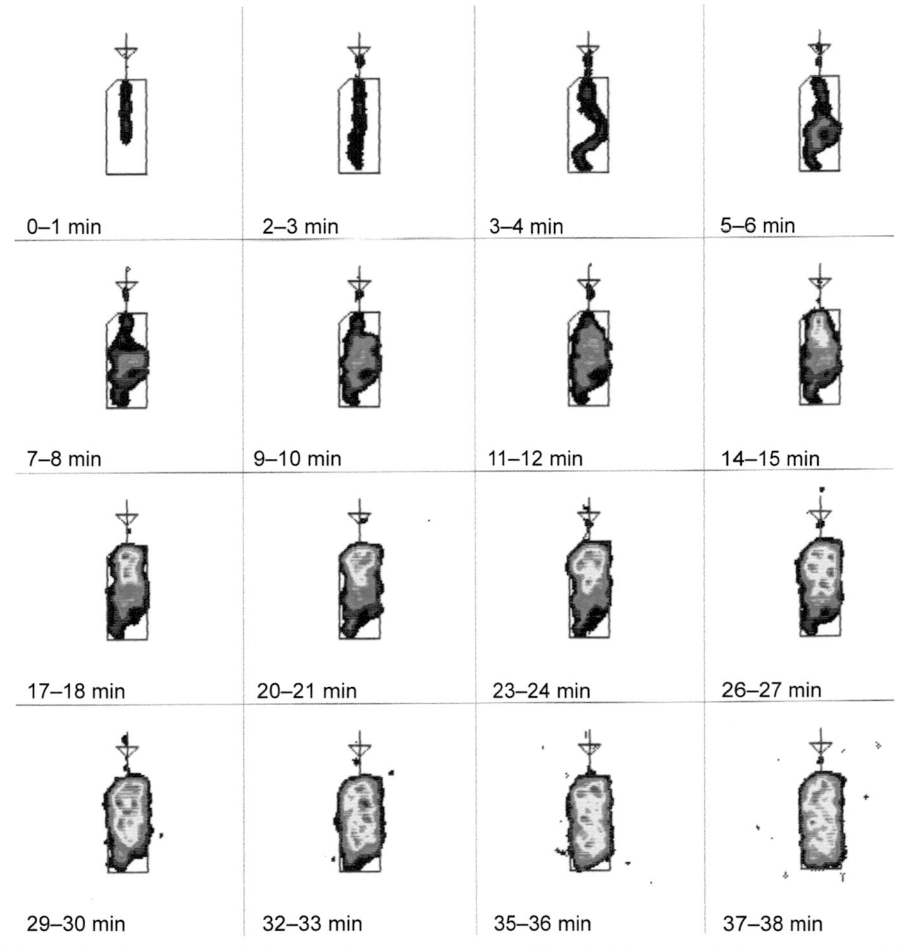

Figure 7.3 Selected 1 min images showing extrusion of labeled flour dough into a metal mold.

Figure 7.4 Sequence of images (central slice through 3D positron emission tomography image) showing mixing of a small amount of labeled powder with the unlabeled bulk powder in a pharmaceutical blender.
From Parker, D. J., Leadbeater, T. W., Fan, X., Hausard, M. N., Ingram, A., & Yang Z. (2008). Positron imaging techniques for process engineering: Recent developments at Birmingham. *Measurement Science and Technology, 19*, 094004.

(Lagrangian measurements), one can build up Eulerian time-averaged density and velocity maps for the labeled species along with information on dispersion.

Two separate techniques for tracking radioactive particles have been developed. Radioactive particle tracking (RPT) determines the source position by comparing count rates of high-energy gamma rays detected by different detectors within an array of individual detectors, whereas positron emission particle tracking (PEPT) uses triangulation of the pairs of back-to-back annihilation photons from positron–electron annihilation. These two techniques are described below.

7.3.1 Radioactive particle tracking

In RPT, the location of the tracer is found by comparing the gamma ray intensities (count rates) measured in a number of small detectors, which are strategically positioned around the system under study. The intensity falls off with distance from the tracer according to the inverse square law, modified by the effects of gamma ray attenuation and scattering within the system. For a particular geometry and material layout, the actual relationships must be determined via a calibration procedure; for a given set of count rates, least-squares fitting then provides an estimate of tracer position. The technique has mainly been used with sources that emit high-energy gamma rays (above about 1 MeV) so as to minimize attenuation. For practical reasons, sources with long half-lives are also preferred: examples include ^{46}Sc (84 days, 889 and 1121 keV gamma rays) and ^{60}Co (5.3 years, 1173 and 1332 keV) and source activity is typically 100 μCi.

The concept of tracking a radioactive particle can be traced back at least to the 1960s. The first modern RPT system was developed by Lin, Chen, and Chao at Illinois (1985). They used 12 2-inch NaI(Tl) scintillator detectors, mounted in three staggered rings, to track a 0.5 mm encapsulated particle of ^{46}Sc moving in a gas-fluidized bed of 0.5 mm glass beads, and obtained detailed velocity fields and autocorrelation functions.

This technique was refined by Dudukovic's group at St Louis, who reported a computer automated radioactive particle tracking (CARPT) facility (Moslemian, Devanathan, & Dudukovic, 1992) comprising 16 detectors operating at a sampling rate of 33 Hz. Over a vertical range of 60 cm in a 30 cm-diameter fluidized bed, this system determined the tracer location with a mean error of 9 mm. This system was extensively used to study gas–solid fluidized beds and the behavior of the liquid phase in bubble columns.

The original CARPT technique assumed that for each detector the count rate depended only on the distance from the source to the center of the detector, and not on direction or varying material content. Subsequently in Montreal, Larachi, Chaouki, and coworkers developed an improved approach: the volume of interest is divided into a fine grid and the expected count rates are determined for each grid location using Monte Carlo modeling which incorporates effects such as detector geometry and attenuation in the rig. During subsequent tracking, the experimental count rates in the various detectors are compared to the modeled rates, searching for the location that best matches the measured values (Larachi, Chaouki, & Kennedy, 1995). A neural network approach has also been used (Godfroy, Larachi, Kennedy, Grandjean, & Chaouki, 1997). A problem with all these methods is that they are unable to correct for temporal variations in attenuation due to moving parts or changes in bed holdup.

Variants on the RPT technique have been used by a number of groups to study particle or fluid motion especially in fluidized beds and bubble columns. As an alternative to using long-lived tracers, Vieira, Pinto, Lopes, and de Souza (2020) report a technique for manufacturing hybrid PMMA La_2O_3 microspheres (diameters ranging from 100 to 900 μm) which are then activated using thermal neutrons to generate ^{140}La (half-life 40 h).

7.3.2 Positron emission particle tracking

The PEPT technique was conceived and first developed at the University of Birmingham, UK, in the late 1980s. Since positron−electron annihilation generates pairs of almost back-to-back photons, if a single positron-emitting source is present in the field of view of a PET scanner, then in principle it should be possible to estimate its location by triangulation using just two pairs of photons whose lines of response both pass close to this location (Fig. 7.5). In practice, the data contain a significant number of corrupt events—random coincidences or events where one or both photons has scattered prior to detection—so that it is necessary to start with a larger dataset, typically 100 pairs. It should then be possible to distinguish the cluster of genuine events, which converge on the tracer location, from the broad background of corrupt events.

The standard PEPT algorithm (Parker, Broadbent, Fowles, Hawkesworth, & McNeil, 1993) uses an iterative approach. Given a set of coincidence events, each of which defines a line of response, the algorithm calculates a centroid, defined as the point that minimizes the sum of distances to all the lines of response. The lines furthest from the point are then discarded, and the centroid is recalculated using the remaining set of events. Iteration continues until only a specified fraction f of the original events remains; these are considered genuine events. The optimum value of f depends on the geometry and on the amount of scattering material through which the 511 keV photons have to pass. In contrast to RPT, calibration is not required, and because the PEPT algorithm efficiently removes scattered events, accurate location is possible even when over 95% of the detected data is corrupt.

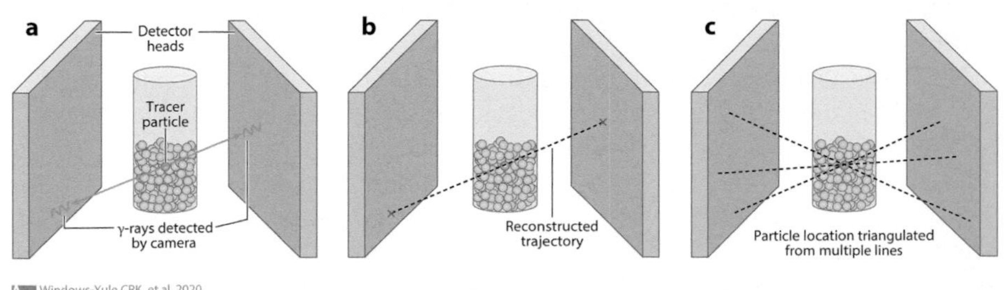

Figure 7.5 Principle of positron emission particle tracking.

PEPT was first developed at Birmingham using the positron camera already mentioned, which consisted of a pair of rectangular gas-filled multiwire proportional chambers, $60 \times 30\ cm^2$ in area. The efficiency of each chamber for detecting an incident 511 keV photon was just 7%, meaning that the overall efficiency for detecting a pair of photons, both of which struck the chambers, was just 0.5% (0.07^2). This low efficiency meant that the true coincidence data were always accompanied by a significant contribution of random coincidences, and in practice the useful data rate was limited to 3000 coincidence events per second. In 1999, this positron camera was replaced by one comprising a pair of sodium iodide gamma camera heads (Forte, manufactured by ADAC Laboratories, Milpitas, CA, USA), giving an order of magnitude improvement in efficiency (Parker, Forster, Fowles, & Takhar, 2002). Using a tracer of optimum activity, the Forte can detect 10^5 coincidence events per second. If it takes 100 events to locate the tracer accurately, then this can be achieved every millisecond. For a slow-moving tracer, the location accuracy obtained using an initial sample of N events can be estimated as $\frac{w}{\sqrt{fN}}$, where w represents the spatial resolution of the positron camera (around 8 mm). Taking $N = 100$ and $f = 0.25$, this formula predicts a location accuracy of around 1.6 mm. The location accuracy deteriorates if the tracer moves significantly during the location interval; for the example cited, this occurs at speeds above a few m/s. At lower speeds, more accurate location can be achieved by increasing the sampling interval.

PEPT has been used at Birmingham to study a very wide range of systems including stirred tanks, rotary kilns, gas-fluidized beds and vibrofluidized granular gases. The ADAC Forte positron camera is still used for many of these studies, and is popular due to its open geometry. To complement this system, a series of modular positron cameras have been developed, using the detector blocks and electronics from redundant medical PET scanners. As well as giving greater geometrical flexibility, these modular cameras have the advantage of being transportable, so that PEPT studies can be performed "in the field." For example, during 2006, a modular camera was used to perform a PEPT study of a large fluidized bed at BP's Hull site (Ingram et al., 2007). Modular cameras have also been used to follow the motion of small refractory inclusions during the casting of liquid metals (molten steel or aluminum). Fig. 7.6 shows a camera consisting of 64 BGO blocks (8 banks, each of 8 detector blocks) being used to study flow of molten aluminum along a channel containing a weir: tracking was possible over a length of 600 mm. In some cases, these modular cameras can generate data rates of over 1 MHz, enabling very accurate high-speed tracking.

PEPT has also been performed by Hoffman's group using a ring scanner: in 2011 they reported tracking of 500 μm particles in a hydrocyclone (Chang, Ilea, Aasen, & Hoffmann, 2011) located every millisecond with an accuracy of better than 200 μm. In 2009, the University of Cape Town established a PEPT facility at iThemba Labs in South Africa. PEPT Cape Town is equipped with a ring scanner (ECAT EXACT3D with an axial field of view of 23.4 cm) and more recently with an ADAC Vertex positron camera (similar to the ADAC Forte but with a slightly different gantry), and specializes in applications of PEPT in minerals processing (Buffler, Cole, Leadbeater, & van Heerden, 2018).

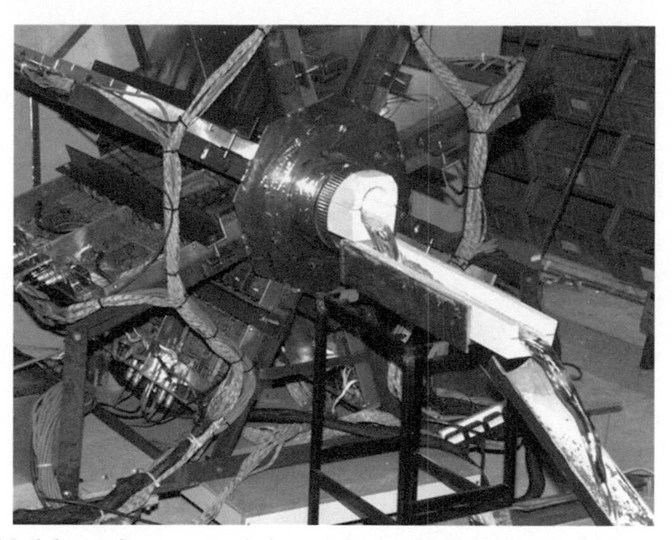

Figure 7.6 Modular positron camera being used to study flow of molten aluminum along a channel containing a weir.

Several groups have explored the possibility of using PEPT to track several labeled particles simultaneously. The Birmingham group extended the basic PEPT algorithm to locate multiple particles in turn (Yang, Parker, Fryer, Bakalis, & Fan, 2006) and are currently investigating alternative methods. The Tennessee group reported tracking multiple particles using a preclinical scanner (Langford, Wiggins, Tenpenny, & Ruggles, 2016), while the Cape Town group offer a different approach (Bickell, Buffler, Govender, & Parker, 2012). To date, most publications on multiparticle tracking are demonstrations of feasibility rather than actual studies, but it seems likely that future work will increasingly exploit these techniques.

As TOF PET systems become more widely available, the capability of PEPT to locate particles using small numbers of events should improve even more.

References

Anger, H. O. (1964). Scintillation camera with multichannel collimators. *Journal of Nuclear Medicine, 5*, 515–531.

Barth, T., Ludwig, M., Kulenkampff, J., Gründig, M., Franke, K., Lippmann-Pipke, J., et al. (2014). Positron emission tomography in pebble beds. Part 1: Liquid particle deposition. *Nuclear Engineering and Design, 267*, 218–226.

Bickell, M., Buffler, A., Govender, I., & Parker, D. J. (2012). A new line density tracking algorithm for PEPT and its application to multiple tracers. *Nuclear Instruments and Methods A, 682*, 36–41.

Buffler, A., Cole, K., Leadbeater, T. W., & van Heerden, M. R. (2018). Positron emission particle tracking: A powerful technique for flow studies. Proceedings of the 2017 international conference on applications of nuclear techniques. *International Journal of Modern Physics: Conference series, 48*, 1860113.

Chang, Y.-F., Ilea, C. G., Aasen, O. L., & Hoffmann, A. C. (2011). Particle flow in a hydrocyclone investigated by positron emission particle tracking. *Chemical Engineering Science, 66*, 4203–4211.

Godfroy, L., Larachi, F., Kennedy, G., Grandjean, B. P. A., & Chaouki, J. (1997). On-line flow visualization in multiphase reactorsusing neural networks. *Applied Radiation and Isotopes, 48*, 225–235.

Hawkesworth, M. R., Parker, D. J., Fowles, P., Crilly, J. F., Jefferies, N. L., & Jonkers, G. (1991). Nonmedical applications of a positron camera. *Nuclear Instruments and Methods in Physics Research Section A, 310*, 423–434.

Ingram, A., Hausard, M., Fan, X., Parker, D. J., Seville, J. P. K., Finn, N., et al. (2007). Portable positron emission particle tracking (PEPT) for industrial use. In F. Berruti, X. Bi, & T. Pugsley (Eds.), *The 12th international conference on fluidization – new horizons in fluidization engineering. ECI Symposium Series* (Vol. RP4). http://dc.engconfintl.org/fluidization_xii/60.

Kulenkampff, J., Grundig, M., Zakhnini, A., & Lippmann-Pipke, J. (2016). Geoscientific process monitoring with positron emission tomography (GeoPET). *Solid Earth, 7*, 1217–1231.

Langford, S., Wiggins, C., Tenpenny, D., & Ruggles, A. (2016). Positron emission particle tracking (PEPT) for fluid flow measurements. *Nuclear Engineering and Design, 302*, 81–89.

Larachi, F., Chaouki, J., & Kennedy, G. (1995). 3-D mapping of solids flow fields in multiphase reactors with RPT. *AIChE Journal, 41*, 439–443.

Legoupil, S., Pascal, G., Chambellan, D., & Bloyet, D. (1997). An experimental single photon emission computed tomograph method for dynamic 2D fluid flow analysis. *Applied Radiation and Isotopes, 48*, 1507–1514.

Lin, J. S., Chen, M. M., & Chao, B. T. (1985). A novel radioactive particle tracking facility for measurement of solids motion in gas fluidized beds. *AIChE Journal, 31*, 465–473.

Moslemian, D., Devanathan, N., & Dudukovic, M. P. (1992). Radioactive particle tracking technique for investigation of phase recirculation and turbulence in multiphase systems. *Review of Scientific Instruments, 63*, 4361–4372.

Parker, D. J., Broadbent, C. J., Fowles, P., Hawkesworth, M. R., & McNeil, P. A. (1993). Positron emission particle tracking – a technique for studying flow within engineering equipment. *Nuclear Instruments and Methods in Physics Research Section A, 326*, 592–607.

Parker, D. J., Forster, R. N., Fowles, P., & Takhar, P. S. (2002). Positron emission particle tracking using the new Birmingham positron camera. *Nuclear Instruments and Methods in Physics Research Section A, 477*, 540–545.

Parker, D. J., Leadbeater, T. W., Fan, X., Hausard, M. N., Ingram, A., & Yang, Z. (2008). Positron imaging techniques for process engineering: Recent developments at Birmingham. *Measurement Science and Technology, 19*, 094004.

Park, J. G., Jung, S.-H., Kim, J. B., Moon, J., Han, M. C., & Kim, C. H. (2014). Development of advanced industrial SPECT system with 12-gonal diverging-collimator. *Applied Radiation and Isotopes, 89*, 159–166.

Park, J. G., Jung, S.-H., Kim, J. B., Moon, J., & Kim, C. H. (2015). Development of a SPECT system for industrial process flow measurement using diverging collimators. *Nuclear Technology, 192*, 133–141.

Perret, J., Prasher, S. O., Kantzas, A., Hamilton, K., & Langford, C. (2000). Preferential solute flow in intact soil columns measured by SPECT scanning. *Soil Science Society of America Journal, 64*, 469–477.

Surti, S. (2015). Update on time-of-flight PET imaging. *Journal of Nuclear Medicine, 56*(1), 98.

Vieira, W. S., Pinto, J. C., Lopes, R. T., & de Souza, M. N. (2020). Development of hybrid microspheres for assessment of multiphase processes. *Applied Radiation and Isotopes, 158*, 109035.

Yang, Z., Parker, D. J., Fryer, P. J., Bakalis, S., & Fan, X. (2006). Multiple-particle tracking – an improvement for positron particle tracking. *Nuclear Instruments and Methods A, 564*, 332–338.

Ultrasound tomography

Nicholas James Watson
Faculty of Engineering, University of Nottingham, Nottingham, United Kingdom

8.1 Introduction

Ultrasound is a branch of acoustics which utilizes mechanical pressure waves above 20 KHz. This frequency relates to the approximate uppermost frequency detected by humans. Acoustic waves are oscillations of pressure which elastically propagate through a medium. These pressure oscillations generate density and small thermal variations. The origins of study into acoustic waves date back to Pythagoras in Greek times and the theory has advanced thanks to the aid of distinguished scientists such as Helmholtz, Newton, Raleigh, and Stokes. Acoustics now has applications in a wide range of disciplines such as medical imagining, detecting fatigue, and flaws in mechanical components and mapping the ocean floors. There are many different classifications of acoustic and ultrasound techniques and these will briefly be presented before discussing the techniques relevant to industrial tomography.

The first classification is whether a system is passive or active. A passive system is defined as one that only detects acoustic waves, such as human hearing or microphones used to detect gas leaking from a pipe. In comparison, an active system is defined as one that transmits and receives acoustic waves through a medium. The second common classification of ultrasound is low power or high power. Low power ultrasound generally has an intensity below 100 mW/cm^2 where high power is between 10 and 10,000 W/cm^2 (Martini, 2013). Low power waves are elastic in nature and make no changes to the medium in which they are propagating. Low power ultrasound can be used to measure material properties of a substance, detects flaws or inhomogeneities, and images samples. In contrast, high power ultrasound is nonelastic and is used to disrupt and change a system. High power ultrasound can be used to mix or homogenize fluid samples, modify the crystallization process, sterilize, and degas samples and has numerous additional processing applications (Awad, Moharram, Shaltout, Asker, & Youssef, 2012).

As ultrasound tomography is used to monitor and image processes, it is classified as a low power active system. For these types of ultrasound systems, there are three necessary components: A transmitter, a medium for the wave to propagate through, and a receiver. Often the transmitter can act as the receiver if the acoustic wave is reflected from an interface within the system. Tomography is defined as the 2D or 3D imaging of a sample or process; therefore, ultrasound tomography is imaging using ultrasonic waves. Ultrasound tomography was initially developed for medical applications in the 1970s (Greenleaf, 1970, pp. 125–136) and in the early 1980s the first

Industrial Tomography. https://doi.org/10.1016/B978-0-12-823015-2.00010-8

systems were developed to study industrial processes such as two-phase flows (Fincke, 1980) and identify solid objects in a fluid (Satti & Szilard, 1983). Ultrasound tomography systems are also being developed to operate in solid materials and have been used to image material loss in aircraft components (Hay, Royer, Gao, Zhao, & Rose, 2006), image stress fields in sandstone (Scott & Roegiers, 1994), and detect decay in trees (Liang, Wang, Wiedenbeck, Cai, & Fu, 2007).

The key benefits of using ultrasound for a tomography system include the following:

- Nondestructive
- Noninvasive
- Can operate online
- Can operate in near real time
- Can be used on opaque systems
- Operate on a wide range of length scales
- No moving parts
- No radioactive or other dangerous components
- often less expensive than other techniques

Although these benefits make ultrasound appear to be an ideal topographic technique, there are challenges which must be overcome, such as transmitting the waves through heterogeneous samples and interpreting the received signals. An additional issue which must always be considered when using ultrasound is the wave propagation speed. For water at ambient temperature, sound waves travel at approximately 1500 m/s; this is significantly slower than electromagnetic waves which travel at around 300,000,000 m/s in the same medium (Hoyle, 1996). This relatively slow propagation speed can result in a major issue if the area to be images is large or the process is changing in a short period of time.

The remainder of this chapter will be structured as follows. Section 8.2 will present fundamental ultrasound/acoustic theory; this will include how acoustic waves are characterized and what happens when they interact with boundaries and changing physical properties within a system. This theory can then be used to explain what can be measured and imaged using ultrasound techniques. As well as covering all of the techniques currently used in ultrasound tomography systems, a more extensive explanation of the uses of ultrasound in process monitoring will be included. This is to enable to the reader to imagine what additional uses ultrasound could have in future process tomography applications. Section three will present the different components required for an ultrasound tomography systems and discuss image reconstruction techniques and how the specifications of the system can be calculated. Section 8.4 will discuss the applications of ultrasound tomography in process monitoring and the chapter will end with sections on future trends, further reading, and a summary.

8.2 Ultrasound theory

There are different types of ultrasound waves which can propagate in a medium; these are longitudinal, shear, surface (Rayleigh), and plate (Lamb). Generally speaking, only longitudinal waves propagate in fluids due to their densities being too low to support shear waves and no solid boundaries for surface or plate waves. The focus of this chapter is ultrasound tomography for process monitoring which is generally a pipe-bound fluid system, so the theory will only be presented for longitudinal waves. For more details of acoustic propagation in solids, the reader can refer to the books of Krautkrämer and Krautkrämer (1990) and Rose (1999). Ultrasound tomography systems have been developed utilizing surface and plate waves with an example being the imaging of defects in the floor of a fuel tank (Mazeika, Kazys, Raisutis, Demcenko, & Sliteris, 2006).

This theoretical section will firstly present the theory for a propagating longitudinal pressure wave and then discuss how it interacts with the material it propagates through and any interface or inhomogeneities within it. It is important to understand these interactions as they are what make an ultrasound system capable of delivering information on a sample or process. For further information on acoustic and ultrasound theory in fluid systems, the reader should refer to Challis, Povey, Mather, and Holmes (2005), Cheeke (2002), Dukhin and Goetz (2002), and Povey (1997),

8.2.1 Acoustic propagation wave theory

Longitudinal or compression waves are defined as waves where the particle motion is in the same direction in which the wave is propagating. The oscillations in pressure are sinusoidal in nature and characterized by their frequency, amplitude, and wavelength (Fig. 8.1).

The frequency (f) of a wave is related to the period (T) of a single cycle by the following equation:

$$f = \frac{1}{T}$$

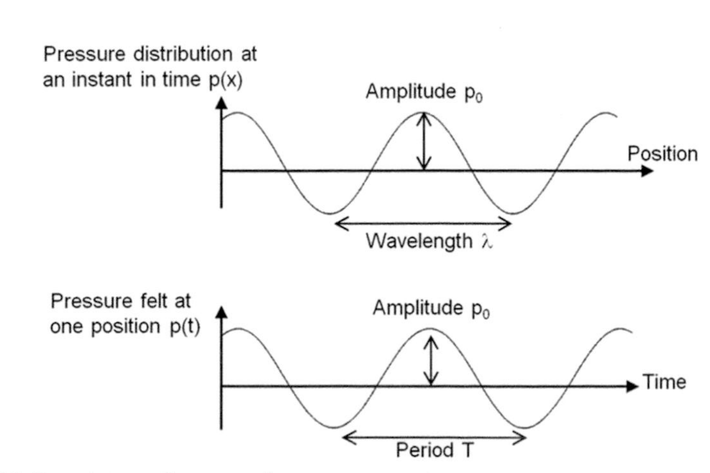

Figure 8.1 Descriptors of an acoustic wave.

with the unit of frequency being the Hz (s^{-1}). This frequency can be described as an angular or radial frequency (ω):

$$\omega = 2\pi f$$

In addition, we must define the phase (ζ) of the wave which is always relative to a specified point.

$$\zeta = k_c' x,$$

k_c' is the real part of the acoustic wavenumber

$$k_c' = \frac{2\pi}{\lambda} = \frac{\omega}{v_p} = \frac{2\pi f}{v_p},$$

$$\boldsymbol{k_c} = k_c' + ik_c'' = k_c' + i\alpha$$

Utilizing these equations, we can express the propagating pressure wave as follows:

$$\Delta p = \Delta p_0 \exp\{i(\omega t - \zeta + i\alpha x)\}$$

where Δp is the instantaneous pressure and Δp_0 is the maximum pressure deviation, respectively; $i - \sqrt{-1}$, k', and k'' are the real and imaginary parts of the complex wave vector $\mathbf{k}$; v_p is the phase velocity of the wave; λ is wavelength; and α is the attenuation coefficient.

Often an acoustic wave will travel as a pulse consisting of multiple frequency components. The Group Velocity v of a pulse is the velocity of the envelope of the wave. This contains all cycles in the pulse and is given by the following:

$$v = \frac{d\omega}{dk}$$

The velocity of the different frequencies contained within the pulse is each given a phase velocity v_p given by the following:

$$v_p = \frac{\omega}{k} = f\lambda$$

where λ is the wavelength at the specific frequency. Pressure is related to amplitude or displacement ξ through the following:

$$p \sim \xi^2$$

The attenuation of the wave can then be defined in terms of measureable pressure differences as follows:

$$-\propto = \frac{1}{x}\ln\frac{\xi}{\xi_0}\ \text{Neper/m}$$

Or

$$-\propto = \frac{1}{x}20\log\frac{\xi}{\xi_0}\ \text{dB/m}$$

The decibel (dB) and Neper (N) are called dimensionless units, because they are ratios. The attenuation is normally expressed in terms of pressure rather than displacement and in this case

$$-\alpha\left[\text{dBm}^{-1}\right] = \frac{1}{x}10\log\frac{p}{p_0} = \frac{1}{x}20\log\frac{\xi}{\xi_0}.$$

8.2.2 Acoustic impedance

An acoustic wave will continue to propagate through a medium until it encounters an interface in the elastic distribution. These interfaces are parameterized through the characteristic acoustic impedance:

$$Z = \rho v$$

where ρ is the density. Consider plane sound waves in medium 1 (Z_1, v_1, and ρ_1) at normal incidence to an interface with medium 2 (Z_2, v_2, and ρ_2). The reflection coefficient R, the ratio of the reflected pressure amplitude (p_R) to the incidence pressure amplitude (p_I), is (Krautkrämer & Krautkrämer, 1990) given by the following equation:

$$R = \frac{Z_2 - Z_1}{Z_2 + Z_1}$$

The above equation indicates that reflection increases with the size of the impedance mismatch. However, strictly, R is a function of incident angle θ and can change markedly with θ, e.g., due to critical angles at which complete internal reflection or surface wave generation can occur. Broadly speaking, the interaction is a reflection when the length scale of the boundary/bounded object l is much greater than the sound wavelength λ. Meanwhile, when $l \leq \lambda$, the feature acts as an inhomogeneity and scatters the sound. Scattering itself has a spectrum of behavior, ranging from the mid-frequency regime ($l \sim \lambda$), where the scattering is sensitive to shape and size resonances (Uberall, 1992), up to the far limit of Rayleigh scattering where the scattering becomes insensitive to shape ($l >> \lambda$) (Morse & Ingard, 1968). To determine whether reflection or scattering will dominate when an incident wave hits the interface of a spherical object, we define the value *Ka:*

$$ka = k_c' a = \frac{2\pi a}{\lambda}$$

where a is the radius of the sphere. When $ka \gg 1$, reflection will dominate and when $ka \ll 1$, scattering will dominate (Fig. 8.2).

8.2.3 Speed of sound

The frequency and amplitude of the propagating pulse are governed by the transducer and excitation source and these can be selected depending on the measurements required. However, the speed v at which the pulse propagates in a fluid is a function of the materials adiabatic compressibility (Woods, 1941):

$$v = \frac{1}{\sqrt{\kappa \rho}}$$

where κ is the adiabatic compressibility of the material and ρ is the materials density. The fact that the speed of sound is dependent on the material properties of the medium within which it propagates can be used to gain information on this medium. The speed of sound varies for different materials and is fastest in solids and slowest in gases. Table 8.1 indicates the speed of sound for some common materials:

8.2.4 What can ultrasound measure in industrial processes?

Acoustic waves can be used to gain information on a system in which they are propagating by measuring:

1. How the waves propagates in a single phase system
2. What happens at an interface in a multiphase system

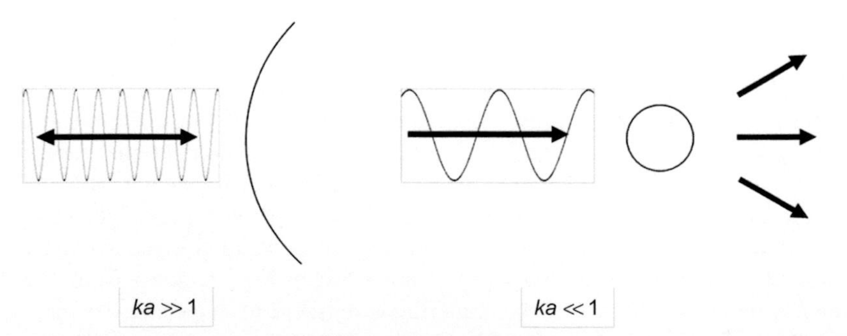

Figure 8.2 When an incident wave reaches an interface, either reflection ($ka \gg 1$)or scattering ($ka \ll 1$) dominates.

Table 8.1 Speed of sound in some common materials.

Material	Speed of sound (v) (m/s)
Iron	5170
Lead	2160
Water	1464
Oxygen	316
Air	331
Ethanol	1207
Methanol	1103
Castor oil	1477

Adapted from Dukhin, A. S., & Goetz, P. J. (2002). *Ultrasound for characterizing colloids: Particle sizing, zeta potential, rheology*. Oxford: Elsevier.

It is these two principles which are the focus of all ultrasound tomography systems. Therefore, the following section will discuss what information can be gained from each of them.

8.2.4.1 Wave propagation in a single phase

For a single phase homogenous fluid, the speed of sound in its simplest form is a function of the density and the compressibility of the medium. Therefore, any changes in these material properties will affect the speed at which the wave propagates through the medium and by regularly measuring the speed of sound any spatial or temporal changes in material properties such as density can be measured. The speed of a wave in a medium is recorded by sending an acoustic pulse through the medium from one transducer to another. If the distance (d) between the transducers is known and the time of flight (tof) of the pulse is recorded v can be calculated from

$$v = \frac{d}{\text{tof}}$$

The temperature of a fluid is known to have a significant effect on the speed of sound and attenuation of a wave propagating through it. This makes ultrasound techniques an ideal tool for monitoring temperature changes in a fluid although prior knowledge of the speed of sound across the temperature range to be monitored must be known. Fig. 8.3 displays the speed of sound and attenuation of water from 0 to 100°C. Temperature fluctuations due to process or environmental factors can be one of the biggest challenges when using ultrasound techniques to monitor industrial processes. It is important to constantly monitor the temperature where the measurements are being made and ensure that an understanding of the temperature effects on the particular fluid or system is known.

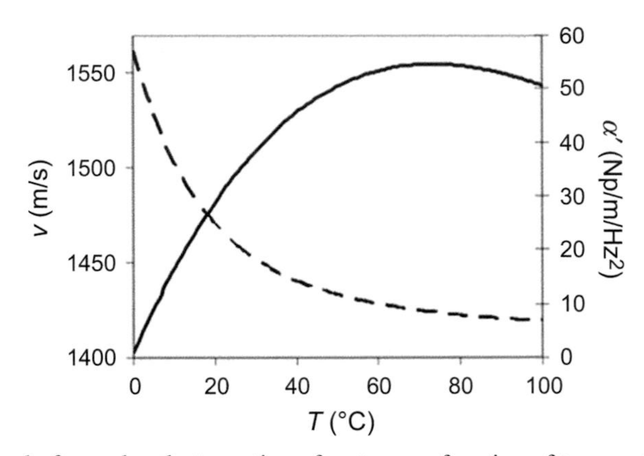

Figure 8.3 Speed of sound and attenuation of water as a function of temperature.
Adapted from Povey, M. J. W. (1997). *Ultrasonic techniques for fluids characterization*. San Diego: Academic Press.

The velocity of sound is also sensitive to the pressure of the fluid and for water can be calculated from the following equation (Pierce, 1981):

$$v = 1402.39 + 5.03711T - 0.0580852T^2 + 3.33420 \times 10^{-4}T^3 - 1.47800 \times 10^{-6}T^4$$
$$+ 3.14643 \times 10^{-9}T^5 + 1.6 \times 10^{-6}(p_0 - 10^5)$$

where T is the temperature in degrees Celsius and p_0 is the absolute pressure in Pascals. The concentration of mixtures can be determined from the sound velocity within them. In the simplest form, the speed of sound of two miscible fluids is a function of the volume averaged densities and compressibility's of the two fluids:

$$v = \frac{1}{\sqrt{k_{VA}\rho_{VA}}}$$

$$k_{VA} = \phi k_2 + (1 - \phi)k_1$$

$$\rho_{VA} = \phi \rho_2 + (1 - \phi)\rho_1$$

where ϕ is the volume fraction of fluid 2 in fluid 1. These equations have being adapted further to measure concentrations of particular substances in water, such as sugar (Contreras montes de Oca, Fairley, McClements, & Povey, 1992) and alcohol (Povey, 1997).

Phase transitions can be monitored using ultrasound techniques in emulsions, dispersions, and homogenous fluids. Accurate measurements of the temperature can be made at which the phase transition occurs. In a solid/liquid or liquid/solid phase

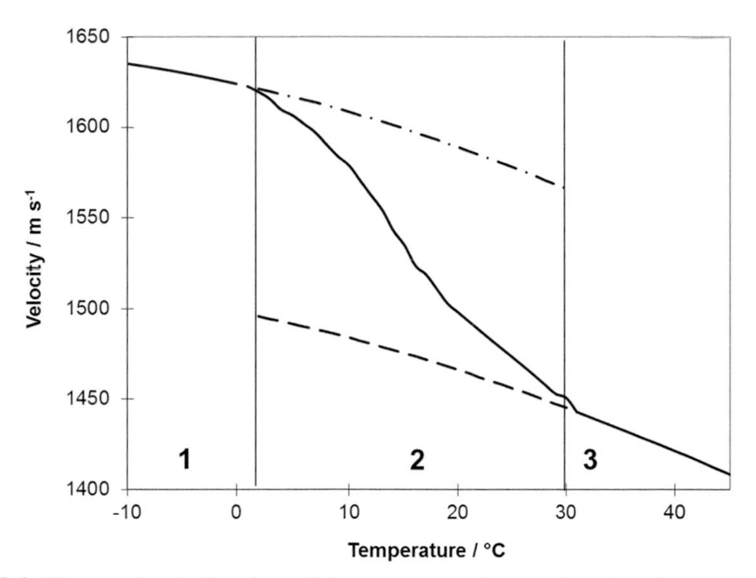

Figure 8.4 Ultrasound velocity of an oil-in-water margarine as a function of temperature. In region 1 the margarine is solid, in region 3 it is liquid, and in region 2 it is a mixture of solid and liquid.
Adapted from McClements, D. J., & Povey, M. J. W. (1992). Ultrasonic analysis of edible fats and oils. *Ultrasonics*, *30*(6), 383−388.

transition, a large change in compressibility occurs and this has an effect on the speed of sound (recall speed of sound is largest in solids). An example of using speed of sound to detect phase changes is presented in Figs. 8.4 and 8.5 where an oil-in-water emulsion goes from a liquid dispersed phase to a solid dispersed phase. Note the crystallization temperature of 2.5 Degrees Celsius and the melting temperature of 15 Degrees Celsius and the associated changes on the speed of sound.

8.2.4.2 What happens at an interface in a multiphase system

When a propagating acoustic wave reaches an interface characterized by a difference in acoustic impedance, reflection and/or scattering occurs. Whether reflection or scattering dominates depends on the size and shape of the scattering object in relation to the wavelength of the incoming wave. The majority of current ultrasound tomography systems operate on the premise that the system is operating in the short wavelength limit $(ka \gg 1)$, that is that the scatterers are significantly larger than the wavelength of the incident wave. These techniques assume that the majority of the incident wave is reflected at this interface and the location of the interface can be found by measuring the time it takes the transmitted ultrasound pulse to travel to the interface, be reflected, and then propagate to the receiving transducers.

When ka is equal to or less than 1, acoustic scattering will occur. There are different techniques available to measure the size distribution of scattering particles in a fluid using ultrasound techniques, such as ultrasound spectroscopy (Alba, 1992).

This technique measures the velocity and/or attenuation of the ultrasound as a function of frequency and acoustic path length and uses mathematical inversion models such as ECAH (Allegra & Hawley, 1972; Foldy, 1945, Lloyd & Berry, 1967) to calculate the particle size distribution. Ultrasound spectroscopy requires several measurements at one position so would not be suited to ultrasound tomography due to the many measurements required to construct tomographs. However, other more simple models such as the modified Urick equation (Pinfield, Povey, & Dickinson, 1995) exist which can calculate size from a single measurement but do assume a monodispersed size distribution where the wavelength is much larger than the radius of the spherical scattering objects. When using ultrasound to measure particle size, it is important to understand the size distribution and localized concentrations of the scatterers as often the systems may be outside the operating realms of the inversion models. An example being where simultaneous multiple scattering occurs.

Care should always be taken when using ultrasound techniques to measure bubble size distributions as the incident wave will cause the bubbles to oscillate and resonate at a frequency dependent upon their radii (Leighton, Phelps, Ramble, & Sharpe, 1996). Although sizing the bubbles is possible, this resonance can cause complication in the interpretation of the acoustic signals and make data interpretation more difficult.

8.2.5 Limitations on ultrasound systems

Although ultrasound appears to be an attractive process monitoring tool, it does have several limitations. These must always be considered before deciding if it is the most suitable process tomography system to use. These limitations include the following:

- slow propagation velocity of sound waves
- difficult to understand signals in complex systems (e.g., large range of inhomogeneity size and concentration)
- temperature must remain constant or constantly be monitored
- Inversion techniques must improve

8.3 Equipment and techniques

So far this chapter has focused on the theory of ultrasound propagation and discussed what information can be gained from a measurement in a single position with an individual or pair of transducers. Tomography requires measurements to be made from multiple positions (projections) so that 2D and 3D spatial images can be constructed where the image contrast is governed by the property of a system under investigation (for example, density or void fraction). This section will introduce the main components required for an ultrasound tomography system and discuss the different setups which can be used.

8.3.1 Equipment setups

Ultrasound tomography instruments consist of various different hardware components and software to control the hardware and reconstruct the image from the many recorded projections. Although numerous different instrument setups exist, the primary components are as follows: Acoustic transducers, Pulser receiver device, control and signal processing hardware, and a PC to control and generate the images.

8.3.1.1 Transducers

The ultrasound transducer is one of the most critical components of the tomography systems as it is responsible for transmitting and receiving the pulse and has a significant effect of the performance parameters of the systems. The most common types of transducers are piezoelectric ones. These transducers utilize the piezoelectric effect which is when a piezoelectric material deforms when an electrical voltage is applied across it. By applying an oscillating voltage, the material will vibrate and emit a pressure wave. The inverse of this occurs when a pressure wave hits a piezoelectric material and a voltage is produced. Design and construction of piezoelectric transducers is a well-documented area and the reader can refer to Silk (1984) for more details. Piezoelectric transducers are not the only transducers available and other types include capacitance transducers (Wright, Schindel, Hutchins, Carpenter, & Jansen, 1998) and hydrophones (Martin, Beesle, & Myers, 1995).

How the acoustic wave propagates from the transducer face is an important property of the ultrasound tomography system and transducers can be built which emit relatively straight waves, focusing waves, and waves which spread out like a fan. As ultrasound tomography systems wish to construct a 2D image of a flowing process, transducers with a fan beam (Fig. 8.5) are often used as these cover the most area in two dimensions so can generate the most spatial information for a single transmitted

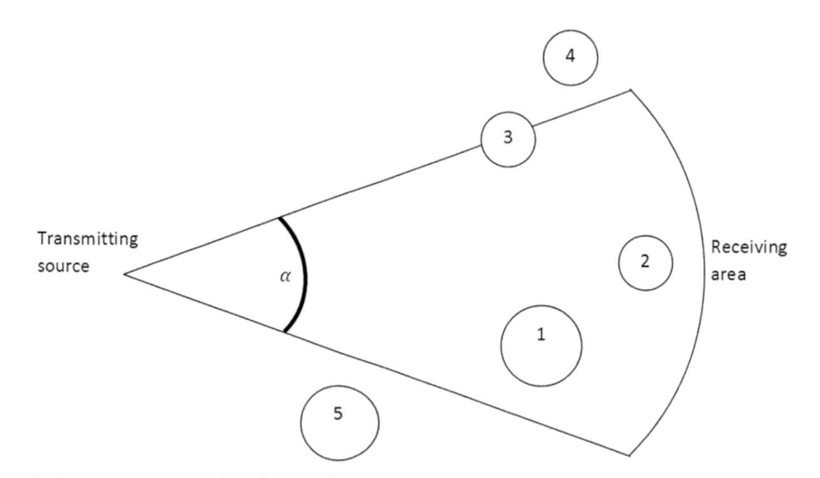

Figure 8.5 Wave propagation from a fan-based transducer. For the beam spread angle α, the scatterers 1, 2, and 3 will have an effect on the propagating wave.

pulse. The beam spread of the fan-based transducer is characterized by the beam angle (α) at which the waves propagates. The selection of beam angle should be considered carefully as although a wider angle will cover a greater area, this will have a negative effect on the spatial resolution of the system (Schlaberg, Yang, & Hoyle, 1998). A fan-shaped beam can be produced by careful design of the size and shape of the transducer face or by adding a cylindrical lens (Steiner & Podd, 2006).

Ultrasound transducers are characterized by their frequency and bandwidth and it should always be remembered that the frequency of a propagating wave is a characteristic of the emitting transducer. The selection of suitable frequency is critical in any ultrasound system as the frequency affects the propagation distance of the wave and the spatial resolution of the system. Acoustic attenuation increases with the frequency squared of the propagating wave, so lower frequencies would be more desirable in highly attenuating systems although lower frequencies have lower spatial resolution. It is obvious that a trade-off exist, so the selection of operating frequency should be selected based on the requirements of the system and the medium the wave will be propagating in. Fig. 8.6 displays the trade-off in spatial resolution and attenuation as a function of frequency for a transducer with a focused beam. In general, ultrasound tomography systems operate between 18 KHz and 20 MHz (Fazalul Rahiman, Abdul Rahim, Abdul Rahim, & Ayob, 2012) with the most commonly used transducer being around 2 MHz (Abdul Rahim, Fazalul Rahiman, & Mohd Taib, 2005).

The bandwidth of the transducer is a description of the frequency range over which is capable of emitting waves. This is usually characterized by a central frequency and the frequency range at which the frequency-dependent output reduces by 6 dB. In most

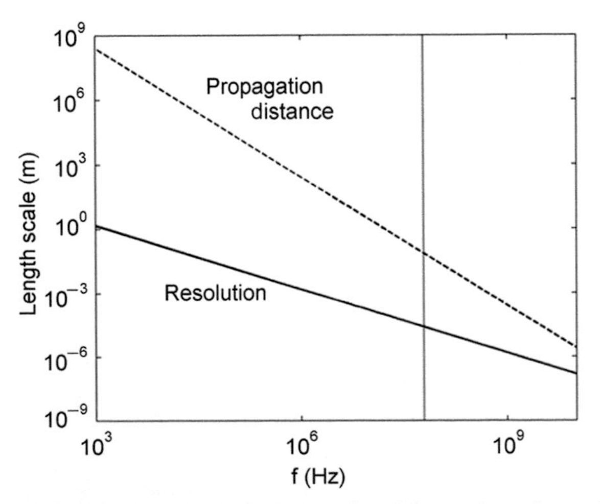

Figure 8.6 Trade-off between frequency and operating distance for a focused transducer in water.
Adapted from Parker, N. G., & Povey, M. J. W. (June 23–24, 2009). Scanning acoustic microscopy for mapping the microstructure of soft materials. *What where when multi-dimensional advances for industrial process monitoring, Leeds, UK.*

ultrasound tomography systems, each transducer will emit at a single frequency to minimize difficulties in deconvoluting received signals. However, it may be desirable to transmit across a range of frequencies for applications such as particle sizing, as well as transducer frequency and bandwidth other properties need to be considered such as the operating temperature and physical size of the transducers.

The different transducer configurations used for ultrasound tomography can be categorized as follows:

- Single or pair of transducers attached to a mechanical system and rotated around pipe
- An array of emitting or receiving transducers
- An array of transmitting and receiving transducer

Generally, multiple transducers will be equally spaced around the domain to be imaged which is usually a circular pipe cross-section. The ultrasound tomography instrument begins by pulsing the first transmitting transducer and recording the detected signal at each receiving transducers. After all propagation from this transmitter have faded beyond a detectable level, the next transducer will be excited. This process continues until a pulse has been transmitted from all transmitting transducers. It may seem ideal to have as many transducers as possible as the more projections recorded, the greater the level of detail in the reconstructed image. However, the number of transducers and projections per image affects the time required to construct an image (Yang, Schlaberg, Hoyle, Beck, & Lenn, 1999).

How the transducers are attached to the system under investigation also requires thought. Some systems use transducers placed in the flowing system; these would benefit from a simpler propagation problem but could interfere with the flow so were not classified as noninvasive and transducers could suffer damage from the flowing medium. Other ultrasound tomography systems attach the transducers to the outside of the pipe removing these issues (Steiner & Podd, 2006). When coupling transducers to the outside of the pipe, it should be considered that the waves must propagate through the pipe wall (which may not have uniform thickness) and a coupling medium is required between the transducer and the outside of the wall due to the large acoustic impedance mismatch between air and solid materials.

8.3.1.2 Pulser, receiver, and signal processing

For an ultrasound transducer to generate an acoustic wave, it must be excited by some form of electrical voltage. These can be provided by standalone pulsing devices or by USB operated signal generators. Pulsers operate by either applying a high voltage spike of known voltage, a square wave of known duration and voltage, or a sinusoidal pulse of known voltage, frequency, and duration. The excited voltage is selected as one that will result in a pulse which can propagate through the system under inspection and be detected by the receiving transducers. High voltage excitations are required for highly attenuating systems although the downside of high voltage excitations is that they can distort the transmitted wave introducing error into the signal analysis and image reconstruction process.

The signal received at the transducers must be saved before being processed and used to generate a tomograph. For earlier systems, this would have being performed using an oscilloscope, but more recently this will be integrated into the software system. It is important that the excitation pulse is synchronized with the data acquisition process as often time of flight measurements are required to calculate speed of sound in the system or to locate inhomogeneities or boundaries in space.

For airborne ultrasound or systems operating on processes with high attenuation, it may be necessary to have some form of signal amplification for the transmitted and/or received signals. Other devices often utilized in an ultrasound tomography system include transducer positioning equipment and temperature monitoring or controlling equipment.

8.3.2 Different types of systems

The two most popular techniques are transmission and reflection ultrasound tomography systems. Transmission systems work by propagating a sound wave through a material and gaining information on the systems by the time and amplitude at which the pulse is received by transducers at different radial locations. Reflection devices operate by sending a pulse which is reflected at an interface with a different acoustic impendence before being detected by the transducers. Fig. 8.7 highlights the difference in operation between a transmission and reflection system. Reflection systems are generally used to study multiphase flows where a liquid and gas exist as the acoustic impedance is large between liquids and gases (Al-Aufi et al., 2019). Transmission systems are more suited to detecting localized changes in material properties such as

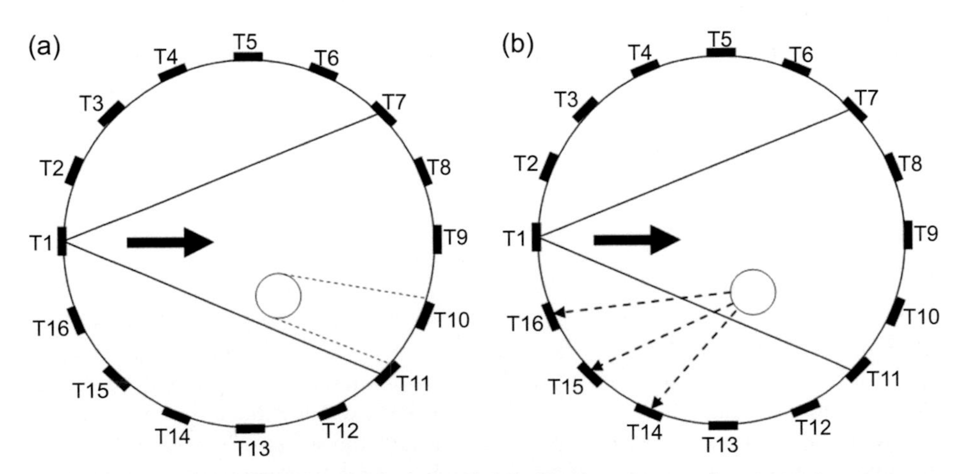

Figure 8.7 Operation of (a) Transmission and (b) reflection ultrasound tomography systems.

density or temperature but can also be used to detect inhomogeneities as these will block the transmitted waves. Examples of transmission systems include the work of Fazalul Rahiman et al. (2006) and Holstein et al. (2004), whereas examples of reflection systems include the work of Chen and Sanderson (1996) and Schlaberg et al. (1998).

It should be noted that often ultrasound tomography systems operate by utilizing the transmission and reflection of acoustic waves. These systems have the benefit of acquiring more process information for each transmitted pulse but add additional complexity to the signal processing and inversion processes. Ultrasound tomography can also be performed on nonfluid pipe flows; for example, acoustic microscopy inspection of electronic components and airborne ultrasound for foreign body detection in foods. In these systems, a single or pair of transducers is used. The transducers will be scanned over the sample under inspection and a two- or three-dimensional picture can be constructed where each pixel represents a single pulse through a sample. A single transducer setup would be operating in reflection mode, whereas two transducers placed either side of the sample are required for transmission mode.

8.3.3 Image reconstruction

Efficient and accurate image reconstruction algorithms are required to accurately produce images of the system property under investigation from the recorded ultrasound data. The majority of image reconstruction work for ultrasound tomography is focused on medical applications (Lasaygues, Mensah, Guillermin, Rouyer, & Franceshini, 2012; Steiner & Watzenig, 2008). The requirements of medical ultrasound tomography systems are often different to those in industrial processes, which include flowing fluids with rapidly changing properties (Gai, Li, Plaskowski, & Beck, 1989). These flows will also often include gas or solid inhomogeneities. The image reconstruction algorithms used for industrial ultrasound tomography do not differ dramatically from the ones used for X-ray, ERT, or ECT techniques, with the most popular method being the linear back projection method (Wright & O'sullivan, 2003). This is used primarily due to the low computational cost (Fazalul Rahiman, Abdul Rahim, Abdul Rahim & Ayob, 2012). The linear back projection method can be split into the following stages (Fig. 8.8):

1. Split the domain into a determined number of pixels. For a 2D image, this could be, for example, a 64 by 64 array giving 4096 pixels
2. Create a sensitivity map which is the expected response at each receiver from a given transmitter
3. Measure the desired property at the receiving transducer (amplitude or time of flight)
4. Back project each measured value
5. Calculate the concentration map for each pixel using a suitable method
6. Represent each pixel as a color level

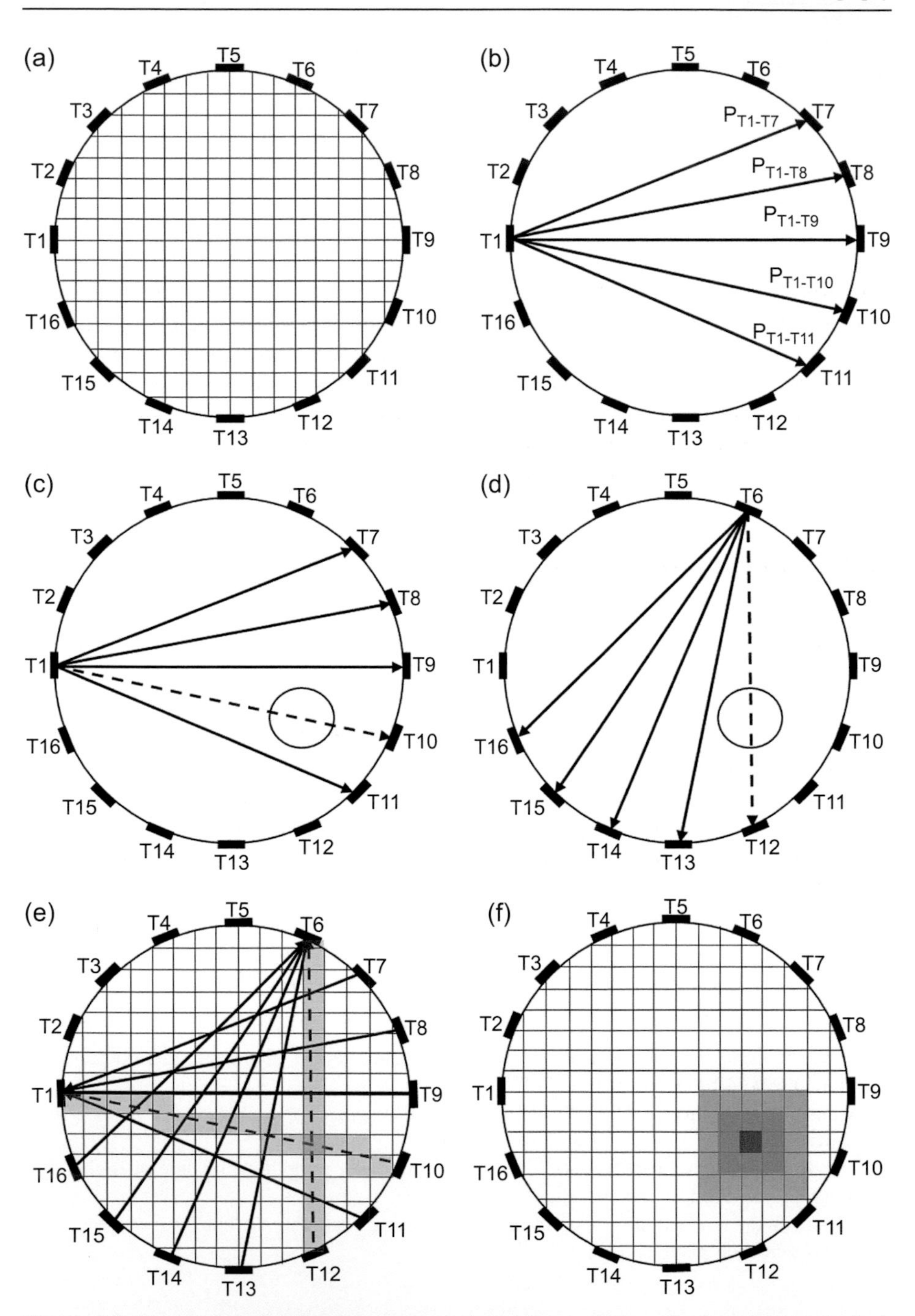

Figure 8.8 Image reconstruction for a transmission system utilizing a back projection method. (a) The cross-sectional area is split into pixels. (b) Sensitivity map created for each possible projection. (c) Projections from transducer 1; note how projection $T_1{-}T_{10}$ is affected by the circular inhomogeneity. (d) Projection from transducer 6; note how the projection $T_6{-}T_{12}$ is affected by the circular inhomogeneity. (e) Back projection method. (f) Pixel intensity value.

For example, where the recorded value is the received voltage, the voltage distribution for a pixel with coordinates x and y can be found using the following (Fazalul Rahiman, Abdul Rahim, Abdul Rahim & Ayob, 2012):

$$V_{\text{LBP}}(x,y) = \sum_{Tx=1}^{m} \sum_{Rx=1}^{n} S_{Tx,Rx} \times \overline{M}_{Tx,Rx}(x,y)$$

where $V_{\text{LBP}}(x,y)$ is the voltage distribution from the LBP algorithm in the concentration matrix and $S_{Tx,Rx}$ is the sensor loss voltage for the corresponding projection from Tx to Rx and $\overline{M}_{Tx,Rx}(x,y)$ is the normalized sensitivity matrices.

Fazalul Rahiman, Abdul Rahim, Abdul Rahim and Ayob (2012) reported that the LBP method is not the most accurate reconstruction method as it can smear out details and loose definition at interface boundaries. These and other researchers have adapted the back projection technique to include thresholding methods (Fazalul Rahiman, Abdul Rahim, Fazalul Rahiman, & Tajjudin, 2006; Opielinski & Gudra, 2006). This is where a pixel is given, either one of two values based on the value of the pixel relative to some predetermined threshold. These thresholding methods have been found to give an improvement in the classification of two-phase flows (Abdul Rahim, Fazalul Rahiman, Chan, & Nawawi, 2007; Yang et al., 1999).

To identify the location of interfaces or reflectors in a system the following equations can be used:

$$\text{tof} = \frac{1}{v}(s_1 + s_2) = \frac{1}{v}\left(\sqrt{(x_i - x_t)^2 + (y_i - y_t)^2} + \sqrt{(x_r - x_i)^2 + (y_r - y_i)^2}\right)$$

where tof is the time of flight between the transmit and received pulse, s_1 is the distance from the transmitter to the interface and s_2 is the distance from the interface to the receiver, and v is ultrasonic velocity in the fluid. X and Y are the coordinates in Cartesian space of the transmitter (t), receiver (r), and interface (i), respectively. These equations do of course assume that the acoustic pulse follows a straight line.

Other techniques used to reconstruct the image include parameterized contour model (Steiner & Watzenig, 2008), Fourier techniques (Mensah & Franceschini, 2006), sparse signal representation (Jovanovic, Hormati, Sbaiz, & Vetterli, 2007), and wave-based reconstruction (Roy, Jovanovic, Hormati, Parhizkar, & Vetterli, 2010). For further information on different reconstruction techniques for ultrasound tomography, the reader can refer to the thesis of Jovanovic (2008). To aid the understanding of the accuracy and effect of different image reconstruction algorithms, often computer simulations are performed to access their effectiveness for a particular application. Examples of such simulations include those of Chen and Sanderson (1996), Fazalul Rahiman, Abdul Rahim, and Zakaria (2008), Li and Hoyle (1997), Liu, Tan and Dong (2020) and Norton and Linzer (1979).

8.3.4 Ultrasound tomography linked with additional techniques

Ultrasound tomographic techniques have also being linked with other modalities to increase the sensing capability of a system. Hoyle et al. (2001) describe a system which combines electrical resistance, electrical capacitance, and ultrasound tomography. They discuss the challenges in relation to hardware and data processing. Deinhammer, Steiner, Sommer, and Watzenig (2008) combine ultrasound and electrical capacitance techniques to monitor multiphase flows. They conclude that ultrasound techniques are more suited to detecting interfaces between materials, whereas the ECTs are better at characterizing the individual phases. Liu et al. (2011) combine ultrasound spectroscopy and electrical resistance tomography to monitor an emulsification process. The combination of techniques is used to monitor the droplet size distribution in the emulsion. One issue with using ultrasound spectroscopy is the number of inputs required in the inversion model, so ERT is used to image the concentration profile and remove this unknown from the ultrasound inversion problem.

It should be remembered when using multiple sensing technologies any propagating radiation does not interact with any other in an undesirable way and that the effect of additional measurements on the system specifications (such as temporal resolution) is known.

8.3.5 Calculating system resolution

For any industrial monitoring or imaging system, it is important to be able to understand and have methods to calculate the operating specification of the system. For tomography systems, the two most crucial are spatial and temporal resolution.

8.3.5.1 Spatial resolution

Spatial resolution is defined by the size of inhomogeneities that can be detected and the accuracy of spatial results in the reconstructed pixelated image. As ultrasound tomography systems operate on the premise that reflection dominates at an interface ($ka \gg 1$), it is possible to calculate the size of the smallest scattering object that can be detected. Assuming that the transmitted pulse is 2 Mhz and the fluid is water at 25°C, the wavelength of the acoustic wave will be approximately 0.75 mm. To operate in the long wavelength limit, the scattering objects must have a radius of approximately 1 mm or larger. Increasing the transmitted frequency will result in the ability to detect smaller inhomogeneities without having to consider scattering but does come at the cost of increased signal attenuation.

For further discussion and methods for calculating spatial resolution of ultrasound tomography systems, the reader can refer to Xu and Xu (1998) and Yang et al. (1999). Xu and Xu (1998) found that their ultrasound tomography system had a spatial resolution equal to 1% of the cross-sectional area and highlighted the fact that spatial resolution was not the same at different radial locations in this circular area.

8.3.5.2 Temporal resolution

Temporal resolution is the time to take the multiple measurements of the cross-section and then reconstruct the image. This time is important for process tomography systems as often changes can occur within the system in a space of time which is shorter than the one required to record the measurements. This is further hampered by the fact that acoustic waves have a relative slow propagation speed. For example, acoustic waves propagate as approximately 1500 m/s in water at 25°C compared to 300,000,000 m/s for electromagnetic waves. Abdul Rahim, Fazalul Rahiman et al. (2007) developed a system capable of making measurements at 300 frames per second but found that the image reconstruction process limited them to an imaging capability of 10 frames per second. The time to acquire a single 2D image for a reflection system in a pipe can be calculated from the following (Yang et al., 1999):

$$t = n\frac{2d}{v}(1+p)$$

where n is the number of transducers, d is the diameter of the pipe, c is the speed of sound in the medium, and p is an additional percentage of time for the ultrasonic wave to attenuate below a detectable level. For a transmission system, the 2 is removed from the equation. For a system with 32 transducers in a pipe of diameter 100 mm, assuming a fluid with $v = 1500$ m/s and $p = 0.3$, typical acquisition times are around 6.2 ms. However, the time to reconstruct an image is much longer as each pulse from a transducer will be detected by multiple receiver resulting in many projections. Image reconstruction can be improved by more efficient algorithms, more powerful computers, and parallel processing.

8.4 Industrial applications

Section 8.2.4 described the different properties of a system in which information can be gained by utilizing ultrasound techniques. This section will focus on reviewing the ultrasound tomography systems presented in the literature and will be split into two sections. The first will focus on single phase system and the seconds will focus of multiphase systems.

8.4.1 Characterization of single phase systems

One of the earliest reported uses of ultrasound tomography to monitor single phase systems was the work of Martin et al. (1995). In this work temperature, variations were imaged in a fluid using multiple hydrophones. Wright and coworkers have

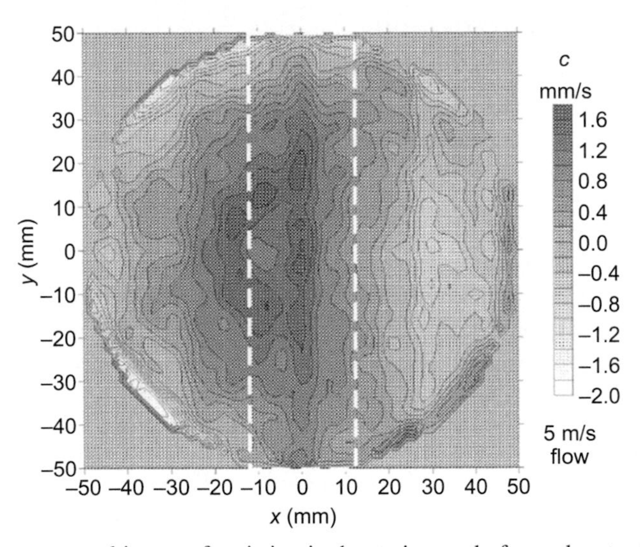

Figure 8.9 Reconstructed image of variation in the static speed of sound c at a flow speed of 5 m/s. The dashed white line indicates where a bluff body was located downstream of the flow. Reproduced with permission Wright, W. M. D., & O'Sullivan, I. J. (2003). Ultrasonic tomographic imaging of air flow in pipes using an electrostatic transducer array. *Review of Quantitative Nondestructive Evaluation*, 22.

imaged the flow velocity and spatial temperature differences in air using electrostatic ultrasound transducers (Ingleby & Wright, 2002; Wright et al., 1998; Wright & O'Sullivan, 2003) (Fig. 8.9). During this work, they have used different transducer configurations. Some work involved transducers which were moved to record the different projections, whereas other work had transducers fastened around a pipe section. All of their systems operated in the transmission mode and they recorded time of flight and received voltage amplitude data. Ingleby and Wright (2002) were also able to identify solid objects in the air flows using their ultrasound tomography system. Holstein et al. (2004) imaged air flow velocity and spatial temperature distribution using a transmission system as did Jovanovic (2008). li and Zhou (2006) used an ultrasound tomography systems based on speed of sound measurements to monitor single phase gas flow in a furnace.

8.4.2 Characterization of multiphase flows

The majority of published work in the literature using process ultrasound tomography systems is for imaging multiphase flows. Some of the first work in the 1980s was Kolbe, Turko, and Leskovar (1986) who used 20 transducers in a reflection system to detect bubbles in a liquid flow and Wolf (1988) who used a transmission system to also image bubbles in a liquid flow.

In the 1990s, research using ultrasound tomography systems continued; Brown, Reilly, and Mills (1996) used a transmission system with 18 transducers to detect solid objects in a gas system. Chen and Sanderson (1996) used a reflection system and imaged air bubbles in water using a fixed transmitter and a position able receiver. Brown and Reilly (1996) also used repositionable transducers, but in a transmission, systems and were capable of imaging solid objects in air. Xu and Xu (1997), Xu, Han, Xu, and Yang (1997), used a transmission system to detect bubbles in a liquid flow and were capable of using their ultrasound tomography system to classify different types of two phase flows such as bubble; slug; annular; and stratified flow. Yang et al. (1999) used a reflection-based ultrasound tomography system to image bubbles and solid objects in a liquid flow. Warsito, Ohkawa, Kawata, and Uchida (1999) developed a transmission system with six moveable transducers to image gas and particle holdups in a slurry bubble common using ultrasonic velocity and attenuation measurements. Utomo, Warsito, Sakai, and Uchida (2001) used the same system to then image gas and TiO_2 holdups in a liquid flow.

Research using ultrasound tomography systems continued into the 2000s. Schlaberg, Podd, and Hoyle (2000) and West, Jia and Williams (2000) were able to use a reflection ultrasound tomography system to study air liquid flows in a hydrocyclone. Deinhammer et al. (2008) coupled reflection ultrasound tomography with electrical capacitance tomography to detect air regions in a flowing liquid. Opielnski and Gudra (2006) used a transmission system with two pairs of moveable transmitters and receivers to detect multiple solid objects of varying shape in a gas system. The Malaysia researchers Abdul Rahim and Fazalul Rahiman have many published articles on a transmission ultrasound tomography system using a fan beamed transducer to image bubbles in flow and classify different types of liquid gas flows such as plug, stratified, and annular (Abdul Rahim et al., 2005, Abdul Rahim, Fazalul Rahiman et al., 2007; Abdul Wahab, Abdul Rahim, Fazalul Rahiman, & Ahmad, 2011; Fazalul Rahiman et al., 2006; Fazalul Rahiman, Abdul Rahim, Abdul Rahim, & Ayob, 2012; Fazalul Rahiman, Abdul Rahim, Abdul Rahim, Muji, & Mohamad, 2012) (Fig. 8.10). This ultrasound tomography system was also capable of imaging gas and solids in the same fluid continuum (Abdul Wahad et al., 2011) and was use to image the quantity of oil and water flowing in a pipe by measuring the received signal voltage (Abdul Rahim, Nyap, Fazalul Rahiman, & San, 2007).

As we move into 2020s, research has continued exploring the use of ultrasound tomography systems for industrial multiphase flow applications. Rahiman et al. (2016) used high-speed imaging to determine the effectiveness of an ultrasound tomography system to study a bubble column and Langener et al. (2016) studied the effect of measurement rate on reconstructed image quality of air bubbles in fluid systems. For the more widespread deployment of ultrasonic technologies in industrial applications, it is important that they can operate effectively on industrially relevant pipe wall material such as steel. The effect of pipe wall material has been studied by several researchers who identified that system properties such as transducer frequency and excitation voltages should be adjusted for different pipe wall materials (Cailly et al., 2020; Goh et al., 2017).

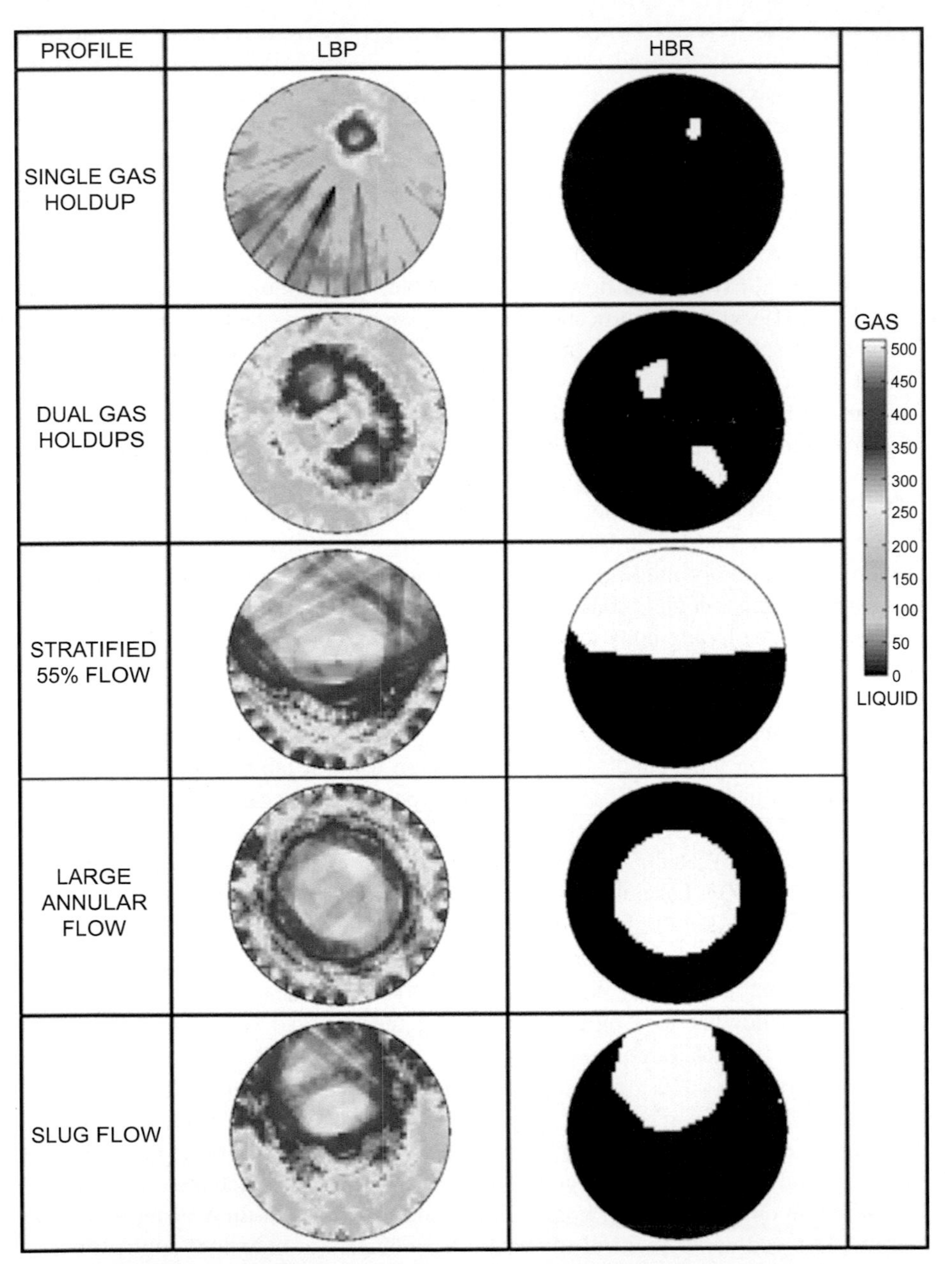

Figure 8.10 Image reconstruction of liquid gas flows using the linear back projection and hybrid binary reconstruction techniques.
Reproduced with permission Abdul Rahim, R., Fazalul Rahiman, M. H., Chan, K. S., & Nawawi, S. W. (2007). Non-invasive imaging of liquid/gas flow using ultrasonic transmission-mode tomography. *Sensors and Actuators*, 135, 337–345.

8.5 Summary

This chapter has focused on ultrasound process tomography instruments. Section 8.2 introduced the basic ultrasound theory and described what happens when an acoustic wave interacts with an interface. As the majority of industrial processes are fluid based, the theory was only presented for ultrasound and acoustics in fluids. This chapter has included what measurements are made using ultrasound methods so the reader can understand what can be imaged using the technique. This section not only included what measurements are made currently with ultrasound process tomography systems, but included an overview on all applications of ultrasound for process monitoring so the reader could understand the techniques full potential. Section 8.3 described the main components for an ultrasound tomography system and discussed the different types of systems and how the images are reconstructed and system specifications calculated. Section 8.4 presented the different systems used for ultrasound process tomography systems and was split into two sections, the first for imaging single phase flows and the second for imaging multiphase flows. This chapter should have given the reader a detailed insight into how ultrasound techniques can be used for process imaging although a powerful technique with many applications and a significant potential in the future of process imaging the original limitations such as slow wave propagation speed will always remain. These issues should always be considered when deciding if it is the most suitable process imaging technique to use.

8.6 Future trends

There was a large body of work performed by various different research groups in the 1990s developing ultrasound tomography systems mainly with the purpose to monitor flow or temperature profiles or characterize multiphase flows in pipes. However, since this time, there has only been a modest amount of research in this area with advances focused on the challenges associated with industry deployment. One area of growing research interest is the use of industrial digital technologies and data-driven modeling methods such as artificial intelligence and machine learning. These methods are been utilized routinely within industry under the headings of digital manufacturing or Industry 4.0. Tomography techniques are aligned with this digital revolution as they capture the data required to develop these models. Ultrasonic tomography methods will also benefit from the use of new data-driven modeling techniques. Recent research has shown how machine learning methods can be combined with ultrasonic measurements to monitor industrial processes such as mixing and fouling removal (Bowler, Bakalis & Watson, 2020; Escrig et al., 2020a,b) and analyze data acquired from ultrasound tomography systems to identify different multiphase flow regimes (Fang et al., 2020; Figueiredo et al., 2016). As the adoption of digital technologies grows, we would expect to see a greater adoption of tomography systems. Below summarizes some trends in ultrasound techniques which should have a positive benefit on future ultrasound tomography systems for industrial processes:

- Constant Improvement in design of ultrasound transducers
- Novel design and construction of airborne ultrasound transducers
- Reduction in cost and size of ultrasound transducers and other hardware components
- Combination with other digital technology research areas such as the Internet of Things and machine learning
- Improvement in image reconstruction techniques and use of deep learning methods
- Improvements in computer technologies, e.g., parallel processing
- Development of process tomography systems using ultrasound and other imaging modalities

8.7 Source of further information and advice

For detailed reading on acoustic theory in solids, the reader should refer to the textbooks of Krautkrämer and Krautkrämer (1990) and Rose (1999). These books also include techniques for surface waves and waves in thin plates which have applications in ultrasound tomography. For ultrasound theory and techniques in fluid and colloidal systems, the reader can refer to the books of Dukhin and Goetz (2002) and Povey (1997). For an overview of ultrasound techniques in industrial processes, the reader can refer to Chapter 9 in Ultrasonics, Fundamental technologies, and applications (Ensminger & Bond, 2012). For further reading specifically on ultrasound process tomography, the reader can refer to specific references within this chapter or Chapter 7 of Process Tomography, principles, techniques, and applications (Williams & Beck, 1995). Ultrasound microscopy generates 2D and 3D images using ultrasound technology. These techniques may also be relevant to ultrasound process tomography, so for further information, the reader should refer to the books of Briggs and Kolosov (2010) and Maev (2008).

References

Abdul Rahim, R., Fazalul Rahiman, M. H., Chan, K. S., & Nawawi, S. W. (2007). Non-invasive imaging of liquid/gas flow using ultrasonic transmission-mode tomography. *Sensors and Actuators, 135*, 337–345.

Abdul Rahim, R., Fazalul Rahiman, M. H., & Mohd Taib, M. N. (2005). Non-invasive ultrasonic tomography: Liquid/gas flow visualization. In *1st international conference on computers, communications, & signal processing with special track on biomedical engineering. Kuala Lumpur, Malaysia, November 14–16.*

Abdul Rahim, R., Nyap, N. W., Fazalul Rahiman, M. H., & San, C. K. (2007). Determination of water and oil flow composition using ultrasound tomography. *Elektrika, 9*(1), 19–23.

Abdul Wahab, Y., Abdul Rahim, R., Fazalul Rahiman, M. H., & Ahmad, M. A. (2011). Application of transmission-mode ultrasonic tomography to identify multiphase flow regime. In *International conference on electrical, control and computer engineering, June 21–22.*

Al-Aufi, Y. A., et al. (2019). Thin film thickness measurements in two phase annular flows using ultrasonic pulse echo techniques. *Flow Measurement and Instrumentation, 66*. https://doi.org/10.1016/j.flowmeasinst.2019.02.008

Alba, F. (1992). *Method and apparatus for determining particle size distribution and concentration in a suspension using ultrasonics*. U.S. Patent no. 5121629.

Allegra, J. R., & Hawley, S. A. (1972). Attenuation of sound in suspensions and emulsions theory and experiments. *Journal of the Acoustical Society of America, 51*, 1545–1564.

Awad, T. S., Moharram, H. A., Shaltout, O. E., Asker, D., & Youssef, M. M. (2012). Applications of ultrasound in analysis, processing and quality control of food: A review. *Food Research International, 48*, 410–427.

Bowler, A. L., Bakalis, S., & Watson, N. J. (2020). Monitoring mixing processes using ultrasonic sensors and machine learning. *Sensors, 20*(7), 1813. https://doi.org/10.3390/s20071813. Multidisciplinary Digital Publishing Institute.

Briggs, G. A. D., & Kolosov, O. (2010). *Acoustic microscopy*. New York: Oxford Science Publications.

Brown, G. J., & Reilly, D. (1996). Ultrasonic tomographic imaging of solid objects in air using an array of fan-shaped-beam electrostatic transducers. *Ultrasonics, 34*, 111–115.

Brown, G. J., Reilly, D., & Mills, D. (1996). Development of an ultrasonic tomography system for application in pneumatic conveying. *Measurement Science and Technology, 7*, 396–405.

Cailly, W., et al. (2020). Pipe two-phase flow non-invasive imaging using ultrasound computed tomography: A two-dimensional numerical and experimental performance assessment. *Flow Measurement and Instrumentation*, 74. https://doi.org/10.1016/j.flowmeasinst.2020.101784. Elsevier Ltd.

Challis, R. E., Povey, M. J. W., Mather, M. L., & Holmes, A. K. (2005). Ultrasound techniques for characterizing colloidal dispersions. *Reports on Progress in Physics, 68*, 1541–1637. https://doi.org/10.1088/0034-4885/68/7/R01

Cheeke, J. N. (2002). *Fundamentals and applications of ultrasonic waves*. Boca Raton: CRC Press.

Chen, Z. X., & Sanderson, M. L. (1996). Ultrasonic tomography for process measurement. In *IEEE instrumentation and measurement technology conference, Brussels, Belgium, June 4–6*.

Contreras montes de Oca, N. I., Fairley, P., McClements, D. J., & Povey, M. J. W. (1992). Analysis of the sugar content of fruit juices and drinks using ultrasonic velocity measurements. *International Journal of Food Science and Technology, 27*, 515–529.

Deinhammer, C., Steiner, G., Sommer, M., & Watzenig, D. (2008). Electromechanical flow imaging using ultrasound and electrical capacitance data. *IEEE Sensors Conference*.

Dukhin, A. S., & Goetz, P. J. (2002). *Ultrasound for characterizing colloids: Particle sizing, zeta potential, rheology*. Oxford: Elsevier.

Ensminger, D., & Bond, J. (2012). *Ultrasonics: Fundamentals, technologies, and applications* (3rd ed.). Taylor and Francis.

Escrig, J., et al. (2020a). Monitoring the cleaning of food fouling in pipes using ultrasonic measurements and machine learning. *Food Control, 116*. https://doi.org/10.1016/j.foodcont.2020.107309

Escrig, J. E., et al. (2020b). Ultrasonic measurements and machine learning for monitoring the removal of surface fouling during clean-in-place processes. *Food and Bioproducts Processing, 123*, 1–13. https://doi.org/10.1016/j.fbp.2020.05.003

Fang, L., et al. (2020). Identification of two-phase flow regime using ultrasonic phased array. *Flow Measurement and Instrumentation, 72*, 101726. https://doi.org/10.1016/j.flowmeasinst.2020.101726. Elsevier Ltd.

Fazalul Rahiman, M. H., Abdul Rahim, R., Abdul Rahim, H., & Ayob, N. M. N. (2012). Design and development of ultrasonic process tomography. In D. Santos (Ed.), *Ultrasonic waves*.

Fazalul Rahiman, M. H., Abdul Rahim, R., Abdul Rahim, H., Muji, S. Z. M., & Mohamad, E. J. (2012). Ultrasonic tomography – image reconstruction algorithms. *International Journal of Innovative Computing, Information and Control, 8*(1B), 527–538.

Fazalul Rahiman, M. H., Abdul Rahim, R., Fazalul Rahiman, M. H., & Tajjudin, M. (2006). Ultrasonic transmission-mode tomography imaging for liquid/gas two-phase flow. *IEEE Sensors Journal, 6*, 6.

Fazalul Rahiman, M. H., Abdul Rahim, R., & Zakaria, Z. (2008). Design and modelling of ultrasonic tomography for two-component high-acoustic impedance mixture. *Sensors and Actuators A: Physical, 147*, 409–414.

Figueiredo, M. M. F., et al. (2016). The use of an ultrasonic technique and neural networks for identification of the flow pattern and measurement of the gas volume fraction in multiphase flows. *Experimental Thermal and Fluid Science, 70*, 29–50. https://doi.org/10.1016/j.expthermflusci.2015.08.010. Elsevier Inc.

Fincke, J. R. (1980). *The application of reconstructive tomography to the measurement of density distribution in two phase flow proc. 26th int. instrumentation symp* (pp. 235–243). Research Triangle Park, NC: ISA.

Foldy, L. L. (1945). The multiple scattering of waves - I. General theory of isotropic scattering b randomly distributed scatterers. *Physics Review, 67*.

Gai, H., Li, Y. C., Plaskowski, A., & Beck, M. S. (1989). Ultrasonic flow imaging using time-resolved transmission-mode tomography. In *Proc. IEE 3rd international conference on image processing and its applications* (pp. 237–241). Warwick: Warwick University Press.

Goh, C. L., et al. (2017). Simulation and experimental study of the sensor emitting frequency for ultrasonic tomography system in a conducting pipe. *Flow Measurement and Instrumentation, 54*, 158–171. https://doi.org/10.1016/j.flowmeasinst.2017.01.003. Elsevier Ltd.

Greenleaf, J. F. (1970). Introduction to computer ultrasound tomography. In *Computed aided tomography and ultrasonics in medicine, North-Holland*.

Hay, T. R., Royer, R. L., Gao, H., Zhao, X., & Rose, J. L. (2006). A comparison of embedded sensor lamb wave ultrasonic tomography approaches for material loss detection. *Smart Materials and Structures, 15*, 946–951.

Holstein, P., Raabe, A., Muller, R., Barth, M., Mackenzie, D., & Starke, E. (2004). Acoustic tomography on the basis of travel-time measurement. *Measurement Science and Technology, 15*, 1420–1428.

Hoyle, B. S., Jia, X., Podd, F. J. W., Schlaberg, H. I., Tan, H. S., Wang, M., et al. (2001). Design and application of a multi-modal process tomography system. *Measurement Science and Technology, 12*, 1157–1165.

Hoyle, B. S. (1996). Process tomography using ultrasonic sensors. *Measurement Science and Technology, 7*, 272–280.

Ingleby, P., & Wright, W. M. D. (2002). Ultrasonic imaging in air using fan-beam tomography and electrostatic transdcuers. *Ultrasonics, 40*, 507–511.

Jovanovic, I. (2008). *Inverse problems in acoustic tomography: Theory and applications* (Ph.D. thesis). Serbia: University of Belgrade.

Jovanovic, I., Hormati, A., Sbaiz, L., & Vetterli, M. (2007). Efficient and stable acoustic tomography using sparse reconstruction methods. In *19th international congress on acoustics, Madrid, Spain, September 2–7.*

Kolbe, W. F., Turko, B. T., & Leskovar, B. (1986). Fast ultrasonic imaging in a liquid filled pipe. *IEEE Transactions on Nuclear Science, 33*(1), 715–722.

Krautkrämer, J., & Krautkrämer, H. (1990). *Ultrasonic testing of materials*. Berlin: Springer-Verlag.

Langener, S., et al. (2016). *A real-time ultrasound process tomography system using a reflection-mode reconstruction technique*. https://doi.org/10.1016/j.flowmeasinst.2016.05.001

Lasaygues, P., Mensah, S., Guillermin, R., Rouyer, J., & Franceshini, E. (2012). Non-linear inversion modeling for ultrasound computer tomography transition from soft to hard tissues imaging. *Medical Imaging 2013: Ultrasonic Imaging, Tomography and Therapy.*

Leighton, T. G., Phelps, A. D., Ramble, D. G., & Sharpe, D. A. (1996). Comparison of the abilities of eight acoustic techniques to detect and size a single bubble. *Ultrasonics, 34*, 661–667.

Liang, S., wang, X., Wiedenbeck, J., Cai, Z., & Fu, F. (2007). Evaluation of acoustic tomography for tree decay detection. In *Proceedings of the 15th international symposium on non destructive testing of wood. Duluth, Minnesota, USA. 10–12 September.*

Li, W., & Hoyle, B. S. (1997). Ultrasonic process tomography using multiple active sensors for maximum real-time performance. *Chemical Engineering Science, 52*(13), 2161–2170.

Liu, L., Li, R. F., Collins, S., Wang, X. Z., Tweedie, R., & Primrose, K. (2011). Ultrasound spectroscopy and electrical resistance tomography for online characterisation of concentrated emulsions in crossflow membrane emulsifications. *Powder Technology, 213*, 123–131.

Liu, H., Tan, C., & Dong, F. (2020). Absolute reconstruction of ultrasonic tomography for oil-water biphasic medium imaging using modified ray-tracing technique. *Measurement: Sensors, 7–9*, 100023. https://doi.org/10.1016/j.measen.2020.100023. Elsevier BV.

Li, Y. Q., & Zhou, H. C. (2006). Experimental study on acoustic vector tomography of 2-D flow field in an experiment-scale furnace. *Flow Measurement and Instrumentation, 17*, 113–122.

Lloyd, P., & Berry, M. V. (1967). Wave propagation through an assembly of spheres. IV. Relations between different multiple scattering theories. In *Proceedings of the physical society, London* (Vol. 91, pp. 678–688).

Maev, R. G. (2008). *Acoustic microscopy: Fundamentals and applications*. Weinheim: Wiley-VCH.

Martin, P. D., Beesle, M., & Myers, P. E. (1995). Ultrasound imaging in large gas and liquid processing vessels. *The Chemical Engineering Journal, 56*, 183–185.

Martini, S. (2013). Sonocrystallization of fats. In *Springer briefs in food, health, and nutrition*. Springer.

Mazeika, L., Kazys, R., Raisutis, R., Demcenko, A., & Sliteris, R. (2006). Long-range ultrasonic testing of fuel tanks. In *EC NDT conference, Berlin, September 25-29.*

Mensah, S., & Franceschini, E. (2006). Near-field ultrasound tomography. *Journal of the Acoustical Society of America, 121*(3), 1423–1433.

Morse, P. M., & Ingard, K. U. (1968). *Theoretical acoustics*. New York: McGraw Hill.

Norton, S. J., & Linzer, M. (1979). Ultrasonic reflectivity tomography: Reconstruction with circular transducer arrays. *Ultrasonic Imaging, 1*, 154–184.

Opielinski, K. J., & Gudra, T. (2006). Recognition of external object features in gas media using ultrasound transmission tomography. *Ultrasonics, 44*, 1069–1076.

Pierce, A. D. (1981). Temperature and pressure dependence of the sound velocity in distilled water. In *Acoustics: An introduction to its physical principles and applications*. New York: McGraw-Hill.

Pinfield, V. J., Povey, M. J. W., & Dickinson, E. (1995). The application of the modified forms of the Urick equation to the interpretation of ultrasound velocity is scattering systems. *Ultrasonics, 33*(3), 243–251.

Povey, M. J. W. (1997). *Ultrasonic techniques for fluids characterization*. San Diego: Academic Press.

Rahiman, M. H. F., et al. (2016). An evaluation of single plane ultrasonic tomography sensor to reconstruct three-dimensional profiles in chemical bubble column. *Sensors and Actuators, 246*, 18–27. https://doi.org/10.1016/j.sna.2016.04.058

Rose, J. L. (1999). *Ultrasonic waves in solid media*. Cambridge University Press.

Roy, O., Jovanovic, I., Hormati, A., Parhizkar, R., & Vetterli, M. (2010). Sound speed estimation using wave-based ultrasound tomography: Theory and GPU implementation. *Medical Imaging 2010: Ultrasonic Imaging, Tomography, and Therapy.*

Satti, A. M. H., & Szilard, J. (July 1983). Computerised ultrasonic tomography for testing solid propellant rocket motors. *Ultrasonics*, 162–166.

Schlaberg, H. I., Podd, F. J. W., & Hoyle, B. S. (2000). Ultrasound process tomography systems for hydrocyclones. *Ultrasonics, 38*, 813–816.

Schlaberg, H. I., Yang, M., & Hoyle, B. S. (1998). Ultrasound reflection tomography for industrial processes. *Ultrasonics, 36*, 297–303.

Scott, T. E., & Roegiers, J.-C. (1994). Acoustic tomographic difference imaging of dynamic stress fields. In *SPE/ISRM rock mechanics in petroleum engineering conference. Delft, The Netherlands, 29–31 August.*

Silk, M. G. (1984). *Ultrasonic transducers for non-destructive testing*. Bristol: Hilger.

Steiner, G., & Podd, F. (2006). A non-invasive and non-intrusive transducer array for process tomography. In *XVIII IMEKO world congress, Rio de Jaeiro, Brazil, September 17–22.*

Steiner, G., & Watzenig, D. (2008). A bayesian filtering approach for inclusion detection with ultrasound reflection tomography. *Journal of Physics: Conference Series, 124.*

Uberall, H. E. (1992). *Acoustic resonance scattering*. Philadelphia: Gordon & Breach Science.

Utomo, M. B., Warsito, W., Sakai, T., & Uchida, S. (2001). Analysis of distributions of gas and TiO_2 particles in slurry bubble column using ultrasonic computed tomography. *Chemical Engineering Science, 56*, 6073–6079.

Warsito, Ohkawa, M., Kawata, N., & Uchida, S. (1999). Cross-sectional distributions of gas and solid holdups in slurry bubble column investigated by ultrasonic computed tomography. *Chemical Engineering Science, 54*, 4711–4728.

West, R. M., Jia, X., & Williams, R. A. (2000). Parametric modelling in industrial process tomography. *Chemical Engineering Journal, 77*, 31–36.

Wood, A. B. (1941). *A textbook of sound*. London: Bell and Sons.

Williams, R. A., & Beck, M. S. (1995). *Process tomography: Principles, techniques and applications*. Oxford, Uk: Butterworth – Heinemann.

Wolf, J. (1988). Investigation of bubbly flow by ultrasonic tomography. *Particle and Particle Systems Characterization, 5*, 170–173.

Wright, W. M. D., & O'Sullivan, I. J. (2003). Ultrasonic tomographic imaging of air flow in pipes using an electrostatic transducer array. *Review of Quantitative Nondestructive Evaluation, 22.*

Wright, W. M. D., Schindel, D. W., Hutchins, D. A., Carpenter, P.,W., & Jansen, D. P. (1998). Ultrasonic tomographic imaging of temperature and flow fields in gases using air-coupled capacitance transdcuers. *Journal of the Acoustical Society of America, 104*, 6.

Xu, L. J., & Xu, L. A. (1998). Gas/liquid two-phase flow regime identification by ultrasonic tomography. *Flow Measurement and Instrumentation, 8*(3/4), 145−155.

Xu, L., Han, Y., Xu, L. A., & Yang, J. (1997). Application of ultrasonic tomography to monitoring gas/liquid flow. *Chemical Engineering Science, 52*(13), 2171−2183.

Yang, M., Schlaberg, H. I., Hoyle, B. S., Beck, M. S., & Lenn, C. (1999). Real-time ultrasound process tomography for two-phase flow imaging using a reduced number of transducers. *IEEE Transactions on Ultrasonics, Ferroelectrics, and Frequency Control, 46*, 3.

Spectro-tomography

Brian S. Hoyle [1] *and Yanlin Zhao* [2]
[1]University of Leeds, Leeds, United Kingdom; [2]China University of Petroleum, Beijing, China

9.1 Introduction

Industrial processes may be classified using many descriptors. Perhaps the most generic and pervasive are terms such as *efficient*, *productive*, and *sustainable*. Their designers and operators must seek to achieve increasingly demanding objectives in these key business parameters. Processes have an almost unlimited variety of forms. Many are concerned with the transformation of raw materials into products in a closed space where distribution is difficult to measure. For a simple process with homogeneous materials, it may be satisfactory to locate a 1D sensor at one convenient sampling point to sense a relevant parameter. This may also suffice for the detection of a specific state in a process trajectory, or when continuous process materials have little variation. For many processes, such assumptions are false, due to wide-ranging operating parameters, or with variations in materials and recipes. Process models may not be available, or their range may not extend over all relevant materials, or recipes. Many processes have multiple components, and wide operating envelopes. In such cases, an explicit recognition of the multidimensional nature of processes in the widest sense is essential in the design of process sensing technology.

Industrial Process Tomography (IPT) is based upon injecting excitation energy into a process and sensing a set of spatially distributed responses, usually termed a *projection* (Hoyle, McCann, & Scott, 2005). Other chapters of Part One provide many examples. Some energy modes only require a single measurement per projection, e.g., of attenuation narrow beam X-ray. Other modes require several measurements per projection, e.g., of peripheral voltage distribution in a conduction field created by a constant current between two points. In medical X-ray applications, many thousands of projections can be taken to gain high spatial resolution data, as the subject is practically stationary. In process applications, dynamic constraints typically impose limits on sampling time to measure projection data, inherently limiting both spatial and temporal resolution. IPT *reconstruction* is the general term for this *inverse problem* which aims to estimate the corresponding unknown internal distribution of the process from a projection dataset. Many methods are available for this task which is explored in depth in other chapters of Part Two.

In order to define general objectives for development, it is first useful to consider the application background in which IPT typically has two application role requirements (other requirements are considered later):

Industrial Tomography. https://doi.org/10.1016/B978-0-12-823015-2.00013-3

Process design and pilot stage: confirm or develop a process model in which 2D or 3D spatial plus temporal process data can be used to quantify or verify model parameters; and

Process operation stage: derive gains from the increased information arising from its tomographic data.

As exemplified in other Part One Chapters, excitation energy may be X-ray, optical, electrical, etc. An X-ray instrument may feature a constant energy level; an electrical conduction instrument features a constant voltage (or current). Although the designed projection data collection may be satisfactory in terms of temporal and spatial resolution, the quality of actual sensed response data (in terms of the above needs) is critically determined by the complex interaction that takes place within the sensing region between process materials and the excitation energy. The simple term *Contrast* is commonly used to summarize this complex linkage. A simple illustration arises in a classic medical attenuation-based X-ray tomography (XRT) case. X-ray energy passes with minimal attenuation through low density soft body tissues, but is strongly attenuated by higher density cartilage and bone (resulting in a measurable energy difference). Thus, high density objects, in a field of low-density material, have *high contrast* and are easily imaged by XRT. Similarly, for an electrical energy mode, Electrical Resistance Tomography (ERT) current excitation flows in low-resistance material result in a low (measurable potential difference) voltage; but in high resistance, material results in a high voltage. Thus, high resistance objects, in a field of low-resistance material, have *high contrast* and are easily imaged by ERT. It is evident that, in terms of the above application role requirements, design of tomographic measurement requires careful attention to spatial and temporal resolutions, but critically must also consider the added dimension(s) of mode and level of excitation energy(s), and include consideration of their interaction(s) with process materials.

It is useful to consider IPT instrumentation in this wider context. IPT instruments may be classified in terms of space and time resolution, but also of modes and levels of excitation energy; and may also usefully take advantage of any axis of symmetry (Hoyle, 2016). Thus, a *First Generation* IPT (1G-IPT) instrument may be defined as having a single energy level and mode, with a complete abbreviation: 1G-SE-SM-IPT; e.g., constant energy X-ray, or fixed current ERT instruments. 1G-IPT units have the advantage of simple practical realization both in electronic systems and computational capability but are only truly viable for an unambiguous case, where only one material in the sensing zone has significant contrast with differing concentrations, that may then be easily imaged, showing on the material concentration as a *gray* scale. Here any background material must have a low contrast and hence will not be visible in the tomograph image. Of course, a monochrome scale may be replaced typically by a visible spectrum color scale for increased graphical effect for human viewers.

1G-IPT instrument products have been developed for applications in both process pilot study and in-line monitoring roles. A typical example is the Industrial Tomography Systems (ITS) P2+ (ERT) product (Industrial Tomography Systems, 2021a), which supports an array of up to 128 electrodes, mountable around the circumference of the pipe or vessel, or on a baffle, or dip-pipe. This would, for example, support data

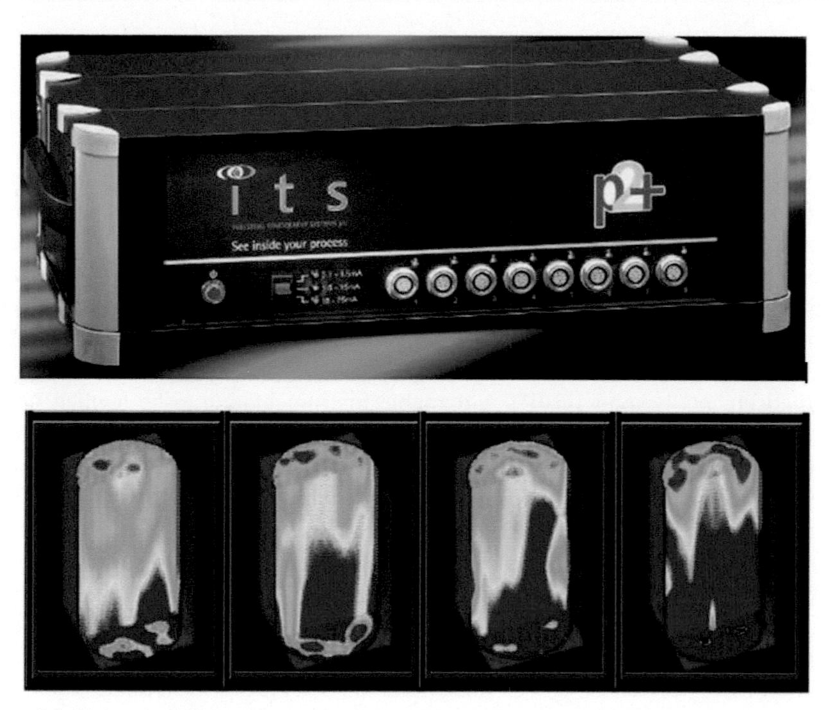

Figure 9.1 Example IPT instrument (upper); and gas–liquid mixing application example (lower) showing increasing gas injection from left to right.
Courtesy of Industrial Tomography Systems.

capture in a cylindrical vessel, such as a process mixer, from 8 vertical planes each having 16 electrodes. Of course, in such cases the fixed current level may be changed to suit materials. Data are processed to provide a conductivity map of the sensing zone, under the assumption that tomographs directly represent process distribution. Fig. 9.1 shows the instrument and typical results for an application in gas–liquid mixing used in this 3D mode.

Many applications have been supported using 1G-IPT, providing major, and otherwise inaccessible, internal distribution data, even though this may feature inaccuracies. Early adoption included key sectors such as Oil and Gas (Dickin et al., 1992) and Pharmaceuticals (Ricard, Brechtelsbauer, Xu, & Lawrence, 2005). ITS cite a wide range of application sectors for this product: biotechnology; bulk and speciality chemicals; environmental processes such as material transport; fast-moving consumer goods such as personal products; mining separation processes; nuclear processing monitoring; oil and gas in oilfield flows; pharmaceutical product monitoring and control; and pulp and paper processing (Industrial Tomography Systems, 2021b). Part Three Chapters provide a wide landscape of applications, with an overview in Chapter 17 of several 1G-IPT deployments, which despite limitations have made valuable contributions, as IPT is one of only a small number of methods able to reveal internal operational process information.

9.2 Multidimensional process sensing

Where the combination of process materials is incompatible with the singular energy form and level limitations of the 1G-IPT, case results may be ambiguous. For processes of two or more materials, confusion is likely in identification of specific component features. For example, when a *contrast profile* (defined in relevant terms) of two components P and Q is similar, a specific *gray level* could indicate a low P concentration or a high Q concentration. Increased processing of 1G-IPT raw data using iterative methods, such as that proposed by Wang (2002), can offer improvements at modest computational cost. However, it is clearly preferable to enhance the quality of raw IPT data.

Enhancements beyond 1G-IPT will enable instrumentation to be applied in more complex and high value processes, and where process models are not available, or where the model range does not extend to a wide range of candidate materials or process recipes. Such applications require *Second Generation* tomography (2G-IPT) approaches for *multidimensional* IPT with *dimensionality* having the forms outlined below. In general terms, there is a sensing need to exploit increased *contrast opportunities*, so that the raw data are richer in terms of the intrinsic information of temporal and spatial distributions but also of identified material components. As introduced above, we assume *dimensions* of Space and Time but add consideration of detectable excitation energy effects on process material(s) as other key critical *dimensions*. In engineering terms, these are also linked to the synchronization, detection, and processing of the sensor data. Taking an electrical tomography example, energy may be applied over a certain *range* to suit a broad *contrast* response function. From this multidimensional viewpoint, key engineering requirements for IPT may be expressed as follows:

Spatial resolution: Sufficient to yield information of interest;

Time resolution: Sufficient to reveal dynamics of interest;

Component and material identification: Sufficient to identify 3D distributions of process components of interest.

The third requirement relates critically to these additional dimensions. These requirements may be usefully combined with the above introductory application requirements.

The view of IPT as a multidimensional sensing technology includes *data fusion* of datasets from multiple sensing dimensions, to deliver enhanced information for the design, or operation of the application process. For example, to deliver operational data, a specific goal may be to estimate the current state in the process trajectory for a batch reactor. Fig. 9.2 provides a schematic illustration of the hypothetical attributes of input data; a state estimation algorithm; and a corresponding state trajectory illustrating the synthesis of product *C*, from reagents *A* and *B*, with a proposed optimal endpoint, before the reaction begins to reverse.

Input data in Fig. 9.2 offer the obvious spatial and temporal data available from the sensing system, and importantly, information on the mode(s) of excitation energy deployed, to enable sensor interpretation to reveal estimated process states. These must integrate data from multiple sources that are likely to feature different sensing

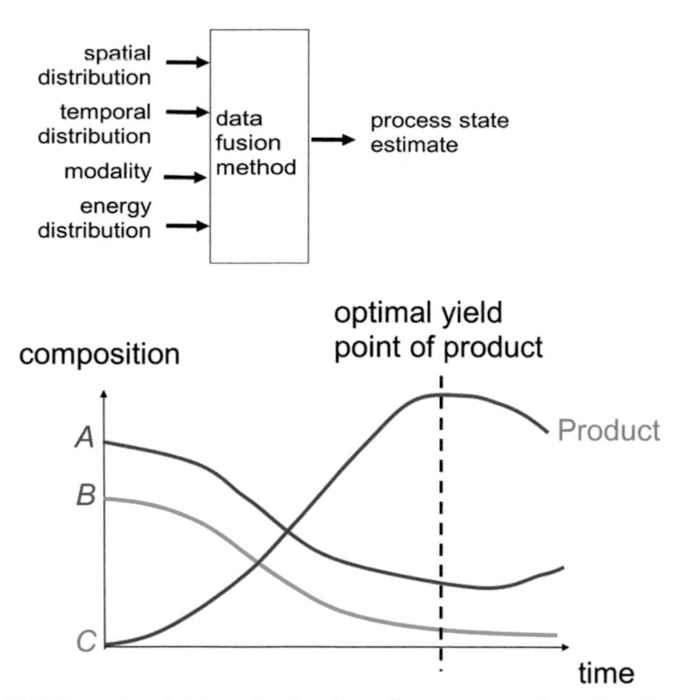

Figure 9.2 Multidimensional data and estimation of process state: (upper) attributes and data fusion; (lower) process state trajectory.

regions and dimensionality. Data from different modalities, perhaps offering insights into different process components, may be fused through a multicomponent, composition model. 2G-IPT systems have been designed to support data collection and fusion from sensors of different energy modes. For example, a generic multimodal system design allows the selection of specific in-built fusion methods to suit various combinations of multimodal modules (Hoyle et al., 2001). This design was implemented by ITS as the M3M IPT instrument. One example description of its use features the fusion of Electrical Resistance and Electrical Capacitance mode data in the study and measurement of oil/water/gas flows (Qui, Hoyle, & Podd, 2007). Other variants in this application area have used gamma ray sensing to locate gas structures with electrical capacitance tomography. Multimodal IPT systems are explored in depth in Chapter 17.

The input data of Fig. 9.2 include the obvious spatial and temporal data extracted from the sensing system and the excitation energy mode which may include a range or set of levels; the specific level that relates to the sensed data is also a key factor in the data fusion interpretation method. This section began with an introduction to the notion of the type and level of excitation energy considered as an extra dimension in any IPT study. *Spectroscopy* provides a classic analysis of the relationship between this dimension and materials.

9.3 Spectroscopic sensing

Electromagnetic wave excitation induces microscopic motion in materials at the molecular and atomic levels which may be sensed electrically providing information of material structure, composition, and chemical change. Such spectroscopic sensing may be classified in terms of the electromagnetic spectrum as illustrated in Fig. 9.3, which identifies ranges for common measurement methods: Electrical Impedance Spectroscopy (EIS), Magnetic Resonance Imaging (MRI), Raman Spectroscopy, and Infrared Spectroscopy.

Tomographic process applications of some of these techniques are developed in other Part One chapters: EIS in Chapter 2, MRI in Chapter 4, and Infrared sensing in Chapter 5. Chapter 8 adds the use of acoustic and ultrasound sensing, which rely upon sound propagation in a material.

Spectroscopic sensing offers opportunities to gain industrial process material information. Electrical excitation, with level expressed in frequency terms, applied over a range results in a spectroscopic response that may be used to infer specific material properties and identification.

In homogeneous materials, bulk properties can be sampled at a convenient access point. A sample may be excited with electrical energy over a range of frequencies in discrete steps. At each step, a measured single frequency response value provides a discrete spectral impedance characteristic. Materials generally exhibit complex impedance characteristics whose parameters are complex quantities representing energy dissipation (resistive) and storage (reactive) parts. Characteristics are typically described by graphical frequency-based functions, either as a complex value (with real and imaginary parts) or a *phasor* (with magnitude and phase-shift parts). Fig. 9.4 illustrates the complex characteristic of a hypothetical material in its real and imaginary parts.

Chemical compounds exhibit a variety of *spectral fingerprints*, based primarily upon their chemical structure. Electrical excitation produces responses arising from

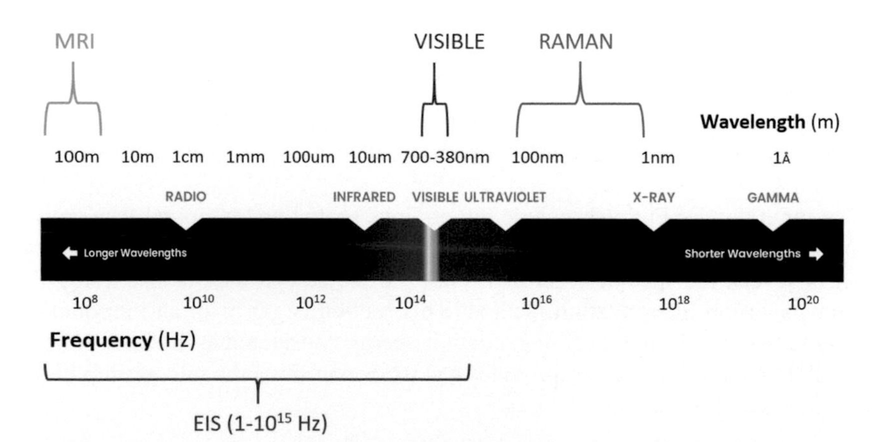

Figure 9.3 Electromagnetic energy spectrum and associated sensing methods.

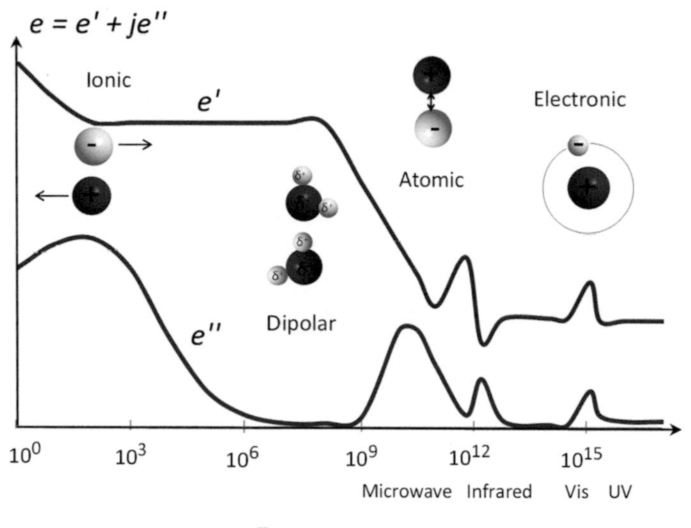

Figure 9.4 Electro-chemical effects over electrical excitation spectrum.

interaction with specific molecular and atomic features. Based upon the energy levels expressed in frequency terms, corresponding resonance features may be classified as illustrated in Fig. 9.4. Ionic features in the material typically present spectral characteristics at lower frequencies; dipole structure resonances become evident at frequencies from 1 MHz to 1 GHz; and atomic and electronic resonances appear at higher frequencies.

The general term *Spectroscopy* may be refined for specific objectives. *Dielectric Spectroscopy* specifically addresses the polarization effect (resulting from the electric field) expressed in terms of frequency-dependent complex permittivity and conductivity. *Electrical Impedance Spectroscopy* (EIS), noted above, addresses electro-kinetic properties of materials and their interfaces (Barsoukov & Macdonald 2018) and is more relevant here. Analogous principles may also be described for nonelectromagnetic acoustic energy propagation and its resulting spectral characteristics.

EIS data can be analyzed using a suitable physical model (e.g., an equivalent circuit) to describe physical or chemical processes in the material/interface. EIS measurement involves applying alternating excitation $V(t) = V_m \sin(\omega t)$, at frequency $f = \omega/2\pi$, to the material and measuring resulting response current $I(t) = I_m \sin(\omega t + \theta)$. Conventional impedance is defined as $Z(\omega) = V(t)/I(t)$. The impedance magnitude (or modulus) is $|Z(\omega)| = V_m/I_m(\omega)$, with phase angle θ. The complex impedance in real part and imaginary parts is defined as follows:

$$Z = Z' + jZ'' \tag{9.1}$$

where $j = \sqrt{-1}$, Z' is the real part, and Z'' is imaginary part.

This may be defined in phasor terms as follows:

$$\mathrm{Re}(Z) = Z' = |Z|\cos(\theta); \mathrm{Im}(Z) = Z'' = |Z|\sin(\theta) \tag{9.2}$$

where $\theta = \tan^{-1}(Z''/Z')$ and $|Z| = \left[(Z')^2 + (Z'')^2\right]^{1/2}$.

Impedance Z can also be expressed in resistance (R) and reactance (X) terms: $Z = R + jX$.

For example, electrical process measurement can provide major insight into the many materials that are based upon particles in liquids. In more general terms, a measured spectral characteristic can be considered a *spectral fingerprint* of a specific material. Even if this is not unique, it is likely to support discrimination between several distinct component materials in a specific process. For example, Fig. 9.5 illustrates the spectral fingerprint of L-Glutamic Acid (LGA), a commonly used simulant for crystallizing pharmaceutical materials (Zhao, Wang, & Hammond, 2011).

Many processes comprise also particles in a liquid and are subject to specific effects which are critical to material interactions. For example, for a two-phase system in which one phase is dispersed in a continuous phase, particles of the dispersed phase in the effective electrolyte solution may become charged (e.g., due to the dissociation of the surface groups). Charges created are balanced by the presence of ions of opposing sign absorbed on the particle surface, and a layer of counter charges can be formed. Thus, an electric double layer exists around each particle as two parts: an inner region (Stern layer), where ions are strongly bound, and an outer region (diffuse layer) where they are less firmly associated. This diffuse layer, where counter charges are carried by ions, has a notional boundary known as the slipping plane within which the particle acts as a single entity. The electrostatic potential difference

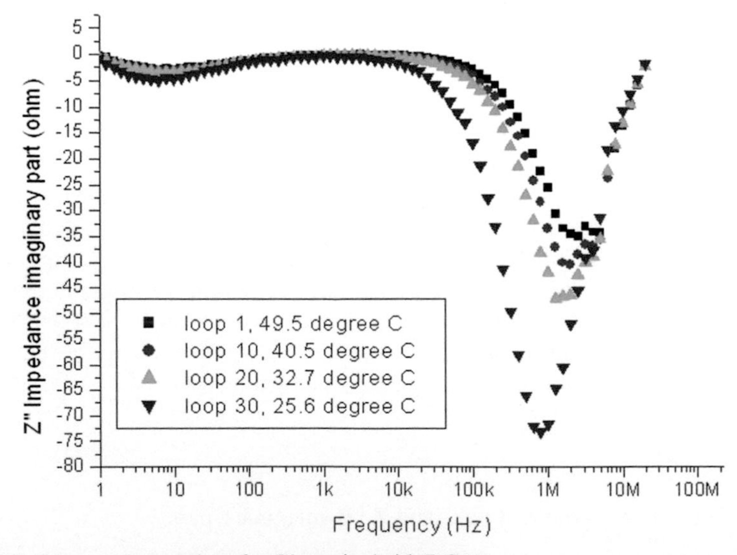

Figure 9.5 Spectral fingerprint of L-Glutamic Acid (LGA).

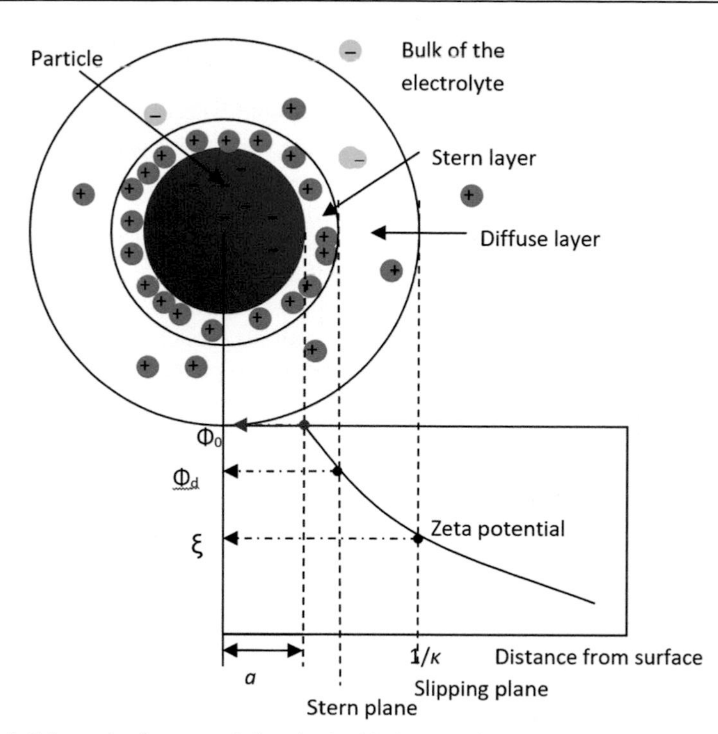

Figure 9.6 Schematic diagram of electric double layer and zeta potential, where Φ_0 is the potential at the particle surface, Φ_d is the potential at the Stern plane, and ξ is the zeta potential at the slipping plane.

between Stern and diffuse layers is known as the Zeta potential. Fig. 9.6 shows the electric double layer and zeta potential in schematic form.

The charged particles (with surface electric double layer) can move through a continuous medium under the influence of an external electric field (Hunter, 2001). For an alternating electric field, both particle and ions will be set in motion, with particle and counterions moving in opposite directions, as illustrated by Fig. 9.7. This results in the double layer distorting in an alternating fashion giving rise to an

Without applied electric field

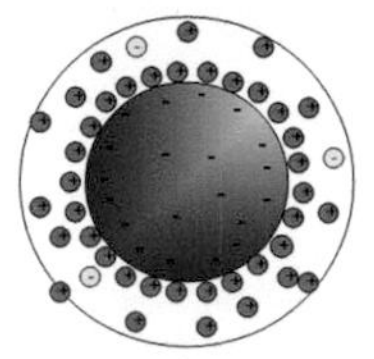

With applied electric field

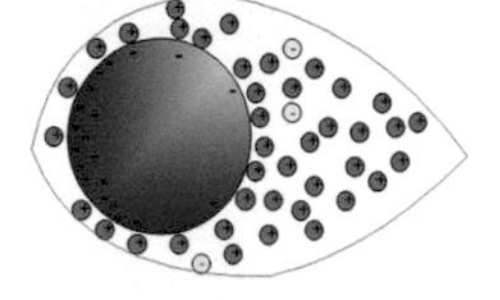

Equilibrium double layer

Nonequilibrium double layer

Figure 9.7 Schematic diagram of charged particles and double layer showing effect of applied electric field.

electric field which at large distances from the particle has the same form as that of an alternating electric dipole (O'Brien, 1982). Polarization of the electric double layer reverses periodically, with a phase lag to the electric field. This can be monitored by EIS to obtain complex conductivity, impedance, and permittivity parameters, which are related to particle size and shape, particle concentration, electrolyte properties (ionic concentration, ionic species), zeta potential, etc.

Impedance data obtained from electric impedance spectra may be related to particle size (Zhao & Wang, 2015). For example, Fig. 9.8 illustrates the variation of complex impedance with various particle sizes (14, 91, 190, 200, and 385 nm) in silica particle suspensions, showing imaginary part changes with the particle size. For a particulate suspension, the impedance imaginary part represents the phase lag between the applied voltage excitation and the response current signal. Here, polarization of the electric double layer around each particle reverses with excitation, but with a phase lag due to the finite time diffusion of ions along the particle surface trajectory, resulting in the observed imaginary part changes in terms of particle size.

Many important industrial processes feature component spatial distributions which are critical in terms of efficiency and product quality. IPT offers a powerful insight into these distributions, but 1G-IPT instruments, limited to a single energy mode and level with a resulting single gray scale tomograph, are likely to fail to reveal the rich process detail illustrated in Figs. 9.5–9.8.

Multimodal IPT may be employed, where each mode has contrast sensitivity to a specific component material. But each mode then typically has a single energy level, in effect for two modes: a *red scale* and a *green scale*, etc., where discrimination may not support reliable material identification in mixtures. However, an augmentation of

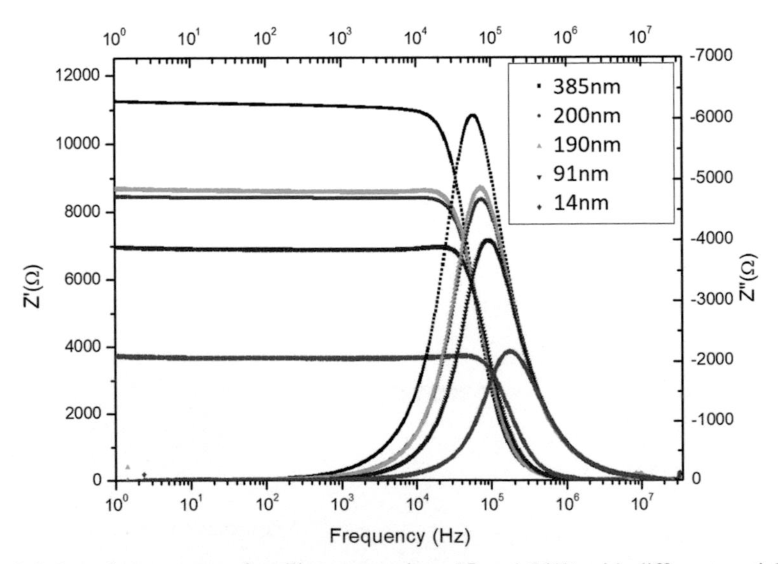

Figure 9.8 Impedance spectra for silica suspensions ($\Phi = 4.76\%$) with different particle size (14, 91, 190, 200, and 385 nm); real and imaginary part versus frequency.

IPT with a spectroscopic dimension has the potential to jointly estimate the spatial distribution of several distinct components explicitly identified using spectroscopic data. In terms of the introductory discussion of Section 9.1, this provides a major increase in the utilization of the contrast opportunity.

For real-time viability, sensing operations must be carried out within a time window that delivers temporal data and reliably tracks the process dynamics of interest. Where process dynamics are significant, such as a stirred vessel, conventional spectroscopic sensing, which relies upon a discrete stepped set of single frequency excitations, is likely to be unsatisfactory. The sensing excitation frequency range is clearly critical to explore the spectral fingerprint of the materials in a process, as illustrated for LGA in Fig. 9.5, showing an overall bandwidth from 10 Hz to 50 MHz.

A composite excitation signal is required, which includes frequencies throughout the band of interest, coupled with joint processing of the target process response set to enable spectroscopic identification of materials. This combination of IPT with integrated fast real-time spectroscopy and material identification is called *Spectro-tomography* (S-T).

9.4 Spectro-tomography principles

9.4.1 Simple process model

To illustrate the requirement and potential of S-T, we return to the typical batch process trajectory illustrated in Fig. 9.2, in which two reagents A and B combine to form a product C, where the key sensing challenge is to determine the optimal yield point.

We can investigate this challenge using a 1G-IPT approach. Fig. 9.9 provides a schematic pictorial view of three states from this trajectory: (1) an early initial state when only the two reagents A and B are present; (2) an intermediate state in which some product C has formed; and (3) a final state when practically all the material present is transformed to product C.

Fig. 9.10 provides a spectral characteristic for a process with hypothetical *spectral fingerprints* for three components, arbitrarily color coded for clarity. It also shows

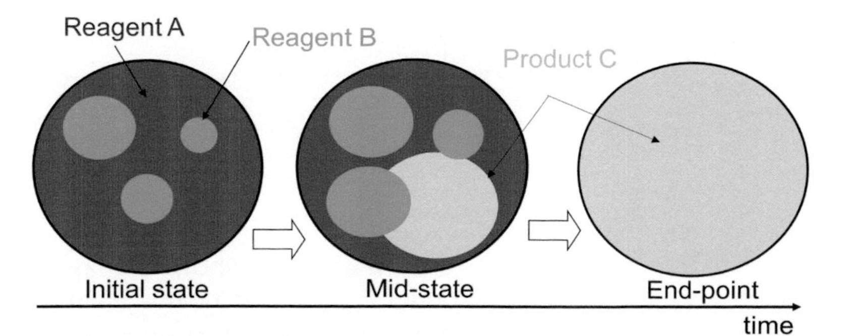

Figure 9.9 Pictorial view of three states within example process trajectory.

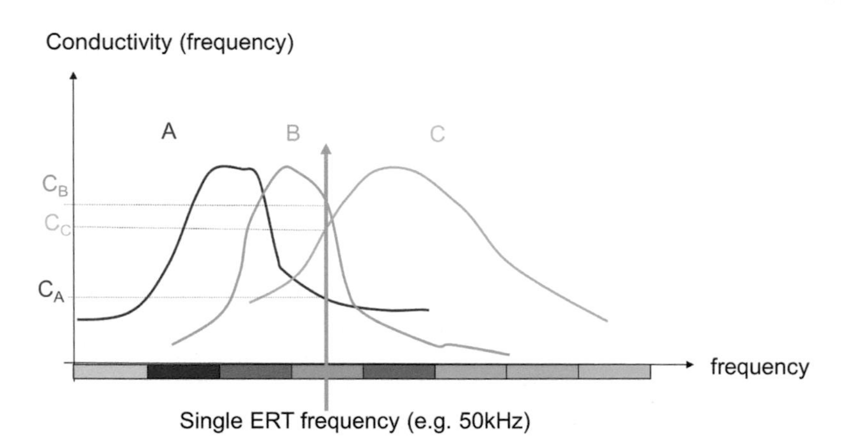

Figure 9.10 Example spectral fingerprints for reagents A, B, and product C.

three possible single excitation frequencies: 20, 50, and 80 kHz also using colors to indicate frequency steps. If the single excitation frequency (energy) of 50 kHz is used in a 1G-IPT trial, then, as illustrated, reagent B has the highest conductivity, followed closely by product C with reagent A having much lower conductivity. Resulting three-step time-lapse tomographs generated using a 1G-IPT unit with conventional reconstruction methods are shown using a *green scale* (for 50 Hz) in Fig. 9.11. Since conductivities C_B and C_c are similar in value, reagent B and product C appear at a similar scale value, while reagent A has a lower value. In such a multicomponent mixture, although some contrast is available, it is unlikely that components can be discriminated reliably. Perhaps major postprocessing may deliver marginal improvement, but it would clearly be preferable if richer raw data were available.

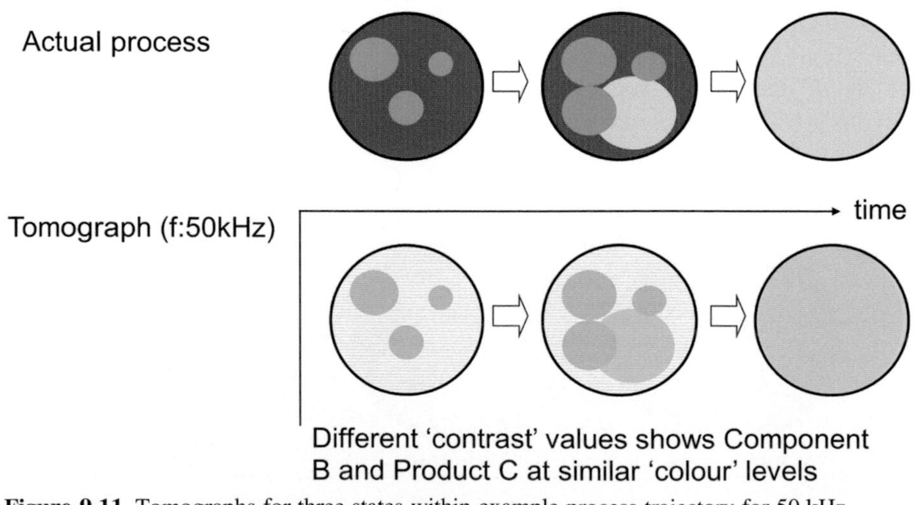

Figure 9.11 Tomographs for three states within example process trajectory for 50 kHz excitation.

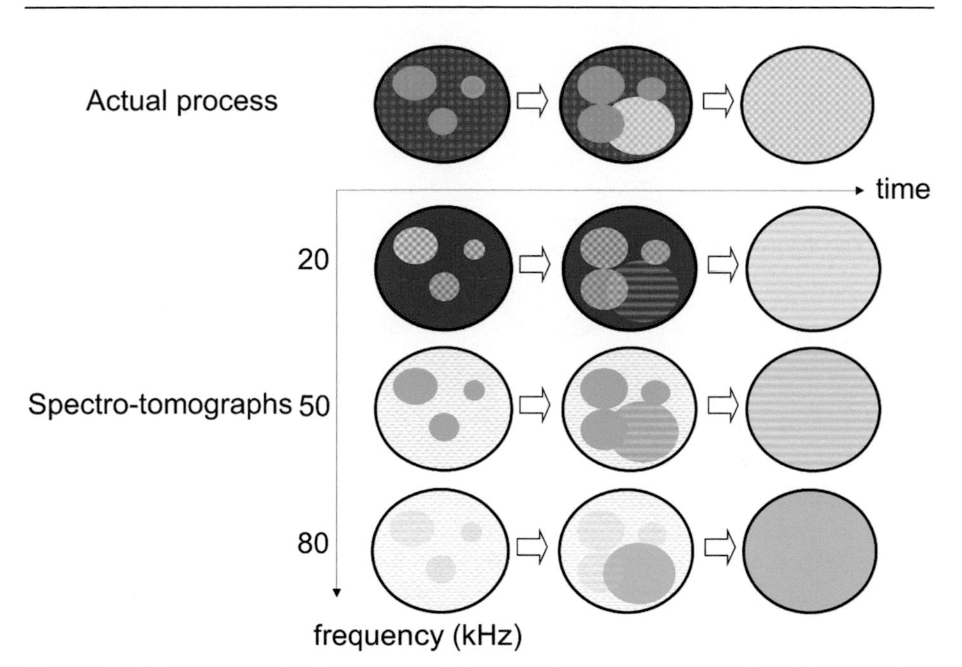

Figure 9.12 Tomographs for three states within example process trajectory for triple excitation.

To extend the raw data, we explore use of the three frequency excitations: 20, 50, and 80 kHz shown in Fig. 9.10. For simplicity, we assume that three excitations and corresponding measurements are carried out sequentially using a 1G-IPT unit, although in the real-world, due to process dynamics, it is unlikely that there will be sufficient sample time. Sequential IPT response data are extracted for each frequency, shown in red (dark gray in print version), green (light gray in print version), and blue (gray in print version)-scale forms, in the corresponding set of three tomographs shown in Fig. 9.12.

9.4.2 Process component identification

To complete the sensing approach for material identification, we examine the (3 in this case) corresponding pixels (or possibly groups of pixels) across the three tomographs. By comparing the three pixel values with the known material spectral fingerprints, such as those of Fig. 9.10, we can assign an *interpreted* material identification for the pixel and indicate this in the schematic example using a red, green, or blue material identification color. If we repeat this for all pixels (or all selected groups of pixels) across the three tomographs, we obtain the *Spectro-tomograph* illustrated in Fig. 9.13. Three-point identification may be a minimum for estimation, but further spectral-line data will clearly enhance this estimation, especially where data for all spectral lines can be collected within a response data time slice. The estimation process identifies the best-fit candidate from the possible materials to form a single resulting *Spectro-tomograph* for each time-lapse state.

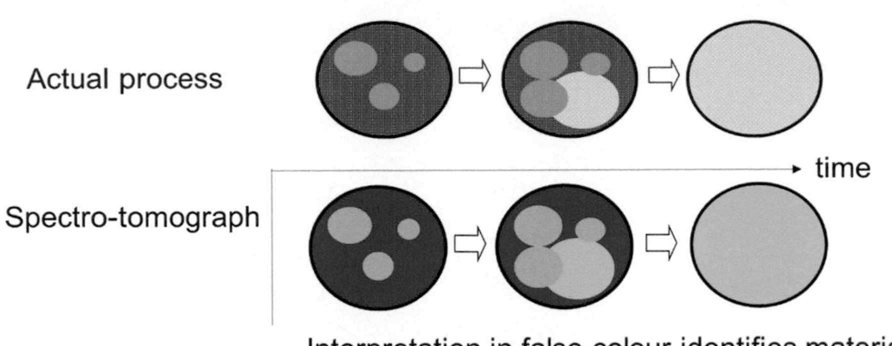

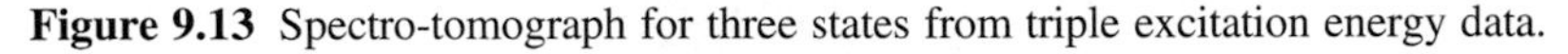

Figure 9.13 Spectro-tomograph for three states from triple excitation energy data.

This data fusion estimation process may be founded upon specific, known data model spectral fingerprints. Where a data model is not available, or useable for a given process, experimental tests may be performed on process material samples at relevant operating conditions to create an empirical model (Nahvi & Hoyle, 2009a, 2009b). The specific material identification exploits these data, either from a *Look-up Table* (LUT) of spectral fingerprints of known materials or from the computational fitting of the sensed form to a mathematical model (Nahvi & Hoyle, 2010, pp. 1309–1312). The resulting spectro-tomograph from this estimation process, based upon the raw tomograph dataset of Fig. 9.12, is shown in Fig. 9.13. Here colors represent the identified materials from the three components.

9.5 Spectro-tomography system implementation

9.5.1 System requirements

The ability to estimate the presence, location, and concentration of a specific material promises to overcome major barriers to effective process management. To realize this need requires in-line sensing (rather than laboratory sampling), to satisfy real-time process trajectory constraints. The S-T method offers a new approach, based upon electrical methods, to provide these data and fulfill operational requirements. Fast sensing of wideband electrical impedance in a material is required. A suggested implementation exploits an electrical impedance method, where electrodes are in contact with the process materials, and hence is directly suited to liquid-based processes. Analogous implementations could be developed using different energy modes and excitation methods suited to other process types. For example, for dry solids processes, a method based upon capacitive coupled excitation, could be used. Implementation of the method outlined above in principle in Section 9.4 may be considered in three stages:

Stage 1 – Wideband process excitation spanning the required frequency range, within a single sampling interval, is applied to the process. This may be conveniently synthesized using a compressed frequency signal format to enhance overall dynamic performance. Excitation will be designed flexibly to suit the spectral fingerprints of the expected materials in the process which may be selective to specific bands.

Stage 2 – Resulting wideband response data are processed to extract relevant energy (frequency) bands providing raw data for computation of energy (frequency)-banded tomographs that correspond to parts of the spectral range of interest. For an optimum solution, the format of the excitation is likely to be linked to the method used in the response processing, e.g., a compressed excitation format may be linked to a suitable response decomposition method.

Stage 3 – Data fusion process is used to estimate the material spectral fingerprint for each region of interest (ROI) across the set of banded tomographs. Here an ROI may be scaled from a single pixel to a zone corresponding to a spatial resolution objective. This process may be based upon a database of spectral fingerprints for expected materials of interest developed either through models or empirical data. For materials in solutions, these data may need to support concentration levels.

9.5.2 Process excitation

To deliver a required real-time performance, the S-T method excitation and response processing must be coherently aligned. The use of multifrequency excitation coupled with analogue or digital filtering of response processing could be used, but this is likely to lead to difficulties in effective frequency coverage and in ease of band selection. Multiscale methods hold promise in providing flexible management of real-time excitation and processing.

A wide range of viable compressed signal types are available, each having distinct properties (Nahvi & Hoyle, 2008b). In terms of simplicity, ease of selection of frequency band, and relatively flat spectral energy distribution, a linear frequency modulated *chirp* signal, provides ideal properties. This signal form provides the basic design parameters expressed in Eq. (9.3).

$$B/T = (f_2 - f_1)/T \tag{9.3}$$

where B is the design bandwidth, between the upper frequency bound, f_2, and the lower frequency bound, f_1, and T is the duration of the signal.

Based upon the design parameters of Eq. (9.3), the process real-time sampling regime can be carefully controlled. The design methodology can simply specify the frequency endpoints, leading directly to the total period, and selection of an appropriate sampling interval. Fig. 9.14 illustrates the general form of the chirp signal scaled to a unit-time value.

9.5.3 Response processing

The wideband excitation is expected to produce a synchronized wideband response which must be detected for all relevant physical tomographic projections for the process. The wideband response is then decomposed into segments for specific selected

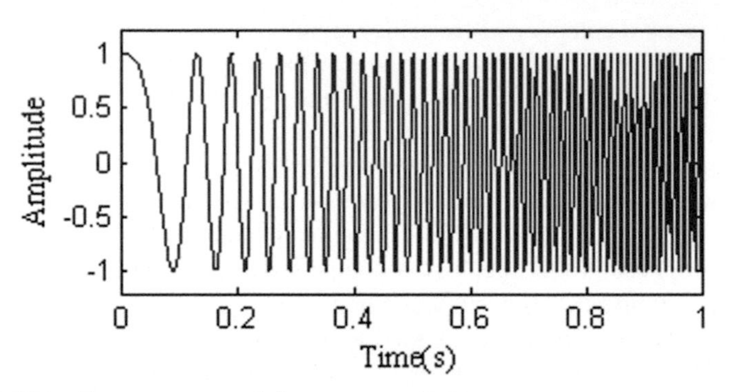

Figure 9.14 Chirp linear compressed frequency excitation.

bands within the total bandwidth, B. The specific organization of this selection is dependent upon the nature of the subject process. In some applications, materials may have specific narrow-band spectral fingerprint features, for example, a specific oscillation at a known narrow range of frequencies, as previously explored and illustrated in Fig. 9.4. In other processes, material fingerprints may present various characteristics. Each designed segment will be used to produce a separate tomograph for that band. To provide a multiscale match to a chirp linear compressed frequency excitation, a wavelet decomposition may be selected. This exploits the Wavelet Transform (WT), which provides an efficient intrinsic ability to extract spectral data. By scaling a selected mother wavelet, time resolution can be traded for frequency resolution, and vice versa. This scaling property provides variable time–frequency resolution making the WT an efficient tool to analyze wideband response datasets. The width and amplitude adaptation of the wavelet to a required frequency band is illustrated in Fig. 9.15. Where wideband data are required, a set of frequency steps can be simply generated, as described in detail by Nahvi and Hoyle (2007, 2008a). Since the underlying response data correspond to the excitation bandwidth, an iterative process can be designed in which specific bands are computed in a controlled sequence.

9.5.4 Data fusion processing

The final processing stage aims to deliver reduced and focused high-level process information useful for plant monitoring and control. As illustrated in the schematic example of Fig. 9.12, the intermediate result for each temporal sampling instant (here shown at three example instants) is a sequential set of spectral tomographs spanning the requisite energy (frequency) range (here shown for three values). As introduced above, these multibanded data are matched to spectral signatures, for example, via a LUT, to obtain material identification, as illustrated in schematic terms of Fig. 9.13. From these data, it is clearly possible to obtain instantaneous time–domain estimation of component fraction data leading to the complete estimation of the process trajectory, as illustrated in Fig. 9.2. Once this reduced information is available, a trend detection algorithm can be deployed to search for changes, and for the optimum yield point illustrated in Fig. 9.2.

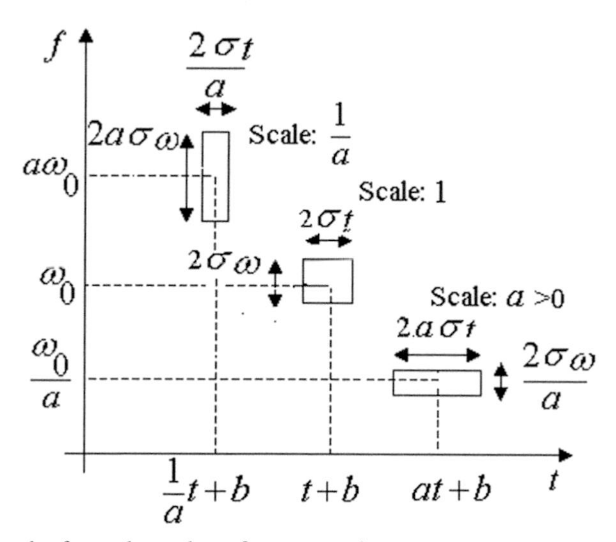

Figure 9.15 Graph of wavelet scale to frequency−time.

In the overview of Spectroscopic Sensing of Section 9.3, the spectral characteristic of a class of liquid pharmaceutical compounds exemplified by the material LGA is illustrated in Fig. 9.5. Corresponding manufacturing processes can clearly be investigated using the S-T method, where the emergence of LGA product material could be identified in a batch reaction vessel. Further detailed spectroscopic insights to EIS characterization in particulate products are discussed in Section 9.3 and illustrated in Figs. 9.6−9.8. Fig. 9.8 illustrates a spectroscopy opportunity to sense particle size in a specific silica suspension with particle sizes over the range 13−384 nm. Here use of the S-T method to extract a relevant spectral signature at each pixel (or ROI) may clearly be deployed to estimate corresponding point particle size distributions for the monitoring and control of production processes.

It is important to note in this overall method that the set of tomographs computed from data acquired within a given sampling interval incurs no relative evolution error due to process dynamics. Any and all computer energy (frequency) band tomographs are obtained from one wideband response acquired within the same sampling instant. All tomographs share the same spatial mesh. For such a tomograph set, any given pixel (or voxel for 3D tomographs) is identical in time and space to the corresponding pixel in all other tomographs computed from the wideband data. A vector of contrast intensity values taken at a specific spatial (pixel or region) position from all tomographs will vary only in terms of the extracted response to the energy (frequency) from each source tomograph.

9.5.5 Data capture and processing architecture

To support the above data acquisition and processing needs, an electronic system architecture must include appropriate control synchronization for the wideband

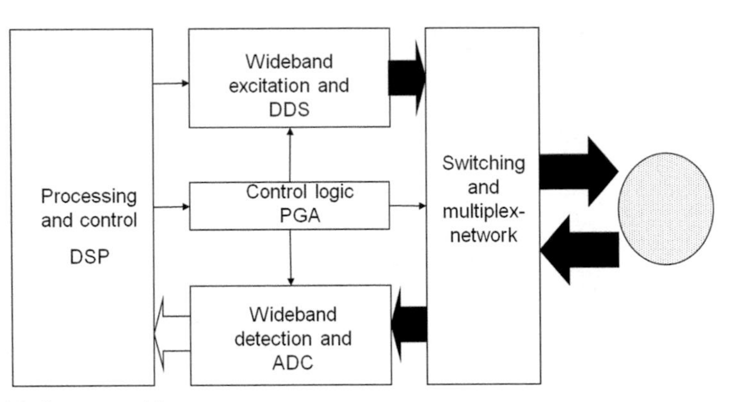

Figure 9.16 System architecture.

excitation, detection, and the extended signal and data fusion processing requirement. Fig. 9.16 illustrates a proposed schematic system architecture for a pilot trial system. The various functions are implemented through suggested hardware features: Direct Digital Synthesis unit, to produce the chirp excitation; switching and multiplexing logic to route and synchronize data streams; detection and analogue to digital conversion (ADC), with logic control implemented using a Programmable Gate Array (PGA); and overall control and processing using a Digital Signal Processor (DSP). A multistage processing algorithm defines the key steps from the specification of the excitation signal to the interpretation of the process features, as reviewed above. This may involve a set of preset and computed key steps that are ideal for execution using a DSP (Nahvi & Hoyle, 2009b, 2010, pp. 1309−1312).

9.5.6 Implementation design

An S-T design aims to deliver a temporal, spatial, and material identification view of a process, able to be interpreted in terms of the user requirements for the process (Hoyle, Nahvi, Ye, & Wang, 2012). This can take many forms, varying primarily with the mode of tomographic excitation, and ideally be driven by process *contrast opportunities*. It may be considered in terms of a modified conventional multichannel IPT instrument which provides estimated internal target process distribution as a time series of tomograph sets, from which materials (or states such as particle size) may be identified followed by interpretation.

Contrast opportunity assessment is concerned critically with process materials and available sensor technology. This chapter has focused upon exemplar trial processes which exhibit electrochemical variations, and hence can deploy simple electrical excitation. The variety of modes reviewed briefly in Section 9.3 offer the capability to vary excitation energy level (typically expressed as frequency) to capture spectroscopic data. For example, a direct amplitude and impedance analogy can be found with acoustic energy but relies upon availability of wide band actuators (which generally offer a dual sensor role as detectors). In this case, speed of sound and temperature are also critical parameters (Chapter 8). Optical (infrared) lasers and detectors have been used with two frequencies for combustion exhaust gas assessment (Chapter 5).

For optimum estimation using S-T, it is preferable to obtain spectral fingerprints of known process materials, with calibrations for concentration ranges. This typically requires simple one-time laboratory tests. As also described in Section 9.3, such spectral fingerprints may also be defined and obtained for specific material properties of interest such as material particle size. Spectral fingerprints are stored for comparative detection of nominated material content within each specified spatial ROI (with the ultimate resolution of a single spatial voxel/pixel in the Spectro-tomograph). Where processes have material composition variations, or where spectral properties vary with component mixtures, fingerprints may be represented in terms of LUTs, used, for example, to identify a "20/80% mixture of A in B." LUTs are easily incorporated in instrument set-up procedures. In processes where no reliable a priori data are available, rule sets may be defined to support empirical annotation of constituent materials, providing useful spectro-tomographic user information.

Process dynamics are often critical in terms of the time available for a total projection of data over spatial, temporal, and spectral dimensions. A specific issue is the delivery of requisite spectral energy excitation within limits from process dynamics. As described above, an optimal form features composite wideband excitation delivered efficiently as a linear chirp, coupled with a matched response decomposition, providing smooth step-free, frequency ramp excitation. Simpler, suboptimal forms may also be viable in some applications.

An electronically simple multienergy level *sequential* excitation method may deploy blocks of single-frequency (*spectral line*) sinusoidal energy to span selected frequencies; providing the requisite number of blocks can be applied within the maximum sampling interval arising from process dynamic limitations. In practice, each block produces an onset transient; necessitating a delay before a response can be reliably measured, which must be applied at each block *step*. An ERT demonstration using this method for Gas, Liquid, Solid mixing by Oguh, Hall, Bolton, Simmons, and Stitt (2012) used two sequential block frequencies (of 0.3 and 9.6 kHz) to investigate gas dispersion and solids suspensions.

Where the total sinusoidal excitation time is limited, an additive, multifrequency, *parallel* sinusoidal excitation having orthogonal properties (to facilitate easier detection) may be viable and has the advantage that (like the chirp) it produces a single onset transient. An Ultrasonic Tomography (UST) demonstration (where speed of sound is 10^5 slower than light) investigated by Li and Hoyle (1995) deployed several summed parallel sinusoids to facilitate efficient spectral and spatial response detection to form interpreted spectro-tomographs.

As detailed above, more optimal S-T implementations may use composite wideband excitation forms (such as a linear chirp) with matched detection. In simple terms, such efficient S-T wideband response data can be captured within the approximate single-frequency sampling interval of a corresponding 1G-IPT system, as higher frequency components add little to the capture time.

9.6 Trial demonstrations

Experimental trials described below demonstrate specific key steps of the composite method. For simplicity, they exploit convenient physical simulants of industrial materials, pseudostationary physical models, and computer simulation data.

9.6.1 Basic wideband excitation and response extraction

This first demonstration aims to illustrate the basic method described above in Section 9.5. This includes excitation, using a wideband chirp signal, extraction of the wideband response function, and direct comparison with the independently measured response of the known material. This focuses upon the basic spectroscopic response rather than a 2D or 3D distribution. This demonstration is also useful to illustrate the principles that would be exploited for the pixel/voxel interpretation processing for material identification in distributions.

The simple experimental process arrangement is shown in both schematic and image forms in Fig. 9.17. The excitation port and detection port are respectively driven and sampled using a synchronized digital signal synthesizer and digital storage oscilloscope. As summarized in Eq. 9.3, the chirp excitation is defined as chirp amplitude 5 mA; bandwidth 10–500 kHz; and signal (and test) duration 5 ms. For each 5 ms trial excitation, a calibration measurement was carried out using a commercial Spectrum Analyzer, which applied a sequence of sinusoidal excitations, taking about 2°min (120°s) to acquire data for 21 frequency steps. The ratio of test times of 24,500 provides an interesting indication of the temporal efficiency of the chirp excitation.

The material in the small process vessel shown in Fig. 9.17 is the commonly used simulant for pharmaceutical materials, LGA, as discussed above in Section 9.3. Following the 5 ms excitation period and following data acquisition, the basic S-T method is deployed using a wavelet set for the required number of spectral points.

Fig. 9.18 shows comparative data for the two methods: the 21-point spectra from the spectrum analyzer, with the smooth spectra synthesized from several thousand

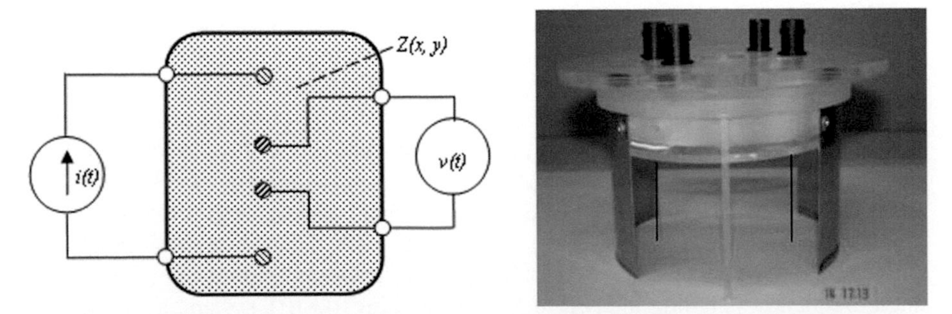

Figure 9.17 Experimental process (left) schematic; (right) physical arrangement—prior to insertion in flask with temperature-controlled water jacket.

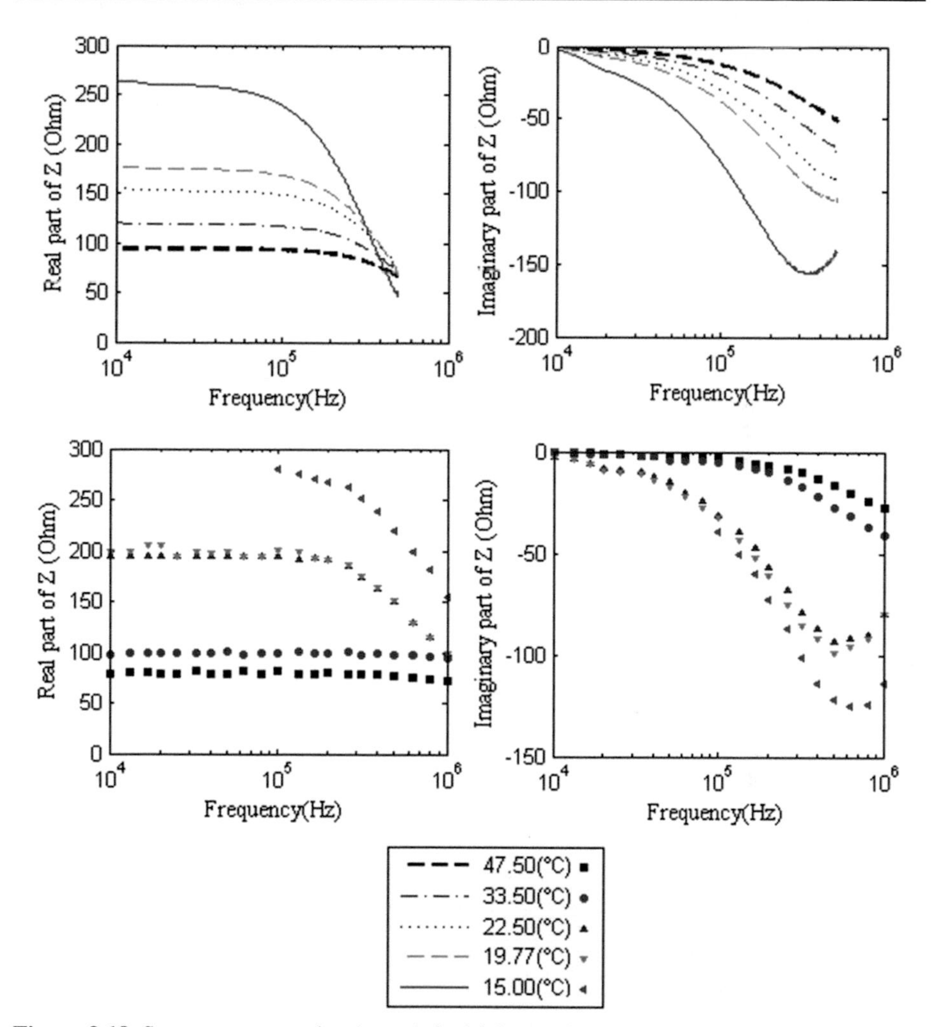

Figure 9.18 Spectroscopy results: (upper) for high-speed method (5 ms data capture); (lower) commercial spectrum analyzer (2 min data capture), with temperature key.

points computed from the single 5 ms wideband response. This demonstrates the fast identification of the electrical impedance spectroscopic characteristics of the LGA material during crystallization. As shown, tests were carried out at each of five temperature steps. Results show the expected impedance spectroscopic behavior that crystallization increases at lower temperatures, as illustrated also in direct test results of Fig. 9.5. The demonstration confirms the viability of the basic wideband excitation and wavelet frequency extraction method.

9.6.2 Energy frequency tomograph set processing

This next demonstration aims to illustrate multiple wideband process excitation, response detection, and the formation of a tomographic projection dataset. This can then be used, with an appropriate wavelet set, to deliver the raw data from which to compute the corresponding set of energy frequency tomographs. This demonstrates the second key stage of the method, prior to interpretation.

Time domain measurement issues have been avoided by the simple design use of a simulated process formed of a shallow dish containing a pseudostationary conductive gel having two regions with different sample slices of cucumber and banana, and a third region of saline water. This was fitted with a 16-electrode sensor array as illustrated in Fig. 9.19. The parallel data collection typical of an industrial IPT system was simulated for this exercise by superposition, by simply repeating the excitation for a sequence of separate measurements. Synchronized excitation and digital data acquisition units were used to collect projection data sequentially in time through manual connections.

The data fusion method described in Section 9.5.4 is then deployed to extract a range of energy frequency response datasets, from the single wideband response dataset, in the range of 20–200 kHz. In this application case expected differences highlighted are at lower frequencies, as shown in results of Fig. 9.20. The images show spectral variations, mainly due to the complex fruit regions. The demonstration confirms the viability of the formation of a set of energy frequency tomographs based upon wideband excitation, response detection, and wavelet frequency extraction method.

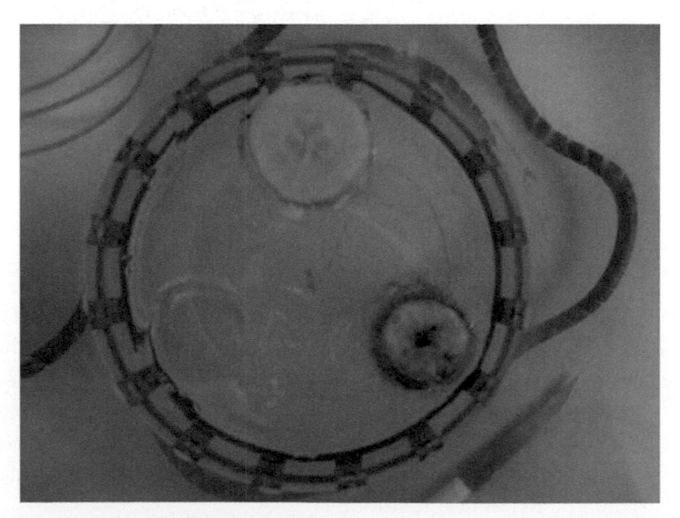

Figure 9.19 Pseudostationary 2D physical process simulation with 16-electrode sensing array.

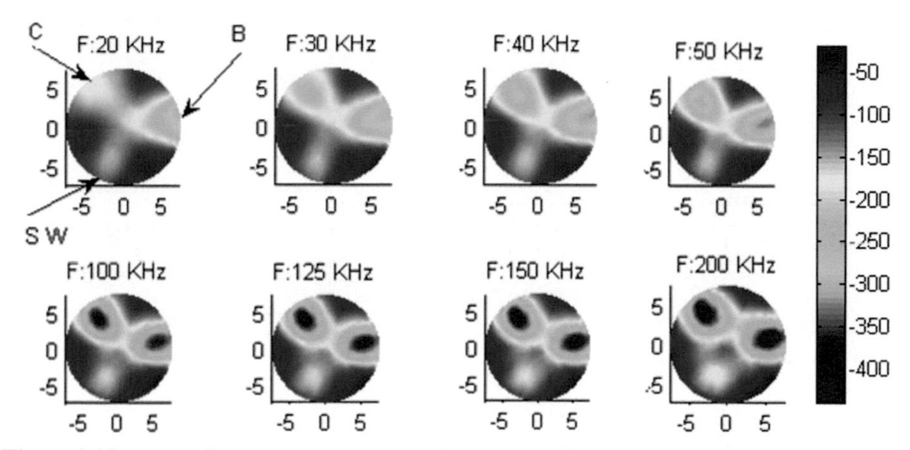

Figure 9.20 Energy frequency tomographs of cucumber (C), banana (B) and saline water (SW) objects in flat spectrum gel; for frequency steps from 20 to 200 kHz.

9.6.3 *Overall processing with interpretation*

To demonstrate the overall operation from excitation to interpretation, a computer-based simulated process distribution was synthesized using a distributed 2D mesh circuit model (Nahvi & Hoyle, 2008a). This provides a known and fixed comparative base for evaluation, with definitive electrical impedance values at key ROIs. The simulated process is used in a standard electrical network simulation tool to perform excitation and response acquisition. The resulting data are then processed, as demonstrated in above in Section 9.6.2, through to the data fusion stage to produce energy frequency tomographs, which can be explored using the interpretation processes described in Section 9.5.4.

The simulated experimental process conditions are based upon those used in Section 9.6.2 above, here using two ROIs in a background that does not exhibit major spectral variation. Fig. 9.21 illustrates a range of 8 energy frequency tomographs; (uppe-r) directly from the original simulated model data; and (lower) the corresponding data generated from the S-T method.

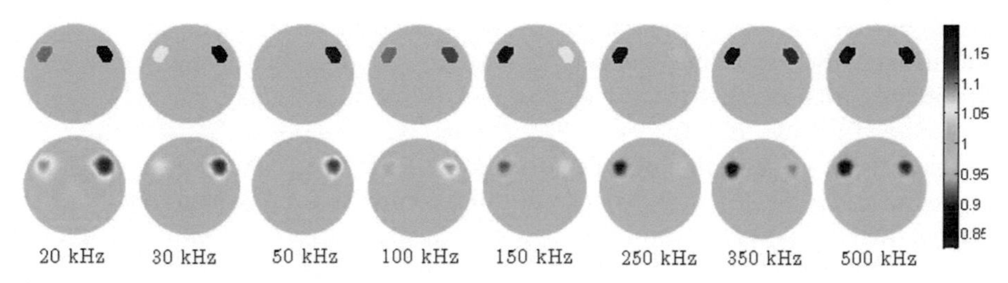

Figure 9.21 (Upper row) Simulation process distribution with two Regions of Interest in flat spectrum background; (lower row) Corresponding energy frequency tomographs, for frequency steps from 20 to 500 kHz.

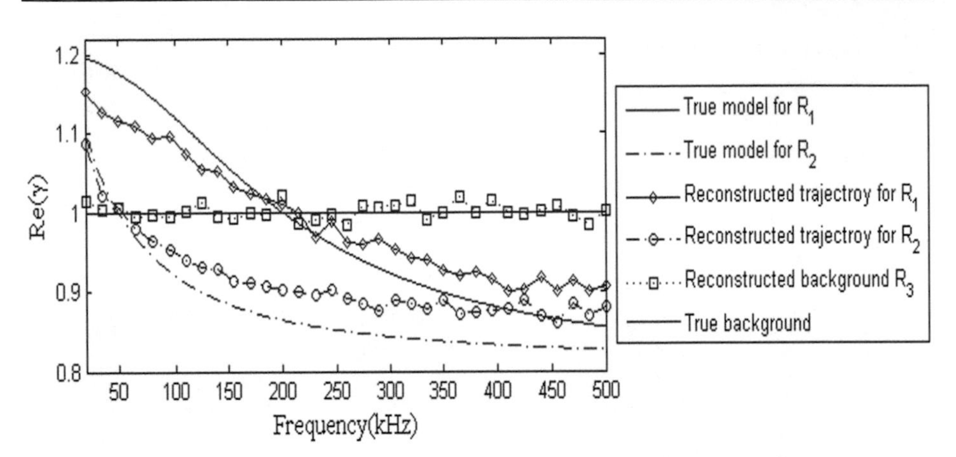

Figure 9.22 Comparison of Region of Interest R_1, R_2, and R_3 trajectory graphs.

As explored in Section 9.5.4, a range of interpretation variants can be applied. Where distinct regions are likely to be present, a simple image processing operation across the set can easily identify these. This variant is demonstrated here. The identified ROIs are described as R1—upper left side; R2—upper right side; R3—background. In this case, the two ROIs and the background area may then be compared with a selected set of spectral fingerprints to identify three materials using an LUT or an analytic model. As described, the set of tomographs represent the same sampling instant and hence there is no movement of materials across the set. Fig. 9.22 illustrates the comparison process in graphical terms. ROI R1 is the upper curve with the marked continuous line as the known model from the original simulated data. The corresponding R1 line reconstructed from the energy frequency tomograph set is seen as a reasonable fit. The constant R3 model and its reconstructed form are the middle lines. The lower lines are the R2 data. It is apparent that a spectral fingerprint fitting operation would successfully identify the three candidates.

9.6.4 General system design conclusions

This review and demonstrations have explored *real-time* performance and demonstrated that an S-T system can capture raw response data, including an added spectral dimension, with minor time overhead. It also provides a major step-up in data precision in industrial settings, where data are typically sparse and noisy, due to hard limits on time and number of projections. The simple 1G-IPT (1G-SE-SM-IPT) form has intrinsic single-energy, single-mode limitations. Following data acquisition, image reconstruction is the next step to obtain useful information from the limited raw tomographic data. Chapter 11 provides a comprehensive overview of reconstruction methods, with details in other Part Two chapters. These range from simple and fast, to complex and compute intensive (e.g., with time series filtering or optimization); but all provide *gray-scale* (or *false color*) 1D intensity line pixel data.

In terms of precision, 1% pixel gray-scale values appear practically demanding, particularly for low sensitivity regions. Hence, a practicable upper limit range of 100 levels would generally be adequate. In terms of energy level (frequency), results for the simple trials of Section 9.6.1 (in Fig. 9.18) show continuous characteristics for fast linear chirp implementation. This arises as the linear chirp excitation stimulates a continuous response; from which any frequency response value may be extracted with computation, justifying a 1% precision. This contrasts with the spectrum analyzer comparison limited to the specific selected (above 20) discrete frequencies. Thus, overall, the S-T method offers a 2D, 100×100 pixel estimation space in comparison to the 100 points offered by 1G-IPT. This demonstrates the major step-up in potential estimation precision gained for a small temporal overhead, when optimal S-T excitation and processing is deployed.

9.7 Future trends for spectro-tomography

Spectro-tomography is an advanced instrumentation solution to the need for comprehensive internal process information in a range of application sectors. This engineering review has shown that the S-T method adds the major key dimension of *what it is* (material), to the *where it is* (space), and *when it is* (time) information obtained from conventional tomographic instrumentation.

9.7.1 System development trends

As introduced at a generic level in Section 9.2, the S-T multidimensional approach forms a radical IPT research and development direction, seen within the landscape as *Multi-Energy Level* (ME): 2G-ME-IPT (Hoyle, 2016). A *Multi-Mode* (MM) system may also be considered for processes that require more than one excitation mode. A *Third Generation* capability features a combination of two or more modes, each of which provides multienergy level capability: 3G-MM-ME-IPT.

Future S-T in-process design will exploit modern digital systems, where control and in-line processing are straightforward to realize from large-scale integrated electronics, with on-chip subsystems, able to synthesize powerful, small-scale instrumentation for in-line use.

9.7.2 Industrial application trends

This chapter provides a rationale for the integration of fast joint sensing of spatial distribution and spectroscopic material properties. It compares this capability with basic 1G-IPT instrumentation, noted as suitable only for investigations where a single material of interest offers a high contrast to selected excitation, in a background material having a low contrast.

Processes in several business sectors feature containments, such as vessels and pipes, whose internal contents and behavior are of interest but cannot comply with 1G-IPT limitations. These include many processes used to synthesize important materials that typically involve several raw, or intermediate, feedstock materials and often also feature two or even three phases. In principle, such processes are perfect candidates for the S-T concept, defined above as 2G-ME-IPT, which as exemplified may be realized with various base excitation modes. For example, in complex mixtures found in oil and gas applications, or in mineral separation, composition may be measured using S-T based upon ECT excitation. Similarly, mixing reactions with aqueous compounds may deploy conductive EIT excitation. S-T is applicable to a wide range of complex multicomponent processes.

A classification of process types is useful to form an overview based upon spatial distribution features for IPT sensing (Hoyle, 2005), augmented by system design terms for symmetry considerations for circular and linear sensing arrays (Hoyle, 2016). It is useful to compare more sophisticated 2G-IPT and 3G-IPT forms presented against the classic (lower cost and simpler) 1G-SE-SM-IPT form, with its single-energy, single-mode limitations. Examples are presented in three groups based upon process mobilization:

Group 1 – pressure or gravity driven motion predominantly in one direction

Examples: many processes include materials constrained as flows. Chapters 18–21, and 24 provide example details. Specific examples: Pipe flow, Bubble column, Pneumatic conveying, Hopper flow, and Tank sedimentation.

Sensing need: monitor cross-sectional instantaneous composition. Most applications deploy cross-sectional 2D view sensor arrays, where electrodes or devices are often mounted internal to a pressure containment. An appropriate means of flow rate sensing is often included; this may involve two sensor rings if flow regime can be used to estimate flow profile. Acoustic or electromagnetically coupled sensors may be mounted externally.

1G-IPT applications: composition measurement in single material transport flows (where a propellant has small relative contrast and is not of sensing interest), such as single-mineral solids fraction in pneumatic flow or single-liquid fraction in liquid/gas flows.

2G, 3G-IPT applications: joint metering of 2–3 components in material transport flows using 2G-ME-IPT. Especially suited to high-precision needs of fiscal metering. Where contrast opportunities are diverse, 3G-ME-MM-IPT may be deployed.

Group II – Forced agitation or stirring within a closed vessel

Examples: many processes include materials contained in a vessel or reactor, while various actions are taken to encourage physical and/or chemical transformations. Chapters 25 and 26 provide example details. Specific examples: Stirred tank reactor, Crystallizer, Powder blender, and Fluidized bed.

Sensing need: monitor and track whole vessel composition in time for batch or continuous process, to inform quality assessment and/or control needs. Applications may deploy a single 2D cross-sectional (often circular) sensor array (to infer whole vessel from model); or multiple cross-sectional arrays with interpolation; or where cylindrical stirring is employed justifying circular symmetry (Hoyle, 2016), a single, vertical, radial planar view sensor array may be used synchronized to rotation position sensing.

1G-IPT applications: monitoring of single product compound where contrast opportunity allows it to be discriminated in a background which may involve several reagents.

2G, 3G-IPT applications: joint metering of 2–3 components in vessels using 2G-ME-IPT and where higher precision particularly in regional detail can aid enhanced control. Where contrast opportunities are diverse, 3G-ME-MM-IPT may be deployed.

Group III – Forced motion of materials through structured matrix of fixed elements

Examples: many processes feature passage of materials through a vessel with specific critical structured design to encourage physical and/or chemical transformations. Chapters 25 and 26 provide example details. Specific examples: Packed columns, Plate towers, Distillation columns, Evaporators, Heat exchangers, Leaching processes, Flotation separators, and Precipitators.

Sensing need: monitor and track passage of materials through active areas or parts to inform quality assessment and/or control needs to ensure optimal efficient use of active zones, for example, for catalysis. Applications may deploy a single 2D cross-sectional sensor array at a strategic location, e.g., at header gravity feed plane or at several selected check planes.

1G-IPT applications: monitoring of single product compound where contrast opportunity allows it to be discriminated in a background if this involves several reagents to check efficiency of material passage.

2G, 3G-IPT applications: monitoring multiple materials to monitor rate of transformation of input material and resulting product at test cross-sections for product quality assessment and/or rate material inlet flow control.

Consideration of these aspects for all processes offers design rationalization opportunities for "packaged" IPT hardware and software solutions which promise future enhanced product quality and reduction of waste and emissions. The S-T aspects provide a critical capability to sense material distributions in many processes that feature multiple components, and which have a critical need to accurately track internal state.

References

Barsoukov, E., & Macdonald, J. R. (2018). *Impedance spectroscopy: Theory, experiment, and applications* (3rd ed.), ISBN 978-1-119-33318-0.

Dickin, F. J., Hoyle, B. S., Hunt, A., Huang, S. M., Ilyas, O., Lenn, C., et al. (1992). Tomographic imaging of industrial process equipment – techniques and applications. *Proceedings of the IEEE Part G, 139*(1), pp. 72–82. https://doi.org/10.1049/ip-g-2.1992.0013

Hoyle, B. S. (2005). Schema for generic process tomography. *IEEE Sensors Journal, 5*(2), 117–124.

Hoyle, B. S. (2016). IPT in industry – application need to technology design. In *ISIPT 7th world congress in industrial process tomography, Iguazu Falls, Brazil* (p. 7), ISBN 978 0 85316 3497. Sept 2016.

Hoyle, B. S., Jia, X., Podd, F. J. W., Schlaberg, H. I., Wang, M., West, R. M., et al. (2001). Design and application of a multi-modal process tomography system. *Measurement Science and Technology, 12*(8), 1157–1165.

Hoyle, B. S., McCann, H., & Scott, D. M. (2005). Process tomography. In D. M. Scott, & H. McCann (Eds.), *Process imaging for Automatic control* (pp. 85–126). Taylor and Francis.

Hoyle, B. S., Nahvi, M., Ye, J., & Wang, M. (2012). Systems engineering for electrical spectro-tomography. In *6th International Symposium on process tomography, cape Town, South Africa, ISIPT, OR19, 1-10*, ISBN 978-0-620-53039-2.

Hunter, R. J. (2001). *Foundations of colloid science*. Oxford University Press, ISBN 9780198505020.

Industrial Tomography Systems. (2021a). *P2+ IPT Instrument*. http://www.itoms.com/products/p2-electrical-resistance-tomography/ (Accessed 28 January 2021).

Industrial Tomography Systems. (2021b). https://www.itoms.com/industries-and-applications/ (Accessed 30 January 2021).

Li, W., & Hoyle, B. S. (April 1995). Spectral discrimination in multiple active real-time ultrasonic tomography. In *Proceedings of European concerted action on process tomography, Bergen* (pp. 187–195).

Nahvi, M., & Hoyle, B. S. (2007). Process impedance spectroscopy through chirp waveform excitation. In *5th world congress on industrial process tomography, Bergen, Norway, 2007* (pp. 630–637).

Nahvi, M., & Hoyle, B. S. (2008a). Wideband electrical impedance tomography. *Measurement Science and Technology, 19*. https://doi.org/10.1088/0957-0233/19/9/094011,094011-094020

Nahvi, M., & Hoyle, B. S. (2008b). Wideband excitation signals for electrical impedance industrial process tomography. In *5th International Symposium on process tomography. Zakopane, Poland* (p. 6). August 2008.

Nahvi, M., & Hoyle, B. S. (2009a). Electrical impedance spectroscopy sensing for industrial processes. *IEEE Sensors Journal, 9*(12), 1808–1816.

Nahvi, M., & Hoyle, B. S. (2009b). Data fusion for electrical spectro-tomography. In *IEEE imaging systems and techniques conference, Shenzen, China, 2009b* (pp. 229–234).

Nahvi, M., & Hoyle, B. S. (November 2010). Spectro-tomography interpretation for integrated sensing of process component identification and distribution. In *IEEE Sensors Conference, Hawaii, US*.

O'Brien, R. W. (1982). The response of a colloidal suspension to an alternating electric-field. *Advances in Colloid and Interface Science, 16*, 281–320.

Oguh, U. I., Hall, J. F., Bolton, G. T., Simmons, M. J. H., & Stitt, E. H. (2012). Characterisation of gas/liquid/solid mixing using using tomographic electrical resistance spectroscopy. In , *PO21. 6th International Symposium on process tomography, cape Town, South Africa* (pp. 1–8), ISBN 978-0-620-53039-2.

Qiu, C., Hoyle, B. S., & Podd, F. J. W. (2007). Engineering and application of a dual-modality process tomography system. *Flow Measurement and Instrumentation, 18*, 247–254.

Ricard, F., Brechtelsbauer, C., Xu, X., & Lawrence, C. (2005). Monitoring of multiphase pharmaceutical processes by electrical resistance tomography. *Chemical Engineering Research and Design, 83*(A7), 794–805.

Wang, M. (2002). Inverse solutions for electrical impedance tomography based on conjugate gradient methods. *Measurement Science and Technology, 13*, 101–117.

Zhao, Y., & Wang, M. (2015). Experimental study on dielectric relaxation of SiO_2 nano-particle suspensions for developing a particle characterization method based on electrical impedance spectroscopy. *Powder Technology, 281*, 200–213.

Zhao, Y., Wang, M., & Hammond, R. B. (2011). Characterization of crystallisation processes with electrical impedance spectroscopy. *Nuclear Engineering and Design, 241*, 1938–1944.

Electron tomography

Sean M. Collins
University of Leeds, Leeds, United Kingdom

10.1 Introduction

A central objective of modern materials engineering is the control and manipulation of the position and chemical interactions of atoms within solids. Functional properties arising from surfaces and interfaces in materials are directly derived from the atomic structure and composition at these nanoscale features, whether in heterogeneous catalysis, the formulation of complex products, separations technologies, or in industrial metallurgy and semiconductor manufacturing. The underlying measurement challenge has been termed the "nanostructure problem" (Billinge & Levin, 2007). Characterization of macroscopic volumes inevitably evaluates the ensemble average of many local, nanoscale features. While tomographic measurements seek to invert this problem, in most cases, the spatial resolution is limited, precluding access to the fundamental nanometer to atomic-scale information at surfaces and buried interfaces, connected pore or particle networks, and localized defects.

Direct imaging of nanoscale features is possible using electron microscopy and selected other microscopy techniques. Today, it is possible to image individual atoms with picometer precision in transmission electron microscopy (TEM) and scanning transmission electron microscopy (STEM). These techniques grant access to fundamental, atomic-scale features but remain limited to integration through a volume of the material (TEM and STEM) or to the exposed surface of the material (scanning probe microscopy) or are inherently destructive (atom probe tomography). Tomography in the electron microscope (TEM and STEM) offers a nondestructive, fully three-dimensional solution to this limitation, functioning in a similar way to a computed tomography (CT) scan of a nanoscale sample volume. Through advances in TEM and STEM capabilities, it is now possible to recover atomic resolution information and quantify chemical composition in three-dimensions at the nanoscale. These capabilities have seen substantial application in the characterization of heterogeneous catalysts and catalyst supports from the earliest development of electron tomography in materials applications (Midgley et al., 2004; Weyland, Midgley, & Thomas, 2001), as well as wide application in materials for the semiconductor and metals industries.

While nanoscale electron tomography offers detailed insights into individual local atom configurations, the technique is often plagued by poor sampling statistics relative to the macroscopic volumes of interest in many industrial processes. However, electron tomography is now poised to address this challenge with emerging trends in improved statistical sampling via increased automation and routine integration of electron tomography with multiscale characterization workflows. Moreover, electron

Industrial Tomography. https://doi.org/10.1016/B978-0-12-823015-2.00012-1

tomography is increasingly becoming possible for a wider class of materials and sample types including complex products containing organic as well as inorganic fractions (including pharmaceutical and battery materials among many others), as well as tracking the evolution of materials in time and under applied heating or biasing, and in liquid or under exposure to gas.

In the medical and biotechnology arenas, there has been a "resolution revolution" in atomic resolution imaging of proteins and viruses using cryogenic electron microscopy and tomographic approaches (Bai et al., 2015), with, for example, rapid characterization of coronavirus particles within the first several months of the COVID-19 pandemic. In structural biology, a key feature has been a degree of similarity in experimental conditions repeated for tens or hundreds of thousands of copies of identical or near-identical samples. Electron tomography for industrial materials presents a wider range of experiments, but also, concomitantly, rich opportunities in quantitative chemical analysis with capabilities for simultaneous spectroscopy, diffraction (crystallographic analysis), and ultimate spatial resolution. For industrial materials and process engineering, a similar revolution will be possible with increased automation and streamlined integration for on-line and off-line process monitoring in tandem with improvements in quantitative analysis at the nano- and atomic scale.

10.2 Tomography in the electron microscope

Tomography in the electron microscope, in general terms, consists of recording a series of two-dimensional images which are then fed to an algorithm which performs reconstruction of the three-dimensional volume. For electron tomography, the electron microscope is operated in transmission mode, either in what is termed TEM or STEM mode of operation (Fig. 10.1). The TEM mode is characterized by a highly parallel beam of electrons which form a "broad beam," i.e., the beam is spread out across the sample. In TEM mode, the distribution of electrons covers an area matched to or larger than the targeted field of view for imaging. Images are formed by magnifying the illuminated area at the sample onto a two-dimensional electron sensitive detector or

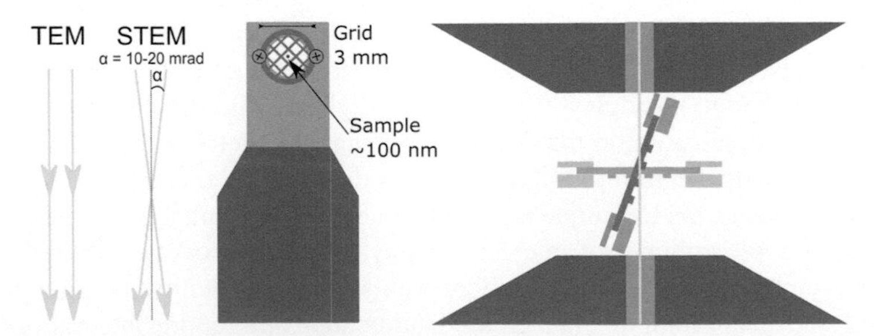

Figure 10.1 Schematic illustration of major operating modes of an electron microscope for tomography, a diagram of a sample mounted on a grid on a side-entry holder, and an illustration of an idealized position and rotation of a sample within the objective lens.

camera. The STEM mode is characterized instead by a "focused beam" where the electron distribution is brought to a spot at the sample. The spot or electron probe is then rastered across the sample in a two-dimensional scan pattern and one or more intensity signals are recorded on a detector or detectors. An image is formed by plotting the signal intensity as a function of probe position.

In order to achieve transmission of electrons through a sample, the electrons are accelerated to a significant fraction of the speed of light, typically to energies of 80–300 keV, corresponding to ca. $0.5c$ to $0.7c$ where c is the speed of light in vacuum. The electrons have a short de Broglie wavelength at these energies corresponding to <5 p.m. While these values may suggest a diffraction limit of a few picometers, in practice, electron microscope resolution is limited by the optics and optical aberrations of the round magnetic lenses and angle-limiting apertures used in setting up the electron beam. For STEM mode operation, the incident angles used in the focused probe are typically of the order of 10–20 mrad (~1 degrees). Even for these high energy electrons, due to the strong interactions of electrons with matter, samples must be thin to enable transmission imaging. In practice, samples should be on the order of 100 nm or less in thickness along the beam direction. As a result of thin samples and small angles involved, the electron beam is typically considered as an effectively parallel beam within the sample volume.

In addition to the requirement for electron transparency, the sample environment in the electron microscope is generally at high vacuum to minimize electron scattering by gases which would otherwise distort imaging conditions. Samples are introduced to this setting by mounting thin sections or microscopic particles onto a grid (most commonly 3 mm diameter) which is then placed in a holder (~30 cm scale). The combined sample grid assembly is mounted on a microscope stage which enables movement of the sample vertically (z) and horizontally (x, y) and a goniometer controls the primary tilt axis of the stage and sample.

To obtain TEM or STEM images, the sample is positioned in the electron beam by moving the stage. The grid is held within the objective lens of the microscope, which consists of two round magnetic lenses, with one soft iron pole piece above and one below the sample. This design provides for a homogeneous magnetic field for optimal image formation and focusing of the electron beam. The electron beam is typically aligned to a fixed position and focal plane where the electron optics are optimal for the design of the microscope. The vertical motion of the stage (z) is used to bring the sample close to this imaging plane, and the electron optics are adjusted minimally to finely focus the beam on the sample. The horizontal motion (x, y) enables selection of a sample site within the 3 mm grid.

In order to obtain a series of two-dimensional images in the electron microscope, the sample is tilted through a set of angles θ defined from 0 degrees with the grid positioned horizontally. In order to achieve high-quality electron imaging, the gap in the objective lens pole piece assembly is ideally kept as small as possible. For tomography, however, tilting through ± 90 degrees is ideal to capture images from all orientations of the sample. Due to these opposing requirements, microscopes typically used for electron tomography restrict tilting to approximately ± 70 degrees in practice with a conventional 3 mm grid, to avoid contact between the holder and the pole pieces.

Notably, the bars of the grid itself may block the electron beam at lower tilt angles or the sample may be too thick and consequently out of focus or not sufficiently electron transparent when tilted beyond a certain angle, further restricting the available tilt range.

First, an idealized reconstruction process over all angles will be presented followed by discussion of the effects of limited tilt range. A useful starting point in electron tomography is the Radon transform of an object f:

$$\mathcal{R}\{f\} = \int_L f(\boldsymbol{R}'(\theta), s)ds \tag{10.1}$$

where the integral is taken over all possible lines L along directions s with coordinates in the perpendicular plane to the direction determined by the tilt angle as $\boldsymbol{R}' = (x', y')$. The full Radon transform includes all possible line integrals and therefore suggests an infinitesimally stepped tilt angle. In practice, discrete projection images are acquired for a finite number of tilt angles. A single projection may be written as follows:

$$P_z = \int_{-\infty}^{\infty} f(\boldsymbol{R}, z)dz \tag{10.2}$$

where now the integral is along the fixed beam direction aligned with z. A full set of experimental projection images can then be defined in operator notation as follows:

$$\widehat{P}f(\boldsymbol{R}, z) = \Gamma^{\text{exp}} \tag{10.3}$$

where the projection operator takes into account projections of the object along a finite set of tilt angles to return the experimental dataset Γ^{exp}. This equation expresses the data generation process and casts the tomographic reconstruction as a set of equations constituting a general inverse problem. This general type of problem is ill-posed; that is, there may be any number of objects $f(\boldsymbol{R}, z)$ that match the experimental data. Additional properties of the sample and projection operation mean that reconstruction algorithm is not entirely unconstrained, but it should be clear that in cases where there is limited and often noisy experimental data, it may not be possible to arrive at a unique solution. Advanced tomography algorithms (Section 10.4) seek to address these challenges.

A cornerstone assumption for electron tomography is the projection requirement. The projection requirement is satisfied when the recorded data vary monotonically with the object quantity of interest, often the density, composition, or otherwise the "amount" of the object. Monotonic variation means that the experimental signal increases with an increasing amount of the object along the beam direction (projection direction). A simple mock object or "phantom" is shown in Fig. 10.2 consisting of three circles. Each arrow indicates a projection direction. For a two-dimensional image, which may be considered as a single yz slice from a three-dimensional object with a tilt axis along x (out of the page), the projections are linear "shadows" of the

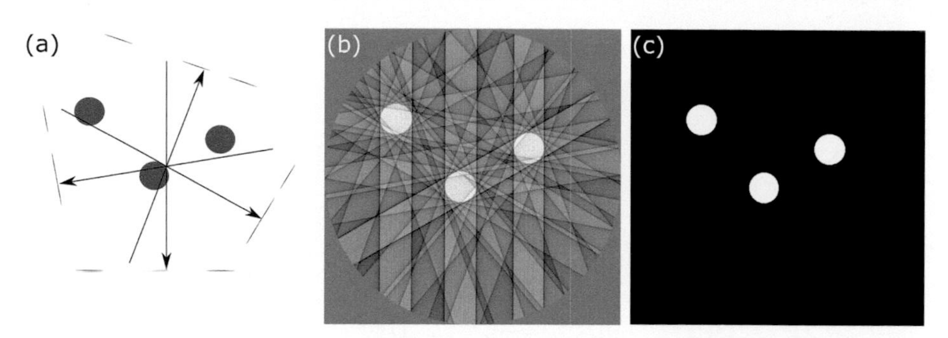

Figure 10.2 (a) Phantom object consisting of three circles shown with four linear shadow images or projections. (b) A back-projection reconstruction from 10 projections compared with (c) the ground truth.

phantom object. A simple back-projection reconstruction, where intensity at each tilt angle is effectively smeared back along the beam direction, is depicted in Fig. 10.2B using 10 projections. The back-projection approach effectively articulates that there is uncertainty associated with where the projected intensity originates, so the algorithm distributes that intensity along the beam trajectory. Additional projections constrain this uncertainty (or the number of possible positions in the volume to place the intensity). Even with very few projections, it is apparent where the circles are located. However, the reconstruction error is still quite significant in that the volume contains many intensities that are not approaching a flat background. Additional projections will flatten the background, leading to a reconstruction increasingly resembling the ground truth.

For some electron microscopy images, the beam is attenuated when passing through the sample, in what is known as "bright field" imaging. In this case, an image without any sample present would be white (flat intensity) and the sample decreases the intensity recorded resulting in darkening of the image proportional to the density of the material. Alternatively, electron microscopy images are instead captured in a "dark field" modality where the scattering of the electron beam is recorded by a detector and the features are bright on an otherwise black background when the electron beam is not scattered onto the detector by the sample. These two representations can be considered equivalent, provided the signal is monotonic with respect to a quantity of interest. The same spheres shown in this dark field representation are depicted in Fig. 10.2C.

The series of projections (one-dimensional projections) can be plotted as a function of angle in what is termed a "sinogram" (Fig. 10.3). In order to consider the sampling of information about the underlying object (two-dimensional for the phantom or the three-dimensional volume as a series of such slices), it is useful to consider the Fourier transform representation of the objects. The Fourier slice theorem, which links the sinogram (real space sampling) and the Fourier domain representation, states that each projection for tilt angle θ corresponds to a radial slice through the Fourier transform at the same angle. This is schematically represented with high density "full" sampling of the tilt angles and the Fourier transform. When the tilt step increment is

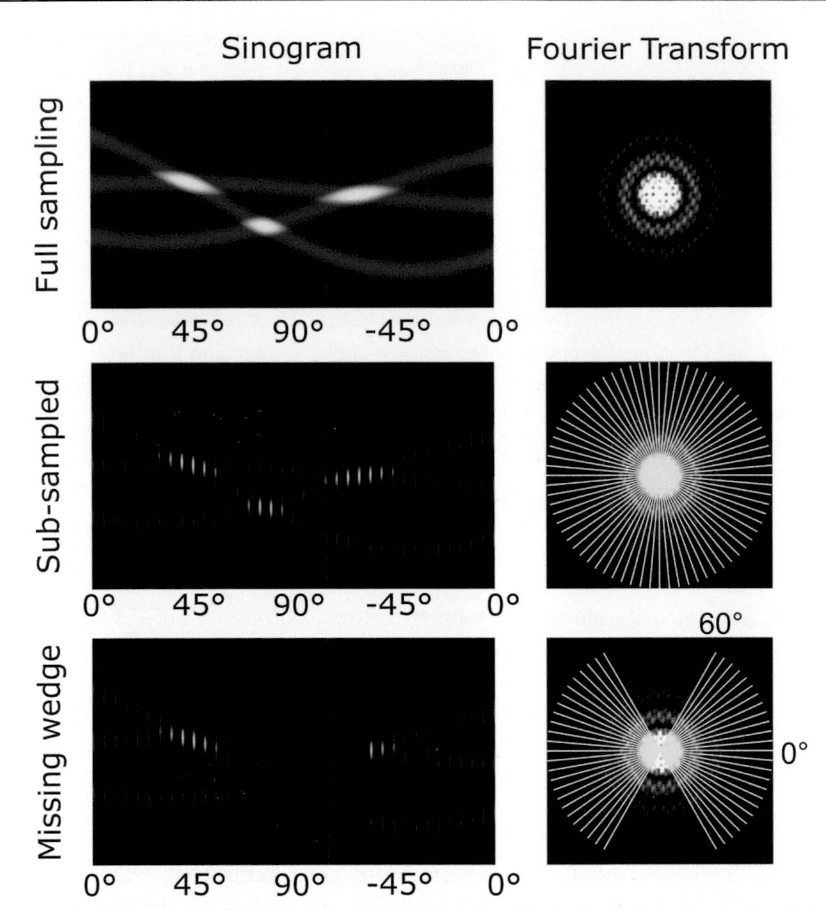

Figure 10.3 Illustration of a sinogram with full sampling (shown as 1 degree steps) and changes with finite step sizes (5 degrees steps) and with a missing wedge (±60 degrees).
The corresponding Fourier space representations are depicted highlighting the radial slices captured in the Fourier domain.

increased, gaps appear in the sinogram and between radial slices in the Fourier domain. If there is a limit to the maximum tilt range, a gap appears in the middle of the sinogram and a "missing wedge" appears in the Fourier transform representation. That is, there is no information captured about the Fourier components at these points. The missing wedge is particularly prominent in that it leaves a significant and highly anisotropic gap in the information that is recorded about the object. In practice, the missing wedge means that there is highly limited information recorded about the features along the beam direction. In the context of the three-circle phantom, there would be very few back-projected bands to define the top and bottom of the circles, resulting in a blurring of these features. This is observed widely in experimental electron tomography, resulting in an elongation of features in the missing wedge direction. This elongation is an extension of the concept of uncertainty in where to place intensity in the reconstruction volume, i.e., the intensity is poorly constrained in the missing wedge direction.

Collectively, gaps in the information between projections at different tilt angles and due to the missing wedge can be termed undersampling or "subsampling." In a simplified representation, if the positions of the circles in the phantom were the targeted feature, only three projections should be required in that there would be six unknowns (three unknown pairs of coordinates), resolved by a single known coordinate and a known angle for each projection. In fact, this "triangulation" approach can be applied to reconstruct line, edge, and point features from far fewer projections than in conventional tomography. Electron tomography, however, seeks the intensities and not just the positions, giving rise to many more unknowns. As such, particularly in experimental electron tomography where constraints on acquisition time and sample stability during the data acquisition (changes to the sample volume due to significant accumulation of ionization-induced damage preclude conventional reconstruction approaches as these assume the reconstruction volume is constant) mean that electron tomography data are usually highly undersampled.

A variety of reconstruction algorithms have been applied in electron tomography of industrial materials. The Fourier slice theorem establishes that one possibility is to perform the reconstruction in Fourier space and invert the Fourier transform. Due to poor sampling at high spatial frequencies (outer edge of the Fourier representation), Fourier space inversion techniques can suffer from noise amplification. The mapping of the sinogram to Fourier space also presents some challenges in requiring nonuniform Fourier transforms, or alternatively, the experimental acquisition can be adjusted to improve the sampling (e.g., the equally sloped tomography approach (Miao, Förster, & Levi, 2005)). Weighted or filtered back-projection algorithms account for this issue by reweighting the contributions to the reconstruction. Iterative update algorithms, such as the simultaneous iterative reconstruction technique (SIRT), show improved robustness to noise and limited angle sampling and are widely used in materials science and engineering materials applications. SIRT can be cast as a least-squares problem where a volume is sought that minimizes the discrepancies between its reprojection and the experimental data:

$$f(\boldsymbol{R}, z) = \arg\min \left\{ \sum_j \left(\widehat{P}_j f(\boldsymbol{R}, z) - \Gamma^{\exp} \right)^2 \right\} \tag{10.4}$$

Iterative refinements can also be incorporated into Fourier space reconstruction techniques, such as for iteratively updating the interpolation of Fourier space (Pryor et al., 2017) or for incorporation with compressed sensing, regularized reconstruction, or model-based approaches (see Section 10.4). However, there is a significant penalty in computational time for iterative approaches and convergence of the algorithm must be carefully considered. SIRT, for example, has a well-known semiconvergence property, meaning that the reconstruction improves for a certain number of iterations but then degrades as the algorithm "overfits" to noise in the experimental data.

Additional constraints can improve the performance of the algorithm. Most reconstruction algorithms operate on the assumption that the quantity of interest is entirely contained in a volume defined by the size of the projection images, further constraining the inverse problem. However, if the sample is not contained within the field of view, there will be inconsistencies in the amount of the object projected through relative to the reconstruction volume. This is a common and sometimes unavoidable issue in electron tomography giving rise to what may be termed "truncation artifacts" when the reconstruction volume is smaller than the true volume contributing to the projection images. Nonnegative projection operators or constraints to impose nonnegativity on the reconstruction volume can also significantly improve reconstructions as this reduces spurious negative intensity oscillations, particularly prominent otherwise at edges and sharp boundaries. Constraining the reconstruction to nonnegative values is a physically motivated assumption for electron tomography signals that satisfy the projection requirement. In this case, zero or near-zero intensity pixels immediately constrain the reconstruction volume along the ray corresponding to that pixel to near-zero along the entire ray. Such zero-intensity features significantly constrain the reconstruction as the algorithm must then place the intensity in the fewer remaining volume elements (or voxels) (Collins, Leary et al. 2017).

10.3 Practical electron tomography

The modern TEM/STEM can be considered as an optical table for electron beams. Many designs are intended as a platform for flexible applications from routine imaging in a mixed-user facility through to bespoke optical configurations for advanced experiments. For the purposes of a practical discussion of routine and advanced tomography, this section will focus on the STEM mode. Principally, imaging in STEM more readily satisfies the projection requirement for electron tomography. STEM, particularly using electrons scattered onto a high angle annular dark field (HAADF) detector, offers a chemically sensitive signal which is dependent on the average atomic number density at the probe position with minimal contributions from Bragg diffraction at low angles. TEM imaging often incorporates signals related to the crystallographic orientation of the specimen, due to strong Bragg scattering at low angles from crystalline samples. Still, many reported examples of TEM-based tomography illustrate its utility for industrial materials, and some early work used energy-filtered TEM where incoherent inelastic scattering is selected to satisfy the projection requirement (Weyland & Midgley, 2001).

Fig. 10.4 presents a diagrammatic overview of a STEM column. The column is not unique to STEM mode operation, and, by changing lens excitations by altering the current passing in the coils of the cylindrical lens assembles, many microscopes operate flexibly between TEM and STEM mode. In either case, electrons are extracted from an electron source or gun and are accelerated to the selected operating energy. The beam then passes through a series of condenser lenses which serve to demagnify the source and set up the angular spread of the STEM probe. Round magnetic lenses in the

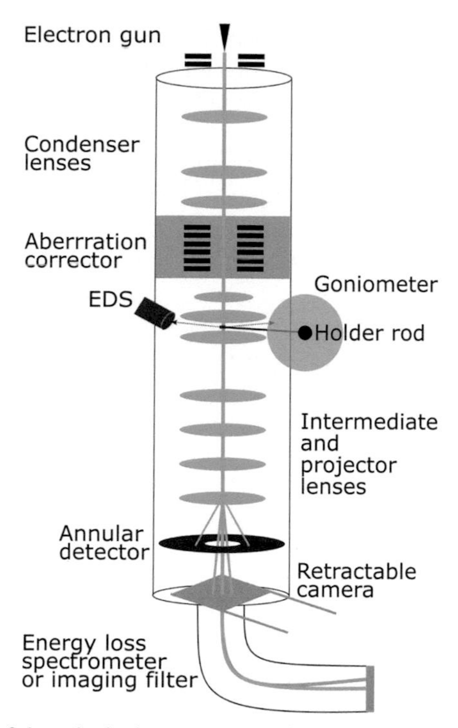

Figure 10.4 Diagram of the principal components of a STEM column.

electron microscope suffer from severe aberrations arising from the geometry of the lens. These aberrations result in a blurring of the STEM probe and a reduction in the attainable spatial resolution for imaging. However, by breaking the symmetry of the cylindrical lens, these aberrations can be compensated. A series of hexapolar or quadrupolar magnetic lenses can be used to offset the aberrations of the cylindrical lenses preceding the sample. Today, this enables imaging with a larger convergence semiangle which increases the numerical aperture of the lens and consequently the achievable spatial resolution, with demonstrated spatial resolution <60 pm reported.

The sample is sandwiched within the objective lens assembly, with a side-entry holder inserted via an air-lock illustrated in Fig. 10.4. The electron beam passes through the sample and is collected by a further series of lenses to project the resulting scattering onto appropriate detectors. By changing the excitations of these lenses, the angle subtended by a fixed annular detector can be adjusted to achieve HAADF imaging. Inelastic interactions with the sample cause ionization of valence and core electrons. The electrons transmitted through the sample that lose energy as a result of these ionizations can be dispersed in energy to record electron energy loss spectroscopy (EELS) at each probe position in the image. Characteristic X-rays are also emitted as a result of core electron ionizations, and these can be collected by detectors typically positioned just above the sample to record an X-ray energy dispersive spectroscopy (EDS) signal. Finally, the diffraction pattern associated with each probe position can also be acquired on a retractable camera (4D-STEM), either for use in crystallographic analysis or for sophisticated imaging modalities such as ptychography that recover information on the phase shift induced in the electron beam upon passing through the sample.

HAADF, EELS, and EDS signals have seen the widest adoption in electron tomography applications, primarily as these signals satisfy or can be processed to satisfy the projection requirement relatively straightforwardly. Examples using diffraction patterns recorded at each position for tomography have also been reported (Eggeman, Krakow, & Midgley, 2015).

Electron tomography datasets are acquired within integrated microscope control software, either using semiautomated acquisition of the tilt-series or, equivalently, manual acquisition of each frame. In the ideal situation for tomography, the holder rod is perfectly aligned with the tilt axis of the goniometer, and the sample is rotated on axis and remains at optimal focus and centered in the field of view. This arrangement is termed the "eucentric" position in the column. Notably, the eucentric position relative to the goniometer may not be perfectly colocated with the optimal focal plane which is also sometimes referred to as the eucentric height. In practice there are several nonidealities that introduce shifts in the x, y, and z directions of the sample and may modify the true tilt angle relative to the angle recorded during at the acquisition time. Shifts in the (x,y) directions of the lab frame (not the sample stage shifts; when tilted, these no longer correspond to pure in-plane shifts) do not present an intrinsic limitation to tomographic reconstruction process, provided the images can be aligned relative to a common tilt axis.

Vertical shifts present a more substantial challenge because the focal plane is set by the optics. Significant changes to the electron optics to compensate for vertical shifts of the sample will reduce the resolution of the images and may modify the magnification between images. It is therefore necessary to bring the sample back to near focus using the stage and minimize adjustments of the focusing optics. Adjustments to the stage hardware provide the preferred route to minimize stage shifts in electron tomography but generally do not completely remove shifts and may only optimize the stage position to near-eucentricity with the goniometer for a single holder. Automated approaches to acquisition involve a combination of calibrated stage shifts, image tracking, and automatic focusing, building in time for the stage to settle at each tilt angle. Tracking is most often executed using cross-correlation—based approaches either on the active sample area or on adjacent material. Automatic focusing relies on iteratively adjustments while monitoring the image intensities (e.g., contrast, or other image figures of merit). These routines are not always robust and may require intermittent manual intervention during tilt-series acquisition.

Acquisition software typically allows preselecting image acquisition settings including the frequency and procedures for auto-focusing and image tracking, stage settling time, selection of the magnification used for each of these procedures, and definition of the detector settings such as brightness and contrast. Within semiautomated acquisition, these parameters are automatically set to preselected values and all files are written to a single output file with logged information about the acquisition. However, most currently available acquisition software tools have a limited number and type of signals they can acquire and so less standard techniques like spectroscopic tomography or customized acquisition routines may require manual acquisition.

When first setting up an acquisition, it can be useful to check the sample will be unobstructed across the angular tilt range of interest. This check can be carried out by tilting through the intended angular range and checking for overlap with other sample features, obstruction by grid bars, or limitations to the accessible angular range due to the proximity of the pole piece at high tilt angles. The accessible range may be significantly reduced if the selected grid square is far from the center of the holder or if the feature of interest is in a narrow section of the grid (Fig. 10.5). Once the sample is selected and the tilt range verified, the sample is tilted to high tilt. Reversing the direction of the mechanical components of the goniometer results in an instability referred to as "backlash" errors which can be reduced by exceeding the intended tilt range marginally and changing directions to the initialize the intended tilt direction. Once started, the tilt direction should generally be maintained to minimize jumps in the stage position and to ensure precise relative angular increments.

A final major consideration for tomography acquisition is the tilt angle sampling scheme (Fig. 10.5). As suggested in Fig. 10.1, the grid containing the sample imposes a finite angular range resulting in a missing wedge. Some holders enable rotation of the

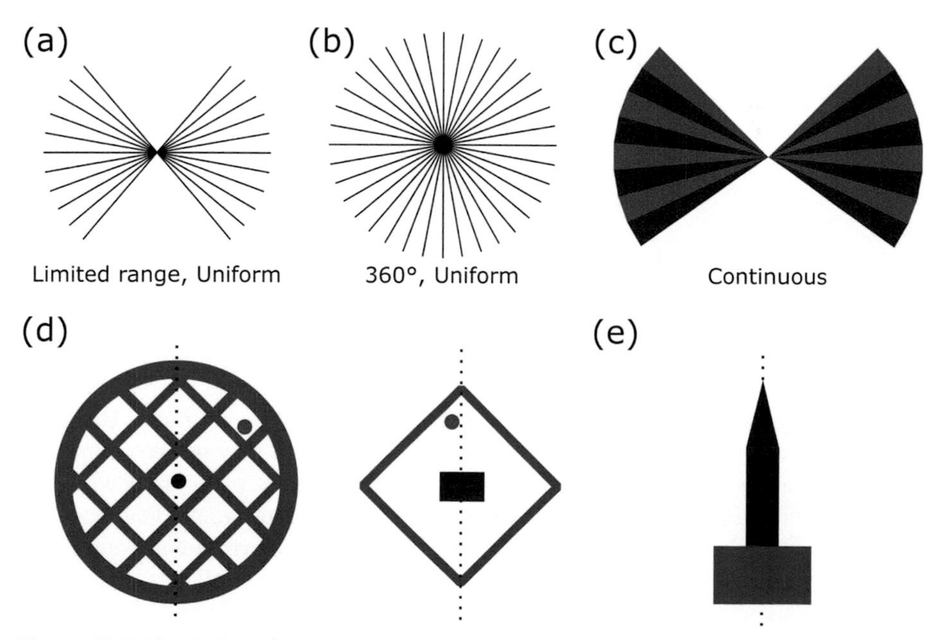

Figure 10.5 Illustration of several common sampling schemes (a)–(c). The colors in (c) indicate an integrated range of angles captured in a single acquisition frame, exaggerated in size for visual clarity. (d) considerations for position of the sample within a grid and a grid square. The magenta (light gray in print version) dots indicate nonideal sample positions. (e) An illustration of a pillar or needle sample as prepared by FIB. The dotted line illustrates idealized position of the tilt axis.

grid for reorientation and collection of data along a second tilt axis. This approach, termed "dual axis tomography," can reduce the missing wedge to a missing pyramid although in practice the added value of the information must be balanced against any errors introduced from alignments of the two tilt axes and the additional electron exposure. For a single tilt axis, a commonly used sampling scheme is uniform sampling. That is, the tilt angle is stepped through equally spaced angles throughout the tilt range. This approach is easily implemented and recovers information relatively isotropically in the Fourier domain within the tilt range. Samples that are slab-like or significantly wider at zero tilt than the sample thickness have an intrinsic anisotropy in terms of the information distribution in Fourier space, and so nonuniform sampling as a function of a cosine function of the tilt angle has been proposed to sample more finely at high tilt and more coarsely at low tilt angles (Saxton, Baumeister, & Hahn, 1984).

Sample geometry often drives the choice of sampling scheme. Focused ion beam (FIB) sample preparation can be used to prepare "lamellae" which can be transferred to standard grids or mounted on specialized "half grids" containing a set of posts rather than grid. Attachment to a post generally requires consideration to avoid shadowing by the support grid. FIB sample preparation techniques also enable the fabrication of "needle" samples (Fig. 10.5) which can be positioned on special grids to reduce shadowing by grid bars and can be used with specialized holders to allow for complete 180 degrees rotation, achieved by enabling a 90 degree fixed rotation of the grid using a mechanism on the holder in combination with the goniometer.

Finally, continuous rotation of the sample is also possible, though this places strict conditions on the quality of the alignment to enable minimal adjustments during the rotation process. In this configuration, the sample is rotated without interruption from the starting tilt angle to the final tilt angle and images are recorded rapidly over small angular ranges, e.g., for fast tomography of dynamic processes (see Section 10.6).

Given small shifts and inconsistent positioning in the field of view during acquisition, an alignment procedure is required and typically constitutes the first major post-processing step. There are broadly two approaches: (i) the use of fiducial markers and (ii) alignment by image registration techniques. Fiducial markers often consist of monodisperse gold nanoparticles deposited on the sample. Tracking the nanoparticles enables alignment when sample features are faint in the presence of noise. However, the particles will contribute to the reconstruction volume and may not be compatible with many sample geometries (needles, lamellae) or chemistries, particularly where the signal from the fiducial markers is difficult to distinguish from sample features.

Alignment without fiducial markers generally consists of cross-correlation type registration approaches. These approaches assume that under small tilts, changes to the projections will be small and allow for sequential image-to-image alignment. It is also possible to calculate the shifts for small subregion of the image and apply the shifts to the original full image stack. When aligned correctly, features in the image should shift only along directions perpendicular to the tilt axis. If the sample itself moves or reorients during acquisition, there may be some irretrievable errors in this type of image alignment. Large angular increments may be necessary for some samples due to restrictions on the acquisition time, e.g., for spectroscopic signal acquisition which tends to be slower than conventional imaging, or due to limited stability of the sample. In these and other cases, center of mass alignments have been shown to be effective although the sample must be contained within the field of view with near-zero background contribution from a supporting film for the simplest application. Additional procedures for refining alignments using multiframe acquisitions during initial data gathering (to reduce effects of sample drift) as well as automated and iterative updates to the alignment parameters during reconstruction have shown improvements and have made particular contributions to electron tomography at atomic resolution.

Once aligned as a stack of images, the tilt-axis must be aligned (Fig. 10.6). The tilt axis alignment consists of adjustments of the shift as well as any rotation of the tilt axis. Generally, this procedure is done manually with iterative adjustment and inspection of a rapid tomographic reconstruction of a few slices through the volume, positioned near the middle and near the edges of the sample or field of view. Misalignment of the tilt axis, assumed to be vertical and at the center of the image in Fig. 10.6, results in "banana artifacts" in the reconstruction. For shift misalignments, arcs will appear in the same direction for all reconstruction slices. Rotation misalignments will result in arcs in opposing directions above and below the center of the image. Fig. 10.6 shows these artifacts for a sample of lanthanum-doped ceria (Collins, Fernandez-Garcia et al., 2017), a materials family widely used in catalysis. This example highlights that in

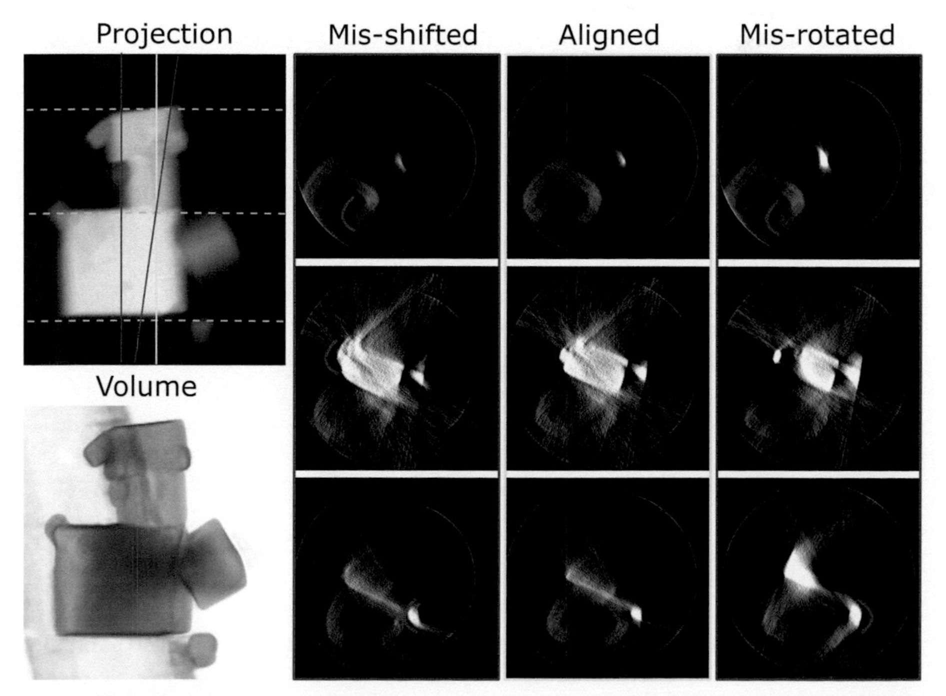

Figure 10.6 Key alignment steps and commonly observed artifacts during correction of the tilt-axis shift and rotation.

many real samples, the arcs may be difficult to distinguish from features of the sample, such as an oblique slice through a cube or curled piece of carbon support films. This particular sample used a nonflat support film to reduce background signals in spectroscopic measurements on the sample in tandem with the tomography data. The dataset also exhibited some nonmonotonic diffraction contrast giving rise to streaking in the depicted weighted back-projection reconstruction. Iterative refinement and evaluation of multiple slices enables choice of corrections to the image shift and rotation to minimize errors insofar as possible for a given dataset.

The goal of electron tomography, ultimately, is the extraction of physically relevant parameters from the final three-dimensional reconstruction. Postprocessing of the volume is an integral step and is typically tailored to the specific science or engineering objective for the sample. Broadly, visualization of the volume can consist of extracting single slices or planar cuts through the volume, extracting surfaces, or by rendering the entire volume by plotting the voxels with a transparent intensity proportional to the intensity at that point in space. Simple visualization can directly enable measurement of feature sizes. However, a great deal of quantitative structural information can also be extracted by considered image processing. These steps usually require a voxel classification process ("segmentation") where the voxels are labeled according to their intensity. The simplest approach is to set a threshold to classify voxels based on intensity above or below a specified intensity value, but more sophisticated approaches to analyze the intensities are often required to group and classify pixels. Image processing techniques, e.g., filtering, morphological, and watershed operations among many others, can be used successfully in complex workflows to achieve high-quality separation (Leary et al., 2012). In general, higher quality reconstructions enable more straightforward postprocessing emphasizing the importance of high-quality data acquisition and reconstruction.

Processing the volume gives access to physical parameters including particle, void, and pore sizing and connectivity (Biermans, Molina, Batenburg, Bals, & Van Tendeloo et al., 2010; Leary et al., 2017), facet analysis (Livi et al., 2017), particle distribution analysis including nearest neighbor analysis (Swearer et al., 2017), as well as details on surface and interface quality, morphology, and abruptness (Herzing, Henry, & Steel, 2012; Saghi et al., 2011). These parameters are not readily discernible or are ambiguous in two-dimensional imaging and bulk measurements but play key roles in determining the transport of liquids, gases, and ions through materials, access or exposure of catalytic sites, or the functional optical, mechanical, and electronic properties of devices and structural materials.

10.4 Advanced electron tomography

Several persistent challenges in electron tomography arising from limited sampling, incomplete data, or nonmonotonic signals place limitations on the achievable resolution and quantitative analysis in conventional electron tomography. However, electron microscopy data are also far richer in information content than is otherwise suggested by these problematic characteristics. There is also substantial prior knowledge both about the nature of the samples as well as about the physics of the signal generation. Moreover, for STEM tomography in particular, there are multiple signal channels available which together comprise greater information than for a single signal. Together, these properties of electron tomography enable advanced approaches to data acquisition and reconstruction.

Advanced electron tomography approaches can be collectively treated in terms of a generalized model-based approach (Collins & Midgley, 2017) where the reconstruction problem is recast as follows:

$$f(R_0, z) = \arg\min\left\{ \left\| \widehat{\Phi} f(\boldsymbol{R}, z) - \Gamma^{\exp} \right\|_{\ell}^{p} + \lambda \left\| \widehat{\Psi} f(\boldsymbol{R}, z) \right\|_{\ell} \right\} \tag{10.5}$$

where $\widehat{\Phi}$ is now a generalized projection operator, which can incorporate additional refined parameters of the alignment process or additional physics of the signal generation. The $||\cdot||$ notation refers to a choice of norms determined by the ℓ parameter and combined with the power p. The least-squares formulation sets these as the ℓ_2-norm and $p = 2$, which makes assumptions about the distribution of errors (e.g., from detector noise or Poissonian electron counting). The additional term weighted by the factor λ serves as a regularization term in this formulation of the reconstruction as a minimization problem. A second operator $\widehat{\Psi}$ can be used to transform the reconstruction volume to a domain where some property of the object is well known. For example, if the sample is known to consist of regions of homogeneous density with sharp boundaries, a total variation (TV) norm might be selected for this regularization term which resembles an ℓ_1-norm (sum of absolute values rather than sum of squares). TV- and ℓ_1-norms promote sparsity in solutions, i.e., the number of nonzero intensities is small, for the transform domain selected. For first-order TV, this means the gradient of the volume is captured in a small number of voxels, promoting piece-wise constant solutions which are defined by a small number of gradients at the boundaries between domains of constant intensity. This generalized formulation allows for consideration of other approaches as well, such as choosing solutions from a dictionary of known features (e.g., dictionary-learning approaches) or, for example, restricting the reconstruction to a discrete number of intensities (e.g., the discrete algebraic reconstruction technique, DART).

One large family of reconstruction techniques using this approach is compressed sensing electron tomography (CS-ET) (Leary, Saghi, Midgley, & Holland, 2013) grounded in established mathematical theories. Compressed sensing uses the sparsity promotion of TV- and ℓ_1-norms together with an incoherence between the projection and the sparsifying transform to enable reconstructions from highly undersampled data. CS-ET has been used in a large number of signals in electron tomography while building in relatively general prior knowledge of the sample. The information CS-ET uses to constrain the reconstruction can be considered as promoting particular solutions from a collection of possible solutions. In undersampled tomography or where there is information missing (e.g., the missing wedge), standard tomography blurs features because of the uncertainty in the voxel assignments of the intensity distribution. CS-ET methods do not directly fill in the missing information but identify the most plausible solution for the assumptions made about the sample. For example, this undesirable blurring in the missing wedge direction can be reduced where solutions are promoted that have a sharp rather than diffuse boundary. The data-fidelity term (first term) is used as the driving force for the reconstruction process. CS-ET is particularly noteworthy in making possible high-fidelity reconstructions from sometimes 10−20 projections rather than a more conventional >100 projections. This capability has particularly enabled electron tomography using analytical (spectroscopic) signals which are generally slow to acquire and require significant electron dose per frame (and associated potential for sample damage).

Analytical signals are almost always collected in parallel with an HAADF signal, and so there are also significant gains in combining multiple signals. Two key approaches for combining these signals have been presented, one referred to as HAADF-EDS bimodal tomography (HEBT) (Zhong, Goris, Schoenmakers, Bals, & Batenburg, 2017) and a second using total nuclear variation (TNV) (Zhong, Palenstijn, Adler, & Batenburg, 2018). HEBT incorporates a physical model relating the HAADF and EDS characteristic X-rays, whereas the TNV approach uses a framework akin to TV-norm formulation. Both approaches show improvements in the noisy but chemically specific EDS tomography when combined with HAADF tomography signals.

Two major areas of key interest in industrial electron tomography are tomography techniques for recovering quantitative intensities and atomically resolved electron tomography. These two objectives are interlinked, as the first three-dimensional imaging at atomic resolution in electron microscopy relied on quantitative atom counting (Van Aert, Batenburg, Rossell, Erni, & Van Tendeloo, 2011). Quantification of atomic resolution images in electron microscopy is largely beyond the scope of this chapter, but broadly consists of two approaches: statistical extraction of atom counts by making use of the discrete number of atoms and calibration of intensities for atom counting. The latter is a more general case and applies to quantitative tomography at all spatial resolutions. Conversely, exact calibration of intensities is not essential for general atomic resolution tomography provided the intensities are self-consistent throughout.

Quantification of intensities has a key role to play in physical parameter extraction. Whereas qualitative intensities still enable quantitative analysis of the structure of a sample, the intensities are known to relate to specific materials properties such as atomic number density (HAADF), chemical composition (EDS), or in some cases chemical bonding information per voxel (EELS) akin to recovering an X-ray absorption spectrum at each voxel. Fig. 10.7 presents an example of the results possible using absolute quantification, counting the number of atoms per unit area in two-dimensional projections, and in return the number of atoms per unit volume in the reconstruction. Here, the samples consisted of two metal-organic framework glasses (Collins et al., 2019). The samples differed in the composition and glass formation process: the first exhibits a multicomponent domain structure, whereas the second was the result of a flux-melting process where one component dissolves the other in the melt. CS-ET was applied with a second-order TV-norm to enable recovery of piece-wise linear solutions from ~15 projection images. The intensities were quantified by separately

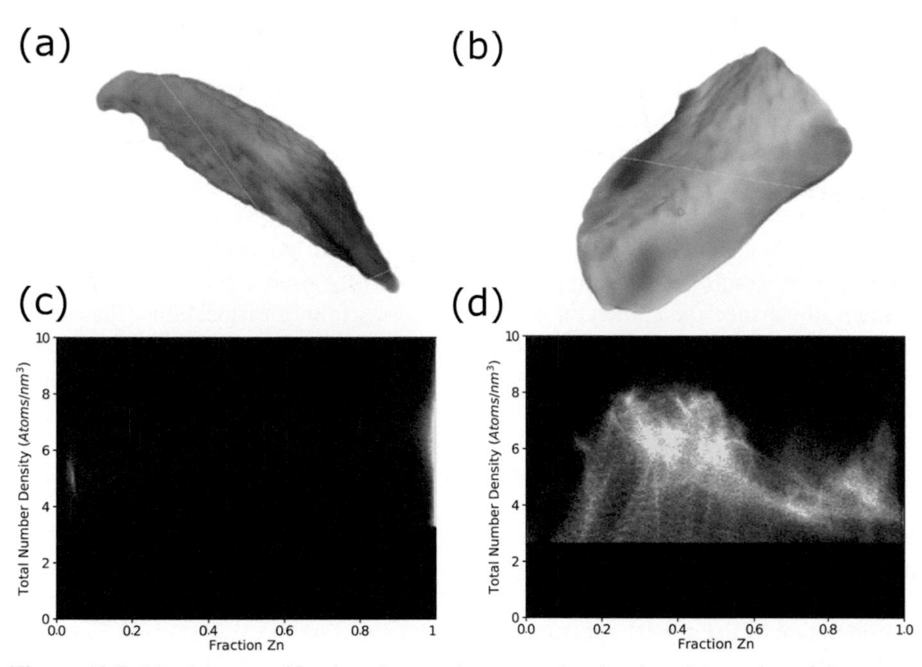

Figure 10.7 Absolute quantification elemental tomography showing the number of Co (red [dark gray in print version]) and Zn (blue [gray in print version]) atoms per unit volume for a metal-organic framework glass (a) blend and (b) flux-melted glass. (c)–(d) The corresponding density-composition phase diagrams extracted from the volumetric composition.
Adapted from Collins, S. M., MacArthur, K.E., Longley, L., Tovey, R., Benning, M., Schönlieb, C.-B., et al. (2019). Phase diagrams of liquid-phase mixing in multi-component metal-organic framework glasses constructed by quantitative elemental nano-tomography. *APL Materials, 7*(9), 091111), CC-BY https://creativecommons.org/licenses/by/4.0/.

calibrating single-component reference samples with known composition, with sample thickness likewise determined by electron tomography for the reference samples. This approach enabled extraction of both density and composition information and the construction of phase diagrams for the mixing processes in the two different glass formation mechanisms.

Other examples of quantitative HAADF tomography have enabled compositional segmentation without a spectroscopic signal for monitoring compositional changes in, e.g., bimetallic metal nanoparticles (Slater et al., 2014). EELS voxel tomography has also enabled recovery of bonding information per voxel (Haberfehlner, Orthacker, Albu, Li, & Kothleitner, 2014). For quantitative tomography, calibration is typically required using standards or reference samples so that the raw intensities can be converted to the physical quantity of interest for a given detector and microscope setting. Some measurements may be avoided where theoretical scattering cross-sections can be used reliably or compared with secondary tomographic measurements (Einsle et al.,

2018), but a workflow incorporating experimental calibrations provides the lowest uncertainties.

The development of electron tomography at atomic resolution has emerged from atom quantification but has also relied on significant improvements in microscope stability and achievable resolution through the use of aberration correctors. Some of the first atomic resolution electron tomography examples achieved partial reconstruction of thin edges of nanoparticle samples (Scott et al., 2012), small single-element rods (Goris et al., 2012), used averaging over many near-identical particles (Azubel et al., 2014), combined modeling (Li et al., 2008), or used filtering in the Fourier domain to enhance the extraction of atom positions from a dislocation (Chen et al., 2013). In 2015, the positions of thousands of atoms with tens of picometers precision were reported for a tungsten needle sample (Xu et al., 2015) to enable strain and point defect imaging. This advance was made possible through careful multiframe acquisition, full 180 degrees sampling, and state-of-the-art aberration correction. Notably, the needle was oriented with a set of planes approximately perpendicular to the tilt-axis which provided a structural constraint arising from the data acquisition strategy for placing atom intensity within these planes (gaps between planes near zero will drive the reconstruction in this way). These provided a "low entropy" information set (Collins, Leary et al. 2017) enabling high-quality reconstructions even without direct consideration of the crystal structure in the reconstruction.

Fig. 10.8 presents a reconstruction of the atomic positions from the tungsten needle sample, using a CS-ET algorithm applied to the underlying data available as a dataset for electron tomography development (Levin et al., 2016). Whereas the originally published reconstructions used an equally sloped algorithm, this treatment does not appear to be essential for recovery of atomically resolved features. However, significant post-processing is also required to retrieve atom coordinates for analysis (Xu et al., 2015).

Additional techniques such as fitting three-dimensional Gaussian functions in combination with an SIRT reconstruction have enabled atom position extraction for analysis of strain in twinned nanoparticles (Goris et al., 2015). While the earliest examples

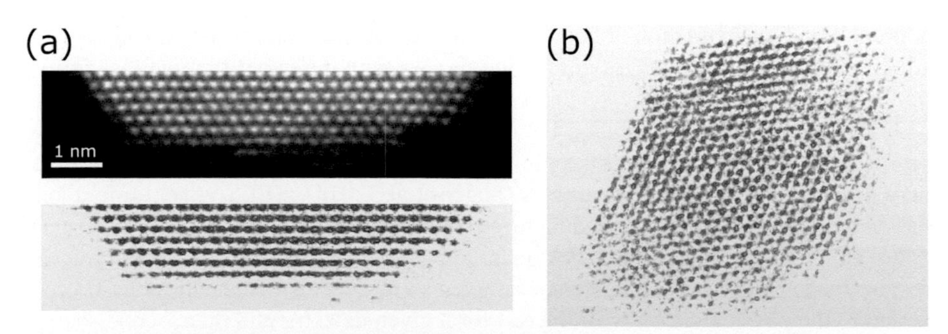

Figure 10.8 (a) Comparison of the zero-tilt projection and the reconstructed atoms shown as a volume rendering. (b) Second orientation of the sample oriented with the needle axis approximately out of the page showing the atomic resolution in this second high symmetry direction.

have been single metals, and therefore relatively radiation resistant, bimetallic examples have also been reported leveraging the atomic number sensitivity of the HAADF signal (Goris et al., 2013; Haberfehlner et al., 2015; Yang et al., 2017), as have nanoparticles within a vacuum-compatible liquid cell by using the tumbling and reorientation of the particles to provide a tilt axis (Kim et al., 2020; Park et al., 2015).

These techniques reveal fundamental properties such as atomic displacements from perfect crystalline lattices due to strain and may yet find application in amorphous materials. The dose limited resolution for many materials (Egerton, 2014), a concept combining the balance between signal for atomic resolution and damage during electron beam irradiation, is likely too low for many materials. Fortunately, the physical properties of interest such as strain and short-range order can also be measured using techniques at lower spatial resolution and significantly lower doses through, e.g., scanning electron diffraction (Tovey et al., 2021) and other 4D-STEM techniques or signatures recorded spectroscopically. These areas are expected to see expanded development in electron tomography, especially for determining defects and disorder in diverse classes of beam-sensitive industrial materials.

10.5 Off-line electron tomography

Applications of electron tomography have been reviewed thoroughly in a variety of articles and textbooks, in both general and application specific terms. Undoubtedly, heterogeneous catalysis applications have dominated work in industrial materials, linked to the detailed information provided by electron tomography on support and catalyst interactions, particle and pore size and shape analysis, and determination of connectivity and exposed crystallographic facets involved in catalytic performance (Friedrich, de Jongh, Verkleij, & de Jong, 2009; Hungría, Calvino, & Hernández-Garrido, 2019). Electron tomography has also seen extensive application in semiconductor materials and devices (Kübel et al., 2005) for examination of buried interfaces as well as in metallurgical samples for microstructure and composition analysis. Combined with the growing development of lithium ion batteries, electron tomography has also seen increased application to batteries and to component materials. Rather than a review of these application areas, this section and the subsequent section will outline relevant procedures for the incorporation of electron tomography procedures in industrial processes.

Due to the constraints on the sample size and compatibility with a vacuum environment (or a cell or holder compatible with the microscope), one of the major routes for incorporating three-dimensional electron microscopy with industrial materials and processes is via an off-line approach where a piece of material or a device component is subjected to a treatment prior to electron tomography or where samples are extracted at intervals from the ongoing process with the state of the material "quenched" at these intermediate stages.

Fig. 10.9 presents electron tomography results for bimetallic nanoparticles consisting of Co and Ni after the particles were subjected to an oxidation process

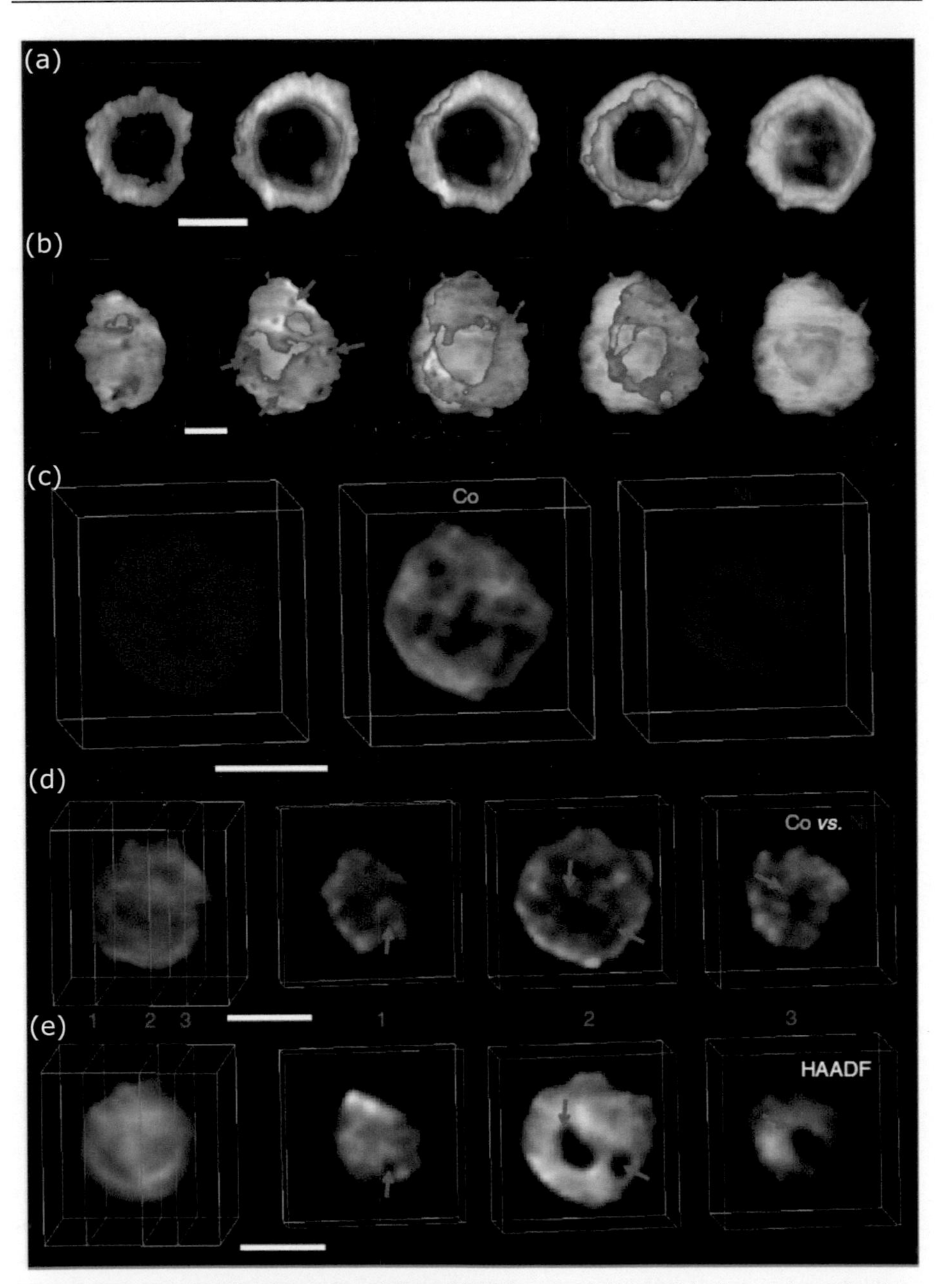

Figure 10.9 3D visualizations of particles with (a) large and (b) small hollow volume fractions. (c) Elemental tomography by EELS tomography. (d) Co and Ni distributions and (e) structural HAADF tomography. The scale bars are 50 nm.
Adapted from Han, L., Meng, Q., Wang, D., Zhu, Y., Wang, J., Du, X., et al. (2016). Interrogation of bimetallic particle oxidation in three dimensions at the nanoscale. *Nature Communications, 7*(1), 13335, CC-BY http://creativecommons.org/licenses/by/4.0/.

(Han et al., 2016). These measurements, together with in situ observations of the oxidation process as well as *ex situ* measurements for samples processed at different temperatures and degrees of oxidation, provided an insight into the structures and compositions present during and following the oxidative treatment. In situ measurements were carried out with two-dimensional imaging and compositional mapping given insufficient time to complete tomography in the time intervals targeted. Postoxidation analysis in 3D using both structural as well as EELS signals provided specific compositional detail on changes in composition at the surface and at pore structures. These details are hidden in two-dimensional compositional mapping where the beam trajectory integrates signal through the top and bottom surfaces as well as the internal volume and any associated interfaces.

Examples highlighting tomography at discrete stopping points in materials processing can provide further stepwise detail within a sequence of tilt-series datasets. Electron tomography of a zeolite precursor followed by tomography of the material after swelling and acidification and finally again after thermal treatment has revealed morphological features as a function of processing step (Arslan et al., 2015). Extraction of material at process interruption points has also enabled electron tomography at intervals, e.g., throughout battery cycling. This approach was applied to an anode comprised of nickel oxide nanosheets by incorporating the electron microscope grid into a coin cell battery assembly (Lin et al., 2014).

A limitation of a simple off-line approach is that different material is examined by electron tomography at each point in the process due to independent extraction of sample material. This constraint has been circumvented by using careful grid and sample registration approaches for identical location electron tomography. This approach enabled potential cycling of a Pt-NbO_x catalyst with a sample prepared on an electron microscopy grid and electron tomography data acquired before and after cycling to track 3D changes in individual particles (Rossouw, Chinchilla, Kremliakova, & Botton, 2017). Where the device structure can be accommodated on an electron microscope grid, or where a model process can be applied to a microscopic quantity (e.g., of catalyst), these approaches are particularly feasible. Bulk samples or devices which are incompatible with electron microscopy likely require additional sample extraction and sample preparation (e.g., polishing, thinning, or FIB milling) to provide off-line electron tomography solutions. Many of these sample preparation processes can increasingly be integrated with standardized and reproducible workflows.

10.6 Electron tomography of dynamic processes

Many processes of interest are not static or cannot be quenched adequately or with sufficient time resolution to provide needed 3D nanoscale structural or compositional information. Through advances in microelectronics and microfluidic devices, a variety of liquid, gas, heating, and biasing electron microscopy holders are now available, offering alternatives for direct monitoring of dynamic processes at second to millisecond time resolution. Notably, the miniaturization of the process as well as interactions of the electron beam with the in situ cell environment must also be considered and may not always reflect bulk processing conditions. However, these methods offer a critical route to nanoscale analysis and provide a closer match to process conditions than *ex situ* measurements can offer in isolation. Some of these holders preclude significant tilting due to additional component size or other limitations, and much of the initial work on in situ electron microscopy has focused on two-dimensional imaging. A further challenge for integration with electron tomography is that traditional reconstruction approaches assume a static sample throughout the tilt-series acquisition. This restricts the time resolution (and ideally the timescale of any structural or compositional evolution) to the time to acquire a complete tilt-series of the sample.

This challenge has been overcome in several key examples, termed "fast tomography," by reducing the acquisition time by applying a continuous rotation approach. The time-savings arise from elimination of the recentring and refocusing steps associated with standard electron tomography acquisitions. These changes to the standard workflow, however, require careful adjustment of the stage hardware to minimize travel of the stage and optimize eucentricity. Using a more parallel beam with a high depth of field as well as a large field of view at each tilt angle can relax some of these requirements. This approach has been applied in both TEM mode to track the evolution of zeolite particles supporting metal catalysts (Migunov et al., 2015; Roiban, Li, Aouine, Tuel, Farrusseng, & Epicier, 2018) as well as in STEM mode to metal nanoparticles (Skorikov et al., 2019; Vanrompay et al., 2018). *Operando* measurements including gas have been provided through environmental TEM techniques reducing the requirement for an in situ cell although combined heating still requires a specialist heating holder (Koneti et al., 2019; Roiban et al., 2018).

Fig. 10.10 presents an example from a gold "star" nanoparticle with tomography acquired at three time intervals and, for different particles, at different temperatures (Vanrompay et al., 2018). Morphological changes as a function of temperature can be quantified (Vanrompay et al., 2018), and this approach has further been extended using quantitative STEM intensities to track diffusion in a bimetallic nanoparticle (Skorikov et al., 2019). In many of these cases, it is possible to decouple the acquisition from a much slower alignment, reconstruction, and postprocessing workflow. For back-projection reconstruction algorithms, it is also possible to reduce the reconstruction speed due to the independence of reconstructions at each voxel; reconstructing only a subset of voxels to produce a limited number of orthogonal slices from the volume can provide "live" tomographic processing in electron tomography (Vanrompay et al., 2020).

10.7 Future trends

Electron microscopy and electron tomography are in a state of acceleration toward a wider range of materials and processes previously deemed inaccessible by electron beam techniques due to prohibitive vacuum, electron beam sensitivity, or sample preparation limitations. A general trend in the field is a shift toward extracting maximal

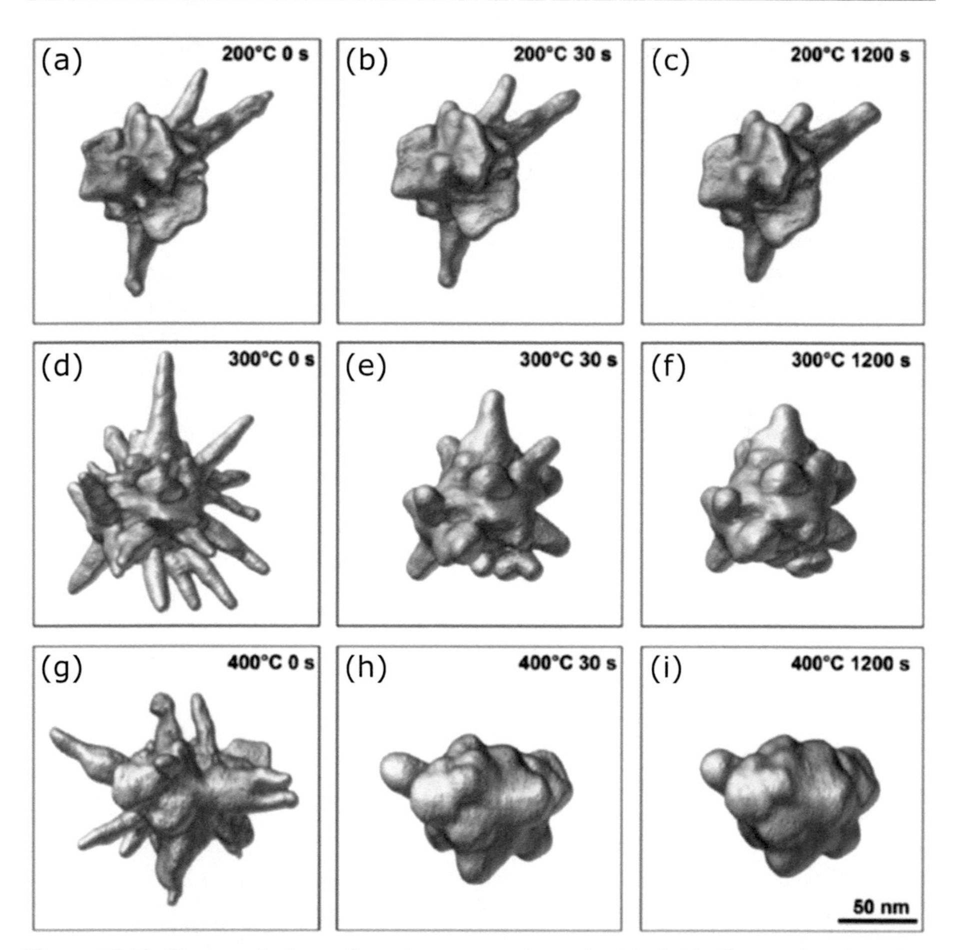

Figure 10.10 The morphology of metal nanostars determined individually as a function of applied temperature using fast electron tomography.
Adapted from Vanrompay, H., Bladt, E., Albrecht, W., Béché, A., Zakhozheva, M., Sánchez-Iglesias, A., et al. (2018). 3D characterization of heat-induced morphological changes of Au nanostars by fast in situ electron tomography. *Nanoscale, 10*(48), 22792–22801. Published by The Royal Society of Chemistry, CC-BY 3.0.

information per incident electron. This concept spans a focus on imaging, diffraction, and spectroscopy at low electron fluence and lower dose exposures through to rapid data acquisition for improved time resolution. It also encompasses a push toward rich information extraction through the use of multiple simultaneous signal acquisitions as well as more sophisticated analyses of signals through the use of multidimensional signal acquisition and analysis coupled with advances in modeling and data processing. The rise of direct electron detectors has enabled counting electrons which has opened opportunities to use lower electron currents due to lower detector contributions to noise. These advances have particularly seen application in 4D-STEM signal acquisition schemes but also in EELS detection. Additional improvements in X-ray detector technology for STEM-EDS have meant that many signals can be acquired with increasing speeds. The combination of these signals for higher spatial resolution, detailed chemical insights, and lower dose experiments enables electron microscopy of organic materials such as organic molecular crystals, polymers, metal-organic frameworks, and hybrid and composite materials for applications in pharmaceuticals, plastics, and complex products.

Simultaneously, a variety of advances in data science have emerged including concepts around compressed sensing, machine learning and artificial intelligence, and automation of acquisition and data handling steps. These trends intersect with tomography with demonstrated improvements in alignment and reconstruction procedures already manifest. These developments will likely see increased application in high-speed multiframe (video) acquisition schemes and alternative scanning strategies in STEM or detector readout in TEM to enable faster imaging at lower electron fluence to further knock down barriers in sample damage and time resolution. Increases in automation are in high demand not only to reduce beam exposure when possible and to enable "smarter" data acquisition schemes but also for high-throughput electron tomography, offering increased statistics for the characterization of nanoscale quantities of material.

The structural biology electron microscopy community has already seen high levels of adoption of automation for imaging and tomography, functioning increasingly akin to automated scanning electron microscopy platforms used in particle analysis workflow such as in the pharmaceuticals and automotive sectors. Electron tomography of industrial materials has begun to see adoption of techniques from the structural biology community including particle averaging (Wang et al., 2019) but will likely increasingly borrow strategies for handling noisy data. The direct application of cryogenic TEM techniques, based on the sample damage reductions from working at cryogenic temperatures as well as the ability to work with vitrified liquid and hydrated soft matter samples (Allen et al., 2015; Nudelman, With, & Sommerdijk, 2010) under static conditions and including time-series snapshots of dynamic processes, is also on the rise. Materials from emulsions to materials requiring cryogenic transfer to stabilize mobile species, e.g., lithium ion battery materials, benefit from stable, reproducible cryogenic sample preparation.

These approaches are also needed in combination with multiscale and multitechnique tomographic characterization, including in a cryo-transfer pipeline in some cases, which is becoming possible via integration with compatible FIB systems. FIB systems, including common Gallium ion systems as well as increasingly available plasma FIB systems using noble gas or other ion beams, can be used to bridge sample preparation for electron tomography with X-ray tomography or FIB slice-and-view tomography at larger length scales. FIB capabilities for site-specific sample extraction from a device or at features identified by means of a coarser tomographic modality will enable integration with industrial tomography across length scales.

10.8 Sources of further information

A number of reviews and books covering principles of electron tomography have been published. Key recent examples include Bals, Van Aert, & Van Tendeloo, 2013; Ercius, Alaidi, Rames, & Ren, 2015; Leary & Midgley, 2019.

References

Allen, F. I., Comolli, L. R., Kusoglu, A., Modestino, M. A., Minor, A. M., & Weber, A. Z. (2015). Morphology of hydrated as-cast Nafion revealed through cryo electron tomography. *ACS Macro Letters, 4*(1), 1–5.

Arslan, I., Roehling, J. D., Ogino, I., Batenburg, K. J., Zones, S. I., Gates, B. C., et al. (2015). Genesis of delaminated-zeolite morphology: 3-D characterization of changes by STEM tomography. *Journal of Physical Chemistry Letters, 6*(13), 2598–2602.

Azubel, M., Koivisto, J., Malola, S., Bushnell, D., Hura, G. L., Koh, A. L., et al. (2014). Electron microscopy of gold nanoparticles at atomic resolution. *Science, 345*(6199), 909–912.

Bai, X., McMullan, G., & Scheres, S. H. W. (2015). How cryo-EM is revolutionizing structural biology. *Trends in Biochemical Sciences, 40*(1), 49–57.

Bals, S., Van Aert, S., & Van Tendeloo, G. (2013). High resolution electron tomography. *Current Opinion in Solid State & Materials Science, 17*(3), 107–114.

Biermans, E., Molina, L., Batenburg, K. J., Bals, S., & Van Tendeloo, G. (2010). Measuring porosity at the nanoscale by quantitative electron tomography. *Nano Letters, 10*(12), 5014–5019.

Billinge, S. J. L., & Levin, I. (2007). The problem with determining atomic structure at the nanoscale. *Science, 316*(5824), 561–565.

Chen, C.-C., Zhu, C., White, E. R., Chiu, C.-Y., Scott, M. C., Regan, B. C., et al. (2013). Three-dimensional imaging of dislocations in a nanoparticle at atomic resolution. *Nature, 496*(7443), 74–77.

Collins, S. M., Fernandez-Garcia, S., Calvino, J. J., & Midgley, P. A. (2017). Sub-nanometer surface chemistry and orbital hybridization in lanthanum-doped ceria nano-catalysts revealed by 3D electron microscopy. *Scientific Reports, 7*(1), 5406.

Collins, S. M., Leary, R. K., Midgley, P. A., Tovey, R., Benning, M., Schönlieb, C.-B., et al. (2017). Entropic Comparison of atomic -ResolutionElectron tomography of crystals and amorphous materials. *Ph*ysical Review Letters, *119*(16), 166101.

Collins, S. M., MacArthur, K. E., Longley, L., Tovey, R., Benning, M., Schönlieb, C.-B., et al. (2019). Phase diagrams of liquid-phase mixing in multi-component metal-organic framework glasses constructed by quantitative elemental nano-tomography. *APL Materials, 7*(9), 091111.

Collins, S. M., & Midgley, P. A. (2017). Progress and opportunities in EELS and EDS tomography. *Ultramicroscopy, 180*, 133–141.

Egerton, R. F. (2014). Choice of operating voltage for a transmission electron microscope. *Ultramicroscopy, 145*, 85–93.

Eggeman, A. S., Krakow, R., & Midgley, P. A. (2015). Scanning precession electron tomography for three-dimensional nanoscale orientation imaging and crystallographic analysis. *Nature Communications, 6*(1), 7267.

Einsle, J. F., Eggeman, A. S., Martineau, B. H., Saghi, Z., Collins, S. M., Blukis, R., et al. (2018). Nanomagnetic properties of the meteorite cloudy zone. *Proceedings of the National Academy of Sciences, 115*(49), E11436–E11445.

Ercius, P., Alaidi, O., Rames, M. J., & Ren, G. (2015). Electron tomography: A three-dimensional analytic tool for hard and soft materials research. *Advanced Materials, 27*(38), 5638–5663.

Friedrich, H., de Jongh, P. E., Verkleij, A. J., & de Jong, K. P. (2009). Electron tomography for heterogeneous catalysts and related nanostructured materials. *Chemical Reviews, 109*(5), 1613–1629.

Goris, B., Bals, S., Van den Broek, W., Carbó-Argibay, E., Gómez-Graña, S., Liz-Marzán, L. M., et al. (2012). Atomic-scale determination of surface facets in gold nanorods. *Nature Materials, 11*(11), 930–935.

Goris, B., De Backer, A., Van Aert, S., Gómez-Graña, S., Liz-Marzán, L. M., Van Tendeloo, G., et al. (2013). Three-dimensional elemental mapping at the atomic scale in bimetallic nanocrystals. *Nano Letters, 13*(9), 4236–4241.

Goris, B., De Beenhouwer, J., De Backer, A., Zanaga, D., Batenburg, K. J., Sánchez-Iglesias, A., et al. (2015). Measuring lattice strain in three dimensions through electron microscopy. *Nano Letters, 15*(10), 6996–7001.

Haberfehlner, G., Orthacker, A., Albu, M., Li, J., & Kothleitner, G. (2014). Nanoscale voxel spectroscopy by simultaneous EELS and EDS tomography. *Nanoscale, 6*(23), 14563–14569.

Haberfehlner, G., Thaler, P., Knez, D., Volk, A., Hofer, F., Ernst, W. E., et al. (2015). Formation of bimetallic clusters in superfluid helium nanodroplets analysed by atomic resolution electron tomography. *Nature Communications, 6*(1), 8779.

Han, L., Meng, Q., Wang, D., Zhu, Y., Wang, J., Du, X., et al. (2016). Interrogation of bimetallic particle oxidation in three dimensions at the nanoscale. *Nature Communications, 7*(1), 13335.

Herzing, A., Henry, K., & Steel, E. (2012). Three-dimensional structural and chemical characterization via STEM of a candidate standard reference material for atom probe tomography. *Microscopy and Microanalysis, 18*(S2), 580–581.

Hungría, A. B., Calvino, J. J., & Hernández-Garrido, J. C. (2019). HAADF-STEM electron tomography in catalysis research. *Topics in Catalysis, 62*(12), 808–821.

Kim, B. H., Heo, J., Kim, S., Reboul, C. F., Chun, H., Kang, D., et al. (2020). Critical differences in 3D atomic structure of individual ligand-protected nanocrystals in solution. *Science, 368*(6486), 60–67.

Koneti, S., Roiban, L., Dalmas, F., Langlois, C., Gay, A.-S., Cabiac, A., et al. (2019). Fast electron tomography: Applications to beam sensitive samples and in situ TEM or operando environmental TEM studies. *Materials Characterization, 151*, 480–495.

Kübel, C., Voigt, A., Schoenmakers, R., Otten, M., Su, D., Lee, T.-C., et al. (2005). Recent advances in electron tomography: TEM and HAADF-STEM tomography for materials science and semiconductor applications. *Microscopy and Microanalysis, 11*(5), 378–400.

Leary, R. K., & Midgley, P. A. (2019). Electron tomography in materials science. In P. W. Hawkes, & J. C. H. Spence (Eds.), *Springer handbook of microscopy* (pp. 1279–1329). Springer International Publishing.

Leary, R., Parlett, C., Barnard, J., Pena, F. de la, Isaacs, M., Beaumont, S., et al. (2017). Multidimensional multi-functional catalytic architecture: A selectively functionalized three-dimensional hierarchically ordered macro/mesoporous network for cascade reactions analyzed by electron tomography. *Microscopy and Microanalysis, 23*(S1), 2042–2043.

Leary, R., Saghi, Z., Armbrüster, M., Wowsnick, G., Schlögl, R., Thomas, J. M., et al. (2012). Quantitative high-angle annular dark-field scanning transmission electron microscope (HAADF-STEM) tomography and high-resolution electron microscopy of unsupported intermetallic $GaPd_2$ catalysts. *Journal of Physical Chemistry C, 116*(24), 13343–13352.

Leary, R., Saghi, Z., Midgley, P. A., & Holland, D. J. (2013). Compressed sensing electron tomography. *Ultramicroscopy, 131*, 70–91.

Levin, B. D. A., Padgett, E., Chen, C.-C., Scott, M. C., Xu, R., Theis, W., et al. (2016). Nanomaterial datasets to advance tomography in scanning transmission electron microscopy. *Scientific Data, 3*, Article 160041.

Lin, F., Nordlund, D., Weng, T.-C., Zhu, Y., Ban, C., Richards, R. M., et al. (2014). Phase evolution for conversion reaction electrodes in lithium-ion batteries. *Nature Communications, 5*(1), 3358.

Livi, K. J. T., Villalobos, M., Leary, R., Varela, M., Barnard, J., Villacís-García, M., et al. (2017). Crystal face distributions and surface site densities of two synthetic goethites: Implications for adsorption capacities as a function of particle size. *Langmuir, 33*(36), 8924–8932.

Li, Z. Y., Young, N. P., Di Vece, M., Palomba, S., Palmer, R. E., Bleloch, A. L., et al. (2008). Three-dimensional atomic-scale structure of size-selected gold nanoclusters. *Nature, 451*(7174), 46–48.

Miao, J., Förster, F., & Levi, O. (2005). Equally sloped tomography with oversampling reconstruction. *Physical Review B, 72*(5), 052103.

Midgley, P. A., Thomas, J. M., Laffont, L., Weyland, M., Raja, R., Johnson, B. F. G., et al. (2004). High-resolution scanning transmission electron tomography and elemental analysis of zeptogram quantities of heterogeneous catalyst. *The Journal of Physical Chemistry B, 108*(15), 4590–4592.

Migunov, V., Ryll, H., Zhuge, X., Simson, M., Strüder, L., Batenburg, K. J., et al. (2015). Rapid low dose electron tomography using a direct electron detection camera. *Scientific Reports, 5*(1), 14516.

Nudelman, F., With, G. de, & Sommerdijk, N. A. J. M. (2010). Cryo-electron tomography: 3-dimensional imaging of soft matter. *Soft Matter, 7*(1), 17–24.

Park, J., Elmlund, H., Ercius, P., Yuk, J. M., Limmer, D. T., Chen, Q., et al. (2015). 3D structure of individual nanocrystals in solution by electron microscopy. *Science, 349*(6245), 290–295.

Pryor, A., Yang, Y., Rana, A., Gallagher-Jones, M., Zhou, J., Lo, Y. H., et al. (2017). GENFIRE: A generalized Fourier iterative reconstruction algorithm for high-resolution 3D imaging. *Scientific Reports, 7*(1), 10409.

Roiban, L., Li, S., Aouine, M., Tuel, A., Farrusseng, D., & Epicier, T. (2018). Fast 'Operando' electron nanotomography. *Journal of Microscopy, 269*(2), 117–126.

Rossouw, D., Chinchilla, L., Kremliakova, N., & Botton, G. A. (2017). The 3D nanoscale evolution of platinum–niobium oxide fuel cell catalysts via identical location electron tomography. *Particle & Particle Systems Characterization, 34*(7), 1700051.

Saghi, Z., Holland, D. J., Leary, R., Falqui, A., Bertoni, G., Sederman, A. J., et al. (2011). Three-dimensional morphology of iron oxide nanoparticles with reactive concave surfaces. A compressed sensing-electron tomography (CS-ET) approach. *Nano Letters, 11*(11), 4666–4673.

Saxton, W. O., Baumeister, W., & Hahn, M. (1984). Three-dimensional reconstruction of imperfect two-dimensional crystals. *Ultramicroscopy, 13*(1), 57–70.

Scott, M. C., Chen, C.-C., Mecklenburg, M., Zhu, C., Xu, R., Ercius, P., et al. (2012). Electron tomography at 2.4-ångström resolution. *Nature, 483*(7390), 444–447.

Skorikov, A., Albrecht, W., Bladt, E., Xie, X., van der Hoeven, J. E. S., van Blaaderen, A., et al. (2019). Quantitative 3D characterization of elemental diffusion dynamics in individual Ag@Au nanoparticles with different shapes. *ACS Nano, 13*(11), 13421–13429.

Slater, T. J. A., Macedo, A., Schroeder, S. L. M., Burke, M. G., O'Brien, P., Camargo, P. H. C., et al. (2014). Correlating catalytic activity of Ag–Au nanoparticles with 3D compositional variations. *Nano Letters, 14*(4), 1921–1926.

Swearer, D. F., Leary, R. K., Newell, R., Yazdi, S., Robatjazi, H., Zhang, Y., et al. (2017). Transition-metal decorated aluminum nanocrystals. *ACS Nano, 11*(10), 10281–10288.

Tovey, R., Johnstone, D. N., Collins, S. M., Lionheart, W. R. B., Midgley, P. A., Benning, M., et al. (2021). Scanning electron diffraction tomography of strain. *Inverse Problems, 37*, 015003.

Van Aert, S., Batenburg, K. J., Rossell, M. D., Erni, R., & Van Tendeloo, G. (2011). Three-dimensional atomic imaging of crystalline nanoparticles. *Nature, 470*(7334), 374–377.

Vanrompay, H., Bladt, E., Albrecht, W., Béché, A., Zakhozheva, M., Sánchez-Iglesias, A., et al. (2018). 3D characterization of heat-induced morphological changes of Au nanostars by fast in situ electron tomography. *Nanoscale, 10*(48), 22792–22801.

Vanrompay, H., Buurlage, J.-W., Pelt, D. M., Kumar, V., Zhuo, X., Liz-Marzán, L. M., et al. (2020). Real-time reconstruction of arbitrary slices for quantitative and in situ 3D characterization of nanoparticles. *Particle & Particle Systems Characterization, 37*(7), 2000073.

Wang, Y.-C., Slater, T. J. A., Leteba, G. M., Roseman, A. M., Race, C. P., Young, N. P., et al. (2019). Imaging three-dimensional elemental inhomogeneity in Pt–Ni nanoparticles using spectroscopic single particle reconstruction. *Nano Letters, 19*(2), 732–738.

Weyland, M., & Midgley, P. A. (2001). 3D-EFTEM: Tomographic reconstruction from tilt series of energy loss images. *Conference Series - Institute of Physics, 168*, 239–242.

Weyland, M., Midgley, P. A., & Thomas, J. M. (2001). Electron tomography of nanoparticle catalysts on porous supports: A new technique based on rutherford scattering. *The Journal of Physical Chemistry B, 105*(33), 7882–7886.

Xu, R., Chen, C.-C., Wu, L., Scott, M. C., Theis, W., Ophus, C., et al. (2015). Three-dimensional coordinates of individual atoms in materials revealed by electron tomography. *Nature Materials, 14*(11), 1099–1103.

Yang, Y., Chen, C.-C., Scott, M. C., Ophus, C., Xu, R., Pryor, A., et al. (2017). Deciphering chemical order/disorder and material properties at the single-atom level. *Nature, 542*(7639), 75–79.

Zhong, Z., Goris, B., Schoenmakers, R., Bals, S., & Batenburg, K. J. (2017). A bimodal tomographic reconstruction technique combining EDS-STEM and HAADF-STEM. *Ultramicroscopy, 174*, 35–45.

Zhong, Z., Palenstijn, W. J., Adler, J., & Batenburg, K. J. (2018). EDS tomographic reconstruction regularized by total nuclear variation joined with HAADF-STEM tomography. *Ultramicroscopy, 191*, 34–43.

Part Two

Tomographic image reconstruction and data fusion

Mathematical concepts for image reconstruction in tomography

Sin Kim[1], *Anil Kumar Khambampati*[2] *and Kyung Youn Kim*[2]
[1]School of Energy Systems Engineering, Chung-Ang University, Seoul, South Korea;
[2]Department of Electronic Engineering, Jeju National University, Jeju, South Korea

11.1 Introduction

Tomography intends to image the cross-section of an object without altering it. From the image, we want to know the internal configuration of the substances comprising the object. For this purpose, we must use "something" that should infiltrate into the object, interact with internal substances, and escape out of the object without altering the object significantly. Based on the change of "something" due to the interaction, the internal configuration is imaged. Of course, the change should be measurable.

Any particles or waves could be used for tomography if they do not alter the object significantly while passing within and interacting with the object. Tiny particles like elementary particles such as neutrons, protons, and electrons could be considered as particles for tomography. Among these, charged particles are less attractive unless the object is very small because their range, or average penetration depth, is very short. Neutrons are electrically neutral and highly penetrative; however, their generation, treatment, and measurement are, in general, less practical. Photons are used as common particles for tomography; their stream is an electromagnetic wave, which is a self-propagating transverse oscillating wave of electric and magnetic fields. The electromagnetic wave or stream of photons is characterized according to its frequency or wavelength, as shown in Fig. 11.1. As its wavelength is shorter or its frequency is larger, it becomes more energetic and more penetrative. In other words, it resembles a solid particle more, and its path becomes straighter. However, as its wavelength is longer or its frequency is smaller, it resembles a wave more and behaves totally differently from the high-frequency electromagnetic waves. Another type of wave for tomography is the pressure wave or sound. For tomography, an electric (or magnetic) field can be directly applied to the object to be imaged. Electric signals are imposed on the object, and the induced electric signals, which are dependent on the internal distribution of the electrical properties, are measured.

The pattern of interaction between penetrating waves and the object can be classified into three groups. When the penetrating wave is strongly penetrative and behaves like a stream of particles, it can be assumed that the wave brings the information on the substances along a straight path connecting the source and the detector. However, if an electric (or magnetic) field is used for tomography, any part of the object can affect the field; namely, the entire domain of the object is the path. When the penetrating wave

Industrial Tomography. https://doi.org/10.1016/B978-0-12-823015-2.00028-5

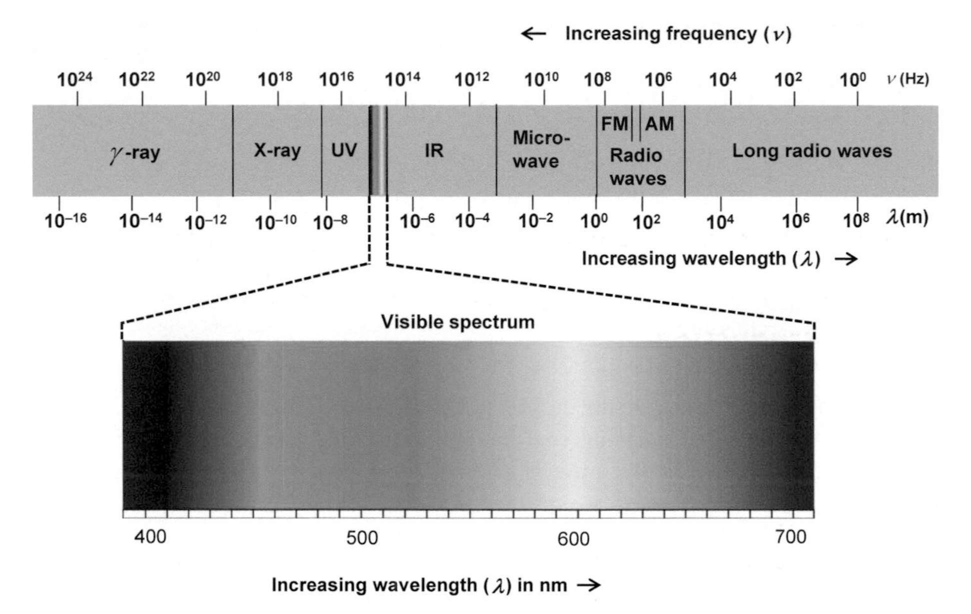

Figure 11.1 Spectrum of electromagnetic wave.

behaves just as a wave, on the other hand, its path will be bent due to diffraction, refraction, and reflection. Thus, the path cannot be modeled as a straight line anymore. Depending on the interaction pattern, the tomography can be categorized into three groups as transmission, electrical, and diffraction tomography (DT). For each tomography, the mathematical description has a different form, and the following sections will introduce the mathematical concepts for each tomography.

11.2 Transmission tomography

Radiation is a stream of energetic particles or energetic electromagnetic waves. An electromagnetic wave is often regarded as a particle called a photon when it is highly energetic. Photon can have a wavelength so short as to be comparable to or less than the size of an atom, for example, X-rays and γ-rays. Radiation loses its particles or energy upon interaction with matter. That is, radiation attenuates, resulting in a decrease of intensity as it passes through matter. The ways of interaction can be roughly classified into absorption and scattering. In the case of absorption, the energy of particles is entirely absorbed by the matter (or radiation particles get disappeared). Often the absorption may result in the production of other radiations. On the other hand, the radiation particle may collide with the matter, and its momentum is altered. After the scattering, of course, the direction of the radiation is changed. The extent of attenuation depends on the sort of radiation, and the matter interacted with. Also, it is a function of the radiation energy.

11.2.1 Mathematical formulation of transmission tomography

For a mathematical formulation of the relationship between the intensity and the material property, suppose an infinite slab of thickness a is placed in a monodirectional, and a mono-energetic plane beam of intensity I_0 and a detector is behind the slab as shown in Fig. 11.2.

Let $I(x)$ be the intensity at x, where the intensity is defined as the power transferred per unit area or the number of radiation particles per unit time passing through a unit area. It is assumed that in traveling an additional distance dx, the intensity decrease is proportional to $I(x)$ and dx. If there is no scattering, then this assumption is exact. In this, the proportional constant μ termed as the *attenuation coefficient* or *linear absorption coefficient* is introduced:

$$dI(x) = -\mu I(x)dx. \tag{11.1}$$

The attenuation coefficient is a physical property of the matter for given radiation. Strictly speaking, μ is a function of radiation energy, but in practice, energy dependency is usually neglected. Integrating this equation gives

$$I(x) = I_0 \exp(-\mu x). \tag{11.2}$$

The intensity measured by the detector I_d is

$$I_d = I(a) = I_0 \exp(-\mu a). \tag{11.3}$$

When the slab is multilayered, and each layer has a different attenuation coefficient, then the intensity measured by the detector becomes

$$I_d = I_0 \exp\left(-\sum_n \mu_n a_n\right), \tag{11.4}$$

where μ_n and a_n are the attenuation coefficient and the thickness of the n-th layer, respectively. It should be noted that, in the one-dimensional infinite slab, the above

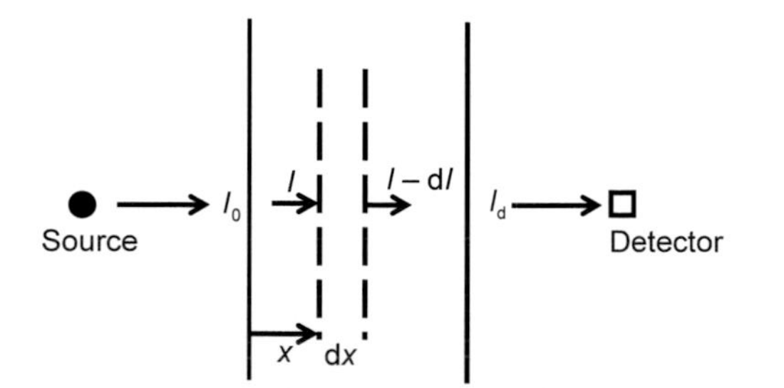

Figure 11.2 Intensity variation when radiation is passing through a slab.

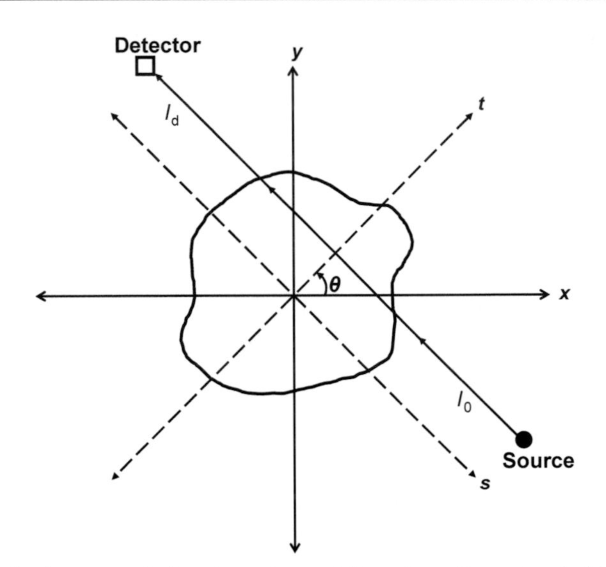

Figure 11.3 Radiation transmission through an arbitrary two-dimensional object.

derivation is valid even when the attenuation is solely caused by the scattering because the contribution to the detector by the scattering is uniform. However, for a point source or when the target size is finite, as the scattering contributes more, the above model becomes worse.

For an arbitrary two-dimensional object, as shown in Fig. 11.3, Eq. (11.4) can be generalized to

$$I_d(t;\theta) = I_0 \exp\left(- \int_{L(t;\theta)} \mu(x,y)ds \right), \tag{11.5}$$

where $L(t;\theta)$ is the line (ray) connecting the source and the detector, (s,t) is the coordinate rotated by $\theta \in [0,\pi)$ from (x,y) to align s-axis parallel to the ray, and $I_d(t;\theta)$ is again the intensity measured by the detector. If $I_d(t;\theta)$ is measurable and distinguishable in terms of measurement uncertainty and $\mu(x,y)$ is invertible from $I_d(t;\theta)$, we can expect that the radiation can be used for the reconstruction of the cross-sectional image because the attenuation coefficient distribution corresponds to the material distribution.

11.2.2 Radon transform and direct back-projection

The basic idea to reconstruct the original function from a series of projected functions was given by an Austrian mathematician Johann Radon in 1917 and realized by R.N. Bracewell (1956). When the rotated coordinate is expressed as (see Fig. 11.3)

$$t = x \cos\theta + y \sin\theta,\ s = -x \sin\theta + y \cos\theta, \tag{11.6}$$

the Radon transform of $\mu(x, y)$ along a ray parallel to s-axis is defined as a line integral or a ray sum

$$p(t;\theta) = \widehat{R}_s[\mu(x,y)](t;\theta) = \int_{-\infty}^{\infty} \mu(x,y)ds. \tag{11.7}$$

Hence, the Radon transform of the attenuation coefficient is directly related to the intensity ratio, which can be measured (Deans, 2007):

$$p(t;\theta) = \ln\frac{I_0}{I_d(t;\theta)}. \tag{11.8}$$

11.2.2.1 Direct back-projection

At each angle, the attenuated intensity is measured, and it can be readily converted to the sum of the attenuation coefficients of the pixels along a ray. But we do not know the contribution of each pixel. The simplest estimation of the contribution is to assume that each pixel along the ray contributes equally. Hence, the simplest form of the back-projection to the pixel at (x, y) will be the sum-up of all the projections crossing (x, y) and can be expressed as (Kak & Slaney, 1988; Natterer, 1986; Stark, Woods, Paul, & Hingorani, 1981)

$$\mu_{\mathrm{DBP}}(x,y) = \int_0^{\pi} p[t(x,y;\theta);\theta]d\theta. \tag{11.9}$$

Fig. 11.4 shows the back-projected image using three projections. If the contribution is expressed in a gray level, the target can be captured with its thicker darkness. However, as can be seen in the figure, with only a few projections, a star-like pattern around the true target remains inevitably. As the number of projections is increased, the star-like pattern is smeared out, and we can get an improved image. Fig. 11.5a

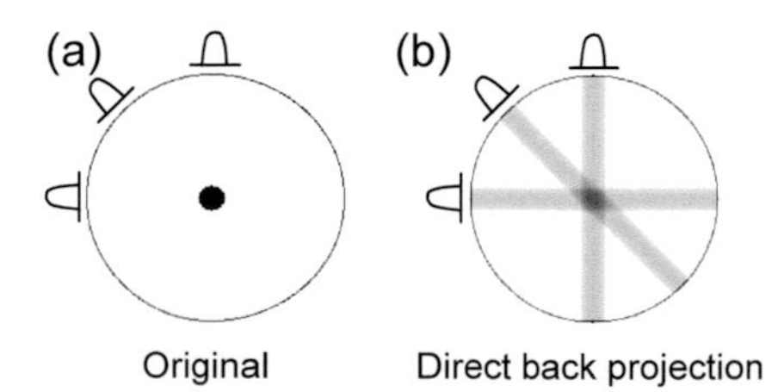

Figure 11.4 Image reconstruction using projections (a) original image (b) image from direct back-projection with three projections.

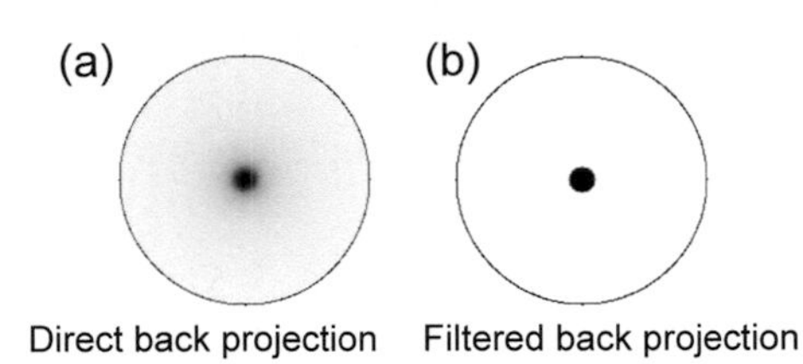

Figure 11.5 Comparison of reconstructed images from the projections (a) direct back-projection, (b) filtered back projection.

is reconstructed by 180 projections covering a full rotation from 0 to 180 degrees. With increasing the number of projections, unfortunately, the object image, as well as the star-like pattern, is blurred at the same time. Even with more projections, spatial resolution is not improved further.

11.2.3 Fourier transform and filtered back-projection

The direct back-projection is not the inversion of the Radon transform, but just a back-projection of the Radon transform of the attenuation coefficient distribution. Before preceding the inversion of the Radon transform, let us review the relationship between the Radon transform and Fourier transform (Bracewell, 1965; Feldkamp, Davis, & Kress, 1984; Jain, 1989, pp. 431–475; Katsevich, 2004; Schafer, Borgert, Rasche, & Grass, 2006). The two-dimensional Fourier transform $\mu(x, y)$ is defined as

$$M(u, v) = \widehat{F}_{(x,y)}[\mu(x, y)](u, v) = \int_{-\infty}^{\infty}\int_{-\infty}^{\infty} \mu(x, y)\exp[-j(ux + vy)]dxdy, \tag{11.10}$$

and its inverse is

$$\mu(x, y) = \widehat{F}^{-1}_{(u,v)}[M(u, v)](x, y) = \frac{1}{(2\pi)^2}\int_{-\infty}^{\infty}\int_{-\infty}^{\infty} M(u, v)\exp[j(ux + vy)]dudv. \tag{11.11}$$

Noting that the one-dimensional Fourier transform of $p(t; \theta)$ for a given θ can be written as

$$P(\omega; \theta) = \widehat{F}_t[p(t; \theta)](\omega; \theta) = \int_{-\infty}^{\infty} p(t; \theta)\exp(-j\omega t)dt, \tag{11.12}$$

introducing $(u, v) = (\omega \cos\theta, \omega \sin\theta)$ we can easily derive the relationship

$$M(\omega\cos\theta,\omega\sin\theta)=\int_{-\infty}^{\infty}\int_{-\infty}^{\infty}\mu(x,y)\exp[-j\omega(x\cos\theta+y\sin\theta)]dxdy$$
$$=\int_{-\infty}^{\infty}\int_{-\infty}^{\infty}\mu(x,y)\exp(-j\omega t)dxdy=\int_{-\infty}^{\infty}\int_{-\infty}^{\infty}\mu(x,y)\exp(-j\omega t)\frac{\partial(x,y)}{\partial(s,t)}dtds$$
$$=\int_{-\infty}^{\infty}\left[\int_{-\infty}^{\infty}\mu(x,y)ds\right]\exp(-j\omega t)dt=P(\omega;\theta),$$

or

$$M(\omega\cos\theta,\omega\sin\theta)=P(\omega;\theta). \tag{11.13}$$

This is called the Fourier slice theorem (or projection-slice theorem), which explains that the one-dimensional Fourier transform of a projection of a two-dimensional function onto a line $P(\omega;\theta)$, is equal to the slice parallel to the projection line passing through an origin of the two-dimensional Fourier transform of the same function, $M(\omega\cos\theta,\omega\sin\theta)$ (Bracewell, 1965; Herman, 1979, 1980; Jain, 1989, pp. 431−475; Levoy, 1992).

If $p(t;\theta)$ can be measured for all θ, $M(\omega\cos\theta,\omega\sin\theta)$ can be obtained from $P(\omega;\theta)$, the Fourier transform of $p(t;\theta)$. Finally, $\mu(x,y)$ can be reconstructed by the inverse Fourier transform $M(\omega\cos\theta,\omega\sin\theta)$. The inversion process is as follows. At first, the inverse Fourier transform is written in a polar coordinate

$$\mu(x,y)=\frac{1}{(2\pi)^2}\int_{-\infty}^{\infty}\int_{-\infty}^{\infty}M(u,v)\exp[j(ux+vy)]dudv$$
$$=\frac{1}{(2\pi)^2}\int_{0}^{\pi}\int_{-\infty}^{\infty}M(\omega\cos\theta,\omega\sin\theta)\exp[j\omega(x\cos\theta+y\sin\theta)]|\omega|d\omega d\theta,$$

and recalling the Fourier slice theorem and $t=x\cos\theta+y\sin\theta$ we have

$$\mu(x,y)=\frac{1}{(2\pi)^2}\int_{0}^{\pi}\int_{-\infty}^{\infty}P(\omega;\theta)\exp[j\omega(x\cos\theta+y\sin\theta)]|\omega|d\omega d\theta$$
$$=\frac{1}{2\pi}\int_{0}^{\pi}\left[\frac{1}{2\pi}\int_{-\infty}^{\infty}P(\omega;\theta)|\omega|\exp(j\omega t)d\omega\right]d\theta=\frac{1}{2\pi}\int_{0}^{\pi}\widehat{F}_{\omega}^{-1}[P(\omega;\theta)|\omega|](t;\theta)d\theta$$

or

$$\mu(x,y)=\frac{1}{2\pi}\int_{0}^{\pi}\widehat{F}_{\omega}^{-1}[P(\omega;\theta)|\omega|](t;\theta)d\theta. \tag{11.14}$$

Comparing with the direct back-projection, $\frac{1}{2\pi}\widehat{F}_{\omega}^{-1}[P(\omega;\theta)|\omega|](t;\theta) = \frac{1}{2\pi}\widehat{F}_{\omega}^{-1}\left[|\omega|\widehat{F}_t[p(t;\theta)](\omega;\theta)\right](t;\theta)$ can be regarded as a filtered version of the projection function $p(t;\theta)$. The inverse Fourier transform of the Fourier transform of $p(t;\theta)$ multiplied by $|\omega|$ instead of $p(t;\theta)$ only, in practice, implies that a high-pass filter is introduced (Pan & Kak, 1983). Fig. 11.5 illustrates the comparison between the images reconstructed by a direct back-projection and by a filtered back projection. The filtered back-projection reconstructs the edge of the object sharply, while the direct back-projection fails.

11.2.4 Algebraic reconstruction technique

In the radiation tomography mentioned above, we have seen that the main relationship between the unknowns to be imaged and the measured projection data has a form of

$$p(t;\theta) = \widehat{R}_s[\mu(x,y)](t;\theta) = \int_{-\infty}^{\infty} \mu(x,y)ds, \tag{11.7a}$$

the unknown is the attenuation coefficient $\mu(x,y)$, and the measured data is

$$p(t;\theta) = \ln\frac{I_0}{I_d(t;\theta)}. \tag{11.8a}$$

This relationship in the integral form will be converted to a discrete algebraic form. The ray is assumed to be a strip with a width rather than a line. In reality, the radiation detector has a finite window size, not a point. The domain to be imaged is discretized into small N pixels, and each pixel width is approximately equal to the ray width. In each pixel, the attenuation coefficient is set to be constant. The relationship can be expressed as an algebraic form as (Gordon, Bender, & Herman, 1970; Kak & Slaney, 1988)

$$p_i = \sum_{j=1}^{N} w_{ij}\mu_j, \; i = 1, 2, \cdots, M, \tag{11.15}$$

where p_i is the projection of the i-th ray, M the number of rays or projections, μ_j the attenuation coefficient of the j-th pixel, and w_{ij} the weighting factor representing the contribution of the j-th pixel to the i-th projection. Solving the algebraic Eq. (11.15) to reconstruct μ_j based on the projection data p_i is called the algebraic reconstruction technique (ART) (Fleischmann & Boas, 2011; Gordon, 1974; Gordon et al., 1970).

When the projection line (ray) is straight, the evaluation of the weighting factors w_{ij} is straightforward, w_{ij} is equal to the area of the j-th pixel intercepted by the i-th ray strip. Therefore, the matrix w_{ij} is very sparse. Once w_{ij} are known, then the system of linear Eq. (11.15) can be solved easily using any well-known available solvers. In practice, the size of the system of linear equations is not small, and direct matrix inversion methods are not feasible. In the case of 1% spatial resolution, for example, $N = 100 \times 100$ and M will have a similar size, then the size w_{ij} is approximately $10^4 \times 10^4$, which requires iterative methods. This is why ART is classified as an iterative method.

One of the popular methods to solve Eq. (11.15) is to apply Kaczmarz's method, which is efficient in handling sparse matrices like those in the ART. In image reconstruction, often, Kaczmarz's method is known as the ART. Kaczmarz's method is a sequence of processes to project the previous solution $\boldsymbol{\mu}^{(k-1)}$ on the hyperplane constructed by the i-th Eq. (11.15) to find the updated solution $\boldsymbol{\mu}^{(k)}$, where $\boldsymbol{\mu} = (\mu_1, \mu_2, \cdots, \mu_N)^T$ and the superscript (k) denotes the k-th iteration step. Hence, Kaczmarz's method is classified into a projection method for solving large linear systems of equations (Herman, 1979, 1980; Strohmer & Vershynin, 2009). This process is performed for all $i = 1, 2, \cdots, M$ in a predetermined sequence. If the solution does not converge, then other cycles are repeated until convergence, when $M = N = 2$ this process can be easily understood, as shown in Fig. 11.6.

This process can be expressed in a mathematical form as

$$\boldsymbol{\mu}^{(k)} = \boldsymbol{\mu}^{(k-1)} + \mathbf{d}^{(k-1)} \text{ and } \mathbf{d}^{(k-1)} = \mathbf{w}_i^T \frac{p_i - \mathbf{w}_i \left(\boldsymbol{\mu}^{(k-1)}\right)^T}{\mathbf{w}_i \mathbf{w}_i^T}, \tag{11.16}$$

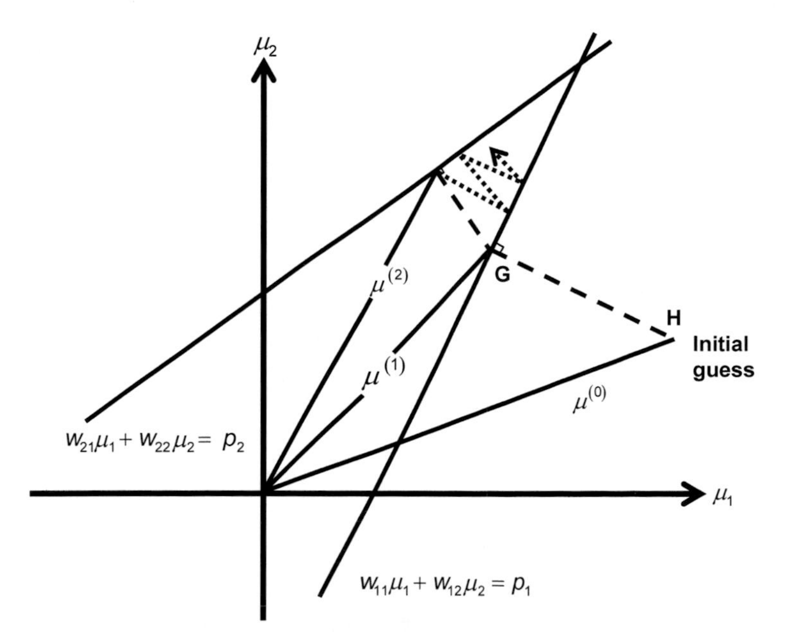

Figure 11.6 Conceptual description of Kaczmarz's method.

where $\mathbf{w}_i = (w_{i1}, w_{i2}, \cdots, w_{iN})$ is the i-th row vector of the matrix $[w_{ij}]$. Often, to control the speed of convergence, the relaxation parameter λ $(0 < \lambda < 2)$ is introduced as $\boldsymbol{\mu}^{(k)} = \boldsymbol{\mu}^{(k-1)} + \lambda \mathbf{d}^{(k-1)}$ and it can be regarded as a variant of the successive overrelaxation (SOR) method. In the ART method, each pixel contains several rays that pass through and cross; each iteration and satisfying a single equation results in a stripe along the particular ray corresponds to that equation. Updating the pixels in the grid for each ray thus causes salt and pepper characteristic in the ART reconstructed image. To counter this problem, a simultaneous iterative reconstruction technique (SIRT) is developed where the pixels are not updated for each ray, and instead, the pixels are corrected simultaneously after all equations are considered at least once (Gilbert, 1972; Jiang & Wang, 2003). The SIRT algorithm is represented as

$$\mu_j^{(k)} = \mu_j^{(k-1)} + \lambda \frac{\sum_i w_{ij} \frac{p_i - \sum_{n=1}^{N} w_{in} \mu_n^{(k)}}{\sum_{n=1}^{N} w_{in}}}{\sum_i w_{ij},} \tag{11.17}$$

where i represents the projections for every ray per angle. The error correction terms used for updating the pixels are the average value of all changes computed for that pixel. Although the SIRT image produced smooth images, the convergence is slow and requires a lot of iterations for the final image. As an improvement to ART methods discussed above, the simultaneous algebraic reconstruction technique (SART) that combines the features of ART and SIRT is introduced by Anderson and Kak (1984a) and has been the most popular iterative reconstruction methods that produce reconstructed images of good quality and accuracy in one iteration. In SART, the superior reconstructions are obtained by using a model of the forward projection process that is more accurate than what can be obtained by choice of pixel basis functions. This is done by using bilinear elements, which are the simplest higher-order basis functions. Instead of sequentially updating pixels on a ray-by-ray basis, we simultaneously apply to a pixel the average of the corrections generated by all the rays in a projection. This entire process is repeated for all projection angles. The SART algorithm is described as (Anderson & Kak, 1984a; Mueller & Yagel, 2000)

$$\mu_j^{(k)} = \mu_j^{(k-1)} + \lambda \cdot \frac{\sum_{i \in I_\theta} w_{ij} \frac{p_i - \sum_{n=1}^{N} w_{in} \mu_n^{(k)}}{\sum_{n=1}^{N} w_{in}}}{\sum_{i \in I_\theta} w_{ij},} \tag{11.18}$$

where I_θ are the projections for a given angle θ.

11.2.5 Maximum likelihood expectation maximization for transmission tomography

An alternative way is to consider the statistical approach that models the physics of transmission in a more practical way that can produce efficient reconstruction with lower patient radiation dose. The maximum likelihood (ML) estimation algorithm is one of the essential statistical approaches that estimate the parameters by maximizing the log-likelihood function. The ML method was first proposed by Rockmore and Macovski (1976) to image reconstruction from projections. Shepp and Vardi (1982) introduced the expectation maximization (EM) approach for computing the ML estimates of pixel concentrations in emission tomography, and later Maximum likelihood expectation maximization (ML-EM) was developed for positron, single-photon emission tomography, and transmission tomography by Lange and Carson (1984). Here, we summarize the ML-EM algorithm for transmission tomography.

Let us consider that from the X-ray source, a total of W_i photons is injected as an infinitely thin beam for each projection i, and due to attenuation of the medium, Y_i photons are detected. The probability for each of these injected photons W_i reaching the detector is the same and is given by

$$p_j = \exp\left(-\sum_{j\in I_i} l_{ij}\mu_j\right), \tag{11.19}$$

where μ_j is the attenuation coefficient for pixel j, I_i is the set of pixels contributing to projection i, l_{ij} is the length of i-th projection intersecting pixel j. Each of the detected photons Y_i has a Poisson distribution with mean described by the product of mean of W_i, d_i times the probability, i.e.,

$$E[Y_i] = d_i E[W_i] p_j = d_i E[W_i] \exp\left(-\sum_{j\in I_i} l_{ij}\mu_j\right), \tag{11.20}$$

where $d_i = \Delta t_i \alpha_i$. Here Δt_i is the time in which the projection i is collected and α_i is the source intensity of projection i. The log-likelihood function of the measured projection data is represented as (Browne & Holmes, 1992; Lange & Carson, 1984)

$$\ln g(Y,\mu) = \sum_i \left\{ -d_i \exp\left(-\sum_{j\in I_i} l_{ij}\mu_j\right) - Y_i \sum_{j\in I_i} l_{ij}\mu_j + Y_i \ln d_i - \ln Y_i! \right\}. \tag{11.21}$$

The standard procedure of finding the maximum of log-likelihood is to compute the partial derivative of Eq. (11.21) with respect to μ_j. However, Eq. (11.21) gives rise to equations that are not trackable. The EM algorithm can be used to find the maximum log-likelihood estimates by applying an iterative procedure. The EM algorithm hypothesizes the complete unobserved data X_{ij} that contains photons that enter pixel j along with projection i. The incomplete data Y_i are the photons that are measured, i.e.,

$$Y_i = \sum_{j \in I_i} X_{ij}. \tag{11.22}$$

For simplicity, consider a single projection where the pixels are labeled as 1 to $m-1$ where pixel 1 is adjacent to the source and pixel $m-1$, adjacent to the detector. Let us consider the number of photons entering pixel j is X_j, where $1 \leq j \leq m-1$. For the first pixel, $X_1 = W$ the number of photons leaving the source and $X_m = Y$ the number of photons detected. The number of photons leaving pixel j depends on the photons that enter X_j, photons that leave pixel X_{j+1} and product of projection path length, attenuation coefficient. Using the above assumptions, the conditional expectation of full data X_j conditional on partial data X_m can be written as (Browne & Holmes, 1992; Lange & Carson, 1984)

$$E(X_j | X_m) = X_m + E(X_j) - E(X_m). \tag{11.23}$$

Now considering all projections i and defining M_{ij}, N_{ij}as the total number of photons entering and leaving pixel j, the E step has the form (Lange & Carson, 1984)

$$E(X_j | X_m, \mu^n) = \sum_i \sum_{j \in I_i} \{ N_{ij} \ln(\exp(-l_{ij}\mu_j)) + (M_{ij} - N_{ij}) \ln(1 - \exp(-l_{ij}\mu_j)) \} + R, \tag{11.24}$$

where R contains all the terms that do not include parameter μ_j. The terms are excluded when computing the M step as the conditional expectation of these terms is not relevant. The M step is obtained by computing the partial derivative of the above conditional expectation Eq. (11.24) with respect to μ_j and equating it to zero

$$0 = \sum_{i \in J_j} N_{ij} l_{ij} + \sum_{i \in J_j} (M_{ij} - N_{ij}) \frac{l_{ij}}{\exp(l_{ij}\mu_j) - 1}, \tag{11.25}$$

where $i \in j_j$ are the set of projections to which the pixel j contributes. In most applications where the region to be imaged contains many small pixels such that $s = l_{ij}\mu_j \ll 1$ and the last term can be approximated using the Taylor series, and an approximate solution can be obtained as

$$\mu_j^{(k)} = \frac{\sum_{i \in J_j} (M_{ij} - N_{ij})}{\frac{1}{2} \sum_{i \in J_j} (M_{ij} - N_{ij})\, l_{ij}.} \tag{11.26}$$

11.2.5.1 Gradient type algorithm for transmission tomography

Using an analogy with emission tomography (Vardi, Shepp, & Kaufman, 1985), a gradient type algorithm for transmission tomography is devised as

$$\mu^{(k)} = \mu^{(k-1)} + D\left(\mu^{(k-1)}\right)\frac{\partial}{\partial \mu}L\left(\mu^{(k-1)}\right), \tag{11.27}$$

where L is a log-likelihood function $\ln\, g(Y,\mu)$ defined in Eq. (11.21); $D\left(\mu^{k}\right)$ is the j-th diagonal matrix having the j-th diagonal element as

$$D\left(\mu^{(k-1)}\right) = \frac{\mu_j^{(k-1)}}{\sum_{i \in J_j} Y_i l_{ij},} \tag{11.28}$$

where $i \in J_j$ are the set of projections to which pixel j contributes, and substituting Eq. (11.28) in Eq. (11.27), the solution for j-th entry can be expressed as

$$\mu_j^{(k)} = \mu_j^{(k-1)} + \frac{\mu_j^{(k-1)}}{\sum_{i \in J_j} Y_i l_{ij}}\frac{\partial}{\partial \mu_j}L\left(\mu_j^{(k-1)}\right) = \mu_j^{(k-1)}\frac{\sum_{i \in J_j} d_i \exp\left(-\sum_{j \in I_i} l_{ij}\mu_j^{(k-1)}\right) l_{ij}}{\sum_{i \in J_j} Y_i l_{ij}.} \tag{11.29}$$

11.2.5.2 Maximum likelihood expectation maximization algorithm using prior models

To improve the convergence of the EM algorithm for transmission tomography, gamma distribution priors are imposed on each pixel (Lange, Bahn, & Little, 1987). Assuming the attenuation coefficients μ_j are independent with prior probability density

$$\frac{1}{\Gamma(\alpha_j)}\left(\frac{\alpha_j}{\beta_j}\right)^{\alpha_j}\mu_j^{\alpha_j - 1}\exp\left(-\frac{\alpha_j \mu_j}{\beta_j}\right), \tag{11.30}$$

where the gamma density has mean β_j, variance $\frac{\beta_j^2}{\alpha_j}$. Moreover, α_j is the coefficient of variation such that $\alpha_j > 1$. The log prior is shown below without the constant terms that do not depend on μ_j

$$S(\mu) = \sum_j \left[(\alpha_j - 1)\ln\left(\mu_j\right) - \alpha_j \mu_j / \beta_j\right]. \tag{11.31}$$

The log prior Eq. (11.31) is added to the complete likelihood given in Eq. (11.24) and incomplete log-likelihood data $\ln\, g(Y,\mu)$ described in Eq. (11.21). The formation of E step is the same as in Eq. (11.24) and in M step using approximation as in Eq. (11.25), the Bayesian EM transmission algorithm using gamma prior is represented as

$$\mu_j^{(k)} = \frac{\sum_{i\in J_j}\left(M_{ij} - N_{ij}\right) + \alpha_j - 1}{1/2\sum_{i\in J_j}\left(M_{ij} - N_{ij}\right) l_{ij} + \alpha_j/\beta_j}. \tag{11.32}$$

Similarly, the Bayesian extension of gradient type algorithm using gamma prior is described as

$$\mu_j^{(k)} = \mu_j^{(k-1)} \frac{\sum_{i\in J_j} d_i \exp\left(-\sum_{j\in I_i} l_{ij}\mu_j^{(k-1)}\right) + \alpha_j - 1}{\sum_{i\in J_j} Y_i l_{ij} + \alpha_j/\beta_j}, \tag{11.33}$$

where $j\in I_i$ are the set of pixels that contribute to projection i, $i\in J_j$ are the set of projections that contribute to pixel j.

11.3 Electrical tomography

11.3.1 Mathematical formulation of electrical tomography

The electric field over a domain of interest depends on the geometry of the domain, the distribution of electrical properties of the medium, and the excitation sources (voltage or current) imposed on the domain boundary. For the given domain under the known boundary conditions, the measured electric signals are determined only by the electrical property distribution over the domain. Hence, if we can measure the excited electric signals (current or voltage) on the boundary, we may consider the use of electric signals to reconstruct the tomographic image or the electrical property distribution. In practice, when the medium is composed of substances with a sufficiently distinct electrical property, it is possible to estimate the distribution of the substances using the relationship between the electric signals and the distribution of electrical properties. This imaging modality is called *electrical tomography* (ET).

It is well-known that the electric field always induces a magnetic field and is affected by it also. Even when excitation sources are electrical, strictly speaking, the electric field cannot exist solely and should be analyzed with a magnetic field simultaneously. Hence, for the mathematical formulation of ET, we will start from Maxwell's equations governing the electromagnetic field. Maxwell's equations in the matter are composed of four equations:

$$\text{Gauss's law } \nabla \cdot \mathbf{D} = \rho \tag{11.34}$$

$$\text{Gauss's law for magnetism } \nabla \cdot \mathbf{B} = 0 \tag{11.35}$$

$$\text{Faraday's law of induction } \nabla \times \mathbf{E} = -\frac{\partial \mathbf{B}}{\partial t} \tag{11.36}$$

$$\text{Ampére's circuital law } \nabla \times \mathbf{H} = \mathbf{J} + \frac{\partial \mathbf{D}}{\partial t} \tag{11.37}$$

where $\mathbf{E}$ is the electric field [V/m] = [N/C], $\mathbf{D}$ the electric displacement field [C/m^2], $\mathbf{B}$ the magnetic induction [T] = [Ns/Cm], and $\mathbf{H}$ the magnetic field [A/m]. ρ and $\mathbf{J}$ are the charge density [C/m^3] and the current density [A/m^2] not including any induced polarization. If the medium is linear isotropic, the following relations are valid:

$$\mathbf{D} = \varepsilon \mathbf{E}, \; \mathbf{B} = \mu \mathbf{H}, \; \text{and } \mathbf{J}_f = \sigma \mathbf{E}, \tag{11.38}$$

where ε is the permittivity [F/m], μ the permeability [Tm/A] = [Vs/Am], and σ the electric conductivity [1/Ωm].

Combining the above relations and assuming that the injected electric signals (voltage or current) are time harmonic with an angular frequency ω, i.e., $\mathbf{E} = \mathbf{E}_0 e^{j\omega t}$, $\mathbf{B} = \mathbf{B}_0 e^{j\omega t}$, the Faraday's and Ampére's laws become (Doerstling, 1994; Sommersalo, Isaacson, & Cheney, 1992b)

$$\text{Faraday's law of induction } \nabla \times \mathbf{E}_0 = -j\omega\mu \mathbf{H}_0 \tag{11.39}$$

$$\text{Ampére's circuital law } \nabla \times \mathbf{H}_0 = (\sigma + j\omega\varepsilon)\mathbf{E}_0. \tag{11.40}$$

Further simplification is made in ET. Due to Gauss's law for magnetism, there exists a magnetic vector potential of $\mathbf{B}$ such as

$$\mathbf{B} = \nabla \times \mathbf{A}. \tag{11.41}$$

From the vector identity, hence, the electric field can be expressed in general as

$$\mathbf{E} = -\nabla u - \frac{\partial \mathbf{A}}{\partial t}, \tag{11.42}$$

where u is the electric potential. Again, if we assume a time-harmonic field for the magnetic vector potential as $\mathbf{A} = \mathbf{A}_0 e^{j\omega t}$ and the electric potential as $u = u_0 e^{j\omega t}$,

$$\mathbf{E}_0 = -\nabla u_0 - j\omega \mathbf{A}_0. \tag{11.43}$$

Here, the magnitude of $\mathbf{A}_0$ may be estimated as

$$|\mathbf{B}_0| = |\mu \mathbf{H}_0| = |\nabla \times \mathbf{A}_0| \text{ and } |\mathbf{A}_0| \sim \mu L_c |\mathbf{H}_0|, \tag{11.44}$$

where L_c is the characteristic length of the magnetic field. On the other hand, from Ampére's law, the scale relation will be

$$|\mathbf{E}_0| = |\nabla u_0 + j\omega \mathbf{A}_0| \sim \frac{|\mathbf{H}_0|/L_c}{|\sigma + j\omega\varepsilon|}. \tag{11.45}$$

The relative contribution of the magnetic vector potential to the electric field is estimated as

$$\frac{|\omega \mathbf{A}_0|}{|\nabla u_0 + j\omega \mathbf{A}_0|} \sim \omega\mu L_c^2 |\sigma + j\omega\varepsilon|. \tag{11.46}$$

If the relative contribution is much smaller than 1, i.e.,

$$\omega\mu L_c^2 |\sigma + j\omega\varepsilon| << 1, \tag{11.47}$$

then the effect of magnetic induction inducing electric field can be neglected (Nunez, 1981). Consider a pipe of diameter 1 m filled with water. For the frequency of 100 kHz, the physical properties of water, $\varepsilon \sim 80 \times 10^{-9}/36\pi$ F/m, $\mu \sim 4\pi\times 10^{-7}$ Vs/Am, and $\sigma \sim 5.5 \times 10^{-6}$ 1/Ωm yields $\omega\mu L_c^2|\sigma + j\omega\varepsilon| \sim 3.5 \times 10^{-4} << 1$. Hence, the above simplification is valid.

Using the vector identity, readily, we have the governing equation for the forward problem of the ET:

$$\nabla \cdot (\sigma + j\omega\varepsilon)\nabla u = 0 \text{ or } \nabla \cdot \gamma \nabla u = 0. \tag{11.48}$$

Here, for simplicity, the subscript "0" is dropped. The coefficient $\gamma = \sigma + j\omega\varepsilon$ is called the admittivity (Cheney, 1999).

According to the ratio of $\omega\varepsilon/\sigma$, when the resistance or the capacitance term is dominant to the other, the governing equation is simplified further (Baker, 1989; Barber & Brown, 1984; Webster, 1990):

$$\nabla \cdot \sigma \nabla u = 0 \text{ for } \omega\varepsilon/\sigma << 1 \text{ (ERT)}, \tag{11.49}$$

$$\nabla \cdot \varepsilon \nabla u = 0 \text{ for } \omega\varepsilon/\sigma >> 1 \text{ (ECT)}, \tag{11.50}$$

where ERT (Electrical Resistance Tomography) means the resistance component is used for the tomography, while ECT (Electrical Capacitance Tomography) uses the capacitance component. ET combining ERT and ECT is called EIT (Electrical Impedance Tomography). It should be noted that ERT is frequently known as EIT.

11.3.2 Image reconstruction based on transmission tomography algorithms

As can be seen from the governing equations, the electric potential satisfies Eq. (11.48) or Eqs. (11.49) and (11.50). Since the equations are elliptic electric potential, they can be altered by any variation of the electrical properties at any location within the domain and by any variation of the boundary conditions. This means that the electric field may not be straight, and the image reconstruction concepts for EIT should be quite different from transmission tomography concepts. In transmission tomography, only the substances in the radiation paths can contribute to the attenuation of the radiation, and the path is assumed to be a straight line between the source and the detector. Nevertheless, if we can estimate the path of electric current, which may be curved, for the given geometry and boundary conditions, the path can be approximated as an electric circuit unaffected by the substances out of the path and the use of transmission tomography concepts for EIT may be feasible. In this, the referenced electric current paths may be approximately determined with known electrical property distribution, usually with a homogeneous medium of background substance. Whenever the distortion of the electric field due to any inclusion of anomaly against the background substance is not significant, we may adopt transmission tomography concepts for ET.

As an application of the transmission tomography concept to ET, the linear back-projection algorithm in ECT is introduced. In practice, this was successfully applied to gas—solid or gas—liquid systems and extended to air—water systems (Dyakowski, Jeanmeure, & Jaworski, 2000; Warsito & Fan, 2001; Xie et al., 1992). After the discretization of the problem domain and linearization of the relationship between capacitance and the permittivity, we can have the expression as follows:

$$\mathbf{C}_{(M\times 1)} = \mathbf{S}_{(M\times N)}\mathbf{g}_{(N\times 1)}. \tag{11.51}$$

In this, $\mathbf{C}$ is the capacitance vector whose entities are independent capacitance data of electrode pairs, M is the number of independent capacitance data, $\mathbf{g}$ is the gray level vector which denotes the normalized permittivity distribution in a discrete number (N) of pixels, and $\mathbf{S}$ is the sensitivity matrix (or projection matrix) which represents the mapping of the gray level vector into the capacitance vector. The above expression called "forward projection" implies that the capacitance measured between a certain electrode pair is the sum of the capacitance for that particular electrode pair contributed by the permittivity in each pixel.

Xie et al. (1989) suggested a "back-projection" algorithm inverting the forward projection to obtain $\mathbf{g}$:

$$\mathbf{g}_{(N\times 1)} = \mathbf{S}^T_{(M\times N)}\mathbf{C}^*_{(M\times 1)}. \tag{11.52}$$

Here, $C_{i,j}$ means that the measured capacitance data between i-th and j-th electrodes, $\mathbf{C}^*_{(M\times 1)}$, are normalized capacitance data computed as

$$C^*_{i,j} = \frac{C_{i,j} - C_{i,j(\mathrm{low}-\varepsilon)}}{C_{i,j(\mathrm{high}-\varepsilon)} - C_{i,j(\mathrm{low}-\varepsilon)}}, \tag{11.53}$$

and pseudoinverse $\mathbf{S}^{-1}_{(M\times N)} = \mathbf{S}^{T}_{(M\times N)}$ is used. The sensitivity matrix is obtained as follows. At first, they defined sensitivity areas $A_{i,j}$ such that the permittivity change of the pixels in $A_{i,j}$ effect normalized capacitance $C^*_{i,j}$ and the permittivity change of the pixels out of $A_{i,j}$ contribute little to $C^*_{i,j}$. Determining which pixel belongs to $A_{i,j}$ could be done by evaluating the change $C^*_{i,j}$ due to the permittivity change of that pixel. If the change is greater than a certain threshold, that pixel is assumed to belong to $A_{i,j}$. The change $C^*_{i,j}$ could be obtained from numerical simulations or experiments by inserting a small anomaly with high permittivity into that pixel in a homogeneous domain with low permittivity. Mathematically, the gray level for the n-th pixel for an eight-electrode system can be calculated as

$$g(n) = \sum_{i=1}^{7} \sum_{j=i+1}^{8} S_{i,j}(n) C^*_{i,j} \tag{11.54}$$

where the sensitivity matrix is evaluated as

$$S_{i,j}(n) = \frac{\delta_{i,j}(n)}{\sum_{i=1}^{7} \sum_{j=i+1}^{8} \delta_{i,j}(n)} \tag{11.55}$$

and the characteristic function $\delta_{i,j}(n)$ is 1 if the pixel n belongs to $A_{i,j}$, and 0 otherwise. This back-projection is conceptually very simple and straightforward; however, many assumptions are engaged in the algorithm and the image reconstruction performance may not be so plausible. To improve the image quality, some modifications of the sensitivity matrix have been suggested (Kim, Choi, & Kim, 2006; Reinecke & Mewes, 1997; Villares et al., 2012).

11.3.3 Linear algorithms

In the above subsection, the transmission tomography concept is applied to ECT. Namely, it is assumed that the pixels influencing the capacitance between a certain electrode pair can be extracted and the area composed of those pixels is considered as the path that radiation passes along in the transmission tomography. In reality, the change of electrical property in any pixel affects the electric signals measured at the boundary electrodes. If we abandon the transmission tomography concept, we can have an exact expression of the sensitivity matrix. For more concrete derivation, let us start from the governing equation for the ECT. The problem domain Ω enclosed by $\partial\Omega$ is filled with substances whose permittivity is $\varepsilon(x)$. Along $\partial\Omega$, electrodes e_l ($l = 1, 2, \cdots, L$) are mounted and the gaps between electrodes are insulated (Yang, 1996). When an electric potential U_l is imposed on each e_l, the electric potential distribution $u(x)$ satisfies

$$\nabla \cdot \varepsilon(x)\nabla u(x) = 0, \; x \in \Omega, \tag{11.56}$$

subject to the boundary condition

$$u(x) = U_l, \; x \in e_l, \tag{11.57a}$$

and

$$\frac{\partial u(x)}{\partial \nu} = 0, \; x \in \partial\Omega / \cup_{l=1}^{L} e_l. \tag{11.57b}$$

where ν is the outward normal unit vector. To reconstruct the permittivity distribution, the domain is discretized into small pixels Ω_n $(n = 1, 2, \cdots, N)$ and in each pixel, the permittivity is assumed to be constant. Namely, the number of unknowns will be the number of pixels. For the estimation of the permittivity of each pixel, we need the equations of the potential excitation and the induced capacitances as many as possible. It means that for the same object we should measure capacitance data sets for a number of boundary condition sets that are independent each other.

Assume that u^p is the potential subject to the p-th boundary condition set $(p = 1, 2, \cdots, P)$ for the given permittivity distribution ε, while u_0^q is the potential subject to the qth boundary condition set for a reference permittivity distribution ε_0 (Fig. 11.7). Then,

$$\nabla \cdot \varepsilon \nabla u^p = 0 \text{ and } \nabla \cdot \varepsilon_0 \nabla u_0^q = 0. \tag{11.58}$$

Here, multiplying each equation by u_0^q and u^p each, subtracting, integrating over Ω, and applying the divergence theorem yields

$$\int_{\partial\Omega} \left(u_0^q \varepsilon \frac{\partial u^p}{\partial \nu} - u^p \varepsilon_0 \frac{\partial u_0^q}{\partial \nu} \right) dS = \int_{\Omega} (\varepsilon - \varepsilon_0) \nabla u^p \cdot \nabla u_0^q d\Omega. \tag{11.59}$$

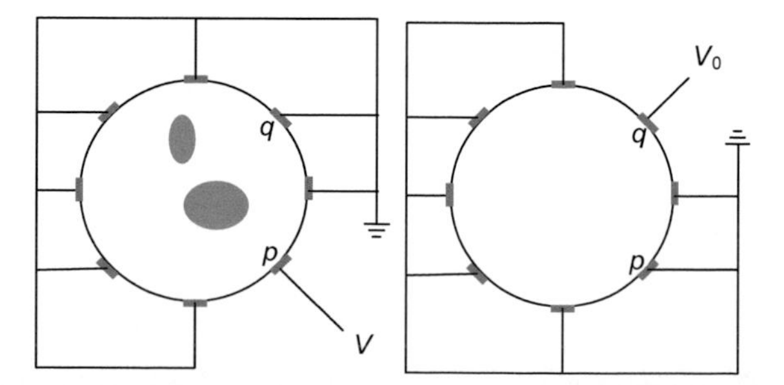

Figure 11.7 Diagram showing measurement configuration for the ECT problem.

The left-hand side collects the electrical excitations and responses on the electrodes. Recalling the boundary conditions, we have

$$\sum_{l=1}^{L}\left[U_{0,l}^{q}\int_{e_l}\varepsilon\frac{\partial u^{p}}{\partial\nu}dS-U_{l}^{p}\int_{e_l}\varepsilon_0\frac{\partial u_0^{q}}{\partial\nu}dS\right]=\int_{\Omega}(\varepsilon-\varepsilon_0)\nabla u^{p}\cdot\nabla u_0^{q}d\Omega. \tag{11.60}$$

As a simple case, consider $U_l^p = V\delta_{pl}$ and $U_{0,l}^q = V_0\delta_{ql}$, where δ_{ij} is the Kronecker delta and $p,q = 1,2,\cdots,L$. And assume that $V, V_0 > 0$.

Then,

$$-V_0Q_q^p + VQ_{0,p}^q = \int_{\Omega}(\varepsilon-\varepsilon_0)\nabla u^p\cdot\nabla u_0^q d\Omega \text{ for } p\neq q \tag{11.61a}$$

or

$$C_{p,q} - C_{0,q,p} = -\int_{\Omega}(\varepsilon-\varepsilon_0)\frac{\nabla u^p}{V}\cdot\frac{\nabla u_0^q}{V_0}d\Omega \text{ for } p\neq q \tag{11.61b}$$

where $Q_l^p = -\int_{e_l}\varepsilon\frac{\partial u^p}{\partial\nu}dS$ is the charge at the l-th electrode and $C_{p,q} = \frac{Q_q^p}{V}$ is the capacitance between the p-th and q-th electrodes when the p-th electrode is activated by electric potential V, while other electrodes are grounded (see Fig. 11.7). Note that from the reciprocity theorem, $C_{p,q} = C_{q,p}$. Also, by applying the divergence theorem to the integration of the governing equation over Ω, we can see that $C_{p,p} = -\sum_{i\neq p}^{L} C_{p,i}$. Hence, the equation for $p = q$ is not independent of other equations for $p \neq q$ and the number of independent data is $M = L(L-1)/2$.

Let us introduce the normalized capacitance as

$$C_{p,q}^{*} = \frac{C_{p,q} - C_{p,q(\text{low}-\varepsilon)}}{C_{p,q(\text{high}-\varepsilon)} - C_{p,q(\text{low}-\varepsilon)}} = \frac{\int_{\Omega}(\varepsilon-\varepsilon_{\text{low}})\frac{\nabla u^p}{V}\cdot\frac{\nabla u_{\text{low}}^q}{V_{\text{low}}}d\Omega}{(\varepsilon_{\text{high}}-\varepsilon_{\text{low}})\int_{\Omega}\frac{\nabla u_{\text{high}}^p}{V_{\text{high}}}\cdot\frac{\nabla u_{\text{low}}^q}{V_{\text{low}}}d\Omega} \text{ for } p\neq q. \tag{11.62}$$

In this, the subscripts "high" and "low" mean that the problem domain is filled with "high" and "low" conductivity (or permittivity) material only, respectively. Note that $\nabla u_{\text{high}}^p / V_{\text{high}} = \nabla u_{\text{low}}^p / V_{\text{low}}$. If the domain is decomposed into N small elements as $\Omega = \cup_{n=1}^{N}\Omega_n$ and in each Ω_n the permittivity is as $\varepsilon = \varepsilon_n$ then

$$C^*_{p,q} = \sum_{n=1}^{N} \varepsilon^*_n S^{p,q}_n \text{ for } p \neq q, \tag{11.63}$$

where ε^*_n is the normalized permittivity

$$\varepsilon^*_n = \frac{\varepsilon_n - \varepsilon_{\text{low}}}{\varepsilon_{\text{high}} - \varepsilon_{\text{low}}} \tag{11.64}$$

and $S^{p,q}_n$ is the sensitivity matrix

$$S^{p,q}_n = \frac{\int_{\Omega_n} (\nabla u^p / V) \cdot \left(\nabla u^q_{\text{low}} / V_{\text{low}}\right) d\Omega}{\int_{\Omega} \left(\nabla u^p_{\text{high}} / V_{\text{high}}\right) \cdot \left(\nabla u^q_{\text{low}} / V_{\text{low}}\right) d\Omega} \text{ for } p \neq q. \tag{11.65}$$

In an algebraic form,

$$\begin{pmatrix} C^*_{1,2} \\ C^*_{1,3} \\ \vdots \\ C^*_{2,3} \\ \vdots \\ C^*_{(L-1),L} \end{pmatrix} = \begin{bmatrix} S^{1,2}_1 & S^{1,2}_2 & \cdots & S^{1,2}_N \\ S^{1,3}_1 & S^{1,3}_2 & \cdots & S^{1,3}_N \\ \vdots & \vdots & \ddots & \vdots \\ S^{2,3}_1 & S^{2,3}_2 & \cdots & S^{2,3}_N \\ \vdots & \vdots & \ddots & \vdots \\ S^{(L-1),L}_1 & S^{(L-1),L}_2 & \cdots & S^{(L-1),L}_N \end{bmatrix} \begin{pmatrix} \varepsilon^*_1 \\ \varepsilon^*_2 \\ \vdots \\ \vdots \\ \vdots \\ \varepsilon^*_N \end{pmatrix} \tag{11.66a}$$

or

$$\mathbf{C}^* = \mathbf{S}\boldsymbol{\varepsilon}^*, \tag{11.66b}$$

where $\mathbf{C}^* \in \Re^{L(L-1)/2 \times 1}$ is the normalized capacitance vector, $\boldsymbol{\varepsilon}^* \in \Re^{N \times 1}$ is the normalized permittivity vector, and $\mathbf{S} \in \Re^{L(L-1)/2 \times N}$ is the sensitivity matrix. The normalized capacitance vector is a measurable quantity so that if the sensitivity matrix is available, the permittivity distribution can be reconstructed. Recalling the definition of the sensitivity matrix, unless ∇u^p under the given $\boldsymbol{\varepsilon}$ is known in advance, $\mathbf{S}$ cannot be evaluated. If there are only two substances with $\varepsilon_{\text{high}}$ and ε_{low}, it is possible to evaluate the sensitivity matrix exactly. Consider a domain with $\varepsilon(x) = \varepsilon_{\text{low}} + \left(\varepsilon_{\text{high}} - \varepsilon_{\text{low}}\right)\chi_{\Omega_n}(x)$, where $\chi_{\Omega_n}(x)$ is the characteristic function defined as

$$\chi_{\Omega_n}(x) = \begin{cases} 1 & \text{for } x \in \Omega_n \\ 0 & \text{otherwise} \end{cases}.$$

Then, we have

$$S_n^{p,q} = \frac{C_{p,q}^{(n)} - C_{p,q(\text{low}-\varepsilon)}}{C_{p,q(\text{high}-\varepsilon)} - C_{p,q(\text{low}-\varepsilon)}} = C_{p,q}^{*(n)} \text{ for } p \neq q. \tag{11.67}$$

In this, $C_{p,q}^{(n)}$ is the capacitance for $\varepsilon(x) = \varepsilon_{\text{low}} + (\varepsilon_{\text{high}} - \varepsilon_{\text{low}})\chi_{\Omega_n}(x)$. These can be obtained by solving the governing equation.

It should be noted that the above formulation is valid for the ERT when the conductivity and the conductance are substituted for the permittivity and the capacitance, respectively.

11.3.4 Gradient-based algorithms

In general, the problem domain may contain more than two substances with distinct electrical properties or miscible mixture with continuously varying electrical properties. Then, the above linear algorithms may not be feasible since the electric field for a given electrical property distribution cannot be known in advance. However, if the deviation of the electrical property distribution from the reference one, we can try a linearization about the reference. For example, if the permittivity is perturbed as $\varepsilon(x) = \varepsilon_{\text{low}} + (\delta\varepsilon)\chi_{\Omega_n}(x)$, then both the potential field and the capacitance are perturbed, and we have

$$\delta C_{p,q}^* = \delta\varepsilon^* \frac{\int_{\Omega_n} (\nabla u_{\text{low}}^p / V_{\text{low}}) \cdot (\nabla u_{\text{low}}^q / V_{\text{low}}) d\Omega}{\int_{\Omega} \left(\nabla u_{\text{high}}^p / V_{\text{high}}\right) \cdot (\nabla u_{\text{low}}^q / V_{\text{low}}) d\Omega} + O(\delta\varepsilon^{*2}) \text{ for } p \neq q \tag{11.68}$$

or the Jacobian, the partial derivative of $C_{p,q}^*$ with respect to the permittivity of the n-th subdomain Ω_n, can be expressed as

$$J_n^{p,q} = \frac{\partial C_{p,q}^*}{\partial \varepsilon_n^*} = \frac{\int_{\Omega_n} (\nabla u_{\text{low}}^p / V_{\text{low}}) \cdot (\nabla u_{\text{low}}^q / V_{\text{low}}) d\Omega}{\int_{\Omega} \left(\nabla u_{\text{high}}^p / V_{\text{high}}\right) \cdot (\nabla u_{\text{low}}^q / V_{\text{low}}) d\Omega} \text{ for } p \neq q. \tag{11.69}$$

Linearizing the normalized capacitance about the homogeneous configuration with ε_{low}, it can be approximated as

$$\mathbf{C}^* = \mathbf{J}\boldsymbol{\varepsilon}^*, \tag{11.70}$$

in which the evaluation of the Jacobian $\mathbf{J}$ requires much fewer computations compared with the previous one. However, since this is a first-order approximation, it is expected that the more the permittivity distribution deviates from ε_{low}, the worse the reconstruction performance becomes.

Now, let us find general approaches to arbitrary situations. For a given domain, if we know the electrical property (conductivity and/or permittivity) distribution and the imposed boundary conditions, we can calculate the potential distribution by solving the governing equation, Eq. (11.48) or Eqs. (11.49) and (11.50). Also, the measured electric signals should be very close to the calculated ones with the known information. Note that they cannot be the same as each other since the measurement error and the modeling error cannot be avoided. Hence, the EIT image reconstruction can be regarded as a proccss to properly assume the electrical property distribution such that the difference between the measured electric signals and the calculated ones with the assumed distribution is minimized. In other words, the EIT is an optimization problem to find the electrical property distribution minimizing the difference between the measured and the calculated electric signals (Vauhkonen, 1997; Yorkey, Webster, & Tompkins, 1987).

Let us consider EIT more strictly ERT problem. In this, it is common to inject electric currents through the electrodes attached along the boundary discretely for excitation and to measure excited voltages on the electrodes. The domain is discretized into small pixels as $\Omega = \cup_{n=1}^{N}\Omega_n$ and the conductivity distribution is assumed to be a pixel-wise constant function such that $\sigma(x) = \sum_{n=1}^{N} \sigma_n \chi_{\Omega_n}(x)$, where the conductivity vector $\boldsymbol{\sigma} = [\sigma_1, \sigma_2, \cdots, \sigma_N]^T \in \Re^{N\times 1}$ is the unknown to be estimated by minimizing the cost function that is the sum of squares of the voltage difference $\mathbf{f}(\boldsymbol{\sigma}) = \mathbf{U}(\boldsymbol{\sigma}) - \mathbf{V}$

$$\Phi(\boldsymbol{\sigma}) = \frac{1}{2}[\mathbf{U}(\boldsymbol{\sigma}) - \mathbf{V}]^T[\mathbf{U}(\boldsymbol{\sigma}) - \mathbf{V}] = \mathbf{f}^T\mathbf{f}. \tag{11.71}$$

The measured voltage vector is denoted as $\mathbf{V} \in \Re^{LM\times 1}$ and the calculated voltage vector based on the assumed conductivity vector $\boldsymbol{\sigma}$ is $\mathbf{U}(\boldsymbol{\sigma}) \in \Re^{LM\times 1}$ (Vauhkonen, 1997; Woo, Hua, Webster, & Tompkins, 1994). In this, it is assumed that voltages are measured at all the electrodes e_l $(l = 1, 2, \cdots, L)$ and the number of current injection patterns, or the boundary condition sets, is M. According to the data collection strategy, the voltages may be measured at some selected electrodes and not at all the electrodes. Let the number of voltage data P, i.e., $\mathbf{f}, \mathbf{V}, \mathbf{U}(\boldsymbol{\sigma}) \in \Re^{P\times 1}$. The calculated voltage $\mathbf{U}(\boldsymbol{\sigma})$ can be obtained by solving the governing equation

$$\nabla \cdot \sigma \nabla u = 0 \tag{11.72}$$

subject to the boundary condition

$$\int_{e_l} \frac{\partial u(x)}{\partial \nu} dS = I_l, \; x \in e_l, \tag{11.73a}$$

and

$$\frac{\partial u(x)}{\partial \nu}=0,\ x\in \partial\Omega/\cup_{l=1}^{L}e_l. \tag{11.73b}$$

Also, two more constraints should be imposed:

$$\sum_{l=1}^{L} I_l=0 \text{ and } \sum_{l=1}^{L} U_l=0. \tag{11.74}$$

The former is to satisfy Kirchhoff's law, and the latter is introduced to fix the potential distribution since only the Neumann conditions are given along the boundary (Somersalo, Cheney et al., 1992). For the latter, another form of constraint is possible as long as the potential can be fixed, e.g., $U_1 = 0$. For the current injection boundary condition, the current itself, as given in Eq. (11.73a), cannot be used as the boundary condition. The current density, instead of the current, should be specified. A simple approach is to assume a uniform current density:

$$\frac{\partial u(x)}{\partial \nu}=\frac{I_l}{|e_l|},\ x\in e_l, \tag{11.75}$$

which is called the average gap model (Cheng, Isaacson, Newell, & Gisser, 1989). On the other hand, we need to obtain the electrode voltages from the potential distribution. As electrodes, very conductive metals are used, and the potential distribution within an electrode can be assumed to be a constant, which will be the electrode voltage. Hence, we may consider an approximation that the potential of the substance at the points contacting the electrode may be set to electrode voltage. This is called the shunt model.

$$u(x)=U_l,\ x\in e_l, \tag{11.76}$$

However, in reality, the current density may not be uniform, and there is contact impedance between the electrode and the substance. As a more realistic model, the complete electrode model considers the contact impedance and the shunt effect at the same time (Cheng et al., 1989; Vauhkonen, 1997):

$$u(x)+z_l\sigma\frac{\partial u(x)}{\partial \nu}=U_l,\ x\in e_l, \tag{11.77}$$

where z_l is the effective contact impedance at the l-th electrode. The disadvantages of the complete electrode model are that z_l is unknown in general, and we should estimate in advance through separate experiments or simultaneously estimate along with the conductivity distribution, which increases the number of unknowns and degrades the image reconstruction performance. For the solution of the governing equation with the boundary conditions, the constraints, and the complete electrode model, usually, the finite element method (FEM) is preferred since it can easily handle the pixel-wise conductivity distribution (Brenner & Scott, 1994; Hinton & Owen, 1979; Vauhkonen, 1997). The finite element formulation is well established, and many references and computer programs are available (Adler et al., 2009; Kim et al., 2006; Vauhkonen, 1997).

One of the most common approaches to the above nonlinear least squares problems is the Gauss−Newton algorithm. It can be regarded as a modification of Newton's method for finding a minimum of a function. The Gauss−Newton algorithm iteratively finds the minimum of the sum of squares (Adler, Dai, & Lionheart, 2007; Kohn & McKenney, 1990; Polydorides, Lionheart, & McCann, 2002; Vauhkonen, 1997; Yorkey et al., 1987). To find the minimum, the derivative of the cost functional with respect to $\boldsymbol{\sigma}$, denoted by prime $'$, is set to zero

$$\Phi'(\boldsymbol{\sigma}) = \mathbf{J}^T(\boldsymbol{\sigma})\mathbf{f}(\boldsymbol{\sigma}) = \mathbf{0},$$

and from the first-order Taylor series expansion, its recurrence relation is written as

$$\mathbf{0} = \Phi'\left(\boldsymbol{\sigma}^k\right) + \Phi''\left(\boldsymbol{\sigma}^k\right)\left(\boldsymbol{\sigma}^{k+1} - \boldsymbol{\sigma}^k\right)$$

or

$$\boldsymbol{\sigma}^{k+1} = \boldsymbol{\sigma}^k - \left[\Phi''\left(\boldsymbol{\sigma}^k\right)\right]^{-1}\Phi'\left(\boldsymbol{\sigma}^k\right).$$

Here, $\mathbf{J}(\boldsymbol{\sigma})$ is the Jacobian matrix of $\mathbf{U}(\boldsymbol{\sigma})$ defined as

$$[\mathbf{J}(\boldsymbol{\sigma})]_{ij} = [\mathbf{f}'(\boldsymbol{\sigma})]_{ij} = \left[\frac{\partial \mathbf{f}(\boldsymbol{\sigma})}{\partial \boldsymbol{\sigma}}\right]_{ij} = \left[\frac{\partial \mathbf{U}(\boldsymbol{\sigma})}{\partial \boldsymbol{\sigma}}\right]_{ij} = \frac{\partial U_i(\boldsymbol{\sigma})}{\partial \sigma_j} \in \Re^{P\times N}, \tag{11.78}$$

and $\Phi''(\boldsymbol{\sigma})$ is called the Hessian matrix $\mathbf{H}$ of Φ expressed as

$$\begin{aligned}\Phi''(\boldsymbol{\sigma}) &= \mathbf{H}(\boldsymbol{\sigma}) = [\mathbf{f}']^T[\mathbf{f}'] + [\mathbf{f}'']^T[\mathbf{I}_{n\times n} \otimes \mathbf{f}] \\ &= \mathbf{J}^T\mathbf{J} + \left[\nabla^2_{N\times N} \otimes \mathbf{f}^T\right][\mathbf{I}_{N\times N} \otimes \mathbf{f}] \in \Re^{N\times N}.\end{aligned} \tag{11.79}$$

$\nabla^2_{N\times N}$ is the second-order partial differential operator whose entity is $\left[\nabla^2_{N\times N}\right]_{ij} = \frac{\partial^2}{\partial\sigma_i\partial\sigma_j}$, $\mathbf{I}_{N\times N} \in \Re^{N\times N}$ is the identity matrix, and $\otimes$ is the Kronecker product. In the Gauss−Newton method, the second-order derivative terms $\frac{\partial^2 f_k}{\partial\sigma_i\partial\sigma_j} f_k$ are ignored. That is, the Hessian is approximated as

$$\Phi''(\boldsymbol{\sigma}) = \mathbf{H} \approx \mathbf{J}^T\mathbf{J} \in \Re^{N\times N}. \tag{11.80}$$

The convergence of the Gauss–Newton method may not be guaranteed in all cases unless

$$\left|\frac{\partial^2 f_k}{\partial\sigma_i\partial\sigma_j}f_k\right| << \left|\frac{\partial f_k}{\partial\sigma_i}\frac{\partial f_k}{\partial\sigma_j}\right|,$$

This approximation is valid if the voltage difference f_k is small in magnitude, at least around the minimum, or the nonlinearity is not severe, so that $\frac{\partial^2 f_k}{\partial\sigma_i\partial\sigma_j}$ is relatively small in magnitude. Consequently, the rate of convergence of the Gauss–Newton algorithm is at most quadratic.

In general, the number of unknowns in EIT is larger than the number of the current–voltage data sets, i.e., $N > M$, which means that the problem is ill-posed. Also, the sensitivity of the boundary voltages with respect to the change of the conductivity in a certain pixel may be highly nonuniform, and it leads to an ill-conditioning of the Hessian matrix. Hence, in practice, regularization is introduced to mitigate the ill-posed and ill-conditioned nature of the problem. The regularized cost functional is expressed as

$$\Phi(\boldsymbol{\sigma}) = \frac{1}{2}[\mathbf{U}(\boldsymbol{\sigma}) - \mathbf{V}]^T[\mathbf{U}(\boldsymbol{\sigma}) - \mathbf{V}] + \frac{\alpha}{2}[\mathbf{R}(\boldsymbol{\sigma} - \boldsymbol{\sigma}^*)]^T[\mathbf{R}(\boldsymbol{\sigma} - \boldsymbol{\sigma}^*)], \tag{11.81}$$

where α is the regularization parameter and $\mathbf{R} \in \Re^{N\times N}$ is the regularization matrix. $\boldsymbol{\sigma}^* \in \Re^{N\times 1}$ is the prior information for the conductivity distribution if available (Hua, Webster, & Tompkins, 1988; Hua, Woo, Webster, & Tompkins, 1991; Vauhkonen, 1997).

Starting with an initial guess $\boldsymbol{\sigma}^0$ for the minimum, the Gauss–Newton method proceeds by the iterations

$$\boldsymbol{\sigma}^{k+1} = \boldsymbol{\sigma}^k - \left[\mathbf{J}^T\mathbf{J} + \alpha\mathbf{R}^T\mathbf{R}\right]^{-1}\left[\mathbf{J}^T\left(\mathbf{U}\left(\boldsymbol{\sigma}^k\right) - \mathbf{V}\right) + \alpha\mathbf{R}^T\mathbf{R}(\boldsymbol{\sigma} - \boldsymbol{\sigma}^*)\right] \tag{11.82}$$

with $\left[\mathbf{J}\left(\boldsymbol{\sigma}^k\right)\right]_{ij} = \frac{\partial U_i(\boldsymbol{\sigma}^k)}{\partial\sigma_j^k}$.

The Jacobian can be obtained from the following identity:

$$\sum_{l=1}^{L}\left[U_{0,l}^q\int_{e_l}\sigma\frac{\partial u^p}{\partial\nu}dS - U_l^p\int_{e_l}\sigma_0\frac{\partial u_0^q}{\partial\nu}dS\right] = \int_{\Omega}(\sigma - \sigma_0)\nabla u^p\cdot\nabla u_0^q d\Omega, \tag{11.83}$$

which can be derived in a similar way used for Eq. (11.59) in ECT. From the boundary condition of the current injection, Eq. (11.83) becomes

$$\sum_{l=1}^{L}\left[U_{0,l}^q I_l^p - U_l^p I_l^q\right] = \int_{\Omega}(\sigma - \sigma_0)\nabla u^p\cdot\nabla u_0^q d\Omega. \tag{11.84}$$

Note that even with the complete electrode model, the above identity is the same. Assume that we apply the same boundary condition for both electric fields $u_0(x)$ under $\sigma_0(x)$ and $u(x)$ under $\sigma(x)$. If $\sigma(x) = \sigma_0(x) + \delta\sigma_n \chi_{\Omega_n}(x)$ is perturbed, the potential and the electrode voltages are perturbed as $u^p = u_0^p + \delta u^p$ and $U_l^p = U_{0,l}^p + \delta U_l^p$. The perturbed equation is written as

$$\sum_{l=1}^{L} \left[U_{0,l}^q I_l^p - U_{0,l}^p I_l^q - \delta U_l^p I_l^q \right] = \delta\sigma_n \int_{\Omega_n} \nabla u_0^p \cdot \nabla u_0^q d\Omega + O(\delta\sigma_n^2).$$

Considering $\sum_{l=1}^{L} \left[U_{0,l}^q I_l^p - U_{0,l}^p I_l^q \right] = 0$, we can have the equation for the Jacobian matrix:

$$\sum_{l=1}^{L} \frac{\partial U_l^p}{\partial \sigma_n} I_l^q = - \int_{\Omega_n} \nabla u_0^p \cdot \nabla u_0^q d\Omega. \tag{11.85}$$

The right-hand side can be calculated for the reference conductivity distribution or the conductivity distribution at the previous iteration step. In practice, the right-hand side is calculated during the process of the finite element calculation for solving the governing equation so that extra computations are not required.

A similar method as in Eq. (11.82) is proposed, which is called a one-step Gauss–Newton approach, also called Newton's one-step error reconstructor (NOSER). In the NOSER method, the regularization matrix is formulated by considering the sensitivity of each element, i.e., the regularization matrix $\mathbf{R}^T\mathbf{R}$ is taken as diagonal of the matrix $\mathbf{J}^T\mathbf{J}$ (Cheney et al., 1990). The Jacobian $\mathbf{J}$ and forward solution is computed based on the homogeneous distribution σ_0 for which an analytical solution can easily be obtained. It is assumed that the inhomogeneities in the domain vary a little from the background; therefore, the problem can be considered as linear, and only one step is needed to get the desired solution.

In EIT, different types of regularization methods are adopted depending on available prior information of conductivity distribution. Generally, in most cases, the regularization matrix is based on Tikhonov methods, which is an L_2 type regularization. If the solution is bounded and fluctuating, the penalty term is set to $\|\mathbf{R}\sigma\|^2$ and $\mathbf{R}^T\mathbf{R} = \mathbf{I}(\text{Identitymatrix})$. If the solution is continuous, the penalty term is represented by a first-order difference method, i.e., $\|\mathbf{R}\nabla\boldsymbol{\sigma}\|^2$ (Kim, Kim, Kim, & Lee, 2001; Vauhkonen, 1997; Webster, 1990). The gradient of the conductivity for the i-th element is approximated as the difference of the conductivity of i-th element and the neighboring elements that share the common face (Kim et al., 2001; Vaukhonen, 1997)

$$\mathbf{R}_i = [0, \cdots, 0, -1, 0, \cdots, 0, -1, 0, \cdots, 0, 3, 0, \cdots, 0, -1, 0, \cdots, 0], \tag{11.86}$$

here, the i-th column has weight 3, and neighboring elements of i-th element are assigned value -1. Weight 3 corresponds to the absolute sum value of the weight -1s. If i-th element is a boundary element, then there are only two elements adjoining the element. Hence, the weight 3 is replaced with 2. Apart from the above, there is Levenberg–Marquardt regularization, similar to the one used in the NOSER, $\mathbf{R}^T\mathbf{R} = \text{diag}\left(\mathbf{J}^T\mathbf{J}\right)$. L_2 type regularization methods enforce a certain degree of smoothness to obtain stability of the solution; therefore, the resulting reconstructed images are smooth and have blurred edges. In many situations, edges contain practical significance and are desired when the imaged domain contains several discontinuities (Wang et al., 2012). To image the discontinuous conductivity profile, the L_1 type of regularization is preferred which is represented as $\|\mathbf{R}(\sigma)\|^1$ (Dai & Adler, 2008). Another popular method is the total variation (TV) method, which is also an L_1 type of regularization method. The TV of conductivity is defined as (Borsic, Graham, Adler, & Lionheart, 2007; Dai & Adler, 2008)

$$\text{TV}(\boldsymbol{\sigma}) = \int_{\Omega} \|\nabla\sigma\| d\Omega \tag{11.87}$$

Let us consider the discretized version of Eq. (11.87) and assuming that the conductivity is described by piecewise constant elements, the TV of the 2D image can be expressed as the sum of the TV of each of the k edges, with each edge weighted by its length (Borsic et al., 2007)

$$\text{TV}(\boldsymbol{\sigma}) = \sum_k l_k |\sigma_{m(k)} - \sigma_{n(k)}| = \sum_k |\mathbf{L}_k \boldsymbol{\sigma}|, \tag{11.88}$$

where l_k is the length of the k-th edge in the mesh, $m(k)$ and $n(k)$ are the indices of the two elements on opposite sides of the k-th edge, and the index k ranges over all the edges. In Eq. (11.88), $\mathbf{L}$ is a sparse matrix where each row represents an edge in the mesh. The row vector $\mathbf{L}_k$ for a k-th edge is represented as $\mathbf{L}_k = [0, \ldots, 0, l_k, 0, \ldots\ldots, 0, -l_k, 0, \ldots, 0]$ where $\mathbf{L}_k$ has two nonzero elements in the columns $m(k)$ and $n(k)$. The EIT inverse problem with the discretized TV regularization method can be written as

$$\Phi(\boldsymbol{\sigma}) = \frac{1}{2}\|\mathbf{U}(\boldsymbol{\sigma}) - \mathbf{V}\|^2 + \alpha \sum_k |\mathbf{L}_k \boldsymbol{\sigma}| \tag{11.89}$$

The above minimization is called the primal problem, and it is solved using iterative methods. The primal-dual interior point method is a popular method that is used to solve the TV minimization problem (Borsic et al., 2007).

In absolute imaging, the solution is highly dependent on the accuracy of the forward solution and measured voltages. Apart from the measurement noise, there are uncertainties due to the modeling errors associated with imprecise knowledge of electrodes position, the unknown shape of outer boundary, and contact impedances of electrodes (Kolehmainen, Lassas, & Ola, 2005). One way to tackle the modeling errors is by considering the uncertainty parameters as additional state variables and estimate simultaneously along with the conductivity distribution (Dardé, Hyvönen, Seppänen, & Staboulis, 2013; Heikkinen, Vilhunen, West, & Vauhkonen, 2002; Vilhunen, Kaipio, Vauhkonen, Savolainen, & Vauhkonen, 2002). The approximation error approach introduced by Kaipio and Somersalo (2007) considers the modeling errors as an auxiliary error along with the measurement noise. The Bayesian framework is then used to model the statistical parameters based on prior distribution models for conductivity and auxiliary parameters. The most often and commonly used method is the linear difference imaging (LDI) approach, which is robust to modeling errors to an extent. Difference imaging estimates the conductivity change $\delta\boldsymbol{\sigma}$ based on the measurements $\mathbf{V}_1$ and $\mathbf{V}_2$ that associated with the data before and after the conductivity change at different time intervals

$$\mathbf{V}_i = U(\boldsymbol{\sigma}_i) + \mathbf{e}_i,\ i = 1, 2, \tag{11.90}$$

where $\mathbf{e}_i$ is the Gaussian distributed measurement noise. Linearizing Eq. (11.90) and then subtracting, we have the expression for difference voltage data as

$$\mathbf{V}_2 - \mathbf{V}_1 = \delta\mathbf{V} = \mathbf{J}\delta\boldsymbol{\sigma} + \delta\mathbf{e}, \tag{11.91}$$

where $\delta\mathbf{V} = \mathbf{V}_2 - \mathbf{V}_1$, $\mathbf{U}(\boldsymbol{\sigma}_2) - \mathbf{U}(\boldsymbol{\sigma}_1) \cong \mathbf{J}(\boldsymbol{\sigma}_2 - \boldsymbol{\sigma}_1) = \mathbf{J}\delta\boldsymbol{\sigma}$, and $\delta\mathbf{e} = \mathbf{e}_2 - \mathbf{e}_1$. The resistivity change from the difference voltage data can be estimated by minimizing the cost function

$$\Phi(\boldsymbol{\sigma}) = \operatorname{argmin}\left\{\|\delta\mathbf{V} - \mathbf{J}\delta\boldsymbol{\sigma}\|^2 + \alpha\|\mathbf{R}\delta\boldsymbol{\sigma}\|^2\right\}, \tag{11.92}$$

and assuming no prior knowledge about the conductivity, the solution for conductivity change is obtained as

$$\delta\boldsymbol{\sigma} = \left(\mathbf{J}^T\mathbf{J} + \alpha\mathbf{R}^T\mathbf{R}\right)^{-1}\mathbf{J}^T\delta\mathbf{V} \tag{11.93}$$

Difference imaging is commonly used in medical applications to study the temporal phenomena; for example, the impedance changes during respiration for lung imaging. LDI using the difference data measured from the same system can compensate for part of the modeling errors. The linearity assumption in LDI is only satisfied for a small change in conductivity, and if the conductivity contrast is high, then the LDI will produce unsatisfactory performance. Nonlinear difference imaging (NDI), which is introduced recently, estimates the initial conductivity and the change simultaneously. NDI method exhibits improved reconstruction performance and is robust to modeling errors (Khambampati, Liu, Konki, & Kim, 2018; Liu, Kolehmainen, Siltanen, & Seppanen, 2015a,b). The conductivity parameterization restricts the conductivity change to a region of interest (ROI) inside the object and allows modeling independently background and ROI regions (Liu et al., 2015a)

$$\boldsymbol{\sigma}_2 = \boldsymbol{\sigma}_1 + \kappa\delta\boldsymbol{\sigma}_{\mathrm{ROI}}, \tag{11.94}$$

where κ is the mapping operator such that

$$\kappa\delta\boldsymbol{\sigma}_{\mathrm{ROI}} = \begin{cases} \delta\boldsymbol{\sigma}_{\mathrm{ROI}}, & x \in \Omega_{\mathrm{ROI}} \\ 0, & x \in \Omega/\Omega_{\mathrm{ROI}} \end{cases}. \tag{11.95}$$

Based on the parameterization using Eq. (11.94), the measurement $\mathbf{V}_2$ can be written as

$$V_2 = U(\sigma_1 + \kappa\delta\sigma) + e_2 \tag{11.96}$$

Augmenting the measurements $\mathbf{V}_1$ and $\mathbf{V}_2$, we have the new observation model for NDI

$$\overline{\mathbf{V}} = \overline{\mathbf{U}}(\overline{\boldsymbol{\sigma}}) + \overline{\mathbf{e}} \tag{11.97}$$

where $\overline{\boldsymbol{\sigma}} = \begin{bmatrix} \boldsymbol{\sigma}_1 \\ \delta\boldsymbol{\sigma} \end{bmatrix}$ is the initial and change in resistivity to be estimated and it is obtained by minimizing the cost functional given below

$$\widehat{\overline{\boldsymbol{\sigma}}} = \operatorname{argmin}\left\{ \left\| \overline{\mathbf{V}} - \overline{\mathbf{U}}(\overline{\sigma}) \right\|^2 + \mathbf{R}(\overline{\boldsymbol{\sigma}}) \right\} \tag{11.98}$$

where the regularization function $\mathbf{R}(\overline{\boldsymbol{\sigma}})$ has the form $\mathbf{R}(\overline{\boldsymbol{\sigma}}) = \mathbf{R}(\boldsymbol{\sigma}_1) + \mathbf{R}(\delta\boldsymbol{\sigma}_{\mathrm{ROI}})$.

11.3.5 Dynamic algorithms

For EIT image reconstruction, a number of boundary condition sets are applied, and the resulting electric signals are measured sequentially. To get one frame of an image, this sequential process of data collection is needed, and it consumes time. If the internal configuration is stationary during the data collection, the measured data sets are for the same object. In process imaging, however, the internal configuration may be varying during the data collection since the process may proceed. Then, the measured data at each boundary condition cannot be said for the same internal configuration, and it is expected that the image reconstruction performance becomes worse as the characteristic time of the process becomes shorter than the data collection time (Bar-Shalom & Li, 1983; Kaipio, Karjalainen, Somersalo, & Vauhkonen, 1999; Kim Kim, Kim, Lee, & Vauhkonen, 2001). Compared with the static imaging where the internal configuration is stationary during the data collection time required to get one frame of an image, the dynamic imaging where the internal configuration varies within the data collection time requires different approaches.

For the dynamic imaging, the Kalman filter algorithm and its variants are plausible (Khambampati, Rashid, Kim, Soleimani, & Kim, 2009; Kim, Kim, Kim, Lee, & Vauhkonen, 2001; Kim et al., 2006; Trigo, Gonzalez-Lima, & Amato, 2004; Vauhkonen, Karjalainen, & Kaipio, 1998). The Kalman filter, well known in the control theory, was devised to estimate the state of a discrete-time controlled process governed by the linear stochastic difference equation with measurements. The Kalman filter provides a recursive procedure to estimate the state based on the previous states and the measurements (Kim Kim, Kim, Lee, & Vauhkonen, 2001, Kim et al., 2006). In the sense of dynamic process, the conductivity (or permittivity) is considered as a state variable evolving in time. And, at each time step (or data collection), a new set of measured data is available. Hence, the state transition model for the conductivity $\boldsymbol{\sigma}\in \Re^{N\times 1}$ in EIT can be written as (Kaipio & Somersalo, 1999; Kim Kim, Kim, Lee, & Vauhkonen, 2001; Seppanen, Vauhkonen, Somersalo, & Kaipio, 2001a)

$$\boldsymbol{\sigma}_k = \mathbf{F}_{k-1}\boldsymbol{\sigma}_{k-1} + \mathbf{w}_{k-1} \tag{11.99}$$

with the measurement vector, or measured electrode voltage vector, $\mathbf{V}\in \Re^{P\times 1}$ such that

$$\mathbf{V}_k = U_k(\boldsymbol{\sigma}_k) + \mathbf{v}_k. \tag{11.100}$$

The subscript "k" denotes the time step. The system matrix $\mathbf{F}_k\in \Re^{N\times N}$ and the observation model U_k are determined from the physics of the problem considered. In the EIT problem, the conductivity evolution is governed by the motion of the substances within the problem domain. If we are imaging multiphase flows or mixing process, the flow field or the concentration distribution evolves according to physical laws, and proper physical models describing the evolution can be used as the transition model (Seppanen et al 2001b; Vauhkonen et al 1998; Kaipio and somersalo 1999). In case such a model is not available, a so-called random work model is used; $\mathbf{F}_k$ is set to the identity matrix, and it means that the state evolves randomly. The observation model can be the numerical solution of the governing equation. The random variables $\mathbf{w}_k\in \Re^{N\times 1}$ and $\mathbf{v}_k\in \Re^{P\times 1}$ denote the process and the measurement noise, respectively. They are assumed to be white noises and independent of each other with normal probability distributions.

$$p(\mathbf{w}_k) \sim N(\mathbf{0}, \mathbf{Q}_k) \text{ and } p(\mathbf{v}_k) \sim N(\mathbf{0}, \mathbf{R}_k), \tag{11.101}$$

where $\mathbf{Q}_k\in \Re^{N\times N}$ and $\mathbf{R}_k\in \Re^{P\times P}$ are the process and the measurement noise covariance, respectively. Usually, it is assumed that both state variables and measured data are independent of each other, and both covariance matrices are diagonal. Now, let us derive a recursive procedure to estimate $\boldsymbol{\sigma}_k$ with the previous states $\boldsymbol{\sigma}_1, \cdots, \boldsymbol{\sigma}_{k-1}$ and the measurements $\mathbf{V}_1, \cdots, \mathbf{V}_k$. Set $\boldsymbol{\sigma}_{k|k-1}\in \Re^{N\times 1}$ to be a priori state estimate at step k given knowledge of the process up to step $k-1$, and $\boldsymbol{\sigma}_{k|k}\in \Re^{N\times 1}$ to be a posteriori state estimate at step k given measurement $\mathbf{V}_k$. There are two different types of Kalman filter formulation, depending on how the nonlinear measurement Eq. (11.100) is linearized. First is the extended Kalman filter (EKF) where the observation model is linearized about a priori state estimate at step k, $\boldsymbol{\sigma}_{k|k-1} \in \Re^{N \times 1}$:

$$\begin{aligned}\mathbf{V}_k &= U_k\left(\boldsymbol{\sigma}_{k|k-1}\right) + \frac{\partial U_k\left(\boldsymbol{\sigma}_{k|k-1}\right)}{\partial \boldsymbol{\sigma}_k}\left(\boldsymbol{\sigma}_k - \boldsymbol{\sigma}_{k|k-1}\right) + \mathbf{v}_k + \mathbf{HOT} \\ &= \mathbf{V}_{k|k-1} + \mathbf{J}_k\left(\boldsymbol{\sigma}_{k|k-1}\right)\left(\boldsymbol{\sigma}_k - \boldsymbol{\sigma}_{k|k-1}\right) + \mathbf{v}_k + \mathbf{HOT},\end{aligned} \tag{11.102}$$

where **HOT** denotes higher-order terms. Let us define the pseudomeasurement $\mathbf{y}_k \in \Re^{P \times 1}$:

$$\mathbf{y}_k \equiv \mathbf{V}_k - U_k\left(\boldsymbol{\sigma}_{k|k-1}\right) + \mathbf{J}_k\left(\boldsymbol{\sigma}_{k|k-1}\right)\boldsymbol{\sigma}_{k|k-1} = \mathbf{J}_k\left(\boldsymbol{\sigma}_{k|k-1}\right)\boldsymbol{\sigma}_k + \widetilde{\mathbf{v}}_k, \tag{11.103}$$

where $\widetilde{\mathbf{v}}_k \in \Re^{P \times 1}$ is the pseudomeasurement noise $\widetilde{\mathbf{v}}_k = \mathbf{v}_k + \mathbf{HOT}$ with zero mean and known covariance $\widetilde{\mathbf{R}}_k \in \Re^{P \times P}$:

$$E\left[\widetilde{\mathbf{v}}_k\right] = \mathbf{0} \text{ and } \widetilde{\mathbf{R}}_k = \operatorname{cov}(\mathbf{v}_k + \mathbf{HOT}). \tag{11.104}$$

The Jacobians $\mathbf{J}_k \in \Re^{P \times N}$ will be defined as

$$\mathbf{J}_k\left(\boldsymbol{\sigma}_{k|k-1}\right) = \left.\frac{\partial U_k(\boldsymbol{\sigma}_k)}{\partial \boldsymbol{\sigma}_k}\right|_{\boldsymbol{\sigma}_k = \boldsymbol{\sigma}_{k|k-1}}. \tag{11.105}$$

The second variant of nonlinear Kalman filter is the linearized Kalman filter (LKF), where the observation model is linearized about a reference or nominal state ($\boldsymbol{\sigma}_0$), i.e.,

$$\begin{aligned}\mathbf{V}_k &= U_k(\boldsymbol{\sigma}_0) + \frac{\partial U_k(\boldsymbol{\sigma}_0)}{\partial \boldsymbol{\sigma}_k}(\boldsymbol{\sigma}_k - \boldsymbol{\sigma}_0) + \mathbf{v}_k + \mathbf{HOT}, \\ &= U_k(\boldsymbol{\sigma}_0) + \mathbf{J}_k(\boldsymbol{\sigma}_0)(\boldsymbol{\sigma}_k - \boldsymbol{\sigma}_0) + \mathbf{v}_k + \mathbf{HOT}.\end{aligned} \tag{11.106}$$

Using Eq. (11.106), the pseudomeasurement equation for LKF has the form

$$\mathbf{y}_k \equiv \mathbf{V}_k - U_k(\boldsymbol{\sigma}_0) + \mathbf{J}_k(\boldsymbol{\sigma}_0)\boldsymbol{\sigma}_0 = \mathbf{J}_k(\boldsymbol{\sigma}_0)\boldsymbol{\sigma}_k + \widetilde{\mathbf{v}}_k, \tag{11.107}$$

where the Jacobian in Eq. (11.107) is defined as

$$\mathbf{J}_k(\boldsymbol{\sigma}_0) = \left.\frac{\partial U_k(\boldsymbol{\sigma}_k)}{\partial \boldsymbol{\sigma}_k}\right|_{\boldsymbol{\sigma}_k = \boldsymbol{\sigma}_0}. \tag{11.108}$$

In LKF, the Jacobian $\mathbf{J}_k$ is constant at each iteration step, whereas, in EKF, the Jacobian is evaluated with respect to a priori state estimate $\boldsymbol{\sigma}_{k|k-1}$ and updated at each iteration step. Therefore, in situations when the reference state is not known with good accuracy, EKF has better estimation performance as compared to LKF. Here, we derive the EKF formulation to estimate the state variable $\boldsymbol{\sigma}_k$ from measurements $\mathbf{y}_k$. Let us consider the a priori state estimate at step k, $\boldsymbol{\sigma}_{k|k-1}$, and the pseudomeasurement estimate at step k, $\mathbf{y}_{k|k}$, will be

$$\begin{aligned}\boldsymbol{\sigma}_{k|k-1} &= E[\boldsymbol{\sigma}_k|\mathbf{y}_{k-1}] = E[\mathbf{F}_{k-1}\boldsymbol{\sigma}_{k-1} + \mathbf{w}_k|\mathbf{y}_{k-1}] = \mathbf{F}_{k-1}E[\boldsymbol{\sigma}_{k-1}|\mathbf{y}_{k-1}], \\ &= \mathbf{F}_{k-1}\boldsymbol{\sigma}_{k-1|k-1}\end{aligned} \tag{11.109}$$

$$\mathbf{y}_{k|k} = E[\mathbf{y}_k|\boldsymbol{\sigma}_k] = E\left[\mathbf{J}_k\boldsymbol{\sigma}_k + \widetilde{\mathbf{v}}_k \middle| \boldsymbol{\sigma}_k\right] = \mathbf{J}_k\boldsymbol{\sigma}_{k|k}. \tag{11.110}$$

The predicted error $\mathbf{e}_{k|k-1} \in \Re^{N\times 1}$ and its covariance $\mathbf{C}_{k|k-1} \in \Re^{N\times N}$ will be

$$\mathbf{e}_{k|k-1} = \boldsymbol{\sigma}_k - \boldsymbol{\sigma}_{k|k-1} = \mathbf{F}_{k-1}\mathbf{e}_{k-1|k-1} + \mathbf{w}_{k-1}, \tag{11.111}$$

$$\mathbf{C}_{k|k-1} = \mathrm{cov}\left(\mathbf{e}_{k|k-1}\right) = \mathbf{F}_{k-1}\mathbf{C}_{k-1|k-1}\mathbf{F}_{k-1}^T + \mathbf{Q}_{k-1}. \tag{11.112}$$

Let the updated state estimate $\boldsymbol{\sigma}_{k|k}$ be the state maximizing the probability density $\boldsymbol{\sigma}_k$ given $\mathbf{y}_k$

$$p(\boldsymbol{\sigma}_k|\mathbf{y}_k) \sim p(\boldsymbol{\sigma}_k|\mathbf{y}_{k-1})p(\mathbf{y}_k|\boldsymbol{\sigma}_k) \tag{11.113}$$

or minimizing the following cost function that corresponds to the exponent of the probability density function

$$\Phi(\boldsymbol{\sigma}_k) = \left\|\boldsymbol{\sigma}_k - \boldsymbol{\sigma}_{k|k-1}\right\|_{\mathbf{C}_{k|k-1}^{-1}} + \|\mathbf{y}_k - \mathbf{J}_k\boldsymbol{\sigma}_k\|_{\widetilde{\mathbf{R}}_k^{-1}}, \tag{11.114}$$

where $\|x\|_R = x^T Rx$. Incase regularization is considered, the cost function will be of the form

$$\Phi(\boldsymbol{\sigma}_k) = \left\|\boldsymbol{\sigma}_k - \boldsymbol{\sigma}_{k|k-1}\right\|_{\mathbf{C}_{k|k-1}^{-1}} + \|\mathbf{y}_k - \mathbf{J}_k\boldsymbol{\sigma}_k\|_{\widetilde{\mathbf{R}}_k^{-1}} + \alpha\|\mathbf{R}(\boldsymbol{\sigma}_k - \boldsymbol{\sigma}^*)\|, \tag{11.115}$$

where α is the regularization parameter and $\mathbf{R} \in \Re^{N\times N}$ is the regularization matrix. $\boldsymbol{\sigma}^* \in \Re^{N\times 1}$ is the prior information for the conductivity distribution if it is available. If we define the augmented pseudomeasurement $\widetilde{\mathbf{y}}_k \in \Re^{(P+N)\times 1}$ and the augmented pseudoobservation matrix $\mathbf{H}_k \in \Re^{(P+N)\times N}$ as

$$\widetilde{\mathbf{y}}_k = \begin{pmatrix} \mathbf{y}_k \\ \sqrt{\alpha}\mathbf{R}\boldsymbol{\sigma}^* \end{pmatrix} \text{ and } \mathbf{H}_k = \begin{pmatrix} \mathbf{J}_k\left(\boldsymbol{\sigma}_{k|k-1}\right) \\ \sqrt{\alpha}\mathbf{R} \end{pmatrix}, \tag{11.116}$$

then the cost function can be rearranged as

$$\Phi(\boldsymbol{\sigma}_k) = \left\|\boldsymbol{\sigma}_k - \boldsymbol{\sigma}_{k|k-1}\right\|_{\mathbf{C}_{k|k-1}^{-1}} + \left\|\widetilde{\mathbf{y}}_k - \mathbf{H}_k\boldsymbol{\sigma}_k\right\|_{\boldsymbol{\Gamma}_k^{-1}}. \tag{11.117}$$

The augmented measurement model can be written as

$$\widetilde{\mathbf{y}}_k = \mathbf{H}_k\boldsymbol{\sigma}_k + \gamma_k, \tag{11.118}$$

where the augmented measurement noise $\gamma_k \in \Re^{(P+N)\times 1}$ is assumed to have zero mean and its covariance $\boldsymbol{\Gamma}_k \in \Re^{(P+N)\times(P+N)}$ is a block diagonal matrix if the pseudomeasurement noise covariance $\widetilde{\mathbf{R}}_k$ is a diagonal matrix:

$$E[\gamma_k] = \mathbf{0} \text{ and } \boldsymbol{\Gamma}_k \equiv \text{cov}(\gamma_k) = \text{Blockdiag}\left[\widetilde{\mathbf{R}}_k, \mathbf{I}_{N\times N}\right]. \tag{11.119}$$

Then, by differentiating $\Phi(\boldsymbol{\sigma}_k)$ with respect to $\boldsymbol{\sigma}_k$ and setting it to zero

$$\frac{\partial\Phi(\boldsymbol{\sigma}_k)}{\partial\boldsymbol{\sigma}_k} = \mathbf{C}_{k|k-1}^{-1}\left(\boldsymbol{\sigma}_k - \boldsymbol{\sigma}_{k|k-1}\right) - \mathbf{H}_k^T\boldsymbol{\Gamma}_k^{-1}\left(\widetilde{\mathbf{y}}_k - \mathbf{H}_k\boldsymbol{\sigma}_k\right) = \mathbf{0}, \tag{11.120}$$

the updated state estimate $\boldsymbol{\sigma}_{k|k}$ is obtained as

$$\boldsymbol{\sigma}_{k|k} = \left(\mathbf{H}_k^T\boldsymbol{\Gamma}_k^{-1}\mathbf{H}_k + \mathbf{C}_{k|k-1}^{-1}\right)^{-1}\left(\mathbf{H}_k^T\boldsymbol{\Gamma}_k^{-1}\widetilde{\mathbf{y}}_k + \mathbf{C}_{k|k-1}^{-1}\boldsymbol{\sigma}_{k|k-1}\right). \tag{11.121}$$

On the other hand, the augmented measurement error $\boldsymbol{\varepsilon}_k \in \Re^{P\times 1}$ and its covariance $\mathbf{S}_k \in \Re^{P\times P}$ are

$$\boldsymbol{\varepsilon}_k = \widetilde{\mathbf{y}}_k - \widetilde{\mathbf{y}}_{k|k-1} = \widetilde{\mathbf{y}}_k - \mathbf{H}_k\boldsymbol{\sigma}_{k|k-1} = \mathbf{H}_k\mathbf{e}_{k|k-1} + \gamma_k, \tag{11.122}$$

$$\mathbf{S}_k = \text{cov}(\boldsymbol{\varepsilon}_k) = \mathbf{H}_k\mathbf{C}_{k|k-1}\mathbf{H}_k^T + \boldsymbol{\Gamma}_k. \tag{11.123}$$

If we set the posterior estimate as a linear combination of the prior estimate $\boldsymbol{\sigma}_{k|k-1}$ and the weighted measurement residual of $\boldsymbol{\varepsilon}_k$

$$\boldsymbol{\sigma}_{k|k} = \boldsymbol{\sigma}_{k|k-1} + \mathbf{K}_k\boldsymbol{\varepsilon}_k = \boldsymbol{\sigma}_{k|k-1} + \mathbf{K}_k\left(\widetilde{\mathbf{y}}_k - \mathbf{H}_k\boldsymbol{\sigma}_{k|k-1}\right), \tag{11.124}$$

where $\mathbf{K}_k \in \Re^{N \times P}$ is the Kalman gain. The Kalman gain can be determined by minimizing the a posteriori error covariance of the a posteriori estimate error:

$$\begin{aligned}\mathbf{e}_{k|k} &= \boldsymbol{\sigma}_k - \boldsymbol{\sigma}_{k|k} = \boldsymbol{\sigma}_k - \boldsymbol{\sigma}_{k|k-1} - \mathbf{K}_k\left(\widetilde{\mathbf{y}}_k - \widetilde{\mathbf{y}}_{k|k-1}\right) \\ &= \mathbf{e}_{k|k-1} - \mathbf{K}_k\left(\mathbf{H}_k\mathbf{e}_{k|k-1} + \gamma_k\right) = (\mathbf{I} - \mathbf{K}_k\mathbf{H}_k)\mathbf{e}_{k|k-1} - \mathbf{K}_k\gamma_k.\end{aligned} \tag{11.125}$$

Its covariance will be

$$\mathbf{C}_{k|k} = (\mathbf{I} - \mathbf{K}_k\mathbf{H}_k)\mathbf{C}_{k|k-1}(\mathbf{I} - \mathbf{K}_k\mathbf{H}_k)^T + \mathbf{K}_k\boldsymbol{\Gamma}_k\mathbf{K}_k^T = (\mathbf{I} - \mathbf{K}_k\mathbf{H}_k)\mathbf{C}_{k|k-1}, \tag{11.126}$$

so that from $\frac{\partial \mathbf{C}_{k|k}}{\partial \mathbf{K}_k} = \mathbf{0}$, we can have the Kalman gain as

$$\mathbf{K}_k = \mathbf{C}_{k|k-1}\mathbf{H}_k^T\left(\mathbf{H}_k\mathbf{C}_{k|k-1}\mathbf{H}_k^T + \boldsymbol{\Gamma}_k\right)^{-1} = \mathbf{C}_{k|k-1}\mathbf{H}_k^T\mathbf{S}_k^{-1}. \tag{11.127}$$

With the Kalman gain, the covariance becomes

$$\mathbf{C}_{k|k} = (\mathbf{I} - \mathbf{K}_k\mathbf{H}_k)\mathbf{C}_{k|k-1}, \tag{11.128}$$

Finally, a full set of recursive equations to estimate the state variable is as follows:
Time Update Equations

$$\text{Predicted (a priori) state}: \ \boldsymbol{\sigma}_{k|k-1} = \mathbf{F}_{k-1}\boldsymbol{\sigma}_{k-1|k-1} \tag{11.129}$$

$$\text{Predicted (a priori) error covariance}: \ \mathbf{C}_{k|k-1} = \mathbf{F}_{k-1}\mathbf{C}_{k-1|k-1}\mathbf{F}_{k-1}^T + \mathbf{Q}_{k-1} \tag{11.130}$$

$$\text{Jacobian}: \ \mathbf{J}_k = \frac{\partial U_k\left(\boldsymbol{\sigma}_{k|k-1}\right)}{\partial \boldsymbol{\sigma}_k} \tag{11.131}$$

$$\text{Augmented observation model } \widetilde{\mathbf{y}}_k = \begin{pmatrix} \mathbf{y}_k \\ \sqrt{\alpha}\mathbf{R}\boldsymbol{\sigma}^* \end{pmatrix}, \ \mathbf{H}_k = \begin{pmatrix} \mathbf{J}_k\left(\boldsymbol{\sigma}_{k|k-1}\right) \\ \sqrt{\alpha}\mathbf{R} \end{pmatrix} \tag{11.132}$$

$$\text{Augmented measurement noise covariance}: \ \boldsymbol{\Gamma}_k = \text{Blockdiag}\left[\widetilde{\mathbf{R}}_k, \mathbf{I}_{N\times N}\right] \tag{11.133}$$

$$\text{Predicted measurement error}: \ \boldsymbol{\varepsilon}_k = \widetilde{\mathbf{y}}_k - \mathbf{H}_k \boldsymbol{\sigma}_{k|k-1} \tag{11.134}$$

$$\text{Predicted measurement error covariance}: \ \mathbf{S}_k = \mathbf{H}_k \mathbf{C}_{k|k-1} \mathbf{H}_k^T + \boldsymbol{\Gamma}_k \tag{11.135}$$

Measurement Update Equations

$$\text{Kalman gain}: \ \mathbf{K}_k = \mathbf{C}_{k|k-1} \mathbf{H}_k^T \mathbf{S}_k^{-1} \tag{11.136}$$

$$\text{Updated (a posteriori) state}: \ \boldsymbol{\sigma}_{k|k} = \boldsymbol{\sigma}_{k|k-1} + \mathbf{K}_k \boldsymbol{\varepsilon}_k \tag{11.137}$$

$$\text{Updated (a posteriori) error covariance}: \ \mathbf{C}_{k|k} = (\mathbf{I} - \mathbf{K}_k \mathbf{H}_k) \mathbf{C}_{k|k-1} \tag{11.138}$$

11.4 Diffraction tomography

11.4.1 Mathematical formulation of diffraction tomography

Diffraction is related to the spreading of waves upon interaction with an object in the medium. DT is an inverse scattering technique to reconstruct the object characteristics from the scattered field data measured at receiver locations. DT considers both scattering as well as absorption effects, whereas, in transmission tomography, the absorption effect is assumed to be dominant. The power or energy of X- or γ-ray used in transmission tomography techniques is very high; therefore, the ray path can be regarded as a straight line. Ultrasound and microwave imaging involves transmitting low-power nonionizing waves to propagate through the medium; therefore, the path may not be straight, and they come under the category of DT (Fig. 11.8). Ultrasound imaging uses a transducer that converts the exciting electrical signal into mechanical disturbance (pressure wave), and vice versa. The attenuation coefficient, sound speed, and reflection coefficient are estimated from the scattered field measurements in ultrasound. Microwave imaging consists of an antenna that converts the electrical signal

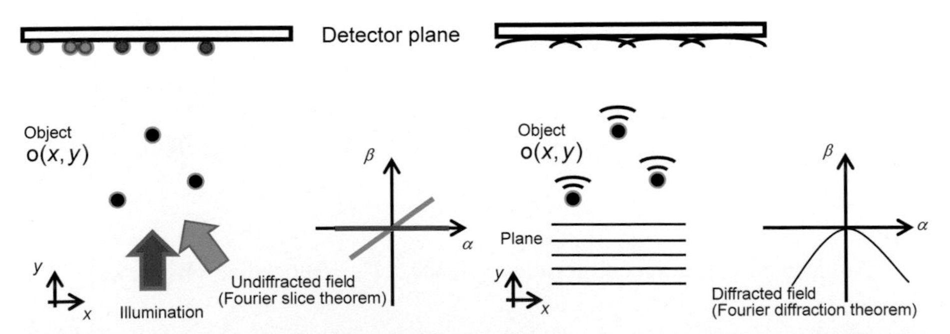

Figure 11.8 Schematic diagram that explains the difference between projection and diffraction tomography.

into low power electromagnetic microwaves and vice versa. Complex permittivity distribution is reconstructed from the received scattered microwaves that give us the spatial distribution within the medium. As ultrasound and microwaves get refracted and diffracted while passing through the objects, more sophisticated algorithms as compared to transmission tomography are required. Here, the reconstruction algorithms that are usually used in DT.

If the object is relatively large and has a small refractive index, then it can be assumed that the waves generated from ultrasound or microwave travel in straight lines after interaction with the object. In this case, the Fourier slice theorem that was discussed in Section 11.2.3 can be used to reconstruct the refractive index (Greenleaf, Johnson, Lee, Herman, & Wood, 1974, pp. 591–603). However, for the case when the transmitted waves undergo refraction and diffraction, methods based on ray tracing (Andersen & Kak, 1982; Mueller, Kaveh, & Wade, 1979) and solving wave equations are usually employed. Ray tracing methods that consider only refraction of waves are not so accurate, especially when the change in refractive index is more than 10% (Andersen & Kak, 1984b). Methods based on solving wave equations are more accurate and are considered in solving DT problem.

Consider a plane wave insonified on the object such that the propagation of waves can be modeled with Helmholtz's wave equation. For a loss-free object with constant density, the waves transmitted with field magnitude $u(\mathbf{r})$ and wave number $k(\mathbf{r})$ satisfy Helmholtz's equation

$$\left[\nabla^2 + k^2(\mathbf{r})\right]u(\mathbf{r}) = 0, \tag{11.139}$$

where the wavenumber $k(\mathbf{r})$ with respect to position $\mathbf{r}$ is defined as

$$k(\mathbf{r}) = \frac{2\pi}{\lambda} = \frac{2\pi\omega}{c}, \tag{11.140}$$

where ω is the temporal frequency, λ, c are the field wavelength, and velocity, respectively. For acoustic or ultrasound tomography, the field $u(\mathbf{r})$ is the pressure field, and for microwave tomography, it is a complex amplitude of the electric field. In the homogeneous conditions, the complex wavenumber is constant, i.e., $k(\mathbf{r}) = k_0$, therefore the wave equation in Eq. (11.139) becomes

$$\left[\nabla^2 + k_0^2\right]u(\mathbf{r}) = 0. \tag{11.141}$$

Rewriting Eq. (11.139) as

$$\left[\nabla^2 + k_0^2\right]u(\mathbf{r}) = -\left[k^2(\mathbf{r}) - k_0^2\right]u(\mathbf{r}). \tag{11.142}$$

In electromagnetic case, if the polarization effects are neglected, the wavenumber is related to the complex refractive index $n(\mathbf{r})$ given by

$$k(\mathbf{r}) = k_0 n(\mathbf{r}) = k_0[1 + n_\delta(\mathbf{r})], \tag{11.143}$$

where $n_\delta(\mathbf{r})$ is the deviation from the average refractive index. The electromagnetic refractive index is defined as (Kak, 1984)

$$n(\mathbf{r}) = \sqrt{\frac{\mu(\mathbf{r})\varepsilon(\mathbf{r})}{\mu_0 \varepsilon_0}}, \tag{11.144}$$

where $\mu(\mathbf{r})$, $\varepsilon(\mathbf{r})$ are the spatially varying magnetic permeability and permittivity inside the object and μ_0, ε_0 are corresponding to the homogeneous medium. Also, in ultrasound, the complex refractive index is defined as the ratio of the propagation velocity of medium in which object is immersed to the spatially varying velocity in the object $c(\mathbf{r})$, i.e., (Kak, 1984)

$$n(\mathbf{r}) = \frac{c_0}{c(\mathbf{r})}. \tag{11.145}$$

The general wave equation that can be modeled for both ultrasound and microwave tomography can be written as

$$\left[\nabla^2 + k_0^2\right] u(\mathbf{r}) = -k_0^2\left[n(\mathbf{r})^2 - 1\right] u(\mathbf{r}). \tag{11.146}$$

Substituting $n(\mathbf{r}) = [1 + n_\delta(\mathbf{r})]$ in the right-hand side of Eq. (11.146) and neglecting the higher-order terms considering $n_\delta(\mathbf{r}) << 1$ we have

$$\left[\nabla^2 + k_0^2\right] u(\mathbf{r}) = -2k_0^2 n_\delta(\mathbf{r}) u(\mathbf{r}). \tag{11.147}$$

Defining the object function, $o(\mathbf{r}) = -2k_0^2 n_\delta(\mathbf{r})$, the above equation is simplified as

$$\left[\nabla^2 + k_0^2\right] u(\mathbf{r}) = -o(\mathbf{r}) u(\mathbf{r}). \tag{11.148}$$

Let us consider the field $u(\mathbf{r})$ is composed of incident field $u_0(\mathbf{r})$ without any inhomogeneities and scattered field $u_s(\mathbf{r})$ due to the presence of inhomogeneities

$$u(\mathbf{r}) = u_0(\mathbf{r}) + u_s(\mathbf{r}). \tag{11.149}$$

Substituting Eq. (11.149) into Eq. (11.148), we have the inhomogeneous wave equation for scattered field

$$\left[\nabla^2 + k_0^2\right] u_s(\mathbf{r}) = -o(\mathbf{r}) u(\mathbf{r}). \tag{11.150}$$

The inhomogeneous Helmholtz wave equation given above cannot be solved directly for the scattered field rather; its solution can be obtained in terms of Greens function (Morse & Feshbach, 1953)

$$\left[\nabla^2 + k_0^2\right] g(\mathbf{r}|\mathbf{r}') = -\delta(\mathbf{r} - \mathbf{r}'). \tag{11.151}$$

The solution of the two-dimensional Helmholtz equation can be expressed in terms of zero-order Henkel function of the first kind, i.e.,

$$g(\mathbf{r}|\mathbf{r}') = \frac{j}{4} H_0^1(k_0 R). \tag{11.152}$$

with

$$R = |\mathbf{r} - \mathbf{r}'|. \tag{11.153}$$

The forcing function on the right-hand side of Eq. (11.150) can be considered as the summation of an array of impulses

$$o(\mathbf{r})u(\mathbf{r}) = \int o(\mathbf{r}')u(\mathbf{r}')\delta(\mathbf{r} - \mathbf{r}')d\mathbf{r}'. \tag{11.154}$$

Performing simple convolution, the total radiation from all sources on the right-hand side of Eq. (11.148) is given by the following superposition

$$u_s(\mathbf{r}) = \int g(\mathbf{r} - \mathbf{r}')o(\mathbf{r}')u(\mathbf{r}')d\mathbf{r}'. \tag{11.155}$$

The solution to the scattered field given in Eq. (11.155) is not straight forward as it is nonlinear. Approximation methods based on perturbation methods are usually used and are discussed in the succeeding subsections.

11.4.2 Born and Rytov approximations for weakly scattered objects

Here, approximation methods called Born and Rytov are formulated that provide the solution for the scattered field (Iwata & Nagata, 1974; Kaveh, Soumekh, & Muller, 1982). They serve as the basis for the Fourier diffraction theorem that will be discussed in the next subsection. At first, let us formulate the solution for the scattered field using Born approximation. Recalling that the definition of the total field which is the sum of the incident field and scattered field and substituting in Eq. (11.155), we have (Born & Wolf, 1999; Kak & Slaney, 1988; Oppengeim & Schafer, 1975; Wolf, 1969)

$$u_s(\mathbf{r}) = \int g(\mathbf{r} - \mathbf{r}')o(\mathbf{r}')u_0(\mathbf{r}')d\mathbf{r}' + \int g(\mathbf{r} - \mathbf{r}')o(\mathbf{r}')u_s(\mathbf{r}')d\mathbf{r}'. \tag{11.156}$$

If the scattered field effect is very small compared to the incident field, then the above equation for the scattered field is called Born scattered field, which is given by

$$u_s(\mathbf{r}) \cong u_B(\mathbf{r}) = \int g(\mathbf{r}-\mathbf{r}')o(\mathbf{r}')u_0(\mathbf{r}')d\mathbf{r}'. \tag{11.157}$$

This Born approximation is valid when the magnitude of the scattered field is less compared to the incident field (Born & Wolf, 1999; Newton, 1966).

In formulating a solution using Rytov approximation, at first let us consider that the total field is represented by a complex phase Ψ (Ishimaru, 1978)

$$u(\mathbf{r}) = e^{\Psi(\mathbf{r})} \tag{11.158}$$

and the total phase is equal to incident phase function and scattered phase function

$$\Psi(\mathbf{r}) = \Psi_0(\mathbf{r}) + \Psi_s(\mathbf{r}), \tag{11.159}$$

where

$$u_0(\mathbf{r}) = e^{\Psi_0(\mathbf{r})}. \tag{11.160}$$

Using the above relation in the wave equation, we can get

$$\left[\nabla^2 + k_0^2\right]u_0\Psi_s = -u_0\left[(\nabla\Psi_s)^2 - o(\mathbf{r})\right]. \tag{11.161}$$

The solution to this equation can be represented in terms of Greens function as before to have

$$u_0\Psi_s = \int g(\mathbf{r}-\mathbf{r}')u_0\left[(\nabla\Psi_s)^2 - o(\mathbf{r}')\right]d\mathbf{r}'. \tag{11.162}$$

Using the Rytov approximation, the terms in the brackets are approximated as

$$(\nabla\Psi_s)^2 - o(\mathbf{r}) \cong -o(\mathbf{r}). \tag{11.163}$$

Under the above approximation, the first-order Rytov approximation for scattered phase is represented as

$$\Psi_s(\mathbf{r}) = \frac{1}{u_0(\mathbf{r})}\int g(\mathbf{r}-\mathbf{r}')u_0(\mathbf{r}')o(\mathbf{r}')d\mathbf{r}'. \tag{11.164}$$

From the Born approximation for scattered field Eq. (11.157), the above Eq. (11.164) can have the form

$$\Psi_s = \frac{u_B\left(\vec{r}\right)}{u_0\left(\vec{r}\right)}. \tag{11.165}$$

Although it appears both the Born and the Rytov solutions look similar, they are obtained using different approximations.

It is to be noted that Born approximation produces a better estimate of the scattered amplitude for large deviations in the refractive index for objects small in size. On the other hand, the Rytov approximation gives a more accurate estimate of the scattered phase for large-sized objects with small deviations in the refractive index (Andersen & Kak, 1984b; Keller, 1969). If the object is small and the refractive index of the object has less deviation from the background, both Born and Rytov approximations result in the same solution.

11.4.3 Fourier diffraction theorem

Fourier diffraction theorem relates the measured diffracted projections to the Fourier transform of the object function. Fourier diffraction theorem serves as the main basis for the formulation of the reconstruction methods in DT. The theorem states that when an object is illuminated with a plane wave, the Fourier transform of the forward scattered data measured at receiver plane TT′ gives us a 2D Fourier transform of object function $O(\omega_1, \omega_2)$ over a semicircular arc in the frequency domain. The object function can then be recovered from the Fourier inversion. It is only valid for weakly scattering objects. Here, the theoretical procedure to formulate the Fourier diffraction theorem is discussed.

At first, let us consider the scattered field of the object due to the incidence of a plane wave from Born approximation, which is given by

$$u_s(\mathbf{r}) \cong u_B(\mathbf{r}) = \int g(\mathbf{r}-\mathbf{r}')o(\mathbf{r}')u_0(\mathbf{r}')d\mathbf{r}'. \tag{11.166}$$

Let us define the Fourier transform pairs for terms in the above equation

$$\widehat{F}[o(\mathbf{r})] = O(\mathbf{\Lambda}),\ \widehat{F}[g(\mathbf{r})] = G(\mathbf{\Lambda}),\ \text{and}\ \widehat{F}[u(\mathbf{r})] = U(\mathbf{\Lambda}). \tag{11.167}$$

The scattered field from the first-order Born equation can be considered as the convolution of Greens function $g(\mathbf{r}-\mathbf{r}')$ and the product of object function $o(\mathbf{r}')$ and incident field $u_0(\mathbf{r}')$. The Born equation for the scattered field can be written in terms of the Fourier transform pairs as

$$U_S(\mathbf{\Lambda}) = G(\mathbf{\Lambda})\{O(\mathbf{\Lambda}) * U_0(\mathbf{\Lambda})\}, \tag{11.168}$$

where $\mathbf{\Lambda} = (\alpha, \gamma)$ and (α, γ) are the spatial frequencies along x and y directions, respectively. Considering a plane wave of the incident field $u_0(\mathbf{r}) = e^{j\mathbf{K}\cdot\mathbf{r}}$ and recalling

$$G(\mathbf{\Lambda}|\mathbf{r}') = \frac{e^{-j\mathbf{\Lambda}\cdot\mathbf{r}'}}{\Lambda^2 - k_0^2}, \tag{11.169}$$

the scattered field in Eq. (11.168) becomes

$$U_S(\mathbf{\Lambda}) = 2\pi\frac{O(\mathbf{\Lambda} - \mathbf{K})}{\Lambda^2 - k_0^2}. \tag{11.170}$$

From Eq. (11.152), it can be noticed that the Greens function has a singularity at the origin in the space domain; therefore, the function is approximated by a two-dimensional average of values near the singularity. Also, the Fourier transform representation of the Greens function can be interpreted such that the point scatter acts as both source and sink waves. If the outgoing waves are only considered, then it is considered as transmission DT, and in other cases for the waves traveling inward are related to reflection DT.

To evaluate the measured scattered field at the receiver line l_0, let us consider a plane wave traveling along the positive y-direction, i.e., $\mathbf{K} = (0, k_0)$. The scattered field measured at the receiver line $y = l_0$ is computed from the inverse Fourier transform:

$$u_s(x, y = l_0) = \frac{1}{4\pi^2}\int_{-\infty}^{\infty}\int_{-\infty}^{\infty} U_s(\mathbf{\Lambda})e^{j\mathbf{\Lambda}\cdot\vec{r}}\,d\alpha\,d\gamma$$
$$= \frac{1}{4\pi^2}\int_{-\infty}^{\infty}\int_{-\infty}^{\infty}\frac{O(\alpha, \gamma - k_0)}{\alpha^2 + \gamma^2 - k_0^2}e^{j(\alpha x + \beta l_0)}\,d\alpha\,d\gamma. \tag{11.171}$$

Integration of above expression Eq. (11.171) with γ for a given α has singularity at

$$\gamma = \pm\sqrt{k_0^2 - \alpha^2}. \tag{11.172}$$

Performing contour integration along the circular arc with respect γ results in

$$u_s(x, y) = \frac{1}{2\pi}\int \Gamma_1(\alpha, y)e^{j\alpha x}d\alpha + \frac{1}{2\pi}\int \Gamma_2(\alpha, y)e^{j\alpha x}d\alpha, \tag{11.173}$$

where

$$\Gamma_1(\alpha, y) = \frac{jO(\alpha, \sqrt{k_0^2 - \alpha^2} - k_0)}{2\sqrt{k_0^2 - \alpha^2}}e^{j\sqrt{k_0^2 - \alpha^2}l_0}, \tag{11.174}$$

$$\Gamma_2(\alpha, y) = \frac{-jO\left(\alpha, \sqrt{k_0^2 - \alpha^2} - k_0\right)}{2\sqrt{k_0^2 - \alpha^2}} e^{-j\sqrt{k_0^2 - \alpha^2} l_0}. \tag{11.175}$$

It is to be noted that only the waves that satisfy the condition $\alpha^2 + \gamma^2 = k_0^2$ are considered. In the above expression, the solution $\Gamma_1(\alpha, y)$ represents the solution for transmission DT for outgoing waves

$$u_s(x, y) = \frac{1}{2\pi} \int \Gamma_1(\alpha, y) e^{j\alpha x} d\alpha, \; y > \text{object}. \tag{11.176}$$

If the receiver location is less than the y coordinates of the object, then the solution $\Gamma_2(\alpha, y)$ represents the solution for reflection DT

$$u_s(x, y) = \frac{1}{2\pi} \int \Gamma_2(\alpha, y) e^{j\alpha x} d\alpha, \; y < \text{object}. \tag{11.177}$$

This is called Fourier diffraction theorem or generalized Fourier slice theorem (Devaney, 2012; Kak & Slaney, 1988; Mueller, 1979; Soumekh, Kaveh, & Mueller, 1984). Using the Fourier diffraction theorem, it is possible to estimate the object function distribution from the measured scattered field. In the following subsection, some reconstruction methods using the Fourier diffraction theorem are discussed.

11.4.4 Image reconstruction methods for weakly scattering objects

Using the Fourier diffraction theorem, the diffracted projections are related to the Fourier transform of the object along the semicircular arc $\left(\gamma = \pm\sqrt{k_0^2 - \alpha^2}\right)$. To estimate the object function from the measured scattered fields, there are two interpolation methods using frequency and space domain. Frequency domain methods are fast and use direct Fourier inversion to estimate the object function (Devaney, 1982; Mueller, 1979; Pan & Kak, 1983; Soumekh et al., 1984). Space domain methods are similar to back-projection techniques for transmission tomography, but they are computationally intensive (Devaney, 1982; Pan & Kak, 1983; Rouseff & Winters, 1994).

11.4.5 Image reconstruction methods for highly scattering objects

DT, which involves diffraction mostly, provides good reconstruction using first-order Born or Rytov approximations for weakly scattering objects. The use of first-order Born or Rytov approximations based on perturbation theory is not plausible when the size of the object and change in the refractive index is large. Methods based on perturbation only can be used when the change in the refractive index is less than 10%.

However, in general, for industrial and medical applications, the contrast between the objects may be high, therefore, involves more scattering than diffraction. Hence, the problem of reconstructing the properties for high scattering object is termed as an inverse scattering problem. The formulation of a direct method to solve for the exact Helmholtz wave equation could allow imaging of objects with large attenuation and change in the index of refraction. The relationship between the scattered field and scattering object is nonlinear, and multiple scattering increases the nonlinearity of the problem.

Let us consider the expression of the field again:

$$u(\mathbf{r}) = u_0(\mathbf{r}) + \int g(\mathbf{r}-\mathbf{r}')o(\mathbf{r}')u(\mathbf{r}')d\mathbf{r}'. \tag{11.178}$$

To solve the above equation for a high scattering problem usually, numerical methods are employed, and the method of moment (MOM) is commonly used (Colton & Kress, 1998; Conte & de Boor, 1980; Gibson, 2008). In MOM, the unknowns are approximated as a linear combination of known basis functions $f_n(\mathbf{r})$ with unknown coefficients such as (Gibson, 2008; Harrington, 1968)

$$u(\mathbf{r}) = \sum_{n=1}^{N} u_n f_n(\mathbf{r}) \text{ and } o(\mathbf{r}) = \sum_{n=1}^{N} o_n f_n(\mathbf{r}). \tag{11.179}$$

Two known functions, $u_0(\mathbf{r})$ and $u_s(\mathbf{r})$, are also approximated in the same way. Usually, due to the nonlinearity, the solution is obtained in two steps, and in each step, the equation is assumed to be linear:

$$\begin{aligned} u_0(\mathbf{r}) &= u(\mathbf{r}) - \int g(\mathbf{r}-\mathbf{r}')o(\mathbf{r}')u(\mathbf{r}')d\mathbf{r}' \\ u_s(\mathbf{r}) &= \int g(\mathbf{r}-\mathbf{r}')o(\mathbf{r}')u(\mathbf{r}')d\mathbf{r}'. \end{aligned} \tag{11.180}$$

Evaluating the above equations at discrete M locations, we can construct two sets of linear equations (Devaney, 2012; Haddadin & Ebbini, 1996; Wang & Chew, 1989)

$$\mathbf{u}_0 = \mathbf{P}\mathbf{u} \tag{11.181}$$

$$\mathbf{u}_s = \mathbf{Q}\ \mathbf{o} \tag{11.182}$$

where

$$P_{m,n} = \delta_{mn} - \int g(\mathbf{r}_m - \mathbf{r}')o(\mathbf{r}')f_n(\mathbf{r}')d\mathbf{r}',\ m,n = 1,\cdots,N \tag{11.183}$$

$$Q_{m,n} = \int g(\mathbf{r}_m - \mathbf{r}')f_n(\mathbf{r}')u(\mathbf{r}')d\mathbf{r},\ m = 1, 2, \cdots, M;\ n = 1, 2, \cdots, N. \quad (11.184)$$

One of the simplest ways to solve the incident and scattered field is the Born iterative method (BIM) (Wang & Chew, 1989; Haddadin & Ebbini, 1996; Tsihrintzis & Devaney, 2000), in which with the initial guess for the total field, $u(\mathbf{r}) = u_0(\mathbf{r})$ the object function $o(\mathbf{r})$ is evaluated by solving $\mathbf{u}_s = \mathbf{Q}\ \mathbf{o}$, then $\mathbf{P}$ is computed from the estimated object function $o(\mathbf{r})$ and total field coefficients $\mathbf{u}$ are obtained by inverting $\mathbf{P}$ in Eq. (11.181). This procedure is repeated until convergence. The BIM method is robust to noise; however, the performance is affected by high scattering media and inhomogeneous background. An extension of the BIM method is the distorted BIM (DBIM), where along with the total field $\mathbf{u}$, the equation that contains the Green's function is corrected. At each iteration step, the latest estimate of scattering function is used to compute the scattered field for all the field points, which are further used in the forward problem Eq. (11.181) to compute the total field. The object function $o(\mathbf{r})$ is then solved using Eq. (11.142) based on computed total field and scattered field (Chew & Wang, 1990; Haddadin & Ebbini, 1996). DBIM has better reconstruction performance and has faster convergence as compared to BIM, but this method is not robust to measurement noise. An improved method called modified DBIM that combines both BIM and DBIM is introduced, which is robust to measurement noise and has improved reconstruction performance (Haddadin & Ebbini, 1996). Also, the problem can be converted into a minimization problem by defining the cost function as

$$\Phi = \frac{1}{2}\|\mathbf{Q}\ \mathbf{o} - \mathbf{u}_s\|^2, \quad (11.185)$$

and any optimization algorithms used for ET can be used to solve the field (Caorsi, Gragnani, & Pastorino, 1990; Franchois & Pichot, 1997; Haddadin & Ebbini, 1996; Joachimowicz, Pichot, & Hugonin, 1991; Tsihrintzis & Devaney, 2000; Wang & Chew, 1989). A nonlinear model based on DBIM that considers multiple scattering is developed by Kamilov, Liu, Mansour, and Boufounos (2016). The method developed can be related to a feed-forward neural network where each layer corresponds to the number of interactions the wave undergoes. In neural network representation, consider the number of measurements as M, number of nodes in input and hidden layers as N, and for $k = 1, \ldots, K$ where K is the scattering order, the discretized version of Eq. (11.180) can be represented as

$$\begin{aligned} \mathbf{u}^k &= \mathbf{u}^0 + \mathbf{G}\left(\mathbf{u}^{k-1} \odot \mathbf{o}\right), \\ \mathbf{z} &= \mathbf{H}\left(\mathbf{u}^K \odot \mathbf{o}\right), \end{aligned} \quad (11.186)$$

where o is the object function, $\mathbf{z} \in \Re^M$ is the predicted scatter field u_s from the M sensors located around the object, $\mathbf{u}^0 \in \Re^N$ is the incident field u_0 discretized inside the domain Ω, $\mathbf{H} \in \Re^{M\times N}$ and $\mathbf{G} \in \Re^{N\times N}$ correspond to the discretization of Green's function at sensor location and inside Ω, respectively. Here, $\mathbf{u}^k \in \Re^N$ represents the internal field discretized after k-th scattering. Moreover, the symbol $\odot$ symbolizes element-wise multiplication. Eq. (11.186) can be represented as a feed-forward network where nodes are related to object function; **H**, **G** are the weights across each edge of the network. The inverse scattering problem is formulated as a minimization problem, and computing the derivative of Eq. (11.186) with respect to object function **o** and rearranging the expressions, we have the error back propagation algorithm. The predicted and measured scattered fields are then compared, and the error is back propagated to compute the updated weights and efficiently estimate the objection function.

11.5 Future trends

Tomography helps us to know about the distribution of the substance properties in the domain of interest; thus, it provides valuable information about the actual process. Various tomography technologies existing in the market use different interaction phenomena between penetrating wave and medium to generate tomograms. For the same physical process, using different tomography techniques would result in the distribution of different properties which depend on the type of interaction. The different property distributions together would give us a better understanding of the process. The fusion of two or more different techniques has attracted the scientific community and is the current hot topic for imaging these days. However, for the application of these systems, they require efficient mathematical formulations that depict the process phenomenon.

In imaging objects that are varying in space with respect to time, if their kinematic behavior can be represented using available constraints and used in the mathematical formulation, it can give us additional prior information about the process. Also, in the tomography reconstruction problem, there is a lot of prior information needed to get satisfactory results. However, in real situations, the prior information available is limited; therefore, still there is a need to develop efficient inverse algorithms that tackle the ill-posed nature. Moreover, the existing tomography systems that are developed for industry use static image reconstruction algorithms. Although data acquisition is very fast with the current measurement hardware, in some applications that involve fast transient behavior such as two- or multiphase flows, there is a need for dynamic algorithms that use limited data to reconstruct the image. Although the dynamic algorithms are studied for many years, they are not used in image reconstruction in real-time monitoring due to the challenges they pose in computational complexity.

In most of the tomography methods, the forward and inverse problem is associated with nonlinear behavior. To solve the nonlinear forward problem, numerical methods based on the FEM and boundary element method are often used. There are also meshless methods that do not require any mesh or boundary integration that can be an alternative. Many of the tomography reconstruction methods use linear based methods to solve inner material properties. However, the linear methods do not reflect the true behavior of the medium interaction; therefore, the accuracy is limited. Furthermore, prior knowledge is often used to have a stable, satisfactory solution. In many situations, however, such prior knowledge is not available. This limits the development and improvement of the mathematical formulation of algorithms. Recently, conventional neural network approaches that made great growth in the areas of vision, speech, and text processing are applied by many researchers to solve the nonlinear tomography forward and inverse problem. Due to limited data and computational challenges, implementing deep neural methods for different tomography applications requires further study.

11.6 Source of further information

This chapter primarily contains mathematical concepts for three modes of tomography techniques. Radon transform forms the main idea in realizing the tomography imaging (Radon, 1917). Fourier transform is used to reconstruct the projected data (Bracewell & Riddle, 1967). Fourier slice theorem relates the projected data to the unknown field (Kak & Slaney, 1988). Computerized tomography is also discussed in an easy way (Natterer, 1986). The filtered back-projection method offers better reconstruction performance compared to direct projection (Kak & Slaney, 1988; Natterer, 1986). Matrix-based methods known as the ART are used to solve a large set of linear equations (Gordon et al., 1970; Herman, 1980; Strohmer & Vershynin, 2009). Improvements to ART namely SIRT and SART are developed for better accuracy and computational efficiency (Anderson & Kak, 1984a; Gilbert, 1972). In the case of low dose radiation, ML approaches can be used to have good reconstruction (Lange & Carson, 1984; Lange et al., 1987; Vardi et al., 1985).

Maxwell equations are used to derive the governing equations for ET (Holder, 2004; Webster, 1990). Different electrode models used in ET are discussed in Cheng (1989). The forward problem is usually solved using the FEM (Vauhkonen, 1997). Different types of regularization and priors are used to reduce the ill-posedness (Vauhkonen, 1997). Static imaging is done using the whole frame of data (Yorkey et al., 1987), and in dynamic imaging, few measurements are only available (Khambampati et al., 2009; Kim et al., 2001; Vaukhonen et al., 1998).

In DT, wave equations are solved using perturbation methods based on Born and Rytov for weakly scattered media (Born & Wolf, 1999; Iwata & Nagata, 1974). Fourier diffraction theorem and filtered back propagation theorem are similar in concept with transmission tomography (Devaney, 1982). For highly scattering medium, numerical methods based on the MOM are used (Gibson, 2008; Harrington, 1968). The inverse scattering problem is solved using iterative methods such as the BIM and the DBIM (Haddadin & Ebbini, 1996). Also, the inverse scattering problem can be formulated and solved in the same way as the ET problem (Chew & Wang, 1990).

References

Adler, A., Arnold, J. H., Bayford, R., Borsic, A., Brown, B., Dixon, P., et al. (2009). Greit: A unified approach to 2D linear EIT reconstruction of lung images. *Physiological Measurement, 30*, S35.

Adler, A., Dai, T., & Lionheart, W. R. B. (2007). Temporal image reconstruction in electrical impedance tomography. *Physiological Measurement, 28*, S1–S11.

Andersen, A. H., & Kak, A. C. (1982). Digital ray tracing in two dimensional refractive fields. *Journal of the Acoustical Society of America, 72*, 1593–1606.

Andersen, A. H., & Kak, A. C. (1984a). Simultaneous algebraic reconstruction technique (SART): A superior implementation of the art algorithm. *Ultrasonic Imaging, 6*, 81–94.

Andersen, A. H., & Kak, A. C. (1984b). *The application of ray tracing towards a correction for refracting effects in computed tomography with diffracting sources, TR-EE 84-14*. School of Electrical Engineering, Purdue University.

Baker, L. E. (1989). Principles of the impedance technique. *IEEE Engineering in Medicine and Biology Magazine, 8*, 11–15.

Bar-Shalom, Y., & Li, X. R. (1993). *Estimation and tracking- principles, techniques, and software*. Norwood, MA: Artech House, Inc.

Barber, D. C., & Brown, B. H. (1984). Applied potential tomography. *Journal of Physics E: Scientific Instruments, 17*, 723–733.

Born, M., & Wolf, E. (1999). *Principles of optics: Electromagnetic theory of propagation, interference and diffraction of light* (7th ed.). Cambridge: Cambridge University Press.

Borsic, A., Graham, B. M., Adler, A., & Lionheart, W. R. (2007). *Total variation regularization in electrical impedance tomography*.

Bracewell, R. N. (1956). Strip integration in radio astronomy. *Australian Journal of Physics, 9*, 198–217.

Bracewell, R. N. (1965). *The Fourier transform and its applications*. New York: McGraw-Hill.

Bracewell, R. N., & Riddle, A. (1967). Inversion of fan-beam scans in radio astronomy. *The Astrophysical Journal, 150*, 427–434.

Brenner, S. C., & Scott, L. R. (1994). *The mathematical theory of finite element methods*. Springer.

Browne, J. A., & Holmes, T. J. (1992). Developments with maximum likelihood X-ray computed tomography. *IEEE Transactions on Medical Imaging, 11*, 40–52.

Caorsi, S., Gragnani, G. L., & Pastorino, M. (1990). Two-dimensional microwave imaging by a numerical inverse scattering solution. *IEEE Transactions on Microwave Theory and Techniques, 38*(8), 980e981.

Cheney, M., Isaacson, D., & Newell, J. C. (1999). Electrical impedance tomography. *SIAM Review, 41*, 85–101.

Cheney, M., Isaacson, D., Newell, J. C., Simke, S., & Goble, J. (1990). Noser: An algorithm for solving the inverse conductivity problem. *International Journal of Imaging Systems and Technology, 2*(2), 66–75.

Cheng, K. S., Isaacson, D., Newell, J. C., & Gisser, D. J. (1989). Electrode models for electric current computed tomography. *IEEE Transactions on Biomedical Engineering, 36*, 918–924.

Chew, W. C., & Wang, Y. M. (1990). Reconstruction of two-dimensional permittivity distribution using the distorted Born iterative method. *IEEE Transactions on Medical Imaging, 9*(2), 218–225.

Colton, D., & Kress, R. (1998). *Inverse acoustic and electromagnetic scattering theory* (2nd ed.). New York: Springer Verlag.

Conte, S. D., & de Boor, C. (1980). *Elementary numerical analysis: An algorithmic approach*. New York: McGraw-Hill.

Dai, T., & Adler, A. (2008). Electrical Impedance Tomography reconstruction using ℓ1 norms for data and image terms. In *2008 30th annual international conference of the IEEE engineering in medicine and biology society* (pp. 2721–2724). IEEE.

Dardé, J., Hyvönen, N., Seppänen, A., & Staboulis, S. (2013). Simultaneous recovery of admittivity and body shape in electrical impedance tomography: An experimental evaluation. *Inverse Problems, 29*(8), 085004.

Deans, S. R. (2007). *The Radon transform and some of its applications*. Courier Corporation.

Devaney, A. J. (1982). A filtered backpropagation algorithm for diffraction tomography. *Ultrasonic Imaging, 4*, 336–350.

Devaney, A. J. (2012). *Mathematical foundations of imaging, tomography and wavefield inversion*. Cambridge: Cambridge Press.

Doerstling, B. H. (1994). *A 3-D reconstruction algorithm for the linearized inverse boundary value problem for Maxwell's equations* (Ph.D. thesis). Troy, New York: Rensselaer Polytechnic Institute.

Dyakowski, T., Jeanmeure, L. F. C., & Jaworski, A. J. (2000). Applications of electrical tomography for gas-solids and liquid-solids flows — a review. *Powder Engineering, 112*, 174–192.

Feldkamp, L. A., Davis, L. C., & Kress, J. W. (1984). Practical cone-beam algorithm. *Journal of the Optical Society of America A, 1*, 612–619.

Fleischmann, D., & Boas, F. E. (2011). Computed tomography—old ideas and new technology. *European Radiology, 21*, 510–517.

Franchois, A., & Pichot, C. (1997). Microwave imaging-complex permittivity reconstruction with a Levenberg-Marquardt method. *IEEE Transactions on Antennas and Propagation, 45*(2), 203–215.

Gibson, W. C. (2008). *The method of moments in electromagnetics*. London: Chapman & Hall/CRC.

Gilbert, P. (1972). Iterative methods for the reconstruction of three dimensional objects from their projections. *Journal of Theoretical Biology, 36*, 105–117.

Gordon, R. (1974). A tutorial on ART (Algebraic Reconstruction Techniques). *IEEE Transactions on Nuclear Science, 21*, 78–93.

Gordon, R., Bender, R., & Herman, G. T. (1970). Algebraic reconstruction techniques (ART) for three-dimensional electron microscopy and x-ray photography. *Journal of Theoretical Biology, 29*, 471–481.

Greenleaf, J. F., Johnson, S. A., Lee, S. L., Herman, G. T., & Wood, E. H. (1974). Algebraic reconstruction of spatial distributions of acoustic absorption within tissue from their two dimensional acoustic projections. In *Acoustical holography*. Plenum Press.

Haddadin, O. S., & Ebbini, E. S. (1996). Adaptive regularization of a distorted Born iterative algorithm for diffraction tomography. In *Proceedings of 3rd IEEE international conference on image processing* (Vol. 2, pp. 725–728). IEEE.

Harrington, R. F. (1968). *Field computation by moment methods*. New York: Macmillan.

Heikkinen, L. M., Vilhunen, T., West, R. M., & Vauhkonen, M. (2002). Simultaneous reconstruction of electrode contact impedances and internal electrical properties: II. Laboratory experiments. *Measurement Science and Technology, 13*(12), 1855.

Herman, G. T. (1979). *Topics in applied physics: Vol. 32. Image reconstruction from projections: Implementation and applications*. Berlin: Springer-Verlag.

Herman, G. T. (1980). *Image reconstruction from projections — the fundamentals of computerised tomography*. New York: Academic Press.

Hinton, E., & Owen, D. R. (1979). *An introduction to finite element computations*. Wales: Pineridge Press.

Holder, D. S. (Ed.). (2004). *Electrical impedance tomography: Methods, history and applications*. CRC Press.

Hua, P., Webster, J. G., & Tompkins, W. J. (1988). A regularized electrical impedance tomography reconstruction algorithm. *Clinical Physics and Physiological Measurement, Supplement, 9*, 137–141.

Hua, P., Woo, E. J., Webster, J. G., & Tompkins, W. J. (1991). Iterative reconstruction methods using regularization and optimal current patterns in electrical impedance tomography. *IEEE Transactions on Medical Imaging, 10*, 621–628.

Ishimaru, A. (1978). *Wave propagation and scattering in random media*. New York: Academic Press.

Iwata, K., & Nagata, R. (1974). Calculation of refractive index distribution from interferograms using the Born and Rytov's approximation. *Japanese Journal of Applied Physics, 14*, 379.

Jain, A. K. (1989). *Fundamentals of digital image processing, chapter: Image reconstruction from projections*. New York: Prentice-Hall.

Jiang, M., & Wang, G. (2003). Convergence of the simultaneous algebraic reconstruction technique (SART). *IEEE Transactions on Image Processing, 12*, 957–961.

Joachimowicz, N., Pichot, C., & Hugonin, J. P. (1991). Inverse scattering: An iterative numerical method for electromagnetic imaging. *IEEE Transactions on Antennas and Propagation, 39*, 1742–1753.

Kaipio, J. P., Karjalainen, P. A., Somersalo, E., & Vauhkonen, M. (1999). State estimation in time-varying electrical impedance tomography. *Annals of the New York Academy of Sciences, 873*, 430–439.

Kaipio, J. P., & Somersalo, E. (1999). Nonstationary inverse problems and state estimation. *Journal of Inverse and Ill-Posed Problems, 7*, 273–282.

Kaipio, J. P., & Somersalo, E. (2007). Statistical inverse problems: Discretization, model reduction and inverse crimes. *Journal of Computational and Applied Mathematics, 198*(2), 493–504.

Kak, A. C. (1984). Tomographic imaging with diffracting and non-diffracting sources. In A. Metherell (Ed.), *Acoustical imaging*. Plenum Press.

Kak, A. C., & Slaney, M. (1988). *Principles of computerized tomographic imaging*. New York: IEEE.

Kamilov, U. S., Liu, D., Mansour, H., & Boufounos, P. T. (2016). A recursive born approach to nonlinear inverse scattering. *IEEE Signal Processing Letters, 23*(8), 1052–1056.

Katsevich, A. (2004). An improved exact filtered backprojection algorithm for spiral computed tomography. *Advances in Applied Mathematics, 32*, 681–697.

Kaveh, M., Soumekh, M., & Muller, R. K. (1982). Tomographic imaging via wave equation inversion. *The International Conference on Acoustics, Speech, & Signal Processing (ICASSP), 82*, 1553–1556.

Keller, J. B. (1969). Accuracy and validity of the Born and Rytov approximation. *Journal of the Optical Society of America, 59*, 1003.

Khambampati, A. K., Rashid, A., Kim, S., Soleimani, M., & Kim, K. Y. (2009). Unscented Kalman filter approach to track moving interfacial boundary in sedimentation process using three-dimensional electrical impedance tomography. *Philosophical Transactions of the Royal Society A, 367*, 3095–3120.

Khambampati, A. K., Liu, D., Konki, S. K., & Kim, K. Y. (2018). An automatic detection of the ROI using Otsu thresholding in nonlinear difference EIT imaging. *IEEE Sensors Journal, 18*(12), 5133–5142.

Kim, J. H., Choi, B. Y., & Kim, K. Y. (2006). Novel iterative image reconstruction algorithm for electrical capacitance tomography: Direct algebraic reconstruction technique. *IEICE Transactions on Fundamentals of Electronics, Communications and Computer Sciences, 89*, 1578–1584.

Kim, M. C., Kim, S., Kim, K. Y., & Lee, Y. J. (2001). Regularization methods in electrical impedance tomography technique for the two-phase flow visualization. *International Communications in Heat and Mass Transfer, 28*(6), 773–782.

Kim, K. Y., Kim, B. S., Kim, M. C., Lee, Y. J., & Vauhkonen, M. (2001). Image reconstruction in time-varying electrical impedance tomography based on the extended Kalman filter. *Measurement Science and Technology, 12*, 1032.

Kohn, R. V., & McKenney, A. (1990). Numerical implementation of a variational method for electrical impedance tomography. *Inverse Problems, 6*, 389–414.

Kolehmainen, V., Lassas, M., & Ola, P. (2005). The inverse conductivity problem with an imperfectly known boundary. *SIAM Journal on Applied Mathematics, 66*(2), 365–383.

Lange, K., Bahn, M., & Little, R. (1987). A theoretical study of some maximum likelihood algorithms for emission and transmission tomography. *IEEE Transactions on Medical Imaging, 6*(2), 106–114.

Lange, K., & Carson, R. (1984). EM reconstruction algorithms for emission and transmission tomography. *Journal of Computer Assisted Tomography, 8*, 306–316.

Levoy, M. (1992). *Volume rendering using the Fourier projection-slice theorem* (Vol. 94, pp. 305–405). California: Computer Systems Laboratory Departments of Electrical Engineering and Computer Science Stanford University Stanford.

Liu, D., Kolehmainen, V., Siltanen, S., & Seppanen, A. (2015a). Estimation of conductivity changes in a region of interest with electrical impedance tomography. *Inverse Problems, 9*, 211–229.

Liu, D., Kolehmainen, V., Siltanen, S., & Seppänen, A. (2015b). A nonlinear approach to difference imaging in EIT; assessment of the robustness in the presence of modelling errors. *Inverse Problems, 31*(3), 035012.

Morse, P.,M., & Feshbach, H. (1953). *Method of theoretical physics*. New York: McGraw Hill Book Company.

Mueller, R. K., Kaveh, M., & Wade, G. (1979). Reconstructive tomography and applications to ultrasonics. *Proceedings of the IEEE, 67*, 567.

Mueller, K., & Yagel, R. (2000). Rapid 3-D cone-beam reconstruction with the simultaneous algebraic reconstruction technique (SART) using 2-D texture mapping hardware. *IEEE Transactions on Medical Imaging, 19*, 1227–1237.

Natterer, F. (1986). *The mathematics of computerized tomography*. New York: Wiley.

Newton, R. G. (1966). *Scattering theory of waves and particles*. New York: McGraw-Hill.

Nunez, P. L. (1981). *Electric fields of the brain: The neurophysics of EEG*. New York: Oxford University Press.

Oppengeim, A. V., & Schafer, R. W. (1975). *Digital signal processing*. NJ: Prentice-Hall, Englewood Cliffs.

Pan, S. X., & Kak, A. C. (1983). A computational study of reconstruction algorithms for diffraction tomography: Interpoltion vs. filtered backpropagation. *IEEE Transactions on Acoustics, Speech, & Signal Processing, 10*, 1262–1275.

Polydorides, N., Lionheart, W. R. B., & McCann, H. (2002). Krylov subspace iterative techniques: On the detection of brain activity with electrical impedance tomography. *IEEE Transactions on Medical Imaging, 21*, 596–603.

Radon, J. (1917). On the determination of functions from their integral values along certain manifolds. English translation 1986 *IEEE Transactions on Medical Imaging, 5*, 170–176.

Reinecke, N., & Mewes, D. (1997). Investigation of the two-phase flow in trickle-bed reactors using capacitance tomography. *Chemical Engineering and Science, 52*, 2111–2138.

Rockmore, A. J., & Macovski, A. (1976). A maximum likelihood approach to emission image reconstruction from projections. *IEEE Transactions on Nuclear Science, 23*, 1428–1432.

Rouseff, D., & Winters, K. B. (1994). Two-dimensional vector flow inversion by diffraction tomography. *Inverse Problems, 10*(3), 687.

Schafer, D., Borgert, J., Rasche, V., & Grass, M. (2006). Motion-compensated and gated cone beam filtered back-projection for 3-D rotational X-ray angiography. *IEEE Transactions on Medical Imaging, 25*, 898–906.

Seppanen, A., Vauhkonen, M., Somersalo, E., & Kaipio, J. P. (2001a). State space models in process tomography – approximation of state noise covariance. *Inverse Problems in Engineering, 9*, 561–585.

Seppanen, A., Vauhkonen, M., Vauhkonen, P. J., Somersalo, E., & Kaipio, J. P. (2001b). State estimation with fluid dynamical evolution models in process tomography— an application to impedance tomography. *Inverse Problems, 17*, 467.

Shepp, L. A., & Vardi, Y. (1982). Maximum likelihood reconstruction for emission tomography. *IEEE Transactions on Medical Imaging, 1*, 113–122.

Sommersalo, E., Cheney, M., & Isaacson, D. (1992). Existence and uniqueness for electrode models for electric current computed tomography. *SIAM Journal on Applied Mathematics, 52*, 1023–1040.

Sommersalo, E., Isaacson, D., & Cheney, M. (1992). A linearized inverse boundary value problem for Maxwell's equations. *Journal of Computational and Applied Mathematics, 42*, 123–136.

Soumekh, M., Kaveh, M., & Mueller, R. K. (1984). Fourier domain reconstruction methods with application to diffraction tomography. *Acoustical Imaging, 13*, 17–30.

Stark, H., Woods, J. W., Paul, I., & Hingorani, R. (1981). An investigation of computerized tomography by direct Fourier inversion and optimum interpolation. *IEEE Transactions on Biomedical Engineering, 28*, 496–505.

Strohmer, T., & Vershynin, R. (2009). A randomized Kaczmarz algorithm for linear systems with exponential convergence. *Journal of Fourier Analysis and Applications, 15*, 262–278.

Trigo, F. C., Gonzalez-Lima, R., & Amato, M. B. P. (2004). Electrical impedance tomography using the extended Kalman filter. *IEEE Transactions on Biomedical Engineering, 51*, 72–81.

Tsihrintzis, G. A., & Devaney, A. J. (2000). Higher-order (nonlinear) diffraction tomography: Reconstruction algorithms and computer simulation. *IEEE Transactions on Image Processing, 9*(9), 1560–1572.

Vardi, Y., Shepp, L. A., & Kaufman, L. (1985). A statistical model for positron emission tomography. *Journal of the American Statistical Association, 80*, 8–37.

Vauhkonen, M. (1997). *Electrical impedance tomography and prior information* (Ph.D thesis).

Vauhkonen, M., Karjalainen, P. A., & Kaipio, J. P. (1998). A Kalman filter approach to track fast changes in electrical impedance tomography. *IEEE Transactions on Biomedical Engineering, 45*, 486–493.

Vilhunen, T., Kaipio, J. P., Vauhkonen, P. J., Savolainen, T., & Vauhkonen, M. (2002). Simultaneous reconstruction of electrode contact impedances and internal electrical properties: I. Theory. *Measurement Science and Technology, 13*(12), 1848.

Villares, G., Begon-Lours, L., Margo, C., Oussar, Y., Lucas, J., & Holé, S. (2012). Non-linear model of sensitivity matrix for electrical capacitance tomography. *Electrostatic Joint Conference*, R4.

Wang, Y. M., & Chew, W. C. (1989). An iterative solution of the two-dimensional electromagnetic inverse scattering problem. *International Journal of Imaging Systems and Technology, 1*(1), 100–108.

Wang, H., Tang, Q., & Zheng, W. (2012). L1-norm-based common spatial patterns. *IEEE Transactions on Biomedical Engineering, 59*(3), 653–662. https://doi.org/10.1109/TBME.2011.2177523

Warsito, W., & Fan, L. S. (2001). Measurement of real-time flow structures in gas–liquid and gas–liquid–solid flow systems using electrical capacitance tomography(ECT). *Chemical Engineering Science, 56*, 6455–6462.

Webster, J. G. (1990). *Electrical impedance tomography*. Adam Hilger.

Wolf, E. (1969). Three-dimensional structure determination of semitransparent objects from holographic data. *Optics Communications, 1*, 153–156.

Woo, E. J., Hua, P., Webster, J. G., & Tompkins, W. J. (1994). Finite element method in electrical impedance tomography. *Medical, & Biological Engineering & Computing, 32*, 530–536.

Xie, C. G., Huang, S. M., Beck, M. S., Hoyle, B. S., Thorn, R., Lenn, C., et al. (1992). Electrical capacitance tomography for flow imaging-system model for development of image reconstruction algorithms and design of primary sensors. *IEE Proceedings G, 139*, 89–98.

Xie, C. G., Plaskowski, A., & Beck, M. S. (1989). 8-electrode capacitance system for two-component flow identification. Part 1: Tomographic flow imaging. *IEE Proceedings A (Physical Science, Measurement and Instrumentation, Management and Education), 136*(4), 173–183. https://doi.org/10.1049/ip-a-2.1989.0031

Yang, W. Q. (1996). Hardware design of electrical capacitance tomography systems. *Measurement Science and Technology, 7*, 233–246.

Yorkey, T. J., Webster, J. G., & Tompkins, W. J. (1987). Comparing reconstruction algorithms for electrical impedance tomography. *IEEE Transactions on Biomedical Engineering, 34*, 843–852.

Direct image reconstruction in electrical tomography and its applications

Zhang Cao and Lijun Xu
School of Instrumentation and Opto-Electronic Engineering, Beihang University, Beijing, China

12.1 Introduction

Electrical tomography (ET) has been developed to reconstruct the spatial permittivity, conductivity, or admittivity distributions of target materials by using the measured data. Compared with the X-ray tomography, ET possesses the advantages of no radiation, fast response, robustness, and simple structure. It is considered as a promising technology for many industrial applications, such as multiphase flow measurement in pipelines, choking transition in gas–solid risers (Du, Warsito, & Fan, 2006a), flames in porous media (Liu, Chen, Xiong, Zhang, & Lei, 2008), mass flow measurement of pneumatically conveyed solids (Sun, Liu, Lei, & Li, 2008), solids distribution measurement in cyclone separator (Wang, Liu, Jiang, & Yang, 2004) and the analysis of dynamic processes in fluidized beds (Du, Warsito, & Fan, 2006b), etc. For example, as one of the typical imaging modalities of ET, electrical capacitance tomography (ECT) reconstructs the permittivity distribution from a set of electrical capacitances measured on the boundary of the region of interest (ROI) (Wang & Yang, 2020).

Unlike the so called "hard field" property of the X-ray, the distribution of the electrical field can be distorted by the objects placed inside. The distortion stems from the "soft field" property of the electromagnetic field. The property indicates that any of the electrodes has the ability to collect the information of changes in any part of the whole sensing region, so the ability of information capture for any electrode is global, while that of the sensing unit in X-ray tomography is local, as only the information along the X-ray radiation lines can be detected. As a result, besides the image reconstruction algorithms that can be inherited from the X-ray tomography, such as the sensitivity matrix–based algorithms, the ET has its own boundary map–based algorithms, which are also termed as direct algorithms (Cao, Xu, & Wang, 2009; Muller, Mueller, & Mellenthin, 2017). There usually exist two different boundary maps, i.e., the Dirichlet-to-Neumann map and the Neumann-to-Dirichlet map, which can be physically interpreted as the voltage-to-current density map and the current density-to-voltage map, respectively, and the discretized form of the maps can be constructed from the measured mutual capacitances (Cao & Xu, 2013a, 2013b, 2013c).

Industrial Tomography. https://doi.org/10.1016/B978-0-12-823015-2.00018-2

In the literature, most of the image reconstruction algorithms of ET were implemented based on the sensitivity theorem (Geselowitz, 1971; Lehr, 1972), which is a kind of linearization method using the perturbation principle (Soleimani, Yalavarthy, & Dehghani, 2010). However, ET possesses the "soft field" property, as the sensing field is distorted by objects in the imaging area. The sensitivity distribution changes with the spatial permittivity distribution. In recent years, iterative image reconstruction with an updated sensitivity matrix has been investigated (Fang & Cumberbatch, 2005; Li & Yang, 2008; Smolik & Radomski, 2009), but they are very time consuming and the property of convergence of the iteration has not been reported. Some existing noniterative methods are usually simplifications of the iterative ones, e.g., preiteration method (Wang, Wang, & Yin, 2004), linear back projection (Yang & Peng, 2003). These noniterative methods based on the sensitivity or Jacobian matrix are not direct algorithms. The permittivity/conductivity/admittivity perturbations at all positions in the ROI can only be reconstructed simultaneously.

In recent years, several direct algorithms have been introduced to ET by using the Calderon's method (Boverman, Kao, Isaacson, & Saulnier, 2009; Cao, Xu, & Wang, 2009; Mueller & Bikowski, 2008), the dbar method (Cao et al., 2009a; Hamilton, Herrera, Mueller, & Von Herrmann, 2012; Hamilton & Mueller, 2013; Knudsen, Mueller, & Siltanen, 2004; Zhao, Xu, & Cao, 2019), the complex geometrical optics method (Astala, Mueller, Päivärinta & Siltanen, 2010; Astala, 2006), and the factorization method (Cao, Xu, Fang, & Wang, 2011; Cao & Xu, 2013c; Gebauer & Hyvönen, 2007; Harrach, 2013; Schmitt, 2009). Both the dbar method and the complex geometrical optics method are originated from the Calderon's method (Calderon, 2006), which is practically suitable for the low-contrast material distribution (Cao et al., 2009b). Among these methods, the factorization method is suitable to reconstruct the disturbed distributions with a high-contrast background, as it is a method based on the principle of the value range identification. The factorization method is usually suitable to reconstruct the disturbed distributions with a connected background, e.g., multirods. If the region to be reconstructed is of a ring shape, for example, the factorization method fails to work (Schmitt, 2009). The dbar method has no such requirement on the shape of the ROI. However, the sensor configurations used in electrical impedance tomography (EIT) are different from those applied in ET. The dbar method that was reported and implemented in EIT requires a sinusoidal wave-like current pattern, in which all the values of the current injected at each electrode generate a discrete approximation of the sine function (Knudsen et al., 2004). As the ET sensor has an earthed screen and each electrode is usually either excited or grounded for hardware implementation, the dbar method successfully implemented in EIT cannot be directly applied to ECT.

Moreover, the sensors and associated reconstruction techniques in ET are often of and for circular shapes. There exists a demand for non-circular sensors, e.g., the cross-sections of the pipes in the power industry or chemical reactors are square or rectangular (Wang, Yang, Senior, Raghavan, & Duncan, 2008). However, direct calculation of the scattering transform from measured capacitances in the case of a noncircular sensor, e.g., an elliptical region, is very complicated (Murphy, Mueller, & Newell, 2007; Murphy & Mueller, 2009). Also, the factorization method requires the calculation of the Neumann-to-Dirichlet map, i.e., the current density-to-voltage map. The test function was constructed specifically for the noncircular regions (Hakula & Hyvönen, 2009), which is very computationally complicated.

In this chapter, we also provide a new treatment to the direct methods, e.g., Calderon's method, dbar method, and factorization method, when applied to the noncircular sensors by using the conformal transformation after the reconstruction. The direct methods summarized here can also be used to a circular sensor, which is a simplified case of non-circular sensors. Also, iterative Calderon's method is also introduced to improve the image quality, by using a PID controller in a closed control loop in case of ECT. Three cases of image reconstruction were carried out to show the effectiveness of these methods. Finally, dynamical flames were captured by dual modality ET in form of admittivity distributions to extract the temperature evolutions.

12.2 Invariant property of the governing equation via conformal transformation

In ET, the governing equation of the sensing field in a two-dimensional region inside its sensor is

$$\nabla \cdot [\gamma(z) \nabla \varphi(z)] = 0 \tag{12.1}$$

where $z = x + \mathrm{j}y$ is the point denoted by coordinates (x, y), and j is the imaginary unit. $\gamma(z)$ and $\varphi(z)$ are, respectively, the admittivity and electric potential at z.

The image reconstruction in two dimensions can be simplified through using the conformal transformation. The case of the sensor with any simply connected region can be equivalent to the case of a sensor with circular cross-section, and the convenience and effectiveness have been well demonstrated in our previous work (Cao et al., 2009a; Cao & Xu, 2013b, p03004). However, no rigorous proof has been reported yet to support the conformal transform in ET, and only the intuitive physical interpretation was utilized. Here, a mathematical proof was provided for the first time to show that for the isotropic admittivity distribution, the reconstructed result of any simply connected region can be equivalent to the circular sensor by using conformal transformation. As a result, it could greatly reduce the computation complexity for the image reconstruction in the noncircular sensor. The process of proof is written as follows.

According to Riemann mapping theorem, the unit disc $\Omega := \{ \leq 1\} z : |z| \leq 1\} \leq 1\}$ can be uniquely mapped conformally onto any simply connected domain E. As Fig. 12.1 shows, let w be a point in E, then there exists a unique function

$$w = f(z) \tag{12.2}$$

that transforms a point denoted by z in Ω conformally onto the point w in E, where $w = u + \mathrm{j}v$ represents the point denoted by coordinates (u, v) in the region E.

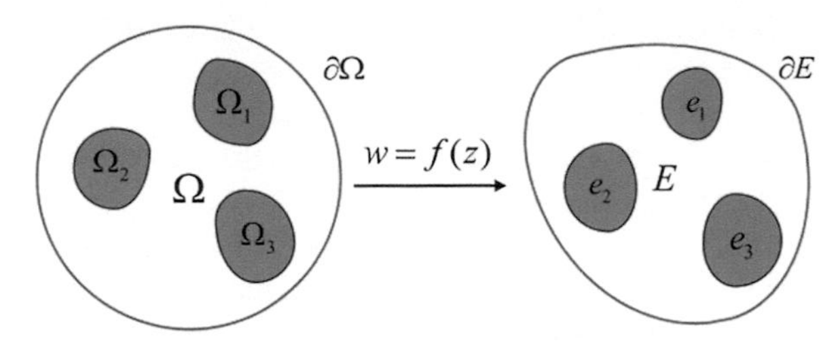

Figure 12.1 Conformal map from a unit disc to a simply connected region.

Accordingly, any continuous function defined in Ω holds its values after the conformal mapping onto E, and relates the boundary $\partial\Omega := \{z : |z| = 1\}$ of Ω to the boundary ∂E of E, for example. Meanwhile, the subdomains of Ω, such as Ω_1, Ω_2, and Ω_3, are conformally mapped onto the subdomains of E, i.e., e_1, e_2, and e_3, respectively.

If the electrical potential at w in E is denoted as $V(w)$ and the related potential value at the z in Ω is denoted as $\varphi(z)$, then they satisfy

$$V(w) = \varphi(z) = \varphi\left[f^{-1}(w)\right] \tag{12.3}$$

And the electrical potential in E satisfies

$$\nabla \cdot \left[\gamma(w)\nabla V(w)\right] = 0 \text{ in } E \tag{12.4}$$

where $\gamma(w)$ is the admittivity at w and satisfies

$$\gamma(w) = \left.\frac{A(z)\varepsilon(z)[A(z)]^{\mathrm{T}}}{\det[A(z)]}\right|_{z=f^{-1}(w)}, \quad w \in E \tag{12.5}$$

and $A(z)$ is Jacobian matrix for the conformal map in equation, namely

$$A(z) = \begin{pmatrix} \dfrac{\partial y}{\partial x} & \dfrac{\partial y}{\partial u} \\[2ex] \dfrac{\partial v}{\partial x} & \dfrac{\partial v}{\partial u} \end{pmatrix} \tag{12.6}$$

The map used in conformal transformation is harmonic, and satisfies the Cauchy–Riemann Equations (Gong & Gong, 2007)

$$\begin{cases} \dfrac{\partial y}{\partial x} = \dfrac{\partial v}{\partial u} \\[2ex] \dfrac{\partial y}{\partial u} = -\dfrac{\partial v}{\partial x} \end{cases} \tag{12.7}$$

It can be obtained that

$$\gamma(w)=\frac{A(z)\varepsilon(z)[A(z)]^{\mathrm{T}}}{\det[A(z)]}\bigg|_{z=f^{-1}(w)}=\frac{\varepsilon(w)\left[\left(\frac{\partial y}{\partial x}\right)^{2}+\left(\frac{\partial y}{\partial u}\right)^{2}\right]\begin{pmatrix}1&0\\0&1\end{pmatrix}}{\left(\frac{\partial y}{\partial x}\right)^{2}+\left(\frac{\partial y}{\partial u}\right)^{2}}=\varepsilon(w)\begin{pmatrix}1&0\\0&1\end{pmatrix} \tag{12.8}$$

The results in Eq. (12.8) show that in the region E after conformal transformation, the form of the permittivity remains unchanged. In other words, the conclusion obtained from a simply connected region of any shape is consistent with that obtained from the unit disc after conformal transformation. Therefore, the reconstruction in the region E can be firstly performed in the unit disc Ω, then transform the gray values onto the mapped coordinates in E though conformal transformation to obtain the solution in E. As a result, direct calculations on the noncircular region can be avoided for image reconstruction in noncircular sensors.

Since the positions of pixels in the reconstructed image are transformed from the unit disk to the noncircular region in advance, the image reconstruction is performed in two steps. Firstly, the image data are calculated using the reconstruction method of the circular sensor. Secondly, the data are displayed in the noncircular region in a point-to-point mapping manner from the circular region to the noncircular region through the conformal transformation. In the two steps, the scattering transform only needs to be calculated for the circular sensor before the conformal transformation that transforms the points in the noncircular region onto those in the unit disk (Cao et al., 2011a).

If the admittivity distribution is $\gamma(z)$, denote the voltage-to-current density map, i.e., the Dirichlet-to-Neumann map, as

$$\Lambda_{\gamma}:\ \varphi(z)|_{\partial\Omega}\rightarrow\gamma(z)\frac{\partial\varphi(z)}{\partial n}\bigg|_{\partial\Omega} \tag{12.9}$$

here, $\Lambda_{\gamma}(\cdot)$ denotes the voltage-to-current density map when the sensing field Ω contains the admittivity distribution $\gamma(z)$. In special, $\Lambda_{1}(\cdot)$ denotes the voltage-to-current density map when Ω contains constant permittivity of one. The functional $\Lambda_{\delta\gamma}(\cdot)$ represents the difference between the two Dirichlet-to-Neumann maps $\Lambda_{\gamma}(\cdot)$ and $\Lambda_{1}(\cdot)$, namely

$$\Lambda_{\delta\gamma}(\cdot)=(\Lambda_{\gamma}-\Lambda_{1})(\cdot) \tag{12.10}$$

where $\gamma=1+\delta\gamma$.

Similarly, the current density-to-voltage map, i.e., the Neumann-to-Dirichlet map, can be denoted as

$$R_\gamma:\ \gamma(z)\frac{\partial\varphi(z)}{\partial n}\bigg|_{\partial\Omega} \rightarrow \varphi(z)|_{\partial\Omega} \tag{12.11}$$

The functional $R_{\delta\gamma}(\cdot)$ represents the difference between the two Neumann-to-Dirichlet maps R_γ and R_1, namely

$$R_{\delta\gamma}(\cdot) = \left(R_\gamma - R_1\right)(\cdot) \tag{12.12}$$

where $\gamma = 1 + \delta\gamma$.

For the same sensor configuration, the Dirichlet-to-Neumann maps $\Lambda_\varepsilon(\cdot)$ and the Neumann-to-Dirichlet map $R_\varepsilon(\cdot)$ satisfy

$$\begin{cases} \Lambda_\gamma R_\gamma(\cdot) = I(\cdot) \\ R_\gamma \Lambda_\gamma(\cdot) = I(\cdot) - P(\cdot) \end{cases} \tag{12.13}$$

where $I(\cdot)$ and $P(\cdot)$ mean the identity operator and a nonzero projection operator, respectively (Siltanen et al., 2001). It should be noted that $\Lambda_\gamma(\cdot)$ and $R_\gamma(\cdot)$ are not the inverse operator of each other, as the dimension of the operator is N-1 if there exists N electrodes in a sensor. In the following sections, the symbols for the operators are rewritten for convenience, e.g., $\Lambda_\gamma(\cdot)$ and $R_\gamma(\cdot)$ are denoted by Λ_γ and R_γ, respectively.

12.3 Typical direct algorithms for electrical tomography

12.3.1 Calderon's method in a circular ET sensor

Calderon's method is a linearized method to deal with the inverse problem in the form of the admittivity Eq. (12.1). The main idea of Calderon's method can be simply expressed into two steps. Firstly, assume that the admittivity distribution is nearly constant, e.g., one. The distribution of electrical potential can be approximated analytically in this case. Secondly, the two-dimensional Fourier transform can be applied to reconstruct the perturbations of the admittivity distribution.

The divergence theorem relates integrals of the spatially varying admittivity to voltage and current measurements on the boundary, i.e.,

$$\int_\Omega v(z)[\nabla\cdot\gamma(z)\nabla\varphi(z)]\ dz = \int_{\partial\Omega} v(z)\gamma(z)\frac{\partial\varphi(z)}{\partial n}dl - \int_\Omega \nabla v(z)\cdot[\gamma(z)\nabla\varphi(z)]\ dz = 0 \tag{12.14}$$

where $v(z)$ is an arbitrary continuous function in L^2-Space. $dL = |dz|$ denotes the measured length of arc on $\partial\Omega$.

Eq. (12.14) can be rewritten as

$$\int_{\Omega} \gamma(z)\nabla v(z)\cdot\nabla\varphi(z)\ dz = \int_{\partial\Omega} v(z)\Lambda_{\gamma}[\varphi(z)]dL \tag{12.15}$$

For a disturbed admittivity $\gamma = 1 + \delta\gamma$, if the perturbation is strictly within the whole domain Ω, we obtain

$$\int_{\Omega} \nabla v(z)\cdot[1+\delta\gamma(z)]\nabla\varphi_{1+\delta\gamma}(z)\ dz = \int_{\partial\Omega} v(z)\Lambda_{1+\delta\gamma}[\varphi_1(z)]dL \tag{12.16}$$

and

$$\int_{\Omega} \nabla v(z)\cdot\nabla\varphi_1(z)\ dz = \int_{\partial\Omega} v(z)\Lambda_1[\varphi_1(z)]dL \tag{12.17}$$

As the excitation voltage is applied to an electrode or several electrodes of the ET system, $\varphi_{\gamma}(z) = \varphi_1(z)$ strictly holds on the boundary of the whole domain Ω. If we assume $\varphi_{\gamma}(z) \approx \varphi_1(z)$ in the whole sensing field, subtracting Eq. (12.16) by Eq. (12.17) results in

$$\int_{\partial\Omega} v(z)\Lambda_{\gamma}[\varphi_1(z)]dL - \int_{\partial\Omega} v(z)\Lambda_1[\varphi_1(z)]dL \approx \int_{\Omega} \delta\gamma(z)\ \nabla v(z)\cdot\nabla\varphi_1(z)\ dz \tag{12.18}$$

i.e.,

$$\int_{\partial\Omega} v(z)\left(\Lambda_{\gamma} - \Lambda_1\right)[\varphi_1(z)]dL \approx \int_{\Omega} \delta\gamma(z)\ \nabla v(z)\cdot\nabla\varphi_1(z)\ dz \tag{12.19}$$

The approximation in Eq. (12.19) holds when the disturbed admittivity is small. Let $\varphi_1(z) = e^{jkz}$ and $v(z) = e^{j\overline{kz}}$. The upper bar notation denotes the conjugate, $k = k_1 + jk_2$ is a complex number, while k_1 and k_2 are real numbers. $\varphi_1(z)$ and $v(z)$ can be treated as a linear combination of the trigonometric functions, e.g., sine and cosine functions. They are real-valued functions and can be realized on the hardware system using different voltage excitation patterns. Denote the left side of Eq. (12.19) as t(k), i.e.,

$$t(k) = \int_{\partial\Omega} e^{j\overline{kz}}\Lambda_{\delta\gamma}\left(e^{jkz}\right)dL \tag{12.20}$$

Then, Eq. (12.20) can be rewritten as

$$t(k) \approx -2\left(k_1^2 + k_2^2\right) \iint_{\Omega} \delta\gamma \; e^{-\mathrm{j}(-2k_1, 2k_2)\cdot(x,y)} \; dxdy \tag{12.21}$$

Using the inverse Fourier transform (Cao et al., 2009b), we obtain the distribution of the disturbed permittivity as follows

$$\delta\gamma(x, y) = \delta\gamma(z) \approx \frac{1}{2\pi^2} \iint_{R^2} \frac{t(k_1 + jk_2)}{k_1^2 + k_2^2} e^{\mathrm{j}(-2k_1 x + 2k_2 y)} dk_1 dk_2 \tag{12.22}$$

where $k = k_1 + jk_2$ is a complex number, and k_1 and k_2 are real numbers.

It has been proved that the reconstruction error of $\delta\varepsilon(x, y)$ approaches zero when $\delta\varepsilon(x, y)$ is close to zero (Mueller & Bikowski, 2008). Since Calderon's method is a simplification of the dbar method (Knudsen, Lassas, Mueller, & Siltanen, 2008), the theoretical scattering transform $t(k)$ has a general form

$$t(k) = \int_{\partial\Omega} e^{\mathrm{j}\overline{kz}} \Lambda_{\delta\gamma}[\psi(\cdot, k)] dL \tag{12.23}$$

where $\psi(\cdot, k)$ is the exponentially growing solution to equation (1), the upper bar notation denotes conjugate. On the boundary, the following equation holds

$$A_{\delta\gamma}[\psi(\cdot, k)]\big|_{\partial\Omega} = e^{\mathrm{j}kz} \tag{12.24}$$

where

$$A_{\delta\gamma}(\cdot) = [I + S_0 \Lambda_{\delta\varepsilon}](\cdot) \tag{12.25}$$

Here, $I(\cdot)$ means the identity operator. The Green function S_0 satisfies

$$(S_0\varphi)(z_0) = -\frac{1}{2\pi} \int_{\partial\Omega} \log|z_0 - z| \varphi(z) dL, \quad z_0 \in \partial\Omega \tag{12.26}$$

When the sensing field Ω is circular, Eq. (12.22) can be rewritten in the polar coordinates of parameters r and θ, namely,

$$\delta\gamma(x, y) \approx \frac{1}{2\pi^2} \int_0^{R_0} \int_{-\pi}^{\pi} \frac{t\left(re^{\mathrm{j}\theta}\right)}{r} e^{\mathrm{j}2r(-x\cos\theta + y\sin\theta)} d\theta dr \tag{12.27}$$

where R_0 is the radius of the region defined in the numerical dual integral.

The reconstruction error of $\delta\gamma(x, y)$ approaches zero, once the absolute values of $\delta\gamma(x, y)$ are sufficiently small (Calderon, 2006). However, if the admittivity disturbance is deviated from zero, Calderon's method tends to produce only rough estimates. That is why in the following sections, the closed-loop control principle was introduced to compensate the deviation caused by the complicated admittivity disturbances.

12.3.2 Iterative Calderon's method based on the closed-loop control

The flow diagram of the proposed iterative reconstruction algorithm based on Calderon's method for ECT is shown as Fig. 12.2. Initial permittivity distribution was obtained from measured data using Calderon's method in Eq. (12.27). The gray value reconstructed by the Calderon's method was normalized according to the actual permittivity contrast, i.e., in the range from lower to higher permittivity. The normalized distribution was used to construct the stiffness matrix in the finite element model. The floating potential in the model was achieved in a novel operation on the matrix for the first time. The Dirichlet-to-Neumann map can subsequently be realized from the matrix directly and neither excitation nor measurement was involved. The map was used as the negative feedback to the measured data.

In the design of the PID controller here, the Calderon's method was treated as the subject to be controlled. In our previous work, Calderon's method has been verified to be effective in cases of low-contrast distributions (Cao et al., 2009b). Usually, the image quality of the distributions reconstructed by using Calderon's method is degraded if the dielectrics contrast in the distributions to be reconstructed is increasing. In the typical PID control, the controller is designed to regulate the distortion introduced by the inexact model of the control subject, i.e., Calderon's method here, and that is why the closed-loop PID control is introduced in this work. The deviation between measured and calculated capacitances was used as the input of a PID controller to reduce the distortion introduced by the Calderon's method. The image reconstruction algorithm was taken as the control subject. The parameters of controller can be tuned, to make the whole closed loop be convergent. As a result, a new iterative algorithm was achieved in this work to reduce the reconstructed image error and improve the reconstruction quality.

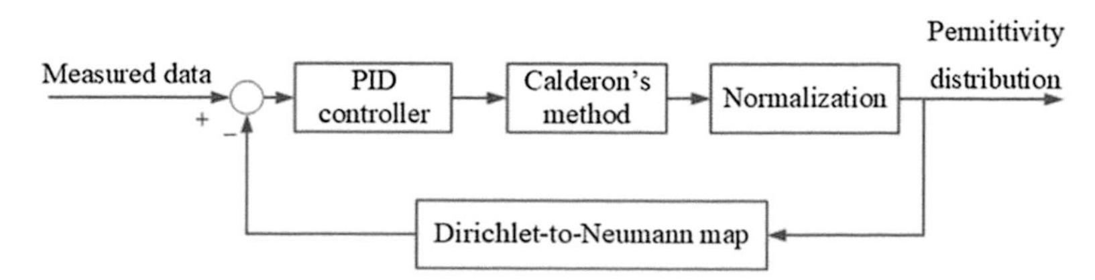

Figure 12.2 Flow diagram of the iterative reconstruction algorithm based on Calderon's method for electrical capacitance tomography.

The typical form of a digital PID controller is

$$x_2(i) = x_2(i-1) + \left(K_p + K_i + K_d\right)x_1(i) - \left(K_p + 2K_d\right)x_1(i-1) + K_d x_1(i-2) \tag{12.28}$$

where i represents the number of iterations. $x_1(i)$ and $x_2(i)$ are the input and output of PID controller at the i-th iteration, respectively. The parameters K_p , K_i , K_d are the coefficients for proportional, integral, and derivatives components, respectively. These values can be adjusted to make the closed-loop control be convergent. Here, the data normalization was carried out by Eq. (12.29),

$$x^* = (a-b)\cdot\frac{x - min(x)}{max(x) - min(x)} + b \tag{12.29}$$

where a and b are upper and lower limits of the normalization, respectively, and K_p is set to be one, then Eq. (12.29) can be simplified as

$$x_2(i) = x_2(i-1) + (1 + K_i + K_d)x_1(i) - (1 + 2K_d)x_1(i-1) + K_d x_1(i-2) \tag{12.30}$$

In other words, adjusting the normalized parameter is equivalent to the effect of K_p, which can affect the response speed of the system.

Fig. 12.3 depicts the flowchart of the proposed iterative reconstruction algorithm to sum up the steps described above. The whole process of the proposed algorithm can be summarized in five steps, as follows.

Step 1: Measure the capacitance data $\Lambda^N_{\delta\gamma}$ from N-electrode ECT sensor.

Step 2: Select the parameters K_p , K_i , K_d of PID controller and the threshold p for the stop criterion. Reconstruct initial permittivity distribution from measured data via using Calderon's method. Normalize the gray values according to the prior information of the actual permittivity contrast.

Step 3: Calculate the i-th Dirichlet-to-Neumann map $\Phi^N_{\delta\gamma}(i)$ by using normalized distribution and the calculation method introduced in section 0.

Step 4: Calculate the output $x^N_2(i)$ of the PID controller from the i-th deviation $x^N_1(i)$ between data $\Lambda^N_{\delta\gamma}$ and negative feedback data $\Phi^N_{\delta\gamma}(i)$. Update permittivity distribution $G(i)$ using Calderon's method. Normalize the gray value according to the actual permittivity contrast.

Step 5: When $|G(i) - G(i-1)|_2 < p$, the iteration terminates and output reconstructed results. If not, return to step 3 and continue the iterations.

12.3.3 Dbar method in a circular ET sensor

When the distribution of permittivity is varied from a uniformly distributed permittivity, as $\gamma\ (z)\ = 1$, the scattering transform $t(k)$ is defined as

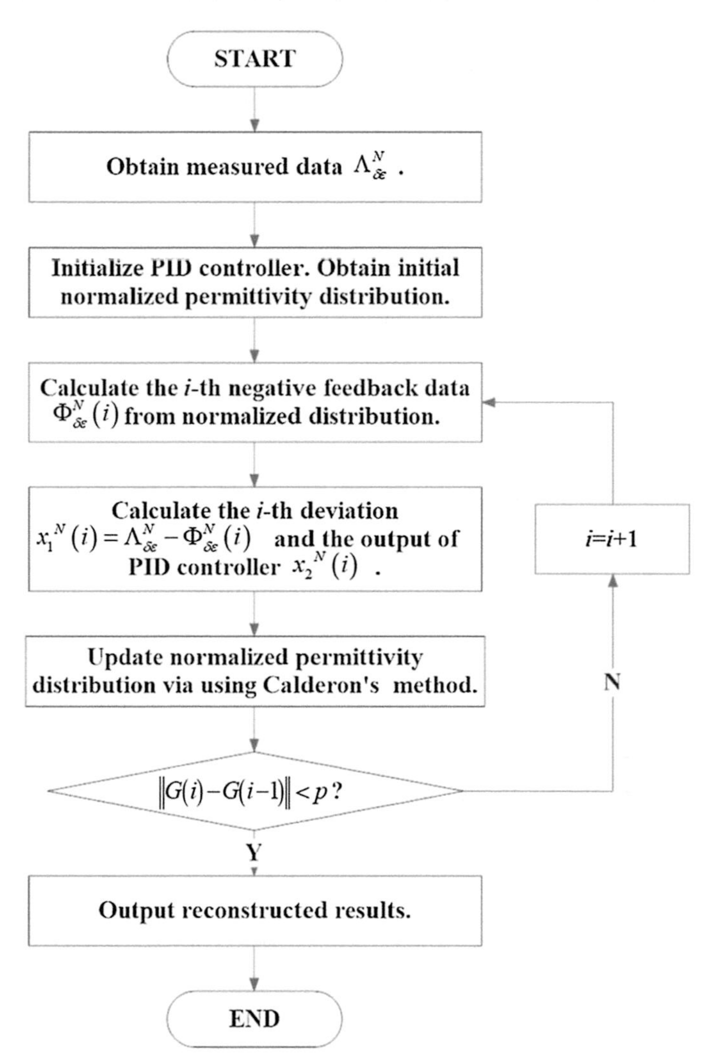

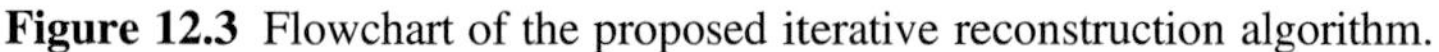

Figure 12.3 Flowchart of the proposed iterative reconstruction algorithm.

$$t(k) = \int_{\partial\Omega} e^{\mathrm{j}\overline{kz}} \Lambda_{\delta\gamma}\left(e^{\mathrm{j}kz}\right) dL \tag{12.31}$$

where $k = k_1 + \mathrm{j}k_2$ represents a complex number as k_1 and k_2 are real numbers. The values of $\gamma\ (z)$ can be obtained as it satisfies

$$\sqrt{\gamma(z)} = \lim_{k\to 0} \mu(z, k) \tag{12.32}$$

where $\mu(z,k)$ is the solution of the weakly singular Fredholm integral equation of the second kind (Nachman, 1996), i.e.,

$$\mu(z,s) = 1 + \frac{1}{(2\pi)^2}\iint\limits_{R^2} \frac{t(k)}{(s-k)\overline{k}} e^{-j\left(kz+\overline{kz}\right)}\overline{\mu(z,k)}dk_1dk_2 \tag{12.33}$$

In image reconstruction for noncircular sensors, direct calculations on the noncircular region can be avoided through using the conformal transformation (Cao et al., 2009a).

12.3.4 Factorization method in a circular ET sensor

The factorization method (Kirsch & Grinberg, 2007) is based on a characterization using the range of difference between two Neumann-to-Dirichlet maps (Harrach, 2013). If the subdomain, e.g., Ω_1 contains a different material other than the background in Ω, then we have the following results:

$z\in\Omega_1$ if and only if $\Phi_z|_{\partial\Omega_1}$ is in the range of the operator $\left|R_\gamma - R_1\right|^{1/2}$, where R_1 corresponds to Neumann-to-Dirichlet map when the conductivity $\gamma\ (x) = 1$ and Φ_z is the solution of

$$\begin{cases} \Delta_x\Phi_z(x) = d\cdot\nabla_x\delta_z(x) \\ \left.\dfrac{\partial}{\partial\nu}\Phi_z\right|_{\partial B} = 0 \\ \displaystyle\int\limits_{\partial\Omega}\Phi_z(x)dx = 0 \end{cases} \tag{12.34}$$

where d is a unit vector and $\delta_z(x)$ is the Dirac delta function at z. Especially, for the unit circle in the two dimensions, the test function Φ_z can be written as

$$\Phi_z(x) = \frac{1}{\pi}\frac{(z-x)\cdot d}{|z-x|^2} \quad \text{for all } x\in\partial\Omega \tag{12.35}$$

Since there exists only a limited number of electrodes in a real ET sensor, we have an approximate to the map, i.e., $|R_\varepsilon - R_1|^{1/2}$ that denoted by A. Let $Av_k = \lambda_k v_k$ be the spectral decomposition of the operator A, then it is yielded according to the Picard criterion that

$$\Phi_z|_{\partial\Omega_1}\in\text{Range}\left(A^{\frac{1}{2}}\right) \tag{12.36}$$

if and only if

$$f(z) := \frac{1}{\Phi_z|_{\partial\Omega}} \sum_{k=1}^{\infty} \frac{(\Phi_z|_{\partial\Omega}, v_k)^2}{\lambda_k} \tag{12.37}$$

is in the range of the operator $|R_\gamma - R_1|^{1/2}$.

Using the singular value decomposition of the discrete approximation $A_d \in C^{N-1 \times N-1}$,

$$A_d v_k^d = \lambda_k^d u_k^d,\ k = 1, 2, \cdots, N-1 \tag{12.38}$$

where N denotes the number of electrodes. With nonnegative $\{\lambda_k^d\} \subset R$ (sorted in descending order) and orthonormal bases $\{u_k^d\}, \{v_k^d\} \subset C^N$, the function $f(z)$ can be approximated by

$$f_d(z) := \frac{\sum_{k=1}^{m} \frac{\left|\Phi_{d_z^*} v_k^d\right|^2}{\lambda_k^d}}{\sum_{k=1}^{m} \left|\Phi_{d_z^*} v_k^d\right|^2} \tag{12.39}$$

where $\Phi_{d_z} \in C^N$ is the value of $\Phi_z|_{\partial\Omega_1}$ on each electrode.

To obtain a numerical criterion telling whether a point z belongs to the unknown inclusion Ω or not, one now has to decide whether the approximate value $f_d(z)$ is greater than a large positive number, e.g., a threshold denoted by C_∞, or not. The following indicator function can be used, i.e.,

$$\text{Ind}(z) := \log[f_d(z)]^{-1} \tag{12.40}$$

Then a reconstruction of the whole sensing region is then obtained by evaluating $f_d(z)$ and saying that all points with $f_d(z) < C_\infty$ belong to the inclusion (Hyvönen, 2005; Schmitt, 2009).

12.3.5 Direct methods implementation in a noncircular ET sensor

Without loss of generosity, here, the direct methods are evaluated in cases of ECT. To evaluate the performance of the direct methods and the conformal transformation, a square electrical capacitance tomographic sensor was constructed and used as an example to implement and validate the methods for a noncircular sensor.

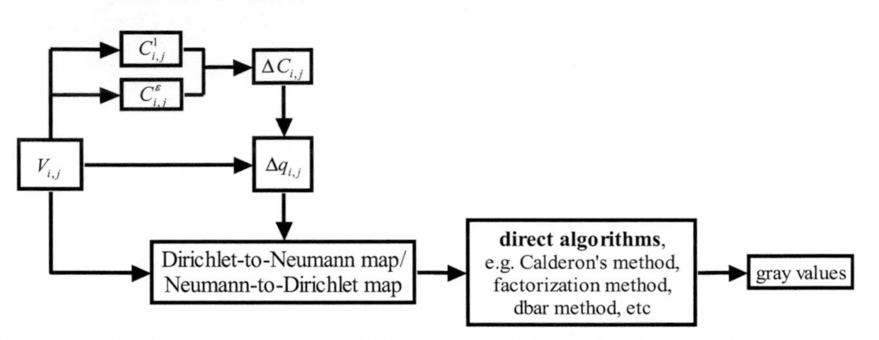

Figure 12.4 Implementation process of the direct methods in electrical capacitance tomography.

The implementation of the direct methods can be depicted in Fig. 12.4. It can be summarized into three steps. Firstly, the traditional excitation and measurement scheme was applied to the N-electrode sensor in order to obtain the $N\,(N\text{-}1)/2$ independent measurements of capacitance changes, i.e., $\Delta C_{i,j}$, the $N\times N$ transform matrix ΔC was constructed and the change of electric charge on each electrode, i.e., $\Delta q_{i,j}$ that is associated with the applied voltage excitations, i.e., $V_{i,j}$. Secondly, the discretized form of the Neumann-to-Dirichlet map or Dirichlet-to-Neumann map was obtained. Finally, the values of the indicator function at each pixel were calculated by using a direct algorithm and conformally transformed to the associated point in the simply connected region for reconstructed images.

12.4 Dirichlet-to-Neumann/Neumann-to-Dirichlet maps

For the direct algorithms mentioned above, the construction of the map on the boundary links the measurement to the reconstructed gray values. The boundary maps can be obtained in several methods, e.g., the capacitance/conductance matrix, matrix transformation methods, as well as the inner product method. In the section, the first two methods will be involved; more details in other methods can be found in the literature (Cao & Xu, 2013a, 2013b, p03004; Hamilton et al., 2012).

12.4.1 Construction of the Dirichlet-to-Neumann map

As the ET sensor has an earthed screen that introduces stray capacitances, the voltage patterns required by Eq. (12.9) cannot be explicitly implemented on the electrodes. Without loss of generosity, the map construction is detailed in cases of ECT. The electric charge on each electrode is given by the general expression q = C V, in matrix form as (Yang, Spink, Gamio, & Beck, 1997)

$$\begin{bmatrix} q_1 \\ q_2 \\ q_3 \\ \vdots \\ q_N \end{bmatrix} = \begin{bmatrix} C_{1,1} & -C_{1,2} & -C_{1,3} & \cdots & -C_{1,N} \\ -C_{2,1} & C_{2,2} & -C_{2,3} & \cdots & -C_{2,N} \\ -C_{3,1} & -C_{3,2} & C_{3,3} & \cdots & -C_{3,N} \\ \vdots & \vdots & \vdots & \ddots & \vdots \\ -C_{N,1} & -C_{N,2} & -C_{N,3} & \cdots & C_{N,N} \end{bmatrix} \begin{bmatrix} V_1 \\ V_2 \\ V_3 \\ \vdots \\ V_N \end{bmatrix} \tag{12.41}$$

where q_s is the electric charge on the s-th electrode, V_t the voltage applied to the t-th electrode, and $C_{s,t}$ the capacitance between electrode pair s-t ($s \neq t$). The self-capacitance of the s-th electrode, i.e., $C_{s,s}$, is defined as the sum of the capacitances between this electrode and all other electrodes plus the capacitance between the electrode and the earthed screen, i.e.,

$$C_{s,s} = \sum_{\substack{t=1 \\ t \neq s}}^{N} C_{s,t} + C_s^e \tag{12.42}$$

where C_s^e is the stray capacitance between the s-th electrode and the earthed screen, $C_{s,t}$ the capacitance between electrodes s and t ($s \neq t$).

When the interval between each two neighboring electrodes is small, the effect of the variation in capacitance caused by the earthed screen can be neglected (Cao et al., 2007, 2008). The relationship between $\Delta C_{s,s}$ and variations in capacitance on all N electrodes can be given by

$$\Delta C_{s,s} = \sum_{j=1, j \neq s}^{N} \Delta C_{s,j} \tag{12.43}$$

where $\Delta C_{s,s}$ is the variation of the self-capacitance of the s-th electrode, $\Delta C_{s,j}$ the variation of the mutual capacitance between the electrode pair s-j ($s \neq j$). When the voltage V_j^t in the t-th voltage pattern is applied to the j-th electrode, the variations of the electric charge Δq_j^t ($j = 1, 2, \ldots, N$) on each of the N electrodes can be expressed as

$$\Delta q_j^t = \Delta C_{j,j} V_j^t - \sum_{s \neq j} \Delta C_{s,j} V_s^t \tag{12.44}$$

Here, $\Delta C_{s,j}$ can be measured through the traditional excitation scheme, i.e., the s-th electrode is excited with all other electrodes grounded. The N variations of the electric charge, i.e., Δq_j^t ($j = 1, 2, \ldots, N$) under the t-th voltage pattern are obtained by using the transformation described in formula Eq. (12.44). Finally, the discretized form of the voltage-to-current density map, i.e., the Dirichlet-to-Neumann map $\Lambda_{\delta\gamma} = \Lambda_\gamma - \Lambda_1$ can be constructed as follows, namely,

$$\Lambda_{\delta\gamma} = \begin{bmatrix} \Delta C_{1,1} & -\Delta C_{1,2} & -\Delta C_{1,3} & \cdots & -\Delta C_{1,N} \\ -\Delta C_{2,1} & \Delta C_{2,2} & -\Delta C_{2,3} & \cdots & -\Delta C_{2,N} \\ -\Delta C_{3,1} & -\Delta C_{3,2} & \Delta C_{3,3} & \cdots & -\Delta C_{3,N} \\ \vdots & \vdots & \vdots & \ddots & \vdots \\ -\Delta C_{N,1} & -\Delta C_{N,2} & -\Delta C_{N,3} & \cdots & \Delta C_{N,N} \end{bmatrix} \tag{12.45}$$

According to Eq. (12.45), $\Delta C_{i,j}$ can be measured from the traditional scheme, i.e., the i-th electrode is excited while all other electrodes grounded. The resulted electric charges of the voltage patterns can be obtained by multiplying the matrix given in Eq. (12.45) with the voltages on the electrodes.

12.4.2 Construction of the Neumann-to-Dirichlet map

There is another method to construct the Dirichlet-to-Neumann map as well as the Neumann-to-Dirichlet map. The Neumann-to-Dirichlet map $R_{\delta\gamma} = R_\gamma - R_1$ can be constructed by using the inner product through a method similar to adjacent excitation measurement (Mueller & Isaacson, 2004, pp. 121–128). Here we provide an alternative method to calculate the Neumann-to-Dirichlet map.

If the excitation voltage on the j-th electrode during the i-th scanning can be denoted as $V_{i,j}$, the varied electric charges after the distribution distorted from 1 to $\gamma(z)$ at the same electrode are $\Delta q_{i,j}$ ($1 \le i \le N-1 \;\; 1 \le j \le N-1$), then the excitations and measurements can be constructed in the forms of matrices,

$$\begin{bmatrix} V_{1,1} & V_{1,2} & V_{1,3} & \cdots & V_{1,N-1} \\ V_{2,1} & V_{2,2} & V_{2,3} & \cdots & V_{2,N-1} \\ V_{3,1} & V_{3,2} & V_{3,3} & \cdots & V_{2,N-1} \\ \vdots & \vdots & \vdots & \ddots & \vdots \\ V_{N-1,1} & V_{N-1,2} & V_{N-1,3} & \cdots & V_{N-1,N-1} \end{bmatrix}$$

$$= \begin{bmatrix} \Delta q_{1,1} & -\Delta q_{1,2} & -\Delta q_{1,3} & \cdots & -\Delta q_{1,N-1} \\ -\Delta q_{2,1} & \Delta q_{2,2} & -\Delta q_{2,3} & \cdots & -\Delta q_{2,N-1} \\ -\Delta q_{3,1} & -\Delta q_{3,2} & \Delta q_{3,3} & \cdots & -\Delta q_{3,N-1} \\ \vdots & \vdots & \vdots & \ddots & \vdots \\ -\Delta q_{N-1,1} & -\Delta q_{N-1,2} & -\Delta q_{N-1,3} & \cdots & \Delta q_{N-1,N-1} \end{bmatrix} \frac{R_{\delta\gamma}}{A_E} \tag{12.46}$$

Here, the values of $V_{i,j}$ in each excitation are normalized to have a zero mean, and then a unique expression of $R_{\delta\gamma}$ without loss of generality can be obtained. It should be noted that the rank of the operator is only N-1 for a sensor with N electrodes.

Eq. (12.46) can be rewritten as

$$V = q\frac{R_{\delta\gamma}}{A_E} \tag{12.47}$$

According to Eq. (12.46), values of electric charges in Eq. (12.44) can be obtained. Using Gram-Schmidt orthonormalization, we have a set of orthonormalized basis for V, e.g., denoted by V_o, which satisfies

$$V = V_0 G_s \tag{12.48}$$

where G_s is the transfer matrix resulted from the orthonormalization, $V_o^T V_o = I$. Inserting Eq. (12.48) into Eq. (12.47) leads to the discretized form of the Neumann-to-Dirichlet map, i.e.,

$$R_{\delta\gamma} = A_E\left(G_S^{-1} V^T q\right)^{-1} \tag{12.49}$$

12.4.3 Fast calculation of the Dirichlet-to-Neumann map from the stiffness matrix in the finite electrode model

For a given ET sensor, if the distribution in the sensing region and electrode configuration is determined, the Dirichlet-to-Neumann map uniquely exists, no matter whether its electrodes are excited or not. In our previous work, a fast calculation method of the Dirichlet-to-Neumann map is derived from the finite element model, i.e., the stiffness matrix (Cao, Ji, & Xu, 2018). It is direct and efficient and can be achieved from the node voltage method derived from the systematic application of Kirchhoff's laws. In the literature, the Dirichlet-to-Neumann map can be constructed from stiffness matrix, if each electrode in the model can be represented by a single node in the stiffness matrix assembling. However, in most cases, it is unlikely to achieve the stiffness matrix like that, especially, when the commercial finite element analysis software was used to provide the stiffness matrix. In the commercial software, the meshing can be achieved adaptively and efficiently, and stiffness matrix can be assembled automatically. But at the same time, more nodes can be found on each electrode, and the methods mentioned in the literature failed to work directly in this case. Here, the boundary constraints of the floating potential on each electrode were introduced on the finite element model for the first time; it can make the electrical potential on all nodes in each electrode be exactly equal. By applying the Kirchhoff's current law, the node voltage method can be written as linear equations, i.e.,

$$A\begin{bmatrix} U_e \\ \\ U_o \end{bmatrix} = \begin{bmatrix} I_e \\ \\ I_o \end{bmatrix} \tag{12.50}$$

where A is the matrix denoting the Dirichlet-to-Neumann map on all nodes in the finite element model, and in the matrix, all the nodes for each electrode are listed together for further operation. U_e and I_e are, respectively, the electrical potential and current on the nodes on the electrodes, and, U_o and I_o are the electrical potential and current on the other nodes in the model, respectively. For the writing convenience, the nodes not on the electrodes are denoted as "others," and the matrix A was rewritten in the form of four submatrices, namely

$$A = \begin{bmatrix} A_{ee} & A_{oe} \\ \\ A_{eo} & A_{oo} \end{bmatrix} \tag{12.51}$$

where A_{ee}, A_{oe}, A_{eo}, and A_{oo} mean the maps of the electrical potential to current from electrodes-to-electrodes, others-to-electrodes, electrodes-to-others, and others-to-others, respectively.

And Eq. (12.50) can be rewritten as

$$\begin{cases} A_{ee}U_e + A_{oe}U_o = I_e \\ A_{eo}U_e + A_{oo}U_o = I_o \end{cases} \tag{12.52}$$

For the finite element model of an ET sensor, the excitation was applied only on the electrodes on the boundary. No excitation and no current or voltage source exists in its interior, namely

$$I_o = 0 \tag{12.53}$$

Combining Eqs. (12.52) and (12.53), the voltage and current on the nodes on the boundary can be related as

$$\left(A_{ee} - A_{eo}A_{oo}^{-1}A_{oe}\right)U_e = I_e \tag{12.54}$$

From the definition of the Dirichlet-to-Neumann map in Eq. (12.9), the map can be derived from Eq. (12.54), and be expressed as

$$\Lambda_\varepsilon = A_{ee} - A_{eo}A_{oo}^{-1}A_{oe} \tag{12.55}$$

For an ET sensor with N electrodes, however, the map cannot be applied directly by using Eq. (12.55). In fact, the nodes on each electrode have the same potential, and this prior information should be incorporated to achieve the precise map. Take the case on the first electrode as an example. If there are m nodes on the first electrode, and the model consists of M nodes, then the matrix A can be written as

$$A = \begin{bmatrix} a_{1,1} & \cdots & a_{1,m} & a_{1,m+1} & \cdots & a_{1,M} \\ \vdots & \ddots & \vdots & \vdots & \ddots & \vdots \\ a_{m,1} & \cdots & a_{m,m} & a_{m,m+1} & \cdots & a_{m,M} \\ a_{m+1,1} & \cdots & a_{m+1,m} & a_{m+1,m+1} & \cdots & a_{m+1,M} \\ \vdots & \ddots & \vdots & \vdots & \ddots & \vdots \\ a_{M,1} & \cdots & a_{M,m} & a_{M,m+1} & \cdots & a_{M,M} \end{bmatrix}_{M\times M} \tag{12.56}$$

In the matrix A, the nodes for N electrodes are listed as N groups. For the first electrode, the elements of the m nodes are listed in the first m rows and arrays. In the same way, the elements of the electrodes for the second electrode followed, and then all the elements of the electrodes can be arranged in the matrix for convenience. Since all the m nodes on the first electrode have the same potential, the first electrode can be represented by any node in all the m nodes, and the matrix A for the first electrode can be rewritten as

$$A_1 = \begin{bmatrix} \sum_{l=1}^{m}\sum_{k=1}^{m} a_{l,k} & \sum_{l=1}^{m} a_{l,m+1} & \sum_{l=1}^{m} a_{l,m+2} & \cdots & \sum_{l=1}^{m} a_{l,M} \\ \sum_{k=1}^{m} a_{m+1,k} & a_{m+1,m+1} & a_{m+1,m+2} & \cdots & a_{m+1,M} \\ \sum_{k=1}^{m} a_{m+2,k} & a_{m+2,m+1} & a_{m+2,m+2} & \cdots & a_{m+2,M} \\ \vdots & \vdots & \vdots & \ddots & \vdots \\ \sum_{k=1}^{m} a_{M,k} & a_{M,m+1} & a_{M,m+2} & \cdots & a_{M,M} \end{bmatrix}_{(M+1-m)\times(M+1-m)} \tag{12.57}$$

Similarly, the simplified version of the matrix A can also be obtained by incorporating the equipotential constraints on the nodes of all the other electrodes, and the matrix can be reduced to be smaller size of $(M - N_0 + N) \times (M - N_0 + N)$; here, N_0 is the total number of all the nodes in all electrodes. As a result, the Dirichlet-to-Neumann map can be calculated from the reduced matrix by using Eq. (12.57), and it is easy to operate to make the numerical operations.

12.5 Calculation of the scattering transforms

Theoretically, the ideal scattering transform $t(k)$ in Eqs. (12.20) and (12.23) requires infinite number of electrodes. However, only a limited number of electrodes are available in a real ET system. Two approximations of the scattering transform $t(k)$ can be used. One is the approximation, referred to as t^{exp}, when $\psi(\cdot, k)$ is replaced by the exponential solution (Knudsen, Lassas, Mueller, & Siltanen, 2007). The other is deduced from the famous Faddeev Green function that is widely used in the inverse scattering theory (Nachman, 1996). It is termed as $t^b(k)$, which needs to solve a weakly singular Fredholm integral equation of the second kind on the boundary. In fact, through a linear conformal transformation, any simply connected region is equivalent to the unit disk; we assume that the radius of the sensor is one.

12.5.1 Approximation of $t(k)$ using t^{exp}

Expand e^{jkz} into a Fourier series with $z = e^{j\theta}$ to obtain

$$e^{jkz} = \sum_{n=-\infty}^{\infty} a_n(k) e^{jn\theta} \tag{12.58}$$

where $a_n(k) = \frac{(jk)^n}{n!}$.

If we assume $A_{\delta\varepsilon}(\cdot)$ to be the identity operator, then the scattering transform in Eq. (12.5) can be simplified as

$$t(k) = \int_{\partial\Omega} e^{j\overline{kz}} (\Lambda_\gamma - \Lambda_1)(e^{jn\theta_n}) dL \tag{12.59}$$

According to Eq. (12.59), the scattering transform can be rewritten as

$$t^{exp}(k) = \sum_{m=0}^{\infty} a_m(\overline{k}) \sum_{n=0}^{\infty} a_n(k) \int_{\partial\Omega} e^{-jm\theta_m} \Lambda_{\delta\gamma}(e^{jn\theta_n}) dL \tag{12.60}$$

Let A_E denote the area of an electrode, N the number of the electrodes in the ET sensor. Assume that the s-th electrode is centered at angle $\theta_s = \frac{2\pi s}{N}$ (Isaacson, Mueller, Newell, & Siltanen, 2004). For an N-electrode ET sensor, the values of $t(k_1 + jk_2)$ can be well approximated from the following equation, namely

$$t^{exp}(k) = t(k_1 + jk_2) \approx A_E \frac{2\pi}{N} \overline{E} \left(\Lambda_\gamma^N - \Lambda_1^N \right) E^T \tag{12.61}$$

Here, A_E represents the area of an electrode, Λ_γ^N and Λ_1^N are Dirichlet-to-Neumann maps in the form of $N \times N$ matrices obtained from measured. If the center of s-th electrode is located at angle $\theta_s = \frac{2\pi s}{N}$, the vector E can be determined as $E = \left[e^{j(k_1+jk_2)e^{j\theta_1}}, e^{j(k_1+jk_2)e^{j\theta_2}}, \ldots, e^{j(k_1+jk_2)e^{j\theta_N}} \right]$. $\overline{E}$ and E^T, respectively, represent the conjugate and transpose of the vector.

12.5.2 Approximation of $t(k)$ using $t^b(k)$

If we use $A_{\delta\varepsilon}(\cdot)$ defined in Eqs. (12.24) and (12.25), then

$$\psi(\cdot,k)|_{\partial\Omega} = \left[I + S_0\Lambda_{\delta\gamma}\right]^{-1}\left(e^{jkz}\right) \tag{12.62}$$

For an N-electrode square ET sensor, $t^b(k)$ can be obtained using the following equation

$$t^b(k) \approx A_E\frac{2\pi}{N}\overline{E}\Lambda_{\delta\gamma}\left[I + S_0\Lambda_{\delta\gamma}\right]^{-1}E^T \tag{12.63}$$

Apparently, if $S_0\Lambda_{\delta\gamma} \approx 0$, then $t^b(k) = t^{exp}$. From this point of view, t^{exp} is a simplification of $t^b(k)$. Meanwhile, Eq. (12.63) is ill posed and theoretically solvable; the operator that must be inverted to solve the equation needs to be regularized. In particular, t^{exp} is a regularized form of $t^b(k)$, the reconstructed images by using t^{exp} are similar to those by using $t^b(k)$ (Cao et al., 2011a).In the following sections, only t^{exp} will be used to calculate the scattering transforms.

12.6 Applications in image reconstruction

12.6.1 Numerical cases

The conformal transformation from a square to the unit disk in the complex plane is

$$w = \frac{sn\left(\sqrt{2}\,z, \frac{1}{\sqrt{2}}\right)}{\sqrt{2}dn\left(\sqrt{2}\,z, \frac{1}{\sqrt{2}}\right)} \tag{12.64}$$

where $sn(\cdot,\cdot)$ and $dn(\cdot,\cdot)$ are the Jacobi elliptic functions. z and w are the complex numbers that represent the locations in the unit disk and the square, respectively. In the calculations, the precise integration method was used to calculate the elliptic functions, as it generated the results in the machine precision (Cao et al., 2008; Zhong & Yao, 2004). The square sensor used here was optimized in the sense of information entropy. The size and interval of each electrode in the square sensor were associated with those of the evenly positioned electrode configuration in the circular sensor through a unique conformal transformation, see Fig. 12.5.

Six phantoms with different permittivity distributions were constructed as shown in Fig. 12.6. The distributions of permittivity were of binary values. The electrical permittivity in the gray and white regions are one (representing air) and three (representing solids), respectively. The phantoms are designed to represent three typical distributions, i.e., one disturbed region (phantoms (a) and (f)), two disturbed regions (phantoms (b), (c), and (e)), and more than two disturbed regions (phantom (d)). In the case of one disturbed region, phantoms (a) and (f) represent a simply connected and a multiply connected region, respectively. Phantoms (b), (c), and (e) provide the

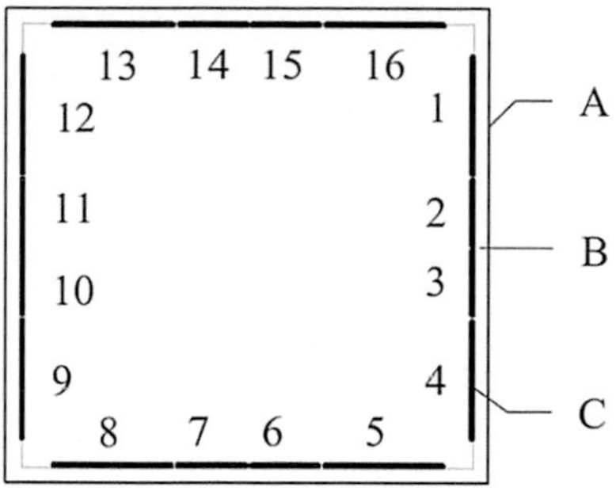

13 14 15 16
12 1
11 2
10 3
9 4
8 7 6 5
A
B
C

Figure 12.5 Cross-section of the optimized square sensor with 16 electrodes. A. copper shielding, B. plastic frame, C. measurement electrode.

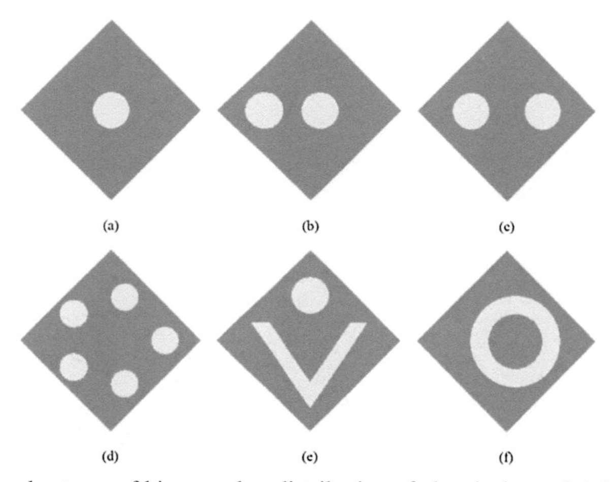

Figure 12.6 Six phantoms of binary-value distribution of electrical permittivity.

comparisons of different locations and shapes. Phantom (d) can be used to test the performance of the proposed method for multirod distributions. The configuration of 16 electrodes is depicted in Fig. 12.5. The forward problem was solved by using the finite element analysis software.

For the convenience of comparison, the gray values of pixels in all the reconstructed images are normalized to the range from one to three. The reconstructed results in Figs. 12.7–12.9 show the reconstructed images by using Calderon's method, dbar method, and factorization method, respectively. In the Calderon's method, the *priori* information of the distributions was used to minimize the artifacts, i.e., all the negative real parts of $\delta\varepsilon(x,y)$ were set to be zero (Cao et al., 2009b). The image quality in Fig. 12.7 appears to be better than those in Fig. 12.8, which were produced by using dbar method. However, it should be noted that the results produced by using dbar method are all positive; as a result, no priori information was used. Fig. 12.9 shows that factorization method fails to work when the background is not a connected region,

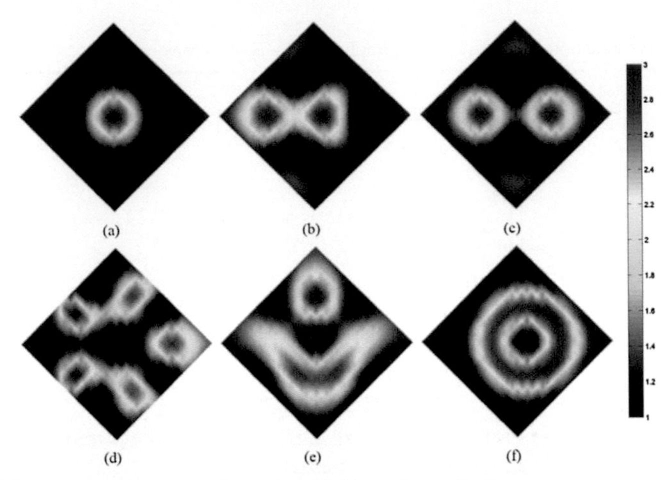

Figure 12.7 Images reconstructed by using Calderon's method.

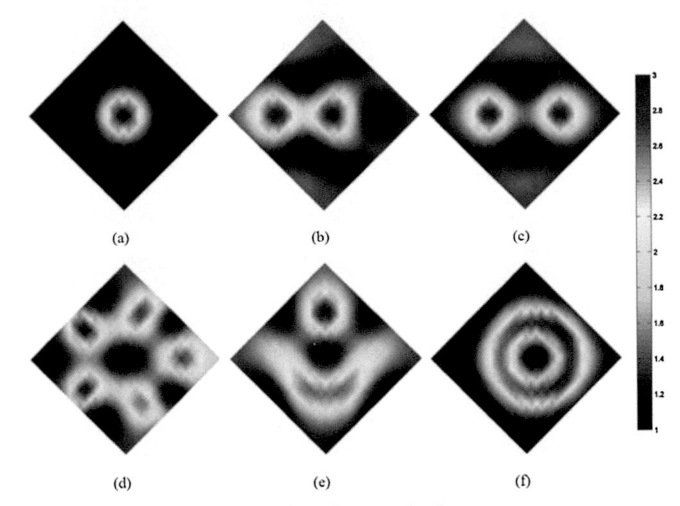

Figure 12.8 Images reconstructed by using dbar method.

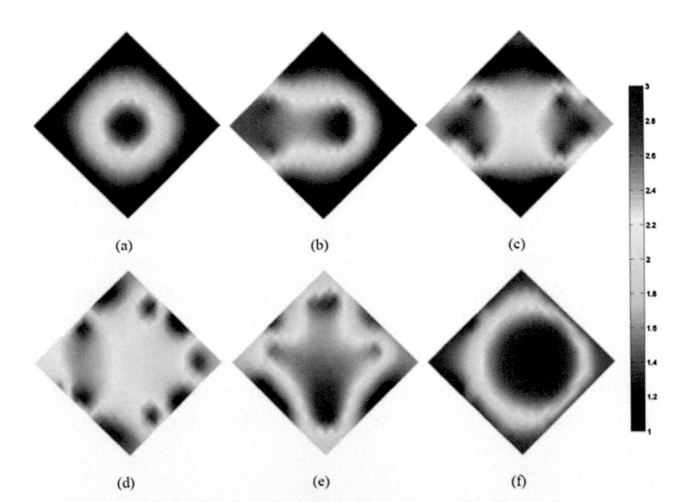

Figure 12.9 Images reconstructed by using factorization method.

e.g., Fig. 12.9(f). Compared with the results shown in Figs. 12.7 and 12.8, the spatial resolution of reconstructed images in Fig. 12.9 was low. However, the factorization method takes about 0.03 s to reconstruct one image, while the time spent for one image by using Calderon's method and dbar method were about 0.7 and 56.8 s, respectively. Moreover, the reconstructed value at each pixel of the images was obtained directly and independently.

Performance of the proposed algorithm was verified by using both simulation and experimental data in this section. Four phantoms in circular region and two phantoms in square region with different binary-value distributions were chosen for comparisons. The real distributions are depicted in Fig. 12.10. The region in white stands for higher permittivity material and the permittivity value was set to be three. The region in black means lower permittivity material and the permittivity was set to be one. The phantoms include the single object and more objects, and objects with complicated shapes, e.g., V-shape and cross. Also, in the case of a noncircular sensor, e.g., square sensor, two typical distributions were selected to evaluate the performance of the proposed algorithm, as illustrated in Fig. 12.10. The reconstruction of noncircular region can be summarized as following two steps. First, reconstruction results were calculated by using the proposed iterative reconstruction algorithm in the unit disc. Second, the results were displayed in the noncircular region through the conformal transformation from the unit disc to noncircular region.

The conformal transformation from the unit disc to the square in the complex plane can be formulated as

$$w_s = f(z) = \int_0^z \frac{dt}{\sqrt{1 - t^4}} \tag{12.65}$$

where z and w_s represent the complex numbers denoting points in the unit disc and the square, respectively.

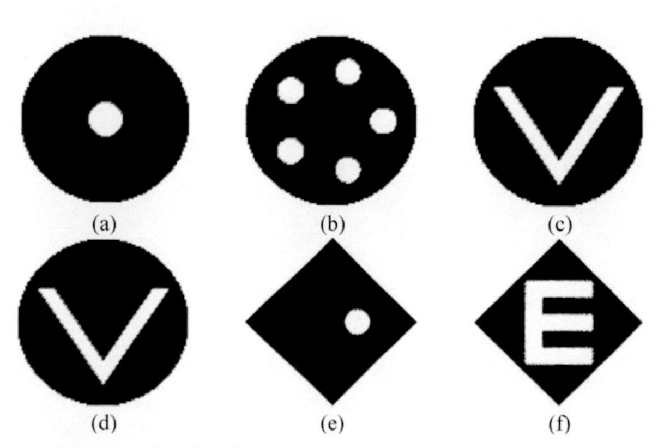

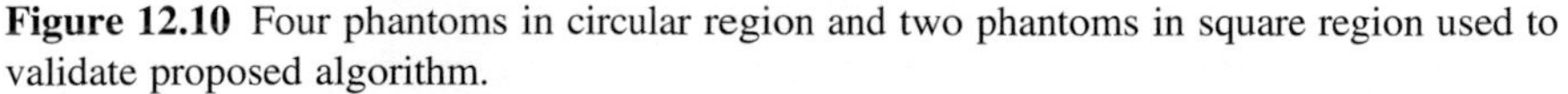
Figure 12.10 Four phantoms in circular region and two phantoms in square region used to validate proposed algorithm.

Two criteria were used to evaluate the quality of reconstructed images. One was to evaluate the visual appearance of reconstructed images, that is, qualitative way. The other was quantitative and reconstructed images are assessed in the aspect of the image error (Yang & Peng, 2003), namely

$$I_e = \frac{\|\widehat{g} - g\|}{\|g\|} \tag{12.66}$$

where g and $\widehat{g}$ are the real permittivity distribution of sensing region and the reconstructed result, respectively. However, the image error formulated in Eq. (12.66) has its own limitation in evaluating the image quality. The image is a three-dimensional vector, while image error is just a scalar parameter. The image error cannot reflect all the information of reconstructed images. Therefore, two methods will both be used to compare the quality of reconstructed images.

The simulated data in cases of both circular and square ET sensor with 16 electrodes that are commonly used in ET were used to reconstruct the images firstly to retrieve the real distributions in Fig. 12.10. The number of pixels of the reconstructed image is 7668. The degree of freedom of finite element model is about 3000. All the computation were carried out in the MALAB environment on a PC that has a Core i7 processor of 3.60 GHz and 8-Gbyte RAM, and it took about 3–4 s for each iteration.

Fig. 12.11 shows the initial images and images after 120 iterations of phantoms (a)–(d) with circular sensor and phantoms (e)–(f) with square sensor. The initial

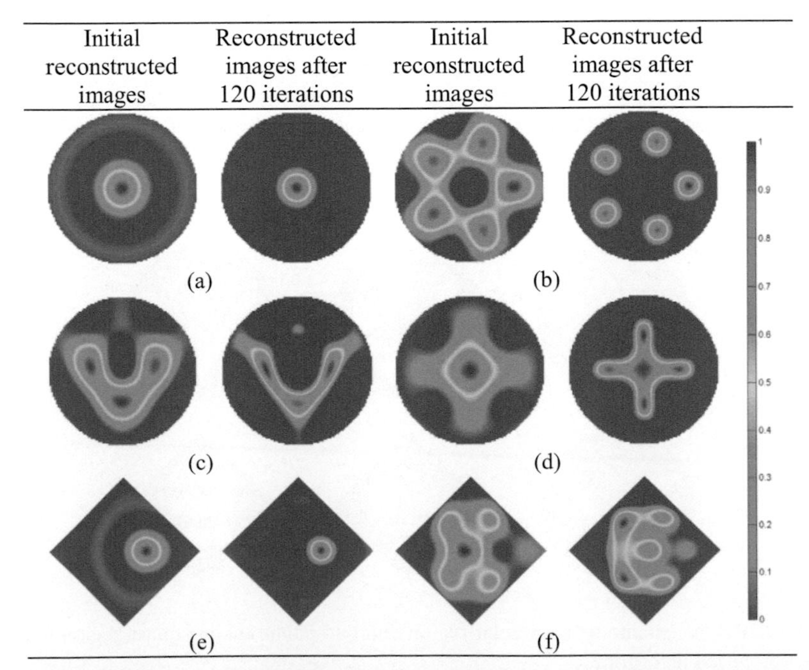

Figure 12.11 Initial reconstructed images and reconstructed images after 120 iterations of phantoms (a)–(f) by using noise-free data.

images were reconstructed by applying the Calderon's method. The reconstructed images after 120 iterations were obtained from the proposed iterative algorithm. When implementing the proposed algorithm, the numerical integration radius R_I was set as five through trial-and-error method. If the numerical integration radius R_I in Eq. (12.27) gets too small, too much useful information will be truncated. If R_I gets too large, the reconstruction will become unstable because of the high frequency noises. The values of the PID parameters were respectively selected as $K_p = 1$, $K_i = 100$, and $K_d = 0$ through various numerical trial and error simulations.

For all the reconstructed images, both Calderon's method and the proposed algorithm give acceptable reconstructions about the position of the objects. The proposed algorithm performs better in reconstructing the size of objects, see Figs. 12.11(a) and 12.12(e). When multiple objects existed in the sensing region, the proposed algorithm can well distinguish each object, as shown in Fig. 12.11(b). For other phantoms with complicated permittivity distributions, as shown in Fig. 12.11(c)-(e), the iterative algorithm reconstructs the shape of the objects more accurately. In cases of both circular and square sensors, the iterative Calderon's algorithm achieves reconstructions in better agreement with real distributions than Calderon's method.

For quantifiable evaluation, image errors of phantoms (a)–(f) are plotted in Fig. 12.12. During the iterative process, the image error curves of some phantoms have concussion, but eventually tends to be stable and less than initial image error. This shows that with the proper PID controller parameters, the whole closed loop has good convergence. The number of iterations required to reach convergence varies for different phantoms. For example, the reconstruction for phantoms (a) and (e) only takes less than 10 iterations, while those for phantoms (b), (c), and (f) requires for

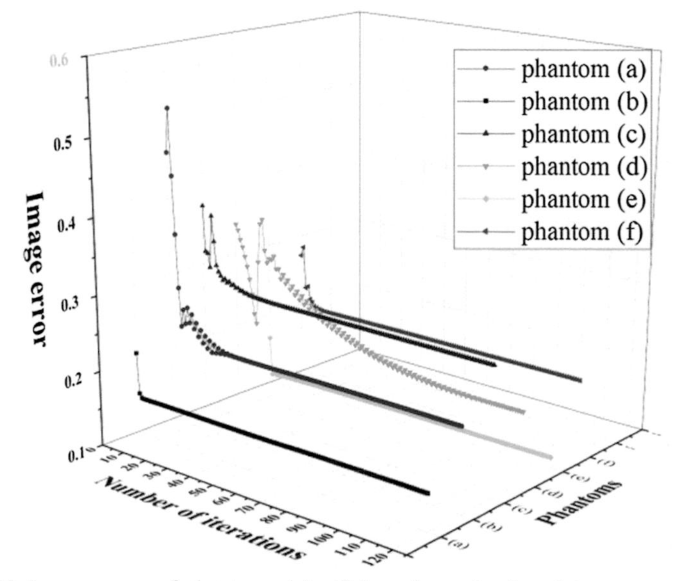

Figure 12.12 Image errors of phantoms (a)–(f) by using noise-free data.

about 30 iterations. In the case of the phantom (d), more iterations are required, and it stabilizes after about 100 iterations. In general, the distributions with simpler shapes stabilize faster than more complicated distributions. Even if the number of iterations continues to increase, the proposed algorithm would not diverge for these phantoms. Therefore, once the proposed algorithm reaches convergence, the iteration can be stopped and output reconstructed results. Here, the iterations terminate when $|G(i) - G(i-1)|_2 < p$, and the threshold was set as $p = 5.0 \times 10^{-2}$.

In the aspect of the visual appearance of reconstructed images and image errors, the proposed algorithm shows good performance to improve image reconstructed quality.

12.6.2 *Static phantoms*

Two cases of permittivity distributions were constructed for experimental test, i.e., two plastic rods and one plastic pipe. The reconstructed images of the two plastic rods in the square sensor were depicted in Fig. 12.13(c),(e) and (g). With the same process, the reconstructed image of a ring-shape plastic pipe (see Fig. 12.13(b)) was also obtained as shown in Fig. 12.13(d),(f) and (h). Like the results from simulated data, the reconstructed images from experimental data by using both Calderon's method and dbar

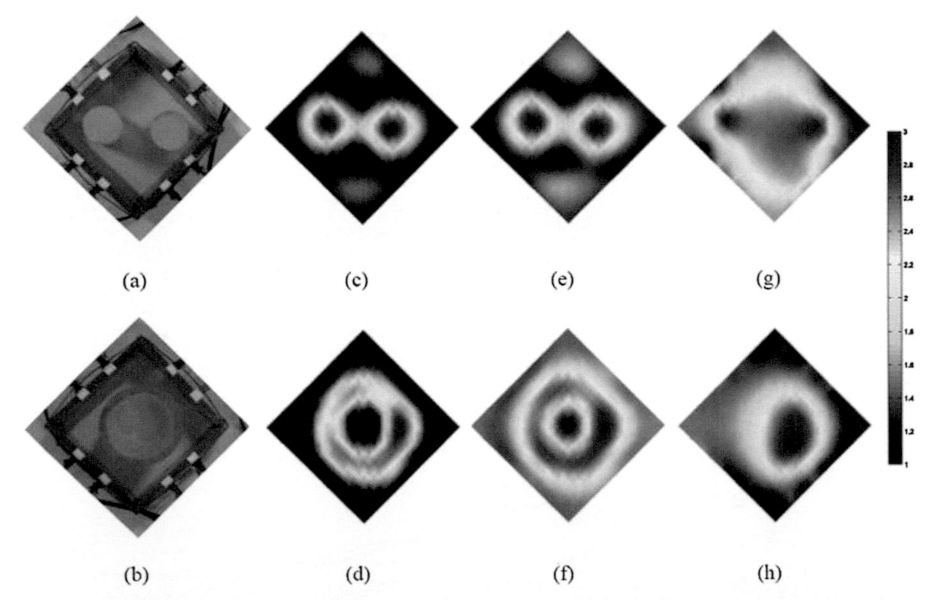

Figure 12.13 Experiment results obtained by using direct methods. (a) and (b): phantoms of two plastic rods and a ring-shape plastic pipe, respectively. (c) and (d): reconstructed images of (a) and (b) by using Calderon's method, respectively. (e) and (f): reconstructed images of (a) and (b) by using dbar method, respectively. (g) and (h): reconstructed images of (a) and (b) by using factorization method, respectively.

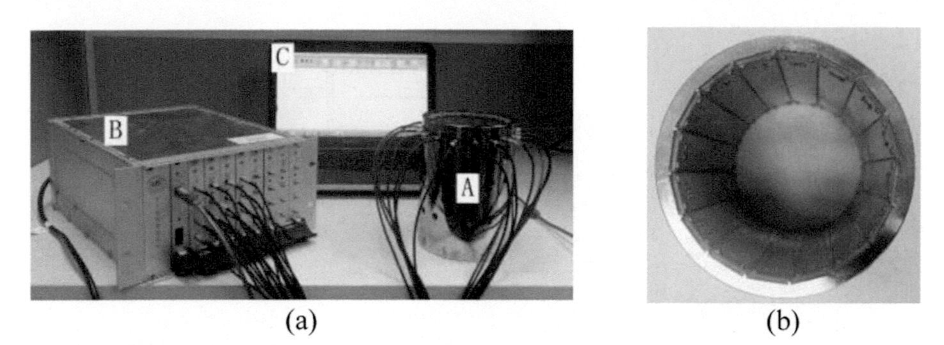

Figure 12.14 System and sensors for electrical capacitance tomography. (a) System of electrical capacitance tomography with a circular sensor. A: sensor, B: data acquisition system, C: computer. (b) Top view of the circular sensor in (a).

method are of high contrast in the sense of visual appearance. The factorization method produced images with low resolution, as it is a kind of sampling method (Gebauer et al., 2008).

In this section, experimental data was acquired to assess the performance of the proposed iterative reconstruction algorithm in practical applications. The data was acquired from the 16-electrode ET system, shown as Fig. 12.14(a), it utilized digital recursive demodulation, and worked in a parallel way (Huang, Cao, Sun, Lu, & Xu, 2019). The working frequency of this ET system ranges from 10 kHz to 1 MHz. The SNR of the capacitance measurement varies from about 50 to 80 dB for different channels. Each electrode was set to one of the following three operation modes, namely, for excitation, for measurement, and grounded. The excitation frequency was selected as 100 kHz and the excitation voltage amplitude was set to be 16V in the experiment. The 16-electrode ET sensor used in the experiments is shown in Fig. 12.14(b). The inner diameter of the ET sensor is 100 mm.

The photo and cross-sectional view of real distributions are illustrated in Fig. 12.15. The ET system measured the changes in capacitance of nylon rod as high permittivity material and air as low permittivity material in phantoms (a)−(d). The radius of nylon rod is 15 mm. Phantoms (e)−(h) in Fig. 12.15 were used to test the effect of sand with different holdup. The holdup of sand was about 16%, 28%, 50%, and 70%, respectively.

When implementing the proposed algorithm, the value of numerical integration radius R_I in Eq. (12.27) was selected to be 4.5 for phantoms (a)−(d) and two for phantoms (e)−(h) by using a trial-and-error method, respectively.

Fig. 12.16 depicts the reconstructed results with Calderon's method and the proposed algorithm by using experimental data. The initial images were reconstructed by applying Calderon's method. The reconstructed images after iterations were obtained from the proposed algorithm. It can be seen that the shape of nylon rods reconstructed by two methods in Fig. 12.16(a)-(d) has certain deformation. The proposed

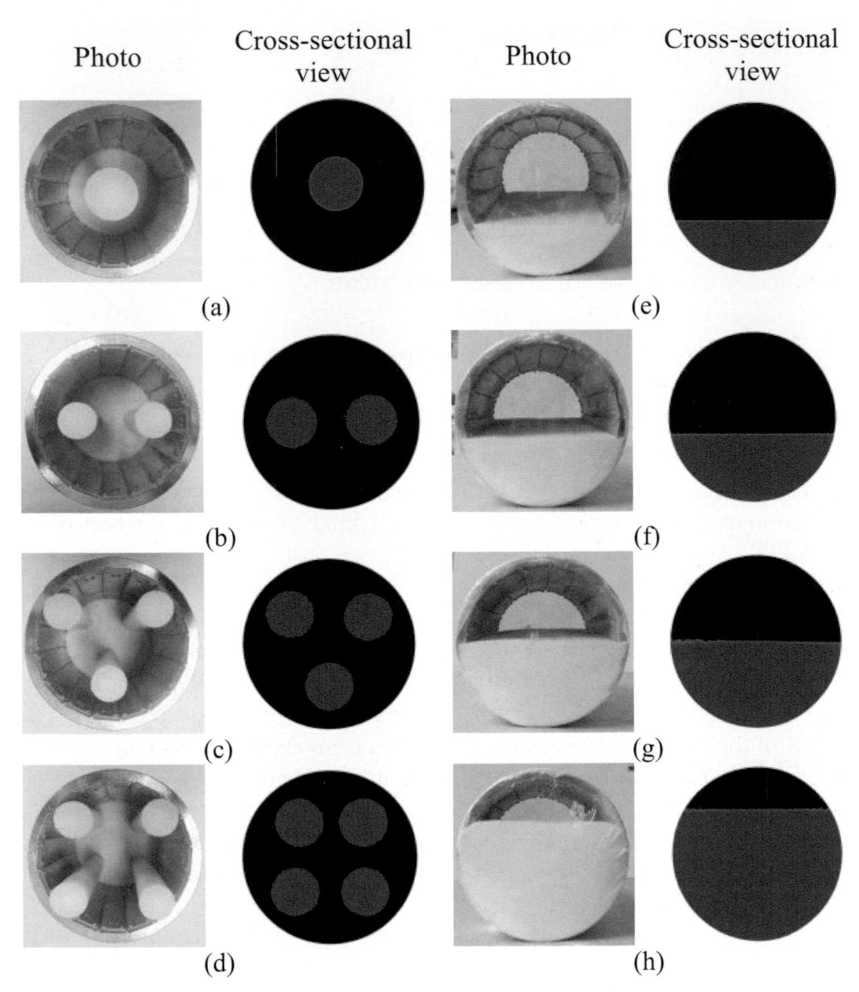

Figure 12.15 Real distributions with circular sensor in the experiment.

algorithm performs better in reconstructing the size of nylon rods and distinguishing each rod in the region, especially for four rods close to each other in Fig. 12.16(d). For the distribution (e)−(h) in Fig. 12.15, the reconstructed images after iterations have stronger contrast compared to initial reconstructed images. In order to analyze the relationship between the real holdup and the reconstructed holdup, four additional experiments were added, for which the holdup of sand were 8%, 33%, 62%, and 100%, respectively. The relationship between the real holdup and the reconstructed holdup is depicted in Fig. 12.17, where S and S' are the area of circular region and the sand region. ε and ε' are the number of pixels and the sum of reconstructed gray value normalized within the range of [0,1]. In the ideal case, the relationship is linear, i.e.,

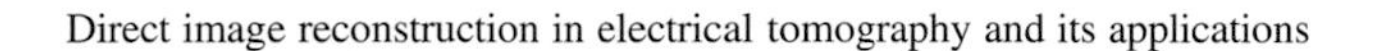

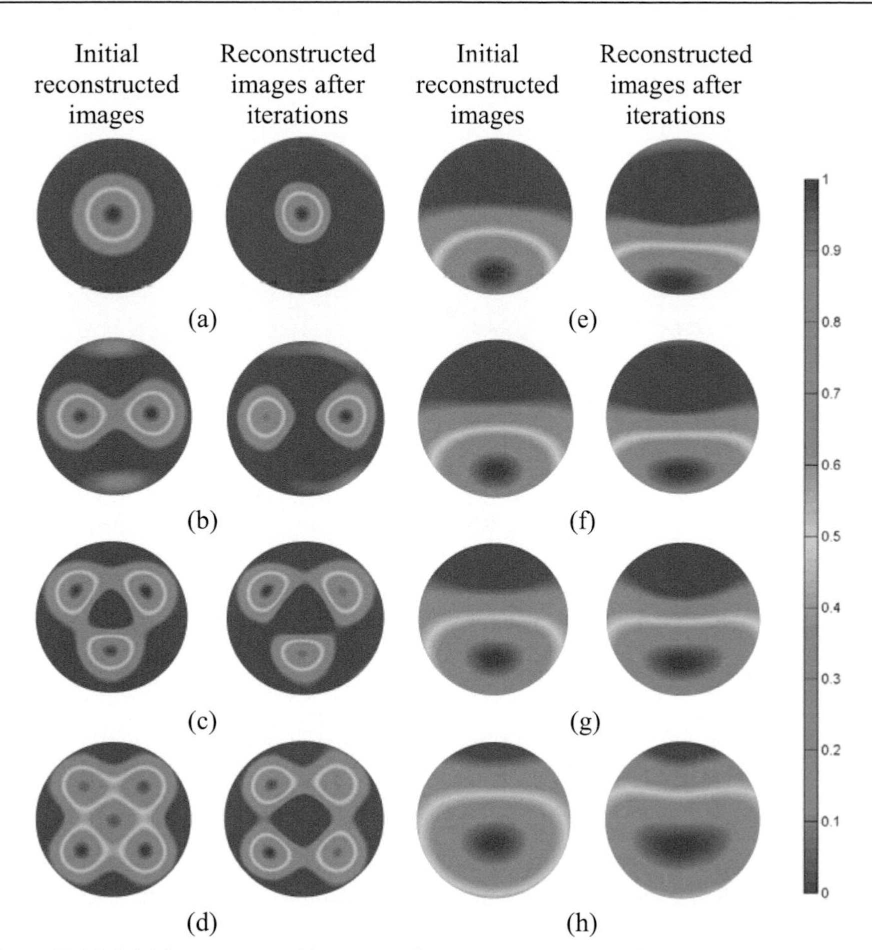

Figure 12.16 Initial reconstructed images and reconstructed images after iterations with circular sensor by using experimental data.

$$\frac{\varepsilon'}{\varepsilon} \propto \frac{S'}{S} \tag{12.67}$$

And the values of correlation coefficient r^2 should be one ideally and reflected in Fig. 12.17 as dotted line. The values of r^2 calculated by using Calderon's method and the proposed algorithm are 0.964 and 0.986, respectively, which indicates that the proposed algorithm generates better reconstruction. Moreover, in the visual appearance, the line provided by the proposed method is closer to the ideal one. These experimental results validate the robustness of the proposed algorithm in cases of real measurements.

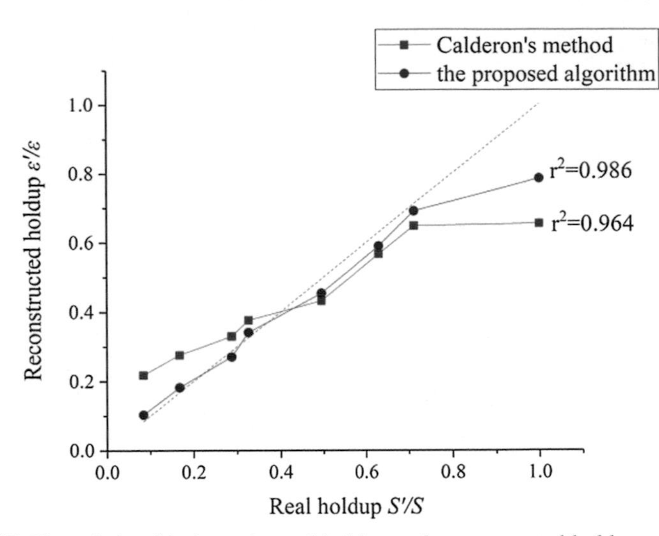

Figure 12.17 The relationship between real holdup and reconstructed holdup with eight different holdups of sand in the air.

12.7 Dynamic flame monitoring

The ignition and blowout processes of an alcohol lamp flame were monitored by the ET system, and then the variations in temperature field were estimated (Hu et al., 2020). In the first stage of experiment, the alcohol lamp was ignited with a lighter. When the alcohol lamp burned gradually and the flame stabilized, the flame was suddenly blown out. In the second stage, the flame burned in an enclosed space, so the alcohol lamp flame went out gradually with the decrease of oxygen. Each dynamic process lasted for 10 s, during which over 20,000 images were obtained. The whole combustion process was compressed into 200 frames, and the temperature images mapped from admittivity data were displayed in Fig. 12.18.

As shown in Fig. 12.18(a) and (c) represented three-dimensional temperature variations along the vertical axis of time in the first and second stages, respectively, where the time increased along t-axis from bottom to top. 3D temperature distribution intuitively reflected the position of flame and the approximate range of temperature in the combustion process. The longitudinal sections of (a) and (c) were shown in (b) and (d), respectively, which clearly reflected the variation of temperature along time in the flame field.

In the first stage of the experiment, at the bottom of Fig. 12.18(a) and (b), the temperature increased a little, which reflected the existence of the flame of lighter when the lighter was used to ignite the alcohol lamp. After that, there was a small truncation in the temperature distribution above it, because the lighter extinguished when the alcohol lamp flame was small at beginning. Then during the combustion process of the alcohol lamp, (a) and (b) showed that with the increase of combustion time, the temperature of flame increased gradually. When the flame burned steadily, it was

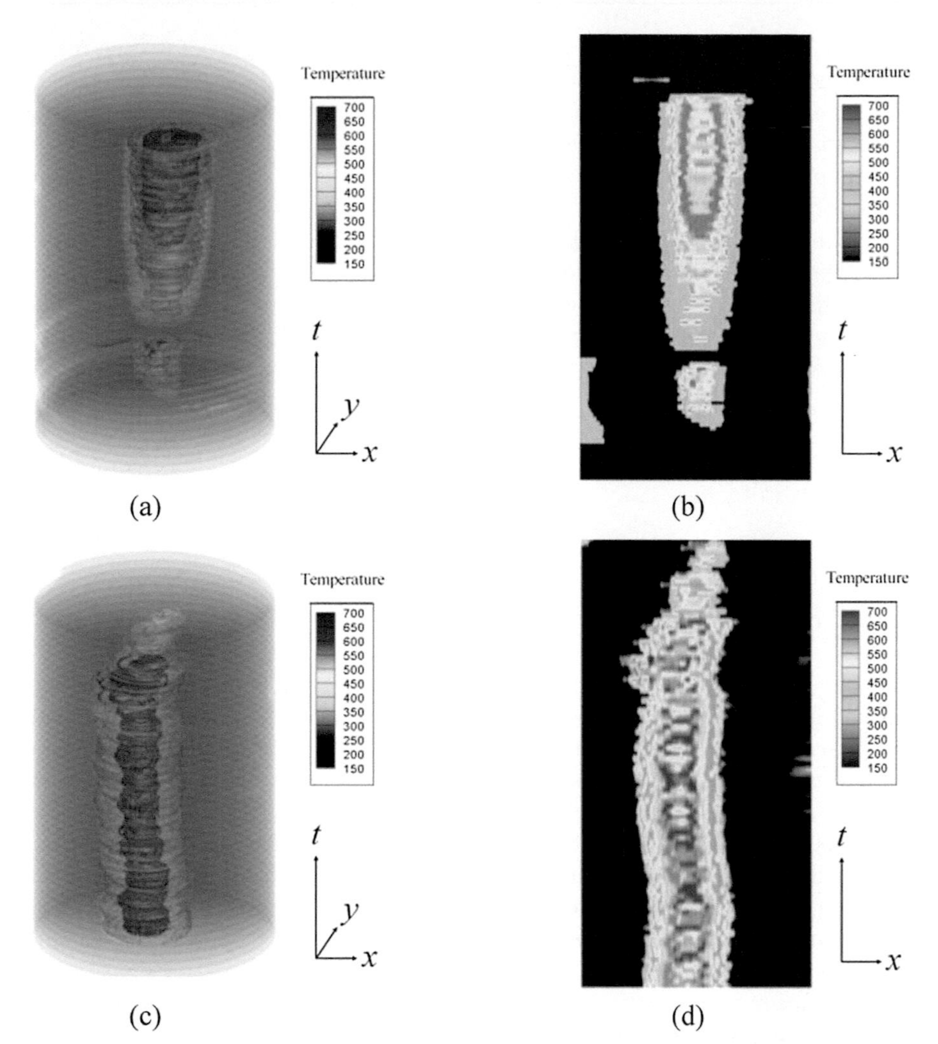

Figure 12.18 Variation of combustion temperature field during ignition and blowout processes of an alcohol lamp flame with time (vertical axis). (a) 3D view of temperature field variation during ignition and sudden blowout process, (b) longitudinal section of (a), (c) 3D view of temperature field variation during slow blowout process, (d) longitudinal section of (c).

observed from (b) that the temperature distribution inside the flame was obviously ring-shaped, which was consistent with the prior knowledge. Finally, temperature in (a) and (b) decreased quickly, which reflected that the alcohol lamp flame was suddenly blown out by a strong wind.

In the second stage of experiment, as shown in the lower part of (c) and (d), the range of high temperature distribution was basically stable at the beginning. It could be observed that the alcohol lamp flame swung a little, so the ring-shaped area of the highest temperature was not clear. As shown in the upper part of (c) and (d), the highest temperature gradually decreased, as the alcohol lamp flame went out gradually after the burning space was enclosed. At the same time, the distribution range of high temperature decreased and shifted to one side, which reflected that during the slow blowout process, the flame burned eccentrically. Finally, the alcohol lamp completely extinguished, and the temperature dropped rapidly, as shown in the top part of (c) and (d). In summary, the results validated the proposed method for estimating the combustion temperature field and monitoring dynamic temperature variations in flame.

For different burners, flowrates of fuel and air, or environmental conditions including the temperature and pressure, the parameters of the mapping model varied, but the tendency that higher combustion rate causes higher temperature and admittivity should be consistent. Therefore, the admittivity to temperature mapping can be obtained by using the same method for other burner types and combustion conditions. In other words, this method should be helpful for noncontact estimation of temperature in other types of flame.

12.8 Future trends

Currently, most algorithms are based on a simplified linear mathematical model that is deduced from the sensitivity theorem. It should be noted that in the algorithms based on the sensitivity or Jacobian matrix, the permittivity perturbations for the whole domain must be reconstructed simultaneously. The direct methods mentioned here provide a new view for image reconstruction in electrical tomography, and there exist three major trends for the direct algorithms. Firstly, the methods are generalized from the two-dimensional distributions to the three-dimensional case, which is very time consuming for the nondirect algorithms. Secondly, these methods are also developed to deal with the multimodality reconstruction in an integrated way, in which the sensitivity matrix-based methods usually reconstruct each modality separately. Finally, partial boundary measurements will be required to retrieval the material distribution by using the direct methods, as in some applications, only measurements on part of the boundary are available.

12.9 Further information

The authors gratefully acknowledge the financial support from the National Natural Science Foundation of China (No. 61827802, 61620106004, 61961130393), Newton Advanced Fellowship (NAF191193), and Program for Changjiang Scholar and Innovative Research Team in University (IRT1203).

References

Astala, K.和P. I. (2006). Calderón's inverse conductivity problem in the plane. *Annals of Mathematics, 163*, 265–299.

Astala, K., Mueller, J. L., Päivärinta, L., & Siltanen, S. (2010). Numerical computation of complex geometrical optics solutions to the conductivity equation. *Applied and Computational Harmonic Analysis, 29*, 2–17.

Boverman, G., Kao, T.-J., Isaacson, D., & Saulnier, G. J. (2009). An implementation of Cal deron's method for 3-D limited-view EIT. Medical imaging. *IEEE Transactions on, 28*, 1073–1082.

Calderon, A. P. (2006). On an inverse boundary value problem (reprint). *Computational and Applied Mathematics, 25*, 133–138.

Cao, Z., Ji, L., & Xu, L. (2018). Iterative reconstruction algorithm for electrical capacitance tomography based on Calderon's method. *IEEE Sensors Journal, 18*, 8450–8462.

Cao, Z., Wang, H., & Xu, L. (2008). Electrical impedance tomography with an optimized calculable square sensor. *Review of Scientific Instruments, 79*, 103710.

Cao, Z., Wang, H., Yang, W., & Yan, Y. (2007). A calculable sensor for electrical impedance tomography. *Sensors and Actuators A: Physical, 140*, 156–161.

Cao, Z., & Xu, L. (2013a). Direct recovery of the electrical admittivities in 2D electrical tomography by using Calderon's method and two-terminal/electrode excitation strategies. *Measurement Science and Technology, 24*, 074007.

Cao, Z., & Xu, L. (2013b). 2D image reconstruction of a human chest by using Calderon's method and the adjacent current pattern. *Journal of Instrumentation, 8*. P03004–P03004.

Cao, Z., & Xu, L. (2013c). Direct image reconstruction for 3-D electrical resistance tomography by using the factorization method and electrodes on a single plane. *IEEE Transactions on Instrumentation and Measurement, 62*, 999–1007.

Cao, Z., Xu, L., Fang, W., & Wang, H. (2011b). 2D electrical capacitance tomography with sensors of non-circular cross sections using the factorization method. *Measurement Science and Technology, 22*, 114003.

Cao, Z., Xu, L., Fan, W., & Wang, H. (2009a). Electrical capacitance tomography with a non-circular sensor using the dbar method. *Measurement Science and Technology, 21*, 015502.

Cao, Z., Xu, L., Fan, W., & Wang, H. (2011a). Electrical capacitance tomography for sensors of square cross sections using Calderon's method. *IEEE Transactions on Instrumentation and Measurement, 60*, 900–907.

Cao, Z., Xu, L., & Wang, H. (2009b). Image reconstruction technique of electrical capacitance tomography for low-contrast dielectrics using Calderon's method. *Measurement Science and Technology, 20*(10), 104027. https://doi.org/10.1088/0957-0233/20/10/104027

Du, B., Warsito, W., & Fan, L.-S. (2006a). Imaging the choking transition in Gas–Solid risers using electrical capacitance tomography. *Industrial & Engineering Chemistry Research, 45*, 5384–5395.

Du, B., Warsito, W., & Fan, L.-S. (2006b). Behavior of the dense-phase transportation regime in a circulating fluidized bed. *Industrial & Engineering Chemistry Research, 45*, 3741–3751.

Fang, W., & Cumberbatch, E. (2005). Matrix properties of data from electrical capacitance tomography. *Journal of Engineering Mathematics, 51*, 127–146.

Gebauer, B., Hanke, M., & Schneider, C. (2008). Sampling methods for low-frequency electromagnetic imaging. *Inverse Problems, 24*, 015007.

Gebauer, B., & Hyvönen, N. (2007). Factorization method and irregular inclusions in electrical impedance tomography. *Inverse Problems, 23*, 2159–2170.

Geselowitz, D. B. (1971). An application of electrocardiographic lead theory to impedance plethysmography. *IEEE Transactions on Biomedical Engineering BME, 18*, 38–41.

Gong, S., & Gong, Y. (2007). *Concise complex analysis* (revised edition). Singapore: World Scientific Publishing Company.

Hakula, H., & Hyvönen, N. (2009). On computation of test dipoles for factorization method. *BIT Numerical Mathematics, 49*, 75–91.

Hamilton, S. J., Herrera, C. N. L., Mueller, J. L., & Von Herrmann, A. (2012). A direct D-bar reconstruction algorithm for recovering a complex conductivity in 2D. *Inverse Problems, 28*, 095005.

Hamilton, S. J., & Mueller, J. L. (2013). Direct EIT reconstructions of complex admittivities on a chest-shaped domain in 2-D. *IEEE Transactions on Medical Imaging, 32*, 757–769.

Harrach, B. (2013). Recent progress on the factorization method for electrical impedance tomography. *Computational and Mathematical Methods in Medicine, 2013*, 425184.

Huang, A., Cao, Z., Sun, S., Lu, F., & Xu, L. (2019). An agile electrical capacitance tomography system with improved frame rates. *IEEE Sensors Journal, 19*, 1416–1425.

Hu, D., Tian, Y., Chang, L., Sun, S., Sun, J., Cao, Z., et al. (2020). Estimation of combustion temperature field from the electrical admittivity distribution obtained by electrical tomography. *IEEE Transactions on Instrumentation and Measurement, 69*, 6271–6280.

Hyvönen, N. (2005). Application of a weaker formulation of the factorization method to the characterization of absorbing inclusions in optical tomography. *Inverse Problems, 21*, 1331–1343.

Isaacson, D., Mueller, J. L., Newell, J. C., & Siltanen, S. (2004). Reconstructions of chest phantoms by the D-bar method for electrical impedance tomography. *IIEEE Transactions on Medical Imaging, 23*(7), 821–828. https://doi.org/10.1109/TMI.2004.827482

Kirsch, A., & Grinberg, N. (2007). *The factorization method for inverse problems*. Oxford University Press. https://doi.org/10.1093/acprof:oso/9780199213535.001.0001

Knudsen, K., Lassas, M., Mueller, J. L., & Siltanen, S. (2007). D-Bar method for electrical impedance tomography with discontinuous conductivities. *SIAM Journal on Applied Mathematics, 67*, 893–913.

Knudsen, K., Mueller, J., & Siltanen, S. (2004). Numerical solution method for the dbar-equation in the plane. *Journal of Computational Physics, 198*(2), 500–517. https://doi.org/10.1016/j.jcp.2004.01.028

Knudsen, K., Lassas, M., Mueller, J., & Siltanen, S. (2008). Reconstructions of piecewise constant conductivities by the D-bar method for electrical impedance tomography. *Journal of physics. Conference series, 124*, Article 012029. https://doi.org/10.1088/1742-6596/124/1/012029

Lehr, J. (1972). A vector derivation useful in impedance plethysmographic field calculations. *IEEE Transactions on Biomedical Engineering BME-, 19*, 156–157.

Liu, S., Chen, Q., Xiong, X., Zhang, Z., & Lei, J. (2008). Preliminary study on ECT imaging of flames in porous media. *Measurement Science and Technology, 19*, 094017.

Li, Y., & Yang, W. (2008). Image reconstruction by nonlinear Landweber iteration for complicated distributions. *Measurement Science and Technology, 19*, 094014.

Mueller, J., & Bikowski, J. (2008). 2D EIT reconstructions using Calderon's method. *Inverse Problems and Imaging, 2*, 43–61.

Mueller, J. L., & Isaacson, D. (2004). *Regularization of the computed scattering transform for the D-bar method for electrical impedance tomography*. SPIE.

Muller, P. A., Mueller, J. L., & Mellenthin, M. M. (2017). Real-time implementation of calderón's method on subject-specific domains. *IEEE Transactions on Medical Imaging, 36*, 1868–1875.

Murphy, E. K., & Mueller, J. L. (2009). Effect of domain shape modeling and measurement errors on the 2-D D-bar method for EIT. *IEEE Transactions on Medical Imaging, 28*, 1576–1584.

Murphy, E. K., Mueller, J. L., & Newell, J. C. (2007). Reconstructions of conductive and insulating targets using the D-bar method on an elliptical domain. *Physiological Measurement, 28*, S101–S114.

Nachman, A. I. (1996). Global uniqueness for a two-dimensional inverse boundary value problem. *Annals of Mathematics, 143*, 71–96.

Schmitt, S. (2009). The factorization method for EIT in the case of mixed inclusions. *Inverse Problems, 25*, 065012.

Siltanen, S., Mueller, J., & Isaacson, D. (2001). Reconstruction of high contrast 2-D conductivities by the algorithm of A. Nachman. *Contemporary Mathematics, 278*, 241–254.

Smolik, W. T., & Radomski, D. (2009). Performance evaluation of an iterative image reconstruction algorithm with sensitivity matrix updating based on real measurements for electrical capacitance tomography. *Measurement Science and Technology, 20*, 115502.

Soleimani, M., Yalavarthy, P. K., & Dehghani, H. (2010). Helmholtz-type regularization method for permittivity reconstruction using experimental phantom data of electrical capacitance tomography. *IEEE Transactions on Instrumentation and Measurement, 59*, 78–83.

Sun, M., Liu, S., Lei, J., & Li, Z. (2008). Mass flow measurement of pneumatically conveyed solids using electrical capacitance tomography. *Measurement Science and Technology, 19*, 045503.

Wang, H. G., Liu, S., Jiang, F., & Yang, W. Q. (2004). Using electrical capacitance tomography for measuring the solids distribution in cyclone separator. *Proceedings of the Chinese Society for Electrical Engineering, 24*, 174–179.

Wang, H., Wang, C., & Yin, W. (2004). A pre-iteration method for the inverse problem in electrical impedance tomography. *IEEE Transactions on Instrumentation and Measurement, 53*, 1093–1096.

Wang, H., & Yang, W. (2020). Application of electrical capacitance tomography in circulating fluidised beds – a review. *Applied Thermal Engineering, 176*, 115311.

Wang, H. G., Yang, W. Q., Senior, P., Raghavan, R. S., & Duncan, S. R. (2008). Investigation of batch fluidized-bed drying by mathematical modeling, CFD simulation and ECT measurement. *AIChE Journal, 54*, 427–444.

Yang, W. Q., & Peng, L. (2003). Image reconstruction algorithms for electrical capacitance tomography. *Measurement Science and Technology, 14*, R1–R13.

Yang, W. Q., Spink, D. M., Gamio, J. C., & Beck, M. S. (1997). Sensitivity distributions of capacitance tomography sensors with parallel field excitation. *Measurement Science and Technology, 8*, 562–569.

Zhao, J., Xu, L., & Cao, Z. (2019). Direct image reconstruction for electrical capacitance tomography using shortcut D-bar method. *IEEE Transactions on Instrumentation and Measurement, 68*, 483–492.

Zhong, W., & Yao, Z. (2004). Precise integration method for elliptic functions. *Advances in Applied Mechanics(in Chinese)*, 106–111.

Machine learning process information from tomography data

Brian S. Hoyle [1], *Thomas D. Machin* [2] *and Junita Mohamad-Saleh* [3]
[1]University of Leeds, Leeds, United Kingdom; [2]Stream Sensing, Manchester, United Kingdom; [3]Universiti Sains Malaysia, Penang, Malaysia

13.1 Introduction

Industrial Process Tomography (IPT) has become a well-established method of gaining a deep multidimensional insight into a wide range of industrial processes. *Machine Learning* (ML), also widely known by the term *Artificial Intelligence* (AI), offers significant advantages for the processing of IPT data. Due to the ill-posed nature of its intrinsic inverse process, and the typical sparse data due to sampling time constraints, results from classical IPT methods are often disappointing, even when sophisticated analytical or statistical modeling is used. In contrast, an ML approach does not rely upon modeling, but instead exploits actual (or simulated) process performance data through a learning network able to crystallize trends and relationships between measurement data and selected process state estimation parameters, from image pixels to interpretation data such as component fractions. Having learned from this, data networks can be implemented with major parallelism to deliver rolling state estimation for fast processes dynamics.

To contrast classical IPT and ML methods, we consider two examples: (i) a batch process to mix several reagents to form a product; and (ii) a continuous process to transport gas/liquid materials. IPT technology typically embodies three core tomographic steps, plus an optional step 4:

S1 – Sensing energy excitation of space/time material distribution of interest;

S2 – Detection of corresponding response;

S3 – Inversion of response data to estimate reconstructed material distribution, often as an image;

S4 – Interpretation of the distribution to identify relevant process characteristics.

Pilot process studies are useful for both optimization of plant design and in trials of new formulations prior to scale-up for mass production. Detailed image data of the internal process state may be valuable to plant and process designers. Many IPT applications have time constraints due to material and process dynamics. In engineering terms, the temporal performance of IPT instrumentation must ensure that Step 1 excitation and Step 2 response detection are completed within the limits of the process dynamics. If response data can be stored for later, Step 3 processing (and any required

Industrial Tomography. https://doi.org/10.1016/B978-0-12-823015-2.00033-9

Step 4) may be carried out after the test completion. Data are typically sparse and ill-posed. As explored in detail in other chapters, a wide range of iterative reconstruction techniques can be applied in the S3 inversion process, ideally exploiting a priori knowledge of process characteristics. Sophisticated techniques such as regularization and optimization methods require considerable computation and typically are more suited for use on stored data, following a pilot scale test.

Operating production processes are focused on efficiency and product quality. For a batch process, this may be on minimizing cycle-time to reduce costs and enhance plant utilization. For a continuous process, the key requirement may be on tracking and controlling the process state within desired levels. In such cases, the temporal performance of IPT instrumentation is critical to deliver timely data for all Steps for quality monitoring and control. For example, continuous in-line monitoring using IPT can remove the need for periodic sampling for laboratory quality checks, whose time lag may be costly in continued production of out of specification product.

In both application cases, S3 can be used to synthesize an image-based dataset with S4 providing further interpreted process information. In the pilot process case, the composite information may be used for design verification. In the operating process case, an image dataset may not be required, as direct estimation of process state for automated monitoring and control is achievable; in effect S3 and S4 are integrated to directly yield process information.

ML offers potential advantages in both pilot and production process cases, through the ability to amass and structure prior process experience data and interpret and generalize this to recognize both prior and newly presented data. ML and AI can be implemented by specific electronic devices, but also for design through software using a conventional digital computer.

In the pilot application case, this may provide enhanced performance; for example, under some conditions the process may exhibit a graded interface between materials, which can be distorted by a conventional IPT regularization algorithm, into a false hard boundary. In the operating production process, the prestored and structured data can, with appropriate hardware, offer fast direct interpretation for automatic process monitoring and control.

13.2 Machine learning methods for information needs

ML methods have developed strongly over the past 20 years, inspired originally of course by biological neural systems; and foundational concepts are now widely used. Early IPT research using ML methods, for example, Nooralahiyan, Hoyle, and Bailey (1993), demonstrated considerable promise, but was not developed further due to major skepticism that estimation results were not verifiable in analytic or statistical terms. As discussed briefly below, ML methods do rest on rigorous foundations, and now have wide ranging science and technology that spans many areas but remain challenging to verification by *reverse engineering*. Due to their impressive performance, they have more recently gained widespread trust, in some cases beyond sensible expectation. Interestingly, recent IPT meetings have included several new ML developments, notably at the ISIPT ninth World Congress 2018, by: Johansen et al., Lähivaara et al., Machin et al., Yan and Mylvaganma, and by Zheng and Peng.

The following subsections present a short introduction to relevant ML theory and methods focusing primarily upon IPT realization of systems able to meet the demanding monitoring and control case discussed above. To provide a comparative link, the analytical basis of simple ML architecture is contrasted with a corresponding Linear Back-Projection (LBP) case, commonly used due to its implementation simplicity and speed. Although foundations and speed appear comparable, ML solutions have greater potential due to their ability to learn and generalize subtle data interactions, such as the *soft field* effect. These topics are further illustrated in depth using two contrasting design case studies: for complex multiphase flow measurement; and in the sophisticated application of multifaceted IPT sensors for the direct, in-process measurement of rheological properties of stream flows.

Finally, promising new approaches to ML IPT systems are considered. The term *Deep Learning* describes a newer range of ML techniques in which the analytics of data evaluation, organization, and preprocessing is also subject to ML creating a *deeper* learning process. Disordered data are preexamined for subtle relationships which can be exploited to speed and simplify later neural processing structures. The final section also provides a short overview of current design tools and implementation systems.

We include foundational references and others mainly focused on IPT focused ML work. Readers interested in more general ML science and technology are directed to the expansive general literature.

13.2.1 Outline of neural computing

Many texts on ML begin with foundations of neural simulation, based loosely upon a biological neuron, and commonly termed an *Artificial Neuron* (AN), in which the capability to perform a perceptive operation is determined in terms of settable coefficients in its structure. It must be stated for clarity that this is a minimal imitation of the major complexity of a biological neuron. A more detailed introduction centered upon instrumentation applications, coinciding with early IPT applications, is provided by Bishop (1994).

A collection of ANs forming an *Artificial Neural Network* (ANN) provides a variety of processing configurations, typically divided into the classes illustrated in Fig. 13.1, that includes their common name, sometimes based upon the name of their first proposer.

The primary classification separates configurations into two groups: Unsupervised ANNs, designed to self-learn without explicit training, and Supervised ANNs, which must be provided with an explicit training configuration procedure and data. Supervised Configurations are further divided into Static and Dynamic forms. In the Static case, a specific learning configuration uses experiential training data, representing key variations the network will be expected to recognize, and as a basis for generalization.

SUPERVISED		**UNSUPERVISED**
Feedforward	**Feedback**	
Radial Basis Function	Jordon Simple Recurrent	Kohonen
Perceptron	Mozer Partially Recurrent	Adaptive Resonance
Multilayer Perceptron	Hopfield Fully Recurrent	
Convolutional		
Time Delay		

Figure 13.1 Common configurations of artificial neural network.

13.2.2 Artificial neural network—perceptron

Here we focus upon a Static ANN: a single neuron; a single layer of neurons; and multiple cascaded layers. IPT application networks often exploit ANNs in a configuration commonly called a *Perceptron*, first proposed by Rosenblatt (1962). This combines detailed interpretive performance and straightforward (if extensive, one-time) training. A single Perceptron node is illustrated in Fig. 13.2 to define notation and abbreviations, based here upon biological neuron names for input dendrites (d), and output axon (a).

The Perceptron has a single bias input, d_0; and M measurement data inputs: $d_{1 \ldots M}$, abbreviated as the vector, $\boldsymbol{d}$. All inputs have an applied multiplicative *Weight* value, w_m. The bias weight, w_0, effectively provides a fixed offset value when the bias input is set (conventionally) to constant 1. The resulting weighted sum, s, is generated at the node. This sum is passed to a threshold (or *activation*) function, giving a composite output, a defined by the following:

$$a = f(s) \text{ where } s = \sum_{m=0}^{M} w_m \, d_m \tag{13.1}$$

The function $f(s)$ may have many forms. A classic threshold *step* function causes the neuron to trigger at the step level. For IPT applications, the Perceptron is often used

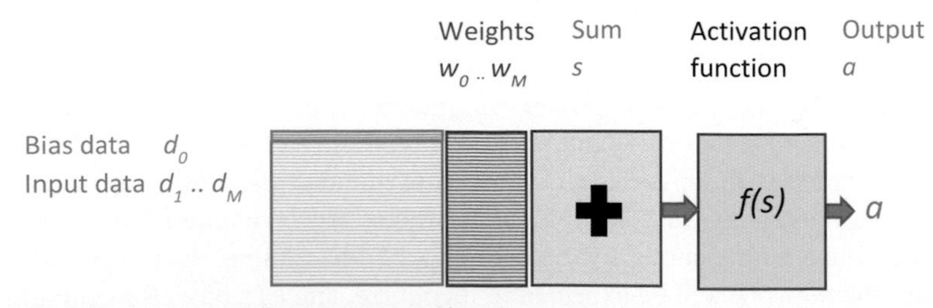

Figure 13.2 Perceptron artificial neuron node.

with a *sigmoid* function. This offers sensitivity to small signals and provides gain control of high values, and is defined as follows:

$$f(s) = \frac{1}{1 + e^{-K_s}} \tag{13.2}$$

where K is a *slope spread* constant, e.g., 0: linear; 0.5: shallow; 5: steep; infinity: step.

This function is nonlinear but is also continuously differentiable, an important property for training procedures discussed below. From Eqs. (13.1) and (13.2), the derivative is the following:

$$f'(s) = a' = \frac{Ke^{-K_s}}{(1 + e^{-K_s})^2} = Kf(s)\,(1 - f(s)) = Ka(1 - a) \tag{13.3}$$

A Network of Nodes, an ANN, consists of multiple nodes arranged in a sequential and parallel format. In an IPT context we may assume an ANN produces a corresponding output instantly, with respect to typical process dynamics. Fig. 13.3 illustrates a corresponding Perceptron ANN with N nodes and M measurement inputs.

Bias and measurement inputs to each node from the *first* to the final Mth input are shown in abbreviated *databus* form. As illustrated, each of the 1 … N nodes has a bias input and, as is usually the case, all M measurement inputs are *fully connected* to each Node, meaning that each input is connected to every node in the layer. For any given task, ANN node and connection structure are determined by design; and the internal settings of the bias and input (1+M) weight values are determined by a learning process using training data. Optimal design can be complex. Obvious key starting points are the number of inputs, and the required outputs. It may also be advantageous to pre-process training set data, to enhance orthogonality and reduce correlated data,

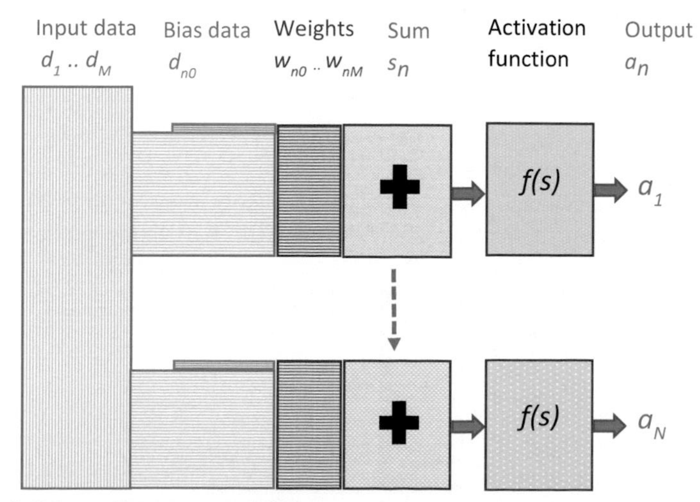

Figure 13.3 Schematic perceptron ANN.

providing a simpler and more efficient ANN; training data are then processed in advance of network design.

For tomographic applications, we infer that an IPT sensor is specified to measure appropriate raw data. Early chapters present a complete range of single and multiple sensing modes. A sensor typically provides a *projection* of all unique measurements at each sample interval. For example, a 12-electrode ECT sensor (Chapter 1, Eq. 1.1) provides a vector of 66 (*M*) measurement inputs, which is combined with the bias input. Various interpretations could be estimated in parallel from raw IPT data. For example, a scalar value of solids fraction in a liquid and binary status flags for quality control limits and fault alarms. It may also be useful to encode possible output values, even to the extent of creating an image matrix synthesized using an output node for each estimated pixel. A useful design starting point is therefore to specify ANN output nodes to suit these requirements.

Diagrams involving ANNs where each neuron is depicted quickly become complex, even for small networks. A corresponding compact form, depicting essential design information in *databus* and *process block* form, is shown in Fig. 13.4, and is used in all following sections.

Here data and process flows are from Left to Right. Bias blocks are outlined in Green; Data blocks are outlined in Blue; and Operator blocks are outlined in Red. Names or descriptions of data items in blocks are shown in Blue text. Symbols or values are included at the top and bottom of a block in Black text: the relevant variable at the top; and the data name, or number of data at the bottom. All layers are assumed to be fully connected. Here the 1 bias input and *M* measurement inputs are combined in a weighted summation followed by the sigmoid activation function with output, *a*.

Having defined the basic ANN Perceptron, we may consider its operation and limitations. If it is possible to separately classify data presented as a sequence of inputs, Rosenblatt (1962) proved that a Perceptron can identify any specific classification. ANN weights embody interaction data, allowing the network to recognize previously presented datasets, and importantly to generalize when presented with new data, based upon the learning experience. This provides the capability to interpret complex object behavior, for example, electrical tomography *soft field* distortion, which arises from interaction between sensing energy and the object field. As significant effects feature during learning, the ANN will compute a result that embodies this effect. Hence, the ANN is a powerful estimation engine.

For comparison, a classical process estimation usually involves an initial (one-time) *model-driven* selection of a physical model and for a dynamic process, a rolling

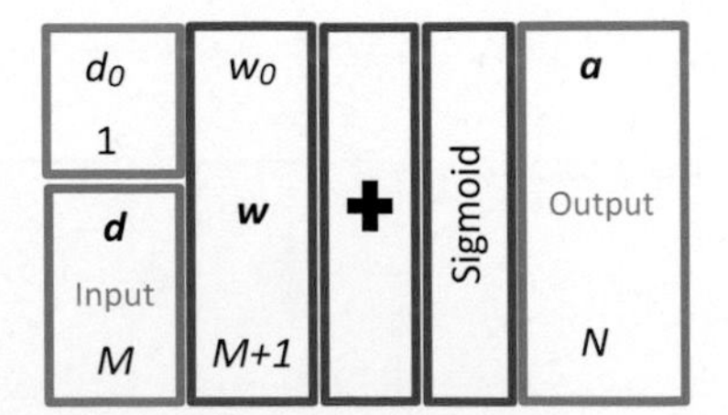

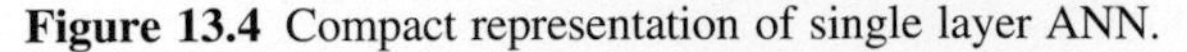
Figure 13.4 Compact representation of single layer ANN.

temporal cycle of measuring process data, followed by a conventional digital computed algorithm to estimate process *output* values. For example, LBP inversion of tomographic measurements to create an image, which may be used to further estimate component fraction values in a mixture.

The ANN method also involves an initial (one-time) *data-driven* stage but does not employ any assumed analytic or statistical model. Instead it exploits the alternate computing paradigm that *learns* process characteristics in a *training* process in terms of the network structure and the set of network bias and weight values, from data sampled from wide ranging actual plant measurements (or from simulations). Rolling estimation in a dynamic process again proceeds in a time cycle of measuring process data, but the trained ANN then directly transforms measurements to component fraction values.

13.2.3 Artificial network training

The foundational Perceptron *back-propagation* learning process attributed to Rumelhart is based upon gradient descent optimization. A major review is beyond the scope of this IPT contextual overview, and for more details, readers are referred to Bishop (1994), and a collection of key works by Anderson and Rosenfeld (1988, 1990). In summary, Rumelhart describes the M weight values at each ANN node (including the bias weight) in terms of an M-dimension hyperspace. For any linearly separable set of input *patterns*, a single-input, single-output layer ANN can extract a corresponding output solution pattern as an $M-1$ dimensional hyperplane. Where the set of input patterns are not linearly separable, Rumelhart found that one or more intermediate *hidden* layers must be used, forming a *multilayer* ANN, to enable a solution.

The availability of adequate training data for the learning process is critical for a successful search for an optimum set of weights and biases for all layers and nodes, in which ANN data are input for feed forward estimation, followed by the back-propagation weight update step. During training, each IPT measurement vector must be accompanied by a corresponding synchronized and independent measurement of the *target* process output, usually measured using temporary supplementary instrumentation fitted to the process plant. An integrated process and sensing model—based simulation may also be used to generate process (measurement with target) data.

One, or several combined, sources may be used to create a *Design Dataset*. A training strategy will typically randomly select three specific datasets from the collected Design Dataset: (i) a *Training Dataset* to be *learned* by the ANN; (ii) a *Test Dataset* to assess error levels during training; and (iii) a *Validation Dataset* of wide-ranging data used to assess if training is complete (and hence avoid *over-training*). The Training set is the only data for which the specific *Training Mode* which features back-propagation is deployed. *Estimation Mode* is the normal application use for a trained network. During training, the Test and Validation datasets are applied to the network in Estimation Mode, but with an added operation that error values are still computed using the included target data. At this stage, network training is incomplete as network structure and weight values are not finalized. For clarity, the Training dataset is the sole *learned* data; the Test and Validation datasets consist of *unlearned* data but is not used in training. This is important to assess generalization and interpolation capabilities.

It is also common to *preprocess* input data to enhance orthogonality and remove duplication, increasing ANN efficiency to learn and generalize. This is explored in detailed in later sections, particularly via a major example in Section 13.4.3.

To illustrate the back-propagation learning mode, we consider a single-layer ANN having M-inputs vector $\boldsymbol{d}$, and N nodes, with output vector $\boldsymbol{a}$, and corresponding target vector $\boldsymbol{a_t}$. This presents an optimization surface having a singular minimum global Mean Square Error (MSE), E. The error value E_n, at each node, from Eq. (13.1) and resulting expression for E are as follows:

$$E_n = a_{nt} - a_n = a_{nt} - f\left(\sum_{m=1}^{M}(w_{nm}d_{nm})\right) \tag{13.4}$$

$$E = \frac{1}{N}\left\{\sum_{n=1}^{N}(a_{nt} - f(s_n))^2\right\} = \frac{1}{N}\left\{\sum_{n=1}^{N}\left(a_{nt} - f\left(\sum_{m=1}^{M}(w_{nm}d_{nm})\right)\right)^2\right\} \tag{13.5}$$

To seek the minimum MSE point in terms of design weight values, we consider the partial derivative of input pattern error with each weight. Analysis of this derivative (Anderson & Rosenfeld, 1988) for change in error for each training pattern leads to the following expression:

$$\delta_n = -\frac{1}{a_n}\frac{\partial E_d}{\partial w_{nm}} \tag{13.6}$$

where δ represents a weight change coefficient.

Eq. (13.6) indicates that making weight changes proportional to δ will reduce error. Rationalizing from Eq. (13.3) shows that the change in node output error may then be stated as follows:

$$\delta_n = f_n'(s_n)(a_{nt} - a_n) = K\, a_n(1 - a_n)(a_{nt} - a_n) \tag{13.7}$$

This result is used to update the M weight (t) values for each node for the next $(t+1)$ iteration:

$$w_{nm}(t+1) = w_{nm}(t) + R\,\delta_n\, a_n \tag{13.8}$$

where R is a *Learning Rate* constant, for control of search stability and convergence.

Initialization for training begins by setting all bias threshold weights, w_{0n}, and all input weights, w_{nm} (m:1 ... M, n:1 ... N) to small random values in the interval [−0.5 ... +0.5]. The iteration should then use a randomly ordered sequence of data patterns from the Training Dataset. Training is normally paused at regular intervals to apply Test Datasets to assess error levels. When errors are acceptable, the Validation Dataset is used to assess when training is complete, to avoid overtraining.

Step 1. Apply input pattern vector, $\boldsymbol{d}$, with bias $d_0 = 1$ and measurements d_1, d_2, ... d_M.

Step 2. Compute node output vector $\boldsymbol{a}$, for target vector $\boldsymbol{a_t}$ from Eqs. (13.1) and (13.2).

Step 3. Compute error change vector, $\boldsymbol{\delta}$ from Eq. (13.7).

Step 4. Back-propagate node for each weight: get next vector: $\boldsymbol{w}$ $(t+1)$ from Eq. (13.8).

Step 5. IF [Test/Validate needed?] THEN [Execute]; ELSE [Return to Step 1].

This simple gradient decent process is used to locate the expected single minimum error point. It is extended to multilayer ANNs by addition of update steps for hidden layers, where back-propagation is executed beginning from the network output. The optimization process for a multilayer ANN is expected to feature local minima and an appropriate search strategy is then required to minimize a global MSE criterion. For brevity, this is not covered in this conceptual introduction; Anderson and Rosenfeld (1988) offer a detailed review. Other training strategies may also provide faster and more efficient design optimization for multilayer networks. For example, an approximation of Newton's method by Levenberg–Marquardt can increase speed but requires large memory. The Momentum method is shown to ignore small local minima in the error surface; related methods also exploit an adaptive Learning Rate.

ANN design and training typically use a computer workstation which features a digital network simulation. In Training Mode, long compute times may be required, but this is a one-time process, and has little relation to the speed of the trained ANN when later used in Estimation Mode, where feedback training features are not required.

IPT instrumentation applications for both pilot plants and operating processes have been outlined in Section 13.1. A design could retain a digital computer implementation if incorporated into an experimental pilot plant configuration, where new ANN configurations can be trained and trialled. In contrast, an operating process configuration could deploy specific purpose-designed hardware to increase speed, reduce instrument costs, and increase plant reliability. Modern ANN design tools (exemplified below in Section 13.6.2) include sophisticated ANN simulation and training facilities. When the design is complete, they include support for the transfer of the ANN structure including all bias and weight values to dedicated Estimation Mode hardware for forming a rolling process estimation instrument.

13.3 Artificial neural networks for IPT applications

To interpret IPT data, we first consider an ANN design to estimate an object image, in effect realizing steps S1, S2, and S3 of Section 13.1 above. As discussed, image information is primarily suited to pilot applications, or in plant diagnostics, where a human observer has an interest in the current process state. However, it is also useful here for direct comparison with the classical IPT method outlined in Section 13.1. The ANN will examine an input vector of M normalized tomographic measurement values that make up a complete projection, corresponding to a specific time-slice within the dynamics of the process of interest.

13.3.1 ECT example

To illustrate principles, we take the specific IPT example of an Electrical Capacitance Tomography (ECT) sensor having 12-electrodes (Chapter 1, Eq. 1.1). This provides a unique 66-value normalized capacitance measurement vector, $\boldsymbol{c}$, for each tomographic projection of a circular section object space, such as a pipeline transporting liquid/gas. We wish to transform these data to a grayscale image pixel vector, $\boldsymbol{g}$, where pixel positions are assigned using an object space grid, often based upon a Finite Element Model (FEM). We do not require any electro-physical model, or sensitivity distribution. Instead, we simply require training data, sampled from the process; or from a simulation, delineated using the same grid positions.

To estimate an image, we deploy a simple ANN with a single layer for the direct processing of input data to yield the required image output. For simplicity, and recognizing the limited spatial resolution of ECT, we specify an image of 100 pixels. Our Perceptron ANN design thus has 100 nodes, each having a bias input, 66 measurement inputs, and $(1 + 66 = 67)$ weights as shown in Fig. 13.5.

As described above, the realization of ANN performance is critically dependent upon an iterative training process. This ECT example features 100 bias values plus (66×100) measurement input values, a total of 6700 input/bias weight values. Training data will typically include a variety of subtle behavior, for example, including soft field effects.

For later comparison, in terms of processing load during the feed-forward estimation operation, this includes 6700 multiplications; plus 100 summations; and activation computations.

13.3.2 ANN versus linear back-projection

It is interesting to compare the above ANN design with a classical LBP approach, where an electro-physical model is used to generate sensitivity distribution data,

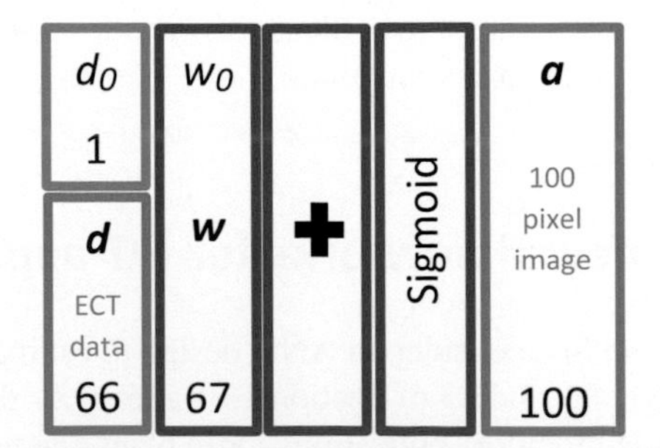

Figure 13.5 ANN for ECT 66 measurement direct 100-pixel image generation.

$s(p)$ which, using the same capacitance measurements described above, enables computation of each positioned grayscale pixel in an image:

$$g(p) = W(p) \sum_{m=1}^{M} c_m \, s_m(p), \text{ where } W(p) = \left[\sum_{m=1}^{M} s_m(p) \right]^{-1} \tag{13.9}$$

This is a simple process to implement and therefore offers practical advantages of speed, but produces results which are nonlinear, and due to soft field and dynamic range issues, typically includes artifacts and distortion. A postprocessing dynamic thresholding operation is also usually required.

Taking the previous ECT example case data, each image estimate requires 100×66 multiplications (6600); plus 100 scaling additions each of 66 values (6600), plus processing for typical thresholding.

In simple terms, the LBT processing load (without an allowance for threshold processing) is almost double that required to execute the ANN Estimation Mode solution.

To gain speed, both approaches will benefit from parallelized processing hardware. In terms of the result quality, the ANN approach is very likely to out-perform an LBP estimation, as it incorporates learned properties of soft field distortions and thresholding and has faster direct computation speed. Of course, the ML data-driven method requires a major investment in its one-time training.

13.3.3 Multilayer example

For simplicity, we add a further stage to the ANN shown above in Fig. 13.5. The additional hidden layer has inputs from the previous image output layer; and a new final layer now provides an *interpretation* of the image data, for example, to identify the flow configuration. In practice, if an image of the pipeline flow is not required, the interpretation output may be achievable directly by a single-layer ANN with only 66 inputs and outputs for each interpretation state. A design addition to the above network is used to identify which of four common flow states is present:

Stratified flow - Bubble flow - Core flow - Annular flow

Using the simple design rule, an additional output stage will have four nodes, one for each state, each fully connected and having 100 inputs. In this revised model, the image layer now becomes a *hidden* layer to compute the *internal image*. If the revised ANN has twin purposes, it is possible to simply train the added layer using a training set incorporating image configuration data. The revised Multi-Layer Perceptron (MLP) ANN network is shown in Fig. 13.6:

The revised ANN includes the previous 100 (bias) + 100×66 (inputs), totaling 6700 weight values, with the addition of a further 4 (bias) + 4×100 (inputs), totaling 404 weight values; totaling 7104 weight values. It could be deployed for this dual function of providing an image plus a flow regime interpretation. If the only requirement was flow regime identification, we would be likely to find a simpler ANN would suffice, as we next explore.

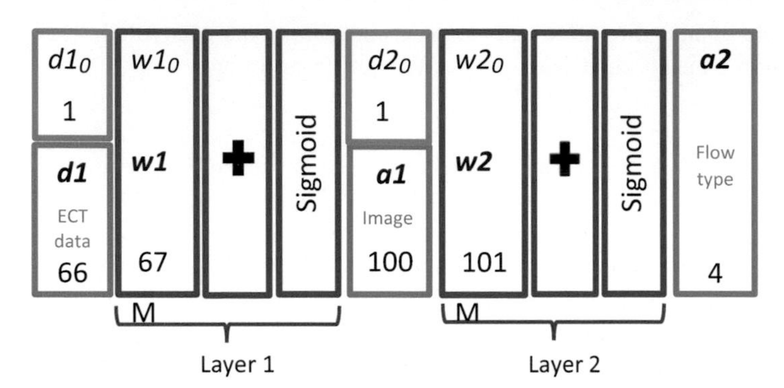

Figure 13.6 ANN for ECT 66 measurement direct image and flow configuration.

Part 3, Chapter 16, also provides a following supplementary application review of AI in process control.

13.4 Case study A—estimating multiphase flow parameters

Multiphase flows arise in several industry sectors, for example in Oil and Gas processing, where volume rate measurements are important. This Case Study addresses an ML approach to the key enabling measurement of component fraction, defined as the area occupied by gas, oil, or water, relative to a total normalized unit area.

Direct flow parameter estimation methods, which are not reliant on reconstructed tomographic images, have been proposed for component fraction and concentration estimation. Early pioneering proposals were reported to be flow regime dependent; more recently proposals by Yu, Yang, Wang, and Yu (2019) are limited to a subset of flow states. Many instruments are limited to two component flows. A direct flow measurement capability that does not require extensive image reconstruction processing would be valuable for instrumentation (Zainal-Mokhtar & Mohamad-Saleh, 2013).

13.4.1 Overview and measurement requirements

IPT ECT sensors have often been deployed for estimating flow parameters within a pipe cross-section based on image reconstructions, as detailed in Chapter 1. Their accuracy is critically affected by the nonuniform sensitivity of the sensing area and the *soft field* effect. Enhancements gained through extensive processing can improve accuracy, but with limitations to certain flow conditions. They are also expensive in processing cost and time further limiting application in dynamic processes such as transport pipelines.

Hence, ANN-based solutions should be flow regime independent and able to directly and speedily transform ECT measurements directly to cross-sectional, component fraction estimates for two-component gas—oil flows, and three-component gas—oil—water flows. Fig. 13.7 shows the schematic flow regimes investigated, categorized as Stratified, Annular, Bubble, or Homogeneous, where 100% content is assumed be of air, oil, or water.

13.4.2 ECT sensor and simulated ECT measurements

The core sensing unit deployed in this simulation case study is a 12-electrode ECT sensor, as in the introductory example of Section 13.3, for which capacitance measurements were simulated. The simulated pipe has normalized inner, outer pipe, and screen wall radii of 1, 1.2, and 1.4 units, respectively, to facilitate scaled future adaptation. Sensor and guard electrodes have subtended angles of 22 degrees and 2.5 degrees, respectively, resulting in gaps of 8 degrees between pairs of sensor electrodes. Capacitance values between electrode pairs are annotated as c_{ij} (where, i,j: 1 … 12).

For convenience, measurements are normalized with respect to the value when the pipe is filled with the low permittivity material, and scaled with respect to the value range when filled with high and low permittivity material, defined as follows:

$$c_{ijn} = \frac{c_{ij} - c_{ij}\mathrm{low}}{c_{ij}\mathrm{high} - c_{ij}\mathrm{low}} \tag{13.10}$$

The fraction for a component is then calculated simply as follows:

$$f_{\mathrm{component}} = \frac{A_{\mathrm{component}}}{A_{\mathrm{pipe}}} \tag{13.11}$$

where $A_{\mathrm{component}}$ is the area covered by the component of interest (either oil or water), and A_{pipe} is the area of the entire pipe cross-section. Hence, component fraction ranges from 0 (empty pipe) to 1 (pipe full of oil or water).

A two-dimensional FEM ECT simulator by Spink (1996) was used to obtain target flow regime data from electric field lines for corresponding permittivity distributions. Its *decomposed geometry* feature was used to generate sets of 66 ECT projection vectors, $\boldsymbol{c}$ (Mohamad-Saleh & Hoyle, 2002). Design Datasets were generated for all flow patterns of interest for network training and evaluation and included relevant Target parameter values:

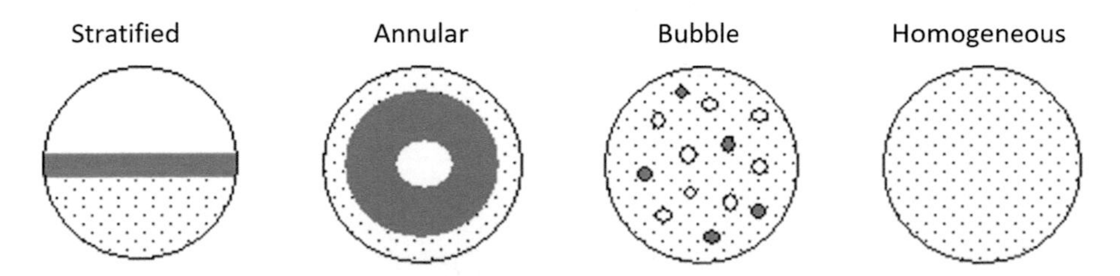

Figure 13.7 Schematic model of the target flow regimes.

Oil–Gas: 3185 flow patterns each with calculated component fractions: 2 simple patterns, 1 empty pipe, 1 full pipe flow, 2520 stratified flows, 636 bubble flows, and 27 annular flows.

Oil–Gas–Water: 4932 flow patterns with calculated oil and water fractions: 3 simple patterns, 1 full gas, 1 full oil and 1 full water, 2730 stratified flows, 2028 bubble flows, and 171 annular flows.

13.4.3 Network design and training for flow estimation

Network design centered upon a computer-based ANN Design Toolsuite, as outlined in Section 13.2.3, based upon the **MATLAB** Neural Network Toolbox (Mohamad-Saleh & Hoyle 2008). Execution speed was not critical as trial data were stored for later evaluation. This supported switching between Training Mode and (simulated) Estimation Mode and was used to design optimum networks for the noted flow regimes for two and three-component flows for parameters: number of hidden layers, all weight and bias values. The aim was to facilitate application deployment, either for a pilot-plant demonstration, or for embedding into purpose-designed hardware forming part of on-line instrumentation for a process monitoring.

The design process used the above combined Dataset of 3185 patterns for Oil/Gas flows and 4932 patterns for Oil/Gas/Water flows. This was randomly divided into 67% Training, 13% Test and 20% Validation datasets. As outlined in Section 13.2.3, Training Datasets form the possible *learned* data used only in Training Mode; Test Datasets were used in simulated Estimation Mode to assess error levels in training; and wide-ranging *Validation* datasets were used in to assess when training was complete.

Preprocessing of input data is commonly used to scale data values and refine data to enhance network efficiency. Here datasets are first preprocessed to normalize values to a convenient range. Next, a further valuable preprocessing operation is used to enhance orthogonality, by removing highly correlated and repeated data. Training Datasets are processed using the Singular Value Decomposition method to realize a Principal Component Analysis (PCA), which ranks orthogonality of data elements. The percentage value of total variation in the transformed Principal Component (PC) dataset is referred to as the PC-Variance. Fig. 13.8 illustrates *Training Mode Preprocessing Operations* for a candidate input Training Dataset.

As illustrated, the M elements of every capacitance measurement vector, $\boldsymbol{c}$, are normalized to produce vector $\boldsymbol{c_n}$, based upon zero mean and unity variance across the Training Dataset. The PCA transformation eliminates measurements which contribute less than a specified fraction (here 5%) of the total Training Dataset variation. The operations produce two outputs: (i) a composite PCA transformation matrix, $\boldsymbol{t}$, generated for the Training Dataset and stored for later corresponding Estimation Mode use; and (ii) the resulting PCA preprocessed Training Dataset, $\boldsymbol{c_{np}}$, of L ($L \leq M$) orthogonal, uncorrelated measurement vectors available for Training Mode operations, in which each vector also includes a time-synchronized target parameter value for error computation.

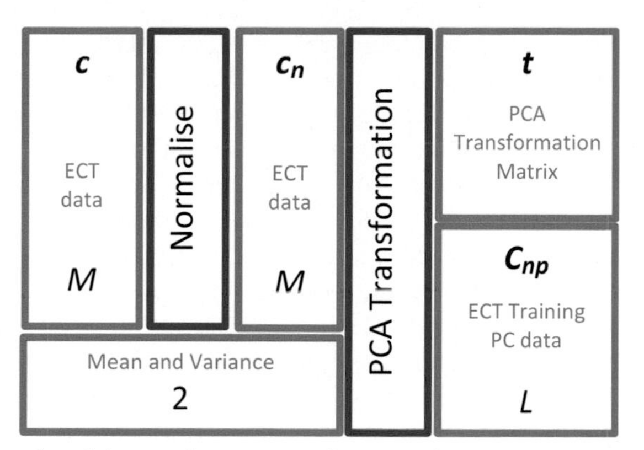

Figure 13.8 Network training mode preprocessing operations.

The Test Dataset is used periodically as training progresses; and the Validation Dataset is used to assess when to cease training. All datasets must embody corresponding preprocessing operations. We also require the same efficiency features to be applied to in-line application networks. Hence, a further preprocessing operation is needed for two roles: (1) to allow the transformation of both Test and Validation Datasets within the network design process and (2) to facilitate preprocessing of in-line process data in pure Estimation Mode when the trained network is fed with process plant ECT vector data. For both preprocessing needs, optimum PCA values have been determined, and for an application network, the design is complete. Hence, this preprocessing operation need only normalize (against preset norms), and then apply the prestored PCA transformation matrix, $\boldsymbol{t}$. The resulting preprocessing operations are illustrated in Fig. 13.9, which has similar nomenclature to Fig. 13.8, but where input vector $\boldsymbol{c_n}$ and preprocessed result vector, $\boldsymbol{c_{np}}$, refer to the specific role, as noted above as (1) and (2).

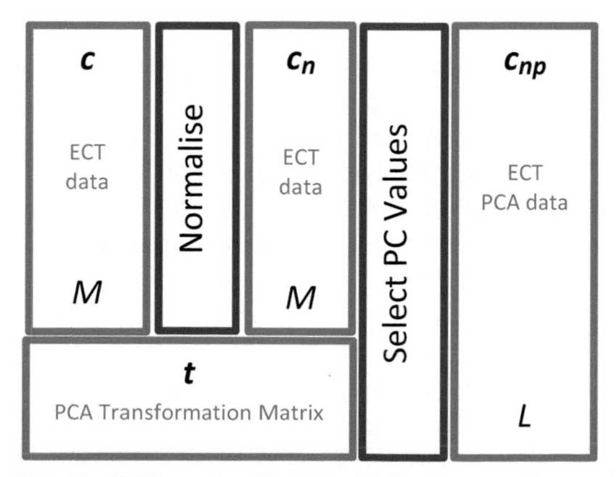

Figure 13.9 Network test/validate and estimation mode preprocessing operations.

For role (1) Training Mode deployment, both preprocessing operations of Figs. 13.8 and 13.9 are single-use operations executed on a one-time basis to transform a complete Dataset. For role (2), Fig. 13.9 operations must be integrated into the front-end of the Estimation Mode ANN design, as illustrated later.

An important ANN design stage is the determination of the number of hidden neurons and the optimum PC variance value in a training process. This can begin firstly with the MLP trained with zero hidden neurons and using the normal practice of initializing weights with small random values, as described in Section 13.2.3.

An *inner iteration* using a fixed PC variance value of preprocessed pattern data is used with a Bayesian Regularization search from the initial weights. Validation patterns are used to assess performance, and if this improves, weights are updated and used in the next training cycle. Training is suspended when there is no improvement. A single hidden neuron having a sigmoid activation function is then added and training is resumed. The iteration of adding a further hidden neuron at each stage continues if performance improves. This *network-growing* process is terminated when there is no further improvement, when the optimum number of hidden neurons is assumed to be attained. The corresponding state information of all current training weights is stored for use in the testing stage. 30 training trials (each with different starting locations on the network error surface) are performed to ensure a global minimum is located.

An *outer iteration* of the network growing process is repeated over a range of PC variance values, using a Mean Absolute Error (MAE) criterion, to locate the global optimum design of PC variance and number of hidden neurons. Here, the MAE criterion is preferred to the MSE criterion of Section 13.2.3.

For this flow estimation application, separate MLP designs are used for each of three component fraction estimation tasks. A linear activation function provides component estimation outputs in the range [0 … 1]. The variants, typically with differing PCA and hidden neuron configurations, must be optimized for their selected flow estimation application role. Hence, the overall system features three subunits each with Estimation Mode Preprocessing Operations and MLP for a specific role:

F1 - Oil fraction from Gas/Oil flows;

F2 - Oil fraction from Gas/Oil/Water flows;

F3 - Water fraction from Gas/Oil/Water flows.

To explore the optimal design of the F1 ANN, measurement data for Gas/Oil Flows MAE results for a range of PC variance values are shown in Fig. 13.10.

This figure presents data for hidden neuron configurations from 2 to 8. A PC variance of 0.05% having the lowest MAE corresponds to an optimum 27 inputs. Use of all 66 measurements (0% PC variance) is shown to produce the highest MAE; likely to arise from correlated components in the full measurement set. This also demonstrates that the network generalizes better when input data are optimal for learning. The figure reveals that the optimum network configuration with five hidden neurons has a test MAE of about 0.32%, which increases when either six or four hidden neurons are used, demonstrating overfitting and underfitting, respectively.

To explore the optimal design of the related F2 and F3 ANNs, measurement data for Gas/Oil/Water Flows MAE results for a range of PC variance values are shown in Figs. 13.11 and 13.12, respectively.

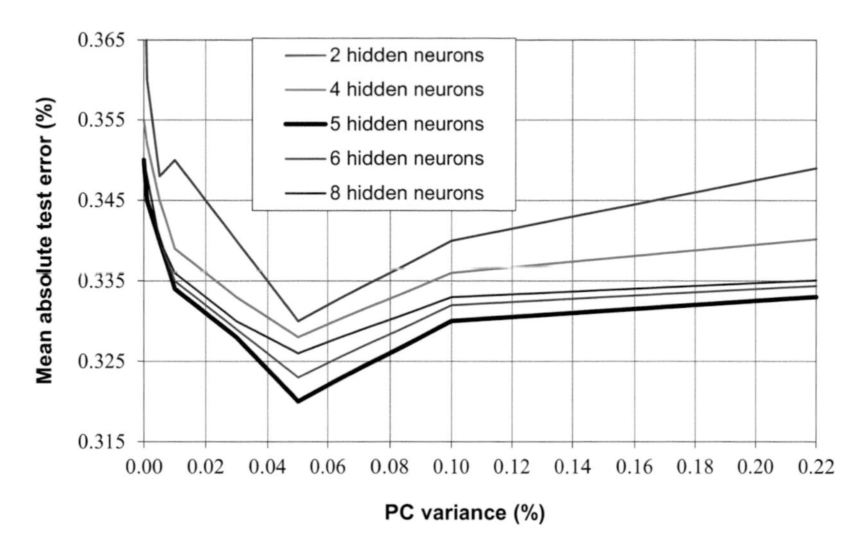

Figure 13.10 Performance with various hidden neurons trained using different PC variance values to estimate oil fractions in Gas/Oil flows.

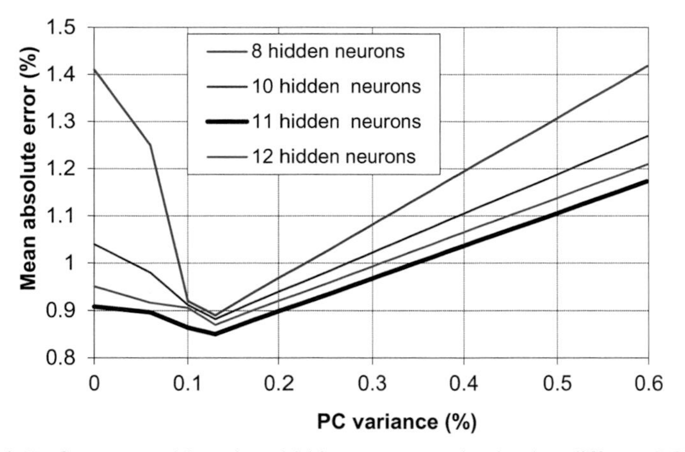

Figure 13.11 Performance with various hidden neurons trained using different PC variance values to estimate oil fractions in Gas/Oil/Water flows.

The F2 and F3 networks are separately trained to estimate oil fraction, and water fraction, from 3-component gas—oil—water flows. Figs. 13.11 and 13.12 show their performance at various PC variance values in the estimation of oil and water fractions from Gas/Oil/Water flows, respectively. In both cases, the least MAE occurs at a PC variance of about 0.13, corresponding to 29 inputs. The F2 oil fraction estimator performs best with 11 hidden neurons, with an MAE of 0.85%. The F3 water fraction estimator performs best with five hidden neurons with MAE of 0.72%. In both cases, MAE increases when overfitting or underfitting occurs.

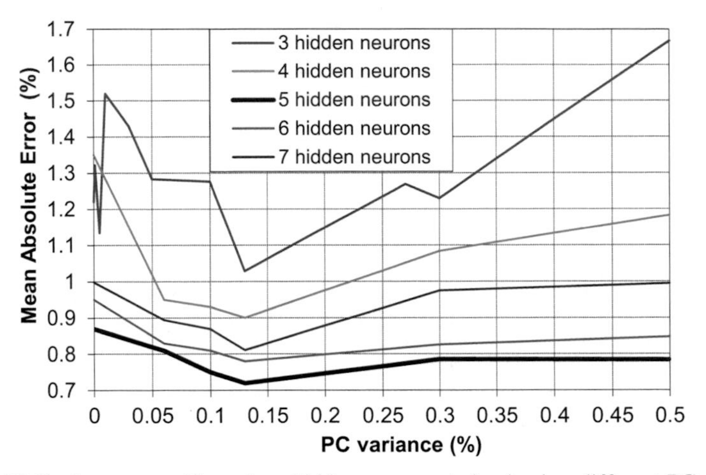

Figure 13.12 Performance with various hidden neurons trained using different PC variance values to estimate water fractions in Gas/Oil/Water flows.

Schematic block diagrams for each of the three optimal ANN designs are illustrated below, including the Estimation Mode Pre-Processing Operations represented here as a single operation in each part of Fig. 13.13.

These clearly indicate the common forms of the networks and the implicit potential to standardize upon a hardware approach to realize instrumentation that fulfills these related application requirements.

13.4.4 Network trials

Trials were carried out on both target flow types: Oil/Gas (F1) and on Oil/Gas/Water (F2 and F3).

Trial performance results for the F1 2-component case, using the ANN of Fig. 13.13a, are classified by flow regimes in Table 13.1.

Results indicate the largest average absolute error occur for annular flow patterns, expected for gas bubbles located in the pipe center, where ECT sensitivity is low. Stratified flows present a simpler estimation of the location of the gas−oil interface, which presents at the ECT sensor boundary where sensitivity is high. The absolute difference in error levels and standard deviations is small, indicating good overall performance. To assess worst case annular data, Fig. 13.14a provides example calibration data of actual and estimated values for 33 annular test patterns, providing a regression fit coefficient of 0.998. These results offer a high confidence level that oil fraction estimation in Oil/Gas flows using the ANN estimation is accurate and not flow regime dependent.

Trials performance results for the F2 3-component case, using the ANN of Fig. 13.13b, are classified by flow regimes in Table 13.2.

Results follow the trends for F1 for the same reasons. Overall errors are slightly higher, but 3-component flows are more challenging. To assess worst case annular

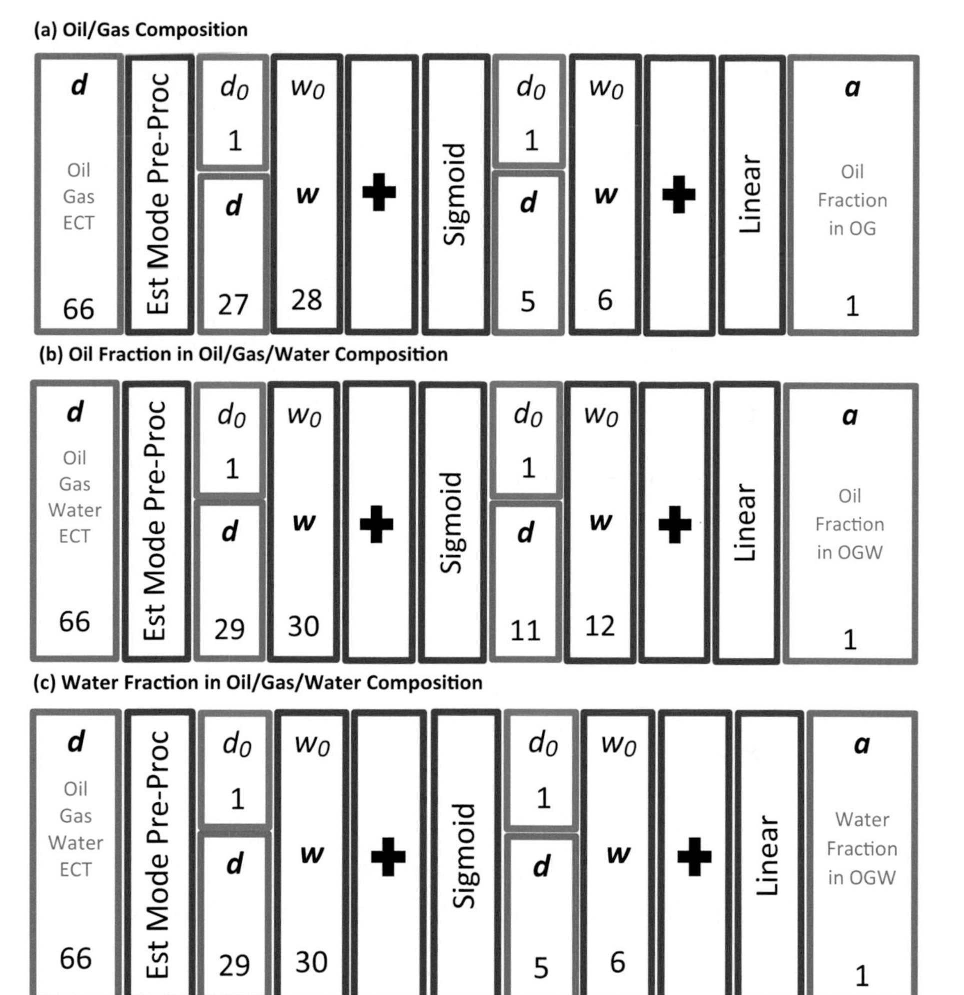

Figure 13.13 MLP ANN - Direct fraction estimations — (A) F1- Oil Fraction in Gas/Oil Composition; (B) F2 - Oil Fraction in Gas/Oil/Water Composition; (C) F3 - Water Fraction in Gas/Oil/Water Composition.

Table 13.1 Oil fraction errors for annular, bubble, and stratified Oil/Gas flows.

	Annular	**Bubble**	**Stratified**
Mean absolute test error (%)	0.20%	0.15%	0.026%
Standard deviation of test error (%)	±0.15%	±0.12%	±0.023%

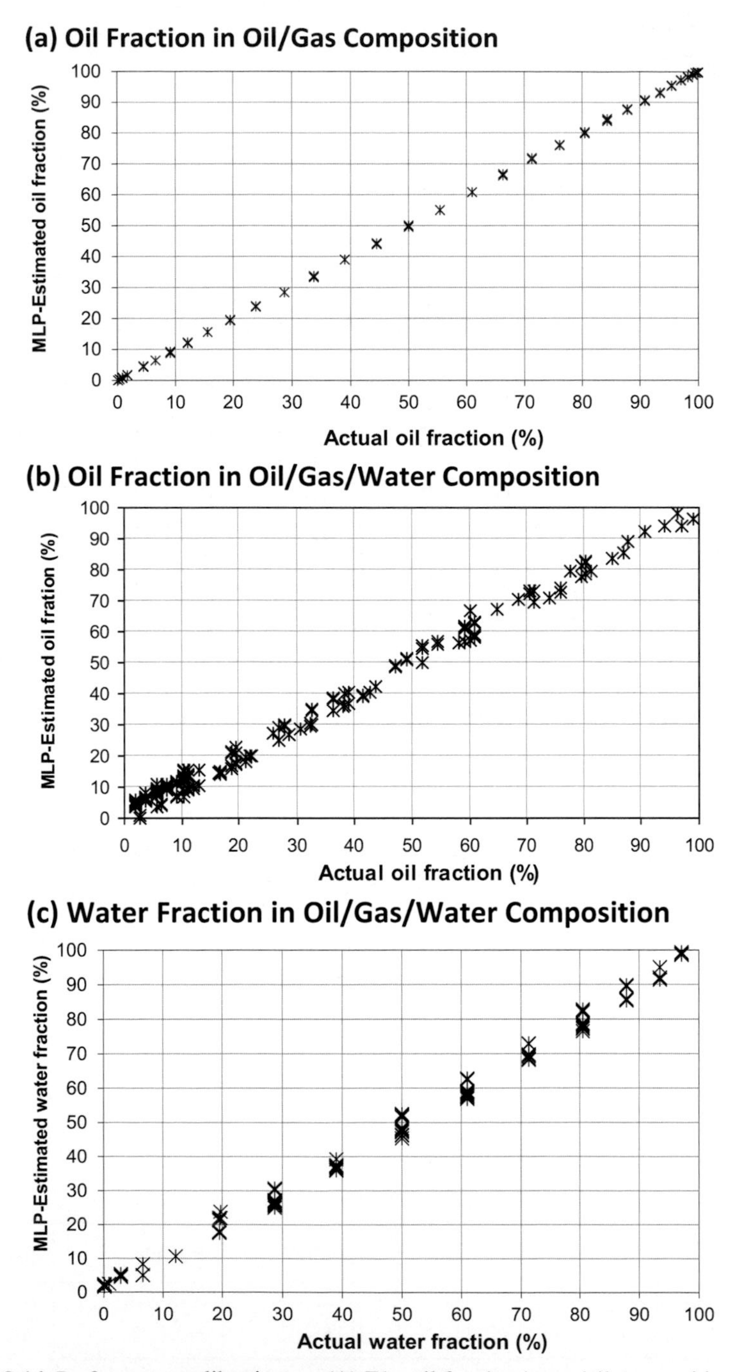

Figure 13.14 Performance calibrations — (A) F1 - oil fraction in gas/oil composition; (B) F2 - oil fraction in Oil/Gas/Water composition; (C) F3 - water fraction in gas/oil/water composition.

Table 13.2 Oil fraction errors for annular, bubble, and stratified regimes of Oil/Gas/Water flows.

	Annular	**Bubble**	**Stratified**
Mean absolute test error (%)	2.3%	0.85%	0.15%
Standard deviation of test error (%)	±0.75%	±0.37%	±0.085%

data, Fig. 13.14b provides example calibration data of actual and estimated values and provide a regression fit coefficient of 0.982. These results offer a high confidence level that oil fraction estimation in Oil/Gas/Water flows using an ANN estimation system is accurate and not flow regime dependent.

Trials performance results for the F3 3-component case, using the ANN of Fig. 13.13c, are classified by flow regimes in Table 13.3.

Trial results again follow the general trends of F1 and F2 results for the same reasons. Overall errors are slightly lower than those for F2. Again, to assess worst case annular data, Fig. 13.14c provides example calibration data of actual and estimated values and provide a regression fit coefficient of 0.980. These results offer a high confidence level that water fraction estimation in OGW flows using an ANN estimation system is accurate and not flow regime dependent.

Further details of these ML flow estimation performance results are available (Mohamad-Saleh & Hoyle 2002).

13.4.5 Evaluation of ANN performance of flow component sensing

This case study has shown that it is feasible to use MLP ANNs to estimate component fractions from 2-component Gas/Oil flows and 3-component Gas/Oil/Water flows. Performance and calibration trial results indicate the success of the method through very low MAE for 2-component, oil−gas flows, with more complex 3-component Gas/Oil/Water flows only slightly increased MAE. The 3-component water fraction estimations have lower error compared to oil fractions. This probably arises as capacitance measurements are calibrated based on materials with highest and lowest permittivity values; in this case, water and gas, respectively. Hence, measurement values are more dominant when the sensing area is filled with water, reinforcing learning associated to water content. However, oil fraction estimation errors are

Table 13.3 Water fraction errors for annular, bubble, and stratified regimes of Oil/Gas/Water flows.

	Annular	**Bubble**	**Stratified**
Mean absolute test error (%)	2.0%	0.7%	0.13%
Standard deviation of test error (%)	±0.61%	±0.31%	±0.1%

slightly higher; they are small in absolute terms offering high confidence that ML estimation is feasible in 3-component flows. Importantly, results indicate that the trained ANNs are robust to different flow regimes and measurements meet the primary aim of flow regime independence.

13.5 Case study B—estimating inline rheology properties of product flow

Rheology, the way a material flows, impacts upon every aspect of our daily lives from the texture of food to the paint on the walls. Not only does rheology affect the final product quality, but it also governs in-process efficiency and is critical in all chemical and physical processing (Barnes, Hutton, &Walters, 2001).

13.5.1 Overview and measurement requirements

The measurement of rheology, termed *rheometry*, is typically conducted off-line through careful sampling of a process stream. Rheological properties from off-line measurements are often considered, with assumptions, as directly applicable to real process flows. However, they can only provide a retrospective characterization of a fluid sample, which may not be representative of the structure as a function of the time-shear history received during processing (Machin, Wei, Greenwood, & Simmons, 2018a,b), and is therefore often considered unsatisfactory.

In situ measurements in a flow environment overcome this deficiency, and also remove the possibility of sample degradation. Due to the critical nature of rheology in processing, the development of an in-line rheometer can advance rheometry from a process end-point, quality control aid to a capability for control, and optimize processes and material structures. Hence, there is an increasing interest in the development of in-line rheometry, as the majority of industrial fluids exhibit complex non-Newtonian behavior.

To address this demand, Machin et al. (2018a,b) developed a novel, in-pipe, tomographic measurement which obtains real-time rheological information of process fluids, termed *Electrical Resistance Rheometry* (ERR). The ability of ERR to estimate rheological parameters is enhanced by the use of ML algorithms and has been demonstrated and validated at an industrial pilot plant of a multinational manufacturer. The trial focused upon the characterization of a wide range of industrial personal and homecare products including shampoos, fabric washes, conditioners, and body washes. Its aim was to compare ERR results against off-line rotational rheometry to determine suitability for in-line quality control.

ERR enables the characterization of fluid rheological properties within a pipe under steady state, incompressible, laminar flow conditions. To achieve such conditions, the Reynolds number of the flow must be less than 2000 (Paul, Atiemo-Obeng, & Kresta, 2004). By cross-correlating fluctuations of computed conductivity pixels across and along a pipe, using noninvasive, microelectrical tomography sensors, rheometric data are obtained through the direct measurement of velocity profile. The use of microelectrical tomography sensors means that the fluid must be relatively conductive, in the range of 0.5−50 mS/cm. The approach is extremely robust, noninvasive, and able to interrogate process fluids, without limitation due to optical opacity. The wide-ranging applicability and simple implementation of microelectrical tomography sensors to industrial processes is an ideal platform for the development of an in-line rheometer. ERR is derived from Electrical Impedance Tomography (EIT) and is capable of simultaneously sensing mixing performance using the cross-sectional impedance map. Thus, ERR combines two significant engineering quality and process control concepts in a powerful characterization instrument and provides an understanding of the inhomogeneity and nonuniformity of a process.

13.5.2 Sensing and processing system

The ERR instrument used is supplied by Stream Sensing Ltd. and utilizes four sensor arrays of eight electrodes, consisting of two linear and two circular arrays, as shown in Fig. 13.15. This novel configuration provides a complete velocity profile with high sensitivity data near to the pipe-wall boundary provided by the linear arrays, and data from center of the pipe by the circular arrays.

This sensor is linked to a v5r EIT system, supplied by Stream Sensing Ltd., capable of operating a sensor with a maximum of 32 electrodes. Here, the arrays consist of eight electrodes to provide high temporal resolution for velocity precision. The EIT instrument is controlled using v5r software, in Rheology Mode, and to capture a measurement every 30 s. The modified sensitivity Back-Projection algorithm from Wang, Mann, and Dickin (1996) is utilized to reconstruct both linear and circular sensor

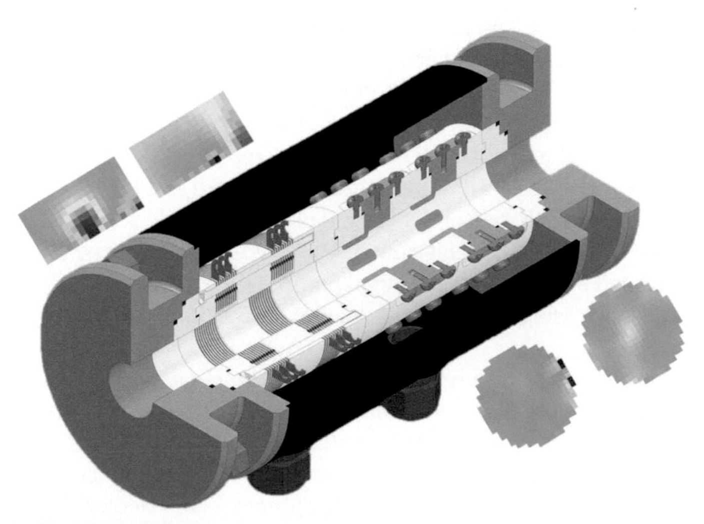

Figure 13.15 Electrical resistance rheometer electrode configuration.
Courtesy of Stream Sensing.

tomograms. These are segmented into zones of interest to assign several radial velocity measurement positions in the pipe cross-section.

A heat pulse is generated within the sensor which causes a minor increase in temperature, in turn producing a conductivity perturbation. This pulse generates uniform heating of the cross-section; is 95% efficient; removes the requirement for a tracer; provides negligible degradation of thermosensitive products; and is automated (Knirsch et al., 2010). Thermal degradation is reduced further as the required temperature rise is only 0.7°C for 0.5 s.

For each zone, the sensed conductivity is averaged and normalized. A cross-correlation algorithm is employed to tag the fluid motion across and within multiple measurement arrays. As the distance between the arrays is fixed and known, the extracted time delay can be converted into two- and three-dimensional velocity profiles. However, due to the axisymmetric nature of velocity profiles in laminar flow, there is only a requirement for a two-dimensional velocity profile in the determination of rheology (Bergman, Lavine, Incropera, & DeWitt, 2019), which reduces computational requirements.

Rheological properties strongly influence velocity profile shape in laminar pipe flow, due to a shear rate response of the fluid. Raw velocity measurements provide a *fingerprint* aid to predict rheological parameters and can be utilized to infer rheological behavior exhibited by complex fluids systems exhibiting shear-banding, wall depletion, and shear-induced phase migration. The velocity profile may be coupled with measurement of differential pressure to obtain the linear shear stress profile in steady state, incompressible laminar flow. If the rheological behavior is known to be described by conventional rheological models, a parametric fitting of the velocity profile can be utilized to directly output the desired rheological properties (Machin et al., 2018a,b). This can be used to extract rheological parameters of fluids modeled by simple, specific constitutive equations; however, as their complexity is increased, parametric fitting becomes increasingly ill-conditioned with large discrepancies. A number of industrial fluids do not comply with such models.

ML methods are preferred in order to provide robust sensing for all materials. To validate this novel approach, several industrial fluids have been investigated at a pilot plant of a multinational manufacturer of homecare and personal care products. The industrial ERR sensor utilized within this study is depicted in Fig. 13.16.

The 25.4 mm Dia. ERR sensor was installed in the simple recirculation flow loop pipeline, shown in Fig. 13.17. This contains a 200 L jacketed vessel, agitated by an anchor impeller and a controlled centrifugal pump. Differential pressure was measured across a 1 m distance using an Endress + Hauser PMD55 meter, with logged flow rate and pressure.

The jacket heater was used to maintain the fluid temperature at 25°C, monitored by a probe at the vessel outlet, before the fluid was circulated at three selected flow rates of 500, 750, and 1000 kg/h. The experiment was repeated at 30°C and again at 35°C. Three ERR measurements were performed for each experimental condition.

A wide range of personal and homecare products were selected for the industrial trial at the pilot scale: shampoos; body washes; fabric washes; and hand washes.

Figure 13.16 Electrical resistance rheometer sensor.
Courtesy of Stream Sensing.

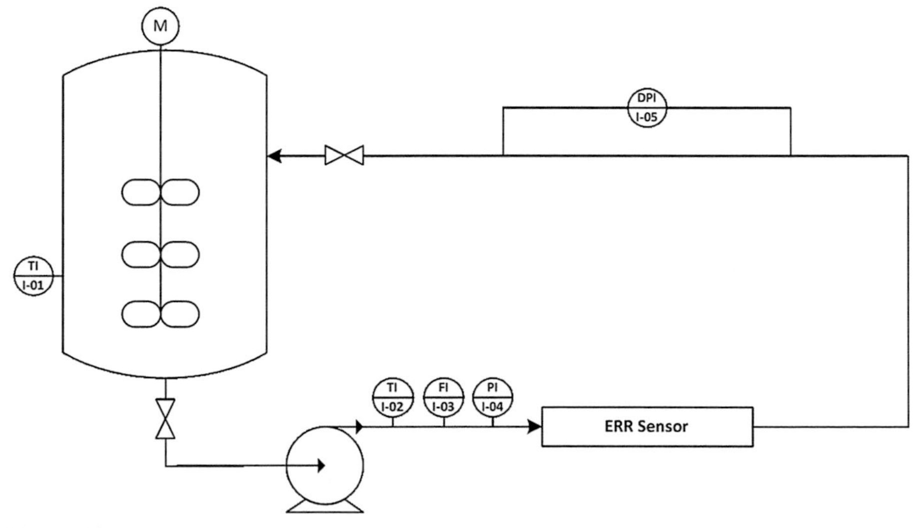

Figure 13.17 Rheology trial flow loop.

To provide comparison of off-line measurements, an HAAKE Rheostress 1 Rotational Rheometer (Thermo Fischer Scientific, USA) equipped with a smooth-walled, 1 degree stainless steel cone and plate geometry was used. A Peltier plate was utilized to maintain each sample at $25 \pm 0.1°C$, $30 \pm 0.1°C$, and $35 \pm 0.1°C$. To assess the impact on structure from the time-shear history during processing, samples were collected, and off-line measurements performed, at each experimental flow rate. To minimize such effects further, a new sample was loaded onto the rheometer for each experimental condition. Upon loading, a logarithmic ramp was applied with a shear rate range of $0.1-1000\ s^{-1}$, with five points per decade over 5 min. The resulting data were analyzed using a nonlinear least squares regression, using the HAAKE

RheoWin software, to extract rheological parameters. This analysis was performed across the shear rate range of 0.1−100 s^{-1}, reflective of the subsequent flow experiments. Two specific shear rates of 4.83 and 20.7 s^{-1} were then selected for comparison with viscosities obtained from both in-line and off-line measurements.

13.5.3 Machine learning approach

A supervised two-layer ANN was selected to estimate the viscosity at the noted shear rates. Inputs include 34 velocity data points from the ERR sensor, plus the differential pressure across the sensor. Input values are acquired over approximately 30 s to ensure that dynamic changes in the process are captured. A sigmoidal transfer function was employed for the 30-neuron hidden layer. The output layer is a single neuron with a linear transfer function to provide an estimated viscosity output. The ANN is shown in Fig. 13.18.

For each shear rate, an independent model was applied with target data and viscosity obtained from off-line rotational rheometry. Considering all test material types, viscosity varied by up to three orders of magnitude. To ensure measurement sensitivity for small viscosity values, two models were trained for each shear rate. Due to the importance of rheology measurements in processing, there is a requirement to obtain a high level of accuracy and thus Bayesian Regularization was selected as the training algorithm. Training was performed using the MATLAB ML toolbox, with the data split into 75% training and 25% test groups to verify the performance.

13.5.4 Network trials

Rheological properties of a fluid strongly influence the velocity profile shape in laminar pipe flow, due to a shear rate fluid response, as highlighted in Fig. 13.20 below. The displayed velocity profiles were obtained using the method outlined by Machin et al. (2018a,b), utilizing a parametric fit of ERR data to a power law rheological model. It is evident that velocity profiles of shampoo, Fig. 13.19d, feature increased blunting, compared to the hand dish wash, Fig. 13.19b, due to increase in shear thinning behavior. This is also reflected in the rheology data obtained from off-line measurements, in Fig. 13.19a and c.

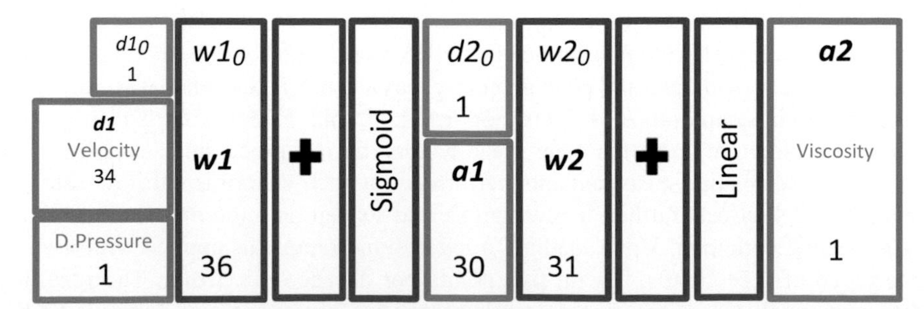

Figure 13.18 Two-layer feed forward ANN to estimate rheological properties.

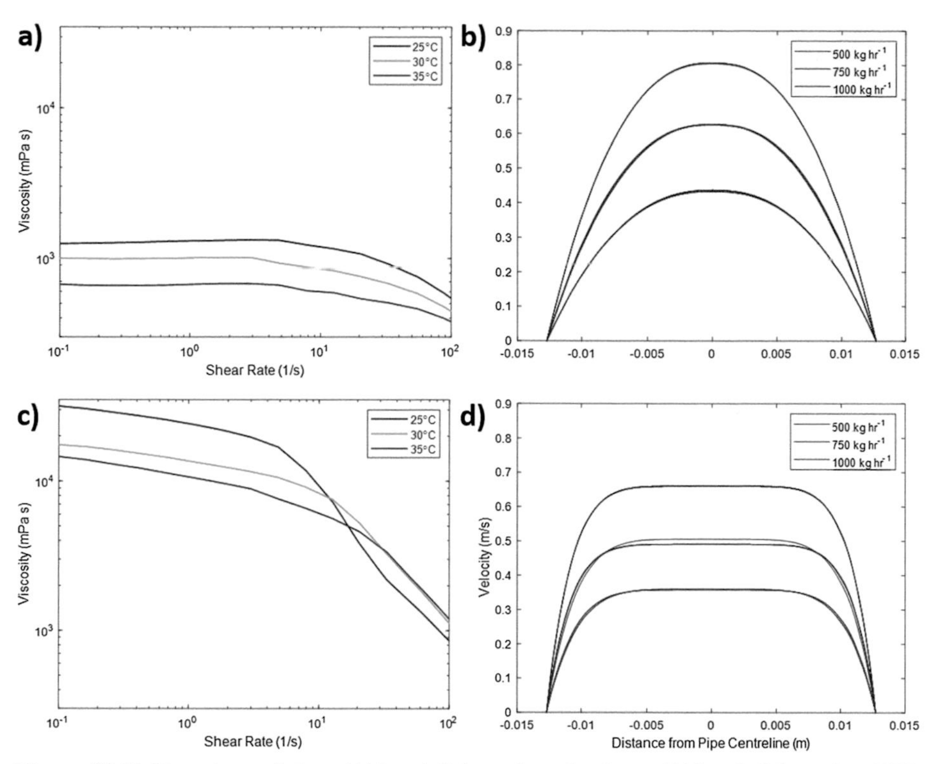

Figure 13.19 Experimental data: (A) hand dish wash – rheology; (B) hand dish wash – ERR velocity profiles; (C) shampoo – rheology; (D) shampoo – ERR velocity profiles.

It is acknowledged that the parametric fitting approach is not ideal, as selected fluids do not adhere to the power law rheological model. Therefore, to capture the increased complexity of the fluid structure, an ML algorithm was developed for in-line quality control. The estimation of in-line viscosity was seen to correspond to that of the off-line rheometer, with results outlined in Fig. 13.20. When considering low viscosity fluids, at 4 s^{-1}, the correlation coefficients between the measured and estimated viscosity were 0.999, 0.998, and 0.999 for the training, test, and complete datasets, respectively. High correlation is also found in the corresponding root-mean squared error (RMSE) of 2.50, 30.6, and 11.87 mPa s. This ensures that the application of an ANN to ERR provides a highly accurate estimation of in-line viscosity.

The high performance has been attained using a relatively small dataset and is likely to be improved further, with the performance of algorithms trained using Bayesian Regularization enhanced monotonically with increasing size of the dataset, as discussed by Neal (1996). The ability of the high viscosity model to predict viscosity was also seen to have high correlation with correlation coefficients of 0.999, 0.997, and 0.999 obtained for the training, test, and complete dataset, respectively.

To estimate rheological behavior, the viscosity of the system must be estimated at multiple shear rates and hence a second shear rate of 20.7 s^{-1} was selected. As the

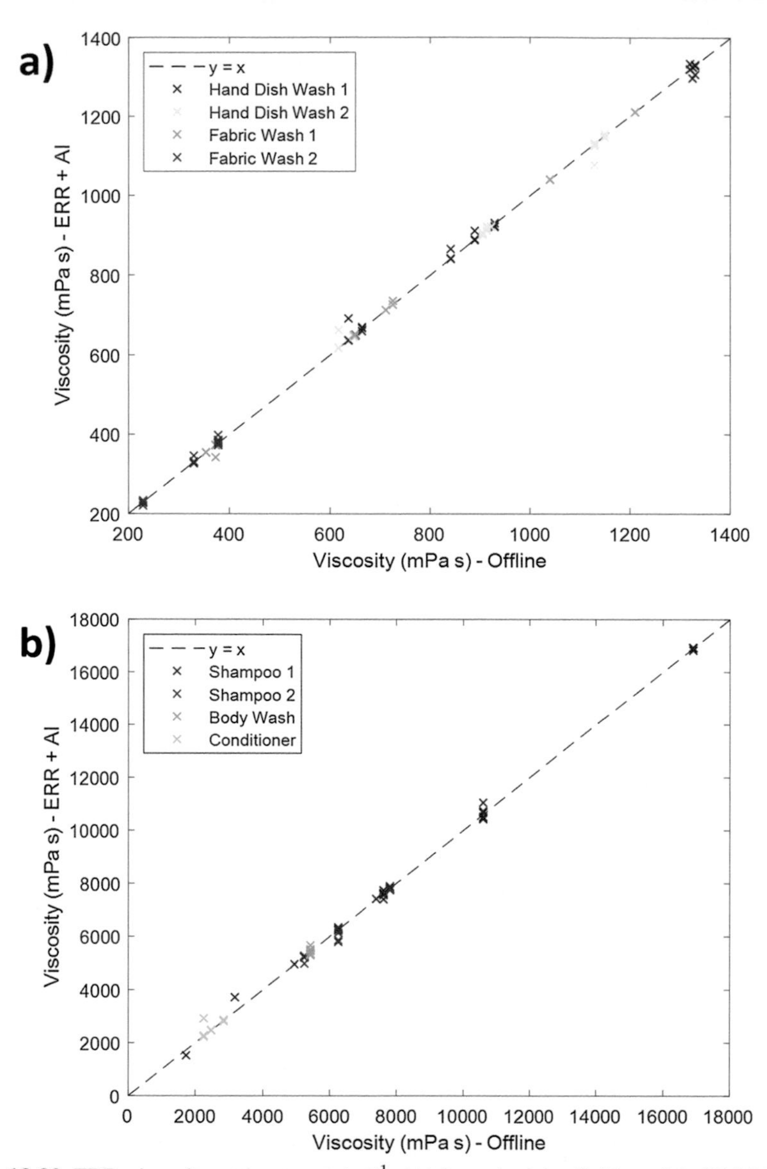

Figure 13.20 ERR viscosity estimates at 4 s^{-1}: (A) low viscosity fluid model; (B) high viscosity fluid model.

shear rate was increased, a minor decrease in the viscosity estimate error was observed. The RMSE for the low viscosity fluid model reduced from 30.6 to 21.8 mPa s; the test data for all conditions are displayed in Fig. 13.21. Similarly, the RMSE in the high viscosity model decreased by 59% with ERR able to accurately estimate the viscosity at multiple shear rates and ultimately rheological properties of a wide range complex industrial fluids. The RMSE error obtained for all test data conditions is below the desired error for an in-line rheology measurement ensuring high in-plant applicability.

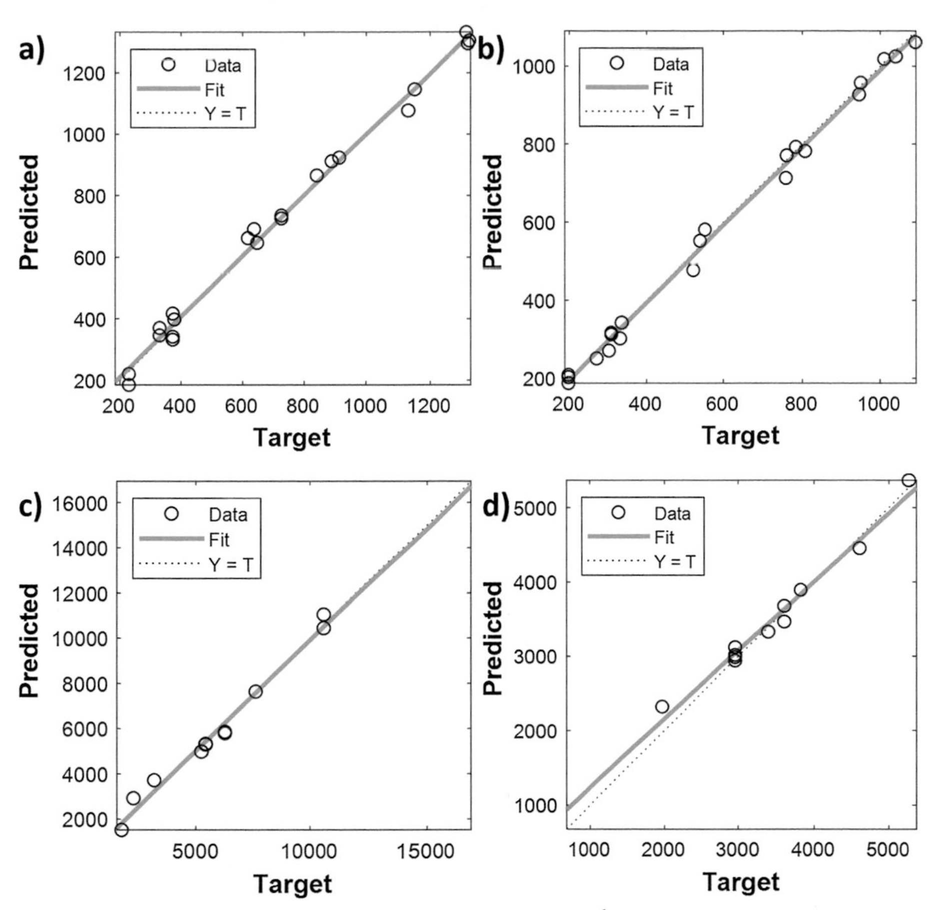

Figure 13.21 Test data: (A) low viscosity fluid model — 4 s^{-1}, (B) high viscosity fluid model — 4 s^{-1}, (C) low viscosity fluid model — 21 s^{-1}, (D) high viscosity fluid model — 21 s^{-1}.

Although not demonstrated in this study, if the fluid adheres to a specific constitutive equation, AI may be employed to estimate parameters of the model and determine the rheological behavior across the entire shear rate ranges of Fig. 13.20. The use of an ANN to estimate parameters of complex rheological models reduces the extent of ill-conditioning when compared to the parametric fitting approach.

13.5.5 Evaluation of ML performance for rheology sensing

This study has demonstrated that ML algorithms can be successfully applied to ERR measurements to accurately estimate in-line rheological properties across a range of industrial personal and home care products. AI methods are also able extend the capability of ERR to capture complex fluid behaviors often observed in industrial fluids, such as shear-banding and wall depletion, greatly enhancing in-plant applicability.

The application of AI algorithms to ERR transforms rheometry from a critical, off-line control tool, to a powerful on-line capability to support control and optimization of processes.

13.6 Future trends

This chapter has introduced ML from first principles for applications to IPT, in both pilot scale and embedded instrumentation; and illustrated this with two contrasting, detailed Case Studies providing detailed insights into design. They demonstrate that sampled experiential process data can be transformed into effective and accurate ANN designs, which are then able to transform IPT measurement data directly into process information.

It is interesting to project potential contributions of ML-IPT to industrial development for a cleaner, greener, and more efficient and productive world. Industrial strategies such as Industry 4.0 (Schwab, 2017) seek to exploit major in-depth information, from individual processes to complete manufacturing plants, to enhance performance, impacting business, economic, and environmental goals. ML methods hold great promise in the direct conversion of spatial measurement data to meaningful information, based upon prior deep knowledge of the process. This requires compatible in-line sensors and diagnostic methods for high response process analytics. Tomographic measurements can interrogate major internal process information. Accurate timely and smart information, typically in digital form, such as that available from IPT, is an essential prerequisite to the success of Industry 4.0.

It may be said that IPT is currently hampered in wider application, by an apparent *manual need* to find models and *tune* them into a design approach for any new candidate process. As demonstrated above this technically demanding and time-consuming process can be largely replaced by a simpler exercise of sampling and training data selection. In Section 13.6.1 next section, we overview the future reality of also removing much of this process through new trends in deep learning.

Tools for investigation and design continue to develop to further enable the lifecycle process of instrumentation design from concept to products. Section 13.6.2 reviews leading products and systems through to the deep learning stage noted above.

Final conclusions are noted in Section 13.6.3.

13.6.1 Deep learning futures for ANNs for IPT

A key theme of this chapter is data-driven design, represented by the essential training requirements for many ML system architectures, particularly the supervised forms noted in Fig. 13.1. An essential element is the selection and formation of training data. Section 13.2.3 introduced the possibility of preprocessing training data for efficiency in representing an optimal variety of the process experience. Section 13.4.3 has provided a detailed exemplar in preprocessing ECT measurement data for ANN Training Mode to select, and then in Estimation Mode deploy, optimum orthogonalized and uncorrelated data. This avoids overtraining which is both inefficient and importantly reduces generalization capabilities. In these examples ANN Training Mode is presented as a pseudomanual design process, where evidence is obtained for network design decisions.

More recent deep learning methods take further lessons from biological systems, in their ability to prestructure complex information, inspired initially from vision processing. For example, we learn to recognize familiar shapes even when they are oriented differently (or *shifted*). These powerful capabilities are modeled in the Convolutional Neural Network (CNN), which like a conventional MLP has input, output, and multiple hidden layers. Its structure was originally proposed by Fukushima (1980) for pattern recognition. Its capabilities are provided by convolutional layers that convolve multiple data with a multiplicative, or other dot product operation, to reveal hidden spatial patterns, such as repeated patterns from nearby measurements, identified by intrinsic shift and compare operations. The activation function is commonly a Rectifier-Linear unit which features a dead-band that responds linearly from a significant feature detection level. Additional convolutions such as data pooling layers, which share selected convolution results; fully connected layers; and normalization layers, are all referred to as hidden layers because their inputs and outputs are masked by the activation function and final convolution.

The shift and compare operations can also be temporal, realized through a further CNN variant, the Time Delay Neural Network, which provides temporal shift invariance identification. This variant was first proposed by Waibel et al. (1989, p. pp328) and has recently been deployed by Johansen, Østby, Dupré, and Mylvaganam (2018) for flow regime estimation from ECT data.

In general terms, a composite raw data training set may be considered as a tensor that can be operated upon jointly by a set of spatial and/or temporal convolution operations, to automatically reveal repeated or correlated data, and to then encode it in compressed form for processing by later layers. We see *deep learning* commonly applied in many application areas where copious data exist. IPT offers raw data. Clear steps and research are needed to join-up these developments, so that, perhaps based upon temporary supplementary instrumentation, deep learning strategies can automate ANN learning in industrial processes delivering fast application realization.

13.6.2 Practical future steps forward in machine learning for IPT

A systematic approach to IPT instrumentation must consider both design and implementation. In principle, a classical approach is *model driven*, whereas the neural approach is *data driven*. Although any project can usefully utilize known process information, a neural computing approach must primarily consider from the outset if, and how, adequate operational data for training can be obtained. Where a classical design will focus upon the system hardware and software needs, a neural design will focus on ANN architecture and its training, with major process investigation and investment to acquire training data. Based upon wider examples, this one-time investment in process data is worthwhile in a major enhancement of performance.

There are currently several tools for ANN design, which typically include ANN simulation and training support useable on any ubiquitous design workstation. All major computing systems manufacturers offer major support to link application design to fast highly parallel devices and subsystems that can be integrated into process instrumentation design. This technology area is fast changing.

Here we offer a brief summary of current provision from supplier through five examples.

Major global microsystem manufacturer ARM (2021) currently offers CMSIS-NN, a library of efficient kernels able to maximize performance and minimize memory requirements for ANN applications using ARM Cortex-M processors. A trained prototype can be optimized using ARM tools for compatibility with the CMSIS-NN library. High-level functions from the kernels can then be simply integrated into high-level application code.

As deployed in case studies in this chapter, MathWorks (2021) offer a range of mathematical tools which include specific support for ML and may be suited especially for early pilot scale investigations. Automatic machine learning includes feature selection, model selection, and hyperparameter tuning. Automated generation of code for embedded and high-performance applications is provided. Integration with Simulink as native or MATLAB Function blocks and supports embedded deployment or simulations and popular classification, regression, and clustering algorithms for supervised and unsupervised learning with faster execution than open source on most ML computations.

Multichannel Graphics Processing Unit microsystem manufacturer NVIDIA (2021) is the world leading supplier of products used for high-performance graphics applications. Due to their featured major parallelization, these devices have been deployed for ANN applications. The TensorRT software development toolset for application development may be used to optimize ANN models trained in all major frameworks and also supports deployment to embedded hardware products such as process instrumentation.

Systems solutions provider NXP (2021) offers its eIQ ML development environment for NXP MCUs, i.MX RT crossover processors, and i.MX family *System on Chip* (SoC) devices. The environment also supports ARM CMSIS-NN and ARM NN SDK, TensorFlow Lite, and OpenCV inference engines.

Major configurable systems provider Xilinx (2021) offers its Deep Neural Network Development Kit (DNNDK) for porting ANN configurations to Xilinx devices. The Deep Learning Processing Unit (DPU) is a configurable computation engine optimized for CNNs (see Section 13.6). The degree of parallelism utilized in the engine is a design parameter and can be selected according to the target device and application. It includes a set of highly optimized instructions and supports most CNNs. The DPU, a hardware platform for Xilinx FPGAs, is scalable for use on to fit various Xilinx Zynq-7000 and Zynq UltraScale+ MPSoC devices. The DNNDK accepts models from Caffe and/or Tensorflow and maps them into DPU instructions. The DPU has a runtime engine supporting generic software and APIs to facilitate efficient deployment and testing of ANNs.

13.6.3 Predicting the future of ML in process sensing

This chapter has presented encouraging steps that promise to liberate the exploitation of IPT technology which has been severely limited by the current need to follow a complex path of modeling, trial, and individual interpretation design in which every application must be designed as an expensive and time consuming special case.

We know that the optical data presented to the retina of a human eye are a blur of color and intensity. The human brain learns the massively complex capability we call *sight* to make sense of these data to create powerful and complex visual information purely by *practicing*.

We can expect deep learning CNN-based interpretation to directly learn to design and train ANN structures to solve the much simpler task of processing raw IPT measurement data, to gain major high-value, multidimensional process information to realize many facets of Industry 4.0.

ML is waiting to supercharge IPT!

References

Anderson, J. A., & Rosenfeld, E. (Eds.). (1988). *Neurocomputing: Foundations of research* (Vol. 1). MIT Press, ISBN 978 0 26251 0486.

Anderson, J. A., & Rosenfeld, E. (Eds.). (1990). *Neurocomputing: Foundations of research* (Vol. 2). MIT Press, ISBN 978 0 26251 0752.

ARM. (2021). https://www.arm.com/solutions/artificial-intelligence/machine-learning. (Accessed 29 January 2021).

Barnes, H. A., Hutton, J. F., & Walters, K. (2001). *An introduction to rheology* (3rd ed.). Amsterdam: Elsevier, ISBN 978 1 49330 2611.

Bergman, T. L., Lavine, A. S., Incropera, F. P., & DeWitt, D. P. (2019). *Fundamentals of heat and mass transfer* (8th ed.). New York: Wiley, ISBN 978 1 119-53734 2.

Bishop, C. (1994). Neural networks and their applications. *Review of Scientific Instruments, 65*(6), 1803–1831. https://doi.org/10.1063/1.1144830

Fukushima, K. (1980). Neocognitron: A self-organizing neural network model for a mechanism of pattern recognition unaffected by shift in position. *Biological Cybernetics, 36*(4), 193–202. https://doi.org/10.1007/BF00344251

Johansen, R., Østby, T. G., Dupré, A., & Mylvaganam, S. (2018). Long short-term memory neural networks for flow regime identification using ECT. In *ISIPT 9th world congress in industrial process tomography, Bath UK* (pp. 135–142), ISBN 978 0 85316 3497.

Knirsch, M., dos Santos, C., de Oliveira Soares Vicente, A., & Vessoni Penna, C. (2010). Ohmic heating, a review. *Trends Food Science, 21*(9), 436–441.

Machin, T., Wei, K., Greenwood, R. W., & Simmons, M. J. H. (2018a). Electrical resistance rheometry – the application of multi-scale tomography sensors to provide in-pipe rheology in complex processes. In *ISIPT world congress in IPT-9, Bath UK* (pp. 143–154), ISBN 978 0 85316 3497.

Machin, T. D., Wei, K., Greenwood, R. W., & Simmons, M. J. H. (2018b). In-pipe rheology and mixing characterisation using electrical resistance sensing. *Chemical Engineering Science, 187*, 327–341. https://doi.org/10.1016/j.ces.2018.05.017

MathWorks. (2021). https://www.mathworks.com/solutions/machine-learning.html. (Accessed 29 January 2021).

Mohamad-Saleh, J., & Hoyle, B. S. (2002). Determination of multi-component flow process parameters based on electrical capacitance tomography data using artificial neural networks. *Measurement Science and Technology, 13*(12), 1815–1821. https://doi.org/10.1088/0957-0233/13/12/303

Mohamad-Saleh, J., & Hoyle, B. S. (2008). Improved neural network performance using principal component analysis on matlab. *International Journal of Computer Integrated Manufacturing, 16*(2), 1–8. ISSN 0858-7027.

Neal, R. (1996). *Bayesian learning for neural networks*. New York: Springer, ISBN 978 0 387 94724 2.

Nooralahiyan, A. Y., Hoyle, B. S., & Bailey, N. J. (1993). Application of a neural network in image reconstruction for capacitance tomography. In *Proc. European concerted action on process tomography* (pp. 50–53).

NVIDIA. (2021). https://www.nvidia.com/en-us/ai-data-science. (Accessed 29 January 2021).

NXP. (2021). https://www.nxp.com/applications/enabling-technologies/ai-and-machine-learning:MACHINE-LEARNING. (Accessed 29 January 2021).

Paul, T. D., Atiemo-Obeng, V., & Kresta, S. (2004). *Handbook of industrial mixing: Science and practice*. New York: Wiley-Interscience, ISBN 978 0 47126 9199.

Rosenblatt, F. (1962). *Principles of neurodynamics: Perceptrons and the theory of brain mechanisms*. Washington DC: Spartan Books, ISBN 978 3 642 70913 5.

Schwab, K. (2017). *The fourth industrial revolution*. New York: Crown Publishing Group, ISBN 978 1 52475 8875.

Spink, D. M. (1996). Direct finite element solution for the capacitance, conductance or inductance, and force in linear electrostatic and magnetostatic problems. *International Journal for Computation and Mathematics in Electrical and Electronic Engineering, 15*(3), 70–84.

Waibel, A., et al. (1989). Phoneme recognition using time-delay neural networks. *IEEE Transactions on Acoustics, Speech, & Signal Processing, 37*(3), 328–339.

Wang, M., Mann, R., & Dickin, F. J. (1996). A large scale tomographic sensing system for mixing processes. In *Proc. third international workshop on image and signal processing* (pp. 647–650), ISBN 978 0 44482 5872.

XILINX. (2021). https://www.xilinx.com/applications/megatrends/machine-learning.html. (Accessed 29 January 2021).

Yu, H., Yang, G., Wang, Y., & Yu, C. (2019). Design of capacitive sensor for phase concentration measurement in two-phase flow. In *31st Chinese control and decision conference (2019 CCDC), Nanchang, China, China* (pp. 3221–3226).

Zainal Mokhtar, K., & Mohamad-Saleh, J. (2013). An oil fraction neural sensor developed using electrical capacitance tomography sensor data. *Sensors, 13*(9), 11385–11406.

Further reading

Lähivaara, T., Kärkkäinen, L., Huttunen, J. M. J., & Hesthaven, S. (2018). Estimation of porous material parameters using ultrasound tomography and deep learning. In *ISIPT 9th world congress in industrial process tomography, Bath UK* (pp. 283–288), ISBN 978 0 85316 3497.

Wang, J., Liang, J., Cheng, J., Guo, Y., & Zeng, L. (2020). Deep learning based image reconstruction algorithm for limited-angle translational computed tomography. *PLoS One, 15*(1), 1–20. e0226963.

Wang, G., Zhang, Y., Ye, X., & Mou, X. (2019). *Machine learning for tomographic imaging*. IOP Publishing, ISBN 978 0 7503 2214 0.

Yan, R., & Mylvaganma, S. (2018). Flow regime identification with single plane ECT using deep learning. In *ISIPT 9th world congress in industrial process tomography, Bath UK* (pp. 289–298), ISBN 978 0 85316 3497.

Zheng, J., & Peng, L. (2018). Deep learning based image reconstruction for electrical capacitance tomography. In *ISIPT 9th world congress in industrial process tomography, Bath UK* (pp. 337–346), ISBN 978 0 85316 3497.

Advanced electrical tomography visualisation

Qiang Wang
University of Edinburgh, Edinburgh, United Kingdom

14.1 Introduction

Electrical tomography (ET) has broad applications in many areas for qualification and quantification of multiphase flow (Wang, 2015), where phase distribution is reconstructed in terms of physical properties, such as impedance for electrical impedance tomography. Since values in the distribution are in general dimensionless, a predominant approach to reflect those scalar values is known as color mapping (Hansen & Johnson, 2011), i.e., transforming reconstructed values to colors. Color mapping is the simplest and most straightforward way to visualize the reconstructed images, but it also has its limitations. For example, nonlinear distribution of the colors in commonly used RGB color mapping is hardly able to fully represent phase distribution in a linear range. Therefore, advanced visualization technologies are in great demand for more insightful information into multiphase flow for both human and machine perception.

Section 14.2 reviews commonly observed flow patterns in horizontal and vertical pipelines, and multiphase flow visualization technologies. Advanced technologies for multiphase flow visualization using ET are addressed in Section 14.3. Multidimensional data fusion for two- and three-phase flow with multimodality ET is discussed in Section 14.4. In Section 14.5, some remarks are made regarding the remaining challenges and future trends of multiphase flow visualization.

14.2 Background

14.2.1 Multiphase flow patterns

Flow patterns/regimes refer to the geometrical distribution of involving phases in a multiphase flow (Brennen, 2005). Due to the complex interactions among flow phases and physical properties of the surrounding environment, there usually are an infinite number of flow patterns which can be observed. To simplify the study, we assume that observed flow is fully developed and in a steady state. Consequently, flow regimes can be estimated using a so-called flow regime map, where superficial velocity of gas and liquid phase is used for the construction of the map.

In the horizontal pipeline, there are seven commonly inspected flow regimes (Corneliussen et al., 2005). Stratified flow is characterized by a complete separation

Industrial Tomography. https://doi.org/10.1016/B978-0-12-823015-2.00024-8

of gas and liquid with a stable horizontal interface. Wavy stratified flow is similar to the stratified, but with an unstable interface due to the difference in interfacial velocity. At slug flow, the interface becomes turbulent and liquid finally reaches the top wall of the pipe to form slug bubbles which can be extremely large (low-frequency slug flow). For plug flow, the gas phase is split into relatively small bubbles at the top of the pipe, with a diameter smaller than the pipe diameter. There is usually a thin film between the bubbles and the top wall of the pipe. With high liquid superficial velocity and low gas superficial velocity, the gas phase becomes fully dispersed, in terms of small bubbles of different diameters and thus, bubbly flow forms. Annular flow is generated with a very high gas flowrate, where it is fully enclosed by liquid. In mist flow, gas becomes dominant, with homogeneously distributed liquid droplets inside the gas.

Flow regimes in the vertical pipeline have distinctive structures (Corneliussen et al., 2005). Generally, four common flow regimes can be found, including bubbly, slug, churn, and annular flow. In bubbly flow, gas at low flowrate is in the form of small individual bubbles, surrounded by liquid. With the increase in gas flowrate, small bubbles merge into large bubbles of bullet shape, with a diameter close to the pipe diameter, and slug flow is produced. Further increasing gas flowrate, large bubbles become twisted and collapse into smaller, yet completely irregular, bubbles, and churn flow appears. When the gas flowrate is beyond a certain threshold, liquid is pushed toward the pipe wall to form a liquid film, with gas surrounding the liquid inside, and annular flow is produced. There may also be some liquid droplets in the gas core (Fig. 14.1).

14.2.2 Multiphase flow visualisation technologies

In general, multiphase flow can be visualized using two different approaches: simulation and experimental methodologies. Computational Fluid Dynamics (CFD) is the most applied way to visualize multiphase flow. CFD numerically solves equations which reflect physical problems with respect to fluid flow (Brennen, 2005; Prosperetti & Tryggvason, 2009), with the assistance of the extreme computational power of contemporary computers. The results generated by CFD are manifold, but the most common usages include concentration and velocity field, where the concentration distribution can be visualized by conventional color mapping (Fig. 14.1) and the velocity

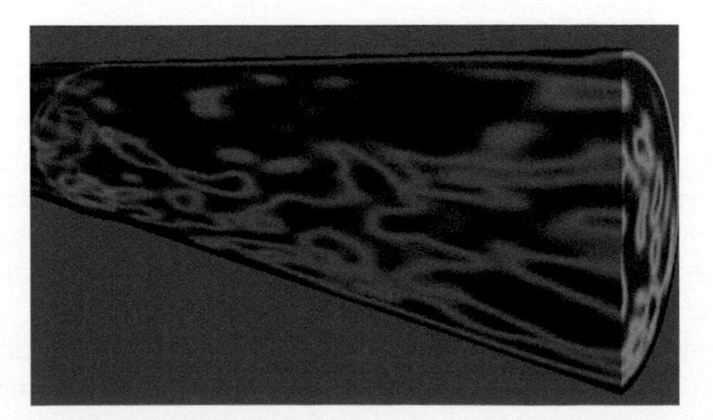

Figure 14.1 Concentration distribution of gas–liquid flow by CFD (Wang, 2017).

field can be visualized using vector-suitable approaches (Prosperetti & Tryggvason, 2009; Hansen & Johnson, 2011). Due to the requirements when numerically solving the equations, in computer graphics (CG), some researchers are also interested in physically based visualization of multiphase flow. The principle behind it is similar to CFD, where the Navier–Stokes equation is approximated with various approaches to simulate multiphase flow (Tan & Yang, 2009). However, those approaches concentrate more on vivid, yet plausible representation of multiple phases in a fluid flow by revealing clear boundaries among the phases, rather than quantification of multiphase flow.

Experimental methodologies visualize multiphase flow with the assistance of instruments. The simplest yet most straightforward way is by using a high-speed camera to capture phase distribution via a transparent chamber. However, such an observation chamber is not always available, particularly in real-world industrial cases. In addition, physical properties and phase proportion also affect the observation. It is reported that in a gas–water flow, this approach is impractical when gas concentration is beyond 10% (Prasser et al., 2001). Tomography is another set of technologies capable of "seeing" the inside of multiphase flow. Fig. 14.2 exemplifies visualization of gas–water upward flow with X-ray (Fig. 14.2(a)), wire-mesh sensor (WMS) (Fig. 14.2(b)), and electrical impedance tomography (Fig. 14.2(c)). Note that all images in Fig. 14.2 have been postprocessed with advanced visualization technologies to emphasize individual bubbles, except for the first images in Fig. 14.2(c).

A common characteristic of X-ray and WMS is that they are able to generate reconstructed images with high spatial resolution enough to reconstruct individual bubbles

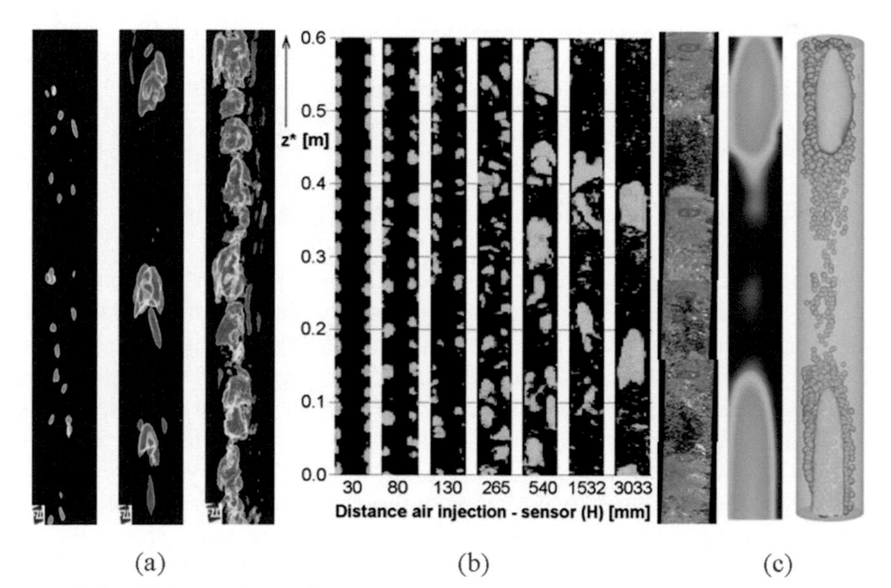

Figure 14.2 Advanced visualization of gas–water vertical flow based on (a) X-ray (Fischer & Hampel, 2010), (b) wire-mesh sensor (Prasser, Scholz, & Zippe, 2001), and (c) electrical impedance tomography (Wang, Jia, & Wang, 2019).

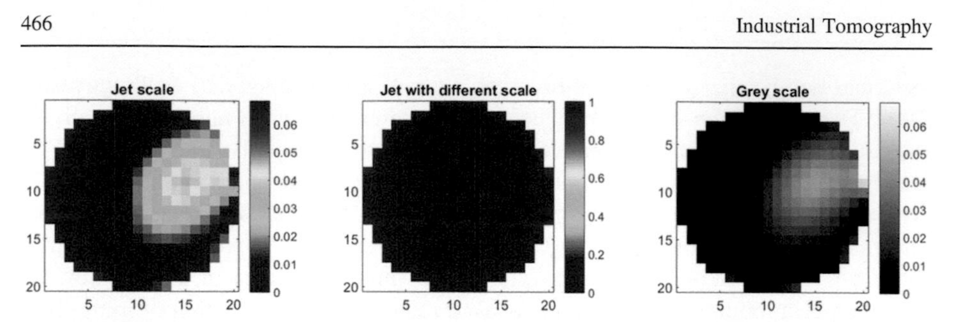

Figure 14.3 Electrical concentration tomograms using (a) Jet look-up table and normal scale, (b) Jet look-up table with scale [0, 1], and (c) gray look-up table with normal scale (Wang, 2017).

beyond certain size. In contrast, ET, including impedance and capacitance, is unable to produce reconstructed images with sharp boundaries between gas and liquid, due to the inhomogeneity of its sensitivity distribution (Wang, 2015). The middle image in Fig. 14.2(c) depicts a gas—water upward slug flow, where slug bubbles can be roughly positioned but small bubbles disappear. Apart from relatively low spatial resolution, another important reason is that the transformation using color mapping is nonlinear, since the scalar values from tomography are discrete. A major disadvantage of the nonlinearity is that small changes in tomography may be suppressed by a wide range of tomographic values. In this case, a limited number of colors are assigned to a large number of values in a narrow range. Fig. 14.3 demonstrates the impact of using different color mappings to an electrical tomogram on the visualization results, where the tomograms are unable to deliver sufficient information on flow dynamics, e.g., bubble size and distribution, for human/machine perception. Therefore, it is of great importance that advanced technologies should be applied to ET to provide more insightful information into the multiphase flow investigated.

14.3 Advanced visualization of multiphase flow

Multiphase flow is, by its nature, a four-dimensional (4D) physical phenomenon, i.e. space and time, and thus, 4D presentation with clear interfaces among available phases is ideal for human and machine perception with little ambiguity. Due to the limitations of ET and conventional presentation, e.g., using color mapping, electrical tomograms are unable to deliver sufficient flow dynamics. Before reviewing the possible advanced visualization technologies suitable for ET, the characteristics of electrical tomograms are briefly discussed here.

Although electrical tomograms for multiphase flow are in a multidimensional format, they are very different from those in medical imaging. Cross-sectional electrical industrial tomograms reflect both spatial and temporal information. Let V be a sequence of electrical tomograms (x, y, t, v), where an interested property v locates at space (x, y) and time t. The property v can be an electrical property reconstructed, such as resistance for ERT, or dispersed-phase concentration converted using

Maxwell's equation (Maxwell, 1982). To correctly present physical distributions of multiphase flow, a transformation from (x, y, t, v) to (x, y, z, v') is preferred. A possible resolution is introduced in Wang et al. (2019), where the conversion is performed by integrating information from the velocity field and pipe diameters, along with the assumption that flow is fully developed. Since the retrieval of the phase velocity field is extremely challenging, it is approximated by local velocity derived via cross-correlation algorithm, or by superficial velocity if local velocity is not derivable in certain flow regimes, e.g. stratified flow.

Another predominant aspect is the geometric structure of the underlying grid represented by electrical tomograms. Fig. 14.4 illustrates an example of commonly used grids in electrical tomograms, where the grid can be either regular

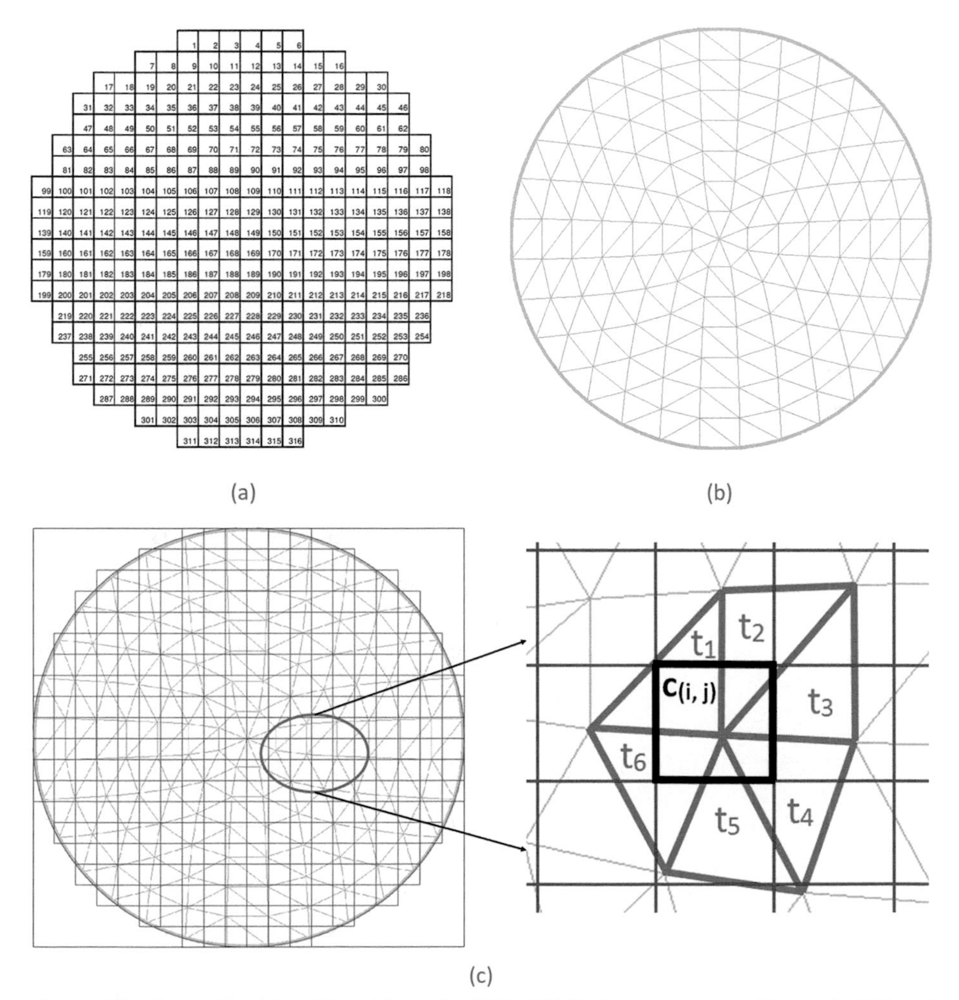

Figure 14.4 Example of (a) a 20 × 20 regular (ITS, 2009), (b) an irregular grid used by ET, and (c) converting the irregular grid to the regular one by averaging (Wang, 2017).

(Fig. 14.4(a)) or irregular (Fig. 14.4(b)). Since advanced visualization technologies are usually grid independent, the underlying grid of ET needs to be considered when choosing the technologies. For example, marching squares, a 2D analogue of marching cubes (Lorensen & Cline, 1987), can be used on the grid in Fig. 14.4(a). Note that an irregular grid can generally be converted into a regular one, as depicted in Fig. 14.4(c), and therefore, an advanced visualization approach on a regular grid can be performed after converting irregular electrical tomograms to regular ones.

14.3.1 Three-dimensional volume rendering

Volume rendering is considered as the simplest 3D visualization technology to retrieve information from scalar datasets, and thus, it has been broadly utilized in the visualization of volumetric data, such as medical imaging (Bankman, 2008) and multiphase flow visualization (Reinecke, Petritsch, Schmitz, & Mewes, 1998; Wang, Marashdeh, Fan, & Williams, 2009; Ye & Yang, 2012; Yao & Takei, 2017). Fig. 14.5 illustrates the flow regime visualization of a gas—water flow in a horizontal pipeline by volume rendering of a stack of cross-sectional ERT tomograms. Compared with 2D cross-sectional electrical tomograms, 3D representations apparently reflect more information regarding multiphase flow dynamics.

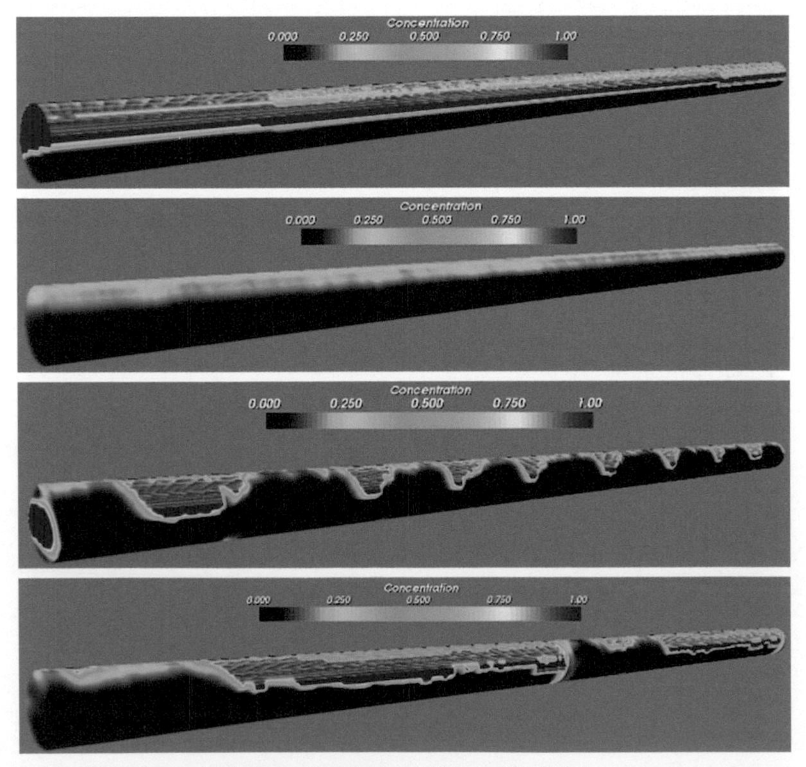

Figure 14.5 3D volume rendering of multiphase flow with ERT.

With the support of contemporary hardware, particularly graphics processing units, transformed 3D electrical tomograms can be rendered on-line to provide real-time monitoring and perception of multiphase flow. However, as shown in Fig. 14.5, an obvious disadvantage of volume rendered multiphase flow with ET is that the interiors are visually ambiguous to be interpreted. Although large bubbles are identifiable, to a large extent, the boundaries between large bubbles and liquid phase are not clear. Meanwhile, small bubbles are still missing, due to the limitations of electrical tomograms, e.g., spatial resolution (Wang, Dickin, & Mann, 1999).

14.3.2 Isosurface

When using color mapping-based visualization, the human eye is able to distinguish similar colors mapped from similar scalar values, where the areas with similar colors usually represent boundaries between those areas. For example, the red (light gray in print version) areas in Fig. 14.5 reflect large bubbles, and the blue (dark gray in print version) areas reflect liquid. Consequently, a particular boundary can be identified between the red (light gray in print version) and the blue (dark gray in print version), where the interface of the gas and liquid phase is derived. Given a set of transformed electrical tomograms (x, y, z, v'), a 3D surface S can be defined as follows:

$$S(v') = t \tag{14.1}$$

where t is a threshold value, called isovalue or contour value, and the surface defined in Eq. (14.1) is called isosurface (Wenger, 2013). On the surface, all points have an identical value t. Two regions, where v' $> t$ and v' $< t$, are usually referred to as inside and outside of the surface. An isoline is a 2D analogue of an isosurface.

Isoline/Isosurface has various applications in ET-based multiphase flow characterization, which can be traced back to a few decades ago, such as multivalue isosurface to present gas distribution in a gas–liquid mixing with ERT (Wang, 2002). Fig. 14.6 depicts two examples of isosurfacing applications in visualization. Fig. 14.6(a) illustrates an application of isoline on cross-sectional electrical tomograms reconstructed by an advanced algorithm (Wang, 2002) to demonstrate the asymmetrical distribution of solid along a horizontal pipe in a swirling flow (Wang et al., 2003). Another example is the bubble mapping approach (Wang et al., 2019) in Fig. 14.6(b), where the matching cubes approach is applied to reconstruct the interface between large bubbles and liquid with an isovalue 0.4. Note that the small bubbles in Fig. 14.6(b) are not generated by isosurfacing algorithms, which will be reviewed in the following section.

When applying isosurface extraction algorithms to ET-based multiphase flow visualization, a few aspects need to be taken into account. As shown in Fig. 14.4, the tomogram grid can be regular or irregular, and hence, targeting algorithms need to be compatible with the grid. In addition, a common mistake when isosurfacing multiphase flow with ET is the lack of a physical boundary, particularly when a bubble directly contacts a physical boundary, e.g., the pipe wall. For example, the blue (dark gray in print version) part on the right side of the isosurface rendering in Fig. 14.6(a) is open, which does not reflect the actual physical phenomenon.

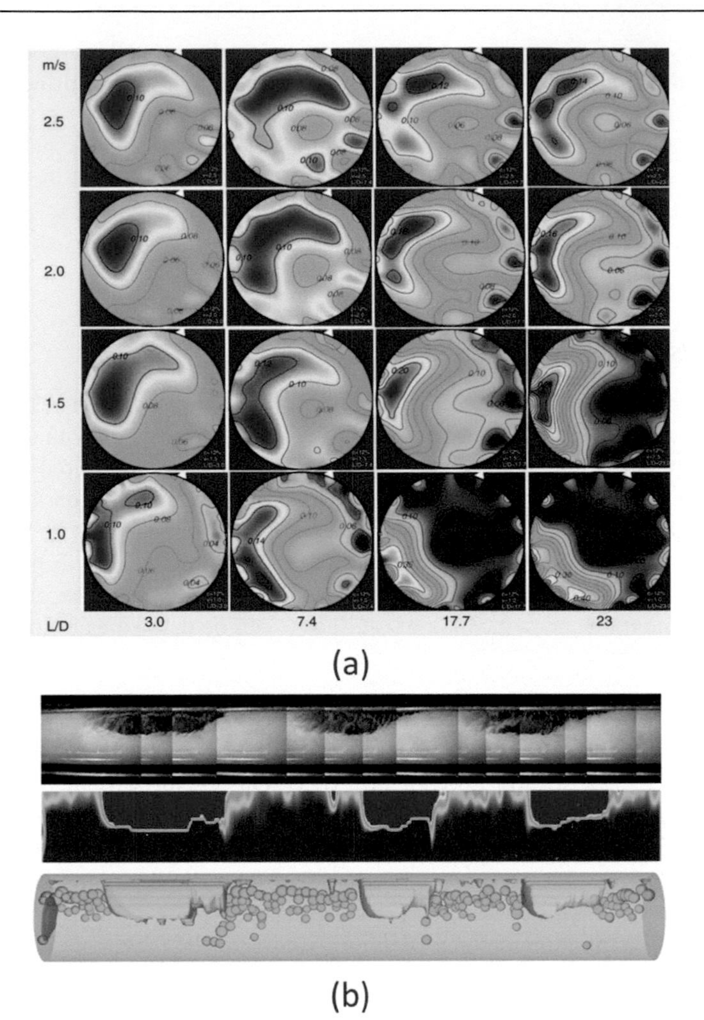

Figure 14.6 Isoline/isosurface visualization of multiphase flow with ET (a) isolines for a swirling flow (Wang, Jones, & Williams, 2003) and (b) marching cubes-based isosurface extraction (bottom image) of electrical tomograms (mid image) to reconstruct large bubbles in a horizontal slug flow (Wang et al., 2019).

Last, but not least, the issue is the determination of isovalue for realistic visualization of multiphase flow with ET, where reconstructed bubble size and distribution should be plausible and, ideally, mean concentration should be conserved.

Like volume rendering, isosurface extraction is expensive in time and computational resource. However, with the support of modern hardware and advanced extraction algorithms, isosurface can be constructed in real time (Wenger, 2013). Compared with volume rendering, isosurface explicitly reflects the interface between dispersed and continuous phase, and therefore, can deliver more flow dynamics, such as large bubble size and distribution. Due to the relatively low spatial resolution of ET, unfortunately, small bubbles below certain sizes are still missing.

14.3.3 Bubble reconstruction

In order to maximize flow dynamics reflected by electrical tomograms, many methodologies have been proposed to extract individual bubbles of various sizes and shapes. Generally, those methodologies can be categorized into two groups: one is direct reconstruction and the other is indirect reconstruction. The former concentrates on generating bubbles during image reconstruction, whereas the latter applies advanced technologies to reconstruct bubbles by postprocessing of reconstructed tomographic images.

Conventional pixel-based image reconstruction generates tomograms with high temporal but low spatial resolution. To retrieve individual bubbles, extra constraints need to be integrated into the reconstruction process. For example, Kim, Khambampati, Hong, Kim and Kim (2013) applied multithresholding values into the iterative Gauss−Newton method to separate objects from their background with sharp boundaries. Comparing with pixel-based image reconstruction algorithms, shape-based advanced methodologies are able to reconstruct bubbles directly. Soleimani, Lionheart and Dorn (2006) proposed a level set-based image reconstruction approach for shape reconstruction. The boundary element method was also applied to estimate the free interface in two-phase flow with ERT (Khambampati, Kim, Lee & Kim, 2016; Ren, Dong, Tan, & Xu, 2012). Tan, Lv, Dong, and Takei (2018) applied convolutional neural networks for image reconstruction in ERT to achieve high-quality reconstructed images with individually identifiable objects. The primary advantage of those advanced image reconstructions is that the quality of the reconstructed electrical tomograms is satisfied to identify boundaries between involving phases, and hence, little postprocessing is required. However, the major issue is that they are usually time and computational resource expensive. In addition, most of them were evaluated by simulation or simple experimental setup, thereby the suitability for real-case multiphase flow is still uncertain.

As discussed in Section, 14.3.2, isosurfacing reconstructed electrical tomograms make it possible to identify the interface between large bubbles and continuous phase. However, it is not possible to extract small bubbles below certain sizes, even with multiple isovalues, due to the low spatial resolution and distinguishability of the tomograms. In order to reconstruct bubbles of different sizes and shapes, advanced technologies are needed. Wang et al. (2019) proposed bubble mapping for 3D visualization, namely BM3D, of multiphase flow with bubbles. In their method, large bubbles are generated based on Marching Cubes, and small bubbles are approximated per averaged concentration in an interrogation cell. Fig. 14.7 depicts the application of BM3D to gas−liquid vertical flow, which clearly demonstrates that there is little ambiguity in visual recognition of flow regimes. Iacovides, Wang and Wang (2018) applied BM3D to investigate transient Taylor bubble flows. In BM3D, the threshold used for generating large bubbles is based on empirical knowledge, although several different values were evaluated to demonstrate its invariance to flow regime

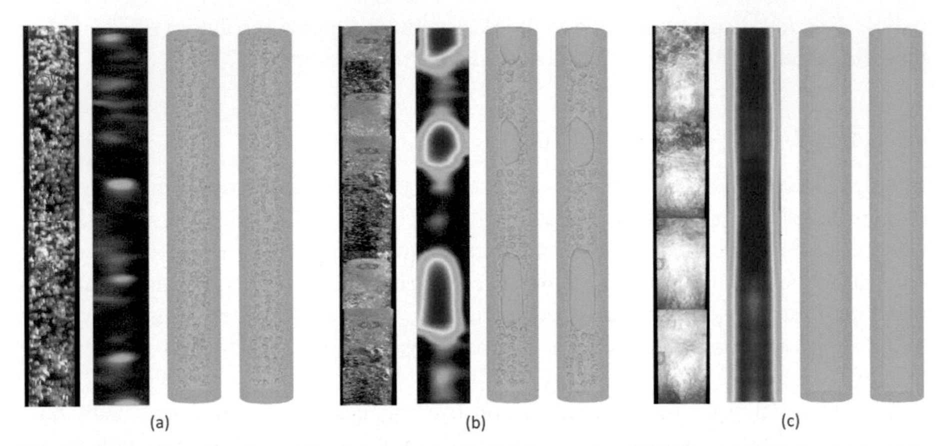

Figure 14.7 Visualization of horizontal gas−liquid flow using BM3D and SPA for (a) bubbly flow, (b) slug flow, and (c) annular flow (Li, Wang, & Wan, 2019). For each flow, a stacked image by high-speed camera (first plot), an axially stacked ERT tomogram (second plot), BM3D-based visualization (third plot), and SPA-based visualization (fourth plot) are presented.

recognition. In Li et al. (2019c), the authors eliminated the arbitrariness by proposing a size projection algorithm (SPA), where an optimal threshold is determined by minimizing the projection error between step-wise estimated boundary data and the measured boundary data. SPA was successfully applied to visualize large bubbles in gas−liquid two-phase flow (Li, Wang, Wang, & Han, 2019), and also integrated to BM3D for large bubble determination (Li et al., 2019), and the result on vertical flow is presented in Fig. 14.7.

14.4 Multidimensional data fusion

Multimodality tomographic systems (MMTS), i.e., an integration of multiple tomographic systems, have broad applications in multiphase flow visualization and measurement (Dias et al., 2020; Hoyle et al., 1999, 2001; Qiu, Hoyle, & Podd, 2007; Wang Wang, Wei, & Qiu, 2017; Zhang, Wang, Wang, Zhang, & Li, 2014). Formation of MMTS can be individually mounted along pipeline (Hoyle et al., 2001; Qiu et al., 2007; Wang et al., 2017), or integrated together but with separate sensors, such as ERT and ultrasound tomography (UT) (Zhang, Tan & Dong, 2020), or with integrated sensors, e.g., ERT−ECT (Ji et al., 2016; Rodriguez-Frias & Yang, 2020). The utilization of MMTS is usually for two purposes. One is to improve the spatial resolution of reconstructed images in two-phase flow (Liang, Ren, Zhao, & Dong, 2019; Steiner, 2006; Zhang et al., 2020), and the other is to discriminate individual phases in multiphase flow with three or more components (Dias et al., 2020; Qiu et al., 2007; Wang et al., 2017).

With single-modality tomographic systems, multidimensional data are generated, including, but not limited to, space, time, and frequency (Hoyle & Wang, 2012). With MMTS, at least extra dimensional data are introduced from the individual systems. Consequently, it is critical to sort out how to integrate those data to maximize the information retrieved from MMTS for insightful perception by humans and machines on multiphase flow dynamics. Ideally, reconstructed tomograms by individual modality should have identical spatial and temporal resolution, i.e., signals are collected at the same time for each system and the reconstructed tomograms have the same underlying grid. Unfortunately, this is not always the case. Therefore, multidimensional data fusion is required. Spatial integration requires corresponding tomograms from all systems are geometrically aligned, i.e., all corresponding pixels in the tomograms from individual tomography represent the same physical location. Temporal integration needs all corresponding tomograms to be generated for the same physical phenomenon at a certain time, so that multiphase flow dynamics can be precisely reflected. Frequency integration is usually demanded for multifrequency tomographic systems, where multiple tomograms are generated from multiple frequency and to be used for decomposing properties of interest (Hoyle & Wang, 2012).

Moreover, data fusion can also be performed at various levels, including, but not limited to, signal, reconstruction, and image level (West & Williams, 1999; Hoyle & Wang, 2012; Wang et al., 2017). Fig. 14.8 illustrates a conceptual scheme of the fusion at multiple levels for dual-modality tomographic systems (DMTS) on multiphase flow. Signal-level fusion is featured by combining the measurements from the sensors of all modalities, which usually needs further processing to reflect physical distribution of interested objects. In order to derive optimal results, signal-level fusion is desired, where lost information during reconstruction or postprocessing is minimized (Steiner, 2006; West & Williams, 1999). However, little effort has been made on signal-level fusion (Hjertaker, Maad, & Johansen, 2011; Wang et al., 2017).

Fusion during reconstruction utilizes reconstructed information from one modality as a priori knowledge into the reconstruction process of the other modality. Fusion at the reconstruction level is possible for MMTS on two-phase flow since different modalities are dedicated to distinguishing one component from the other, such as UT with

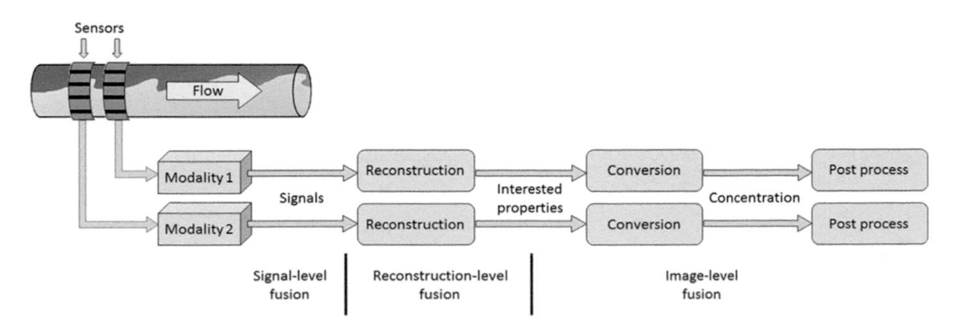

Figure 14.8 Data fusion at various levels for dual-modality tomographic systems on multiphase flow (Wang, 2017).

EIT (Liang et al., 2019) or with ECT (Steiner, 2006). In contrast, this is hardly feasible for three-phase flow, since each modality is responsible for differentiating one distinctive component from the mixture.

Image-level fusion is performed on the reconstructed images, which can be further categorized into different levels, such as pixel- and feature-level fusion (Mitchell, 2012). Image-level fusion is feasible for two- and three-phase flow. Due to the relatively low spatial resolution of ET, features retrievable from general images are usually challenging to be extracted from ET tomograms, and hence, feature-level fusion is not universally applicable for ET-involved MMTS. A more common practice of image-level fusion with MMTS for multiphase flow is at pixel level, such as Qiu et al. (2007); Sun and Yang (2015); and Wang et al. (2017).

Fig. 14.9 shows an example of spatial and temporal alignment for dual-modality ERT−ECT systems, where the ERT has a spatial resolution of 20 × 20 and temporal resolution of 62.5 frames per second (fps), and the ECT has a spatial resolution of

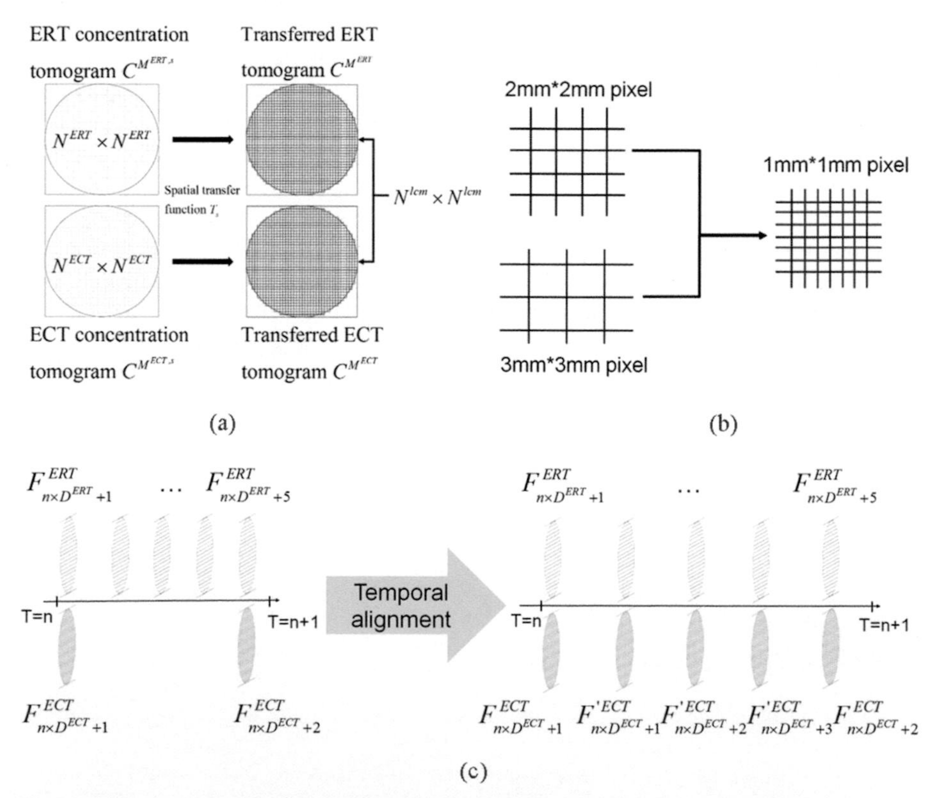

Figure 14.9 An example of data alignment for multidimensional data fusion by linear interpolation with separated dual-modality electrical tomography. (a) spatial alignment for 20 × 20 ERT and 30 × 30 ECT tomograms with different underlying grids, (b) physical meaning of the spatial alignment for the systems, and (c) temporal alignment due to different temporal resolutions of the modalities (Wang et al., 2017).

30 × 30 and temporal resolution of 12.5 fps. In order to make all corresponding pixels in the system represent identical physical positions, the tomograms are spatially transformed to a 60 × 60 grid (Fig. 14.9(a) and 14.9(b)), and temporally to 62.5 fps (Fig. 14.9(c)). To simplify the transformation, linear interpolation was employed. Although it was proposed for ERT–ECT systems, it can also be extended to other MMTS having two or more different modalities. In addition, it is also applicable for MMTS-based two-phase flow characterization.

14.4.1 Two-phase flow

As discussed in the previous section, applying MMTS to two-phase flow visualization aims to improve the quality of resultant images, which can be conducted at both reconstruction and image level. Fig. 14.10 shows typical scenarios of data fusion at reconstruction and image.

In reconstruction-level data fusion for two-phase flow, the reconstructed information from one modality is usually deployed as a priori knowledge for the reconstruction of the other modality (Fig. 14.10(a)). Steiner (2006) investigated reconstruction-level fusion, or so-called sequential fusion, for gas–solid two-phase flow visualization using ECT and UT. In the study, the UT identified region was used as the background with fixed permittivity values, leaving the remaining area to be determined by ECT. Dyakowski et al. (2006) employed γ-ray (GCT) to constrain the ECT reconstruction process, with an extra statistical analysis on the ECT information per the k-nearest neighbor rule. Zhang et al. (2014) proposed a data fusion for ECT and GCT on oil–gas two-phase flow, where the measured data from ECT and GCT are linearly combined and normalized for the reconstruction (fifth column in Fig. 14.11). Liang, Ren, and Dong (2017a, 2017b) performed a similar strategy but using UT with ERT, where UT was responsible for position measurement as prior information utilized by EIT for free-interface reconstruction. Zhang et al. (2020) proposed a different approach, namely ERT and UTT projection sorting, to fuse UT and ERT data during the

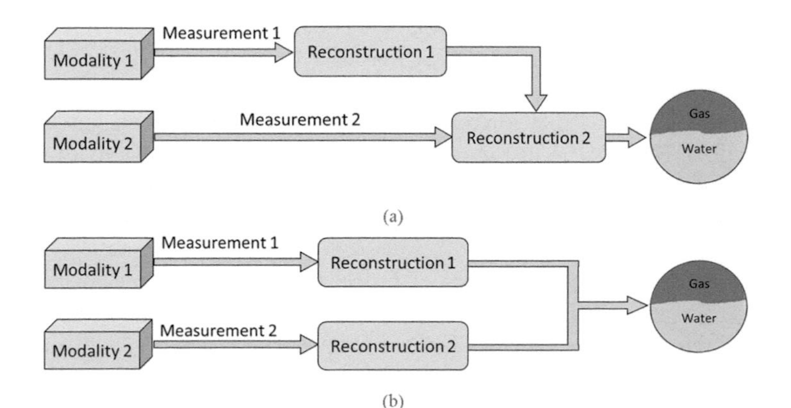

Figure 14.10 Illustrative data fusion for two-phase flow by DMTS at (a) reconstruction and (b) image level.

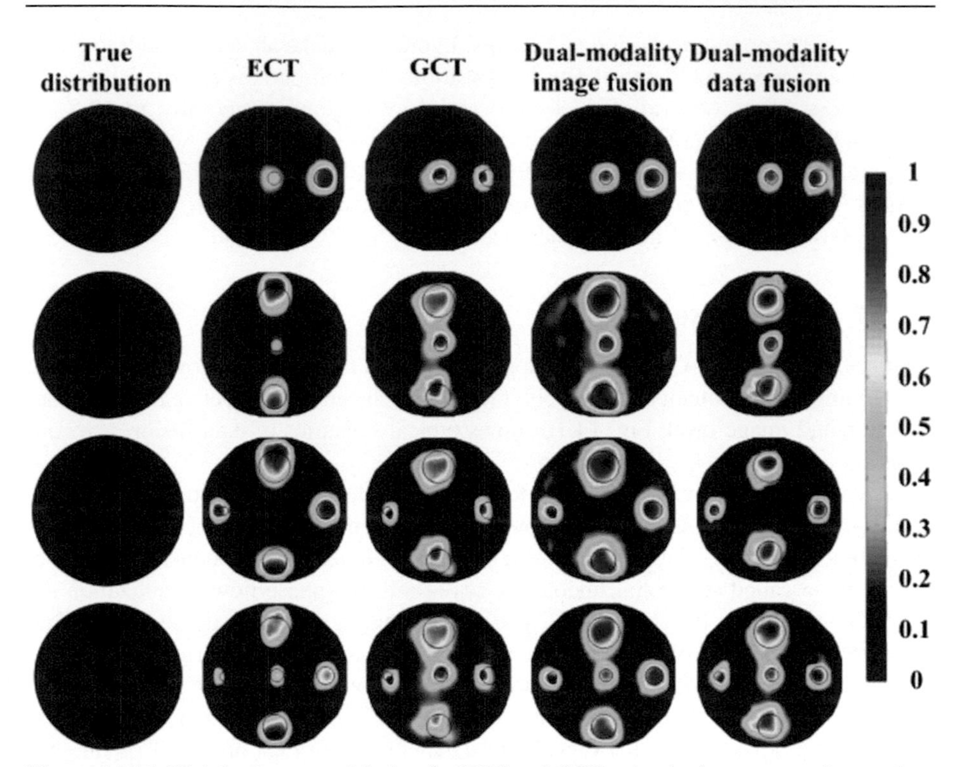

Figure 14.11 Weighted averaged fusion for ECT and GCT using both reconstruction- and image-level fusion (Zhang et al., 2014).

reconstruction. In their approach, UT results were not used as a prior constraint. Instead, both UT and ERT information were examined pixel-wise for deriving maximal contribution from either modality as the resultant value.

Data fusion for two-phase flow with MMTS can also be performed at image level, where reconstruction by individual modalities is carried out in parallel, and fusion is done on reconstructed images from each modality (Fig. 14.10(b)). Yue et al. (2013) applied fuzzy clustering to integrate EIT and ECT images for object discrimination, as well as EIT images from adjacent and opposite driven patterns on gas−liquid two-phase flow. Fig. 14.12 illustrates the advantage of the fused results over an individual approach. In Zhang et al. (2014), they also proposed a weighted average for pixel-level fusion of ECT and GCT, and their results are depicted in Fig. 14.11 (fourth column).

Apparently, data fusion, regardless of fusion levels, is promising to improve the quality of resultant images, with higher spatial resolution than those by individual modality. From Fig. 14.11, we can also find that image-level fusion is visually superior to reconstruction-level fusion, but with the cost of more computational time (Zhang et al., 2014). Surprisingly, fused reconstruction even costs less than the ET reconstruction, probably because fused reconstruction converges faster with complementary

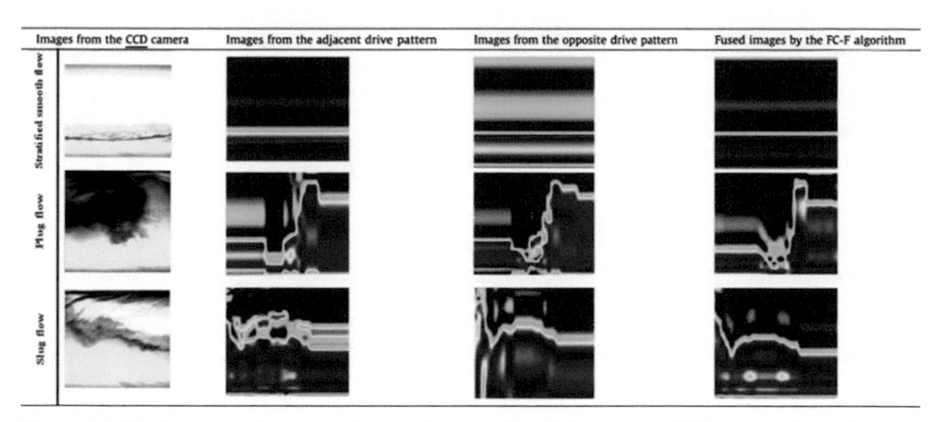

Figure 14.12 K-means clustering fusion for gas–liquid flow using EIT images from the adjacent and opposite drive patterns (Yue, Wu, Pan, & Wang, 2013).

constraints. In contrast, image-level fusion requires much more time than the one at reconstruction level, where the overhead is due to the postprocessing of the individually reconstructed images.

14.4.2 Three-phase flow

Various DMTS have been proposed for three-phase flow characterization, and the formation has evolved from separated modalities to integrated sensors (Ji et al., 2016; Pusppanathan et al., 2017; Qiu et al., 2007; Rodriguez-Frias & Yang, 2020; Sun & Yang, 2015; Xia, Cui, Zhai, & Wang, 2020b). They usually have a distinctive objective, compared with those for two-phase flow. That is, they aim to decompose individual phase, rather than improving the quality of reconstructed images. However, as far as data fusion is concerned, the level of the fusion can also be at reconstruction and image level, as depicted in Fig. 14.10. Due to the complexity of three-phase flow and limitations of existing reconstruction algorithms, there is little research on pure reconstruction-level fusion for three-phase flow. As a result, image-level fusion methods play a dominant role in MMTS-based three-phase flow visualization and measurement.

A broadly utilized pixel-level image fusion technology in MMTS-based three-phase flow visualization is thresholding, where an individual phase is distinguished by threshold values, usually based on empirical knowledge. One of the pioneering works has been done by Qiu et al. (2007) to DMTS for three-phase flow visualization, where a stratified stationary gas–oil–water flow in a horizontal pipe was qualified and quantified, with a simple threshold-based pixel-wise fusion approach (Fig. 14.13(a)). Sun and Yang (2015) also proposed a similar method to fuse an integrated dual-modality ERT–ECT system for stratified gas–oil–water flow. However, each phase in their results was presented in separate images, rather than one with fused results Fig. 14.13(c). In Wang et al. (2017), pixel-level image fusion for a gas–oil–water horizontal pipeline flow was systematically studied, with in-depth discussion of how to spatially and temporally align ERT and ECT tomograms for effective pixel-wise image

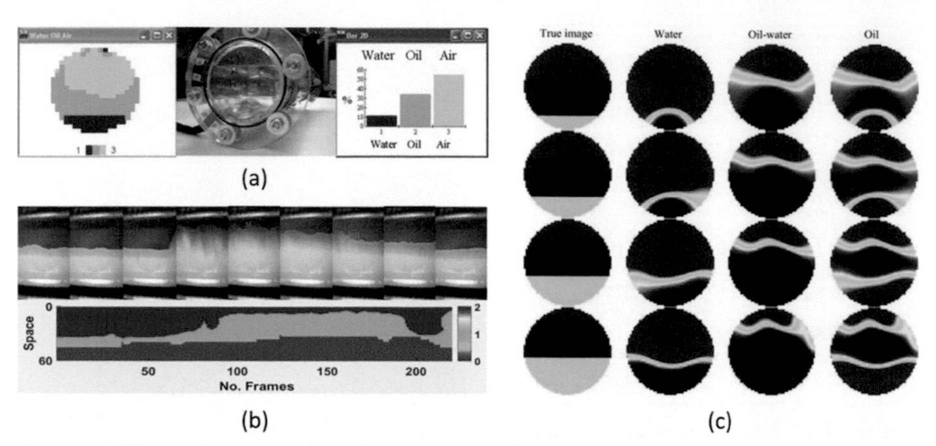

Figure 14.13 (a) characterization of a stratified stationary gas–oil–water flow in a horizontal pipe (Qiu et al., 2007), (b) threshold-based visualization of a slug gas–oil–water flow in horizontal pipeline with ERT–ECT systems (Wang et al., 2017), and (c) threshold-based pixel-level image fusion for gas–oil–water stratified flow (Sun and Yang, 2015).

fusion. In addition, their approach was experimentally evaluated on six commonly observed gas–liquid flow patterns in horizontal pipe, rather than on simulation or simple flow regimes. Fig. 14.13(b) shows visualization results on a slug flow. Later, their approach was applied to thoroughly evaluate the performance of dual-modality ERT–ECT systems on gas–oil–water horizontal flow on an industrial-scale flow facility at TUV-NEL, and the results demonstrated the effectiveness and efficiency of the approach on three-phase flow characterization (Wang et al., 2018).

Besides threshold-based fusion methods, some effort has been made on advanced fusion technologies to improve the results. Zhang, Ma and Soleimani (2015) integrated magnetic inductance tomography (MIT) and ECT to distinguish gas–oil–water flow, where the conductivity distribution derived by MIT was utilized as a priori knowledge into ECT for generating a sensitivity matrix and resolving the inverse problem, with a pixel-wise image fusion approach for differentiating individual phases in gas–oil–water mixture. Pusppanathan et al. (2017) investigated a fuzzy logic–based image fusion approach for a single-plane dual-modality ECT–UT system for stratified gas–oil–water flow. Due to the fuzzy logic technology, their approach does not require empirical threshold values, but still had promising results. In Wang et al. (2020), the authors proposed combining fuzzy logic with a decision tree to overcome the issues caused by empirical threshold values in Wang et al., 2017, where fuzzy logic was to fuse ERT and ECT tomograms, and the decision tree was to decompose the fused results to discriminate individual phases (Fig. 14.14(a)). The resultant images demonstrated the feasibility and robustness of the approach on real-case industrial gas–oil–water flow. Xia et al. (2020b) applied machine learning (ML) technologies to ECT and electromagnetic tomography (EMT), where a least-squares support-vector machine-based regression approach was utilized to generate fused results per the measured data and reconstructed images from ECT and EMT. Xia et al. (2020b)

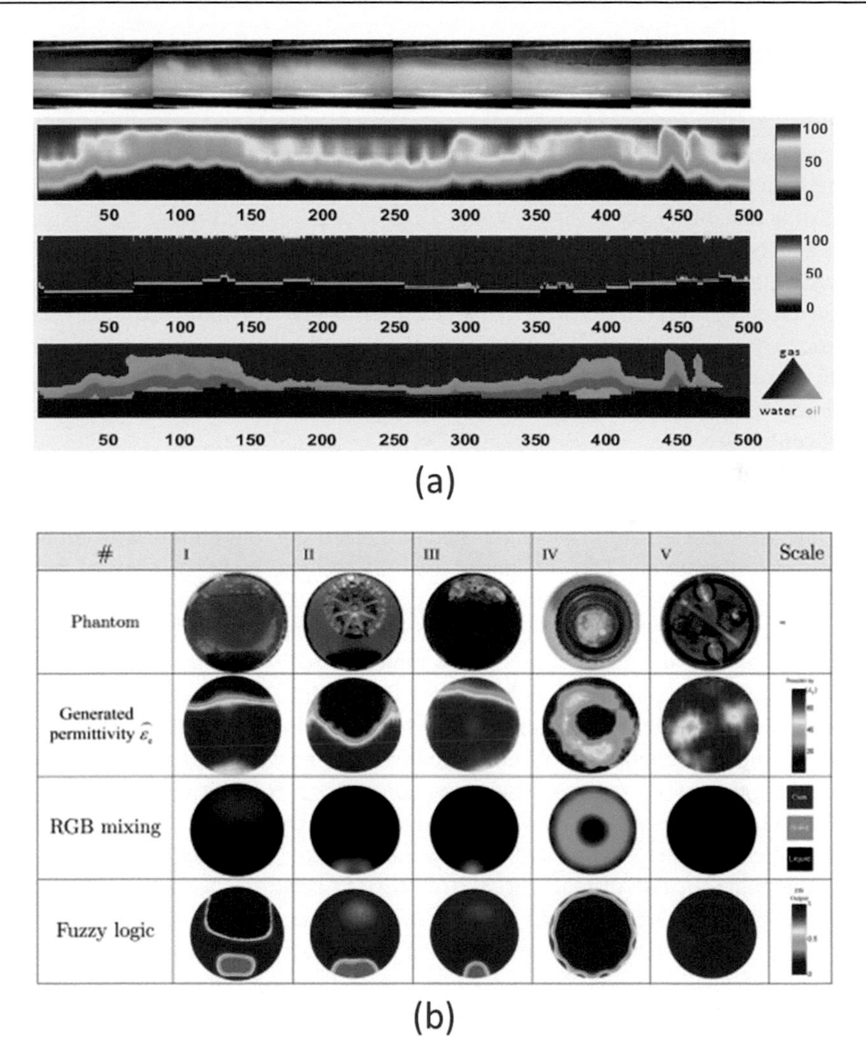

Figure 14.14 (a) fuzzy logic-based data fusion for a gas–oil–water slug flow with dual-modality ERT–ECT (Wang, Jia, & Wang, 2020) and (b) GAN-based data fusion for gas–liquid–solid flow with dual-modality ECT–EMT (Xia, Cui, Chen, Hu, & Wang, 2020).

explored the fusion using Generative Adversarial Networks (GANs) (Goodfellow et al., 2014). In their proposal, the training data, containing the electromagnetic properties, measurements, and reconstructed images, were derived from CFD simulation results. During the training, the fusion outcomes were generated from the measurements and the reconstructed images, using the electromagnetic properties as the ground truth. In contrast, during the testing stage, the data from real measurements were used as the input to the GAN to produce the fused images. Fig. 14.14(b) demonstrates the results by their proposed approach, along with the ones by RGB mixing and fuzzy logic.

14.5 Future trends

ET is well known for its inexpensiveness, noninvasiveness, and submillisecond temporal resolution. Due to the limited number of measurements and the nature of electromagnetic fields, electrical tomograms suffer from low spatial resolution, around 5% and distinguishability about 20% (Wang, 2015). As far as multiphase flow visualization is concerned, a direct impact is the blurred boundaries between bubbles and liquid, as well as the absence of small bubbles below certain sizes, resulting in ambiguous interpretation of multiphase flow dynamics, particularly for inexperienced users. In addition, conventional color mapping approaches also introduced some ambiguity on the perception of electrical tomograms. In order to minimize the ambiguity and maximize the information retrieved from electrical tomograms, various attempts have been made.

Some attempts applied advanced methodologies in CG, such as volume rendering and isosurface, to overcome the limitations of electrical tomograms. Those methodologies are particularly helpful to single-modality electrical tomograms with little complementary information. The challenges are still realistic reconstruction of bubbles. Large bubbles can be effectively identified by an isovalue from either empirical knowledge (Wang et al., 2019) or an iterative numerical solution (Li et al., 2019b). However, small bubbles below certain sizes are extremely challenging to retrieve, although Wang et al. (2019) approximated them via a bubble mapping approach. To improve the resultant small bubbles, supplementary information is required. For example, since the BM3D method requires local velocity to transform the original tomograms from (x, y, t, v) to (x, y, z, v'), a more precise derivation of local velocity will be helpful, e.g., the best correlation approach (Mosorov, Sankowski, & Dyakowski, 2002). In addition, small bubbles may be reconstructed via the integration of experimental results with numerical simulation. For instance, Ren et al. (2015) proposed a simple yet effective method to simulate bubbles based on a volume fraction representation, where bubble deformation and motion at a subcell level can be approximated. Another potential solution could be ML technologies, particularly deep learning (DL), where correlations between blurred electrical tomograms and actual bubble size and distribution could be "learned" by DL technologies.

As far as multidimension data fusion for MMTS is concerned, various solutions have been proposed for two- and three-phase flow at different levels. The majority of fusions for two-phase are at the reconstruction level, rather than at image level, due to reconstruction-level integration usually requiring less time and computational resources, compared to those at image level, Zhang et al. (2014). Due to the complementary information from other modalities, the reconstruction of the primary modality tends to numerically converge quicker and produce better results. Therefore, reconstruction-level fusion would be a promising direction for MMTS-based two-phase flow characterization. Conventionally, the complementary information is utilised as constraints, e.g., background separation (Steiner, 2006). Recent research demonstrates that measured data from individual modalities can be fused to generate a new sensitivity matrix and measured data for the reconstruction (Liang et al., 2017a,

2017b; Zhang et al., 2014, 2020). In this case, effective and efficient fusion for optimal reconstruction results is critical, where ML methodologies may be a potential solution for optimal fusion.

Data fusion for three-phase flow has a distinctive objective, i.e., each modality is responsible for decomposition of one phase from the rest. As a result, reconstruction-level fusion is extremely challenging for MMTS, unless more than two modalities are deployed, and hence, most of existing data fusion proposals for the flow is at image level, particularly pixel-wise fusion, Qiu et al. (2007); Sun and Yang (2015); Wang et al. (2017); Pusppanathan et al. (2017).

Conventional threshold-based methods are simple yet effective for the fusion, but the determination of the threshold values is usually empirical. Therefore, advanced technologies would be beneficial to minimize the uncertainty introduced by the empirical threshold, such as fuzzy logic by Pusppanathan et al. (2017) and Wang et al. (2020). In addition, fusion at higher level, such as feature and decision level, would be also helpful. A potential challenge is whether effective features could be retrieved from the tomograms involved. Another potential solution is the DL-based application in image fusion, where DL technologies have proved to be efficient for image fusion (Liu et al., 2018). It is also worth noting that the aforementioned advanced technologies in CG are also applicable to the fused results to further enhance the presentation of the flow under investigation.

References

Bankman, I. (2008). *Handbook of medical image processing and analysis*. Academic Press.

Brennen, C. E. (2005). *Fundamentals of multiphase flows*. Cambridge University Press.

Corneliussen, S., Couput, J. P., Dahl, E., Dykesteen, E., Froysa, K. E., Malde, E., et al. (2005). *Handbook of multiphase flow metering*. http://nfogm.no/wp- content/uploads/2014/02/MPFM_Handbook_Revision2_2005_ISBN-82-91341-89-3.pdf.

Dias, F. D., dos Santos, E. N., da Silva, M. J., Schleicher, E., Morales, R. E., Hewakandamby, B., et al. (2020). New algorithm to discriminate phase distribution of gas-oil-water pipe flow with dual-modality wire-mesh sensor. *IEEE Access, 8*, 125163−125178.

Dyakowski, T., Johansen, G. A., Hjertaker, B. T., Sankowski, D., Mosorov, V., & Wlodarczyk, J. (2006). A dual modality tomography system for imaging gas/solids flows. *Particle & Particle Systems Characterization, 23*, 260−265. https://doi.org/10.1002/ppsc.200601057

Fischer, F., & Hampel, U. (2010). Ultra-fast electron beam x-ray computed tomography for two-phase flow measurement. *Nuclear Engineering and Design, 240*, 2254−2259.

Goodfellow, I., Pouget-Abadie, J., Mirza, M., Xu, B., Warde-Farley, D., Ozair, S., et al. (2014). Generative adversarial nets. In *Advances in neural information processing systems* (pp. 2672−2680).

Hansen, C. D., & Johnson, C. R. (2011). *Visualization handbook*. Elsevier Science.

Hjertaker, B., Maad, R., & Johansen, G. (2011). Dual-mode capacitance and gamma-ray tomography using the landweber reconstruction algorithm. *Measurement Science and Technology, 22*, 104002.

Hoyle, B. S., Jia, X., Podd, F. J. W., Schlaberg, H. I., Tan, H. S., Wang, M., et al. (2001). Design and application of a multi-modal process tomography system. *Measurement Science and Technology, 12*, 1157.

Hoyle, B. S., Podd, F., Schlaberg, H., Wang, M., Williams, R. A., & York, T. (1999). Multi-sensor process tomography system design: Part 1—systems and hardware engineering. In *Proc. Of the 1st world congress on industrial process tomography* (pp. 323—327).

Hoyle, B. S., & Wang, M. (2012). Multi-dimensional opportunities and data fusion in industrial process tomography. In *Instrumentation and measurement technology conference (I2MTC), 2012 IEEE international* (pp. 916—920). https://doi.org/10.1109/I2MTC.2012.6229323

Iacovides, S., Wang, M., & Wang, Q. (2018). Bubble mapping method for transient taylor bubble flows. In *9th world congress on industrial process tomography (WC-IPT-9)* (pp. 871—878). International Society for Industrial Process Tomography.

ITS. (2009). *ITS system 2000 version 7.0 p2+ electrical resistance tomography system - user's manual. Industrial tomography systems plc. Speakers house, 39 deansgate, manchester M3 2BA*.

Ji, H., Tan, W., Gui, Z., Wang, B., Huang, Z., Li, H., et al. (2016). A new dual-modality ect/ert technique based on c4d principle. *IEEE Transactions on Instrumentation and Measurement, 65*, 1042—1050. https://doi.org/10.1109/TIM.2016.2526758

Khambampati, A. K., Kim, K. Y., Lee, Y. G., & Kim, S. (2016). Boundary element method to estimate the time-varying interfacial boundary in horizontal immiscible liquids flow using electrical resistance tomography. *Applied Mathematical Modelling, 40*, 1052—1068.

Kim, B. S., Khambampati, A. K., Hong, Y. J., Kim, S., & Kim, K. Y. (2013). Multiphase flow imaging using an adaptive multi-threshold technique in electrical resistance tomography. *Flow Measurement and Instrumentation, 31*, 25—34. https://doi.org/10.1016/j.flowmeasinst. 2012.11.003. special Issue IWPT-4 in Flow Measurement and Instrumentation

Liang, G., Ren, S., & Dong, F. (2017). An adaptive local weighted image reconstruction algorithm for eit/utt dual-modality imaging. In *2017 IEEE international instrumentation and measurement technology conference (I2MTC)* (pp. 1—6).

Liang, G., Ren, S., & Dong, F. (2017). Ultrasound guided electrical impedance tomography for 2d free-interface reconstruction. *Measurement Science and Technology, 28*, 074003.

Liang, G., Ren, S., Zhao, S., & Dong, F. (2019). A Lagrange-Newton method for eit/ut dual-modality image reconstruction. *Sensors, 19*, 1966.

Liu, Y., Chen, X., Wang, Z., Wang, Z. J., Ward, R. K., & Wang, X. (2018). Deep learning for pixel- level image fusion: Recent advances and future prospects. *Information Fusion, 42*, 158—173. https://doi.org/10.1016/j.inffus.2017.10.007

Li, K., Wang, Q., & Wang, M. (2019a). Three-dimensional visualisation of gas-water two-phase flow based on bubble mapping method and size projection algorithm. *Flow Measurement and Instrumentation, 69*, 101590. https://doi.org/10.1016/j.flowmeasinst.2019.101590

Li, K., Wang, Q., Wang, M., & Han, Y. (2019b). Imaging of a distinctive large bubble in gas—water flow based on a size projection algorithm. *Measurement Science and Technology, 30*, 094004. https://doi.org/10.1088/1361-6501/ab16b0

Li, K., Yang, N., Wang, J., Han, Y., Nie, P. F., & Zhang, M. (2019c). Size projection algorithm: Optimal thresholding value selection for image segmentation of electrical impedance tomography. *Mathematical Problems in Engineering, 2019*, 1368010.

Lorensen, W. E., & Cline, H. E. (1987). Marching cubes: A high resolution 3d surface construction algorithm. *SIGGRAPH Comput. Graph., 21*, 163—169.

Maxwell, J. C. (1982). *A treatise on electricity and magnetism* (3rd ed., Vol. 1). Oxford University Press.

Mitchell, H. B. (2012). *Data fusion: Concepts and ideas*. Springer Science & Business Media.

Mosorov, V., Sankowski, D., & Dyakowski, T. (2002). The 'best-correlated pixels' method for solid mass flow measurements using electrical capacitance tomography. *Measurement Science and Technology, 13*, 1810.

Prasser, H. M., Scholz, D., & Zippe, C. (2001). Bubble size measurement using wire-mesh sensors. *Flow Measurement and Instrumentation, 12*, 299–312. https://doi.org/10.1016/S0955-5986(00)00046-7

Prosperetti, A., & Tryggvason, G. (2009). *Computational methods for multiphase flow*. Cambridge University Press.

Pusppanathan, J., Abdul Rahim, R., Phang, F. A., Mohamad, E. J., Nor Ayob, N. M., Fazalul Rahiman, M. H., et al. (2017). Single-plane dual-modality tomography for multiphase flow imaging by integrating electrical capacitance and ultrasonic sensors. *IEEE Sensors Journal, 17*, 6368–6377.

Qiu, C., Hoyle, B. S., & Podd, F. J. W. (2007). Engineering and application of a dual-modality process tomography system. *Flow Measurement and Instrumentation, 18*, 247–254. https://doi.org/10.1016/j.flowmeasinst.2007.07.008. process Tomography and Flow Visualization

Reinecke, N., Petritsch, G., Schmitz, D., & Mewes, D. (1998). Tomographic measurement techniques– visualization of multiphase flows. *Chemical Engineering & Technology, 21*, 7–18.

Ren, S., Dong, F., Tan, C., & Xu, Y. (2012). A boundary element approach to estimate the free surface in stratified two-phase flow. *Measurement Science and Technology, 23*, 105401.

Ren, B., Jiang, Y., Li, C., & Lin, M. C. (2015). A simple approach for bubble modelling from multiphase fluid simulation. *Computational Visual Media, 1*, 171–181.

Rodriguez-Frias, M. A., & Yang, W. (2020). Dual-modality 4-terminal electrical capacitance and resistance tomography for multiphase flow monitoring. *IEEE Sensors Journal, 20*, 3217–3225.

Soleimani, M., Lionheart, W. R. B., & Dorn, O. (2006). Level set reconstruction of conductivity and permittivity from boundary electrical measurements using experimental data. *Inverse Problems in Science and Engineering, 14*, 193–210. https://doi.org/10.1080/17415970500264152

Steiner, G. (2006). Sequential fusion of ultrasound and electrical capacitance tomography. *International Journal Of Information and Systems Sciences, 2*, 487–497.

Sun, J., & Yang, W. (2015). A dual-modality electrical tomography sensor for measurement of gas–oil–water stratified flows. *Measurement, 66*, 150–160. https://doi.org/10.1016/j.measurement.2015.01.032

Tan, C., Lv, S., Dong, F., & Takei, M. (2018). Image reconstruction based on convolutional neural network for electrical resistance tomography. *IEEE Sensors Journal, 19*, 196–204.

Tan, J., & Yang, X. B. (2009). Physically-based fluid animation: A survey. *Science in China - Series F: Information Sciences, 52*, 723–740. https://doi.org/10.1007/s11432-009-0091-z

Wang, M. (2002). Inverse solutions for electrical impedance tomography based on conjugate gradients methods. *Measurement Science and Technology, 13*, 101.

Wang, M. (Ed.). (2015) (1st ed.)*Woodhead publishing series in electronic and optical materialsIndustrial tomography: Systems and applications*. Woodhead Publishing. https://doi.org/10.1016/C2013-0-16466-5

Wang, Q. (2017). *A data fusion and visualisation platform for multi-phase flow by electrical tomography*. Ph.D. thesis. University of Leeds.

Wang, M., Dickin, F. J., & Mann, R. (1999). Electrical resistance tomography sensing systems for industrial applications. *Chemical Engineering Communications, 175*, 49–70. https://doi.org/10.1080/00986449908912139

Wang, Q., Jia, X., & Wang, M. (2019). Bubble mapping: Three-dimensional visualisation of gas–liquid flow regimes using electrical tomography. *Measurement Science and Technology, 30*, 045303. https://doi.org/10.1088/1361-6501/ab06a9

Wang, Q., Jia, X., & Wang, M. (2020). Fuzzy logic based multi-dimensional image fusion for gas–oil- water flows with dual-modality electrical tomography. *IEEE Transactions on Instrumentation and Measurement, 69*, 1948–1961.

Wang, M., Jones, T., & Williams, R. (2003). Visualization of asymmetric solids distribution in horizontal swirling flows using electrical resistance tomography. *Chemical Engineering Research and Design, 81*, 854–861. https://doi.org/10.1205/026387603322482095. particle Technology

Wang, F., Marashdeh, Q., Fan, L. S., & Williams, R. A. (2009). Chapter 5 electrical capacitance, electrical resistance, and positron emission tomography techniques and their applications in multi-phase flow systems. In J. Li (Ed.), *Of advances in chemical engineering: Vol. 37. Characterization of flow, particles and interfaces* (pp. 179–222). Academic Press. https://doi.org/10.1016/S0065-2377(09)03705-3

Wang, Q., Polansky, J., Wang, M., Wei, K., Qiu, C., Kenbar, A., et al. (2018). Capability of dual-modality electrical tomography for gas-oil-water three-phase pipeline flow visualisation. *Flow Measurement and Instrumentation, 62*, 152–166. https://doi.org/10.1016/j.flowmeasinst.2018.02.007

Wang, Q., Wang, M., Wei, K., & Qiu, C. (2017). Visualization of gas–oil–water flow in horizontal pipeline using dual-modality electrical tomographic systems. *IEEE Sensors Journal, 17*, 8146–8156.

Wenger, R. (2013). *Isosurfaces: Geometry, topology, and algorithms* (1st ed.). A K Peters/CRC Press.

West, R. M., & Williams, R. A. (1999). Opportunities for data fusion in multi-modality tomography. In *Proc. 1st world congress on industrial process tomography* (pp. 195–200).

Xia, Z., Cui, Z., Chen, Y., Hu, Y., & Wang, H. (2020a). Generative adversarial networks for dual-modality electrical tomography in multi-phase flow measurement. *Measurement.* https://doi.org/10.1016/j.measurement.2020.108608, 108608.

Xia, Z., Cui, Z., Zhai, L., & Wang, H. (2020b). Dual-modality tomography for gas-liquid-solid three phase flow imaging: A simulation study. In *2020 IEEE international instrumentation and measurement technology conference (I2MTC)* (pp. 1–6). IEEE.

Yao, J., & Takei, M. (2017). Application of process tomography to multiphase flow measurement in industrial and biomedical fields: A review. *IEEE Sensors Journal, 17*, 8196–8205.

Ye, L., & Yang, W. (2012). Real-time 3d visualisation in electrical capacitance tomography. In *2012 IEEE international conference on imaging systems and techniques proceedings* (pp. 40–44). IEEE.

Yue, S., Wu, T., Pan, J., & Wang, H. (2013). Fuzzy clustering based et image fusion. *Information Fusion, 14*, 487–497. https://doi.org/10.1016/j.inffus.2012.09.004

Zhang, M., Ma, L., & Soleimani, M. (2015). Dual modality ect-mit multi-phase flow imaging. *Flow Measurement and Instrumentation, 46*(Part B), 240–254. https://doi.org/10.1016/j.flowmeasinst.2015.03.005. special issue on Tomography Measurement & Modelling of Multiphase Flows

Zhang, W., Tan, C., & Dong, F. (2020). Dual-modality tomography by ert and utt projection sorting algorithm. *IEEE Sensors Journal, 20*, 5415–5423.

Zhang, R., Wang, Q., Wang, H., Zhang, M., & Li, H. (2014). Data fusion in dual-mode tomography for imaging oil-gas two-phase flow. *Flow Measurement and Instrumentation, 37*, 1–11. https://doi.org/10.1016/j.flowmeasinst.2014.03.003

Part Three

Tomography applications

Applications of electrical resistance tomography to chemical engineering

Mohadeseh Sharifi and Brent Young
Department of Chemical & Materials Engineering, University of Auckland, Auckland, New Zealand

15.1 Introduction

The need to retrieve information and knowledge from inside of objects, which may be difficult to gain access to, is common within various industries from medical diagnosis of body parts, to identifying leakages and cracks in underground pipes, to investigating processes within chemical reactors (Sharifi & Young, 2013a,b,c; Stanley & Bolton, 2008). This need brought about the concept of tomography, the possibility to expose the internal intricacies of an object without needing to invade it. It was initially discovered by a Norwegian physicist, Abel, for an axisymmetrical object and then nearly 100 years later, by an Austrian mathematician, Radon, for arbitrary shaped objects. Since then, continuous advances in tomography and its applications have made it widely applied as a medical diagnostic tool and more recently as a reliable tool for imaging industrial applications including oil and gas, pharmaceuticals, agrochemical, and fine chemical manufacture (Aw, Rahim, Rahiman, Yunus, & Goh, 2014).

The chemical engineering industry is a world of pipes, tanks, reactors, and various other vessels which are usually opaque and generally inaccessible, commonly containing various forms of chemicals in continuous processes which require monitoring and control. For this purpose, continuous knowledge and information of the internal state of these vessels, although required, are hard, and in many cases impossible to attain. The idea of remote noninvasive interior inspection through volume scanning in tomography provides the solution to this problem.

There are different tomography modalities or data collection methods depending on the knowledge required and the attributes of the process under examination, such as radiation, acoustic and electrical. Electrical resistance tomography (ERT), a specific case of Electrical Impedance Tomography (EIT), was initially invented in the 1980s and has now become one of the most widespread and adapted modalities of tomography and is a particularly promising technique for monitoring and analyzing various industrial vessels. ERT's potential is attracting widespread focus in terms of applications and research due to its numerous distinct advantages, such as high speed, low cost, suitability for various sizes of pipes and vessels, having no radiation hazard, and being nonintrusive (Aw et al., 2014).

Industrial Tomography. https://doi.org/10.1016/B978-0-12-823015-2.00032-7

Thorough evaluation of all related literature shows that the core of research on ERT to date has been largely on Chemical and Geotechnical Engineering. A limited number of papers have utilized ERT in Biomedical Engineering applications such as medical body imaging. The reported literature in the Geotechnical Engineering field utilized ERT commonly with the purpose of detecting, monitoring and controlling ground water, subsurface properties, superficial deposits, soil characteristics, seismic properties, buried structural remains, fluid or moisture distribution, structural cracks or faults, degradation of wall foundations, water leakages, flood banks, and shallow reservoirs (Sharifi & Young, 2013a,b,c).

A large amount of research has focused on the various applications of ERT to Chemical Engineering. This chapter aims on reviewing and classifying all such studies which have applied ERT to numerous vessels/media in the Chemical Engineering industry for a number of application purposes. The idea is to provide a broad review of the proven advantages and applications of ERT in Chemical Engineering.

Generally research on ERT and all other process tomography techniques can be divided into two areas of focus. A vast amount of research has been focused on the development and enhancement of the related hardware and software to expand the abilities of the technique. The application of the technique to different subjects and for various purposes and the development of methodologies required for these novel applications are another emphasis of the available literature. The numerous application purposes in which ERT has been utilized for in the Chemical Engineering arena are classified and described in this chapter.

15.2 Applications of ERT

ERT has various advantages such as being a nonintrusive, safe, fast, simple, and low cost technology, making it a suitable and favorable choice for a wide range of applications (Aw et al., 2014), and due to these advantages, considerable research has also been done in order to enhance speed of data acquisition, sensitivity, flexibility, and noise immunity which therefore extends its application to more diverse situations and subjects (Sharifi & Young, 2013a,b,c).

As shown in Fig. 15.1, careful review of all published literature on ERT applications has revealed that ERT measurements have been applied for numerous purposes in the Chemical Engineering field. Mixing and flow investigations are the most common with more than 100 research articles on each. Phase hold up; i.e., Liquid, gas, and solid phase hold ups are next in line in terms of number of dedicated research papers. A number of other application purposes such as solid particle suspension, dissolution and precipitation, phase separation, phase boundary detection, concentration monitoring, cleaning-in-place (CIP), malfunction detection, and process control applications have also been researched. Each of these divisions and subdivisions are discussed in detail in this chapter.

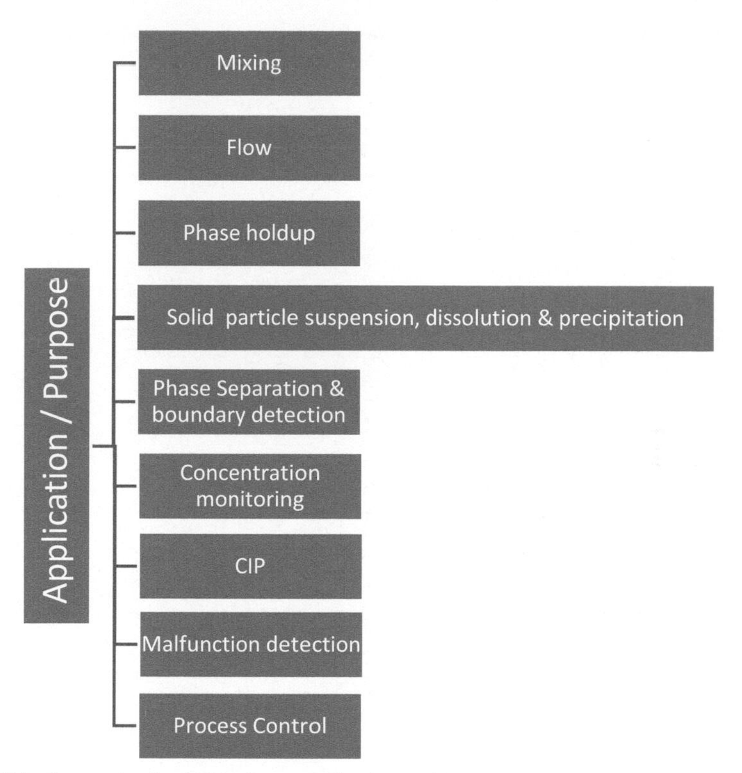

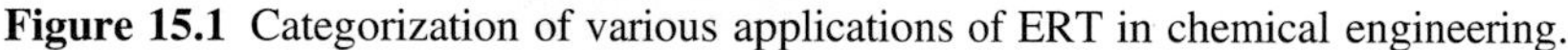

Figure 15.1 Categorization of various applications of ERT in chemical engineering.

15.2.1 Mixing investigations

Mixing and storage vessels are one of the most common unit operations which are utilized in almost every process line for the purpose of mixing and blending different process materials, heating and cooling, holding, and reactions. As these vessels are usually opaque and inaccessible, gaining knowledge of the state of the material/phases, degree of mixing and phase distribution inside, although essential, is often difficult to achieve. ERT has the ability to provide real-time three-dimensional images of the internal state of mixing and storage vessels for various purposes. Once ERT measurements are made on mixing or storage vessels, the resulting multidimensional conductivity data can be utilized for the investigation of numerous characteristics of the mixing process. Mixing quality/performance, mixing time/index, evaluation of the effect of mixer and baffle characteristics on quality of mixing, and three-dimensional qualitative mixing visualization are some of the main types of information acquired through ERT measurements to date, as can be seen in Table 15.1.

Table 15.1 Main applications of ERT to mixing evaluations.

Mixing quality, mixing time/index investigations	
Mann et al. (1996)	Patel et al. (2013a)
Williams, Jia and McKee (1996)	Patel et al. (2013b)
Li and Wei (1999)	Patel et al. (2013c)
West, Jia and Williams (1999)	Patel, Ein-Mozaffari and Mehrvar (2014)
Williams and Wang (2000)	Montante and Paglianti (2015)
Wang and Yin (2001)	Hashemi, Ein-Mozaffari, Upreti and Hwan (2016)
Forte, Alberini, Simmons and Stitt (2019)	Carletti, Montante, De Blasio and Paglianti (2016)
Yenjaichon, Grace, Lim and Bennington (2011)	Kazemzadeh, Ein-Mozaffari, Lohi and Pakzad (2016a)
Mann et al. (1997)	Kazemzadeh, Ein-Mozaffari, Lohi and Pakzad (2016b)
Wang, Dorward, Vlaev and Mann (2000)	Kazemzadeh, Ein-Mozaffari, Lohi and Pakzad (2016c)
Spear (2003)	Kennedy, Bhattacharjee, Eshtiaghi and Parthasarathy (2016)
Wabo, Kagoshima and Mann (2004)	Mishra and Ein-Mozaffari (2016)
Ricard, Brechtelsbauer, Xu and Lawrence (2005)	Jegatheeswaran, Ein-M and Wu (2017)
Stanley, Hristov, Mann and Primrose (2005)	Kazemzadeh et al. (2017a)
Kim, Nkaya and Dyakowski (2006)	Kazemzadeh et al. (2017b)
Pakzad et al. (2008a)	Khalili, Jafari Nasr, Kazemzadeh and Ein-Mozaffari (2017)
Pakzad et al. (2008b)	Low, Eshtiaghi, Shu and Parthasarathy (2017)
Zhao, Mehrvar and Ein-Mozaffari (2008)	Paglianti, Carletti, Busciglio and Montante (2017)
Hui, Bennington and Dumont (2009)	Paglianti, Carletti, and Montante (2017)
Li et al. (2009)	Jegatheeswaran, Ein-M and Wu (2018)
Rodgers, Cooke, Siperstein and Kowalski (2009)	Low, Allitt, Eshtiaghi and Parthasarathy (2018)
Bhole, Hui, Gomez, Bennington and Dumont (2011)	Malik and Pakzad (2018)
Liu et al. (2011)	Mihailova et al. (2018)
Rodgers, Gangolf, Vannier, Parriaud and Cooke. (2011)	Naghavi-Anaraki, Turcotte and Ein-Mozaffari (2018)
Tahvildarian et al. (2011)	Duan et al. (2019)
Yenjaichon et al. (2011)	Forte et al. (2019)
Yenjaichon, Pageau, Bhole, Bennington and Grace (2011)	Jamshed, Cooke and Rodgers (2019)
Hamood-ur-Rehman, Dahman and Ein-Mozaffari (2012)	Khajeh Naeeni and Pakzad (2019)
Yenjaichon et al. (2012)	Mirshekari and Pakzad (2019)
Pakzad, Ein-Mozaffari, Upreti and Lohi (2013a)	Montante, Carletti, Maluta and Paglianti (2019)
Pakzad, Ein-Mozaffari, Upreti and Lohi (2013b)	Harrison, Kotsiopoulos, Stevenson and Cilliers (2020)
Pakzad, Ein-Mozaffari, Upreti and Lohi (2013c)	Maluta, Montante and Paglianti (2020)
Pakzad, Ein-Mozaffari, Upreti and Lohi (2013d)	Khalili, Jafari Nasr, Kazemzadeh and Ein-Mozaffar (2017)
Pakzad, Ein-Mozaffari, Upreti and Lohi (2013e)	Kagoshima and Mann (2005)

Mixer and baffle characteristics investigations	
Mann et al. (1999)	Kazemzadeh et al. (2016)
Kaminoyama, Taguchi, Misumi and Nishi (2005)	Lassaigne, Blais, Fradette and Bertrand (2016)
Pakzad et al. (2008a)	Mishra and Ein-Mozaffari (2016)
Pakzad et al. (2008b)	Busciglio, Opletal, Moucha, Montante and Paglianti (2017)
Hui et al. (2009)	Kazemzadeh et al. (2017a)
Bhole and Bennington (2010)	Kazemzadeh et al. (2017b)
Hosseini, Patel, Ein-Mozaffari and Mehrvar (2010)	Hashemi, Ein-Mozaffari, Upreti and Hwan (2018)
Bhole et al. (2011)	Jamshed, Cooke, Ren and Rodgers (2018)
Gumery, Ein-Mozaffari and Dahman (2011)	Kazemzadeh, Elias, Tamer and Ein-Mozaffari (2018)
Tahvildarian et al. (2011)	Malik and Pakzad (2018)
Pakzad, Ein-Mozaffari, Upreti and Lohi (2013a)	Mihailova et al. (2018)
Pakzad, Ein-Mozaffari, Upreti and Lohi (2013b)	Naghavi-Anaraki, et al. (2018)
Pakzad, Ein-Mozaffari, Upreti and Lohi (2013c)	Jamshed et al. (2019)
Pakzad, Ein-Mozaffari, Upreti and Lohi (2013d)	Kazemzadeh, Ein-M and Lohi (2019)
Pakzad, Ein-Mozaffari, Upreti and Lohi (2013e)	Khajeh Naeeni and Pakzad (2019)
Patel et al. (2013a)	Mirshekari and Pakzad (2019)
Patel et al. (2013b)	Harrison et al. (2020)
Patel et al. (2013c)	Kazemzadeh, et al. (2020a)
Patel et al. (2014)	Kazemzadeh et al. (2020b)
Montante and Paglianti (2015)	Maluta et al. (2020)
Kazemzadeh et al. (2016)	Hashemi et al. (2016)
Three-Dimensional mixing visualizations	
Mann et al. (1996)	Bolton et al. (2006)
Mann, Williams, Dyakowski, Dickin and Edwards (1997)	Simmons et al. (2009)
Holden, Wang, Mann, Dickin and Edwards (1998)	Bhole and Bennington (2010)
Stanley, Mann and Primrose (2002)	Rodgers and Kowalski (2010)
Hume (2003)	Sharifi and Young (2010)
	Sharifi and Young (2011)

15.2.1.1 Mixing quality and mixing time/index investigations

One of the most common processes in industrial applications is mixing, but due to its complication, theoretical methodologies are very restricted. Qualitative observations and accurate quantification of mixing are of great significance from the practical point of view and for the confirmation of theoretical models as well (Kim, Nkaya, & Dyakowski, 2006). In this category of research, ERT has been implemented to evaluate mixing quality both qualitatively and quantitatively.

Research in this area was initiated in 1996 with ERT being used for the purpose of development of mixing models (Mann et al., 1997; Williams et al., 1996). ERT was also later applied for the validation of Computational Fluid Dynamics (CFD) (Spear, 2003).

Measurement of mixing time/quality and also evaluation of cavern dimensions in agitation of pseudoplastic fluids such as Xanthan gum, under various mixing conditions and using a variety of mixers has attracted a lot of attention (Jegatheeswaran, et al., 2017, 2018; Kazemzadeh et al., 2016a, 2016b, 2016c, 2017a, 2017b; Pakzad et al. 2008a,b, 2013a,b,c,d,e; Patel et al., 2013a,b,c, 2014).

Studying mixing time/quality of various types of media such as a non-Newtonian highly viscous surfactant (sodium laureth sulfate) in water (Rodgers et al., 2009), or nanoparticle suspensions (Wang et al., 2009a,b) and also mixing in a variety of vessels such as an agitated multilamp photoreactor (Zhao et al., 2008) or vessels with aspect ratios greater than one (Rodgers et al., 2011) are also other areas of research focus in this category.

15.2.1.2 Mixer and baffle characteristics investigations

In mixing processes, impeller type, speed, rotation mode, off-bottom clearance, presence, and characteristics of baffles have crucial effects on the quality of mixing. ERT's ability to provide interior insight into mixing processes has provided researchers a tool to evaluate such effects in detail.

This research was initiated in 1999 to study imperfect mixing and improve impeller configuration (Mann et al., 1999). The effect of impeller characteristics on mixing and cavern dimensions in pulp fiber suspensions have also been investigated (Bhole & Bennington, 2010; Bhole et al., 2011; Hui et al., 2009). Effect of impeller type, speed, rotation mode, and off-bottom clearance on different dispersion regimes in solid—liquid mixing (Hosseini et al., 2010), on micron-sized polymeric particles in a slurry (Tahvildarian et al., 2011), on high solids concentrations in viscous liquids (Lassaigne et al., 2016), or on highly concentrated slurries of large particles using RSM (Kazemzadeh et al., 2019, 2020a,b) have been some focuses of research in this area.

15.2.1.3 3-Dimensional mixing visualizations

Providing qualitative insight into otherwise opaque and visibly inaccessible vessels is one of the main benefits of ERT. Such qualitative insight provides the ability to provide vast knowledge from within the mixing process, such as visualization of phase distributions or cavern geometry and boundaries.

Evaluation of plant scale mixing was initiated by Mann et al. in 1996 and later further enhanced for the purpose of vortex detection, miscible, and gas–liquid mixing (Holden et al., 1998; Mann et al., 1996, 1997, 1999; Wang et al., 2000).

ERT was later applied for the purpose of visualization of cavern boundaries in Newtonian (Bhole & Bennington, 2010; Simmons et al., 2009) and non-Newtonian media (Kennedy et al., 2016), phase inversion (Kaminoyama et al., 2010), physical stability and phase separation (Kowalski, Davidson, Flanagan, & York, 2010), and visualization of significant level change (Rodgers & Kowalski, 2010).

It was also utilized for quantitative auditing and attaining informative data such as total solids content in milk powder processing (Sharifi & Young, 2010) and also overall homogeneity or nonhomogeneity, total solids content monitoring, adulteration, cream separation, aeration, and object/powder lump detection which are parameters requiring monitoring and control in the current milk processing industry (Sharifi & Young, 2011).

Application of ERT for three-dimensional visualization of various vessels has also been an area of research interest such as visualization of mixing in draft-tube airlift reactors (Gumery et al., 2011) and cross-flow membrane emulsifications (Liu et al., 2011).

15.2.2 Flow investigations

Acquiring various forms of information from media in different flow situations is the next most common application of ERT to Chemical Engineering processes. Single, two-phase, or multiphase flows can very commonly be seen in numerous locations in a process line, such as pipes, columns, and various types of reactors. Similar to mixing and storage vessels, these vessels are also usually visibly obscure and inaccessible. Therefore, obtaining essential information of the state of the material/phases, degree of mixing, phase distribution, and flow and velocity profiles is often complicated to attain. The rapid data collection capability of ERT facilitates observation of such dynamic processes. Using ERT, cross-sectional profiles of the flowing media inside the vessels can be obtained from which otherwise unattainable knowledge such as qualitative flow visualization or quantitative measurements such as phase distribution, phase flowrate, and velocity profiles can be achieved.

Flow and phase distribution visualization and velocity and flow profile measurements are two main categories of research into applications of ERT to flow investigations as can be seen in Table 15.2.

15.2.2.1 Flow and phase distribution visualizations

When multiphase media is in a flowing condition, whether inside a pipe or any other vessel such as bubble column or a digester, being able to actually visualize the flow of each phase in relation to the other, and generally the distribution of the various phases within the vessel, is essential for monitoring and control of the process. In solid–liquid or liquid–gas flow, understanding whether the solid particles are in suspension or

Table 15.2 Main applications of ERT to flow investigations.

Flow and phase distribution visualizations	
Wang, Jones and Williams (2003)	Annamalai, Pirouzpanah, Gudigopuram and Morrison (2016)
Vlaev and Bennington (2004)	Shi, Wang and Shen (2016)
Norman and Bonnecaze (2005)	Jia, Wang and Millington (2017)
Vlaev and Bennington (2005)	Liu, Deng, Zhang, Yu and Li (2017)
Stevenson, Harrison, Miles and Cilliers (2006)	Low et al. (2017)
Vlaev and Bennington (2006)	Singh, Quiyoom and Buwa (2017)
Lee and Bennington (2007)	Son et al. (2017)
Regner, Henningsson, Wiklund, Östergren and Trägårdh (2007)	Wang, Wang, Wei and Qiu (2017)
Xu, Wang and Cui (2009)	Liu, Zhang, Yang and Xu (2018)
Lee and Bennington (2010)	Shen, Tan, Dong, Smith and Escudero (2018)
Razzak, Barghi, & Zhu (2010a,b)	Son et al. (2018)
Alaqqad, Bennington and Martinez (2012)	Wang et al. (2018)
Faraj and Wang (2012)	Zbib, Ebrahimi, Ein-Mozaffari and Lohi (2018a)
Hamood-ur-Rehman et al. (2012)	Zbib, Ebrahimi, Ein-Mozaffari and Lohi (2018b)
Harrison, Stevenson and Cilliers (2012)	Zhang, Li, Ma, Chen and Xu (2018)
Sharifi and Young (2012a,b)	Fang et al. (2019)
Wang, Meng, Huang, Ji and Li (2012)	Vadlakonda, Kopparthi, Mukhurjee and Mangadoddy (2019)
Yenjaichon et al. (2013a)	Wu, Hutton and Soleimani (2019)
Yenjaichon et al. (2013b)	Yan et al. (2019)
Yenjaichon et al. (2013c)	Rodriguez-Frias and Yang (2020)
Yenjaichon et al. (2013d)	Wang et al. (2020)
Vakamalla, Kumbhar, Gujjula and Mangadoddy (2014)	Zhou, Xu, Cao, Hu and Liu (2012)
Wang, Jia and Wang (2020)	Tan, Ren and Dong (2015)
Zhao et al. (2015)	

Velocity and flow profile measurement investigations	
Bolton, Hooper, Mann and Stitt (2004)	Zhao et al. (2015)
Toye et al. (2005)	Annamalai et al. (2016)
Wu, Li, Wang and Williams (2005)	Jin et al. (2016)
Henningsson, Östergren and Dejmek (2006)	Shi et al. (2016)
Lee and Bennington (2007)	Tan, Wang and Dong (2016)
Razzak, Barghi and Zhu (2007a)	Liu et al. (2017)
Razzak, Barghi and Zhu (2007b)	Low et al. (2017)
Ruzinsky and Bennington (2007)	Ren, Kowalski and Rodgers (2017)
Vijayan, Schlaberg and Wang (2007)	Wang and Yang (2017)
Giguère, Fradette, Mignon and Tanguy (2008a)	Zhang et al. (2017)
Razzak et al. (2008)	Shen et al. (2018)
Giguère, Fradette, Mignon and Tanguy. (2009)	Vauhkonen, Hänninen, Jauhiainen and Lehtikangas (2019)
Razzak et al. (2009)	Wang, Li, Wang, Qin and Liu (2019)
Razzak, Zhu and Barghi (2009a,b)	Faraj and Wang (2012)
Han, Yang, Jin and Zhu (2010)	Wu et al. (2005)
Lee and Bennington (2010)	Liu, Wang, Huang and Li (2007)
Parvareh, Rahimi, Alizadehdakhel and Alsairafi (2010)	Dong, Xu, Zhang and Ren (2012)
Faraj and Wang (2012)	Tan et al. (2016)
Hamood-ur-Rehman et al. (2012)	Wang and Yang (2017)
Sharifi and Young (2012a,b)	Wang et al. (2020)
Sharifi and Young (2013a,b,c)	Sun, Cui, Yue and Wang (2016)
Huang, Ding, Li and Wu (2015)	Wang et al. (2019)
	Wu et al. (2019)

settling or where the bubbles are located within the flow is essential process information which can be attained by the application of ERT.

This group of applications was introduced with the investigation of application of ERT to phase distribution visualizations inside various types of pipes. Solids distribution in a slurry flow inside pipes (Norman & Bonnecaze, 2005) and generally phase distribution in swirling two-phase flows (Stevenson et al., 2006; Wang et al., 2003; Zhang et al., 2018) were initially investigated. As a particular example of this category of ERT application purposes, the displacement of yogurt by water in a pipe setting or in other words, the water and yogurt phase distributions was visualized using ERT (Regner et al., 2007).

Later research broadened and application of ERT to multiphase distributions inside pipes (Faraj & Wang, 2012; Li & Yang, 2009) was also examined. Phase distribution inside other types of vessels was later commenced such as phase distribution inside a gas−liquid−solid circulating fluidized bed (Razzak et al., 2010a, 2010b). Further progress enabled visualization of fluid−particle and particle−particle interactions in a liquid−solid fluidized bed (Zbib et al., 2018a, 2018b).

Visualization of liquor flow and phase dispersion in pulp digesters using ERT has attracted the attention of many researchers in which ERT presented great potential with providing accurate results in comparison with CFD and other mathematical models or other visualization techniques (Alaqqad et al., 2012; Vlaev & Bennington, 2004; Yenjaichon et al., 2013a, 2013b, 2013c, 2013d, 2013e).

15.2.2.2 Velocity and flow profile measurement investigations

Flow rate and velocity of moving media have been measured for decades using various types of flow meters. With increasing complexity of processes and the presence of multiphase media, quantitative analysis of the velocity distribution and flow profiles of individual phases are essential knowledge which are otherwise unattainable. ERT pixel by pixel reconstructed conductivity measurements have the ability of providing the essential flow process knowledge, phase propagation velocity, by cross-correlation between the data obtained from a pair of planes of ERT, one upstream, and one downstream.

Utilizing ERT for velocity profile and flow regime recognition of various types of media flowing in pipes is the main area of research in this group. This application was investigated for slurries or gas−liquid flows in pipes (Giguère et al., 2008, 2009; Parvareh et al., 2010; Wang et al., 2012; Wu et al., 2005), for yogurt (Henningsson et al., 2006; Kabengele, Sutherland & Fester, 2017), for milk (Sharifi & Young, 2012, 2013), for oil/water two-phase flows (Huang et al., 2015; Jia et al., 2017; Liu et al., 2017; Shi, Wang, & Shen, 2016; Tan et al., 2016; Wang &Yang, 2017), for shampoo (Ren et al., 2017), and for three-phase gas−liquid−solid flow (Rodriguez-Frias & Yang, 2020; Vadlakonda & Mangadoddy, 2018; Wang et al., 2018, 2020).

ERT has also been utilized for flow distributions and velocity measurements in various other types of process vessels such as a radial flow fixed bed reactor, in which ERT measurements were also used for validation of CFD (Bolton et al., 2004), an external loop airlift reactor (Hamood-ur-Rehman et al., 2012; Han et al., 2010), and an airlift inner loop reactor (Jin et al., 2016). Flow patterns inside a bubble column were also investigated through application of ERT (Singh et al., 2017; Vijayan et al., 2007; Yan et al., 2019; Zhang et al., 2017).

Utilizing ERT for the measurement of velocity distributions inside various types of fluidized beds is another common area of research. Liquid–solid circulating fluidized beds (Razzak et al., 2007, 2009; Zbib et al., 2018) and gas–liquid–solid three-phase circulating fluidized beds (Razzak et al., 2007, 2008, 2009; Razzak et al., 2009a, 2009b; Xu et al., 2009) have been investigated in detail.

Similar to the application of ERT to qualitative visualization of flow and phase distributions, quantitative measurements of flow patterns inside model digesters, in which a fluid is flowing through a stationary bed of uncooked chips (Lee & Bennington, 2007, 2010; Ruzinsky & Bennington, 2007; Vlaev & Bennington, 2005, 2006), are also a common area of research attracting the attention of many research groups.

15.2.3 Phase holdup investigations

ERT applications to phase holdup investigations are a broad group of research in itself, although similar to the previous ERT application purpose, phase distributions. Understanding phase holdup or distribution, whether it's gas, solid, or liquid holdup, inside a process vessel is critical information essential for process monitoring and control. ERT as a noninvasive technique based on conductivity measurements of the continuous phase can also provide color-coded cross-sectional views of phases at high frequencies. The local conductivity measured by a number of electrodes located at the periphery of the plane is then further converted into a local phase concentration distribution based on Maxwell's relation. Since ERT is applicable only to conductive phase(s), e.g., liquid phase, another technique such as optical fiber probes needs to be utilized at the same time to quantify all three phases (Razzak et al., 2007, 2008).

Applications of ERT to phase holdup investigations are divided into three subcategories as can be seen in Table 15.3. Gas holdup, which is also referred to as void fraction, is the most common area of research in this group, which also covers bubble behavior (size, rise velocity, and dispersion) investigations. The second subcategory, solids holdup, is also a common area of research. Liquid holdup has had much less focused interest, as liquid is usually the continuous main phase.

ERT has been utilized for evaluation of phase holdup on a range of media and in various types of vessels. Gas–liquid mixing in a stirred vessel has been broadly investigated for gas holdup using the application of ERT (Busciglio et al., 2017; Forte et al., 2019; Hashemi et al., 2016a, 2016b, 2018; Kazemzadeh et al., 2018; Khalili et al., 2018; Montante & Paglianti, 2015; Sardeshpande et al., 2017). Similarly, three-phase mixing in a stirred vessel has been researched for phase holdup evaluations (Abdullah et al., 2011; Forte et al., 2019). Water holdup has been measured using the application of ERT in oil–water two-phase flow (Wang et al., 2016) and also liquid holdup distribution has been measured in a trickle bed reactor (Singh et al., 2019).

Gas holdup or void fraction in two-phase flow has been investigated generally (Dong et al., 2005), and also with specific focus such as gas dispersion in pulp fiber suspension flows (Yenjaichon et al., 2013a, 2013b, 2013c), or gas holdup in non-Newtonian fluids with the presence of a mixer (Jegatheeswaran & Ein-Mozaffari, 2020).

Table 15.3 Applications of ERT to phase holdup investigations.

Gas holdup/void fraction/bubble behavior	
Miettinen, Laakhonen and Aittamaa (2003)	Rakesh, Kumar Reddy and Narasimha (2014)
Abdullah, Dave, Nguyen, Cooper and Adesina (2011)	Babaei, Bonakdarpour and Ein-Mozaffari (2015a)
Montante and Paglianti (2015)	Babaei, Bonakdarpour and Ein-Mozaffari (2015b)
Hashemi et al. (2016)	Ridzuan Aw et al. (2015)
Abdullah and Adesina (2017)	Dong, Xu, Qiao, Xu and Xu (2005)
Busciglio et al. (2017)	Wang et al. (2012)
Sardeshpande, Gupta and Ranade (2017)	Hernandez-Alvarado et al. (2018)
Hashemi et al. (2018)	Fang et al. (2019)
Jamshed et al. (2018)	Jin et al. (2007)
Fransolet, Crine, Marchot and Toye (2005)	Hamood-ur-Rehman et al. (2013)
Jin, Wang and Williams (2006)	Jin, Lian, Qin, Yang and He (2013)
Razzak et al. (2007)	Qin, Jin and Yang (2013)
Vijayan et al. (2007)	Khalili et al. (2017)
Razzak et al. (2008)	Hashemi et al. (2018)
Han and Jin (2009)	Khalili et al. (2018)
Razzak et al. (2009)	Bobade, Evans and Eshtiaghi (2019)
(Razzak et al. (2009a, 2009b)	Forte et al. (2019)
(Razzak et al. (2009a, 2009b)	Hamzah et al. (2019)
Han et al. (2010)	Jegatheeswaran, Kazemzadeh and Ein-Mozaffari (2019)
Ishkintana and Bennington (2010)	Suard et al. (2019)
Jin, Yang, He, Wang and Williams (2010)	Jegatheeswaran and Ein-Mozaffari (2020)
Razzak et al. (2010a, 2010b)	Vadlakonda and Mangadoddy (2020)
Razzak et al. (2010a, 2010b)	Yang, Zou, Ma, Wang and Xu (2020)
Razzak et al. (2010)	Dong et al. (2005)
Yenjaichon et al. (2013a)	Liu, Wang and Li (2018)
Yenjaichon et al. (2013b)	Yang et al. (2020)
Yenjaichon et al. (2013c)	Toye et al. (2005)
Hashemi et al. (2016)	Jin, Wang and Williams (2007)
Jin et al. (2016)	Wang and Cilliers (1999)
Vadlakonda and Mangadoddy (2017)	Aw et al. (2014)
Zhang et al. (2017)	Ridzuan Aw et al. (2015)
Vadlakonda and Mangadoddy (2018)	Liu et al. (2018)
Vadlakonda et al. (2019)	
Yan et al. (2019)	
Aw et al. (2014)	
Jin et al. (2014)	
Solids holdup	
Razzak et al. (2007a)	Razzak et al. (2009a, 2009b)
Razzak et al. (2007b)	Han et al. (2010)
Razzak et al. (2008)	Razzak et al. (2010a, 2010b)
Han and Jin (2009)	Razzak et al. (2010a, 2010b)
Razzak et al. (2009a)	Razzak et al. (2010)
Razzak et al. (2009b)	Vadlakonda and Mangadoddy (2018)
Razzak et al. (2009a, 2009b)	Vadlakonda and Mangadoddy (2020)
Liquid holdup	
Razzak et al. (2010a, 2010b)	Singh, Jain and Buwa (2019)
Razzak et al. (2010a, 2010b)	Wang, Tan and Dong (2016)

Bubble columns are a common vessel used in process lines in which various phases come into contact. Measurement of phase holdup in these vessels by applying ERT has been a popular area of research, such as measurement of gas holdup in bubble columns in general (Babaei et al., 2015; Hernandez-Alvarado et al., 2018; Jin et al., 2013, 2014, 2010; Singh et al., 2017; Zhang et al., 2017), and in situations with non-Newtonian fluids as the liquid phase (Fransolet et al., 2005; Ishkintana & Bennington, 2010).

ERT has been applied for the purpose of phase holdup measurements in various other types of process vessels as well, such as in liquid–solid circulating fluidized beds (Razzak et al., 2007, 2009; Razzak et al., 2009a, 2009b), in gas–liquid–solid circulating fluidized beds (Razzak et al., 2008, 2009, 2010a, 2010b; Razzak et al., 2010), in three-phase external loop airlift reactors (Hamood-ur-Rehman et al., 2013; Han & Jin, 2009; Han et al., 2010), and in airlift inner-loop columns (Jin et al., 2016). Air-core size analysis in hydrocyclones has also been performed using the application of ERT (Rakesh et al., 2014; Vakamalla et al., 2014).

Application of ERT, other than just quantitative measurement of phase holdup, also provides the ability to evaluate bubble flow behavior such as rise velocity of gas bubble swarms and mean bubble size. Application of ERT for this purpose was evaluated in bubble columns (Bobade et al., 2019; Jin et al., 2007; Qin et al., 2013). Also the effect of sparger geometry on gas bubble flow behavior has been evaluated (Jin et al., 2006; Vijayan et al., 2007). Breakup, coalescence, and migration regularity of bubbles under gas–liquid swirling flow in a gas–liquid cylindrical cyclone have been another new area of research (Yang et al., 2020).

15.2.4 Solid particles suspension, dissolution, and precipitation

The ability of ERT to provide knowledge from inside vessels comes from injecting current to sensors located around the periphery of a process vessel and measuring the resulting voltage between all electrode pairs. This data is then reconstructed to the electrical conductivity distribution within the sensing domain. Therefore, conductivity variation within the sensing domain is critical in providing the required interior knowledge. Such conductivity variation is commonly seen in process vessels in which solid particles are in contact with a conductive liquid phase, and the solid particles are either in suspension in the liquid phase (slurry flows) or in the process of dissolution or precipitation (crystallization). Accurate information of the state of such processes which is critical for the purpose of process monitoring and control can be acquired by ERT utilization and has been a popular area of research in literature.

This area of research was initiated by detecting and quantifying nonuniformities in solid suspensions in mixing vessels in three dimensions using ERT (Mann et al., 1997) and monitoring the extent of solid deposition during slurry conveying in a pipe (Fangary, Williams, Neil, Bond, & Faulks, 1998). Later this area of research was extended by utilizing other techniques in combination with ERT such as dynamic light scattering (DLS) and ultrasound spectroscopy (USS) in online monitoring of nanoparticle suspensions (Wang et al., 2009), and MRI in monitoring partial suspension of various glass and nylon particles (Stevenson et al., 2010). ERT as a nonintrusive/noninvasive visualization technique was also utilized to compute the degree of solid suspension within a mixing vessel to assess the mixing performance of a Maxblend impeller (Mishra & Ein-Mozaffari, 2016).

The liquid mixing time and suspension in solid–liquid systems is difficult to determine by quantitative and nonintrusive methods, particularly at high solid loadings, but this issue has been resolved with the application of ERT. The suspension of nondilute concentrations of spherical particles in viscous liquids has been investigated using ERT (Lassaigne et al., 2016). Various methods for measuring mixing time have been considered for the postprocessing of raw conductivity data collected by ERT in a baffled stirred tank in the presence of high solid loadings (Paglianti et al., 2017). ERT has also been applied to analyze the suspension of large solid particles and extent of homogeneity in highly concentrated slurries (Kazemzadeh et al., 2019). The mathematical models of a turbulent solid–liquid stirred vessel with high solids loading have also been compared to the predicted solids concentration profiles with experimental data measured by ERT (Maluta, Paglianti, & Montante, 2019).

The influence of reactor scale, impeller tip speed, and specific power on the overall homogeneity (multiphasic mixing) of similar stirred tanks has been compared using two-dimensional data collected from ERT for particle suspensions (Harrison et al., 2020). Also the effect of impeller type on mixing of highly concentrated slurries of large particles has also been evaluated (Kazemzadeh et al., 2020a, 2020b).

Food process industries are more progressively using nonsoluble powders with high protein content in their recipes. These nonsoluble powders are complicated to hydrate and it is always considered essential to know when their hydration is finalized. ERT has been applied to monitor the rates of dispersion, dissolution, and settling of such powders (Shirhatti, Wang, & Williams, 2005). The dissolution of salt particles in turbulent stirred vessels has also been investigated using the application of ERT (Montante et al., 2019).

Investigating the reaction process of precipitation using ERT was commenced for various media such as barium sulfate precipitation (Kagoshima & Mann, 2005; Stanley, 2006; Stanley et al., 2005), heavy metal precipitation (Bolton et al., 2006, 2007), and aluminum hydroxide precipitation (Edwards, Axon, Barigou, & Stitt, 2009). Recently, a 2-D ERT system with a low conductivity sensor unit was designed and tested with demineralized water, tap water, and industrial food grade saturated sucrose solution. Nonconducting phantom and sugar crystals were observed within the saturated sucrose solution using a Bayesian reconstruction algorithm (Rao, Aghajanian, Koiranen, Wajman, & Jackowska-Strumiłło , 2020).

Other research with similar purposes in this area has been evaluation of flow patterns for coarse particles transported in non-Newtonian carrier fluids using ERT (Kabengele et al., 2017), measurements of local solids concentration distribution at steady-state conditions and of liquid homogenization in the presence of dispersed particles at transient conditions (Paglianti et al., 2017), and using ERT for monitoring settling slurry pipe flow and location of bed interface (Adler, Sutherland, & Kotze, 2019).

15.2.5 Monitoring separation and phase boundaries

ERT has been successfully applied to monitor a radioactive waste separation process by estimating the boundary between two distinct waste streams in a rotating separator as well as providing the conductivity value of each stream (Park, Moon, Lee, & Kim, 2008). ERT has also been applied to assist in the investigation of inlet bubble size distribution, gas phase concentration, and swirling intensity in the swirling flow field of a vane-type separator and the results have been verified with numerical studies (Liu et al., 2018). ERT has been used to estimate the separation efficiency and mean mixing index of a liquid–liquid cyclone reactor under different conditions and again the results have been compared with numerical studies (Duan et al., 2019).

Flow of two immiscible stratified liquids in a pipe is observed in many industrial applications. Estimating the interfacial boundary between the immiscible liquids inside the pipeline can provide valuable information about volume fraction which is essential in monitoring the flow process and can be provided by the application of ERT (Khambampati, Hong, Kim, & Kim, 2013; Khambampati, Kim, Lee, & Kim, 2016). ERT has also been applied in the detection of the location of the interface in a settling slurry in a pipe (Adler et al., 2019).

Other research studies with similar objectives in this area are application of ERT for the determination of the phase inversion phenomenon, from water–in–oil to oil–in–water (Kaminoyama, Kato, Misumi, & Nishi, 2010), applying ERT for early characterization of physical stability in liquid compositions, and to reveal behavior which is indicative of instability prior to visually discernible effects (Kowalski et al., 2010). ERT, when utilized together with electrical capacitance tomography (ECT), has also been presented as a liquid level detection device which can detect many different layers (Liu et al., 2018).

15.2.6 Concentration monitoring

The application of ERT to concentration monitoring was initiated by quantitative measurement of solids concentration in multiphase flows (Giguère et al., 2008). Later various studies were conducted in which ERT was applied for quantitative auditing and attaining informative data such as total solids content of various milk solutions in the processing of milk to produce milk powder (Sharifi & Young, 2010, 2011, 2012, 2013). A high-performance dual-plane ERT system was also used for the measurement of mean local solids concentration and solids volumetric concentration profile of a settling slurry in a pipeline (Faraj & Wang, 2012).

15.2.7 Cleaning-in-place

In a broad range of processes such as multifunctional food and detergent production process lines, precise detection of ending point of the cleaning process for the previous product is vital to guarantee product integrity. Monitoring and controlling the CIP process and detecting the end-point is a novel, promising and very intriguing purpose for the application of ERT which is being investigated in recent years.

Monitoring CIP of multiphase formulated products by ERT (Uppal et al., 2014), monitoring the removal of a non-Newtonian soil (shampoo) by water from a CIP circuit containing different pipe geometries using ERT (Hou, Martin, Uppal, & Kowalski, 2016), and using ERT with dynamic references for monitoring CIP (Wang & Yin, 2016) are some of the research studies conducted in this area. In order to improve quality control and avoid contamination, inline measurement of biofilms could deliver an instrumental technology for water, food, and bioprocessing industries. For this purpose, ERT has been applied to detect the removal of biofilms in a pipe setting (Díaz De Rienzo, Hou, & Martin, 2018). The application of ERT for visualization of surface deposits during a CIP process, which is a common problem in various industries, has also been evaluated. For this purpose, a pilot-scale cleaning rig with a purposely designed ERT test section was used for monitoring the prerinse CIP stage of cleaning milk paste adhering to a test section (Ren et al., 2019).

15.2.8 Malfunction detection

As the application of ERT to any process vessel provides remote visual insight to the interior of the vessel, any variation from the required and desired state can be easily detected. Therefore, detection of any sort of malfunction or displacement can be made possible using ERT.

This purpose was initially investigated by utilizing ERT in a stirred vessel for detecting (1) a misplaced gas sparger, (2) inadvertent solids accumulation, and (3) a displaced feed point behind a baffle (Holden et al., 1999). Later, ERT was applied to a solid–liquid filtration process and provided the ability to detect movement of the liquid level, any tilt of the filter assembly, and any pathological behavior due to malfunction or accidental displacement of the filter support plate. Moreover, illustrative distortions of a filter cake formed by artificial surface depressions could also be readily observed (Vlaev, Wang, Dyakowski, Mann, & Grieve, 2000). ERT has also been applied in milk holding tanks for the purpose of detecting any form of nonhomogeneity, adulteration, cream separation, aeration, and presence of object/powder lump which are parameters requiring control in the current milk processing industry (Sharifi & Young, 2011; Sharifi, Yu, & Young, 2014). The capability of ERT in detecting faults or malfunctions can also be seen across other areas of research such as damage detection in concrete (Zhao et al., 2016).

15.2.9 Process control

Being able to utilize the realtime multidimensional data provided by the application of ERT to process vessels for the purpose of process control is the ultimate goal of such research. A few researchers have been able to make this possible by applying process tomography data for multivariate statistical process control (MSPC) (Boonkhao, Li, Wang, Tweedie, & Primrose, 2011). Multiple on-line sensors including ERT, DLS, and USS have been utilized for real-time characterization of process operations processing emulsions and nanoparticle slurries and developing MSPC strategies (Li et al., 2009).

Later Principal Component Analysis has been applied for the purpose of reducing the high dimensionality of the ERT-produced data from various situations to lower dimensions holding most of the information. The reduced set of information is then used for the detection of whole and skim milk inhomogeneity, aeration, and external object detection in milk holding tanks (Sharifi et al., 2014).

15.3 Conclusions

ERT has the ability to deliver real-time three-dimensional conductivity distribution measurements from inside a given process plant. Such information can provide time evolving multidimensional information from the process vessel which often improves essential process understanding while advancing the design and function of the process equipment (Stanley & Bolton, 2008). Due to these abilities, ERT is attaining progressively more recognition and therefore developing into an established measurement system for examination and observation of pilot plant vessels.

ERT has been shown through research to be a successful measurement tool for the evaluation of a variety of unit operations, numerous process media, and for various purposes in the Chemical Engineering world. It has been applied on process vessels such as mixing/holding tanks, various pipe geometries, a variety of reactors, and fluidized beds, and on several media such as water, saline, pseudoplastic fluids, dairy products, and other aqueous media. It has been demonstrated that ERT measurements can be useful in the qualitative monitoring and quantitative computation of various crucial process parameters such as mixing time/indexes, flow regimes, flow rates, phase holdups, solids suspension, dissolution and precipitation, phase separation and boundaries, concentrations, CIP, and detection of pathologies. The use of other measurement techniques such as USS, Pressure Transducers, Optical Fiber probes, and ECT, in conjunction with ERT, has enhanced its capabilities (Sharifi & Young, 2013a,b,c).

Due to the numerous parameters of interest in each process sector in the endlessly growing industry of Chemical Engineering, the wide variety of process vessels and mediums available, and the fact that there is still scope for further improvement through advances in sensor design, measurement electronics and protocol, computing hardware reconstruction algorithms, and also through engagement of other measurement techniques, ERT's capabilities remain to be explored in the future.

References

Abdullah, B., & Adesina, A. A. (2017). Evaluation of gas–liquid mass transfer in gas-induced stirred tank reactor using electrical resistance tomography. *Journal of Chemical Technology and Biotechnology, 92*(8), 2123–2133.

Abdullah, B., Dave, C., Nguyen, T. H., Cooper, C. G., & Adesina, A. A. (2011). Electrical resistance tomography-assisted analysis of dispersed phase hold-up in a gas-inducing mechanically stirred vessel. *Chemical Engineering Science, 66*(22), 5648–5662.

Adler, A., Sutherland, A., & Kotze, R. (2019). *Location of bed interface in settling slurry pipe flow using ert.*

Alaqqad, M., Bennington, C. P. J., & Martinez, D. M. (2012). An estimate of the axial dispersion during flow through a compressible wood-chip bed. *Canadian Journal of Chemical Engineering, 90*(6), 1602–1611.

Annamalai, G., Pirouzpanah, S., Gudigopuram, S. R., & Morrison, G. L. (2016). Characterization of flow homogeneity downstream of a slotted orifice plate in a two-phase flow using electrical resistance tomography. *Flow Measurement and Instrumentation, 50*, 209–215.

Aw, S. R., Rahim, R. A., Rahiman, M. H. F., Yunus, F. R. M., Fadzil, N. S., Zawahir, M. Z., et al. (2014). Application study on bubble detection in a metallic bubble column using electrical resistance tomography. *Jurnal Teknologi (Sciences and Engineering), 69*(8), 19–25.

Aw, S. R., Rahim, R. A., Rahiman, M. H. F., Yunus, F. R. M., & Goh, C. L. (2014). Electrical resistance tomography: A review of the application of conducting vessel walls. *Powder Technology, 254*, 256–264.

Babaei, R., Bonakdarpour, B., & Ein-Mozaffari, F. (2015a). Analysis of gas phase characteristics and mixing performance in an activated sludge bioreactor using electrical resistance tomography. *Chemical Engineering Journal, 279*, 874–884.

Babaei, R., Bonakdarpour, B., & Ein-Mozaffari, F. (2015b). The use of electrical resistance tomography for the characterization of gas holdup inside a bubble column bioreactor containing activated sludge. *Chemical Engineering Journal, 268*, 260–269.

Bhole, M. R., & Bennington, C. P. J. (2010). Performance of four axial flow impellers for agitation of pulp suspensions in a laboratory-scale cylindrical stock chest. *Industrial & Engineering Chemistry Research, 49*(9), 4444–4451.

Bhole, M. R., Hui, L. K., Gomez, C., Bennington, C. P. J., & Dumont, G. A. (2011). The effect of off-wall clearance of a side-entering impeller on the mixing of pulp suspensions in a cylindrical stock chest. *Canadian Journal of Chemical Engineering, 89*(5), 985–995.

Bobade, V., Evans, G., & Eshtiaghi, N. (2019). Bubble rise velocity and bubble size in thickened waste activated sludge: Utilising electrical resistance tomography (ERT). *Chemical Engineering Research and Design, 148*, 119–128.

Bolton, G. T., Bennett, M., Wang, M., Qiu, C., Wright, M., Primrose, K. M., et al. (2007). Development of an electrical tomographic system for operation in a remote, acidic and radioactive environment. *Chemical Engineering Journal, 130*(2–3), 165–169.

Bolton, G. T., Bennett, M., Wang, M., Qiu, C., Wright, M., & Rhodes, D. (2006). *On the development of an electrical tomographic system for monitoring the performance of a heavy metal precipitation step during nuclear fuel reprocessing.*

Bolton, G. T., Hooper, C. W., Mann, R., & Stitt, E. H. (2004). Flow distribution and velocity measurement in a radial flow fixed bed reactor using electrical resistance tomography. *Chemical Engineering Science, 59*(10), 1989–1997.

Boonkhao, B., Li, R. F., Wang, X. Z., Tweedie, R. J., & Primrose, K. (2011). Making use of process tomography data for multivariate statistical process control. *AIChE Journal, 57*(9), 2360–2368.

Busciglio, A., Opletal, M., Moucha, T., Montante, G., & Paglianti, A. (2017). Measurement of gas hold-up distribution in stirred vessels equipped with pitched blade turbines by means of electrical resistance tomography. *Chemical Engineering Transactions, 57*, 1273–1278.

Carletti, C., Montante, G., De Blasio, C., & Paglianti, A. (2016). Liquid mixing dynamics in slurry stirred tanks based on electrical resistance tomography. *Chemical Engineering Science, 152*, 478–487.

Díaz De Rienzo, M. A., Hou, R., & Martin, P. J. (2018). Use of electrical resistance tomography (ERT) for the detection of biofilm disruption mediated by biosurfactants. *Food and Bioproducts Processing, 110*, 1–5.

Dong, F., Xu, Y., Qiao, X., Xu, L., & Xu, L. (2005). Void fraction measurement for two-phase flow using electrical resistance tomography. *Canadian Journal of Chemical Engineering, 83*(1), 19–23.

Dong, F., Xu, C., Zhang, Z., & Ren, S. (2012). Design of parallel electrical resistance tomography system for measuring multiphase flow. *Chinese Journal of Chemical Engineering, 20*(2), 368–379.

Duan, S., Meng, X., Zhang, R., Liu, H., Xu, J., Du, W., et al. (2019). Experimental and computational investigation of mixing and separation performance in a liquid-liquid cyclone reactor. *Industrial & Engineering Chemistry Research, 58*(51), 23317–23329.

Edwards, I., Axon, S. A., Barigou, M., & Stitt, E. H. (2009). Combined use of PEPT and ERT in the study of aluminum hydroxide precipitation. *Industrial & Engineering Chemistry Research, 48*(2), 1019–1028.

Fangary, Y. S., Williams, R. A., Neil, W. A., Bond, J., & Faulks, I. (1998). Application of electrical resistance tomography to detect deposition in hydraulic conveying systems. *Powder Technology, 95*(1), 61–66.

Fang, L., Wang, P., Zeng, Q., Li, M., Li, X., Wang, M., et al. (2019). Measurement of interphase forces based on dual-modality ERT/DP sensor in horizontal two-phase flow gas-water. *Measurement: Journal of the International Measurement Confederation, 136*, 703–717.

Fang, Z., Zhaolu, H., Li, J., Tang, Z., & Xu, C. (2019). Experimental investigation on suppression of microbubbles to slug flow in vertical pipeline. *Dongnan Daxue Xuebao (Ziran Kexue Ban)/Journal of Southeast University (Natural Science Edition), 49*(3), 527–534.

Faraj, Y., & Wang, M. (2012). *ERT investigation on horizontal and vertical counter-gravity slurry flow in pipelines*.

Forte, G., Albano, A., Simmons, M. J. H., Stitt, H. E., Brunazzi, E., & Alberini, F. (2019). Assessing blending of non-Newtonian fluids in static mixers by planar laser-induced fluorescence and electrical resistance tomography. *Chemical Engineering & Technology, 42*(8), 1602–1610.

Forte, G., Alberini, F., Simmons, M. J. H., & Stitt, E. H. (2019). Measuring gas hold-up in gas–liquid/gas–solid–liquid stirred tanks with an electrical resistance tomography linear probe. *AIChE Journal, 65*(6).

Fransolet, E., Crine, M., Marchot, P., & Toye, D. (2005). Analysis of gas holdup in bubble columns with non-Newtonian fluid using electrical resistance tomography and dynamic gas disengagement technique. *Chemical Engineering Science, 60*(22), 6118–6123.

Giguère, R., Fradette, L., Mignon, D., & Tanguy, P. A. (2008a). Characterization of slurry flow regime transitions by ERT. *Chemical Engineering Research and Design, 86*(9), 989–996.

Giguère, R., Fradette, L., Mignon, D., & Tanguy, P. A. (2008b). ERT algorithms for quantitative concentration measurement of multiphase flows. *Chemical Engineering Journal, 141*(1–3), 305–317.

Giguère, R., Fradette, L., Mignon, D., & Tanguy, P. A. (2009). Analysis of slurry flow regimes downstream of a pipe bend. *Chemical Engineering Research and Design, 87*(7), 943–950.

Gumery, F., Ein-Mozaffari, F., & Dahman, Y. (2011). Macromixing hydrodynamic study in draft-tube airlift reactors using electrical resistance tomography. *Bioprocess and Biosystems Engineering, 34*(2), 135–144.

Hamood-ur-Rehman, M., Dahman, Y., & Ein-Mozaffari, F. (2012). Investigation of mixing characteristics in a packed-bed external loop airlift bioreactor using tomography images. *Chemical Engineering Journal, 213*, 50–61.

Hamood-ur-Rehman, M., Ein-Mozaffari, F., & Dahman, Y. (2013). Dynamic and local gas holdup studies in external loop recirculating airlift reactor with two rolls of fiberglass packing using electrical resistance tomography. *Journal of Chemical Technology and Biotechnology, 88*(5), 887–896.

Hamzah, A. A., Ruzairi, A. R., Takriff, M. S., Mohamad, E. J., Pusppanathan, J., Azman, I. N., et al. (2019). Application of electrical resistance tomography in an Oscillatory Baffled column for Gas-Liquid two-phase flow. *International Journal of Integrated Engineering, 11*(6), 119–125.

Han, Y. H., & Jin, H. B. (2009). Experimental study on phase holdup in three-phase external loop airlift reactors using electrical resistance tomography. *Guocheng Gongcheng Xuebao/The Chinese Journal of Process Engineering, 9*(3), 431–436.

Han, Y. H., Yang, S. H., Jin, H. B., & Zhu, J. H. (2010). Distribution characteristics of phase holdups and flow structure of gas-liquid in a three-phase external loop airlift reactor. *Guocheng Gongcheng Xuebao/The Chinese Journal of Process Engineering, 10*(5), 862–867.

Harrison, S. T. L., Kotsiopoulos, A., Stevenson, R., & Cilliers, J. J. (2020). Mixing indices allow scale-up of stirred tank slurry reactor conditions for equivalent homogeneity. *Chemical Engineering Research and Design, 153*, 865–874.

Harrison, S. T. L., Stevenson, R., & Cilliers, J. J. (2012). Assessing solids concentration homogeneity in Rushton-agitated slurry reactors using electrical resistance tomography (ERT). *Chemical Engineering Science, 71*, 392–399.

Hashemi, N., Ein-Mozaffari, F., Upreti, S. R., & Hwang, D. K. (2016a). Analysis of mixing in an aerated reactor equipped with the coaxial mixer through electrical resistance tomography and response surface method. *Chemical Engineering Research and Design, 109*, 734–752.

Hashemi, N., Ein-Mozaffari, F., Upreti, S. R., & Hwang, D. K. (2016b). Analysis of power consumption and gas holdup distribution for an aerated reactor equipped with a coaxial mixer: Novel correlations for the gas flow number and gassed power. *Chemical Engineering Science, 151*, 25–35.

Hashemi, N., Ein-Mozaffari, F., Upreti, S. R., & Hwang, D. K. (2016c). Experimental investigation of the bubble behavior in an aerated coaxial mixing vessel through electrical resistance tomography (ERT). *Chemical Engineering Journal, 289*, 402–412.

Hashemi, N., Ein-Mozaffari, F., Upreti, S. R., & Hwang, D. K. (2018). Hydrodynamic characteristics of an aerated coaxial mixing vessel equipped with a pitched blade turbine and an anchor. *Journal of Chemical Technology and Biotechnology, 93*(2), 392–405.

Henningsson, M., Östergren, K., & Dejmek, P. (2006). Plug flow of yoghurt in piping as determined by cross-correlated dual-plane electrical resistance tomography. *Journal of Food Engineering, 76*(2), 163–168.

Hernandez-Alvarado, F., Kleinbart, S., Kalaga, D. V., Banerjee, S., Joshi, J. B., & Kawaji, M. (2018). Comparison of void fraction measurements using different techniques in two-phase flow bubble column reactors. *International Journal of Multiphase Flow, 102*, 119–129.

Holden, P. J., Wang, M., Mann, R., Dickin, F. J., & Edwards, R. B. (1998). Imaging stirred-vessel macromixing using electrical resistance tomography. *AIChE Journal, 44*(4), 780–790.

Holden, P. J., Wang, M., Mann, R., Dickin, F. J., & Edwards, R. B. (1999). On detecting mixing pathologies inside a stirred vessel using electrical resistance tomography. *Chemical Engineering Research and Design, 77*(8), 709–712.

Hosseini, S., Patel, D., Ein-Mozaffari, F., & Mehrvar, M. (2010). Study of solid-liquid mixing in agitated tanks through electrical resistance tomography. *Chemical Engineering Science, 65*(4), 1374–1384.

Hou, R., Martin, P. J., Uppal, H. J., & Kowalski, A. J. (2016). An investigation on using electrical resistance tomography (ERT) to monitor the removal of a non-Newtonian soil by water from a cleaning-in-place (CIP) circuit containing different pipe geometries. *Chemical Engineering Research and Design, 111*, 332–341.

Huang, Y. T., Ding, J., Li, H., & Wu, Y. X. (2015). Study of oil/water two-phase flow pattern characteristics using electrical resistance tomography technology. *Shuidonglixue Yanjiu yu Jinzhan/Chinese Journal of Hydrodynamics Ser. A, 30*(1), 70–74.

Hui, L. K., Bennington, C. P. J., & Dumont, G. A. (2009). Cavern formation in pulp suspensions using side-entering axial-flow impellers. *Chemical Engineering Science, 64*(3), 509–519.

Hume, C. (2003). Tomography create a stir at Thorp. *The Chemical Engineer*, (750), 36.

Ishkintana, L. K., & Bennington, C. P. J. (2010). Gas holdup in pulp fibre suspensions: Gas voidage profiles in a batch-operated sparged tower. *Chemical Engineering Science, 65*(8), 2569–2578.

Jamshed, A., Cooke, M., Ren, Z., & Rodgers, T. L. (2018). Gas–liquid mixing in dual agitated vessels in the heterogeneous regime. *Chemical Engineering Research and Design, 133*, 55–69.

Jamshed, A., Cooke, M., & Rodgers, T. L. (2019). Effect of zoning on mixing and mass transfer in dual agitated gassed vessels. *Chemical Engineering Research and Design, 142*, 237–244.

Jegatheeswaran, S., & Ein-Mozaffari, F. (2020). Investigation of the detrimental effect of the rotational speed on gas holdup in non-Newtonian fluids with scaba-anchor coaxial mixer: A paradigm shift in gas-liquid mixing. *Chemical Engineering Journal, 383.*

Jegatheeswaran, S.,F., Ein, M., & Wu, J. (2017). Efficient mixing of yield-pseudoplastic fluids at low Reynolds numbers in the chaotic SMX static mixer. *Chemical Engineering Journal, 317*, 215–231.

Jegatheeswaran, S.,F., Ein, M., & Wu, J. (2018). Process intensification in a chaotic SMX static mixer to achieve an energy-efficient mixing operation of non-Newtonian fluids. *Chemical Engineering and Processing: Process Intensification, 124*, 1–10.

Jegatheeswaran, S., Kazemzadeh, A., & Ein-Mozaffari, F. (2019). Enhanced aeration efficiency in non-Newtonian fluids using coaxial mixers: High-solidity ratio central impeller with an anchor. *Chemical Engineering Journal, 378.*

Jia, J., Wang, H., & Millington, D. (2017). Electrical resistance tomography sensor for highly conductive oil-water two-phase flow measurement. *IEEE Sensors Journal, 17*(24), 8224–8233.

Jin, H., Lian, Y., Liu, X., Lin, J., Yang, S., & He, G. (2016). Hydrodynamic parameters in an airlift inner-loop column using electrical resistance tomography. *Zhongnan Daxue Xuebao (Ziran Kexue Ban)/Journal of Central South University (Science and Technology), 47*(11), 3935–3939.

Jin, H., Lian, Y., Qin, Y., Yang, S., & He, G. (2013). Distribution characteristics of holdups in a multi-stage bubble column using electrical resistance tomography. *Particuology, 11*(2), 225–231.

Jin, H., Lian, Y., Qin, L., Yang, S., He, G., & Guo, Z. (2014). Parameters measurement of hydrodynamics and CFD simulation in multi-stage bubble columns. *Canadian Journal of Chemical Engineering, 92*(8), 1444–1454.

Jin, H., Wang, M., & Williams, R. A. (2006). Effect of sparger geometry on gas bubble flow behaviors using electrical resistance tomography. *Chinese Journal of Chemical Engineering, 14*(1), 127–131.

Jin, H., Wang, M., & Williams, R. A. (2007). Analysis of bubble behaviors in bubble columns using electrical resistance tomography. *Chemical Engineering Journal, 130*(2–3), 179–185.

Jin, H., Yang, S., He, G., Wang, M., & Williams, R. A. (2010). The effect of gas-liquid counter-current operation on gas hold-up in bubble columns using electrical resistance tomography. *Journal of Chemical Technology and Biotechnology, 85*(9), 1278–1283.

Kabengele, K., Sutherland, A., & Fester, V. (2017). *Effect of coarse solids on non-Newtonian settling slurry flow regime transitions: Using pressure gradients and electrical resistance tomography.*

Kagoshima, M., & Mann, R. (2005). Interactions of precipitation and fluid mixing with model validation by electrical tomography. *Chemical Engineering Research and Design, 83*(7 A), 806–810.

Kaminoyama, M., Kato, K., Misumi, R., & Nishi, K. (2010). Measurements of the phase inversion phenomenon in a suspension polymerization reactor with an electrical resistance tomography system. *Journal of Chemical Engineering of Japan, 43*(1 Suppl. L), 52–55.

Kaminoyama, M., Taguchi, S., Misumi, R., & Nishi, K. (2005). Monitoring stability of reaction and dispersion states in a suspension polymerization reactor using electrical resistance tomography measurements. *Chemical Engineering Science, 60*(20), 5513–5518.

Kazemzadeh, A.,F., Ein, M., & Lohi, A. (2019). Mixing of highly concentrated slurries of large particles: Applications of electrical resistance tomography (ERT) and response surface methodology (RSM). *Chemical Engineering Research and Design, 143*, 226–240.

Kazemzadeh, A.,F., Ein, M., & Lohi, A. (2020a). Effect of impeller type on mixing of highly concentrated slurries of large particles. *Particuology, 50*, 88–99.

Kazemzadeh, A.,F., Ein, M., & Lohi, A. (2020b). Hydrodynamics of solid and liquid phases in a mixing tank containing high solid loading slurry of large particles via tomography and computational fluid dynamics. *Powder Technology, 360*, 635–648.

Kazemzadeh, A., Ein-Mozaffari, F., Lohi, A., & Pakzad, L. (2016a). Effect of the rheological properties on the mixing of Herschel-Bulkley fluids with coaxial mixers: Applications of tomography, CFD, and response surface methodology. *Canadian Journal of Chemical Engineering, 94*(12), 2394–2406.

Kazemzadeh, A., Ein-Mozaffari, F., Lohi, A., & Pakzad, L. (2016b). Investigation of hydrodynamic performances of coaxial mixers in agitation of yield-pseudoplasitc fluids: Single and double central impellers in combination with the anchor. *Chemical Engineering Journal, 294*, 417–430.

Kazemzadeh, A., Ein-Mozaffari, F., Lohi, A., & Pakzad, L. (2016c). A new perspective in the evaluation of the mixing of biopolymer solutions with different coaxial mixers comprising of two dispersing impellers and a wall scraping anchor. *Chemical Engineering Research and Design, 114*, 202–219.

Kazemzadeh, A., Ein-Mozaffari, F., Lohi, A., & Pakzad, L. (2017a). Effect of impeller spacing on the flow field of yield-pseudoplastic fluids generated by a coaxial mixing system composed of two central impellers and an anchor. *Chemical Engineering Communications, 204*(4), 453–466.

Kazemzadeh, A., Ein-Mozaffari, F., Lohi, A., & Pakzad, L. (2017b). Intensification of mixing of shear-thinning fluids possessing yield stress with the coaxial mixers composed of two different central impellers and an anchor. *Chemical Engineering and Processing - Process Intensification, 111*, 101–114.

Kazemzadeh, A., Elias, C., Tamer, M., & Ein-Mozaffari, F. (2018). Hydrodynamic performance of a single-use aerated stirred bioreactor in animal cell culture: Applications of tomography, dynamic gas disengagement (DGD), and CFD. *Bioprocess and Biosystems Engineering, 41*(5), 679–695.

Kennedy, S., Bhattacharjee, P. K., Eshtiaghi, N., & Parthasarathy, R. (2016). Cavern formation in non-Newtonian media in a vessel agitated by submerged recirculating liquid jets. *Industrial & Engineering Chemistry Research, 55*(40), 10771–10781.

Khajeh Naeeni, S., & Pakzad, L. (2019). Experimental and numerical investigation on mixing of dilute oil in water dispersions in a stirred tank. *Chemical Engineering Research and Design, 147*, 493–509.

Khalili, F., Jafari Nasr, M. R., Kazemzadeh, A., & Ein-Mozaffari, F. (2017). Hydrodynamic performance of the ASI impeller in an aerated bioreactor containing the biopolymer solution through tomography and CFD. *Chemical Engineering Research and Design, 125*, 190–203.

Khalili, F., Jafari Nasr, M. R., Kazemzadeh, A., & Ein-Mozaffari, F. (2018). Analysis of gas holdup and bubble behavior in a biopolymer solution inside a bioreactor using tomography and dynamic gas disengagement techniques. *Journal of Chemical Technology and Biotechnology, 93*(2), 340–349.

Khambampati, A. K., Hong, Y. J., Kim, K. Y., & Kim, S. (2013). A boundary element method to estimate the interfacial boundary of two immiscible stratified liquids using electrical resistance tomography. *Chemical Engineering Science, 95*, 161–173.

Khambampati, A. K., Kim, K. Y., Lee, Y. G., & Kim, S. (2016). Boundary element method to estimate the time-varying interfacial boundary in horizontal immiscible liquids flow using electrical resistance tomography. *Applied Mathematical Modelling, 40*(2), 1052–1068.

Kim, S., Nkaya, A. N., & Dyakowski, T. (2006). Measurement of mixing of two miscible liquids in a stirred vessel with electrical resistance tomography. *International Communications in Heat and Mass Transfer, 33*(9), 1088–1095.

Kowalski, A., Davidson, J., Flanagan, M., & York, T. (2010). Electrical resistance tomography for characterisation of physical stability in liquid compositions. *Chemical Engineering Journal, 158*(1), 69–77.

Lassaigne, M., Blais, B., Fradette, L., & Bertrand, F. (2016). Experimental investigation of the mixing of viscous liquids and non-dilute concentrations of particles in a stirred tank. *Chemical Engineering Research and Design, 108*, 55–68.

Lee, Q. F., & Bennington, C. P. J. (2007). Measuring flow velocity and uniformity in a model batch digester using electrical resistance tomography. *Canadian Journal of Chemical Engineering, 85*(1), 55–64.

Lee, Q. F., & Bennington, C. P. J. (2010). Liquor flow in a model kraft batch digester. *Chemical Engineering Journal, 158*(1), 51–60.

Li, R. F., Liu, L., Wang, X. Z., Tweedie, R., Primrose, K., Corbett, J., et al. (2009). Multivariate statistical control of emulsion and nanoparticle slurry processes based on process tomography, dynamic light scattering, and acoustic sensor data. *Computer Aided Chemical Engineering, 27*, 1317–1322.

Liu, Y., Deng, Y., Zhang, M., Yu, P., & Li, Y. (2017). Experimental measurement of oil-water two-phase flow by data fusion of electrical tomography sensors and venturi tube. *Measurement Science and Technology, 28*(9).

Liu, L., Li, R. F., Collins, S., Wang, X. Z., Tweedie, R., & Primrose, K. (2011). Ultrasound spectroscopy and electrical resistance tomography for online characterisation of concentrated emulsions in crossflow membrane emulsifications. *Powder Technology, 213*(1), 123–131.

Liu, T., Wang, B., Huang, Z., & Li, H. (2007). Measurement circuits and image reconstruction techniques of electrical resistance tomography system for two-phase flow measurement. *Huagong Xuebao/Journal of Chemical Industry and Engineering (China), 58*(4), 862–868.

Liu, S., Wang, H., & Li, Y. (2018). *Level measurement of separator by using electrical tomography sensor.*

Liu, S., Yang, L. L., Zhang, D., & Xu, J. Y. (2018). Separation characteristics of the gas and liquid phases in a vane-type swirling flow field. *International Journal of Multiphase Flow, 107*, 131–145.

Liu, S., Zhang, D., Yang, L. L., & Xu, J. Y. (2018). Breakup and coalescence regularity of non-dilute oil drops in a vane-type swirling flow field. *Chemical Engineering Research and Design, 129*, 35–54.

Li, L., & Wei, J. (1999). Three-dimensional image analysis of mixing in stirred vessels. *AIChE Journal, 45*(9), 1855–1865.

Li, Y., & Yang, W. (2009). *Measurement of multi-phase distribution using an integrated dual-modality sensor.*

Low, S. C., Allitt, D., Eshtiaghi, N., & Parthasarathy, R. (2018). Measuring active volume using electrical resistance tomography in a gas-sparged model anaerobic digester. *Chemical Engineering Research and Design, 130*, 42–51.

Low, S. C., Eshtiaghi, N., Shu, L., & Parthasarathy, R. (2017). Flow patterns in the mixing of sludge simulant with jet recirculation system. *Process Safety and Environmental Protection, 112*, 209–221.

Malik, D., & Pakzad, L. (2018). Experimental investigation on an aerated mixing vessel through electrical resistance tomography (ERT) and response surface methodology (RSM). *Chemical Engineering Research and Design, 129*, 327–343.

Maluta, F., Montante, G., & Paglianti, A. (2020). Analysis of immiscible liquid-liquid mixing in stirred tanks by Electrical Resistance Tomography. *Chemical Engineering Science, 227.*

Maluta, F., Paglianti, A., & Montante, G. (2019). RANS-based predictions of dense solid–liquid suspensions in turbulent stirred tanks. *Chemical Engineering Research and Design, 147*, 470–482.

Mann, R., Dickin, F. J., Wang, M., Dyakowski, T., Williams, R. A., Edwards, R. B., et al. (1997). Application of electrical resistance tomography to interrogate mixing processes at plant scale. *Chemical Engineering Science, 52*(13), 2087–2097.

Mann, R., Wang, M., Dickin, F. J., Dyakowski, T., Holden, P. J., Forrest, A. E., et al. (1996). Resistance tomography imaging of stirred vessel mixing at plant scale. *Chemical Engineering Science.*

Mann, R., Wang, M., Forrest, A. E., Holden, P. J., Dickin, F. J., Dyakowski, T., et al. (1999). Gas-liquid and miscible liquid mixing in a plant-scale vessel monitored using electrical resistance tomography. *Chemical Engineering Communications, 175*, 39–48.

Mann, R., Williams, R. A., Dyakowski, T., Dickin, F. J., & Edwards, R. B. (1997). Development of mixing models using electrical resistance tomography. *Chemical Engineering Science, 52*(13), 2073–2085.

Miettinen, T., Laakhonen, M., & Aittamaa, J. (2003). Comparison of various flow visualisation techniques in a gas-liquid mixed tank. *Computer Aided Chemical Engineering, 14*, 773–778.

Mihailova, O., Mothersdale, T., Rodgers, T., Ren, Z., Watson, S., Lister, V., et al. (2018). Optimisation of mixing performance of helical ribbon mixers for high throughput applications using computational fluid dynamics. *Chemical Engineering Research and Design, 132*, 942–953.

Mirshekari, F., & Pakzad, L. (2019). Mixing of oil in water through electrical resistance tomography and response surface methodology. *Chemical Engineering & Technology, 42*(5), 1101–1115.

Mishra, P., & Ein-Mozaffari, F. (2016). Using tomograms to assess the local solid concentrations in a slurry reactor equipped with a Maxblend impeller. *Powder Technology, 301*, 701–712.

Montante, G., Carletti, C., Maluta, F., & Paglianti, A. (2019). Solid dissolution and liquid mixing in turbulent stirred tanks. *Chemical Engineering & Technology, 42*(8), 1627–1634.

Montante, G., & Paglianti, A. (2015). Gas hold-up distribution and mixing time in gas-liquid stirred tanks. *Chemical Engineering Journal, 279*, 648–658.

Naghavi-Anaraki, Y., Turcotte, G., & Ein-Mozaffari, F. (2018). Characterization of mixing and yield stress of pretreated wheat straw slurries used for the production of biofuels through tomography technique. *Bioprocess and Biosystems Engineering, 41*(9), 1315–1328.

Norman, J. T., & Bonnecaze, R. T. (2005). Measurement of solids distribution in suspension flows using electrical resistance tomography. *Canadian Journal of Chemical Engineering, 83*(1), 24–36.

Paglianti, A., Carletti, C., Busciglio, A., & Montante, G. (2017). Solid distribution and mixing time in stirred tanks: The case of floating particles. *Canadian Journal of Chemical Engineering, 95*(9), 1789–1799.

Paglianti, A., Carletti, C., & Montante, G. (2017). Liquid mixing time in dense solid-liquid stirred tanks. *Chemical Engineering & Technology, 40*(5), 862–869.

Pakzad, L., Ein-Mozaffari, F., & Chan, P. (2008a). Measuring mixing time in the agitation of non-Newtonian fluids through electrical resistance tomography. *Chemical Engineering & Technology, 31*(12), 1838–1845.

Pakzad, L., Ein-Mozaffari, F., & Chan, P. (2008b). Using electrical resistance tomography and computational fluid dynamics modeling to study the formation of cavern in the mixing of pseudoplastic fluids possessing yield stress. *Chemical Engineering Science, 63*(9), 2508–2522.

Pakzad, L., Ein-Mozaffari, F., Upreti, S. R., & Lohi, A. (2013a). Characterisation of the mixing of non-Newtonian fluids with a scaba 6SRGT impeller through ert and CFD. *Canadian Journal of Chemical Engineering, 91*(1), 90–100.

Pakzad, L., Ein-Mozaffari, F., Upreti, S. R., & Lohi, A. (2013b). Evaluation of the mixing of non-Newtonian biopolymer solutions in the reactors equipped with the coaxial mixers through tomography and CFD. *Chemical Engineering Journal, 215–216*, 279–296.

Pakzad, L., Ein-Mozaffari, F., Upreti, S. R., & Lohi, A. (2013c). Experimental and numerical studies on mixing of yield-pseudoplastic fluids with a coaxial mixer. *Chemical Engineering Communications, 200*(12), 1553–1577.

Pakzad, L., Ein-Mozaffari, F., Upreti, S. R., & Lohi, A. (2013d). A novel and energy-efficient coaxial mixer for agitation of non-Newtonian fluids possessing yield stress. *Chemical Engineering Science, 101*, 642–654.

Pakzad, L., Ein-Mozaffari, F., Upreti, S. R., & Lohi, A. (2013e). Using tomography to assess the efficiency of the coaxial mixers in agitation of yield-pseudoplastic fluids. *Chemical Engineering Research and Design, 91*(9), 1715–1724.

Park, B. G., Moon, J. H., Lee, B. S., & Kim, S. (2008). An electrical resistance tomography technique for the monitoring of a radioactive waste separation process. *International Communications in Heat and Mass Transfer, 35*(10), 1307–1310.

Parvareh, A., Rahimi, M., Alizadehdakhel, A., & Alsairafi, A. A. (2010). CFD and ERT investigations on two-phase flow regimes in vertical and horizontal tubes. *International Communications in Heat and Mass Transfer, 37*(3), 304–311.

Patel, D., Ein-Mozaffari, F., & Mehrvar, M. (2013a). Characterization of the continuous-flow mixing of non-Newtonian fluids using the ratio of residence time to batch mixing time. *Chemical Engineering Research and Design, 91*(7), 1223–1234.

Patel, D., Ein-Mozaffari, F., & Mehrvar, M. (2013b). Using tomography technique to characterize the continuous-flow mixing of non-Newtonian fluids in stirred vessels. *Chemical Engineering Transactions, 32*, 1465–1470.

Patel, D., Ein-Mozaffari, F., & Mehrvar, M. (2013c). Using tomography to characterize the mixing of non-Newtonian fluids with a Maxblend impeller. *Chemical Engineering & Technology, 36*(4), 687–695.

Patel, D., Ein-Mozaffari, F., & Mehrvar, M. (2014). Using tomography to visualize the continuous-flow mixing of biopolymer solutions inside a stirred tank reactor. *Chemical Engineering Journal, 239*, 257–273.

Qin, Y. J., Jin, H. B., & Yang, S. H. (2013). Local bubble parameters in a pressurized gas-liquid bubble column reactor measured by electrical resistance tomography. *Gao Xiao Hua Xue Gong Cheng Xue Bao/Journal of Chemical Engineering of Chinese Universities, 27*(3), 372–379.

Rakesh, A., Kumar Reddy, V. T. S. R., & Narasimha, M. (2014). Air-core size measurement of operating hydrocyclone by electrical resistance tomography. *Chemical Engineering & Technology, 37*(5), 795–805.

Rao, G., Aghajanian, S., Koiranen, T., Wajman, R., & Jackowska-Strumiłło, L. (2020). Process monitoring of antisolvent based crystallization in low conductivity solutions using electrical impedance spectroscopy and 2-D electrical resistance tomography. *Applied Sciences, 10*(11).

Razzak, S., Barghi, S., & Zhu, J. (2007a). *Measurement of holdup and propagation velocities in a liquid-solid circulating fluidized bed using Electrical Resistance Tomography.*

Razzak, S. A., Barghi, S., & Zhu, J. X. (2007b). Electrical resistance tomography for flow characterization of a gas-liquid-solid three-phase circulating fluidized bed. *Chemical Engineering Science, 62*(24), 7253–7263.

Razzak, S. A., Barghi, S., & Zhu, J. X. (2008). *Local holdups and phase propagation velocity measurement in GLSCFB riser using electrical resistance tomography and optical fibre probe.*

Razzak, S. A., Barghi, S., & Zhu, J. X. (2009a). Application of electrical resistance tomography on liquid-solid two-phase flow characterization in an LSCFB riser. *Chemical Engineering Science, 64*(12), 2851–2858.

Razzak, S. A., Barghi, S., & Zhu, J. X. (2010a). Axial hydrodynamic studies in a gas-liquid-solid circulating fluidized bed riser. *Powder Technology, 199*(1), 77–86.

Razzak, S. A., Barghi, S., & Zhu, J. X. (2010b). Phase distributions in a gas-liquid-solid circulating fluidized bed riser. *Canadian Journal of Chemical Engineering, 88*(4), 579–585.

Razzak, S. A., Barghi, S., Zhu, J. X., & Mi, Y. (2009b). Phase holdup measurement in a gas-liquid-solid circulating fluidized bed (GLSCFB) riser using electrical resistance tomography and optical fibre probe. *Chemical Engineering Journal, 147*(2–3), 210–218.

Razzak, S. A., Zhu, J. X., & Barghi, S. (2009c). Particle shape, density, and size effects on the distribution of phase holdups in an LSCFB riser. *Chemical Engineering & Technology, 32*(8), 1236–1244.

Razzak, S. A., Zhu, J. X., & Barghi, S. (2009d). Radial distributions of phase holdups and phase propagation velocities in a three-phase gas-liquid-solid fluidized bed (GLSCFB) riser. *Industrial & Engineering Chemistry Research, 48*(1), 281–289.

Razzak, S. A., Zhu, J. X., & Barghi, S. (2010c). Effects of particle shape, density, and size on a distribution of phase holdups in a gas-liquid-solid circulating fluidized bed riser. *Industrial & Engineering Chemistry Research, 49*(15), 6998–7007.

Regner, M., Henningsson, M., Wiklund, J., Östergren, K., & Trägårdh, C. (2007). Predicting the displacement of Yoghurt by water in a pipe using CFD. *Chemical Engineering & Technology, 30*(7), 844–853.

Ren, Z., Kowalski, A., & Rodgers, T. L. (2017). Measuring inline velocity profile of shampoo by electrical resistance tomography (ERT). *Flow Measurement and Instrumentation, 58*, 31–37.

Ren, Z., Trinh, L., Cooke, M., De Hert, S. C., Silvaluengo, J., Ashley, J., et al. (2019). Development of a novel linear ERT sensor to measure surface deposits. *IEEE Transactions on Instrumentation and Measurement, 68*(3), 754–761.

Ricard, F., Brechtelsbauer, C., Xu, X. Y., & Lawrence, C. J. (2005). Monitoring of multiphase pharmaceutical processes using electrical resistance tomography. *Chemical Engineering Research and Design, 83*(7 A), 794–805.

Ridzuan Aw, S., Abdul Rahim, R., Fazalul Rahiman, M. H., Mohamad, E. J., Mohd Yunus, F. R., Wahab, Y. A., et al. (2015). Simulation study on electrical resistance tomography using metal wall for bubble detection. *Jurnal Teknologi, 73*(6), 31–35.

Rodgers, T. L., Cooke, M., Siperstein, F. R., & Kowalski, A. (2009). Mixing and dissolution times for a cowles disk agitator in large-scale emulsion preparation. *Industrial & Engineering Chemistry Research, 48*(14), 6859–6868.

Rodgers, T. L., Gangolf, L., Vannier, C., Parriaud, M., & Cooke, M. (2011). Mixing times for process vessels with aspect ratios greater than one. *Chemical Engineering Science, 66*(13), 2935–2944.

Rodgers, T. L., & Kowalski, A. (2010). An electrical resistance tomography method for determining mixing in batch addition with a level change. *Chemical Engineering Research and Design, 88*(2), 204–212.

Rodriguez-Frias, M. A., & Yang, W. (2020). Dual-modality 4-terminal electrical capacitance and resistance tomography for multiphase flow monitoring. *IEEE Sensors Journal, 20*(6), 3217–3225.

Ruzinsky, F., & Bennington, C. P. J. (2007). Aspects of liquor flow in a model chip digester measured using electrical resistance tomography. *Chemical Engineering Journal, 130*(2–3), 67–74.

Sardeshpande, M. V., Gupta, S., & Ranade, V. V. (2017). Electrical resistance tomography for gas holdup in a gas-liquid stirred tank reactor. *Chemical Engineering Science, 170*, 476–490.

Sharifi, M., & Young, B. (2010). *The potential utilisation of Electrical Resistance Tomography (ERT) in milk powder processing for monitoring and control.*

Sharifi, M., & Young, B. (2011). 3-Dimensional spatial monitoring of tanks for the milk processing industry using electrical resistance tomography. *Journal of Food Engineering, 105*(2), 312–319.

Sharifi, M., & Young, B. (2012a). Milk total solids and fat content soft sensing via electrical resistance tomography and temperature measurement. *Food and Bioproducts Processing, 90*(4), 659–666.

Sharifi, M., & Young, B. (2012b). Qualitative visualization and quantitative analysis of milk flow using electrical resistance tomography. *Journal of Food Engineering, 112*(3), 227–242.

Sharifi, M., & Young, B. (2013a). Electrical resistance tomography (ert) applications to chemical engineering. *Chemical Engineering Research and Design, 91*(9), 1625–1645.

Sharifi, M., & Young, B. (2013b). Electrical resistance tomography (ERT) for flow and velocity profile measurement of a single phase liquid in a horizontal pipe. *Chemical Engineering Research and Design, 91*(7), 1235–1244.

Sharifi, M., & Young, B. (2013c). Towards an online milk concentration sensor using ERT: Correlation of conductivity, temperature and composition. *Journal of Food Engineering, 116*(1), 86–96.

Sharifi, M., Yu, W., & Young, B. (2014). Towards fault detection of the operation of dairy processing industry tanks using electrical resistance tomography. *Food Control, 38*(1), 192–197.

Shen, Y., Tan, C., Dong, F., Smith, K., & Escudero, J. (2018). *Gas-water two-phase flow pattern recognition based on ERT and ultrasound Doppler.*

Shirhatti, V. S., Wang, M., & Williams, R. A. (2005). *Visualisation of dispersion, Dissolution and settling of powders in a stirred mixing vessel by electrical resistance tomography.*

Shi, Y., Wang, M., & Shen, M. (2016). Characterization of oil-water two-phase flow in a horizontal pipe with multi-electrode conductance sensor. *Journal of Petroleum Science and Engineering, 146*, 584–590.

Simmons, M. J. H., Edwards, I., Hall, J. F., Fan, X., Parker, D. J., & Stitt, E. H. (2009). Techniques for visualization of cavern boundaries in opaque industrial mixing systems. *AIChE Journal, 55*(11), 2765–2772.

Singh, B. K., Jain, E., & Buwa, V. V. (2019). Feasibility of electrical resistance tomography for measurements of liquid holdup distribution in a trickle bed reactor. *Chemical Engineering Journal, 358*, 564–579.

Singh, B. K., Quiyoom, A., & Buwa, V. V. (2017). Dynamics of gas–liquid flow in a cylindrical bubble column: Comparison of electrical resistance tomography and voidage probe measurements. *Chemical Engineering Science, 158*, 124–139.

Son, Y., Kim, G., Lee, S., Kim, H., Min, K., & Lee, K. S. (2017). Experimental investigation of liquid distribution in a packed column with structured packing under permanent tilt and roll motions using electrical resistance tomography. *Chemical Engineering Science, 166*, 168–180.

Son, Y., Lee, S., Han, S., Yang, D., Min, K., & Lee, K. S. (2018). Liquid distribution model based on a volume cell for a column with structured packing under permanent tilt and roll motions. *Chemical Engineering Science, 182*, 1–16.

Spear, M. (2003). Validating CFD with tomography. *Process Engineering, 84*(3), 15–16.

Stanley, S. J. (2006). Tomographic imaging during reactive precipitation in a stirred vessel: Mixing with chemical reaction. *Chemical Engineering Science, 61*(24), 7850–7863.

Stanley, S. J., & Bolton, G. T. (2008). A review of recent electrical resistance tomography (ERT) applications for wet particulate processing. *Particle & Particle Systems Characterization, 25*(3), 207–215.

Stanley, S. J., Hristov, H., Mann, R., & Primrose, K. (2005). Reconciling electrical resistance tomography (ERT) measurements with a fluid mixing model for semi-batch operation of a stirred vessel. *Canadian Journal of Chemical Engineering, 83*(1), 48–54.

Stanley, S. J., Mann, R., & Primrose, K. (2002). Tomographic imaging of fluid mixing in three dimensions for single-feed semi-batch operation of a stirred vessel. *Chemical Engineering Research and Design, 80*(8), 903–909.

Stanley, S. J., Mann, R., & Primrose, K. (2005). Interrogation of a precipitation reaction by electrical resistance tomography (ERT). *AIChE Journal, 51*(2), 607–614.

Stevenson, R., Harrison, S. T. L., Mantle, M. D., Sederman, A. J., Moraczewski, T. L., & Johns, M. L. (2010). Analysis of partial suspension in stirred mixing cells using both MRI and ERT. *Chemical Engineering Science, 65*(4), 1385–1393.

Stevenson, R., Harrison, S. T. L., Miles, N., & Cilliers, J. J. (2006). Examination of swirling flow using electrical resistance tomography. *Powder Technology, 162*(2), 157–165.

Suard, E., Clément, R., Fayolle, Y., Alliet, M., Albasi, C., & Gillot, S. (2019). Electrical resistivity tomography used to characterize bubble distribution in complex aerated reactors: Development of the method and application to a semi-industrial MBR in operation. *Chemical Engineering Journal, 355*, 498–509.

Sun, B., Cui, Z., Yue, S., & Wang, H. (2016). *An ECT/ERT dual-modality system based on an internal electrode sensor*.

Tahvildarian, P., Ng, H., D'Amato, M., Drappel, S., Ein-Mozaffari, F., & Upreti, S. R. (2011). Using electrical resistance tomography images to characterize the mixing of micron-sized polymeric particles in a slurry reactor. *Chemical Engineering Journal, 172*(1), 517–525.

Tan, C., Ren, S. J., & Dong, F. (2015). Reconstructing the phase distribution within an annular channel by electrical resistance tomography. *Heat Transfer Engineering, 36*(12), 1053–1064.

Tan, C., Wang, N. N., & Dong, F. (2016). Oil–water two-phase flow pattern analysis with ERT based measurement and multivariate maximum Lyapunov exponent. *Journal of Central South University, 23*(1), 240–248.

Toye, D., Fransolet, E., Simon, D., Crine, M., L'Homme, G., & Marchot, P. (2005). Possibilities and limits of application of electrical resistance tomography in hydrodynamics of bubble columns. *Canadian Journal of Chemical Engineering, 83*(1), 4–10.

Uppal, H. J., Rodgers, T. L., Cooke, M., Trinh, L., Kowalski, A. J., & Martin, P. J. (2014). *Monitoring cleaning-in-place (CIP) of multiphase formulated products by electrical resistance tomography (ERT)*.

Vadlakonda, B., Kopparthi, P., Mukhurjee, A. K., & Mangadoddy, N. (2019). *Investigation of column flotation hydrodynamics using electrical resistance tomography coupled with pressure transducers*.

Vadlakonda, B., & Mangadoddy, N. (2017). Hydrodynamic study of two phase flow of column flotation using electrical resistance tomography and pressure probe techniques. *Separation and Purification Technology, 184*, 168–187.

Vadlakonda, B., & Mangadoddy, N. (2018). Hydrodynamic study of three-phase flow in column flotation using electrical resistance tomography coupled with pressure transducers. *Separation and Purification Technology, 203*, 274–288.

Vadlakonda, B., & Mangadoddy, N. (2020). *Measurement of gas–solid dispersion characteristics in a slurry flotation column using ERT technique*. Transactions of the Indian Institute of Metals.

Vakamalla, T. R., Kumbhar, K. S., Gujjula, R., & Mangadoddy, N. (2014). Computational and experimental study of the effect of inclination on hydrocyclone performance. *Separation and Purification Technology, 138*, 104–117.

Vauhkonen, M., Hänninen, A., Jauhiainen, J., & Lehtikangas, O. (2019). Multimodal imaging of multiphase flows with electromagnetic flow tomography and electrical tomography. *Measurement Science and Technology, 30*(9).

Vijayan, M., Schlaberg, H. I., & Wang, M. (2007). Effects of sparger geometry on the mechanism of flow pattern transition in a bubble column. *Chemical Engineering Journal, 130*(2–3), 171–178.

Vlaev, D. S., & Bennington, C. P. J. (2004). Using electrical resistance tomography to image liquor flow in a model digester. *Journal of Pulp and Paper Science, 30*(1), 15–21.

Vlaev, D. S., & Bennington, C. P. J. (2005). Flow uniformity in a model digester measured with electrical resistance tomography. *Canadian Journal of Chemical Engineering, 83*(1), 42–47.

Vlaev, D. S., & Bennington, C. P. J. (2006). Flow of liquor through wood chips in a model digester. *Chemical Engineering Communications, 193*(7), 879−890.

Vlaev, D., Wang, M., Dyakowski, T., Mann, R., & Grieve, B. D. (2000). Detecting filter-cake pathologies in solid-liquid filtration: Semi-tech scale demonstrations using electrical resistance tomography (ERT). *Chemical Engineering Journal, 77*(1−2), 87−91.

Wabo, E., Kagoshima, M., & Mann, R. (2004). Batch stirred vessel mixing evaluated by visualized reactive tracers and electrical tomography. *Chemical Engineering Research and Design, 82*(9 SPEC. ISS.), 1229−1236.

Wang, M., & Cilliers, J. J. (1999). Detecting non-uniform foam density using electrical resistance tomography. *Chemical Engineering Science, 54*(5), 707−712.

Wang, M., Dorward, A., Vlaev, D., & Mann, R. (2000). Measurements of gas-liquid mixing in a stirred vessel using electrical resistance tomography (ERT). *Chemical Engineering Journal, 77*(1−2), 93−98.

Wang, Q., Jia, X., & Wang, M. (2020). Fuzzy logic based multi-dimensional image fusion for gas-oil-water flows with dual-modality electrical tomography. *IEEE Transactions on Instrumentation and Measurement, 69*(5), 1948−1961.

Wang, M., Jones, T. F., & Williams, R. A. (2003). Visualization of asymmetric solids distribution in horizontal swirling flows using electrical resistance tomography. *Chemical Engineering Research and Design, 81*(8), 854−861.

Wang, X. Z., Liu, L., Li, R., Tweedie, R. J., Primrose, K., Corbett, J., et al. (2009a). Online monitoring of nanoparticle suspensions using dynamic light scattering, ultrasound spectroscopy and process tomography. *Computer Aided Chemical Engineering, 26*, 351−356.

Wang, X. Z., Liu, L., Li, R. F., Tweedie, R. J., Primrose, K., Corbett, J., et al. (2009b). Online characterisation of nanoparticle suspensions using dynamic light scattering, ultrasound spectroscopy and process tomography. *Chemical Engineering Research and Design, 87*(6), 874−884.

Wang, P., Li, Y. B., Wang, M., Qin, X. B., & Liu, L. (2019). An adaptive electrical resistance tomography sensor with flow pattern recognition capability. *Journal of Central South University, 26*(3), 612−622.

Wang, B., Meng, Z., Huang, Z., Ji, H., & Li, H. (2012). Voidage measurement of air-water two-phase flow based on ert sensor and data mining technology. *Chinese Journal of Chemical Engineering, 20*(2), 400−405.

Wang, Q., Polansky, J., Wang, M., Wei, K., Qiu, C., Kenbar, A., et al. (2018). Capability of dual-modality electrical tomography for gas-oil-water three-phase pipeline flow visualisation. *Flow Measurement and Instrumentation, 62*, 152−166.

Wang, N., Tan, C., & Dong, F. (2016). *Water holdup measurement of oil-water two-phase flow based on KPLS regression.*

Wang, Q., Wang, M., Wei, K., & Qiu, C. (2017). Visualization of gas-oil-water flow in horizontal pipeline using dual-modality electrical tomographic systems. *IEEE Sensors Journal, 17*(24), 8146−8156.

Wang, Y., & Yang, Y. (2017). *Refined composite multivariate multiscale fuzzy entropy analysis of horizontal oil-water two-phase flow.*

Wang, M., & Yin, W. (2001). Measurements of the concentration and velocity distribution in miscible liquid mixing using electrical resistance tomography. *Chemical Engineering Research and Design, 79*(8), 883−886.

Wang, S., & Yin, W. (2016). *Monitoring Cleaning-In-Place by electrical resistance tomography with dynamic references.*

West, R. M., Jia, X., & Williams, R. A. (1999). Quantification of solid-liquid mixing using electrical resistance and positron emission tomography. *Chemical Engineering Communications, 175*, 71–97.
Williams, R. A., Jia, X., & McKee, S. L. (1996). Development of slurry mixing models using resistance tomography. *Powder Technology, 87*(1), 21–27.
Williams, R. A., & Wang, M. (2000). Dynamic imaging of process plant reactors and separators using electrical process tomography. *Oil and Gas Science and Technology, 55*(2), 185–186.
Wu, C., Hutton, M., & Soleimani, M. (2019). Smart water meter using electrical resistance tomography. *Sensors, 19*(14).
Wu, Y., Li, H., Wang, M., & Williams, R. A. (2005). Characterization of air-water two-phase vertical flow by using electrical resistance imaging. *Canadian Journal of Chemical Engineering, 83*(1), 37–41.
Xu, Y., Wang, H., & Cui, Z. (2009). Phase information extraction of gas/liquid two-phase flow in horizontal pipe based on independent component analysis. *Huagong Xuebao/CIESC Journal, 60*(12), 3012–3018.
Yang, L., Zou, L., Ma, Y., Wang, J., & Xu, J. (2020). Breakup, coalescence, and migration regularity of bubbles under gas-liquid swirling flow in gas-liquid cylindrical cyclone. *Industrial & Engineering Chemistry Research, 59*(5), 2068–2082.
Yan, P., Jin, H., He, G., Guo, X., Ma, L., Yang, S., et al. (2019). CFD simulation of hydrodynamics in a high-pressure bubble column using three optimized drag models of bubble swarm. *Chemical Engineering Science, 199*, 137–155.
Yenjaichon, W., Grace, J. R., Jim Lim, C., & Bennington, C. P. J. (2013a). Characterisation of gas mixing in water and pulp-suspension flow based on electrical resistance tomography. *Chemical Engineering Journal, 214*, 285–297.
Yenjaichon, W., Grace, J. R., Jim Lim, C., & Bennington, C. P. J. (2013b). Gas dispersion in horizontal pulp-fibre-suspension flow. *International Journal of Multiphase Flow, 49*, 49–57.
Yenjaichon, W., Grace, J. R., Lim, J. C., & Bennington, C. P. J. (2011). *Characterization of liquid-pulp fiber suspension mixing at tee mixers based on electrical resistance tomography*.
Yenjaichon, W., Grace, J. R., Lim, C. J., & Bennington, C. P. J. (2012). In-line jet mixing of liquid-pulp-fibre suspensions: Effect of concentration and velocities. *Chemical Engineering Science, 75*, 167–176.
Yenjaichon, W., Grace, J. R., Lim, C. J., & Bennington, C. P. J. (2013c). Gas dispersion in pulp-suspension flow in the presence of an in-line mechanical mixer. *Chemical Engineering Science, 93*, 22–31.
Yenjaichon, W., Grace, J. R., Lim, C. J., & Bennington, C. P. J. (2013d). In-line jet mixing of liquid-pulp-fiber suspensions: Effect of fiber properties, flow regime, and jet penetration. *AIChE Journal, 59*(4), 1420–1430.
Yenjaichon, W., Grace, J. R., Lim, C. J., & Bennington, C. P. J. (2013e). Pilot-scale examination of mixing liquid into pulp fiber suspensions in the presence of an in-line mechanical mixer. *Industrial & Engineering Chemistry Research, 52*(1), 485–498.
Yenjaichon, W., Pageau, G., Bhole, M., Bennington, C. P. J., & Grace, J. R. (2011). Assessment of mixing quality for an industrial pulp mixer using electrical resistance tomography. *Canadian Journal of Chemical Engineering, 89*(5), 996–1004.
Zbib, H., Ebrahimi, M., Ein-Mozaffari, F., & Lohi, A. (2018a). Comprehensive analysis of fluid-particle and particle-particle interactions in a liquid-solid fluidized bed via CFD-DEM coupling and tomography. *Powder Technology, 340*, 116–130.

Zbib, H., Ebrahimi, M., Ein-Mozaffari, F., & Lohi, A. (2018b). Hydrodynamic behavior of a 3-D liquid-solid fluidized bed operating in the intermediate flow regime - application of stability analysis, coupled CFD-DEM, and tomography. *Industrial & Engineering Chemistry Research, 57*(49), 16944–16957.

Zhang, J. Z., Li, H., Ma, J. W., Chen, X. P., & Xu, J. Y. (2018). Study of the swirling flow field induced by guide vanes using electrical resistance tomography and numerical simulations. *Chemical Engineering Communications, 205*(10), 1351–1364.

Zhang, B., Qin, Y., Jin, H., Yang, S., He, G., Luo, G., et al. (2017). CFD simulation of pressurized gas-liquid bubble column and experimental verification based on electrical resistance tomography. *Huaxue Fanying Gongcheng Yu Gongyi/Chemical Reaction Engineering and Technology, 33*(4), 335–342.

Zhao, T., Eda, T., Achyut, S., Haruta, J., Nishio, M., & Takei, M. (2015). Investigation of pulsing flow regime transition and pulse characteristics in trickle-bed reactor by electrical resistance tomography. *Chemical Engineering Science, 130*, 8–17.

Zhao, Z. F., Mehrvar, M., & Ein-Mozaffari, F. (2008). Mixing time in an agitated multi-lamp cylindrical photoreactor using electrical resistance tomography. *Journal of Chemical Technology and Biotechnology, 83*(12), 1676–1688.

Zhao, J., Su, C., Ren, H., Qin, L., Dong, B., & Xing, F. (2016). *Application of electrical resistance tomography to damage detection in concrete.*

Zhou, H., Xu, L., Cao, Z., Hu, J., & Liu, X. (2012). Image reconstruction for invasive ERT in vertical oil well logging. *Chinese Journal of Chemical Engineering, 20*(2), 319–328.

From process understanding to process control—Applications in industry

16

Ken Primrose[1,2]
[1]Industrial Tomography Systems Ltd, Manchester, United Kingdom; [2]Stream Sensing Ltd, Manchester, United Kingdom

16.1 Introduction

Process instruments are designed to measure variables within a volume over time during the physical or chemical transformation of a material. This section will not address instruments applied to model industrial processes in a laboratory research environment or sectors such as oil and gas which are covered elsewhere in this book. Process instrumentation is normally differentiated from laboratory instrumentation by virtue of its operating in a manufacturing environment, either at full or pilot scale. In this context, scale effects are much more significant—and hence the need to measure at different places through technologies such as process tomography.

16.1.1 Process instrumentation levels

While Technology Readiness Levels (European Commission, 2014, p. 4995) are often used as a measure of technology development. Fig. 16.1 shows a simpler categorization of process instrumentation levels; these measurements are normally used at different levels of complexity, affecting both the design of the instruments and also the interpretation of their measurements:

- Level 1: developing a fundamental understanding of the process.
- Level 2: optimizing a process; this is frequently a one-off exercise. Level 2 also encompasses process scale-up and validation of process models.

  Once the process is optimised it may be controlled through a timed sequence of process steps or simpler apparatus than that used for detailed process understanding.
- Level 3: process analytics, where process variables are continually measured, such as for quality control or regulatory compliance. However, Level 3 process instrumentation does not include a real-time feedback for process control.
- Level 4: process control, where the process instrumentation feeds into a mechanism for process control. There are two subgroups of process control; where measurements are continuously used or periodic (for either safety or to initiate a process step); and where feedback is either through a human interaction or automated.

Industrial Tomography. https://doi.org/10.1016/B978-0-12-823015-2.00023-6

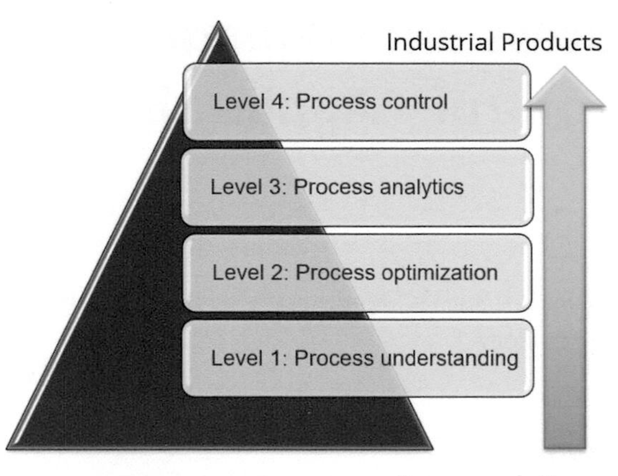

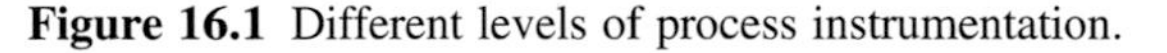
Figure 16.1 Different levels of process instrumentation.

As applications move from Level 1 (process understanding) to Level 4 (automated process control), there is a change in the nature of the instrumentation.

For **Level 1 (process understanding)**, instrumentation is generally operated by a highly trained engineer experienced in the particular instrument, where the engineer will often fine tune settings; interpret data, and interact with the instrument. The installation of the instrument will usually be temporary and changes or adaptation to the process can be introduced ("tolerated") for the expected benefits to be derived from the process improvements.

Data outputs may be recorded on a PC; calibration can be applied by the engineer attending the instrument or postprocessed. Once recorded, measurements may be correlated with other process variables (temperature, time, yield, etc.) and analysis carried out to understand the process conditions in question.

In contrast, **Level 4 (process control)**, instrumentation needs to be considerably more robust and consistently reliable. While there may be skilled instrument engineers to hand, these people will usually be responsible for a wide range of instruments at that particular facility.

The requirement for robustness and reliability place a different set of requirements on the process instrumentation in a control environment: Level 4 instruments will usually be

- operate to industrial standards; this may include site or process-specific standards such as operating in explosive environments (such as ATEX) and to elevated pressures and temperatures
- automated in terms of data collection and able to auto calibrate to changing process conditions
- transmit data in real time using a range of digital and analog industrial protocols
- carry remote diagnostics, making it easier for on-site engineers to resolve any measurement issues.
- standardized, rather than bespoke, with full supporting documentation

This chapter will review examples of industrial tomography instrumentation at all four levels.

16.1.2 Distinctiveness of process tomography in industrial applications

The distinctiveness of process tomography lies in its nature of taking data.

The simplest class of process instrumentation takes data at a single point, with typical examples as temperature, which measures energy at the location of a detector and pressure where it measures average force at the point of the sensor based on an average within the process volume. Similar to pressure are electro-magnetic flow instruments which measure the average current produced by movement of a conducting fluid through a field (although interestingly there are some localized EMF detection methods which are similar to tomography, but failed to be effectively commercialized).

A second class of process instrumentation takes measurements through a chord of the process volume—for example, laser based and spectroscopic instruments. Related to these are nucleonic which measure across a fan beam between the nuclear source and detector array. However, these measurements are usually aggregated and averaged. This class also includes ultrasound detection methods which are able to measure a wide range of material properties based on the transmission and detection of high frequency pulses.

The distinctiveness of process tomography instrumentation is that they use an array of excitation and detection points. These are then used to determine spatially distributed properties as described elsewhere.

The basis of detection and measurement sets out the categories of properties which can be measured and hence the industrial processes most amenable to process tomography. This book describes a wide range of process tomography measurement principles—high and low frequency electrical tomography; gamma tomography; X-ray tomography; optical tomography; ultrasound; and so on. However, to be amendable to effective installation in an industrial process, as described above, then industrial process tomography most favors electrical techniques.

Applications of electrical techniques in oil and gas flow are described in chapters 26 and 27. This chapter will describe capacitance and resistance tomography applications in other industrial sectors, looking at each of the above categories in greater detail.

16.2 Applications to improve process understanding

Two case studies are discussed here, both in the burgeoning space sector. The number of satellites being launched globally is growing rapidly: 314 were launched in 2018, up by 26% from the previous year (Satellite Industry Association, 2019). Each satellite remains costly and the launch costs and risks remain high; ECT can contribute directly to addressing these issues. The first case study below shows how ECT can be used to give direct measurements of combustion in low-cost rocket launchers, while the next shows how ECT can be used to give accurate propellant gauging in spacecraft and direct information on the impact of liquid movement in various gravity environments on the spacecraft and its maneuvering capability. These applications show significant promise of accurate and valuable measurements, and trials are being supported by several large organizations including Orbital-ATK (Northrop-Grumman) (UK Space Agency, 2019), the German Space Agency (DLR), and Ariane Group (Ariane Press Release).

While space-borne applications of ECT systems need to be engineered for flight, it is clear that the essential elements of the system, the electrode array, interconnections, and electronics can be made very lightweight and compact.

16.2.1 Case study 1 - monitoring rocket-motor combustion

Monitoring of flames and combustion using ECT has been reported previously (Qi, Jia, Mao, & Ju, 2016; Munson, 2011), but this is a relatively novel area of application and in particular calibration is difficult. Working on low-cost, expendable launch vehicles (Baker, 2018; Marlow, 2016), Kingston University (KU) have utilized ECT to help in monitoring rocket motor combustion. Rocket propulsion is widely acknowledged as the main cost driver for any launch vehicle to space, stemming from complexity, need for extensive testing, use of exotic materials, and use of turbomachinery. Tackling these challenges with a low-cost solution has been a goal of the KU rocketlab since its formation in 2010 and they have developed a novel approach to cooling small bipropellant rocket engines using internal pressure gradients to generate a hot core and a cool wall.

Optimization of this complex fluid mechanics process requires novel approaches to measurement and control and ECT offers the opportunity to image the cross-sectional structure of the flows in the combustion chamber at high speed.

A slide-on tomographic sensor array was developed to enable the glass combustion chamber to be imaged in two sections—at the entry of the propellant and at the exit to the nozzle (see Fig. 16.2 below). The sensor is constructed primarily in titanium, and has two rings of eight electrodes, effectively separating the 70 mm diameter chamber into two halves. This sensor has been used to noninvasively characterize the flow field at nearly 4000 frames per second using a commercial Atout APL-C-900 ECT research

Figure 16.2 Left: sensor mounted on full test assembly, flame issues to right of chamber; right photo of sensor.

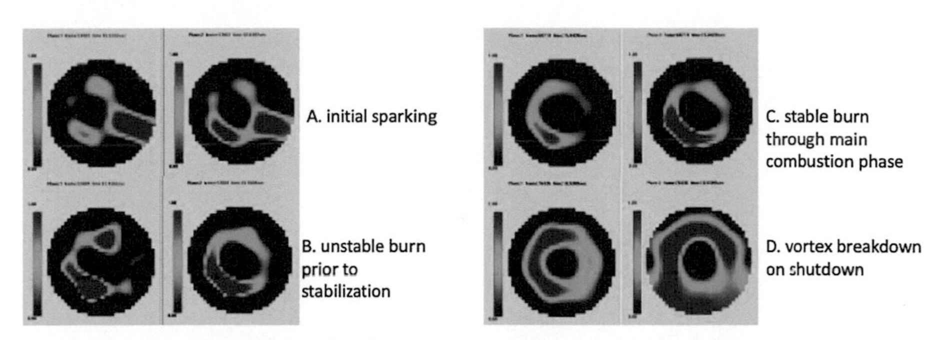

Figure 16.3 Images of combustion from upstream (left of each pair) and downstream (right of each pair). Images are normalized on a scale from blue (dark gray in print version) - 0 (air relative permittivity 1.0) to red (gray in print version) - 1 (relative permittivity >3.0).

system. Details of the combustion process are evident, and in particular, the stability of the combustion at start-up and shut down can be studied in detail.

The equivalent electrical permittivity of a flame is not a straightforward measurement, being related to temperature, composition, ionic content, conductivity, and charge, but it has been shown to be a strong function of burn temperature (Chen, 2016).

The images indicate that there is a burn structure which is ring shaped, but with predominant combustion in the "south-west" sector of the cross-section. This is consistent with the expected annular vortex mixing interface and location of the injection inlets. Fig. 16.3 shows a selection of images from the video available in Ref 8. Each pair of images comes from a different section of the combustion process. Initially (A and B), there is sparking and burn instability; during the main burn phase, the images are stable (C) for a long period, but when the flow of propellant is stopped, the vortex structure breaks down (D). It should be noted that in (A) and (D) there is a flash to the wall which can be considered a fault condition putting the combustion chamber at risk from high surface temperatures. ECT can contribute directly to understanding the physics of the process and, in the future, real-time control of propellant flows to stabilize the burn and avoid such fault conditions.

16.2.2 Case study 2 - smart tanks for space

Most satellites are launched into space carrying liquid propellant constituting a significant fraction of the mass of the spacecraft. This propellant has to last the entire life of the mission, which may be many years, and so maintaining accurate measurements of propellant mass is critical. The measurement problem is severely complicated by the fact that the tank is subjected to zero gravity conditions allowing the liquid to be distributed throughout the tank and invalidating most normal techniques for measuring mass or volume. Existing techniques in use include 'book-keeping' where the burn rate of thrusters is monitored and incremental use is recorded, and radio-frequency (RF) and ultrasonic techniques (US) which are used to monitor absorption of radiation into the mass present. Book-keeping inherently accumulates significant errors, while RF and US techniques offer nonunique solutions for mass due to the complex and nonuniform radiation field interacting with the wide and unpredictable distribution of the propellant. Errors on all current techniques may accumulate to an uncertainty

Figure 16.4 Satellite tank instrumented with ECT for slosh testing.

of + -12 months on a 15-year lifetime of a satellite meaning that satellites must be decommissioned when they may in fact have 5% of propellant reserves (van Benthem, van Es, van Put, & Matthijsen, 2013).

ECT offers a potential accurate solution to the problem of propellant gauging in zero-gravity (Hunt, 2015). Furthermore, the movement of the large mass of propellant during maneuvering can impose significant slosh forces on the spacecraft and compromise position accuracy and propellant usage (Fries, 2012). Because ECT can be used to measure mass at all points in a section of the tank, the center of mass and hence slosh forces can be estimated from nonintrusive measurements (Hunt, 2018).

Fig. 16.4 above shows a satellite tank instrumented with ECT mounted inside a bund on a hexapod at the German Space Agency in Bremen, Germany, while Fig. 16.5, below shows the bunded system mounted ready for dynamic slosh testing.

Typical output results for mass measurement during severe sloshing are shown in Fig. 16.6, above where it can be seen that at a fill level of 10% by mass, typical of near end-of-life conditions in a commercial satellite, the mass of fluid in either half

Figure 16.5 Hexapod system at DLR, Bremen.

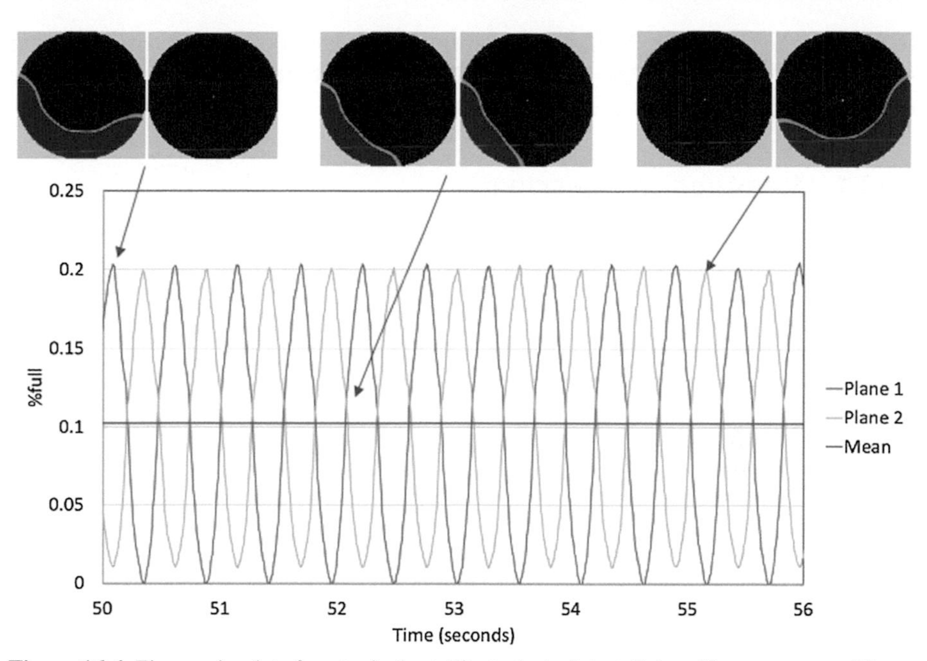

Figure 16.6 Time series data for a typical satellite tank slosh test. Pairs of images at top of figure show plane 1 and plane 2 at three different times: red (gray in print version) - liquid present, blue (dark gray in print version) -no liquid.

of the tank may vary from 0% to 20% as the fluid sloshes around the tank between the two halves. Unlike other techniques, ECT effectively interrogates the whole volume and the resultant mean fill level is stable, as shown by the blue (dark gray in print version) line. Further work has scaled-up this testing to 1m diameter and proven the accuracy and resolution of the technique for commercial application (Behruzi et al., 2020).

16.3 Process modeling and optimization

A significant number of electrical resistance tomography applications are based on the characterization and optimization of mixing. This includes gas–liquid, liquid–liquid, and solid–liquid systems.

In the case the Washington River Protection Program1,2, a high-level waste slurry is formed through mixing approximately 720,000 L of settled solids on the bottom of double shelled storage tanks with the remaining tank volume—up to 3,407,000 L (approximately 1m gallons) (Townson, 2011).

There is some uncertainty surrounding the ability of the mixer pump system to adequately suspend and homogenously distribute the solid particles within the tank. The design basis assumes each staged HLW feed tank is homogenously mixed and delivered in consistent feed delivery batches of 570,000 L (150,000 gallons).

Consistent, as used here, is intended to mean that the first 570,000 L batch has the same solids composition as the last 570,000 L batch (i.e., approximately eight batches).

16.3.1 Measurement objective

The primary objectives of this phase of demonstrations were to test scale models of the large mixing tanks and the scale up methodologies such that there was confidence in the overall mixing program.

The detailed objectives were to:

- Demonstrate equivalent tank mixing behavior can be achieved in the two sizes of scaled tanks by
- Demonstrating equipment performance and the ability of the scaled tanks to meet performance objectives
- Demonstrating instrumentation performance and the ability to measure particulate movement within the tank
- Collect fluid velocity measurements for input into a Computational Fluid Dynamic model. Data will be used to calibrate and validate the model.
- Demonstrate a range of tank operating parameters that define the edges of mixing performance to include:
  - Mixer pump flow rate
  - Mixer pump rotation rate
- Provide the framework to move forward to the next phase of batch transfer and sampling testing
- Support the batch transfer testing by identifying the range of parameters, simulants, and instrumentation to be used during that testing.

A range of test equipment was deployed including ITS p2+ resistance tomography system; the test vessels were equipped with radial tomography arrays and linear probes. This provided real-time, 3-D visualization of the solids' distribution within the vessels, so areas of good and less good mixing could be identified qualitatively.

In addition, a Coriolis flow meter, FBRM (fixed beam reflection microscopy for studying chord lengths of suspended particles), video and visual observations, and direct sampling were also used.

16.3.2 Design considerations

The main components of the platform (Fig. 16.7) are the flush vessel (11,000 L), test vessel TK-201 (110 cm dia., 420 L), and test vessel TK-301 (305 cm dia., 8877 L), both were fitted with 4 circular 16-electrode tomography arrays.

In addition, linear tomography probes were used to study mixing in the z-axis (i.e., solids concentration as a function of height).

16.3.3 Implementation

The circular tomography arrays were hitherto the largest deployed in a mixing rig and while calculations had been carried out, it was not certain how these would perform.

Figure 16.7 Mixing tank.

Group 1: jet starts from top

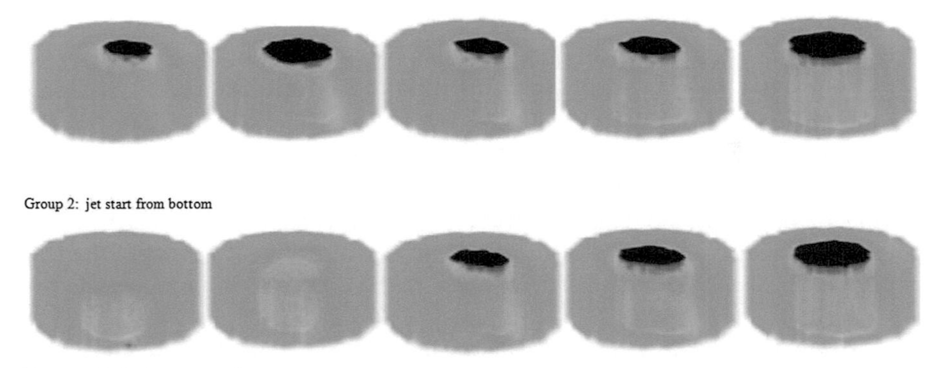

Figure 16.8 Large tank tomograms.

As it transpired, the circular arrays performed beyond expectations and were able to characterize the full mixing of the vessel such that the linear probes were not required. Preliminary results are set out in Fig. 16.8 below.

ERT – Bi-Modal Al(OH)3 and SiC Planes (110 cm Tank)

Using this instrument on both test vessels allowed each vessel to be set up so that the mixing in each was equivalent, and this information, combined with later testing, will be used to derive scaling correlations for this type of fluidically mixed system. The testing demonstrated that the two scales of test vessel could be set up to show almost identical mixing patterns as observed via the ERT system.

16.4 Process analytics

16.4.1 Vortex finder—monitoring of semicontinuous crystallization

A further application of resistance tomography is worthy of more detailed coverage as it represented an early example of implementation of ERT for the monitoring and improved performance of precipitation process under extremely challenging conditions (Bolton, 2006).

16.4.1.1 Process

The process is based on the precipitation of radioactive salts contained in a glass vessel with two overhead feeds and an overflow at the side.

The first feed contains a metal salt in nitric acid (typically 3.5M) and the other contains an organic acid. The contents of the vessel (from the two feeds) are continuously mixed using a rotating magnetic stirrer bar at the base of the vessel to maintain a reaction. The reaction products leave the vessel through the overflow at the side. The effect of the magnetic stirrer bar is to swirl the fluids and create a vortex at the center of the vessel and produce a centrifugal force to enable the precipitate to leave through the overflow.

The size and shape of the vortex depend upon the rotational speed of the stirrer bar.

Access to the vessels is restricted and visual inspection is difficult due to the opaque nature of the tank contents and the presence of a heating jacket.

Prior to the installation of tomographic monitoring, the status of the vessel is by means of a conductivity probe located in the central air space caused by the vortex.

Under normal operation, the conductivity probe should not be in contact with the process liquor. If the vortex fails, the liquid level will rise until it reaches the level of the overflow and the liquors make contact with the conductivity probe and an alarm is signaled.

However, false alarms were created due to splashing of liquor onto the conductivity probe. In addition, very little can be deduced about the internal flow structure within the precipitator vessel (such as size and shape of the vortex).

16.4.1.2 Measurement objective

The measurement objective was to develop an alternate method of determining the vortex which was not vulnerable to false positives should a leak form in the vessel (resulting in the conductivity probe being in air, but process conditions faulty); nor false negatives, when the conductivity probe is splashed (resulting in apparently faulty condition when process is actually running to specification).

A linear tomography probe was proposed which would be sensitive to the vortex depth and free of the limitations of the single probe.

Design considerations.

The design considerations were the following:

- Process environment was acidic
  The acidity limited the selection of materials. In addition, the high acidity also meant the continuous medium was highly conducting (in excess of brine at 600 mS/cm). The high conductivity had an impact on electrode size and also meant that a high current would be required to achieve the necessary signal: noise ratio required to reconstruct images.
- Process environment was radioactive
  This consideration carries across all applications in nuclear waste management and limits the selection of materials.
- Process unit was remote (operating in glove box in different location to control room)
  The impact of the remote siting of the glove box was there and would need to be long cables from sensor to instrument; in addition, it limits the access to the sensor/process making installation more demanding.
- The process itself was a technically validated operation.
  This consideration arises in a number of different sectors (for example, pharmaceuticals). It means that the sensor should not alter process operations, either through the hydrodynamic impact of the probe or through changing operating procedures (such as through taking an online reference).

16.4.1.3 Implementation

In 2003, Industrial Tomography Systems was asked to provide instrumentation to address this challenge and a solution was developed using the p2000 ERT instrument. The University of Leeds was asked to validate the instrumentation with project management provided through BNFL Research (now the UK's National Nuclear Laboratory).

The successful implementation of the sensor was achieved addressing each of the above requirements. Key aspects of this included the following:

- development of signal amplification to optimize sensitivity. The figures below show the performance with and without amplifier. As can be seen, the measurements without the amplifier are below zero (−3.5 mA) which is below the minimum threshold necessary for effective reconstruction.
  With the amplified signal, all measurements are positive and the data set can be reconstructed.
  In addition, the injection current was optimized at 75 mA and measurement frequencies of 1200, 2400, 4800, 9800, and 19,600 Hz were tested.
  These test results showed that the stability of the voltage signature in the absence of a vortex deteriorated with increasing frequency. However, the tests at injection current frequencies of 1200 and 2400 Hz failed to effectively discriminate between the absence and presence of a vortex. On this basis, the best results were obtained with a frequency of 4800 Hz and this value was recommended for all future work.
- design of thin, blade-like probe which was hydrodynamically tested to have minimal impact on the vortex. The main body of the sensor was constructed from polyvinylidene fluoride (PVDF) which possesses good chemical resistance and offers good properties in terms of radiological resistance. The electrodes were manufactured from stainless steel grade 316L, and the design was such that only PVDF and stainless steel were in contact with the process liquor.

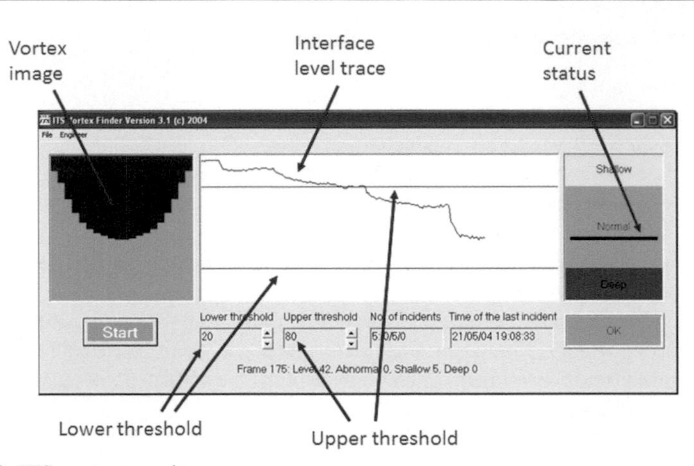

Figure 16.9 ITS vortex monitor.

- development of virtual referencing methodology to enable reference to be taken in presence of vortex. A method of processing the raw image was used to provide an image to characterize the vortex. This was achieved by the average of the pixel values in each of the 10 columns of imaging pixels being calculated and these are plotted as a bar chart. Due to the presence of the vortex, the average pixel value for each column decreases from left to right across the image. This average value (which is normalized between 0 and 1) was assumed to be proportional to the volume fraction of liquor and then used to define the interface of the liquor and gas. It is assumed that liquor exists below this point and gas exists above this point.

A mirror image is used to produce a representation of the V-shaped vortex which was familiar to plant operatives.

A software user interface was developed to acquire the measurements, analyze the data, and display the results as shown in Fig. 16.9. The user interface displays the processed image of the vortex and an historical trace of the vortex depth. Lower and upper thresholds can be defined for the vortex depth and an alarm is registered if the depth reaches either of these values.

A simplified start up routine was programmed so a user could initiate operations with a single mouse click (based on established reference).

Acceptance tests were carried out at the University of Leeds in May 2004 and the system installed at THORP in autumn 2004 and switched on for use in November of the same year.

16.4.2 Rheology

Rheology, the way a material flows, impacts upon every aspect of our daily lives from the texture of food, to the paint on the walls. Not only does rheology of a fluid system affect the final product quality, but it also governs in-process efficiency and hence is critical in all chemical and physical processing.

Conventionally, the measurement of such properties is conducted off-line with careful sampling from the product stream required. The fluid rheology is often considered, with assumptions, as applicable to flows in real processes. However, this approach is in the majority unsatisfactory since off-line measurements only afford a retrospective characterization of the sample structure which may not be representative of the time shear history received during processing (Machin, 2018). In situ measurements are inherently conducted within the flow environment. The in-line measurement of rheology affords a number of additional advantages: provides a basis for real-time product release; improvement in quality control responsiveness; optimization of processes and formulation control; and reductions in waste. Due to the critical nature of rheological behavior in processing, the development of an in-line rheometer possesses the capability to elevate rheometry from a quality control tool, at process end-point, to one which is able to control and optimize processes and materials structures (Rees, 2014). There is thus an ever-increasing demand for the development of in-line rheometry as the majority of industrial fluids exhibit complex non-Newtonian behavior.

Aligned with this demand, Machin et al. developed an in-pipe, tomographic technique capable of obtaining real-time rheological information of process fluids within pipe flows, termed Electrical Resistance Rheometry (ERR). This novel technology is an example of how tomographic measurements can be applied in process control and optimization. This system has been developed by Stream Sensing Ltd. and verified at both laboratory and pilot scales using industrial fluids. A pilot-scale study has been conducted at a multinational manufacturer of home and personal care products; the results of this study are presented in Chapter 20. An image of the ERR sensor, supplied by Stream Sensing Ltd., is depicted in Fig. 16.10 below.

16.4.2.1 Development of electrical resistance rheometry

ERR is an in-line technique capable of obtaining rheological information of process fluids, in situ, based upon electrical resistance sensing. By cross-correlating of computed conductivity pixels across and along a pipe, using microelectrical tomography sensors, rheometric data are obtained through the direct measurement of the radial velocity profile. Under laminar pipe flow conditions, the rheological properties of a

Figure 16.10 ERR sensor.

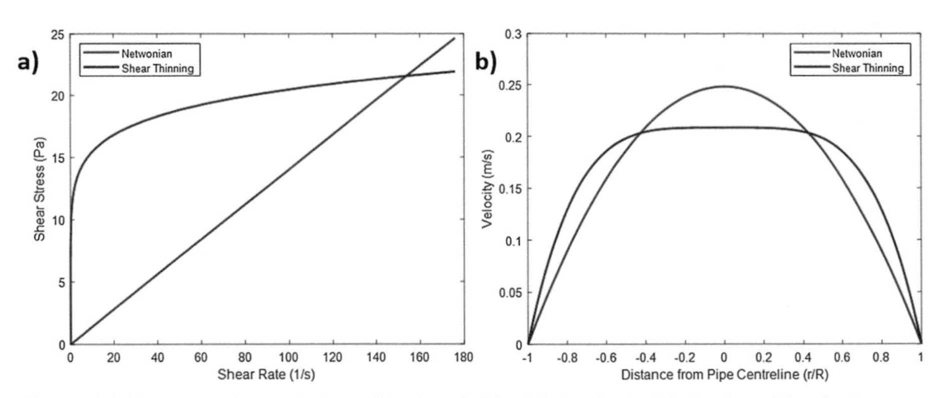

Figure 16.11 Newtonian and shear-thinning fluids: (a) rheological behavior; (b) velocity profile.

fluid strongly influence the shape of the velocity profile due to a shear rate response of the fluid. This is demonstrated in Fig. 16.11 below, for Newtonian fluid and a shear thinning fluids.

To obtain a complete velocity profile, a novel electrical sensing was developed, which consists of four arrays of eight electrodes. The ERR sensor, supplied by Stream Sensing, consists of two linear arrays, which are able to specifically target tomographic information near to the pipe wall with two circular arrays to interrogate the center of the pipe (Machin, 2018). A cross-section of the ERR sensor is depicted in Fig. 16.12 below. Using the modified standard back projection algorithm (Wang, 2002), four tomograms may be reconstructed; two rectangular tomograms and two circular tomograms. The inclusion of circular sensors enables the visualization of cross-sectional

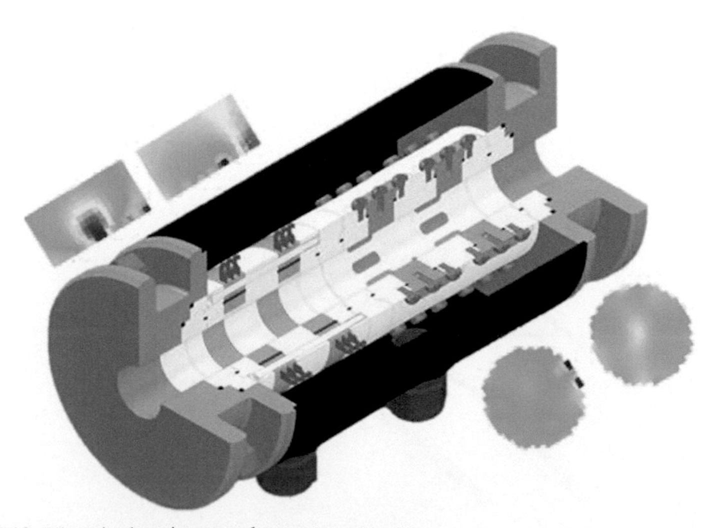

Figure 16.12 Electrical resistance rheometry sensor.

impedance map to simultaneously interrogate localized and global mixing behavior, in-line. Thus, ERR is able to combine two significant engineering quality and control concepts of rheology and mixing to understand inhomogeneity and nonuniformity within a process to form a powerful process characterization tool (Machin, 2018).

These tomograms may then be segmented into zones, to extract the velocity at a number of different radial measurement points and hence the velocity profile. A brief, nonintrusive, thermal stimulus is applied within the sensor, which acts as a small conductivity perturbation as it delivers a small increase in temperature over a short time period. The subsequent implementation of a cross-correlation algorithm to tag the motion of the fluid (Papoulis, 1962, pp. 244–253) can then yield both two- and three-dimensional velocity profiles.

The velocity profile enables the shear rate within the pipe to be determined. To determine the linear shear stress profile within the pipe under laminar flow conditions, the differential pressure is required to be measured (Wilkinson, 1960, pp. 1–38). Combining the velocity profile from the ERR sensor and differential measurement enables the in-line rheological properties to be extracted. If rheological behavior of the fluid is known to adhere to conventional rheological models, i.e., power law, the velocity profile shape can be related to the constitutive equation of the fluid, and the desired parameters can be directly outputted via a parametric fitting (Machin, 2018). This approach is the focus of this section; however, often industrial fluids do not adhere to simple rheological models and as the complexity of the model increases, the parametric fittings become increasingly ill-conditioned. To overcome this, artificial intelligence algorithms may also be utilized to extract rheological behavior in-line from complex fluids, with a pilot plant case study presented in Chapter 16.

This section presents the validation of ERR at the laboratory scale using independent techniques, across a range of model Newtonian and non-Newtonian fluids.

16.4.2.2 Velocity profile

Three model aqueous-based fluids were selected which exhibit the behavior of three rheological constitutive models, namely solutions of

- 75 wt%, 85 wt%, and 92.5 wt% glycerol (Darrant Chemical, UK)
- 0.1 wt%, 0.5 wt%, and 1.0 wt% xanthan gum, *from Xanthomonas Campestris* (Sigma–Aldrich, UK)
- 0.1 wt% and 1.0 wt% Carbopol 940 (Lubrizol), with 0.1 M Sodium Hydroxide (Sigma–Aldrich) added to alter pH to 7.

A simple recirculating pipeline was setup, which consisted of an agitated 60 L vessel and a control positive displacement pump operated by a personal computer to circulate the fluid at a desired flow rate. A 1-inch ERR sensor (Stream Sensing Ltd., UK) was then placed in series with the differential pressure monitored across the length of the sensor. Each of the fluids was circulated across the flow rate range of 40–350 L/h, and operated to capture a single in-line rheometer measurement in 30 s. This is an additional advantage over conventional off-line rheometry techniques which can often take 10 min per measurement; this does not consider the time taken for sampling.

When inspected, the ERR velocity profiles mirrored the expected trend across all fluids, with the Newtonian glycerol observing parabolic profiles in all experiments. When changing the fluid to a shear thinning power law fluid xanthan gum, the increased blunting of the velocity profile, highlighted in Fig. 16.13, was observed. Similar behavior has observed when Carbopol 940 solutions were interrogated; Carbopol may be modeled with the Herschel–Bulkley constitutive equation.

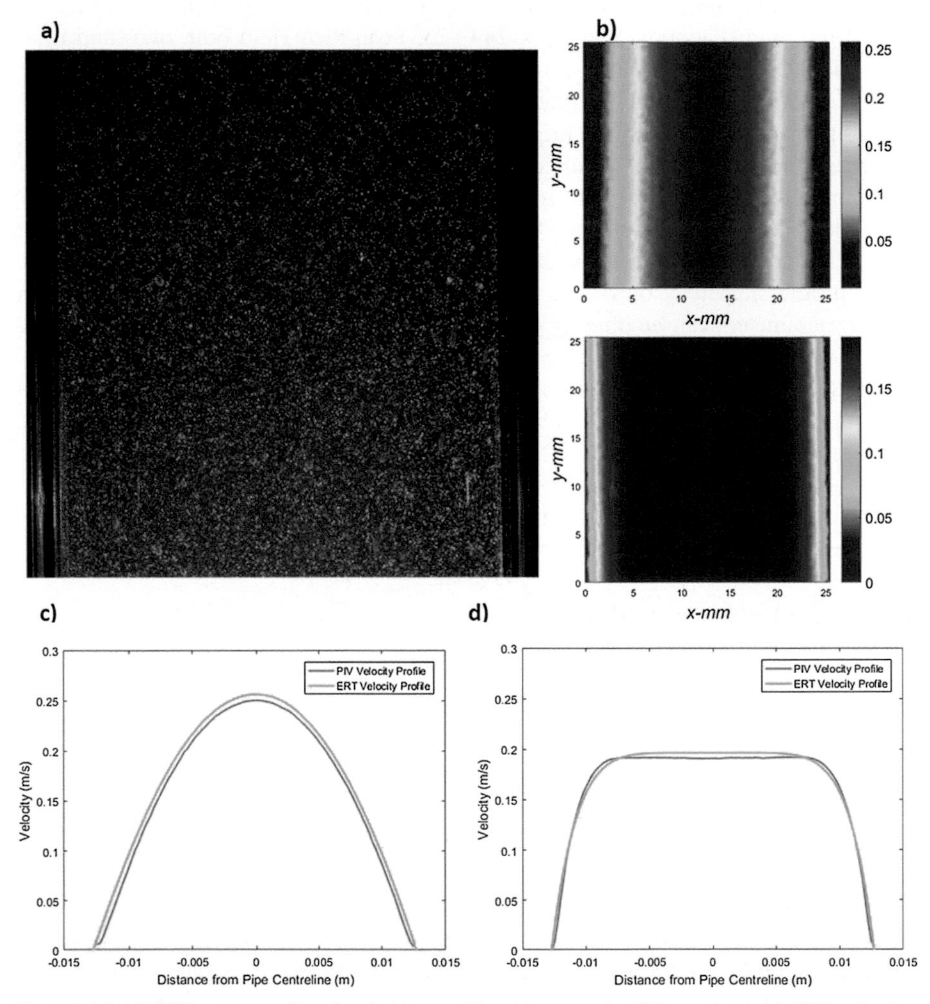

Figure 16.13 PIV cross-sectional velocity profile: (a) example PIV capture; (b) PIV velocity magnitude – glycerol 240 L/h[1] (top), 1.0 wt% Carbopol 940, pH 7 at 260 L/h (bottom); (c) radial averaged PIV and ERR/ERT velocity profiles – glycerol; (d) radial averaged PI. **Note**: This figure has been sourced from Machin et al.

To assess the repeatability of ERR, 20 in-line rheology measurements were conducted for each experimental condition. Across all fluids and flow rates, the mean standard error across the 20 measurements was found to be just 0.0043 m/s. Moreover, this error was seen to be consistent at all radial positions, allowing accurate measurement at the boundary and is able capture complex rheological behavior, i.e., wall-slip. The 95% confidence interval was also found to be ±2.93% when compared to the mean. Therefore, ERR can be considered to be a robust measurement which possesses little measurement uncertainty.

The accuracy of the velocity profile was validated using Particle Image Velocimetry (PIV). The application of PIV allows the measured velocity profile which has been extensively studied velocimetry technique of superior resolution. As PIV is a laser-based technology which is only amenable to transparent fluids and requires the fluid to be seeded with silver nanoparticles, it is not able to be implemented in-plant, safely. Due to the opaque nature of xanthan gum solutions, these were not able to be interrogated with a setup similar to that described by Alberini adapted for pipe flow (Alberini 2014).

When comparing the average PIV velocity profile to the ERR profile, a correlation coefficient of 0.99 was observed for both glycerol and Carbopol solutions. The root-mean squared errors between the ERR and PIV profiles were found to be just 0.0064 m/s and 0.0048 m/s, for the glycerol and Carbopol solutions, respectively. Not only can ERR be considered as repeatable but also accurate.

16.4.2.3 Rheology

The ultimate comparison of ERR is to off-line rotational rheometry. The rheological properties of the fluids were analyzed using an AG R2 rotational rheometer (TA Instruments, USA) equipped with a smooth-walled, 4.006 degrees, stainless steel cone and plate geometry. The sample was held at 23 ± 0.1 °C using a Peltier plate with a logarithmic shear rate ramp applied across the shear rate range of 0.01−1500/s.

Fig. 16.14 demonstrates the comparison of ERR and off-line rotational rheometry outputs. Using ERR, when predicting the single rheology parameter, of viscosity, for the Newtonian fluid, the rheology has been measured with excellent reliability with a 95% confidence interval of 2.5% of the mean value. In addition to such repeatability, the measured viscosity displays good agreement with the accepted off-line technique in industrial processing, a rotational rheometer, with an average accuracy of 99.1%. As the complexity of the model is increased to a power law model, a minor decrease in the accuracy of the descriptive rheological parameters is decreased only slightly to 97.4% and 98.5% for the consistency and power indexes, respectively. However, when comparing the rheology curves, the average correlation coefficient was found to be 0.99.

Overall. ERR coupled with using a nonlinear least square fitting accurately and repeatedly outputs the desirable rheological properties and is able to mimic conventional rotational rheometry. ERR can therefore be utilized as an effective, robust, in-line rheometer. An industrial case study of the technology is demonstrated in Chapter 20.

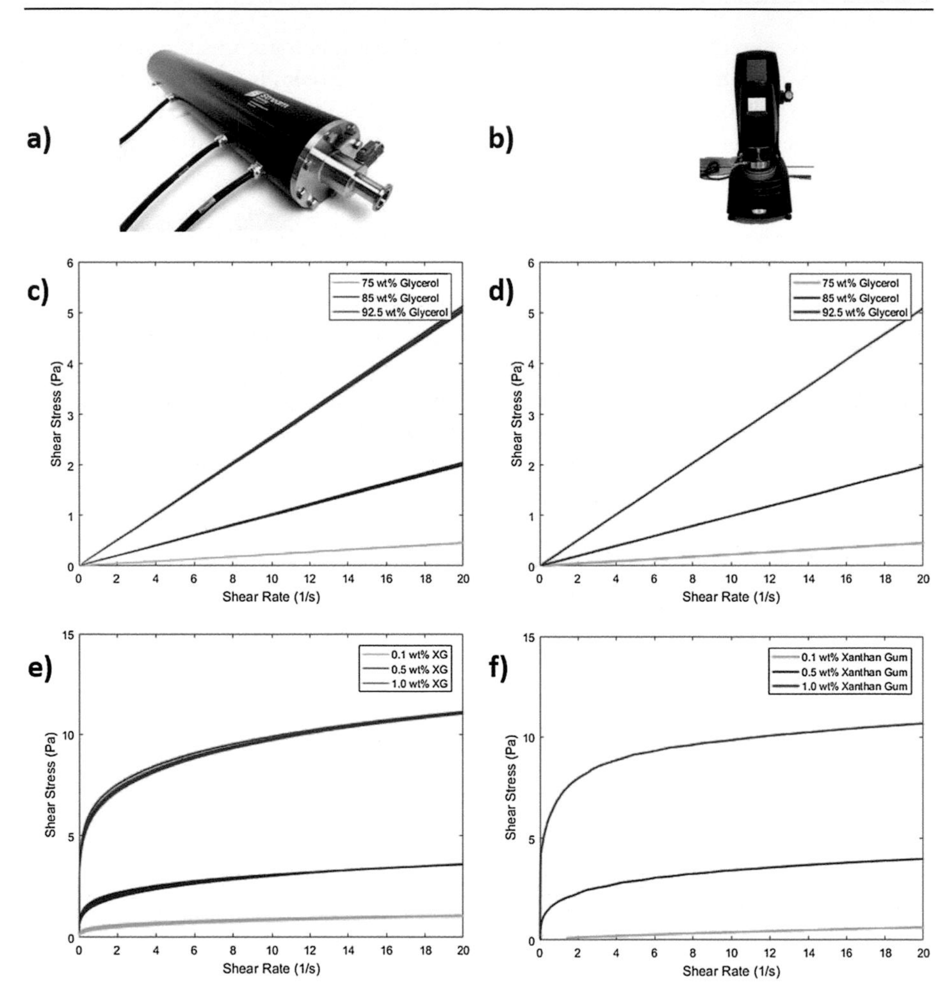

Figure 16.14 (a) ERR in-line rheometer; (b) rotational rheometer (RR); (c) ERR – glycerol; (d) RR – glycerol; (e) ERR – xanthan gum; (f) RR – xanthan gum.

16.5 Process monitoring for process control

16.5.1 Dense phase hydraulic conveying

The measurement of slurry density—along with bulk flow velocity—is a fundamental parameter for the efficient operation of modern dredgers (Miedema & Ramsdell, 2016). Many dredgers will use these two control parameters to ensure that solids are mobilized and are present at a sufficient level in the pipeline for effective transport, but are not at too high a level to endanger operations through settling and subsequent blocking of pipelines.

Gamma densitometers are nearly universally used as the basis of in-line density measurement. Nuclear sources have been in use for decades as a measuring tool and the use of a radioactive source on board a vessel is often seen as a technique of last resort due to the inherent perceived and actual concerns of such materials.

Issues that arise with nuclear instrumentation include the following (Primrose 2016):

- regulatory requirements for the
  - transport
  - maintenance and,
  - disposal of radioactive sources
- Restrictions on access around the nuclear source
- Frequency calibration; while different systems require different procedures, frequently a full calibration will require off-line measurements with standardized plates. As well as normal calibration of instrumentation, recalibration is required due to wear of the pipe-wall and natural depletion of the source.
- Single, fan-beam measurements place restrictions on either the orientation of the sensor or limits to its accuracy. This is due slurry falling outside the fan beam not being measured. If flows are heterogonous, this will bring an inherent error/uncertainties to the resultant density measurement
- Differing absorptions levels of different elemental species in the solid slurry lead to uncertainties in the density data
- Partially filled pipes lead to uncertainties over the density of the slurry as the single measurement of density is a composite of air, water, and solids

In addition to the above, the gamma densitometer is not able to provide information on the distribution of solids and/or flow regime. Nor does it provide information on flow velocities.

Of the above issues, the regulatory burden over the operation of nuclear sources is the most significant.

It is frequently very challenging to transport a replacement source to particular countries, requiring dredgers to be moved to a separate location, at significant cost. In addition, a small number of countries (for example, Russia or Nigeria) limit or prohibit the use of nuclear instrumentation by overseas dredging operators.

In addition, it should be noted that nuclear sources present a possible terrorist threat, with reports in 2015 of Cesium sources being trafficked (NY Post 2015). Such events and increased tension are likely to increase the regulatory burden on the transport, use, and disposal of sources for legitimate purposes.

For the above reasons, there has been a long-standing interest in the development of alternative to gamma densitometers. Standard density measurements such as Coriolis flow meters and vibrating forks are not suitable to dense phase slurries. A range of other techniques have been tried; however, no alternative has effectively met the market's requirements. For example:

- ultrasound/acoustic methods are limited by the high levels of attenuation and scattering of such ultrasound presenting difficulties. The signal can also be highly dependent on the positioning of the transducer which can also present issues with respect to pipe orientation.
- microwave sensors based on phase difference have been deployed in the pulp and paper sector; however, the issues associated with large pipe sizes and process variabilities in dredging have prevented the successful deployment of these sensors

- mass flow – a load-cell sensor was presented a number of years ago, and while some tests appeared promising, the bulkiness of the sensor, limitations with respect to its installation, and other factors prevented it from being a commercially viable alternative

The limitations of nuclear sources and difficulties with the above methods have meant that there remains an ongoing requirement to develop an alternative to nuclear based density meters.

In 2005, a dredging research group, through Van Oord, approached Industrial Tomography Systems to see if electrical resistance tomography could be used as the basis for a nuclear-free density meter and the development path of the technology is illustrative of the issues around the application of emerging lab-based measurement techniques to industrial processes.

In the decade leading up to 2019, there have been many advances to develop the technology, the most recent of which has been integrating the electrical resistance sensor array with a standard EMF (electromagnetic flow) meter to provide the world's first fully integrated nonnuclear production meter for dense phase hydraulic conveying (Primrose, 2019).

Fig. 16.15 above shows a system installed inside a dredger and the key components are described in more detail below:

16.5.1.1 Sensor

Fig. 16.16 shows an example of an integrated DN700 production meter. The liner is built using ceramic tiles to ensure maximum lifetime from abrasion of slurries. These are often pumped at velocities of 10/ms.

Figure 16.15 Integrated production meter (DN800) installed in dredger.

Figure 16.16 Integrated DB700 production meter.

The tomography electrodes can be seen forming a ring in the center of the pipe and close inspection of the image shows one of the EMF electrodes at 3'o'clock. The top of the meter shows the junction box for both ERT and EMF signals.

16.5.1.2 Electronics

The tomography and EMF electronics are installed in an IP68 enclosure which can be mounted in the ship's pump room or on deck. The industrial processor produced by Bachmann is able to process both flow and density data and out output analogue and digital data across a variety of standard formats. The system is able to receive control data from the ship's automation control software and has an autocalibration functionality that can sustain performance under a variety of conditions. Units have been taking uninterrupted data over many years, demonstrated the robustness and reliability of the in-field instruments.

16.5.1.3 Software

Fig. 16.17 shows the graphical user interface for the production meter as developed by Industrial Tomography Systems Ltd.

The graph at the top sets out a number of single performance variables which can be logged over time.

Lower left shows the production tomograph of instantaneous solids concentration and average solids.

Lower right shows the "cross-hairs" display commonly used in dredging. The two lines show liquid velocity and concentration; where they intersect shows volume production. The standard performance objective is for the control engineer to get the intersection as high up as possible. The lower colored strip shows solids flow profile over time and can be used to ensure the continual flow of solids. A build-up of solids leading to a blocked pipe is extremely expensive.

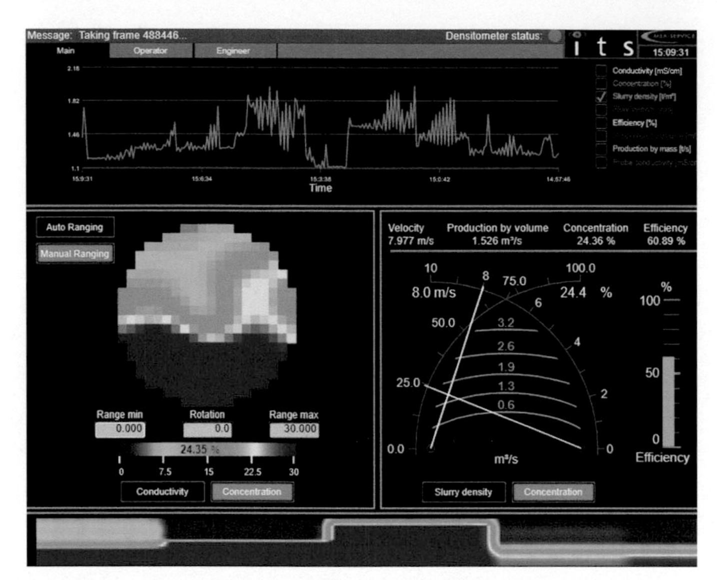

Figure 16.17 Dredging density meter graphic user interface (GUI).

16.5.1.4 Performance

In comparative tests, the tomography density meter has shown equivalent measurements with gamma density meters, as well as other comparisons such as integrated production based on ship's displacement. The EMF meter performance is also equivalent to other independent flow measurements[26].

Ship's crews have been very supportive of the technology as it is considerably more responsive than gamma (which use moving averages), allowing the vessel to respond more quickly to changing production conditions.

As the tomography density measurements are based on Maxwell's equation, reliable conductivity data of the water and solids are essential to good accuracy. Operationally, this has presented some challenges, particularly if the conductivity of the water in the pipe changes during operation (such as in an estuary). Through years of operation, a number of strategies have been deployed which have led to an optimized installation and operation delivering reliable performance to extent that a number of vessels now rely on these instruments for their operations.

16.5.2 Wider application of dense phase hydraulic conveying

Extending beyond the initial application of electrical resistance tomography to dredging, Industrial Tomography Systems Ltd. has installed density meters in other applications:

- Mineral processing

  Gamma density meters are used across a wide range of applications in mineral processing and ITS has installed meters in mining installations.

Work has previously been reported on the application of electrical tomography instrumentation being applied in a number of mineral processing operations where close correspondence between electrical tomography and gamma density measurements were demonstrated.

- Tunneling

  Similarly, slurry-based tunnel boring machines (TBMs) need to accurately measure the solids excavated during operation to ensure safe operation. Hitherto this has been carried out using measuring the difference between solids in the input and output slurry lines by using two gamma density meters.

  Tomography density meters have demonstrated encouraging performance on TBMs and are currently under evaluation prior across a range of operating conditions ahead of being deployed as an alternative to gamma density meters as shown in Fig. 16.19 below.

- Deep sea mining

  Deep sea mining also requires measurement of production through density and flow. As the flows are relatively slow, ITS has developed a tomography meter which is able to measure both flow and concentration of nodules for a remote-operated vehicle (ROV) designed to operate at depths of over 4,000 m. After trials in 2019, this unit is expected to be deployed in the Pacific in 2021.

  Fig. 16.18 above shows the ROV. In addition to measuring production, it is also installed with tomography units designed to measure suspended solids concentrations at the aft of the ROV to monitor the environmental impact of operations.

Figure 16.18 Deep sea (4,000m) ROV fitted with ERT-based density meter and additional tomography senors.

Figure 16.19 Slurry density meter fitted in TBM.

16.6 Conclusions and future trends

This chapter has set out how Industrial Tomography Instruments are applied to improve the performance of process applications. It has done this by setting out the different bases in how process instrumentation is applied:

- Developing process understanding
- Process optimization
- Process monitoring
- Process control

In moving through these different classes of application, the nature of the instrumentation moves from *bespoke instruments*—often the result of university research projects—which require dedicated supervision and fine tuning for their operation and while data are available in real time, outputs are normally postprocessed—to *standard products*—typically produced by instrumentation companies, which are robust, reliable, and repeatable within a much narrower scope of operation.

The transition from bespoke to standard is usually accompanied by a move of processor from Microsoft-based PC to a lower cost processor with connectivity to standard industrial control systems. Such standardization requires investment and can only be commercially justified on the basis of significant unit volumes.

The applications described in this chapter and others described elsewhere in this book are now justifying higher levels of investment which is leading to lower cost, industrially robust tomography platforms.

The accompanying trend of Industry 4.0—in which blue chip process companies are seeking out more data and higher levels of process knowledge—is leading to increased demand for process tomography instrumentation.

Overall process tomography has come a long way in the last 2 decades—indeed even since the first edition of this publication—led by the teamwork of pioneering, entrepreneurial companies, patient academics, and enthusiastic product champions in large process companies who all share a vision of improved productivity through more complete process monitoring with process tomography.

Thanks and Acknowledgments

The author would like to thank and acknowledge the contributions of Prof Andy Hunt (who prepared much of the material on process optimization), Ms Changhua Qiu (who has been at the heart of all the instruments developed by Industrial Tomography Systems, in particular the *Densitometer*), Dr. Kent Wei (who has developed much of the control software applied by Industrial Tomography Systems), and Dr. Tom Machin (who has contributed to the in-line rheology instrumentation of Stream Sensing).

References

Alberini, F., Simmons, M. J. H., Ingram, A., & Stitt, E. H. (2014). Use of an areal distribution of mixing intensity to describe blending of non-Newtonian fluids in a Kenics KM static mixer using PLIF. *AIChE Journal, 60*, 332−1242.

Baker, A. M., Claus, M., Augousti, A., Marlow, J. J., Hunt, A., & Foster-Turner, R. (2018). *Vortex-cooled rocket engine with electrical capacitance tomography, Paper 535, in proc. Space Propulsion 2018.*

Behruzi, P., Hunt, A., & Foster-Turner, R. (2020). *Evaluation of liquid sloshing using electrical capacitance tomography, AIAA propulsion and energy forum, August 24-28 2020.* https://doi.org/10.2514/6.2020-3804

Bolton, G., Bennett, M., Qiu, C., Wang, M., Wright, M., Primrose, K., et al. (2006). *Development of an electrical tomographic system for operation in a remote, acidic and radioactive environment, to be published in Chemical Engineering Journal in October 2006, online on 20th August 2006.*

European Commission. (2014). *"Horizon 2020: Work program 2014-2015 general Annexes." G. Technology readiness levels (TRL), extract from Part 19 - commission decision C.*

Fries, N., Behruzi, P., Arndt, T., Winter, M., Netter, G., & Renner, U. (2012). *Modelling of fluid motion in spacecraft propellant tanks − sloshing, Space Propulsion 2012 Conference, Bordeaux, May 2012.*

Hunt, A., Drury, R., & Foster-Turner, R. (2018). Propellant slosh force and mass measurement. *International Journal of Aerospace Engineering, 2018*. Article ID 3026872, 9 pages, 2018.

Hunt, A., & Foster-Turner, R. (2015). *Real time multi-phase mass and density measurement of propellent in zero G, UK space propulsion workshop, 29 October 2015*. London: Imperial War Museum.

Machin, T. D. (2018). In-pipe rheology and mixing characterisation using electrical resistance sensing. *Chemical Engineering Science, 187*, 327−341.

Marlow, J. J., et al. (2016). *Low-cost rocket engine development at Kingston university London, reinventing space conference BIS-RS-2016-15.*

Miedema, S., & Ramsdell, R. (2016). *"Slurry transport, fundamentals, a historical overview & the delft head loss & limit deposit velocity framework" WODCON XXI continuing education short course on slurry transport.*

Munson, S. M., et al. (2011). *Development of a low-cost vortex-cooled thrust chamber using hybrid fabrication techniques, Joint Propulsion Conference. AIAA 2011-5835.*

NY Post. (2015, October 7). *The smugglers trying to help ISIS make a dirty bomb*. Retrieved from http://nypost.com/2015/10/07/the-smugglers-trying-to-help-isis-make-a-dirty-bomb/.

Papoulis, A. (1962). *The fourier integral and its applications*. New York: McGraw-Hill.

Primrose, K., Boer, A., Bosman, F., Maingay, D., Qiu, C., & Wei, H. Y. (2019). Real Time Production efficiency based on combination of non-nuclear density and magnetic flow instrumentation. In *Dredging summit and expo 2019 proceedings.*

Primrose, K., McCormack, D., Qiu, C., & Wei, H. Y. (2016). Development and deployment of a non-nuclear densitometer, based on electrical resistance tomography. *Wodcon XXI Proceedings.*

Qi, C., Jia, Y., Mao, X., & Ju, Y. (2016). Direct measurements of permittivity of plasma-assisted combustion using electrical capacitance tomography. *IEEE Transactions on Plasma Science, 44*(12).

Rees, J. (2014). Towards online, continuous monitoring for rheometry of complex fluids. *Advances in Colloid and Interface Science, 206*, 294−302.

Satellite Industry Association. (2019). *State of the satellite industry*. https://sia.org/wpcontent/uploads/2019/11/SSIR19-2-Pager-May-7th.pdf.

Townson, P., & Thien, M. (2011). In *Design and fabrication of test facilities for the demonstration of jet mixer performance in the Hanford double shell tanks WM2011 conference, February 27 - March 3 (Phoenix, AZ)*.

UK Space Agency. (2019). *Smart tanks for space, national space technology programme brochure v3, September 2019*. www.gov.uk/government/organisations/ukspace-agency.

van Benthem, R. C., van Es, J., van Put, P., & Matthijsen, R. (2013). *Accuracy analysis of propellant gauging systems, 43rd international conference on environmental systems (ICES), vail, CO, USA, July 2013*.

Wang, M. (2002). Inverse solutions for electrical impedance tomography based on conjugate gradients methods. *Measurement Science and Technology, 13*, 101–117.

Wilkinson, W. L. (1960). *Non-Newtonian fluids – fluid mechanics*. London: Mixing and Heat Transfer Pergamon Press Ltd.

Applications of tomography in oil–gas industry—Part 1

Cheng-gang Xie
Schlumberger, Singapore

17.1 Introduction

Applications of tomography in the upstream oil–gas industry are widespread, covering from kilometer reservoir-scale, submeter borehole scale, to submm rock-core pore scale. This chapter illustrates the use of different seismic tomography modalities in hydrocarbon exploration, reservoir monitoring, and production monitoring.

In Section 17.1, the range of physical parameters of reservoir rocks and fluids that can be explored by different tomographic-sensing methods, such as seismic velocity, resistivity (conductivity), and density, will be introduced.

The basic principle of seismic (traveltime) tomography and depth imaging, widely used for hydrocarbon exploration, is briefly described in Section 17.2.

The potential of multicomponent time-lapse (space-time four dimensional) seismic surveys in monitoring reservoir changes during production and some typical EOR processes is then discussed in Sections 17.3 and 17.4.

In Section 17.5, various borehole seismic imaging techniques are introduced. Some application examples of time-lapse cross-borehole seismic for reservoir EOR monitoring are given.

In Section 17.6, we will make some remarks of the future trend of seismic tomography imaging in oil–gas industry.

17.2 Seismic tomography in hydrocarbon exploration and reservoir characterization

A petroleum reservoir is a subsurface body of rock having sufficient porosity and permeability to store and transmit fluids. Sedimentary rocks are the most common reservoir rocks because they have more porosity than most igneous and metamorphic rocks. A reservoir is a critical component of a complete petroleum system. Subsurface rocks containing hydrocarbon oil and/or gas and connate waters are composed of grains, pores, and cementing materials. Porosity (the percentage of pore volume or void space) provides the fluids storage capacity of the rock. Permeability is a rock's ability to transmit fluids, typically measured in darcies or millidarcies (mD). Porosity and permeability influence a reservoir's ability to accumulate and produce hydrocarbons. A petroleum trap is a configuration of rocks suitable for containing hydrocarbons

Industrial Tomography. https://doi.org/10.1016/B978-0-12-823015-2.00031-5

and is sealed by a relatively impermeable formation (a barrier) through which hydrocarbons will not migrate laterally or upward, allowing oil–gas to accumulate. A trap is an essential component of a petroleum system, and may be a structural trap (in deformed strata such as folds and faults, Fig. 17.1) or a stratigraphic trap (in areas where rock types change, such as unconformities, pinch-outs, and reefs).

Various geophysical exploration techniques have been used in the upstream oil–gas industry to help identify petroleum reservoir prospects and plan drilling operations, such as the gravity method and the seismic method. The gravity method measures local gravity-field differences due to variations in density (ρ) in the subsurface; the effects of such density variations are very small (in the order of 0.1–1 part per million) compared with those of the Earth's (global) gravity field (Seidel & Lindner, 2007, p. 185). A gravity survey may be used to detect, for example, lithological and structural changes and fractures in consolidated rock, or lithology in unconsolidated rock. However, the use of the measurements from the gravity (potential) field alone, due to the principle of equivalence, cannot yield unique information about the depth, size, and/or density of a subsurface gravity anomaly. An integrated geophysical survey including additional data (such as seismic velocity data) can reduce the ambiguity when interpreting gravity survey data.

In the quest to find hydrocarbon resources, electromagnetic surveys are also used to identify subsurface features by electrical resistivity contrast. Near-surface (shallow) electromagnetic surveys may be conducted by deploying an inductive loop at the earth surface (Dawoud, Hallinan, Hermann, & van Kleef, 2009). Two different deep-reading EM technologies, magnetotelluric (MT) and controlled-source electromagnetic (CSEM), provide distinctly different insights into the earth subsurface that are complementary to seismic data. The MT fields generated by passive (atmospheric, plane-wave vertically incident) sources are sensitive to large (basin-scale) conductive features and

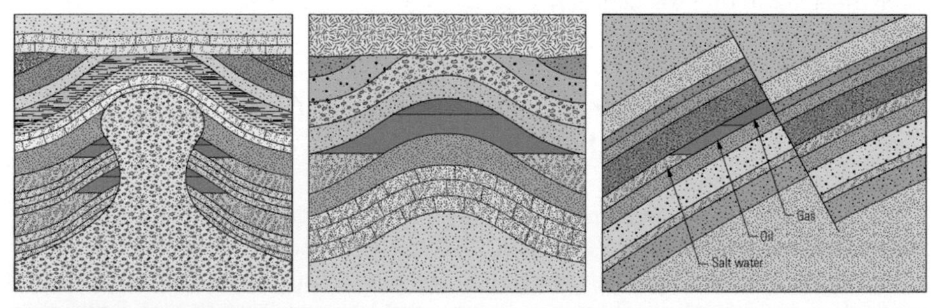

Figure 17.1 Structural traps. The weight of overlying sediments causes salt layers to plastically deform, creating diapirs. As diapirs evolve, sediments onlap their margins, forming traps that typically accommodate hydrocarbons (left). Where the strata have deformed to form an anticline (center), oil (green [dark gray in print version]) and gas (red [gray in print version]) may become trapped under a seal. Faulting may also trap hydrocarbons (right) by sealing the undip margin of a reservoir.
From Barriol, Y., Glaser, K. S., Pop, J., Bartman, B., Corbiell, R., Eriksen, K. O. et al. (2005). The pressure of drilling and production. *Oilfield Review 17*(3), 22–41. Graphic copyright Schlumberger. Used with permission. *Courtesy:* Lisa Stewart (former executive editor of Oilfield Review) & Schlumberger.

are used to detect structure and lithology, such as large salt, basalt, and carbonate bodies due to their contrasts with the conductive surroundings (Brady et al., 2009). The complex electrical impedance (Z) of the subsurface formation (hence the bulk formation resistivity ρ_a) can be obtained by taking the ratio of the horizontal electric field in one direction (E_x) to the horizontal magnetic field in orthogonal direction (H_y), i.e., $Z = E_x/H_y = (i\ \rho_a\ \omega\mu)^{1/2}$ (where ω is the angular frequency, μ the magnetic permeability). Recent interest in MT measurements has focused on evaluation of marine or subsea environments, driven by the increasing cost of deep-water drilling and the complexity of imaging below salt and basalt.

However, the attenuation of the MT fields with depth makes them insensitive to resistivity contrasts of thin reservoir formations. The marine CSEM method imposes an appropriate EM signal from a transmitter source comprising a long horizontal dipole, making it more sensitive in detecting resistivity layers (possibly hydrocarbon-bearing formations) against a conductive background (water-bearing formations). The same receivers can be used for both CSEM and MT measurements. The typical frequency range of the CSEM signal is between 0.05 and 5 Hz; the effective upper limit for marine MT is 1 Hz. Major obstacles to marine MT and CSEM surveys are the high cost and the low data-collection efficiency. Seismic measurements covering large three-dimensional areas can be performed efficiently by deploying vessel-towed multiple receiving streamers and source array (Brady et al., 2009).

The seismic exploration industry was revolutionized in the early 1990s by the transition of imaging from two to three dimensions, from time to depth, and from poststack to prestack. The industry replaced interactive normal moveout (NMO) velocity analysis with efficient methods that can meet the new challenges in large data volume and in growing spatial resolution expectations (In Section 17.2.3, the meanings of poststack, prestack, and NMO will be described). By the late 1990s, the industry was implementing ray-based, prestack depth migration tomography methods to perform full migration velocity analysis (Woodward, Nichols, Zdraveva, Whitfield, & Johns, 2008). Over the last decade, ray-based postmigration grid tomography has become the standard model-building tool for seismic depth imaging. The problems it can solve have become bigger, driven by seismic exploration demands and enabled by 1000-fold increases in computing power. The three main areas of change are as follows (Woodward et al., 2008):

- Standard model spatial resolution has improved 10-fold, from a few thousand meters to a few hundred meters. This has been attributed to high-quality multiple-parameter residual moveout data picked as densely as 25–50 m vertically and horizontally, and to a strategy of working down from long-wavelength to short-wavelength solutions.
- More seismic data sets are being acquired along multiple azimuths for improved illumination and suppression of multiples. In order to prevent short-wavelength velocity heterogeneity from being mistaken for azimuthal anisotropy, high-resolution velocity tomography must solve for all azimuths simultaneously.
- Models have shifted from largely isotropic to largely anisotropic, both vertical transverse isotropic (VTI) and TTI (tilted TI). With four-component data, anisotropic grid tomography can be used to build models that tie PZ and PS images in depth (see Section 17.3).

17.2.1 Basics of seismic waves

The basic principle of all seismic methods is the controlled generation of elastic waves by a seismic source to obtain an image of the subsurface properties. Surface and/or borehole seismic energy sources (e.g., explosives, vibrator, airgun) produce the following type of strain−energy wave pulses that propagate in solids and fluids (Schuck & Lange, 2007, p. 337):

- Body wave that travels in all directions and
- Surface wave that travels along/near the surface (usually considered a source of noise). (Dispersion of surface waves has been used to solve for near-surface velocity variations, Bagaini, Bunting, El-Emam, Laake, & Strobbia, 2010; Strobbia et al., 2009.)

There are two types of body waves:

- *P*-waves (primary, longitudinal, or compressional waves) with particle motion parallel to the direction of propagation and
- *S*-waves (secondary, shear, or transverse waves) with particle motion perpendicular to the direction of propagation (and referred to as *SV*-waves and *SH*-waves with particle motion in the vertical plane and in the horizontal plane, respectively).

In marine seismic surveys, hydrophones are often used to record pressure (stress) P directly. In a medium of density ρ and compressibility (or bulk modulus) κ, the following wave equation can be derived based on the theory of elastic continua, which relates deformation (strain) and stress (Chapman, 2004, p. 104):

$$\frac{\nabla^2 P}{\partial t^2} = \kappa \nabla \cdot \left(\frac{1}{\rho} \nabla P \right) - \kappa \nabla \cdot \left(\frac{1}{\rho} \mathbf{F} \right). \tag{17.1}$$

For a medium of constant density, the above is reduced to

$$\frac{\nabla^2 P}{\partial t^2} = c^2 \nabla^2 P - c^2 \nabla \cdot \mathbf{F}, \tag{17.2}$$

where the acoustic wave velocity c is defined by

$$c = \sqrt{\kappa/\rho} \tag{17.3}$$

The source is modeled as a force distribution $\mathbf{F}$ (e.g., representing an underwater airgun explosion "equivalent force," Chapman, 2004, p. 104). Outside the source region $F = 0$, Eq. (17.2) is reduced to the Helmholtz wave equation. For a wave of harmonic time dependence of the form $P(\mathbf{r}, t) = P(\mathbf{r},\omega)\, e^{-i\omega t}$, the frequency-domain solution for homogeneous media is

$$c^2 \nabla^2 P + \omega^2 P = 0. \tag{17.4}$$

The general solution is of the form $g(|\mathbf{r}| - ct)$. A special wave is the harmonic wave where g is a sinc, cosine, or more generally:

$$P(\mathbf{r}, t) = P(\mathbf{k}, \omega)e^{i(\mathbf{k}\cdot\mathbf{r}-\omega t)}, \tag{17.5}$$

where $P(\mathbf{k}, \omega)$ is the amplitude of the harmonic wave; the wavenumber vector $\mathbf{k}$ in the direction of wave propagation and angular frequency ω are related by the dispersion relationship $|\mathbf{k}| = \omega/c$. A more general wave shape can be formed by summing up individual harmonic waves in a Fourier series. For a system of homogeneous layers, the homogeneous solution can still be applied within each layer. Ray approximation can be used to predict wave propagation in smoothly varying inhomogeneous media (Nolet, 2008, p. 20).

For a homogeneous isotropic elastic medium of very small deformations, according to Hooke's law, strain (change of volume and shape) is proportional to stress (force per unit area). For anisotropic media, a general elastic tensor containing up to 21 independent elastic constants is used to describe the relations between stress and strain. In Table 17.1, these elastic constants are listed versus each other, P- and S-wave velocities (V_p and V_s), and bulk density ρ. For the simplest case of seismic wave propagation

Table 17.1 Relationships between elastic constants, P- and S-wave velocities, and bulk density ρ for isotropic media.

		λ, μ	κ, μ	E, σ	V_p^2, V_s^2
Lamé's constant	λ	λ	$\kappa - \frac{2}{3}\mu$	$\frac{E\sigma}{(1+\sigma)(1-2\sigma)}$	$\rho\left(V_p^2 - 2V_s^2\right)$
Shear modulus	μ	μ	μ	$\frac{E}{2(1+\sigma)}$	ρV_s^2
Bulk modulus	κ	$\lambda + \frac{2}{3}\mu$	κ	$\frac{E}{3(1-2\sigma)}$	$\rho\left(V_p^2 - \frac{4}{3}V_s^2\right)$
Young's modulus	E	$\mu\left(\frac{3\lambda+2\mu}{\lambda+\mu}\right)$	$\frac{9\kappa\mu}{3\kappa+\mu}$	E	$\rho V_p^2\left(\frac{3V_p^2-4V_s^2}{V_p^2-V_s^2}\right)$
Poisson's ratio	σ	$\frac{\lambda}{2(\lambda+\mu)}$	$\frac{3\kappa-2\mu}{6\kappa+2\mu}$	σ	$\frac{\frac{V_p^2}{V_s^2}-2}{2\frac{V_p^2}{V_s^2}-2}$
P-wave velocity	V_p^2	$\frac{\lambda+2\mu}{\rho}$	$\frac{\kappa+\frac{4}{3}\mu}{\rho}$	$\frac{E(1-\sigma)}{\rho(1+\sigma)(1-2\sigma)}$	V_p^2
S-wave velocity	V_s^2	$\frac{\mu}{\rho}$	$\frac{\mu}{\rho}$	$\frac{E}{2\rho(1+\sigma)}$	V_s^2
P- to S-wave velocity ratio	$\left(\frac{V_p}{V_s}\right)^2$	$\frac{\lambda+2\mu}{\mu}$	$\frac{\kappa+\frac{4}{3}\mu}{\mu}$	$\frac{2(1-\sigma)}{1-2\sigma}$	$\left(\frac{V_p}{V_s}\right)^2$

Adapted from Table 4.6-1 of Schuck, A., & Lange, G. (2007). Seismic methods. In: J. Knodel, G. Lange, G., & H.-J. Voigt (Eds.), Environmental Geology – handbook of field methods and case studies, (p. 344). Berlin: Springer, with kind permission from Springer Science + Business Media.

in an isotropic medium, there exist only two elastic constants, Lamé's constant λ (lambda constant) and μ (shear modulus); the wave equation solution results in two different velocities $V_p = \sqrt{(\lambda + 2\mu)/\rho}$ and $V_s = \sqrt{\mu/\rho}$ (Nolet, 2008, p. 28). Because fluids do not support shear stress, *S*-waves cannot propagate in fluids. The velocities of seismic waves are the most fundamental attributes in the seismic method (obviously $V_s < V_p$); they vary with lithology, mineral content (Table 17.2), porosity, pore fluid saturation, and degree of compaction.

As an illustrative example of the dependence of velocity on lithology, we consider shaly sandstones and shales, which are a major component of sedimentary basins and of prime relevance to oil reservoirs. Experimental studies of 80 saturated shaly sandstone samples (at confining pressure 40 MPa and pore pressure 1 MPa) have found that both *P*-wave velocity (V_p) and *S*-wave velocity (V_s) decrease with increasing porosity (ϕ). A clay volumetric content term (C_{clay}) is also needed in order to fit the experimental data (Nur, 1987, p. 205):

$$V_p[km/s] = 5.59 - 6.93\phi - 2.18C_{clay},$$

$$V_s[km/s] = 3.52 - 4.91\phi - 1.89C_{clay}$$

The velocity ratio V_p/V_s is a useful seismic attribute in the determination of rock properties. For water-saturated shaly sandstones, the increase of V_p/V_s with increasing porosity and/or increasing clay content was found by least-square regression (Nur, 1987, p. 207):

$$V_p/V_s = 1.55 + 0.56\phi + 0.43C_{clay}.$$

Sandstones with high clay content have velocity ratios and Poisson's ratios similar to those of carbonate rocks. The combined use of velocity and velocity ratio may provide discrimination of lithology.

Seismic waves are reflected, refracted, or diffracted at subsurface boundaries between layers of different properties. Reflection seismology is the most important method to prospect for oil and gas at great depths. The recording of seismic waves returning from the subsurface to surface and/or borehole receivers (e.g., geophones, hydrophones) and subsequent processing allow inference of structures and lithological composition of the subsurface. With measurements of traveltimes and amplitude attenuations of seismic waves, velocities of subsurface layers can be determined and a geological model of the subsurface reservoir can be constructed.

For a planar wave incident normal to a reflecting interface, the contrast in acoustic impedance (ρV) across the interface is expressed as the reflection coefficient or reflectivity R:

$$R = (\rho_2 V_2 - \rho_1 V_1)/(\rho_2 V_2 + \rho_1 V_1). \tag{17.6}$$

The corresponding transmission coefficient T is

Table 17.2 Typical densities and resistivities of selected sedimentary, metamorphic and igneous rocks, and fluids.

	Density ρ (g/cm)	Resistivity (Ω m)	*P*-wave velocity V_p (m/s)	*S*-wave velocity V_s (m/s)
Sediments				
Fine sand, dry	1.5 (1.3–1.7)	> 10^4	100–600	100–500 (sand)
Fine sand, wet	1.8 (1.7–2.0)	50 (water saturated)	200–2000; 1300–1800 (saturated); 1740 (fine)	
Medium sand	1.7 (1.5–2.0)		1835 (coarse)	
Gravel, dry	1.7 (1.7–2.0)	> 10^4	180–1250;	
Gravel, wet	2.0 (1.8–2.3)	50 (water saturated)	750–1250 (wet)	
Silt	1.7 (1.2–2.0)	20–50		
Clay, dry	1.7 (1.3–1.8)	>1000	500–2800	110–1500 (clay)
Clay, wet	1.9 (1.7–2.2)	5–30		
Sandstone	2.35	<50 (wet,	800–4500	320–2700
Loose	(1.6–2.7);	jointed);	1500–2500	575–1101
Compacted	1.80–2.40;	>10^5	1800–4300	672–1023
Siliceous	2.22–2.69	(compact)	2200–2400	
Limestone	2.55	100 (wet, jointed);	2000–6250	1800–3800
Dolomite	(1.9–2.8)	>10^5	2000–6250	2900–3740
Gypsum	1.75–2.88	(compact)	1500–4600	750–2760
Anhydrite	2.31–2.33 2.15–2.44		4500–6500	750–3600
Halite (rock salt)	2.2 (2.1–2.6)	30 (wet); >10^6 (compact)	4500–6500	2250–3300
Metamorphic and igneous rocks				
Granite	2.65 (2.5–2.7)	<100 (weathered, wet); >10^6 (compact)	4500–6000[a]	2500–3300[a]
Granodiorite	2.7 (2.6–2.7)	- ditto-		
Basalt	3.1	- ditto-	5000–6000[a]	2800–3400[a]
Gneiss	2.7 (2.6–2.8)	- ditto-	4400–5200[a]	2700–3200[a]
Fluids				
Air (gas)	1.29×10^{-3}		310–340	
Water	0.98–1.01	10–300	1430–1590	
Sea water	1.02	0.25	1400–1600	
Brine	Up to ~1.2	<0.15	1400–1600	
Petroleum	0.92–1.07	>10^5	1035–1370	

[a]https://pangea.stanford.edu/courses/gp262/Notes/8.SeismicVelocity.pdf (accessed August 6, 2013).
Adapted from Table 4.2-1 of Seidel, K., & Lindner, H. (2007). Gravity methods, In J. Knodel, G. Lange, & H.J. Voigt (Eds.), Environmental geology – handbook of field methods and case studies, (p. 188). Berlin: Springer; Table 4.3-2 of Seidel, K., & Lange, G. (2007). Direct current resistivity methods, In: J. Knodel, G. Lange, & H.-J. Voigt (Eds.), Environmental geology – handbook of field methods and case studies, (p. 213). Berlin: Springer; Table 4.6-2 of Schuck, A., & Lange, G. (2007). Seismic methods, In J. Knodel, G. Lange, & H.-J. Voigt (Eds.), Environmental geology – handbook of field methods and case studies, (p. 346). Berlin: Springer. With kind permission from Springer Science + Business Media.

$$T = 1 - R = 2\rho_1 V_1/(\rho_2 V_2 + \rho_1 V_1). \tag{17.7}$$

In general, incidence angle−dependent reflection and transmission coefficients for elastic plane waves at a horizontal boundary are given by the Zoeppritz equations (see references in Schuck & Lange, 2007, p. 349). The angle of the reflected and transmitted wave at oblique incidence can be easily obtained from the Snell's law:

$$p = \sin(i_\alpha)/V_\alpha = \text{constant}, \tag{17.8}$$

where p is the ray parameter (slowness), i_α the angle of the incident, reflected, or transmitted ray, and V_α is the P- or S-wave velocity of the wave-propagating medium. The transmitted wave is refracted with a change in its propagation direction if there is a velocity contrast at the interface; the reflection and refraction angles are independent of bulk density.

For the case of P-wave propagation shown in Fig. 17.2, four waves will be generated: reflected P-wave and SV-wave, transmitted (refracted) P-wave and SV-wave. The angles of reflected and refracted rays can be calculated as follows by using Snell's law:

$$\sin(i_{p1})/V_{p1} = \sin(i_{p2})/V_{p2} = \sin(i_{s1})/V_{s1} = \sin(i_{s2})/V_{s2} = p = \text{constant}$$

The above calculations are applicable only to planar and relatively large interfaces, and are not valid for regions around faults or fracture zones if the scale of such a structure is equal to or less than the wavelength ($= 2\pi V/\omega$); at such interfaces, the wave

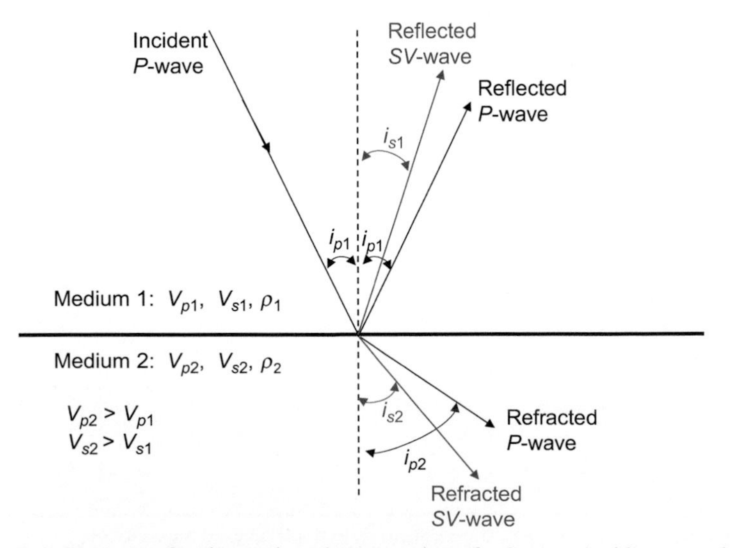

Figure 17.2 Reflection, refraction and mode conversion of a P-wave incident at a planar interface between Medium 1 and Medium 2 with different P- and S-wave velocities V_p and V_s (and bulk densities ρ).

energy will be diffracted instead (as a secondary spherical source according to Huygens' principle). Since the reflected, refracted, and diffracted waves interfere with each other, identifying diffracted waves on seismic record may be a challenge (Wielandt, 1987).

17.2.2 Traveltime tomography

In Eq. (17.8), the (horizontal) slowness $p = dT/dX$ (measured in s/m) is known as the ray parameter. At an extended interface with a discontinuity in seismic velocity V, the slowness remains constant, implying a jump in incidence angle i by refraction. The horizontal distance X traveled by a ray with slowness p is (Nolet 2008, pp. 22,23)

$$X(p) = \int dx = \int \tan i dz = \int \sin i/\sqrt{1-(\sin i)^2} dz$$
$$= \int V \cdot p/\sqrt{1-V^2p^2} dz, \quad (17.9)$$

and the corresponding traveltime is

$$T(p) = \int dt = \int V^{-1} ds = \int V^{-1}/\sqrt{1-V^2p^2} dz \quad (17.10)$$

For a general three-dimensional geological model, the traveltime for a ray is a function of the velocity profile $V(\mathbf{r})$ and the ray path trajectory S. For a number of seismic sources ($j = 1, \ldots, M$) placed at the surface, the classical problem of "traveltime tomography" is to derive the velocity model $V(\mathbf{r})$ of the interior of the Earth from a number of traveltimes measured by a number of receivers ($i = 1, \ldots, N$) at the surface (Nolet, 1987, p. 9):

$$T_{i,j} = \int_{S_{i,j}} V^{-1}(\mathbf{r})\, ds \quad i = 1, \ldots, N; \; j = 1, \ldots, M. \quad (17.11)$$

If there are a sufficient number of ray paths, analytic solutions to $V(\mathbf{r})$ are available (Chapman, 1987). In general, this is not the case and more general numerical methods have to be used (Nolet, 1987).

To solve a traveltime tomography problem numerically, the image domain can be partitioned into pixels with the slowness p specified pixel wise. The resulting matrix equation is as follows:

$$\mathrm{t} = \mathrm{S} \cdot \mathrm{p} \quad (17.12)$$

where $\mathbf{t}$ is the traveltime vector for each ray path, $\mathbf{p}$ the slowness vector for each pixel, and $\mathbf{S}$ the ray path matrix with S_{ij} = arc length of segment of ray path i in pixel j.

Traveltime tomography can be obtained by solving Eq. (17.12) using, for example, simultaneous iterative reconstructive technique (van der Sluis & van der Vorst, 1987), with ray path information obtained from ray tracing (with Snell's law obeyed at interfaces) through the current estimate of the seismic-velocity field (Červeny, 1987). For 2D traveltime tomography problem, Guo et al. (2019) applied supervised descent machine-learning method that helps the inversion to skip local minima and achieve fast convergence. The numerical results show that the accuracy and speed of the inversion based on supervised descent method can be enhanced in comparison with classical gradient-based methods.

The intrinsic amplitude attenuation when a wave propagates through a rock medium can be expressed in terms of effective Q, defined as the integration of the Q effects along the propagation ray path S. If the amplitude attenuation varies linearly with frequency f and if Q effects are frequency independent, the log ratio between a propagated signal of amplitude A and the reference A_o may be approximated as follows (Cavalca & Fletcher, 2008):

$$ln(A/A_o) = -\pi f \int_S Q^{-1}(S)V^{-1}(S)ds = -\pi fT/Q_{eff} + c \tag{17.13}$$

where T is the propagating wave traveltime; c captures remaining elastic effects such as geometrical spreading or transmission losses and is assumed frequency independent.

For attenuated traveltime tomography, an equivalent absorption coefficient α can be deduced from Eq. (17.13) as

$$\alpha = -\pi f/(QV). \tag{17.14}$$

Eq. (17.13) can then be rewritten in the following discretized form for a varying α:

$$-ln(A_i/A_o) = \sum_j \alpha_j S_{ij} \tag{17.15}$$

A_i is the measured amplitude after traversing ray path i, α_j the absorption coefficient in pixel j, and S_{ij} the length of ray path i within pixel j. Leggett, Goulty, and Kragh (1993) applied an appropriate correction to the amplitude A_i in Eq. (17.15) to account for the effects of directivity, geometric spreading, and elastic transmission losses (source-receiver coupling factors).

Traveltimes or amplitudes of propagating waves can be inverted to obtain a distribution of seismic velocity V or absorption coefficient α of the surveyed domain (Tarantola, 1987). For example, in Leggett et al. (1993), the resulting attenuated traveltime tomography locates the waterflood zone in a simulated EOR process. Time-lapse "cross-well" seismic amplitude tomography showed the flood zone as a region of higher absorption. Amplitude absorption tomograms are found to complement velocity tomograms in imaging heterogeneous reservoirs because absorption and velocity respond differently to changes in liquid or gas saturations in reservoir rocks.

17.2.3 Seismic depth imaging—stacking and migration

In structurally complex reservoirs, especially where faulting and salt intrusions necessitate complicated seismic-velocity models, conventional (travel) time-domain processing gives misleading results. Seismic depth imaging reveals the true location and shape of subsurface structures and hence improves the success rates of oil–gas companies in challenging exploration and reservoir delineation projects (Albertin et al., 2002).

Similar to time-domain imaging, depth imaging consists of two main processing aspects: stacking and migration. *Stacking* increases the signal-to-noise ratio (SNR) by summing records obtained from several seismic shots reflecting at the same point (Fig. 17.3). Traces from several source–receiver pairs, centered on a common midpoint (CMP, at the same reflection point but with different offsets), are gathered together. The shape of the two-way traveltimes (T_x) plotted against the offset (X) defines a hyperbola (Dutta, 2002):

$$(T_x)^2 = (T_0)^2 + (X/V)^2, \tag{17.16}$$

where T_0 is the time at zero offset ($X = 0$), the normal incidence reflection time, V the (stacking) velocity of the layer.

The variation of reflected wave traveltime with offset due to an oblique wave path, in comparison with the normal incidence trace, is called NMO. So from Eq. (17.16)

$$NMO = T_x - T_0 = \sqrt{(T_0)^2 + (X/V)^2} - T_0 \tag{17.17}$$

For any sample in a trace in the CMP gather, T_0 and the offset X are known. If V is also known, NMO can be calculated for each trace, and the reflections can be shifted backwards in time (from recorded time T_x to T_0 at zero offset) and then stacked (summed) across the gather to form a single enhanced trace with an improved SNR (by a factor of $\sqrt{n}$ if n traces are stacked). In stacking, the trace amplitude is increased (by a factor of n if n traces are summed).

The velocity V in Eq. (17.17) is initially unknown; a process can be used to find the particular velocity value (called the stacking velocity V_{stack}) that makes the best NMO correction for each reflection and stacks the reflection to maximum amplitude. An approximation of the NMO is as follows (Ashcroft, 2011, p. 44; Dutta, 2002):

$$NMO \approx X^2 / \left(2V_{\text{stack}}^2 \, T_0\right). \tag{17.18}$$

For flat-lying strata comprising m layers, stacking velocity at short range (called V_{NMO}) is a good approximation to the root-mean-square velocity V_{rms}:

$$V_{\text{rms}} = \sqrt{\sum \left(V_i^2 t_i\right) / \sum (t_i)}, \; (i = 1, \dots m) \tag{17.19}$$

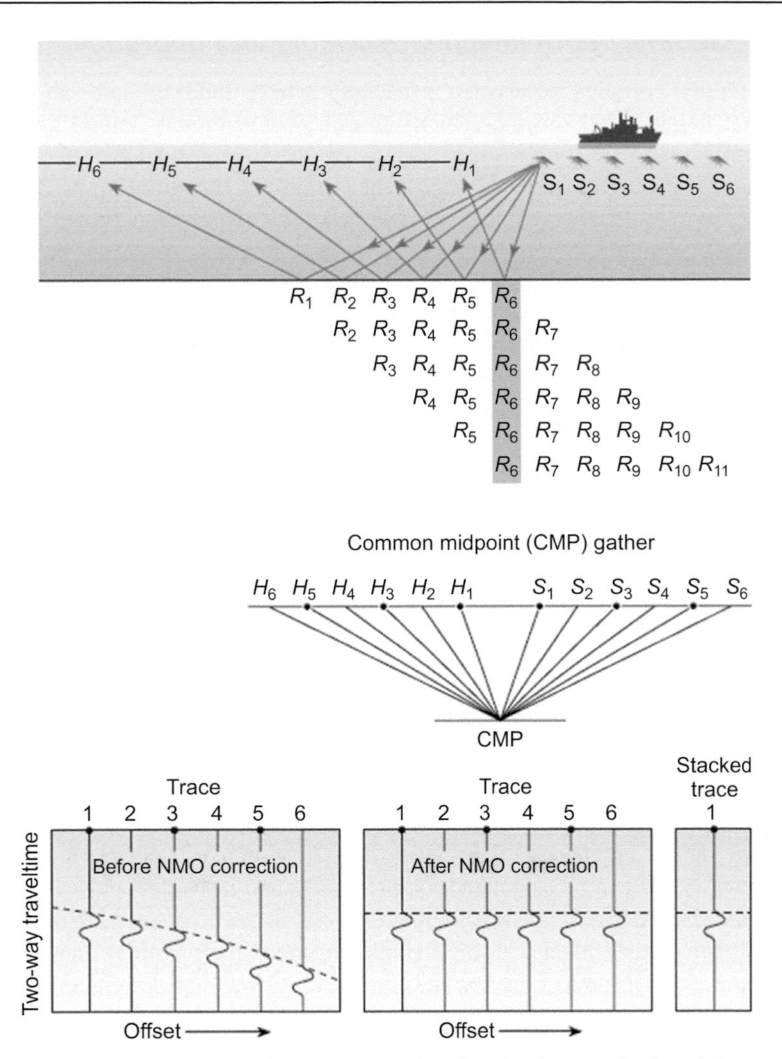

Figure 17.3 Stacking basics. Stacking enhances signal and reduces noise by adding several traces together. In this example, a marine seismic vessel acquires traces at many offsets from every source (top). S numbers represent sources, R numbers represent reflection points, and H numbers represent hydrophones. Stacking first gathers traces from all available source−receiver offsets that reflect at a common midpoint (CMP) (middle). Because arrivals from longer offsets have traveled farther, a time correction, called normal moveout (NMO) correction, is applied to each gather to flatten the arrivals (bottom left). The flattened traces are averaged (bottom right) to produce one stacked trace that represents the normal incidence (zero-offset) trace, with improved signal-to-noise ratio.
From Barclay, F., Bruun, A., Rasmussen, K. B., Alfaro, J. C., Cooke, A., Cooke, D. et al. (2008). Seismic inversion: reading between the lines. *Oilfield Review. 20*(1), 42−63. Graphic copyright Schlumberger. Used with permission. *Courtesy:* Lisa Stewart (former executive editor of Oilfield Review) & Schlumberger.

with the i-th layer interval time and interval velocity being

$$t_i = (T_{i+1} - T_i)/2,$$

$$V_i = 2(D_i - D_{i-1})/(T_i - T_{i-1}),$$

where D_i is the depth of the ith layer from the surface.

Importantly, V_{rms} can be used to derive the desired interval velocity V_i from the following Dix equation (Ashcroft, 2011, p. 42; Dutta, 2002):

$$V_i = \sqrt{(V_2^2 T_2 - V_1^2 T_1)/(T_2 - T_1)}, \tag{17.20}$$

where V_1, T_1, V_2, and T_2 are the V_{rms} and TWT (two-way time) values for the reflections at the top and the bottom of a layer. For strata that are not flat, the Dix velocity of Eq. (17.20) may be used to establish a rough velocity structure of an area with moderate dips.

Once the interval velocity is obtained, it is straightforward to calculate the layer thicknesses (and thus the depths and average velocities) for key reflections in the seismic section. From CMP stacking, useable velocities can hence be derived from the seismic data alone. The suppression of troublesome surface multiple reflections also benefits from the process of CMP stacking (Ashcroft, 2011, p. 44).

Migration, the second aspect in depth imaging, uses a velocity model to redistribute reflected seismic energy from its assumed position at the midpoint to its true position (lateral reposition of the reflection, Fig. 17.4). The horizontal shift X_m is (Ashcroft, 2011, p. 46, p. 46)

$$X_m = \frac{1}{4} V^2 T (dT / dX), \tag{17.21}$$

where V is the velocity down to the reflector, $T =$ TWT at the CMP location, and dT/dX is the gradient of the reflection at the CMP location (on the unmigrated time section). It can be observed that lateral shift is much bigger for deep reflections (at greater T and V^2).

Migration can be performed by solutions to elastic wave propagation equations (Ashcroft, 2011, p. 48), in time or in depth, and either before or after stacking (Fig. 17.5). Time migration can be used to solve certain imaging problems, but more complex ones need depth migration. Time migration is applicable to most sedimentary basins worldwide, where the seismic velocity increases with traveltime (depth) and varies gradually horizontally. For a velocity model having large horizontal or vertical contrasts, depth migration is applied when steeply dipping faults, folds, or intrusions juxtapose layers with significantly different elastic properties (Albertin et al., 2002).

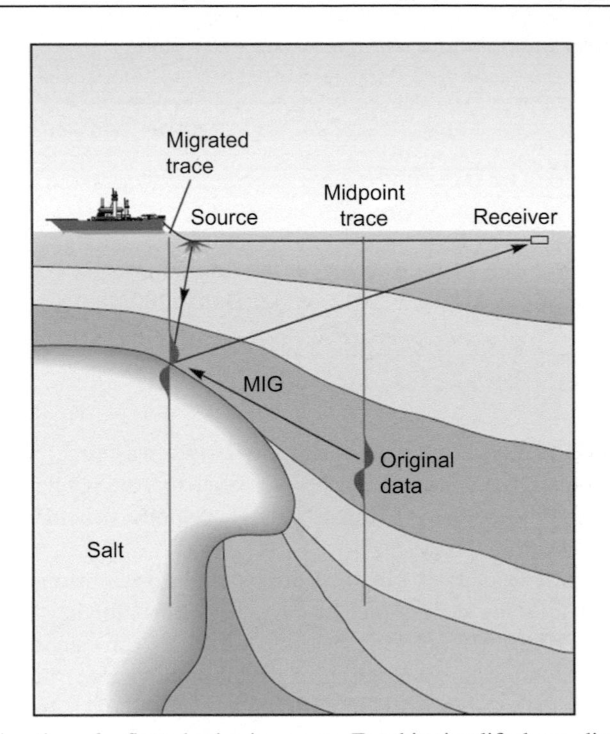

Figure 17.4 Migration of reflected seismic energy. For this simplified two-dimensional example, migration repositions the data trace from its recorded position at the source–receiver midpoint to its true position (MIG) using a velocity model. In 3D cases, reflections may be redistributed to and from positions outside the plane containing the sources and receivers. From Albertin, U., Kapoor, J., Randall, R., Smith, M., Brown, G., Soufleris, C. et al. (2002). The time for depth imaging. *Oilfield Review. 14*(1), 2–15. Graphic copyright Schlumberger. Used with permission. *Courtesy:* Lisa Stewart (former executive editor of Oilfield Review) & Schlumberger.

Stacking reduces the number of processed traces typically by an order of magnitude; migration applied poststacking is much faster than that applied prestacking. Prestack migration requires fewer assumptions than poststack in the variations in the lithology and fluid content over the span of the gathered traces; prestack migration is hence used to handle complex structures and velocity fields, thanks to high-performance computing (Vesnaver, 2008).

The data processing workflow for prestack depth migration (such as for subsalt depth imaging) is a complex interplay of several steps. Building the velocity model itself requires iterations of prestack depth migration to define the velocity and the geometric boundaries of each layer (Albertin et al., 2002). The key step of updating the velocity model is mostly based on *tomographic* inversion that uses traveltime information derived from seismic data to refine velocity models (Dutta, 2002).

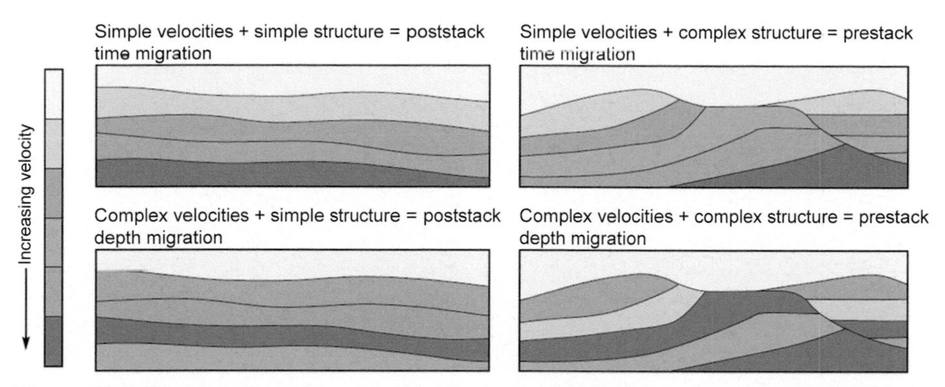

Figure 17.5 Simple and complex seismic velocity models (with two-way traveltime TWT as vertical axis) and structures treated by four seismic migration classes—time (top), depth (bottom), prestack (right), and poststack (left). For time migration, the velocity model may have smooth variations, but only with depth, and only monotonically increasing with depth. Depth migration is required for more complex velocity models, such as those with lateral variation or decreases of velocity with depth. Poststack migration works with models with low structure complexity. Prestack can handle the most complex models.
From Albertin, U., Kapoor, J., Randall, R., Smith, M., Brown, G., Soufleris, C. et al. (2002). The time for depth imaging. *Oilfield Review. 14*(1), 2–15. Graphic copyright Schlumberger. Used with permission. *Courtesy:* Lisa Stewart (former executive editor of Oilfield Review) & Schlumberger.

Classical reflection tomography uses the differences between the predicted and observed reflection traveltimes. Ray tracing is used to predict the arrival times of reflections on CMP gathers at control points. On each gather, the actual arrival time of the shallowest reflector is compared with the predicted arrival times, and the velocity that best flattens the actual arrival times is used to update the model. This time-consuming step derives a final velocity model that optimally satisfies the data at all control points (Albertin et al., 2002). The next step applies depth migration using the updated velocity model. The migrated traces are gathered again and arrival flatness checked. For subsalt imaging, if the preliminary time migration shows that the top of salt is smooth (or structurally simple), the velocities of the overburden can be used in a poststack depth migration to obtain an image of the top of the salt. If this image is rough (or structurally complex), prestack depth migration should be applied. After the top of the salt is interpreted from the image, the velocity model is updated by filling the volume below the salt top with a uniform salt velocity. With this new velocity model, the volume is again prestack depth migrated; the image of the bottom of the salt comes into focus (Fig. 17.6) (Albertin et al., 2002).

In reality, much of the subsurface is anisotropic in some physical property, such as in elasticity, permeability, or electromagnetic properties (for the effect of kerogen on

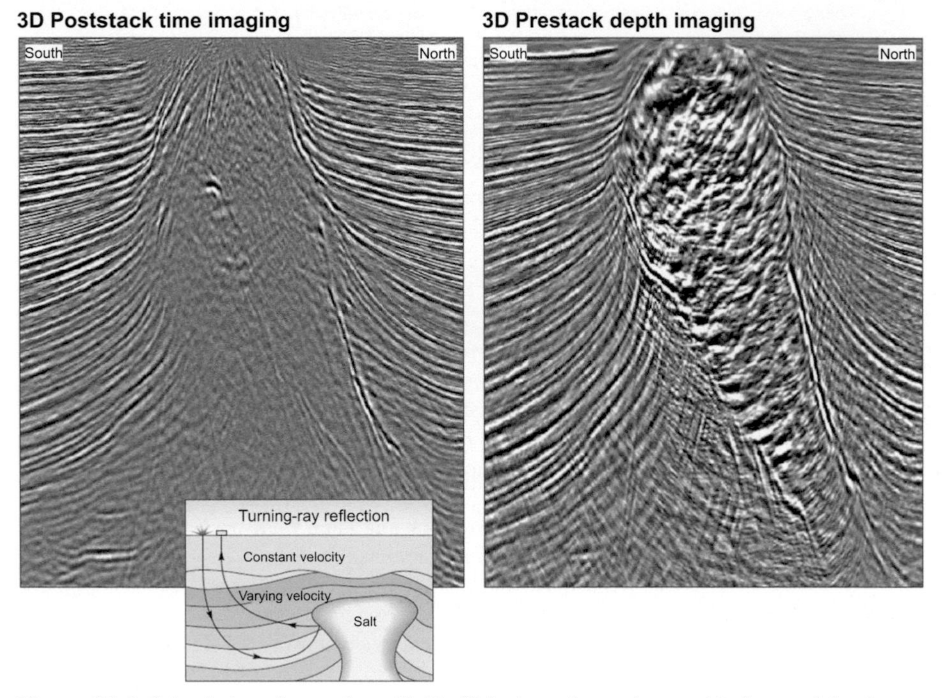

Figure 17.6 Seismic imaging under a Gulf of Mexico salt overhang with time and depth migration. Poststack time migration (left) manages to image the north side of a salt diapir, but the southern side is lost in a shadow created by an overhang. By including turning rays (inset) and rays that pass through the salt, prestack depth migration (right) images the steeply dipping layers and the overhang on the south side of the intrusion.
From Albertin, U., Kapoor, J., Randall, R., Smith, M., Brown, G., Soufleris, C. et al. (2002). The time for depth imaging. *Oilfield Review. 14*(1), 2−15. Graphic copyright Schlumberger. Used with permission. *Courtesy:* Lisa Stewart (former executive editor of Oilfield Review) & Schlumberger.

the elastic anisotropy of organic-rich shales, see Sayers, 2013). Elastic anisotropy is one of the factors that can complicate the seismic-velocity model-building process, and it can be incorporated into the prestack depth migration process. Ignoring anisotropy can lead to vertically and horizontally unfocussed and mispositioned structures. The effects of anisotropy can be seen as a nonhyperbolic shape in the arrivals from a flat reflector (Fig. 17.7). Traces from long offsets arrive earlier than predicted by an isotropic velocity model as they travel longer in the faster, horizontal direction (Albertin et al., 2002).

The use of *P*- and *PS* converted wave reflection moveout for estimating VTI parameters has been reviewed by Li and Zhang (2011).

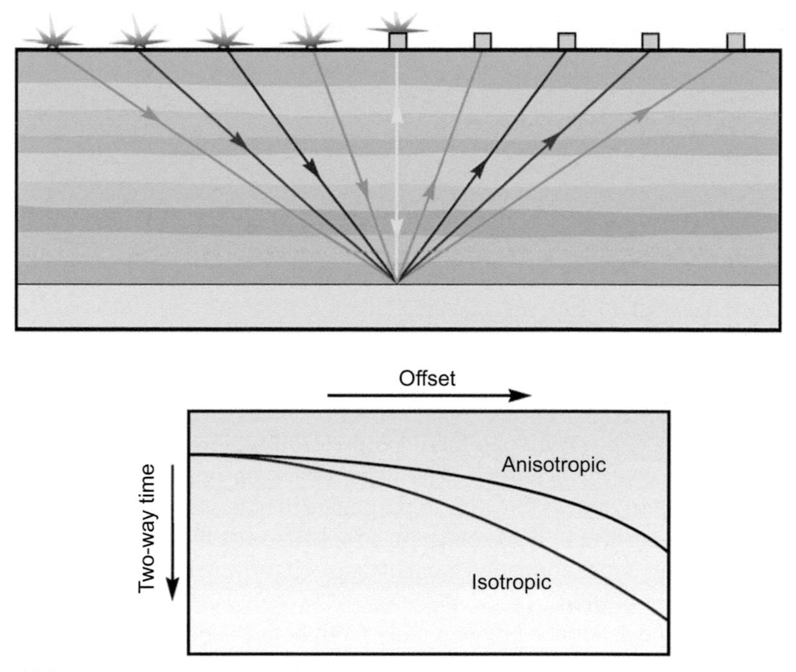

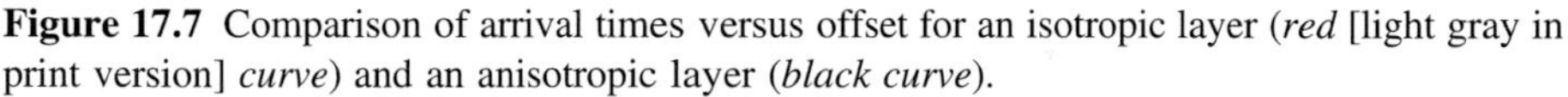

Figure 17.7 Comparison of arrival times versus offset for an isotropic layer (*red* [light gray in print version] *curve*) and an anisotropic layer (*black curve*).
From Albertin, U., Kapoor, J., Randall, R., Smith, M., Brown, G., Soufleris, C. et al. (2002). The time for depth imaging. *Oilfield Review*. *14*(1), 2–15. Graphic copyright Schlumberger. Used with permission. *Courtesy:* Lisa Stewart (former executive editor of Oilfield Review) & Schlumberger.

17.3 Multicomponent seismic data for reservoir characterization

In some cases, the acoustic impedance contrast at the interface between two lithologies may be too small to generate a normal incidence reflection. For example, a shale with low density and high *P*-wave velocity might have nearly the same acoustic impedance as an oil-filled sandstone with high density and low *P*-wave velocity. Without an acoustic impedance contrast, such oil reservoirs are extremely difficult to detect using traditional *P*-wave surface seismic acquisition and processing.

Multicomponent seismic data from both *P*-waves and *S*-waves can improve reservoir characterization and reduce exploration risk (Barkved et al., 2004; Landrø & Kvam 2002). Hydrocarbon reservoirs formed from subtle lithological changes, such as stratigraphic traps, may be delineated from changes in *P*- and *S*-wave velocities and impedances. Variations in the seismic attributes derived from multicomponent data can provide vital information about fluid type and distribution (Li & Zhang, 2011).

From Table 17.1, the velocities of *P*-waves and *S*-waves are as follows:

$$V_p = \sqrt{\left(\kappa + \frac{4}{3}\mu\right)/\rho} \tag{17.22}$$

$$V_s = \sqrt{\mu/\rho} \tag{17.23}$$

The rock density (ρ) can be expressed as a function of the porosity (ϕ), fluid, and rock matrix densities (ρ_{fluid}, ρ_{matrix}), viz.

$$\rho = \phi\ \rho_{fluid} + (1 - \phi)\ \rho_{matrix} \tag{17.24}$$

Bulk modulus (κ) is sensitive to fluid compressibility; this makes *P*-waves highly sensitive to fluid content in a rock. The dependence on both fluid compressibility and shear modulus (μ) allows *P*-waves to propagate in both solids and liquids. A shear wave is almost insensitive to fluid content in a rock; its velocity and reflectivity remain unchanged whether a rock formation contains gas, oil, or water. *S*-waves can originate and propagate only in solids.

Lithology can be determined more readily from both *P*- and *S*-wave velocities (V_p, V_s) than from *P*-wave data alone. The V_p/V_s ratio, the common seismic attribute derived from multicomponent seismic data, can be used to predict rock type. Typically, different lithologies may give rise to different velocity ratios. As shown in Fig. 17.8, hydrocarbon-bearing rocks are often characterized by a low velocity ratio (Li & Zhang, 2011).

Comparison of *P*-wave with *S*-wave behavior at a reflector can distinguish lithology changes from fluid changes: a lateral change in *P*-wave reflection amplitude along a

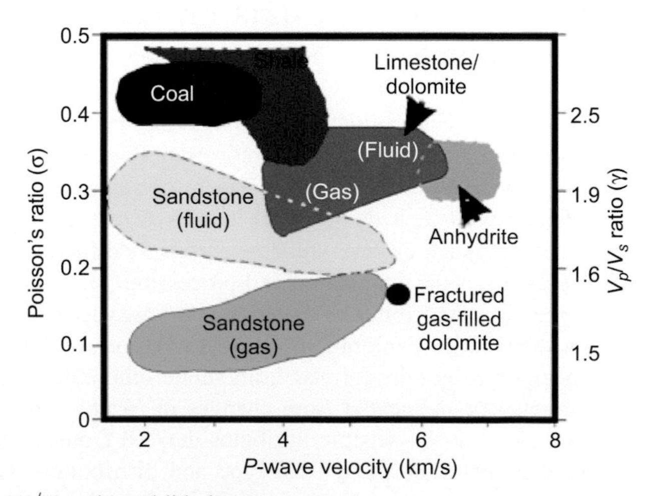

Figure 17.8 V_p/V_s ratio and lithology.
From Li, X.-Y., & Zhang, Y.-G. (2011). Seismic reservoir characterization: how can multicomponent data help? *Journal Geophysics and Engineering, 8*, 128. Copyright IOP Publishing. Used with permission.

layer boundary may indicate either a lithology change or a fluid change, but if *S*-wave reflection amplitude at the same boundary also changes, the variation is more likely due to a lithology change. Consistent *S*-wave reflectivity at the same reflector indicates a change in the fluid type is more probable (Barkved et al., 2004).

In *P*-waves, particle motion is parallel to the direction of propagation, while in *S*-waves, particle motion is perpendicular to the direction of propagation (Fig. 17.9, top). In land-based surface seismic data acquisition, *P*-waves can be recorded by single-component geophones that detect vertical motion in the rock formation. In typical marine seismic surveys, *P*-waves can also be detected by towed-streamer hydrophones (or pressure sensors) surrounded by water. The field of *S*-waves is three dimensional; therefore, three-component sensors (with three geophone accelerometers in orthogonal orientation in each sensor) are required to fully characterize it (Fig. 17.9, bottom). Processing of land multicomponent *S*-wave surveys is often problematic because inhomogeneity of near-surface layers causes large traveltime variations for the *S*-waves (Barkved et al., 2004).

Recording *S*-waves in a marine survey requires deployment of multicomponent sensors on the seafloor. It still is impractical to deploy a shear-wave source on seabed; in a multicomponent survey, the surface source generated *P*-waves undergo partial conversion to *S*-waves at subsurface interfaces and the resulting converted *PS*-waves (or *C*-waves) can be detected by ocean-bottom cables (OBCs) (De Freitas, 2011). These cables contain an array of four-component sensors—three geophones and a hydrophone—spaced at intervals determined by the survey requirements (Fig. 17.9, bottom). The geophones detect the multicomponents of *S*-wave motion; the hydrophone detects *P*-wave signals, designated as *PP* arrivals. The *P*-wave is also detected by the geophones, mainly on the vertical component, giving rise to *PZ* signals.

The combination of *P*- and *S*-waves improves development drilling. For example, in a gas-condensate field in the UK sector of the North Sea, delineating the major reservoir fault is difficult because migrated gas above the structural crest disturbs *P*-wave propagation. A *PP*-image from the conventional 3D towed-streamer survey indicates the imaging problems typical of low-velocity gas clouds (Fig. 17.10, left). Converted *PS*-waves recorded by a 3D multicomponent survey gave a clear picture of the gas-condensate field structure (Fig. 17.10, right). The clearly resolved fault on the *PS*-image made it possible to locate development wells and to guide the drilling of a good production well (Barkved et al., 2004).

In addition to the velocity ratio, *P*- and *S*-wave impedances are also commonly used seismic attributes and may be derived from *P*-wave data using prestack amplitude variation with offset (AVO) (Fig. 17.11) inversion. However, the inversion can be better constrained using multicomponent data (Li & Zhang, 2011; Mathewson et al., 2012).

The AVO for a reservoir interface can be calculated theoretically. The amplitude is defined by the reflection coefficient *R* of the interface; it depends on the angle of incidence at the interface and on the contrast in the *P*-wave velocity, *S*-wave velocity, and density across the interface (Fig. 17.2). For angles of incidence no more than 50 degrees, the *PP*- and *PS*-wave reflection coefficients are derived as follows (Ashcroft, 2011, p. 131; Li & Zhang, 2011):

$$R_{pp}(\theta) \approx \frac{1+(\tan\theta)^2}{2}R_p - 4\gamma^2(\sin\theta)^2R_s - \left(\frac{1}{2}(\tan\theta)^2 - 2\gamma^2(\sin\theta)^2\right)R_d \tag{17.25}$$

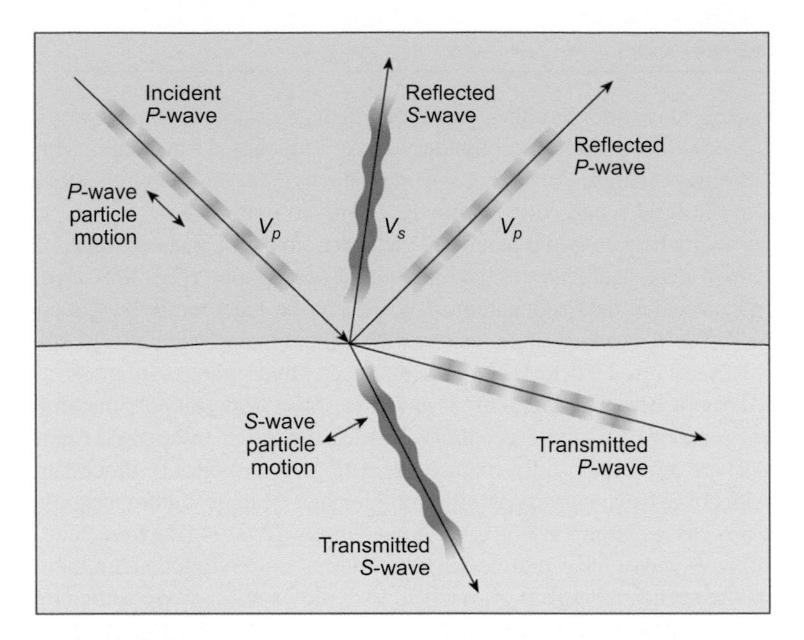

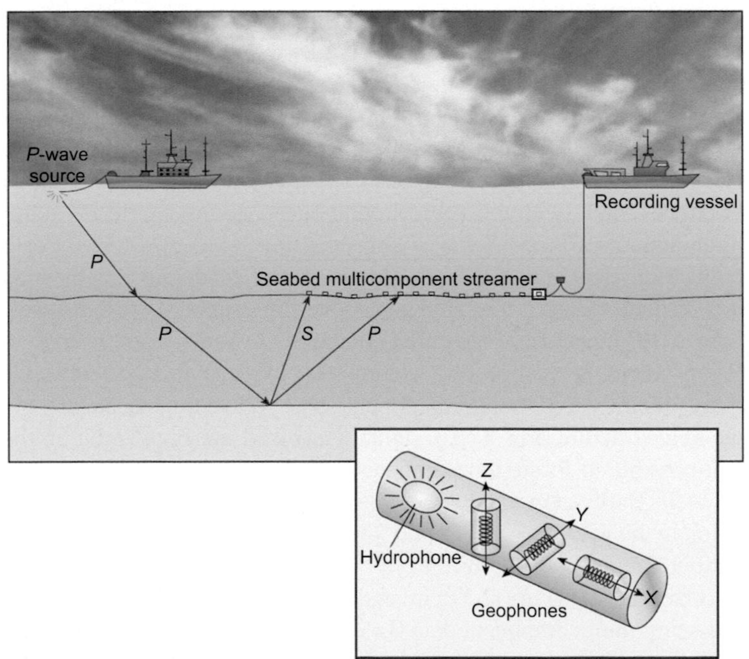

Figure 17.9 Top: Particle motion and propagation of compressional (*P*-) and shear (*S*-) waves. Bottom: Detecting converted waves by seabed sensors. At a subsurface interface, incident *P*-waves reflect and transmit as *P*-waves and also are partially converted to *S*-waves. Upgoing *S*-waves can be detected by seabed receivers sensitive to multiple components of motion. The four receiver components consist of one hydrophone and three orthogonally oriented geophones or accelerometers—X, Y, and Z (inset).
From Barkved, O., Bartman, B., Compani, B., Gaiser, J., Van Dok, R., Johns, T. et al. (2004). The many facets of multicomponent seismic data. *Oilfield Review 16*(2), 42–56. Graphic copyright Schlumberger. Used with permission. *Courtesy:* Lisa Stewart (former executive editor of Oilfield Review) & Schlumberger.

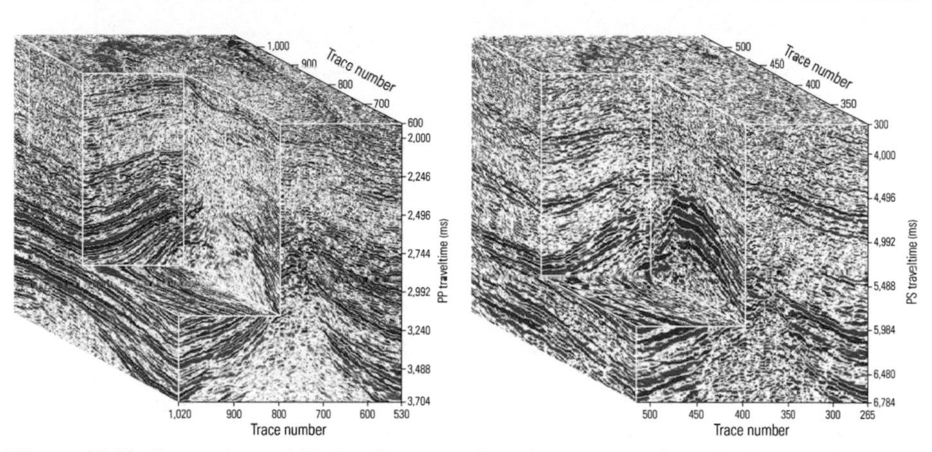

Figure 17.10 Comparison of 3D data from towed-marine survey and a seabed survey. In the *PP* image from the towed survey (left), reflections at the crest of a North Sea gas-condensate field structure are obscured by shallow (low-velocity) gas. The *PS* converted wave image from the seabed survey (right) clearly resolves the large fault passing through the structure at its crest, and fully illuminates the structure with high-amplitude reflections.
From Barkved, O., Bartman, B., Compani, B., Gaiser, J., Van Dok, R., Johns, T. et al. (2004). The many facets of multicomponent seismic data. *Oilfield Review 16*(2), 42−56. Graphic copyright Schlumberger. Used with permission. *Courtesy:* Use of clients data.

$$R_{ps}(\theta,\phi) \approx \frac{\tan\phi}{2\gamma}\left[\left(4(\sin\phi)^2 - 4\gamma\cos\theta\cos\phi\right)R_s - \left(1+2(\sin\phi)^2 - 2\gamma\cos\theta\cos\phi\right)R_d\right] \tag{17.26}$$

where

$\theta = \frac{\theta_2+\theta_1}{2}$, average of *P*-wave angles of incidence and transmission across the interface;

$\phi = \frac{\phi_2+\phi_1}{2}$, average of *S*-wave angles of reflection and transmission across the interface;

$\gamma = V_s/V_p$, velocity ratio;

$V_p = \frac{V_{p2}+V_{p1}}{2}$, $\Delta V_p = V_{p2} - V_{p1}$, *P*-wave average velocity and velocity contrast;

$V_s = \frac{V_{s2}+V_{s1}}{2}$, $\Delta V_s = V_{s2} - V_{s1}$, *S*-wave average velocity and velocity contrast;

$\rho = \frac{\rho_2+\rho_1}{2}$, $\Delta\rho = \rho_2 - \rho_1$, average density and density contrast;

$R_d = \frac{\Delta\rho}{\rho}$, density (reflection) coefficient;

$R_p = \frac{\rho_2 V_{p2} - \rho_1 V_{p1}}{\rho_2 V_{p2} + \rho_1 V_{p1}} \approx \frac{1}{2}\left(\frac{\Delta V_p}{V_p} + \frac{\Delta\rho}{\rho}\right)$, *P*-wave normal incidence reflection coefficient;

$R_s = \frac{\rho_2 V_{s2} - \rho_1 V_{s1}}{\rho_2 V_{s2} + \rho_1 V_{s1}} \approx \frac{1}{2}\left(\frac{\Delta V_s}{V_s} + \frac{\Delta\rho}{\rho}\right)$, *S*-wave normal incidence reflection coefficient.

From multicomponent seismic data, *PP* and *PS* reflectivities (R_{pp}, R_{ps}) are assumed known, and density is assumed to follow Gardner's relation ($\rho = 0.23V^{1/4}$). For a given velocity model, Eqs. (17.25) and (17.26) can be used to solve the two unknowns R_p and R_s. This process is often called joint *PP*- and *PS*-wave AVO inversion.

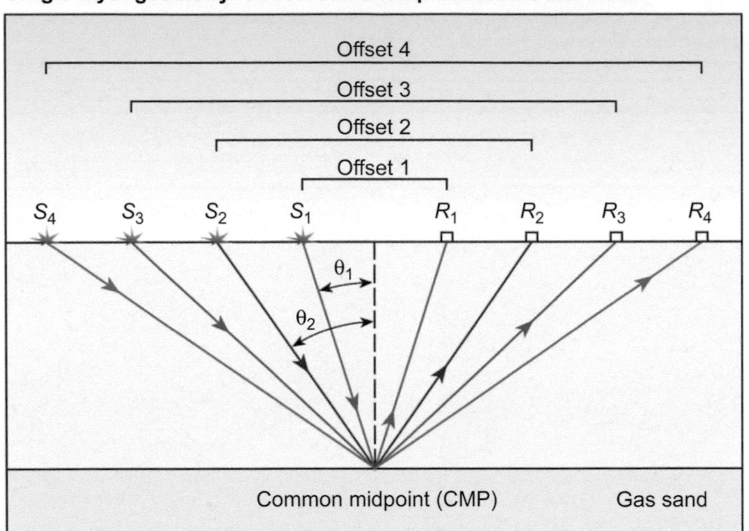

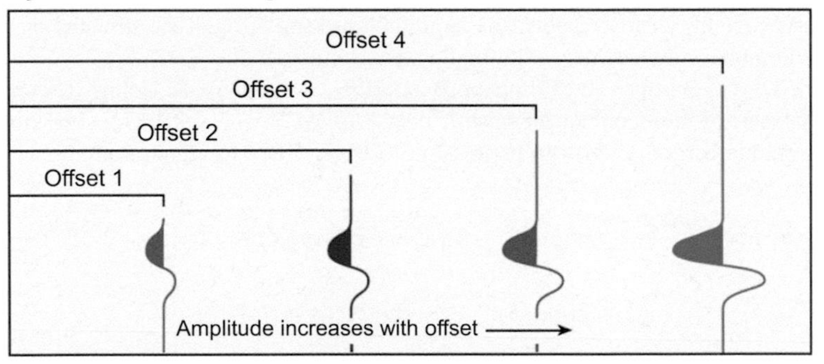

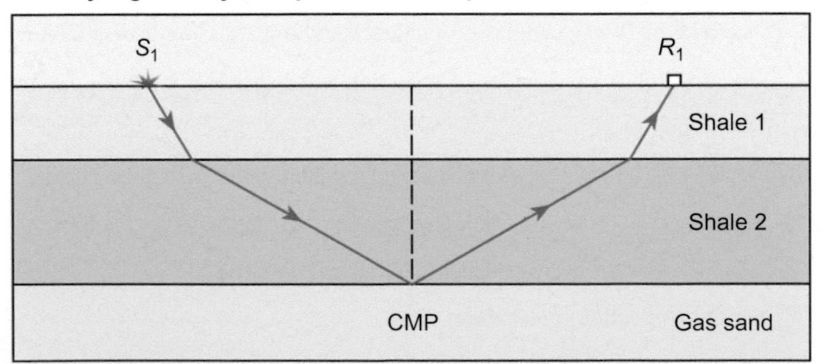

Figure 17.11 Amplitude variation with offset (AVO). In steps similar to preparation for stacking, traces reflecting at a common midpoint are gathered and sorted by offset (top), then the arrivals are flattened using a normal moveout velocity model while preserving the amplitude information (middle). Clearly, in this case, averaging the four traces would produce a trace that does not resemble the zero-offset trace; in other words, stacking would not preserve amplitudes. The offset versus angle (θ) relationship is determined by ray tracing (bottom). From Barclay, F., Bruun, A., Rasmussen, K. B., Alfaro, J. C., Cooke, A., Cooke, D. et al. (2008). Seismic inversion: reading between the lines. *Oilfield Review.* *20*(1), 42–63. Graphic copyright Schlumberger. Used with permission. *Courtesy:* Lisa Stewart (former executive editor of Oilfield Review) & Schlumberger.

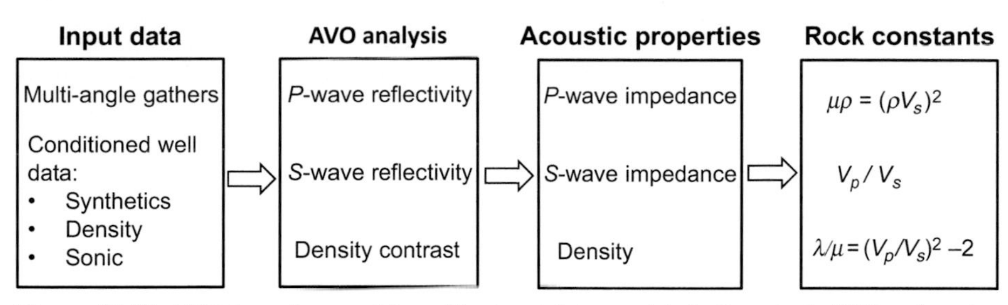

Figure 17.12 AVO inversion workflow. The input data consisted of prestack AVO gathers in, e.g., 7 degrees offset ranges, along with sonic and density well logs (*left*). The first step, three-parameter AVO inversion, produced estimates of *P*-wave and *S*-wave reflectivities and density contrast. These volumes were inverted for *P*-wave and *S*-wave impedances and density. The final step extracted rock properties, in the form of $\mu\rho$, V_p/V_s, and λ/μ.
Adapted from Barclay, F., Bruun, A., Rasmussen, K. B., Alfaro, J. C., Cooke, A., Cooke, D. et al. (2008). Seismic inversion: reading between the lines. *Oilfield Review 20*(1), 42–63. *Courtesy:* Lisa Stewart (former executive editor of Oilfield Review) & Schlumberger.

The resulting attributes can then be used directly for lithological interpretation (Li & Zhang, 2011).

Fig. 17.12 demonstrates the three-parameter AVO inversion workflow applied to a case study in one area of the central North Sea (Barclay et al., 2008). An operating company intended to distinguish the clean sands from the surrounding shales by improving seismic characterization of injected reservoir sands in the Balder interval that were particularly difficult to image from the existing surface seismic data. A new marine seismic survey was designed to map the distribution and thickness of the reservoir pay zone, delineate the geometry of individual sand wings, and assess reservoir connectivity. Data acquisition with the new marine seismic survey system enabled accurate streamer cable positioning, fine spatial sampling and calibration of sources and receivers. These capabilities are all important for achieving a successful inversion, facilitated by precision imaging, improved noise attenuation, increased bandwidth, and preservation of amplitude and phase information.

The prestack seismic input data were divided into seven angle stacks, each containing reflections in a 7 degree range of incidence angles out to 49 degrees. Three-parameter AVO inversion generated estimates of *P*- and *S*-wave reflectivities and density contrast. These data volumes were inverted for *P*- and *S*-wave impedances and density, leading to determination of rock constants such as $\mu\rho = (\rho V_s)^2$, $\lambda/\mu = (V_p/V_s)^2-2$ (see Table 17.1).

Log data from three wells intersecting the reservoir were analyzed for correlations between *P*- and *S*-wave velocities, ρ, μ, λ, lithology, and fluid saturation. For example, cross-plotting V_p/V_s with the product $\mu\rho$, and color coding by lithology, showed that high sand probability correlated with low V_p/V_s and high $\mu\rho$ values. These relationships were then applied to V_p/V_s calculated from seismic inversion to map high sand content throughout the seismic volume. This characterization of the extent and quality of the injected sand bodies helped optimize development of these complex features (Barclay et al., 2008).

In Fig. 17.13, to further illustrate the value of the *PS* data from multicomponent OBC sensors, simultaneous inversion (see next Section 17.4) of the *PZ* data alone is compared with simultaneous inversion of the combined *PZ* and *PS* datasets. The acoustic impedance and density derived from the *PZ* and *PS* reflection amplitudes are much better resolved and match the well-logging values better than those calculated from *PZ* arrivals alone.

17.4 Simultaneous inversion of time-lapse seismic surveys for reservoir monitoring

Most of the examples shown so far are the results from inversion techniques based on trace-by-trace reflectivity methods. A simultaneous inversion method has been developed to examine all traces at once to invert for a model of rock properties, globally optimized based on simulated annealing (Barclay et al., 2008). Simultaneous inversion enhances resolution and accuracy by making use of the full bandwidth (low and high frequencies) of the seismic signal. The underlying global optimization cost function has four penalty terms that are collectively minimized to deliver the best solution. Modification of these terms can be made to incorporate requirements of more complex data types into the simultaneous inversion algorithm (Table 17.3), such as time-lapse surveys to highlight time-lapse changes in rock and fluid properties (Albright et al., 1994; Aronsen et al., 2004) and shear-wave multicomponent data (Fig. 17.13).

In an attempt to help increase oil recovery factor, advanced time-lapse (or 4D) marine seismic surveys have been tested on the Norne field in the North Sea operated by Equinor (formerly Statoil) (Barclay et al., 2008). The Norne field has high-quality sandstone reservoirs, with porosities ranging from 25% to 32% and permeabilities from 200 to 2000 mD; changes in fluid saturation and pressure result in noticeable differences in seismic amplitudes and elastic impedances. The Norne field is therefore conducive to successful time-lapse monitoring. Its first advanced marine seismic survey was acquired in 2001, forming the baseline for the 2003, 2004, and 2006 subsequent monitor surveys. Early time-lapse monitoring delivered vital information for optimizing field development. Differences in the AVO inversions of the 2001 and 2003 surveys revealed changes in acoustic impedance that was interpreted as increases in water saturation; the trajectory of a planned well was modified to avoid a high water saturation zone (Barclay et al., 2008).

To optimize reservoir drainage and injection strategies and to understand the effects of continued field production on changes in effective stress, a simultaneous inversion project incorporated all available seismic data, log data from seven wells and production data from reservoir model. The simultaneous inversion estimated baseline values and changes in acoustic impedance and Poisson's ratio from the time-lapse seismic data (Fig. 17.14). To compensate for the lack of low-frequency information in the seismic bandwidth—needed to determine absolute elastic properties—background models were constructed. For the (time-lapse) baseline survey, the background model was derived by propagating borehole (sonic logging) values of elastic properties throughout the zone of interest, constrained by key interpreted horizons and the seismic interval velocities.

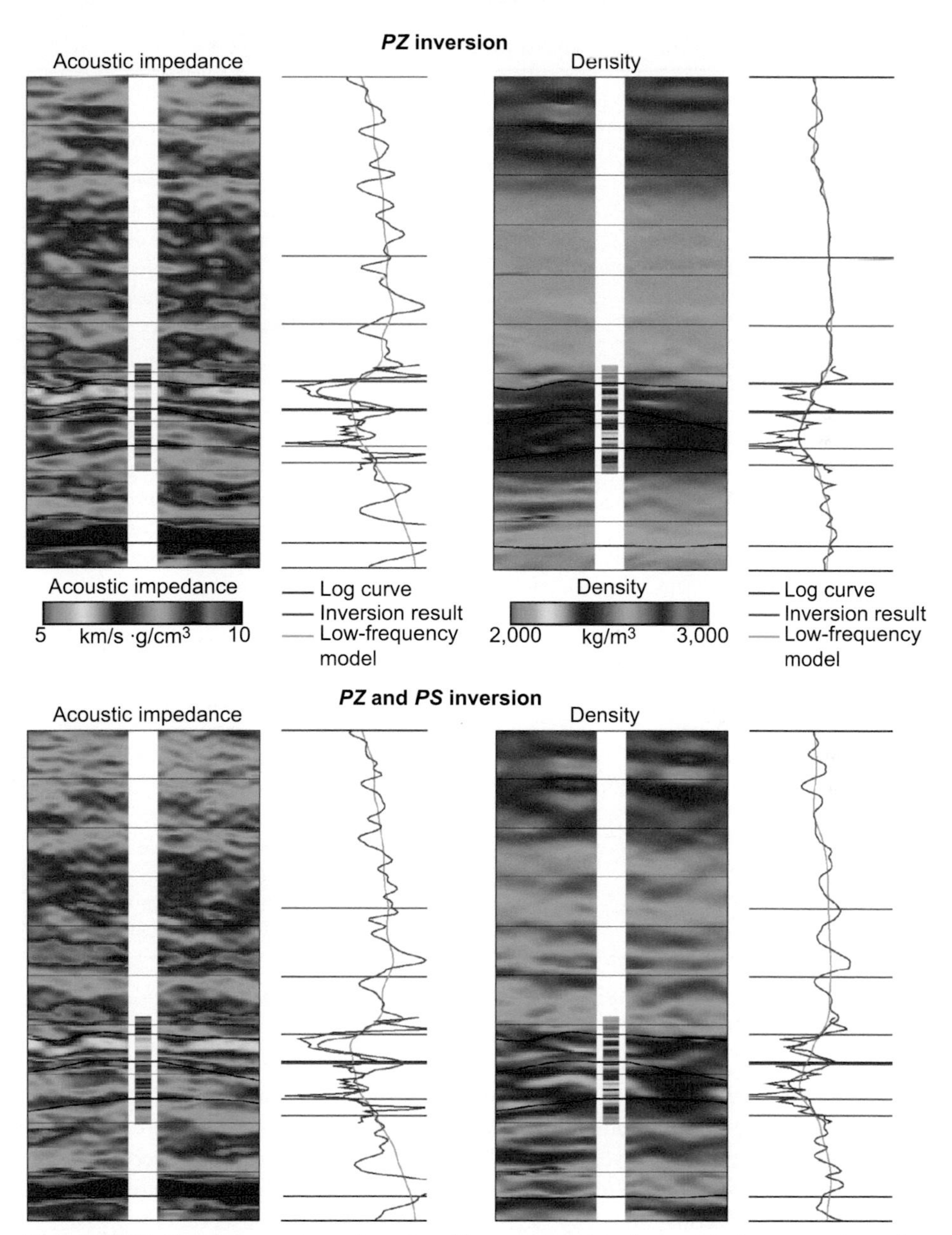

Figure 17.13 Simultaneous inversion of multicomponent data from seabed sensors. Acoustic impedance (left) and density (right) from inversion using only PZ data (top) lack the resolution and continuity of the results of inversion using PZ and PS data (bottom). In particular, compared with PZ inversion, the densities predicted by inversion of PZ and PS data showed much better correlation with log values. In the panels showing inversion results, the nearly horizontal black lines are interpreted horizons.
From Barclay, F., Bruun, A., Rasmussen, K. B., Alfaro, J. C., Cooke, A., Cooke, D. et al. (2008). Seismic inversion: reading between the lines. *Oilfield Review 20*(1), 42–63. Graphic copyright Schlumberger. Used with permission. *Courtesy:* Use of clients data.

Table 17.3 Applications of simultaneous inversion.

Data type	Physical properties
Full-stack data	*P*-wave impedance
Partial-stack AVO data	*P*-wave impedance, V_p/V_s (or *S*-wave impedance) and density, from which Poisson's ratio, λ and μ can be estimated.
Intercept and gradient AVO data	Acoustic impedance from the intercept data; shear impedance from shear seismic data calculated from the intercept and gradient data
Multicomponent full-stack data (*P*-to-*P* and *P*-to-*S* conversions)	*P*-wave impedance and *S*-wave impedance
Multicomponent partial-stack AVO data	*P*-wave impedance, V_p/V_s (or *S*-wave impedance), and density
Borehole seismic (VSP) data	Acoustic impedance from the *PP* data; shear impedance from the *PS* data
Time-lapse full-stack data (which may include multicomponent full-stack data)	Simultaneous time-lapse inversion for baseline *P*-wave impedance and the changes in *P*-wave impedance for each time interval: for multicomponent data, inversion will also output baseline *S*-wave impedance and changes in *S*-wave impedance over the time interval.
Time-lapse partial-stack AVO data (which may include multicomponent partial-stack AVO data)	Simultaneous time-lapse inversion for baseline properties and the changes: for example, for partial-stack data, inversion can determine baseline *P*-wave impedance, V_p/V_s (or *S*-wave impedance) and density and the changes in these properties over the time interval.

Adapted from Barclay, F., Bruun, A., Rasmussen, K.B., Alfaro, J.C., Cooke, A., Cooke, D. et al. (2008). Seismic inversion: reading between the lines, *Oilfield Review 20*(1), 42–63.

For the time-lapse low-frequency models, estimates of elastic properties were obtained from the reservoir simulator in three steps: reservoir properties were converted from depth to time using the velocity model, then converted to elastic property changes using rock physics models. Finally, the spatial and temporal distributions of the property changes were constrained by seismic velocity changes observed in time-lapse traveltime differences. This unique combination of time-converted reservoir properties with seismic-derived traveltime changes delivered accurate changes in elastic properties consistent with the reservoir simulation (Barclay et al., 2008).

The reservoir management team can use these time-lapse results to track the movement of the waterflood front, evaluate the progress of water and gas injection, estimate the pressure distribution, and update the reservoir model.

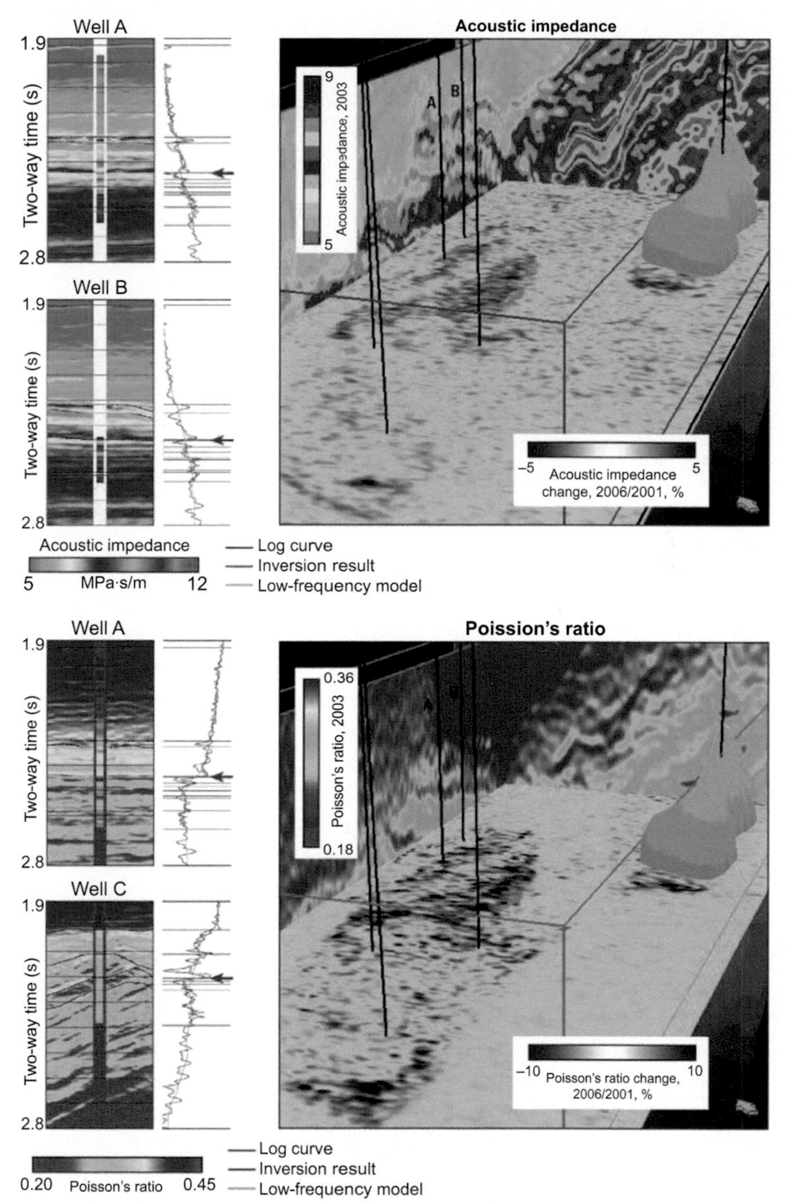

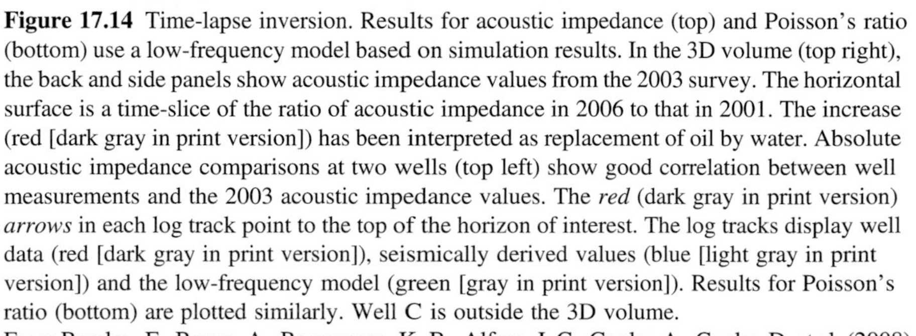

Figure 17.14 Time-lapse inversion. Results for acoustic impedance (top) and Poisson's ratio (bottom) use a low-frequency model based on simulation results. In the 3D volume (top right), the back and side panels show acoustic impedance values from the 2003 survey. The horizontal surface is a time-slice of the ratio of acoustic impedance in 2006 to that in 2001. The increase (red [dark gray in print version]) has been interpreted as replacement of oil by water. Absolute acoustic impedance comparisons at two wells (top left) show good correlation between well measurements and the 2003 acoustic impedance values. The *red* (dark gray in print version) *arrows* in each log track point to the top of the horizon of interest. The log tracks display well data (red [dark gray in print version]), seismically derived values (blue [light gray in print version]) and the low-frequency model (green [gray in print version]). Results for Poisson's ratio (bottom) are plotted similarly. Well C is outside the 3D volume.
From Barclay, F., Bruun, A., Rasmussen, K. B., Alfaro, J. C., Cooke, A., Cooke, D. et al. (2008). Seismic inversion: reading between the lines. *Oilfield Review 20*(1), 42–63. Graphic copyright Schlumberger. Used with permission. *Courtesy:* Use of clients data.

17.5 Borehole seismic surveys

Borehole seismic surveys have an advantage over their surface seismic counterparts in that they record direct signals in a low-noise environment. Traditional vertical seismic profiles (VSPs) consisted of receivers deployed in a vertical borehole to record the most basic signals from a seismic source at the surface to perform time-depth correlation analysis (Fig. 17.15).

Modern VSPs delivered innovations by recording more information and expanding survey geometries (Fig. 17.16) with improved acquisition tools, including 3D VSPs (Fig. 17.17), VSPs acquired while drilling (Fig. 17.18), and walkaround VSP (Fig. 17.19). From these data, exploration and production companies derive important information about reservoir depth, extent, and heterogeneity, as well as fluid content, rock-mechanical properties, pore pressure, enhanced oil-recovery progress, elastic anisotropy, induced-fracture geometry, and natural-fracture orientation and density (Blackburn et al., 2007).

The main types of waves generated and recorded in borehole seismic surveys are body waves emitted by point sources or frequency-sweep sources (airguns, vibrating trucks, or dynamite sources), and consist of *P*-waves (and *PP*, *PZ*) and *S*-waves (and *PS*). Most modern downhole hardware for recording VSPs consists of clamped, calibrated three-component (3C) geophones.

Hydraulically induced fractures can also be monitored using borehole seismic methods. While the fracture is being created in the treatment well, a multicomponent receiver array in a monitor well records the microseismic activity generated by the fracturing process (Fig. 17.20). Locating hydraulically induced microseismic events requires an accurate velocity model (Blackburn et al., 2007). Mapping in real-time the extent of the fracture with time helps monitor the progress of well stimulation treatments, allows comparison between actual and planned fractures, and optimizes reservoir treatments by allowing engineers to modify pumping rates when observed fractures differ from plan.

Another borehole seismic technology, called *passive seismic monitoring* or *microseismic monitoring*, characterizes fractures by recording microseismic signals generated when fluid is produced from or injected into a naturally fractured reservoir. When fluid injection (e.g., hydraulic fracturing) and production modify the stress state enough to cause seismic events, the resulting acoustic emissions can be recorded for long periods of time in nearby monitoring wells by arrays of multicomponent borehole receivers. The microseismic events can be plotted in space and time to identify the fractures that are responding to the change in stress state. Receiver arrays may be installed permanently to record for extended periods (Blackburn et al., 2007). Microseismic monitoring has provided crucial information to help improve the modeling and making decisions on well placement for drilling, well completion design, and hydraulic-fracturing operations that have revolutionized the exploitation of tight and unconventional oil and gas reservoirs (Le Calvez et al., 2016).

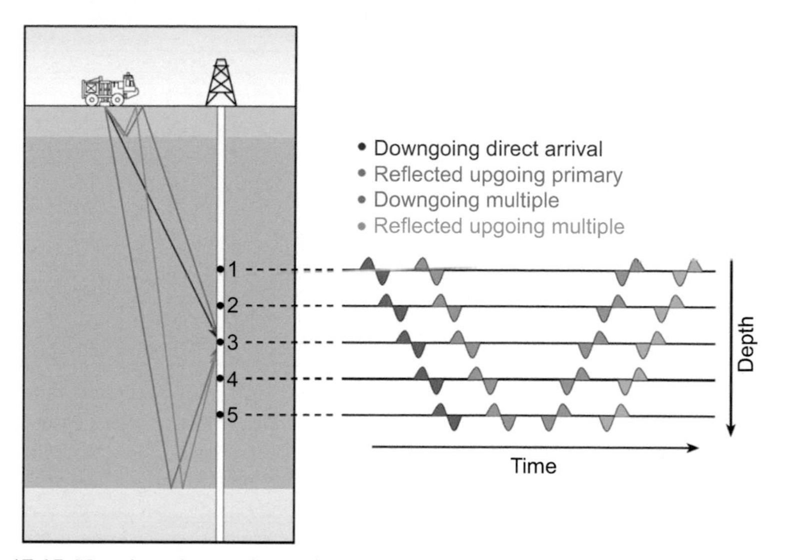

Figure 17.15 Upgoing, downgoing, primary, and multiple arrivals. Upgoing waves reflect at interfaces below the receiver and then travel upward to be recorded (blue [gray in print version] and green [light gray in print version]). Downgoing waves arrive at the receivers from above (red [dark gray in print version] and orange [gray in print version]). A wave that arrives at the receiver without reflecting is called the direct arrival (red [dark gray in print version]). Waves that reflect only once are called primaries. The reflected upgoing primary (blue [gray in print version]) is the arrival that is desired for imaging reflections. Both upgoing and downgoing signals can contain interfering multiples. Downgoing signals can be used to distinguish multiples from (multiples-free) primary arrivals, and to enable more reliable processing of the surface seismic upgoing wavefield. Together with P- and S-waves propagating from a near-surface source to the receiver, source-generated noise tube waves also arise. Another form of noise that sometimes contaminates recordings is casing ringing.
From Blackburn, J., Daniels, J., Dingwall, S., Hampden-Smith, G., Leaney, S., Le Calvez, J. et al. (2007). Borehole seismic surveys: beyond the vertical profile. *Oilfield Review 19*(3), 20–35. Graphic copyright Schlumberger. Used with permission. *Courtesy:* Lisa Stewart (former executive editor of Oilfield Review) & Schlumberger.

In *cross-well seismic surveys* (Figs. 17.21–17.23), downhole seismic sources, such as downhole vibrators, are deployed at selected depths in one borehole, shooting to a receiver array in another borehole (Hu, Abubakar, & Habashy, 2008; Rao & Wang 2011). Deep-reading cross-well seismic survey technology can combine traveltime tomography with reflection imaging between wells (Dong, Marion, Meyer, Xu, & Xu, 2005).

Fig. 17.24 is an example of utilization of such reflection imaging data, based on shooting raypaths above the reservoir as shown in the dotted paths of Fig. 17.22. Recorded data are processed to extract information about the velocities in the interwell region. A limitation of the cross-well method is the maximum allowable distance

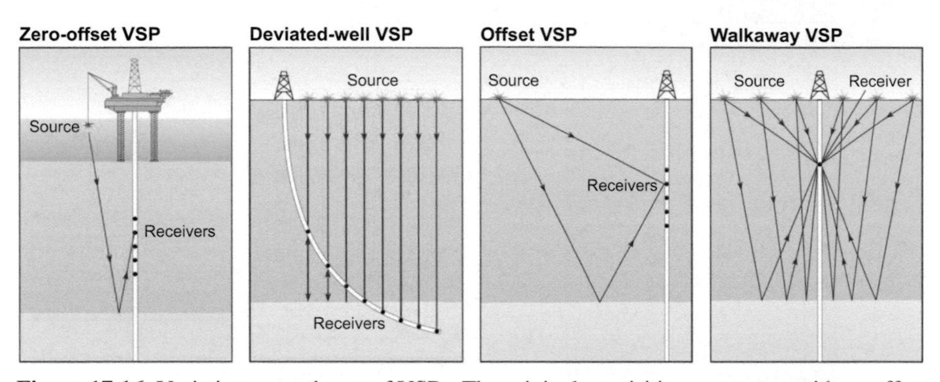

Figure 17.16 Variations on a theme of VSPs. The original acquisition geometry, with no offset between source and wellbore, creates a zero-offset VSP. Seismic waves travel essentially vertically down to a reflector and up to the receiver array. Processing yields velocities of formations at different depths, which can be tied to well log properties and interpreted for detection and prediction of overpressured zones. The velocity model can also be used to generate synthetics to identify multiples in surface seismic processing. Another normal incidence, or vertical incidence, VSP is acquired in deviated wells with the source always vertically above each receiver tool. This is known as a deviated-well, or walkabove VSP; this survey acquires a 2D image of the region below the borehole. The additional benefits of a walkabove VSP are good lateral coverage and fault and dip identification beneath the well. In an offset VSP, an array of seismic receivers is clamped in the borehole at a wide range of depths and a seismic source is placed some distance away, producing a 2D image. The non-vertical incidence can give rise to P- to S-wave conversion. The offset increases the volume of subsurface imaged and maps reflectors at a distance from the borehole that is related to the offset and subsurface velocities. The added volume of illumination enhances the usefulness of the image for correlation with surface seismic images, and for identification of faulting and dip laterally away from the borehole. In addition, because the conversion of P-waves to S-waves increases with offset, an offset VSP allows shear-wave, AVO, and anisotropy analysis. In walkaway VSPs, a seismic source is activated at numerous positions in a line on the surface. The range of offsets acquired in a walkaway VSP is particularly useful for studying shear-wave, AVO, and anisotropy effects. And, because they can illuminate a large volume of subsurface, offset and walkaway VSPs are useful elements in the design of surface seismic surveys. All these survey types may be acquired onshore or offshore.
From Blackburn, J., Daniels, J., Dingwall, S., Hampden-Smith, G., Leaney, S., Le Calvez, J. et al. (2007). Borehole seismic surveys: beyond the vertical profile. *Oilfield Review 19*(3), 20–35. Graphic copyright Schlumberger. Used with permission. *Courtesy:* Lisa Stewart (former executive editor of Oilfield Review) & Schlumberger.

between boreholes—a few thousand feet is typical—which varies with rock type, attenuation, and source strength and frequency content (Blackburn et al., 2007).

Many of the borehole seismic surveys mentioned above can be acquired at different stages in the life of a reservoir. Offset VSPs, walkaways, 3D VSPs, and cross-well surveys can also be acquired in time-lapse fashion, before and after production. Time-lapse surveys can reveal changes in the position of fluid contacts (Albright et al., 1994, p. 8), changes in fluid content, and/or front movement, such as in steamflooding

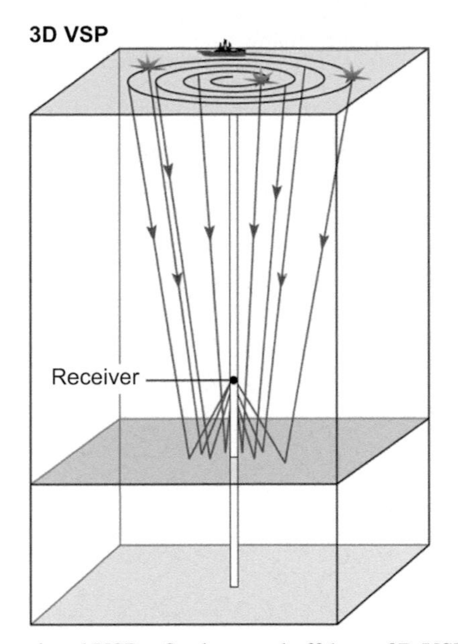

Figure 17.17 Three-dimensional VSPs. Onshore and offshore, 3D VSPs tend to borrow surface seismic acquisition geometries. On land, source positions usually follow lines in a grid. Offshore, source positions can be laid out in lines or in a spiral centered near the well. Ray-trace modeling prior to acquisition ensures proper coverage and illumination of the target. 3D VSPs deliver high-resolution subsurface imaging for exploration and development applications, and require detailed prejob modeling and planning. 3D VSPs can fill in areas that cannot be imaged by surface seismic surveys because of interfering surface infrastructure or difficult subsurface conditions, such as shallow gas, which disrupts propagation of P-waves.
From Blackburn, J., Daniels, J., Dingwall, S., Hampden-Smith, G., Leaney, S., Le Calvez, J. et al. (2007). Borehole seismic surveys: beyond the vertical profile. *Oilfield Review 19*(3), 20–35. Graphic copyright Schlumberger. Used with permission. *Courtesy:* Lisa Stewart (former executive editor of Oilfield Review) & Schlumberger.

(Fig. 17.23) and in CO_2 flooding (Fig. 17.24) in an EOR process, and other variations, such as pore pressure (Fig. 17.25), rock stress (Cook et al. 2007), and temperature.

The detection of high pore pressure (geopressure) zones is important in oil–gas exploration, especially for early detection ahead of the drill bit to prevent blowout due to insufficient mud weight (Dutta, 2002). Both *P*-wave velocity (V_p) and *S*-wave velocity (V_s) have been shown to be anomalously low in a zone of high pore pressure, in a sharp contrast with a gas-bearing zone where there is a marked low velocity in V_p, but not in V_s (Nur, 1987, p. 232). Anomalous pore pressure detection can hence be realized by the combined use of compressional and shear wave velocity data by discriminating between a gas zone and an overpressured zone.

Conventional seismic data smooth out velocity fluctuations and have velocity resolution too coarse for accurate pore pressure prediction for well-planning purposes (Barriol et al., 2005). Seismic reflection tomography based on a ray-tracing modeling–based approach refines the velocity model, leading to a better understanding of the

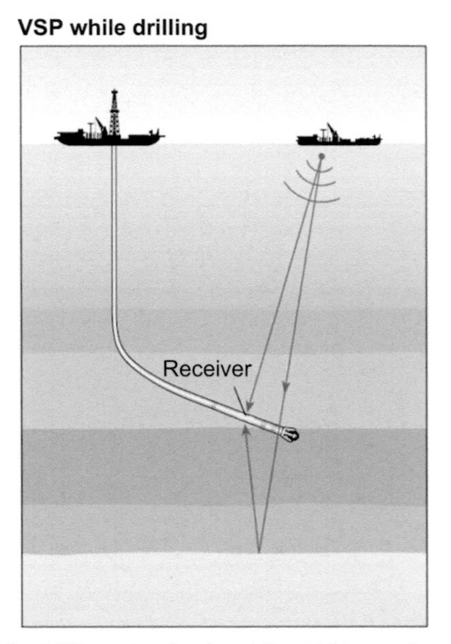

Figure 17.18 A VSP while drilling. A seismic-while-drilling tool positioned near the drill bit receives signals generated by a seismic source at the surface. Signals are transmitted to the surface for real-time, time-depth information. A seismic-while-drilling process can help reduce uncertainty in time-depth correlation without having to stop the drilling process. This technology uses a conventional seismic source at the surface, a logging-while-drilling (LWD) tool containing seismic sensors in the drillstring, and a high-speed mud-pulse telemetry system to transmit data to the surface. Availability of real-time seismic waveforms allows operators to look thousands of feet ahead of the bit to safely guide well drilling. Seismic source activation and signal measurement must take place during quiet periods, when drilling has paused for making drillpipe connections. The drill bit can act as a downhole source, generating vibrations that are detected by sensors deployed at surface or on marine cables. The processed recordings can provide critical answers in time for decisions to be made while drilling, such as changing mud weight or setting casing.
From Blackburn, J., Daniels, J., Dingwall, S., Hampden-Smith, G., Leaney, S., Le Calvez, J. et al. (2007). Borehole seismic surveys: beyond the vertical profile. *Oilfield Review 19*(3), 20−35. Graphic copyright Schlumberger. Used with permission. *Courtesy:* Lisa Stewart (former executive editor of Oilfield Review) & Schlumberger.

amplitude and spatial distribution of pore pressure, reducing the uncertainty in its predictions (Fig. 17.25).

Discovered hydrocarbon reservoirs have increasing complexity in spatial heterogeneity. We have established some understanding of the complex relation between the seismic properties of reservoir and related rocks, and their petrophysical properties in relation to production (porosity, permeability) and state (mineralogy, saturation, pore pressure). The unlocking of the interconnections between seismic attributes and petrophysical parameters is of key importance for the evaluation of stratigraphic

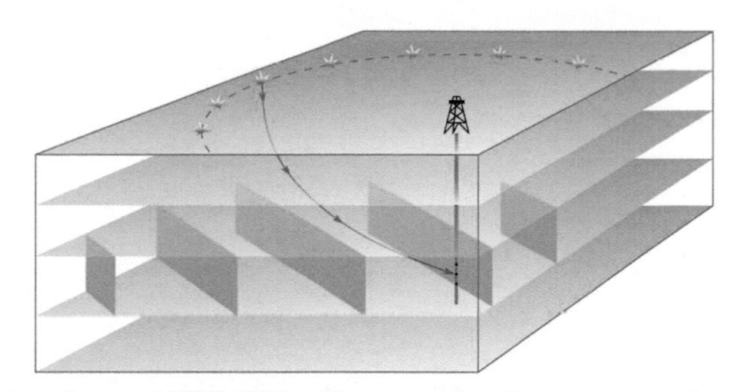

Figure 17.19 A walkaround VSP. With offset source locations spanning a large circular arc to probe the formation from a wide range of azimuths, this survey can be used to characterize direction and magnitude of anisotropy caused by aligned natural fractures. Walkaround VSPs can help understand fractures and fracture systems, both natural and hydraulically induced. From Blackburn, J., Daniels, J., Dingwall, S., Hampden-Smith, G., Leaney, S., Le Calvez, J. et al. (2007). Borehole seismic surveys: beyond the vertical profile. *Oilfield Review 19*(3), 20–35. Graphic copyright Schlumberger. Used with permission. *Courtesy:* Lisa Stewart (former executive editor of Oilfield Review) & Schlumberger.

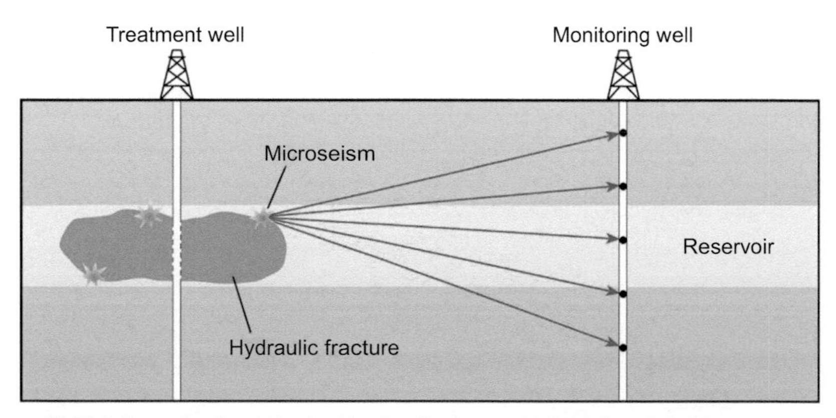

Figure 17.20 Microseismic method of hydraulic fracture monitoring. Sensitive multicomponent sensors in a monitoring borehole record microseismic events, or acoustic emissions, caused by hydraulic fracturing. Data processing determines event location, and visualization allows engineers to monitor the progress of stimulation operations. From Blackburn, J., Daniels, J., Dingwall, S., Hampden-Smith, G., Leaney, S., Le Calvez, J. et al. (2007). Borehole seismic surveys: beyond the vertical profile. *Oilfield Review 19*(3), 20–35. Graphic copyright Schlumberger. Used with permission. *Courtesy:* Lisa Stewart (former executive editor of Oilfield Review) & Schlumberger.

traps, fracture detection, and the spatial distribution of porosity and permeability (Grochau, 2009; Grochau & Gurevich, 2008; Nur, 1987; Scott, Zaman, & Roegiers, 1998; Shapiro & Gurevich, 2002). Some applications of seismic velocity (and attenuation) imaging in the description of reservoirs and the monitoring of reservoir recovery (Nur, 1987, p. 235) have been demonstrated over the years. Some related references are given in Table 17.4.

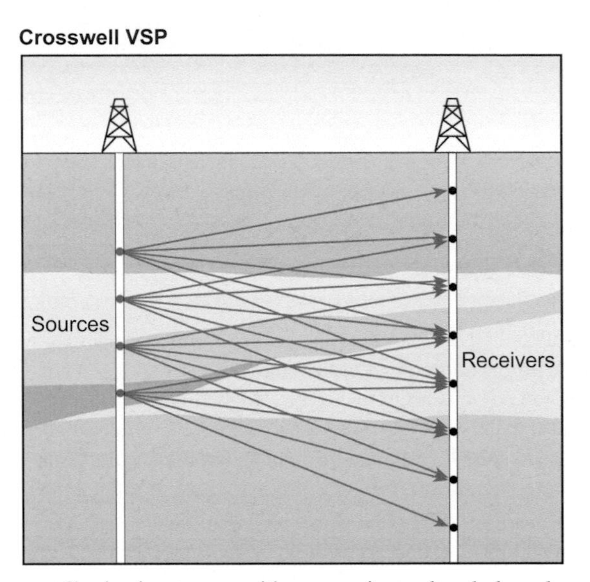

Figure 17.21 Cross-well seismic surveys, with sources in one borehole and receivers in another. Because raypaths are at large angles to any formation interfaces, little energy is reflected; most energy recorded by the receivers comes from direct arrivals. These data reveal information about formation velocities in the interwell volume. The repeatable survey geometry makes cross-well seismic surveys useful for time-lapse monitoring of steam injection, for example.
From Blackburn, J., Daniels, J., Dingwall, S., Hampden-Smith, G., Leaney, S., Le Calvez, J. et al. (2007). Borehole seismic surveys: beyond the vertical profile. *Oilfield Review 19*(3), 20–35. Graphic copyright Schlumberger. Used with permission. *Courtesy:* Lisa Stewart (former executive editor of Oilfield Review) & Schlumberger.

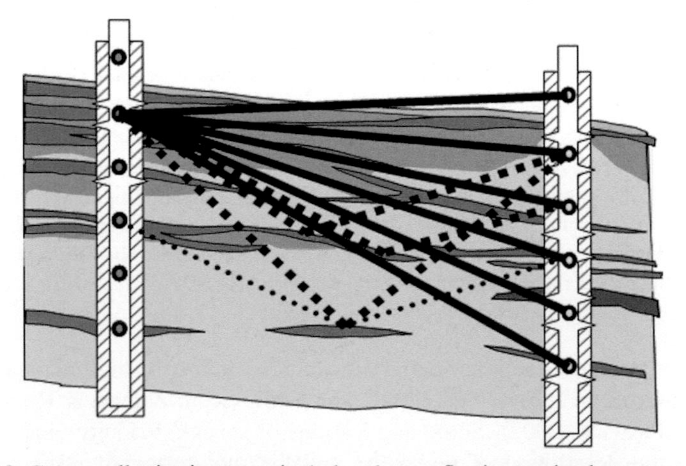

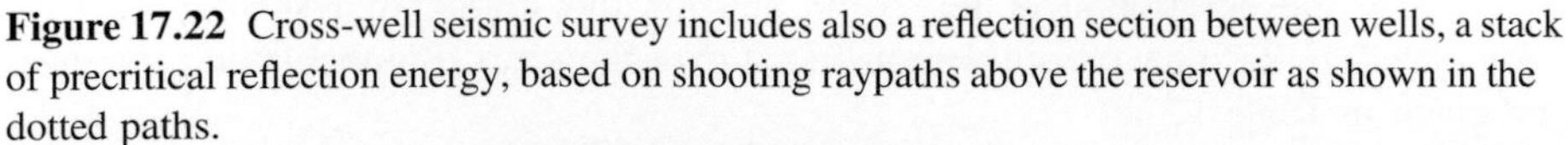

Figure 17.22 Cross-well seismic survey includes also a reflection section between wells, a stack of precritical reflection energy, based on shooting raypaths above the reservoir as shown in the dotted paths.
Courtesy: Bruce Marion (former advisor of Schlumberger).

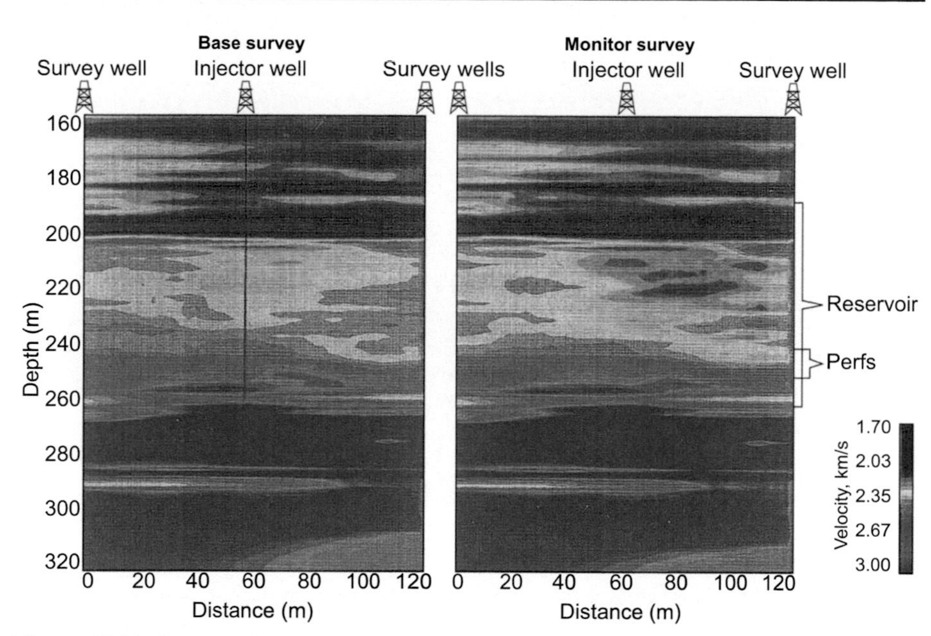

Figure 17.23 Cross-well seismic-velocity images before and after steam injection in the Athabasca tar sands, Alberta, Canada. The base survey (left) was acquired before steam injection. The injection well is halfway between the survey wells. The monitor survey (right) was run 3 months after steam injection began. The reservoir interval is 190–260 m, with a gas cap from 197 to 205 m. Perforation interval is 240–250 m. As seen from the velocity decrease—more orange (gray in print version) and red (dark gray in print version)—the steam has risen and taken off to the right.
From Albright, J., Cassell, B., Dangerfield, J., Deflandre, J.-P., Johnstad, S., & Withers, R. (1994). Seismic surveillance for monitoring reservoir changes. *Oilfield Review 6*(1), 4–14. Graphic copyright Schlumberger. Used with permission. *Courtesy:* Use of clients data.

17.6 Future trends

There have been new developments in full azimuth (FAZ) marine seismic imaging. One of the new data acquisition techniques, in which a single vessel acquires FAZ 3D seismic data by sailing in circles, can deliver for better illumination and more accurate and reliable subsurface images than conventional 3D methods in areas of complex geology. A multivessel implementation of such a new technique has been developed to address subsalt imaging challenges in deepwater areas. A typical FAZ tomography workflow is illustrated in Fig. 17.26 (Brice et al., 2013).

Another FAZ acquisition technique improves marine-seismic cross-line (between-streamer) data quality by combining wavefield–pressure measuring hydrophones with a 3-component (3C) microelectromechanical system (MEMS) unit that contains three orthogonal accelerators to measure full 3D motion of the recorded wavefield. The use of arrays of 3C MEMS units enables the estimation of 3D wavefield around the

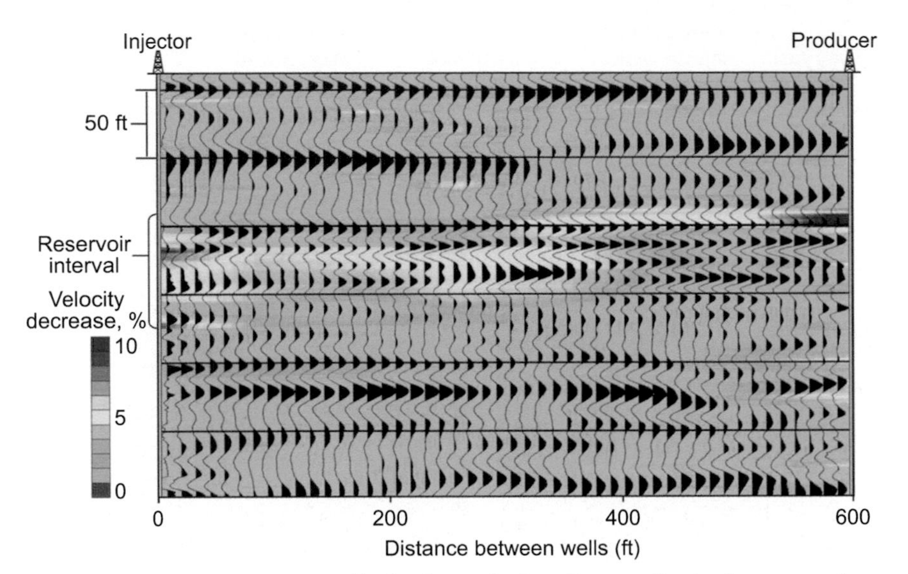

Figure 17.24 - Time-lapse cross-well seismic monitoring. Cross-well seismic tomography was used for mapping miscible CO_2 between an injector and producer in an enhanced oil recovery program. The display shows high-resolution cross-well seismic reflection data (black traces) overlying color-coded velocity-difference data. This color scale helps geoscientists track changes in seismic velocity between a baseline preinjection survey and a later survey conducted 9 months after CO_2 injection commenced. The zone in the center (yellow [very light gray in print version] and red [dark gray in print version]) experienced a greater velocity decrease than surrounding layers (blue [gray in print version] and green [light gray in print version]), indicating a potential CO_2 accumulation. In this case, the operator concluded the CO_2 sweep was not uniform between the injector and producer wells.
From Al-Ali, Z. A., Al-Buali, M. H., AlRuwaili, S., Ma, S. M., Marsala, A. F., Alumbaugh, D. et al. (2009). Looking deep into the reservoir. *Oilfield Review 21*(2), 38–47. Graphic copyright Schlumberger. Used with permission. *Courtesy:* Use of clients data.

reduced number of streamers in all directions—inline, cross-line, and vertical, sharpening the images of deep targets (Assev et al., 2013).

Further improvements are expected from the implementation of innovative acquisition configurations, advanced processing technologies, and new workflows that will extract more information from seismic measurements to enhance our understanding of the subsurface.

Improving seismic imaging resolution requires extending the range of useable signal frequencies at both high and low ends. The preservation of lower frequencies is important not only for imaging deep or steep targets but also for building high-resolution velocity models using full waveform inversion (FWI) (Assev et al., 2016; ten Kroode et al., 2013; Robertsson, Ronen, Singh, & van Borselen, 2013; Vigh, Jiao, & Watts, 2012).

New processing including 3D prestack acoustic FWI uses a two-way wave equation method, to build high-resolution velocity models. FWI performs forward modeling to

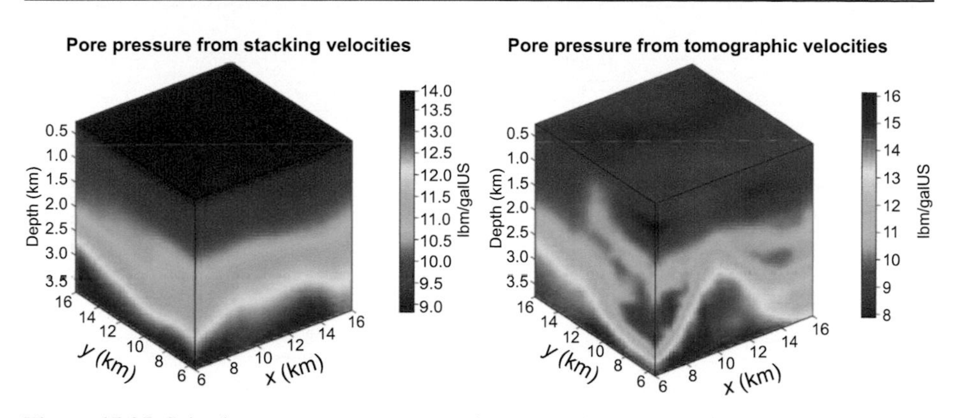

Figure 17.25 Seismic pore pressure tomography. In previous (conventional) methods, interpreters stacked seismic velocities to improve resolution; from this, they generated a pore pressure cube representing pore pressures across a given area (left). Now, tomographic techniques dramatically improve pore pressure resolution (right), reducing uncertainty and increasing accuracy in well planning.
From Barriol, Y., Glaser, K. S., Pop, J., Bartman, B., Corbiell, R., Eriksen, K. O. et al. (2005). The pressure of drilling and production. *Oilfield Review 17*(3), 22–41. Graphic copyright Schlumberger. Used with permission. *Courtesy:* Lisa Stewart (former executive editor of Oilfield Review) & Schlumberger.

Table 17.4 Applications of seismic imaging in reservoir description and monitoring.

Applications listed (Nur 1987)	Some references
Porosity and permeability mapping	Barclay et al. (2008)
Anomalous pore pressure detection	Landrø and Kvam (2002)
Fracture detection	Blackburn et al. (2007), Le Calvez et al. (2016)
Tracking thermal fronts	Zadeh, Srivastava, Vedanti and Landrø (2010)
Gap cap movement	Albright et al. (1994), Barkved (2004)
Water flooding	Leggett et al. (1993)
Steam reservoir boundaries	Justice et al. (1989), Albright et al. (1994), Brzostowski and McMechan (1995)
Distribution of permafrost	

compute the differences between the acquired seismic data and the current model and carries out a process similar to reverse time migration (RTM) on the residual dataset to compute a gradient volume and to update the velocity model (RTM is a prestack two-way wave equation migration algorithm suited to accurate imaging in and below areas with structural and velocity complexities). When combined with imaging using RTM, data-driven velocity model building with FWI improved the final product because consistent wavefield solutions were applied throughout the depth imaging workflow (Aseev et al., 2016; Brice et al., 2013).

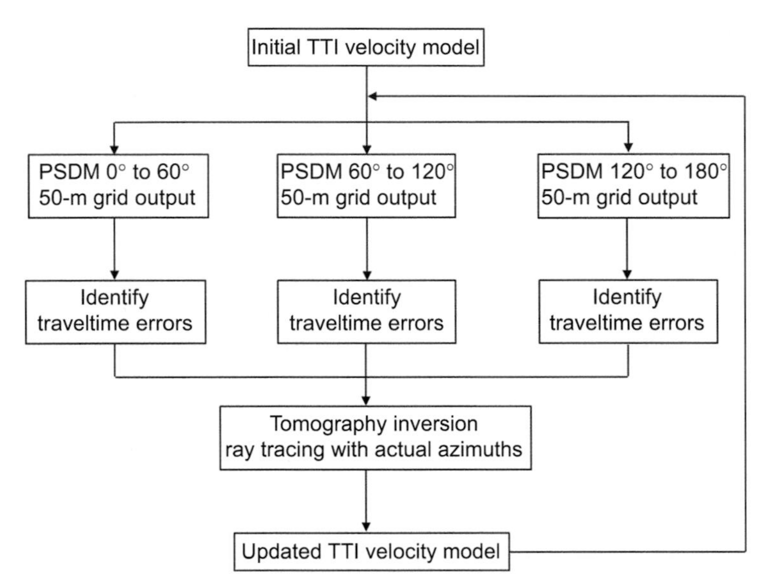

Figure 17.26 Azimuthal tomography workflow. A tilted transverse isotropic (TTI) velocity model derived from time domain processing formed the starting model. The new marine seismic acquisition dataset was divided into three azimuthal groups for prestack depth migration (PSDM). Traveltimes from the PSDM are compared with those predicted by the model using tomography inversion and ray tracing techniques, leading to an updated velocity model. Adapted from Brice, T., Buia, M., Cooke, A., Hill, D., Palmer, E., Khaled, N. et al. (2013). Developments in full azimuth marine seismic imaging. *Oilfield Review 25*(1), 42–55. Used with Schlumberger permission. *Courtesy:* Lisa Stewart (former executive editor of Oilfield Review) & Schlumberger.

Moseley, Nissen-Meyer, and Markham (2019) proposed the use of deep neural networks as fast alternatives to numerical methods (such as finite difference) for the simulation of seismic wave response at multiple locations in complex media. The need to model iteratively the time-dependent seismic wavefield is avoided, resulting in a massive reduction in computation and facilitating real-time seismic simulation and inversion based on forward modeling, such as FWI.

Deep learning methods have the potential to further automate and enhance seismic tomography and the related interpretation workflows (which are highly demanding in the numerical computational solutions to the wave equations for subsurface velocity model building and in the domain experts). To overcome these resources-intensive challenges, deep learning approach (based on, e.g., 3D convolutional neural network) has been used to provide a direct, accurate, and much faster reconstruction of a subsurface predictive earth model from the seismic data (Araya-Polo, Adler, Farris, & Jennings, 2020; Yuan, Zhang, Jia, & Zhang, 2020).

Deep-reading resistivity tomography obtained through electromagnetic (EM) induction cross-well surveys described in Chapter 18 can provide better petrophysical control in building reservoir models than surface seismic surveys. The capability to identify and monitor fluid fronts using cross-well EM measurements can be complemented by cross-well seismic imaging technique, which can delineate structure and monitor fluid movement in the reservoir. The detailed reservoir-level information obtained from combined deep-reading measurements gives petrophysicists and reservoir engineers greater insight into how fluids move and interact, which will ultimately determine the decision-making of oilfield management and production (Al-Ali et al., 2009; Maver, Hulme, McCallum, & Nalonnil, 2009; Dell'Aversana, 2019).

17.7 Source of further information and advice

- The Society of Exploration Geophysicists (SEG) Journals:
  - Geophysics
  - The Leading Edge
- The European Association of Geoscientists and Engineers (EAGE) Journals:
  - Geophysical Prospecting
  - First Break
- Other Journals or Magazines:
  - Petroleum Geoscience
  - Journal of Petroleum Geology
  - Geo Expro

Acknowledgments

I would like express my gratitude to Schlumberger for permission to publish this book chapter, and to several Schlumberger clients for permission to reuse some figures containing their data. Lisa Stewart, former executive editor of Schlumberger *Oilfield Review*, helped in the preparation (of the first edition) of this book chapter. Bruce Marion, former Schlumberger advisor, reviewed the first-edition draft of the manuscript and provided invaluable suggestions for improving the section related to cross-well seismic imaging.

References

Al-Ali, Z. A., Al-Buali, M. H., AlRuwaili, S., Ma, S. M., Marsala, A. F., Alumbaugh, D., et al. (2009). Looking deep into the reservoir. *Oilfield Review, 21*(2), 38–47.

Albertin, U., Kapoor, J., Randall, R., Smith, M., Brown, G., Soufleris, C., et al. (2002). The time for depth imaging. *Oilfield Review, 14*(1), 2–15.

Albright, J., Cassell, B., Dangerfield, J., Deflandre, J.-P., Johnstad, S., Withers, R., & January. (1994). Seismic surveillance for monitoring reservoir changes. *Oilfield Review, 6*(1), 4–14.

Araya-Polo, M., Adler, A., Farris, S., & Jennings, J. (2020). Fast and accurate seismic tomography via deep learning. In W. Pedrycz, & S. M. Chen (Eds.), *Deep Learning: Algorithms and applications. Studies in computational intelligence* (Vol. 865, pp. 129–156). Springer.

Aronsen, H. A., Osdal, B., Dahl, T., Eiken, O., Goto, R., Khazanehdari, J., et al. (2004). Time will tell: New insights from time-lapse seismic data. *Oilfield Review, 16*(2), 6–15.

Aseev, A., Chandola, S. K., Foo, L. C., Cunnell, C., Francis, M., Gupta, S., et al. (2016). Marine imaging in three dimensions: Viewing complex structures. *Oilfield Review, 28*(2), 4–15.

Ashcroft, W. (2011). *A petroleum geologist's guide to seismic reflection*. Chichester: Wiley-Blackwell.

Bagaini, C., Bunting, T., El-Emam, A., Laake, A., & Strobbia, C. (2010). Land seismic techniques for high-quality data. *Oilfield Review, 22*(2), 28–39.

Barclay, F., Bruun, A., Rasmussen, K. B., Alfaro, J. C., Cooke, A., Cooke, D., et al. (2008). Seismic inversion: Reading between the lines. *Oilfield Review, 20*(1), 42–63.

Barkved, O., Bartman, B., Compani, B., Gaiser, J., Van Dok, R., Johns, T., et al. (2004). The many facets of multicomponent seismic data. *Oilfield Review, 16*(2), 42–56.

Barriol, Y., Glaser, K. S., Pop, J., Bartman, B., Corbiell, R., Eriksen, K. O., et al. (2005). The pressure of drilling and production. *Oilfield Review, 17*(3), 22–41.

Blackburn, J., Daniels, J., Dingwall, S., Hampden-Smith, G., Leaney, S., Le Calvez, J., et al. (2007). Borehole seismic surveys: Beyond the vertical profile. *Oilfield Review, 19*(3), 20–35.

Brady, J., Campbell, T., Fenwick, A., Ganz, M., Sandberg, S. K., Buonora, M. P. P., et al. (2009). Electromagnetic sounding for hydrocarbons. *Oilfield Review, 21*(1), 4–19.

Brice, T., Buia, M., Cooke, A., Hill, D., Palmer, E., Khaled, N., et al. (2013). Developments in full azimuth marine seismic Imaging. *Oilfield Review, 25*(1), 42–55.

Brzostowski, M. A., & McMechan, G. A. (1995). Tomographic imaging of 3D seismic low-velocity anomalies to simulate monitoring of enhanced oil recovery. *International Journal of Imaging Systems and Technology, 5*(1), 62–72.

Cavalca, M., & Fletcher, R. P. (2008). Deriving 3D Q models from surface seismic data using attenuated traveltime tomography. In *70th EAGE conference & exhibition, Rome, Italy, 9 - 12 June 2008*. Paper H003.

Červeny, V. (1987). Ray tracing algorithms in three-dimensional laterally varying layered structures. In G. Nolet (Ed.), *Seismic tomography* (pp. 99–133). D. Reidel Publishing Co.

Chapman, C. (1987). The Radon transform and seismic tomography. In G. Nolet (Ed.), *Seismic tomography* (pp. 25–47). D. Reidel Publishing Co.

Chapman, C. (2004). *Fundamentals of seismic wave propagation*. Cambridge: Cambridge University Press.

Cook, J., Frederiksen, R. A., Hasbo, K., Green, S., Judzis, A., Martin, J. W., et al. (2007). Rock matters: Ground truth in geomechanics. *Oilfield Review, 19*(3), 36–55.

Dawoud, M., Hallinan, S., Hermann, R., van Kleef, F., & Spring. (2009). Near-surface electromagnetic surveying. *Oilfield Review, 21*(1), 20–25.

De Freitas, J. M. (2011). Recent developments in seismic seabed oil reservoir monitoring applications using fibre-optic sensing networks. *Measurement Science and Technology, 22*, 052001.

Dell'Aversana, P. (2019). *A global approach to data value maximization: Integration, machine learning and multimodal analysis*. Cambridge Scholars Publishing.

Dong, Q., Marion, B., Meyer, J., Xu, Y., & Xu, D. (2005). Imaging complex structure with crosswell seismic in Jianghan oil field. *The Leading Edge, 24*, 18.

Dutta, N. C. (2002). Geopressure prediction using seismic data: Current status and the road ahead. *Geophysics, 67*(6), 2012–2041.

Grochau, M. H. (2009). *An integrated approach to improve time-lapse seismic interpretation: Investigation of pressure and saturation effects on elastic parameters*. Lambert Academic Publishing.

Grochau, M. H., & Gurevich, B. (2008). Investigation of core data reliability to support time-lapse interpretation in Campos Basin, Brazil. *Geophysics, 3*(2), E59–E65.

Guo, R., Li, M., Yang, F., Xu, S., & Abubakar, A. (2019). First arrival traveltime tomography using supervised descent learning technique. *Inverse Problems, 35*(10).

Hu, W., Abubakar, A., & Habashy, T. (2008). A local contrast source inversion algorithm for cross-well time-lapse seismic applications. In *SEG Las Vegas 2008 annual meeting*.

Justice, J. H., Vassiliou, A. A., Singh, S., Logel, J. D., Hansen, P. A., Hall, B. R., et al. (1989). Acoustic tomography for monitoring enhanced oil recovery. *The Leading Edge, 8*(2), 12–19.

ten Kroode, F., Bergler, S., Corsten, C., de Maag, J., Strijbos, F., & Tijhof, H. (2013). Broadband seismic data — the importance of low frequencies. *Geophysics, 78*(2), WA3–WA14.

Landrø, M., Kvam, Ø., & September. (2002). Pore pressure estimation – what can we learn from 4D? *CSEG Recorder*, 82–87.

Le Calvez, J., Malpani, R., Xu, J., Stokes, J., Williams, M., & May. (2016). Hydraulic fracturing insights from microseismic monitoring. *Oilfield Review, 28*(2), 16–33.

Leggett, M., Goulty, N. R., & Kragh, J. E. (1993). Study of traveltime and amplitude time-lapse tomography using physical model data. *Geophysical Prospecting, 41*(5), 599–619.

Li, X.-Y., & Zhang, Y.-G. (2011). Seismic reservoir characterization: How can multicomponent data help? *Journal of Geophysics and Engineering, 8*, 123–141.

Mathewson, J. C., Evans, D., Leone, C., Leathard, M., Dangerfield, J., & Tonning, S. A. (2012). Improved imaging and resolution of overburden heterogeneity by combining amplitude inversion with tomography. In *SEG Las Vegas 2012 annual meeting*.

Maver, K. G., Hulme, C., McCallum, M., Nalonnil, A., & December. (2009). Improved imaging with deep reading technologies. *Geo Expro, 6*(6), 28–34.

Moseley, B., Nissen-Meyer, T., & Markham, A. (2019). Deep learning for fast simulation of seismic waves in complex media. *Solid Earth Discuss*, 1–23. https://doi.org/10.5194/se-2019-157

Nolet, G. (1987). Seismic wave propagation and seismic tomography. In G. Nolet (Ed.), *Seismic tomography* (pp. 1–23). D. Reidel Publishing Co.

Nolet, G. (2008). *A breviary of seismic tomography – imaging the interior of the Earth and Sun*. Cambridge: Cambridge University Press.

Nur, A. (1987). Seismic rock properties for reservoir description and monitoring. In G. Nolet (Ed.), *Seismic tomography* (pp. 203–237). D. Reidel Publishing Co.

Rao, Y., & Wang, Y. (2011). Crosshole seismic tomography including the anisotropy effect. *Journal of Geophysics and Engineering, 8*, 316–321.

Robertsson, J. O. A., Ronen, S., Singh, S., & van Borselen, R. (2013). Broadband seismology in oil and gas exploration and production — Introduction. *Geophysics, 78*(2), WA1–WA2.

Sayers, C. (2013). The effect of kerogen on the elastic anisotropy of organic-rich shales. *Geophysics, 78*(2), D65–D74.

Schuck, A., & Lange, G. (2007). Seismic methods. In J. Knodel, G. Lange, & H.-J. Voigt (Eds.), *Environmental geology – handbook of field methods and case studies* (pp. 337–402). Berlin: Springer.

Scott, T. E., Jr., Zaman, M. M., & Roegiers, J.-C. (1998). Acoustic-velocity signatures associated with rock-deformation processes. *Journal of Petroleum Technology, 50*(6), 70–74.

Seidel, K., & Lindner, H. (2007). Gravity methods. In J. Knodel, G. Lange, & H.-J. Voigt (Eds.), *Environmental geology – handbook of field methods and case studies* (pp. 185–204). Berlin: Springer.

Shapiro, S. A., & Gurevich, B. (2002). Seismic signatures of fluid transport—Introduction. *Geophysics, 67*(1), 197–198.

van der Sluis, A., & van der Vorst, H. A. (1987). Numerical solution of large, sparse linear algebraic systems arising from tomographic problems. In G. Nolet (Ed.), *Seismic tomography* (pp. 49–83). D. Reidel Publishing Co.

Strobbia, C. L., Glushchenko, A., Laake, A., Vermeer, P., Papworth, S., & Ji, U. (2009). Arctic near surface challenges: The point receiver solution to coherent noise and statics. *First Break, 27*(2), 69–76.

Tarantola, A. (1987). Inversion of travel times and seismic waveforms. In G. Nolet (Ed.), *Seismic tomography* (pp. 135–157). D. Reidel Publishing Co.

Vesnaver, A. (2008). Yardsticks for industrial tomography. *Geophysical Prospecting, 56*, 457–465.

Vigh, D., Jiao, K., & Watts, D. (2012). Elastic full-waveform inversion using 4C data acquisition. In *SEG Las Vegas 2012 annual meeting*.

Wielandt, E. (1987). On the validity of the ray approximation for interpreting delay times. In G. Nolet (Ed.), *Seismic tomography* (pp. 85–98). D. Reidel Publishing Co.

Woodward, M. J., Nichols, D., Zdraveva, O., Whitfield, P., & Johns, T. (2008). A decade of tomography. *Geophysics, 73*(5), VE5–VE11.

Yuan, C., Zhang, X., Jia, X., & Zhang, J. (2020). Time-lapse velocity imaging via deep learning. *Geophysical Journal International, 220*(2), 1228–1241.

Zadeh, H. M., Srivastava, R. P., Vedanti, N., & Landrø, M. (2010). Seismic monitoring of in situ combustion process in a heavy oil field. *Journal of Geophysics and Engineering, 7*(1), 16–29.

Applications of tomography in oil–gas industry—Part 2

Cheng-gang Xie [1], *Michael Wilt* [2] *and David Alumbaugh* [2]
[1]Schlumberger, Singapore; [2]Earth & Environmental Sciences Area – Berkeley Lab, Berkeley, CA, United States

18.1 Introduction

Applications of tomography in the upstream oil–gas industry are widespread, covering from kilometer reservoir-scale, sub-meter borehole scale to submm rock-core pore scale. Continuing with the theme of Chapter 17 on the applications of seismic tomography in the oil–gas industry, this chapter illustrates the use of other tomography modalities in hydrocarbon exploration, reservoir monitoring and production monitoring.

In Section 18.2, cross-well (cross-borehole) tomography for reservoir monitoring is demonstrated by way of electromagnetic (EM) (induction) resistivity imaging of water flood and steam flood processes.

Complex oil–gas–water multiphase flows are often encountered in hydrocarbon production pipelines, and the accurate measurement of their flow rates over a wide range of flow regimes is still a challenge facing the oil–gas industry. In Section 18.3, some tomography modalities investigated for multiphase flow imaging and measurement will be described, and some results from electrical capacitance tomography (ECT) demonstrated.

In Section 18.4, we will make some remarks of the future trend of industrial tomography in oil–gas industry.

18.2 Cross-well electromagnetic tomography in hydrocarbon reservoir monitoring

Seismic surveys described in the last chapter can image reservoir volumes but tend to lack sufficient vertical resolution and sensitivity to fluid type and fluid distribution within a reservoir (seismic surveys are more sensitive to rock matrix). To better monitor movements of hydrocarbons fluid lying between wells at a reservoir scale, fluid saturation surveys based on cross-well EM induction tomography have been performed (Alumbaugh & Morrison, 1995; Wilt et al., 1995; DePavia et al., 2008; Marsala et al. 2008; Wilt, Donadille, AlRuwaili, Ma, & Marsala, 2008. Surface-to-borehole EM uses an array of borehole receivers and a movable surface source, Colombo, McNeice, Cuevas, & Pezzoli, 2018; Marsala et al. 2011; Pardo, Torres-Verdín, & Zhang, 2008). The cross-well EM

Industrial Tomography. https://doi.org/10.1016/B978-0-12-823015-2.00011-X

induction tomography approach produces a resistivity distribution between wells and offers a better resolution than surface seismic surveys (Al-Ali et al., 2009). As indicated in Fig. 18.1, cross-well EM surveys investigating the interwell region fill an intermediate role between high-resolution near-wellbore logs (e.g., nuclear magnetic resonance logging, Allen et al., 2000) and lower resolution surface seismic or electrical resistivity surveys (Ramirez, Daily, Owen, Chesnut, & LaBrecque, 1992; Seidel & Lange, 2007; Tsourlos, Ogilvy, Papazachos, & Meldrum, 2011). High-resolution analyses of rock-core sample are carried out in laboratories by using magnetic resonance imaging (Mitchell, Chandrasekera, Holland, Gladden, & Fordham, 2013; Mitchell, Staniland, et al., 2013), X-ray tomography (Fig. 18.2), or micro-X-ray tomography (Kayser, Knackstedt, & Ziauddin, 2006; Feali et al., 2012; Iglauer, Wang, & Rasouli, 2011; Ziauddin & Bize, 2007).

Induction measurements respond primarily to electrical conductivity (the reciprocal of resistivity); in the downhole environment, high conductivity values generally indicate the presence of salt water. The presence of hydrocarbons may be indicated by low conductivity (high resistivity) values. However, the measurement is influenced by temperature, porosity, pore-fluid resistivity, and fluid saturations. Cross-well EM resistivity imaging is used to distinguish the electrical contrast between resistive oil and conductive saline water. Interwell resistivity data are also useful for mapping variations in reservoir properties (see Table 17.2 in Chapter 17) for heterogeneity or connectivity between wells.

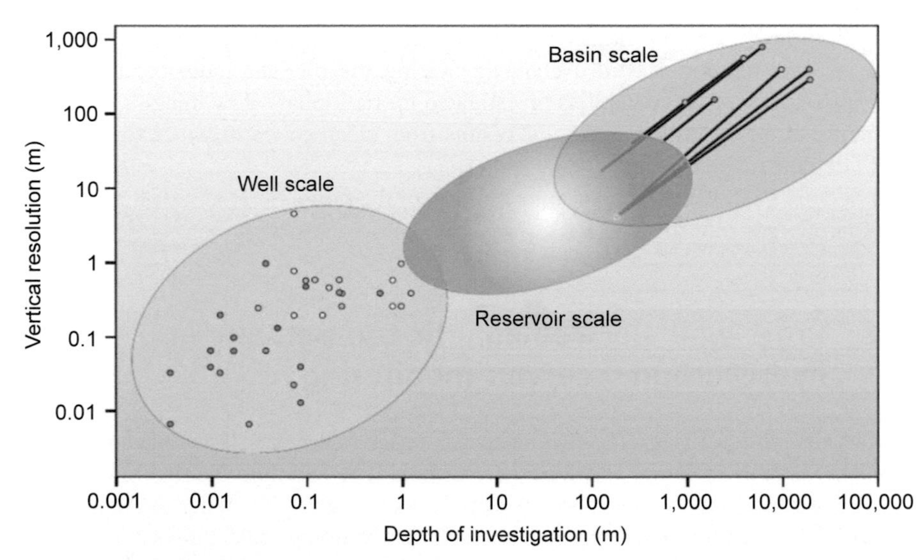

Figure 18.1 Bridging the resolution gap. Most oilfield measurements fall at the ends of a spectrum; they are obtained by probing either the near-wellbore vicinity at high resolution or a basin-wide area at lower resolution. Cross-well measurements sample at an intermediate depth of investigation and resolution to help geoscientists characterize and image the reservoir within the interwell region.

From Al-Ali Z. A., Al-Buali M. H., AlRuwaili S., Ma S. M., Marsala A. F., Alumbaugh D., et al. (2009). Looking deep into the reservoir. *Oilfield Review, 21*(2), 38–47. Graphic copyright Schlumberger. Used with permission. Courtesy: Lisa Stewart (former executive editor of Oilfield Review) & Schlumberger.

Figure 18.2 Computed tomography (CT) scans of a rock core indicate heterogeneity. Graphic copyright Schlumberger. Used with permission. Courtesy: Lisa Stewart (former executive editor of Oilfield Review) & Schlumberger.

18.2.1 Principles of EM induction cross-well tomography

Illustrated in Fig. 18.3, in a cross-well EM survey, a magnetic dipole transmitter is placed in one well and an array of magnetometer receivers in another well to investigate the interwell area. By transmitting from one well to another at frequencies in the Hz to kHz range, signals over a distance up to 1 km can be propagated, depending on wellbore casing and formation characteristics (Table 18.1) (Al-Ali et al., 2009).

EM cross-well data are collected for tomographic resistivity imaging by lowering the receiver array into a wellbore to a specified depth. While this receiver array remains stationary, the EM transmitter in the other well continuously broadcasts as it is moved between selected depths. Once the signals from a complete transmitter traverse are obtained at a depth, the receiver tool is repositioned (typically uphole) to another depth, and the process is repeated. Thus, as the transmitter and receivers are moved across the logging interval, the interwell target zone is interrogated from multiple angles (Fig. 18.3).

The EM transmitter generates a magnetic field, usually a monochromatic sine wave, that is sent into the formation at frequencies between 5 Hz and 1 kHz. This magnetic field, known as the primary field, attenuates with increasing distance from the transmitter. The primary field induces a current in conductive formations (Fig. 18.4). This current generates an opposing secondary EM field, such that the total field decreases in amplitude with decreasing formation resistivity and changes its phase. An array of sensitive induction coil receivers can be stationed in the receiver well to detect the primary EM field generated by the transmitter and the secondary EM field generated by the induced currents.

For a vertically oriented (z-direction) magnetic dipole source in a homogeneous isotropic earth, Maxwell's equations can be combined to yield a second-order equation for the vertical magnetic field H_z:

$$\frac{\partial^2 H_z}{\partial x^2} + \frac{\partial^2 H_z}{\partial y^2} + \frac{\partial^2 H_z}{\partial z^2} + k^2 H_z = m_z \tag{18.1}$$

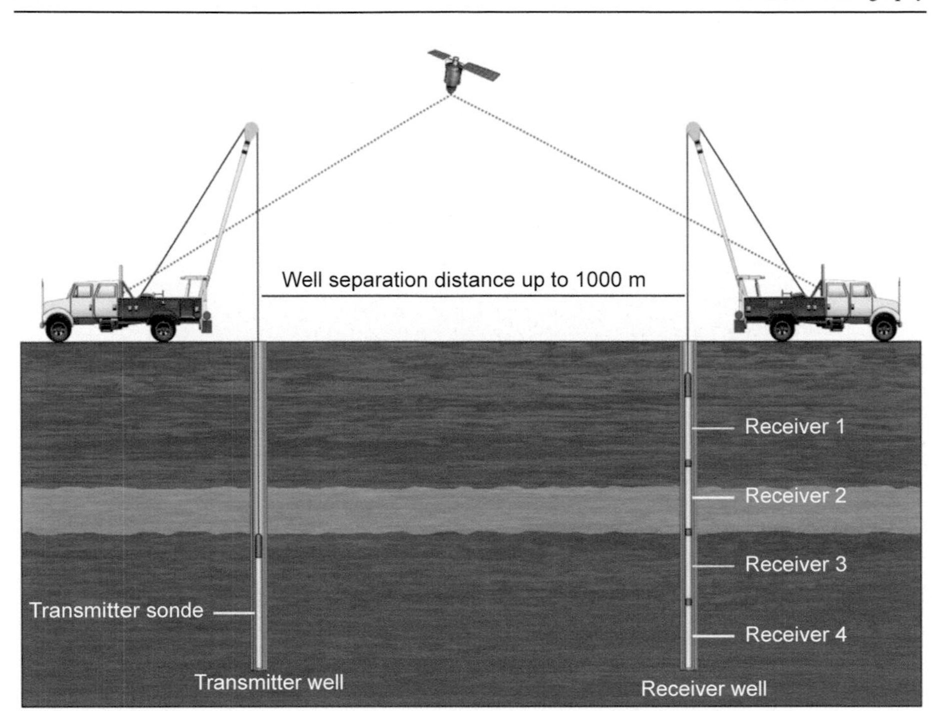

Figure 18.3 Cross-well EM acquisition system. A magnetic dipole transmitter is lowered into the first well (left). In the second well (right), an array of four receivers is deployed. Magnetometers in the receiver array can measure magnetic fields 10 orders of magnitude smaller than the Earth's static magnetic field. The transmitter and receiver components are linked wirelessly between wells and spaced up to 1 km apart.
From Al-Ali Z. A., Al-Buali M. H., AlRuwaili S., Ma S. M., Marsala A. F., Alumbaugh D., et al. (2009). Looking deep into the reservoir. *Oilfield Review, 21*(2), 38–47. Graphic copyright Schlumberger. Used with permission. Courtesy: Lisa Stewart (former executive editor of Oilfield Review) & Schlumberger.

Table 18.1 Distance versus casing. EM induction cross-well surveys are constrained by casing type and formation conditions.

Transmitter well	Receiver well	Maximum well spacing
Open hole	Open hole	1000 m
Fiberglass casing	Fiberglass casing	1000 m
Open hole	Chromium steel casing	500 m
Open hole	Carbon steel casing	450 m
Chromium steel casing	Chromium steel casing	350 m

Adapted from Al-Ali Z. A., Al-Buali M. H., AlRuwaili S., Ma S. M., Marsala A. F., Alumbaugh D., et al. (2009). Looking deep into the reservoir. *Oilfield Review, 21*(2), 38–47.

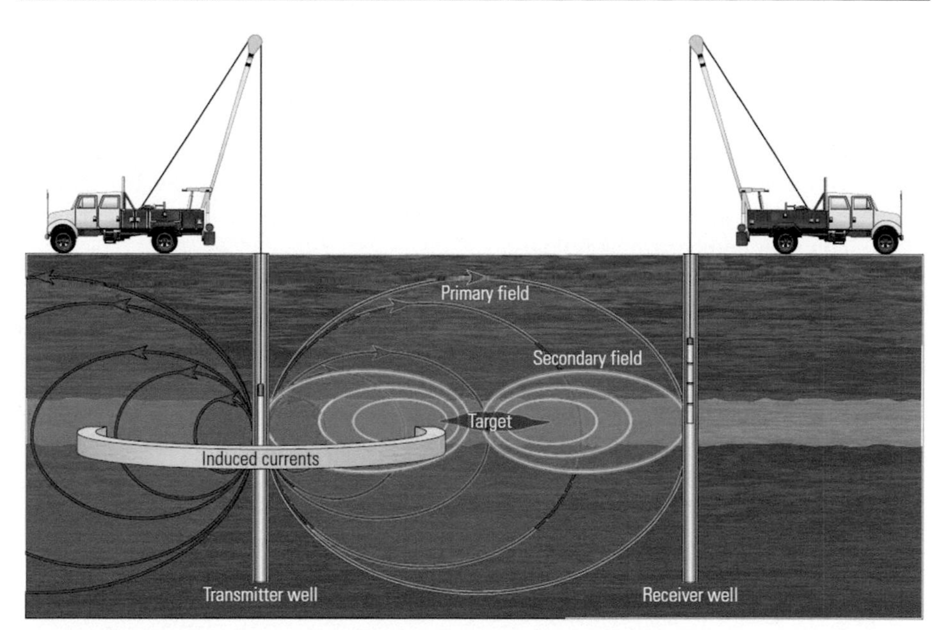

Figure 18.4 Primary and secondary fields. An alternating current excites the magnetic dipole transmitter coil to send an electromagnetic field into the formation. This primary field induces eddy currents that, in turn, generate a secondary alternating electromagnetic field whose strength is inversely proportional to formation resistivity. The secondary electromagnetic field is detected at the receiver array along with the primary field.
From Al-Ali Z. A., Al-Buali M. H., AlRuwaili S., Ma S. M., Marsala A. F., Alumbaugh D., et al. (2009). Looking deep into the reservoir. *Oilfield Review, 21*(2), 38–47. Graphic copyright Schlumberger. Used with permission. Courtesy: Lisa Stewart (former executive editor of Oilfield Review) & Schlumberger.

For a multiturn loop (or coil), the magnetic dipole moment in the above equation is $m = \mu_r NAI$, representing the strength of the source (where A is the cross-sectional area, N the number of windings, I the current flowing in the windings, and $\mu_r = \mu/\mu_o$ the effective magnetic permeability of the core relative to free space).

Eq. (18.1) is the Helmholtz wave equation commonly associated with a propagating energy field such as a seismic field (Eqn 17.4). For low-frequency EM waves where the dielectric constant can be essentially ignored, the propagation wavenumber, $k = \sqrt{i\sigma\omega\mu}$, is complex, leading to the diffusive nature of the fields (here $\sigma = 1/\rho$ is the formation conductivity, $\omega = 2\pi f$ is the angular frequency (with frequency f), μ the magnetic permeability, and $i = \sqrt{-1}$). The diffusive nature means that the waveform changes in shape and amplitude as it propagates through the medium, with higher frequency fields attenuating more rapidly than the lower ones, and the phase shifts as the energy moves away from the source.

For a receiver at the same depth level as the source, the solution to Eq. (18.1) is given by the sum of primary and secondary fields:

$$H_z = H_{z,\text{primary}} + H_{z,\text{secondary}} = \frac{m}{4\pi r^3}\left(k^2 r^2 - ikr - 1\right)e^{-ikr} \tag{18.2}$$

The total field is, thus, a function of the complex propagation constant k. To illustrate how the earth conductivity affects the field, we need to separate the in-phase (real) and out-of-phase (imaginary or quadrature) field components, and to combine terms. The quantity of plane-wave skin depth, $\delta = \sqrt{2/(\sigma\omega\mu)}$, can be used to represent the distance that a propagating plane wave attenuates to $1/e$ of its initial value, and experiences a phase-shift of one radian (57.3 degrees). Eq. (18.2) can be written in terms of the induction number r/δ as follows:

$$H_z = H_{z,\text{primary}} + H_{z,\text{secondary}} = \frac{m}{4\pi r^3}\left(2i\frac{r^2}{\delta^2} - (1+i)\frac{r}{\delta} - 1\right)e^{-ir/\delta}e^{-r/\delta} \quad (18.3)$$

At low frequency, low conductivity or close well spacing (i.e., $r/\delta < 0.01$), the last term in the brackets dominates, and Eq. (18.3) yields the primary field. Data collected under these conditions are useful for determining the positions of the transmitters and receivers but insensitive to the formation conductivity. At high frequency, high conductivity or large well spacing ($r/\delta > 10$), the exponential decay function ($e^{-r/\delta}$) dominates, and the field is attenuated to below system noise levels. In this case, the induced currents shield the transmitter, and the fields are invisible to the receiver.

The intermediate range ($0.4 < r/\delta < 10.0$) is where cross-well EM measurements are most useful (the sweet spot). At the low end of this range, the formation signal is roughly 10% of the total received signal; but at the high end, the fields are exponentially decaying due to the formation conductivity and, thus, very sensitive to the interwell conductivity structure. In cross-well EM tomography, it is a general rule to operate at the highest frequency where one can still achieve a good signal/noise ratio.

In addition to the attenuation term ($e^{-r/\delta}$), Eq. (18.3) also has the oscillatory component given in the complex exponential $e^{-ir/\delta} = \cos(r/\delta) + i\sin(r/\delta)$. The field, therefore, changes in phase as a function of the formation conductivity, a very useful attribute for interwell imaging (e.g., $r/\delta = \pi/2$ means the phase of the signal has changed by 90 degrees during interwell propagation, a very significant change if we can detect phase to within 1 degree).

Fig. 18.5 shows the cross-well vertical magnetic field in-phase and out-of-phase (or quadrature) with respect to the transmitter current as a function of the induction number for a source and receiver at the same depth level. Note that here we plot the total field (primary plus secondary) and that the primary field is in-phase with the transmitter. The shaded region is the sweet spot for cross-well EM surveys, where the fields are sensitive to the formation conductivity but still large enough to be accurately measured.

The conductivity of subsurface with fully or partially saturated sediments (σ_m) is determined by the porosity (ϕ), conductivity of the pore fluid (σ_w) and water saturation (S), and for shale-free, clean, sandy sediments it is defined by the following relation (Archie's law):

$$\sigma_m = (\phi^m / a)\sigma_w S^n \quad (18.4)$$

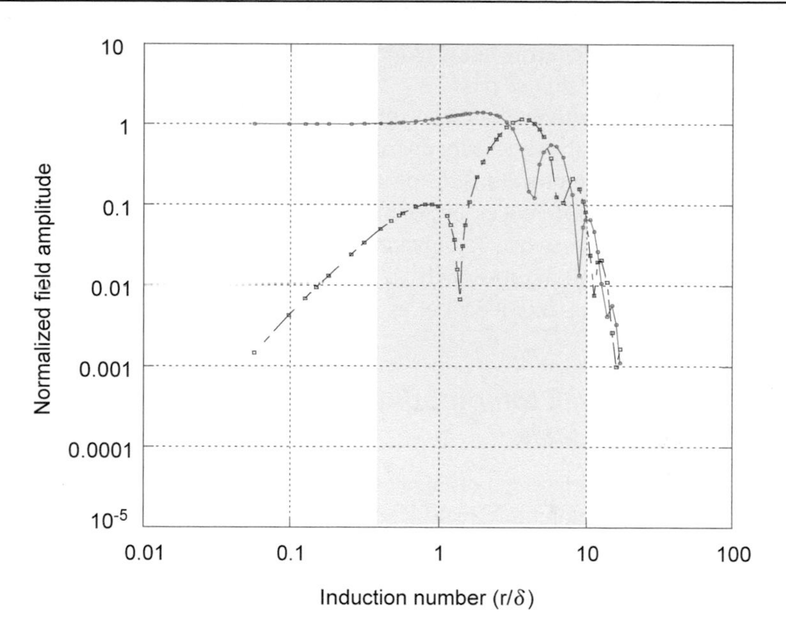

Figure 18.5 The in-phase (real) (red [gray in print version]) and quadrature (imaginary) (blue [black in print version]) cross-well field in a homogeneous whole space.

where n is saturation exponent ($n \cong 2$), m is the cementation exponent (for most sedimentary rocks $1.3 < m < 2.5$), and a is an empirical parameter ($0.5 < a < 1$, depending on the type of sediment) (Lange & Seidel, 2007).

An EM cross-well tomography survey can help map variations in the resistivity distribution within a reservoir, hence reflecting changes in the underlying resistivity-dependent properties such as porosity, fluid saturations, and temperature. The detection of these changes can be valuable for hydrocarbon reservoir management. For instance, changes in porosity may be indicative of subsidence within a reservoir. Changes in saturation often suggest the advancing front of a water flood, whereas static saturation measurements may indicate bypassed pay. Temperature variations can help evaluate the effectiveness of a steam flood. During fluid injection, differences in resistivity distribution over time provide information about fluid movement, which in turn may provide clues to permeability distribution. EM cross-well tomography surveys can hence be utilized for a variety of oilfield applications, such as monitoring reservoir sweep efficiency, identifying bypassed pay, planning infill drilling locations, and improving effectiveness of reservoir simulations (Al-Ali et al., 2009).

An EM survey often constitutes several thousand measurements spanning a depth interval roughly equal to the well spacing. Data are collected using stationary receivers in a multiple sensor array and a continuously moving transmitter. Data are collected while the transmitter is moving and the received signals are averaged over several hundred cycles per logging depth. Transmitter logging speeds may range from 50 to 600 m/h, and it may require 12–30 h of field recording for a vertical section of 300 m with logging depths every 5 m. The depths covered by the multiple evenly spaced coils in the receiver array helps reduce logging times by covering multiple logging depths for each transmitter pass.

Once the actual field survey data are acquired after careful prejob planning (Al-Ali et al., 2009), an image of the resistivity distribution between wells is generated through a nonlinear data inversion process. This process employs an iterative procedure to update and minimize the differences between the measured data and the modeled responses in a least-squares sense (Abubakar et al., 2005; Li, Abubakar, Li, Pan, & Habashy, 2010). The resulting resistivity distributions are used to track interwell fluid movements (Mieles et al., 2009; Wilt et al., 2012).

18.2.2 EM cross-well tomography case 1: imaging fluid flow on a reservoir scale

To understand how uneven sweep of injected water was affecting production, an operator in UAE needed to monitor subsurface water flood movements between wells to help improve reservoir modeling in predicting water front dimensions. The monitoring method should provide better lateral coverage than well logging and a higher vertical resolution than seismic surveys (Al-Ali, Al-Buali et al., 2009; Al Ali, Vahrenkamp, et al., 2009).

Deep-reading EM cross-well profiles of well pairs were obtained among three dedicated observer wells having chromium steel completions (with maximum interwell distance being 350 m for such casings, Table 18.1). The EM survey conducted before injection established the baseline to enable tracking of changes in the interwell reservoir space. Time-lapse survey images obtained 6 months after the baseline indicated effective water flooding from the horizontal injector well in the direction of observer Wells 1 and 3, with a large volume of water swept along a preferential flow direction (Fig. 18.6), confirmed by well logs from a downhole reservoir saturation tool. The results of subsequent time-lapse EM surveys provided more details on the advancement of the water flood. From this information, engineers and geologists developed an improved reservoir engineering flow simulation model and devised strategy for placement of new injectors and infill development wells while fine-tuning the performance of existing injectors for optimal recovery.

18.2.3 EM cross-well tomography case 2: water flood monitoring

In an oil field in Saudi Arabia, seawater was pumped into peripheral water injectors for several years to maintain reservoir pressure. The injected water replaces produced oil and has broken through in some of the producer wells. The distribution of injected water is variable, because of heterogeneities in this carbonate reservoir (Wilt et al., 2008). In one part of this giant field, resistivity measurements obtained by conventional wireline logs indicated two producing wells were nearly watered out. Because the resistivity distribution between the wells was unknown, the fluid saturation in the interwell region could not be determined.

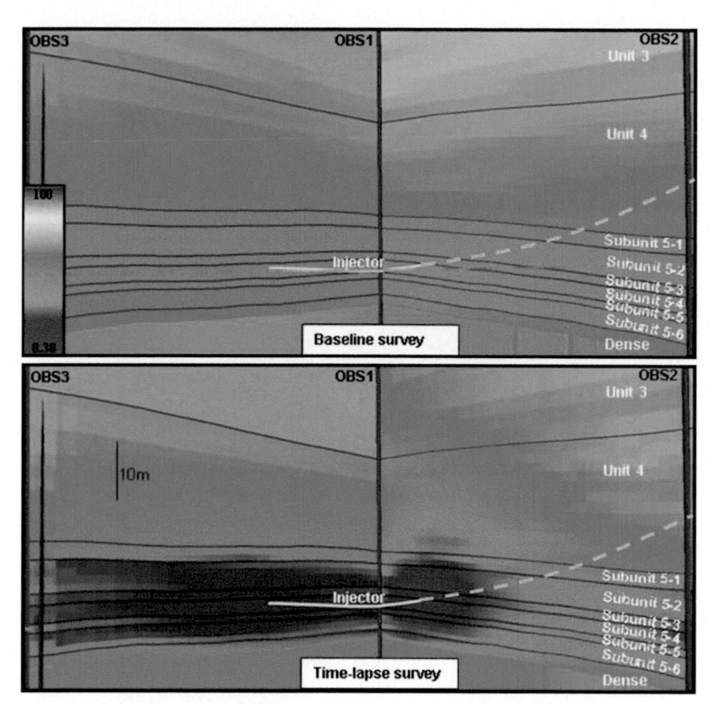

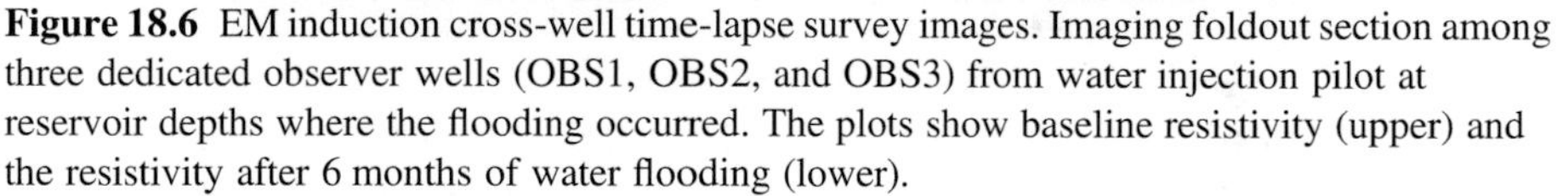

Figure 18.6 EM induction cross-well time-lapse survey images. Imaging foldout section among three dedicated observer wells (OBS1, OBS2, and OBS3) from water injection pilot at reservoir depths where the flooding occurred. The plots show baseline resistivity (upper) and the resistivity after 6 months of water flooding (lower).
From Al-Ali Z. A., Al-Buali M. H., AlRuwaili S., Ma S. M., Marsala A. F., Alumbaugh D., et al. (2009). Looking deep into the reservoir. *Oilfield Review, 21*(2), 38–47. Reproduced with permission of SPE.

Water cut measurements at the producer wells have been used to track the progression of the water front over time; but this method does not delineate the shape of the front between the wells. The injected water is obtained from seawater or recycled produced water, with salinity being around 60 k-ppm. Fluid-swept zones can exhibit a drastic change in formation resistivity, from 30 to 50 Ω m to less than 2 Ω m, depending on the saturation and porosity. There are also corresponding changes in density, but only minor changes in seismic velocity. A high resistivity contrast between injected water and oil in the reservoir would be apparent in deep-reading EM imaging profiles. Cross-well survey results would also facilitate the interpretation of formation resistivity to fluid saturation (Marsala et al., 2008).

A deep-reading EM cross-well survey was therefore conducted between producing wells 845 m apart on the west flank of the reservoir (Marsala et al., 2008), yielding several thousand data points at an operating frequency of 100 Hz. The processed survey indicated that the resistivity distribution differs significantly from the interpolated baseline cross-section based solely on presurvey well logs (Fig. 18.7). Although

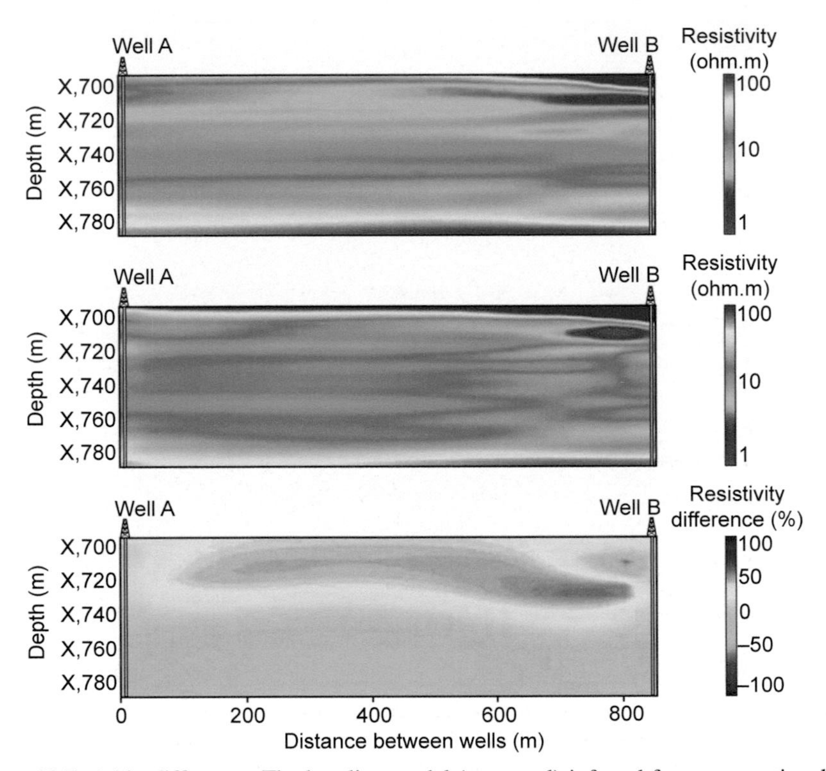

Figure 18.7 A big difference. The baseline model (top panel) inferred from conventional resistivity logs seems to indicate the reservoir interval between the wells is watered out, and the conductive injected fluid (blue [gray in print version] area, top of the section) is sweeping the reservoir. By contrast, the inverted deep-reading EM resistivity image (middle panel) shows a higher-resistivity range (green [dark gray in print version] to yellow [light gray in print version]) in the same section, indicative of oil. The resistivity differences between the baseline and inverted data (bottom panel) indicate the efficiency of the water flood is variable in an area known for higher porosity.
From Al-Ali Z. A., Al-Buali M. H., AlRuwaili S., Ma S. M., Marsala A. F., Alumbaugh D., et al. (2009). Looking deep into the reservoir. *Oilfield Review, 21*(2), 38–47. Graphic copyright Schlumberger. Used with permission. Courtesy: Use of clients data.

conventional resistivity logs indicated the near-wellbore section was essentially watered out, deep-reading EM survey revealed areas of variable sweep efficiency between the wells (Al-Ali et al., 2009). The final results of this pilot field study have been described by Marsala et al. (2008).

18.2.4 EM cross-well tomography case 3: imaging steam fronts

In a heavy-oil field in the San Joaquin Valley of California, cyclic steam injection technique is used to enhance production in several reservoirs. The steam-injected well is shut in for several days to allow hot steam to disperse and thin the oil by reducing its viscosity. The well is then opened and the thinned oil is produced.

The steeply dipping and complex geology of this large field (52 km^2) made it a challenging task for the operator to determine the vertical and radial distributions of steam around each injector and, hence, to have an overall better understanding of how heat is distributed through the reservoir after steam injection.

The proposed approach was to image the reservoir's resistivity response to steam injection. Induction well logging data show that reservoir resistivity values typically decrease by 35%–80% after steam injection. This response is due to the fact that as temperature increases, rocks become less resistive; the response will also be seen when lower resistivity salt water replaces high-resistivity oil. The sensitivity to resistivity changes made deep-reading EM cross-well tomography an obvious candidate technique for tracking the injected steam distribution and identifying swept zones in the reservoirs. Using cross-well EM resistivity tomography, the operator would be able to infer structure, temperature distribution, and residual saturation of steam-affected reservoir volumes.

During the planning stage, a baseline model for EM tomographic inversion was constructed. This model incorporated conventional resistivity logs obtained before the wells were cased or subjected to steam injection. Deep-reading EM transmitters and receivers were deployed in two wells once the feasibility of the survey had been determined for mapping the steam-affected zone between wells. The resulting resistivity image from the survey clearly revealed depleted and unswept zones within the reservoir (Fig. 18.8). Data from this cross-well survey enabled the operator to map the effects of steam injection, guide field development plans, and improve reserves estimates (Al-Ali et al., 2009).

18.3 Potential of tomography in hydrocarbon production monitoring

The previous section focused on the oilfield use of EM tomographic techniques for reservoir fluid movement monitoring on a large scale ($\sim$100 m–1 km). In this section, the status on the research and development of various process tomography techniques relevant to the understanding of multiphase flows in pipes (diameter $<$1 m) and to the online measurement of oil–gas production will be described.

The advent of new surface multiphase flowmeter (MPFM) technology is fundamentally changing the production monitoring of complex flows from oil–gas production wells. This transformation is driven by new technology that accurately measures rapid variations in oil–water–gas multiphase flows that previously were difficult to quantify. The capability to measure multiphase flow in real time increases operational efficiency, saving both time and cost. It is now possible to allocate hydrocarbon production without conventional phase separation and to overcome processing bottlenecks in surface facilities (Fig. 18.9). Accurately quantifying individual fluid phases in a production stream allows operators to make more informed decisions about well

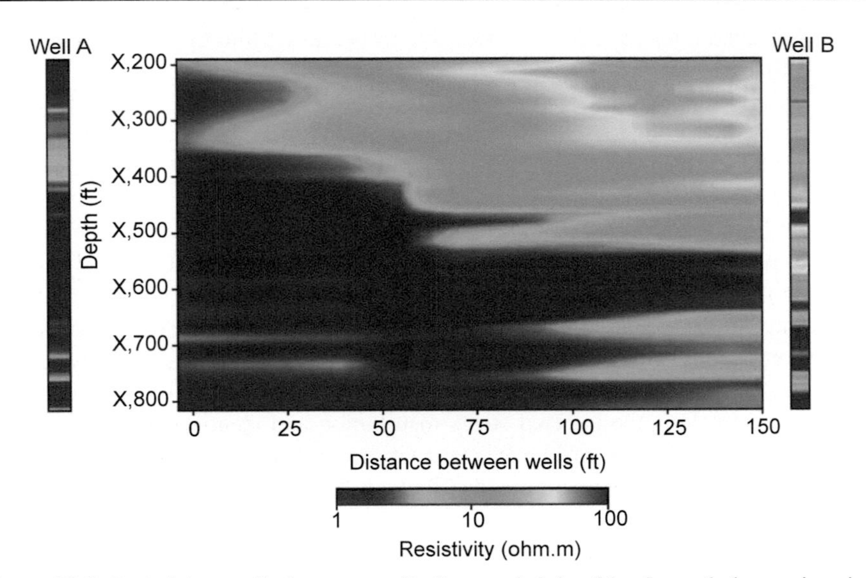

Figure 18.8 Resistivity profile between wells. Low resistivity (blue [very dark gray in print version] shading) is a typical characteristic of shale layers and steam-swept zones; at the other end of the spectrum, higher resistivities (orange [light gray in print version] to red [dark gray in print version]) are a characteristic of unswept oil sands. Intermediate resistivities (green [gray in print version]) represent transition zones between steam and oil. This deep-reading EM survey imaged an abrupt boundary midway between wells—the resistivity changes from 2 Ω m in the depleted zone to more than 50 Ω m in the unswept zone. The low resistivity most probably results from replacement of oil by formation water and steam condensate injected from a well to the left of this survey.
From Al-Ali Z. A., Al-Buali M. H., AlRuwaili S., Ma S. M., Marsala A. F., Alumbaugh D., et al. (2009). Looking deep into the reservoir. *Oilfield Review, 21*(2), 38−47. Graphic copyright Schlumberger. Used with permission. Courtesy: Use of clients data.

performance, to better identify, understand and remediate problematic wells, optimize artificial lift operations, and build dynamic reservoir models (Pinguet, 2012; Xie, Atkinson, & Lenn, 2007).

Unlike conventional test separators, an MPFM continuously measures gas, oil, and water flow rates without physically separating the flow stream into individual fluid phases, giving measurements within minutes of being placed in operation. The pressure drop across an MPFM is significantly less than for conventional separators, allowing wells to be tested close to actual producing conditions. In permanent metering applications, these devices have minimal footprints at surface locations or on offshore platforms.

In the subsea environment, an MPFM provides substantial cost savings through down-scaling or elimination of surface well testing facilities and subsea test lines. Continuous measurement with an MPFM can also help identify potential flow assurance risks−blockages in the subsea production systems (e.g., hydrates, asphaltenes,

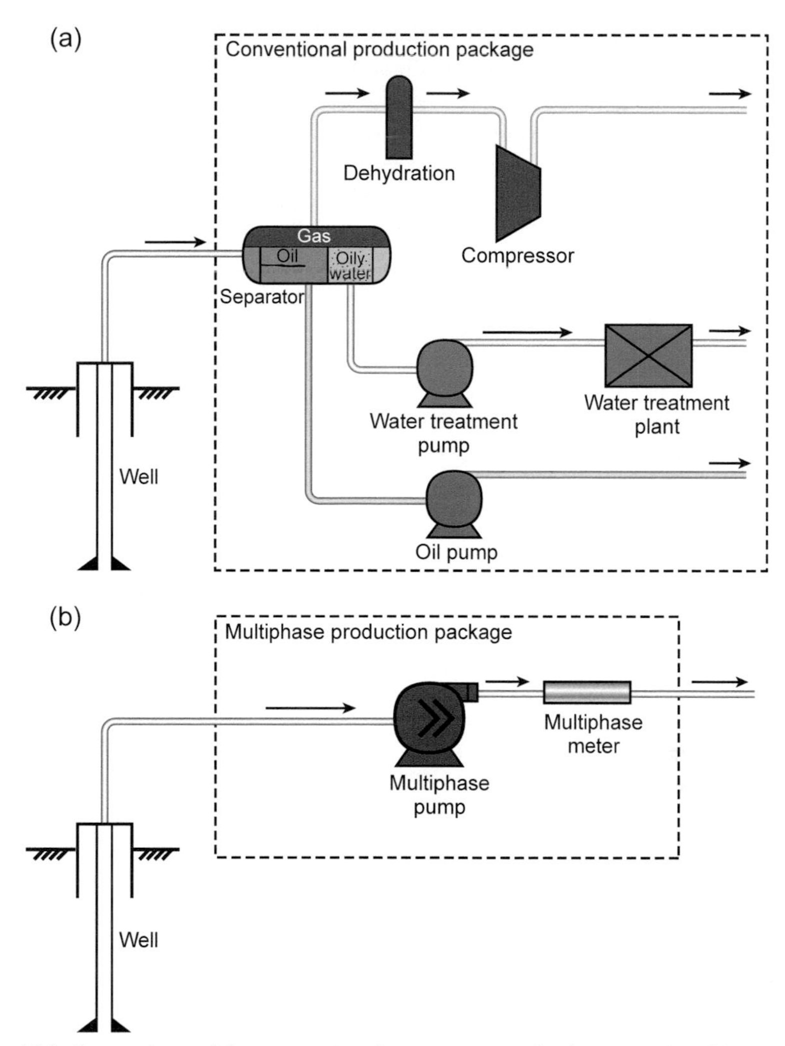

Figure 18.9 Comparison of the conventional permanent production scenario with a separator, associated equipment, and pipelines (a) with the cost-saving scenario (b) that only needs a multiphase pump and meter.
From Pinguet B. (2012). *Fundamentals of multiphase metering*. 09-TS-0022 (2nd ed.) Schlumberger, Fundamentals of Multiphase Metering, second Edition, Schlumberger. p. 11; used with permission. Courtesy: Lisa Stewart (former executive editor of Oilfield Review) & Schlumberger.

wax, sand, scale), or breakthrough of first formation or injection water. This MPFM capability enables oil–gas operators to make appropriate risk mitigation decisions, for example, injection of hydrate and/or corrosion inhibitors in time and at an appropriate injection rate (Couput, 2011; Hansen, Pedersen, & Durdevic, 2019; Pinguet et al., 2014).

Oil–gas–water multiphase flows produced from oil–gas wells are very complex in nature. The accurate measurement of the individual phase flow rates demands a flowmeter with a combination of multiple measurement principles for phase fraction and phase velocity measurements (Fig. 18.10), such as EM, single- or dual-energy gamma ray and Venturi differential pressure (for recent modeling and flow visualization work of multiphase flow through a Venturi, see Gajan, Salque, Couput, & Berthiaud, 2013; Jing, Yuan, Duc, Yin, & Yin, 2019; Paladinoa & Maliskab, 2011; Pan et al., 2019; Wu et al., 2020). To reduce the uncertainty of flow regime variations in the accuracy of multiphase flow rate measurements, most of commercial MPFMs are installed vertically to tackle flow regimes with more axi-symmetry than horizontal flows (Fig. 18.11, Dahl, 2005). MPFM measurement accuracy of individual phase volumetric flow rates is generally within ±5 to ±10% of reading for gas volume fractions (GVFs) up to about 95%, depending on the sensor technologies used (for accuracy comparison of industrial MPFMs, see Hansen, Pedersen, & Durdevic, 2019). Higher-GVF wet gas flow metering often requires different techniques to achieve good accuracy in liquid as well as gas flow rate measurements (Xie et al., 2007).

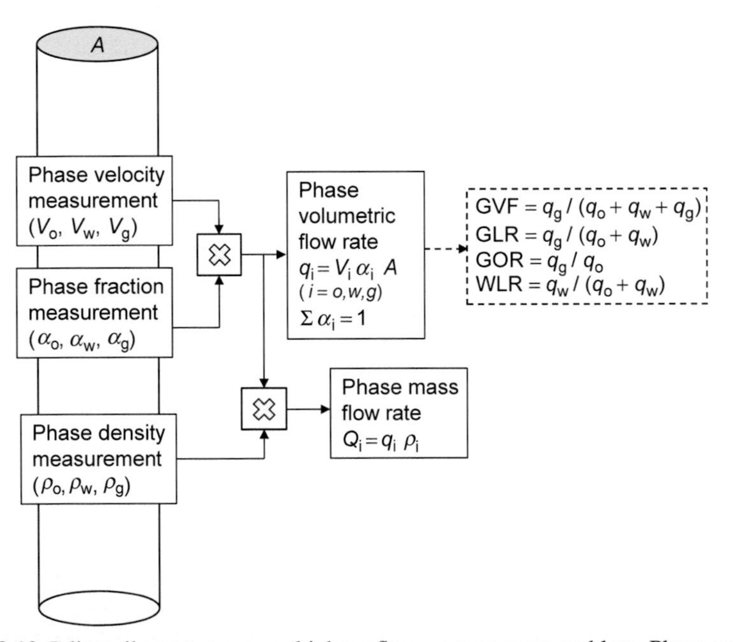

Figure 18.10 Inline oil–water–gas multiphase flow measurement problem. Phase volumetric flow rate is inferred from phase fraction and phase velocity measurements; further measurement or calculation of phase density is needed to determine phase mass flow rate (pressure and temperature sensors are often needed for the determination of phase density; not shown here). Multiphase flow parameters such as gas volume fraction (GVF), gas liquid ratio (GLR), gas oil ratio (GOR), and water liquid ratio (WLR) can be derived. In some of the multiphase flow meters, WLR is measured together with the total liquid flow rate $q_L = q_w + q_o$; the oil and water flow rates are then derived from $q_w = WLR\ q_L$ and $q_o = (1\text{-}WLR)\ q_L$.

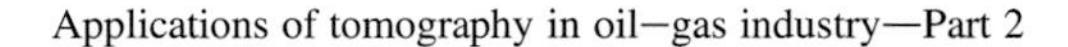

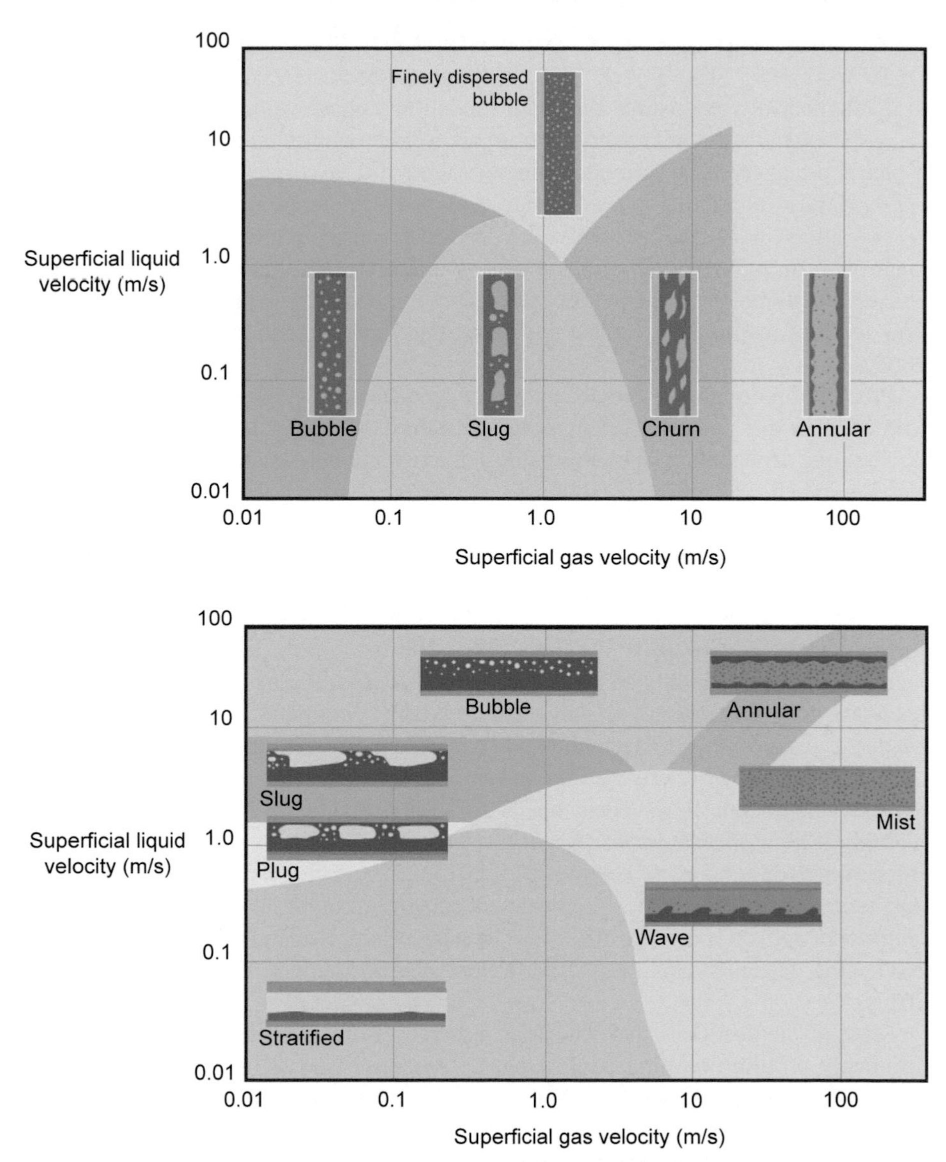

Figure 18.11 Vertical (top) and horizontal (bottom) flow patterns.
From NFOGM & Tekna (2005). *Handbook of Multiphase Metering*; used with permission.

Further descriptions of various commercial MPFMs and the measurement strategies can be found in an industrial monograph (Pinguet, 2012), text book (Falcone, Hewitt, & Alimonti, 2010), or review paper (Hansen, Pedersen, & Durdevic, 2019; Thorn, Johansen, & Hjertaker, 2012).

Process tomography has been conceived to have the potential of measuring dynamic multiphase processes such as multiphase flows of complex regimes through a pipeline (Chaouki, Larachi, & Dudukovic, 1997; Plaskowski, Beck, Thorn, & Dyakowski, 1995; Thorn et al. 1990; Williams & Beck, 1995). The basic concept is to mathematically reconstruct the phase holdup and phase velocity profiles at a sufficient spatial and temporal resolution from appropriate multiple measurements made at a pipe's periphery. The individual phase volumetric flow rates are then derived by appropriately integrating phase holdup and phase velocity profiles over the pipe cross-section (individual mass rates are derived given the phase densities). The challenge with multiphase flow measurement is that both the phase distribution and the phase velocity profile vary significantly in time and space, as manifested by various flow regimes mapped for vertical and horizontal pipe inclinations (Fig. 18.11, Dahl, 2005).

Process tomography, either as a standalone system or as an add-on to an existing MPFM, has the potential to improve multiphase metering accuracy by reducing flow regime dependency. With this being achieved, it would enable an MPFM to operate over much wider flow conditions, with less dependency on flow models. Some level of tomographic (phase fraction) measurements based on multiview microwave, electrical impedance, or electrical capacitance sensing have appeared in surface MPFMs (Brandt, Tol, & Lars, 2009; Scheers & Wee, 2011) and in downhole production logging tools (Hallundbaek & Kjærsgaard-Rasmussen, 2014; Kjærsgaard-Rasmussen & Meyer, 2011; Xie et al., 2005, 2007).

It is yet to be demonstrated that tomography-based techniques can measure individual phase flow rates more accurately than the existing commercially available MPFMs. Magnetic resonance is the technique that can potentially map the 3D flow field (fraction and velocity) of relatively slow-moving fluid in complex geometries. Considerable developments have been made to apply magnetic resonance to oilfield multiphase flow measurement (Appel, Freeman, & Pusiol et al. 2011; Hogendoorn, Boer, Appel, de Jong, & de Leeuw, 2013, 2015, 2019; Pusiol et al., 2012). A few other magnetic resonance flow measurement concepts or prototypes have been conceived or under development (Fordham 2013; Lakshmanan, Maru, Holland, Mantle, & Sederman, 2017; Ong, Chen, & Bussear, 2016; O'Neill, Klotz, Stanwix, Fridjonsson, & Johns, 2017).

Most of the academic and industrial research efforts have been devoted to more accurately profiling multiphase flow phase fraction, based on electrical tomography (RF/microwave, impedance, capacitance, resistance, magnetic, electrostatic, wire mesh sensors; Table 18.2a), gamma-ray and X-ray tomography (Table 18.2b), or their combinations. Dispersed phase velocity profiling is done mainly by cross-correlating axially spaced dual-plane phase fraction segment images, using fast X-ray, electrical capacitance, and conductance tomographic sensors (Table 18.2a and b). EM flowmeters using single- or multiple-pair of potential-sensing electrodes have been investigated for (water) continuous phase velocity measurement or profiling (Leeungculsatien & Lucas, 2013; Ma, McCann, & Hunt, 2017; Wang, Tian, & Lucas, 2007). Other tomography modalities, such as ultrasound (Black, Ahmad, & Noui-Mehidi, 2017; Goh, Rahim, Rahiman, Cong, & Wahad, 2017; Langener, Vogt, Ermert, & Musch, 2017), optical, and thermal, have also been investigated (Table 18.2c). Cross-validation of the accuracy of tomography-based multiphase flow measurement system is also examined (Table 18.2d).

Table 18.2a Electrical tomography techniques for multiphase flow measurement and research.

Modality	Physics parameter(s)	Phase fraction	Phase velocity	References
RF/microwave tomography	Complex permittivity $\varepsilon^* = \varepsilon' - j\sigma/(\varepsilon_o\omega)$ profile from amplitude–attenuation and/or phase-shift data	Gas-, oil-, and water-continuous flows: water fraction, WLR, (with γ-ray liquid fraction input)	(1) Venturi	(1) MPFM based on parametric tomography Scheers & Wee (2011). Taherian & Habashy (1996). Wu et al. (2009), Mallach et al. (2017), Wu & Wang (2017)
Electrical impedance tomography (EIT)	Complex permittivity $\varepsilon^* = \varepsilon' - j\sigma/(\varepsilon_o\omega)$ profile from $Y = G + j\omega C$ data (capacitance C; conductance G)	Gas-, oil-, and water-continuous flows: (1) water fraction, WLR, (2) and with γ-ray liquid fraction input	(3) $V_{dispersed\text{-}phase}$ by cross-correlation, (4) Venturi	(2) (4) (3) Brandt et al. (2009) (MPFM with 6-electrode rotating near-wall regions sensing), (1) (4) Xie (2018b)
Electrical capacitance tomography (ECT)	Permittivity ε profile from capacitance (C) data	Gas- or oil-continuous flows: water fraction, (2) WLR and liquid fraction, (3) solids fraction (in gas)	(1) Venturi (4) $V_{dispersed\text{-}phase}$ (profile) by dual-plane cross-correlation	Yang, Stott, Beck, and Xie (1995), Yang et al. (2004), Hasan and Azzopardi (2007), (1) Huang, Xie, Zhang, and Li (2005), (2) Li et al. (2013), (1) (2) Xie (2018a), (3) (4) Yang and Liu (2000), Hunt, Pendleton, and Byars (2004), Saoud et al. (2017), Wang and Yang (2020).

Continued

Table 18.2a Continued

Modality	Physics parameter(s)	Phase fraction	Phase velocity	References
Electrical resistance tomography (ERT)	Conductivity σ profile from conductance (G) data	Water-continuous flows: water fraction, (4) insulated electrode sensor design, (5) flow-regime identification	(1) $V_{\text{dispersed-phase}}$ (profile) by dual-plane cross-correlation, (2) Venturi, (3) $V_{\text{continuous-phase}}$ by (two-electrode) electromagnetic flow meter	(1) Li et al. (2009), Dong, Xu, Xu, Hua, and Qiao (2005), (2) Meng et al. (2010), (3) Deng, Li, Wei, Yan, and Yang (2011), Faraj et al. (2015), Wang et al. (2015), (4) Wang, Zhang, Huang, Ji, and Li (2013), (5) Tan, Dong, and Wu (2007)
ERT and ECT dual modality	Composite permittivity ε and conductivity σ profiles from conductance and capacitance data.	Oil- and water-continuous flows: water fraction, WLR (separate C and G electrodes)		Qiu, Hoyle, and Podd (2007), Glen and Ross (2010), Primrose et al. (2010), Wang et al. (2018)
Electrical magnetic tomography (EMT) (Magnetic induction tomography - MIT)	Conductivity σ profile from water-induced eddy current data (magnetic coils excitation and pick-up)	Oil- and water-continuous flows: water fraction	(1) $V_{\text{continuous-phase}}$ by electromagnetic velocity tomography (EVT)	Liu et al. (2013), Ma et al. (2015), (1) Ma et al. (2017)
Electromagnetic velocity profiler (multielectrode) (electromagnetic velocity tomography)	Flow velocity—induced potential difference profile measured by electrodes at boundary	Water-continuous flows: (liquid fraction input from ERT)	$V_{\text{continuous-phase}}$ (profile)	Wang et al. (2007), Wang et al. (2015), Leeungculsatien and Lucas (2013), Ma et al. (2017), Webilor et al. (2018)

Electrostatic charge tomography and ECT dual modality	Electrostatic charge–induced voltage profile at boundary	Solids fraction profile (in gas)	($V_{dispersed\text{-}phase}$ by cross-correlation)	Zhou and Zhang (2013), Zhou et al. (2013), Zhang, Hu, Dong, and Yan (2012)
Wire-mesh sensor (WMS) - capacitance	Direct mapping of permittivity ε profile	Gas fraction (profile) Bubble size distribution		Da Silva, Thiele, Abdulkareem, Azzopardi, and Hampel (2010)
Wire-mesh sensor (WMS) - conductance	Direct mapping of conductivity σ profile	Gas fraction (profile)	(1) Gas velocity profiles by auto- /cross-correlation (2) Gas velocity (profile) by cross-correlation with fast X-ray	(1) Hoppe, Grahn, and Schütz (2010), (1) Beyer, Lucas, and Kussin (2010), (2) Zhang, Xu, Wu, Li, and Li (2013), Hampel et al. (2009)

Table 18.2b Nucleonic tomography techniques for multiphase flow measurement and research.

Modality	Physics parameter(s)	Phase fraction	Phase velocity	References
γ-ray tomography	High-energy mass/linear attenuation coefficient μ	Liquid (or gas) fraction	(1) Venturi (2) Cross-correlation	(1) (2) Chen, Li, Ye, and Yu (2012), Johansen, Hampel, and Hjertaker (2010), Sætre et al. (2010), Bruvik, Hjertaker, and Hallanger (2010), Johansen et al. (2012), Sætre, Johansen, and Tjugum (2012), Xue, Wang, Yang, and Cui (2013), Wagner, Bieberle, Bieberle, and Hampel (2017), Hjertaker, Tjugum, Hallanger, and Maad (2018)
Dual-energy γ-ray tomography	Dual-energy mass/linear attenuation coefficient μ	Liquid (oil, water) and gas fractions, WLR		Frøystein, Kvandal, and Aakre (2005)
X-ray tomography	Mass/linear attenuation coefficient μ	Liquid (or gas) fraction	(1) Cross-correlation (2) Flow regime identification	(1) Bieberle et al. (2010), Johansen et al. (2010), Hu and Lawrence (2016), (1) (2) Jones and Hu (2019)
Positron tomography	Absorption coefficient (from positron radiation source and γ-ray detector array)	Gas/oil/water fractions		Li and Chen (2015)

Table 18.2c Other tomography techniques for multiphase flow measurement and research.

Modality	Physics parameter(s)	Phase fraction	Phase velocity	References
Ultrasound tomography	Reflection (acoustic impedance) and/or transmission attenuation	Gas fraction, flow regime identification, gas–liquid interface detection	(1) Liquid velocity from range-gated Doppler	Xu and Xu (1998), (1) Murai, Tasaka, Nambu, Takeda, and Roberto Gonzalez (2010), (1) Huang and Xie (2013), Black, Ahmad, and Noui-Mehidi (2017), Goh, Rahim, Rahiman, Cong, and Wahad (2017), Langener, Vogt, Ermert, and Musch (2017)
Optical tomography	Transmission attenuation	Solids fraction (in gas)		Rahim, Yunos, Rahiman, and Rahim (2010), Muji et al. (2013)
Thermal tomography	Thermal conductivity profile	Water fraction profile (of oil–water flow)		Hoffmann, Amundsen, Schulkes, and Schüller (2012)

Table 18.2d Cross-validation of tomography techniques for multiphase flow measurement and research.

Modality 1	Modality 2	Comparative measurement	References
X-ray tomography	ECT	Solids (or gas) fraction	Rautenbach, Mudde, Yang, Melaaen, and Halvorsen (2013)
X-ray tomography	Wire-mesh sensor (conductance)	Gas fraction profile Gas velocity profile (by cross-correlation of X-ray and WMS)	Zhang et al. (2013), Banowski et al. (2017)
X-ray tomography	Thermal sensor array	Water fraction profile (of oil—water flow)	Hoffmann et al. (2012)
Dual-energy γ-ray (traversable)	Single-energy γ-ray (stationary) – nontomographic	Oil—water-gas three-phase fraction profiles (at 100 bar), flow pattern identification	Hoffmann, and Johnson (2011), Kaku Arvoh, Hoffmann, and Halstensen (2012)
Wire-mesh sensor (WMS) - conductance	Air—water flow-loop reference	Time-averaged integrated gas flow rate (from gas fraction and velocity profiles)	Beyer et al. (2010)
Wire-mesh sensor (WMS) - conductance	High speed video image processing	Bubble size and water rate	Nuryadin et al. (2015)
Electrical tomography	Pressure drop	Oil—water flow liquid fraction	Zhang et al. (2013)
Electrical impedance tomography	High speed video image	Gas—water flow image mapping	Li, Wang & Wang (2019)

The key technical challenges are to demonstrate the capability of determining accurately individual phase holdups and phase velocities, for both oil- and water continuous flows, over a wide range of flow regimes, water and gas cuts, and fluid properties. Spatial and/or temporal resolutions inherent with different physical measurement principles can pose limitations to measuring fast varying multiphase flows (sometimes limited by the accuracy and speed in the image reconstruction and the measurement hardware); further development is continuing (Guo, Tong, Lu, & Liu, 2018; Kim, Khambampati, Kim, & Kim, 2013; Liu, Lei, Wang, & Liu, 2013; Liu, Yang, He, & Tan, 2013; Mallach, Gebhardt, & Musch, 2017; Mallach, Gevers, Gebhardt, & Musch, 2018; Sun & Yang, 2013; Wang, Hu, Wang, & Wang, 2012; Wu, & Wang, 2017). Cost, robustness, and reliability are also issues to be addressed in transferring technology to practical industrial applications.

Coined as "tomometry" (Thorn et al., 2012), the information output from a process tomography system can be a few key process parameters, such as multiphase flow pattern, phase fraction, and WLR (Section 18.3.1 below). An MPFM based on multiple electrode impedance measurements and three gamma-ray beams (for flow regime identification to reduce fraction-sensing error) has been used in oilfield downhole applications (Johansen et al., 2012).

18.3.1 Case study: ECT for multiphase flow WLR and liquid fraction measurement

In this section, the measurement results of oil continuous multiphase flows from a research prototype eight-electrode ECT sensor (Fig. 18.12) are briefly described (for further details, see Li et al., 2013). In the prototype, to minimize the variation in horizontal flow regime over a wide range of GVFs (Fig. 18.11), a novel flow-conditioning device is used at the upstream of the ECT sensor (not shown) to form predominantly a gas–liquid annular flow.

Figure 18.12 The eight-electrode ECT sensor tested on TUV-NEL multiphase flow facility. The experimental section instrumented with (from left to right, flow direction) ultrasound and electrical tomography sensors.

Fig. 18.13 shows the data and image inversion steps developed to determine the WLR from the raw capacitance measurements $\boldsymbol{C_m}$ by (i) converting the measured normalized capacitance $\boldsymbol{C_n}$ to mixture permittivity $\boldsymbol{\varepsilon_m}$ and (ii) converting $\boldsymbol{\varepsilon_m}$ to WLR and liquid fraction for a gas–liquid flow based on the dielectric mixing laws shown. The outcome of the key step (i) also permits a more quantitative reconstruction of the mixture permittivity image (Fig. 18.14) based on multiview $\boldsymbol{\varepsilon_m}$ data (rather than based on the normalized capacitance $\boldsymbol{C_n}$) for transient gas–liquid flows.

Working in the mixture permittivity parameter domain also permits the use of appropriate dielectric mixing models (as shown in Fig. 18.13) for the determination WLR and/or liquid fraction of gas–liquid flows. The purpose of using a flow-conditioning device upstream of the ECT sensor is to create a gas-core and a liquid annulus at the ECT measurement section. As can be seen from the instantaneous permittivity images illustrated in Fig. 18.14, a more steady gas core (or smoother liquid layer) is formed at a relatively low GVF (~25%) than at a relatively high GVF (~63%). As expected, the permittivity of an oil/water mixture near the pipe wall region increases when the WLR increases. The real-time images reconstructed by the linear back-projection method (shown in Fig. 18.13) are also useful to indicate gas–liquid flow regimes. Fig. 18.15 shows that the near-wall measurement (1-electrode apart) at low GVFs preferentially measures the liquid layer mixture permittivity, especially during the time intervals with the liquid-rich slugs passing (see Fig. 18.14, GVF $\leq$0.4).

In multiphase flows, the WLR of the (oil/water) liquid is generally much more stable than the liquid fraction over a short measurement interval (say $\Delta T = 10$ s, with one or more liquid-rich slugs passing the sensors); this feature of near-wall localized sensing (in space) and/or dynamic liquid-rich slug mixture permittivity data capturing (in time, at a fast rate say $\Delta t = 10$ ms) can yield a rolling estimate (within each ΔT interval) of a short time-average permittivity of liquid $\varepsilon_{liquid(\Delta T)}$. This then results in a rolling estimate of the liquid WLR by using an oil/water (uniform) mixing model for an oil continuous flow:

$$WLR_{(\Delta T)} = \frac{\varepsilon_{\text{liquid}(\Delta T)} - \varepsilon_{\text{oil}}}{\varepsilon_{\text{liquid}(\Delta T)} + 2\varepsilon_{\text{oil}}} \tag{18.5}$$

From Fig. 18.16 left, it can be seen that by capturing and analyzing the mixture permittivity data of a near-wall liquid-rich slug, the WLR can be estimated to ±5% for GVF up to about 90%, WLR up to ~40% (largely within the permittivity linearity range of near-wall electrode pairs). With respect to liquid fraction estimates, Fig. 18.15 indicates that, at the same WLR and GVF, the mixture permittivity from cross-pipe (i.e., opposite electrode) measurement (which interrogates the entire gas-rich core) is much lower than that of the near-wall one. From instantaneous gas–liquid mixture permittivity measured cross-pipe, an instantaneous and time-average liquid fraction

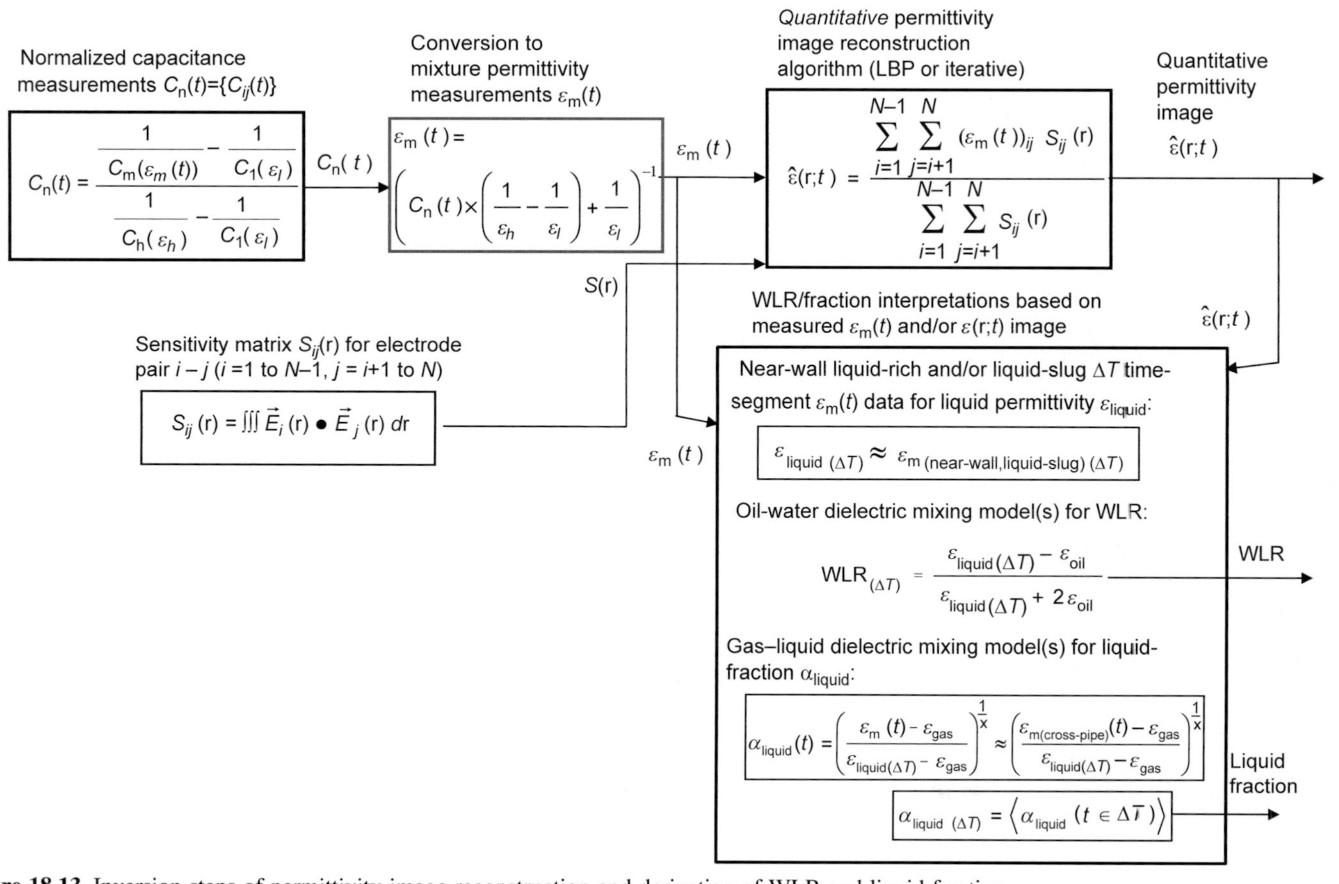

Figure 18.13 Inversion steps of permittivity image reconstruction and derivation of WLR and liquid fraction.

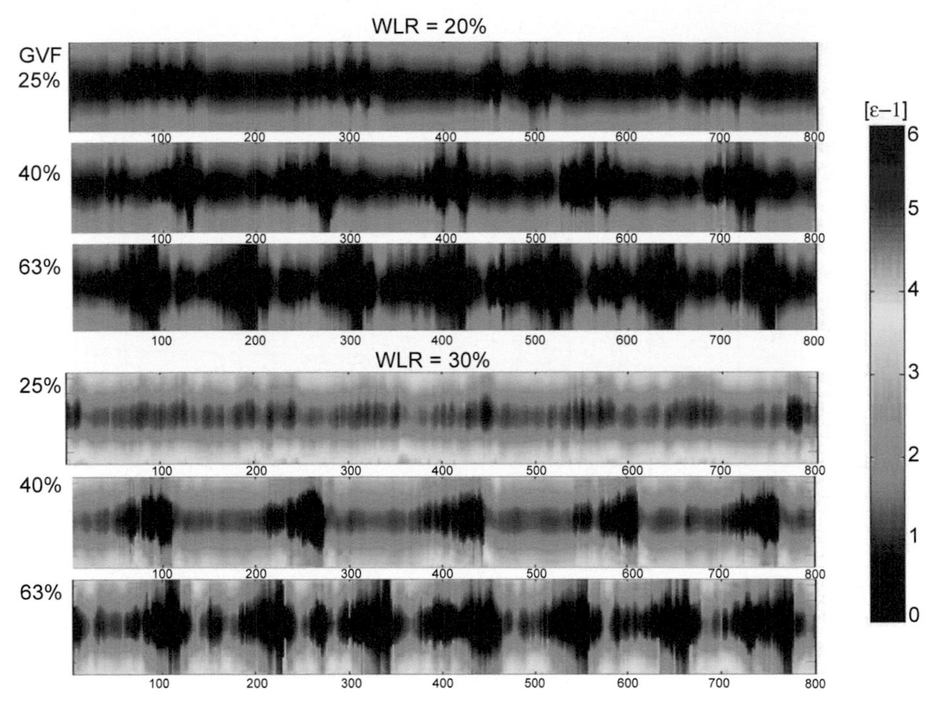

Figure 18.14 Instantaneous longitudinal cross-sectional images stacked with over 8 s (imaging frame rate at 100 frames per second). Permittivity change color scale ε-1 = 0 to 6 as shown on the far right. Image reconstruction based on instantaneous permittivity of gas–liquid mixture for horizontal gas–liquid conditioned flows with total liquid rate $Q_l = 30\ m^3/h$, WLR = 20% and 30%, and GVF = 25%, 40% and 63%.

can be obtained by using an appropriate gas–liquid mixing model as below, with near-wall sensed liquid permittivity as input, viz:

$$\alpha_{\text{liquid}}(t) = \left(\frac{\varepsilon_{\text{mixture}}(t) - \varepsilon_{gas}}{\varepsilon_{\text{liquid}(\Delta T)} - \varepsilon_{gas}}\right)^{\frac{1}{x}} = \left(\frac{\varepsilon_{\text{mixture(cross-pipe)}}(t) - \varepsilon_{\text{gas}}}{\varepsilon_{mixture(\text{near-wall,liquid-slug})(\Delta T)} - \varepsilon_{\text{gas}}}\right)^{\frac{1}{x}}, \tag{18.6}$$

$$\alpha_{\text{liquid}(\Delta T)} = \overline{\alpha_{\text{liquid}}(t \in \Delta T)}, \tag{18.7}$$

where x is an empirical factor for gas–liquid annular-type distribution (including GVF = 0).

The time-averaged liquid fractions derived from the cross-pipe mixture permittivity data given in Fig. 18.16 right show a good agreement with some of the liquid fractions measured by ultrasound sensors (Huang & Xie, 2013).

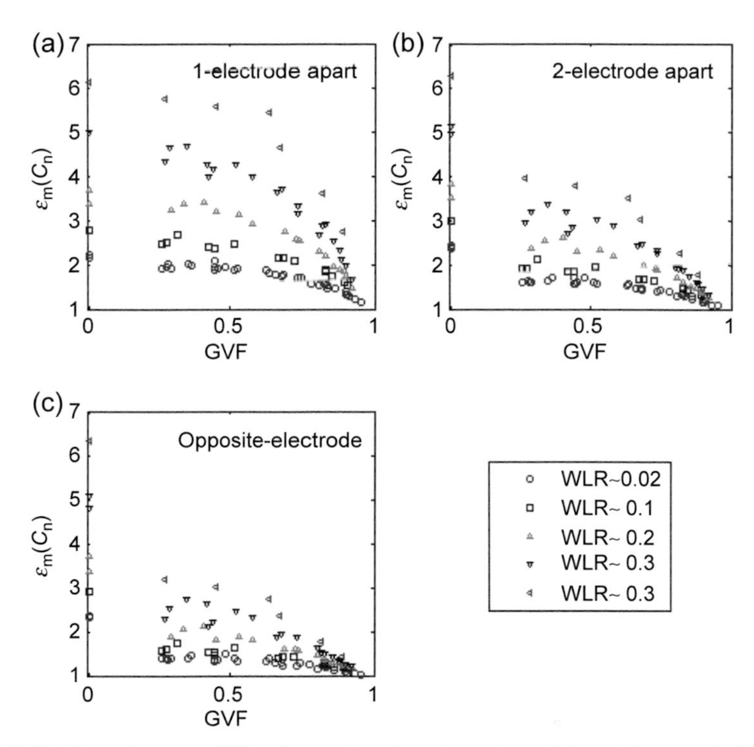

Figure 18.15 Plotted versus GVF of gas–liquid horizontal conditioned flows with WLR up to ∼40%, with permittivity of mixture $\boldsymbol{\varepsilon_m}$ calculated from normalized capacitance $\mathbf{C_n}$ in Fig. 18.13, for (a) 1-electrode apart, (b) 2-electrode-apart, and (c) opposite–electrode. Note some $\boldsymbol{\varepsilon_m}$ of 1-electrode apart are close to saturation for WLR > ∼40%.

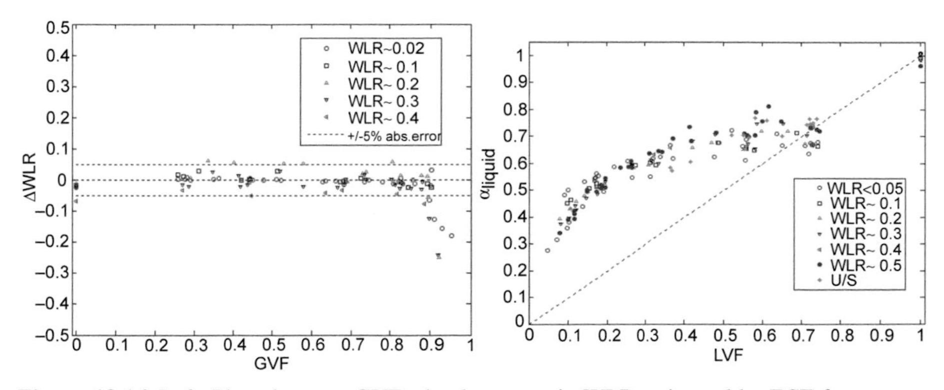

Figure 18.16 Left: Plotted versus GVF, absolute error in WLR estimated by ECT from permittivity of near-wall liquid–slug mixture by using Eq. (18.5) (∼60 s data acquisition interval). Right: Plotted versus LVF (= 1- GVF), time-average liquid fraction α_{liquid} estimated by opposite-electrode mixture permittivity data using Eqs. (18.6) and (18.7). Ultrasound (U/S) sensor measured liquid fraction data are also shown.

18.4 Future trends

Deep-reading resistivity tomography obtained through EM induction cross-well surveys can provide better petrophysical control in building reservoir models than surface seismic surveys. The capability to identify and monitor fluid fronts using cross-well EM measurements can be complemented by cross-well seismic imaging technique, which can delineate structure and monitor gas movement in the reservoir (Göktürkler, 2011; Yu et al., 2008). The detailed reservoir-level information obtained from combined deep-reading measurements gives petrophysicists and reservoir engineers greater insight into how fluids move and interact, which will ultimately determine the decision-making of oilfield management and production (Al-Ali et al., 2009).

Cross-well EM inverse (and forward) problems demand high computational resources. The use of machine-learning (ML) techniques in the form of deep-learning (e.g., 3D convolutional) neural networks for implementing rapid modeling and inversion of EM-based reservoir monitoring (coupled with a dynamic fluid flow simulator) has been explored, with cross-well EM data being generated using an electric source and a multicomponent (electric—magnetic) receiver assemblage (Colombo, Li, Sandoval-Curiel, & McNeice, 2020). ML-based inversion of reservoir saturation (converted from resistivity) has shown to be robust against increased noise levels. ML inversion through deep learning can become an efficient approach to data-driven and physics-constrained real-time reservoir monitoring.

On the oil—gas production monitoring front, with technological innovations, multiphase flow meters are likely to be deployed in higher pressure and temperature environments. This could significantly expand subsea applications for MPFM technology, with additional applications onshore in heavy-oil thermal recovery (Pinguet et al., 2010) and natural gas production (Pinguet et al., 2014).

For permanent-metering applications, MPFMs with minimal calibration requirements are desirable. Research in a novel microwave measurement interpretation has demonstrated a potential multiphase flow WLR measurement system that is robust to changes in water salinity (Xie, 2006). It is shown feasible to characterize the salinity of multiphase flow from a combination of transmission and scattering gamma-radiation measurements (Sætre, Johansen, & Tjugum, 2010). There have been oilfield applications of microwave sensor technology in well-monitoring, flow assurance, or diagnosis, such as detection of formation or injection water breakthrough, of hydrate inhibitor in water concentration, and salinity-independent multiphase flow measurement (Fiore, Xie, & Jolivet, 2019, 2020; Vielliard et al., 2019).

Various ML methods have been used to improve multiphase flow measurement in terms of sensor data interpretation and characterization (Dang et al., 2019; Yan et al., 2018). An accurate solution to the mathematical inverse (and forward) problems of a tomographic imaging modality often demands large computational resources. ML approach has also been explored to develop accurate and fast solver(s) for solving the underlying forward and inverse problems for reconstructing multiphase fluid tomographic images in real time (Fernández-Fuentes, Mera, Gómez, & Vidal-Franco, 2018; Rymarczyk, Kłosowski, Kozłowski, & Tchórzewski, 2019).

With ML-based real-time inversion capability, process tomography has the potential to improve MPFM accuracy by reducing flow regime dependency, identifying (and correcting for the effects of) flow regimes, and measuring directly the individual phase fractions and phase velocities. This improvement would enable accurate measurement of multiphase flow rates, at surface and downhole, over a broader range of flow regimes, water and gas cuts, and fluid properties, with less dependency on flow models than current methods have. Continued collaborative research and joint industrial projects are needed to demonstrate that a process tomography–based MPFM can outperform existing MPFMs, ideally without using a radioactive source, and at an acceptable cost.

18.5 Source of further information and advice

- The Society of Exploration Geophysicists (SEG) Journals: *Geophysics* and *The Leading Edge*
- The European Association of Geoscientists and Engineers (EAGE) Journals: *Geophysical Prospecting* and *First Break*
- Other Journals: *Petroleum Geoscience, Journal of Petroleum Geology, Flow Measurement and Instrumentation, Measurement Science and Technology* and *International Journal of Multiphase Flow*

Acknowledgments

We would like express our gratitude to Schlumberger for permission to publish this book chapter, and to several Schlumberger clients for permission to reuse some figures containing their data. Lisa Stewart, former executive editor of Schlumberger *Oilfield Review*, is thanked for her help in the preparation (of the first edition) of this book chapter. The work described in Section 18.3.1 was cofunded by the UK Technology Strategy Board (TSB); the project team consisted of Schlumberger, the University of Manchester, and TUV-NEL.

References

Abubakar, A., Habashy, T., Druskin, V., Alumbaugh, D., Zhang, P., Wilt, M., et al. (2005). A fast and rigorous 2.5D inversion algorithm for cross-well EM data. *SEG Extended Abstracts, 24*, 534–537.

Al-Ali, Z. A., Al-Buali, M. H., AlRuwaili, S., Ma, S. M., Marsala, A. F., Alumbaugh, D., et al. (2009). Looking deep into the reservoir. *Oilfield Review, 21*(2), 38–47.

Al-Ali, M., Vahrenkamp, V., Elsembawy, S., Bhatti, Z., Reeder, S. L., Clerc, N., et al. (2009). Constraining inter-well water flood imaging with geology and petrophysics: An example from the Middle East. In *SPE Middle East oil and gas show and conference, 15-18 March 2009. Bahrain: Spe 120558.*

Allen, D., Flaum, C., Ramakrishnan, T. S., Bedford, J., Castelijns, K., Fairhurst, D., et al. (2000). Trends in NMR logging. *Oilfield Review, 12*(3), 2–19.

Alumbaugh, D. L., & Morrison, H. F. (1995). Monitoring subsurface changes over time with cross-well electromagnetic tomography. *Geophysical Prospecting, 43*(7), 873–902.

Appel, M., Freeman, J. J., & Pusiol, D. (2011). Robust multi-phase flow measurement using magnetic resonance technology. In *Proceedings SPE Middle East oil and gas show (Manama, Bahrain, September 2011) SPE 141465.*

Banowski, M., Beyer, M., Szalinski, L., Lucas, D., & Hampel, U. (2017). Comparative study of ultrafast X-ray tomography and wire-mesh sensors for vertical gas–liquid pipe flows. *Flow Measurement and Instrumentation, 53A*, 95–106.

Beyer, M., Lucas, D., & Kussin, J. (2010). Quality check of wire-mesh sensor measurements in a vertical air/water flow. *Flow Measurement and Instrumentation, 21*, 511–520.

Bieberle, M., Schleicher, E., Fischer, F., Koch, D., Menz, H.-J., Mayer, H.-G., et al. (2010). Dual-plane ultrafast limited-angle electron beam x-ray tomography. *Flow Measurement and Instrumentation, 21*, 233–239.

Black, M. J., Ahmad, T. J., & Noui-Mehidi, M. N. (2017). *Tomographic imaging of multiphase flows*. Saudi Arabian Oil Co. US patent US9721336B2. August 1, 2017.

Brandt, M., Tol, M., & Lars, A. R. (2009). Improving measurement quality and meeting changing operator challenges with the multiphase meter. Paper 8.2. In *8th south east Asia hydrocarbon flow measurement workshop (Kuala Lumpur, Malaysia, 4-6 March 2009).*

Bruvik, E. M., Hjertaker, B. T., & Hallanger, A. (2010). Gamma-ray tomography applied to hydrocarbon multi-phase sampling and slip measurements. *Flow Measurement and Instrumentation, 21*, 240–248.

Chaouki, J., Larachi, F., & Dudukovic, M. P. (1997). Noninvasive tomographic and velocimetric monitoring of multiphase flows. *Industrial & Engineering Chemistry Research, 36*, 4476–4503.

Chen, J., Li, H., Ye, J., & Yu, H. (2012). *Gamma ray section imaging device, multiphase flow detecting device and detecting method*. Lanzhou Haimo Technologies. PCT patent application WO2012100385. August 2, 2012.

Colombo, D., Li, W., Sandoval-Curiel, E., & McNeice, G. W. (2020). Deep-learning electromagnetic monitoring coupled to fluid flow simulators. *Geophysics, 85*, WA1–WA12.

Colombo, D., McNeice, G. W., Cuevas, N., & Pezzoli, M. (2018). Surface-to-borehole electromagnetics for 3D waterflood monitoring: Results from first field deployment. In *2018 SPE annual technical conference and exhibition (Dallas, USA, 24–26 September 2018). SPE 191544.*

Couput, J.-P. (2011). Subsea multiphase measurements: Where are we and what's next from an oil and gas operator perspective. In *Proceedings 29th int. North Sea flow measurement workshop (Tønsberg, Norway, October 2011).*

Da Silva, M. J., Thiele, S., Abdulkareem, L., Azzopardi, B. J., & Hampel, U. (2010). High-resolution gas–oil two-phase flow visualization with a capacitance wire-mesh sensor. *Flow Measurement and Instrumentation, 21*, 191–197.

Handbook of multiphase flow metering. In Dahl, E. (Ed.), *Oslo: Norwegian society for oil and gas measurement/Norwegian society of chartered technical and scientific professionals* (2nd ed.), (2005) www.nfogm.no Accessed January 2012.

Dang, W., Gao, Z., Hou, L., Lv, D., Qiu, S., & Chen, G. (2019). A novel deep learning framework for industrial multiphase flow characterization. *IEEE Transactions on Industrial Informatics, 15*(11).

Deng, X., Li, G., Wei, Z., Yan, Z., & Yang, W. (2011). Theoretical study of vertical slug flow measurement by data fusion from electromagnetic flowmeter and electrical resistance tomography. *Flow Measurement and Instrumentation, 22*, 272–278.

DePavia, L., Zhang, P., Alumbaugh, D., Levesque, C., Zhang, H., & Rosthal, R. (2008). Next-generation cross-well EM imaging tool. In *SPE annual technical conference and exhibition (Denver, Colorado, USA, 21-24 September 2008). SPE 116344.*

Dong, F., Xu, Y. B., Xu, L. J., Hua, L., & Qiao, X. T. (2005). Application of dual-plane ERT system and cross-correlation technique to measure gas–liquid flows in vertical upward pipe. *Flow Measurement and Instrumentation, 16*, 191–197.

Falcone, G., Hewitt, G. F., & Alimonti, C. (2010). *Multiphase flow metering: Principles and applications*. Amsterdam: Elsevier.

Faraj, Y., Wang, M., Jia, J., Wang, Q., Xie, C. G., Oddie, G., et al. (2015). Measurement of vertical oil-in-water two-phase flow using dual-modality ERT–EMF system. *Flow Measurement and Instrumentation, 46B*, 255–261.

Feali, M., Pinczewski, W. V., Cinar, Y., Arns, C. H., Arns, J.-Y., Turner, M., et al. (2012). Qualitative and quantitative analyses of the three-phase distribution of oil, water, and gas in Bentheimer sandstone by use of micro-CT imaging. *SPE Reservoir Evaluation and Engineering, 15*(6), 706–711.

Fernández-Fuentes, X., Mera, D., Gómez, A., & Vidal-Franco, I. (2018). Towards a fast and accurate EIT inverse problem solver: A machine learning approach. *Electronics, 7*, 422. https://doi.org/10.3390/electronics7120422

Fiore, M., Xie, C. G., & Jolivet, G. (2019). Wider salinity range and lower water detection limit for multiphase flowmeters. In *SPE/IATMI Asia pacific oil & gas conference and exhibition (bali, Indonesia, 29 – 31 October 2019). SPE 196504.*

Fiore, M., Xie, C. G., & Jolivet, G. (2020). Extending salinity operating range and water detection lower limit of multiphase flowmeters. In *12th international petroleum technology conference (Saudi Arabia, 13–15 January 2020). IPTC 19590.*

Fordham, E. (2013). *NMR flowmeter with superconducting polarizer*. Schlumberger Technology. PCT patent application WO2013024456. February 21, 2013.

Frøystein, T., Kvandal, H., & Aakre, H. (2005). Dual energy gamma tomography system for high pressure multiphase flow. *Flow Measurement and Instrumentation, 16*, 99–112.

Gajan, P., Salque, G., Couput, J. P., & Berthiaud, J. (2013). Experimental analysis of the behaviour of a Venturi meter submitted to an upstream air/oil annular liquid film. *Flow Measurement and Instrumentation, 33*, 160–167.

Glen, N., & Ross, A. (2010). Use of dual modality tomography for complex flow visualization. Paper 7.3. In *28th international North Sea flow measurement workshop, 26-29 October 2010.*

Goh, C. L., Rahim, R. A., Rahiman, H. F., Cong, T. Z., & Wahad, Y. A. (2017). Simulation and experimental study of the sensor emitting frequency for ultrasonic tomography system in a conducting pipe. *Flow Measurement and Instrumentation, 54*, 158–171.

Göktürkler, G. (2011). A hybrid approach for tomographic inversion of crosshole seismic first-arrival times. *Journal of Geophysics and Engineering, 8*(1), 99–108.

Guo, G., Tong, G., Lu, L., & Liu, S. (2018). Iterative reconstruction algorithm for the inverse problems in electrical capacitance tomography. *Flow Measurement and Instrumentation, 64*, 204–212.

Hallundbaek, J., & Kjærsgaard-Rasmussen, J. (2014). *Logging tool*. Welltec. AS. US patent US8823379B2, September 2, 2014.

Hampel, U., Otahal, J., Boden, S., Beyer, M., Schleicher, E., Zimmermann, W., et al. (2009). Miniature conductivity wire-mesh sensor for gas-liquid two-phase flow measurement. *Flow Measurement and Instrumentation, 20*, 15–21.

Hansen, L. S., Pedersen, S., & Durdevic, P. (2019). Review multi-phase flow metering in offshore oil and gas transportation pipelines: Trends and perspectives. *Sensors, 19*, 2184.

Hasan, N. M., & Azzopardi, B. J. (2007). Imaging stratifying liquid−liquid flow by capacitance tomography. *Flow Measurement and Instrumentation, 18*, 241−246.

Hjertaker, B. T., Tjugum, S.-A., Hallanger, A., & Maad, R. (2018). Characterization of multiphase flow blind-T mixing using high speed gamma-ray tomometry. *Flow Measurement and Instrumentation, 62*, 205−212.

Hoffmann, R., Amundsen, L., Schulkes, R., & Schüller, R. B. (2012). Measuring phase distribution using external thermal excitation. *Flow Measurement and Instrumentation, 26*, 55−62.

Hoffmann, R., & Johnson, G. W. (2011). Measuring phase distribution in high pressure three-phase flow using gamma densitometry. *Flow Measurement and Instrumentation, 22*, 351−359.

Hogendoorn, J., Boer, A., Appel, M., de Jong, H., & de Leeuw, R. (2013). Magnetic resonance technology: A new concept for multiphase flow measurement. In *31st international North Sea flow measurement workshop, 22-25 October 2013, Tønsberg, Norway.*

Hogendoorn, J., Cerioni, L. M. C., Tromp, R. R., Zoeteweij, M. L., & Bousché, O. J. P. (2019). *Method for operating a nuclear magnetic flowmeter and nuclear magnetic flowmeter*. Krohne. AG. US patent US10393559B2. August 27, 2019.

Hogendoorn, J., van der Zande, M., Zoeteweij, M., Tromp, R., Cerioni, L., Boer, A., et al. (2015). Practical experiences obtained with the magnetic resonance multiphase flowmeter. In *33rd international North Sea flow measurement workshop, 20-23 October 2015, Tønsberg, Norway.*

Hoppe, D., Grahn, A., & Schütz, P. (2010). Determination of velocity and angular displacement of bubbly flows by means of wire-mesh sensors and correlation analysis. *Flow Measurement and Instrumentation, 21*, 48−53.

Huang, S. M., & Xie, C. G. (2013). Ultrasound imaging and measurement of swirling gas-liquid multiphase flow. In *7th world congress on industrial process tomography, 2-5 September 2013, Krakow, Poland.*

Huang, Z., Xie, D., Zhang, H., & Li, H. (2005). Gas−oil two-phase flow measurement using an electrical capacitance tomography system and a Venturi meter. *Flow Measurement and Instrumentation, 16*, 177−182.

Hu, B., & Lawrence, C. (2016). *Method of estimating chordal holdup values of gas, oil and water for tomographic imaging of a three-phase flow through a volume*. Institute of Energy Technology. US patent US9291579B2. March 22, 2016.

Hunt, A., Pendleton, J., & Byars, M. (2004). Non-intrusive measurement of volume and mass using electrical capacitance tomography. In *7th Biennial ASME conference on engineering system design and analysis, Manchester, UK. (July 19-22, 2004)*. ESDA 2004-58398.

Iglauer, S., Wang, S., & Rasouli, V. (2011). X-ray micro-tomography measurements of fractured tight sandstone. In *SPE Asia pacific oil and gas conference and exhibition, 20-22 september, Jakarta, Indonesia: Spe 145960.*

Jing, J., Yuan, Y., Duc, S., Yin, X., & Yin, R. (2019). A CFD study of wet gas metering over-reading model under high pressure. *Flow Measurement and Instrumentation, 69*(101608).

Johansen, G. A., Hampel, U., & Hjertaker, B. T. (2010). Flow imaging by high speed transmission tomography. *Applied Radiation and Isotopes, 68*, 518−524.

Johansen, G. A., Hjertaker, B. T., Tjugum, S.-A., Bruvik, E. M., Maad, R., Sætre, C., et al. (2012). Industrial applications of tomographic gamma-ray methods. In *Proceedings 6th international symposium on process tomography (Cape Town, South Africa, March 2012).*

Jones, W., & Hu, B. (2019). An application of machine learning in multiphase flow regime identification. In *New Advances and key questions in phase-change cooling, 2nd annual ThermaSMART/progress 100 workshop, Kyushu university, Japan*. https://thermasmart.eng.ed.ac.uk/sites/thermasmart.eng.ed.ac.uk/files/attachments/Jones2019.12.02-ThermaSMART-Workshop-Japan-WilJones-FlowCapture.pdf. (Accessed May 2020).

Kaku Arvoh, B., Hoffmann, R., & Halstensen, M. (2012). Estimation of volume fractions and flow regime identification in multiphase flow based on gamma measurements and multivariate calibration. *Flow Measurement and Instrumentation, 23*, 56–65.

Kayser, A., Knackstedt, M., & Ziauddin, M. (Spring 2006). A closer look at pore geometry. *Oilfield Review, 18*(1), 4–13.

Kim, B. S., Khambampati, A. K., Kim, S., & Kim, K. Y. (2013). Improving spatial resolution of ERT images using adaptive mesh grouping technique. *Flow Measurement and Instrumentation, 31*, 19–24.

Kjærsgaard-Rasmussen, J., & Meyer, K. E. (2011). Inside-out electrical capacitance tomography. *Flow Measurement and Instrumentation, 22*, 104–109.

Lakshmanan, S., Maru, W. A., Holland, D. J., Mantle, M. D., & Sederman, A. J. (2017). Measurement of an oil–water flow using magnetic resonance imaging. *Flow Measurement and Instrumentation,* 53*A*, 161–171.

Langener, S., Vogt, M., Ermert, H., & Musch, T. (2017). A real-time ultrasound process tomography system using a reflection-mode reconstruction technique. *Flow Measurement and Instrumentation, 53A*, 107–115.

Lange, G., & Seidel, K. (2007). Electromagnetic methods. In J. Knodel, G. Lange, & H.-J. Voigt (Eds.), *Environmental geology – handbook of field methods and case studies* (pp. 239–281). Berlin: Springer.

Leeungculsatien, T., & Lucas, G. P. (2013). Measurement of velocity profiles in multiphase flow using a multi-electrode electromagnetic flow meter. *Flow Measurement and Instrumentation, 31*, 86–95.

Li, M., Abubakar, A., Li, J., Pan, G., & Habashy, T. (2010). Three-dimensional regularized Gauss-Newton inversion algorithm using a compressed implicit Jacobian calculation for EM applications. *SEG Extended Abstracts, 29*, 4.

Li, H., & Chen, J. (2015). *Positron tomography imaging apparatus and method for multiphase flow*. Lanzhou Haimo Technologies. US patent US9006666B2. April 14, 2015.

Liu, S., Lei, J., Wang, X. Y., & Liu, Q. B. (2013). Generalized multi-scale dynamic inversion algorithm for electrical capacitance tomography. *Flow Measurement and Instrumentation, 31*, 35–46.

Liu, Z., Yang, G., He, N., & Tan, X. (2013). Landweber iterative algorithm based on regularization in electromagnetic tomography for multiphase flow measurement. *Flow Measurement and Instrumentation, 27*, 53–58.

Li, K., Wang, Q., & Wang, M. (2019). Three-dimensional visualisation of gas-water two-phase flow based on bubble mapping method and size projection algorithm. *Flow Measurement and Instrumentation, 69*(101590).

Li, H., Wang, M., Wu, Y. X., & Lucas, G. (2009). Volume flow rate measurement in vertical oil-in-water pipe flow using electrical impedance tomography and a local probe. *Multiphase Science and Technology, 21*, 81–93.

Li, Y., Yang, W. Q., Xie, C. G., Huang, S. M., Wu, Z., Tsamakis, D., et al. (2013). Gas/oil/water flow measurement by electrical capacitance tomography. *Measurement Science and Technology, 24*(074001), 12.

Ma, L., Hunt, A., & Soleimani, M. (2015). Experimental evaluation of conductive flow imaging using magnetic induction tomography. *International Journal of Multiphase Flow, 72*, 198–209.

Mallach, M., Gebhardt, P., & Musch, T. (2017). 2D microwave tomography system for imaging of multiphase flows in metal pipes. *Flow Measurement and Instrumentation, 53*, 80–88.

Mallach, M., Gevers, M., Gebhardt, P., & Musch, T. (2018). Fast and precise soft-field electromagnetic tomography systems for multiphase flow imaging. *Energies, 11*, 1199. https://doi.org/10.3390/en11051199

Ma, L., McCann, D., & Hunt, A. (2017). Combining magnetic induction tomography and electromagnetic velocity tomography for water continuous multiphase flows. *IEEE Sensors Journal, 17*(24), 8271–8281.

Marsala, A. F., Al-Buali, M., Ali, Z., Ma, S. M., He, Z., Tang, B., et al. (2011). First borehole to surface electromagnetic survey in KSA: Reservoir mapping and monitoring at a new scale. In *SPE annual technical conference and exhibition, 30 October-2 November 2011, Denver, Colorado, USA*. 146348-MS.

Marsala, A. F., Ruwaili, S., Ma, S. M., Ali, Z., Buali, M., Donadille, J. M., et al. (2008). Crosswell electromagnetic tomography: From resistivity mapping to inter-well fluid distribution. In *International petroleum technology conference, 3-5 December 2008, Kuala Lumpur, Malaysia*. 12229-MS.

Meng, Z., Huang, Z., Wang, B., Ji, H., Li, H., & Yan, Y. (2010). Air-water two-phase flow measurement using a Venturi meter and an electrical resistance tomography sensor. *Flow Measurement and Instrumentation, 21*, 268–276.

Mieles, L., Darnet, M., Popta, J. V., Singh, M., Wilt, M., & Levesque, C. (2009). Experience with cross-well electromagnetics (EM) for waterflood management in Oman. In *International petroleum technology conference,7-9 December 2009, Doha, Qatar*. IPTC 14080.

Mitchell, J., Chandrasekera, T. C., Holland, D. J., Gladden, L. F., & Fordham, E. J. (2013). Magnetic resonance imaging in laboratory petrophysical core analysis. *Physics Reports, 526*, 165–225.

Mitchell, J., Staniland, J., Chassagne, R., Mogensen, K., Frank, S., & Fordham, E. J. (2013). Mapping oil saturation distribution in a limestone plug with low-field magnetic resonance. *Journal of Petroleum Science and Engineering, 108*, 14–21.

Muji, S. Z. M., Goh, C. L., Ayob, N. M. N., Rahim, R. A., Rahiman, M. H. F., Rahim, H. A., et al. (2013). Optical tomography hardware development for solid gas measurement using mixed projection. *Flow Measurement and Instrumentation, 33*, 110–121.

Murai, Y., Tasaka, Y., Nambu, Y., Takeda, Y., & Roberto Gonzalez, S. A. (2010). Ultrasonic detection of moving interfaces in gas-liquid two-phase flow. *Flow Measurement and Instrumentation, 21*, 356–366.

Nuryadin, S., Ignaczak, M., Lucas, D., & Deendarlianto. (2015). On the accuracy of wire-mesh sensors in dependence of bubble sizes and liquid flow rates. *Experimental Thermal and Fluid Science, 65*, 73–81.

O'Neill, K. T., Klotz, A., Stanwix, P. L., Fridjonsson, E. O., & Johns, M. L. (2017). Quantitative multiphase flow characterisation using an Earth's field NMR flow meter. *Flow Measurement and Instrumentation, 58*, 104–111.

Ong, J. T., Chen, S., & Bussear, T. R. (2016). *Multiphase meter to provide data for production management*. Baker Hughes. US patent US9335195B2. May 10, 2016.

Pan, Y., Hong, Y., Sun, Q., Zheng, Z., Wang, D., & Niu, P. (2019). A new correlation of wet gas flow for low pressure with a vertically mounted Venturi meter. *Flow Measurement and Instrumentation, 70*(101636).

Pardo, D., Torres-Verdín, C., & Zhang, Z. (2008). Sensitivity study of borehole-to-surface and cross-well electromagnetic measurements acquired with energized steel casing to water displacement in hydrocarbon-bearing layers. *Geophysics, 73*(6), F261–F268.

Paladinoa, E. E., & Maliskab, C. R. (2011). Computational modeling of bubbly flows in differential pressure flow meters. *Flow Measurement and Instrumentation, 22*(4), 309–318.

Pinguet, B. (2012). *Fundamentals of multiphase metering*. 09-TS-0022 (2nd ed.). Schlumberger.

Pinguet, B., Gaviria, F., Kemp, L., Graham, J., Coulder, C., Damas, C., et al. (2010). First ever complete evaluation of a multiphase flow meter in SAGD and demonstration of the performance against conventional equipment. In *Proc. 28th int. North sea flow measurement workshop (st Andrews, UK, October 2010).*

Pinguet, B., Smith, M. T., Vagen, N., Merete Alendal, G., Rustad, R., & Xie, C. G. (2014). An innovative liquid detection sensors for wet gas subsea business to improve gas-condensate flow rate measurement and flow assurance issue. In *Offshore technology conference, Kuala Lumpur, 25-28 March 2014*. OTC-25054-MS.

Plaskowski, A., Beck, M. S., Thorn, R., & Dyakowski, T. (1995). *Imaging industrial flows: Applications of electrical process tomography*. Bristol: Institute of Physics Publishing.

Primrose, K. M., Qiu, C., Bolton, G. T., Talmon, A. M., Glen, N., Ross, A., et al. (2010). Visualisation and measurement of 3 phase (air water oil) flow and 2 phase flow (sand water) with electrical tomography. In *Proc. 6th world congress on industrial process tomography (beijing, China, september 2010)* (pp. 108–118).

Pusiol, D., Carpinella, M., Albert, G., Osán, T. M., Ollé, J. M., Freeman, J. J., et al. (2012). *Method for analyzing a multi-phase fluid*. US patent US8212557B2. July 03, 2012. Shell Oil Company.

Qiu, C., Hoyle, B. S., & Podd, F. J. W. (2007). Engineering and application of a dual-modality process tomography system. *Flow Measurement and Instrumentation, 18*, 247–254.

Rahim, R. A., Yunos, Y. M., Rahiman, M. H. F., & Rahim, H. A. (2010). Mathematical modelling of gas bubbles and oil droplets in liquid media using optical linear path projection. *Flow Measurement and Instrumentation, 21*, 388–393.

Ramirez, A. L., Daily, W., Owen, E. W., Chesnut, D. A., & LaBrecque, D. J. (1992). Electrical resistance tomography used to monitor subsurface steam injection. In *SEG annual meeting, October 25-29, new Orleans, Louisiana, USA*. Paper ID: 1992-0492.

Rautenbach, C., Mudde, R. F., Yang, X., Melaaen, M. C., & Halvorsen, B. M. (2013). A comparative study between electrical capacitance tomography and time-resolved X-ray tomography. *Flow Measurement and Instrumentation, 30*, 34–44.

Rymarczyk, T., Kłosowski, G., Kozłowski, E., & Tchórzewski, P. (2019). Comparison of selected machine learning algorithms for industrial electrical tomography. *Sensors, 19*, 1521. https://doi.org/10.3390/s19071521

Saoud, A., Mosorov, V., & Grudzien, K. (2017). Measurement of velocity of gas/solid swirl flow using electrical capacitance tomography and cross correlation technique. *Flow Measurement and Instrumentation, 53A*, 133–140.

Sætre, C., Johansen, G. A., & Tjugum, S. A. (2010). Salinity and flow regime independent multiphase flow measurements. *Flow Measurement and Instrumentation, 21*, 454–461.

Sætre, C., Johansen, G. A., & Tjugum, S. A. (2012). Tomographic multiphase flow measurement. *Applied Radiation and Isotopes, 70*, 1080–1084.

Scheers, L., & Wee, A. (2011). Solutions for reliable and accurate measurement of water production in subsea wet gas fields. In *Subsea controls downunder conference, 17–19 October 2011, Perth, Australia.*

Seidel, K., & Lange, G. (2007). Direct current resistivity methods. In J. Knodel, G. Lange, & H.-J. Voigt (Eds.), *Environmental geology – handbook of field methods and case studies* (pp. 205–237). Berlin: Springer.

Sun, J., & Yang, W. Q. (2013). Fringe effect of electrical capacitance and resistance tomography sensors. *Measurement Science and Technology, 24*, 074002.

Taherian, R., & Habashy, T. (1996). *Microwave device and method for measuring multiphase flows*. Schlumberger Technology. US patent US5485743A. January 23, 1996.

Tan, C., Dong, F., & Wu, M. (2007). Identification of gas/liquid two-phase flow regime through ERT-based measurement and feature extraction. *Flow Measurement and Instrumentation, 18*, 255–261.

Thorn, R., Huang, S. M., Xie, C. G., Salkeld, J. A., Hunt, A., & Beck, M. S. (1990). Flow imaging for multi-component flow measurement. *Flow Measurement and Instrumentation, 1*, 259–268.

Thorn, R., Johansen, G. A., & Hjertaker, B. T. (2012). Three-phase flow measurement in the petroleum industry. *Measurement Science and Technology, 24*(012003), 17.

Tsourlos, P., Ogilvy, R., Papazachos, C., & Meldrum, P. (2011). Measurement and inversion schemes for single borehole-to-surface electrical resistivity tomography surveys. *Journal of Geophysics and Engineering, 8*, 487–497.

Vielliard, C., Hester, K. C., Roccaforte, F., Di Lullo, A. G., Assecondi, L., Elkhafif, H. H., et al. (2019). Real-time subsea hydrate management in the world's longest subsea tieback. In *Offshore technology conference, 6-9 may, Houston, Texas*. OTC 29232.

Wagner, M., Bieberle, A., Bieberle, M., & Hampel, U. (2017). Dynamic bias error correction in gamma-ray computed tomography. *Flow Measurement and Instrumentation, 53*, 141–146.

Wang, H., Hu, H., Wang, L., & Wang, H. X. (2012). Image reconstruction for an electrical capacitance tomography (ECT) system based on a least squares support vector machine and bacterial colony chemotaxis algorithm. *Flow Measurement and Instrumentation, 27*, 59–66.

Wang, M., Jia, J., Faraj, Y., Wang, Q., Xie, C. G., Oddie, G., et al. (2015). A new visualisation and measurement technology for water continuous multiphase flows. *Flow Measurement and Instrumentation, 46B*, 204–212.

Wang, Q., Polansky, J., Wang, M., Wei, K., Qiu, C., Kenbar, A., et al. (2018). Capability of dual-modality electrical tomography for gas-oil-water three-phase pipeline flow visualization. *Flow Measurement and Instrumentation, 62*, 152–166.

Wang, J. Z., Tian, G. Y., & Lucas, G. P. (2007). Relationship between velocity profile and distribution of induced potential for an electromagnetic flow meter. *Flow Measurement and Instrumentation, 18*, 99–105.

Wang, H., & Yang, W. (2020). Application of electrical capacitance tomography in circulating fluidised beds – a review. *Applied Thermal Engineering, 13*(April), 115311. https://doi.org/10.1016/j.applthermaleng.2020.115311

Wang, B., Zhang, W., Huang, Z., Ji, H., & Li, H. (2013). Modeling and optimal design of sensor for capacitively coupled electrical resistance tomography system. *Flow Measurement and Instrumentation, 31*, 3–9.

Webilor, R. O., Lucas, G. P., & Agolom, M. O. (2018). Fast imaging of the velocity profile of the conducting continuous phase in multiphase flows using an electromagnetic flowmeter. *Flow Measurement and Instrumentation, 64*, 180–189.

Williams, R. A., & Beck, M. S. (1995). *Process tomography, principles, techniques and applications*. Oxford: Butterworth-Heinemann.

Wilt, M., Alumbaugh, D. L., Morrison, H. F., Becker, A., Lee, K. H., & Deszcz-Pan, M. (1995). Cross-well electromagnetic tomography: System design considerations and field results. *Geophysics, 60*, 871–885.

Wilt, M., Donadille, J. M., AlRuwaili, S., Ma, S. M., & Marsala, A. (2008). Cross-well electromagnetic tomography in Saudi Arabia: From field surveys to resistivity mapping. In *70th EAGE conference and exhibition, Rome, June 9–12, 2008*.

Wilt, M., Zhang, P., Maeki, J., Netto, P., Queiroz, J., Santos, J., et al. (2012). Monitoring a water flood of moderate saturation changes with cross-well electromagnetics (EM): A case study from Dom Joao Brazil. *SEG Technical Program Expanded Abstracts*, 1–5.

Wu, Z., McCann, H., Davis, L. E., Hu, J., Fontes, A., & Xie, C. G. (2009). Microwave tomographic system for oil and gas multiphase flow imaging. *Measurement Science and Technology, 20*(104026), 8.

Wu, Z., & Wang, H. (2017). Microwave tomography for industrial process imaging: Example applications and experimental results. *IEEE Antennas and Propagation Magazine, 59*(5), 61–71.

Wu, H., Xu, Y., Xiong, X., Mamat, E., Wang, J., & Zhang, T. (2020). Prediction of pressure drop in Venturi based on drift-flux model and boundary layer theory. *Flow Measurement and Instrumentation, 71*(101673).

Xie, C. G. (2006). Measurement of multiphase flow water fraction and water-cut. In *Proceedings 5th international symposium on measurement techniques for multiphase flows, 10-13 December 2006, Macau, China* (Vol. 914, pp. 232–239). AIP Conference Proceedings.

Xie, C. G. (2018a). *Tomography of multiphase mixtures*. Schlumberger Technology. US patent US10132847B2. November 20, 2018.

Xie, C. G. (2018b). *Multiphase flowmeter*. Schlumberger Technology. US patent US9927270B2. March 27, 2018.

Xie, C. G., Atkinson, I., & Lenn, C. (2007). Multiphase flow measurement in oil and gas production. In *5th world congress on industrial process tomography, Bergen, Norway, 3-6 September 2007* (pp. 723–736), ISBN 978 0 85316 265 0.

Xie, C. G., North, R., Wilt, M., Zhang, P., Denaclara, H., & Levesque, C. (2005). Imaging technologies in oilfield applications. *Journal of Zhejiang University - Science, 6A*(12), 1394–1400.

Xue, Q., Wang, H. X., Yang, C. Y., & Cui, Z. Q. (2013). Experimental research on two-phase flow visualization using a low-energy gamma CT system with sparse projections. *Measurement Science and Technology, 24*(074008), 8.

Xu, L. J., & Xu, L. A. (1998). Gas/liquid two-phase flow regime identification by ultrasonic tomography. *Flow Measurement and Instrumentation, 8*, 145–155.

Yang, W. Q., Chondronasios, A., Nattrass, S., Nguyen, V. T., Betting, M., Ismail, I., et al. (2004). Adaptive calibration of a capacitance tomography system for imaging water droplet distribution. *Flow Measurement and Instrumentation, 15*, 249–258.

Yang, W. Q., & Liu, S. (2000). Role of tomography in gas/solids flow measurement. *Flow Measurement and Instrumentation, 11*, 237–244.

Yang, W. Q., Stott, A. L., Beck, M. S., & Xie, C. G. (1995). Development of capacitance tomographic imaging systems for oil pipeline measurements. *Review of Scientific Instruments, 66*, 4326–4332.

Yan, Y., Wang, L., Wang, T., Wang, X., Hu, Y., & Duan, Q. (2018). Application of soft computing techniques to multiphase flow measurement: A review. *Flow Measurement and Instrumentation, 60*, 30–43.

Yu, G., Marion, B., Bryans, B., Carrillo, P., Guo, W., Pang, Y., et al. (2008). Crosswell seismic imaging for deep gas reservoir characterization. *Geophysics, 73*(6), B117–B126.

Zhang, Z., Bieberle, M., Barthel, F., Szalinski, L., & Hampel, U. (2013). Investigation of upward cocurrent gas—liquid pipe flow using ultrafast X-ray tomography and wire-mesh sensor. *Flow Measurement and Instrumentation, 32*, 111—118.

Zhang, J., Hu, H., Dong, J., & Yan, Y. (2012). Concentration measurement of biomass/coal/air three-phase flow by integrating electrostatic and capacitive sensors. *Flow Measurement and Instrumentation, 24*, 43—49.

Zhang, J., Xu, J., Wu, Y. X., Li, D., & Li, H. (2013). Experimental validation of the calculation of phase holdup for an oil—water two-phase vertical flow based on the measurement of pressure drops. *Flow Measurement and Instrumentation, 31*, 96—101.

Zhou, H., Yang, Y., Dong, K., Wu, J., Yan, Y., Qian, X., et al. (2013). Investigation of two-phase flow mixing mechanism of a swirl burner using an electrostatic sensor array system. *Flow Measurement and Instrumentation, 32*, 14—26.

Zhou, B., & Zhang, J. Y. (2013). A novel ECT—EST combined method for gas—solids flow pattern and charge distribution visualization. *Measurement Science and Technology, 24*, 074003.

Ziauddin, M., & Bize, E. (2007). The effects of pore-scale heterogeneities on carbonate stimulation treatments. In *15th SPE Middle East oil & gas show and conference, Bahrain, 11-14 March 2007*. SPE 104627.

Applications of tomography in multiphase transportation

Chao Tan and Feng Dong
Tianjin Key Laboratory of Proces Measurement and Control, School of Electrical and Information Engineering, Tianjin University, Tianjin, China

19.1 Introduction

Slurry flows, gas—solid two-phase flows and oil—gas—water multiphase flows are the frequently encountered processes of industrial multiphase transportations. Measuring their process parameters have been a popular scientific and engineering topic, which is closely involved with many safety and fiscal issues that are discussed worldwide for the past several decades (Thorn, Johansen, & Hjertaker, 2013).

Parameters of each phase are required to describe the process hydrodynamics of custody transportations, including phase distribution and fraction, transport velocity of each phase. Most of the problems encountered in multiphase flow measurements are caused by the increased number of the characteristic flow parameters as compared to single-phase flows, because multiphase flow always involves the interactions between phases, like mass transfer, momentum transfer, and heat transfer. In addition, all the changes and interactions are simultaneous and transient, which make the flow process even more complicated. For example, a typical flow phenomenon in multiphase transportation, the slurry flow, needs the information about phase fraction and velocity to study their dynamic behavior. Thus, slurry concentration profile is an important factor to predict the pressure drop and the frictional pressure losses in pipeline, as well as to understand the hydrodynamics and predict the performance of chemical conversion and heat transfer in chemical reactors. Another rheological change of a multiphase fluid structure in multiphase transportation is named flow regime or flow pattern, which dominates the governing dynamic and consequently the means of measurements conducted.

The difficulty of measuring/monitoring multiphase flow is that both the phase distribution and the velocity profile vary significantly with flow patterns. Lots of measuring techniques are employed to solve this problem; however, no individual method could fully satisfy the field application so far, because of the lack of reliable and nonintrusive measurement methods to obtain concentration and velocity profiles for multiphase transportation (Giguère, Fradette, Mignon & Tanguy, 2008).

To fully understand the multiphase flow and establish the description models, a comprehensive, nonintrusive, and visualized detecting method is required, which falls in the capability of Industrial Process Tomography (IPT) (Yao & Takei, 2017). The IPTs can provide spatial and temporal information of multiphase flow by integrating

Industrial Tomography. https://doi.org/10.1016/B978-0-12-823015-2.00022-4

multisource information at different positions of the measured pipe cross-section. They also provide visual monitoring on the phase distribution within the pipeline in a two-dimensional (2D) or three-dimensional (3D) manner. Therefore, the IPTs are widely studied due to their capability of "looking" inside the fluid mixture where the local phase fraction and velocity can be extracted. In this way, an accurate means of calculating flow rate and eliminating the uncertainty in measurement will be available. Due to these virtues, IPT gains applications widely in multiphase transportation for flow measurement/monitoring, such as the petroleum transportation in oil industry (Ismail, Gamio, Bukhari, & Yang, 2005), fluidized beds (Wang & Yang, 2020), and slurry transportation (Kotze, Adler, Sutherland, & Deba, 2019).

This chapter will give an overview on how to apply the IPTs in measuring the multiphase flow in pipeline transportation. These applications include the process parameter measurement and flow process monitoring, along with the process analysis based on either the reconstructed cross-sectional (or volumetric) images or raw data. The multiphase flow studied in this chapter mainly includes oil—gas—water multiphase flow (in oil industry transportation) and solid—liquid flow (especially slurry flow in multiphase transportation concerning particles). Definition of flow pattern in multiphase flow is introduced in section 19.2 along with flow pattern identification methods with IPTs. Then, flow parameters, mainly the phase fraction and flow velocity, measured with IPTs are discussed in section 19.3, also covering the topics of flow process monitoring and characterization. Because of the complexity of multiphase flow, one sensor may not be able to conduct a comprehensive monitoring or measurement alone, so the IPTs in the frame of multisensor fusion are discussed in section 19.4, covering the topics of multimodality IPTs, and IPTs with other sensors. A closing remark is given in section 19.5.

19.2 Flow pattern and flow pattern identification with IPT

19.2.1 Flow patterns in multiphase transportation

Flow pattern (or flow regime) of multiphase flow is the temporal and spatial distribution of each phase; it reflects the flow conditions and governs nearly all the flow parameters of interests to field applications. The characterization of flow patterns in multiphase transportation is useful not only for the design, the optimization, and the control of a transportation process, but also for research purposes by providing experimental data needed to develop, tune, and validate empirical and numerical multiphase flow models. Flow pattern changes with fluid properties, pipe diameter, pipe inclination, and operation conditions. Therefore, only the horizontal flow patterns are discussed here as an example demonstrating the application of IPTs in flow pattern identification and characterization. The horizontal flow patterns are usually asymmetric in structure, because gravity acts perpendicular to the flow, and the separation of the flow might occur which makes the flow asymmetric and hence more complex.

Four typical flow patterns commonly seen in a horizontal slurry flow can be classified with solid phase concentration where the density of the solid phase is usually higher than the density of the carrier fluid. They are flow with a stationary bed, flow with a moving bed, heterogeneous flow (asymmetric), and pseudohomogenous flow (symmetric) (Giguère, Fradette, Mignon, & Tanguy, 2009). For the oil–gas–water multiphase flow, usually it is defined based on the patterns of gas–liquid (water or oil) two-phase flow and liquid–liquid (oil–water) two-phase flow (Spedding, Donnelly, & Cole, 2005). Due to gravity and buoyancy that simultaneously occur to gas–liquid two-phase flow, the gas phase flows more likely on the top of the pipe. Slug flow and elongated bubble flows can be jointly classified as "intermittent" flow, stratified flow and annular flow are the "separated" flow, and dispersed bubble flow is the "homogeneous" flow. Flow regime of oil–water flows is quite different from that of gas–liquid two-phase flows. The different flow structure is mainly caused by the large liquid/liquid momentum transfer capacity and small buoyancy effects. The lower free energy at the interface allows the formation of shorter interfacial waves and smaller dispersed phase droplet size.

The flow patterns of three-phase flow were investigated by considering the two-liquid phase as a mixture, which simplifies the models and analysis, but also confined its applications within only particular conditions. One of the widely accepted flow regime categorizations is given by Acikgoz Franca and Lahey (1992). They observed and named six different flow regimes, which can be classified into two major groups, oil-based flow and water-based flow, according to the distinct formation of dominated phase.

Besides, solid particle entrained in slurry flow, gas–liquid two-phase flow, or oil–gas–water three-phase flow is frequently encountered in petroleum transportation or chemical process, and these three- or even four-phase flows make the flow pattern much more complicated. Research studies have shown that the multiphase flow regimes transition is jointly governed by many factors, such as liquid viscosity and solid loadings (Roostaei et al., 2020).

19.2.2 Flow pattern identification with IPTs

Radiation-based tomography, such as X-ray Computed Tomography (CT), has been applied on solid–liquid two-phase flow and gas–oil–water three-phase flow pattern identifications (Hu, Langsholt, Liu, Andersson & Lawrence, 2014), because of the linearized projections of CT that reduce the uncertainty of solving inverse problems. Electrical tomography with simple structure and low environmental hazard is also an effective tool for flow pattern identification. For instance, in gas–oil flow, usually the ECT systems are employed (Gamio et al., 2005); when conductive liquid is the continuous phase, such as gas–water two-phase flow, usually electrical resistance tomography (ERT) does the job (Tan, Shen, Smith, Dong & Escudero, 2019). Besides, ultrasound tomography (UT) is considered an effective identification method for the high contrast acoustic impedance distribution medium, for instance, gas–liquid two-phase flow (Xu & Xu, 1997).

Flow regimes can be visually identified either by stacking the reconstructed 2D images, or by 3D reconstructions. However, these observations are highly restricted by the image reconstruction speed and artifacts of the reconstructed images. Therefore, a rapid method of identifying flow regimes by extracting flow features from tomographic measurement data was developed. In this way, the flow features at a certain time interval are further processed with a classification algorithm (or pattern recognition) to separate the features of different flow regimes. This method does not need an image reconstruction procedure, and thus is a time-saving way of identifying flow regimes in a practical application (Tan, Dong, & Wu, 2007).

19.3 Multiphase transportation process measurement and monitoring with IPTs

In multiphase transportations, the most important process parameters of interest are the phase fraction (by volume, cross-section, or mass) and the flow rate (by volume or mass) of each individual phase. Variations of either parameter will bring changes to flow conditions in terms of flow patterns and hydrodynamic models. Therefore, precise estimates of these parameters without disturbing flow are required, and the flow process characterization by the IPTs can be helpful for the understanding the transportation mechanism.

19.3.1 Phase fraction measurement

Phase fraction is a key parameter that reflects flow conditions and is indispensable in multiphase flow models. IPTs can visualize the phase distribution; therefore, have a great perspective in measuring the phase fraction. In general, phase fraction is calculated from the ratio of the area occupied by one phase to the whole cross-sectional area in a reconstructed image. The estimated phase fraction precision is restricted by the gray-level threshold, i.e., how to define the gray level of representing a dispersed phase in the image. The adaptive multithreshold combined the iterative Gauss–Newton method for the multiphase flow imaging using ERT can improve the reconstruction accuracy other than some image processing methods of separating the objects from the background (Kim, Khambampati, Hong, Kim, & Kim, 2013). However, the precision is also restricted by the spatial resolution of the tomography system. Thereby, the method that directly processes the data to extract the phase concentration without image reconstruction was presented (Dong, Jiang, Qiao, & Xu, 2003).

The two most popularly adopted tomographic principles for phase fraction estimation are based on radiation and electrical techniques. Because the radiation beams do not bend at the interface of phases, the accurate phase distribution and further the phase fraction can be obtained. By comparison, the sensing field of electrical tomography is vulnerable to phase distribution and difficult to obtain a phase distribution as precise as radioactive tomography. But electrical tomography is simple in structure and reliable in operation, and the accuracy could reach $\pm 5\%$ (Sharifi & Young, 2013).

The radioactive wave, e.g., X-ray or γ-ray, attenuates when transmits in fluids with an attenuation rate varying with fluid properties. Usually a single-energy γ-ray can measure a two-phase flow system. However, three-phase flow usually requires two independent measurements of phase fractions, then the fraction of the third phase will be calculated from their relation that the summation of the three phase fractions is unity. A second measurement is needed such as capacitance; furthermore, if only radiation-based method is preferred, then two independent measurements can be obtained using the dual-energy technique. One example of measuring the phase fraction of a three-layered oil−gas−water three phase flow by a two-beam (one vertical and one horizontal) X-ray computed tomography is shown in Fig. 19.1. Two independent units of 60 kV and 4 mA with a sampling frequency of 40 Hz were used. The measurement is obtained and processed with the calibration data when the pipe is full of either oil, gas, or water. The results are normalized to reduce the influence of noise and systematic error (Hu et al., 2014).

As with all radiation measurement techniques, a trade-off between measurement speed and accuracy must be made due to the statistical nature of the radiation source. The greater the accuracy required, the longer will be the measurement period. Stronger radiation sources will shorten the measurement period, but much greater safety precautions must be taken. Besides, the vessel size is always not large and hence wall effects usually influence the measurement. An ultrafast electron beam X-ray tomography with fan beam has also been developed for the visualization and measurement of gas−liquid two-phase flow as shown in Fig. 19.2 (Lau, Hampel, & Schubert, 2018). The multibeam γ-ray tomography system is complex and expensive, and yet the component concentration can be measured with an accuracy of a few percent and thus they are being increasingly used in multicomponent flow measurement (Hjertaker, Tjugum, Hallanger, & Maad, 2018).

MRI is a high spatial resolution and noninvasive method for multidimensional visualization of multiphase flow, such as water/hydrate distribution during hydrate formation and methane production (Baldwin et al., 2009). Miscible fluid displacement

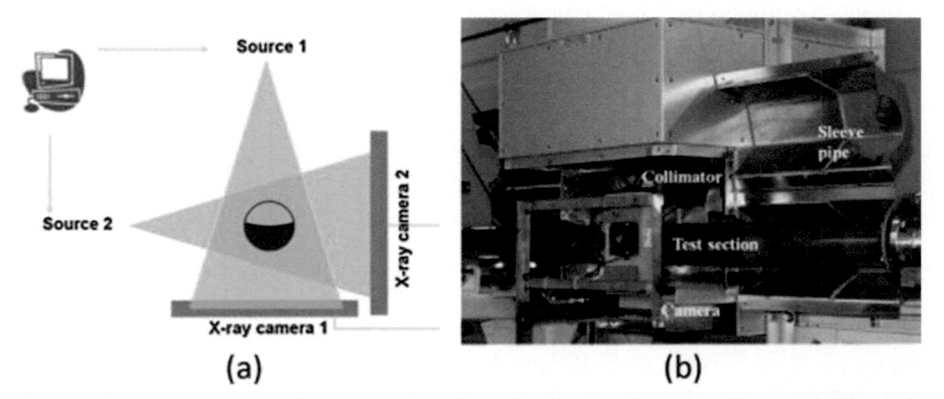

Figure 19.1 X-ray tomography system, (a) schematic diagram, (b) in-line X-ray unit (Hu et al., 2014).

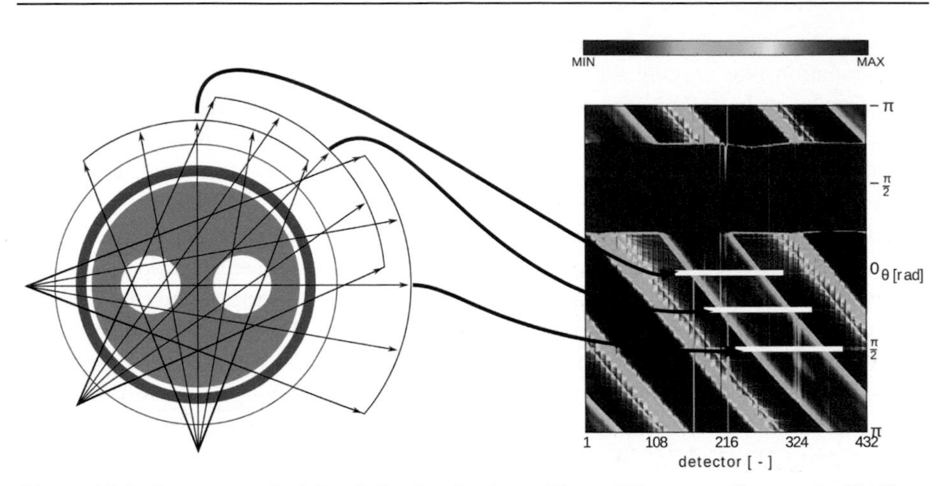

Figure 19.2 Operation principle of ultrafast fan beam X-ray CT scanner (Lau et al., 2018).

always occurs in mineral transport in soil; MRI can monitor the fluid saturation along the length of the sample and the pore size distribution in the sample (Muir, Petrov, Romanenko, & Balcom, 2014).

ERT has been successfully applied in reconstructing phase distribution of multiphase flow (Kotze et al., 2019). Through reconstruction algorithms, the conductivity distribution within the target pipeline could be reconstructed, then the phase fraction can be estimated based on the area each phase occupies, as shown in Fig. 19.3. However, artifacts within reconstructed images bring error to phase fraction measurement. To alleviate artifacts, several measures are proposed, including decreasing the intensity of electrical current injected, increasing the absolute electrical conductivity of the water phase, or designing robust reconstruction algorithms. These methods increase the definition of image reconstruction but decrease the quality of ERT reference measurements.

Another methodology of utilizing ERT measurement on phase fraction estimation is to directly process the measurement data to determine the phase fraction without image reconstruction. For sphere particles evenly distributed in liquid, their volume fraction α_p can be converted from the conductivity distributions with the Maxwell equation:

$$\alpha_p = \frac{2\sigma_l - 2\sigma_m}{\sigma_m + 2\sigma_l}, \tag{19.1}$$

where σ_l and σ_p are the conductivity of the liquid phase and particles, respectively and σ_m is the mixture conductivity distribution.

Razzak used ERT and the Maxwell relationship to obtain the solid fraction distribution of slurry flow (Razzak, Barghi, & Zhu, 2009). The reconstructed phase fraction distribution at superficial liquid velocity, $u_l = 11.2$ cm/s, superficial solids velocity $u_s = 0.75$ cm/s is shown in Fig. 19.4, where the solids concentration was

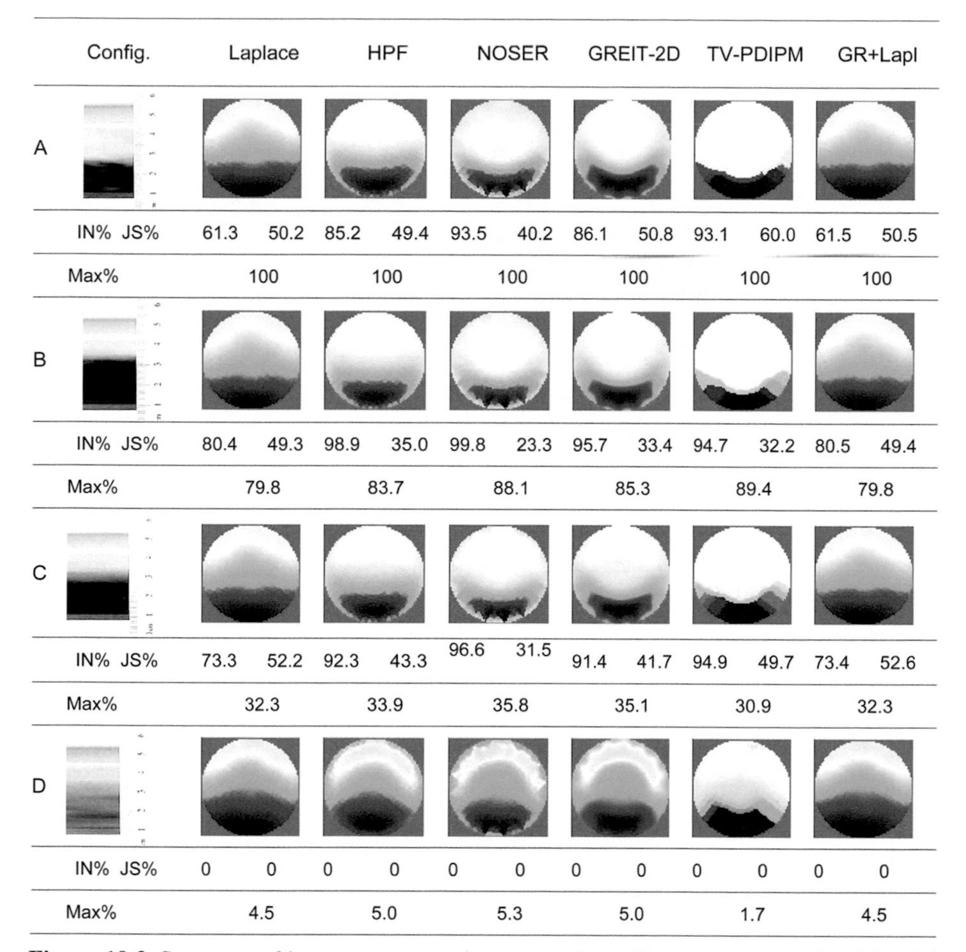

Figure 19.3 Summary of image reconstruction comparisons for measurement of bed levels/flow regime for flow loop experiments A to D (Kotze et al., 2019).

higher near the wall regions and decreased toward the central region. A 3D topographic view of the cross-sectional solids holdup in Fig. 19.5(b) shows the distribution more clearly.

For other electrical tomography modalities, ECT has been used in high pressurized pipe to visualize the different flow regimes of gas–oil flow and the phase volume fraction can be obtained by averaging the local concentration of the images (Perera, Pradeep, Mylvaganam, & Time, 2017). Combining the mixture permittivity measured by the ECT and the dielectric mixing model such as oil–water and gas–water, it is possible to measure phase fraction of oil–water oil–continuous two-phase flows and gas–oil–water three-phase flow oil–continuous flows (Li et al., 2013). Electromagnetic Tomography (EMT) has also been applied for the phase fraction measurement through measuring the conductivity of the fluid; it's preferably used in highly

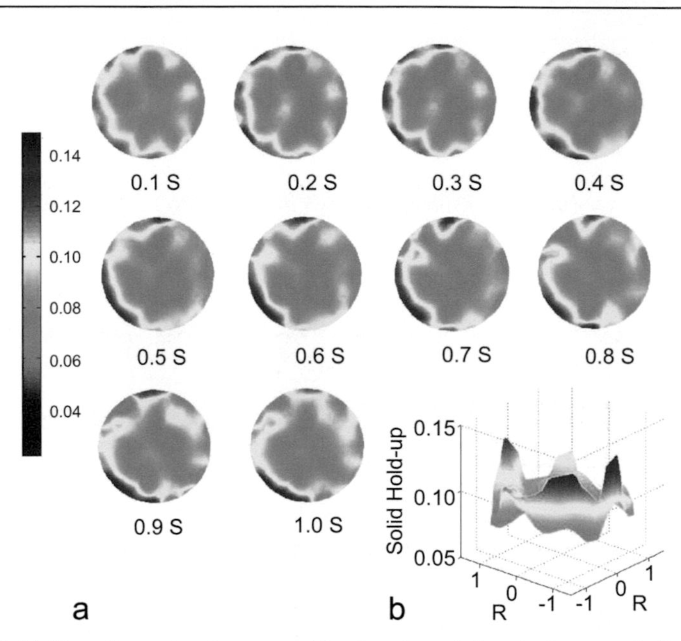

Figure 19.4 (a) Two-dimensional topographic view from 0.1 to 1s frame and (b) three-dimensional topographic view of the cross-sectional solid holdup at superficial liquid velocity, $u_l = 11.2$ cm/s, superficial solids velocity $u_s = 0.75$ cm/s (Razzak et al., 2009).

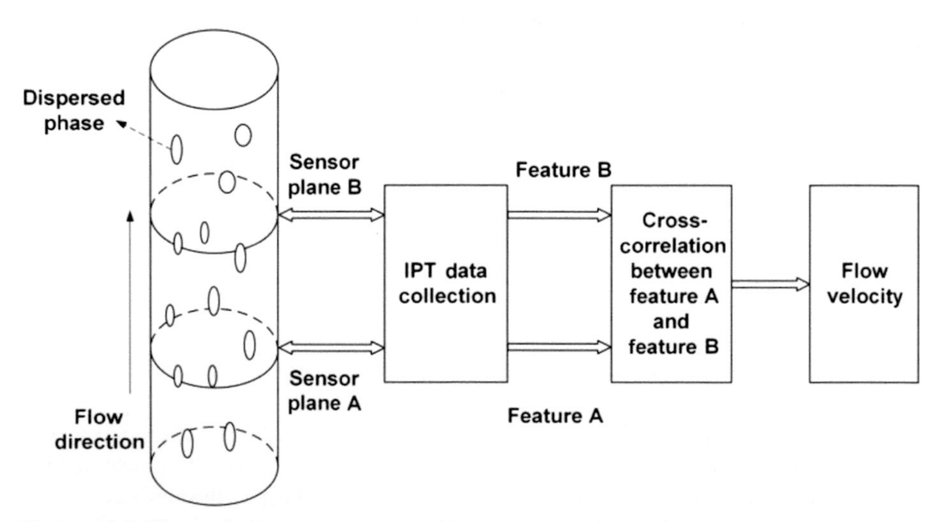

Figure 19.5 Flow velocity measurement with cross-correlation and an IPT.

conductive water-based multiphase flow, such as seawater and oil mixture in pipeline transportation (Watson, Williams, Gough, & Griffiths, 2008).

UT is noninvasive, cheap, and radiation-free for the multiphase flow monitoring and detection. Compared with electrical methods, UT detects the acoustic impedance variations for the multiphase mediums and is capable to overcome the change of conductivity (Goh, Ruzairi, Hafiz, & Tee, 2017). For a better phase fraction measurement, the exciting voltage and frequency should be carefully selected to improve the signal-to-noise ratio. The information obtained by UT is usually the attenuation or the time-of-flight of ultrasound for the cross-section image reconstruction and phase fraction measurement (Tan, Li, Liu, & Don, 2019). Ultrasonic tomography is typically used in gas−liquid two-phase flow, because of the strong reflection at the gas−liquid interface, it could also be extended to liquid−liquid two-phase flow by processing the multiscattering with advanced reconstruction algorithms (Liu, Tan, & Dong, 2019).

19.3.2 Flow velocity measurement

Accurate measurement of flow velocity is broadly needed in multiphase transportation; however, due to the intrinsic temporal and spatial complexity of fluid, as well as the industrial requirement on nonintrusive and low energy loss, it is a challenging topic for decades. Generally, the method that process tomography adopts for flow velocity measurement is cross-correlation, which is to seek the time delay of the flow structure passes from upstream sensors to downstream sensors by capturing turbulence of the fluid or suspended particles moving in the pipe (Plaskowski, Beck, & Krawaczynski, 1987). When applying cross-correlation to process tomography, the upstream and downstream sensors are referred to the tomographic sensor planes mounted on the cross-sections or slices of interest of a pipe (Deng, Dong, Xu, Liu, & Xu, 2001), as shown in Fig. 19.5. This structure is named "Dual-plane Tomography," and its extended version is multiplane tomography which, by definition, employs more than two sensor planes on the pipeline. Taking the ERT cross-correlation method as an example, a multiplane ERT usually has four or eight electrode planes which depends on the scale and the range of interest for flow investigations. For instance, an 8×16 array of resistance tomography sensors for the visualization of the mixing process of a gas−liquid two-phase flow (Wang, Dorward, Vlaev & Mann, 2000).

Basically, two methods of tomography based cross-correlation are employed; one is to correlate the corresponding pixels of reconstructed images on each plane for local velocity measurement (Lucas, Cory, Waterfall, Loh, & Dickin, 1999). In this way, the axial velocity profile can be reconstructed as shown in Fig. 19.6. The pixel-based cross-correlation can also reconstruct the radial velocity distribution by using a dual-plane ERT system to measure local air volume fraction and 3D velocity distribution in gas−liquid flows. A "best-correlated pixels" method was developed for the axial, angular, and radial velocity profiles measurement of swirling flow (Wang, Lucas, Dai, Panayotopoulos, & Williams, 2006). A pixel from upstream sensing plane (plane X) is correlated with axially corresponding pixel and its neighbor's pixels from the downstream sensing plane (plane Y) through the following correlation:

$$R_{x[n,m]y[n+i,m+j]}[p] = \sum_{k=0}^{T-1} x_{[n,m]}[k] y_{[n+i,m+j]}[k+p], \tag{19.2}$$

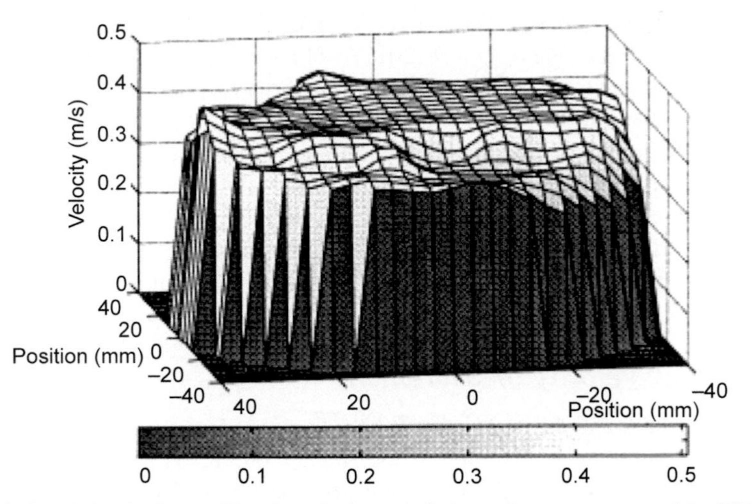

Figure 19.6 Axial velocity profile of vertical upward slurry flow reconstructed by ERT ($Q_s = 1.11$ m^3/h, $Q_w = 6.84$ m^3/h, and $\alpha_m = 0.16$) (Lucas et al., 1999).

where T is the number of the images, for which the cross-correlation is calculated; $p = -(T-1), \ldots 0, 1, 2, 3 \ldots T-1$; $k = 0, 2, 3, 4 \ldots T-1$; n, m are the coordinates of the pixel; x, y are the values of the pixels on plane X and plane Y, respectively; $(i, j) \in B$ is the group of neighboring pixels on plane Y. After calculation of the cross-correlation functions between each pixel, the highest correlation value will be selected as the best-correlated pixel, and thus the velocity between these two pixels is obtained:

$$u_{nm} = \frac{d}{\tau}, \tag{19.3}$$

where d is the separation distance between the corresponding pixels at plane X and plane Y. τ is the transit time of the flow through the two sensing planes and is calculated from the frame number delayed corresponding to the highest correlation value and the known data collection speed.

The other cross-correlation method is to correlate the features extracted from the raw measurement data without image reconstruction. The velocity obtained is the overall flow velocity; it is faster in calculation than the pixel-based cross-correlation but loses details on the velocity profile, thus can be used in fast industrial process measurement. The model of tomography feature cross-correlation is flow regime dependent and needs a priori knowledge on velocity field. Therefore, a priori knowledge of both the flow pattern and the phase fraction is required (Saoud, Mosorov, & Grudzien, 2017). One example of using ERT cross-correlation for oil−water two-phase flow velocity estimation is shown in Fig. 19.7, where the cross-correlation velocity is in line with the reference velocity with a scatter caused by the flow patterns.

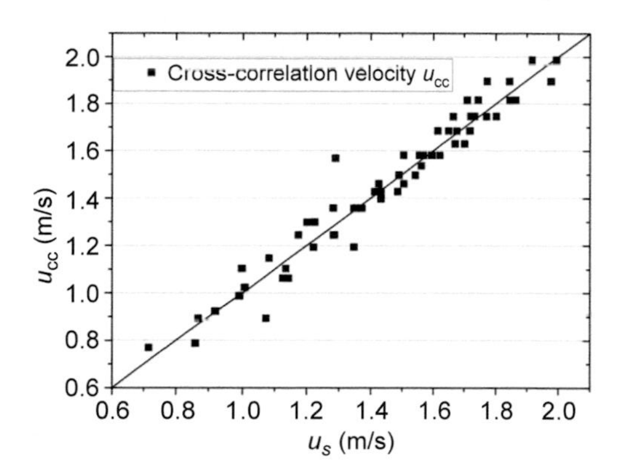

Figure 19.7 Oil–water two-phase flow velocity u_{cc} prediction with cross-correlation method at superficial mixture velocity u_s of ranging from 0.7 m/s to 2.1 m/s.

Many IPTs utilizing the same principle have shown the capability for velocity measurement. UT with cross-correlation can obtain the local gas velocity distribution in a bubbly gas/liquid pipe flow. It should be noted that the cross-correlation velocity measurement with the IPTs does not work properly in the stratified or annular flow as the flow feature to correlate is not obvious like in slug or plug flow; therefore, the cross-correlation calculation would deliver an erroneous results in time delay (Ayob, Rahiman, Zakaria, Yaacob, & Rahim, 2011). Using ECT or X-ray tomography with the cross-correlation method can obtain the angular velocity of the swirl flow (Saoud et al., 2017) or the particulate phase velocity in gas–solid two-phase flow (Barthel, Bieberle, Hoppe, Banowski, & Hampel, 2015).

The accuracy of cross-correlation technique in velocity measurement is jointly determined by the data acquisition rate and the separation between two sensors (arrays). A large separation increases the velocity measurement range while maintaining the time resolution, but the similarity of the signals captured by the two sensing planes decreases due to the transient and complex character of two-phase flow. Besides, a slower data acquisition speed and increased image reconstruction calculation needs a large separation. Therefore, the separation should be carefully selected to match the flow velocity and acquisition rate of the sensor system (Beck, 1981).

An electromagnetic flow meter (EFM) with multiple electrodes can measure flow velocity profiles in multiphase flow (Leeungculsatien & Lucas, 2013). The Helmholtz coil was used to produce a near-uniform magnetic field and an annular electrode array mounted on the pipe wall acquires the induced voltages. The axial flow nonuniform velocity profiles can be reconstructed (Kollár, Lucas, & Zhang, 2014). An EMFT system with 4 excitation coils and 16 measurement electrodes was developed, and reliable velocity field estimates in a laboratory environment with both axisymmetric and asymmetric single phase flow can be obtained (Vauhkonen, Hänninen, & Lehtikangas, 2018). The coil currents of the EMFT were optimized based on the distinguishability approach and minimizing the trace of the posterior covariance matrix (Lehtikangas & Vauhkonen, 2017). Accurate velocity field can be obtained by using the optimal magnetic fields for both axisymmetric and asymmetric flows.

MRI has also been applied to measure the flow velocity. For instance, the flow characteristics of water flow in porous media containing a stagnant immiscible liquid were investigated by MRI and a phase encoding method was proposed for flow velocity measurement (Okamoto, Hirai, & Ogawa, 2001). Recently, a portable MRI-based multiphase flow meter was proposed (Meribout, 2019); a solenoid coil surrounded by several stacks of relatively small permanent cuboid magnet elements was designed and optimized. This structure has the features of lightweight without compromising the intensity and homogeneity of the magnetic field.

19.3.3 Flow process analysis and characterization

Characterization of flow process in multiphase transportation is helpful to the understanding and modeling of multiphase flow. IPT is an ideal tool for flow process analysis owing to the multiple information that is simultaneously provided by the sensors implemented at different locations of the pipe. Similar to flow pattern identification, the methods of flow process analysis and characterization can also be categorized into image-based analysis and data-based analysis.

The image-based methods analyze the flow behavior through the reconstructed images (in 2D or 3D). For example, the phase distribution of a horizontal swirling slurry flow is studied with image-based analysis of a dual-plane ERT (Wang, Jones, & Williams, 2003). Two-dimensional solids volume fraction distributions of slurry with and without the swirl-inducing pipe are shown in Fig. 19.9(a) and (b). The longitudinal stacked views of nonswirling phase distributions are shown in Fig. 19.9 (c) and (d). These stacked views are formed by projecting 150 frames of cross-sectional images, so as to observe the flow history. The two side views in Fig. 19.8(c) are reconstructed by the two electrode planes, the time delay of the same solid volumes can be clearly observed, and the velocity of the solid can be calculated with the cross-correlation method.

The swirling slurry flow can be analyzed with 3D tomogram constructed from four tomograms by a four-plane ERT, through interpolation and thresholding conductivity that defined the bulk—particle/bulk—fluid interface (Stevenson, Harrison, Miles, & Cilliers, 2006). The 3D tomograms of different orientations are shown in Fig. 19.9; these stacked side views are not real 3D reconstruction but are usually referred to 2.5D. A real 3D reconstruction is computationally expensive and time consuming, thus is not practical for the transient flow reconstruction. With the development of the computing technologies, the 3D reconstruction becomes more popular in the academic researches for the dynamic reconstructions. Many similar researches have been done using the ECT for flow characterization (Wang & Yang, 2020). Additionally, combining ERT and ECT sensors can obtain the water-to-liquid ratio (WLR) and gas volume fractions of horizontal oil—gas—water three-phase flow. ERT works in the water continuous flow with WLR higher than 40% and ECT works in the range from 0% to 90% for other conditions. The threshold-based data fusion method was adopted for the three-phase images and it shows well consistency with high-speed camera pictures (Wang et al., 2018). The EMT can also be combined

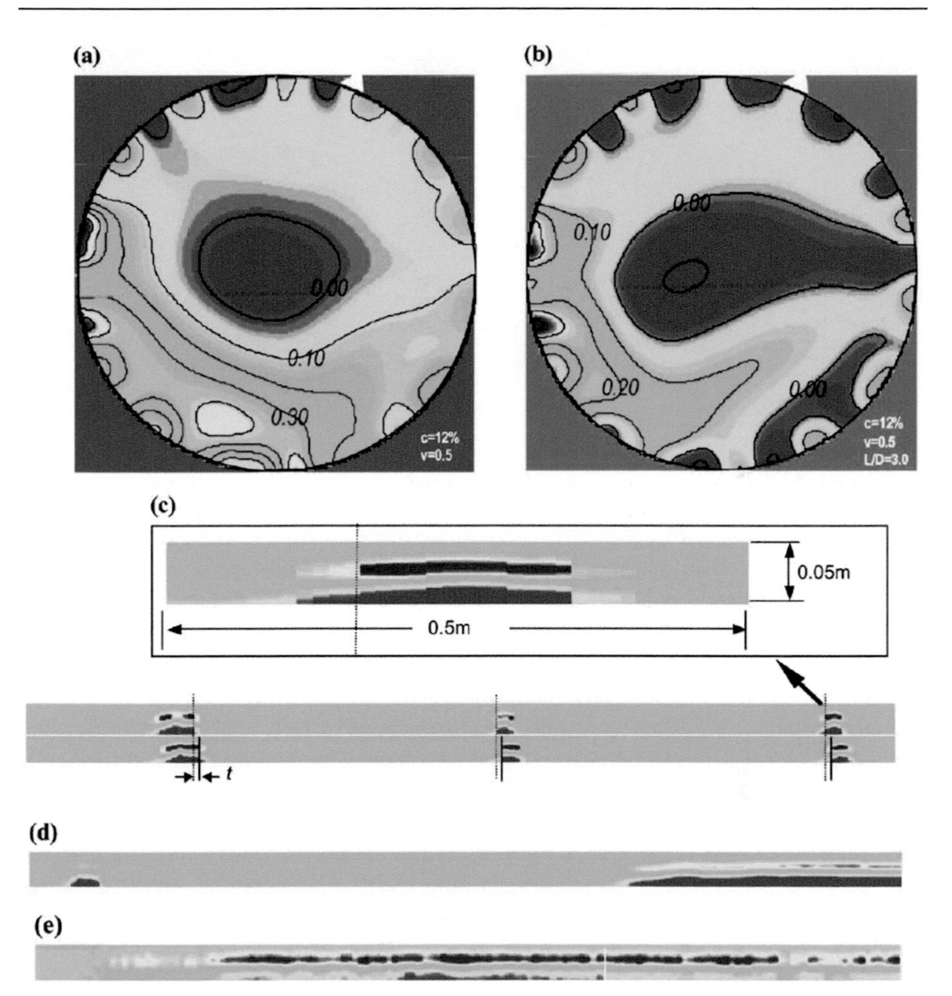

Figure 19.8 Axial solids volume fraction distributions from nonswirling flows with estimated concentrations of 2.1% (a and c) and 8.6% (d), and a swirling flow with a concentration of 8.6% (b and e). The water flow velocities in all cases are 0.5 m/s (Wang et al., 2003).

with EMFT since they both utilize the magnetic coils and the cross-sectional volumetric fraction and the local axial velocity of the water phase could be measured simultaneously (Ma, Spagnul, & Soleimani, 2017).

The X-ray tomography has the advantage of higher sampling rate with better spatial resolution. Flow regimes and quantitative measurements under different flow state are studied with a fast-response X-ray tomography and the multiphase interface behavior is captured and analyzed (Hu et al., 2014). The complex transient gas–liquid two-phase flow and its hydrodynamic characteristics are also investigated by the ultrafast single-slice X-ray tomography system (Schäfer, Meitzner, Lange, & Hampel, 2016). Using the reduced-order description for the flow dynamics characterization, the phase

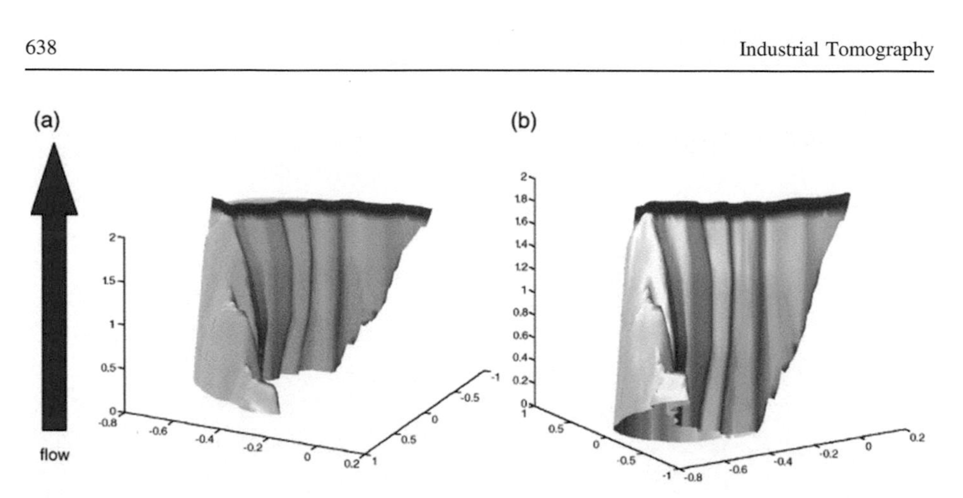

Figure 19.9 3D view of upward vertical swirling flow at a linear velocity of 1.5 m/s and 3% mass solids where the green twisting section represents the bulk fluid–solid interface (Stevenson et al., 2006).

fraction, cross-section phase distributions, and phase interface characteristics of two-phase dispersed and slug flow could be analyzed (Viggiano, SKJæraasen, SCHümann, Tutkun, & Cal, 2018).

Image-based analysis is an intrinsic method utilizing the advantages of IPT, but it suffers from the low spatial and temporal resolution of image reconstruction, and thereby suitable for offline analysis. The time series–based analysis method treats the measurement data of one cross-section as a matrix containing spatial information about phase distribution. With certain matrix dimension reduction or compression, the data matrix is reduced to a set of feature vector or a feature value. A multi/univariate time series is formed if the time history of the feature vector/value is considered. The time series contains rich information regarding the flow process and can be analyzed with a variety of time series analysis methods. For example, Principle Component Analysis, Independent Component Analysis (ICA) and other statistic methods, Wavelet Transform (Tan et al., 2007), and other time-frequency methods, and other methods include nonlinear time series analysis, such as entropy analysis (Tan, Zha & Dong, 2015). Additionally, the multivariate time series could also be studied via their intercorrelation, i.e., the similarity between the local phase fraction fluctuation in the same cross-section but over a certain length of time, as shown in Fig. 19.10 (Tan et al., 2019). This can provide a profound view of discovering the inner mechanism of the structural behavior of multiphase transportation, in a novel manner. By using the above-mentioned analysis, the flow process of different fluctuation scales could be separated for further analysis.

19.4 IPT in multiphase flow measurement with multisensor fusion

19.4.1 Multimodality IPTs

Multiphase flow is a transient process with spatially complex and time-variant structures that are sensitive to the change of phase fraction and velocity, fluid properties,

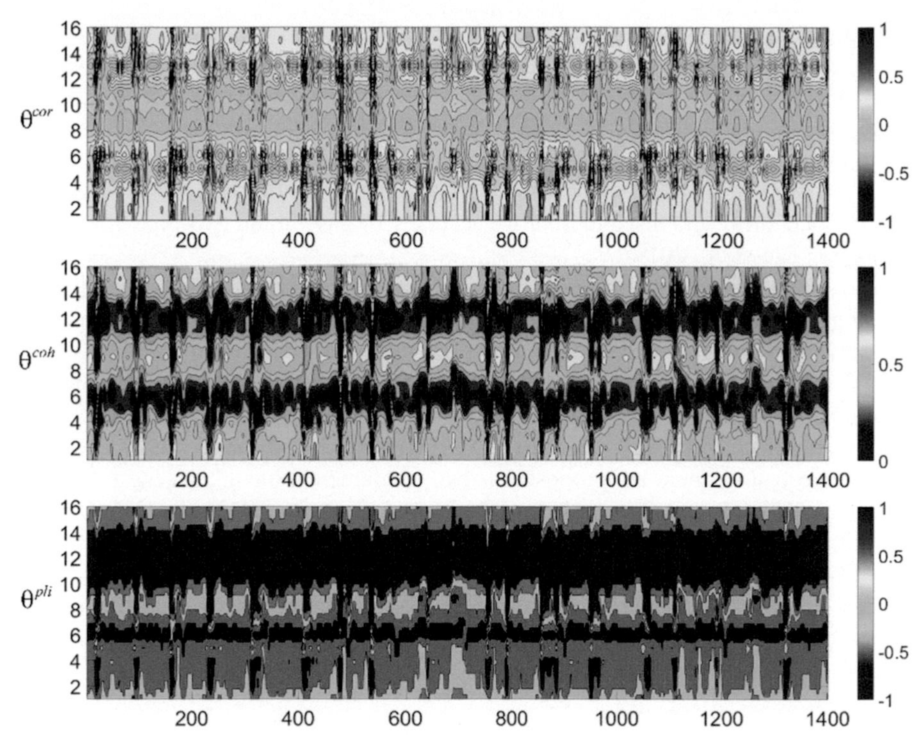

Figure 19.10 Dynamic connectivity of slug flow (Tan et al., 2019).

and ambient conditions. There are usually more than two fluids flowing together, such as oil—gas—water—sand flow, single modality IPT might fail to reconstruct correct distribution of fluids; therefore, multimodality IPT became favorable for its capability of simultaneously sensing different fluid properties by utilizing multiple physical sensing fields. For instance, the electrical tomography and radiation tomography can be combined to compensate their sensing fields to sense different fluids, the electrical capacitance, and resistance tomography can also be combined to collect the conductivity and permittivity information. In three-phase flow, it theoretically solves the failure of ERT under high oil fraction as well as the failure of ECT under high water fraction. With the attempts on integrated dual modality sensor (Wang, Wang, Sun, Cui, & Huang, 2014), the oil—water two-phase and gas—oil—water three-phase flow with water/oil fraction shifting from 0% to 100% could be visualized. Similar to ECT/ERT, there is also a report of using ECT/EMT (Electromagnetic Tomography) combination (Zhang, Ma, & Soleimani, 2015) and ECT/EST (Electrostatic Tomography) (Zhou & Zhang, 2013) to measure the oil—gas—water and solid—liquid—gas three-phase flow. The combinations of ECT and Gamma Densitometry Tomography (GDT) are also studied in three-phase flow measurement (Hjertaker, Maad, & Johansen, 2011). ECT separates conductive water from oil and gas, and GDT recognizes gas from oil and water. This EIT-GDT combination can also measure the gas, liquid, and solid three-phase flow in a vertical pipeline (George, Shollenberger, Torczynski, O'hern, & Ceccio, 2001).

Ultrasound is a potential alternative to X-ray or Gamma-ray tomography for its hard field feature, but with a much lower cost and higher safety. Both time-of-arrival and amplitude attenuation occur during ultrasound propagation. Therefore, the Ultrasonic Reflection Tomography (URT) and Ultrasonic Transmission Tomography (ITT) can be combined with ECT and ERT (Jiang, Soleimani, & Wang, 2019; Pusppanathan et al., 2017). For the dual mode UTT/URT, the data can be collected with one set of transducer array and fused to improve accuracy of gas–liquid two-phase flow measurement (Tan et al., 2019).

The dual-modality data fusion strategy and joint image reconstruction algorithm are also important to the visualization effect. Generally, data fusion can be conducted at three levels: (1) pixel by pixel fusion, (2) image feature fusion, (3) Combination of preclassified data derived from each modality. In pixel-by-pixel image fusion, common solutions are adding threshold and normalization to the images reconstructed by both modalities, and then fuse the image pixel by pixel (Hjertaker et al., 2011). In image feature fusion, there are propositions and implementations like fuzzy fusion and self-metric–based fusion, suitable for different combinations of tomographic modality. In the combination of preclassified data, the extracted feature or reconstructed image of one modality is normally used as input or prior information of another modality as shown in Fig. 19.11 (Zhang et al., 2015). Specifically, the prior information usually comes from hard field modality; thus, the extracted feature is more easily accessed through simple linearization of hard field.

19.4.2 IPTs combined with other sensors

The fundamental purpose of multimodality IPTs is to increase information source to recognize fluids that cannot be seen by single-modality IPT. However, in flow metering applications, the interphase slippage in multiphase transportation causes big difference on flow rate, which will cause instability of flow structure and large metering error. The interphase slip can only be modeled and compensated by acquiring each phase's real-time fraction. IPT is good at measuring phase fraction, while extra flow meters are needed to estimate the flow rate. These kinds of combinations will provide rich information regarding the flow conditions to facilitate the process diagnosis and modeling. One example is the combination of ERT with EFMs to obtain the mixture flow rate (Faraj et al., 2015), in which ERT provides phase distribution and flow pattern, while EFM measures the flow velocity. The combination of ERT with a Venturi meter was presented for individual flow rate estimation of two-phase flow; where ERT provides flow pattern and void fraction, Venturi provides overall flow rate of two-phase flow (Meng, Huang, Wang, Ji, Li, & Yan, 2010). The combination of tomography with other sensors will provide sufficient space for investigation and improvement of data fusion and thus comprehensive measurement of multiphase flow.

The tendency of multimodality IPTs and IPT with extra sensor in multiphase transportation focuses on proposing fusion modalities and integrated sensor configuration with more complementary information, more cooperative workflow, and wider

Position number	Modality	DI water & oil	3%saline solution & oil	5%saline solution & oil
1 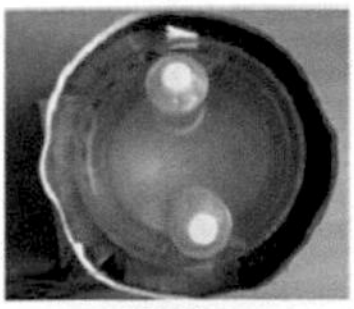 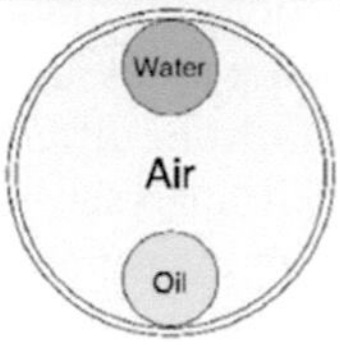	ECT	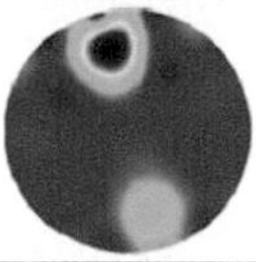	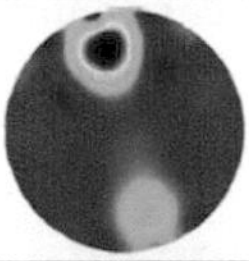	
	MIT-ECT	N/A		
2 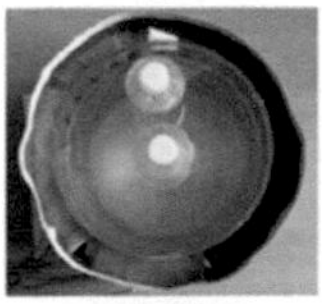 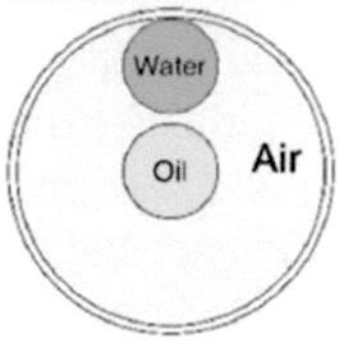	ECT	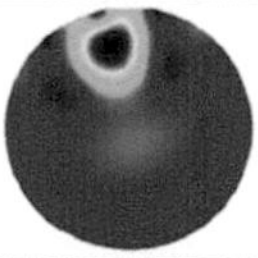	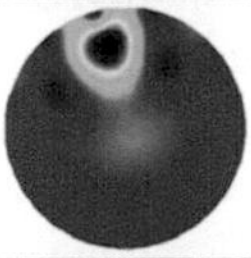	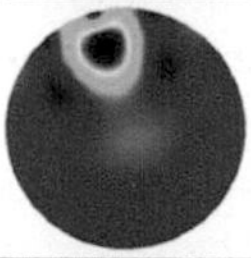
	MIT-ECT	N/A	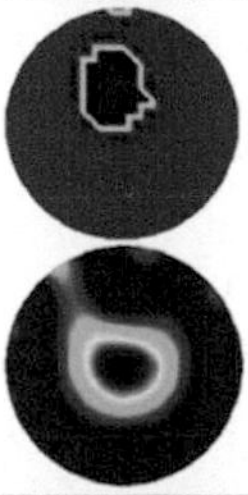	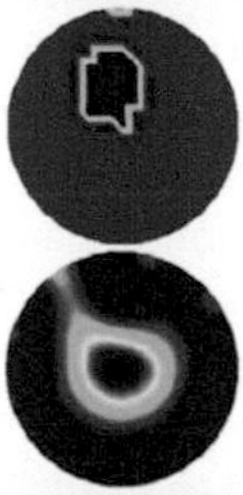
3  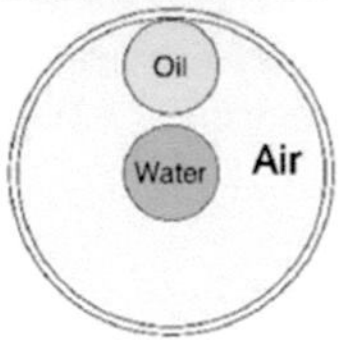	ECT	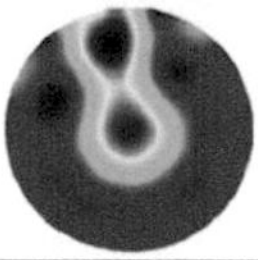	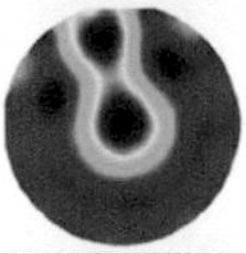	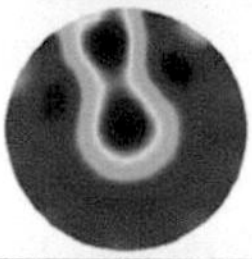
	MIT-ECT	N/A		

Figure 19.11 MIT-ECT dual modality image reconstruction of three-phase distribution (Zhang et al., 2015).

sensing range. The dual-modality fusion strategy will be evolved to a higher level with information decomposition and feature characterization capability, as well as reliability under complex environment and serious signal interference.

19.5 Conclusions and future trends

This chapter reviews the applications of IPT on multiphase transportation, specifically focusing on slurry flow and oil–gas–water multiphase flow. The flow pattern identification methods and flow process parameter measurement with IPTs from different disciplines are discussed. The issue of flow process monitoring and analysis with IPTs is discussed. At last, the multimodality IPTs for multiphase flow measurement and analysis is introduced, and the sensor fusion of IPTs with other flow meters is also investigated with reported examples. IPTs are not only capable of providing visual observations on the internal phase distribution of multiphase flow, but also providing rich information regarding flow process and parameter modeling through the multiple sensors installed on the target vessel.

The future trend of applying IPTs in multiphase transportation will be how to fuse information from multimodality IPTs and with information provided by other flow meters. For the multiphase transportation monitoring in harsh environment, "clamp-on" IPTs will attract more applications due to their noninvasive nature. Besides, the phase distribution and velocity distribution are both very important to the multiphase transportation efficiency, which should be better reconstructed with one set of sensor arrays rather than two sets with data fusion or cross-correlations to form a compact and flexible sensor. Sensing modalities such as broadband and multimode UT are very promising to fit such applications, but they still need intensive research and development to meet such demands.

Acknowledgment

The authors would like to thank the support by the National Natural Science Foundation of China (61903271; 51976137; 61903271), the National Key Research and Development Program of China (Grant No. 2019YFB1504702), and the Natural Science Foundation of Tianjin (No. 19JCZDJC38900).

References

Acikgoz, M., Franca, F., & Lahey, R. T., Jr. (1992). An experimental study of three-phase flow regimes. *International Journal of Multiphase Flow, 18*, 327–336.

Ayob, N. M. N., Rahiman, M. H. F., Zakaria, Z., Yaacob, S., & Rahim, R. A. (2011). Dual-plane ultrasonic tomography simulation using cross-correlation technique for velocity measurement in two-phase liquid/gas flow. *International Journal of Electrical and Electronic Systems Research, 4*, 46–52.

Baldwin, B. A., Stevens, J., Howard, J. J., Graue, A., Kvamme, B., Aspenes, E., et al. (2009). Using magnetic resonance imaging to monitor CH_4 hydrate formation and spontaneous conversion of CH_4 hydrate to CO_2 hydrate in porous media. *Magnetic Resonance in Imaging, 27*, 720–726.

Barthel, F., Bieberle, M., Hoppe, D., Banowski, M., & Hampel, U. (2015). Velocity measurement for two-phase flows based on ultrafast X-ray tomography. *Flow Measurement and Instrumentation, 46*, 196–203.

Beck, M. S. (1981). Correlation in instruments: Cross correlation flowmeters. *Journal of Physics E: Scientific Instruments, 14*, 7–19.

Deng, X., Dong, F., Xu, L. J., Liu, X. P., & Xu, L. A. (2001). The design of a dual-plane ERT system for cross correlation measurement of bubbly gas-liquid pipe flow. *Measurement Science and Technology, 12*, 1024–1031.

Dong, F., Jiang, Z. X., Qiao, X. T., & Xu, L. A. (2003). Application of electrical resistance tomography to two-phase pipe flow parameters measurement. *Flow Measurement and Instrumentation, 14*, 183–192.

Faraj, Y., Wang, M., Jia, J., Wang, Q., Xie, C.-G., Oddie, G., et al. (2015). Measurement of vertical oil-in-water two-phase flow using dual-modality ERT–EMF system. *Flow Measurement and Instrumentation, 46*, 255–261.

Gamio, J. C., Castro, J., Rivera, L., Alamilla, J., Garcia-Nocetti, F., & Aguilar, L. (2005). Visualisation of gas–oil two-phase flows in pressurised pipes using electrical capacitance tomography. *Flow Measurement and Instrumentation, 16*, 129–134.

George, D. L., Shollenberger, K. A., Torczynski, J. R., O'hern, T. J., & Ceccio, S. L. (2001). Three-phase material distribution measurements in a vertical flow using gamma-densitometry tomography and electrical-impedance tomography. *International Journal of Multiphase Flow, 27*, 1903–1930.

Giguère, R., Fradette, L., Mignon, D., & Tanguy, P. A. (2008). Characterization of slurry flow regime transitions by ERT. *Chemical Engineering Research and Design, 86*, 989–996.

Giguère, R., Fradette, L., Mignon, D., & Tanguy, P. A. (2009). Analysis of slurry flow regimes downstream of a pipe bend. *Chemical Engineering Research and Design, 87*, 943–950.

Goh, C. L., Ruzairi, A. R., Hafiz, F. R., & Tee, Z. C. (2017). Ultrasonic tomography system for flow monitoring: A review. *IEEE Sensors Journal, 17*, 5382–5390.

Hjertaker, B. T., Maad, R., & Johansen, G. A. (2011). Dual-mode capacitance and gamma-ray tomography using the Landweber reconstruction algorithm. *Measurement Science and Technology, 22*(10), Article 104002. https://doi.org/10.1088/0957-0233/22/10/104002

Hjertaker, B. T., Tjugum, S. A., Hallanger, A., & Maad, R. (2018). Characterization of multiphase flow blind-T mixing using high speed gamma-ray tomometry. *Flow Measurement and Instrumentation, 62*, 205–212.

Hu, B., Langsholt, M., Liu, L., Andersson, P., & Lawrence, C. (2014). Flow structure and phase distribution in stratified and slug flows measured by X-ray tomography. *International Journal of Multiphase Flow, 67*, 162–179.

Ismail, I., Gamio, J. C., Bukhari, S. F. A., & Yang, W. Q. (2005). Tomography for multi-phase flow measurement in the oil industry. *Flow Measurement and Instrumentation, 16*, 145–155.

Jiang, Y., Soleimani, M., & Wang, B. (2019). Contactless electrical impedance and ultrasonic tomography: Correlation, comparison and complementarity study. *Measurement Science and Technology, 30*(11), Article 114001. https://doi.org/10.1088/1361-6501/ab2292

Kim, B. S., Khambampati, A. K., Hong, Y. J., Kim, S., & Kim, K. Y. (2013). Multiphase flow imaging using an adaptive multi-threshold technique in electrical resistance tomography. *Flow Measurement and Instrumentation, 31*, 25–34.

Kollár, L. E., Lucas, G. P., & Zhang, Z. (2014). Proposed method for reconstructing velocity profiles using a multi-electrode electromagnetic flow meter. *Measurement Science and Technology, 25*(7), Article 075301. https://doi.org/10.1088/0957-0233/25/7/075301

Kotze, R., Adler, A., Sutherland, A., & Deba, C. N. (2019). Evaluation of Electrical Resistance Tomography imaging algorithms to monitor settling slurry pipe flow. *Flow Measurement and Instrumentation, 68*, Article 101572. https://doi.org/10.1016/j.flowmeasinst.2019.101572

Lau, Y. M., Hampel, U., & Schubert, M. (2018). Ultrafast X-ray tomographic imaging of multiphase flow in bubble columns - Part 1: Image processing and reconstruction comparison. *International Journal of Multiphase Flow, 104*, 258–271.

Leeungculsatien, T., & Lucas, G. P. (2013). Measurement of velocity profiles in multiphase flow using a multi-electrode electromagnetic flow meter. *Flow Measurement and Instrumentation, 31*, 86–95.

Lehtikangas, O., & Vauhkonen, M. (2017). Optimal coil currents in electromagnetic flow tomography. *IEEE Sensors Journal, 17*, 8137–8145.

Li, Y., Yang, W., Xie, C.-G., Huang, S., Wu, Z., Tsamakis, D., et al. (2013). Gas/oil/water flow measurement by electrical capacitance tomography. *Measurement Science and Technology, 24*.

Liu, H., Tan, C., & Dong, F. (2019). Continuous-wave ultrasonic tomography for oil/water two-phase flow imaging using regularized weighted least square framework. *Transactions of the Institute of Measurement and Control, 42*(4), 666–679. https://doi.org/10.1177/0142331219853073

Lucas, G. P., Cory, J., Waterfall, R. C., Loh, W. W., & Dickin, F. J. (1999). Measurement of the solids volume fraction and velocity distributions in solids–liquid flows using dual-plane electrical resistance tomography. *Flow Measurement and Instrumentation, 10*, 249–258.

Ma, L., Spagnul, S., & Soleimani, M. (2017). Metal solidification imaging process by magnetic induction tomography. *Scientific Reports, 7*, 14502.

Meng, Z., Huang, Z., Wang, B., Ji, H., Li, H., & Yan, Y. (2010). Air-water two-phase flow measurement using a Venturi meter and an electrical resistance tomography sensor. *Flow Measurement and Instrumentation, 21*, 268–276.

Meribout, M. (2019). Optimal design for a portable NMR- and MRI-based multiphase flow meter. *IEEE Transactions on Industrial Electronics, 66*, 6354–6361.

Muir, C. E., Petrov, O. V., Romanenko, K. V., & Balcom, B. J. (2014). Measuring miscible fluid displacement in porous media with magnetic resonance imaging. *Water Resources Research, 50*, 1859–1868.

Okamoto, I., Hirai, S., & Ogawa, K. (2001). MRI velocity measurements of water flow in porous media containing a stagnant immiscible liquid. *Measurement Science and Technology, 12*, 1465–1472.

Perera, K., Pradeep, C., Mylvaganam, S., & Time, R. W. (2017). Imaging of oil-water flow patterns by electrical capacitance tomography. *Flow Measurement and Instrumentation, 56*, 23–34.

Plaskowski, A., Beck, M. S., & Krawaczynski, J. S. (1987). Flow imaging for multi-component flow measurement. *Transactions of the Institute of Measurement and Control, 9*, 108–112.

Pusppanathan, J., Rahim, R. A., Phang, F. A., Mohamad, E. J., Ayob, N. M. N., Rahiman, M. H. F., et al. (2017). Single-plane dual-modality tomography for multiphase flow imaging by integrating electrical capacitance and ultrasonic sensors. *IEEE Sensors Journal, 17*, 6368–6377.

Razzak, S. A., Barghi, S., & Zhu, J. X. (2009). Application of electrical resistance tomography on liquid–solid two-phase flow characterization in an LSCFB riser. *Chemical Engineering Science, 64*, 2851–2858.

Roostaei, M., Nouri, A., Hosseini, S. A., Soroush, M., Velayati, A., Mahmoudi, M., et al. (2020). A concise review of experimental works on proppant transport and slurry flow. In *SPE international conference and exhibition on formation damage control*. Lafayette, Louisiana, USA: Society of Petroleum Engineers.

Saoud, A., Mosorov, V., & Grudzien, K. (2017). Measurement of velocity of gas/solid swirl flow using Electrical Capacitance Tomography and cross correlation technique. *Flow Measurement and Instrumentation, 53*, 133–140.

Schäfer, T., Meitzner, C., Lange, R., & Hampel, U. (2016). A study of two-phase flow in monoliths using ultrafast single-slice X-ray computed tomography. *International Journal of Multiphase Flow, 86*, 56–66.

Sharifi, M., & Young, B. (2013). Electrical resistance tomography (ERT) applications to chemical engineering. *Chemical Engineering Research and Design, 91*, 1625–1645.

Spedding, P. L., Donnelly, G. F., & Cole, J. S. (2005). Three phase oil-water-gas horizontal Co-current flow. *Chemical Engineering Research and Design, 83*, 401–411.

Stevenson, R., Harrison, S. T. L., Miles, N., & Cilliers, J. J. (2006). Examination of swirling flow using electrical resistance tomography. *Powder Technology, 162*, 157–165.

Tan, C., Dong, F., & Wu, M. M. (2007). Identification of gas/liquid two-phase flow regime through ERT-based measurement and feature extraction. *Flow Measurement and Instrumentation, 18*, 255–261.

Tan, C., Li, X., Liu, H., & Dong, F. (2019a). An ultrasonic transmission/reflection tomography system for industrial multiphase flow imaging. *IEEE Transactions on Industrial Electronics, Online, 66*(12), 9539–9548.

Tan, C., Shen, Y., Smith, K., Dong, F., & Escudero, J. (2019b). Gas-liquid flow pattern analysis based on graph connectivity and graph-variate dynamic connectivity of ERT. *IEEE Transactions on Instrumentation and Measurement, 68*, 1590–1601.

Tan, C., Zhao, J., & Dong, F. (2015). Gas-water two-phase flow characterization with electrical resistance tomography and multivariate multiscale entropy analysis. *ISA Transactions, 55*, 241–249.

Thorn, R., Johansen, G. A., & Hjertaker, B. T. (2013). Three-phase flow measurement in the petroleum industry. *Measurement Science and Technology, 24*, 012003.

Vauhkonen, M., Hänninen, A., & Lehtikangas, O. (2018). A measurement device for electromagnetic flow tomography. *Measurement Science and Technology, 29*(1), Article 015401. https://doi.org/10.1088/1361-6501/aa91dd

Viggiano, B., SKJæraasen, O., SCHümann, H., Tutkun, M., & Cal, R. B. (2018). Characterization of flow dynamics and reduced-order description of experimental two-phase pipe flow. *International Journal of Multiphase Flow, 105*, 91–101.

Wang, M., Dorward, A., Vlaev, D., & Mann, R. (2000). Measurements of gas-liquid mixing in a stirred vessel using electrical resistance tomography (ERT). *Chemical Engineering Journal, 77*, 93–98.

Wang, M., Jones, T. F., & Williams, R. A. (2003). Visualization of asymmetric solids distribution in horizontal swirling flows using electrical resistance tomography. *Chemical Engineering Research and Design, 81*, 854–861.

Wang, M., Lucas, G., Dai, Y., Panayotopoulos, N., & Williams, R. A. (2006). Visualisation of bubbly velocity distribution in a swirling flow using electrical resistance tomography. *Particle & Particle Systems Characterization, 23*, 321–329.

Wang, Q., Polansky, J., Wang, M., Wei, K., Qiu, C., Kenbar, A., et al. (2018). Capability of dual-modality electrical tomography for gas-oil-water three-phase pipeline flow visualisation. *Flow Measurement and Instrumentation, 62*, 152–166.

Wang, P., Wang, H., Sun, B., Cui, Z., & Huang, W. (2014). An ECT/ERT dual-modality sensor for oil-water two-phase flow measurement. *Review of Scientific Instruments, 82*, 124701.

Wang, H., & Yang, W. (2020). Application of electrical capacitance tomography in circulating fluidised beds – a review. *Applied Thermal Engineering, 176.*

Watson, S., Williams, R. J., Gough, W., & Griffiths, H. (2008). A magnetic induction tomography system for samples with conductivities below 10 S m−1. *Measurement Science and Technology, 19*, 045501.

Xu, L., & Xu, L. (1997). Gas-liquid two-phase flow regime identification by ultrasonic tomography. *Flow Measurement and Instrumentation, 8*, 145–155.

Yao, J., & Takei, M. (2017). Application of process tomography to multiphase flow measurement in industrial and biomedical fields: A review. *IEEE Sensors Journal, 17*, 8196–8205.

Zhang, M., Ma, L., & Soleimani, M. (2015). Dual modality ECT-MIT multi-phase flow imaging. *Flow Measurement and Instrumentation, 46*, 240–254.

Zhou, B., & Zhang, J. Y. (2013). A novel ECT–EST combined method for gas–solids flow pattern and charge distribution visualization. *Measurement Science and Technology, 24*(7), Article 074003. https://doi.org/10.1088/0957-0233/24/7/074003

Measurement and characterization of slurry flow using Electrical Resistance Tomography

Yousef Faraj
Department of Chemical Engineering, Faculty of Science and Engineering, University of Chester, Chester, United Kingdom

20.1 Introduction

The presence of solid particles in a carrier liquid forms a complex mixture, which is usually conceived of as slurry. The carrier liquid may be water, Newtonian, or non-Newtonian liquid. However, water is the most commonly used carrier liquid in slurry transportation, which is referred to as "hydraulic transportation." Slurry flows cover a wide spectrum of applications and are the focus of considerable interest in engineering research. It is widely utilized in long distance commodity pipelines, as well as short and medium in plant pipelines such as those in nuclear power plant, chemical, pharmaceutical, and food industry. It has been a progressive technology for transporting a vast amount of different solid materials such as sands, iron concentrates, copper concentrates, phosphate matrix, tailings, limestone and sewage, in different densities, shapes, sizes up to 150 mm through various sizes of pipelines with different orientations (Abulnaga, 2002). In the chemical and allied industries dealing with processing of particulate solid materials, it is quite difficult to find a plant that does not at some stage involve the transport of solid materials in a carrier liquid.

There are two broad classifications for slurry in hydraulic transportation, which are referred to as settling slurry and nonsettling slurry. A settling slurry is usually composed of relatively large solid particles, typically larger than 40 μm, and/or heavy solid particles in a low viscosity carrier liquid. In the absence of sufficient fluid turbulence, the solid particles tend to rapidly settle to the bottom of the transporting pipe, where the particles can remain as a deposit or move along the bottom of the pipe as a sliding bed (Gillies, 1991; Matousek, 2005). Nonsettling or homogeneous slurries, on the other hand, are composed of particles of colloidal dimensions, which are characterized by particle diameters of typically less than 2 μm. They are highly concentrated and maintained in suspension by molecular movement (Brownian motion) within the carrier liquid (Peker, Helvacı, Yener, İkizler, & Alparslan, 2008). Nonsettling slurries are beyond the scope of this chapter; therefore, no further reference will be made to such type of slurry.

Industrial Tomography. https://doi.org/10.1016/B978-0-12-823015-2.00006-6

Industrial slurry flows are influenced by a large number and range of flow variables such as pipeline configuration (horizontal, vertical, and inclined), solid particle size, solids shape, solids density, density of carrier liquid, viscosity of carrier liquid, mean transport velocity, pipe diameter, and flow direction (ascending or descending flow). Therefore, the behavior of slurry flow in pipeline has been systematically investigated from the early work of Blatch (1906) to Howard (1939), Durand (1952), Durand and Condolios (1953), Newitt, Richardson, Abbott, and Turtle (1955) and many others. Flow of settling slurry in horizontal pipeline is enormously complex and the vast majority of literature is devoted to this type of flow orientation. The complexity of this type of flow is due to the influence of gravity, which acts at right angle to the flow and gives rise to various flow patterns (regimes) from pseudohomogenous at high transport velocity to stationary bed at low transport velocity. The emergence of these flow patterns affects the pressure drop and influences pipe wear and other performance characteristics (Brown, 1991; Doron & Barnea, 1993; Wilson & Pugh, 1988). These flow patterns are also encountered in inclined flow with occurrence of backflow (Doron, Simkhis, & Barnea, 1997; Matousek, 2002; Yamaguchi, Niu, Nagaoka, & de Vuyst, 2011). On the other hand, in vertical slurry flow, the flow conditions are straightforward with a homogeneous distribution of solid particles across the pipe cross-section. However, in vertical slurry flows, the gravity acts counter to the dynamic forces and the slippage of the liquid and solid phase occurs (Parvareh, Rahimi, Alizadehdakhel, & Alsairafi, 2010; Shook and Bartosik, 1994). Arguably, most of pipe configurations are horizontal and vertical, inclined flows are unavoidable for certain long-distance overland pipeline, thickener feed systems, pump box feed systems, and so on (Abulnaga, 2002).

The occurrence of separation and slippage of the constituent phases in settling solid–liquid flow in pipelines makes the flow unpredictable and time dependent. Therefore, it is paramount for the operator of these pipelines to monitor and measure the flow continuously, particularly from the local point of view, so as to ensure safe transport and maintaining acceptable control limits. In order to understand the internal structure of such flows, solids volume fraction distribution and solids velocity distribution are of great importance.

The contemporary approach for characterization of slurry flow entails the rapid development of measuring techniques and substantial computational effort, which is a viable approach to describe the complex behavior and nature of settling slurry flow in pipelines (Lahiri, 2009; Matousek, 2005). In the past, several intrusive methods, such as traditional probes, have been used to measure solids volumetric concentration and solids velocity (MacTaggart et al., 1993; Nasr-El-Din, Shook, & Colwell, 1987). The disadvantages of using these probes for solid–liquid flow measurement have been reported (Brown & Heywood, 1991). It is highly unlikely that these devices can survive the harsh condition inside the pipelines due to abrasive nature of slurry. In many cases solids may accumulate around them and cause pipe blockage. It is well known that intrusive devices introduce an undesirable physical disturbance and alter the internal structure of the flow (Heindel, Grayb, & Jensenb, 2008). In order to overcome this limitation, efforts have been made to develop a variety of nonintrusive measurement techniques to interrogate the internal structure of two or multiphase flows, such as optical, electrical, nuclear, and so on. Since slurries are opaque and flow through opaque enclosures, using optical techniques can be extremely difficult if not impossible. Although nuclear techniques provide an accurate measurement, they are highly expensive and suffer from low temporal resolution and environmental issues (Thorn, Johansen, & Hammer, 1997). Among nonintrusive techniques, Electrical Resistance Tomography (ERT) has attracted the interest of many researchers. This is due to the advantages that ERT offers, such as nonintrusive, relatively low cost, no environmental restrictions, providing quantitative and qualitative on-line measurement, fast, and so on. Within the last 2 decades, ERT has seen a significant development and been applied to many industrial processes involving two/multiphase flow systems (Dickin & Wang, 1996; Fangary, Williams, Neil, Bond, & Faulks, 1998; Faraj et al., 2015; Gumery, Ein-Mozaffari, & Dahman, 2011; Jia, Wang, Faraj, & Wang, 2015b; Vijayan, Schlaberg, & Wang, 2007; Wang et al. 2015; Zhao & Lucas, 2011). Particularly, the application of ERT to slurry flow has been reported by a number of investigators (Faraj, Wang, & Jia, 2013, pp. 869–877; Giguere, Fradette, Mignon, & Tanguy, 2008b; Lucas, Cory, Waterfall, Loh, & Dickin, 1999; Marefatallah, Breakey, & Sanders, 2019; Pachowko, Poole, Wang, & Rhodes, 2004; Qureshi et al., 2021; Razzak, Barghi, & Zhu, 2009; Silva, Faia, Garcia, & Rasteiro, 2016; Wang, Jones, & Williams, 2003; Wood & Jones, 2003), and it has been proven to be a promising technique for measurement and visualization of solid–liquid flow in a pipe.

This chapter provides an overview of hydraulic transport and ERT for slurry flow visualization and characterization, while the main emphasis is placed upon settling slurry flow. The physical mechanisms governing solid–liquid flow in a pipe are also described, along with the four typical settling slurry flow regimes in a pipeline. An evaluation of ERT for slurry flow visualization and characterization is presented through reconstructed tomograms and both solids volume fraction distribution and solids axial velocity distribution, with the main focus on stratified flow.

20.2 Physical mechanisms governing hydraulic transport of solid particles

When solid particles in a carrier liquid are transported within a pipe, they are acted upon by several forces, which are resulted from particle–particle interaction, particle–liquid interaction, and particle–pipe wall interaction. The forces, which are resulted from particle–particle interaction, are transmitted as an interparticle stress. However, when a granular bed is formed (i.e., the particles are in continuous contact), the Coulombic stress is created on the particles, while Bagnold stress is created on the surface of the granular bed as a result of sheared granular bed and the particles move in a sporadic fashion. On the other hand, the forces, which are resulted from particle–liquid interaction are buoyancy force, drag force, and lift force, and if the carrier liquid is turbulent, then the turbulent diffusive force is generated (Bagnold, 1956; Matousek, 2002; Wilson, Addie, Sellgren, & Clift, 2006).

A settling slurry flow, in a horizontal or inclined pipe, normally tends to stratify. In other words, at the velocity used during the practical operations, the particle distribution across a pipe cross-section is nonuniform. Under these circumstances, the flow can be fully or partially stratified. If all solid particles form the granular bed (stationary or sliding), the flow is referred to as fully stratified flow, and if a portion of the solid particles form the granular bed (where particles are virtually in permanent contact with each other) and the remaining particles are suspended within the carrier fluid, the flow is referred to as partially stratified.

The description of particle motion was explained by Bagnold (1956), who developed a concept that the solid particles are supported by two major physical mechanisms: fluid suspension and intergranular contact. For the first mechanism, Wilson, Clift, and Sellgren. (2002) observed that the particles that are suspended in the carrier fluid are due to turbulent diffusion. Obviously, industrial settling slurries are composed of fine and coarse particles. The friction behavior of suspended coarse particles differs from that of fine particles. The suspended coarse particles interact with each other and the pipe wall. It is important to mention that the contact is not permanent, but rather sporadic, which is due to turbulent dispersive action and collision dispersive action. In other words, the solid particles are dispersed to all directions by turbulent eddies and also the particles are collided with the other particles of different velocities and impelled in the direction of the wall (Matousek, 2005; Wilson et al., 2006). The fluid suspension mechanism can come to play, when the velocity of turbulent eddies is greater than the settling velocity and by further increasing the mean velocity the more solid particles are suspended by the fluid force (Brown & Heywood, 1991).

In the case of fully stratified flow, the settling velocity of the solid particles is greater than the velocity of the turbulent eddies; therefore, the particles settle and are supported by granular contact rather than fluid suspension. These particles are designated as contact load, in either case stationary or moving bed. The motion of the particles within the contact load can be analyzed by applying the force balance, which was first used by Wilson (1970) and Wilson, Streat and Bantin. (1972), then it was latter used to determine the limit deposition velocity for fully stratified flow.

20.2.1 Slurry flow pattern

Slurry flow patterns are usually described in terms of the distribution of solid particles within the pipe cross-section. In horizontal and inclined flow, the influence of gravity gives rise to several flow patterns, which have been the subject of numerous experimental studies. The identification of these flow patterns is usually based on visual observation of slurry flow, which sometimes can be extremely difficult due to either the system being opaque or high transportation velocity. This implies that they are identified by subjective operator judgment; as a result, they have been given different names and definitions in the literature (Brown, 1991; Doron et al., 1997; Durand & Condolios, 1952, pp. 29–55; Giguere, Fradette, Mignon, & Tanguy, 2008a; Govier & Aziz, 1972; Lazarus & Neilson, 1978; Matousek, 2005, 2009; Newitt et al., 1955; Parvareh et al., 2010; Parzonka, Kenchington, & Charles, 1981; Pashowko, 2004). However, as a refined classification, based on solids concentration profile

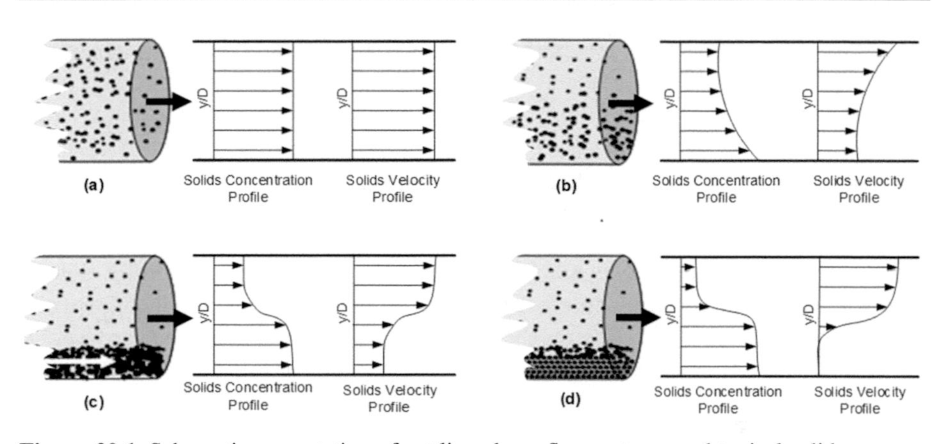

Figure 20.1 Schematic presentation of settling slurry flow patterns and typical solids concentration and velocity profiles: (a) Homogeneous; (b) Heterogeneous; (c) Moving bed; and (d) Stationary bed.
Credit: Drawn by the author.

within the pipe cross-section, there are typically four settling slurry flow patterns: Homogeneous/pseudohomogeneous flow (with all solid particles in suspension), Heterogeneous flow (with all solid particles in suspension), flow with a moving bed (with or without solids in suspension), and flow with stationary bed. The schematic presentation of solid particle distribution within the vertical plane of a horizontal pipe, along with typical solids concentration profile and typical solids axial velocity profile, is shown in Fig. 20.1.

Homogeneous flow is considered as fully suspended flow regime, usually at high velocities well above that used commercially for such slurry (Brown & Heywood, 1991). This type of flow is typical of the fine particles, which are all suspended in the carrier fluid and the slurry property approaches that of a single fluid. It is important to note that the true homogeneous flow rarely occurs in settling slurries. Therefore, several investigators such as Doron and Barnea (1993) and Govier and Aziz (1972) have used different terms for this type of flow pattern such as pseudohomogeneous and symmetric flow pattern. It is worth mentioning that homogeneous differs from pseudohomogeneous flow pattern. When the solid particles are distributed evenly across the pipe cross-section, the flow is homogeneous, whereas when the solid particles approach even distribution, the flow is referred to as pseudohomogeneous or quasihomogeneous. The true homogeneous flow can easily occur in nonsettling slurries, where the solid particles are equally distributed along the vertical plane of the pipe cross-section. In homogeneous (or pseudohomogeneous) flow pattern, solids concentration gradient and solids velocity gradient across the pipe cross-section is absent. As the flow acts as a single phase and generally observed as a vertical line. Therefore, Newitt et al. (1955) used the equivalent fluid model for this type of flow. The schematic presentation of this type of flow regime, along with typical solids volumetric concentration profile and solids axial velocity profile, is shown in Fig. 20.1a.

Heterogeneous flow pattern is the most complex type of flow, as the solid particles are not evenly mixed and a gradient of solids concentration and solids axial velocity exist, as presented in Fig. 20.1B. Heterogeneous flows are encountered in many industries such as mining and dredging applications. Some particles are suspended and supported by fluid turbulence at the top half of the pipe, particularly the fine ones, and the coarse particles are suspended at the bottom half of the horizontal pipe. This phenomenon is due to the occurrence of minimum hydraulic gradient, as the transport velocity is reduced (Brown & Heywood, 1991). In heterogeneous flow, the deposition velocity depends on the particle size, particle density, solids concentration, and the pipe diameter (Shook & Roco, 1991). Heterogeneous flows require a minimum carrier velocity, which is referred to as the critical velocity (Wilson et al., 2006).

In moving bed flow pattern, when the transport velocity is low and a large number of coarse particles exist, the larger particles tend to accumulate at the bottom of the pipe and form a packed layer. This accumulated packed layer, which is referred to as bed, moves along the bottom of the pipe like desert sand dunes. The upper layers of the bed move faster than the lower layers, which is due to the difference in sizes and settling velocities of solid particles forming the bed. The concentration of bed corresponds to maximum, while the upper part of the pipe still contains a heterogeneous mixture, as illustrated in Fig. 20.1C. Many terms and definitions for this type of flow pattern have been introduced in literature such as saltation flow, which means that the fluid above the bed tends to move the finer solid particles by entrainment (Turian & Yuan, 1977).

As the transport velocity drops even further, the required suspension force reduces and the bed thickens; as a result, the lower layer of particles that is in contact with the pipe becomes stationary, as shown in Fig. 20.1D. However, the fluid above the bed tends to move the finer solid particles by entrainment, in which the finer solid particles occupying the upper part of the bed tend to roll and tumble. Flow with saltation and asymmetric suspension occurs just above the velocity of blockage. Therefore, further reduction in the transport velocity increases the pressure drop quite high that is impossible to maintain the flow and consequently the pipe blocks (Matousek, 2005; Wilson et al., 2006).

20.3 Slurry flow characterization with Electrical Resistance Tomography

ERT is used to provide a novel means of nonintrusively interrogating the internal structure of slurry flow in a closed system, such as pipe, without physically looking inside (Seager, Baber, & Brown, 1987; Williams, 1995). Among all the electrical tomography forms (Electrical Impedance Tomography-EIT, Electrical Capacitance Tomography-ECT, Electrical Resistance Tomography-ERT, and Electromagnetic Tomography-EMT), ERT has been found to be the most attractive method to visualize the mechanisms of slurry flow systems and measure its parameters. The ERT's sensing technique provides an excellent time resolution due to fast electrical measurements,

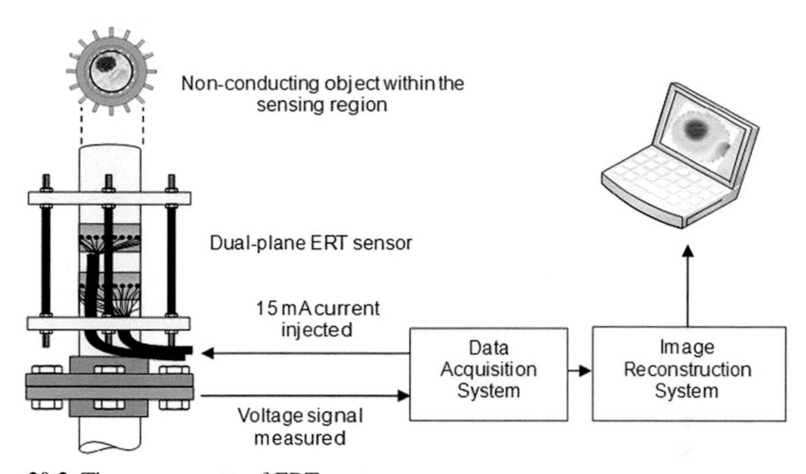

Figure 20.2 The components of ERT system.
From Faraj, Y., Wang, M. (2012). ERT investigation on horizontal and counter-gravity slurry flow in pipelines. *Procedia Engineering*, *42*: 588–606.

which makes it suitable for measurement of dynamic and highly unstable industrial slurry flows (Mann & Wang, 1997; Wang, 2005; Wang & Yin, 2001; Williams et al., 1999).

Fig. 20.2 presents the components of ERT system; the detailed description of each component can be found elsewhere in this book. The working principle of ERT system is to image mixtures and generate color-coded cross-sectional images of the constituent phases (tomogram) inside the sensor, where the continuous phase is conductive and the dispersed phase is insulating or less conductive, such as sand–water mixture.

Among all the data acquisition strategies (adjacent electrode pair strategy, opposite strategy, diagonal or cross strategy, and conducting boundary strategy), the adjacent electrode pair strategy is the most widely used strategy for slurry flow measurement, as it requires less hardware and provides fast image reconstruction (Hosseini, Patel, Ein-Mozaffari, & Mehrvar, 2010; Mann & Wang, 1997; Tapp, Peyton, Kemsley, & Wilson, 2003). Fig. 20.3 presents the adjacent strategy applied on a 16-electrode ERT sensor. Briefly, a low electrical current is injected, typically 15 mA, between adjacent pairs of neighboring boundary of electrodes and measuring the potential difference between the remaining electrodes. This procedure is repeated for the other electrode pairs until the full rotation of the electrical field is completed; then a set of measurements is formed.

The tomograms computed by an image reconstruction algorithm provide valuable information on the process, which can be used for flow monitoring and characterization, intelligent control, or mathematical model verification (Ismail, Gamio, Bukhari, & Yang, 2005). In the case of slurry flow, ERT has been proven to be a powerful tool for online measurement of solids volume fraction and solids velocity distribution, where electrical conductivity difference between the solid phase and liquid phase exists.

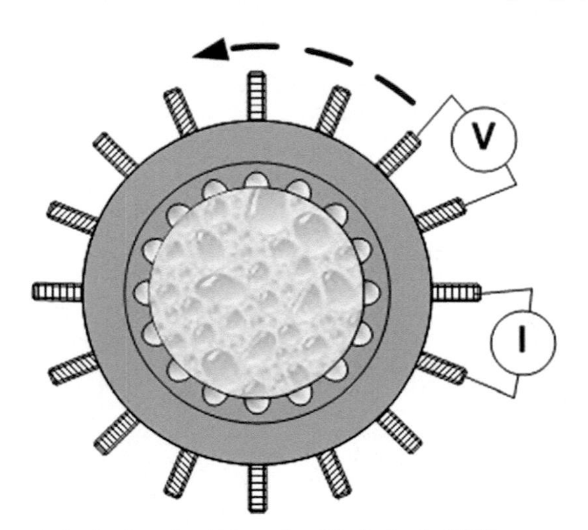

Figure 20.3 Adjacent electrode pair strategy for 16-electrode ERT sensor.
Credit: Drawn by the author.

20.3.1 Solids volume fraction measurement

Since the ERT is able to detect the local changes in the electrical conductivity of a domain, it can be used to characterize the flow of a slurry mixture, where the solid particles have different conductivity from the aqueous continuous phase. This implies that the conductivity of a slurry mixture is dependent on the individual conductivity of each constituent phase. However, ERT can map only the conductivity distribution of a mixture within the pipe cross-section. If solids volume fraction is required, then there must be a relationship to convert the measured conductivity distribution to solids volume fraction.

A number of correlations have been proposed for conversion of conductivity to solids volume fraction of the dispersed phase. In terms of slurry, the most studied correlations are those of Maxwell (1881); Bruggeman (1935); Meredith and Tobias (1962); Prager (1963); and Weissberg (1963), which are shown in Eqs. (20.1) to (20.5).

$$\text{Maxwell}\quad \alpha_c = \frac{2\sigma_1 + \sigma_2 - 2\sigma_m - \left(\dfrac{\sigma_m \sigma_2}{\sigma_1}\right)}{\sigma_m - \left(\dfrac{\sigma_2}{\sigma_1}\right)\sigma_m + 2(\sigma_1 - \sigma_2)} \tag{20.1}$$

$$\text{Bruggeman}\quad \sigma_m = \sigma_1 (1 - \alpha_c)^{1.5} \tag{20.2}$$

$$\text{Meredith and Tobias}\quad \sigma_m = \sigma_1 \left[\frac{8(2 - \alpha_c)(1 - \alpha_c)}{(4 + \alpha_c)(4 - \alpha_c)}\right] \tag{20.3}$$

$$\text{Prager} \quad \sigma_m - \sigma_1 \left[\frac{(3 - \alpha_c)(1 - \alpha_c)}{3} \right] \tag{20.4}$$

$$\text{Weissberg} \quad \sigma_m = \sigma_1 \left[\frac{2(1 - \alpha_c)}{2 - \ln(1 - \alpha_c)} \right] \tag{20.5}$$

where α_c is the solids volume fraction, σ_1 is the conductivity of the continuous phase (liquid phase), σ_2 is the conductivity of thc dispersed phase (solid particles), and σ_m is the local slurry mixture conductivity or the reconstructed conductivity measured by the ERT.

Maxwell's relationship, which relates conductivity distribution of a mixture to particle concentration distribution, was derived for diluted slurry mixtures, and it is expressed in terms of solids (or dispersed phase) volume fraction, α_c (Maxwell, 1881). Bruggeman's equation was derived based on a mixture of nonconductive spherical particles in a conductive media. Meredith and Tobias also generated an equation relating the conductivity and local solid's concentration. Prager's correlation is a generalized diffusion model for the suspension of nonuniform-shaped solid particles having random geometry. Weissberg's correlation, which relates slurry mixture conductivity to solids volume fraction, considers an idealized bed, in which the bed is composed of randomly distributed spherical solid particles.

A question may arise here as to which of the above correlations perform better for converting the conductivity distribution to solids volume fraction. It has been confirmed that the prediction of the mixture conductivity in all the above five correlations is similar, only if the solids volume fraction is less than 30% (Holdich & Sinclair, 1992). Recently, a comparison of the above five correlations was carried out, covering a range of solids volume fraction in a slurry composed of nonconductive solid particles and water as a continuous phase (Pachowko, 2004), the results of which are shown in Fig. 20.4.

It is worth pointing out that Maxwell's correlation, which is the most widely studied, is used in the ERT system to convert the conductivity distribution to solid's volume fraction. Although Maxwell believed that his equation can only be applied on diluted mixtures, it was confirmed that Maxwell's equation can also be used for high solids concentration, even with maximum packing concentration of the slurry (Turner, 1976). If it is assumed that the dispersed particles are nonconductive, then the conductivity of solid particles is considered to be zero and Eq. (20.1) is simplified to Eq. (20.6).

$$\alpha_c = \frac{2\sigma_1 - 2\sigma_m}{\sigma_m + 2\sigma_1} \tag{20.6}$$

Eq. (20.6) is used to calculate the solids volume fraction at 316 locations (pixels) within the pipe cross-section and the resulting reconstruction map is arranged. However, since the ERT conductivity measurements are based on the measurement of the original conductivity of the continuous conductive phase, which is typically

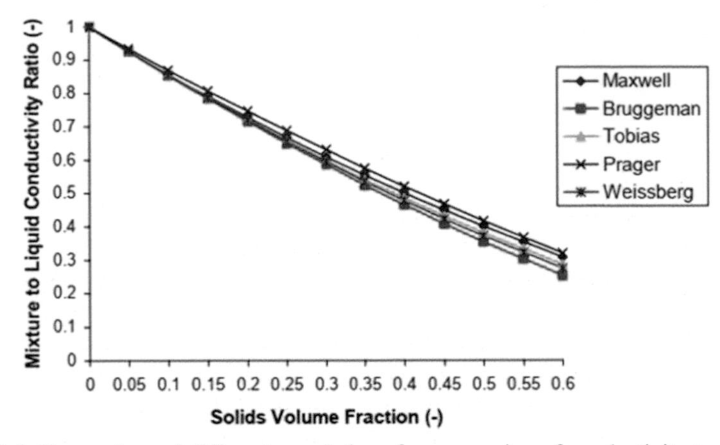

Figure 20.4 Comparison of different correlations for conversion of conductivity to solids volume fraction.
Pachowko, A.D. (2004). *Design and modelling of a coarse particulate slurry handling system* (Doctoral thesis). Leeds: University of Leeds.

referred to as reference measurement, the ERT measurements are valid only in the case of isothermal condition. This implies that any variation in the continuous phase temperature, after taking the reference measurement, results in erroneous measured solids volume fraction values, as the conductivity of continuous phase varies with variations in the continuous phase temperature.

ERT can generate local solids concentration profile across the pipe cross-section. The tomograms reconstructed by ERT can be collected and analyzed to determine the mean solids volume fraction and solids concentration profile across the vertical plane of each tomogram. Each individual tomogram consists of a 20×20 pixel array, which gives 400 spatial elements (Hosseini et al., 2010). The circular image is constructed using 316 pixels from 400 pixels squares grid, as shown in Fig. 20.5. The mean local solids volume fraction can be produced by averaging a block of frames from the concentration map, and the solids concentration profile can be extracted along the vertical centrelines of the concentration tomogram (the shaded area). The size of each pixel is dependent on the pipe diameter, for example, for 50 mm pipe diameter 2.5 by 2.5 mm pixels are generated.

Due to dynamic and instability of slurry flow over time, it is paramount to use an ERT system that can offer an excellent speed in measuring the conductivity of the media and collecting the data in a continuous fashion. Over the last 2 decades, various ERT systems have been developed for process engineering, each with specific features, capabilities, and application areas. Among all the ERT systems, the Fast Impedance Camera System (FICA), which is known as "Leeds FICA System," is a more sophisticated system compared to the other ERT systems. The principle of the hardware operating system and the operation of the control software is described in detail in Schlaberg, Jia, Qiu, Wang, and Hua (2008). The development of FICA system can be considered as a response to the requirement of many industrial processes such as

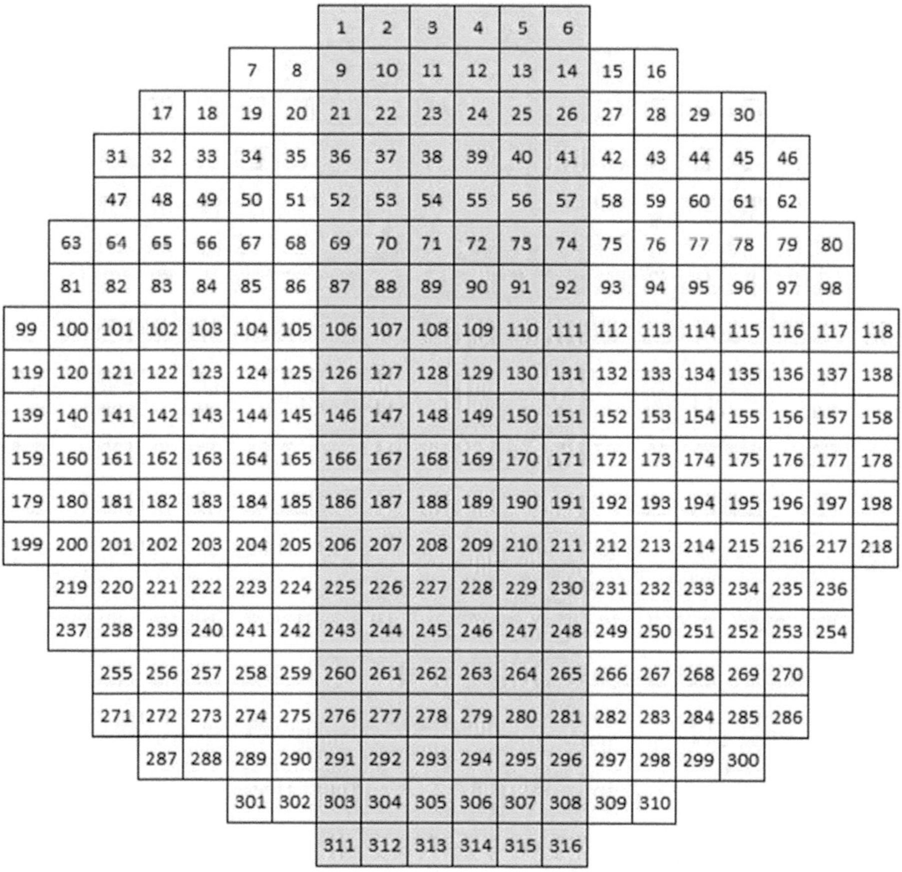

Figure 20.5 316-pixel Image reconstruction grid.
Credit: Drawn by author.

two/multiphase flow, including slurry flow, where a higher (faster) frame rate is required to measure and monitor the flow behavior.

The capability of the ERT FICA system, in measuring the conductivity of solid–liquid mixture and collecting the data with excellent speed, has been reported by Faraj and Wang (2012). They measured the local solids volume fraction in the carrier liquid, the results of which are shown in Fig. 20.6, by collecting a set of block data of 8000 frames at various mixture velocities. Fig. 20.6 shows the profile of solids concentration as a function of transport velocity along the vertical axis of a 50 NB pipe cross-section. Clearly, the local chord concentration can present a clearer picture of the distribution of solid particles and their movement in the vertical plane of the pipe cross-section.

The concentration profiles in Fig. 20.6, generated by the ERT, represent typical concentration profile of settling slurry flow. It can be observed that within the range of higher velocities above the deposition velocity the particles are suspended and

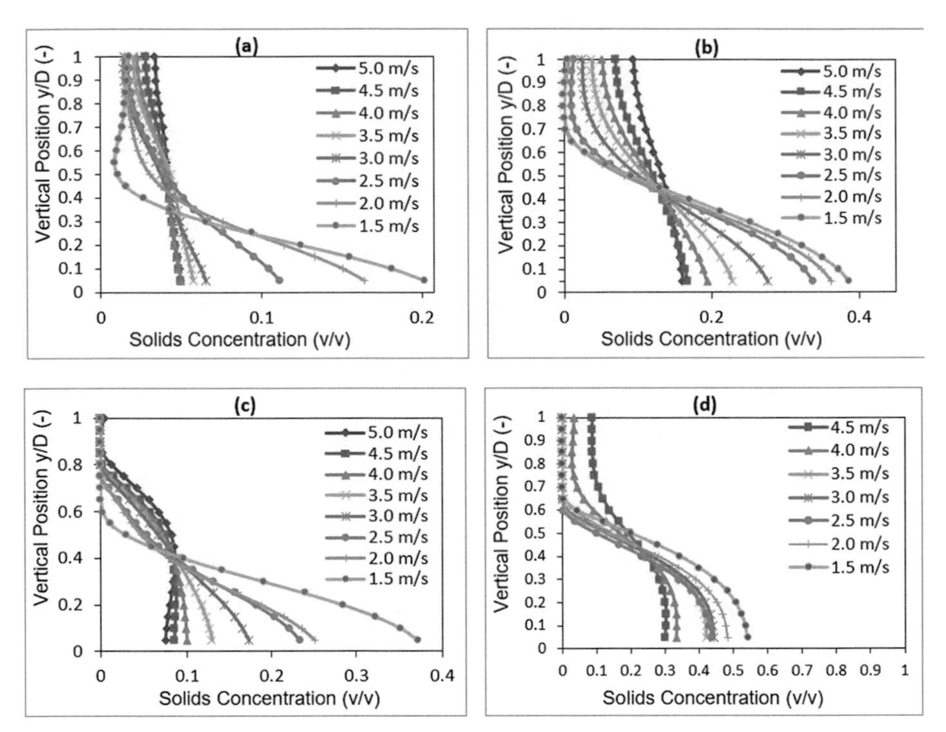

Figure 20.6 Concentration profiles for flowing (a) medium sand at 2% throughput concentration, (b) medium sand at 10% throughput concentration, (c) coarse sand at 2% throughput concentration, and (d) coarse sand at 10% throughput concentration in a horizontal 50 NB pipe as a function of transport velocity.
Faraj, Y., Wang, M. (2012). ERT investigation on horizontal and counter-gravity slurry flow in pipelines. *Procedia Engineering*, *42*, 588–606.

distributed in the carrier liquid, which is due to turbulent eddies formed in the suspending liquid. It can also be noticed that, at velocities below 3.4 m/s, for the two sand types the profiles composed of a concave (left-hand bend) curve in the upper part of the pipe and convex (right-hand side) curve in the lower part of the pipe. This can be explained by the fact that as the transport velocity decreases, the solid particles, mainly the coarse particles, migrate toward the bottom half of the pipe, which leads to an increase in solids concentration at the bottom half of the pipe. The effect of particle size can be manifested in the ERT-generated profiles, as in the case of coarse sand similar trend can be observed as for medium sand with the same throughput concentration, but with higher degree of distortion, which is well captured by the ERT. The effect of solids throughput concentration can also be detected by the ERT. It can be seeing that the profiles of both sand types at 10% throughput concentration have higher degree of distortion than those of 2% throughput concentration at a given transport velocity. This can be attributed to the particle–particle interaction, the effect of which increases with increase of particle concentration.

20.3.2 Solids axial velocity measurement

The solids axial velocity is usually obtained by cross-correlation method in conjunction with ERT measurement. The method of cross-correlation using ERT measurements has been used in a number of studies (Dyakowiski & Williams, 1996; Etuke & Bonnecaze, 1998; Faraj et al., 2015; Henningsson, Ostergren, & Dejmek, 2006; Lucas et al., 1999; Mosorov, Sankowski, Mazurkiewicz, & Dyakowski, 2002; Wang et al., 2015; Wang, Jones & Williams, 2003). The solids axial velocity, using the method of cross-correlation, is obtained from two ERT electrode plane sensors separated by a short distance, as shown in Fig. 20.7. The general concept of cross-correlation is measuring the time between to signals generated by turbulence of the fluid or suspended particles flowing in a pipe through the two sensor planes. In other words, if a fluid flows through the two sensors with a distance L between them and the downstream sensor detects the signal after a certain period T, then the velocity V can be calculated.

It is important that the distance between the two sensing planes is suitably selected so as to realize the cross-correlation. In order to select the distance between the two planes, two important factors must be considered simultaneously: the dynamic behavior of the system (mixture velocity) and the resolution of measurement of the transit time (Dyakowski, Jeanmeure, & Jaworski, 2000). The distance between the two planes could be 50, 75, 85, or 100 mm (Deng, Dong, Xu, Liu, & Xu, 2001). However, it is recommended that the distance to be adjustable, so as to coordinate it with different cases and various flow velocities. The signals can be correlated in different methods, point to point correlation and best correlation pixels (Dai, Wang, Panayotopoulos, Lucas, & Williams, 2004). The detailed discussion of cross-correlation method and its application in ERT can be found in an earlier chapter of this book.

Pixel to pixel correlation method has been widely used, and it was recently developed into a software package (Advanced Imaging and Measurement for Flow, Multiphase Flow, and Complex Flow in the Industrial Plant-AIMFLOW) at the University of Leeds. By importing the conductivity map into the AIMFLOW, the axial solids velocity distribution can be computed, as shown in Fig. 20.8. The right hand-side of

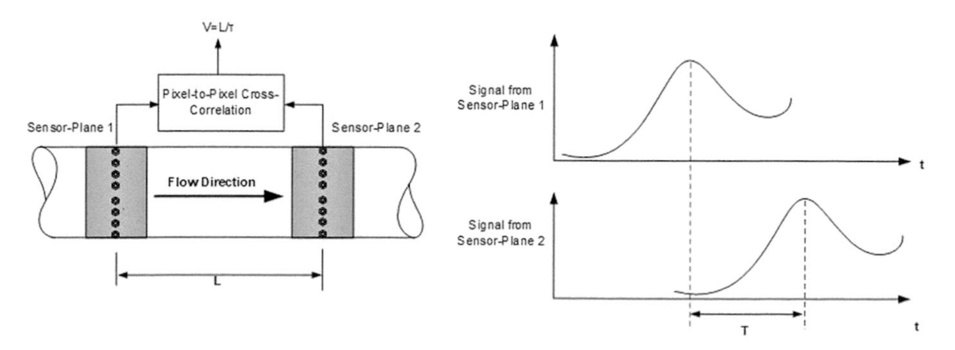

Figure 20.7 The principle of velocity measurement by cross-correlation of ERT signals. Credit: Drawn by the author.

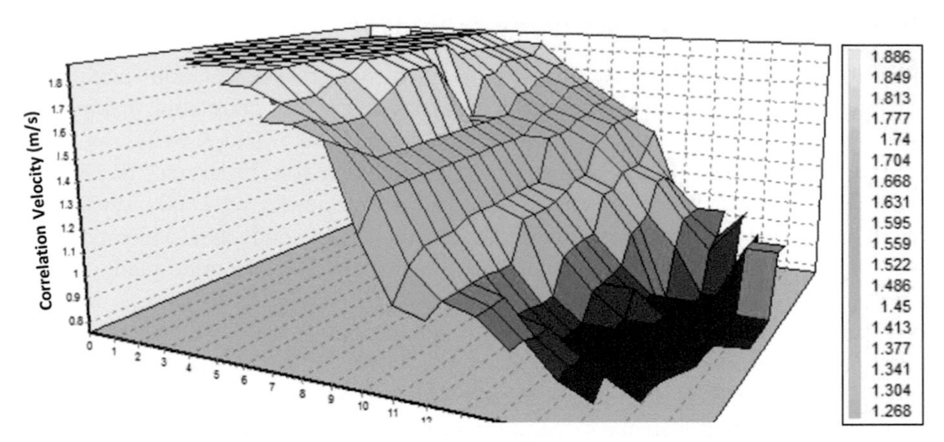

Figure 20.8 Solids velocity profile across a horizontal 50 NB pipe cross-section for coarse sand at 1.5 m/s transport velocity.
Faraj, Y., Wang, M. (2012). ERT investigation on horizontal and counter-gravity slurry flow in pipelines. *Procedia Engineering*, *42*, 588–606.

the profile (blue color) represents the bottom of the pipe, where the lowest solids velocity exists, which is expected for flowing settling slurry at 1.5 m/s transport velocity in a 50 NB pipe. The mean solids local velocity can be determined by averaging the magnitude of solids velocity across 316 pixels in the tomogram.

20.3.3 Solids flow monitoring and visualization

Slurry flow monitoring is paramount in many industrial applications. In order to avoid the formation of a packed stationary bed at the pipe invert, which may eventually lead to pipe blockage, the slurry velocity must be kept sufficiently high to suspend and drag the solids in the stream of a carrier liquid. Doing so results in an unnecessary energy consumption and pipe wear costs. On the other hand, keeping the slurry velocity low results in pipe blockage and undesirable consequences such as pipeline shutdown and production loss, high cost of labor and material used to unblock the pipeline, and environmental pollution caused by the spillage of pipeline contents. Therefore, monitoring the internal structure of a slurry pipeline can prevent pipeline blockage and assist in the optimization of the pump speed and associated costs (Kotze, Adler, Sutherland, & Deba, 2019).

ERT as an image reconstruction scheme can be a promising technique for visualization and monitoring the internal structure of a slurry pipeline, in which the continuous phase is conductive. The ERT tomograms can provide valuable information on dynamics of slurry flow, which can readily be used for the purpose of slurry visualization, as well as an online tool for controlling the transportation process and minimizing energy consumption. ERT has been used in a number of studies for visualization of internal structure of slurry flow (Faraj, Wang & Jia, 2013, pp. 869–877; Giguère et al., 2008a, 2008b, 2009; Kotze et al., 2019; Pachowko, 2004;

Silva et al., 2016). The visualization of tomograms may be beneficial to the operator in industry if certain flow features are sought, such as bubbles, plug flow, and formation of strata in horizontal slurry flow. However, for detailed analysis and evaluation, solids volumetric concentration profile and solids axial velocity profile may be the best tool for characterization and analysis of slurry flow and its prevailing flow patterns.

Since ERT images are reconstructed from measured data via a specific reconstruction algorithm, it is apparent that the choice of algorithm significantly influences the quality and accuracy of the resulting images. Commercially available software Linear Back-Projection (LBP), which offers simplicity and fast image reconstruction, is the most widely used method for the reconstruction of tomograms that can be used for direct visualization of solids concentration distribution across the pipe cross-section (Giguere et al., 2008a). It has been demonstrated that the concentration tomograms reconstructed by noniterative LBP cannot provide an accurate quantitative measurement and the images may contain some artifacts, such as the blurred interface between the bed and the flowing slurry above it, which may be misleading. This is well attributed to the limitations associated with the image reconstruction technique and the hardware capability. These artifacts in the reconstructed images have been observed particularly when the deposited solid particles cover a large portion of the sensing area (electrodes). Several options have been suggested to reduce these artifacts and improve the definition of the interface between the bed and the slurry above it, such as decreasing the magnitude of the injected current, increasing the conductivity contrast of the aqueous continuous phase by adding salt (e.g., NaCl), or modification of the gain used by the hardware (Giguère et al., 2008b; Jia, Wang, & Faraj, 2015).

It is worth mentioning that very few studies regarding the performance, validation, and limitations of quantitative image reconstruction have been conducted. In a bid to overcome the limitations associated with the currently used commercial ERT image reconstruction, quantitative image reconstruction algorithms for ERT have been developed to quantitatively visualize slurry flow in a pipe or any process application. Recently, an iterative image reconstruction method was developed by deriving a Generalized Iterative Algorithm, called GIA, to solve the ERT inverse problem. GIA was devised from the generalization of two image reconstruction techniques, using Landweber iterations and Tikhonov iterations. The reconstructed images of static particle bed (17% (v/v) bulk concentration of 100 μm spherical glass beads) using LBP method and GIA at different combinations of injected current and absolute mixture conductivity (A: 15 mA and 400 μS/cm; B: 15 mA and 1800 μS/cm; and C: 1 mA and 400 μS/cm) are shown in Fig. 20.9.

By observing Fig. 20.9, it is apparent that GIA, compared with LBP, significantly improves the image precision, allowing for better interpretation of the ERT measurements. Furthermore, comparing the GIA set of images, A—B, it can be seen that absolute electrical conductivity of the mixture and current injection significantly influence the ERT results. By injecting too high current or too small mixture conductivity, the corresponding voltage to be measured falls beyond the capability of the hardware, leading to errors in the measurements, hence presence of artifacts in the reconstructed image. This can be seen in case A, in which a high intensity current (15 mA) is injected for a conductive mixture of 400 μS/cm. The reason for this is due to the increase of the

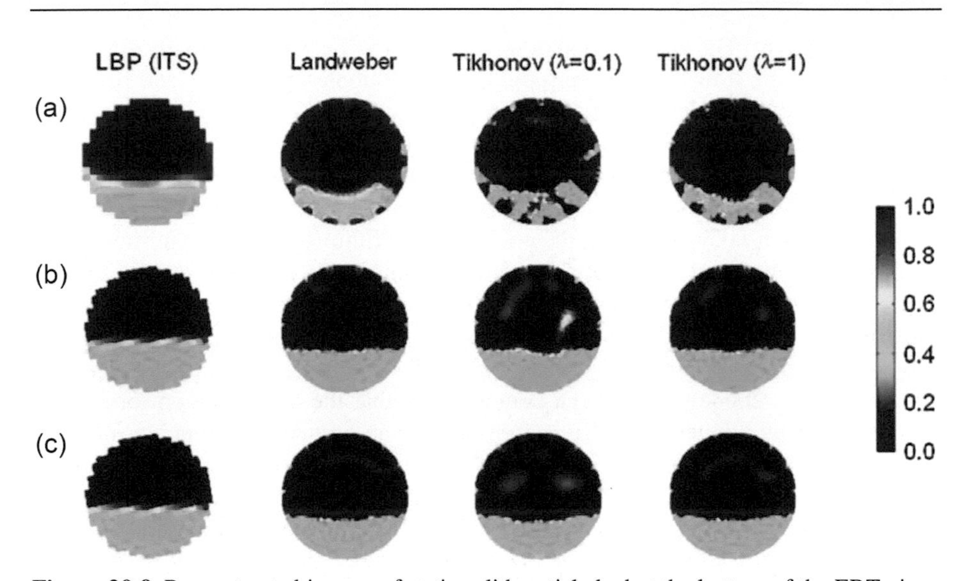

Figure 20.9 Reconstructed images of static solid particle bed at the bottom of the ERT pipe sensor using different image reconstruction algorithms.
Giguere, R., Fradette, L., Mignon, D. Tanguy, P.A. (2008b). ERT algorithms for quantitative concentration measurement of multiphase flows. *Chemical Engineering Journal, 141*, 305–317.

electrical resistance of the materials in the sensor, caused by the presence of nonconductive solid particles, leading to relatively high voltages measurements by ERT. Compared with LBP, which usually overestimates the concentration of solid particles, GIA provides a reasonable prediction of the solids concentration with less than 1% (v/v) error, as shown in Fig. 20.10. On the other hand, the LBP method can provide correct solids concentration for small bed of solid particles 2.5% (v/v), while it overestimates the concentration of solids concentration by 5% (v/v) if higher solid particles are present. Nevertheless, the LBP method suits well for a rapid determination of the size of the bed present at the bottom of a pipe.

Image reconstruction algorithms, which have been originally developed for the purpose of medical EIT applications such as Laplace, HPF Noser, Greit-2D, TV-PDIMP, and GR + Laplace, have also been analyzed for identification of slurry bed at the pipe invert in static and dynamic flow conditions (Kotze et al., 2019). A summary of image reconstruction comparisons for measurement of solid particles bed height, along with photographs of the experimental conditions (A: slowly sliding bed; B: moving bed with some suspended solids; C: moving bed with higher portion of suspended solid particles and D: Fully suspended heterogeneous flow), is shown in Fig. 20.11.

It is apparent that the performance of the six algorithms for imaging coarse particle bed varies significantly. The only two algorithms that provide reasonable interpretation of coarse solid particles in a settled bed at the bottom of the pipe are Laplace and GREIT. By combining GREIT and Laplace, through averaging the outputs of the two algorithms, better representation of the bed can be achieved, perhaps due to synergistic actions of the two algorithms. Although GREIT, Laplace, and GREIT + Laplace perform even

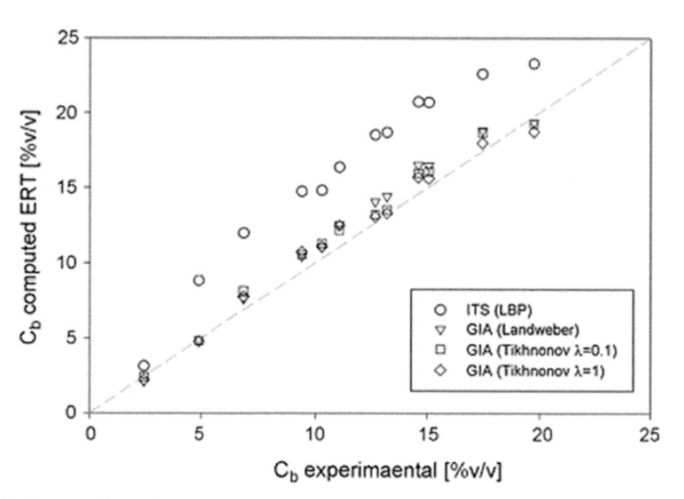

Figure 20.10 Comparison between LBP and GIA in ERT measurements of solid particle bed concentration (C_b).
Giguere, R., Fradette, L., Mignon, D. Tanguy, P.A. (2008b). ERT algorithms for quantitative concentration measurement of multiphase flows. *Chemical Engineering Journal, 141*, 305−317.

better for imaging static object such as imaging an organ in medical applications, they can provide a reasonable interpretation of ERT measurements.

20.3.4 Characterization of flow patterns and analysis of stratified slurry flow by ERT

Characterization of slurry flow patterns is important for design and operation of pipelines conveying settling slurries. ERT tomograms have been utilized for direct interpretation of slurry flow patterns (Giguère et al., 2008a). Although these tomograms can be highly beneficial for rapid monitoring slurry flow within the pipe cross-section, they do not readily provide detailed information regarding the flow and solids distribution across a pipe cross-section. Therefore, the ERT solids concentration profile and/or solids velocity profile may be used as a viable tool for obtaining better information about solids distribution and the prevailing flow patterns within a pipe. These flow patterns can be visualized and characterized through solids concentration profile and axial velocity profile obtained from a dual plane ERT sensor. Fig. 20.12 shows the ERT solids concentration profile (left) and axial velocity profile (right) for coarse sand particles in 50 mm pipe diameter at 2.5 m/s transport velocity.

As the transport velocity is decreased, below the deposit velocity, the suspended solid particles follow a smooth downward trajectory to form a moving bed, which is shown by both concentration and velocity profiles in Fig. 20.12. This gives rise to the emergence of the three zones, which have been described by Wilson and Pugh (1988). These are namely, a turbulent zone in the upper region of the pipe, a shear layer

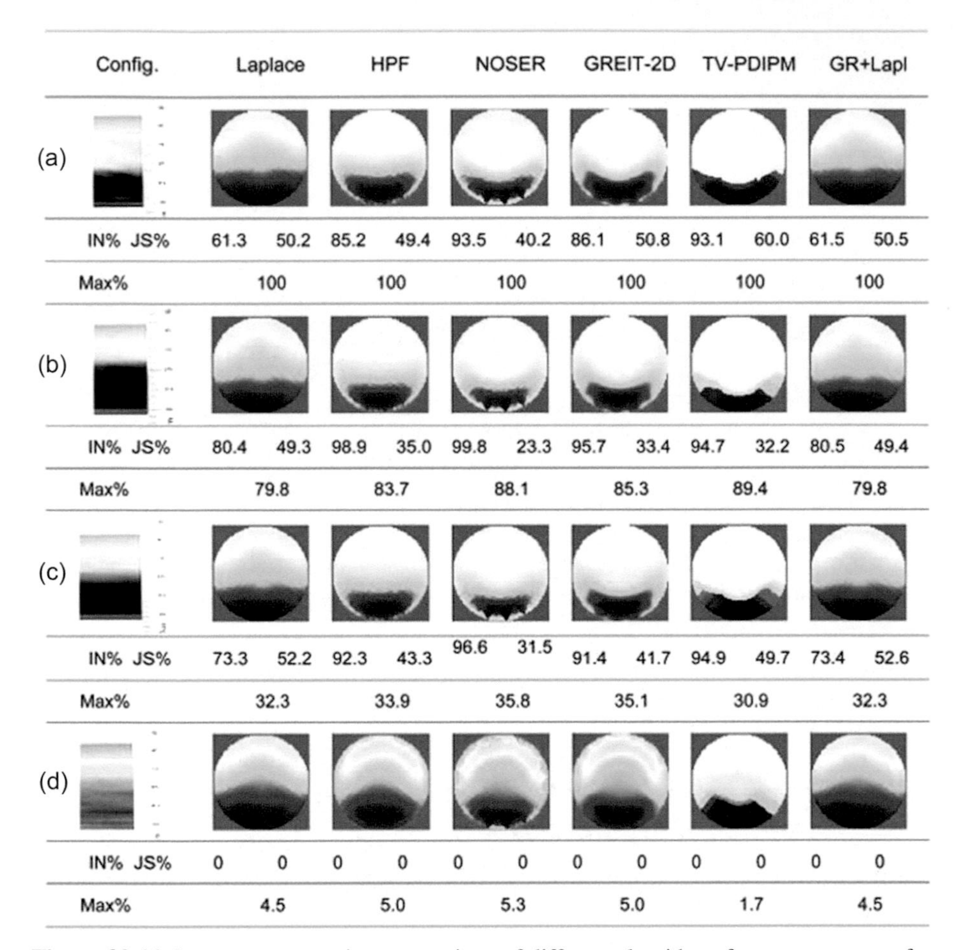

Figure 20.11 Image reconstruction comparison of different algorithms for measurement of dynamic solid particle bed concentration.
Kotze, R., Adler, A., Sutherland, A., Deba, C.N. (2019). Evaluation of Electrical Resistance Tomography imaging algorithms to monitor settling slurry pipe flow. *Flow Measurement and Instrumentation*, *68*, 101572.

above the bed, in which solid particles move in a sporadic fashion, and the sliding bed in the bottom region of the pipe. It is apparent that these zones can be seen in the ERT-generated profiles. There is clearly a good agreement between the inserted real flow photograph into the plot area and the measured profiles by the ERT.

By further observing the ERT profiles, it can be seen that the shear layer is developed, which is associated with the variation of shear stress at the boundary between the shear layer and the uniformly distributed granular moving bed. A wide variation in the velocity gradient can be observed within the shear layer, which is due to the difference in the local bed velocity and the mean slurry velocity above the bed. The high velocity gradient within the lower part of the shear layer causes a chaotic region over the bed,

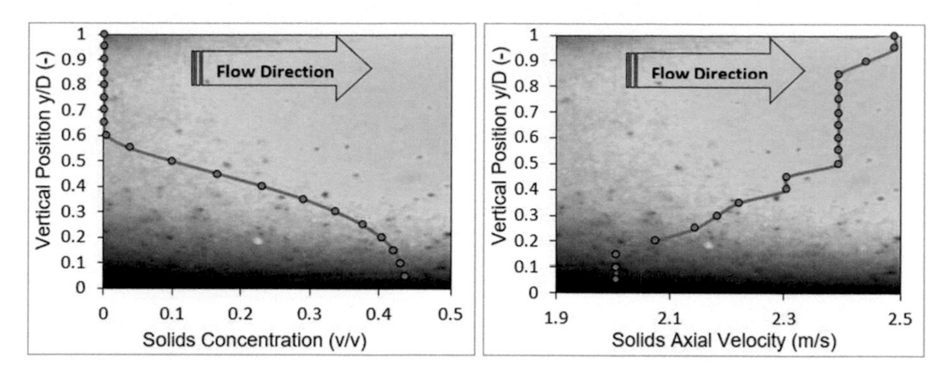

Figure 20.12 Moving bed flow pattern of coarse sand in 50 mm pipe diameter at 2.5 m/s transport velocity: concentration profile (left) and solids velocity profile (right). The actual flow photograph inserted into the plot area.
Faraj, Y., Wang, M., Jia, J. (2013). Application of the ERT for slurry flow regime characterisation. *International Society for Industrial Process Tomography*, pp. 869–877.

where the particles are lifted from the surface of the bed and supported again by the upward impulses of the turbulent eddies. The phenomenon of sporadic movement of particles continuously occurs at the interface between the sliding bed and the upper turbulent region unless the variation in the transport velocity occurs, which has a direct impact on the lifting force. This flow behavior is attributed to an increase of the shear stress at the bottom of the shear layer, by the moving fluid stream in a momentary impulse fashion. The momentary jumping of the bed solid particles into the turbulent stream above the bed may contribute into the magnitude of the mean solids velocity, which in return causes an overestimation of the mean solids axial velocities, particularly those below the deposition velocity (i.e., within the velocity range, in which the bed is present). Another interesting feature can be observed on the velocity profile, where the profile goes through a deformation through its course at the top of the pipe ($y/D = 0.85$). The profile unfolds another region with higher velocity, comparable to the lower layers, and confined by another turbulent particle-rich layer and the pipe wall. The mean velocity of the particles moving within this layer is considerably higher than the rest of the layers flowing below. This is due to absence or very limited particle–particle interaction at the very top of the pipe. The background photograph of the real flow clearly shows a particle-lean layer next to the upper pipe wall.

Slurry flow can be partially or fully stratified, which occur when the intensity of turbulence of a carrier fluid is not sufficient to suspend all solid particles in a pipe. In partially stratified flow, which is common in dredging operations and mining applications, a portion of solids accumulate near the bottom of a pipe and the rest of solids occupy the rest of pipe cross-sectional area in a nonuniform distribution. This type of slurry flow exhibits a considerable solids concentration gradient across a pipe cross-section, whereas in fully stratified flow, there is no sufficient turbulence intensity of the carrier liquid to suspend solid particles in the pipe, and as a result the solid particles form a granular bed, which is either stationary or moving along the bottom of the

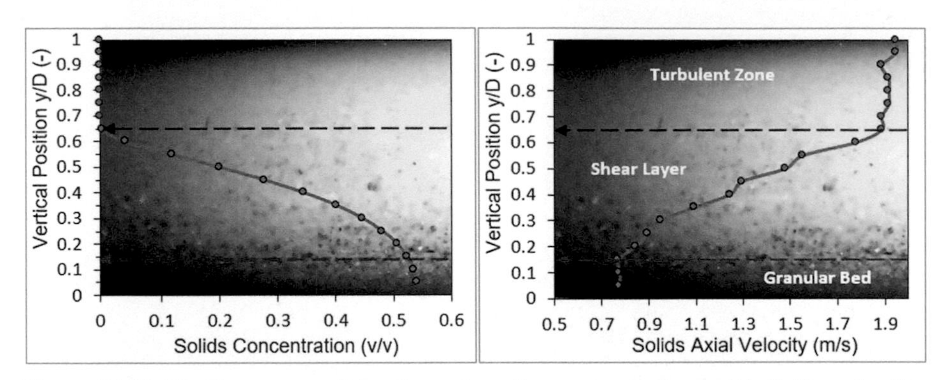

Figure 20.13 The three zones of the distorted profiles, concentration profile (left), and solids axial velocity profile (right), in stratified flow in 50 mm pipe diameter at 1.5 m/s transport velocity.
Faraj, Y., Wang, M., Jia, J. (2013). Application of the ERT for slurry flow regime characterisation. *International Society for Industrial Process Tomography*, pp. 869–877.

pipeline, as shown in Fig. 20.13. If the transport velocity in a pipe is too low, the carrier liquid is not able to keep all solids in motion. The granular bed composed of settled solids at the bottom of the pipe does not slide and as a result an inefficient and potentially dangerous stationary bed flow pattern develops.

Therefore, characterization of stratified slurry flow is paramount for safe slurry transport, slurry flow pipeline design, and optimization. ERT can be utilized for monitoring settling slurry flows and characterization of stratified flow through combination of solids concentration distribution and solids axial velocity distribution. The observation of both profiles and the knowledge of pixel height within the tomogram leads to a reasonable estimation of some parameters related to stratified slurry flow, such as mean granular bed concentration, mean granular bed velocity, the height of granular bed, the height of shear layer, and the height of turbulent zone at the upper part of the pipe.

The height of granular bed can be demarcated at the point where the concave curve (left-hand side of the concentration profile) at the bottom of the pipe marks a sharp gradient in solids concentration distribution. Accordingly, the height of granular bed in Fig. 20.13 (demarcated by a red arrow) can be estimated as 7 mm from the bottom of the pipe (where $y/D = 0.14$, y is the distance from the bottom of the pipe and D is the pipe diameter). This scheme can only be used to measure the height of granular bed if the height of bed is equal or greater than the ratio of inner pipe diameter to the number of the pixels in the vertical centreline of a 316-pixel tomogram. In other words, the height of the settled solid particle bed, moving or stationary, can be estimated if the size of the bed does not exceed the pixel height in a reconstructed tomogram. For example, the pixel size of a reconstructed image of a solid particle bed at the bottom of a 50 mm inner diameter pipe is 2.5 mm (based on 20 pixels in the centreline of the tomogram); in this case, the bed height of only 2.5 mm or greater can be estimated. Once the height of granular bed is estimated, the mean solids concentration occupying the granular bed can be reasonably estimated by averaging the solids concentration within a set of pixel rows covered by the granular bed.

With regard to the mean granular bed velocity, this can readily be estimated from the ERT solids velocity profile. The solids axial velocity within the granular bed can be estimated by demarcating the point where the profile goes through a sharp gradient. In a stationary bed, no sharp gradient of solids velocity is expected throughout a stationary bed, but a uniform velocity distribution. However, the granular bed may be moving or stationary, or in some cases, both moving and stationary layer may exist simultaneously, where a moving layer is sliding over the surface of the stationary layer (Dorn & Barnea, 1993; Matousek, 2002). In this case, a slightly skewed solids velocity may be observed with the granular bed and considered as the height of granular bed, within which the velocities are averaged. Similarly, the height of shear layer (between the red and black arrows) and turbulent zone can be estimated.

20.4 Limitations of ERT in slurry flow measurement and characterization

Although within the last 2 decades ERT has seen a significant development and applications to many industrial processes involving two/multiphase systems, there is still scope for improving some aspects of ERT, such as spatial resolution, conductivity resolution, and the ability of distinguishing a single solid particle.

Spatial resolution helps in identifying a minimum particle size within the flow domain. It has been shown that by varying the particle to vessel or pipe diameter ratio and keeping the mixture conductivity constant with a measurement error less than 1%, a typical ERT spatial resolution of 5% pipe diameter can be achieved (Wang, Mann, & Dickin, 1999). For example, applying ERT on a mixture of solid–liquid flowing through a 50 mm inner pipe diameter, while keeping the mixture conductivity constant and the measurement error is less than 1%, the ERT sensor can identify solid particle size of 2.5 mm or greater. This implies that the ERT cannot be used for determining the physical properties of slurry flow such as particle size; however, it can be considered as an ideal tool for determining the bulk characteristics of slurry flow such as solids volume fraction and solids velocity.

Based on the concept of the ERT measurement, there must be a minimum conductivity difference between the constituent phases. The conductivity resolution determines this minimum conductivity deviation in individual pixels, which enable the image reconstruction algorithm to pick up the change and demonstrate it in the reconstructed images. Regarding this aspect, it has been found that, for a fixed particle to pipe or vessel diameter ratio and measurement error of less than 1%, at least a deviation of 10% of conductivity is required for the object to be imaged (Wang et al., 1999).

Distinguishability of a single solid particle is a parameter that determines the minimum distance between each single particle within the slurry so that ERT can distinguish each of them. Currently, for a measurement error less than 1%, the minimum value for this parameter is 20% of the pipe or vessel diameter (Wang et al., 1999). Taking the previous example of an ERT measurement of a slurry flow in 50 mm pipe diameter, a minimum distance of 10 mm is required between the solid particles, so that they can clearly be seen in the tomograms, otherwise they are seen as a single particle. This suggests that ERT is unable to distinguish every single particle in a highly concentrated slurry flow, even if the flow is homogeneous flow regime. However, this may not have an importance when the contact load of bed is formed.

20.5 Conclusions and future trends

Slurry transport has been a progressive technology for transporting a huge amount and variety of solid materials in both long distance pipelines and short commodity pipelines. Many slurry pipelines already exist, and there will be more to build in all pipeline orientations. In order to ensure safe transport, optimized operation, and reduction of financial costs by avoiding blockage, the operational engineer requires a reliable technique that suits all the conditions in industry. Clearly, the complex nature of settling slurry flow forces the operator to continuously measure the local parameters governing the flow, such as solids concentration and solids velocity, and visualize the internal structure of the flow within the pipeline. Undoubtedly, the measurement of slurry flow parameters and monitoring the motion of solids within the pipe, in fast-evolving processes, requires a fast-responding instrument (i.e., high frame rates of milliseconds). ERT has shown its capability to rapidly interrogate the internal structure of complex and dynamic structure of slurry flow in a nonintrusive fashion. The development of fast-responding ERT system of high-speed imaging in excess of 1000 frames per second, along with the transition of the technology from solids concentration mapping technology to solids velocity mapping technology, can be considered as a direct response to today's slurry flow engineering requirements and a step forward for better controlling the processes and reducing unnecessary financial costs. The current commercially available image reconstruction technique for ERT, LBP, is widely used and provides excellent time resolution for generating qualitative images, which may be sufficient for rapid monitoring slurry flow. Further work on the development of quantitative tomography techniques and efficient strategies of image reconstruction for ERT will be beneficial for detailed characterization of slurry flow and providing a novel insight over the mechanisms of solid−liquid flow. Application of ERT in a wider spectrum of slurry flow, including complex fluid, will further reveal the capabilities of the ERT and pave the way to a reliable and robust technique for characterization of a wide range of slurries and optimization of operation of settling slurry pipelines in industry. Integration of ERT with nonintrusive techniques such as Electrical Capacitance Tomography or ultrasound method will yield an enhanced and versatile multimodality sensor that can be beneficial for a wide range of chemical and allied industries dealing with solid−liquid flow.

20.6 Sources of further information

A recent book by Basu (2018) is a unique in its coverage of multiphase flow and solid−liquid flow with a focus on flow measurement. It also attempts to give a comprehensive discussion regarding the mechanisms governing solid−liquid flow and typical slurry flow patterns in pipeline. A very good review paper by Dyakowski, Jeanmeure, and Jaworski (2000) presents an overview of ERT and its application in slurry flow. The two articles by Wang, Jones and Williams (2003) and Stevenson, Harrison, Miles, and Cilliers (2006) provide interesting discussions and analysis of swirling solid–liquid flow in helical pipe using ERT. An excellent review paper by Sharifi and Young (2013) presents a good coverage of ERT applications to chemical engineering, which includes application of ERT in solid–liquid measurement as part of its coverage. A journey into the articles by Mann and Wang (1997), Fangary et al. (1998), Lucas et al. (1999), Giguere et al. (2008a and 2009) may be beneficial for beginners who are interested in the ERT and its application to solid–liquid flows.

References

Abulnaga, B. (2002). *Slurry systems handbook.* New York, NY: McGraw-Hill.

Bagnold, R. A. (1956). The flow of cohesionless grains in fluids. *Philosophy Transaction of the Royal Society, London, Series A, 249*, 99235–99297.

Basu, S. (2018). *Plant flow measurement and control handbook: Fluid, solid, slurry and multiphase flow*. United Kingdom: Elsevier Science.

Blatch, N. S. (1906). Discussion: Water filtration at Washington D.C. *Transactions of the American Society of Civil Engineers, 57*, 400–408.

Brown, N. P. (1991). Flow regimes of settling slurries in pipes. In N. P. Brown, & N. L. Heywood (Eds.), *Slurry handling design of solid–liquid systems* (pp. 41–52). London: Springer.

Brown, P. B., & Heywood, N. I. (1991). *Slurry handling: Design of solid-liquid systems.* London: Elsevier Applied Science.

Bruggeman, D. A. G. (1935). Calculation of various physical constants of heterogeneous substances, Part 1, constant and conductivity of mixtures of isotropic substances. *Annual Physics, Leipzig, 24*, 636.

Dai, Y., Wang, M., Panayotopoulos, N., Lucas, G., & Williams, A. R. (2004). 3-D visualisation of a swirling flow using electrical resistance tomography. In *4th World congress on industrial process tomography, Aizu, Japan* (pp. 362–369).

Deng, X., Dong, F., Xu, L. J., Liu, X. P., & Xu, L. A. (2001). The design of a dual-plane ERT system for cross correlation measurement of bubbly gas/liquid pipe flow. *Measurement Science and Technology, 12*, 1024–1031.

Dickin, F. J., & Wang, M. (1996). Electrical resistance tomography for process applications. *Measurement Science and Technology, 7*(3), 247–260.

Doron, P., & Barnea, D. (1993). A three-layer model for solid–liquid flow in horizontal pipes. *International Journal of Multiphase Flow, 19*, 1029–1043.

Doron, P., Simkhis, M., & Barnea, D. (1997). Flow of solid–liquid mixtures in inclined pipes. *International Journal of Multiphase Flow, 23*, 313–323.

Durand, R. (1952). Basic relationship of the transportation of solids in pipes-experimental research. In *Proceedings of the Minnesota International hydraulics convention, American society of civil engineers* (pp. 89–103). Minneapolis: University of Minnesota.

Durand, R., & Condolios, E. (1952). *Compterendu des Deuxiemes journees de L'Hydraulique.* Paris: societe Hydrotechnique de France.

Durand, R., & Condolios, E. (1953). The hydraulic transport of coal and solid materials in pipes. In *Proceedings of colloquium on the hydraulic transport of coal* (pp. 39–52). London: National Coal Board.

Dyakowski, T., Jeanmeure, L. F. C., & Jaworski, A. J. (2000). Applications of electrical tomography for gas-solids and liquid-solids flows- a review. *Powder Technology, 112*, 174–192.

Dyakowski, T., & Williams, R. A. (1996). Prediction of high solids concentration regions within a hydrocyclone. *Powder Technology, 87*, 43–47.

Etuke, E. O., & Bonnecaze, R. T. (1998). Measurement of angular velocities using electrical impedance tomography. *Flow Measurement and Instrumentation, 9*, 159–169.

Fangary, Y. S., Williams, R. A., Neil, W. A., Bond, J., & Faulks, I. (1998). Application of electrical resistance tomography to detect deposition in hydraulic conveying systems. *Powder Technology, 95*(1), 61–66.

Faraj, Y., & Wang, M. (2012). ERT investigation on horizontal and counter-gravity slurry flow in pipelines. *Procedia Engineering, 42*, 588–606.

Faraj, Y., Wang, M., & Jia, J. (2013). *Application of the ERT for slurry flow regime characterisation*. International Society for Industrial Process Tomography.

Faraj, Y., Wang, M., Jia, J., Wang, Q., Xie, C., Oddie, G., et al. (2015). Measurement of vertical oil-in-water two-phase flow using dual-modality ERT/EMF system. *Flow Measurement and Instrumentation, 46*(Part B), 255–261.

Giguere, R., Fradette, L., Mignon, D., & Tanguy, P. A. (2008a). Characterization of slurry flow regime transitions by ERT. *Chemical Engineering Research and Design, 86*, 989–996.

Giguere, R., Fradette, L., Mignon, D., & Tanguy, P. A. (2008b). ERT algorithms for quantitative concentration measurement of multiphase flows. *Chemical Engineering Journal, 141*, 305–317.

Giguere, R., Fradette, L., Mignon, D., & Tanguy, P. A. (2009). Analysis of slurry flow regimes downstream of a pipe bend. *Chemical Engineering Research and Design, 87*, 943–950.

Gillies, R. G. (1991). Flow loop studies. In N. P. Brown, & N. I. Heywood (Eds.), *Slurry handling: Design of solid-liquid systems*. Elsevier Applied Science.

Govier, G. W., & Aziz, K. (1972). *The flow of complex mixtures in pipes*. New York, NY: van Nostrand Reinhold Company.

Gumery, F., Ein-Mozaffari, F., & Dahman, Y. (2011). Macromixing hydrodynamic study in draft-tube airlift reactors using electrical resistance tomography. *Bioprocess and Biosystems Engineering, 34*(2), 135–144.

Heindel, T. J., Grayb, J. N., & Jensenb, T. C. (2008). An X-ray system for visualizing fluid flows. *Flow Measurement and Instrumentation, 19*, 67–78.

Henningsson, M., Ostergren, K., & Dejmek, P. (2006). Plug flow of yoghurt in piping as determined by cross-correlated dual-plane electrical resistance tomography. *Journal of Food Engineering, 76*, 163–168.

Holdich, R. G., & Sinclair, I. (1992). Measurement of slurry solids content by electrical conductivity. *Powder Technology, 72*, 77–87.

Hosseini, S., Patel, D., Ein-Mozaffari, F., & Mehrvar, M. (2010). Study of solid-liquid mixing in agitated tanks through electrical resistance tomography. *Chemical Engineering Science, 65*, 1374–1384.

Howard, G. W. (1939). *Transportation of sand and gravel in a four-inch pipe* (Vol. 104, pp. 1334–1348). Transaction of the American Society of Civil Engineers.

Ismail, I., Gamio, J. C., Bukhari, S. F. A., & Yang, W. Q. (2005). Tomography for multi-phase flow measurement in the oil industry. *Flow Measurement and Instrumentation, 16*, 145–155.

Jia, J., Wang, M., & Faraj, Y. (2015). Evaluation of EIT systems and algorithms for handling full void fraction range in two-phase flow measurement. *Measurement Science and Technology, 26*(1), 015305.

Jia, J., Wang, M., Faraj, Y., & Wang, Q. (2015). Online conductivity calibration methods for EIT gas/oil in water flow measurement. *Flow Measurement and Instrumentation, 46*(Part B), 213−217.

Kotze, R., Adler, A., Sutherland, A., & Deba, C. N. (2019). Evaluation of electrical resistance tomography imaging algorithms to monitor settling slurry pipe flow. *Flow Measurement and Instrumentation, 68*, 101572.

Lahiri, S. K. (2009). *Study of slurry flow modelling in pipeline* (Doctoral thesis). India: National Institute of Technology.

Lazarus, J. H., & Neilson, I. D. (1978). A generalized correlation for friction head losses of settling mixtures in horizontal smooth pipelines. In *Proceedings of the fifth International conference on the hydraulic transport of solids in pipes, Hanover, Germany, paper B1* (pp. 1−32).

Lucas, G. P., Cory, J., Waterfall, R. C., Loh, W. W., & Dickin, F. J. (1999). Measurement of the solids volume fraction and velocity distribution in solids-liquid flows using dual-plane electrical resistance tomography. *Flow Measurement and Instrumentation, 10*(4), 249−258.

MacTaggart, R., Nasr-El-Din, H., & Masliyah, J. (1993). A conductivity probe for measuring local solids concentration in a slurry mixing tank. *Separations Technology, 3*(3), 151−160.

Mann, R., & Wang, M. (1997). Electrical process tomography: Simple and inexpensive techniques for process imaging. *Measurement and Control, 30*, 206−211.

Marefatallah, M., Breakey, R. D., & Sanders, S. (2019). Study of local solid volume fraction fluctuations using high speed electrical impedance tomography: Particles with low Stokes number. *Chemical Engineering Science, 203*, 439−449.

Matousek, V. (2002). Pressure drops and flow patterns in sand−mixture pipes. *Experimental Thermal and Fluid Science, 26*, 693−702.

Matousek, V. (2005). Research developments in pipeline transport of settling slurries. *Poweder Technology, 156*(1), 43−51.

Matousek, V. (2009). Predictive model for frictional pressure drop in settling-slurry pipe with stationary deposit. *Powder Technology, 192*, 367−374.

Maxwell, J. C. (1881). *A treatise on electricity and magnetism*. Oxford: Calendar Press.

Meredith, R. E., & Tobias, C. W. (1962). *Advances in electrochemistry and electrochemical engineering* (2nd ed.). New York, NY: Interscience.

Mosorov, V., Sankowski, D., Mazurkiewicz, L., & Dyakowski, T. (2002). The ''best-correlated pixels'' method for solid mass flow measurements using electrical capacitance tomography. *Measurement Science and Technology, 13*, 1810−1814.

Nasr-El-Din, H., Shook, C. A., & Colwell, J. (1987). A conductivity probe for measuring local concentrations in slurry systems. *International Journal of Multiphase Flow, 13*(3), 365−378.

Newitt, D. M., Richardson, J. F., Abbott, M., & Turtle, R. B. (1955). Hydraulic conveying of solids in horizontal pipes. *Transactions of the Institution of Chemical Engineers, 33*, 93−110.

Pachowko, A. D. (2004). *Design and modelling of a coarse particulate slurry handling system* (Doctoral thesis). Leeds: University of Leeds.

Pachowko, A. D., Poole, C., Wang, M., & Rhodes, D. (2004). Measurement of slurry density profiles in horizontal pipes by using electrical resistance tomography. In *Proceedings of the hydrotransport 16th International conference, Santiago, Chile*.

Parvareh, A., Rahimi, M., Alizadehdakhel, A., & Alsairafi, A. A. (2010). CFD and ERT investigations on two-phase flow regimes in vertical and horizontal tubes. *International Communications in Heat and Mass Transfer, 37*, 304–311.

Parzonka, W., Kenchington, J. M., & Charles, M. E. (1981). Hydrotransport of solids in horizontal pipes: Effects of solid concentration and particle size on deposit velocity. *Canadian Journal of Chemical Engineering, 59*, 291–296.

Peker, S. M., Helvacı, Ş.Ş., Yener, H. B., İkizler, B., & Alparslan, A. (2008). *Solid-liquid two phase flow*. Oxford: Elsevier.

Prager, S. (1963). Diffusion and viscous flow in concentrated suspensions. *Physica, 29*, 129–139.

Qureshi, M. F., Moustafa, H. A., Ferroudji, H., Rasul, G., Khan, M. S., Rahman, M. A., et al. (2021). Measuring solid cuttings transport in Newtonian fluid across horizontal annulus using electrical resistance tomography (ERT). *Flow Measurement and Instrumentation, 77*, 101841.

Razzak, S. A., Barghi, S., & Zhu, J. X. (2009). Application of electrical resistance tomography on liquid–solid two-phase flow characterization in an LSCFB riser. *Chemical Engineering Science, 64*, 2851–2858.

Schlaberg, H. I., Jia, J., Qiu, C., Wang, M., & Hua, L. (2008). Development and application of the Fast Impedance Camera - a high performance dual-plane electrical impedance tomography system. In *5th International symposium on process tomography*.

Seager, A. D., Baber, D. C., & Brown, B. H. (1987). Theoretical limits to sensitivity and resolution in impedance imaging. *Clinical Physiological Measurements, 8*(Suppl. A), 13–31.

Sharifi, M., & Young, B. (2013). Electrical resistance tomography (ERT) applications to chemical engineering. *Chemical Engineering Research and Design, 91*(9), 1625–1645.

Shook, C. A., & Bartosik, A. S. (1994). Particle-wall stresses in vertical slurry flows. *Journal of Powder Technology, 81*, 119–134.

Shook, C. A., & Roco, M. C. (1991). *Slurry flow: Principles and practice*. London: Butterworth-Heinemann.

Silva, R., Faia, P. M., Garcia, F. A. P., & Rasteiro, M. G. (2016). Characterization of solid-liquid settling suspensions using electrical impedance tomography: A comparison between numerical, experimental and visual information. *Chemical Engineering Research and Design, 111*, 223–242.

Stevenson, R., Harrison, S. T. L., Miles, N., & Cilliers, J. J. (2006). Examination of swirling flow using electrical resistance tomography. *Powder Technology, 162*(2), 157–165.

Tapp, H. S., Peyton, A. J., Kemsley, E. K., & Wilson, R. H. (2003). Chemical engineering applications of electrical process tomography. *Sensors and Actuators B: Chemical, 92*, 17–24.

Thorn, R., Johansen, G. A., & Hammer, E. A. (1997). Recent developments in three- phase flow measurement. *Measurement Science and Technology, 8*, 691–701.

Turian, R. M., & Yuan, T. Y. (1977). Flow of slurries in pipelines. *American Institute of Chemical Engineers Journal, 23*(3), 232–243.

Turner, J. C. R. (1976). Electrical conductivity of liquid-fluidized bed of spheres. *Chemical Engineering Science, 31*, 487–492.

Vijayan, M., Schlaberg, H. I., & Wang, M. (2007). Effects of sparger geometry on the mechanism of flow pattern transition in a bubble column. *Chemical Engineering Journal, 130*, 171–178.

Wang, M. (2005). Impedance mapping of particulate multiphase flows. *Flow Measurement and Instrumentation, 16*, 183–189.

Wang, M., Jia, J., Faraj, Y., Wang, Q., Xie, C., Oddie, G., et al. (2015). A new visualization and measurement technology for water continuous multiphase flows. *Flow Measurement and Instrumentation, 46*(Part B), 204–212.

Wang, M., Jones, T. F., & Williams, R. A. (2003). Visualisation of asymmetric solids distribution in horizontal swirling flows using electrical resistance tomography. *Transactions on IChemE, 81*(Part A), 854–861.

Wang, M., Mann, R., & Dickin, F. J. (1999). Electrical resistance tomographic sensing systems for industrial applications. *Chemical Engineering Communications, 175*, 49–70.

Wang, M., & Yin, W. (2001). Measurement of the concentration and velocity distribution in miscible liquid mixing using electrical resistance tomography. *Transactions of the Institution of Chemical Engineers, 79*(part A), 883–886.

Weissberg, H. L. (1963). Effective diffusion coefficient in porous media. *Journal of Applied Physics, 34*, 2636–2639.

Williams, R. A. (1995). A journey inside mineral separation processes. *Minerals Engineering, 8*(7), 721–737.

Williams, R. A., Jia, X., West, R. M., Wang, M., Cullivan, J. C., Bond, J., et al. (1999). Industrial monitoring of hydrocyclones operation using electrical resistance tomography. *Minerals Engineering, 12*(10), 1245–1252.

Wilson, K. C. (1970). Slip point of beds in solid-liquid pipeline flow. *Proceedings of the Institution of Civil Engineers, Journal of the Hydraulics Division, 96*, 1–12.

Wilson, K. C., Addie, G. R., Sellgren, A., & Clift, R. (2006). *Slurry transport using centrifugal pumps* (3rd ed.). New York, NY: Springer.

Wilson, K. C., Clift, R., & Sellgren, A. (2002). Operating points of pipelines carrying concentrated heterogeneous slurries. *Powder Technology, 123*, 19–24.

Wilson, K. C., & Pugh, F. J. (1988). Dispersive-force modeling of turbulent suspension in heterogeneous slurry flow. *Canadian Journal of Chemical Engineering, 66*, 721–727.

Wilson, K. C., Streat, M., & Bantin, R. A. (1972). Slip-model correlation of dense two phase flow. In *Proceedings of the second International conference on the hydraulic transport of solids in pipes, BHRA, Coventry, England, Paper B1* (pp. B1-B1–B1-B10).

Wood, R. J. K., & Jones, T. F. (2003). Investigations of sand–water induced erosive wear of AISI 304L stainless steel pipes by pilot-scale and laboratory-scale testing. *Wear, 255*, 206–218.

Yamaguchi, H., Niu, X. D., Nagaoka, S., & de Vuyst, F. (2011). Solid–liquid two-phase flow measurement using an electromagnetically induced signal measurement method. *Journal of Fluids Engineering, 133*, 041302.

Zhao, X., & Lucas, G. P. (2011). Use of a novel dual-sensor probe array and electrical resistance tomography for characterization of the mean and time-dependent properties of inclined, bubbly oil-in-water pipe flows. *Measurement Science and Technology, 22*(10), 104012.

Application of tomography in microreactors

Daisuke Kawashima and Masahiro Takei
Chiba University, Inage-ku, Chiba-shi, Chiba, Japan

21.1 Introduction

This chapter focuses on engineering processes involved in microreactors, generally defined as devices with the internal structural features, most conveniently measured in microns, that are capable of performing, controlling, and processing chemical or biological reactions in a continuous flow manner. A typical microfluidic platform has a lab-on-chip configuration and can manipulate small volumes (from femtoliter to microliter) of fluids constrained within microfluidic channels. A basic microreactor may include enclosed microchannels, simple mixing units, a heat source, and a pumping system to control fluid or gas flow. Devices have been fabricated from a variety of materials including glass, metals, silicon, quartz, polymers, plastics, and ceramics. Fluid pumping and manipulation in the devices can be achieved by hydrodynamic pumping, most commonly using an external syringe or diaphragm pumps to generate flow or electrokinetic flow, which relies on the application of an electric field along the microreactor to induce bulk flow in aqueous or suitable polar solvent systems. Fluid flow in the microreactors is characteristically different from fluid flow in bulk systems and provides a unique environment in which to perform chemical reactions. Microreactors are capable of controlling and transferring small amounts of gases/liquids/particles that allow chemical processes and biochemical assays to be integrated and carried out on a microscale.

A microfluidic platform has been widely used in the fields of biology and chemistry; it plays an important role in the chemical and process industries. Handling systems involving two or more phases are commonplace in areas from processing of fuels and chemicals to the production of feed, food, pharmaceuticals, and specialty materials. Many applications are aimed at improving aspects of critical basic biologic analyses and clinical diagnosis. For example, in biology, the applications of microreactors include immunoassays, polymerase chain reaction amplification, DNA analyses, chemical gradient formation, cell manipulation, separation, and patterning. On the chemistry side, many examples have demonstrated that microreactors are extremely useful in performing a wide variety of chemical reactions with purer products, higher yields, greater selectivities, and shorter reaction times than those obtained in batch-scale reactions. As one of the major attributes of the microfluidic technology, more benign and milder reaction conditions can be found for certain reactions within microreactors. Manufacturing processes to fabricate microreactors are relatively inexpensive and easily amenable to mass production, even with highly elaborate,

Industrial Tomography. https://doi.org/10.1016/B978-0-12-823015-2.00025-X

complicated device designs. Besides, biology and chemistry carried out in integrated microfluidic chips have great promise as a foundation for new chemical technology and processes.

Additionally, microreactors have a series of generic components for introducing and removing multiple reagents and samples within microfluidic channels; efficient mixing of separate reagents; performing reactions with precise temperature control; and carrying out other functionalities, such as purification for chemical syntheses or detection for product characterization. At the microscale, heat and mass transfer are very rapid owing to high-surface area-to-volume ratio and a large heat exchange coefficient; hence, the reaction conditions within such an environment can be precisely controlled. As a consequence, the yield and selectivity of many biological reactions and chemical syntheses carried out in microfluidic channels can be dramatically improved, whereas the consumption of reagents and reaction or process time is significantly reduced. In addition, unlike devices used in the macroscale chemical and biological reactions, microreactors exhibit a high degree of scalability, reproducibility, modularity, and automaticity. The dominance of viscous forces over inertial forces for moving fluids in microreactors generally results in the existence of laminar flow regimes with Reynolds numbers typically less than 10^2.

Although laminar flow is achieved in most microreactors, mixing can in fact be an extremely fast process, provided that the channel diameters (diffusion distances) are less than 100 μm. Consequently, the rates of chemical reactions can be significantly improved compared with standard laboratory methods. The consequence of increased surface area-to-volume ratio is the enhancement of mass transport via rapid diffusion time scales and significantly improved heat transfer either into or out of the microreactor. Furthermore, it is important to note that from concept to working chip, each microreactor takes less than three working days to construct because of the use of soft lithography. When a new reaction protocol is selected, a chip can be rapidly designed by a computer-aided design program on a personal computer, as well as a standardized production scheme, to meet the new conditions. The reconfiguration of microreactors is more flexible and can meet the diverse needs of producing many different types of compounds. Applications of tomography imaging techniques used with microreactor systems are described and summarized in Table 21.1.

21.2 X-ray and γ-ray tomography

The main factors to determine the imaging capabilities of a computed tomography (CT) scanner used in engineering applications are its achievable spatial, temporal, and density resolution. Spatial resolution is the minimum distance that two high-contrast point objects can be separated; temporal resolution refers to the frequency with which the images can be obtained; and density resolution refers to the smallest difference in the mass attenuation coefficients that the system is able to distinguish. The transmission of X-rays or γ-rays through the heterogeneous medium is accompanied by attenuation of the incident radiation, and the measurement of this attenuation

Table 21.1 Summary of the tomography techniques in microreactors.

General measurement	Sensor installed	Size of microreactor/ microchannel tested system	Phase condition	Reference for more details
X-ray microtomography/ transmission CT	Tungsten or molybdenum sources/ionization chamber type or scintillation type detectors	$D = 3.54 \times H = 3.54 \times V = 2.36\ mm^3$	Gas–solid	Grunwaldt and Schroer (2010); Schroer et al. (2003); Smit et al. (2008)
X-ray absorption spectroscopy	Placing a two-dimensional position-sensitive X-ray camera behind the reactor	Quartz glass capillary microreactor ($D = 1$ mm, $W = 20$ mm)	Gas–solid	Hannemann et al. (2006)
X-ray radiography	Sources same as in X-ray CT; detector, sheet of film, or image intensifier camera		Gas–solid	Jamal et al. (1997)
NMRI	External magnetic field gradient and radiofrequency pulses	186 nm, $D = 85$ nm, and $W = 100$ nm	Liquid	Bouchard et al. (2008)
PET	Positron emitter tracers/ position camera as detector	Flat disk shape ($D = 5$ mm, $H = 255$ mm)	Gas–solid	Elizarov, Meinhart et al. (2010)
			Liquid–gas–solid	Gillies et al. (2006a); Gillies et al. (2006b); Watts and Wiles (2007); Elizarov, Michael et al. (2010)

Continued

Table 21.1 Continued

General measurement	Sensor installed	Size of microreactor/microchannel tested system	Phase condition	Reference for more details
		Ring-shaped channel ($W = 100$ μm)		Lee et al. (2005)
		T-shaped channel ($W = 220$ μm × $D = 60$ μm × $L = 14$ mm; $V = \sim 0.2$ mL)		Lu et al. (2004)
		Capillary reactor ($D = 300$ μm, $L = 0.7$ m)		Wester et al. (2009)
		$W = 200$ μm and $H = 45$ μm		Cheng-Lee et al. (2005)
		$D = 100$ μm, $L = 4.0$ m, $V = 31.4$ μL		Lu and Pike (2010)
		$D = 100$ μm, $L = 4.0$ m, $V = 31.4$ μL		Chun et al. (2010)
		$D = 100$ μm, $L = 2.0$ m, $V = 15.6$ μL		Pascali, Nannavecchia, Pitzianti and Salvadori (2011)
		Glass microreactor ($W = 200$ μm × $D = 75$ μm × $L = 5$ m		Miller et al. (2007)
		Glass microreactor: $V = 100$ μL, $W = 600$ μm, $D = 500$ μm and $L = 360$ mm		Van den Broek et al. (2012)
Electrical impedance tomography	Electrodes energized sequentially	$H = 800$ μm, $L = 20$ mm	Solid–Liquid	Yonghong et al. (2013); Liu et al. (2018);

D, Inner diameter; *H*, Height, *L*; Length; *V*, Volume; *W*, Width.

provides a measure of the line integral of the local mass density distribution along the path traversed by beam. The measurement of several such beams at different spatial and angular orientations with respect to test section or volume, followed by an image reconstruction procedure, provides density distribution of phases to a high degree of spatial resolution. Scanners for transmission tomography employ radiation sources, such as an X-ray tube or an encapsulated γ-ray source, positioned on one side of the object to be scanned and a set of collimated detectors arranged on the other side. With X-ray and γ-ray tomography, the data obtained for the concentration distribution of the phases are almost always time-averaged, since a significant period of time is required to obtain the photon count rates for all the projections (necessary to rotate the sensors and the source). Depending on the design of the scanner, the period can range from a few minutes to close to 1 h. This is a basic limitation of these techniques in studies of the time-evolving flow phenomena. However, dynamic information can be obtained when the dispersion rate or flow velocities are sufficiently slow.

For example, a tomographic study of a Cu/ZnO catalyst in a microreactor (ex situ in a closed capillary: after a reduction−oxidation−reduction cycle) was investigated (Hannemann et al., 2006). In this way, a series of absorption tomograms for all energies in the absorption spectrum were obtained. Each of the tomograms was reconstructed individually, yielding a reconstruction of the distribution of the attenuation at a given energy. Together, the reconstructions provided a full X-ray absorption near-edge structure (XANES) spectrum at each location in the tomographic reconstruction. By fitting the spectra of different copper species at every location in the tomographic reconstruction, the distributions of metallic, monovalent, and bivalent copper can be determined together with the distribution of other elements. The Cu(I):Cu(II) ratio is very different within the microreactor. Although this study was not performed in situ, it was done on a sealed capillary after several reduction/reoxidation cycles; thus, it is straightforward to extend this approach to an in situ analysis. Full-field X-ray spectroscopic tomography could be performed, for example, in an alternative approach: not the X-ray absorption spectroscopic data but the X-ray diffraction (XRD) patterns were acquired by the so-called tomographic energy-dispersive diffraction imaging technique. It used an approach to image in three dimensions the distribution and composition in the catalyst bodies. The sample is scanned in the beam by an energy-dispersive detector. The signal contains both the diffraction and the fluorescence peaks from the defined volume element and thus a full scan results in XRD and X-ray fluorescence datasets. In this way, both concentration in a pellet and the individual phases can be identified. The impregnation of alumina pellet with $(NH_4)_6Mo_7O_{244}H_2O$, after calcination at 500°C MoO_3, is found with the resolution in the 100-μm range. Their method (Hannemann et al., 2006) is related to tomographic X-ray absorption spectroscopic imaging of catalysts applied by Schroer et al. (2003). Fig. 21.1 shows in detail the tomographic reconstruction of copper in (a) the metallic (Cu), (b) the monovalent $[Cu(I)_2O]$, and (c) the bivalent state $[Cu(I)_2O]$ on a virtual section through a reactor capillary filled with a Cu/ZnO catalyst. The results show that the bivalent copper is not present in the sample within the detection limit, while the metallic and monovalent copper are distributed slightly differently, as shown in the difference map in (d). The shade of gray outside the specimen corresponds to

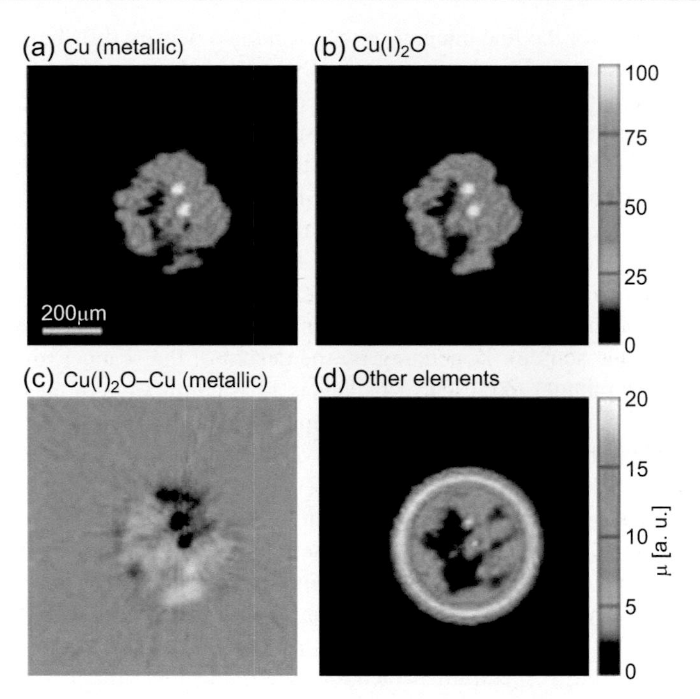

Figure 21.1 Tomographic reconstruction of copper in (a) the metallic (Cu) and (b) the monovalent [Cu(I)2O] on a virtual section through a reactor capillary filled with a Cu/ZnO catalyst. Metallic and monovalent copper are distributed slightly differently, as shown in the difference map in (c). The shade of gray outside the specimen corresponds to zero difference. While brighter areas depict a higher concentration of Cu(I)2O, darker regions show a higher concentration of metallic copper. (d) Attenuation due to other elements.
Reproduced from Schroer, C. G., Kuhlmann, M., Günzler, T. F., Lengeler, B., Richwin, M., Griesebock, B., et al. (2003) Mapping the chemical states of an element inside a sample using tomographic X-ray absorption spectroscopy. *Applied Physics Letters, 82* 3360–3362, with the permission of AIP Publishing. https://doi.org/10.1063/1.1573352.

zero difference. Brighter areas depict a higher concentration of $Cu(I)_2O$, and darker regions show a higher concentration of metallic copper.

Also, Hannemann et al. (2006) investigated the distinct spatial changes of catalyst structure inside a fixed-bed microreactor during partial oxidation of methane over Rh/ Al_2O. The experimental results are shown in Fig. 21.2, where the amount of oxidized Rh-species and the distribution of other elements show a featureless absorption spectrum in the given energy range. Selected examples demonstrate the application of X-ray microscopy and tomography to monitor structural gradients in the catalytic reactors and catalyst preparation with micrometer resolution but also the possibility to follow structural changes in the sub-100 nm regime. In addition, Smit et al. (2008) investigated the nanoscale chemical imaging of a working catalyst by scanning transmission X-ray microscopy. The detailed strategy of their experiment for two-dimensional (2D) mapping of samples on the microscale, experimental setup, and

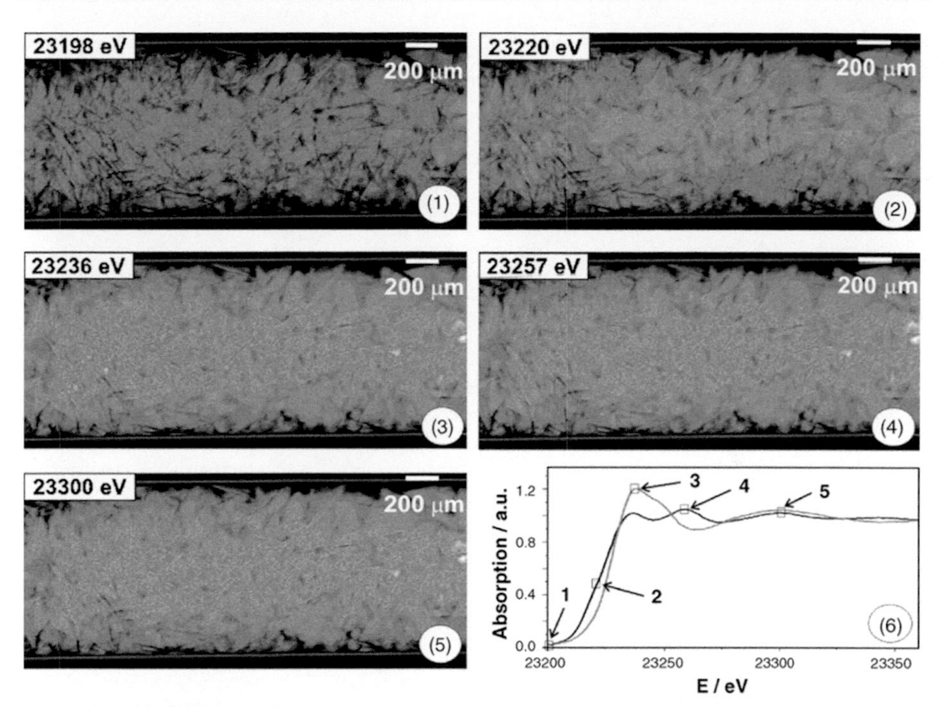

Figure 21.2 Absorption extracted from transmission images (recorded by the CCD camera; flat- and dark-field corrected) of Rh/Al_2O_3 inside the microreactor at 362 C during the CPO of methane at the energies indicated in the corresponding images; center position is 1.8 mm for image (1)−(5); (6) shows the extracted XANES spectra from the flat- and dark-field corrected transmission images at about 0.3 and 3.3 mm at 362 C.
Reprinted with permission from Hannemann et al. (2006). Copyright 2006, Catalyst Today Publishing. https://doi.org/10.1016/j.cattod.2006.08.065.

data acquisition method for soft X-ray scanning microscopy is shown in Fig. 21.3. In order to improve their situation for X-ray absorption spectroscopy, either a continuously scanning monochromator in the so-called QEXAFS mode or dispersive EXAFS is used. In their experiment, by using the QEXAFS mode, the X-rays were focused onto the sample by parabolic refractive X-ray lenses made of beryllium. The sample was scanned through the focus in translation and rotation, recording at each position of the tomographic scan a full near-edge absorption (XANES) spectrum at the Cu K-edge by means of a fast monochromator (sampling frequency 100 kHz, 10 spectra per sample point). In total, about 101 projections over 3601 were recorded, each consisting of 90 translational steps of 10 mm each.

Espinosa-Alonso et al. (2009) approached the real in situ measurements and investigated in a specially designed environmental cell the changes of impregnated $[Ni(en)^3](NO_3)^2/g\text{-}Al_2O_3$ and $[Ni(en)^3]Cl_2/g\text{-}Al_2O_3$ treated in N_2. Despite that the profile was uniform after impregnation with $[Ni(en)^3](NO_3)^2/g\text{-}Al_2O_3$ already during drying at 20°C, the crystallization of large $[Ni(en)^3](NO_3)^2$ was observed, especially

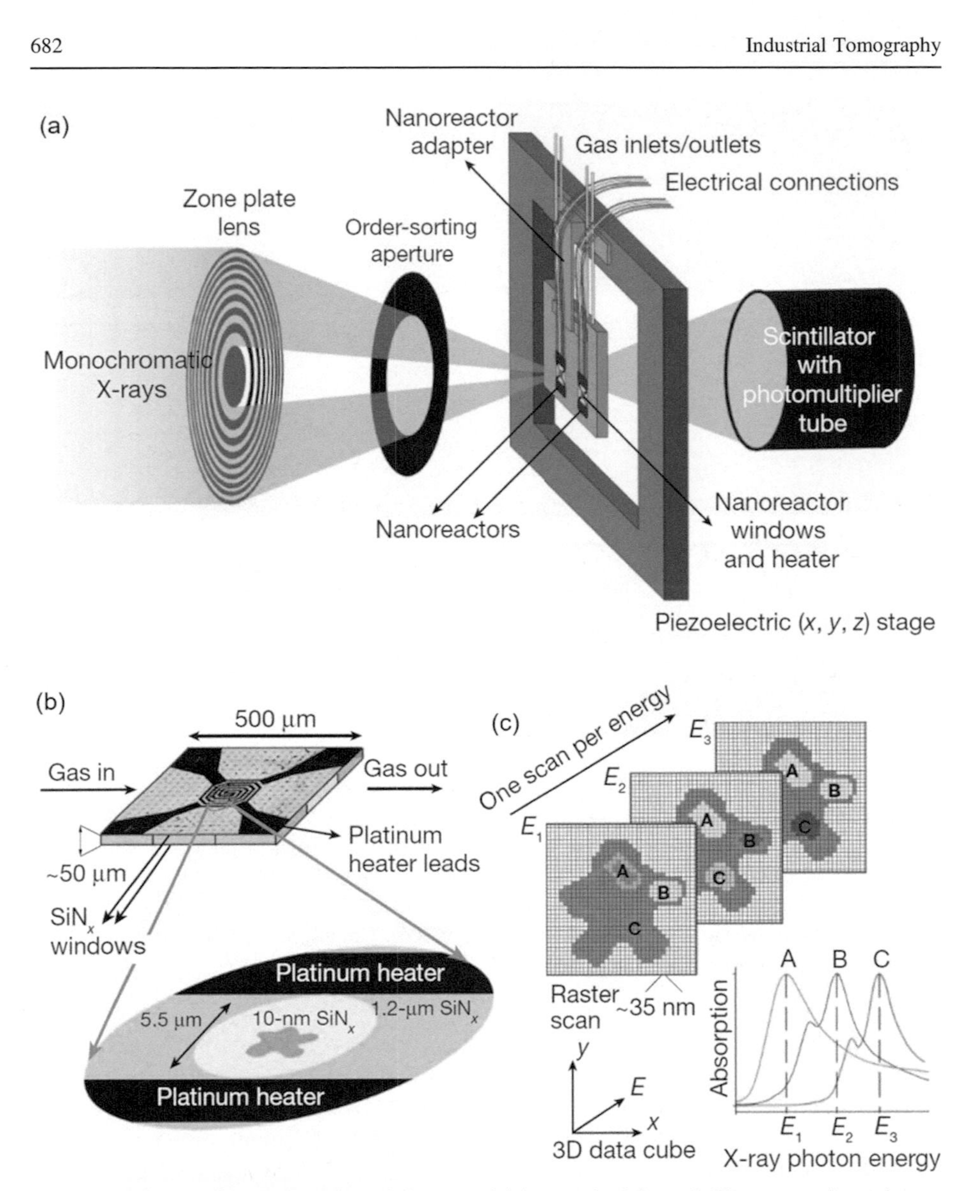

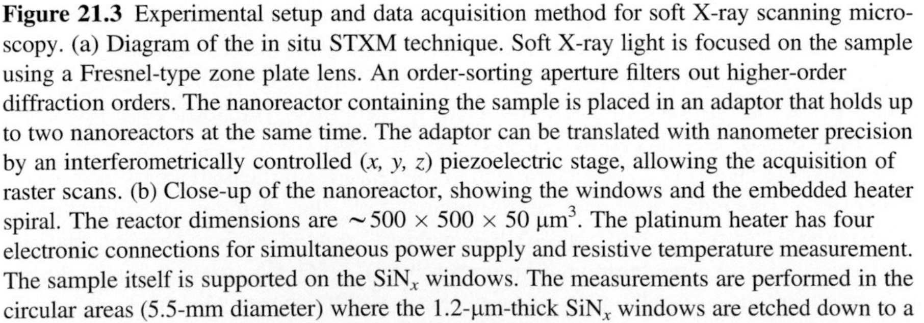

Figure 21.3 Experimental setup and data acquisition method for soft X-ray scanning microscopy. (a) Diagram of the in situ STXM technique. Soft X-ray light is focused on the sample using a Fresnel-type zone plate lens. An order-sorting aperture filters out higher-order diffraction orders. The nanoreactor containing the sample is placed in an adaptor that holds up to two nanoreactors at the same time. The adaptor can be translated with nanometer precision by an interferometrically controlled (*x, y, z*) piezoelectric stage, allowing the acquisition of raster scans. (b) Close-up of the nanoreactor, showing the windows and the embedded heater spiral. The reactor dimensions are $\sim 500 \times 500 \times 50\ \mu m^3$. The platinum heater has four electronic connections for simultaneous power supply and resistive temperature measurement. The sample itself is supported on the SiN_x windows. The measurements are performed in the circular areas (5.5-mm diameter) where the 1.2-μm-thick SiN_x windows are etched down to a

at the outer part of the pellet. In situ 1D scans applied during the calcination process showed the decomposition of the precursor at 190°C. The egg-shell distribution was maintained, after calcinations at 500°C, where 5-nm large particles were detected. These examples demonstrate the importance of imaging the catalyst structure under working conditions and the preparation of catalysts in the millimeter-to-centimeter length scale, where hard X-rays are the optimal choice. Grunwaldt and Schroer (2010) investigated the spatial changes of catalyst structure inside a fixed-bed microreactor during the partial oxidation of methane by using in situ X-ray absorption spectroscopy tomography. They found that in certain cases a variation of catalyst structure can occur inside a catalytic reactor, as a result of temperature or concentration gradients, by using a position-sensitive X-ray camera to record XANES spectra of the catalytic reactor with a spatial resolution on the scale of a few micrometers. Moreover, they also investigated the hard and soft X-ray microscopy and tomography in catalysis, bridging the different time and length scales. They reported that a particularly important aspect is to study catalysts during their preparation and activation and under operating conditions, where X-rays have an inherent advantage due to their good penetration length, especially in a hard X-ray regime. While reaction cell design for hard X-rays is straightforward, smart in situ cells were also reported for the soft X-ray regime. In their study, the axial temperature distribution over a 5.0% Rh/Al_2O_3 catalyst is observed in a quartz microreactor (length 10 mm) upon heating from 300°C to 400°C. The inset gives the thermographic views at furnace temperatures of 300 and 400°C. After ignition, a strong hot spot can be found, connected to strong structural changes.

21.3 X-ray and γ-ray absorption/radiography tomography

X-ray absorption or radiography is a technique based on the same principle as X-ray tomography, but the attenuation of the beam emitted by the X-ray source is registered by sheets of film or an image-intensifier camera. The registered images are recorded on a cine camera or a video recorder and transferred to a computer for processing and analysis. X-ray absorption or radiography has been used for many years to probe the interior of microreactors and enable direct observations of motion. Impregnation of

thickness of ~10 nm. (c) Diagram of a typical STXM data acquisition method. By acquiring images at different X-ray photon energies (for example, *E1*, *E2*, and *E3*), a three-dimensional (3D) data cube with full spectral information at every pixel is obtained. These data can be used to image and distinguish between specific chemical species (for example, species A, B, and C). Reprinted with permission from Smit, E.de., Swart, I., Creemer, J. F., Hoveling, G. H., Gilles, M. K., Tyliszczak, T., et al. (2008). Nanoscale chemical imaging of a working catalyst by scanning transmission X-ray microscopy. *Nature 456*, 222–225 (2008). Copyright 2008, Nature Publishing.

preshaped metal oxide particles is one of the most applied preparation procedures of industrial heterogeneous catalysts. Therefore, different mathematical descriptions of the impregnation step have been developed. For example, Jamal, Larachi, and Dudukovic (1997) reported on the application of X-ray absorption tomography to identify the oxidation state distribution inside a catalytic microreactor. It nondestructively yields information on the microstructure, e.g., grain size structure, of the preshaped catalyst pellets, and the element contrast can be achieved if the X-ray absorption is different. To demonstrate the perspective of their approach, they determined the microstructure of a commercial 0.5% Pd/Al_2O_3 catalyst successfully applied in the alcohol oxidation in supercritical fluids where a Cu-based catalyst with different impregnation depths was prepared using different impregnation times and a similar kind of Al_2O_3 pellet.

Moreover, Elizarov (2009) studied functionalized inorganic monolithic microreactors for high productivity in fine chemical catalytic synthesis. In their study, two striking features of Monosil are a uniform radial distribution of voids and struts by X-ray tomography and a very narrow size distribution of windows and diffusion pores. These characteristics guarantee an extremely flat flow profile and the same residence time for all reactant molecules, as well as a very short diffusion path of 3 μm in the struts. Compared to a microreactor packed with 50 μm particles, the typical diffusion time scale is then 200 times smaller in Monosil. The schematic setup and picture for mapping the oxidation state inside a catalytic reactor in two dimensions under reaction conditions are shown in Fig. 21.4 (Hannemann et al., 2006), and Fig. 21.5 shows the experimental result of the extracted components from the analysis of the 160 dark and flat-field corrected transmission images.

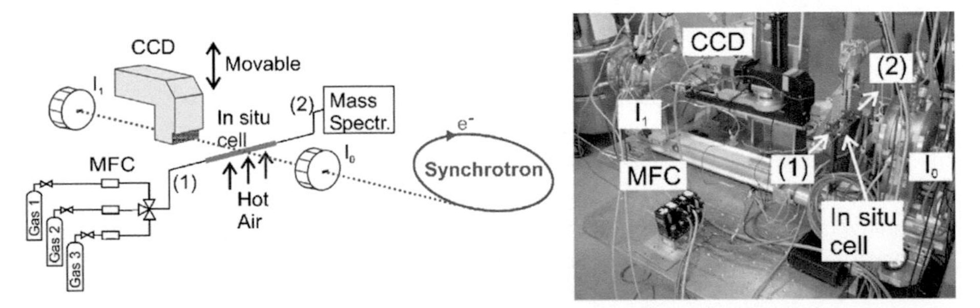

Figure 21.4 Schematic setup and picture for mapping the oxidation state inside a catalytic reactor in two dimensions under reaction conditions; CCD detector (position sensitive detection of the X-ray absorption), ionization chambers ("integral" X-ray absorption spectra) as well as microreactor (in situ cell), oven, and gas supply including MFCs (mass flow controllers) are depicted; (1) denotes the inlet of the in situ cell and (2) the outlet connected to a mass spectrometer.

Reprinted with permission from Hannemann, S., Grunwaldt, J.-D., Vegten, N., Baiker, A., Boye, P., & Schroer Christian, G. (2006). Distinct spatial changes of the catalyst structure inside a fixed-bed microreactor during the partial oxidation of methane over Rh/Al_2O_3. *Catalyst today 126*, 54−63. Copyright 2006, Catalyst Today Publishing.

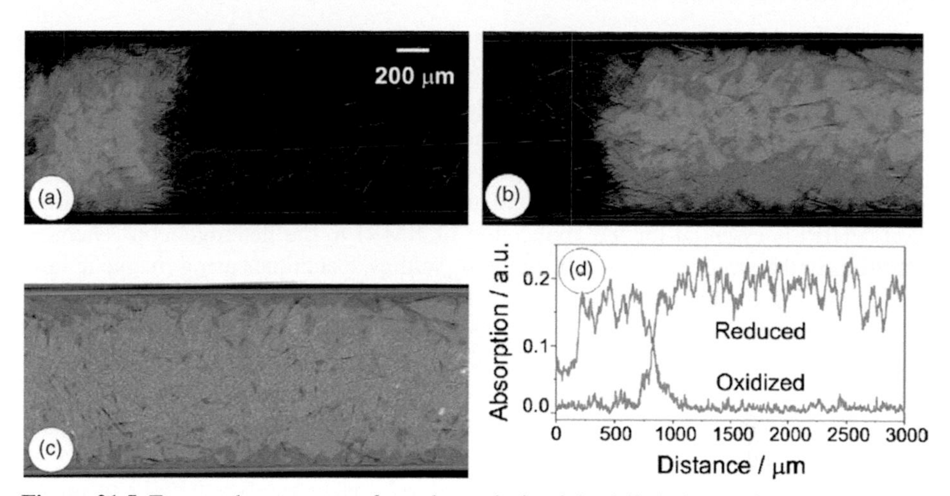

Figure 21.5 Extracted components from the analysis of the 160 dark- and flat-field corrected transmission images: (a) oxidized Rh-species; (b) reduced Rh-species; (c) featureless background; (d) relative concentration of the oxidized (red [dark gray in print version]) and reduced (blue [gray in print version]) Rh-particles in the axis of the fixed-bed (conditions: 362 8C, space velocity 1.9105 h1).

Reprinted with permission from Hannemann, S., Grunwaldt, J.-D., Vegten, N., Baiker, A., Boye, P., & Schroer Christian, G. (2006). Distinct spatial changes of the catalyst structure inside a fixed-bed microreactor during the partial oxidation of methane over Rh/Al_2O_3. *Catalyst today 126*, 54–63. Copyright 2006, Catalyst Today Publishing.

21.4 Nuclear magnetic resonance imaging

Nuclear magnetic resonance imaging (NMRI) has received a great deal of attention, particularly due to the remarkable advances that have been made in the field of medical imaging. The NMRI is a noninvasive method based on the paramagnetic properties of the nuclei. Atomic nuclei are characterized by states that are quantum mechanical in their behavior. Each nucleus has a spin quantum number associated with it and a fixed quantity characterizing its stable ground state. An angular momentum and a magnetic dipole moment, proportional to the angular momentum, are associated with the spin of the nucleus. The constant of proportionality relating the angular momentum and the magnetic dipole moment is known as the gyromagnetic ratio. In NMRI experiments, signals are obtained by having radiofrequency pulses (Callaghan & Xia, 1991) and magnetic field gradient pulses interact with the spin system positioned in a static magnetic field. Imaged velocity and liquid diffusion profiles in laminar flow were investigated through an abrupt contraction and expansion and in capillary flows using spin-echo technique.

The NMRI technique is also applied for catalyst application in microreactor systems. It has proven challenging to correlate the active regions in the heterogeneous catalyst beds with morphology and to monitor multistep reactions within the catalyst bed. Previous applications of NMRI to heterogeneous catalysis included studies of the hydrogenation processes without parahydrogen (pH_2), catalyst morphologies and synthesis technique, and fluid flow through the catalyst bed, as well as monitoring of esterification reactions in situ. All of these applications were based on the nuclear magnetic resonance signal detection of the liquid phase and thus offer a sensitivity that is three orders of magnitude larger as compared with gases, resulting from the difference in density. The sensitivity enhancement offered by pH_2-induced polarization (PHIP) is essential for the application of NMRI to the heterogeneous chemical reactions in the gas phase. From the previous studies, techniques using magnetic resonance imaging and pH_2 polarization were demonstrated that allow direct visualization of gas-phase flow and the density of active catalyst in the packed-bed microreactor, as well as control over the dynamics of the polarized state in space and time to facilitate the study of the subsequent reactions. These procedures are suitable for characterizing reactors and reactions in microreactors; low sensitivity of the conventional magnetic resonance would otherwise be the limiting factor.

To demonstrate the effectiveness of heterogeneously catalyzed PHIP in microreactors, two model catalytic reactors are used. Bouchard et al. (2008) reported that at time, $T = 0$ ms, no isotropic mixing with the polarized product is observed in the catalyst layer. Meanwhile, during $T = 10$ ms after isotropic mixing, the polarized product travels about 5 mm and at $T = 40$ ms after isotropic mixing the product has traveled 10 mm. The schematic of the microreactors, consisting of catalyst layers sandwiched between layers of glass beads for stability, is shown in Fig. 21.6(a), and the result of the density of active catalyst and flow map imaging by NMRI is shown in Fig. 21.6(b).

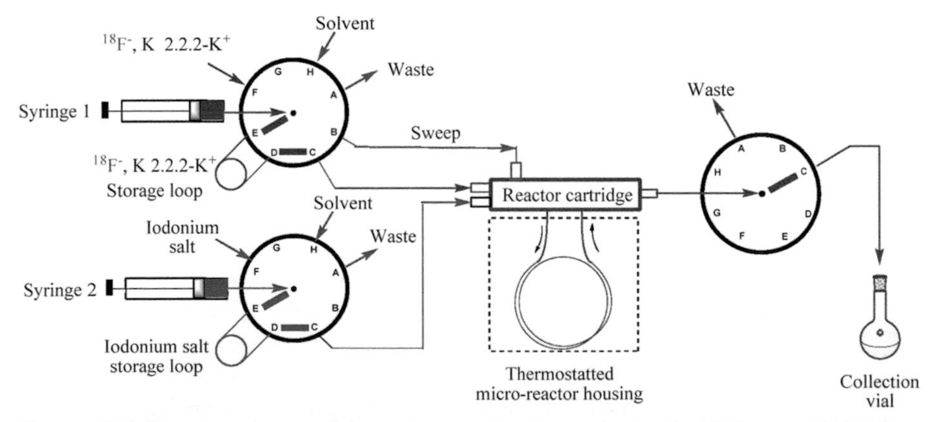

Figure 21.6 Experiment setup: Schematic setup for the synthesis of a PET tracer (FMISO). Reprinted with permission from Chun, J. -H., Lu, S., Lee, Y. -S., & Pike, V. W. (2010). Fast and high-yield microreactor syntheses of ortho-substituted [18F]fluoroarenes from reactions of [18F] fluoride ion with diaryliodonium salts. *Journal of Organic Chemistry 75*, 3332–3338. Copyright 2010, J. Org. Chem Publishing.

21.5 Positron emission tomography

Positron emission tomography (PET) is a very powerful diagnostic technique; it can be used in a wide range of medical applications including oncology, neurology, and cardiology. The basic principle of PET relies on the incorporation of short-lived positron-emitting radioisotopes, most commonly ^{18}F (half-life of 110 min) or ^{11}C (half-life of 20 min), into the radiopharmaceutical; other isotopes that are sometimes used include ^{13}N, ^{15}O, and ^{124}I. The fact that ^{18}F and ^{11}C radioisotopes have such short half-lives is a significant challenge for the synthetic chemist, as the organic reactions used to prepare the molecule of interest must be relatively fast chemical transformations, with the goal being to prepare, purify, and isolate the desired product in one to two half-lives; consequently, for ^{11}C, this means that the chemist must synthesize and isolate the pure target in less than 30 min. The enhanced mass transport and heat transfer in such microreactors should result in shorter reaction times and better radiochemical yields (RCYs) and radiochemical purities (RCPs). As PET radionuclides in the tracer molecules decay, the emitted positrons travel a short distance in a range of about 1–5 mm in tissue, depending on the radionuclide, and collide with the nearby electrons. Widely expected benefits included a superior reaction control, enhanced product yields, enhanced reaction kinetics, production reliability, safety, and facile automation and integration with downstream processing, all essential for good manufacturing practices in production of PET tracers.

The miniaturization of the PET radiosynthesis devices has the potential to address many of the current issues associated with PET probe production. This is because the microfluidic radiochemistry microreactor platforms can offer several advantages, such as reliable production of PET probes with high RCP and RCY, stand-alone and self-shielded devices without the full-sized hot cells, production of various probes using custom-designed microfluidic chips, use of lower starting activity and less probe precursor, and a significant reduction in separation challenge during purifications. Typically, the PET uses radioactive tracers that decay via the emission of a positron. The radioactive decay occurs when the nucleus of these tracers, rich in protons, converts one proton into an uncharged neutron, expelling a positron and the antiparticle of the electron. Any positron expelled from the nucleus will combine with an electron and these can annihilate each other, and their total rest mass energy is converted into two 511-keV back-to-back γ-rays. By detecting γ-rays simultaneously with the positron-sensitive detectors, a line of sight along which the decay must have taken place can be inferred. By detecting many such decays, the distribution of tracer activity can be determined (Benton & Parker, 1997).

In the case of [^{11}C] methylation application in microreactors, Lu et al. (2004) synthesized ^{11}C-labeled carboxylic ester via [^{11}C] methylation using a simple hydrodynamically driven microreactor. It was composed of two inlet ports for reagent delivery: a microfluidic channel for mixing and radiolabeling (with dimensions of 220 μm [width] × 60 μm [height] × 14 mm [length] and total volume of approximately 0.2 μL) and one outlet port for product collection. Precise control of flow rate was attained by syringe pumps, delivering [^{11}C]CH_3I and 3-(3-pyridinyl) propionic acid accurately into the microchannel. From their study, the best RCY, where the RCY is decay corrected of the corresponding [^{11}C]methyl ester, was 88% when the infusion rate of the reagents was set at 1 μL/min at room temperature. The average residence time of the reagents on the chip device was approximately 12 s and gave a total processing time of 10 min for the whole reaction, which is commensurate with the time scale of PET radiolabeling reactions. Their microreactor was further applied to prepare the compound of brain peripheral benzodiazepine receptor ligand via [^{11}C] methylation and the highest RCY was 65% when the infusion rate was at 1 μL/min.

Lee et al. (2005) reported the synthesis of [^{18}F]FDG in a polydimethylsiloxane microreactor consisting of a complex array of channels, typically 200 μm wide and 45 μm deep. Their microreactor was designed to have different temperature zones; the fluorination reaction was conducted at 100°C for 30 s followed by 120°C for 50 s, while acid hydrolysis was conducted at 60°C. Using their approach, [^{18}F]FDG was obtained in 38% RCY but in very high purity, 97%. The total processing time was reduced to 14 min compared with the typical 50 min for current batch protocol. They also designed a chemical reaction circuit with the capacity to synthesize large [^{18}F]FDG doses. The chip has a coin-shaped reactor with a volume of 5 mL equipped with a vacuum vent. It was used to synthesize 1.74 mCi of [^{18}F]FDG, an amount sufficient for several mouse experiments. From the purified and sterilized product, two doses (375 and 272 mCi) were used for micro-PET molecular imaging of two mouse models of cancer. They concluded that a proof of principle for automated multi-step syntheses at the nanogram to microgram scale may be generalized to a range of radiolabeled substrates.

In the case of radiohalogenation of small- and large-molecular weight molecules with a microfluidic device, Gillies et al. (2006a) and Gillies et al. (2006b) used two microreactor devices in series to achieve two steps ([^{18}F] fluorination + deprotection) required for [^{18}F]FDG synthesis. These reactions involved direct radioiodination of apoptosis marker annexin V using ^{124}I, the indirect radioiodination of the anticancer drug doxorubicin from a tinbutyl precursor, and the radiosynthesis of [^{18}F]FDG from a mannose triflate precursor and ^{18}F. They demonstrated the rapid radioiodination of the protein annexin V (40% RCY within 1 min) and the rapid radiofluorination of [^{18}F]FDG (60% RCY within 4 s) using a polymer microreactor chip. In addition, Gillies et al. (2006a) and Gillies et al. (2006b) reported on a three-layer polycarbonate microreactor for radiolabeling the PET isotopes ^{18}F and ^{124}I and SPECT (single-photon emission CT) isotope ^{99m}Tc. In their study, the top plate of their device contains three inlets for reagents; the middle plate is a disc-shaped cavity with a simple vortex mixer, which also acts as an outlet to the bottom disc cavity containing an outlet port. Three separate syringes were respectively loaded with annexin V protein in buffer solution, [^{124}I]NaI in buffer solution and iodogen (an oxidizing agent necessary to convert I− to I^+ for electrophilic oxidation) in acetonitrile, and pumped through the microreactor simultaneously. As a result, after 2 min, the labeling efficiency was about 40% and was comparable with conventional methods over the same time frame. The microreactor was used to conduct the fluorination of mannose triflate, followed by acid hydrolysis of an intermediate to synthesize [^{18}F]FDG; 50% incorporation of the radiolabel was achieved with a residence time of just 4 s.

In other research, Miller et al. (2007) described a simple, low-cost, and effective microreactor employing a reusable silica-supported palladium catalyst packed into a standard Teflon tube for rapid multiphase [^{11}C] carbonylation reactions. It combined three phases: gaseous [^{11}C]CO, substrates in the solution, and a solid-phase catalyst in the [^{11}C] labeling process. The main component of the microreactor is the catalyst-immobilized Teflon microtube with a diameter of 1 mm and a length 45 cm, connected with precision syringe pumps or injectors and a mass flow controller to meter the flow of [^{11}C]CO through a mixing of a T-shaped connector. The active catalyst within the microtube is a palladium phosphine complex attached to the silica support material, where the surface area-to-volume ratio is drastically increased. It is estimated to be about 20,000 times larger than that produced by coating only the catalyst inside the surface of the microchannel. A feature such as this has dual properties in contributing to the improved yields of [^{11}C]-labeled amides: it provides a large active surface area (coated with palladium catalyst) for the catalytic reaction to occur; and, as well, it increases the turbulence of [^{11}C]CO and substrate solution, enhancing the mixing efficiency.

Meanwhile, Steel, O'Brien, Luthra, and Brady (2007) reported the use of a two-stage glass-fabricated microreactor for the automated synthesis of [^{18}F]FDG where a solution of mannose triflate in the anhydrous acetonitrile is reacted with a premade complex from [^{18}F]KF, Kryptofix, and K_2CO_3 in acetonitrile. They reported about 40% conversion for the radiolabeling reaction when a residence time of 2 min was used. A total run time of 10 min was reported from dilute [^{18}F]fluoride ion processed macroscopically to crude [^{18}F]FDG. The automated single device presents a significant advantage, in terms of total processing time, over the two-device method reported; it took about 30–60 min for the same [^{18}F] fluorination/acetyl deprotection steps. Improved RCYs of 40% for [^{18}F]FDG synthesis were reported using different structured microfluidic devices enabling fast-flow rates and [^{18}F] processing within a total processing time of 10 min. In another case, Wester, Schoultz, Hultsch, and Henriksen (2009) used a capillary reactor (inner diameter 300 μm, length 0.7 m) for the production of [^{18}F]FDG. They reported that the reaction had optimum performance at a temperature of 105°C and a residence time of 40 s. The group subsequently conducted a NaOH hydrolysis, to afford [^{18}F]FDG in 88% RCY within a processing time of 7 min. In another instance, Lu and Pike (2007) reported the synthesis and reaction of [^{18}F]XeF_2 in a microreactor. They conducted a detailed study that indicated that the optimum method for the synthesis was to react xenon difluoride with cyclotron-produced [^{18}F]fluoride in either dichloromethane at room temperature or acetonitrile at 80°C, with a residence time of about 1.5 min. With the optimum conditions for generation of [^{18}F]XeF_2 in hand, they subsequently investigated the reaction of the complex. Also, the reaction of a solution of trimethylsilyl silyl enol ether in acetonitrile with [^{18}F]XeF_2 at 80°C for 3.1 min afforded the radiolabeled ketone in 78% RCY.

Also, Lu, Giamis, and Pike (2009) synthesized [^{18}F]fallypride in a microreactor for rapid optimization and multiple production in small doses. A commercial coiled-tube microreactor (NanoTek; Advion) was used as a convenient platform for synthesis of [^{18}F]fallypride in small doses (0.5–1.5 mCi) of brain dopamine subtype-2 receptors in rodents. Each radiosynthesis used low amounts of tosylate precursor and [^{18}F]fluoride ion. Optimization of the labeling reaction in their apparatus, with respect to the effects of precursor amount, reaction temperature, flow rate, and [^{18}F]fluoride ion-to-precursor ratio, was achieved rapidly and the decay-corrected RCY of [^{18}F]fallypride up to 88% was reproducible. From their study, in a 2-m-long reactor, the decay-corrected RCY of [^{18}F]fallypride increased from 0% to 65% over the temperature between 100 and 170°C. The RCYs of [^{18}F]fallypride were quite reproducible and independent of laboratory setting and method of analysis. Each radiosynthesis used 20 or 40 μg (38.7 or 77.4 nmol) of precursor, only 1 or 2% of that used conventionally, and a whole run required only 4 min. By using two of 2-m-long reactors (total length 4 m) connected in series, RCYs reached 88%. In a single 4-m reactor, a longer residence time, achieved by using a slower flow rate from each reactant syringe, moreover increased the RCY. When the volume ratio of [^{18}F]fluoride ion to precursor solution was increased, RCY increased slightly and when this ratio was decreased, RCY decreased slightly with flow rate on the RCYs of [^{18}F]fallypride at 140°C in a 4-m microreactor. The product fraction eluting between 10.9 and 11.9 min was collected for formulation.

Moreover, Chun, Lu, Lee, and Pike (2010) investigated fast and high-yield microreactor syntheses of *ortho*-substituted [^{18}F]fluoroarenes from the reactions of [^{18}F]fluoroarene ion with diaryliodonium salts. In their study, a microreactor was applied to produce the *ortho*-substituted [^{18}F]fluoroarenes from reactions of cyclotron-produced [^{18}F]fluoride ion with $t_{1/2} = 109.7$ min with diaryliodonium salts. The microreactor provided a very convenient means for running sequential reactions rapidly with small amounts of the reagents under well-controlled conditions, thus allowing reaction kinetics to be followed and Arrhenius activation energies to be measured. The schematic of the experimental setup for the synthesis of a PET-tracer (FMISO) is shown in Fig. 21.6. As well, Elizarov et al. (2010) investigated the flow optimization of a batch microfluidics PET-tracer synthesizing device with a coin-shaped microfluidic chip. They used fluorescent imaging techniques for mixing and elution optimization inside a fabricated reaction chamber. For the mixing quality evaluation tests, the reactor was partially filled with an aqueous dye solution (AlexaFluor4 fluorophore dye; 0.1 mg/mL) followed by dead-end injection of pure water and monitoring the quality of mixing over time. Total fluorescence emission within the reactor was monitored to record elution efficiency. As a result, the elution process was considered complete when the total reactor fluorescence measure had decreased by 95% of initial value. Also, Keng et al. (2011) investigated the microchemical synthesis of molecular probes on an electronic microfluidic device. Figs. 21.7 and 21.8 show the electrowetting on dielectric (EWOD) microchip with four concentric heaters with a maximum volume and the sequence of photographic images with corresponding Carenkov images and synthetic scheme of multistep radiosynthesis performed on EWOD.

21.6 Electrical impedance tomography

Electrical impedance tomography (EIT) is a noninvasive technique for imaging the distribution of an electrical property within a medium using electrical measurements from a series of electrodes flush-mounted with the medium surface. Electrical properties employed in EIT include electrical resistance tomography (ERT), electrical

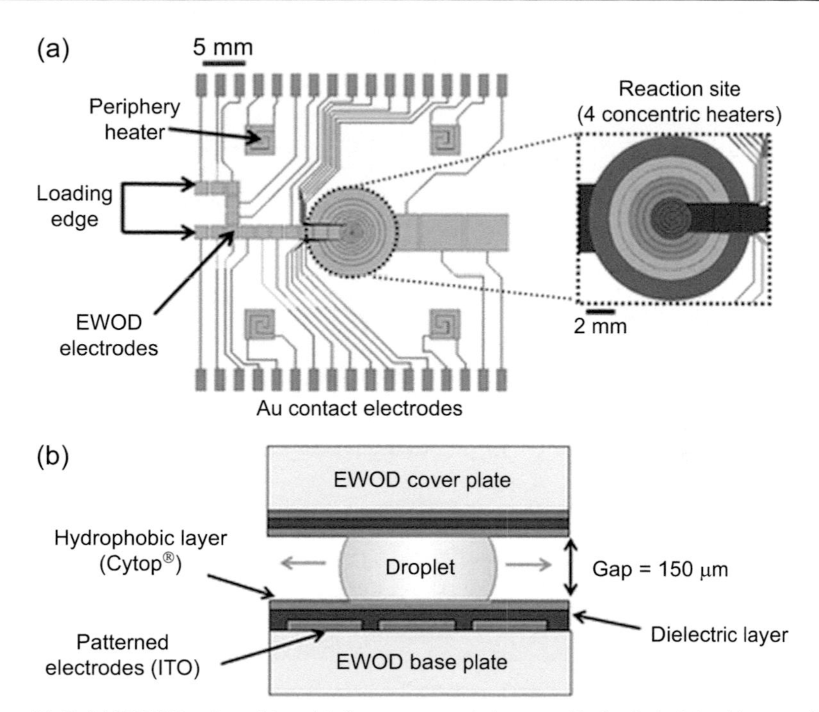

Figure 21.7 (a) EWOD microchip with four concentric heaters (dashed circle) with a maximum volume of 17 μL. Inset shows magnified area of the heater with four concentric individually controlled resistive heating rings. (b) Schematic side view of the EWOD chip sandwiching a reaction droplet between two plates coated with ITO electrodes, a dielectric layer, and a hydrophobic layer of Cytop (not to scale).
Credit: Reprinted with permission from Keng, P. Y., Chen, S., Ding, H., Sadeghi, S., Shah, G. J., & Dooraghi, A. et al. (2011). Micro-chemical synthesis of molecular probes on an electronic microfluidic device. *Proceedings of the National Academy of Sciences of the United States of America* v1−6 (2011). Pnas 1117566109. Copyright 2011, Proceedings of the National Academy of Sciences of the United States of America Publishing.

capacitance tomography (ECT), and electrical magnetic inductance tomography (EMIT). ERT and EMIT are useful for detecting electrically conducting materials or for materials that generate eddy currents or affect magnetic permeability, while ECT is suitable for electrically insulating multiphase systems. In point measurement devices, electrical sensors usually consist of two or more electrodes placed on the opposite sides of the section to be imaged. For tomographic measurements, an array of electrodes around the boundary of the process stream is required. Each measurement can be used to determine the impedance value in one region or pixel in sensed volume.

In general, electrical sensing techniques have the advantage of being faster and safer than hard-field shorter-wavelength nuclear-based tomographies. Also, since the primary signal issued from electrical sensing techniques is itself of an electrical nature, these techniques are easily amenable to automation and digital computer control. As well, the cheaper commercial implementation of EIT and electrical imaging techniques in general accommodate large- and small-vessel geometries with greater flexibility than nuclear-based tomographies. However, even though EIT is powerful in tracing

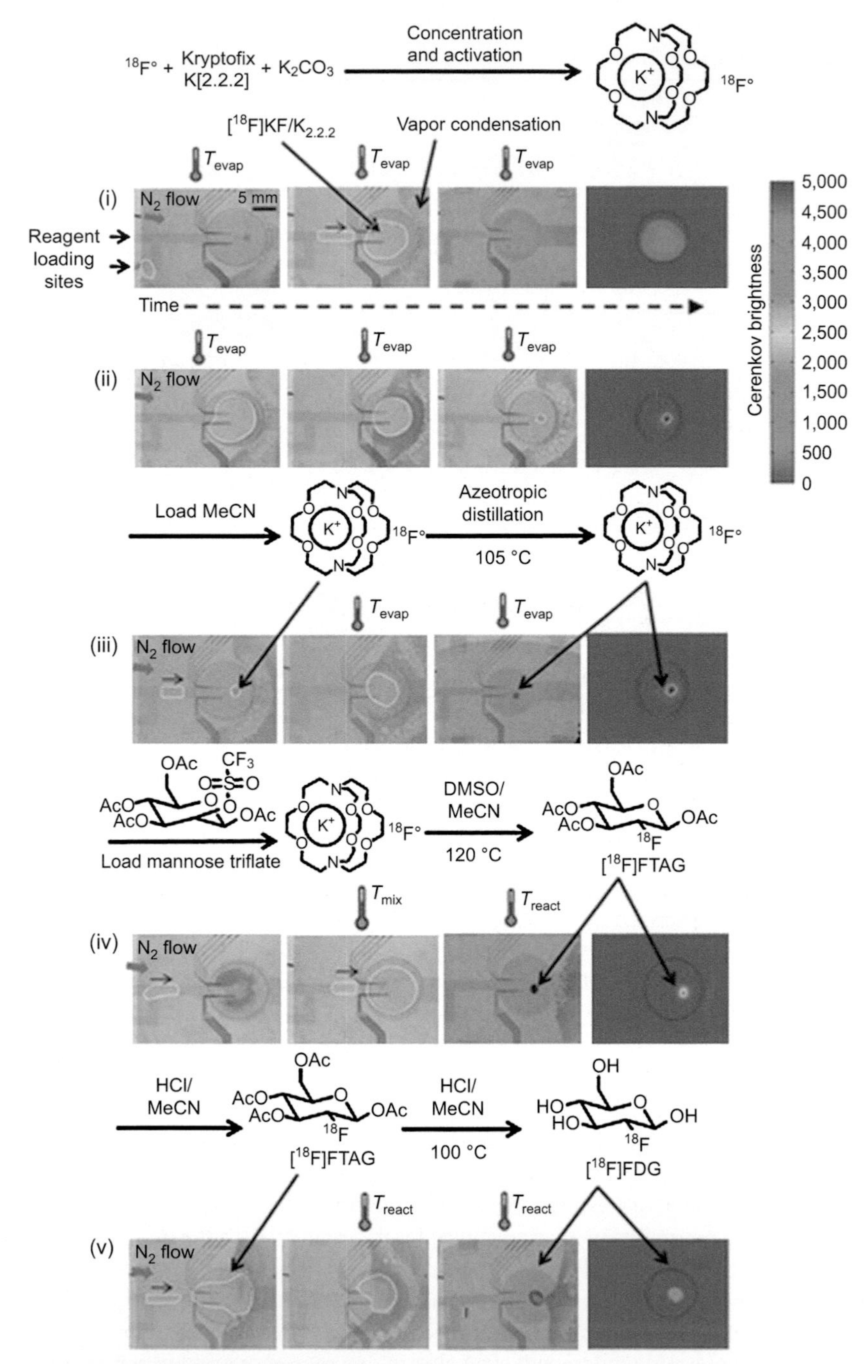

Figure 21.8 Experiment results: Sequence of photographic images with corresponding Carenkov images and the synthetic scheme of the multistep radiosynthesis performed on EWOD.

Credit: Reprinted with permission from Keng, P. Y., Chen, S., Ding, H., Sadeghi, S., Shah, G. J., & Dooraghi, A. et al. (2011). Micro-chemical synthesis of molecular probes on an electronic microfluidic device. *Proceedings of the National Academy of Sciences of the United States of America* v1–6 (2011). Pnas 1117566109. Copyright 2011, Proceedings of the National Academy of Sciences of the United States of America Publishing.

evanescent phenomena and fast-flowing streams exhibiting intense composition fluctuations, the spatial resolution of EIT images is considerably poorer than that of the nuclear-based tomographies. The spatial resolution corresponds to voxels of 1 $cm^2 \times 2.54$ cm (METC ECT, 32 electrodes) and 3.4 $cm^2 \times 10$ cm (UMIST ECT, 12 electrodes) and is limited by uniformity, alignment, and electrode size used to contain and shape electrical flux lines.

Additionally, the fluid flow in a microchannel has emerged as an important area of research that has been motivated by various applications in medical and biomedical use, computer chips, and chemical separations. Recently, EIT techniques were applied to the microchannel, so-called μEIT (Yao & Takei (2017)). Yao, Koder, Obara, Sugawara and Takei (2015) developed an integrated microchannel–electrode system which had five cross-sectional sensor regions with multiple electrodes embed as shown in Fig. 21.9(a). The electrode has 10 μm in thickness and 200 μm in width,

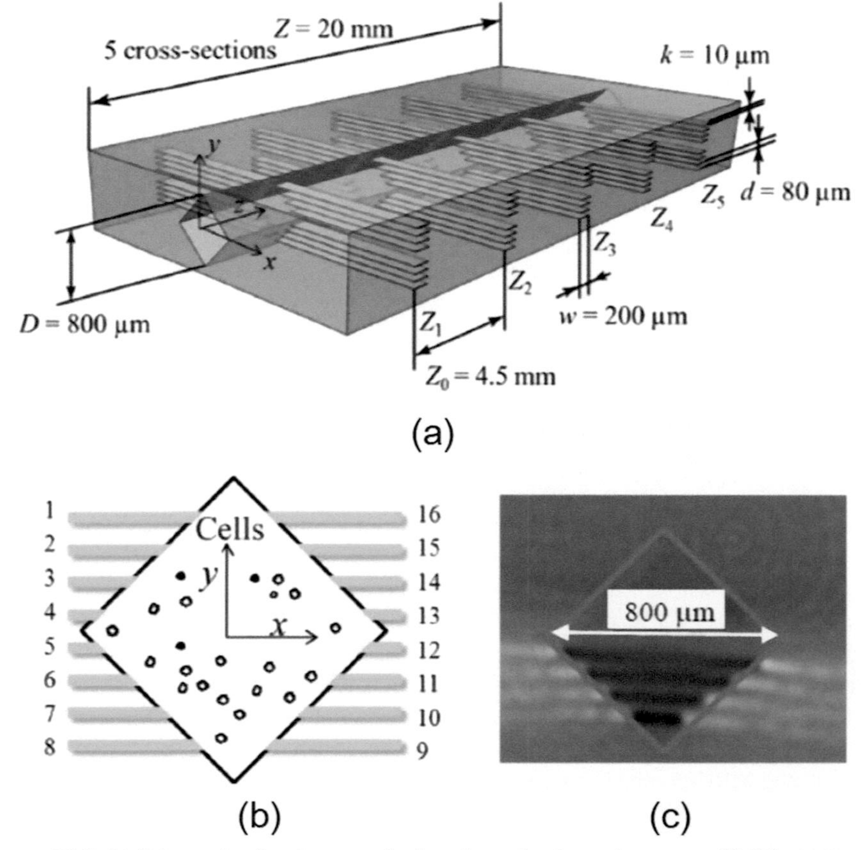

Figure 21.9 (a) Schematic of an integrated microchannel–electrode system. (b) Schematic of a cross-sectional view in the microchannel. (c) Microscopic image of a cross-sectional view in the microchannel.

Reprinted with permission from Yao, J., Kodera, T., Obara, H., Sugawara, M., & Takei M. (2015). Spatial concentration distribution analysis of cells in electrode-multilayered microchannel by dielectric property measurement. *Biomicrofluidics* 9, 044129 (2015). 10.1063/1.4929824. Copyright 2015, Biomicrofluidics Publishing.

and the electrode interval in the cross-section is 80 μm. Fig. 21.9(b) and 21.9(c) show the schematic and microscopic image of the cross-sectional view in microchannel with 16 electrodes. The diagonal length of the cross-section is 800 μm.

On a basis of the system, μEIT can visualize the distribution of chemical particles and biological cells in the microchannel. Yonghong, Xiantao, and Takei (2013) investigated solid–liquid phase flow image reconstruction by μEIT in a microchannel. Fig. 21.10 shows the experimental result of the predicted equipotential in the cross-section from finite element modeling. It gives representative results for the 16-electrode in microchannel, a single particle with 5 μm diameter located near the top right of the diagram, suggesting that the invading solid phase can result in the change of equipotentials distribution. Fig. 21.11 shows voltages measured in the experiment; Figs. 21.11(a)–(c) are corresponding voltages measured on cross-sections I, III, and V. From these results, it has been shown the change with the controlled two-phase flow inside a microchannel with the highest voltage level about 3.5 mV. The results prove that contact resistance can be effectively suppressed, and measured boundary voltages are sensitive to the changes of resistivity distribution in the microchannel under the experimental conditions.

Liu et al., (2018) applied μEIT to visualize the cell distribution in microchannel during cell sedimentation. They performed an experiment using the setup shown in Fig. 21.12(a). They considered the impedance spectral characteristics of contact impedance due to electrical double layer. As shown in Fig. 21.12(b), the contact impedance Z_c in low frequency f is much higher than that in high frequency f. According to their results, the sensor size also influences the contact impedance. The contact impedance in micro electrode (10 × 200 μm) is influenced even in higher frequency region ($f < 100$ kHz) compared to that in macro electrode (4 × 10 mm) ($f < 1$ kHz). Fig. 21.12(c) shows the reconstructed images by μEIT at different frequency in the case of cell sedimentation. According to the images, the image at $f = 1$ MHz shows a better image quality representing the cell sedimentation without the effect of contact impedance.

In addition, Yao, Obara, Sapkota and Takei (2016) and Sato, Yao, Kawashima and Takei (2019) investigated the cell sorting due to dielectrophoresis (DEP) by using the integrated microchannel–electrode system. They developed another structure of cross-sectional sensor whose electrodes were arranged on the opposite walls in microchannel as shown in Fig. 21.13(a), where the channel size is 550 μm in height ×0.550 μm in width. Yao et al. (2016) analyzed four kinds of electrical kinetic force such as DEP, electro-thermal (ET) force, electroosmotic (EO) force, and thermal buoyancy (TB) using dielectric particles under different buffer solution with different conductivity in the case of PBS and water buffer as shown in Fig. 21.13(b). Also, Sato et al. (2019) used MRC-5 cells in sucrose solution to investigate positive DEP which shows particles are attracted to electrode. Fig. 21.14(a) shows the particle tracking of three cells. Those cells were attracted to electrodes and those velocities can be described by the equation described by gradient of electric field $\mathbf{E}$ and applied voltage ϕ_e as shown in Fig. 21.14(b). Their studies can be combined with μEIT to implement the particle distribution monitoring and manipulation system.

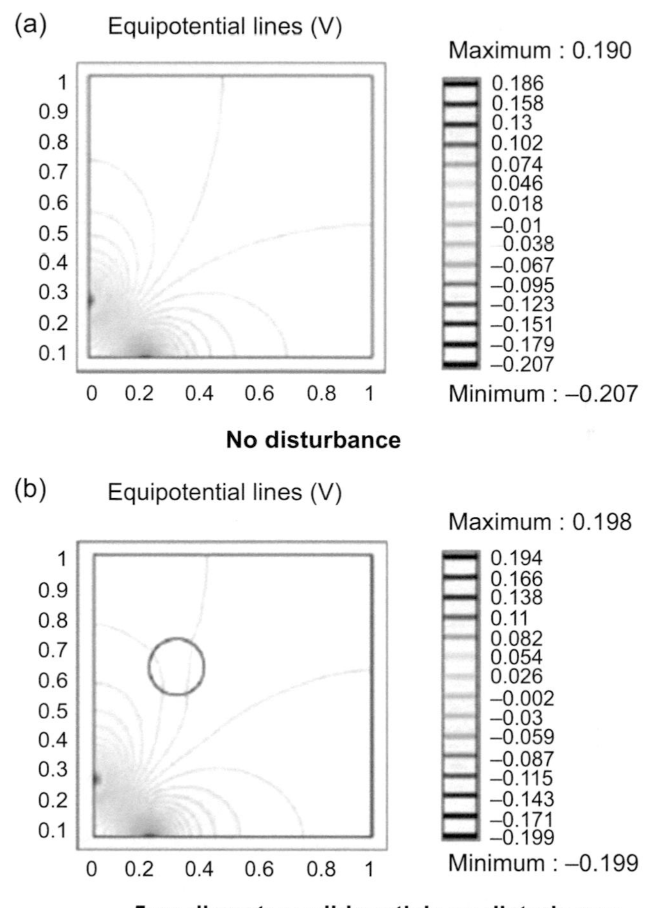

Figure 21.10 Predicted equipotential in the cross-section from finite element modeling. Reproduced from Yao, J., Kodera, T., Obara, H., Sugawara, M., & Takei, M. (2015). Spatial concentration distribution analysis of cells in electrode-multilayered microchannel by dielectric property measurement. *Biomicrofluidics 9*, 044129. 10.1063/1.4929824, with the permission of AIP Publishing. https://doi.org/10.1063/1.4929824.

21.7 Future trends

Microreactor and microchannel technology is a powerful tool and has become indispensable over a wide application range, from organic synthesis to enzymatic controlled reactions. A miniaturization with the possibility of reaction tomography, and thus a significant reduction in costs, will be the next step in the foreseeable future. Many chemical reactions, especially organic and fine chemical synthesis, could already be transferred to continuous microreaction processes. The small geometric dimensions result in an intensified mass and heat transfer, which often leads to increased yield

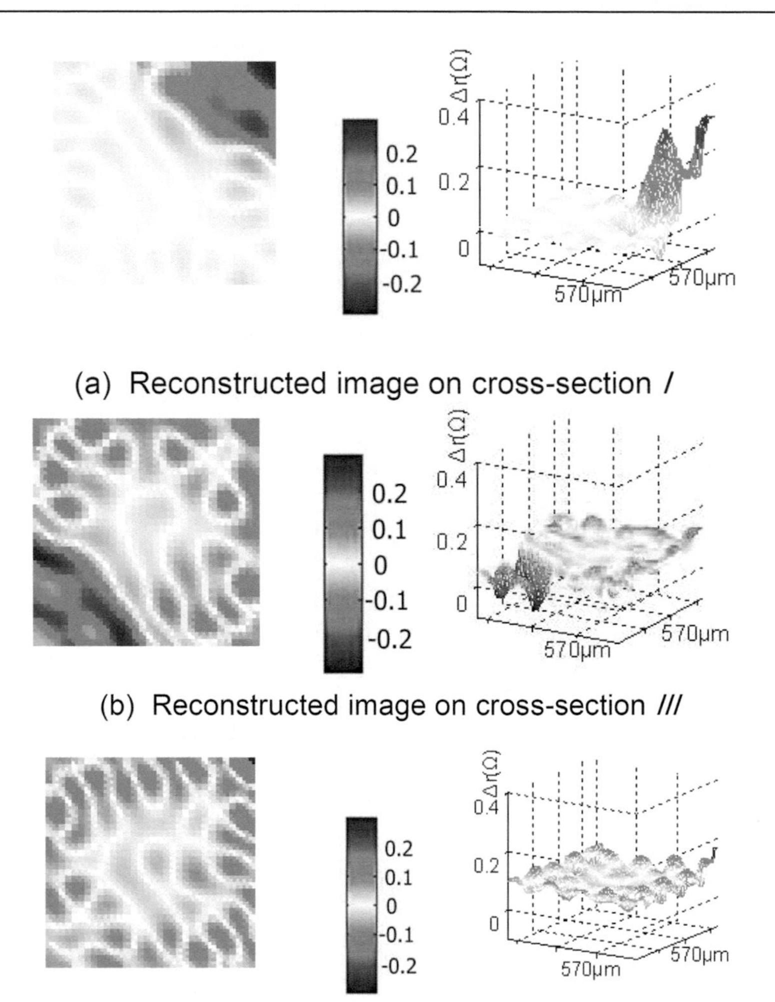

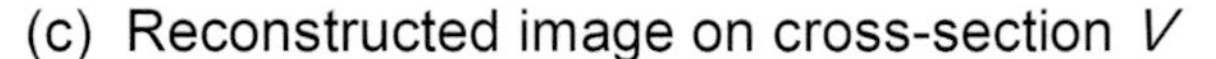

Figure 21.11 2D and 3D distribution on some sample cross-sections. It shows the changes of resistivity distribution in all of the five cross-sections in the microchannel. It can be seen that along the flow direction from cross-section I (a) to cross-section III (b) and cross-section V (c), polymer microsphere suspension and NaCl solution resolve into each other gradually except in cross-section I, NaCl solution just meet each other. The test result also reveals that under certain flow rate, polymer microparticles have already dispersed evenly near cross-section V. Reprinted with permission from Yonghong et al. (2013). Copyright 2013, TELKOMNIKA Publishing.

and selectivity compared to the classical batch approach. However, microreaction technology today is still at the threshold between academia and industry. A crucial factor for the successful implementation of microstructured production processes in industry is a suitable process analytical technology. Time and spatially resolved

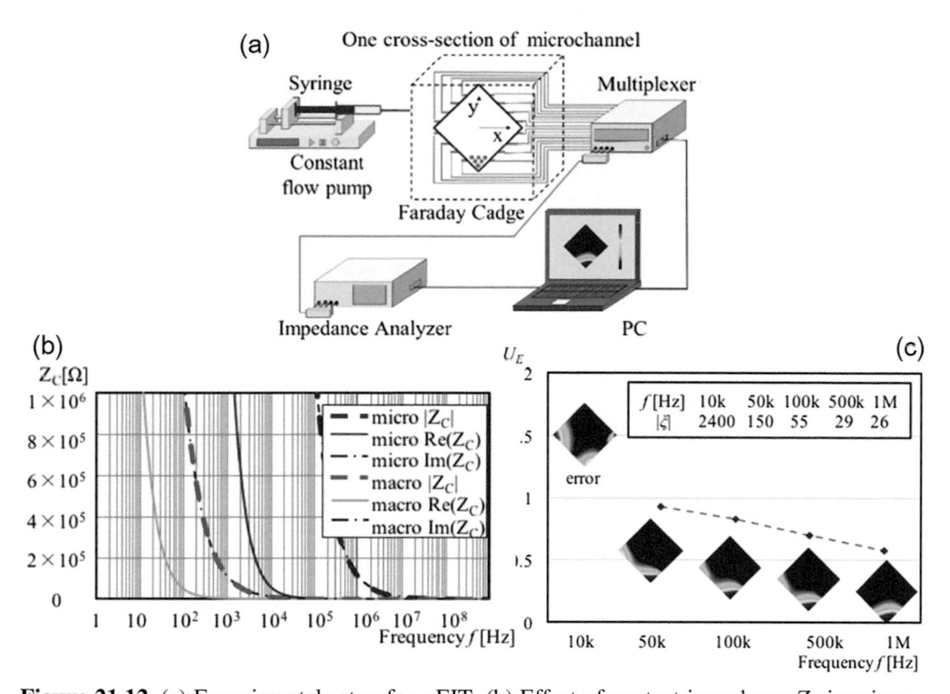

Figure 21.12 (a) Experimental setup for μEIT. (b) Effect of contact impedance *Zc* in micro electrode and in macro electrode. (c) Reconstructed images by μEIT.
Credit: Reprinted with permission from Liu, X., Yao, J., Zhao, T., Obara, H., Cui, Y., &Takei, M. (2018). Image reconstruction under contact impedance effect in micro electrical impedance tomography sensors. *IEEE Transactions on Biomedical Circuits and Systems 12* 623−631. 10.1109/TBCAS.2018.2816946. Copyright 2018, IEEE Transactions on Biomedical Circuits and Systems Publishing.

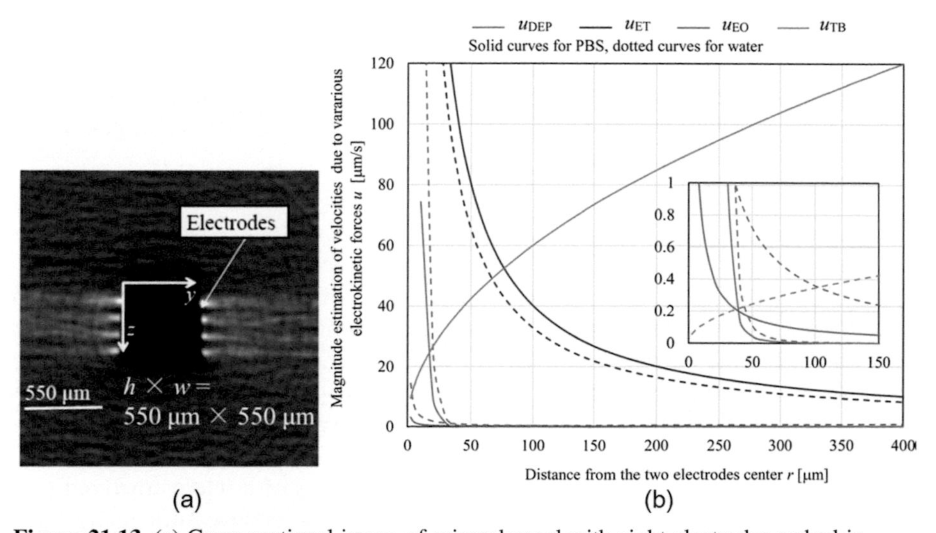

Figure 21.13 (a) Cross-sectional image of microchannel with eight electrodes embed in opposite walls. (b) Estimated velocities due to various electrokinetic forces.
Reprinted with permission from Yao, J., Obara, H., Sapkota, A., Takei, M. (2016). Development of three-dimensional integrated microchannel-electrode system to understand the particles' movement with electrokinetics. *Biomicrofluidics 10*, 024105. 10.1063/1.4943859. Copyright 2016, Biomicrofluidics Publishing.

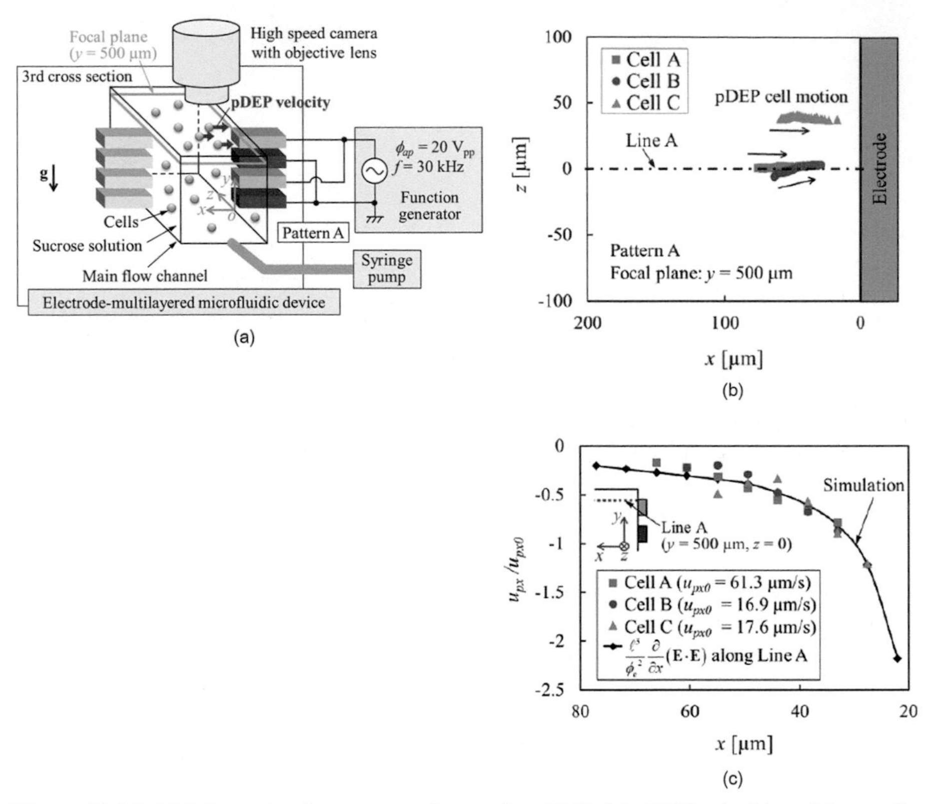

Figure 21.14 (a) Schematic of experimental setup for pDEP. (b) pDEP velocities of three cells. (c) Nondimensional pDEP velocities against position x.
Reproduced from Yao, J., Obara, H., Sapkota A., & Takei M. (2016). Development of three-dimensional integrated microchannel-electrode system to understand the particles' movement with electrokinetics. *Biomicrofluidics 10* 024105, with the permission of AIP Publishing. https://doi.org/10.1063/1.4943859.

online analysis must be implemented directly in the microfluidic channels. Thus, parallel and multiplexed measurement technologies are needed to reduce costs and increase the robustness of information. To date, no commercially available solutions exist. Typical state-of-the-art procedures to study chemical processes use flow cells positioned after microstructured environments or off-line methods.

Moreover, no information about the actual reaction process inside the microreactor and microchannel is generated. However, a disadvantage for the industry is the increasing costs for analytical devices required to assure constant product quality. The production of high-added-value chemicals by means of a microstructured process requires several hundred microreactors in order to produce sufficiently large amounts of final product. Thus, large numbers of flow cells, each attached to a separate spectrometer, are needed to meet these unique requirements. An alternative technology is to analyze several microchannels simultaneously or to analyze a single microchannel spatially resolved along the reaction path using push-broom imaging technology. This type of optical online spectroscopy is an ideal tool to characterize chemical reactions in a fast and reliable way, even on a molecular level. Transmission or reflectance spectra are registered through fixed prism–grating–prism optics with a 2D charge-coupled device (CCD) camera attached to it. Thereby, the x-axis of the CCD array corresponds to the spatial resolution, and the y-axis of the camera provides the full spectrum of the sample.

References

Benton, D. M., & Parker, D. J. (1997). Non-medical applications of positron emission tomography. In J. Chaouki', F. Larachi, & M. P. Duduković (Eds.), *Non-invasive monitoring of multiphase flows* (Vol. 5, p. 161). Amsterdam, the Netherlands: Elsevier.

Bouchard, L. S., Burt, S. R., Anwar, M. S., Kovtunov, K. V., Koptyug, I. V., & Pines, A. (2008). NMR imaging of catalytic hydrogenation in microreactors with the use of para-hydrogen. *Science, 319*(5862), 442–445. https://doi.org/10.1126/science.1151787

Callaghan, P. T., & Xia, Y. (1991). Velocity and diffusion imaging in dynamic NMR microscopy. *Journal of Magnetic Resonance, 91*, 326.

Cheng-Lee, C., Sui, G., Elizarov, A., Shu, C. J., Shin, Y.-S., Dooley, A. N., et al. (2005). Multistep synthesis of a radiolabeled imaging probe using integrated microfluidics. *Science, 310*, 1793–1796.

Chun, J.-H., Lu, S., Lee, Y.-S., & Pike, V. W. (2010). Fast and high-yield microreactor syntheses of ortho-substituted [18F]fluoroarenes from reactions of [18F]fluoride ion with diaryliodonium salts. *Journal of Organic Chemistry, 75*, 3332–3338.

Elizarov, A. M. (2009). Microreactors for radiopharmaceutical synthesis. *Lab on a Chip, 9*, 1326–1333.

Elizarov, A. M., Meinhart, C., Miraghaie, R., Michael, R., Huang, J., Daridon, A., et al. (2010). Flow optimization study of a batch microfluidics PET tracer synthesizing device. *Biomedical Microdevices, 13*, 231–242.

Elizarov, A. M., Michael, R., Shik Shin, Y., Hartmuth, C. K., Henry, C. P., Stout, D., et al. (2010). Design and optimization of coin-shaped microreactor chips for PET radiopharmaceutical synthesis. *Journal of Nuclear Medicine, 51*, 282–287.

Espinosa-Alonso, L., O'Brien, M. G., Jacques, S. B. M., Beale, A. M., de Jong, K. P., Barnes, P., et al. (2009). Tomographic energy dispersive diffraction imaging to study the genesis of Ni nanoparticles in 3D within γ-Al2O3 catalyst bodies. *Journal of the American Chemical Society, 131*, 16932.

Gillies, J. M., Prenant, C., Chimon, G. N., Smethurst, G. J., Perrie, W., Dekker, B. A., et al. (2006). Microfluidic technology for PET radiochemistry. *Journal of Applied Radiation and Isotopes, 64*, 333–338.

Gillies, J. M., Prenant, C., Chimon, G. N., Smethurst, G. J., Perrie, W., Hamblett, I., et al. (2006). Microreactor for the radiosynthesis of PET radiotracers. *Journal of Applied Radiation and Isotopes, 64*, 325–332.

Grunwaldt, J.-D., & Schroer, C. G. (2010). Hard and soft X-ray microscopy and tomography in catalysis: Bridging the different time and length scales. *Chemical Society Reviews, 39*, 4741–4753.

Hannemann, S., Grunwaldt, J.-D., Vegten, N., Baiker, A., Boye, P., & Schroer Christian, G. (2006). Distinct spatial changes of the catalyst structure inside a fixed-bed microreactor during the partial oxidation of methane over Rh/Al_2O_3. *Catalyst today, 126*, 54–63.

Jamal, C., Larachi, F., & Dudukovic, M. P. (1997). Noninvasive tomographic and velocimetric monitoring of multiphase flows. *Industrial & Engineering Chemistry Research, 36*, 4476–4503.

Keng, P. Y., Chen, S., Ding, H., Sadeghi, S., Shah, G. J., Dooraghi, A., et al. (2011). Micro-chemical synthesis of molecular probes on an electronic microfluidic device. *Proceedings of the National Academy of Sciences of the United States of America v1–6*, 1117566109.

Lee, C. C., Sui, G. D., Elizarov, A., Shu, C. J., Shin, Y. S., Dooley, A. N., et al. (2005). Multistep synthesis of a radiolabeled imaging probe using integrated microfluidics. *Science, 310*, 1793–1796.

Liu, X., Yao, J., Zhao, T., Obara, H., Cui, Y., & Takei, M. (2018). Image reconstruction under contact impedance effect in micro electrical impedance tomography sensors. *IEEE Transactions on Biomedical Circuits and Systems, 12*, 623–631. https://doi.org/10.1109/TBCAS.2018.2816946

Lu, S. Y., Giamis, A. M., & Pike, V. W. (2009). Synthesis of [18F]fallypride in a micro-reactor: Rapid optimization and multiple-production in small doses for micro-PET studies. *Current Radiopharmaceuticals, 2*, 49–55.

Lu, S. Y., & Pike, V. W. (2007). Micro-reactors for PET tracer labeling. In A. Schubiger, L. Lehmann, & M. Friebe (Eds.), *PET chemistry: The driving force in molecular imaging* (Vol. 62, pp. 271–287). Berlin: Springer.

Lu, S. Y., & Pike, V. W. (2010). Synthesis of [18F]xenon difluoride as a radiolabeling reagent from [18F]fluoride ion in a micro-reactor and at production scale. *Journal of Fluorine Chemistry, 131*, 1032–1038.

Lu, S. Y., Watts, P., Chin, F. T., Hong, J., Musachio, J. L., Briard, E., et al. (2004). Syntheses of 11C- and 18F-labeled carboxylic esters within a hydrodynamically driven micro-reactor. *Lab on a Chip, 4*, 523–525.

Miller, P. W., Long, N. J., de Mello, A. J., Vilar, R., Audrain, H., Bender, D., et al. (2007). Rapid multiphase carbonylation reactions by using a microtube reactor: Applications in positron emission tomography 11C-radiolabeling. *Angewandte Chemie International Edition, 46*, 2875–2878.

Pascali, G., Nannavecchia, G., Pitzianti, S., & Salvadori, P. A. (2011). Dose-on-demand of diverse 18F-fluorocholine derivatives through a two-step microfluidic approach. *Nuclear Medicine and Biology, 38*, 37–644.

Sato, N., Yao, J., Kawashima, D., & Takei, M. (2019). Numerical study of enhancement of positive dielectrophoresis particle trapping in electrode-multilayered microfluidic device. *IEEE Transactions on Biomedical Engineering, 66*, 2936–2944. https://doi.org/10.1109/TBME.2019.2898876

Schroer, C. G., Kuhlmann, M., Günzler, T. F., Lengeler, B., Richwin, M., Griesebock, B., et al. (2003). Mapping the chemical states of an element inside a sample using tomographic X-ray absorption spectroscopy. *Applied Physics Letters, 82*, 3360–3362.

Smit, E.de, Swart, I., Creemer, J. F., Hoveling, G. H., Gilles, M. K., Tyliszczak, T., et al. (2008). Nanoscale chemical imaging of a working catalyst by scanning transmission X-ray microscopy. *Nature, 456*, 222–225.

Steel, C. J., O'Brien, A. T., Luthra, S. K., & Brady, F. (2007). Automated PET radiosyntheses using microreactor. *Journal of Labelled Compounds and Radiopharmaceuticals, 50*, 308–311.

Van den Broek, S. A. M., Becker, R., Koch, K., & Nieuwland, P. J. (2012). Microreactor technology: Real-time flow measurements in organic synthesis. *Micromachines, 3*, 244–254. https://doi.org/10.3390/mi3020244

Watts, P., & Wiles, C. (2007). Recent advances in synthetic micro reaction technology. *Chemical Communications, 5*, 443–467.

Wester, H. J., Schoultz, B. W., Hultsch, C., & Henriksen, G. (2009). Fast and repetitive incapillary production of F-18 FDG. *European Journal of Nuclear Medicine and Molecular Imaging, 36*, 653–658.

Yao, J., Kodera, T., Obara, H., Sugawara, M., & Takei, M. (2015). Spatial concentration distribution analysis of cells in electrode-multilayered microchannel by dielectric property measurement. *Biomicrofluidics, 9*, 044129. https://doi.org/10.1063/1.4929824

Yao, J., Obara, H., Sapkota, A., & Takei, M. (2016). Development of three-dimensional integrated microchannel-electrode system to understand the particles' movement with electrokinetics. *Biomicrofluidics, 10*, 024105. https://doi.org/10.1063/1.4943859

Yao, J., & Takei, M. (2017). Application of process tomography to multiphase flow measurement in industrial and biomedical fields: A review. *IEEE Sensors Journal, 17*, 8196–8205. https://doi.org/10.1109/JSEN.2017.2682929

Yonghong, L., Xiantao, W., & Takei, M. (2013). Solid-liquid two-phase flow image reconstruction based on ERT technique in microchannel. *TELKOMNIKA, 11*, 173–180. https://doi.org/10.11591/telkomnika.v11i1.1885

X-ray tomography of fluidized beds

Apostolos Kantzas
University of Calgary, Calgary, AB, Canada

22.1 Introduction

Nondestructive, nonintrusive flow visualization techniques have been extensively used in understanding fluid flow phenomena in porous media since the early 1980s (Kantzas, 1990, 1994). The term *process tomography* describes all the techniques used for monitoring and visualizing physical and chemical processes at the laboratory, pilot plant, or plant scale (Williams & Beck, 1995). The techniques used are grouped into two major classes: electrical and electromagnetic. The utilization of X-ray tomography (CT) for monitoring the fluid and solid distribution in laboratory gas-solid fluidized beds was studied extensively by the Kantzas group, as illustrated in this chapter. This technique is the basis for determining bubble diameter, volumetric flow rate, and linear velocity of descending solids. The objective is to determine the limits of X-ray tomography for such predictions and define the boundaries of overlapping electromagnetic techniques used in our laboratory flow visualization and imaging of flow phenomena in fluidized beds. Several other approaches, including radioactive particle tracking (RPT) and particle image velocimetry (Chen & Fan, 1992; Yates & Simons, 1994), were also used for similar purposes.

CT provides a 2D image of X-ray absorption by an object by manipulating a series of one-dimensional projections of the X-ray absorption measured through different angles. This is done through a series of data that is acquired and manipulated using sophisticated reconstruction algorithms. When dealing with chemical reactors, the reconstructed images are interpreted to provide useful information to chemical engineers. The property that describes the X-ray absorption is called the linear attenuation coefficient, and it is a function of density and atomic number (Kantzas, 1990). In the experiments described in this chapter, the atomic number of the system is constant, and the density is the main variable. In the cases where polymer beds are evaluated, the measured density is the bulk density of the voxel, which is the volume average of the polymer particle density and the surrounding gas density. If the density of the polymer is known, then the voidage can be calculated accurately at a predefined resolution.

The voidage measured by the X-ray CT scanner is different from that usually defined as voidage in the literature. In the common definition of voidage, only interspace was considered. With the CT scanner, the sum of the interspace and the intraspace voidage of particles can be mapped. This is especially important when the solid particles are porous, as in the case of polyethylene resins. In the cases of nonporous particles, the intraspace voidage is zero. Therefore, for a dense phase, the voidage

Industrial Tomography. https://doi.org/10.1016/B978-0-12-823015-2.00021-2

measured by the X-ray CT scanner would be higher than the voidage measured with conventional liquid displacement techniques.

Fluidization is the process where solid particles are brought into a fluid-like state through suspension in a gas or liquid. Fluidized beds were first used in the coal gasification industry in 1926 (Kunii & Levenspiel, 2013). Fluidized beds are very important to the chemical industry because fluidization provides an excellent mixing ability, high heat and mass transfer rates, and typically low-pressure drops. Most gas-solid fluidized bed reactors operate at high temperatures, and some, such as those used in the production of polyolefins, also operate at elevated pressures. Due to the multiphase nature of fluidized beds, multiregion models (Behie & Kehoe, 1973) are required to properly account for the hydrodynamics of flow. Since the early 1990s, significant strides have been made on the modeling front. Current models range from the classical approaches (Behie & Kehoe, 1973; Davidson & Harrison, 1963; Kunii & Levenspiel, 2013) to multiphase Eulerian models, which are based on the assumption of an interpenetrating continuum (Gidaspow, 1994), to discrete particle simulations (Hoomans, Kuipers, & van Swaaij, 1998; Kaneko, Shiojima, & Horio, 1999; Tsuji, Kawaguchi, & Tanaka, 1993). Because of computational limitations (Ranade, 2002), the Eulerian models are preferred over discrete particle simulations even though the latter treat particle dynamics more rigorously, which results in better predictions. Computational fluid dynamics (CFD) simulation is becoming more popular in chemical reactor design, scale-up, and modeling. However, reliable experimental data are rare to validate the CFD simulation. It is necessary to carry out experimental measurements of flow behavior in fluidized beds at elevated temperature and pressure. In this chapter, most of the examples presented are based on fluidization of sand or polyethylene.

Many researchers have indicated that the performance of a fluidized bed reactor is strongly influenced by the specific characteristics and behavior of the bubbles. These bubbles are nothing but regions of lower solid density. The main bubble characteristics that affect a fluidized bed are size, rising velocity, and frequency (Cheremisinoff & Cheremisinoff, 1986). Another parameter used by researchers to infer the state of a fluidized bed is the pressure fluctuation measurement. It is the only means of obtaining detailed data on an industrial scale. (Johnsson, Zijerveld, Schouten, van den Bleek, & Leckner, 2000; Peirano, Delloume, Johnsson, Lechner, & Simonin, 2002; Schober, Hulme, Chandrasekaran, & Kantzas, 2003). Pressure fluctuations are believed to be caused by the passage of bubbles in the column and therefore also measure the dynamics of a fluidized bed. Results from the study indicate sensitive variations in the pressure fluctuations with the coefficient of restitution. In the last 30 years, researchers started using pressure fluctuations as another important parameter to evaluate model predictions (Lyczkowski et al., 1993; Peirano et al., 2002). In this chapter, pressure fluctuation analysis is often referred to as a means to either complement or verify modeling data. Also, because of the noninvasive nature of the pressure transducers, the pressure fluctuation analysis can be a complementary tool for tomographic measurements.

Fluidization characteristics detection and fluidization state monitoring are very important for the operation and control of fluidized bed reactors. Statistics of the triboelectric current detect the local fluidization characteristics (Briens et al., 1999).

Coefficient of variation and fractal dimension of conductivity measurements were used to characterize channeling in a liquid fluidized bed (Briens, Briens, Margaritis, Cooke, & Bergougnou, 1997). Hurst's analysis of signals from various measurement probes was used to detect minimum fluidization and gas maldistribution in fluidized beds (Briens, Briens, Hay, Hudson, & Margaritis, 1997). However, among the sensors used in previous investigations, only pressure fluctuation measurements are commonly used in industrial fluidized beds systems. More recently, pressure fluctuations were proposed to monitor and control the state of fluidized beds (Croxford & Gilbertson, 2006; Gheorghiu, Van Ommen, & Coppens, 2003; Schouten & Van Den Bleek, 1998; Van Ommen, Coppens, & Van Den Bleek, 2000). Due to the global characteristic of pressure fluctuation measurement, local flow behavior needs to be measured and compared to pressure measurements, though some local flow behavior can be extracted using statistical or other mathematical methods (Gheorghiu et al., 2003). X-ray CT measurements are known to provide detailed flow patterns of fluidized beds and can be used to monitor fluidization characteristics (Kantzas & Kalogerakis, 1996). It is necessary to investigate fluidized beds using pressure fluctuation and X-ray CT measurement. If a correlation between pressure fluctuation and tomographic data can be found, then pressure fluctuation measurements can be another scale-up tool that will help decipher large column hydrodynamics, where tomography with any reasonable resolution will be unattainable.

22.2 Imaging of fluid beds

Bubble imaging with X-rays goes back to the pioneering work of Rowe, MacGillivray, and Cheesman (1979). However, the recent use of digital fluoroscopy (DF) offers a very powerful option that can accumulate and process massive amounts of data that can be used in fluid bed characterization. X-ray fluoroscopy in conjunction with an image acquisition system provides a means for viewing moving objects. Fig. 22.1 shows a single black-and-white frame from a series of images captured using the X-ray camera. Some recent studies are included as examples only and are by no means exhaustive.

Although X-ray CT is well developed for medical application, the application to fluidized beds only started in the 1990s (Kantzas, 1994). It was first applied to determine the time-average gas holdup in a gas-solid bed (Kantzas, Wright, & Kalogerakis, 1997; Wright et al., 2001). Most of the fluidization experiments in this group's work were conducted using lab-scale Plexiglas columns. Fig. 22.2 shows the experimental setup used to collect column wall pressure measurements. Either tomography or direct imaging was complemented by pressure fluctuation analysis. Pressure transducers were used to sample the pressure oscillations at a frequency of 500 Hz. In later studies, columns made out of aluminum were utilized to run tests at elevated pressures.

An X-ray imaging system was developed by Heindel, Gray, and Jensen (2008) to achieve 3D visualization for fluidized beds. It can also provide time-averaged CT imaging for the fluidized beds (Drake & Heindel, 2012a, 2012b). All the early

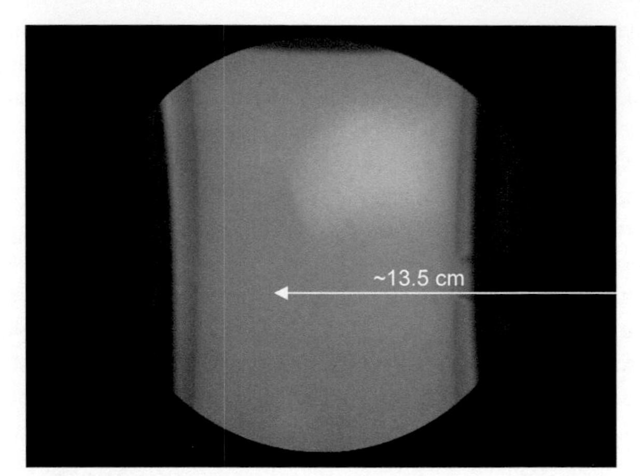

Figure 22.1 A rising bubble acquired with X-ray imaging.
From Chandrasekaran A. (2005). *Verification of the hydrodynamics of a polyethylene fluidized bed reactor using CFD and imaging experiments* (M. Sc. thesis). Calgary, Alberta: University of Calgary.

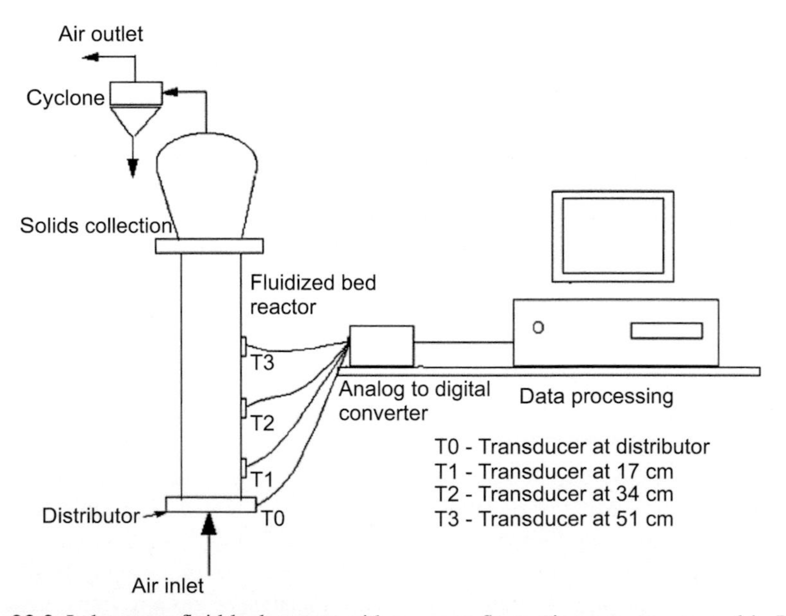

Figure 22.2 Laboratory fluid bed system with pressure fluctuation sensors, as used in X-ray imaging and X-ray CT experimental work at the University of Calgary.
From Hulme I. (2003). *Verification of the hydrodynamics of a polyethylene fluidized bed reactor using CFD and imaging experiments* (M. Sc. thesis). Calgary: University of Calgary.

applications faced a practical limit of X-ray CT: the scanning time of traditional X-ray CT is too long to record the dynamics of the flow of a fluidized bed (Heindel, 2011). Time-averaged data are limited when analyzing the dynamic phenomena of fluidized beds.

For a fluidized bed reactor, the intensity of the X-ray beam after attenuation is calculated based on the Beer-Lambert law, which implies that the bubbles in a fluidized bed appear as bright regions due to less attenuation of the beam compared to the beam that passes through the emulsion. Denser materials, such as glass beads, attenuate X-rays more when compared to polyethylene particles. In other words, glass beads yield a high contrast image compared to that of a bed of polyethylene particles, due to the higher density of the former.

22.3 Computational models and their experimental validation

Measurement alone is not enough. Since the experiments cannot be scaled up to reactor sizes, an intermediate approach must be followed. This is done by CFD modeling. The models can match the images obtained from the experiments, and then the models can be scaled up to reactor conditions.

Most of the available computational models are extended versions of the Navier–Stokes equations for multiphase flow. The various governing equations and constitutive relations used in closing these equations have been a subject of debate for many years. The closure laws range from empirical formulations to the kinetic theory of granular flow (Syamlal, Rogers, & O'Brien, 1993). Because of these reasons, many researchers highlight the importance of validating these models before a specific application. (Boemer, Qi, & Renz, 1998; Kantzas, Wright, Bhargava, Li, & Hamilton, 2001; Kuipers, Duin van, Beckum van, & Swaaij van, 1993; McKeen & Pugsley, 2003; Van Wachem, Schouten, Krishna, Van & Bleek, 1999).

The validation of these models poses significant problems, as the experimental data available must be obtained in a nonintrusive fashion. Most experimental methods for tracking bubbles are either intrusive or do not represent a complete picture of the dynamics of the column.

22.4 Experimental studies

22.4.1 *Fluid bed characterization using computer tomography*

In order to study fluidization behavior, scanners capable of scanning large vertical objects have to be designed. Until recently, this presented a challenge. Instead, many of the presented studies were run using third- or fourth-generation medical X-ray CT scanners that have been modified to perform scans on vertically placed objects. The spatial resolution was as low as 0.2×0.2 mm in cross-section and the temporal resolution as low as 2 s. Data were acquired by scanning the bed using a slice thickness of up to 1 cm at various heights, usually spaced 1 cm along the column and at various superficial velocities.

Zarabi and Kantzas (1998, 2000) tomographically quantified solid and gas distributions as a function of position with high resolution in a series of laboratory fluid beds containing air and polyethylene particles. The resolution used was $0.4 \times 0.4 \times 3$ mm. The laboratory models were Plexiglas columns of 10 cm in diameter, and the settling bed L/D ratios varied between one and three. Large particles (up to 1.5 mm in diameter) of high-density polyethylene and linear low-density polyethylene were used. The superficial gas velocities varied from the minimum fluidization velocity to 50 cm/s. The analysis of fluid bed CAT scanner images was extended to show bubble, emulsion, and dense phase distribution. The analysis was also used to determine bubble diameters and to predict the flow direction of solid particles, as well as the velocity of descending solids. The voidage frequency distributions of a bed at different gas flow rates were compared to each other, and the voidage threshold values corresponding to gas, emulsion, and dense phases were determined. These threshold values were used to prepare ternary images that clearly showed the parts of the bed cross-section corresponding to bubble, emulsion, and dense phases.

Hydrodynamic properties of a three-phase laboratory fluidized bed were characterized using X-ray CT, RPT, and pressure fluctuation analysis (Song et al. (1999)). The solid-liquid mass transfer was measured using benzoic acid as a tracer. Columns 12.7 cm in diameter and 160–200 cm in length were used. The fluids used were water and air, and the solids were spherical polycarbonate particles 0.65 cm in diameter.

The combination of the above techniques provided a unique opportunity to integrate several tools into a single apparatus. Thus, a comprehensive characterization of the transport phenomena in this three-phase fluid bed system was obtained. The application of X-ray CAT scanning was extended in the qualitative and quantitative analysis of fluidized beds. The lower and higher threshold values were introduced for deconvolution of voidage frequency distribution by Gaussian curves. From deconvolution, the area fractions of three distinct phases were calculated (i.e., dense, emulsion, and gas phases). The lower and higher threshold values were defined as the voidage value for minimum fluidization and minimum bubbling, respectively.

These threshold values were used for constructing ternary images showing the phase distributions across the cross-section. Also, the bubble diameters and number of bubbles could be determined from the images. These images are a powerful tool for visualizing the flow pattern and explaining the behavior of fluidized bed. Moreover, it was found that the particle morphology and its size distribution are important parameters in the hydrodynamics of the fluid beds. The analysis was extended to determine the average solid flow rate and solid linear velocity at any height. The results pertaining to the bubble diameters were satisfactory when compared with some correlations. For calculating the solid velocity, the two-phase model and some correlations from the literature were used.

Grassler and Wirth (2000) performed a tomographic measurement to determine the solid concentration in a fluidized bed filled with glass beads of 70 mm. The measuring system consisted of a 60 keV X-ray source and a linear X-ray detector with 1024 sensitive elements and a signal resolution of 8 bit. The integration time was 15 s for each projection and 30 min for a radial concentration pattern. The system could measure the solid concentration up to 20% vol and a minimum spatial resolution of 0.2 mm. The superficial gas velocity was in the range of 2−7 m/s.

The tomography results of the upflow fluidized bed showed a decrease in solid concentration at a higher elevation of the bed. The solid concentration was at a maximum of 45% vol near the wall of the tube at any gas superficial flow and elevation of the bed. The solid concentration profile at a different elevation changed in the downflow bed. A higher concentration of solids in the tube center was observed in the downflow bed. More homogeneous distribution of solid was observed at the lower elevations from the solid-gas distributor.

Bhargava, Kabir, Vaisman, Langford and Kantzas (2004) used X-ray tomography to study the hydrodynamics of an annular dual function photocatalytic reactor operating in absorption (fixed bed) and regeneration (fluid bed) modes. This was a unique application in solid-liquid fluidized bed columns. Degradation experiments of model pollutant (phenol) were conducted with a pilot-scale reactor to evaluate its effectiveness. Adsorption of pollutant to the catalyst and pollutant degradation with respect to various catalyst loadings was investigated.

Wu, Kantzas, Bellehumeur, He and Kryuchkov (2007), Wu, Cheng, Ding, Wei and Jin (2007) used pressure fluctuations, X-ray DF, and X-ray computed tomography (CT) to characterize the flow behavior of gas-solid fluidized bed using polyethylene particles in three Plexiglas columns with diameters of 10, 20, and 30 cm. The particles were linear low-density polyethylene resins with the same mean particle size (750 μm) and particle size distribution (140−1500 μm). The static bed height was 40 cm. Experiments were carried out using air as the gas phase under ambient conditions and at different superficial gas velocities. Pressure fluctuation measurements were obtained from four pressure transducers (Schlumberger Solartron, model 8000 DPD). Dynamic flow behavior of the gas−solids was analyzed using both chaos and wavelet analysis. Characteristics of the flow behavior caused by bubble and particle movement at different scales were extracted from wavelet decomposition of the pressure fluctuation time series. Statistical and chaos analysis methods were further used to distinguish the dynamics behavior at different scales. The CT technique was also used to visualize the bubbling behavior of the polyethylene powder. CT images at different bed heights were obtained over a broad range of gas velocities. Based on the processed images, the time-average voidage distribution, bubble phase area fraction, bubble diameter, and bubble number distribution varying with the bed heights and the superficial velocities were extracted for all three scales of operation. In addition to pressure fluctuation and CT measurements, an X-ray DF system was also used. The bubble diameter, bubble velocity, bubble number distribution, and bubble diameter distribution varying with the bed heights and the superficial velocities were obtained and analyzed in conjunction with that obtained from other techniques, to evaluate the effect of particle size distribution and scale of operation on the fluidization behavior of polyethylene powder. The results extracted from their analysis can be used to validate models and facilitate the implementation of changes for further improvement of hydrodynamic modeling and simulation of gas-solid fluidized bed systems. The results also contribute to the theoretical basis of the design, scale-up, operation optimization, and assessment of new processes.

Particles with different shapes will affect the bed packing properties, the granular flow, and the hydrodynamics of fluidized beds. Zhang, Wu, Li, Bellehumeur and Kantzas (2007) used a CT scanner to investigate the effect of particle shape on the behavior of gas-solid fluidized beds in a 20-cm-diameter column. Particles with different shapes were used as the solid phase, and air was used as the gas phase. An algorithm verified by bed height measurements was developed to remove the artifacts that existed in the CT images. The processed CT images were then used to analyze the time-averaged voidage, voidage distribution, and bubble behavior under different superficial gas velocities and bed heights with different particles.

22.4.2 Fluid bed characterization by X-ray fluoroscopy and pressure measurements

The X-ray DF unit used in these studies was a GE RFXII (which consists of the X-ray tube and the image intensifier) with a GE MPX100 generator. The voltage and current for the experiments conducted in the study were 100 kV and 6 mA, respectively. The apparatus was complemented by software that could analyze the X-ray fluoroscopy images. The program identifies the bubble boundaries, and an excel macro tracked the bubbles from frame to frame. The original image is processed using different techniques to enhance contrast, change to grayscale, remove noise, and get a binary image to obtain area, centroid, equivalent diameter, and boundaries.

Chandrasekara, Van der Lee, Hulme and Kantzas (2005) used power spectral analysis of pressure fluctuations in conjunction with bubble properties obtained from an advanced experimental technique to validate the CFD simulations of an air/polyethylene-fluidized bed. The current Eulerian models based on the kinetic theory of granular flow have been studied extensively for systems that have ideal characteristics for the solid phase. However, the models need to be investigated for particles such as polyethylene that do not have ideal characteristics with respect to shape, size distribution, and elastic character from a simulation standpoint. Unlike previous works, which mostly use average bubble properties for validation, this work utilizes both pressure fluctuations and bubble properties of a high temporal and spatial resolution for the validation process. Results have indicated that significant work is required to adapt the solid phase model CFD code for particles such as polyethylene.

Wu, Briens and Zhu (2006) worked with nonintrusive techniques to characterize the hydrodynamics in gas-solid bubbling fluidized beds using polyethylene powder and glass beads with comparable mean diameters. Pressure fluctuations and X-ray fluoroscopy measurements were performed on a pseudo-2D fluidized bed. Statistical, wavelet, and chaos analyses were applied to the nonstationary pressure signal series to extract and characterize the intrinsic features of the gas-solid fluidized beds. Dominant cycle time was calculated from the approximate coefficient of scale 6 decomposed from cleaned pressure fluctuation signals. The global bubbling behavior of the glass bead system was greatly affected by changes in the superficial gas velocity, while polyethylene powder only significantly varied with the distance from the distributor. Average cycle time, dominant cycle time, Kolmogorov entropy, and wavelet energy were also calculated from detail coefficients of scale 1−6 decomposed from cleaned pressure fluctuation to investigate flow dynamics at micro- and mesoscales. Similarities and differences of bubbling behavior at different scales for glass beads and polyethylene powder systems from pressure fluctuations were verified from X-ray fluoroscopy measurements. Results showed that polyethylene particle systems have different bubble properties compared with glass bead particle systems under comparable operating conditions. The combination of statistical, chaos and wavelet analyses proved to be an effective method to characterize multiscale flow behavior in the gas−solids fluidized bed.

Orta, Wu, Guerrero, Ghods M, Bellehumeu and Kantzas (2011) studied hydrodynamics of a polyethylene-fluidized bed at different operating pressures (191−2908 kPa) and constant temperature of 30°C. In order to extract measurements in systems that can operate in temperatures up to 100°C and pressure up to 3.1 MPag, under X-ray transparent conditions, an aluminum body column was designed and built. Polyethylene particles (Geldart B) were used as the bed material. Minimum fluidization velocity (U_{mf}) decreased with increasing operating pressures. Power spectral density of pressure fluctuation series indicated similar frequency distribution of the bubbling behavior at different superficial velocities. Bubble diameter and bubble velocities were estimated from X-ray fluoroscopy images. Bubble diameter and bubble velocity increase with increasing bed height and superficial gas velocity (Hulme & Kantzas, 2004). At ambient conditions, bubble diameter and bubble velocity increase while increasing bed height and superficial gas velocity due to coalescence and more gas flowing upward under the multiple bubbling regimes in theory. The effect of pressure over the U_{mf} indicates that while increasing the pressure, the minimum fluidization velocity decreases for polyethylene particles (530 μm and 918 kg/m^3).

22.5 Data evaluation

Once the images from the experiments and simulations are obtained, separate image processing algorithms can be used to identify the bubbles and calculate the bubble properties such as bubble diameters and velocities. The important step in this process is bubble identification and tracking. The first step is to select a threshold to define bubble boundaries. For most of the work referenced in this chapter, the bubbles are defined as regions with a void fraction greater than 0.8. The definition was obtained by comparing the predicted bubble diameters with experimentally observed bubble diameters. Extra steps were required during the bubble identification procedure to remove noise. Furthermore, a concept called "local thresholding" was incorporated to identify the bubbles. This is necessary for bubbles in experimental images that do not have well-defined boundaries and are affected by background noise as well as noise created by the X-ray beam.

When this is completed, comparisons between bubble properties of simulation and experiment can be done. Standardization is required and needs to be explicitly stated. Ensemble averaging is sometimes used.

A successful match will imply that both image shapes and pressure fluctuation analysis spectra between experiment and simulation should be the same. The presented literature work indicates that when comparing simulated and experimental pressure drop across the bed at different gas superficial velocities, almost a 10% deviation from the experimental results can be achieved. The simulated and experimental pressure drop decreases to the same extent for increases in gas superficial velocity.

Snapshots of the simulated and experimental polyethylene beds presented by Chandrasekaran (2005) demonstrate that the simulation results often did not show a major change in the size distribution of the bubbles. However, there were specific cases where the opposite was observed. Thus, the argument that the simulation matches the experiment invariably cannot (unfortunately) be made.

Fig. 22.3 shows a comparison of the experimental and simulated bubble frequency distributions with height above the distributor at a gas-superficial velocity of $3U_{mf}$ (Hulme, 2003). It shows that there were more bubbles in the experiment than the simulation except at lower bed heights. This is quite surprising because it was

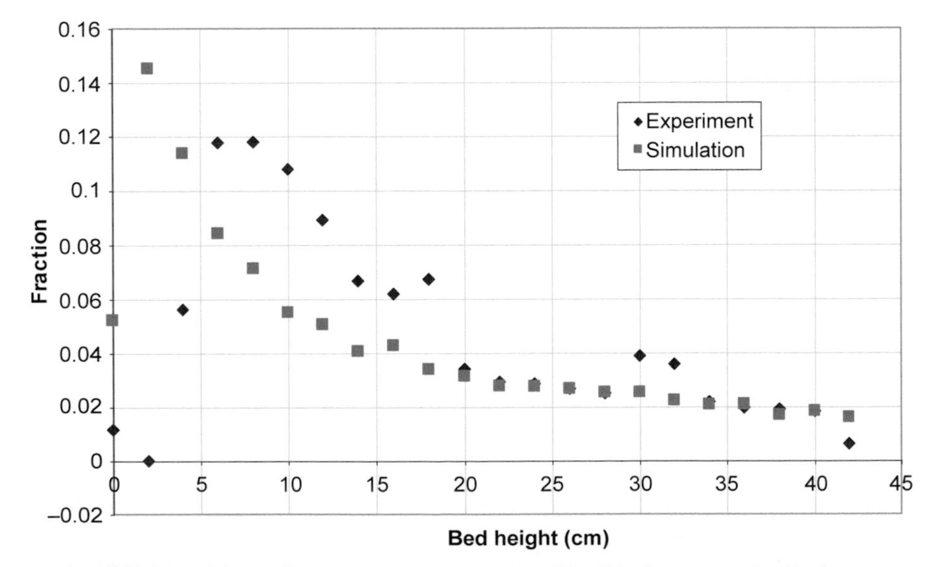

Figure 22.3 Comparison of experimental and simulated bubble frequency distribution as a function of bed height.
From Hulme I. (2003). *Verification of the hydrodynamics of a polyethylene fluidized bed reactor using CFD and imaging experiments* (M. Sc. thesis). Calgary: University of Calgary.

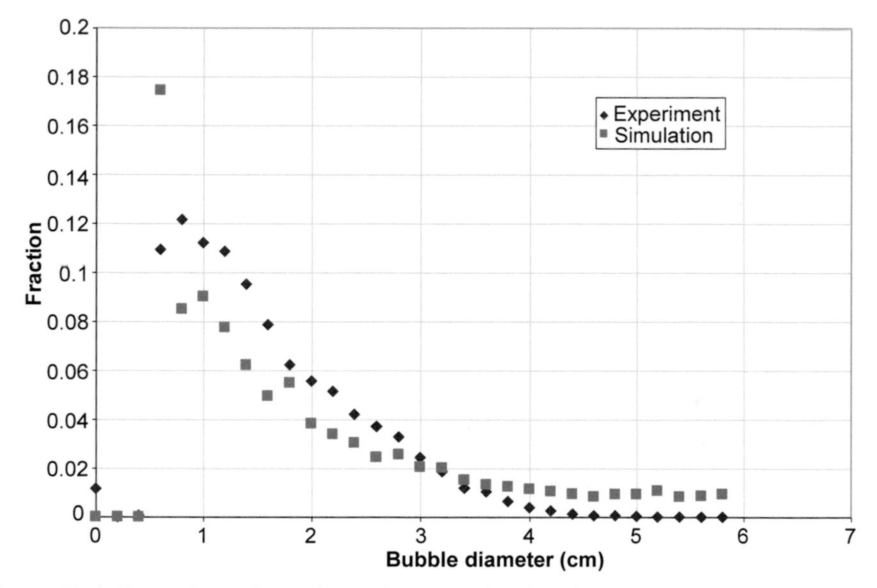

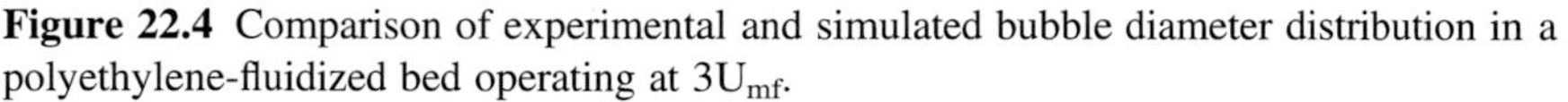

Figure 22.4 Comparison of experimental and simulated bubble diameter distribution in a polyethylene-fluidized bed operating at $3U_{mf}$.
From Hulme I. (2003). *Verification of the hydrodynamics of a polyethylene fluidized bed reactor using CFD and imaging experiments* (M. Sc. thesis). Calgary: University of Calgary.

expected that the experiment should mask smaller bubbles in the wake of larger bubble streams. The figure illustrates that if bubble behavior is to be used as a yardstick for fluid bed reactor performance, then precise and accurate bubble tracking will have to be an important parameter.

Fig. 22.4 (Hulme, 2003) shows that the bubble diameters were distributed similarly for both the experiments and the simulation, with more smaller bubbles than larger bubbles. It should be noted that all the bubble properties from simulation and experiments are time-averaged.

In literature, the average bubble size is invariably used to validate model predictions. As can be observed in Fig. 22.5, there is a considerable deviation between the simulated bubble diameters and experimental results at lower heights. However, at higher bed heights, it appears that the simulations were able to closely predict the experimentally observed results. Similar results were reported consistently by Hulme (2003), Chandrasekaran (2005), and Orta et al. (2011), among others. There was no consistent correlation that matched experimentally derived bubble properties. This adds value to the need for accurate and precise experimental measurements.

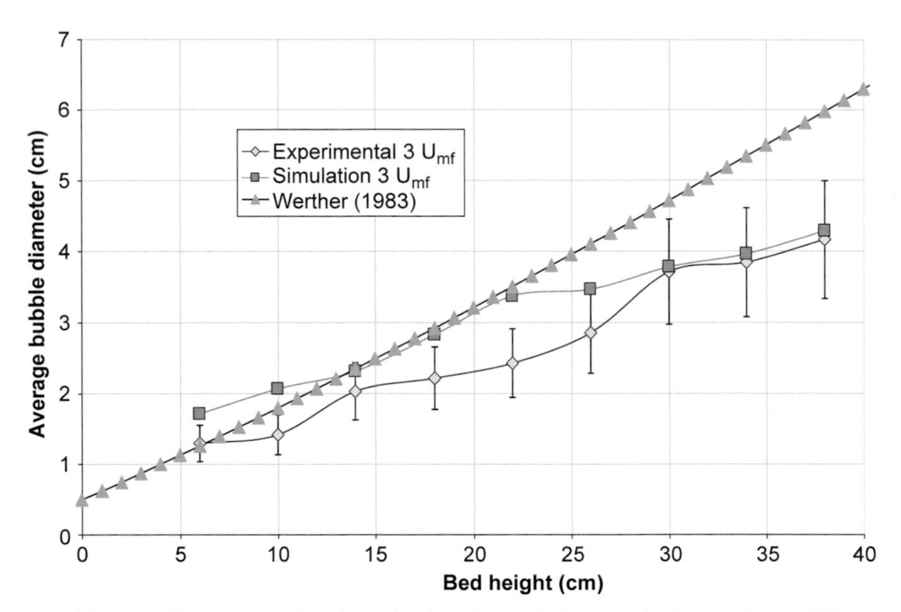

Figure 22.5 Bubble size as a function of height in a polyethylene bed operating at $3U_{mf}$. From Hulme I. (2003). *Verification of the hydrodynamics of a polyethylene fluidized bed reactor using CFD and imaging experiments* (M. Sc. thesis). Calgary: University of Calgary.

22.6 Validation experiments for narrow and wide particle size distribution

One of the major assumptions of multiphase CFD codes is the assumption of uniform particle size and shape. In polymerization processes the size is not controlled, hence there is a wide particle size distribution. Because of this, there is difficulty in capturing the bubble properties with a change in operating conditions. Therefore, to assess the effect of particle size distribution, simulations and experiments of a narrow particle size distribution of polyethylene were conducted.

The simulation bubble diameter predictions for the wide particle size sample showed larger deviations than the narrow size distribution sample, which showed excellent agreement with the experimental average bubble sizes. However, a comparison of the experimental power spectra with the simulation predictions showed significant differences in power spectra magnitude, frequency distribution, and dominant frequencies for both the wide and narrow particle size samples. Subsequent analyses showed that the differences in the bubble size distributions between the simulation and experimental results contributed to the differences in the power spectra.

However, the presence of high-frequency energy in the simulations could not be explained from that study. It is speculated that oscillations in the numerical solution of the governing equations can be one of the contributing factors but remains to be investigated. Typically, in validation studies, the average size distribution is used for the evaluation of the simulation results. This might lead to erroneous conclusions. Various factors contribute to the deviations between the simulation and experiment for the air/polyethylene-fluidized bed. Parallel studies with an ideal particle system, such as glass beads with a narrow size distribution and nearly spherical particles, have shown excellent predictions in both the power spectral and bubble property analyses using CFD packages.

Chandrasekaran, Van der Lee, Hulme and Kantzas (2005) showed that the bubble diameters and pressure fluctuations are very sensitive to the coefficient of restitution for polyethylene particles. This sensitivity increased dramatically for larger particles and almost disappeared for smaller particles. Typically, the bubble diameters would increase with a decrease in the restitution coefficient or an increase in the angle of internal friction. Hence, there will be an error introduced in the simulation results for polyethylene-fluidized bed systems, as experimental values of the solid parameters are currently absent in the literature. This will be vital to the further development of the current granular flow model. The same investigations also highlighted issues with modeling the solid phase stress tensor. The sensitivity of the model to the restitution coefficient is not understood at this time. Preliminary results (Fan, Marchisio and Fox, 2004; Goldschmidt, Kuipers and Van Swaaji, 2001) suggest that a consistent and tested formulation for the solid phase stress tensor will make the difference in understanding nonideal particle fluidization systems such as polyethylene. Multiple solid phases, representative of the various size fractions are currently available for polydisperse powders.

22.7 Comparison between different validation approaches

Hulme (2003) conducted simulation using CFD software and experiments on a scaled-down cold flow model of a gas-solid-fluidized bed reactor used for the production of polyethylene. Using an Eulerian−Eulerian multiphase model based on the two-phase model of Anderson and Jackson (1967) and Jackson (2000), simulations were conducted to determine the effect of several parameters—time step, differencing scheme, frictional stress, and closing equations—on the bubble properties (predominantly bubble diameter). Once the model flexibility was accessed, the next step was to compare these results with those from experiments conducted using X-ray fluoroscopy (Hulme, 2003). X-ray fluoroscopy was used to create a statistically significant data set. Preliminary experiments comparing the bubble diameter and axial bubble velocity were conducted using glass beads. The results demonstrated that, for glass beads, the average bubble diameters were comparable for CFD and experiments.

Fig. 22.6 (Van der Lee, Shepperson & Kantzas, 2005) shows the experimental bubble diameters for the polyethylene bed (wide particle size distribution: 165−1500 μm) at three superficial gas velocities and a static bed height of 40 cm. Fig. 22.6 also shows the bubble diameters predicted from a correlation developed by Cai, Schiavetti, De Michele, Grazzini, and Miccio (1994), and those predicted from pressure measurements taken at 17 and 34 cm above the distributor (indicated by "PF" on the legend). The correlation by Cai et al. (1994) overpredicts bubble diameters at 2.5 times minimum fluidization and underpredicts at 3.5 times minimum fluidization. However, at 3.0 times minimum fluidization, the bubble diameter predictions match closely with the

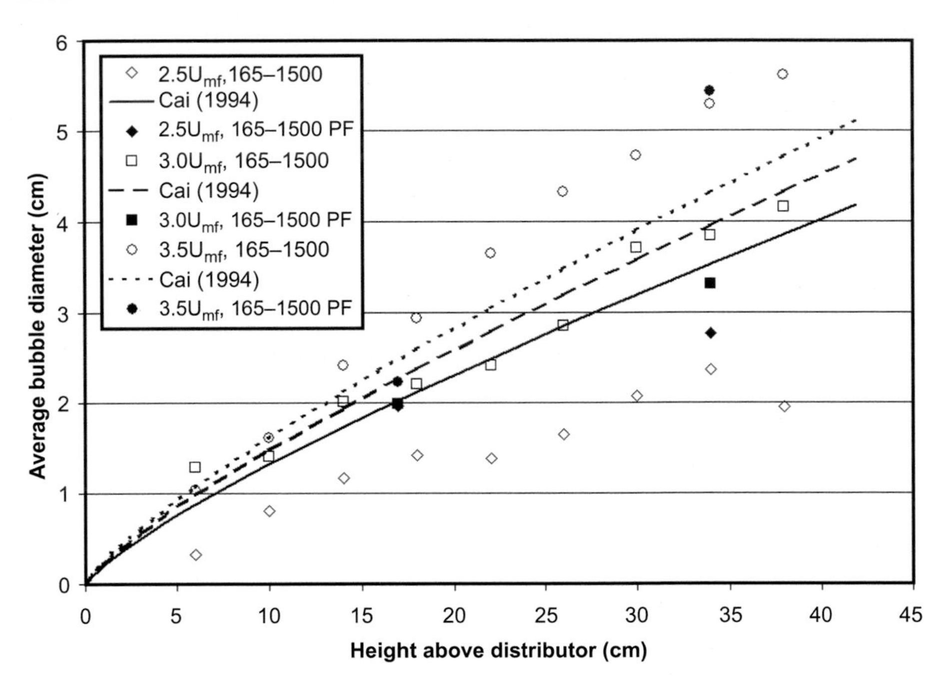

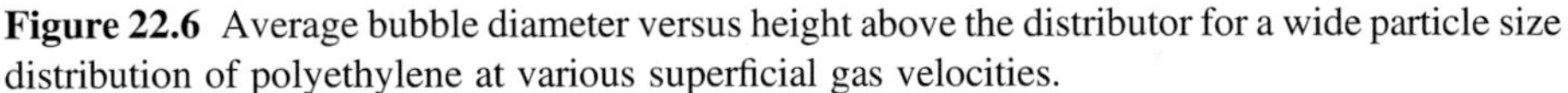

Figure 22.6 Average bubble diameter versus height above the distributor for a wide particle size distribution of polyethylene at various superficial gas velocities.
From Van der Lee, L., Shepperson, J, & Kantzas, A. (2005). A comparison of experimental and predicted bubble characteristics of a linear low density polyethylene. In: *Proceedings, 3rd European conference on the reaction engineering of polyolefins*, Lyon, France, June 20–24. (2005).

experimental values. The comparison of bubble diameters predicted by pressure fluctuations and those measured experimentally changes with measurement height as well as superficial gas velocity.

From the comparison of correlations, it was evident that an incorrect choice of correlation for bubble diameter prediction for polyethylene fluidized beds would result in deviant fluidized bed behavior. For instance, it was found that most correlations grossly overpredicted bubble diameters at a given superficial gas velocity. This type of error could translate into a difficulty such as agglomeration inside an operational fluidized bed.

The average bubble diameters measured while fluidizing the two narrow particle size distributions (<425 μm and 425–707 μm), were also compared with published correlations, and it was found that correlations by Darton, LaNauze, Davidson and Harrison (1977) and Yasui and Johanson (1958) were most suited to predict bubble diameters for the low (<425 μm) and medium (425–707 μm) particle size distributions, respectively. In addition to this, the bubble diameters predicted from pressure fluctuations matched well with the medium (425–707 μm) particle size distribution. Axial velocities were also compared. It was found that a correlation by Geldart (1973) (using calculated bubble diameters from Cai et al. (1994)) predicted the data most closely, except for the smallest particle size distribution (less than 425 μm) where correlations by Wallis and Grace (as cited in Kunii and Levenspiel (2013)) were found to fit best.

22.8 Validation for reactor scale-up

The scale-up of the fluidized bed reactors is difficult due to the complexity of hydrodynamics and chemical kinetics (Rudisuli, Schildhauer, Biollaz & Van Ommen, 2012). The fluidized bed performance is linked to particle interactions, chemical kinetics, and heat and mass transfer in the reactors. As a result, the mathematical modeling of hydrodynamics becomes challenging. Lab-scale cold models can predict the hydrodynamics of the hot scaled-up reactors, but the scaling-up of the cold models is still challenging.

The most common procedures in scaling-up the fluidized bed use dimensionless numbers such as Gliksman, which is a function of Reynolds number, Froude number, gas-to-particle density ratio, sphericity, and size distribution of the particles and bed geometry. Dimensionless scaling-up can be validated using spectral analysis and CFD. However, some of the fluidization phenomena, such as wall effect and particle interactions are neglected in dimensionless scaling-up.

The Kantzas group has focused on developing nonintrusive measurement techniques for cold laboratory-scale polymerization reactors to provide statistically significant data for comparison with CFD models for the production of polyethylene. This is illustrated by the multiple references of this chapter. To model the behavior of these reactors, they have chosen commercial CFD packages. They used the combination of experiment and simulation to study the effect of certain model parameters on the predicted bubble properties. These parameters in simulation include the time step, differencing scheme, solid stress closure equations, and frictional stress. The geometric configuration and conditions specified for these simulations were based on a case run by Syamlal and O'Brien (1987) and Laux and Johansen (1998), which provided a means for quantitative comparison. In addition, simulation results were compared with results from a statistically significant data set obtained from X-ray fluoroscopy experiments, using relatively uniform glass beads as the solid phase.

Due to the difficulty in solving the momentum and continuity equations for every particle in the gas-solid fluidized bed reactor, much effort has been focused on reducing the number of equations to describe this system. The most common method is to apply an averaging technique to the Navier–Stokes equations. In this manner, the two fluids (gas and solid) are treated as two fully interpenetrating continua. The development of the equations used for the continuum models, or two-fluid models, can be found in Anderson and Jackson (1967) and Jackson (2000).

In recent years, due to the controversy surrounding the governing and closing equations used for the simulations, many researchers have been validating proposed models. Of particular note is the work by Boemer, Qi, and Renz (1997), Boemer et al. (1998), Van Wachem et al. (1999), and Van Wachem, Van Der Schaaf, Schouten, Krishna and Van Den Bleek (2001). Boemer's work (1997) determined the closing equation that best predicted spontaneous bubble formation in a fluidized bed. In addition, Boemer et al. (1998) conducted experiments and simulations to verify properties using a 2D bed. Among the properties studied were the bubble wake angle, voidage, and pressure distribution, bubble size, and velocity. In general, these properties found good agreement with the experiments. It should be noted that while their work gave positive results, the amount of data studied for bubble properties is limited. Van Wachem et al. (1999) also conducted a validation of the Eulerian-simulated dynamic behavior of gas-solid fluidized beds. They studied the behavior of Geldart B particles to determine if the dynamic behavior matched the experiments. The choice of closing equations was based on fast numerical convergence and accurate physical results. The closing equation chosen followed the work of Boemer, Qi and Renz (1995), Ding and Gidaspow (1990), and Lun, Savage, Jeffrey and Chepurnity (1984), Syamlal et al. (1993) with, the interphase exchange being that of Syamlal et al. (1993). Van Wachem et al. (1999) mainly examined the power spectral density of the pressure and voidage fluctuations, which they found matched correlations. The paper by Van Wachem et al. (2001) summarizes the governing and closing equations that are more commonly used. Showing results both qualitatively and quantitatively, they were able to get a feel for the range of applicability of simulating the behavior of a bed with glass beads. They also concluded that the choice of solids stress models, or radial distribution functions, plays little role in the final simulation result.

Very little open literature discusses simulations of large commercial-scale or even pilot-scale reactors for the polymerization process. Gobin et al. (2012) is one of the only groups in open literature, studying large-scale polymerization reactors. They describe their simulator's ability to predict the gross flow patterns of a large-scale reactor. Their numerical code, ESTET-ASTRID, developed by Electricité de France for CFB boilers and based on the two fluids model, is used for these simulations. They were able to determine the pressure drop across a fluid bed and determine some of the gross flow patterns, such as solid flow rate and solid volume fraction. They also conducted some preliminary work to examine the credibility of modeling a 3D reactor in two dimensions. While there were some variations, they concluded that for industrial use they were comparable. An additional study by Lettieri, Cammarata, Micale and Yates (2003) concluded that a 2D model provided sufficient accuracy.

Many authors agree that CFD will be a powerful tool in determining the macro- and microscopic phenomena associated with gas–solid–fluidized bed reactors. Yet, it is generally accepted that the next step necessary in providing a reliable, fully predictable model is experimental validation (Gobin et al., 2012). A set of experiments and simulations that were conducted on a glass-bead fluidized bed system were performed by He, Wu, Bellehumeur and Kantzas (2007). Although polyethylene systems are the focus of their research, the nonideality of the particles makes it difficult to access the applicability of the CFD model and the effect of its parameters. By initially using an "ideal" and much-studied system, such as glass beads, they determined the value of simulation results as well as the effect of various parameters by comparison with literature and the experimental data. The glass bead system was also used to validate the image processing software that was developed to analyze the statistically significant amount of quantitative experimental data produced with X-ray fluoroscopy. The glass bead data were used to compare experimental results with correlations presented in the literature (Hulme, 2003; Hulme & Kantzas, 2004). A simple parametric study was conducted to determine the effect of parameters such as time step, differencing scheme, closing equations, and frictional stress.

The vessel was a scaled-down cold model reactor used for the production of polyethylene, with a diameter of 0.1 and 1 m in height. Images showing bubble movement were captured at a frequency of 30 Hz and then analyzed using in-house software to determine the bubble properties. The particles used in this study were glass beads with a narrow particle size distribution (150−250 μm). Two minutes (3600 frames) of data were captured at each set of fluidization conditions. Experiments were repeated three times and found to be statistically valid. More details on the experimental procedure and data analysis can be found in Hulme (2003) and Hulme and Kantzas (2004), with a summary provided above.

Experimental results using glass beads as the solid phase were compared with the results obtained from the simulation. The simulation of the cylindrical laboratory reactor was performed, assuming it could be modeled in two dimensions (Gobin et al., 2012; Lettieri et al. 2003). The first parameter examined was the average fluidized bed height and pressure drop across the bed. The average height for the experimental bed was found to be approximately 42 cm. For the simulations, the average bed height was closer to 46 cm, or approximately 10% greater. Other authors have noted the same discrepancy, and in the case of Gobin et al. (2012), the simulated bed height was 15% higher.

Examination of the distribution of the bubble diameters throughout the bed indicates similarities between the experimental and simulation results. Once the distribution of the bubbles and bubble diameters was deemed comparable, the effort shifted to examining the bubble diameter and axial velocity for binned bed heights.

The diameters of the bubbles for the simulations were determined using a modified version of the program developed for the experiments. The simulation output images show the volume fraction of solids. To analyze the images, a void fraction of solids had to be chosen as a cutoff (similar to the thresholding for experimental images) for determining the bubble boundaries. While some authors have chosen the value of 0.3 (Guenther & Syamlal, 2001), others have chosen void fractions of 0.2 (Boemer et al., 1997; Gidaspow, 1994) and 0.15 (Kuipers, 1990). After comparing with experimental results, a cutoff of 0.2 was chosen. The conclusion by most authors listed above is that an exact definition of the bubble boundary is not necessary. However, after examining the effect of varying the volume fraction cutoff, it becomes apparent that the definition of the bubble boundary is important.

After comparing the simulation results with a statistically significant amount of experimental data, it becomes apparent that the simulations can reasonably predict average bubble diameters. However, the bubble distribution near the top of the bed did not match well, and the absolute axial velocities were off by approximately 10 cm/s. However, they both followed the same trend, which is encouraging.

22.9 Ultrafast X-ray computer tomography

Fast moving objects typically produce poor quality images and motion artifacts in CT scans. The imaging of bubbling gas-solid fluidized beds, where many bubbles may pass through the sensing cross-section in a tomographic measurement, gives such a problem. Previously, one method of dealing with this problem was to remove artifacts from the reconstructed image itself through low-pass filtering. However, this method leads to lost information from the images. When the concentration of bubbles is low, and their shape is close to cylindrically symmetric, analysis of consecutive scans in the raw data sinogram can also provide valuable information about the frequency of bubbles passing and about their individual size, shape, and position (Bennett, Kryuchkov and Kantzas, 2006). Then, utilizing the inverse Abel transform, this information can also be used to contribute to time-averaged properties of the bed's behavior, such as voidage distribution, bubble phase area fraction, and spatial bubble number distribution. The approach is simple but has uncertainties.

There are uncertainties for bubble definition using pressure fluctuation and X-ray tomography methods (Dantas, Melo & dos Santos, 2008). Measurement of pressure fluctuations due to wall effect causes severe inaccuracy. Application of X-ray single projection for bubble distribution measurement will be inaccurate at the presence of several bubbles at one cross-section area of reactor and wall effects (Grassler & Wirth, 1999).

Insufficient X-ray projections cause limited data in tomography, which is a problem in conventional CT algorithms. Instantaneous tomographic measurements of fluidized beds provide an alternative approach, that has great merit and potential.

A high-speed X-ray tomography system has been developed for measuring the bubbling flow in a fluidized bed by Mudde, Alles and van der Hagen (2008). Three X-ray sources are set up at equal angles around the bed to generate projections simultaneously, so that detector arrays receive the attenuated X-ray signals at the same time. A time resolution of 2500 frames/s can be reached. Meanwhile, another fast X-ray system was built by Bieberle et al. (2007) for gas-liquid flow. The frame rate can be as high as 10,000 frames/second (Bieberle et al., 2010). Later, it was also applied in a fluidized bed measurement (Bieberle, Barthel, & Hampel, 2012). Both fast X-ray CT systems have a high enough time resolution to study multiphase fluid flow. The time resolution is improved, but the spatial resolution changed from millimeter-scale to centimeter-scale for a 25-cm diameter fluidized with the current hardware controlled by the limited number of detectors or limited angle.

(Mudde, Schulte, & Van Den Akker (1994)) have developed a fast CT system consisting of three X-ray sources to measure bubble rise velocity in a bubble bed with 23-cm ID. The system can acquire data with 100 frame/s rate. However, small bubbles with less than 3-cm ID were not detected by this system due to signal noise. By increasing the number of X-ray sources to 5, better accuracy could be obtained (Mudde, 2010).

Kai, Misawa, Takahashi, Tiseanu and Ichikawa (2005) were able to produce high-quality images with pixels of 0.15 mm^2 and a shorter measuring time of 4 ms using 18 X-ray sources and 122 detectors.

Measurements of size and velocity of bubbles were also performed using X-ray computer tomography by Brouwer, Wagner, van Ommen and Mudde (2012). They used CT scans to study the effect of fine concentration and operating pressure on bubble size in a fluidized bed. Three X-ray sources with YXLON Y.TU 160-D06 tubes were mounted at 120 degrees around the fluidized bed. Maximum voltage and current were 160 V and 12 mA. The velocity of the detected phantom between the planes has been measured using a high-speed (500 fps) camera.

Bieberle et al. (2012) have applied an ultrafast X-ray tomography technique to assess the gas phase and the particulate structure in fluidized bed columns. An X-ray beam of 150 kV, 12 mA, and a maximum of 10 kW power has been used rotating the object to have limited-angle and full-angle imaging in two heights of 30 and 70 cm. The X-ray tomography result was validated by performing the experiments in columns with the diameter of 50 and 94 mm filled with ellipsoidal shape polyoxymethylene particles with 3 × 2 mm size. The result of ultrafast X-ray CT at the 50-mm-diameter column showed that as the gas flow rate increases, the bubble fraction increases. An increase in bubble fraction changed the flow regime from bubble flow to slug flow. The bubbles and slugs were in similar shape and centered at different heights in the 50-mm column. The formation of gas channels close to the wall was observed in the 94-mm column. They observed that bubbles coalescence in the center at upper regions.

Although virtually all reconstruction problems in CT are ill-posed, this problem in high-speed X-ray CT is serious in high-speed X-ray CT; the number of data points depends on the number of detectors, which is restricted by space and detector size. The traditional CT algorithms, such as filtered back-projection, are seriously influenced by the ill-posed problem. It is easy to produce unknown errors when the order of the image matrix is much larger than the order of the raw data matrix (Natterer, 1986). The simultaneous algebraic reconstruction technique (SART) (Andersen & Kak, 1984) was first applied to solve such a problem. Alternatively, the genetic algorithm (GA) (Holland, 1992) is introduced to tomographic image reconstruction (Abdoli, Ay, Ahmadian, Dierckx & Zaidi, 2010; Di Gesù, Boscoa, Millonzi & Valenti, 2010; Kihm & Lyons, 1996; Valenti, 2008; Venere, Liao & Clausse, 2000; Wu et al., 2007a, 2007b). Although it is proved that GA performed well in reconstructing images from ill-posed and noisy data, a comparison between SART and GA has not been made. At the completion of this document, there was no commercial ultrafast CT system in the literature.

22.10 Future trends

In this chapter, it was demonstrated that CT can play an important role in characterizing laboratory- and pilot-scale fluidized bed systems. New designs in ultrafast scanning can make this vision a reality in the very near future. It was also demonstrated that without supplementary methods such as CFD and pressure fluctuation analysis, it will be impossible to provide information on plant-scale systems. Thus, CT becomes one tool for process monitoring and process design in fluid bed columns (reactors or not). There are numerous problems to be tackled and numerous types of reactors where tomography will be invaluable from dryers to spouted beds to fixed beds. What needs to be addressed in earnest is how this information will translate into real-world information.

Acknowledgments

The author acknowledges the contributions of all the graduate students, postdoctoral fellows, technologists, and research associates that have worked hard over the years to materialize the vision of the tomography and imaging of fluid bed systems. This work would never have been done without the financial contributions of Nova Chemicals, the Natural Sciences and Engineering Research Council of Canada (NSERC), and the Canada Foundation for Innovation (CFI).

References

Abdoli, M., Ay, M., Ahmadian, A., Dierckx, R., & Zaidi, H. (2010). Reduction of dental filling metallic artifacts in CT-based attenuation correction of PET data using weighted virtual sinograms optimized by a genetic algorithm. *Medical Physics, 37*, 6166–6177.

Andersen, A. H., & Kak, A. C. (1984). *The application of ray tracing towards a correction for refraction effects in computed tomography with diffracting sources*. West Lafayette, Ind: Purdue University Research Rep. TR-EE84–14 Purdue University.

Anderson, T. B., & Jackson, R. (1967). A fluid mechanical description of fluidized beds equations of motion. *Industrial & Engineering Chemistry Fundamentals, 6*(1), 527–538.

Behie, L. A., & Kehoe, P. (1973). The grid region in a fluidized bed reactor. *AIChE Journal, 19*(5), 1070–1072.

Bennett, M., Kryuchkov, S., & Kantzas, A. (2006). Interpretation of CT-Scan sinograms to study the dynamic flow properties of a fluidized bed. In *Multiphase flow: The ultimate measurement challenge, proceedings 5th ISMTMF/IWPT-2, Macau, China, December 10-13*.

Bhargava, A., Kabir, M., Vaisman, E., Langford, C. H., & Kantzas, A. (2004). A novel technique to characterize hydrodynamics and analyze performance of a fluidized bed photocatalytic reactor for wastewater treatment. *Industrial & Engineering Chemistry Research, 43*, 980–989.

Bieberle, M., Barthel, F., & Hampel, U. (2012). Ultrafast X-ray computed tomography for the analysis of gas-solid fluidized beds. *Chemical Engineering Journal*, 356–363.

Bieberle, M., Fischer, F., Schleicher, E., Hampel, U., Koch, D., Aktay, K. S. D. C., et al. (2007). Ultrafast limited-angle-type X-ray tomography. *Applied Physics Letters, 91123516*.

Bieberle, M., Schleicher, E., Fischer, F., Koch, D., Menz, H.-J., Mayer, H.-G., et al. (2010). Dual-plane ultrafast limited-angle electron beam X-ray tomography. *Flow Measurement and Instrumentation, 21*(3), 233–239.

Boemer, A., Qi, H., & Renz, U. (1995). Eulerian computation of fluidized bed hydrodynamics – a comparison of physical models. In *Proceedings of the 13th international conference on fluidized bed combustion, Orlando, FL., USA* (pp. 775–786).

Boemer, A., Qi, H., & Renz, U. (1997). Eulerian simulation of bubble formation at a Jet in a two-dimensional fluidized bed. *International Journal of Multiphase Flow, 23*(5), 927–944.

Boemer, A., Qi, H., & Renz, U. (1998). Verification of Eulerian simulation of spontaneous bubble formation in a fluidized bed. *Chemical Engineering Science, 53*(10), 1835–1846.

Briens, C. L., Briens, L. A., Barthel, E., Le Blevec, J. M., Tedoldi, A., & Margaritis, A. (1999). Detection of local fluidization characteristics using the V statistic. *Powder Technology, 102*, 95–103.

Briens, C. L., Briens, L. A., Hay, J., Hudson, C., & Margaritis, A. (1997). Hurst's analysis to detect minimum fluidization and gas maldistribution in fluidized beds. *AIChE Journal, 43*(7), 1904–1908.

Briens, L. A., Briens, C. L., Margaritis, A., Cooke, S. L., & Bergougnou, M. A. (1997). Characterization of channeling in multiphase systems. Application to a liquid fluidized bed of angular biobone particles. *Powder Technology, 91*, 1–9.

Brouwer, G., Wagner, E., van Ommen, J., & Mudde, R. (2012). Effects of pressure and fines content on bubble diameter in a fluidized bed studied using fast X-ray tomography. *Chemical Engineering Journal*, 207-208 711–717.

Cai, P., Schiavetti, M., De Michele, G., Grazzini, G. C., & Miccio, M. (1994). Quantitative estimation of bubble size in PFBC. *Powder Technology, 80*, 99–109.

Chandrasekaran, A. (2005). *Verification of the hydrodynamics of a polyethylene fluidized bed reactor using CFD and imaging experiments (M. Sc. thesis)*. Calgary, Alberta: University of Calgary.

Chandrasekaran, B., Van der Lee, L., Hulme, I., & Kantzas, A. (2005). A simulation and experimental study of the hydrodynamics of a bubbling fluidized bed of linear low density polyethylene using bubble properties and pressure fluctuations. *Macromolecular Materials and Engineering, 290*, 592–609.

Chen, R. C., & Fan, L.-S. (1992). Particle image velocimetry for characterizing the flow structure in three-dimensional gas-liquid-solid fluidized beds. *Chemical Engineering Science, 47*(13/14), 3615–3622.

Chereminsinoff, N. P., & Cheremisinoff, P. N. (1986). *Hydrodynamics of gas-solids fluidization*. Houston, Texas: Gulf Publishing Company.

Croxford, A. J., & Gilbertson, M. A. (2006). Control of the state of a bubbling fluidized bed. *Chemical Engineering Science, 61*, 6302–6315.

Dantas, C. C., Melo, S. B., & dos Santos, V. A. (2008). A study of uncertainty evaluation in transmission measurement in gamma ray tomography. In *12th IMEKO TC1 & TC7 joint symposium on man science & measurement*. France: Annecy.

Darton, R. C., LaNauze, R. D., Davidson, J. F., & Harrison, D. (1977). Bubble growth due to coalescence in fluidized beds. *Transactions of the Institution of Chemical Engineers, 55*, 274–280.

Davidson, J. F., & Harrison, D. (1963). *Fluidized particles*. New York: Cambridge University Press.

Di Gesù, V., Boscoa, G. L., Millonzi, F., & Valenti, C. (2010). A mimetic approach to discrete tomography from noisy projections. *Pattern Recognition, 43*(9), 3073–3082.

Ding, J., & Gidaspow, D. (1990). A bubbling fluidization model using kinetic theory of granular flow. *AIChE Journal, 36*, 523–538.

Drake, J. B., & Heindel, T. J. (2012). Comparisons of annular hydrodynamic structures in 3D fluidized beds using X-ray computed tomography imaging. *Journal of Fluids Engineering-Transactions of the ASME, 134*.

Drake, J. B., & Heindel, T. J. (2012). Local time-average gas holdup comparisons in a cold flow fluidized beds with side-air injection. *Chemical Engineering Science, 68*, 157–165.

Fan, R., Marchisio, D. L., & Fox, R. O. (2004). Gas-solid fluidized beds. *Powder Technology, 139*, 7–20.

Geldart, D. (1973). Types of gas fluidization. *Powder Technology, 7*, 285–292.

Gheorghiu, S., Van Ommen, J. R., & Coppens, M. O. (2003). Power law distribution of pressure fluctuations in multiphase flow. *Physical Review, 67*, 041305-1–041305-7.

Gidaspow, D. (1994). *Multiphase flow and fluidization*. Boston: Academic Press Inc.

Gobin, A., Neau, H., Simonin, O., Llinas, J., Reiling, V., & Selo, J. (2012). Numerical simulation of a gas phase polymerisation reactor. In *European congress on computational methods in applied sciences and engineering ECCOMAS computational fluid dynamics conference*. Wales.

Goldschmidt, M. J. V., Kuipers, J. A. M., & Van Swaaji, W. P. M. (2001). Hydrodynamic modeling of dense gas-fluidized beds using the kinetic theory of granular flow: Effect of coefficient of restitution on bed dynamics. *Chemical Engineering Science, 56*(2), 571.

Grassler, T., & Wirth, K. E. (1999). Computerized tomography and image processing. In *DGZfP proceedings, BB 67-CD. Computerized tomography for industrial applications and image processing in radiology March* (Vols. 15 − 17).

Grassler, T., & Wirth, K. E. (2000). X-ray computer tomography potential and limitation for the measurement of local solids distribution in circulating fluidized beds. *Chemical Engineering Journal, 77*, 65−78.

Guenther, C., & Syamlal, M. (2001). The effect of numerical diffusion on simulation of isolated bubbles in a gas-solid fluidized bed. *Powder Technology, 116*(2−3), 142−154.

Heindel, T. J. (2011). A review of X-ray flow visualization with applications to multiphase flows. *Journal of Fluids Engineering-Transactions of the ASME, 133*.

Heindel, T. J., Gray, J. N., & Jensen, T. C. (2008). An X-ray system for visualizing fluid flows. *Flow Measurement and Instrumentation, 19*, 67−78.

He, Z., Wu, B., Bellehumeur, C., & Kantzas, A. (2007). In *"X-ray fluoroscopy measurements and CFD simulation of hydrodynamics in a two-dimensional gas-solids fluidized bed", fluidization XII*. Vancouver, B.C.: Harrison Hot Springs. May 13-17.

Holland, J. H. (1992). Genetic algorithms. *Scientific American, 267*, 44−50.

Hoomans, B. P. B., Kuipers, J. A. M., & van Swaaij, W. P. M. (1998). Discrete particle simulation of segregation phenomena in dense gas-fluidized beds. In *Fluidization IX, proceedings of the ninth engineering foundation conference on fluidization* (pp. 485−492). Colorado: Durangeo.

Hulme, I. (2003). *Verification of the hydrodynamics of a polyethylene fluidized bed reactor using CFD and imaging experiments (M. Sc. thesis)*. Calgary: University of Calgary.

Hulme, I., & Kantzas, A. (2004). Determination of bubble diameter and axial velocity for a polyethylene fluidized bed using X-ray fluoroscopy. *Powder Technology, 147*(1−3), 20−33.

Jackson, R. (2000). *The dynamics of fluidized bed particles*. New York: Cambridge University Press.

Johnsson, F., Zijerveld, R. C., Schouten, J. C., van den Bleek, C. M., & Leckner, B. (2000). Characterization of fluidization regimes by time-series analysis of pressure fluctuations. *International Journal of Multiphase Flow, 26*, 663−715.

Kai, T., Misawa, M., Takahashi, T., Tiseanu, I., & Ichikawa, N. (February 2005). Observation of 3-D structure of bubbles in a fluidized catalyst bed. *Canadian Journal of Chemical Engineering, 83*(1), 113−118.

Kaneko, Y., Shiojima, T., & Horio, M. (1999). DEM simulation of fluidised beds for gas-phase olefin polymerization. *Chemical Engineering Science, 54*, 5809−5821.

Kantzas, A. (1990). Investigation of physical properties of porous rocks and fluid flow phenomena in porous media using computer assisted tomography. *In Situ, 14*(1), 77−132.

Kantzas, A. (1994). Computation of holdups in fluidized beds and trickle beds by computer assisted tomography. *AIChE Journal, 40*(7), 1254−1261.

Kantzas, A., & Kalogerakis, N. (1996). Monitoring the fluidization characteristics of polyolefin resins using X-ray computer assisted tomography scanning. *Chemical Engineering Science, 51*(10), 1979.

Kantzas, A., Wright, I., Bhargava, A., Li, F., & Hamilton, K. (2001). Measurement of hydrodynamic data of gas phase polymerization reactors using non-intrusive methods. *Catalysis Today, 64*(3−4), 189−203.

Kantzas, A., Wright, I., & Kalogerakis, N. (1997). Quantification of channeling in polyethylene resin fluid beds using X-ray computer assisted tomography (CAT). *Chemical Engineering Science, 52*(13), 2023–2035.

Kihm, K. D., & Lyons, D. P. (1996). Optical tomography using a genetic algorithm. *Optics Letters, 21*(17), 1327–1329.

Kuipers, J. A. M. (1990). *A two-fluid micro Balance model of fluidized bed (Ph.D. thesis).* Enschede, The Netherlands: Twente University.

Kuipers, J. A. M., Duin, van K. J., Beckum, van F. P. H., & Swaaij, van W. P. M. (1993). Computer simulation of the hydrodynamics of a two-dimensional gas-fluidized bed. *Computers & Chemical Engineering, 17*(8), 839–858.

Kunii, D., & Levenspiel, O. (2013). *Fluidization engineering* (2nd ed.). Newton: Butterworth-Heinemann.

Laux, H., & Johansen, S. T. (1998). Computer simulation of bubble formation in a gas-solid fluidized bed. In *Fluidization IX, proceedings of the ninth engineering foundation conference on fluidization* (pp. 517–524). Colorado: Durangeo.

Lettieri, P., Cammarata, L., Micale, G., & Yates, J. (2003). CFD simulation of gas fluidized beds using alternative Eulerian–Eulerian modelling approaches. *International Journal of Chemical Reactor Engineering, 1.*

Lun, C., Savage, S., Jeffrey, D., & Chepurnity, N. (1984). Kinetic theories for granular flow: Inelastic particles in Couette flow and slightly inelastic particles in a general flow-field. *Journal of Fluid Mechanics, 140*, 233–256.

Lyczkowski, R. W., Gamwo1, I. K., Dobran, F., Ai, Y. H., Chao, B. T., Chen, M. M., et al. (1993). Validation of computed solids hydrodynamics and pressure oscillations in a bubbling atmospheric fluidized bed. *Powder Technology, 76*, 65–77.

McKeen, T., & Pugsley, S. T. (2003). Simulation and experimental validation of a freely bubbling bed of FCC catalyst. *Powder Technology, 129*, 139–152.

Mudde, R. F. (2010). Time-resolved X-ray tomography of a fluidized bed. *Powder Technology, 199*, 55–59.

Mudde, R. F., Alles, J., & van der Hagen, T. H. J. J. (2008). Feasibility study of a time-resolving X-ray tomographic system. *Measurement Science and Technology, 19.*

Mudde, R., Schulte, H. B. M., & Van Den Akker, H. (1994). Analysis of a bubbling 2-D gas-fluidized bed using image processing. *Powder Technology, 81*, 149.

Natterer, F. (1986). *The mathematics of computerized tomography.* John Wiley & Sons and B.G. Teubner.

Orta, A., Wu, B., Guerrero, A., Ghods, M., Bellehumeur, C., & Kantzas, A. (2011). Pressure effect on hydrodynamics of a high pressure X-ray transparent polyethylene fluidized bed. *IJCRE, 9*(1).

Peirano, E., Delloume, V., Johnsson, F., Lechner, B., & Simonin, O. (2002). Numerical simulation of the fluid dynamics of a freely bubbling fluidized bed: Influence of the air supply system. *Powder Technology, 122*, 69–82.

Ranade, V. V. (2002). *Computational flow modeling for chemical reactor engineering* (Vol. 5). San Diego, California: Academic Press.

Rowe, P. N., MacGillivray, H. J., & Cheesman, D. J. (1979). Gas discharge from an orifice into a gas fluidized bed. *Transactions of the Institution of Chemical Engineers, 57*, 194–199.

Rudisuli, M., Schildhauer, T. J., Biollaz, S. M. A., & Van Ommen, J. R. (2012). Scale-up of bubbling fluidized bed reactors – a review. *Powder Technology, 217*, 21–38.

Schober, L., Hulme, I., Chandrasekaran, B., & Kantzas, A. (2003). An experimental and simulation study of the hydrodynamics of linear low density polyethylene. In *3rd world congress on industrial process tomography.* Alberta: Banff. September 2-5.

Schouten, J. C., & Van Den Bleek, C. M. (1998). Monitoring the quality of fluidization using the short-term predictability of pressure fluctuation. *AIChE Journal, 44*(1), 48−60.

Song, J., Hyndman, C. L., Jakher, R. K., Hamilton, K., Kryuchkov, S., & Kantzas, A. (1999). Fundamentals of hydrodynamics and mass transfer in a three-phase fluidized bed system. 4th international conference on gas−liquid and gas−liquid−solid reactor engineering, Delft, The Netherlands, August 23−25. *Chemical Engineering Science, 54*(21), 4967−4973.

Syamlal, M., & O'Brien, T. J. (1987). *A generalized drag correlation for multiparticle systems*. Morgantown, WV: Unpublished Report U.S. Department of Energy, Office of Fossil Energy, Morgantown Energy Technology Center.

Syamlal, M., Rogers, W., & O'Brien, T. J. (1993). *MFIX documentation: Volume 1, theory guide*. Springfield, VA: National Technical Information Service.

Tsuji, Y., Kawaguchi, T., & Tanaka, T. (1993). Discrete particle simulation of two-dimensional fluidized bed. *Powder Technology, 77*, 79−87.

Valenti, C. (2008). A genetic algorithm for discrete tomography reconstruction. *Genetic Programming and Evolvable Machines, 9*(1), 85−96.

Van Ommen, J. R., Coppens, M. O., & Van Den Bleek, C. M. (2000). Early warning of agglomeration in fluidized beds by attractor comparison. *AIChE Journal, 46*(11), 2183−2197.

Van Wachem, B. G., Schouten, J. C., Krishna, R., Van, D., & Bleek, C. M. (1999). Validation of the Eulerian simulated dynamic behavior of gas-solid fluidized beds. *Chemical Engineering Science, 54*, 2141−2149.

Van Wachem, B. G. M., Van Der Schaaf, J., Schouten, J. C., Krishna, R., & Van Den Bleek, C. M. (2001). Experimental validation of Lagrangian−Eulerian simulations of fluidized beds. *Powder Technology, 116*, 155−165.

Van der Lee, L., Shepperson, J., & Kantzas, A. (2005). A comparison of experimental and predicted bubble characteristics of a linear low density polyethylene. In *Proceedings, 3rd European conference on the reaction engineering of polyolefins, Lyon, France, June 20−24*.

Venere, M., Liao, H., & Clausse, A. (July 2000). A genetic algorithm for adaptive tomography of elliptical objects. *IEEE Signal Processing Letters, 7*(7), 176−178.

Williams, R. A., & Beck, M. S. (Eds.). (1995). *Process tomography: Principles, techniques and applications*. Butterworth-Heinemann.

Wright, I., Hamilton, K., Kruychkov, S., Chen, J., Li, F., & Kantzas, A. (2001). On the measurement of hydrodynamic properties of an air−polyethylene fluidized bed system. *Chemical Engineering Science, 56*(13), 4085−4097.

Wu, B. Y., Briens, L., & Zhu, J.-X. (2006). Multi-scale flow behaviors in gas-solids two-phase flow systems. *Chemical Engineering Journal*, 187−195.

Wu, C., Cheng, Y., Ding, Y., Wei, F., & Jin, Y. (2007). A novel X-ray computed tomography method for fast measurement of multiphase flow. *Chemical Engineering Science, 62*, 4325−4335.

Wu, B., Kantzas, A., Bellehumeur, C., He, Z., & Kryuchkov, S. (2007). Multiresolution analysis of pressure fluctuations in a gas-solids fluidized bed: Application to glass beads and polyethylene powder systems. *Chemical Engineering Journal, 131*(1−3), 23−33.

Yasui, G., & Johanson, L. N. (1958). Characteristics of gas pockets in fluidized beds. *AIChE Journal, 4*, 445−452.

Yates, J. G., & Simons, S. J. R. (1994). Experimental methods in fluidization research. *International Journal of Multiphase Flow, 20*, 297−330.

Zarabi, T., & Kantzas, A. (October 1998). Predictions on bubble and solids movement in laboratory polyethylene fluid beds as visualized by X-ray computer assisted tomography (CAT) scanning. *Canadian Journal of Chemical Engineering, 76*(5), 853–865.

Zarabi, T., & Kantzas, A. (October 2000). Analysis of gamma-ray imaging data to describe hydrodynamic properties of gas–polyethylene fluidized beds. *Canadian Journal of Chemical Engineering, 78*(5), 849–857.

Zhang, Q., Wu, B., Li, Y., Bellehumeur, C., & Kantzas, A. (2007). Effect of particle shape on fluidized bed hydrodynamics using X-ray tomography and pressure measurement. In *5th world congress on industrial process tomography, Bergen, Norway, September 3–6*.

Applications of tomography in bubble column and fixed bed reactors

Daniel J. Holland
Department of Chemical and Process Engineering, University of Canterbury, Christchurch, New Zealand

23.1 Introduction

Many chemical reactions of industrial interest require contacting of gases, liquids, and solids. Two of the most common reactor configurations used for these reactions are bubble column and fixed bed reactors. Some examples of industrial reactions commonly using bubble column or fixed bed reactors include oxidation, hydrogenation, polymerization, fermentation, hydrodesulphurization, and Fischer−Tropsch synthesis of liquid fuels (Deckwer, 1992; Dudukovic, Larachi, & Mills, 1999; Gianetto & Specchia, 1992; Kantarci, Borak, & Ulgen, 2005; Satterfield, 1975). The efficient design and operation of these systems requires understanding the interaction of the gas and liquid phases. For example, the reaction rate is frequently determined not by the kinetics of the reaction but rather by the rate at which the gas and liquid come into contact, often with a solid catalyst as well. For many years, the design of these reactors has been based on empirical correlations derived from measurements of the inlet and outlet conditions of the reactor, including pressure, temperature, and concentration. However, these conventional measurements provide little insight into the internal operations of the reactor and therefore can be adversely affected by, for example, maldistribution of the fluid entering the reactor, imperfect mixing of the continuous phase, or changes in the flow regime. Without detailed knowledge of these factors, it is necessary to assume that the reactor behaves as if it were following an ideal model whereby each phase is treated as being either in plug flow or well mixed. It is well known that such models are only approximate descriptors of industrial reactors. Indeed, it is common for full-scale industrial reactors designed using these approaches to fail to achieve their design objectives, leading to significant increases in production costs and loss of efficiency (Dudukovic, 2010).

Computational fluid dynamic (CFD) techniques potentially provide significantly improved insight into the design and operation of bubble column and fixed bed reactors, and many researchers are actively exploring these techniques (Kuipers & van Swaaij, 1998; Olmos, Gentric, Vial, Wild, & Midoux, 2001; Sanyal, Vasquez, Roy, & Dudukovic, 1999; Selma, Bannari, & Proulx, 2010; van Sint Annaland, Dijkhuizen, Deen, & Kuipers, 2006). However, such numerical models of multiphase

Industrial Tomography. https://doi.org/10.1016/B978-0-12-823015-2.00020-0

reactors require closure laws to describe the interaction of the gas, liquid, and/or solid phases. These closure laws are typically derived from simplified phenomenological models or experimental measurements of bulk properties such as the hold-up and their accuracy is questionable (Dudukovic, 2010). Accurate specification of closure laws becomes even more complex in the presence of a stationary solid phase. Detailed noninvasive measurements of local characteristics of the hydrodynamics are required to improve our understanding of multiphase flows and to validate the predictions of numerical models. Such information is not available using conventional measurement techniques. Tomographic imaging techniques can provide noninvasive measurements of the local phase fraction, velocity, and composition. Such measurements will enable new science-based approaches to modeling industrial-scale chemical reactors to be developed and therefore will lead to improved designs and increased efficiency of operation.

This chapter explores some recent studies of bubble column and fixed bed reactors using tomographic imaging techniques. The theoretical bases for the tomographic imaging techniques are covered elsewhere in this book and therefore will not be described here. This chapter focuses on describing some of the key properties of the two reactor types considered, and how tomographic techniques have been used to investigate these properties. Owing to the wealth of literature on the topic, this review cannot summarize all of the work in the area. The interested reader is referred to earlier reviews for further discussion of the importance of tomography in understanding multiphase reactors (Al-Dahhan, Larachi, Duduković, & Laurent, 1997; Boyer, Duquenne, & Wild, 2002; Chaouki, Larachi, & Dudukovic, 1997; Dixon & Deutschmann, 2017; Dudukovic, 2002; Dudukovic et al., 1999; Gladden & Sederman, 2017; Hampel et al., 2020; Heindel, 2011; Mudde, 2010a; Reinecke & Mewes, 1996; Schubert, Bieberle, Barthel, Boden, & Hampel, 2011; Sharifi & Young, 2013; Zhivonitko, Svyatova, Kovtunov, & Koptyug, 2018). An emerging area of interest is in exploiting the capability of additive manufacturing (such as 3D printing) to customize the internal structure of these reactors. In this second edition, a new section has been added detailing progress and future developments in the area of additive manufacturing of fixed bed reactors. The chapter concludes with an outlook for expected future developments of tomographic imaging in these systems and some sources of further information.

23.2 Bubble column reactors

Bubble column reactors consist of a vessel containing a liquid or a liquid—solid suspension with a gas distributor at the bottom. Gas is injected into the column through the distributor to form bubbles. A schematic illustration of a bubble column is shown in Fig. 23.1. They are used throughout the chemical and biochemical industries especially for reactions involving oxidation, chlorination, alkylation, polymerization, hydrogenation, Fischer—Tropsch synthesis, fermentation, and wastewater treatment. Bubble columns are an attractive choice owing to their excellent heat and mass transfer

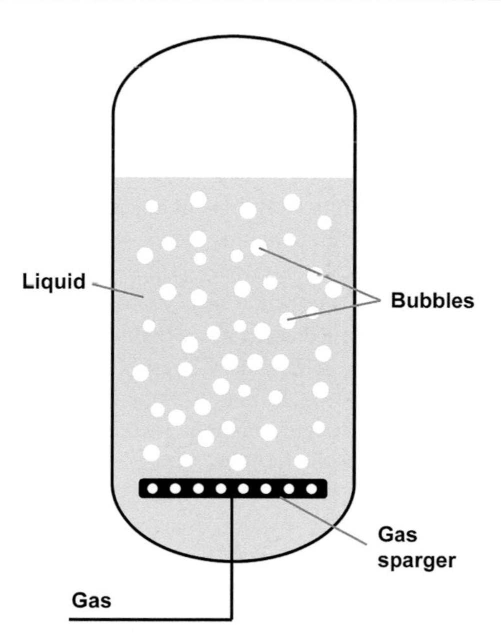

Figure 23.1 Schematic diagram illustrating a bubble column reactor. Gas enters the column through a gas distributor or sparger in the base of the column. Once in the column, gas forms discrete bubbles that rise due to their buoyancy and induce motion of the continuous liquid phase.

characteristics, low operating costs, and durability of any solid material present (Kantarci et al., 2005). A further advantage is the relative ease with which catalyst can be added or removed from the column. However, there are several significant challenges associated with using bubble column reactors.

One of the characteristics of industrial bubble column reactors is their large size. Bubble column reactors used for commodity products typically have a volume on the order of 100 m^3; when used as fermenters or in waste treatment applications, the volumes can be up to two orders of magnitude larger. Owing to the complex interaction of the gas and liquid phases, small changes in the column geometry, internals, distributor, gas flow rate, and solid or liquid phase properties can lead to dramatic changes in the flow regime in a bubble column reactor. The scale of these systems and their sensitivity to the operating conditions makes accurate design and scale-up of bubble columns challenging. There is therefore a need to obtain measurements of the key parameters that determine the operation of bubble column reactors including local gas and liquid fraction, bubble size, and gas and liquid velocity (Kantarci et al., 2005). The following sections describe some of the advanced tomographic measurement techniques that can provide such measurements.

23.2.1 Fluid phase fraction measurements

Perhaps the simplest measurement obtainable from tomographic techniques is a local phase fraction or hold-up measurement. A variety of techniques have been used to characterize the hold-up including electrical capacitance tomography (ECT) (Bennett, West, Luke, Jia, & Williams, 1999; Cai, Shen, Shen, & Dai, 2010; Warsito & Fan, 2001, 2003; Warsito, Ohkawa, Kawata, & Uchida, 1999), electrical resistance tomography (Fransolet, Crine, Marchot, & Toye, 2005; Ishkintana & Bennington, 2010; Okonkwo, Wang, & Azzopardi, 2013; Toye et al., 2005; Vijayan, Schlaberg, & Wang, 2007), ultrasonic tomography (Rahiman et al., 2014, 2016; Supardan, Masuda, Maezawa, & Uchida, 2007; Utomo, Warsito, Sakai, & Uchida, 2001; Warsito et al., 1999), X-ray or γ-ray computed tomography (CT) (Chen et al., 1998; Fischer & Hampel, 2010; Gulati, Behling, Munshi, Luke, & Mewes, 2010; Hubers, Striegel, Heindel, Gray, & Jensen, 2005; Nedeltchev, Shaikh, & Al-Dahhan, 2006, 2007; Ong, Gupta, Youssef, Al-dahhan, & Dudukovic, 2009; Patel & Thorat, 2008; Rabha, Schubert, & Hampel, 2013b, 2013a; Rabha, Schubert, Wagner, Lucas, & Hampel, 2013; Shaikh & Al-dahhan, 2005), and magnetic resonance imaging (MRI) (Daidzic, Schmidt, Hasan, & Altobelli, 2005; Sankey, Yang, et al., 2009; Tayler, Holland, Sederman, & Gladden, 2012a). These measurements have been used to explore a variety of characteristics of bubble column reactor performance. Flow in bubble column reactors is divided into up to five different regimes including homogeneous, slugging, churn, semiannular, and annular flow (Spedding, Woods, Raghunathan, & Watterson, 1998). The hold-up in bubble column reactors is most frequently used to track changes in the flow regime; however, if the temporal and spatial resolution are sufficiently great, tomography can also be used to estimate the bubble size distribution and interfacial area, which are perhaps the most important characteristics required to predict the performance of a bubble column reactor.

CT techniques have high spatial resolution, but typically a low temporal resolution, and therefore have mostly been used to measure time-averaged voidage distributions (Al Mesfer, Sultan, & Al-Dahhan, 2016; Chen, Kemoun, et al., 1999; Shaikh & Al-dahhan, 2005; Shaikh, Taha, & Al-Dahhan, 2020; Sultan, Sabri, & Al-Dahhan, 2018). CT techniques have been used to characterize changes in the flow structure from different sparger designs (Ong et al., 2009; Varma & Al-dahhan, 2007), internals (Chen, Kemoun, et al., 1999; Chen, Li, et al., 1999), and composition (Hubers et al., 2005; Patel & Thorat, 2008) and have also been explored as a means to characterize the flow regime using ideas from chaos analysis (Nedeltchev, 2009; Nedeltchev et al., 2006, 2007). However, these studies have been limited to time-averaged measurements owing to the long time taken to acquire conventional CT measurements.

An exciting development of CT techniques has been the advent of fast CT scanners (Bieberle & Hampel, 2006; Mudde, Alles, & Hagen, 2008). The ultrafast electron beam X-ray tomography system developed in Dresden promises to be particularly beneficial for high resolution, dynamic imaging (Bieberle, Barthel, Menz, Mayer, & Hampel, 2011; Bieberle et al., 2007, 2010; Bieberle & Hampel, 2006; Fischer & Hampel, 2010). This facility is used to obtain two-dimensional images at a rate of up to 7000 frames/s with a spatial resolution of about 1 mm, though in practice, a slower frame rate is often used owing to the large quantity of data produced (Rabha, Schubert, & Hampel, 2013b; Rabha, Schubert, Wagner, et al., 2013). Fig. 23.2 shows an example of the quality of data that can be produced using this facility (Lau, Möller, Hampel, & Schubert, 2018). A clear demarcation between the gas and liquid phases is

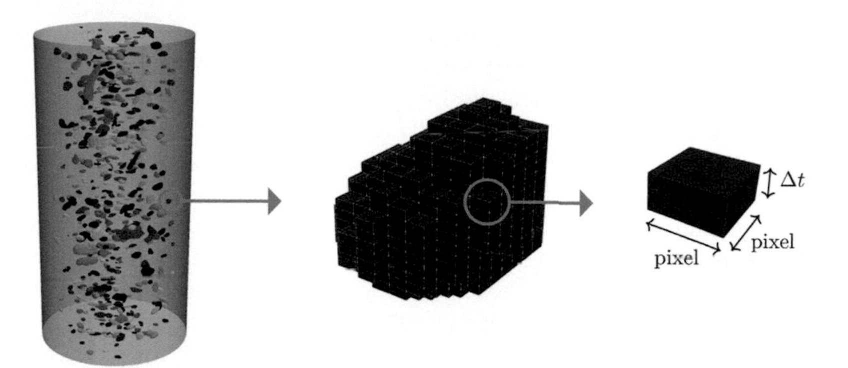

Figure 23.2 Ultrafast X-ray images of a bubble column. Images shown are a virtual projection of the pseudothree-dimensional gas phase structure obtained from a time series of two-dimensional images of the cross-section of the column. Individual bubbles in the image are identified by connected regions of gas phase. The images shown were obtained for an air–water mixture at time intervals of 0.001 s.
Adapted from Lau, Y. M. et al. (2018). Ultrafast X-ray tomographic imaging of multiphase flow in bubble columns – Part 2: Characterisation of bubbles in the dense regime, *International Journal of Multiphase Flow, 104*, pp. 272–285.

obtained with sufficient resolution to identify the shape and size of individual bubbles. The measurements have been used to extract an apparent bubble size distribution, and how this size distribution changes in the presence of internals in the bubble column. Measurements such as those obtained by Lau et al. (2018) provide detailed information on the void structure and will in the future lead to more advanced models of bubble size, coalescence, and break-up than has hitherto been possible. Already these measurements are being used to test CFD simulations of two-phase flow (Rabha, Schubert, & Hampel, 2013b). However, a particular strength of this facility is that it will be integrated into the TOPFLOW facility at Dresden (Fischer & Hampel, 2010), which will permit the characterization of flows at up to 70 bar and nearly 300°C (Prasser et al., 2006). Such measurements will provide invaluable data for improving our understanding of bubble column reactor hydrodynamics at industrially relevant conditions. The first results from these systems are now becoming available. In one recent study, the ultrafast X-ray CT measurements of a gas liquid system operating at 4 bar were presented in comparison with detailed CFD simulations (Banowski, Hampel, Krepper, Beyer, & Lucas, 2018). The system used in that study was capable of operating at up to 70 bar, indicating significant potential for further advances in the near future.

One limitation of this ultrafast X-ray system is the diameter of the column that can be studied, which is apparently limited to about 100 mm at this stage (Kipping, Kryk, & Hampel, 2021). This limitation has been partially overcome by the design used in TUDelft (Mudde, 2010b), though the increase in scale that is achievable comes at the cost of a decrease in the spatial and temporal resolution. The system at TU Delft has been extensively used to study gas–solid fluidized beds, but has had only limited application to study gas–liquid systems (Jahangir, Wagner, Mudde, & Poelma, 2019; Mandalahalli, Wagner, Portela, & Mudde, 2020).

MRI presents another promising technique for the characterization of bubble column reactors (Daidzic et al., 2005; Sankey, Holland, Gladden, & Sederman, 2009; Tayler et al., 2012a). The chemical sensitivity of MRI means that it is possible to measure the gas and liquid phases independently to provide a complete characterization of gas–liquid flows (Adair, Richard, & Newling, 2021; Sankey, Holland, et al., 2009). Conventional MRI is too slow to resolve the structure of gas–liquid flows and thus was initially confined to time-averaged measurements (Daidzic et al., 2005; Sankey, Holland, et al., 2009). Ultrafast MRI techniques such as EPI (Stehling, Turner, & Mansfield, 1991) are available; however, they are sensitive to turbulent motion in the fluid phase and therefore are not well suited to study gas–liquid flows. However, spiral MRI techniques have been adapted to study gas–liquid flows (Tayler et al., 2012a; Tayler, Holland, Sederman, & Gladden, 2011) at a spatial resolution of ~500 μm and a temporal resolution of 55 frames/s. The spatial resolution of these images has since been further increased through the introduction of compressed sensing (Holland, Malioutov, Blake, Sederman, & Gladden, 2010; Tayler, Holland, Sederman, & Gladden, 2012b). Spiral MRI is advantageous as it is robust to motion artifacts and therefore can be used to provide quantitative measurements of the bubble size distribution in bubble columns for a gas hold-up of up to 0.2 and potentially higher using compressed sensing (Tayler et al., 2012b). MRI measurements of bubble columns were used to explore the variation in bubble size distribution and interfacial area with radial and axial position in the column. A particular advantage of MRI-based imaging is that it is possible to acquire a two-dimensional image at any orientation. Thus, Tayler et al. (2012a) demonstrate the acquisition of vertical images in the center of a bubble column and use these to study bubble break-up and coalescence. In the future, measurements such as these could be used to derive kernel functions for the population balance models used to simulate industrial bubble column reactors.

Although spiral MRI shows significant promise for characterizing bubble columns, there are some limitations. MRI generally requires the use of a superconducting magnet and therefore is restricted to columns <50 mm in diameter. An interesting approach that could potentially overcome this limitation is the use of Bayesian analysis (Holland, Blake, Tayler, Sederman, & Gladden, 2011, 2012). This approach has been shown to provide measurements of the bubble size distribution in a bubble column from very simple measurements. Further, the technique has been adapted for use on a low-field permanent magnet system for characterizing the grain size distribution in rock samples at a resolution an order of magnitude or more better than could be achieved with conventional MRI techniques (Holland, Mitchell, Blake, & Gladden, 2013). In the future, such an approach could be used with other low-field instrumentation (see for example Leblond, Javelot, Lebrun, & Lebon, 1998) to permit the measurement of bubble size distributions under conditions of relevance to industrial bubble columns.

Industrial process tomography requires sensors that can be operated on large-scale processes and under industrially relevant conditions. MRI and ultrafast X-ray measurements are typically limited to small-scale systems (perhaps <100 mm diameter) and relatively benign conditions. While some progress is being made to overcome these limitations as noted above, perhaps a more promising approach for industrial process monitoring is to use the electrical and ultrasonic tomography techniques. ECT and ERT have both been used to study bubble column systems for some time. An early example is the work of Bennett et al. (1999) who were able to distinguish between churn flow and homogeneous flow by comparing the differences between ECT measurements of a gas–water flow with and without a surfactant present, as shown in Fig. 23.3. In their work, the homogeneous flow produced in the presence of the surfactant resulted in a uniform decrease in the permittivity throughout the tomogram as the gas fraction increased. By contrast, when no surfactant was present, large bubbles and slugs were formed that were visible in the tomograms. Similarly, Fransolet et al. (2005) observed a transition from homogeneous flow with small bubbles through to heterogeneous flow with large bubbles as the viscosity of the water was increased through the addition of xanthan gum.

In general, ECT, ERT, and ultrasonic tomography techniques are all characterized by a relatively low spatial resolution. Therefore, the major benefit of these types of tomographic imaging techniques is in the extension to larger-scale processes, industrial process conditions, and the characterization of macroscopic heterogeneity arising from, for example, sparger malfunctioning, drift in operating conditions, or flow regime transitions. Full-scale industrial processes are still challenging to access with these techniques owing to limitations of signal to noise and practicality, though several

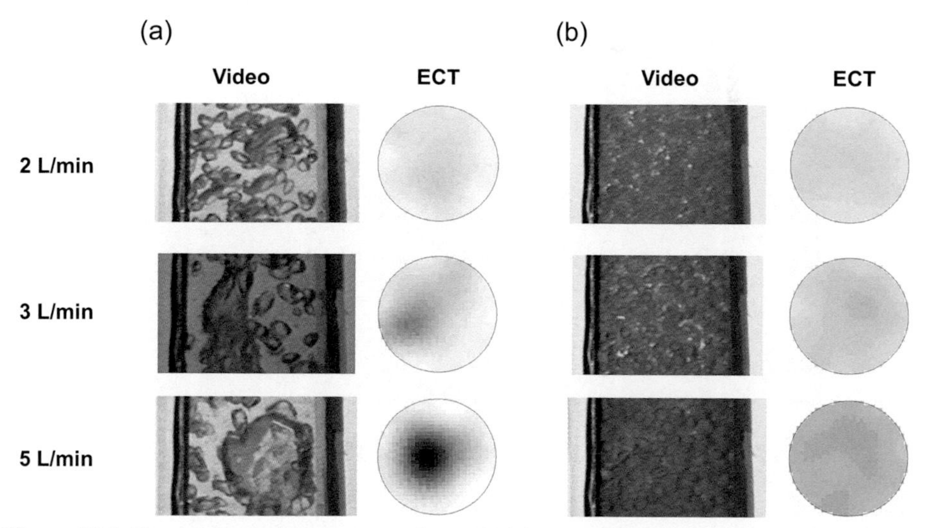

Figure 23.3 These images show a comparison of high speed video (left) and electrical capacitance tomography measurements (right) of gas–liquid flow when the liquid has either (a) no surfactant or (b) 32 ppm of a surfactant frother. In the absence of the surfactant, the column undergoes a regime transition where large bubbles or slugs are formed at high gas flow rates. The presence of the surfactant ensures a homogeneous flow regime and no large bubbles are formed regardless of the flow rate.
Adapted from Bennett, M. A., West, R. M., Luke, S. P., Jia, X., Williams, R. A. (1999). Measurement and analysis of flows in a gas-liquid column reactor. *Chemical Engineering Science, 54*, 5003–5012 with permission.

studies of pilot-scale facilities have been presented (Mohammed, Hasan, Ibrahim, Dimitrakis, & Azzopardi, 2019; Sines et al., 2019; Suard et al., 2019; Wang, Marashdeh, & Fan, 2014; Wang, Marashdeh, Fan, & Warsito, 2010; Wang, Dorward, Vlaev, & Mann, 2000). Linear sensor arrays have been proposed to permit measurements on full-scale processes (Schlaberg et al., 2006); however, these have not yet been applied to study bubble column reactors. One interesting study of bubble column reactors using electrical tomography mapped the transition from bubbling to churn-turbulent flow in pilot-scale columns up to 0.29 m in diameter (Mohammed et al., 2019). Fig. 23.4 shows a visualization of the different flow regimes, as measured by ECT, and compares these with high speed photography images of the flow. The consistency of the flow features identified between the photographs and ECT measurements is clear. These measurements demonstrate that electrical techniques are able to capture the key flow features. However, to be able to use them as a tool in industrial process monitoring, it is necessary to derive simple quantitative information from the data. Cai et al. (2010) have demonstrated such an approach in the bubbling/jetting

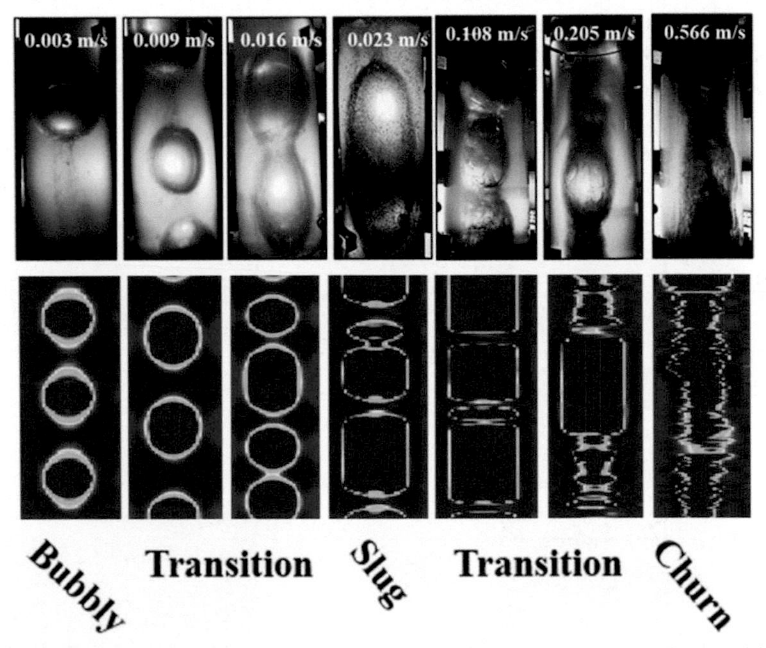

Figure 23.4 Electrical capacitance tomography and optical photographs of the transition from bubbly to churn-turbulent flow in a large diameter bubble column. The top row of images show photographs of the system, while the bottom row of images were generated from a time series of two-dimensional ECT images of the cross-section of the column. In the ECT images, blue (dark gray in print version) corresponds to gas, while red (gray in print version) corresponds to liquid.

Adapted from Mohammed, S. K. et al. (2019) 'Dynamics of flow transitions from bubbly to churn flow in high viscosity oils and large diameter columns', International Journal of Multiphase Flow. 120, 103095.

regime transition. They show that the local void fraction and the standard deviation of the void fraction are linearly related to the gas flow rate in both the bubbling and jetting flow regimes. However, a sharp transition in the slope of this relationship was observed, indicating a simple mechanism for characterizing the change in flow regime. Measurements such as these potentially provide a model for future exploitation of the robustness of electrical-based industrial process tomography sensors. For example, measurements of the local void fraction might be obtained using small electrodes located in the vicinity of an object of interest, such as a nozzle. The limiting case of this approach could perhaps be considered to be the wire mesh sensor.

Wire mesh sensors consist of a network of wires that are placed inside the column to be imaged (Prasser, Böttger, & Zschau, 1998). Initially, the conductivity or capacitance through the fluid between pairs of these wires is measured (Silva, Schleicher, & Hampel, 2007). Wire mesh sensors are advantageous as they are robust and can be applied to study systems at high pressure and temperature (Pietruske & Prasser, 2007). Wire mesh sensors also do not require the solution of an ill-posed, nonlinear inverse problem, as is required for other electrical tomography techniques. Of course, the disadvantage of wire mesh sensors is that they are inherently invasive, and therefore, the application of these devices in certain processes may be problematic. The intrusive effects of wire mesh sensors have been studied by comparison with high speed camera measurements in transparent systems (Fuangworawong, Kikura, Aritomi, & Komeno, 2007; Ito, Prasser, Kikura, & Aritomi, 2011; Wangjiraniran et al., 2003). These studies revealed that the wire mesh can cause bubbles to break up and to either slow or accelerate, depending on the liquid velocity. However, despite this limitation, wire mesh sensors have been used successfully to study the transition between flow regimes, bubble size distribution, and bubble flow patterns (Fuangworawong et al., 2007; Lucas, Krepper, & Prasser, 2005; Pietruske & Prasser, 2007; Prasser, Krepper, & Lucas, 2002; Richter, Aritomi, Prasser, & Hampel, 2002; Wangjiraniran, Aritomi, Kikura, Motegi, & Prasser, 2005). Wire mesh sensors have also been used extensively in validation studies of other tomographic imaging techniques including ultrafast X-ray, ECT, and ERT (Azzopardi et al., 2010; Olerni, Jia, & Wang, 2013; Prasser, Misawa, & Tiseanu, 2005).

Up to this point, the discussion has focused on the characterization of the distribution of the gas and liquid phases within the bubble column. In the following section, the discussion will consider a second critical parameter in the characterization of bubble column reactors, the velocity of each phase.

23.2.2 Fluid velocity measurements

Measurements of the velocity are more difficult to obtain than measurements of the voidage. Wire mesh sensors, ECT, ERT, radioactive particle tracking (RPT, also known as computer-aided radioactive particle tracking, CARPT, or positron emission particle tracking, PEPT, depending on the exact technique used), and MRI have all been used to measure the velocity of one or both phases in bubble column reactors.

Wire mesh sensors, ECT, ERT, and ultrafast X-ray CT have been used to obtain measurements of the velocity of the discrete phase (Barthel, Bieberle, Hoppe,

Banowski, & Hampel, 2015; Gumery, Ein-Mozaffari, & Dahman, 2011; Jin, Wang, & Williams, 2007; Neumann, Bieberle, Wagner, Bieberle, & Hampel, 2019; Richter et al., 2002; Warsito & Fan, 2003). In these examples, the distribution of the two phases is measured by sensors separated by a known distance. A cross-correlation of the phase fraction is obtained from a time series of images, and from the cross-correlation the time-averaged velocity distribution is obtained. Wang et al. (2006) modified the standard cross-correlation approach by calculating a "best-correlated" pixel to allow for the transverse motion of the bubbles in between the two measurement planes. This approach was validated using point measurements obtained with a four-sensor ERT probe and enabled the measurement of the horizontal velocity in a swirling pipe flow. Jin et al. (2007) validate the cross-correlation approach by comparison of estimates of the bubble velocity from ERT and a drift-flux model. Their results demonstrate excellent agreement, provided that the bubble column is operated in a regime consisting of relatively isolated swarms of bubbles, that is, below the transition to slugging flow. The high resolution of ultrafast X-ray CT permits another approach whereby individual bubbles are identified and correlated between the two planes (Barthel et al., 2015). This approach enables measurement of the velocity of individual bubbles, a level of detail not possible with cross-correlation−based imaging. However, none of these techniques are sensitive to the motion of the continuous phase. Such a limitation can be partially overcome through the introduction of a tracer molecule (Gumery et al., 2011; Hamood-ur-rehman, Dahman, & Ein-mozaffari, 2012), although this increases the invasiveness of the measurement.

An alternative method of tracking the motion of the continuous phase is to use RPT. These techniques measure the flow field by tracing the position of a single radioactive tracer particle over time. A map of the distribution of the velocity of the continuous phase is extracted by assuming ergodicity, that is that a long-time average of a single particle is equivalent to a spatial average of many particles, and that the motion of the tracer particle is determined by the motion of the continuous phase (Chen, Kemoun, et al., 1999; Karim, Varma, Vesvikar, & Al-dahhan, 2004; Rados, Shaikh, & Al-Dahhan, 2005; Upadhyay, Pant, & Roy, 2013; Varma & Al-dahhan, 2007). A benefit of RPT is that they are capable of measuring the velocity and velocity fluctuations of the continuous phase of opaque flows at high pressure and temperature (Karim et al., 2004; Rados et al., 2005), and in relatively large-scale systems (Besagni, Inzoli, Zieghenein, & Lucas, 2019). Thus, RPT is able to probe operations at industrially relevant conditions. Such experiments have proven particularly beneficial when combined with CT measurements of the same system, as in this case it is possible to extract both time-averaged density and velocity information. In one study, RPT and CT were combined to examine the effect of changing the gas distribution in an airlift bioreactor (Varma & Al-dahhan, 2007). These measurements demonstrate that a multiorifice distributor improves the gas distribution, as expected. However, in addition, the multiorifice distributor was found to provide enhanced liquid recirculation for the same power input and therefore improved mixing. RPT and CT have also been combined to demonstrate the importance of tomographic measurements for scale-up studies. Shaikh and Al-dahhan (2010) studied bubble columns operating in the same flow regime and with similar overall hold-up, but with different radial distributions of the hold-up as measured by CT. RPT was then used to study the mixing patterns in the bubble columns. Shaikh and Al-dahhan demonstrate that when the radial distribution of the hold-up is the same, similar mixing patterns are established and the columns exhibit hydrodynamic similarity. However, when the radial distribution of the hold-up is not consistent, the mixing patterns are also dissimilar and the reactors are not hydrodynamically similar, even though the overall hold-up and flow regime are similar. These results demonstrate the need for detailed tomographic measurements of multiphase flows in order to develop fundamental scale-up criteria for bubble column reactors.

One disadvantage of RPT techniques for measuring the velocity is that it is necessary to assume that the long-time-averaged motion of the tracer particle provides an accurate estimate of the map of the time-averaged velocity in the continuous phase. This approximation is likely valid in most situations. However, in a bubble column, it is possible that the particle may become trapped at the interface of the gas and liquid and therefore travel with the discrete phase for a period of time. One recent study enables us to examine this issue closely, as they measured the velocity of the gas and liquid phases separately using ultrafast X-ray CT and RPT (Azizi et al., 2017). By combining the measurements, the authors were able to close the mass balance for upward and downward flow in the column. Hence, it is likely the RPT measurements were able to recover the liquid velocity accurately, as desired. These complementary techniques have since been used to develop detailed hydrodynamic correlations from measurements of bubble columns (Azizi et al., 2019).

Another approach to overcome uncertainty with liquid phase characterization is to measure the motion of the continuous phase directly, as in MRI measurements. The chemical sensitivity of MRI enables time-averaged imaging of the gas and liquid phases independently, provided that a suitable gas is used (Adair et al., 2021; Sankey, Holland, et al., 2009). Recently, the adaptation of spiral MRI to study gas–liquid flows has enabled the characterization of the liquid velocity dynamically (Tayler et al., 2011, 2012a, 2012b). Fig. 23.5 provides an example of an MRI image of the liquid velocity distribution in a bubble column (Tayler et al., 2012b). These images were obtained at a rate of 188 frames per second for each velocity component, or 63 frames per second for all three velocity components. The measurements were used to study the onset of turbulence in the continuous liquid phase in bubble columns. It is well known that rising gas bubbles can lead to the formation of turbulent structures in multiphase flows (Balachandar & Eaton, 2010), and indeed, this turbulence is one of the reasons why bubble columns are desirable as industrial reactors as it leads to high heat and mass transfer. However, it is still unclear how this turbulence arises and is difficult to simulate numerically. Tayler et al. (2012b) used MRI to measure the velocity in the wake of single and multiple rising bubbles. They demonstrated that the formation of a pair of secondary vortices in the wake of a single bubble drives the horizontal motion of the bubble as it rises. When multiple bubbles rise together, these vortices were observed to join together to form "vorticity chains": large-scale structures that extend over multiple bubble-length scales. These vorticity structures are hypothesized to be significant in the formation of multiphase-induced turbulence.

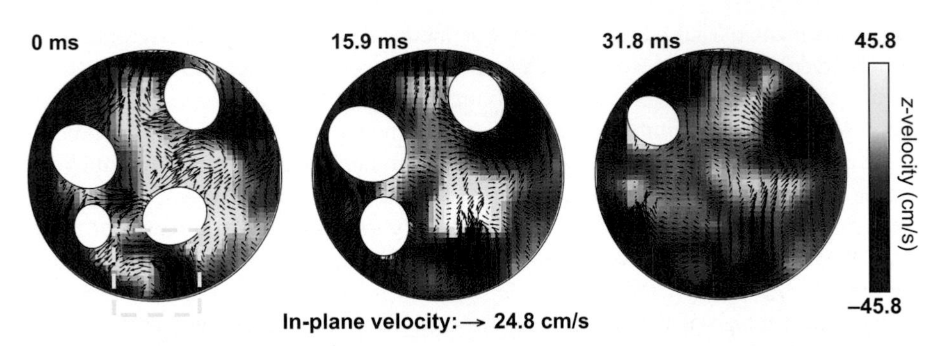

Figure 23.5 MRI measurements of the velocity of the water around freely rising air bubbles. The color-bar gives the through-plane velocity, while the arrows indicate the direction and speed of the in-plane velocity. Three-component velocity images were acquired at a rate of 63 frames per second and a spatial resolution of 390 × 390 μm.
Adapted from Tayler, A. B., Holland, D. J., Sederman A. J., Gladden L. F. (2012). Exploring the origins of turbulence in multiphase flow using compressed sensing MRI. *Physical Review Letter, 108*, 264505.

One of the challenges with tomographic measurements is that the complexity of the measurement and/or image reconstruction can make it difficult to identify the accuracy of the measurement precisely. To overcome this limitation, it is useful to cross-validate measurement techniques by comparing them on the same system (Bieberle, Schubert, Silva, & Hampel, 2010; Hernandez-Alvarado et al., 2018). One recent study found that pressure and gamma-ray densitometry were able to provide robust, accurate results across a range of flow regimes, but that the accuracy of wire mesh sensors, electrical tomography, and optical void probes depended on the system being studied (Hernandez-Alvarado et al., 2018); that study did not consider MRI or X-ray CT techniques.

The research described above presents an introduction to some of the measurements of bubble column reactors performed using tomographic imaging techniques. Low-resolution measurements of the local phase fraction have been obtained using CT, MRI, and electrical tomography techniques, as well as direct imaging of the individual bubbles using high speed CT, MRI, and wire mesh sensors. Measurements of the velocity of the discrete phase have been obtained using electrical tomography, ultrafast CT, and MRI, while RPT and MRI have been used to measure the velocity of the continuous phase. These measurements characterize the key parameters of bubble column reactors. Together, they will lead to improved design and operation of industrial reactors.

23.3 Fixed bed reactors

Fixed bed reactors consist of a vessel filled with solid catalyst over which gas and/or liquid flow. Fixed bed reactors are used throughout the chemical industry, especially for hydrogenation and oxidation reactions in the petrochemicals industry and for chromatographic separation in biochemical engineering. In one classic design, the gas and liquid flow cocurrently downward over a collection of particles that are randomly packed into the column. These reactors are known as trickle bed reactors, a schematic illustration of which is shown in Fig. 23.6. A high-profile example of the use of a trickle bed reactor is at the Pearl Gas-to-Liquid plant in Ras Laffan, Qatar. In other systems, the gas and/or liquid may flow upward, or only a single phase may be present. This review will focus on fixed bed reactors in chemical engineering. Fixed beds are attractive reactors as liquid flow patterns tend to produce high conversions, catalyst losses are minimal, they can be operated at high temperature and pressure, and are scaled up with relative ease. However, there are also several drawbacks of trickle bed reactors. For example, the catalyst effectiveness in fixed bed reactors is low owing to the large particle size needed to achieve a reasonable pressure drop, flow maldistribution can lead to decreased performance, side reactions can lead to fouling, and heat transfer is poor (Gianetto & Specchia, 1992). Industrial fixed bed reactors tend to be large, on the order of 10 m diameter. However, scale-up of these reactors can be

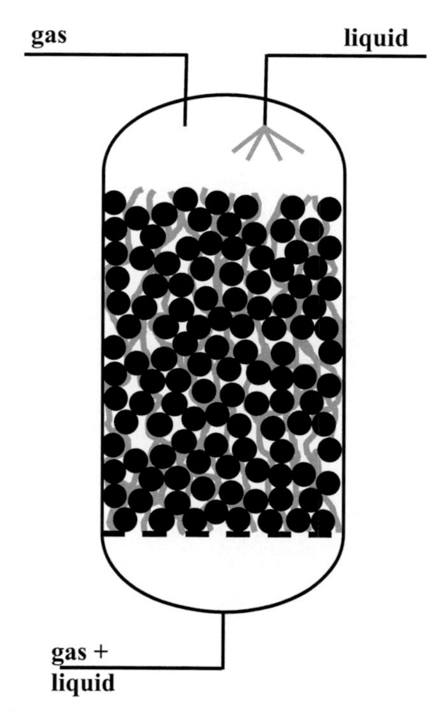

Figure 23.6 Schematic diagram of a trickle bed reactor. Gas and liquid enter the column at the top and flow cocurrently over a fixed bed of solid particles. The solid particles typically contain a catalyst, and the reaction between the gas and liquid occurs inside the particles. A two-phase mixture containing the products and any unreacted material is withdrawn from the bottom of the column. Gas and liquid flow may be stable or unstable depending on the operating conditions of the reactor.

achieved using parallel small diameter columns contained in a single process vessel and hydrodynamic features tend to be dominated by particle-scale phenomena that are largely scale invariant. As a result, the challenge for tomographic measurements of trickle bed reactors is not so much to determine how best to scale-up these reactor designs but rather to identify how to operate the reactors most efficiently, regardless of scale. Research has tended to focus on either hydrodynamic studies in cold-model fixed beds or on the interaction of operating conditions with the chemical conversion.

23.3.1 Hydrodynamic studies

Hydrodynamic studies of fixed bed reactors are challenging owing to the opacity of most granular materials. Optical imaging of fixed bed reactors is possible using refractive index matched materials (Dijksman, Rietz, Lorincz, Van Hecke, & Losert, 2012). Optical imaging is typically able to achieve high temporal (100 Hz) and spatial (12 μm) resolution, and thus has been used to track transient flow features within fixed beds including the transition to turbulence (Khayamyan, Lundström, Gren, Lycksam, & Hellström, 2017; Ziazi & Liburdy, 2019). Fig. 23.7 presents some example results from these PIV measurements at particle Reynolds numbers Re_p of 20 and 240. A clear

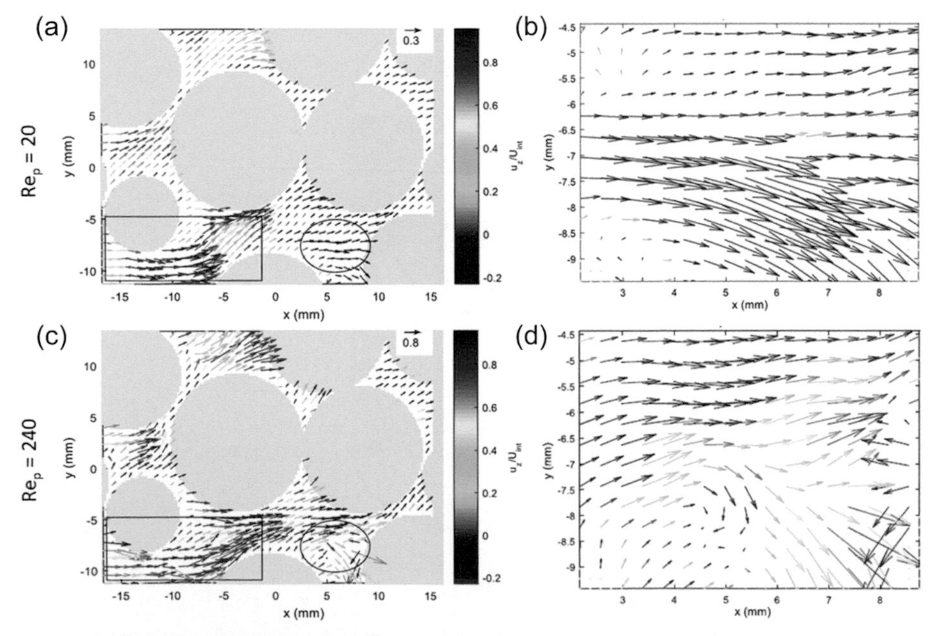

Figure 23.7 PIV measurements of flow within a random packed bed of spheres at (a, b) a Reynolds number of 20 and (c,d) a Reynolds number of 240. PIV is possible because the fluid and particles are refractive index matched. The images in (b) and (d) are of the region enclosed by the ellipse in (a) and (c).
Adapted from Khayamyan, S. et al. (2017). Transitional and turbulent flow in a bed of spheres as measured with stereoscopic particle image velocimetry. *Transport in Porous Media 117*, 45–67.

transition in the flow behavior is seen with relatively stable flow at $Re_p = 20$, to a flow dominated by vortices and inertial flow patterns at $Re = 240$. Optical imaging techniques have significant limitations due to the requirement for refractive index matching. For example, studying gas–liquid flow through a packed bed optically is likely not possible. However, they are significantly cheaper and more accessible than many of the true tomographic imaging techniques described below.

MRI is one technique that has been used widely to study packed and trickle bed reactors (Anadon, Sederman, & Gladden, 2006; Collins et al., 2017; Gladden, Abegão, Dunckley, Holland, Sankey, & Sederman, 2010; Gladden, Lim, Mantle, Sederman, & Stitt, 2003; Kutsovsky, Scriven, Davis, & Hammer, 1996; Sederman & Gladden, 2001, 2005; Suekane, Yokouchi, & Hirai, 2003). MRI can extract the distribution of gas and liquid within the three-dimensional structure of the packing. These measurements provide quantitative, local measurements of the hold-up, wetting efficiency, and velocity (Gladden, Abegão, Dunckley, Holland, Sankey, & Sederman, 2010; Gladden et al., 2003; Sankey, Holland, et al., 2009). MRI is particularly valuable for dynamic studies of, for example, the transition from trickle to pulsing flow. Using real-time measurements of the liquid hold-up in a trickle bed, Anadon et al. (2006) were able to identify the characteristics of the onset of the pulsing regime. At low liquid flow rates, the bed is stable with a uniformly low standard deviation in intensity. As the liquid flow rate increases, isolated pulsing events occur at the pore scale causing local increases in the standard deviation of signal intensity. As the liquid flow rate increases further, these isolated events merge until the entire cross-section of the bed is pulsing. Thus, the transition to pulsing flow occurs over a range of flow rates. In a subsequent study, MRI results were compared with measurements of the pressure and conductance in a trickle bed (Anadon, Sederman, & Gladden, 2008). These results demonstrate that conductance identifies the onset of the macroscopic pulsing regime, whereas pressure fluctuations are sensitive to the onset of isolated pulsing and the point at which the maximum number of liquid pulses occurs. Together, conductance and pressure fluctuations were able to fully characterize the transition from trickle to pulsing flow. Thus, using insights from MRI, they demonstrate that an accurate characterization of the transition from trickle to pulsing flow can be obtained using conventional instrumentation, such as could be installed on many industrial trickle bed reactors.

Hydrodynamic effects are known to affect the chemical conversion in trickle bed reactors. One approach that has been proposed to improve the performance of a trickle bed reactor is to operate in a periodic regime, whereby the flow rate of the gas or liquid feed is increased periodically (Silveston, Hudgins, & Renken, 1995). Numerical models have been presented to describe this periodic operation (Dietrich, Grünewald, & Agar, 2005; Kouris, Neophytides, Vayenas, & Tsamopoulos, 1998); however, such models require an accurate description of the dynamic behavior of flow of liquid and gas over the catalyst particles. MRI studies of periodic operation demonstrate that periodic operation leads to a heterogeneous distribution of liquid contacting in the pores of the reactor (Anadon et al., 2006; Lim, Sederman, Gladden, & Stitt, 2004). Anadon et al. (2006) were able to identify three characteristic pore-scale draining events, each of which was associated with a number of different time scales for the draining events. Using the experimentally measured heterogeneity of the liquid contacting, reactor models can be used to explore changes in operation to improve reactor performance. In one study, the reaction rate was demonstrated to almost double if the distribution of pores that rapidly fill and empty could be increased (Dietrich, Anadon, Sederman, Gladden, & Agar, 2012). These results demonstrate the importance of understanding the local hydrodynamic phenomena, or mesoscale structure, in order to predict the overall behavior of the reactor.

One drawback of MRI measurements of trickle bed reactors is that they are restricted to narrow diameter columns made of nonconducting materials. Thus, the application of MRI to reactors with metallic internals has been challenging. X-ray and γ-ray CT are potentially advantageous in this regard (Bazer-Bachi, Haroun, Augier, & Boyer, 2013; Boyer & Fanget, 2002; Johnson, Levison, Shearing, & Bracewell, 2017; Schubert et al., 2011); one recent facility has even been certified to operate in an explosive environment (Hoffmann & Kögl, 2017). CT techniques have been used in an analogous manner to the MRI techniques discussed above to characterize the distribution of gas and liquid in structured packing materials such as Katapak-SP 12 (Aferka, Crine, Saroha, Toye, & Marchot, 2007) and Mellapak 752Y (Roy et al., 2005). These measurements provide the specific surface area and gas/liquid interfacial area within the packing materials. CT techniques have also been used to explore the distribution of liquid at the entrance of fixed bed columns and catalytic beds (see, e.g., Bazer-Bachi et al., 2013; Fourati, Roig, & Raynal, 2012; Kalaga et al., 2009; Merwe, Nicol, & Beer, 2007; Roy et al., 2005 and references therein). For example, Fig. 23.8 illustrates the distribution of liquid in a trickle bed

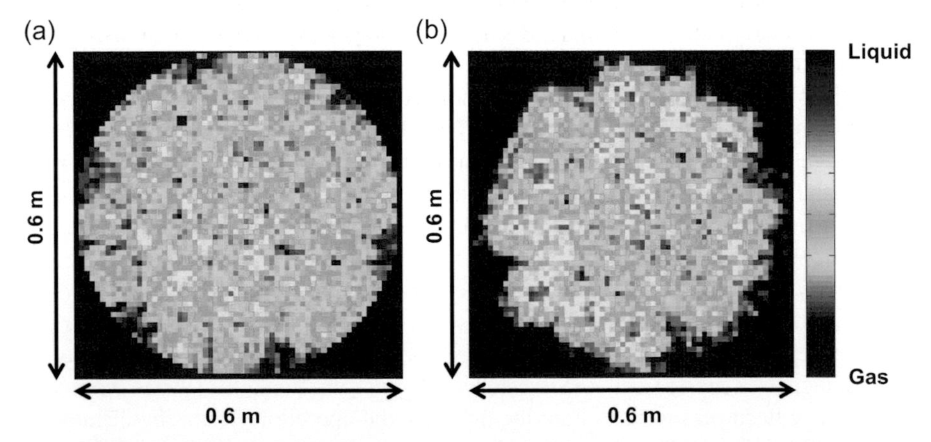

Figure 23.8 γ-Ray tomography measurements of the distribution of liquid in a trickle bed reactor with (a) chimney tray or (b) gas lift distributor designs. The liquid phase was heptane and the gas phase was nitrogen and the column was operated at low pressure (0.1–0.4 bar gauge) and room temperature. The images are shown for a superficial liquid velocity of 2 mm/s and a superficial gas velocity of 20 mm/s.
Adapted with permission from Bazer-Bachi, F., Haroun, Y., Augier, F., Boyer, C. (2013). Experimental evaluation of distributor technologies for trickle-bed reactors. *Industrial and Engineering Chemical Research 52*, 11189–11197. Copyright 2013 American Chemical Society.

reactor of diameter 0.6 m with two types of distributors: a chimney tray with deflectors and a gas lift distributor. Commercial literature suggests that gas lift distributors are beneficial for achieving a homogeneous distribution of liquid in the trickle bed. However, the experimental measurements shown in Fig. 23.8 indicate that the chimney tray gives superior liquid distribution, with no obvious bias of the liquid flow in the vicinity of the drip points.

Although CT techniques can potentially be used on large, metallic vessels, typical CT hardware systems are difficult to apply to industrial processes owing to the hazards of ionizing radiation and the bulk and power requirements of the source of radiation. Therefore, instruments used to study industrial processes typically obtain only very limited measurements and do not reconstruct entire images (Bukur, Daly, & Patel, 1996; Shaikh & Al-Dahhan, 2013; Sharaf et al., 2011; Toukan, Alexander, AlBazzaz, & Al-Dahhan, 2017). However, Kuzeljevic, Dudukovic, and Stitt (2011) note that industrial measurements require only simple diagnostic information, and that such information may be extracted from far fewer measurements than a complete tomographic reconstruction would require. Thus, even these relatively simple devices can provide significant insight into optimizing the operation of industrial reactors. Alternatively, it is possible to devise portable CT systems that can achieve relatively high spatial resolution and full tomographic reconstructions, though these may have other restrictions regarding the size of the equipment on which they can be applied (Bieberle et al., 2013; Kim et al., 2012).

Electrical tomography techniques have also proven beneficial for studying trickle bed reactors owing to the high time resolution that can be achieved (Bieberle et al., 2010; Chen et al., 2017; Häfeli, Hutter, Damsohn, Prasser, & Rudolf von Rohr, 2013; Llamas et al., 2008; Matusiak, Jose, Hampel, & Romanowski, 2010; Reinecke & Mewes, 1997; Singh, Jain, & Buwa, 2019; Son et al., 2017; Wang, Marashdeh, Motil, & Fan, 2014; Zhao et al., 2015). Measurements of the conductivity require the conductive phase to be continuous, which is not often the case in a trickle bed. For this reason, electrical measurements in trickle beds have primarily used invasive wire mesh sensors or capacitance measurements. Wire mesh sensors provide a higher spatial resolution than capacitance tomography measurements; however, they are invasive so not suitable for all applications. Measurements with wire mesh sensors have been compared with capacitance tomography (Matusiak et al., 2010). Both techniques were shown to yield very similar measurements of the average liquid hold-up and were able to identify macroscopic heterogeneity of the liquid distribution. Additionally, the wire mesh sensor was able to measure local variations in the liquid hold-up around individual particles. Such measurements demonstrate the potential for electrical tomography techniques to characterize flow within trickle bed reactors and indicate they will provide a valuable tool for testing models of these reactors in the future.

23.3.2 Reaction and mass transfer studies

One of the major challenges in optimizing the performance of fixed bed reactors is to understand the interplay between the chemical reaction kinetics, mass transfer, and hydrodynamic phenomena. Tomographic imaging of reacting systems has advanced significantly in recent years. Much of this work has used MRI, but recently, X-ray CT has also been used to study reaction and mass transfer. The attraction of MRI is that it can map the local chemical concentration, distribution of phases, and their velocity (Gladden, Abegão, Dunckley, Holland, Sankey, Sederman, et al., 2010; Lysova, Koptyug, & Weckhuysen, 2010). Hence, this review will concentrate on these measurements, but the latest developments using X-ray CT will also be covered.

One of the first examples to demonstrate the importance of understanding the interaction of flow and reaction was a study of the liquid phase esterification of methanol and acetic acid (Yuen, Sederman, & Gladden, 2002). Previous work had illustrated that the velocity of fluid flowing through a fixed bed reactor does not follow an ideal plug flow model, but rather that there are variations in the local fluid velocity of up to an order of magnitude (Sederman, Johns, Alexander, & Gladden, 1998). However, it had been assumed that dispersion over the length of the reactor would lead to a relatively homogeneous distribution of concentration. The study of Yuen et al. (2002) showed fractional variations in the steady-state conversion of approximately 20% arising from the heterogeneity in flow.

MRI has also been used to probe microbial growth and oxygen consumption in biofilms in a fixed bed (Simkins, Stewart, Codd, & Seymour, 2019, 2020; Simkins, Stewart, & Seymour, 2018). In those studies, a combination of ^{19}F and ^{1}H MRI are used to probe the oxygen concentration and flow field, respectively, in an actively growing biofilm. Oxygen concentration is mapped by encapsulating an oxygen sensitive perfluorocarbon in alginate beads. The relaxation rate of the ^{19}F signal from this perfluorocarbon is dependent on the local oxygen concentration. Using these measurements, the authors were able to demonstrate the presence of an asymmetric boundary layer during flow. The boundary layer thickness varied through the system indicating variation in the local mass transfer resistance due to the complex pore structure. These findings may help explain some of the unusual microbial growth patterns observed in biochemical fixed beds.

Much of the work examining reactions has been confined to systems at close to ambient conditions owing to the constraints on the geometry and materials of construction when working with MRI. However, there has been a move toward studying reactions at conditions of industrial relevance (Gladden, Abegão, Dunckley, Holland, Sankey, Sederman, et al., 2010; Koptyug, Khomichev, Lysova, & Sagdeev, 2008; Ulpts, Dreher, Klink, & Thöming, 2015). ^{13}C NMR has been used to explore changes in selectivity and conversion in trickle bed reactors operating at up to 5 barg and 120°C (Akpa, Mantle, Sederman, & Gladden, 2005; Gladden, Abegão, Dunckley, Holland, Sankey, Sederman, et al., 2010; Sederman, Mantle, Dunckley, Huang, & Gladden, 2005; Zheng, Russo-Abegao, Sederman, & Gladden, 2017). These measurements reveal changes in selectivity throughout the reactor. The latest of these studies extracts mass transfer coefficients during reaction (Zheng et al., 2017). The mass transfer coefficients were compared with various correlations that exist in the literature. The correlations predicted mass transfer coefficients that varied by about an order of magnitude. The mass transfer coefficients measured using the MRI experiments showed good agreement with correlations obtained from similar hydrodynamic operating conditions to that used in their study. This work highlights the importance of studying fixed bed reactors under conditions as close to those used industrially as possible.

Many industrial reactions occur at temperatures well in excess of 120°C. Such conditions are challenging to achieve in an MRI system, where the magnet itself is typically superconducting and hence cooled cryogenically to 4K (−269°C). Despite this, some measurements have been reported of reacting systems at over 400°C (Koptyug et al., 2008; Koptyug, Sagdeev, Gerkema, Van As, & Sagdeev, 2005; Lysova, Kulikov, Parmon, Sagdeev, & Koptyug, 2012). For example, Fig. 23.9 shows the temperature distribution in a Pt/γ-Al_2O_3 catalyst pellet when contacted on one side with a mixture of hydrogen and air. The temperature in this system exceeded 400°C (700 K) locally, while a temperature difference of up to 150°C was observed across the catalyst pellet.

More recently, a novel reactor was developed to operate within an MRI system at up to 350°C and 30 barg (Baker et al., 2018; Roberts et al., 2013). The first studies reported using this reactor investigate ethene oligomerization at 110°C and 28 barg. The authors combine a variety of techniques to study the formation of oligomers

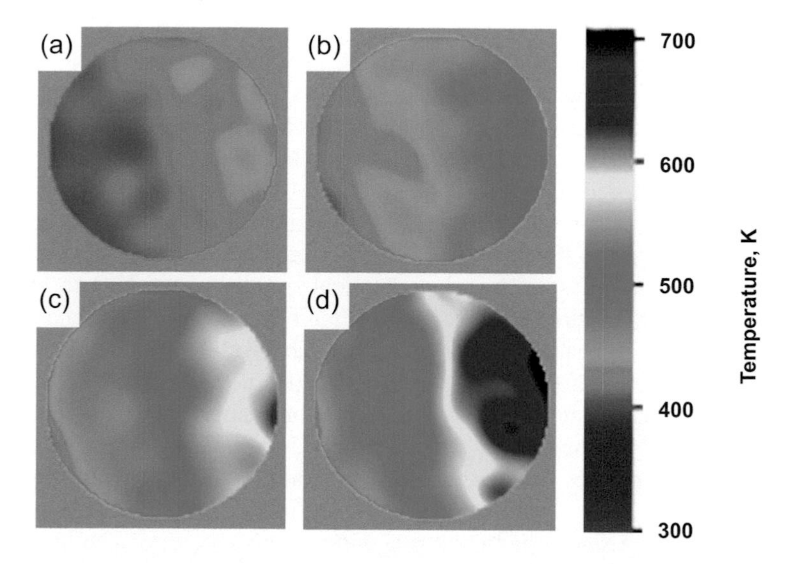

Figure 23.9 MRI measurements of the temperature of a catalyst pellet during oxidation of H_2. The temperature of the catalyst was obtained from the ^{27}Al NMR signal intensity, which is a function of the local temperature of the pellet. Images were detected for hydrogen flow rates of (a) 0.8, (b) 1.4, (c) 2.0, and (d) 2.5 mL/s.

Adapted from Lysova, A.A., Kulikov, A. V, Parmon, V.N., Sagdeev, Z., Koptyug, I. V. (2012). Quantitative temperature mapping within an operating catalyst by spatially resolved 27Al NMR. *Chemical Communication 48*, 5763−5765 with permission from The Royal Society of Chemistry.

with carbon numbers ranging from C4 to over C20. Through careful analysis, the authors demonstrate the formation of long-chain olefins in the pores of the catalyst pellets, with the proportion of long-chain olefins increasing with time on stream. The authors suggest that the low diffusivity of the long-chain olefins blocks the pores of the catalyst and hence influences the production rate and distribution of oligomers formed. In the future, measurements such as these will help design improved catalyst formulations and may lead to new reactor start-up and catalyst reactivation processes.

One challenge of MRI measurements is its relatively low sensitivity. Over the years, many methods have been developed to increase the sensitivity of MRI, particularly those using hyperpolarization methods (Barskiy et al., 2017; Halse, 2016). In the last decade, some of these techniques have begun to be used to study reactions in fixed beds (Burueva et al., 2020; Kovtunov et al., 2019; Svyatova et al., 2020; Zhivonitko et al., 2018). One of the challenges with applying hyperpolarization methods to systems as complex as fixed bed reactors is in obtaining quantitative measurements. Relaxation is known to vary significantly in the presence of solids, often becoming very short. Thus, it can be difficult to ensure changes in signal intensity are related to reaction and not relaxation. One approach to partially overcome this problem is to design model reactors in which the catalyst is deposited on the wall of a tube (Kovtunov et al., 2019). With this approach, it is possible to study some important gas phase reactions such as propene hydrogenation using parahydrogen-induced polarization.

Recently, CT techniques have also begun to be used to study reactions in fixed beds (Andrews & Weckhuysen, 2013; Basile et al., 2010; Price et al., 2015; Vamvakeros et al., 2015). These techniques typically rely on synchrotron radiation to produce sufficiently high-quality data, but they permit characterization of changes in catalyst structure at an unprecedented resolution. Fig. 23.10 presents an example of data acquired using X-ray diffraction (XRD) CT of the activation of a catalyst by contacting it with a 20% H_2/He gas flow at 800°C (Vamvakeros et al., 2018). The diffraction peaks are used to identify the changing chemical structure from NiO to Ni during the reduction. The interaction of the reduction of NiO with the promoters characterized by the CeO_2 and ZrO_2 peaks in Fig. 23.10 is also able to be resolved. These experiments, and others with this emerging technology (Burueva et al., 2020; Kovtunov et al., 2019), demonstrate an entirely new capability of tomographic imaging to resolve changes in the catalysts themselves. This technique is complementary to MRI, which predominantly investigates the reactants and is relatively insensitive to the catalyst itself.

Collectively, these studies represent a significant advance in our ability to study phenomena occurring inside chemical reactors at industrially relevant conditions. X-ray CT, MRI, and ECT have all been used to study the distribution of liquid and gas within trickle bed reactors, though CT and MRI are the only noninvasive techniques that are able to provide the high spatial resolution required to measure the liquid distribution around individual particles. MRI, and now CT, provides direct measurements of the interaction of flow and reaction and how these may affect the catalyst itself. These tomographic imaging techniques are being used to probe the performance benefits that can be achieved using a variety of "process intensification" technologies, including periodic operation (Dietrich et al., 2012; Hamidipour, Chen, &

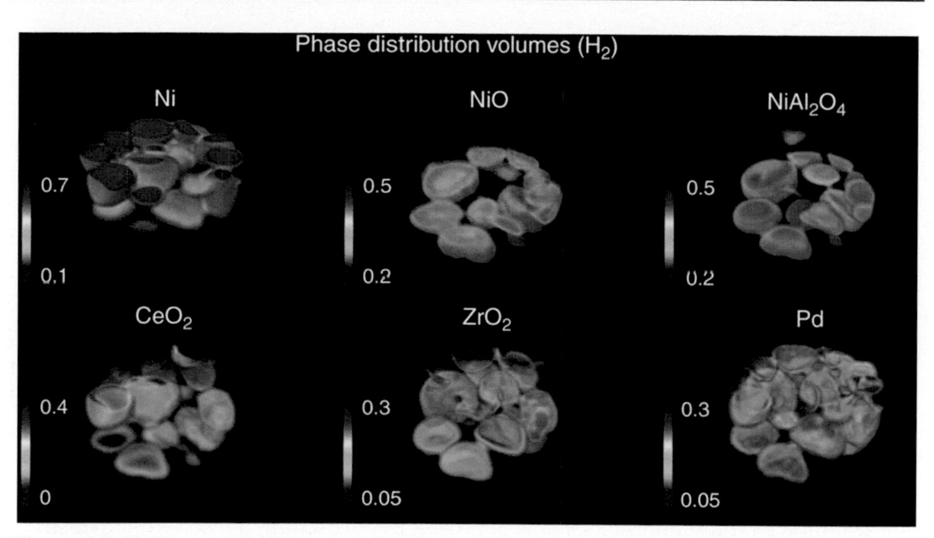

Figure 23.10 X-ray diffraction tomography images of the catalyst activation process. The images show volume renderings of the 3D-XRD-CT data collected at 800°C with a flow of 20% H_2/80% He. The colorbar axes were chosen to achieve the best possible contrast. The catalyst is initially NiO but is reduced on contact with the H_2. The CeO_2 and ZrO_2 are present in a mixed phase as a promotor of the catalyst activation.
Adapted from Vamvakeros, A. et al. (2018). 5D operando tomographic diffraction imaging of a catalyst bed, *Nature Communications. Nature Publishing Group, 9*(1), pp. 1–11. https://doi.org/10.1038/s41467-018-07046-8.

Larachi, 2013; Hamidipour, Larachi, & Ring, 2010; Härting, Bieberle, Lange, Larachi, & Schubert, 2015) and inclined or rotating reactors (Assima, Hamitouche, Schubert, & Larachi, 2015; Groβ et al., 2019; Härting et al., 2015; Motamed Dashliborun, Hamidipour, & Larachi, 2017; Motamed Dashliborun, Härting, Schubert, & Larachi, 2017; Wongkia, Larachi, & Assabumrungrat, 2015). It is anticipated that the insights provided from these studies will enable researchers to further optimize the operating conditions of industrial fixed bed reactors.

23.3.3 Complex porous structures

The studies discussed thus far have mostly considered random packed beds. Monolithic reactors have long been used owing to their low pressure drop and highly regular structure, perhaps the most common example being the catalytic converter used in most petrol engine cars (Williams, 2001). There have been a wide variety of tomographic investigations of such reactors (Al-Dahhan, Kemoun, & Cartolano, 2006; Hamidipour & Larachi, 2010; Mantle et al., 2002; Ramskill et al., 2013; Sederman, Heras, Mantle, & Gladden, 2007; Ulpts, Dreher, Kiewidt, Schubert, & Thöming, 2016), and how insights from tomographic measurements help understand reactor performance (Schubert et al., 2016; Ulpts, Kiewidt, Dreher, & Thöming, 2018). However, random packed beds remain the dominant fixed bed reactor type used industrially.

Over the last decade or so, open-cell foam reactors have emerged as a potential alternative to monoliths and packed beds (Twigg & Richardson, 2007). Open-cell foams achieve lower pressure drop than packed beds, while achieving high rates of heat and mass transfer. These features make the open-cell foam structure a potentially attractive alternative to packed beds and monoliths. Recently, tomographic imaging has begun to be used to gain greater insight into the design and operation of open-cell foam reactors. Ultrafast X-ray CT and MRI have both been used to study foam reactors (Dong, Korup, Gerdts, Roldán Cuenya, & Horn, 2018; Sadeghi, Mirdrikvand, Pesch, Dreher, & Thöming, 2020; Tschentscher et al., 2011; Ulpts et al., 2018; Zalucky, Clauβnitzer, Schubert, Lange, & Hampel, 2017; Zalucky, Möller, Schubert, & Hampel, 2015). The high spatial and temporal resolution of ultrafast CT makes it well suited to study the dynamics of gas–liquid flow through foam structures. On the other hand, the ability of MRI to resolve the motion within the fluid itself makes it ideally suited to characterize single phase flow through these structures.

Recently it has been proposed that additive manufacturing (of which 3D printing is one common example) presents a new approach to developing the solid internal structure of fixed beds (Díaz-Marta et al., 2018; Fee, 2017; Hurt et al., 2017). Additive manufacturing enables the generation of structures with complex internal features that would not be possible to manufacture using conventional technologies. In the last decade the cost, range of materials, and resolution of additive manufacturing technologies have all improved at an astonishing rate (Bikas, Stavropoulos, & Chryssolouris, 2016; Bourell, 2016). The new capability of additive manufacturing potentially enables researchers to design the internal structure of a fixed bed with a geometry that optimizes the heat and/or mass transfer performance for a specific application (Faure, Flin, Gallo, & Wagner, 2018; Kaur & Singh, 2021). Tomographic imaging has been widely used to characterize the quality of additively manufactured materials (du Plessis, Yadroitsava, & Yadroitsev, 2020; Khosravani & Reinicke, 2020). Now, the first tomographic investigations of the flow through these structures have been presented (Clarke, Dolamore, Fee, Galvosas, & Holland, 2021). Fig. 23.11 shows selected maps of the flow through an additively manufactured triply periodic minimal surface measured using MRI. The flow was studied at a Reynolds number of 1.25 and at a Reynolds number of 17.6. At the higher Reynolds number, a change in the recirculation patterns within the porous material is apparent, which is associated with the transition to inertial flow. The measurements were compared with CFD simulations and found to be in excellent agreement. In the future, this work will help design novel internal structures to optimize the flow path for heat and/or mass transfer.

23.4 Future trends

The literature covered in the previous sections has demonstrated the significant potential for tomographic imaging techniques for studying bubble column and fixed bed processes. Tomographic techniques have been shown to provide detailed measurements of the local phase fraction and interfacial area, as well as the velocity distribution

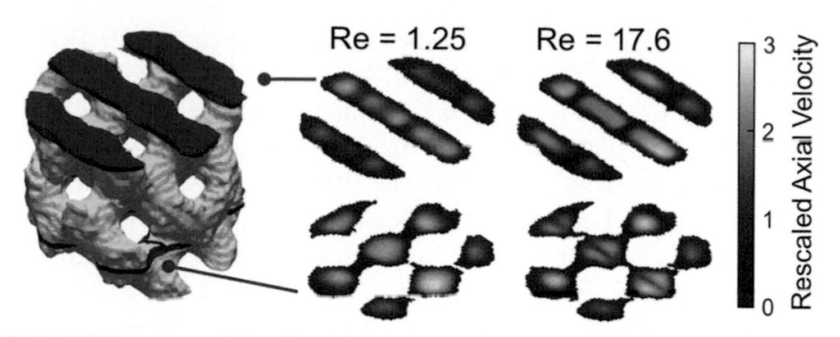

Figure 23.11 MRI images of the flow through an additively manufactured fixed bed. The geometry of the solid phase was defined using a triply periodic minimal surface. The image on the left was obtained from a high-resolution three-dimensional image of the column. Axial velocity maps are shown on the right at two positions within the three-dimensional structure at the given Reynolds numbers.
Adapted from Clarke, D. A. et al. (2021). Investigation of flow through triply periodic minimal surface-structured porous media using MRI and CFD, *Chemical Engineering Science. 231*, 116264.

of both the continuous and discrete phases. Furthermore, various tomographic techniques have been cross-validated against each other to ensure that the measurements obtained are reliable. Thus, tomographic imaging is already influencing the design and operation of industrial-scale processes. Recently, tomographic imaging has begun to be applied at close to industrial conditions, which promises to provide another step change in our understanding of the phenomena that govern the operation of these complex reactors. In the future, such studies will likely influence the design of catalyst pellets, the composition of the reactant mixture, and the geometry of the reactors themselves.

In addition to such fundamental studies, tomographic imaging techniques will also be used to develop computational models of industrial-scale reactors. As noted earlier, the design of industrial reactors has traditionally been based on correlations derived from measurements of the inlet and outlet of small-scale reactors and simple assumptions of the flow patterns in the reactor. Now, increasingly advanced CFD models of reactors are being used to describe the flow patterns in more detail (Alopaeus, Hynynen, Aittamaa, & Manninen, 2006; Atta, Roy, & Nigam, 2007; Delnoij, Kuipers, & van Swaaij, 1997; Jiang, Khadilkar, Al-dahhan, & Dudukovic, 1999, 2002; Kuipers & van Swaaij, 1997; Kuzeljevic & Dudukovic, 2012; Lappalainen, Manninen, & Alopaeus, 2009; Lopes & Quinta-ferreira, 2009; Lopes & Quinta-Ferreira, 2009; Mahr & Mewes, 2008). Closure models are needed to describe, for example, the interaction between fluid phases in these CFD models. These closure models are typically developed from empirical measurements, or from high resolution CFD simulations. However, uncertainties in the correct form and implementation of these closure models remain. Therefore, such simulations need to be validated before they can be used to design industrial systems with confidence (Dudukovic, 2010). Tomographic imaging techniques can provide the validation studies required for these CFD models (Atta, Hamidipour, Roy, Nigam, & Larachi, 2010; Atta, Schubert, Nigam, Roy, & Larachi,

2010; Banowski et al., 2018; Bazmi, Hashemabadi, & Bayat, 2012; Boyer, Koudil, Chen, & Dudukovic, 2005; Dong et al., 2018; Gaeini, Wind, Donkers, Zondag, & Rindt, 2017; Gunjal, Kashid, Ranade, & Chaudhari, 2005; Hamidipour et al., 2013; Khopkar, Rammohan, Ranade, & Dudukovic, 2005; Krepper, Reddy Vanga, Zaruba, Prasser, & Lopez de Bertodano, 2007; Liu et al., 2011; Lopes & Quinta-Ferreira, 2010; Lopes, Quinta-ferreira, & Larachi, 2012; Motamed Dashliborun, Hamidipour, et al., 2017; Rabha, Schubert, & Hampel, 2013a; Rampure, Buwa, & Ranade, 2003; Robbins, El-bachir, Gladden, Cant, & von Harbou, 2012; Wood et al., 2015; Yan et al., 2019; Yin, Afacan, Nandakumar, & Chuang, 2002). In one of these studies, summarized in Fig. 23.12, an Euler–Euler computational model was compared with wire mesh sensor measurements of the time-averaged voidage in the bubble column (Krepper et al., 2007). The computational model was shown to describe the distribution of the average voidage accurately high in the column, where the fluid phase turbulence was not too severe. The model was even able to predict the sharp rise in void fraction at the wall that was seen experimentally. However, in the region near the distributor, the model performed poorly, significantly underestimating the heterogeneity of the voidage distribution. The poor performance of the model near the distributor was attributed to the simple turbulence model used in the numerical study. These results highlight the need to incorporate detailed turbulence models for the fluid phase when predicting the dynamics of bubble column reactors. Advanced measurement techniques are now becoming available that will permit the direct characterization of the closure models required to describe some of the phenomena occurring in these systems. For example, an MRI study presented simultaneous measurements of the gas and liquid velocity distribution in a trickle bed reactor (Sankey, Holland, et al., 2009). Such measurements will provide a means for testing directly the

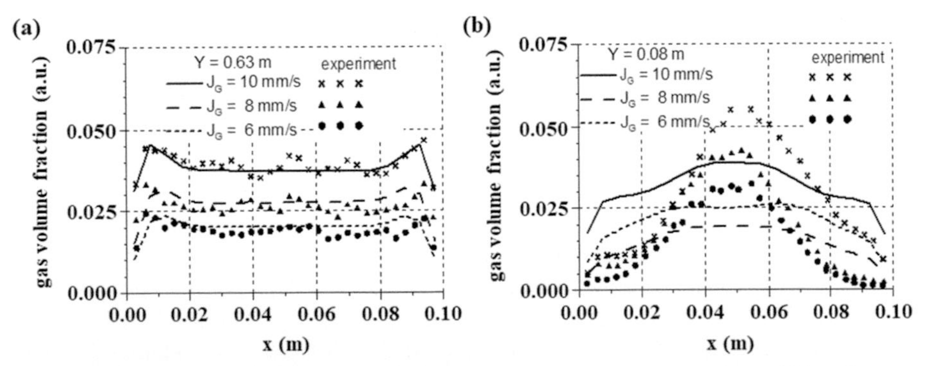

Figure 23.12 Profiles of the gas volume fraction are shown in a rectangular bubble column reactor (a) high in the column and (b) near the distributor. Gas volume fraction profiles are obtained from either experimental measurements using a wire mesh sensor or numerical simulations of the column. The gas was introduced through a stone sparger 0.02 m in width by 0.01 m in depth. The column was 0.10 m in width and 0.02 m in depth. Comparisons are shown for superficial gas velocities of 10, 8, and 6 mm/s.
This figure is adapted from Krepper, E., Reddy Vanga, B. N., Zaruba, A., Prasser, H.-M., Lopez de Bertodano, M. A. (2007). Experimental and numerical studies of void fraction distribution in rectangular bubble columns. *Nuclear Engineering Design 237*, 399–408 with permission.

assumptions that underpin the closure models used for gas–liquid interaction in trickle bed reactors (Attou, Boyer, & Ferschneider, 1999; Iliuta, Larachi, & Al-Dahhan, 2000; Iliuta, Larachi, & Grandjean, 1998; Sáez & Carbonell, 1985). Tomographic measurements are also being used in conjunction with computational simulations to identify the effect of flow maldistribution on reactor performance (Dietrich et al., 2012; Schubert et al., 2016). Future developments of numerical models of bubble column and trickle bed reactors will depend on detailed experimental investigations, such as those described above, in order to establish the correct methods for modeling these complex systems.

One interesting development in the last decade has been the growth of the open data movement (Popkin, 2019). The open data movement encourages researchers to make their raw and analyzed data available to other researchers. It has been around for decades and is well established in disciplines such as astronomy. However, in the last decade or so, it has spread more widely. Tomographic measurements are a perfect example of data that should be made available to all. Tomographic imaging equipment is often expensive and complex to operate. Consequently, relatively few laboratories have access to these techniques. Similarly, high-quality CFD simulations of multiphase flows require significant expertise, even with the availability of modern CFD software packages. Therefore, very few research groups are capable of performing both high-quality tomographic measurements and world-leading CFD simulations. If tomographic imaging researchers made their data widely available, it would enable those working at the forefront of CFD to test new modeling approaches readily. Complete data sets from tomographic imaging studies of bubble columns and fixed bed reactors are now becoming available (Neumann-Kipping, Bieberle, & Hampel, 2020) and this trend is likely to continue due to pressures from government funding agencies. A major challenge with the open data movement is the time and effort required by researchers to make sure their data are available in a suitable format. Furthermore, to be useful for validation studies, it is critical that the physical set up of the system is described clearly. However, the investment in making the data available to the wider community is worthwhile (Piwowar & Vision, 2013). It is therefore anticipated that increasing numbers of tomographic investigations will be made widely available in the coming years.

23.5 Sources of further information

The literature cited in this chapter primarily focuses on publications in the last 20 years. Significant publications describing tomographic imaging of bubble column and trickle bed reactors are not included here as they have been included in earlier review articles (Al-Dahhan et al., 1997; Boyer et al., 2002; Chaouki et al., 1997; Dixon & Deutschmann, 2017; Dudukovic, 2002; Dudukovic et al., 1999; Gladden & Sederman, 2017; Hampel et al., 2020; Heindel, 2011; Mudde, 2010a; Reinecke & Mewes, 1996; Schubert et al., 2011; Zhivonitko et al., 2018). The reader is referred to these articles and the references contained therein for further information. Additionally, the articles cited within this chapter contain relevant background information; the reader is encouraged to explore these for further information.

For details of the fundamental principles of the tomographic techniques discussed in this chapter, the reader is referred to the earlier chapters in this book, as well as the seminal text by Williams, and Beck entitled Process Tomography: Principles, Techniques, and Applications. Butterworth Heinemann, 1995.

For further information regarding the fundamentals of the operation of bubble column and fixed bed reactors and how to estimate the performance of these, the reader is referred to the following material:

Akita, K., & Yoshida, F. (1974). Bubble size, interfacial area, and liquid-phase mass transfer coefficient in bubble columns. *Industrial & Engineering Chemistry Process Design and Developments, 13*(1), 84–91.

Al-Dahhan, M., Larachi, F., Duduković, M. P., & Laurent, A. (1997). High-pressure trickle-bed reactors: a review. *Industrial & Engineering Chemistry Research, 36*(8), 3292–3314.

Deckwer, W. (1992). Bubble column reactors. Chichester: Wiley.

Gianetto, A., & Specchia, V. (1992). Trickle-bed reactors: state of art and perspectives. *Chemical Engineering Science, 47*, 3197–3213.

Kantarci, N., Borak, F., & Ulgen, K. O. (2005). Bubble column reactors. *Process Biochemistry, 40*(7), 2263–2283.

Satterfield, C. N. (1975). Trickle-bed reactors. *American Institute of Chemical Engineers Journal, 21*(2), 209–228.

Shah, Y. T., Kelkar, B. G., Godbole, S. P., & Deckwer, W.-D. (1982). Design parameters estimations for bubble column reactors. *American Institute of Chemical Engineers Journal, 28*(3), 353–379.

For details of the techniques available for numerical modeling of bubble column and trickle bed reactors, the reader is referred to the following material:

Deen, N. G., Mudde, R. F., Kuipers, J. A. M., Zehner, P., Kraume, M., (2010). Bubble columns. In: Ullmann's encyclopedia of industrial chemistry. Wiley Online Library.

Deen, N. G., van Sint Annaland, M., & Kuipers, J.A. M. (2004). Multi-scale modeling of dispersed gas–liquid two-phase flow. *Chemical Engineering Science, 59*(8–9), 1853–1861. https://doi.org/10.1016/j.ces.2004.01.038

Jakobsen, H. A. (2008). Chemical reactor modeling (vol. 3). Berlin, Heidelberg: Springer Berlin Heidelberg.

Joshi, J. B. (2001). Computational flow modeling and design of bubble column reactors. *Chemical Engineering Science, 56*, 5893–5933.

Kuipers, J. A. M., & van Swaaij, W. P. M. (1997). Application of computational fluid dynamics to chemical reaction engineering. *Reviews in Chemical Engineering, 13*, 1–118.

Rafique, M., Chen, P., & Duduković, M. P. (2004). Computational modeling of gas-liquid flow in bubble columns. *Reviews in Chemical Engineering, 20*, 225–375.

Sokolichin, A., Eigenberger, G., & Lapin, A. (2004). Simulation of buoyancy driven bubbly flow: established simplifications and open questions. *American Institute of Chemical Engineers Journal, 50*(1), 24–45.

References

Adair, A., Richard, S., & Newling, B. (2021). Gas and liquid phase imaging of foam flow using pure phase encode magnetic resonance imaging. *Molecules, 26*, 28. https://doi.org/10.3390/molecules26010028

Aferka, S., Crine, M., Saroha, A. K., Toye, D., & Marchot, P. (2007). In situ measurements of the static liquid holdup in Katapak-SP12 packed column using X-ray tomography. *Chemical Engineering Science, 62*, 6076–6080. https://doi.org/10.1016/j.ces.2007.06.025

Akpa, B. S., Mantle, M. D., Sederman, A. J., & Gladden, L. F. (2005). In situ 13C DEPT-MRI as a tool to spatially resolve chemical conversion and selectivity of a heterogeneous catalytic reaction occurring in a fixed-bed reactor. *Chemical Communications*, (21), 2741–2743. https://doi.org/10.1039/b501698c

Al Mesfer, M. K., Sultan, A. J., & Al-Dahhan, M. H. (2016). Impacts of dense heat exchanging internals on gas holdup cross-sectional distributions and profiles of bubble column using gamma ray Computed Tomography (CT) for FT synthesis. *Chemical Engineering Journal, 300*, 317–333. https://doi.org/10.1016/j.cej.2016.04.075

Al-Dahhan, M. H., Kemoun, A., & Cartolano, A. R. (2006). Phase distribution in an upflow monolith reactor using computed tomography. *AIChE Journal, 52*(2), 745–753. https://doi.org/10.1002/aic.10665

Al-Dahhan, M. H., Larachi, F., Duduković, M. P., & Laurent, A. (1997). High-pressure trickle-bed reactors: A review. *Industrial & Engineering Chemistry Research, 36*(8), 3292–3314. https://doi.org/10.1021/ie9700829

Alopaeus, V., Hynynen, K., Aittamaa, J., & Manninen, M. (2006). Modeling of gas-liquid packed-bed reactors with momentum equations and local interaction closures. *Industrial & Engineering Chemistry Research, 45*, 8189–8198.

Anadon, L. D., Sederman, A. J., & Gladden, L. F. (2006). Mechanism of the trickle-to-pulse flow transition in fixed-bed reactors. *AIChE Journal, 52*(4), 1522–1532. https://doi.org/10.1002/aic

Anadon, L. D., Sederman, A. J., & Gladden, L. F. (2008). Rationalising MRI, conductance and pressure drop measurements of the trickle-to-pulse transition in trickle beds. *Chemical Engineering Science, 63*(19), 4640–4648. https://doi.org/10.1016/j.ces.2007.10.033

Andrews, J. C., & Weckhuysen, B. M. (2013). Hard X-ray spectroscopic nano-imaging of hierarchical functional materials at work. *ChemPhysChem, 14*(16), 3655–3666. https://doi.org/10.1002/cphc.201300529

Assima, G. P., Hamitouche, A., Schubert, M., & Larachi, F. (2015). Liquid drainage in inclined packed beds-Accelerating liquid draining time via column tilt. *Chemical Engineering and Processing: Process Intensification, 95*, 249–255. https://doi.org/10.1016/j.cep.2015.06.021

Atta, A., Hamidipour, M., Roy, S., Nigam, K. D. P., & Larachi, F. (2010a). Propagation of slow/fast-mode solitary liquid waves in trickle beds via electrical capacitance tomography and computational fluid dynamics. *Chemical Engineering Science, 65*(3), 1144–1150. https://doi.org/10.1016/j.ces.2009.09.069

Atta, A., Roy, S., & Nigam, K. D. P. (2007). Investigation of liquid maldistribution in trickle-bed reactors using porous media concept in CFD. *Chemical Engineering Science, 62*(24), 7033–7044. https://doi.org/10.1016/j.ces.2007.07.069

Atta, A., Schubert, M., Nigam, K. D. P., Roy, S., & Larachi, F. (2010b). Co-current descending two-phase flows in inclined packed beds: Experiments versus simulations. *Canadian Journal of Chemical Engineering, 88*(October). https://doi.org/10.1002/cjce.20340. n/a-n/a.

Attou, A., Boyer, C., & Ferschneider, G. (1999). Modelling of the hydrodynamics of the cocurrent gas−liquid trickle flow through a trickle-bed reactor. *Chemical Engineering Science, 54*(6), 785−802. https://doi.org/10.1016/S0009-2509(98)00285-1

Azizi, S., Yadav, A., Lau, Y. M., Hampel, U., Roy, S., & Schubert, M. (2017). On the experimental investigation of gas-liquid flow in bubble columns using ultrafast X-ray tomography and radioactive particle tracking. *Chemical Engineering Science, 170*, 320−331. https://doi.org/10.1016/j.ces.2017.02.015

Azizi, S., Yadav, A., Lau, Y. M., Hampel, U., Roy, S., & Schubert, M. (2019). Hydrodynamic correlations for bubble columns from complementary UXCT and RPT measurements in identical geometries and conditions. *Chemical Engineering Science, 208*. https://doi.org/10.1016/j.ces.2019.07.017

Azzopardi, B. J., Abdulkareem, L.a., Zhao, D., Thiele, S., da Silva, M. J., Beyer, M., et al. (2010). Comparison between electrical capacitance tomography and wire mesh sensor output for air/silicone oil flow in a vertical pipe. *Industrial & Engineering Chemistry Research, 49*(18), 8805−8811. https://doi.org/10.1021/ie901949z

Baker, L., Renshaw, M. P., Mantle, M. D., Sederman, A. J., Wain, A. J., & Gladden, L. F. (2018). Operando magnetic resonance studies of phase behaviour and oligomer accumulation within catalyst pores during heterogeneous catalytic ethene oligomerization. *Applied Catalysis A: General, 557*, 125−134. https://doi.org/10.1016/j.apcata.2018.03.011

Balachandar, S., & Eaton, J. K. (2010). Turbulent dispersed multiphase flow. *Annual Review of Fluid Mechanics, 42*(1), 111−133. https://doi.org/10.1146/annurev.fluid.010908.165243

Banowski, M., Hampel, U., Krepper, E., Beyer, M., & Lucas, D. (2018). Experimental investigation of two-phase pipe flow with ultrafast X-ray tomography and comparison with state-of-the-art CFD simulations. *Nuclear Engineering and Design, 336*, 90−104. https://doi.org/10.1016/j.nucengdes.2017.06.035

Barskiy, D. A., Coffey, A. M., Nikolaou, P., Mikhaylov, D. M., Goodson, B. M., Branca, R. T., et al. (2017, January 18). NMR hyperpolarization techniques of gases. *Chemistry - A European Journal, 23*, 725−751. https://doi.org/10.1002/chem.201603884

Barthel, F., Bieberle, M., Hoppe, D., Banowski, M., & Hampel, U. (2015). Velocity measurement for two-phase flows based on ultrafast X-ray tomography. *Flow Measurement and Instrumentation, 46*, 196−203. https://doi.org/10.1016/j.flowmeasinst.2015.06.006

Basile, F., Benito, P., Bugani, S., De Nolf, W., Fornasari, G., Janssens, K., et al. (2010). Combined use of synchrotron-radiation-based imaging techniques for the characterization of structured catalysts. *Advanced Functional Materials, 20*(23), 4117−4126. https://doi.org/10.1002/adfm.201001004

Bazer-Bachi, F., Haroun, Y., Augier, F., & Boyer, C. (2013). Experimental evaluation of distributor technologies for trickle-bed reactors. *Industrial & Engineering Chemistry Research, 52*, 11189−11197.

Bazmi, M., Hashemabadi, S. H., & Bayat, M. (2012). CFD simulation and experimental study of liquid flow mal-distribution through the randomly trickle bed reactors. *International Communications in Heat and Mass Transfer, 39*(5), 736−743. https://doi.org/10.1016/j.icheatmasstransfer.2012.03.005

Bennett, M. A., West, R. M., Luke, S. P., Jia, X., & Williams, R. A. (1999). Measurement and analysis of flows in a gas−liquid column reactor. *Chemical Engineering Science, 54*, 5003−5012.

Besagni, G., Inzoli, F., Zieghenein, T., & Lucas, D. (2019). Experimental study of liquid velocity profiles in large-scale bubble columns with particle tracking velocimetry. *Journal of Physics: Conference Series, 1224*(1). https://doi.org/10.1088/1742-6596/1224/1/012036

Bieberle, M., Barthel, F., Menz, H.-J., Mayer, H.-G., & Hampel, U. (2011). Ultrafast three-dimensional x-ray computed tomography. *Applied Physics Letters, 98*(3), 034101. https://doi.org/10.1063/1.3534806

Bieberle, M., Fischer, F., Schleicher, E., Hampel, U., Koch, D., Aktay, K. S. D. C., et al. (2007). Ultrafast limited-angle-type x-ray tomography. *Applied Physics Letters, 91*(12), 123516. https://doi.org/10.1063/1.2787879

Bieberle, M., & Hampel, U. (2006). Evaluation of a limited angle scanned electron beam x-ray CT approach for two-phase pipe flows. *Measurement Science and Technology, 17*(8), 2057−2065. https://doi.org/10.1088/0957-0233/17/8/001

Bieberle, A., Nehring, H., Berger, R., Arlit, M., Härting, H.-U., Schubert, M., et al. (2013). Compact high-resolution gamma-ray computed tomography system for multiphase flow studies. *Review of Scientific Instruments, 84*(3), 033106. https://doi.org/10.1063/1.4795424

Bieberle, M., Schleicher, E., Fischer, F., Koch, D., Menz, H.-J., Mayer, H.-G., et al. (2010b). Dual-plane ultrafast limited-angle electron beam x-ray tomography. *Flow Measurement and Instrumentation, 21*(3), 233−239. https://doi.org/10.1016/j.flowmeasinst.2009.12.001

Bieberle, A., Schubert, M., Silva, M. J. da, & Hampel, U. (2010a). Measurement of liquid distributions in particle packings using wire-mesh sensor versus transmission tomographic imaging. *Industrial & Engineering Chemistry Research, 49*, 9445−9453.

Bikas, H., Stavropoulos, P., & Chryssolouris, G. (2016). Additive manufacturing methods and modeling approaches: A critical review. *International Journal of Advanced Manufacturing Technology, 83*(1−4), 389−405. https://doi.org/10.1007/s00170-015-7576-2

Bourell, D. L. (2016). Perspectives on additive manufacturing. *Annual Review of Materials Research, 46*(1), 1−18. https://doi.org/10.1146/annurev-matsci-070115-031606

Boyer, C., Duquenne, A., & Wild, G. (2002). Measuring techniques in gas − liquid and gas − liquid − solid reactors. *Chemical Engineering Science, 57*, 3185−3215.

Boyer, C., & Fanget, B. (2002). Measurement of liquid flow distribution in trickle bed reactor of large diameter with a new gamma-ray tomographic system. *Chemical Engineering Science, 57*, 1079−1089.

Boyer, C., Koudil, a., Chen, P., & Dudukovic, M. P. (2005). Study of liquid spreading from a point source in a trickle bed via gamma-ray tomography and CFD simulation. *Chemical Engineering Science, 60*(22), 6279−6288. https://doi.org/10.1016/j.ces.2005.03.049

Bukur, D. B., Daly, J. G., & Patel, S. A. (1996). Application of γ-ray attenuation for measurement of gas holdups and flow regime transitions in bubble columns. *Industrial & Engineering Chemistry Research, 35*(1), 70−80. https://doi.org/10.1021/ie950134z

Burueva, D. B., Pokochueva, E. V., Wang, X., Filkins, M., Svyatova, A., Rigby, S. P., et al. (2020). In situ monitoring of heterogeneous catalytic hydrogenation via 129Xe NMR spectroscopy and proton MRI. *ACS Catalysis, 10*(2), 1417−1422. https://doi.org/10.1021/acscatal.9b05000

Cai, Q., Shen, X., Shen, C., & Dai, G. (2010). A simple method for identifying bubbling/jetting regimes transition from large submerged orifices using electrical capacitance tomography (ECT). *Canadian Journal of Chemical Engineering, 88*(June), 340−349. https://doi.org/10.1002/cjce.20295

Chaouki, J., Larachi, F., & Dudukovic, M. P. (1997). Noninvasive tomographic and velocimetric monitoring of multiphase flows. *Industrial & Engineering Chemistry Research, 36*(11), 4476−4503. https://doi.org/10.1021/ie970210t

Chen, J., Gupta, P., Degaleesan, S., Al-Dahhan, M. H., Duduković, M. P., Toseland, A., et al. (1998). Gas holdup distributions in large-diameter bubble columns measured by computed tomography. *Flow Measurement and Instrumentation, 9*(2), 91–101. https://doi.org/10.1016/S0955-5986(98)00010-7

Chen, J., Kemoun, A., Al-dahhan, M. H., Dudukovic, M. P., Lee, D. J., & Fan, L. (1999a). Comparative hydrodynamics study in a bubble column using computer-automated radioactive particle tracking (CARPT)/computed tomography (CT) and particle image velocimetry (PIV). *Chemical Engineering Science, 54*, 2199–2207.

Chen, J., Li, F., Degaleesan, S., Gupta, P., Al-Dahhan, M. H., Dudukovic, M. P., et al. (1999b). Fluid dynamic parameters in bubble columns with internals. *Chemical Engineering Science, 54*(13–14), 2187–2197. https://doi.org/10.1016/S0009-2509(99)00003-2

Chen, Z., Yang, J., Ling, D., Liu, P., Ilankoon, I. M. S. K., Huang, Z., et al. (2017). Packing size effect on the mean bubble diameter in a fixed bed under gas-liquid concurrent upflow. *Industrial & Engineering Chemistry Research, 56*(45), 13490–13496. https://doi.org/10.1021/acs.iecr.7b00123

Clarke, D. A., Dolamore, F., Fee, C. J., Galvosas, P., & Holland, D. J. (2021). Investigation of flow through triply periodic minimal surface-structured porous media using MRI and CFD. *Chemical Engineering Science, 231*, 116264. https://doi.org/10.1016/j.ces.2020.116264

Collins, J. H. P., Sederman, A. J., Gladden, L. F., Afeworki, M., Douglas Kushnerick, J., & Thomann, H. (2017). Characterising gas behaviour during gas–liquid co-current up-flow in packed beds using magnetic resonance imaging. *Chemical Engineering Science, 157*, 2–14. https://doi.org/10.1016/j.ces.2016.04.004

Daidzic, N., Schmidt, E., Hasan, M., & Altobelli, S. (2005). Gas–liquid phase distribution and void fraction measurements using MRI. *Nuclear Engineering and Design, 235*(10–12), 1163–1178. https://doi.org/10.1016/j.nucengdes.2005.02.024

Deckwer, W. (1992). *Bubble column reactors*. Chichester: Wiley.

Delnoij, E., Kuipers, J.a. M., & van Swaaij, W. P. M. (1997). Computational fluid dynamics applied to gas-liquid contactors. *Chemical Engineering Science, 52*(21–22), 3623–3638. https://doi.org/10.1016/S0009-2509(97)00268-6

Díaz-Marta, A. S., Tubío, C. R., Carbajales, C., Fernández, C., Escalante, L., Sotelo, E., et al. (2018). Three-dimensional printing in catalysis: Combining 3D heterogeneous copper and palladium catalysts for multicatalytic multicomponent reactions. *ACS Catalysis, 8*(1), 392–404. https://doi.org/10.1021/acscatal.7b02592

Dietrich, W., Anadon, L., Sederman, A. J., Gladden, L. F., & Agar, D. W. (2012). Simulation studies on the performance enhancement in periodically operated trickle-bed reactors based on experimental local liquid distribution measurements. *Industrial & Engineering Chemistry Research, 51*(4), 1672–1679. https://doi.org/10.1021/ie200827c

Dietrich, W., Grünewald, M., & Agar, D. W. (2005). Dynamic modelling of periodically wetted catalyst particles. *Chemical Engineering Science, 60*(22), 6254–6261. https://doi.org/10.1016/j.ces.2005.03.054

Dijksman, J. A., Rietz, F., Lorincz, K. A., Van Hecke, M., & Losert, W. (2012). Invited article: Refractive index matched scanning of dense granular materials. *Review of Scientific Instruments, 83*(1). https://doi.org/10.1063/1.3674173

Dixon, A., & Deutschmann, O. (2017). *Spatially resolved operando measurements in heterogeneous catalytic reactors*. Saint Louis, UNITED STATES: Elsevier Science & Technology.

Dong, Y., Korup, O., Gerdts, J., Roldán Cuenya, B., & Horn, R. (2018). Microtomography-based CFD modeling of a fixed-bed reactor with an open-cell foam monolith and experimental verification by reactor profile measurements. *Chemical Engineering Journal, 353*, 176–188. https://doi.org/10.1016/j.cej.2018.07.075

Dudukovic, M. (2002). Opaque multiphase flows: Experiments and modeling. *Experimental Thermal and Fluid Science, 26*(6–7), 747–761. https://doi.org/10.1016/S0894-1777(02)00185-1

Dudukovic, M. P. (2010). Reaction engineering: Status and future challenges. *Chemical Engineering Science, 65*(1), 3–11. https://doi.org/10.1016/j.ces.2009.09.018

Dudukovic, M. P., Larachi, F., & Mills, P. L. (1999). Multiphase reactors – revisited. *Chemical Engineering Science, 54*(13–14), 1975–1995. https://doi.org/10.1016/S0009-2509(98)00367-4

Faure, R., Flin, M., Gallo, P. del, & Wagner, M. (2018). *Add it up! The chemical engineer.*

Fee, C. (2017). November 1). 3D-printed porous bed structures. *Current Opinion in Chemical Engineering, 18*, 10–15. https://doi.org/10.1016/j.coche.2017.07.003

Fischer, F., & Hampel, U. (2010). Ultra fast electron beam X-ray computed tomography for two-phase flow measurement. *Nuclear Engineering and Design, 240*(9), 2254–2259. https://doi.org/10.1016/j.nucengdes.2009.11.016

Fourati, M., Roig, V., & Raynal, L. (2012). Experimental study of liquid spreading in structured packings. *Chemical Engineering Science, 80*, 1–15. https://doi.org/10.1016/j.ces.2012.05.031

Fransolet, E., Crine, M., Marchot, P., & Toye, D. (2005). Analysis of gas holdup in bubble columns with non-Newtonian fluid using electrical resistance tomography and dynamic gas disengagement technique. *Chemical Engineering Science, 60*(22), 6118–6123. https://doi.org/10.1016/j.ces.2005.03.046

Fuangworawong, N., Kikura, H., Aritomi, M., & Komeno, T. (2007). Tomographic imaging of counter-current bubbly flow by wire mesh tomography. *Chemical Engineering Journal, 130*(2–3), 111–118. https://doi.org/10.1016/j.cej.2006.08.033

Gaeini, M., Wind, R., Donkers, P. A. J., Zondag, H. A., & Rindt, C. C. M. (2017). Development of a validated 2D model for flow, moisture and heat transport in a packed bed reactor using MRI experiment and a lab-scale reactor setup. *International Journal of Heat and Mass Transfer, 113*, 1116–1129. https://doi.org/10.1016/j.ijheatmasstransfer.2017.06.034

Gianetto, A., & Specchia, V. (1992). Trickle-bed reactors: State of art and perspectives. *Chemical Engineering Science, 47*, 3197–3213.

Gladden, L. F., Abegão, F. J. R., Dunckley, C. P., Holland, D. J., Sankey, M. H., Sederman, A. J., et al. (2010). MRI: Operando measurements of temperature, hydrodynamics and local reaction rate in a heterogeneous catalytic reactor. *Catalysis Today, 155*(3–4), 157–163. https://doi.org/10.1016/j.cattod.2009.10.012

Gladden, L. F., Lim, M. H. M., Mantle, M. D., Sederman, a. J., & Stitt, E. H. (2003). MRI visualisation of two-phase flow in structured supports and trickle-bed reactors. *Catalysis Today, 79*(80), 203–210. https://doi.org/10.1016/S0920-5861(03)00006-3

Gladden, L. F., & Sederman, A. J. (2017). Magnetic resonance imaging and velocity mapping in chemical engineering applications. *Annual Review of Chemical and Biomolecular Engineering, 8*(1), 227–247. https://doi.org/10.1146/annurev-chembioeng-061114-123222

Groß, K., Bieberle, A., Gladyszewski, K., Schubert, M., Hampel, U., Skiborowski, M., et al. (2019). Analysis of flow patterns in high-gravity equipment using gamma-ray computed tomography. *Chemie Ingenieur Technik, 91*(7), 1032–1040. https://doi.org/10.1002/cite.201800085

Gulati, S., Behling, M., Munshi, P., Luke, A., & Mewes, D. (2010). Tomographic KT-1 signature of phase-fraction distributions in multiphase bubble columns. *Flow Measurement and Instrumentation, 21*(3), 249–254. https://doi.org/10.1016/j.flowmeasinst.2009.12.005

Gumery, F., Ein-Mozaffari, F., & Dahman, Y. (2011). Macromixing hydrodynamic study in draft-tube airlift reactors using electrical resistance tomography. *Bioprocess and Biosystems Engineering, 34*(2), 135–144. https://doi.org/10.1007/s00449-010-0454-2

Gunjal, P. R., Kashid, M. N., Ranade, V. V., & Chaudhari, R. V. (2005). Hydrodynamics of trickle-bed reactors: Experiments and CFD modeling. *Industrial & Engineering Chemistry Research, 44*(16), 6278–6294. https://doi.org/10.1021/ie0491037

Häfeli, R., Hutter, C., Damsohn, M., Prasser, H.-M., & Rudolf von Rohr, P. (2013). Dispersion in fully developed flow through regular porous structures: Experiments with wire-mesh sensors. *Chemical Engineering and Processing: Process Intensification, 69*, 104–111. https://doi.org/10.1016/j.cep.2013.03.006

Halse, M. E. (2016). October 1). Perspectives for hyperpolarisation in compact NMR. *TRAC Trends in Analytical Chemistry, 83*, 76–83. https://doi.org/10.1016/j.trac.2016.05.004

Hamidipour, M., Chen, J., & Larachi, F. (2013). CFD study and experimental validation of trickle bed hydrodynamics under gas , liquid and gas/liquid alternating cyclic operations. *Chemical Engineering Science, 89*, 158–170. https://doi.org/10.1016/j.ces.2012.11.041

Hamidipour, M., & Larachi, F. (2010). Dynamics of filtration in monolith reactors using electrical capacitance tomography. *Chemical Engineering Science, 65*(1), 504–510. https://doi.org/10.1016/j.ces.2009.06.040

Hamidipour, M., Larachi, F., & Ring, Z. (2010). Cyclic operation strategies in trickle beds and electrical capacitance tomography imaging of filtration dynamics. *Industrial & Engineering Chemistry Research, 49*, 934–952.

Hamood-ur-rehman, M., Dahman, Y., & Ein-mozaffari, F. (2012). Investigation of mixing characteristics in a packed-bed external loop airlift bioreactor using tomography images. *Chemical Engineering Journal, 213*, 50–61. https://doi.org/10.1016/j.cej.2012.09.106

Hampel, U., Schubert, M., Alexander, D., Sohr, J., Vishwakarma, V., Repke, J. U., et al. (2020). Recent advances in experimental techniques for flow and mass transfer analyses in thermal separation systems. *Chemie Ingenieur Technik, 92*(7), 926–948. https://doi.org/10.1002/cite.202000076

Härting, H.-U., Bieberle, A., Lange, R., Larachi, F., & Schubert, M. (2015). Hydrodynamics of co-current two-phase flow in an inclined rotating tubular fixed bed reactor - wetting intermittency via periodic catalyst immersion. *Chemical Engineering Science, 128*, 147–158. https://doi.org/10.1016/j.ces.2015.02.008

Heindel, T. J. (2011). A review of X-ray flow visualization with applications to multiphase flows. *Journal of Fluids Engineering, 133*(7), 074001. https://doi.org/10.1115/1.4004367

Hernandez-Alvarado, F., Kleinbart, S., Kalaga, D. V., Banerjee, S., Joshi, J. B., & Kawaji, M. (2018). Comparison of void fraction measurements using different techniques in two-phase flow bubble column reactors. *International Journal of Multiphase Flow, 102*, 119–129. https://doi.org/10.1016/j.ijmultiphaseflow.2018.02.002

Hoffmann, R., & Kögl, T. (2017). An ATEX-proof gamma tomography setup for measuring liquid distribution in process equipment. *Flow Measurement and Instrumentation, 53*, 147–153. https://doi.org/10.1016/j.flowmeasinst.2016.05.009

Holland, D. J., Blake, A., Tayler, A. B., Sederman, A. J., & Gladden, L. F. (2011). A bayesian approach to characterising multi-phase flows using magnetic resonance: Application to bubble flows. *Journal of Magnetic Resonance, 209*(1), 83–87. https://doi.org/10.1016/j.jmr.2010.12.003

Holland, D. J., Blake, A., Tayler, A. B., Sederman, A. J., & Gladden, L. F. (2012). Bubble size measurement using bayesian magnetic resonance. *Chemical Engineering Science, 84*, 735–745.

Holland, D. J., Malioutov, D. M., Blake, A., Sederman, A. J., & Gladden, L. F. (2010). Reducing data acquisition times in phase-encoded velocity imaging using compressed sensing. *Journal of Magnetic Resonance, 203*(2), 236–246. https://doi.org/10.1016/j.jmr.2010.01.001

Holland, D. J., Mitchell, J., Blake, A., & Gladden, L. F. (2013). Grain sizing in porous media using bayesian magnetic resonance. *Physical Review Letters, 110*(1), 018001. https://doi.org/10.1103/PhysRevLett.110.018001

Hubers, J. L., Striegel, A. C., Heindel, T. J., Gray, J. N., & Jensen, T. C. (2005). X-ray computed tomography in large bubble columns. *Chemical Engineering Science, 60*, 6124–6133. https://doi.org/10.1016/j.ces.2005.03.038

Hurt, C., Brandt, M., Priya, S. S., Bhatelia, T., Patel, J., Selvakannan, P., et al. (2017). Combining additive manufacturing and catalysis: A review. *Catal. Sci. Technol., 7*, 3421–3439. https://doi.org/10.1039/C7CY00615B

Iliuta, I., Larachi, F., & Al-Dahhan, M. H. (2000). Double-slit model for partially wetted trickle flow hydrodynamics. *AIChE Journal, 46*(3), 597–609. https://doi.org/10.1002/aic.690460318

Iliuta, I., Larachi, F., & Grandjean, B. P. A. (1998). Pressure drop and liquid holdup in trickle flow reactors: Improved ergun constants and slip correlations for the slit model. *Industrial & Engineering Chemistry Research, 37*(12), 4542–4550. https://doi.org/10.1021/ie980394r

Ishkintana, L. K., & Bennington, C. P. J.Ã. (2010). Gas holdup in pulp fibre suspensions : Gas voidage profiles in a batch-operated sparged tower. *Chemical Engineering Science, 65*(8), 2569–2578. https://doi.org/10.1016/j.ces.2009.12.040

Ito, D., Prasser, H.-M., Kikura, H., & Aritomi, M. (2011). Uncertainty and intrusiveness of three-layer wire-mesh sensor. *Flow Measurement and Instrumentation, 22*(4), 249–256. https://doi.org/10.1016/j.flowmeasinst.2011.03.002

Jahangir, S., Wagner, E. C., Mudde, R. F., & Poelma, C. (2019). Void fraction measurements in partial cavitation regimes by X-ray computed tomography. *International Journal of Multiphase Flow, 120*, 103085. https://doi.org/10.1016/j.ijmultiphaseflow.2019.103085

Jiang, Y., Khadilkar, M. R., Al-dahhan, M. H., & Dudukovic, M. P. (1999). Two-phase flow distribution in 2D trickle-bed reactors. *Chemical Engineering Science, 54*, 2409–2419.

Jiang, Y., Khadilkar, M. R., Al-dahhan, M. H., & Dudukovic, M. P. (2002). CFD of multiphase flow in packed-bed reactors: I . k-fluid modeling issues. *AIChE Journal, 48*(4), 701–715.

Jin, H., Wang, M., & Williams, R. A. (2007a). Analysis of bubble behaviors in bubble columns using electrical resistance tomography. *Chemical Engineering Journal, 130*, 179–185. https://doi.org/10.1016/j.cej.2006.08.032

Jin, H., Yang, S., Wang, M., & Williams, R. A. (2007b). Measurement of gas holdup profiles in a gas liquid cocurrent bubble column using electrical resistance tomography. *Flow Measurement and Instrumentation, 18*, 191–196. https://doi.org/10.1016/j.flowmeasinst.2007.07.005

Johnson, T. F., Levison, P. R., Shearing, P. R., & Bracewell, D. G. (2017). X-ray computed tomography of packed bed chromatography columns for three dimensional imaging and analysis. *Journal of Chromatography A, 1487*, 108–115. https://doi.org/10.1016/j.chroma.2017.01.013

Kalaga, D. V., Kulkarni, A. V., Acharya, R., Kumar, U., Singh, G., & Joshi, J. B. (2009). Some industrial applications of gamma-ray tomography. *Journal of the Taiwan Institute of Chemical Engineers, 40*, 602–612. https://doi.org/10.1016/j.jtice.2009.05.012

Kantarci, N., Borak, F., & Ulgen, K. O. (2005). Bubble column reactors. *Process Biochemistry, 40*(7), 2263–2283. https://doi.org/10.1016/j.procbio.2004.10.004

Karim, K., Varma, R., Vesvikar, M., & Al-dahhan, M. H. (2004). Flow pattern visualization of a simulated digester. *Water Research, 38*, 3659–3670. https://doi.org/10.1016/j.watres.2004.06.009

Kaur, I., & Singh, P. (2021, April 1). Critical evaluation of additively manufactured metal lattices for viability in advanced heat exchangers. *International Journal of Heat and Mass Transfer, 168*, 120858. https://doi.org/10.1016/j.ijheatmasstransfer.2020.120858

Khayamyan, S., Lundström, T. S., Gren, P., Lycksam, H., & Hellström, J. G. I. (2017). Transitional and turbulent flow in a bed of spheres as measured with stereoscopic particle image velocimetry. *Transport in Porous Media, 117*(1), 45–67. https://doi.org/10.1007/s11242-017-0819-y

Khopkar, A. R., Rammohan, A. R., Ranade, V. V., & Dudukovic, M. P. (2005). Gas-liquid flow generated by a rushton turbine in stirred vessel: CARPT/CT measurements and CFD simulations. *Chemical Engineering Science, 60*, 2215–2229. https://doi.org/10.1016/j.ces.2004.11.044

Khosravani, M. R., & Reinicke, T. (2020, December 1). On the use of X-ray computed tomography in assessment of 3D-printed components. *Journal of Nondestructive Evaluation, 39*, 75. https://doi.org/10.1007/s10921-020-00721-1

Kim, J., Jung, S., Moon, J., Guen Park, J., Jin, J., & Cho, G. (2012). Development of transportable gamma-ray tomographic system for industrial application. *Nuclear Instruments and Methods in Physics Research Section A: Accelerators, Spectrometers, Detectors and Associated Equipment, 693*, 203–208. https://doi.org/10.1016/j.nima.2012.07.046

Kipping, R., Kryk, H., & Hampel, U. (2021). Experimental analysis of gas phase dynamics in a lab scale bubble column operated with deionized water and NaOH solution under uniform bubbly flow conditions. *Chemical Engineering Science, 229*. https://doi.org/10.1016/j.ces.2020.116056

Koptyug, I. V., Khomichev, A. V., Lysova, A. A., & Sagdeev, R. Z. (2008). Spatially resolved NMR thermometry of an operating fixed-bed catalytic reactor. *Journal of the American Chemical Society, 130*(32), 10452–10453. https://doi.org/10.1021/ja802075m

Koptyug, I. V., Sagdeev, D. R., Gerkema, E., Van As, H., & Sagdeev, R. Z. (2005). Solid-state 27Al MRI and NMR thermometry for catalytic applications with conventional (liquids) MRI instrumentation and techniques. *Journal of Magnetic Resonance (San Diego, Calif.: 1997), 175*(1), 21–29. https://doi.org/10.1016/j.jmr.2005.03.005

Kouris, C., Neophytides, S., Vayenas, C. G., & Tsamopoulos, J. (1998). Unsteady state operation of catalytic particles with constant and periodically changing degree of external wetting. *Chemical Engineering Science, 53*(17), 3129–3142. https://doi.org/10.1016/S0009-2509(98)00090-6

Kovtunov, K. V., Lebedev, D., Svyatova, A., Pokochueva, E. V., Prosvirin, I. P., Gerasimov, E. Y., et al. (2019). Robust in situ magnetic resonance imaging of heterogeneous catalytic hydrogenation with and without hyperpolarization. *ChemCatChem, 11*(3), 969–973. https://doi.org/10.1002/cctc.201801820

Krepper, E., Reddy Vanga, B. N., Zaruba, A., Prasser, H.-M., & Lopez de Bertodano, M.a. (2007). Experimental and numerical studies of void fraction distribution in rectangular bubble columns. *Nuclear Engineering and Design, 237*(4), 399–408. https://doi.org/10.1016/j.nucengdes.2006.07.009

Kuipers, J. A. M., & van Swaaij, W. P. M. (1997). Application of computational fluid dynamics to chemical reaction engineering. *Reviews in Chemical Engineering, 13*, 1–118. https://doi.org/10.1515/REVCE.1997.13.3.1

Kuipers, J. A. M., & van Swaaij, W. P. M. (1998). Computational fluid dynamics applied to chemical reaction engineering. *Advances in Chemical Engineering, 24*, 227–328.

Kutsovsky, Y. E., Scriven, L. E., Davis, H. T., & Hammer, B. E. (1996). NMR imaging of velocity profiles and velocity distributions in bead packs. *Physics of Fluids, 8*(4), 863–871. https://doi.org/10.1063/1.868867

Kuzeljevic, Z. V., & Dudukovic, M. P. (2012). Computational modeling of trickle bed reactors. *Industrial & Engineering Chemistry Research, 51*(4), 1663–1671. https://doi.org/10.1021/ie2007449

Kuzeljevic, Z., Dudukovic, M., & Stitt, H. (2011). From laboratory to field tomography: Data collection and performance assessment. *Industrial & Engineering Chemistry Research, 50*(17), 9890–9900. https://doi.org/10.1021/ie101759s

Lappalainen, K., Manninen, M., & Alopaeus, V. (2009). CFD modeling of radial spreading of flow in trickle-bed reactors due to mechanical and capillary dispersion. *Chemical Engineering Science, 64*(2), 207–218. https://doi.org/10.1016/j.ces.2008.10.009

Lau, Y. M., Möller, F., Hampel, U., & Schubert, M. (2018). Ultrafast X-ray tomographic imaging of multiphase flow in bubble columns – Part 2: Characterisation of bubbles in the dense regime. *International Journal of Multiphase Flow, 104*, 272–285. https://doi.org/10.1016/j.ijmultiphaseflow.2018.02.009

Leblond, J., Javelot, S., Lebrun, D., & Lebon, L. (1998). Two-phase flow characterization by nuclear magnetic resonance. *Nuclear Engineering and Design, 184*(2–3), 229–237. https://doi.org/10.1016/S0029-5493(98)00199-X

Lim, M. H. M., Sederman, a. J., Gladden, L. F., & Stitt, E. (2004). New insights to trickle and pulse flow hydrodynamics in trickle-bed reactors using MRI. *Chemical Engineering Science, 59*(22–23), 5403–5410. https://doi.org/10.1016/j.ces.2004.07.096

Liu, Y.-J., Li, W., Han, L.-C., Cao, Y., Luo, H., Al-Dahhan, M., et al. (2011). γ-CT measurement and CFD simulation of cross section gas holdup distribution in a gas–liquid stirred standard Rushton tank. *Chemical Engineering Science, 66*(17), 3721–3731. https://doi.org/10.1016/j.ces.2011.03.042

Llamas, J.-D., Pérat, C., Lesage, F., Weber, M., D'Ortona, U., & Wild, G. (2008). Wire mesh tomography applied to trickle beds: A new way to study liquid maldistribution. *Chemical Engineering and Processing: Process Intensification, 47*(9–10), 1765–1770. https://doi.org/10.1016/j.cep.2007.09.017

Lopes, R. J. G., & Quinta-ferreira, R. M. (2009a). Volume-of-Fluid-based model for multiphase flow in high-pressure trickle-bed reactor : Optimization of numerical parameters. *AIChE Journal, 55*(11). https://doi.org/10.1002/aic

Lopes, R. J. G., & Quinta-Ferreira, R. M. (2009b). CFD modelling of multiphase flow distribution in trickle beds. *Chemical Engineering Journal, 147*(2–3), 342–355. https://doi.org/10.1016/j.cej.2008.11.048

Lopes, R. J. G., & Quinta-Ferreira, R. M. (2010). Evaluation of multiphase CFD models in gas–liquid packed-bed reactors for water pollution abatement. *Chemical Engineering Science, 65*(1), 291–297. https://doi.org/10.1016/j.ces.2009.06.039

Lopes, R. J. G., Quinta-ferreira, R. M., & Larachi, F. (2012). ECT imaging and CFD simulation of different cyclic modulation strategies for the catalytic abatement of hazardous liquid pollutants in trickle-bed reactors. *Chemical Engineering Journal, 211–212*, 270–284. https://doi.org/10.1016/j.cej.2012.09.050

Lucas, D., Krepper, E., & Prasser, H.-M. (2005). Development of co-current air–water flow in a vertical pipe. *International Journal of Multiphase Flow, 31*(12), 1304–1328. https://doi.org/10.1016/j.ijmultiphaseflow.2005.07.004

Lysova, A. A., Koptyug, I. V., & Weckhuysen, B. M. (2010). Magnetic resonance imaging methods for in situ studies in heterogeneous catalysis. *Chemical Society Reviews, 39*(12), 4585–4601. https://doi.org/10.1039/b919540h

Lysova, A. A., Kulikov, A. V., Parmon, V. N., Sagdeev, Z., & Koptyug, I. V. (2012). Quantitative temperature mapping within an operating catalyst by spatially resolved 27Al NMR. *Chemical Communications, 48*, 5763–5765. https://doi.org/10.1039/c2cc31260c

Mahr, B., & Mewes, D. (2008). Two-phase flow in structured packings: Modeling and calculation on a macroscopic scale. *AIChE Journal, 54*(3), 614–626. https://doi.org/10.1002/aic

Mandalahalli, M. M., Wagner, E. C., Portela, L. M., & Mudde, R. F. (2020). Electrolyte effects on recirculating dense bubbly flow: An experimental study using X-ray imaging. *AIChE Journal, 66*(1), 1–15. https://doi.org/10.1002/aic.16696

Mantle, M. D., Gladden, L. F., Sederman, a. J., Raymahasay, S., Winterbottom, J. M., & Stitt, E. H. (2002). Dynamic MRI visualization of two-phase flow in a ceramic monolith. *AIChE Journal, 48*(4), 909–912. https://doi.org/10.1002/aic.690480425

Matusiak, B., Jose, M., Hampel, U., & Romanowski, A. (2010). Measurement of dynamic liquid distributions in a fixed bed using electrical capacitance tomography and capacitance wire-mesh sensor. *Industrial & Engineering Chemistry Research, 49*, 2070–2077.

Merwe, W. Van Der, Nicol, W., & Beer, F. De (2007). Trickle flow distribution and stability by X-ray radiography. *Chemical Engineering Journal, 132*, 47–59. https://doi.org/10.1016/j.cej.2007.01.015

Mohammed, S. K., Hasan, A. H., Ibrahim, A., Dimitrakis, G., & Azzopardi, B. J. (2019). Dynamics of flow transitions from bubbly to churn flow in high viscosity oils and large diameter columns. *International Journal of Multiphase Flow, 120*, 103095. https://doi.org/10.1016/j.ijmultiphaseflow.2019.103095

Motamed Dashliborun, A., Hamidipour, M., & Larachi, F. (2017a). Hydrodynamics of inclined packed beds under flow modulation - CFD simulation and experimental validation. *AIChE Journal, 63*(9), 4161–4176. https://doi.org/10.1002/aic.15732

Motamed Dashliborun, A., Härting, H.-U., Schubert, M., & Larachi, F. (2017b). Process intensification of gas-liquid downflow and upflow packed beds by a new low-shear rotating reactor concept. *AIChE Journal, 63*(1), 283–294. https://doi.org/10.1002/aic.15549

Mudde, R. F. (2010a). Advanced measurement techniques for GLS reactors. *Canadian Journal of Chemical Engineering, 88*, 638–647. https://doi.org/10.1002/cjce.20315

Mudde, R. F. (2010b). Double X-ray tomography of a bubbling fluidized bed. *Industrial & Engineering Chemistry Research, 49*(11), 5061–5065. https://doi.org/10.1021/ie901537z

Mudde, R. F., Alles, J., & Hagen, T. H. J. J. Van Der (2008). Feasibility study of a time-resolving x-ray tomographic system. *Measurement Science and Technology, 19*, 085501. https://doi.org/10.1088/0957-0233/19/8/085501

Nedeltchev, S. (2009). Application of chaos analysis for the investigation of turbulence in heterogeneous bubble columns. *Chemical Engineering & Technology, 32*(12), 1974–1983. https://doi.org/10.1002/ceat.200900336

Nedeltchev, S., Shaikh, a., & Al-Dahhan, M. (2006). Flow regime identification in a bubble column based on both statistical and chaotic parameters applied to computed tomography data. *Chemical Engineering & Technology, 29*(9), 1054–1060. https://doi.org/10.1002/ceat.200600162

Nedeltchev, S., Shaikh, A., & Al-Dahhan, M. (2007). Prediction of the Kolmogorov entropy derived from computed tomography data in a bubble column operated under the transition regime and ambient pressure. *Chemical Engineering & Technology, 30*(10), 1445–1450. https://doi.org/10.1002/ceat.200700053

Neumann-Kipping, M., Bieberle, A., & Hampel, U. (2020). Investigations on bubbly two-phase flow in a constricted vertical pipe. *International Journal of Multiphase Flow, 130*, 103340. https://doi.org/10.1016/j.ijmultiphaseflow.2020.103340

Neumann, M., Bieberle, M., Wagner, M., Bieberle, A., & Hampel, U. (2019). Improved axial plane distance and velocity determination for ultrafast electron beam x-ray computed tomography. *Measurement Science and Technology, 30*(8). https://doi.org/10.1088/1361-6501/ab1ba2

Okonkwo, A. D., Wang, M., & Azzopardi, B. (2013). Characterisation of a high concentration ionic bubble column using electrical resistance tomography. *Flow Measurement and Instrumentation, 31*, 69–76. https://doi.org/10.1016/j.flowmeasinst.2012.10.005

Olerni, C., Jia, J., & Wang, M. (2013). Measurement of air distribution and void fraction of an upwards air–water flow using electrical resistance tomography and a wire-mesh sensor. *Measurement Science and Technology, 24*(3), 035403. https://doi.org/10.1088/0957-0233/24/3/035403

Olmos, E., Gentric, C., Vial, C., Wild, G., & Midoux, N. (2001). Numerical simulation of multiphase flow in bubble column reactors. Influence of bubble coalescence and break-up. *Chemical Engineering Science, 56*, 6359–6365.

Ong, B. C., Gupta, P., Youssef, A., Al-dahhan, M. H., & Dudukovic, M. P. (2009). Computed tomographic investigation of the influence of gas sparger design on gas holdup distribution in a bubble column. *Industrial & Engineering Chemistry Research, 48*, 58–68.

Patel, A. K., & Thorat, B. N. (2008). Gamma ray tomography — an experimental analysis of fractional gas hold-up in bubble columns. *Chemical Engineering Journal, 137*, 376–385. https://doi.org/10.1016/j.cej.2007.05.014

Pietruske, H., & Prasser, H.-M. (2007). Wire-mesh sensors for high-resolving two-phase flow studies at high pressures and temperatures. *Flow Measurement and Instrumentation, 18*(2), 87–94. https://doi.org/10.1016/j.flowmeasinst.2007.01.004

Piwowar, H. A., & Vision, T. J. (2013). Data reuse and the open data citation advantage. *PeerJ, 2013*(1), 1–25. https://doi.org/10.7717/peerj.175

du Plessis, A., Yadroitsava, I., & Yadroitsev, I. (2020). February 1). Effects of defects on mechanical properties in metal additive manufacturing: A review focusing on X-ray tomography insights. *Materials and Design, 187*, 108385. https://doi.org/10.1016/j.matdes.2019.108385

Popkin, G. (2019). Setting your data free. *Nature, 569*, 445–447.

Prasser, H.-M., Beyer, M., Carl, H., Manera, A., Pietruske, H., Schütz, P., et al. (2006). The multipurpose thermal hydraulic test facility TOPFLOW: An overview on experimental capabilities, instrumentation and results. *Kernteknik, 71*, 163–173.

Prasser, H.-M., Böttger, A., & Zschau, J. (1998). A new electrode-mesh tomograph for gas–liquid flows. *Flow Measurement and Instrumentation, 9*(1998), 111–119.

Prasser, H.-M., Krepper, E., & Lucas, D. (2002). Evolution of the two-phase flow in a vertical tube—decomposition of gas fraction profiles according to bubble size classes using wire-mesh sensors. *International Journal of Thermal Sciences, 41*(1), 17–28. https://doi.org/10.1016/S1290-0729(01)01300-X

Prasser, H.-M., Misawa, M., & Tiseanu, I. (2005). Comparison between wire-mesh sensor and ultra-fast X-ray tomograph for an air–water flow in a vertical pipe. *Flow Measurement and Instrumentation, 16*(2–3), 73–83. https://doi.org/10.1016/j.flowmeasinst.2005.02.003

Price, S. W. T., Ignatyev, K., Geraki, K., Basham, M., Filik, J., Vo, N. T., et al. (2015). Chemical imaging of single catalyst particles with scanning μ-XANES-CT and μ-XRF-CT. *Physical Chemistry Chemical Physics, 17*(1), 521–529. https://doi.org/10.1039/c4cp04488f

Rabha, S., Schubert, M., & Hampel, U. (2013a). Hydrodynamic studies in slurry bubble columns: Experimental and numerical study. *Chemie Ingenieur Technik, 85*, 1092–1098.

Rabha, S., Schubert, M., & Hampel, U. (2013b). Intrinsic flow behavior in a slurry bubble column: A study on the effect of particle size. *Chemical Engineering Science, 93*, 401–411. https://doi.org/10.1016/j.ces.2013.02.034

Rabha, S., Schubert, M., Wagner, M., Lucas, D., & Hampel, U. (2013). Bubble size and radial gas hold-up distributions in a slurry bubble column using ultrafast electron beam X-ray tomography. *AIChE Journal, 59*(5), 1709–1722. https://doi.org/10.1002/aic

Rados, N., Shaikh, A., & Al-Dahhan, M. H. (2005). Solids flow mapping in a high pressure slurry bubble column. *Chemical Engineering Science, 60*(22), 6067–6072. https://doi.org/10.1016/j.ces.2005.04.087

Rahiman, M. H. F., Rahim, R. A., Rahim, H. A., Green, R. G., Zakaria, Z., Mohamad, E. J., et al. (2016). An evaluation of single plane ultrasonic tomography sensor to reconstruct three-dimensional profiles in chemical bubble column. *Sensors and Actuators, A: Physical, 246*, 18–27. https://doi.org/10.1016/j.sna.2016.04.058

Rahiman, M. H. F., Rahim, R. A., Rahim, H. A., Mohamad, E. J., Zakaria, Z., & Muji, S. Z. M. (2014). An investigation on chemical bubble column using ultrasonic tomography for imaging of gas profiles. *Sensors and Actuators B: Chemical, 202*, 46–52. https://doi.org/10.1016/j.snb.2014.05.043

Rampure, M. R., Buwa, V. V., & Ranade, V. V. (2003). Modelling of gas-liquid/gas-liquid-solid flows in bubble columns: Experiments and CFD simulations. *Canadian Journal of Chemical Engineering, 81*(August), 692–706.

Ramskill, N. P., Gladden, L. F., York, A. P. E., Sederman, A. J., Mitchell, J., & Hardstone, K. A. (2013). Understanding the operation and preparation of diesel particulate filters using a multi-faceted nuclear magnetic resonance approach. *Catalysis Today, 216*, 104–110. https://doi.org/10.1016/j.cattod.2013.06.023

Reinecke, N., & Mewes, D. (1996). Tomographic imaging of trickle-bed reactors. *Chemical Engineering, 51*(10), 2131–2138.

Reinecke, N., & Mewes, D. (1997). Investigation of the two-phase flow in trickle-bed reactors using capacitance tomography. *Chemical Engineering Science, 2509*(97), 2111–2127.

Richter, S., Aritomi, M., Prasser, H.-M., & Hampel, R. (2002). Approach towards spatial phase reconstruction in transient bubbly flow using a wire-mesh sensor. *International Journal of Heat and Mass Transfer, 45*, 1063–1075.

Robbins, D. J., El-bachir, M. S., Gladden, L. F., Cant, R. S., & von Harbou, E. (2012). CFD modeling of single-phase flow in a packed bed with MRI validation. *AIChE Journal, 58*(12), 3904–3915. https://doi.org/10.1002/aic

Roberts, S. T., Renshaw, M. P., Lutecki, M., McGregor, J., Sederman, A. J., Mantle, M. D., et al. (2013). Operando magnetic resonance: Monitoring the evolution of conversion and product distribution during the heterogeneous catalytic ethene oligomerisation reaction. *Chemical Communications, 49*(89), 10519–10521. https://doi.org/10.1039/c3cc45896b

Roy, S., Kemoun, A., Al-Dahhan, M. H., Dudukovic, M. P., Skourlis, T. B., & Dautzenberg, F. M. (2005). Countercurrent flow distribution in structured packing via computed tomography. *Chemical Engineering and Processing, 44*(1), 59–69. https://doi.org/10.1016/j.cep.2004.03.010

Sadeghi, M., Mirdrikvand, M., Pesch, G. R., Dreher, W., & Thöming, J. (2020). Full-field analysis of gas flow within open-cell foams: Comparison of micro-computed tomography-based CFD simulations with experimental magnetic resonance flow mapping data. *Experiments in Fluids, 61*(5), 1–16. https://doi.org/10.1007/s00348-020-02960-4

Sáez, A. E., & Carbonell, R. G. (1985). Hydrodynamic parameters for gas-liquid cocurrent flow in packed beds. *AIChE Journal, 31*(1), 52–62. https://doi.org/10.1002/aic.690310105

Sankey, M. H., Holland, D. J., Gladden, L. F., & Sederman, A. J. (2009a). Magnetic resonance velocity imaging of liquid and gas two-phase flow in packed beds. *Journal of Magnetic Resonance, 196*(2), 142–148. https://doi.org/10.1016/j.jmr.2008.10.021

Sankey, M. H., Yang, Z., Gladden, L., Johns, M. L., Lister, D., & Newling, B. (2009). SPRITE MRI of bubbly flow in a horizontal pipe. *Journal of Magnetic Resonance, 199*(2), 126–135. https://doi.org/10.1016/j.jmr.2009.01.034

Sanyal, J., Vasquez, S., Roy, S., & Dudukovic, M. P. (1999). Numerical simulation of gas-liquid dynamics in cylindrical bubble column reactors. *Chemical Engineering Science, 54*, 5071–5083.

Satterfield, C. N. (1975). Trickle-bed reactors. *AIChE Journal, 21*(2), 209–228. https://doi.org/10.1002/aic.690210202

Schlaberg, H. I., Baas, J. H., Wang, M., Best, J. L., Williams, R. A., & Peakall, J. (2006). Electrical resistance tomography for suspended sediment measurements in open channel flows using a novel sensor design. *Particle & Particle Systems Characterization, 23*(3–4), 313–320. https://doi.org/10.1002/ppsc.200601062

Schubert, M., Bieberle, A., Barthel, F., Boden, S., & Hampel, U. (2011). Advanced tomographic techniques for flow imaging in columns with flow distribution packings. *Chemie Ingenieur Technik, 83*(7), 979–991. https://doi.org/10.1002/cite.201100022

Schubert, M., Kost, S., Lange, R., Salmi, T., Haase, S., & Hampel, U. (2016). Maldistribution susceptibility of monolith reactors: Case study of glucose hydrogenation performance. *AIChE Journal, 62*(12), 4346–4364.

Sederman, A. J., & Gladden, L. F. (2001). Magnetic resonance imaging as a quantitative probe of gas-liquid distribution and wetting efficiency in trickle-bed reactors. *Chemical Engineering Science, 56*, 2615–2628.

Sederman, A. J., & Gladden, L. F. (2005). Transition to pulsing flow in trickle-bed reactors studied using MRI. *AIChE Journal, 51*(2), 615–621. https://doi.org/10.1002/aic.10317

Sederman, A. J., Heras, J. J., Mantle, M. D., & Gladden, L. F. (2007). MRI strategies for characterising two-phase flow in parallel channel ceramic monoliths. *Catalysis Today, 128*(1–2 SPEC. ISS.), 3–12. https://doi.org/10.1016/j.cattod.2007.04.012

Sederman, A. J., Johns, M., Alexander, P., & Gladden, L. F. (1998). Structure-flow correlations in packed beds. *Chemical Engineering Science, 53*(12), 2117–2128.

Sederman, A. J., Mantle, M. D., Dunckley, C. P., Huang, Z., & Gladden, L. F. (2005). In situ MRI study of 1-octene isomerisation and hydrogenation within a trickle-bed reactor. *Catalysis Letters, 103*(1–2), 1–8. https://doi.org/10.1007/s10562-005-7522-2

Selma, B., Bannari, R., & Proulx, P. (2010). Simulation of bubbly flows: Comparison between direct quadrature method of moments (DQMOM) and method of classes (CM). *Chemical Engineering Science, 65*(6), 1925–1941. https://doi.org/10.1016/j.ces.2009.11.018

Shaikh, A., & Al-dahhan, M. (2005). Characterization of the hydrodynamic flow regime in bubble columns via computed tomography. *Flow Measurement and Instrumentation, 16*, 91–98. https://doi.org/10.1016/j.flowmeasinst.2005.02.004

Shaikh, A., & Al-dahhan, M. (2010). A new methodology for hydrodynamic similarity in bubble columns. *Canadian Journal of Chemical Engineering, 88*, 503–517. https://doi.org/10.1002/cjce.20357

Shaikh, A., & Al-Dahhan, M. (2013). A new method for online flow regime monitoring in bubble column reactors via nuclear gauge densitometry. *Chemical Engineering Science, 89*, 120–132. https://doi.org/10.1016/j.ces.2012.11.023

Shaikh, A., Taha, M. M., & Al-Dahhan, M. H. (2020). Phase distribution in Fischer-Tropsch mimicked slurry bubble column via computed tomography. *Chemical Engineering Science, 231*, 116278. https://doi.org/10.1016/j.ces.2020.116278

Sharaf, S., Da Silva, M., Hampel, U., Zippe, C., Beyer, M., & Azzopardi, B. (2011). Comparison between wire mesh sensor and gamma densitometry void measurements in two-phase flows. *Measurement Science and Technology, 22*(10). https://doi.org/10.1088/0957-0233/22/10/104019

Sharifi, M., & Young, B. (2013). Electrical resistance tomography (ERT) applications to chemical engineering. *Chemical Engineering Research and Design, 91*(9), 1625–1645. https://doi.org/10.1016/j.cherd.2013.05.026

Silva, M. J. D., Schleicher, E., & Hampel, U. (2007). Capacitance wire-mesh sensor for fast measurement of phase fraction distributions. *Measurement Science and Technology, 18*(7), 2245–2251. https://doi.org/10.1088/0957-0233/18/7/059

Silveston, P., Hudgins, R. R., & Renken, A. (1995). Periodic operation of catalytic reactors - introduction and overview. *Catalysis Today, 25*, 91–112.

Simkins, J. W., Stewart, P. S., Codd, S. L., & Seymour, J. D. (2019). Non-invasive imaging of oxygen concentration in a complex in vitro biofilm infection model using 19F MRI: Persistence of an oxygen sink despite prolonged antibiotic therapy. *Magnetic Resonance in Medicine, 82*(6), 2248–2256. https://doi.org/10.1002/mrm.27888

Simkins, J. W., Stewart, P. S., Codd, S. L., & Seymour, J. D. (2020). Microbial growth rates and local external mass transfer coefficients in a porous bed biofilm system measured by 19F magnetic resonance imaging of structure, oxygen concentration, and flow velocity. *Biotechnology and Bioengineering, 117*(5), 1458–1469. https://doi.org/10.1002/bit.27275

Simkins, J. W., Stewart, P. S., & Seymour, J. D. (2018). Spatiotemporal mapping of oxygen in a microbially-impacted packed bed using 19F Nuclear magnetic resonance oximetry. *Journal of Magnetic Resonance, 293*, 123–133. https://doi.org/10.1016/j.jmr.2018.06.008

Sines, J. N., Hwang, S., Marashdeh, Q. M., Tong, A., Wang, D., He, P., et al. (2019). Slurry bubble column measurements using advanced electrical capacitance volume tomography sensors. *Powder Technology, 355*, 474–480. https://doi.org/10.1016/j.powtec.2019.07.077

Singh, B. K., Jain, E., & Buwa, V. V. (2019). Feasibility of electrical resistance tomography for measurements of liquid holdup distribution in a trickle bed reactor. *Chemical Engineering Journal, 358*, 564–579. https://doi.org/10.1016/j.cej.2018.10.009

van Sint Annaland, M., Dijkhuizen, W., Deen, N. G., & Kuipers, J.a. M. (2006). Numerical simulation of behavior of gas bubbles using a 3-D front-tracking method. *AIChE Journal, 52*(1), 99–110. https://doi.org/10.1002/aic.10607

Son, Y., Kim, G., Lee, S., Kim, H., Min, K., & Lee, K. S. (2017). Experimental investigation of liquid distribution in a packed column with structured packing under permanent tilt and roll motions using electrical resistance tomography. *Chemical Engineering Science, 166*, 168–180. https://doi.org/10.1016/j.ces.2017.03.044

Spedding, P. L., Woods, G. S., Raghunathan, R. S., & Watterson, J. K. (1998). Vertical two-phase flow part I : Flow regimes. *Transactions of the Institute of Chemical Engineers, 76*(July), 612–619.

Stehling, M. K., Turner, R., & Mansfield, P. (1991). Echo-planar imaging: Magnetic resonance imaging in a fraction of a second. *Science, 254*(5028), 43–50.

Suard, E., Clément, R., Fayolle, Y., Alliet, M., Albasi, C., & Gillot, S. (2019). Electrical resistivity tomography used to characterize bubble distribution in complex aerated reactors: Development of the method and application to a semi-industrial MBR in operation. *Chemical Engineering Journal, 355*, 498–509. https://doi.org/10.1016/j.cej.2018.08.014

Suekane, T., Yokouchi, Y., & Hirai, S. (2003). Inertial flow structures in a simple-packed bed of spheres. *AIChE Journal, 49*(1), 10–17. https://doi.org/10.1002/aic.690490103

Sultan, A. J., Sabri, L. S., & Al-Dahhan, M. H. (2018). Investigating the influence of the configuration of the bundle of heat exchanging tubes and column size on the gas holdup distributions in bubble columns via gamma-ray computed tomography. *Experimental Thermal and Fluid Science, 98*, 68−85. https://doi.org/10.1016/j.expthermflusci.2018.05.005

Supardan, M. D., Masuda, Y., Maezawa, A., & Uchida, S. (2007). The investigation of gas holdup distribution in a two-phase bubble column using ultrasonic computed tomography. *Chemical Engineering Journal, 130*, 125−133. https://doi.org/10.1016/j.cej.2006.08.035

Svyatova, A., Kononenko, E. S., Kovtunov, K. V., Lebedev, D., Gerasimov, E. Y., Bukhtiyarov, A. V., et al. (2020). Spatially resolved NMR spectroscopy of heterogeneous gas phase hydrogenation of 1,3-butadiene with: Para hydrogen. *Catalysis Science and Technology, 10*(1), 99−104. https://doi.org/10.1039/c9cy02100k

Tayler, A. B., Holland, D. J., Sederman, A. J., & Gladden, L. F. (2011). Time resolved velocity measurements of unsteady systems using spiral imaging. *Journal of Magnetic Resonance, 211*, 1−10. https://doi.org/10.1016/j.jmr.2011.03.017

Tayler, A. B., Holland, D. J., Sederman, A. J., & Gladden, L. F. (2012a). Applications of ultrafast MRI to high voidage bubbly flow: Measurement of bubble size distributions, interfacial area and hydrodynamics. *Chemical Engineering Science, 71*, 468−483.

Tayler, A. B., Holland, D. J., Sederman, A. J., & Gladden, L. F. (2012b). Exploring the origins of turbulence in multiphase flow using compressed sensing MRI. *Physical Review Letters, 108*, 264505.

Toukan, A., Alexander, V., AlBazzaz, H., & Al-Dahhan, M. H. (2017). Identification of flow regime in a cocurrent gas − liquid upflow moving packed bed reactor using gamma ray densitometry. *Chemical Engineering Science, 168*, 380−390. https://doi.org/10.1016/j.ces.2017.04.028

Toye, D., Fransolet, E., Simon, D., Crine, M., L'Homme, G., & Marchot, P. (2005). Possibilities and limits of application of electrical resistance tomography in hydrodynamics of bubble columns. *Canadian Journal of Chemical Engineering, 83*(February), 4−10.

Tschentscher, R., Schubert, M., Bierbele, A., Nijhuis, T. A., van der Schaaf, J., Hampel, U., et al. (2011). Tomography measurements of gas holdup in rotating foam reactors with Newtonian, non-Newtonian and foaming liquids. *Chemical Engineering Science, 66*, 3317−3327.

Twigg, M. V., & Richardson, J. T. (2007). Fundamentals and applications of structured ceramic foam catalysts. *Industrial & Engineering Chemistry Research, 46*(12), 4166−4177. https://doi.org/10.1021/ie061122o

Ulpts, J., Dreher, W., Kiewidt, L., Schubert, M., & Thöming, J. (2016). In situ analysis of gas phase reaction processes within monolithic catalyst supports by applying NMR imaging methods. *Catalysis Today, 273*, 91−98. https://doi.org/10.1016/j.cattod.2016.02.062

Ulpts, J., Dreher, W., Klink, M., & Thöming, J. (2015). NMR imaging of gas phase hydrogenation in a packed bed flow reactor. *Applied Catalysis A: General, 502*, 340−349. https://doi.org/10.1016/j.apcata.2015.06.011

Ulpts, J., Kiewidt, L., Dreher, W., & Thöming, J. (2018). 3D characterization of gas phase reactors with regularly and irregularly structured monolithic catalysts by NMR imaging and modeling. *Catalysis Today, 310*(January 2017), 176−186. https://doi.org/10.1016/j.cattod.2017.05.009

Upadhyay, R. K., Pant, H. J., & Roy, S. (2013). Liquid flow patterns in rectangular air-water bubble column investigated with radioactive particle tracking. *Chemical Engineering Science, 96*, 152−164. https://doi.org/10.1016/j.ces.2013.03.045

Utomo, M. B., Warsito, W., Sakai, T., & Uchida, S. (2001). Analysis of distributions of gas and TiO_2 particles in slurry bubble column using ultrasonic computed tomography. *Chemical Engineering Science, 56*, 6073–6079.

Vamvakeros, A., Jacques, S. D. M., Di Michiel, M., Matras, D., Middelkoop, V., Ismagilov, I. Z., et al. (2018). 5D operando tomographic diffraction imaging of a catalyst bed. *Nature Communications, 9*(1), 1–11. https://doi.org/10.1038/s41467-018-07046-8

Vamvakeros, A., Jacques, S. D. M., Middelkoop, V., Di Michiel, M., Egan, C. K., Ismagilov, I. Z., et al. (2015). Real time chemical imaging of a working catalytic membrane reactor during oxidative coupling of methane. *Chemical Communications, 51*(64), 12752–12755. https://doi.org/10.1039/c5cc03208c

Varma, R., & Al-dahhan, M. (2007). Effect of sparger design on hydrodynamics of a gas recirculation anaerobic bioreactor. *Biotechnology and Bioengineering, 98*(6), 1146–1160. https://doi.org/10.1002/bit

Vijayan, M., Schlaberg, H. I., & Wang, M. (2007). Effects of sparger geometry on the mechanism of flow pattern transition in a bubble column. *Chemical Engineering Journal, 130*, 171–178. https://doi.org/10.1016/j.cej.2006.06.024

Wang, M., Dorward, A., Vlaev, D., & Mann, R. (2000). Measurements of gas – liquid mixing in a stirred vessel using electrical resistance tomography (ERT). *Chemical Engineering Journal, 77*, 93–98.

Wangjiraniran, W., Aritomi, M., Kikura, H., Motegi, Y., & Prasser, H.-M. (2005). A study of non-symmetric air water flow using wire mesh sensor. *Experimental Thermal and Fluid Science, 29*(3), 315–322. https://doi.org/10.1016/j.expthermflusci.2004.05.004

Wangjiraniran, W., Motegi, Y., Richter, S., Kikura, H., Aritomi, M., & Yamamoto, K. (2003). Intrusive effect of wire mesh tomography on gas-liquid flow measurement. *Journal of Nuclear Science and Technology, 40*(11), 932–940.

Wang, M., Lucas, G., Dai, Y., Panayotopoulos, N., & Williams, R.a. (2006). Visualisation of bubbly velocity distribution in a swirling flow using electrical resistance tomography. *Particle & Particle Systems Characterization, 23*(3–4), 321–329. https://doi.org/10.1002/ppsc.200601063

Wang, A., Marashdeh, Q., & Fan, L.-S. (2014a). ECVT imaging of 3D spiral bubble plume structures in gas-liquid bubble columns. *Canadian Journal of Chemical Engineering, 92*(12), 2078–2087. https://doi.org/10.1002/cjce.22070

Wang, F., Marashdeh, Q., Fan, L.-S., & Warsito, W. (2010). Electrical capacitance volume tomography: Design and applications. *Sensors, 10*(3), 1890–1917. https://doi.org/10.3390/s100301890

Wang, A., Marashdeh, Q., Motil, B. J., & Fan, L. S. (2014b). Electrical capacitance volume tomography for imaging of pulsating flows in a trickle bed. *Chemical Engineering Science, 119*, 77–87. https://doi.org/10.1016/j.ces.2014.08.011

Warsito, W., & Fan, L. S. (2001). Measurements of real-time flow structures in gas-liquid and gas-liquid-solid flow systems using electrical capacitance tomography (ECT). *Chemical Engineering Science, 56*, 6455–6462.

Warsito, W., & Fan, L. (2003). 3D-ECT velocimetry for flow structure quantification of gas-liquid-solid fluidized beds. *Canadian Journal of Chemical Engineering, 81*(August), 875–884.

Warsito, W., Ohkawa, M., Kawata, N., & Uchida, S. (1999). Cross-sectional distributions of gas and solid holdups in slurry bubble column investigated by ultrasonic computed tomography. *Chemical Engineering Science, 54*, 4711–4728.

Williams, J. L. (2001). Monolith structures, materials, properties and uses. *Catalysis Today, 69*(1–4), 3–9. https://doi.org/10.1016/S0920-5861(01)00348-0

Wongkia, A., Larachi, F., & Assabumrungrat, S. (2015). Hydrodynamics of countercurrent gas-liquid flow in inclined packed beds - a prospect for stretching flooding capacity with small packings. *Chemical Engineering Science, 138*, 256–265. https://doi.org/10.1016/j.ces.2015.08.024

Wood, B. D., Apte, S. V., Liburdy, J. A., Ziazi, R. M., He, X., Finn, J. R., et al. (2015). A comparison of measured and modeled velocity fields for a laminar flow in a porous medium. *Advances in Water Resources, 85*, 45–63. https://doi.org/10.1016/j.advwatres.2015.08.013

Yan, P., Jin, H., He, G., Guo, X., Ma, L., Yang, S., et al. (2019). CFD simulation of hydrodynamics in a high-pressure bubble column using three optimized drag models of bubble swarm. *Chemical Engineering Science, 199*, 137–155. https://doi.org/10.1016/j.ces.2019.01.019

Yin, F., Afacan, A., Nandakumar, K., & Chuang, K. T. (2002). Liquid holdup distribution in packed columns : Gamma ray tomography and CFD simulation. *Chemical Engineering and Processing, 41*, 473–483.

Yuen, E. H. L., Sederman, a. J., & Gladden, L. F. (2002). In situ magnetic resonance visualisation of the spatial variation of catalytic conversion within a fixed-bed reactor. *Applied Catalysis A: General, 232*(1–2), 29–38. https://doi.org/10.1016/S0926-860X(02)00064-9

Zalucky, J., Claußnitzer, T., Schubert, M., Lange, R., & Hampel, U. (2017). Pulse flow in solid foam packed reactors: Analysis of morphology and key characteristics. *Chemical Engineering Journal, 307*, 339–352. https://doi.org/10.1016/j.cej.2016.08.091

Zalucky, J., Möller, F., Schubert, M., & Hampel, U. (2015). Flow regime transition in open-cell solid foam packed reactors: Adaption of the relative permeability concept and experimental validation. *Industrial & Engineering Chemistry Research, 54*(40), 9708–9721. https://doi.org/10.1021/acs.iecr.5b02233

Zhao, T., Eda, T., Achyut, S., Haruta, J., Nishio, M., & Takei, M. (2015). Investigation of pulsing flow regime transition and pulse characteristics in trickle-bed reactor by electrical resistance tomography. *Chemical Engineering Science, 130*, 8–17. https://doi.org/10.1016/j.ces.2015.03.010

Zheng, Q., Russo-Abegao, F. J., Sederman, A. J., & Gladden, L. F. (2017). Operando determination of the liquid-solid mass transfer coefficient during 1-octene hydrogenation. *Chemical Engineering Science, 171*, 614–624. https://doi.org/10.1016/j.ces.2017.04.051

Zhivonitko, V. V., Svyatova, A. I., Kovtunov, K. V., & Koptyug, I. V. (2018). Recent MRI studies on heterogeneous catalysis. *Annual Reports on NMR Spectroscopy, 95*, 83–145. https://doi.org/10.1016/bs.arnmr.2018.06.001

Ziazi, R. M., & Liburdy, J. A. (2019). Vortical structure characteristics of transitional flow through porous media. In *ASME-JSME-KSME 2019 8th joint fluids engineering conference, AJKFluids 2019* (p. 1). https://doi.org/10.1115/AJKFLUIDS2019-5094

Applications of tomography in mixing process

Volodymyr Mosorov
Institute of Applied Computer Science, Lodz University of Technology, Lodz, Poland

24.1 Introduction

Mixing processes are general operations in industrial process engineering. Its main goal is to make heterogeneous physical system homogeneous by using manipulating operations. Generally, efficient mixing can be challenging to achieve especially in the industrial scales. Sometimes industrial mixing is done in batches and then ones can call it as dynamic mixing. However, many industrial applications use static mixers, i.e., reactor position is fixed. The segregation is opposite operation to mixing. The example of such an operation is particle segregation, i.e., segregation of particles according to its size, density, shape, and other properties.

There are two types of patterns produced by typical mixer simultaneously: radial mixing and flow division. In the first case, rotational circulations of a processed material around own center cause its radial mixing. Processed material is intermixed to reduce or eliminate radial gradients in temperature, velocity, and material composition. In case of flow division, a processed material divides at the leading edge of each element of the mixer and follows the channels created by the element shape.

Nowadays, static mixers are used for a wide range of different applications in many different industrial areas such as pharmaceutical, chemical, agricultural, and food industry. General application is mixing n-component powders and particles. Other applications include many different chemical processes, for instance, wastewater treatment. Also static mixers are widely used in the oil and gas refinery, for example, for desalting crude oil. Another example of mixing processes in chemistry is a polymer production, where mixers are applied to facilitate polymerization reactions.

Mixing phenomena have also place in multiphase flows which are widely petroleum and petrochemical industries. In many numerous types of reactors such as airlift loop reactors, bubble columns, packed columns, vessels with combined mechanical agitation occur multiphase flow reactions which can be named mixing. Examples of parameters describing mixing processes in reactor are phase volume fractions, mixing parameters, and mass transfer coefficients. In many cases, these parameters are the critical ones, because they allow understanding operational, geometrical, and physicochemical nature of the reactor and improving its work. Therefore, the mixing operations should be taken into account in terms of process efficiency and ultimately product quality. In mixing applications, the swirl-flow phenomena can be recognized as the main factor (Baker & Sayre, 1974; Hristov, Mann, Lossev, Vlaev, & Seichter, 2001). The swirlers are used in pipes mainly to either enhance convective heat transfer

Industrial Tomography. https://doi.org/10.1016/B978-0-12-823015-2.00008-X

between the fluid and the pipe walls or to intensify mixing, as well as for separation processes. The gas–solid flow in a cyclone separator can be a good example where velocity analysis is required for swirl-flow characterization and metering (Gutmark, Li, & Grinstein, 2002; Liu et al., 2005; Yang & Liu, 2000).

Besides process tomography, there exist several other techniques for monitoring and controlling of mixing processes. Briefly discuss several examples of such techniques. Thus, determining the degree of mixing between components in a mixing process can be done by adding luminescent materials to the mixture see, e.g., Kraft, Launikonis, and Swiegers (2005). Such a luminescent material has a uniquely detectable luminescence emission wavelength. Monitoring of mixing is performed by measuring the emitted luminescence waves coming from mixture. Then, the ratio of luminescence intensities and/or the absolute or relative intensities of luminescence indicative about the degree of mixing between the components.

Another technique is the radioactive particle tracking (RPT) which is based on tracking of the motion of a single radioactive particle in the 3D interesting volume. The radioactive particle should have physical properties identical to the ones of the investigated flow. Nowadays such radioactive chemical elements like Sc-46, Co-60, or another isotopes emitting gamma-ray are widely used to track the motion in one or multiphase systems (Mosorov, 2013). The instantaneous particle positions are identified by monitoring the radiation intensities measured by the detectors arranged in different positions around the reactor. In order to reconstruct the position of the particle, a calibration procedure is needed. The calibration allows to relate the intensities measured at the detector as a function of coordinates of the particle. It is not feasible to carry out the calibration procedure by direct experimental measurements, because of the large number of positions must be considered. Generally, this number depends on the number of the detectors, the dimensional of a volume (vessel) and requirement accuracy although even a simple RPT system requires more than several thousand points for a calibration procedure.

However, the methods mentioned above, comparing with process tomography, have serious disadvantages which limit their use in practice. Thus, they need invention into mixing processes by adding other components. Besides these techniques do not allow on good understanding of the underlying mechanisms and principles of the particular mixing processes, which are often largely dependent on the properties of the components.

Therefore, monitoring based on analysis of the cross-sectional distribution of the material enables “look though” inside the internal behavior of the processes and may be used to improve the economy of operation and stability of technological processes, measuring the degree of mixing between components in commercial mixing processes so as to enable the mixing processes to be monitored or optimized. The objective of this chapter is to show how tomographic techniques can be used to monitor and control mixing processes in order to improve technologies and reduce the power consumption.

24.2 Review of tomographic techniques utilizing for different kinds of mixing processes

Obviously, that a type of the sensor which can be used to monitor the mixing processes depends on the nature of mixing materials. For instance, since electrical resistance tomography (ERT) can detect local changes in conductivity/resistivity, therefore, the technique is usually used for hydrodynamic studies the unsteady mixing dynamics of miscible liquids if the fluids to be blended have different conductivities.

Table 24.1 presents the use of the known sensing principles which can be used to study/monitor of mixing processes. As can been seen from Table 24.1, the following kinds of tomography have been utilized for monitoring/control gas–solids mixing: ECT, Position Emission Technique, and Optical/Near-Infrared tomography, whereas two kinds of electrical tomography are generally applied for gas–liquid mixing monitoring.

One of the tomography types that can be successfully utilized for monitoring of different gas–solids mixing processes is the ECT tomography. During the mixing

Table 24.1 Sensing principles to study/monitor of mixing processes.

Mixing process	Sensing principle		
Gas–solid mixing	ECT tomography (Du, Liang-Shih, Fei, & Warsito, 2002)	Position emission particle (Laurent, Bridgwater, & Parker, 2001)	Optical tomography/ Near-Infrared spectroscopy (Rantanen, Wikstro, Turner, & Taylor, 2005)
Gas–liquid mixing	Electrical resistance tomography (Wang & Yin, 2001; Kagoshima and Mann (2005)	Electrical impedance tomography (West, Meng, Aykroyd, & Williams, 2003)	Ultrasonic transducers (Dyakowski et al., 2003, pp. 761–767)
Liquid–solid mixing	Electrical resistance tomography (Lomtschera et al., 2017)	Positron emission tracking technique (Windows-Yule, Hart-Villamil, Ridout, Kokalova, & Nogueira-Filho, 2020)	
Liquid–liquid mixing	Electrical resistance tomography (Jegatheeswaran, Ein-Mozaffari, & Wu, 2018)	Positron emission particle tracking (Jegatheeswaran et al., 2018)	Magnetic resonance imaging (Jegatheeswaran et al., 2018)

process, the measured capacitance data are changed that could be correlated with the changes in the particle size distribution and the power consumption of the mixer device (see Rimpiläinen et al. 2010, pp. 44–51). Next, the mixing indexes or generally mixing parameters can be calculated from tomograms.

Another kind of electrical tomography is electrical impedance tomography. The example of EIT for investigation of the mixing of chemicals in a turbulent flow by using electrical impedance tomography is described in Kourunen et al. (2007, pp. 803–809). The EIT system (KIT4-system) developed at the Department of Physics, University of Kuopio. The system consists of 16 independent current injection channels and 80 simultaneously sampled measurement channels, real-time controller, and a host PC.

An interesting case of the use of process tomography for mixing processes is the detection of grotto formations which can occur in the mixing of fluids, for instance, production of shampoos, pastes, as well sauces. The grotto patterns (regions) often occur in reactors during non-Newtonian fluid mixing. These weakly mixed regions should be detected because their presentation degrades the quality of the output product. Therefore, process tomography can be an effective tool for detecting such artifacts inside reactors. More details concerning the grotto detection can be found, for instance, in Simmons et al. (2009).

An alternative tomographic technique is the Positron Emission Particle Tracking technique, which is similar to the abovementioned RPT. The technique is based on detection of gamma-rays emitted by positron-emitting tracers. The particle injection unit introduces suspended tracers into the reactor. 2D/3D images of tracer concentration within the reactor are then constructed by computer analysis. Such a technique can be utilized for the study of gas–solid mixing in the case of powder flows, i.e., in the case when material density is relatively low. The mentioned technique allows the macroscopic solids behavior to be analyzed. It can be used to estimate the particle velocity fluctuates both radial and axial positions (Ding, Wang, Wen, & Ghadiri, 2005, pp. 368–374). The main disadvantage of the technique is that time data acquisition in relation to ECT is higher. It limits application of the technique in a case of rapid flows.

Another technique is Near-Infrared (NIR) tomography also known as diffuse optical tomography, where NIR light (650–900 nm) is injected through the optical fibers positioned on the surface of the imaging volume of interest and the emergent light is measured at other locations on the same tissue surface using either other fibers or a detector array, such as a charge coupled device (CCD). Such a system will provide time-varying concentration profile, which will then be cross-correlated to provide a velocity profile over the measurement cross-section. The profile will relate to axial velocities; however, information relating to swirl is available too. For instance, process analysis of three separate phases of high-shear wet granulation (mixing. spraying, and wet massing) using NIR technique was described in Rantanen, Wikstro, Turner, & Taylor. (2005). There are no generally accepted on-line tools to gain insight into such a process in practice; therefore, an NIR tomography can be successful technique to provide corresponding chemical information including the homogeneity of the formulation and the amount of water in wet mass. Thus, NIR spectroscopy could be used to determine the end points of the three subphases of high shear wet granulation and as well provide a fast in-line quality control tool.

Another technique is ultrasonic tomography. The use of the thick-film ultrasonic transducers for deployment within oil and gas extraction plant to measure the composition of heterogeneous mixtures has been reported in Dyakowski et al. (2003, pp. 761–767). The technology of manufacturing the transducers is there described and the preliminary measurements, conducted for mixtures of vegetable oil and saline water, using time-of-flight principle are presented.

24.3 How to extract information about mixing from tomographic images

One of the main problems using of any tomographic technique for investigation of mixing processes is how to extract from tomographic image series information concerning the mixing process.

The tomograms represent temporal images of material distribution in the cross-section only. Moreover, in the case of application of one-plane sensor system, we should assume that the material moves in radial direction with the angular velocity (see Fig. 24.1A). That means that axial moves of material will be ignored. However, in a general case, the material moves in 3D volume. Therefore, analysis of a mixing process in 3D volume requires having at least twin-plane sensor system. Then such a solution will be allowed to analyze the motion of the material both axial and radial direction (see Fig. 24.1B). Angular velocity profiles of a rotating rod measured with two sets of electrodes in a displaced angle (Etuke & Bonnecaze, 1998) and axial

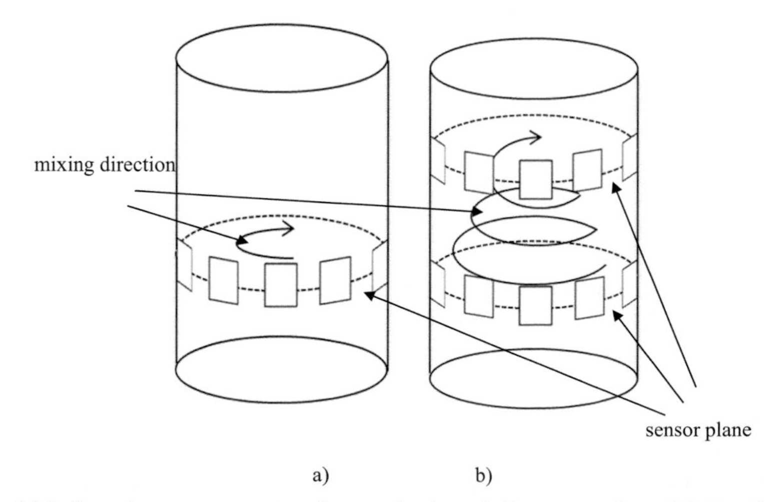

Figure 24.1 One-plane sensor system for monitoring mixing process in radial direction only (A), twin-plane sensors system for monitoring mixing process in radial and axial direction (B).

velocity profiles of solid—liquid flow obtained with dual-plane sensor system (Lucas, Cory, Waterfall, Loh, & Dickin, 1999) have been reported.

In further sections, the use of tomography for two aforementioned cases, i.e., radial and angular mixing will be considered separately.

24.4 Application of one-plane tomography in mixing process

The idea how to analyze a mixing process from the series of tomographic images utilizing one-plane tomographic system is based on an assumption of a strong correlation between neighboring images. The similar idea applying cross-correlation technique to calculate the velocity profiles had been already widely demonstrated in previous chapters. The cross-correlation technique can be also employed for the implementation of mixing analysis. Due to all the data coming from one set of sensor, an autocorrelation method was employed for the implementation of the angular velocity. The basic method to obtain the angular velocity from autocorrelation is the same as the cross-correlation, which is to find the transition time between two profiles with a certain angle where the similarities are most obvious. According to a mathematical description, it intends to find a transition time τ that can make the difference $\varepsilon(\tau)$ minimum. This can be achieved using the least square criterion as given by Eq. (24.1):

$$\varepsilon^2(\tau) = \int_0^T (X_{\alpha 1}(t) - X_{\alpha 2}(t-\tau))^2 \mathrm{dt} \tag{24.1}$$

where $X_{\alpha 1}(t)$ and $X_{\alpha 2}(t)$ are two reconstructed images in a T imaging sequence with starting angle α_1 and α_2, respectively.

Expending Eq. (24.1) and then taking the average values away, the autocorrelation for extracting the angular velocity can be expressed in discrete as shown below:

$$R_{k.\ \Delta\alpha}(n) = \sum_{m=0}^{M-n} X_{k.\alpha_1}(m) X_{k.\alpha_2}(m-n) \tag{24.2}$$

where $R_{k.\ \Delta\alpha}$ is the autocorrelation function at a k discrete radius with an angular difference $\Delta\alpha$, where $(\Delta\alpha = \alpha_1 - \alpha_2)$, M is a sample length, n is correlated sample $(n = 1,\ \ldots, M-1)$.

Unlike the rotated rod in a mixing phantom or the flow regime in a pipeline where the moving feature has no change or a little change during a short transition time, the mixing dynamics of the concentration profile will be diluted or dispersed quickly to a homogeneous distribution. In order to extract the dynamic velocity feature in the miscible liquid mixing, a normalization procedure was applied to data at each imaging ring shown in Fig. 24.2, which gives a same weight factor to all ring profiles. Applying

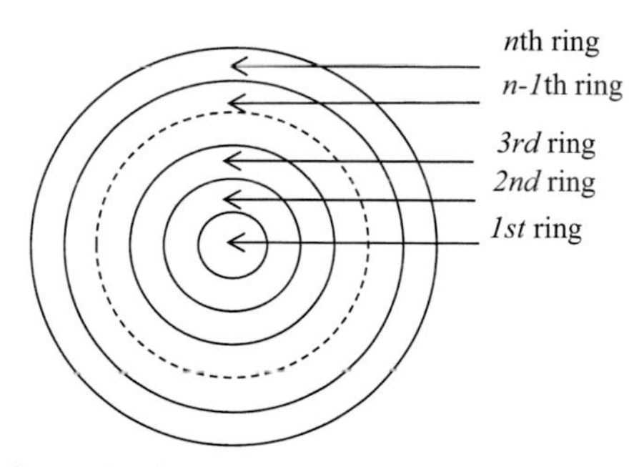

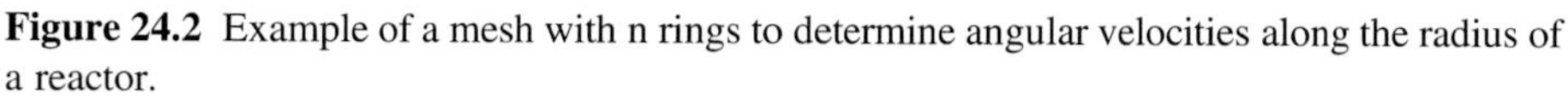

Figure 24.2 Example of a mesh with n rings to determine angular velocities along the radius of a reactor.

the normalization procedure and the autocorrelation method to a data set of 300 tomographic images acquired from the slow stirred mixing phantom, a strong correlation was found as given in Fig. 24.3. It should be noted that the reason for having different correlation angles at these rings in Fig. 24.2 was due to the limited and different pixel

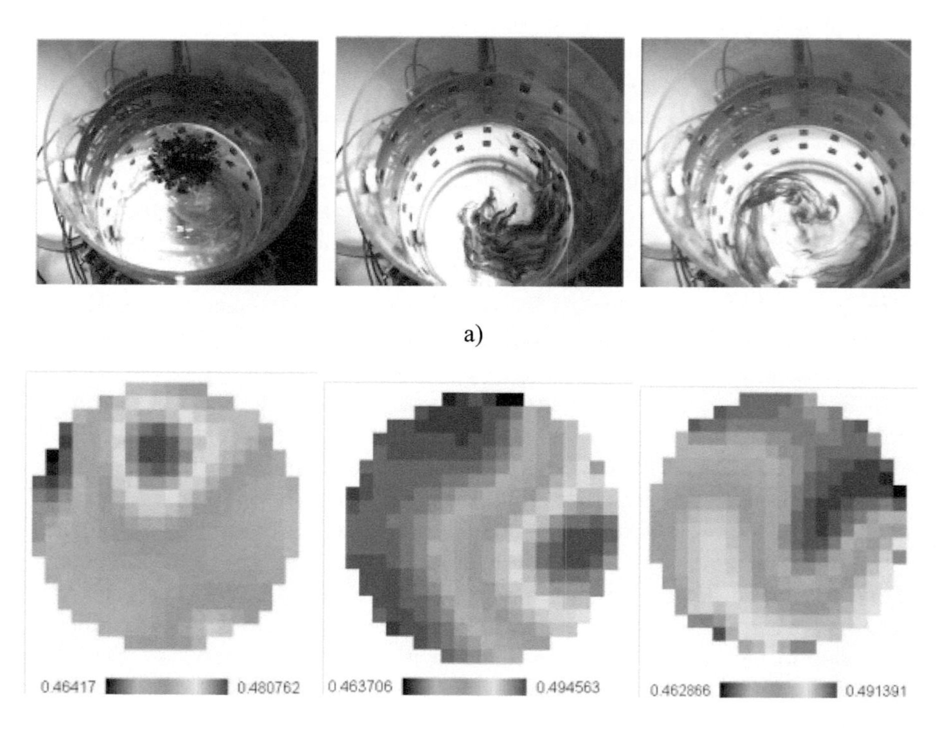

Figure 24.3 Visualization of concentration distribution with the photographic dye distribution (A). The tomographic conductivity distribution (B) (Wang & Yin, 2001, pp. 146−150).

numbers at each ring in the finite element mesh. The conversion from the results to angular velocities is given in Eq. (24.3):

$$\omega = \frac{2\pi}{N_{\text{ring}}} \cdot \frac{N_{\text{corr}}}{\Delta_{\text{time}} \cdot M}; \ [\text{rad}/s] \tag{24.3}$$

where ω denotes angular velocity, N_{ring} is the number of pixels per ring, N_{corr} the number of pixels between a correlated angle, M is the distance (number of images) between two correlated images, Δ_{time} is the time for collecting each frame of data.

24.5 Mixing process monitoring by twin-plane tomographic system

The velocity of flow is often a desirable parameter that provides significant knowledge about the flow state of the controlled process. For laminar flow, where each fluid layer flowing smoothly, parallel to the pipe and without mixing flow structure, the estimation of the velocity profile can be done based only on the analysis of axial velocity. This assumption makes no sense for turbulent flow when the behavior of the flow structure significantly deviating from the laminar flow is observed. Especially, the mixing process (swirl flow) requires an analysis of these three velocity components to determine the swirl angle and to explain the mechanism of the swirl phenomena (Gutmark et al., 2002; Li and Tomita, 2001). The significance of the swirl-flow measurement very broadly confirms the industrial types of the flow phenomena, where the knowledge about the axial, radial, and angular velocity components increases the efficiency of the control system: gas—solid (Li & Tomita, 2001), gas—oil (Li et al., 2013), combustion chambers (Orbay, Nogenmyr, Klingmann, & Bai, 2013), cylindrical separators (Jawarneh, Tlilan, Al-Shyyab, & Ababneh, 2008), and mixing by agitation (Hristov et al. 2001). In these situations, the swirl effects are extensively seen as either the desired result or unavoidable.

As the flow phase distribution changes in space and time for the various flow regimes, it can cause a lot of problems for the flow measurement system. Spatial—temporal changes of the flow characteristics are dependent on a number of conditions, such as flowing material properties, flow rates, or the geometry of the pipe (bends, elbows, control valves, etc.). In order to obtain a high accuracy of the flow measurement system, the flow regime should be considered. The determination of the current flow regimes in pipes is, however, another nontrivial task. On the other hand, an elaboration of the flow pattern by the use of an independent method, which permits the flow velocity field to be unambiguously calculated with higher robustness, is still waiting to be invented.

In order to meet these requirements, the methods of flow velocity measurement considering 3D components will be required. The next subsections describe the velocity measurement methods based on the spatial—temporal cross-correlation technique.

24.5.1 Velocity measurement methods–based cross-correlation technique

One of the possible methods for the velocity calculation is based on the correlation technique. This technique determines a time delay of the measurement signal between two distantly located sensors (Hanus et al., 2021; Mosorov, 2006; Mosorov, Sankowski, Mazurkiewicz, & Dyakowski, 2002; Plaskowski, Beck, Thorn, & Dyakowski, 1995). This delay is interpreted as the time needed for the movement of the material between the first and the second sensor, and this information allows the determination of the flow velocity. Depending on the format of the signal—type of measurement system—a more or less sophisticated data processing method can be applied with the aim to obtain a high accuracy of the velocity measurement. A measurement system providing rich information about the spatial–temporal changes of the flow should be considered for application, in order to conduct a deeper analysis of the flow process character. In this context, tomography systems are very useful and effective for flow measurements. The sequences of images obtained from sensor planes give the possibility of obtaining the velocity profile, as well as for the velocity components of the swirl flow.

Let $x_{[n,m]}(iT)$ and $y_{[n,m]}(iT)n, m = 0, \ldots, N-1$ define the material distribution for nth pixel of the ith tomographic image $X(iT)$ and $Y(iT)$ obtained from the X and Y planes, respectively, with the frame rate resolution T. These changes can be treated as a 1D signal. A high correlation exists between the two corresponding signals of the different planes at a short time lag τ. The classical cross-correlation-based method for determining the lag time τ of the flow propagation through the two planes is related to the assumption that the flow is laminar within a sensor volume. However, the flow material passes between the planes with different velocity directions. Therefore, in this case, the velocity comprises also a radial component. To overcome this drawback, the "best-correlated pixel" method (Mosorov, Sankowski, Mazurkiewicz, & Dyakowski, 2002) was developed for flow velocity profile measurement when the trajectories have no linear character. The "best-correlated pixel" method uses the correlation technique in a different way. The changes within the pixel from the X plane can better correlated with either the corresponding pixel or some different pixel of its neighbors from plane Y. The correlation function between pixel $x_{[n,m]}$ from plane X and the neighborhood $y_{[n+l,m+p]}$ ε B from plane Y can be calculated as follows:

$$R_{x_{[n,m]}y_{[n+l,m+p]}}(kT) = \sum_{l \in B} \sum_{p \in B} \sum_{i=0}^{M-1} x_{[n,m]}(iT)\, y_{[n+l,m+p]}((i-k)T), \tag{24.4}$$

where B is the pixel neighborhoods in plane Y. After calculating the cross-correlation function between each pair of pixels, the best-correlated pair is chosen and the radial velocity is given by the following:

$$Vr = \Delta d / \tau, \tag{24.5}$$

where Δd is the distance between the corresponding pixel and the best one in plane Y, while lag time τ can be determined from the delayed frame number, corresponding to the maximum correlation value.

24.5.2 Spatial cross-correlation for flow angular velocity calculation

In the case of swirl flow, the material changes along the axial direction of the pipe from plane X to plane Y can be considered as a rotated process (Fig. 24.4). Two separated series of images $X(iT)$ and $Y(iT)$ (from planes X and Y, respectively) give the possibility to apply the spatial cross-correlation technique in a new way. The hypothesis can be advanced by proposing that the image from the first plane is highly correlated with some rotated image from the second plane. The idea is to calculate the spatial-time correlation between an image from plane X and a rotated image from plane Y for all possible angles. The highly correlated pair of images determines the spatial difference in the angle between the two images.

The spatial-time correlation function for the chosen time iT can be calculated as follows:

$$R_{X(iT)Y((i-k)T)}(\theta, kT) = \sum_{n=0}^{N-1} \sum_{m=0}^{N-1} x_{[n,m]}(iT) y_{[n,m],\theta}((i-k)T), \quad \theta = 0, 1, \ldots 359^{\circ} \tag{24.6}$$

where θ is the angle of rotation applied to the image $Y_{[n,m]}(iT)$. The lag time is calculated as $\tau = kT$, $k = 1, 2, \ldots K$ corresponding to the number of frame delay. In most cases, the value of k is 1. However, in the case, when the angle velocity is not too

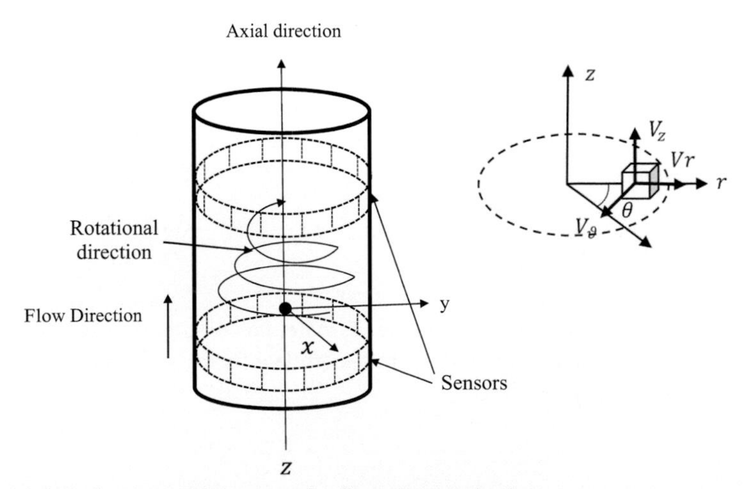

Figure 24.4 Twin plane ECT system for flow characterization.

large, that means that the images of $X(iT)$ and $Y(iT)$ from planes X, Y are very similar, and another value of k can be chosen. The angular velocity V_θ is given by the following:

$$V_\theta = \frac{\Delta\theta}{\tau}. \tag{24.7}$$

The angular displacement $\Delta\theta$ can be calculated as the difference in the angle between images from plane Y for different time instants (kT) when $K = 1, 2, \ldots, K$ (Fig. 24.5). The calculation of the angular velocity requires the condition be fulfilled that

$$\tau \ll T_{\theta\min} \tag{24.8}$$

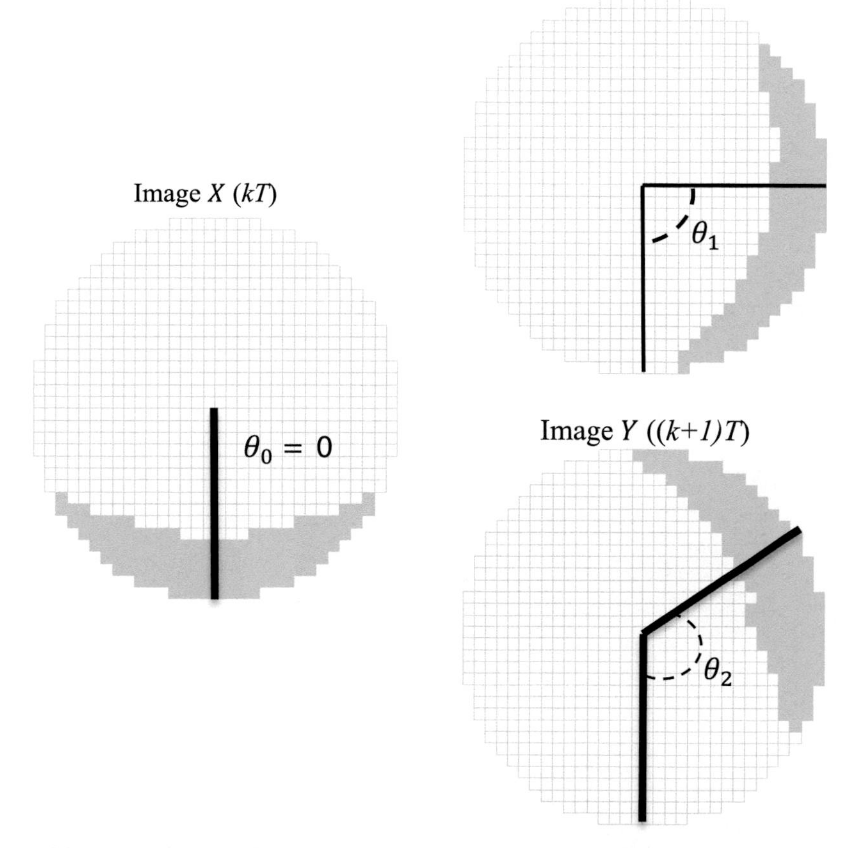

Figure 24.5 The principle of the angular displacement $\Delta\theta$ calculation.

where $T_{\theta min}$ is the minimum period of the material rotation inside the sensor planes. The choice of τ, i.e., the number of frame delayed k, will depend on a specific flow. Thus, k can be set to 1 in the case of rapid material rotation and to a larger one when flow rotation is slower.

24.5.3 Process modeling

To verify the proposed method, a physical model simulating the swirl-flow phenomena can be used. For instance, the physical model consists of a long elastic plastic pipe (filled with dry silica sand—estimated relative permittivity of porosity mixture is approximately in the range of 1.85–2.25) with a diameter of 20 mm and length of 1000 mm wrapped around a cylindrical paper tube (hollow roll with a diameter of 130 mm). The elastic plastic pipe included four loops and the distance between two neighboring loops is 210 mm (see Fig. 24.6C). Such a model represents a high concentration of the rotated material during flow in specified positions. This phantom is inserted in a cylindrical fragment of the pipe made from Plexiglas (permittivity 3.2) with an outer/inner diameter of 150/142 mm and length of 500 mm. The pipe wall thickness is 4 mm. Such a physical phantom can model a spherical bubble-rising process in a spiral motion that takes place in gas–liquid bubble columns or three-phase fluidization systems widely used in chemical and petroleum industries (see Wasito & Fan, 2003). Similar effects of swirl-flow behavior are visible in several types of processes: hydrocyclone/cyclones, separation for gas/solid, or solid–liquid mixing.

In order to visualize the material distribution inside the pipe, a 32-electrode ECT unit (see Fig. 24.6A) was applied. The ECT system (Fig. 24.6A) had specifically designed sensors (Fig. 24.6B), which comprises 32 measuring electrodes, organized in four layers of eight electrodes made of copper with a thickness of 0.2 mm and width of 50 mm for each one, a height of 70 mm for layers 1 and 4, while for layers 2 and 3, the height was 30 mm. Electrodes were mounted on the outer wall of the Plexiglas

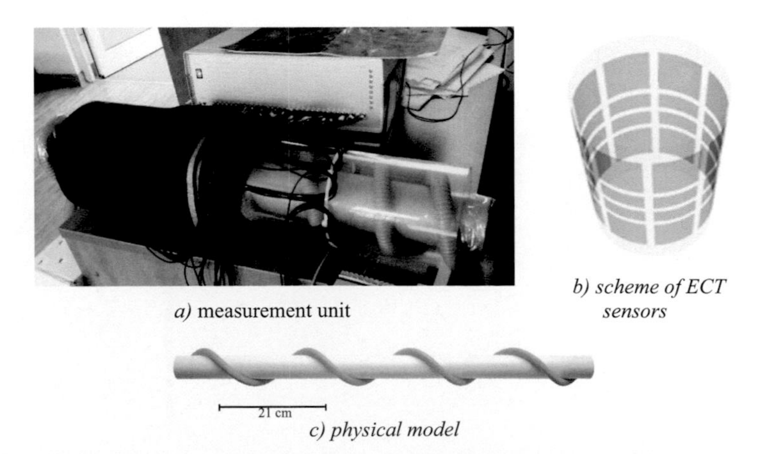

Figure 24.6 ECT Measurement unit and the physical model used in the experiment.

pipe. During the investigation, only layers 1 and 4 were used. The external screen is made of a copper sheet with a thickness of 0.5 mm connected to the grounding measurement system; insulation of the external screen is made of polyurethane foam dielectric (permittivity of 1.6), while the two internal screen boundaries made of copper tape have a thickness of 0.2 mm and height of 25 mm each.

Calibration of the ECT system was conducted with both an empty and full sensor space. In order to calibrate the sensor for full sensor measurement space, a phantom made of Ertalon (relative permittivity 3.8) was used. A series of tests using the proposed physical model were performed. During the measurements, the designed phantom was shifted into the sensor space with an average axial velocity of $V_z = 10$ cm/s. The data were collected with an acquisition time of $F_r = 11$ frames/s.

For known angular V_θ and axial V_z flow velocities, one can determine an angle of phantom rotation along the length of the electrode L as follows

$$\theta = \frac{V_\theta}{V_Z} \cdot L \tag{24.9}$$

Assume that $V_{\theta max} = 4 \text{rad/s}$, $\theta_{max} = 4 \cdot 7/10 = 2.8$ rad. This value is acceptable because it is less than 2π, i.e., the correlation function can be utilized for calculation. In the case of actual flows, a similar consideration can be applied.

The iterative back-projection algorithm with a constant step length (number of iterations: 10 - the value of relaxation factor was equal to 1.2) was used for the 2D image reconstruction (Yang & Peng, 2002). The choice of this reconstructed method was motivated by the need of a swirl-flow estimation algorithm verification for actual laboratory conditions. Such an approach allows to conclude about the robustness of the methods in a much wider aspect than in the case of the use of, e.g., nonlinear methods. The results obtained with the LBP algorithm are of a slightly lower accuracy than in case of the IBP algorithm, although the accuracy difference is not significant. For each plane, there have been 500 images (32 × 32 pixels) reconstructed showing changes of material concentration inside each pixel. Fig. 24.7 shows the ECT

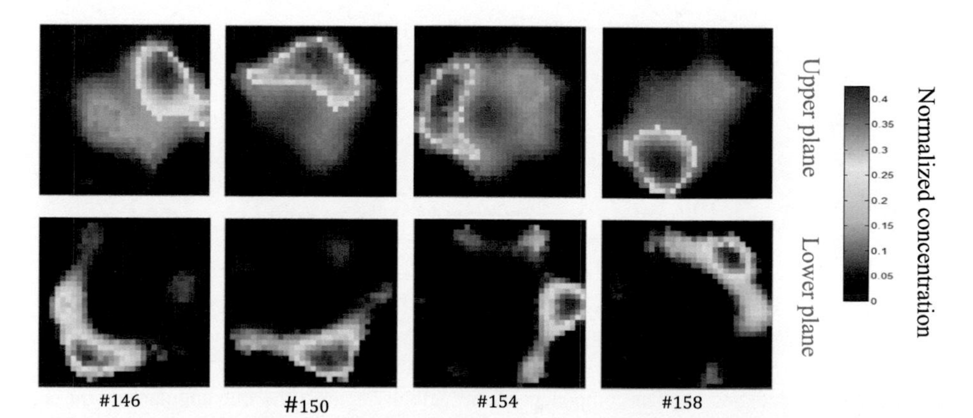

Figure 24.7 ECT normalized tomograms of two planes for chosen time moments.

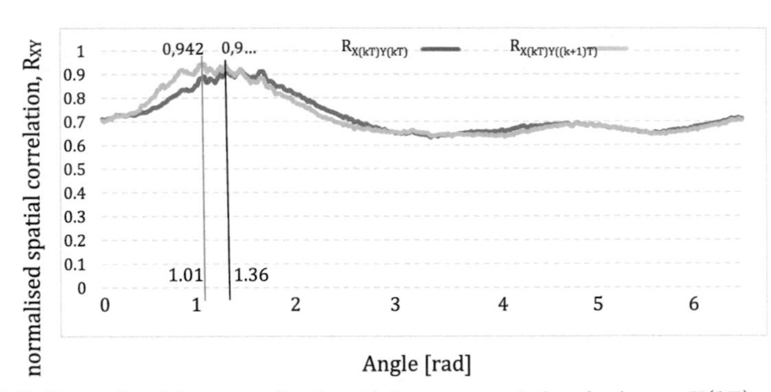

Figure 24.8 Example of the normalized spatial cross-correlation for image $X(kT)$ of the upper plane as well as two images $Y(kT)$ and $Y((k+1)T)$ of the lower plane, where $kT =$ 11.36 s and $(k+1)T =$ 11.45 s.

normalized tomograms for chosen time moments. The reconstructed images demonstrate the rotation of the material inside the sensor space as it takes a place in the case of the swirl flow.

The angular component of the velocity was calculated based on the spatial cross-correlation function (Eq. 24.3). Fig. 24.8 shows an example of calculating the normalized spatial correlation between frame #125 of plane 4 and two frames #125 and #126 of plane 1, respectively. The correlation function $R_{XY}(\theta)$ allows to determine the angular displacement $\Delta\theta$, which is equal to 1.01 and 1.36 rad for two cases $k = 1$ and 2.

Fig. 24.9 presents the calculated angular component of the velocity against the true angular one during flow propagation in the pipe for the conducted experiments. The real angular velocity was extracted from video sequences captured by a CCD camera. The characteristics were estimated based on tomography data sets coming from two planes (planes 4 and 1) and the cross-correlation technique for $k = 1$ and $k = 2$

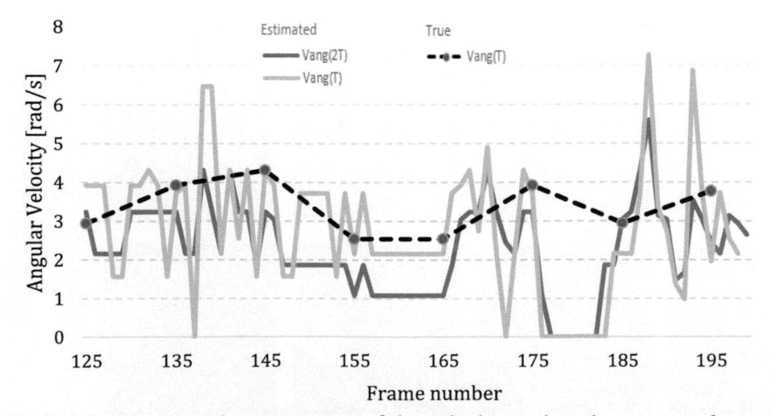

Figure 24.9 Calculated angular component of the velocity against the true angular one during the phantom propagation.

(see Eq. 24.6). The variation of the angular velocity component $V_{ang}(T)$ is between 0 and 7.5 rad/s and one is between 0 and 5.5 rad/s for $V_{ang}(2T)$. The similarity between the two angular velocity variations was noticed. However, there are some unlikely peaks caused by the push procedure of the phantom into the sensors space.

24.5.4 Gravity swirl-drop measurement

In order to show the efficiency of the considered approach on different-flow data sets, it was applied on real data sets derived from the ECT system. To obtain the swirl flow, the gravity drop was rotated by an air injection in the vertical section of the pipe. The schematic of the experimental setup is presented in Fig. 24.10A. The ECT sensor was mounted on the Perspex tube (inner/outer of diameter 60/50 mm. Each sensor includes eight electrodes, 3 cm in length. The two planes of electrodes were fitted on the pipe with a spacing between the center line of the planes of 3 cm. The center of the lower ECT sensor was located approximately 87 cm below the neck of the hopper funnel. The solid phase was a dielectric material: polypropylene beads. The electrodes were numbered anticlockwise when viewed from funnel end of the flow rig. The injected air was also swirled clockwise when viewed from the funnel end of the flow rig. The experiments were carried out for several swirl air-setting control parameters: 0 (no swirl), 20, 30, and 40 (maximum value). The frame rate for each sensor was $F_r = 100$ frames/s. The crosssectional images of the flow propagation from upstream to downstream sensors were obtained for a gravity solid−gas flow. In order to create the swirl phenomena, air was injected to the process as mentioned before. Fig. 24.11

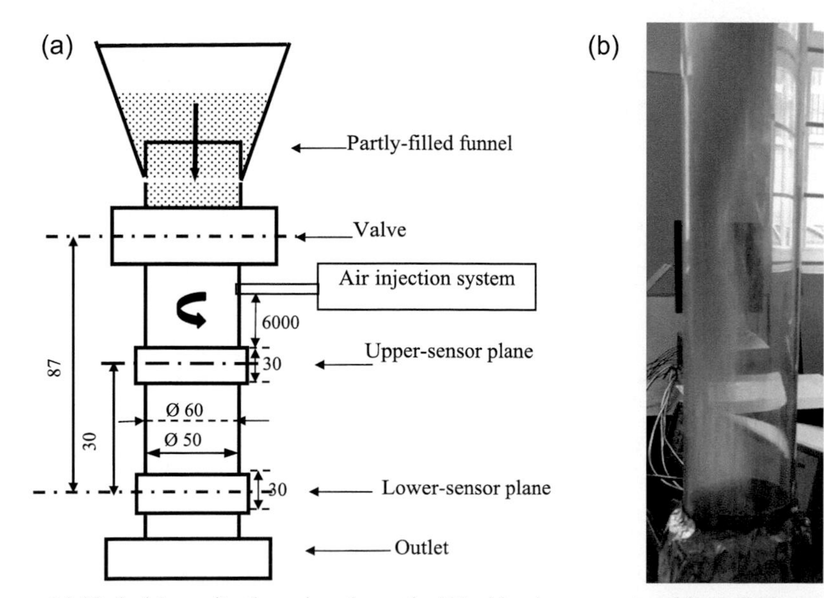

Figure 24.10 Swirl gravity drop rig schematic (A) video image captured by a CCD camera installed next to the transparent pipe section of the swirl gravity drop rig (B).

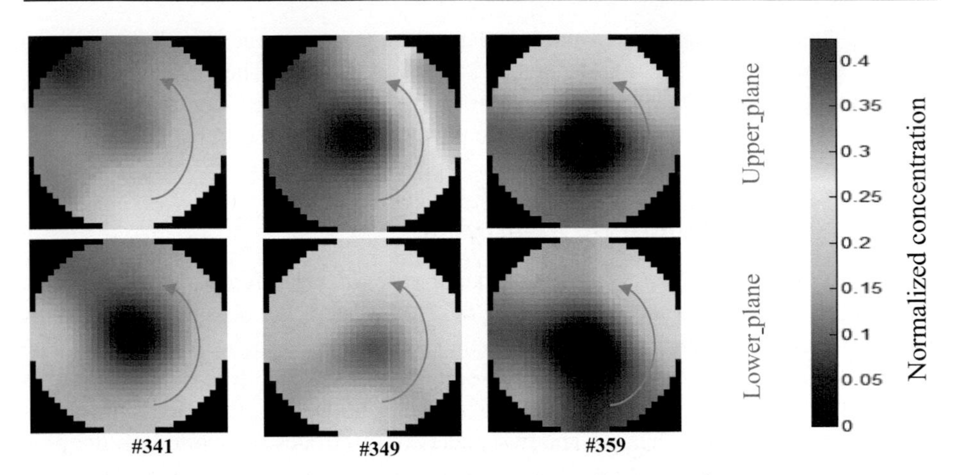

Figure 24.11 ECT normalized tomograms of two planes for chosen time moments.

presents the local solid volume distribution of the material as measured by the ECT system of two planes for chosen time moments. One can observe the swirl-motion movement of the solids inside the pipe. These experiments were carried out with different parameters of air injection control inside the pipe.

The proposed method based on spatial cross-correlation (Eq. 24.3) was employed to evaluate the angular velocity of the considered swirl. Fig. 24.12 presents the variation of the angular velocity during the flow propagation. It clearly shows the difference between the variation of the velocity when the air set control is on and when the air set control is off. The calculated results were compared with results obtained by video images captured by a CCD camera installed next to the transparent pipe section of the swirl gravity drop rig (see Fig. 24.10B). What is worth noting is that the two independent methods yield similar results.

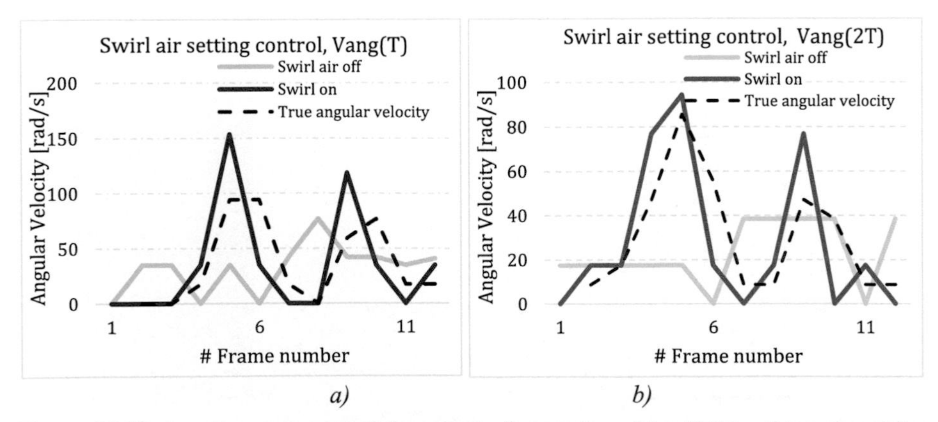

Figure 24.12 Angular component of the velocity for one time delay T (A) and two-time delay $2T$ (B).

For future research, it is necessary to develop a methodology allowing to fit the appropriate time acquisition speed of the measurement system since the one applied in this work is limited to a low sampling rate and it can be insufficient when applied to a higher axial velocity of flow. Another study is necessary to verify the influence of the image reconstruction algorithm on the quality of results. Although the reconstruction method applied in this study is somehow limited, however, the approach introduced shows its applicability for the simulated tomographic data as well as for a range of real dynamic flows.

24.6 Toward to improvement of process measurement

One of the basic operations in Eq. (24.6) is the operation of image rotation. However, at any rotation angle other than a multiple integer of $\pi/2$, the following effects are presented: geometric deformation, e.g., contour and shape changes, nonlinear brightness changes in pixels, etc. This is especially evident in the case of low-resolution images obtained from all process tomography system techniques. The basic reason is a simple fact, that the rotation of a reconstructed image of raster type by an arbitrary angle is done by averaging of adjacent pixels, i.e., the pixel after rotation by an angle which is not a multiple of $\pi/2$ occupies the space corresponding to several pixels. All image rotation techniques, for instance, nearest neighbor, bilinear, and cubic convolution (see Parker, Kenyon, & Troxel, 1983), cause that images obtained by the use of these methods are quite distorted. As a result, the accuracy of determining the rotation angle between two images based on Eq. (24.6) decreases. An additional factor that plays an important role in the accuracy of determining the rotation angle is the low quality of the tomographic images caused by the limitations of reconstruction algorithms. Therefore, in case of the use of image rotation operation, the method of improving the accuracy of angular velocity determination is required.

24.6.1 Method for improving the accuracy of angular velocity determination

The typical resolution of tomographic images is typically 32×32 pixels. Very rarely image resolution can be higher, because number of raw data is strong limited. Thus, number of sensors used in measurements in electrical capacitance tomography number of electrodes rarely exceeds 12, which gives total number of measurements less than 100. Similarly, in gamma-ray industrial tomography, the number of measurements also does not exceed 100. Therefore, such a low image resolution caused that rotation operation will generate significant deformation of image features. Fig. 24.13 shows original image (A) and image rotated 30° by the *cubic convolution method*. As it is seen, there are significant geometric deformations of image contours and shapes as well as blurring of pixel brightness. The reason of such significant deformations is fact, that the pixel's values in the rotated image are not matched with image grid. Each new pixel should be calculated for instance using the 3×3 pixel's block of the original image.

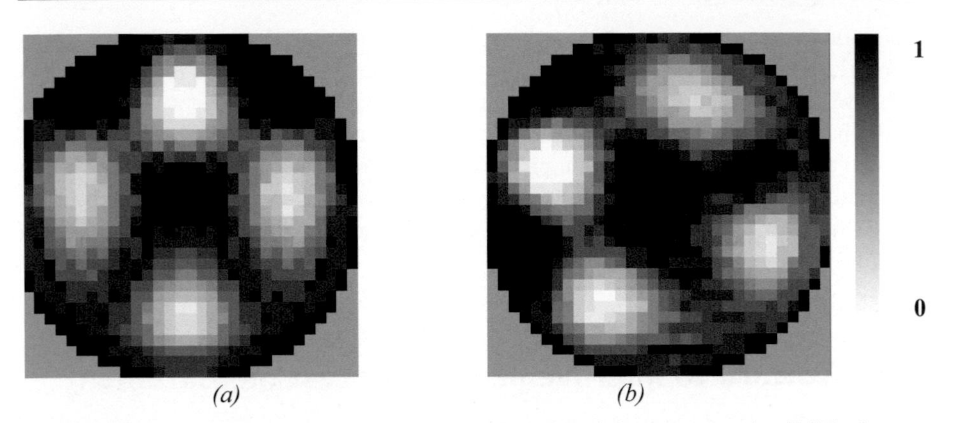

Figure 24.13 Reconstructed 32 × 32 pixel Electrical capacitance tomography (ECT) phantom image (A) the same image rotated by 30° clockwise (B). Grayscale represents normalized concentration.

Since image and its rotated copy will differ, hence, angle determination θ_r of images with low resolution will give inaccurate results.

Improving of the angle determination procedure can be achieved for images with high resolution. For this end, artificial increasing of resolution of tomographic images can be employed. Such a higher resolution artificial image can be constructed by dividing each image pixel of an original image into $k \times k$ subpixels, wherein the intensity of each subpixel should be equal to initial pixel's intensity. Such an artificial increasing of image resolution is enough simple operation and it can be easy done by image processing functions. As result, this artificial increase of image resolution should improve the operation of the image rotation and as consequence improving the accuracy of the rotation angle determination. One can assume that the higher image resolution may lead to higher accuracy of angle determination. However, on the other side, the number of the operations in Eq. (24.6) is growing; therefore, it is important to find appropriate image division ratio k.

24.6.2 Phantom validation

As a previous case, it is worth to check the effectiveness of the proposed approach for relative good quality of reconstructed images of the physical phantoms. Fig. 24.14 shows ECT tomography images reconstructed by the authors' software using the Iterative Least Square Technique algorithm (Isaksen, 1996). Images were obtained for Ø80 polyethylene phantom (density material 0.93 g·cm^{-3}) which was made as cylinder with 4 holes of 20 mm in diameter (Fig. 24.14E). It was rotated at chosen angles to simulate two-phase solid/gas swirl flow. The rotation and resizing operations were performed by original software written in Java language. To determine rotation angle, we need firstly to calculate the spatial correlation function defined by Eq. (24.6) for all possible angles (time moments iT are not considered). Fig. 24.15 shows the examples of calculating the spatial correlation function between 32 × 32 pixel resolution

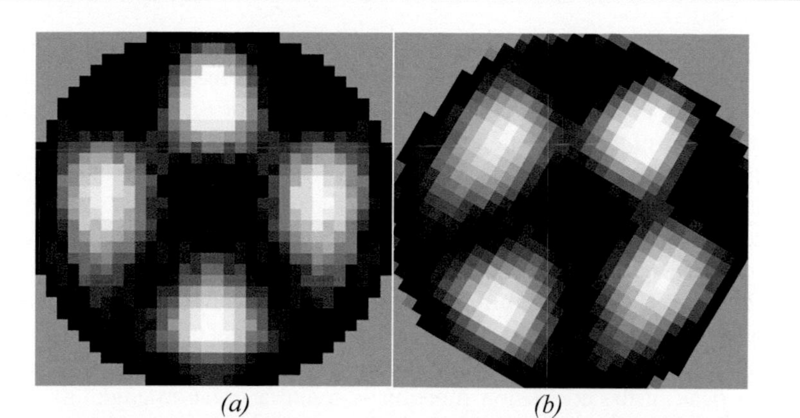

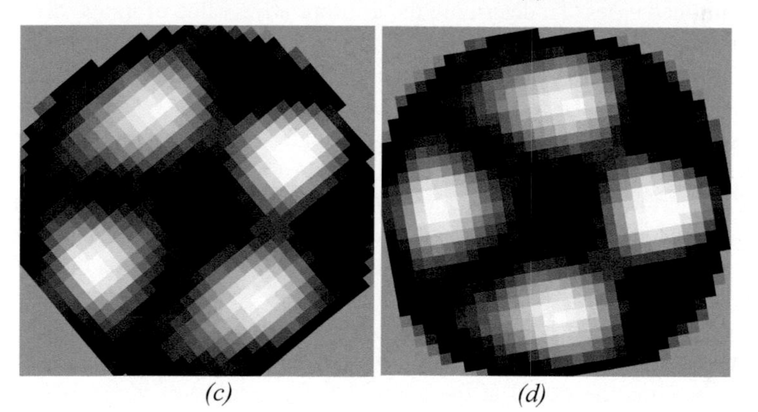

Figure 24.14 Reconstructed 32 × 32 pixel gamma-ray tomography phantom images at chosen rotation angle: (A) 0°, (B) 30°, (C) 50°, and 80° (D) clockwise. Grayscale represents normalized material density.

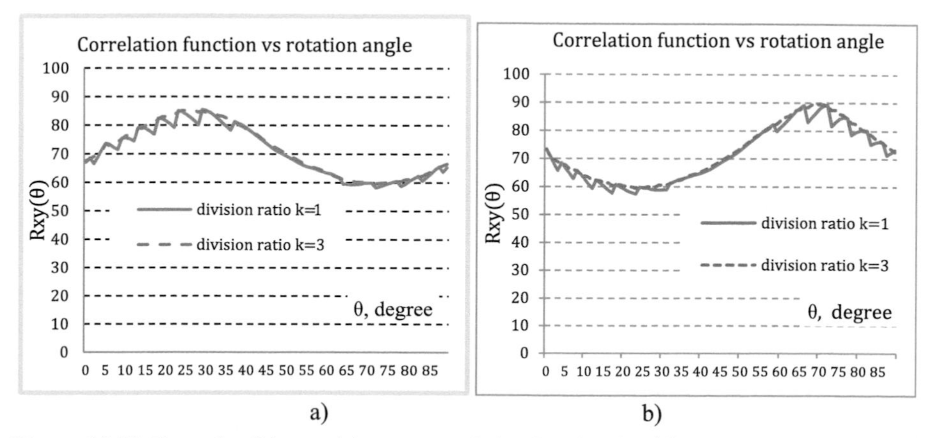

Figure 24.15 Example of the spatial cross-correlation function $R_{XY}(\theta)$ of reconstructed 32 × 32 pixel gamma-ray tomography phantom image (Fig. 24.14A) and rotated image at chosen angle: (A) 30°; (B) 67.5°.

phantom image (Fig. 24.14A) and its copies rotated by 30° and 67.5° angle. Then, maximum of the correlation function $R_{XY}(\theta)$ corresponded to the highest correlated pair of images will be determined by an angular displacement θ_r. As can be noted in Fig. 24.15, the curve of the correlation function is very rough if calculation is based on original low resolution tomographic images. The curve has characteristic "sawtooth" shape especially in neighborhood of its maxima. It is a reason why it is difficult to determine a peak correctly. In the case shown in Fig. 24.15, the peak can be detected at rotated angle 25° or 29° (see Fig. 24.15A) and 66° or 70° for the second case (see Fig. 24.15B). These results differ from true value 30° and 67.5°, respectively. However, the peak can be more accurately detected for the spatial correlation function calculated of artificially increased resolution images. Thus, if chosen image division ratio k is 3, the peak of the correlation function is found at rotation angle 30° and 69° for two mentioned cases. To determine the appropriative value of image division ratio k, the following study is done. Fig. 24.16 shows the dependence between accuracy of the rotation angle determination and image division ratio k, for three different iteration numbers in the image reconstruction procedure. Thus, increasing of image resolution caused that relative error of the rotation angle decreases from 6.7% up to less than 4% for rotated angle 22.5° (see Fig. 24.16A). Additionally, a dependence between the angle determination error and the quality of the reconstructed images, i.e., the iteration number of reconstructed procedure, has been studied. It is shown that the proposed approach can be successfully utilized for lower quality of the reconstructed images and it means that reconstruction time can be reduced too. Thus, the relative errors of rotation angle calculation are very similar for images reconstructed with higher or lower number of iterations. For instance, relative errors are 2.5%, 2.9%, and 3.4% for images rotated at the angle 30° for number of iterations 5, 10, and 20 respectively. This means that even for lower quality of images, the calculated angle of rotation has a similar accuracy.

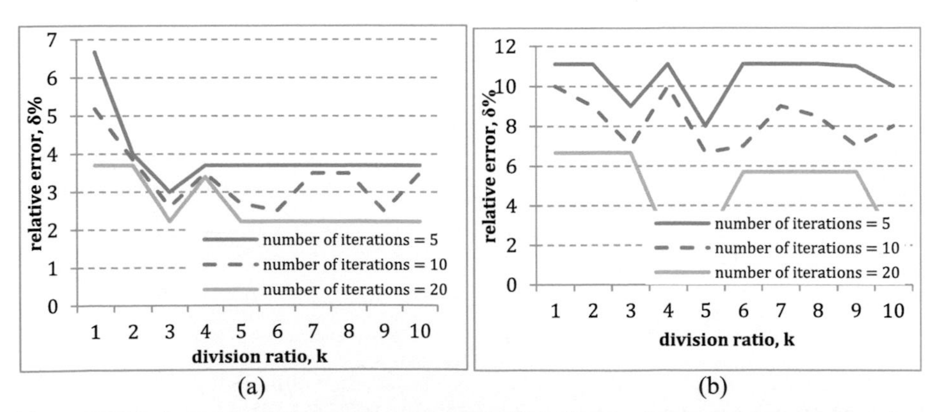

Figure 24.16 Accuracy of determining the angle of rotation vs. image division ratio k and iteration number in image reconstruction procedure (A) angle 30° (B) angle 67.5°.

24.6.3 Real case study

The proposed approach has been applied on real data sets described in previous Section 24.5, i.e., for gravity drop which was rotated by an air injection in the vertical section of the pipe to obtain the swirl flow. Fig. 24.17 presents the variation of the angular velocity during the flow propagation calculated by Eq. (24.6). It clearly shows the difference between the variation of the velocity when the swirl air-setting control parameter was 40. To verify obtained results, they were compared with results obtained by video images captured by a CCD camera. As it can be seen, two independent methods yield enough similar results.

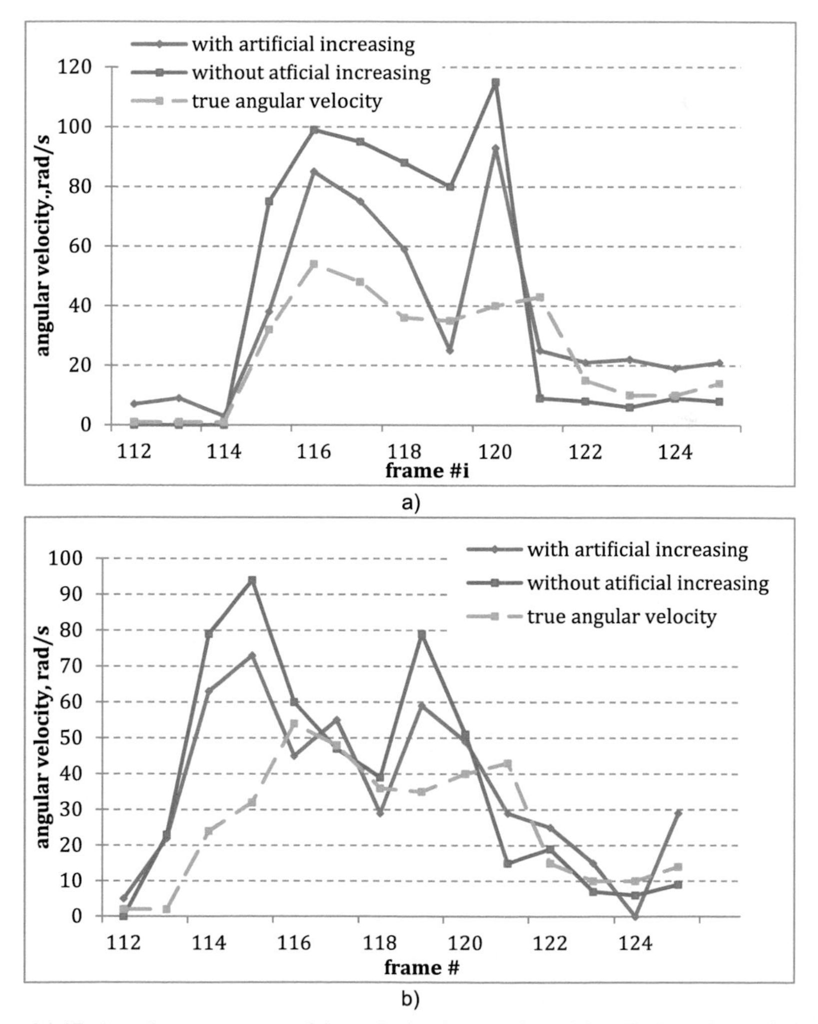

Figure 24.17 Angular component of the velocity for one time delay T (A) and two-time delay $2T$ (B). Division ratio $k = 5$, and swirl air-setting control parameter is 40, iteration number in the reconstruction procedure is 20.

Fig. 24.17 shows that the errors of rotation angle estimation using 'enhancement" of image resolution are smaller comparing to the case when calculation is done for an original image resolution. Thus, at division ratio $k = 5$, the relative errors of angular velocity V_θ calculation are approximately 27% and 19% smaller than without artificial increasing of resolution at delayed frame number the T and $2T$, respectively.

As can be seen in Fig. 24.17B in frames 118–120, there appears a big difference between the actual angular velocities and the one obtained from the proposed method. The probable reason can be that the time shift (parameter rT in Eq. 24.6) between analyzed images captured by two sensor planes is crucial (2T instead of T). This means that changes in instantaneous flow structures have the biggest differences, too, i.e., two images of two planes are less similar, and then angular velocity was calculated with lower accuracy. This is the limitation of the main assumption of all correlation methods calculating a velocity, which assumes that the flow structure is "frozen."

24.7 Future trends

The problem of studies the mixed processes by tomographic means remains still a significant challenge. The main problem is how to extract useful information based on images capturing by corresponding tomographic systems. The main difficulty is not the manner of obtaining the tomogram but the development of efficient processing algorithms that extract information about the mixing process. The known algorithms are based on the use of correlation techniques in the analysis of mixing processes. However, their common drawback is that correlation methods impose certain restrictions on the input data. Thus, for example, input data should be cross-correlated and have significant value changes. Also, there is a problem with recognizing the appropriative patterns of input signals, which can be analyzed by software. Most of the researchers do not focus on the correlability of analyzed data, although without this condition, the correlation method has no sense.

Also, the main problem is how to validate whether the obtained data relating to the mixing process correspond to reality. As mentioned in this chapter, one of the possible solutions is to carry out the verification procedure for a tomographic system by using physical phantoms allowing to model of studied flows. Another approach is the addition of materials like tracers and whose trajectories can be easily identifiable during the mixing. Comparing the results obtained by observing the movement of tracers to results obtained from the tomographic system, one can estimate the correctness of its performance. Developing such methodologies to estimate the correctness can be a main priority in future works. Perhaps it is worth developing some standard physical models, allowing different scientific groups to compare their results. Another important issue is the question of whether 3D tomography can be used to analyze the mixing processes. At present, there is no single answer to this question since such a tomography exists in laboratory versions only.

In the future, other innovative tomographic techniques can appear to provide information on the characterization mixing processes. One of them is hydraulic tomography

(Cho, Zhao, Thomson Neil, & Illman Walter, 2020), allowing the selection of flow parameters that can enhance reagent mixing. A promising approach is to combine industrial tomography with other methodology, for instance, applications of ERT and response surface methodology for the monitoring mixing of highly concentrated slurries (Kazemzadeh, Ein-Mozaffari, & Lohi, 2019).

References

Baker, D., & Sayre, C. (1974). Deacay of swirling turbulent flow of incompressible fluids in long pipes. *Flows — Its Measurement and Control in Science and Industry, 1*, 301—312.

Cho, M. S., Zhao, Z., Thomson Neil, R., & Illman Walter, A. (2020). Use of steady-state hydraulic tomography to inform the selection of a chaotic advection system. *Journal of Contaminant Hydrology, 229*, 103559. https://doi.org/10.1016/j.jconhyd.2019.103559

Ding, Y., Wang, Z., Wen, D., & Ghadiri, M. (2005). Solids motion in a gas-solid two-phase mixture flowing through a packed bed. In *4th World congress on industrial process tomography, Aizu, Japan*.

Du, B., Liang-Shih, F., Fei, W., & Warsito, W. (2002). Gas and solids mixing in a turbulent fluidized bed. *AIChE Journal, 48*(9), 1896—1909.

Dyakowski, T., Hale, J., Jaworski, A., White, N., Harris, N., Nowakowski, A., et al. (2003). Characterisation of heterogeneous mixtures by using thick-film ultrasonic transducers. In *3rd World congress on industrial process tomography, Banff, Canada*.

Etuke, E., & Bonnecaze, R. (1998). Measurement of angular velocities using electrical impedance tomography. *Journal of Flow Measurement and Instrumentation, 9*(3), 159—169.

Gutmark, E. J., Li, G., & Grinstein, F. (2002). Characterization of multiswirling flow. *Engineering Turbulence Modelling and Experiments, 5*, 873—884. https://doi.org/10.1016/b978-008044114-6/50084-3

Hanus, R., Zych, M., Mosorov, V., Golijanek-Jedrzejczyk, A., Jaszczur, M., & Andruszkiewicz, A. (2021). Evaluation of liquid-gas flow in pipeline using gamma-ray absorption technique and advanced signal processing. *Metrology and Measurement Systems, 28*(1). https://doi.org/10.24425/mms.2021.13599

Hristov, H., Mann, R., Lossev, V., Vlaev, S. D., & Seichter, P. (2001). A 3-D analysis of gas-liquid mixing, mass transfer and Bioreaction in a stirred bio-reactor. *Food and Bioproducts Processing, 79*, 232—241. https://doi.org/10.1205/096030801753252306

Isaksen, Ø. (1996). A review of reconstruction techniques for capacitance tomography. *Measurement Science and Technology, 7*(3), 325.

Jawarneh, A. M., Tlilan, H., Al-Shyyab, A., & Ababneh, A. (2008). Strongly swirling flows in a cylindrical separator. *Minerals Engineering, 21*, 366—372. https://doi.org/10.1016/j.mineng.2007.10.012

Jegatheeswaran, S., Ein-Mozaffari, F., & Wu, J. (2018). Laminar mixing of non-Newtonian fluids in static mixers: Process intensification perspective. *Reviews in Chemical Engineering, v36*(3), 423—436.

Kagoshima, M., & Mann, R. (2005). Tomographic visualisation of semi-batch precipitation with multizone mixing model validation. In *4th World congress on industrial process tomography, Aizu, Japan* (pp. 788—793).

Kazemzadeh, A., Ein-Mozaffari, F., & Lohi, A. (2019). Mixing of highly concentrated slurries of large particles: Applications of electrical resistance tomography (ERT) and response surface methodology (RSM). *Chemical Engineering Research and Design, 143*, 226–240.

Kourunen, J., Heikkinen, L. M., Vauhkonen, M., Käyhkö, R., Matula, J., & Käyhkö, J. (2007). Imaging of mixing of two miscible liquids using electrical impedance tomography. In *5th World congress on industrial process tomography, Bergen, Norway*.

Kraft, J., Launikonis, A. L., & Swiegers, G. F. (Aug 22, 2005). *Method of monitoring and controlling of mixing processes CA 2619702 A1*.

Laurent, B. F. C., Bridgwater, J., & Parker, D. J. (2001). Structure of transaxial flow in a powder mixer. In *2nd World congress on industrial process tomography, Hannover, Germany* (pp. 346–352).

Li, H., & Tomita, Y. (2001). Characterization of pressure fluctuation in swirling gas–solid two-phase flow in a horizontal pipe. *Advanced Powder Technology, 12*, 169–185. https://doi.org/10.1163/15685520052385005

Li, Y., Yang, W., Xie, C., Huang, S., Wu, Z., & Tsamakis, D. (2013). Gas/oil/water flow measurement by electrical capacitance tomography. *Measurement Science and Technology, 24*, 074001. https://doi.org/10.1088/0957-0233/24/7/074001

Liu, S., Chen, Q., Wang, H. G., Jiang, F., Ismail, I., & Yang, W. Q. (2005). Electrical capacitance tomography for gas–solids flow measurement for circulating fluidized beds. *Flow Measurement and Instrumentation, 16*, 135–144. https://doi.org/10.1016/j.flowmeasinst.2005.02.013

Lomtschera, A., Jobsta, K., Fogela, S., Rostalski, K., Stempinc, S., & Kraume, M. (2017). Scale-up of mixing processes of highly concentrated suspensions using electrical resistance tomography flow. *Measurement and Instrumentation, 53*, 56–66.

Lucas, G. P., Cory, J., Waterfall, R., Loh, W. W., & Dickin, F. J. (1999). Measurement of the solids volume fraction and velocity distributions in solids-liquid flows using dual-plane electrical resistance tomography. *Journal of Flow Measurement and Instrumentation, 10*(4), 249–258.

Mosorov, V. (2006). A method of transit time measurement using twin-plane electrical tomography. *Measurement Science and Technology, 17*, 753–760. https://doi.org/10.1088/0957-0233/17/4/022

Mosorov, V. (2013). An iterative position reconstruction algorithm for radioactive particle techniques. *Applied Radiation and Isotopes, 79*, 56–61.

Mosorov, V., Sankowski, D., Mazurkiewicz, L., & Dyakowski, T. (2002). The best-correlated pixels method for solid mass flow measurements using electrical capacitance tomography. *Measurement Science and Technology, 13*, 1810–1814. https://doi.org/10.1088/0957-0233/13/12/302

Orbay, R. C., Nogenmyr, K. J., Klingmann, J., & Bai, X. S. (2013). Swirling turbulent flows in a combustion chamber with and without heat release. *Fuel, 104*, 133–146. https://doi.org/10.1016/j.fuel.2012.09.023

Parker, J., Kenyon, R. V., & Troxel, D. E. (1983). Comparison of interpolating methods for image resampling. *IEEE Transactions on Medical Imaging, 2*(1), 31.

Plaskowski, A., Beck, M. S., Thorn, R., & Dyakowski, T. (1995). *Imaging industrial flows: Applications of electrical process tomography*. Bristol: Institute of Physics Publishing.

Rantanen, J., Wikstro, H., Turner, R., & Taylor, L. S. (2005). Use of in-line near-infrared spectroscopy in combination with chemometrics for improved understanding of pharmaceutical processes. *Analytical Chemistry, 77*, 556–563.

Rimpiläinen, V., Poutiainen, S., Heikkinen, L. M., Savolainen, T., Vauhkonen, M., & Ketolainen, J. (2010). Monitoring of high shear mixing and granulation with capacitive measurements and tomography. In *6th World congress on industrial process tomography, Beijing, China.*

Simmons, M., Edwards, I., Hall, J., Fan, X., Parker, D. J., & Stitt, E. (2009). Techniques for visualisation of cavern boundaries in opaque industrial mixing systems. *AIChE Journal, 55*(11), 2765–2772.

Wang, M., & Yin, W. (2001). Measurements of the angular velocity distribution in miscible liquid mixing using electrical resistance tomography. In *2nd World congress on industrial process tomography, Hannover, Germany.*

Warsito, W., & Fan, L.-S. (2003). 3D-ECT Velocimetry for flow structure quantification of gas-liquid-solid fluidized beds. *The Canadian Journal of Chemical Engineering, 81*, 875–884. https://doi.org/10.1002/cjce.5450810372

West, R. M., Meng, S., Aykroyd, R. G., & Williams, R. A. (2003). Spatial-temporal modelling for electrical impedance imaging of a mixing process. In *3rd World congress on industrial process tomography, Banff, Canada* (pp. 226–232).

Windows-Yule, C. R. K., Hart-Villamil, R., Ridout, T., Kokalova, T., & Nogueira-Filho, J. C. (2020). Positron emission particle tracking for liquid-solid mixing in stirred tanks. *Chemical Engineering & Technology, 43*. https://doi.org/10.1002/ceat.202000177

Yang, W., & Liu, S. (2000). Role of tomography in gas/solids flow measurement. *Flow Measurement and Instrumentation, 11*, 237–244. https://doi.org/10.1016/s0955-5986(00)00023-6

Yang, W. Q., & Peng, L. (2002). Image reconstruction algorithms for electrical capacitance tomography. *Measurement Science and Technology, 14*, R1–R13. https://doi.org/10.1088/0957-0233/14/1/201

Further reading

Ashwood, A. C., Vanden Hogen, S. J., Rodarte, M. A., Kopplin, C. R., Rodríguez, D. J., Hurlburt, E. T., et al. (2014). A multiphase, micro-scale PIV measurement technique for liquid film velocity measurements in annular two-phase flow. *International Journal of Multiphase Flow, 67*, 200.

Baker, R. C. (2000). *Flow measurement handbook: Industrial designs, operating principles, performance, and applications*. Cambridge: Cambridge University Press.

Banasiak, R., Wajman, R., Jaworski, T., Fiderek, P., Fidos, H., Nowakowski, J., et al. (2014). Study on two-phase flow regime visualization and identification using 3D electrical capacitance tomography and fuzzy-logic classification. *International Journal of Multiphase Flow, 58*, 1–14.

Giguère, R., Fradette, L., Mignon, D., & Tanguy, P. A. (2009). Analysis of slurry flow regimes downstream of a pipe bend. *Chemical Engineering Research and Design, 87*, 943–950. https://doi.org/10.1016/j.cherd.2009.01.005

Grudzien, K., Chaniecki, Z., Romanowski, A., Niedostatkiewicz, M., & Sankowski, D. (2012). ECT image analysis methods for shear zone measurements during silo discharging process. *Chinese Journal of Chemical Engineering, 20*, 337.

Grudzien, K., Romanowski, A., Chaniecki, Z., Niedostatkiewicz, M., & Sankowski, D. (2010). Description of the silo flow and bulk solid pulsation detection using ECT. *Flow Measurement and Instrumentation, 21*(3), 198.

Hammer, E., Johansen, G. A., Dyakowski, T., Roberts, E. P., Cullivan, J., & Williams, R. (2005). Advanced experimental techniques. In *Multiphase flow handbook*. Boca Raton, FL: Taylor & Francis.

Hartigan, J. A., & Wong, M. A. (1979). A k-means clustering algorithm. *Applied Statistics, 28*(1), 100−108.

Huang, Z., Wang, B., & Li, H. (2003). Application of electrical capacitance tomography to the void fraction measurement of two-phase flow. *IEEE Transactions on Instrumentation and Measurement, 52*, 7−12. https://doi.org/10.1109/tim.2003.809087

Huang, S., & Xie, C. (2013). Ultrasound imaging and measurement of swirling gas-liquid multiphase flow. In *7th World congress on industrial process tomography, WCIPT7, Krakow, Poland* (pp. 1−9).

Jaworski, A. J., & Dyakowski, T. (2001). Tomographic measurements of solids mass flow in dense pneumatic conveying. What do we need to know about the flow Physics?. In *2nd World congress on industrial process tomography. Hannover, Germany* (pp. 353−361).

Krzanowski, W. J. (2000). *Principles of multivariate analysis: A user's perspective*. Oxford: Oxford University Press.

Mosorov, V. (2008). Flow pattern tracing for mass flow rate measurement in pneumatic conveying using twin plane electrical capacitance tomography. *Particle & Particle Systems Characterization, 25*(3), 259.

Mosorov, V. (2015). Applications of tomography in reaction engineering (mixing process). In M. Wang (Ed.), *Industrial tomography: Systems and applications* (1st Edition, pp. 509−528). Elsevier Woodhead Publishing.

Mosorov, V., Sankowski, D., Grudzien, K., & Romanowski, A. (2004). Characterizing solids mixing process within a plug body. In *Bulletin De la Societe Des Sciences Et Des Letters De Lodz, Poland, 2004* (pp. 45−56).

Neuber, D. (2014). Computed tomography for industrial process control. *Chemical Engineering World, 49*(12), 42.

Rodriguez R.N., Tobias R.D. Multivariate methods for process knowledge discovery: The power to know your Process. SAS Institute Inc., Cary, NC. paper 252-26. http://www.sas.com.

Saoud, A., Mosorov, V., & Grudzien, K. (2017). Measurement of velocity of gas/solid swirl flow using Electrical Capacitance Tomography and cross correlation technique. *Flow Measurement and Instrumentation, 53*(1), 133.

Scott, D. M., & McCann, H. (Eds.). (2005). *Process imaging for automatic control*. New York, NY: Marcel Dekker.

Tabe, H. T., Simons, S. J. R., Savery, J., West, R. M., & Williams, R. A. (April 14−17, 1999). Modelling of multiphase processes using tomographic data for optimisation and control. In *1st World congress on industrial process tomography. Buxton. Greater Manchester* (pp. 84−89).

Virdung, T., & Rasmuson, A. (2007). Measurements of continuous phase velocities in solid−liquid flow at elevated concentrations in a stirred vessel using LDV. *Chemical Engineering Research and Design, 85*(2), 193.

Wang, M. (Ed.). (2015). *Industrial tomography: Systems and applications* (p. 744). Elsevier.

Ye, M., & Haralick, R. M. (1999). *Image flow estimation using facet model and covariance propagation. Technical report*. Seattle: University of Washington.

Applications of electrical capacitance tomography in industrial systems

Benjamin Straiton[1], *Shah M. Chowdhury*[2], *Qussai Marashdeh*[1], *Fernando L. Teixeira*[2], *Liang-Shih Fan*[2] *and Aining Wang*[2]
[1]Tech4Imaging LLC, Columbus, OH, United States; [2]The Ohio State University, Columbus, OH, United States

25.1 Introduction

25.1.1 Basic requirements for ECT system

ECT technology requires only a few basic components to operate properly—a passive sensor or electrode array, a data acquisition system (DAS), and software or embedded system which can process the data into an image or extract other metrics.

25.1.1.1 Sensor

ECT sensors are very simple, requiring only a plurality of electrically conductive electrodes between which a capacitance can be detected when excited by the DAS. This simple requirement allows a nearly infinite set of sensor designs which may be optimal for different sensing scenarios. For example, electrodes may be of different shapes such as rectangular, triangular, circular, or helical. These shapes can be placed around the sensing region in different patterns, numbers, and sizes. The electrodes can be built around a pipe with a circular cross-section or a rectangular cross-section. They can also be built around complex shapes such as the tapered bottom of a hopper or a pipe elbow. Each of these shapes would affect how the electric field is distributed throughout the sensing region. Some configurations can simplify the inverse problem to solve for the image, while others can make it more complex. The more uniform the electric field created when the plates are excited, the simpler the inverse problem is to solve. The electric field between an excite and detect plate is most uniform when the plates are large flat surfaces which emulate a parallel plate capacitor. Deviations from this model such as curvature, size, shape, and position of the plates increase the complexity of the problem.

In addition to determining the complexity of the problem to be solved, the sensor design also contributes to the limitations the system will have in regard to resolution and signal-to-noise ratio (SNR). In ECT, larger electrode plates lead to stronger capacitance (signal) and lower noise as long as the distance between the plates remains

Industrial Tomography. https://doi.org/10.1016/B978-0-12-823015-2.00001-7

fixed or increases proportionally less than the area of the plates. The trade-off to increasing SNR with larger plates is that larger plates reduce the spatial sensitivity of the sensor, thus reducing the overall resolution of the final image.

For these reasons, the classical configuration for an ECT sensor is one row of 12 rectangular plates distributed evenly around the sensing region circumference. Conventional wisdom in the ECT community seems to dictate that this is an optimal trade-off between resolution, SNR, and problem complexity. Some of these inherent limitations to ECT, however, were later overcome by fundamental advancements in the field of ECT such as Electrical Capacitance Volume Tomography (ECVT) for measuring 3D space despite increased complexity and Adaptive ECT for increasing resolution without compromising SNR.

25.1.1.2 Data acquisition system

The DAS is a critical component of the ECT system and can often be the limiting component of resolution and stability. The DAS can introduce major drift and noise into the data if it does not properly handle thermal energy dissipation.

The classical ECT DAS follows one of two designs—charge/discharge circuit or AC circuit. In both designs, the standing capacitances of the circuitry tend to be very large, while the changes in capacitance being measured in the sensing region tend to be quite small. For this reason, the standing capacitances of the circuitry are typically canceled by DC offset or AC feedback. Advanced self-balancing techniques have also been developed to increase the dynamic range of the circuitry by eliminating the standing capacitances automatically.

25.1.1.3 Software

Software is required to take the data from the DAS and solve the inverse problem to generate an image. There are many ways to solve the inverse problems such as Linear Back Projection, Singular Value Decomposition, algebraic methods, iterative methods, and neural network methods. ECT image reconstruction is a large field of research unto itself with the goal of overcoming the inherent spatial resolution limitations of ECT without introducing more or different measurements from the hardware. This is different than the Adaptive ECT approach which takes more information from the hardware to more accurately reconstruct the image.

Many of these reconstruction techniques have an inverse relationship between resolution and processing time, which can make a small increase in resolution require a large increase in computational requirement.

25.1.2 Fundamental advantages of ECT in industrial application

ECT has many inherent qualities that make it ideal for industrial application. Recent advancements have proven the ability of this technology to perform reliably in some of the harshest industrial environments.

25.1.2.1 Passive sensor

Because the sensors are passive, all electronics can be placed away from the harsh environment or in a separate controlled environment. This helps to lower the costs of the electronics and reduce typical issues of placing electronics in hot or corrosive environments.

The simplicity of ECT sensors also allows them to be readily adapted into any environment by simply choosing a suitable electrode material and a dielectric material to separate the electrodes from the flow, the outside environment, and each other. For example, Inconel plates can be placed inside a refractory for use in a reactor application which requires up to 1000°C internal temperatures.

25.1.2.2 Noninvasive

The noninvasive nature of ECT allows interrogation of the region of interest without disturbing the process. In addition to allowing more accurate measurement of the true process, this also helps to prevent build up of material on sensing components that would otherwise be required to be inserted into the region of interest.

25.1.2.3 Low power

Because ECT works in a low frequency range (a few kHz to a few MHz), the power demand is relatively low and can often be measured on the order of 10 s of Watts. The low power requirement allows the device to be battery or solar powered in remoted environments. The low power aspect also shows the inherent safety of the technique compared to high powered alternatives such as X-ray and gamma-ray.

25.1.2.4 High speed

Although ECT traditionally has a low spatial resolution, it makes up for this in temporal resolution. Depending on the DAS design, the acquisition rate can reach hundreds or even thousands of frames per second. This is important for imaging very fast moving industrial flows such as pneumatic conveying in which solid particles can reach velocities of over 30 m per second and gas velocities much higher.

25.1.2.5 Cost effective

Due to the low power requirement and the simplicity of the sensors, the system is generally very cost effective, typically only costing a small fraction of hard field tomographic systems such as magnetic resonance imaging (MRI).

25.1.2.6 Compact

The sensors can be extremely compact because they only require electrically conductive surfaces for sensing. This means that the electrodes can be printed on flexible or hard printed circuit boards or even painted on a surface with conductive paint.

The electronics can also be designed to fit into a relatively small space. This makes it ideal for situations in which weight and size are costly such as low earth orbit and deep space applications.

25.1.3 Adapting ECT for industrial applications

With all of the advantages ECT provides for industrial application, it is possible to adapt the technology for use in the harsh conditions of industrial processes including extreme temperatures, high pressures, and vacuum pressures, corrosive environments such as sulfuric acid and steam, radiation hardening for space travel, deep sea conditions, and explosive environments. At the forefront of ECT technological advancements is the packaging of these sensors and electronics to operate in these harsh environments.

Additionally, ECT has been adapted to overcome traditional limitations such as working with saline water and three phase applications.

What follows is an exploration of some of the most recent work in advancing ECT to overcome these limitations and apply the benefits of the technology to a wide variety of industries.

25.2 Two-phase gas–solid systems

25.2.1 Fluidized bed reactor

25.2.1.1 FB background

Fluidization is the process of stationary solid particles brought into a dynamic "fluid-like" state by an upward stream of fluid (gas or liquid) (Fan & Zhu, 2005). The basic components required for a fluidized bed (FB) are a container, a gas distributor, solid powder, and gas. When a gas flow is introduced into the bottom of a bed packed with solid particles, the gas will move upwards through the gaps between the particles. When the gas velocity is low, the drag force on each individual particle is small, and the bed remains in a packed configuration. When the gas velocity exceeds a certain threshold velocity U_{mf} (where mf denotes "minimum fluidization"), the upward aerodynamic drag force becomes large enough to counteract the particle gravity minus buoyancy. In this case, the particles will move away from each other and become suspended within the fluid, causing the bed to expand in volume and exhibit some fluid behavior. As the gas velocity is further increased, the bed will keep expanding and the bulk density will keep dropping, and finally the particles cannot form a bed anymore and will be blown away by the gas.

This process is used extensively as a reactor system in industry. Some of the common uses are described below:

1. Catalytic gas phase reactions. Solid particles serve as catalyst and gas phases react inside the reactor. Solid catalyst materials often combine favorable activity and selectivity

characteristics along with acceptable physical properties such as size, density, and rigidity. The most important application for such kind of reactions is the fluidized catalytic cracking (FCC) process used in petroleum refineries. This process converts the high-molecular weight hydrocarbon fractions of crude oils into gasoline and other products.

2. Gas—solid reactions. In this type of reaction, fluidized particles react with gas to provide useful products such as thermal energy or other types of gas and solids. Typical applications include ore-roasting, which was one of the first usages of fluidized beds; fluidized bed combustion, which can burn coal at lower temperature compared to traditional pulverized coal combustion, units thus reducing nitrogen oxide emissions; and coal gasification process, which is used for producing "syngas" composed primarily of CO and H_2.
3. Other noncatalytic processes. In these processes, the solid particles are inert. The solids can be used as uniform heat medium, or nuclei of product vapor. And very often, the fluidized beds can be used for drying the particulate materials, as well as for agglomeration and granulation of solid particles. These processes are widely used in food and pharmaceutical industries.

25.2.1.2 Tomography techniques for FB

Since gas—solid fluidized beds are very important for industry, many experimental techniques have been developed for such beds. For example, tracers can be used to study the solids residence time; pressure transducers and pilot tubes are employed to monitor pressure fluctuations; and thermal couples are irreplaceable for temperature controlling. In this chapter, those of interest are tomography techniques. Photography and tomography are the main noninvasive techniques used to investigate bubble behaviors, particle motions, and phase distributions in fluidized beds. Below listed are several such types of experimental techniques:

1. Photography. Used in 2D or 3D columns with the aid of transparent reactor walls and external light sources; good for observing near-wall behaviors but difficult to clearly see through the inside of the column.
2. X-ray, gamma-ray, and positron emission tomography. Accurate methods yielding high resolution, but with very high capital costs, and potential environmental and health hazards.
3. Magnetic resonance imaging (MRI). High resolution, but the imaging scale is very limited and also has a high capital cost. Relatively slow acquisition times can also be an issue.
4. Electrical tomography techniques, including electrical capacitance tomography (ECT), electrical impedance tomography (EIT), and electromagnetic inductance tomography (EMT). ECT monitors the permittivity distributions of nonconductive phases; EIT is useful for the reactor with conductive phases inside; and EMT is mainly applied for high conductive fluids that could induce measurable current under a magnetic field, which is not suitable for gas—solid systems. Electrical tomography techniques have been much used in recent years because of their low cost, high imaging speeds, noninvasive design, and safe nature. The most significant drawback for the tomography system is its relatively low spatial resolution. Traditionally, the most advanced electrical tomography system can have a linear resolution of about 3% of the reactor's characteristic length. However, recent advancements in Spatially Adaptive technology have improved this. Because of the advantages of ECT for gas—solids systems, ECT will be the focus of our discussion that follows.

25.2.1.3 ECT verification in FB

The first investigations on capacitance measuring techniques can be traced back to the 1980s in the Morgantown Energy Technology Center of the US Department of Energy (Halow & Nicoletti, 1992). Starting from the early 1990s, ECT systems have been used to investigate multiphase flow phenomena (Halow, Fasching, Nicoletti, & Spenik, 1993; Huang, Plaskowski, Xie, & Beck, 1989). Since then, more advanced ECT systems have been applied in fluidization phenomenon research (Makkawi & Wright, 2004). Besides traditional 2D ECT systems, advanced 3D ECT systems, ECVT systems, and Adaptive ECT systems have been devised and implemented to further expand the capabilities of capacitance tomography.

For any novel measurement technique, the first and maybe the most important feature is the reliability of the technology. For fluidization phenomena, by directly comparing the measured results between ECT systems and some traditional experimental methods, it has been verified that the ECT can be successfully used to measure fluidization phenomena.

In 2003, the results from ECT were compared with those of a fiber optic probe in an operating fluidized bed (Pugsley et al., 2003). Fig. 25.1 shows the diagram of the circulating fluidized bed (CFB) system used in that experiment. The riser is 7 m in height with an inner diameter (ID) of 0.14 m. The CFB riser consists of several interchangeable flanged Plexiglas sections 1.6 m above the riser inlet. One section is equipped with an ECT sensor and the other has an optical fiber probe mounted. By carefully controlling the air and solid flux, flow conditions can be reestablished for comparison. FCC catalyst is used as the solid phase with an average diameter of 89 mm. The particle density is 1550 kg/m^3. The twin-plane ECT system used in the experiment consisted of eight electrode plates in each layer and an iterative linear back-projection algorithm was employed to perform the image reconstruction from the ECT data.

Fig. 25.2 shows the time-averaged radial voidage profiles from both ECT and fiber optic probe measurements under various solid flux conditions. The differences between the two curves from ECT and optical fiber probe are smaller than 10%. In the center core region of the riser, the difference is less than 1%, becoming progressively larger near the wall region. This may be explained by the nonuniformity of the sensitivity distribution and the bias of boundary effect in the ECT system. Basically, the results show good agreement between the ECT data and optical fiber probe results.

Recently, the ECT system was compared with an X-ray tomography system on a fluidized bed with wide particle size distributions (Rautenbach, Mudde, Yang, Melaaen, & Halvorsen, 2013). They compared bubble volume and found good agreement. In 2012, the ECVT system was compared with an MRI tomography system on a spouting bed which also showed satisfactory agreement between the two techniques (Chandrasekera et al., 2012).

All these results corroborate the fact that ECT technique is a reliable experimental technique for fluidization investigations.

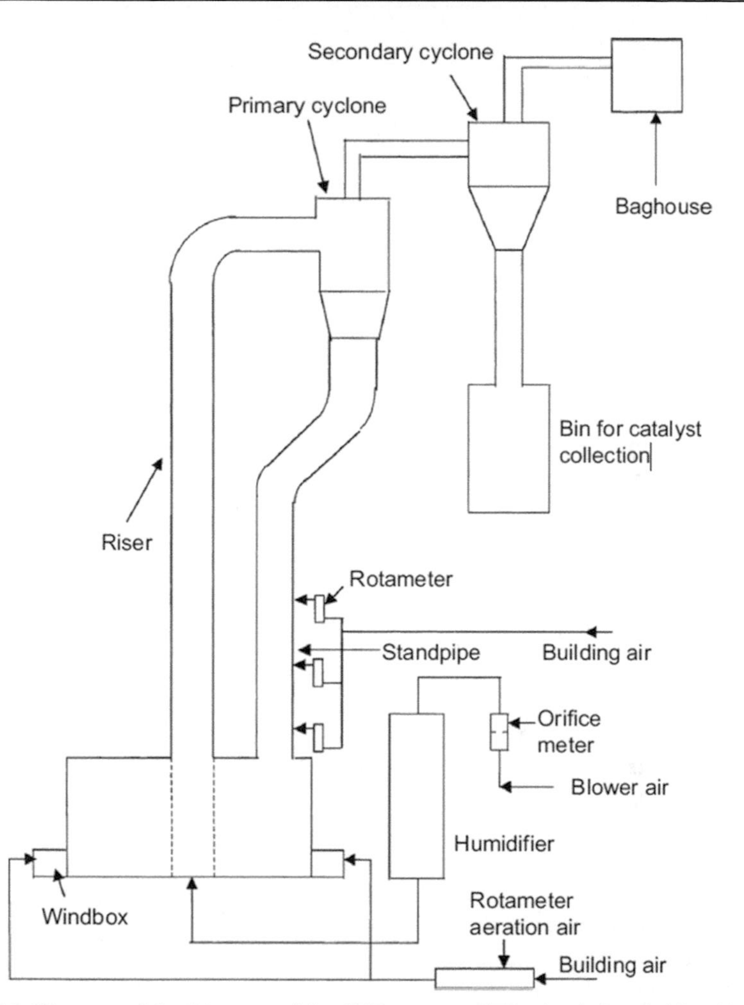

Figure 25.1 Diagram of the Diagram of the CFB system. CFB, circulating fluidized bed. Figure reproduced from Pugsley, T., Tanfara, H., Malcus, S., Cui, H., Chaouki, J., & Winters, C. (2003). Verification of fluidized bed electrical capacitance tomography measurements with a fibre optic probe. *Chemical Engineering Science*, 3923–3934, with permission from Elsevier.

25.2.1.4 ECT application in circulating FB

A gas–solid CFB is a device in which particles are carried upwards by gas in a riser and then separated by a cyclone to return back to the bottom through a downer, forming a recirculation loop for the particles. Major commercial applications of CFBs include fluid catalytic cracking, Fischer–Tropsch synthesis, coal combustion and gasification, and ore processing.

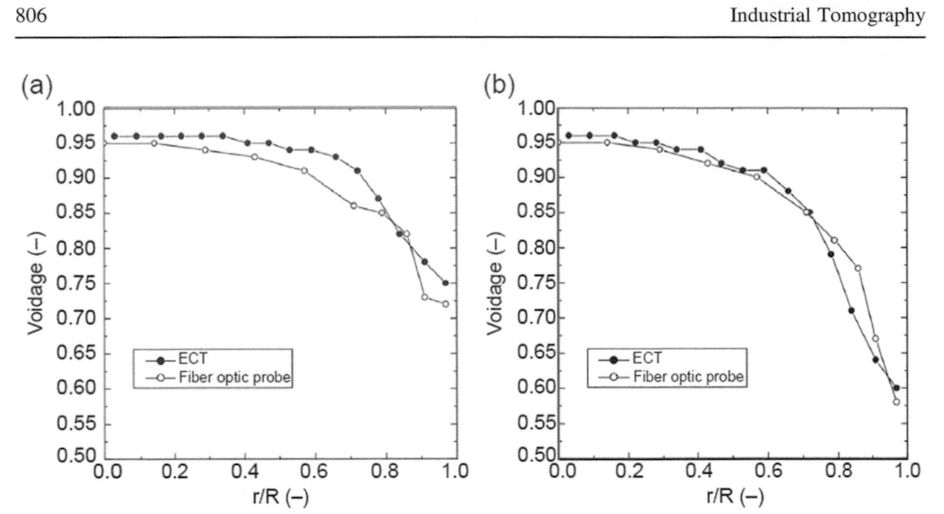

Figure 25.2 Comparison of radial voidage profiles measured with ECT and fiber optic probe in the CFB riser containing FCC catalyst: (A) riser solids mass flux = 148 kg/(m^{2}s), superficial gas velocity = 4.7 m/s; (B) riser solids mass flux = 264 kg/(m^{2}s), superficial gas velocity = 4.7 m/s. *CFB*, circulating fluidized bed; *ECT*, electrical capacitance tomography; *FCC*, fluidized catalytic cracking.
Figure reproduced from Pugsley, T., Tanfara, H., Malcus, S., Cui, H., Chaouki, J., & Winters, C. (2003). Verification of fluidized bed electrical capacitance tomography measurements with a fibre optic probe. *Chemical Engineering Science*, 3923–3934, with permission from Elsevier.

ECT techniques can be used to image flow behaviors and solids distributions in CFBs. Traditional ECT systems have all electrodes placed in one plane and can generate cross-sectional slice images. On the other hand, fully volumetric images can be directly produced using ECVT with several longitudinal layers of sensor plates interacting simultaneously (Wang, Marashdeh, Fan, & Warsito, 2010). Researchers in the Ohio State University have constructed a CFB cold model with ECVT sensors mounted around it. In this section, some experimental results will be discussed to illustrate the application of ECVT to CFBs (Wang, Marashdeh, Wang, & Fan, 2012).

Fig. 25.3 shows the schematic diagram of the CFB system. The Plexiglas CFB consists of a 5 cm ID riser with a height of 2.6 m, a 90° bend exit, a cyclone, a 3.8 cm ID downer, and an L-valve. A porous plate with a mean pore size of 20 mm and 60% fractional free area is used as the gas distributor. FCC particles (Geldart group A, mean diameter 60 mm, density 1400 kg/m^3) are used as the solid phase and the fluidization gas is air. A cylinder and a bend shape ECVT sensor are installed at the riser and bend exit region of the CFB, respectively. Fig. 25.4 is an actual photograph of the CFB system and the ECVT sensors.

From ECVT sensor I, the solids distribution images in the riser are obtained as shown in Fig. 25.5, where time-averaged solids distribution in the riser under different gas superficial velocities can be visualized.

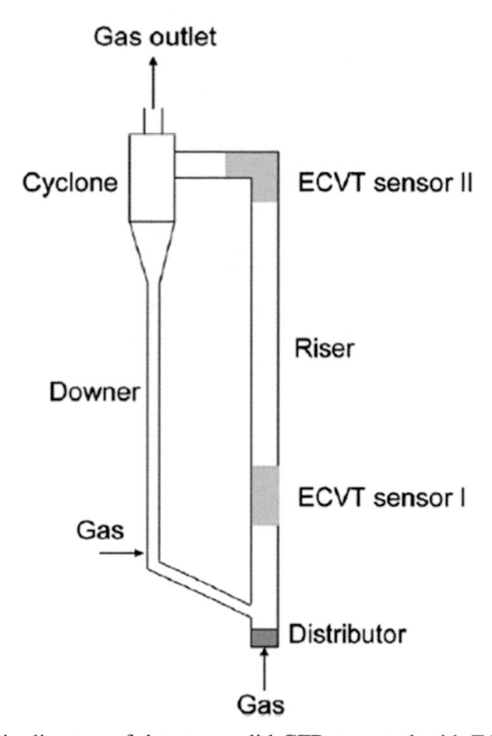

Figure 25.3 Schematic diagram of the gas–solid CFB mounted with ECVT sensors. CFB, circulating fluidized bed; ECVT, electrical capacitance volume tomography.
Figure reproduced from Wang, F., Marashdeh, Q., Wang, A., & Fan, L. S. (2012). Electrical capacitance volume tomography imaging of three-dimensional flow structures and solids concentration distributions in a riser and a bend of a gas-solid circulating fluidized bed. *Industrial & Engineering Chemistry Research*, 10968–10976 with permission from American Chemical Society.

The left figure in each image is the solids distribution along the axial direction, whereas the right image represents three horizontal cross-sectional distributions. The results clearly indicate the core-annular shape lean phase flow in the riser. Most solids are distributed near the wall. In the center of the riser, gas is the dominating phase. The thickness of the annulus and the concentration in the annulus near the wall decrease with the superficial fluidization gas velocity.

The ECVT sensor II mounted at the riser exit is used to monitor the solids distribution at the bend section connecting the riser and the cyclone. Fig. 25.6 illustrates the theoretical prediction that solids move upward through the riser with the fluidization gas, and that, in the bend exit, most solids will turn their moving directions into horizontal and go into the cyclone. Due to the high upwards gas velocity, some solids will be carried to the top corner of the bend and fall back into the riser. Some solids in the horizontal pipe will fall down by gravity and a solids dune will appear as indicated.

Figure 25.4 Photograph of the 0.05 m ID gas–solid CFB mounted with ECVT sensors: (A) CFB; (B) ECVT bend sensor; (C) ECVT riser sensor. *CFB*, circulating fluidized bed; *ECVT*, electrical capacitance volume tomography; *ID*, inner diameter.
Figure reproduced from Wang, F., Marashdeh, Q., Wang, A., & Fan, L. S. (2012). Electrical capacitance volume tomography imaging of three-dimensional flow structures and solids concentration distributions in a riser and a bend of a gas-solid circulating fluidized bed. *Industrial & Engineering Chemistry Research*, 10968–10976 with permission from American Chemical Society.

Fig. 25.7 shows the solids distributions under different gas velocities at the riser bend exit from ECVT sensor II. The solids dune and the entrained solids at the top corner of the bend can be clearly seen from the images. Fig. 25.8 shows the quantitative results of the solids concentrations at the corner of the bend. The solids concentration will increase with the superficial gas velocities.

From these results, the solids distributions in the CFB can be observed directly, which can offer a better understanding of the flow behavior inside the CFB system. The solids distribution information from the bend riser exit section can be used to improve the bending design to reduce the erosion and attrition for the reactor wall as well as the solid particles.

25.2.2 High temperature FB

Many fluidized bed processes such as FCC and Fischer–Tropsch take place at elevated temperatures which can exceed 1000°C. ECVT has been applied to one such experiment to determine how the fluidization behavior changes for iron-oxide particles as temperature increases (Wang et al., 2018). The experiment focused on the relationship between gas velocity and slug frequency at temperatures up to 650°C. The

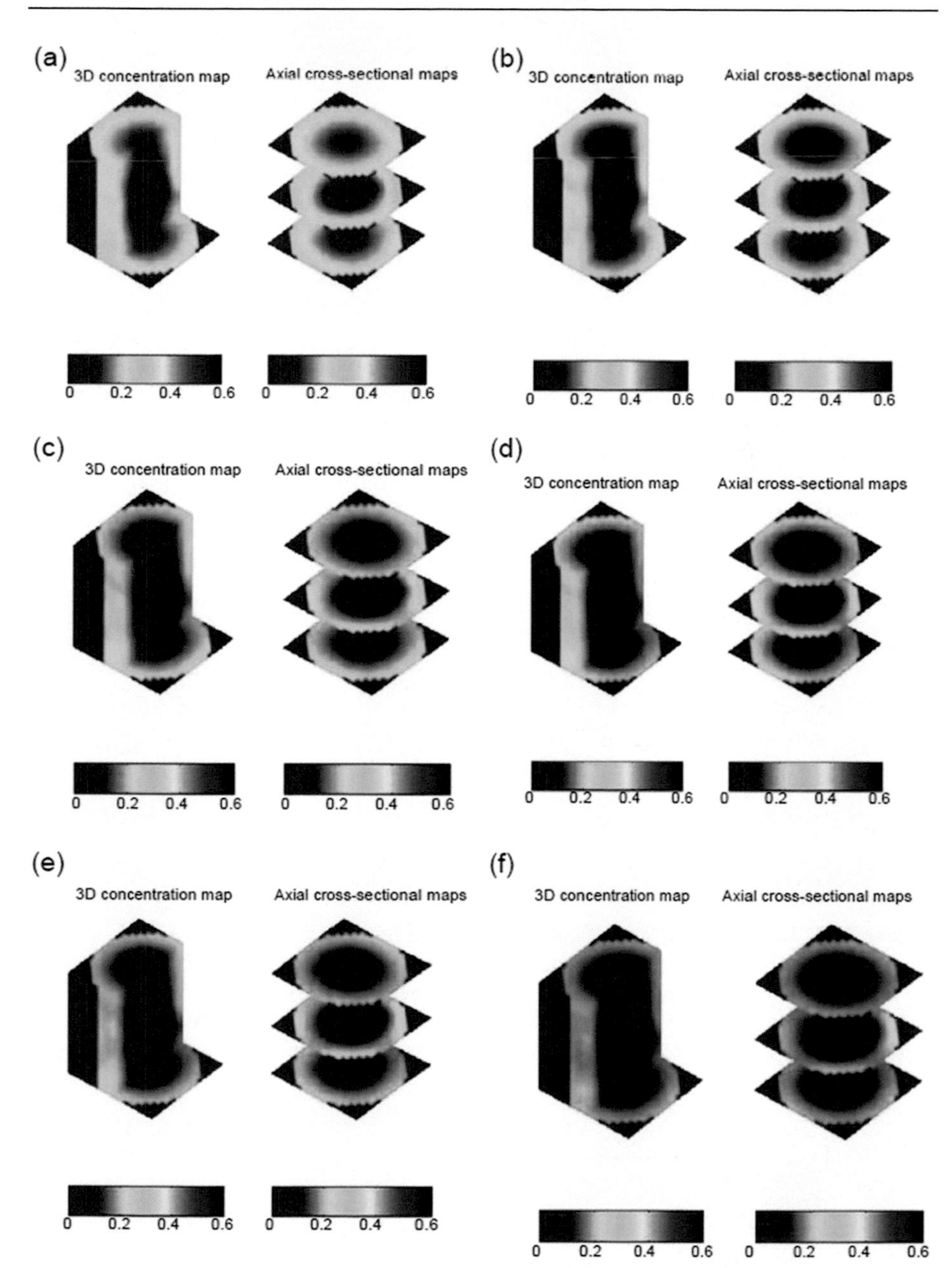

Figure 25.5 Time-averaged solids holdup distribution in the riser at Ug ¼ (A) 0.97; (B) 1.16; (C) 1.36; (D) 1.55; (E) 1.75; (F) 1.94 m/s.
Figure reproduced from Wang, F., Marashdeh, Q., Wang, A., & Fan, L. S. (2012). Electrical capacitance volume tomography imaging of three-dimensional flow structures and solids concentration distributions in a riser and a bend of a gas-solid circulating fluidized bed. *Industrial & Engineering Chemistry Research*, 10968–10976 with permission from American Chemical Society.

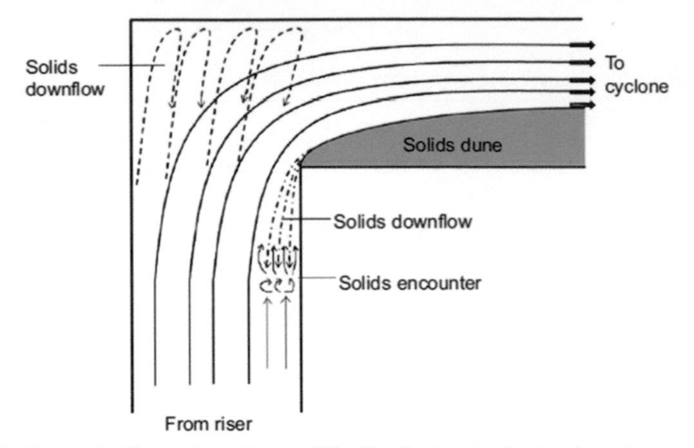

Figure 25.6 Theoretic illustration of the solids distribution in the bend section.
Figure reproduced from Wang, F., Marashdeh, Q., Wang, A., & Fan, L. S. (2012). Electrical capacitance volume tomography imaging of three-dimensional flow structures and solids concentration distributions in a riser and a bend of a gas-solid circulating fluidized bed. *Industrial & Engineering Chemistry Research*, 10968–10976 with permission from American Chemical Society.

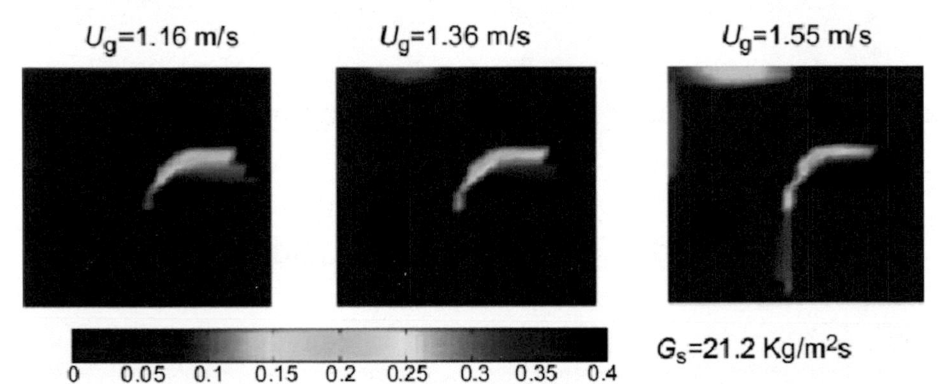

Figure 25.7 Time-averaged solids holdup distribution in the axial plane of the bend.
Figure reproduced from Wang, F., Marashdeh, Q., Wang, A., & Fan, L. S. (2012). Electrical capacitance volume tomography imaging of three-dimensional flow structures and solids concentration distributions in a riser and a bend of a gas-solid circulating fluidized bed. *Industrial & Engineering Chemistry Research*, 10968–10976 with permission from American Chemical Society.

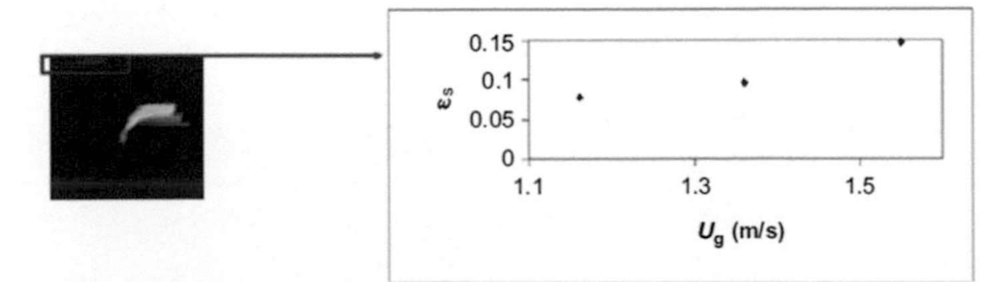

Figure 25.8 Quantitative time-averaged solids holdup distribution in the axial plane of the bend at Gs ¼ 21.2 kg/(m^{2}s): top wall area of the horizontal section.
Figure reproduced from Wang, F., Marashdeh, Q., Wang, A., & Fan, L. S. (2012). Electrical capacitance volume tomography imaging of three-dimensional flow structures and solids concentration distributions in a riser and a bend of a gas-solid circulating fluidized bed. *Industrial & Engineering Chemistry Research*, 10968–10976 with permission from American Chemical Society.

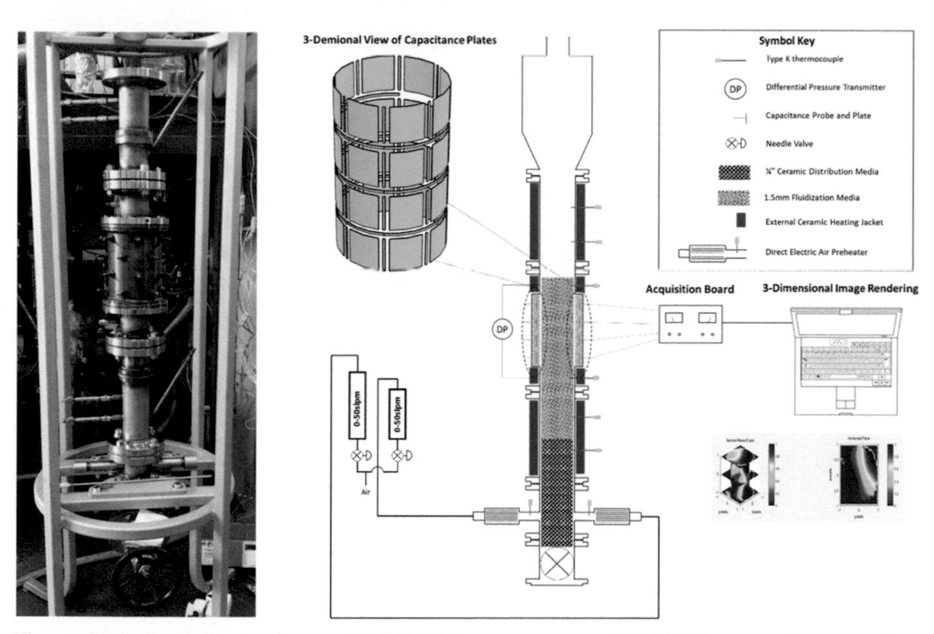

Figure 25.9 (Left) Photo of assembled ECVT test apparatus. (Right) Diagram of process flow vessel and accessories of the ECVT system.
Figure reproduced from Wang, D., Xu, M., Marashdeh, Q., Straiton, B., Tong, A., & Fan, L.-S. (2018). Electrical capacitance volume tomography for characterization of gas–solid slugging fluidization with geldart group D particles under high temperatures. *Industrial and Engineering Chemistry Research*, 2687–2697 with permission from.

experimental set up is shown in Fig. 25.9. An ECVT sensor with four layers of six electrodes was used. The electrodes for the ECVT sensor were embedded in a refractory and placed inline to the fluidized bed using flanges. The ID of the system was 76.2 mm with a height of 0.76 m. The column in use was relatively short to ensure that the particles in the column would be uniformly heated. The system was heated with ceramic semicylindrical heaters. The experiment involved varying the superficial gas velocity between 0.25 and 2.1 m/s above Umf and the temperature between 25°C and 650°C. Over 300 tests were performed in this study.

Pressure drop measurements were taken to verify ECVT measurements. As shown in Fig. 25.10, the measurements track very closely between the two forms of measurement. Low pressure drop corresponded to low volume fraction and high pressure drop corresponded to high volume fraction. The cyclical behavior is indicative of the slugging behavior of the bed. Increasing the temperature increased the Umf and thus the increased slugging frequency as well. Both the pressure drop and ECVT captured this behavior.

Fig. 25.11 shows an image reconstruction of a single slug in the sensing region and the time series volume fraction measurement for each of four layers in the ECVT sensor. The slight delay in signal from the bottom region to the top region shows the time progression of the slug moving from the bottom to the top of the sensor.

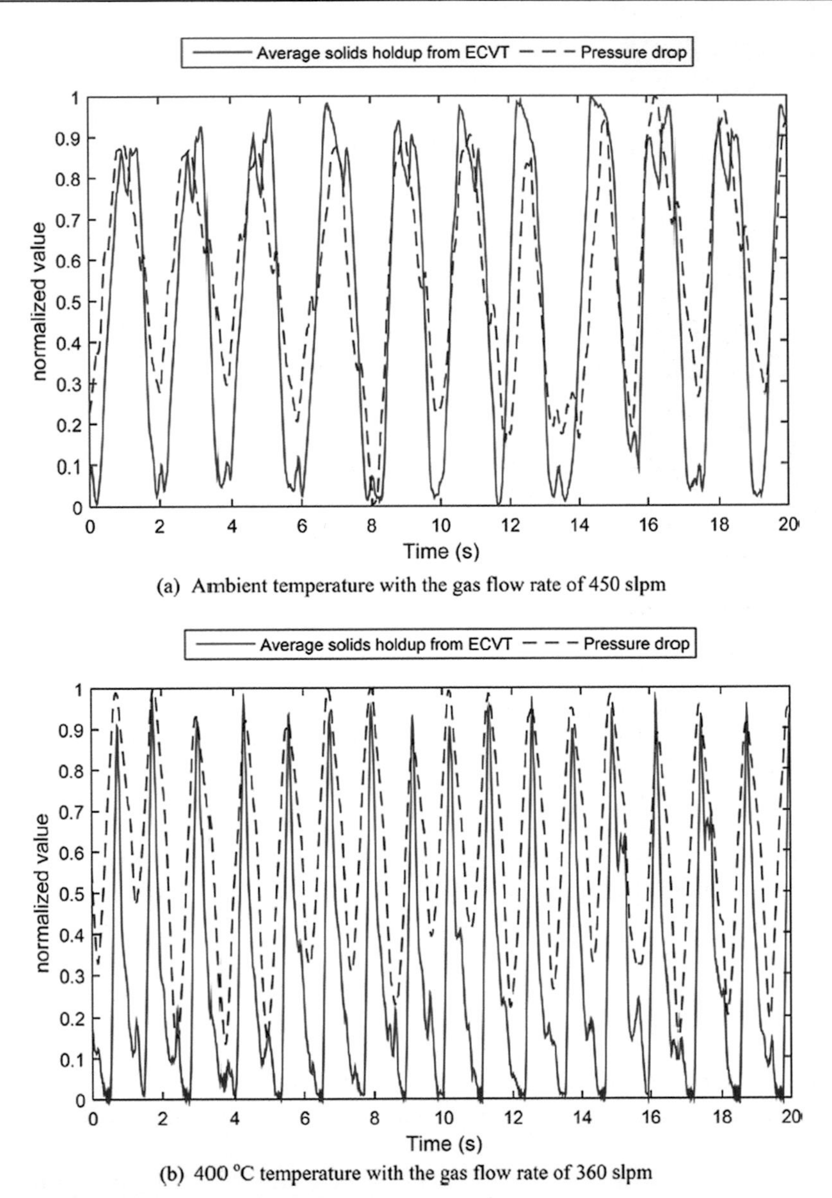

Figure 25.10 Comparison of ECVT measurement and pressure drop profile obtained from pressure transducers.
Figure reproduced from Wang, D., Xu, M., Marashdeh, Q., Straiton, B., Tong, A., & Fan, L.-S. (2018). Electrical capacitance volume tomography for characterization of gas–solid slugging fluidization with geldart group D particles under high temperatures. *Industrial and Engineering Chemistry Research*, 2687–2697 with permission from.

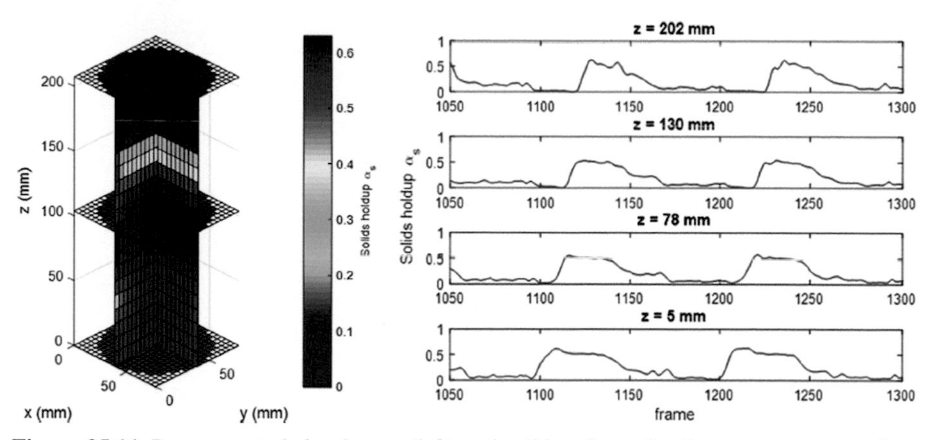

Figure 25.11 Reconstructed slug image (left) and solids volume fraction measurements along the height of the sensor region with frames (right). Solids holdup acquisition frame rate is 80 Hz.
Figure reproduced from Wang, D., Xu, M., Marashdeh, Q., Straiton, B., Tong, A., & Fan, L.-S. (2018). Electrical capacitance volume tomography for characterization of gas–solid slugging fluidization with geldart group D particles under high temperatures. *Industrial and Engineering Chemistry Research*, 2687–2697 with permission from.

The time intervals between successive slugs were measured at different temperatures. Fig. 25.12 shows that a similar slugging interval is evident across all temperatures tested for each superficial gas velocity.

Rising bubble velocity can be obtained using ECVT by investigating the time lag of the signal between successive layers in the sensor. In all temperatures tested, the bubble velocity consistently exhibited three regions of behavior depending on the superficial gas velocity as shown in Fig. 25.13.

These results show the usefulness of ECT technology practically applied to a high temperature industrial application. Important information such as bubble velocity, slugging frequency, and solids holdup can all be obtained in real time. All of this information is useful in optimizing design or operation of such an industrial process.

25.3 Two-phase air–water systems

Gas–liquid bubble column reactors are widely used in industry, especially in petrochemical and biochemical processes. The performance of a bubble column reactor is dictated by the dynamics of the multiphase flow inside the reactor column. Many efforts to observe and quantify the behavior have been made including application of computational fluid dynamics, pitot tubes, flywheel anemometer, heat-pulse

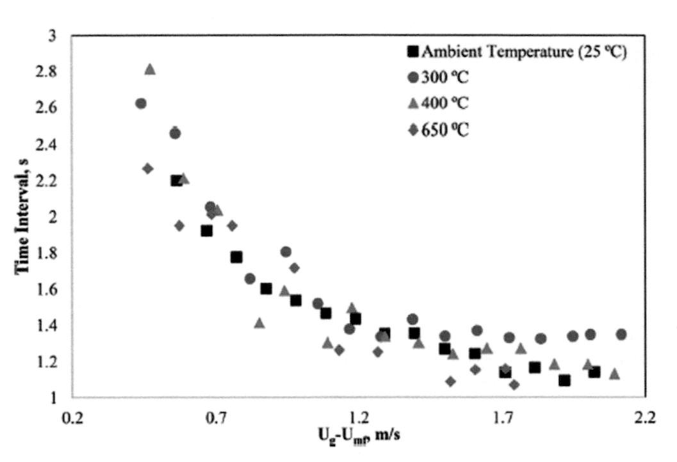

Figure 25.12 Measured time interval between slugs at varying Ug−Umf for 25, 300, 400, and 650 °C.
Figure reproduced from Wang, D., Xu, M., Marashdeh, Q., Straiton, B., Tong, A., & Fan, L.-S. (2018). Electrical capacitance volume tomography for characterization of gas−solid slugging fluidization with geldart group D particles under high temperatures. *Industrial and Engineering Chemistry Research*, 2687−2697 with permission from.

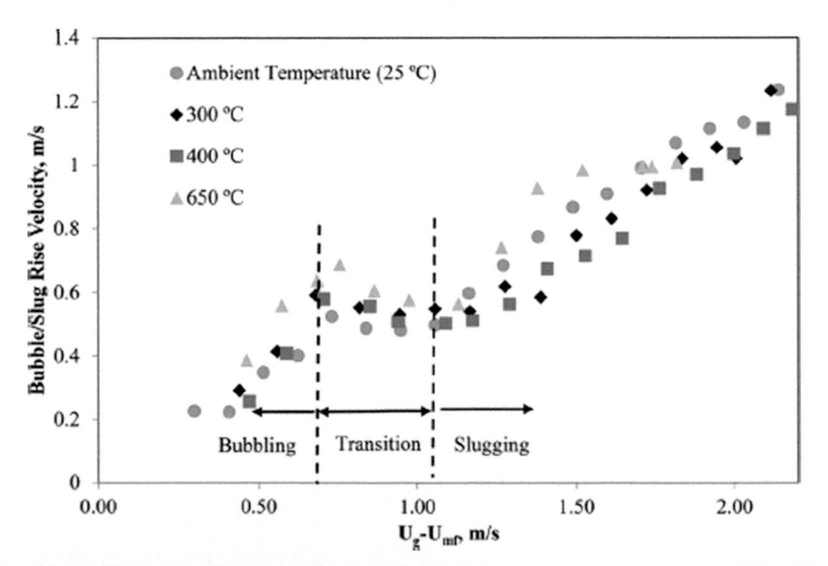

Figure 25.13 Slug velocities at varying Ug−Umf for various temperatures, 25, 300, 400, and 650 °C.
Figure reproduced from Wang, D., Xu, M., Marashdeh, Q., Straiton, B., Tong, A., & Fan, L.-S. (2018). Electrical capacitance volume tomography for characterization of gas−solid slugging fluidization with geldart group D particles under high temperatures. *Industrial and Engineering Chemistry Research*, 2687−2697 with permission from.

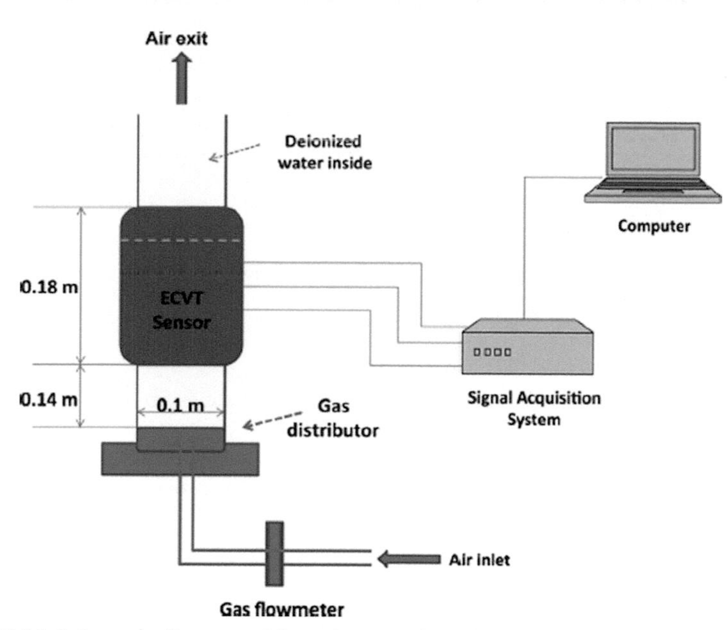

Figure 25.14 Schematic diagram of the column and the location of the ECVT sensor. Figure reproduced from Wang, Marashdeh, & Fan, ECVT Imaging of 3D Spiral Bubble Plume Structures in Gas-Liquid Bubble Columns, 2014 with permission from.

anemometry, hot-film anemometry, optical probes, radioactive particle tracking, ERT, EIT, and ECT. In this section, ECT measurement techniques are explored in application to bubble column reactors, particularly air–water systems.

In one study, an ECVT sensor was used with four layers of six electrodes to examine the bubble flow characteristics of a spiraling bubble plume (Wang, Marashdeh, & Fan, ECVT Imaging of 3D Spiral Bubble Plume Structures in Gas-Liquid Bubble Columns, 2014). Fig. 25.14 shows the experimental set up. The distributor in one portion of the experiment was swapped out for a tapered cylinder which forced the air to the center of the column and created the spiraling behavior. The gas holdup was measured using ECVT and verified by measuring the difference in height of the column from static to bubbling. Fig. 25.15 shows the gas holdup results from ECVT compared to the visual method across different superficial gas velocities.

The gas holdup was also measured across the radius of the vessel and verified with optical probe measurements as shown in Fig. 25.16. Results largely agree within 5%, with more disagreement toward the edge of the vessel.

In this study, when the tapered cylinder was placed inside the column to concentrate the air in the center, a spiraling motion was observed as shown in Fig. 25.17. This snapshot shows the asymmetrical behavior of the plume. Additionally, as shown in Fig. 25.18, the frequency of the spiraling motion of the plume can be obtained and mapped according to the superficial gas velocity.

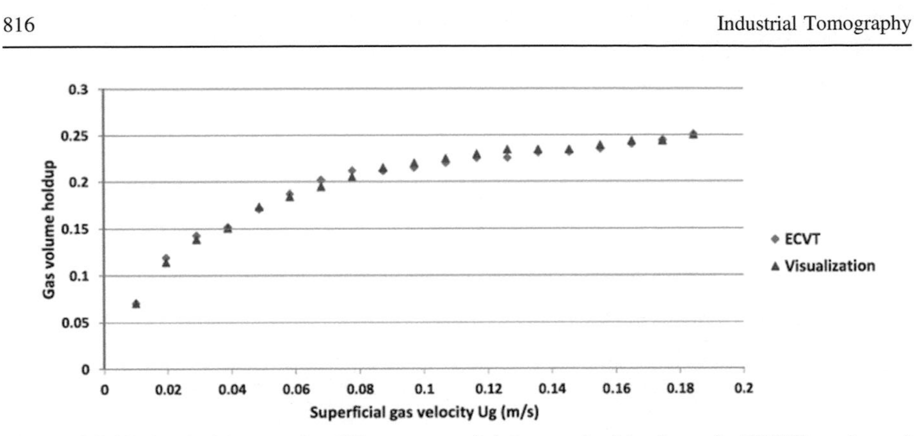

Figure 25.15 Gas holdups under different superficial gas velocities from the ECVT results and the visualization results.
Figure reproduced from Wang, Marashdeh, & Fan, ECVT Imaging of 3D Spiral Bubble Plume Structures in Gas-Liquid Bubble Columns, 2014 with permission from.

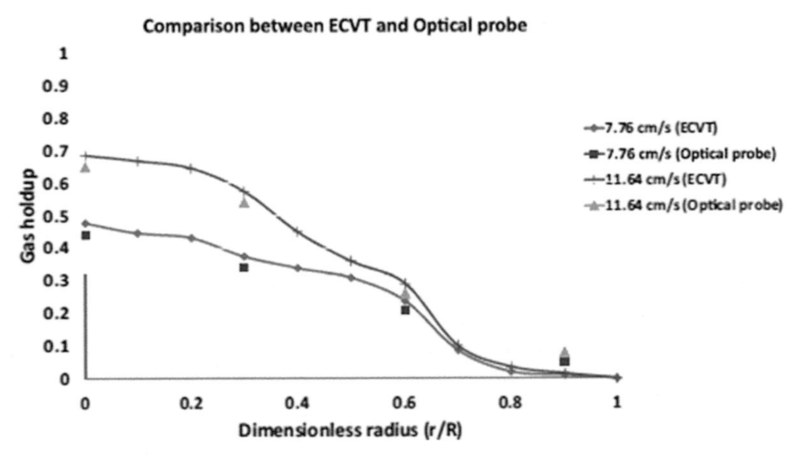

Figure 25.16 Radius gas holdup results from ECVT and Optical fiber probe.
Figure reproduced from Wang, Marashdeh, & Fan, ECVT Imaging of 3D Spiral Bubble Plume Structures in Gas-Liquid Bubble Columns, 2014 with permission from.

25.4 Three-phase systems

25.4.1 Trickle bed reactor

A trickle bed reactor is an important three-phase system for petroleum, biochemical, electrochemical, and water treatment industries in which gas and liquid flow simultaneously through a fixed solid medium. Understanding the complex flow behavior inside this system is vital for optimizing and advancing the trickle bed reactor. Several techniques have been employed for observing this behavior such as MRI and wire mesh. This section will focus on the application of ECT.

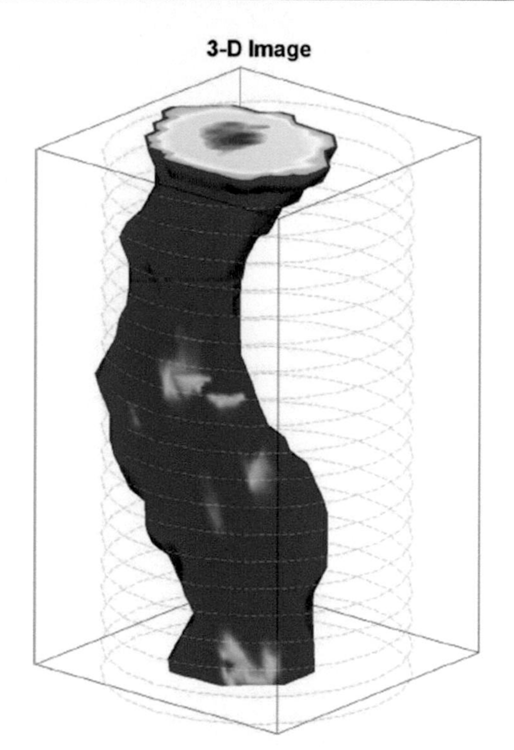

Figure 25.17 A typical 3D spiral locus for a bubble cluster in a cylinder bubble column (Ug¼ 4 cm/s).
Figure reproduced from Wang, Marashdeh, & Fan, ECVT Imaging of 3D Spiral Bubble Plume Structures in Gas-Liquid Bubble Columns, 2014 with permission from.

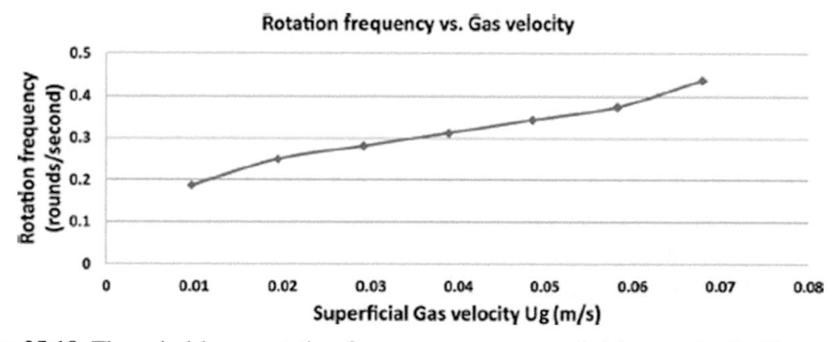

Figure 25.18 The spiral locus rotation frequency versus superficial gas velocity Ug.
Figure reproduced from Wang, Marashdeh, & Fan, ECVT Imaging of 3D Spiral Bubble Plume Structures in Gas-Liquid Bubble Columns, 2014 with permission from.

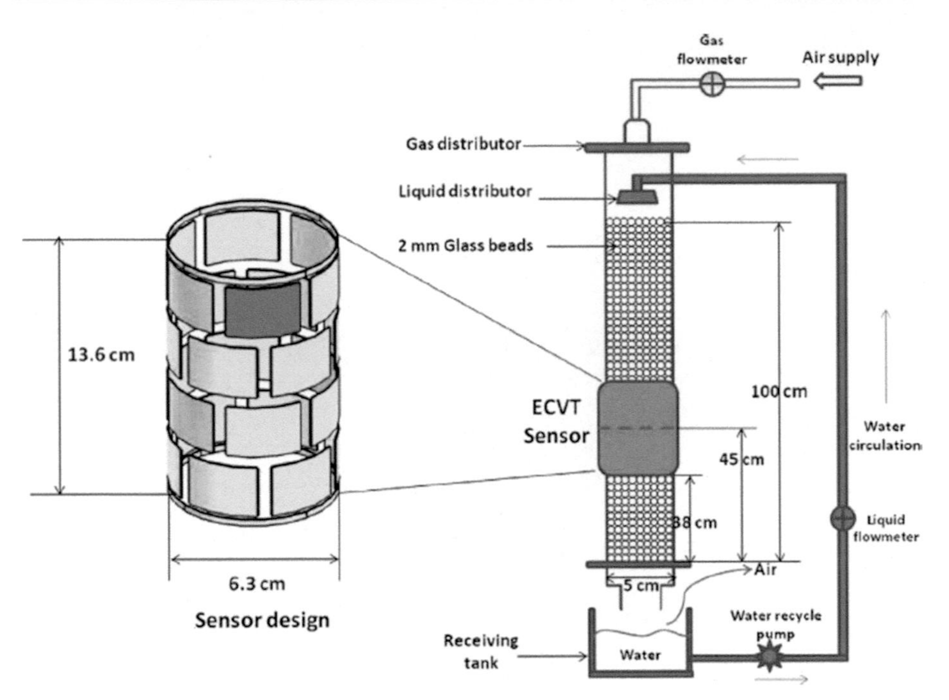

Figure 25.19 Schematic diagram of the experimental setup.
Figure reproduced from Wang, Marashdeh, Motil, & Fan, 2014 with permission from.

In one study, ECVT was applied to a trickle bed which utilized a packed bed of glass beads as the fixed solids medium and air and deionized water as the gas and liquid, respectively (Wang, Marashdeh, Motil, & Fan, 2014). The ECVT sensor consisted of four layers of six plates and was installed on the setup as shown in Fig. 25.19. The gas and water flow rates were set such that the flow regime would be pulsating between gas-rich and water-rich regions. Fig. 25.20 shows an image reconstruction of a pulse in two different viewing planes.

Data were extracted from the images such as gas and water holdup and pulse frequency as they relate to the flow rate of gas and water through the system. Fig. 25.21 shows the mapping of pulse frequency to air and water flow rate. Fig. 25.22 shows the frequency spectrum of the pulse signal for two different flow rates of water and air.

25.5 Future trends

Exciting advancements in recent years are expanding the useful applications from gas—solid to gas—water, and three-phase systems. Advancements in sensor design and manufacturing are lowering barriers to moving ECT technology from the lab into real industrial environments with extreme temperatures, pressures, and corrosive

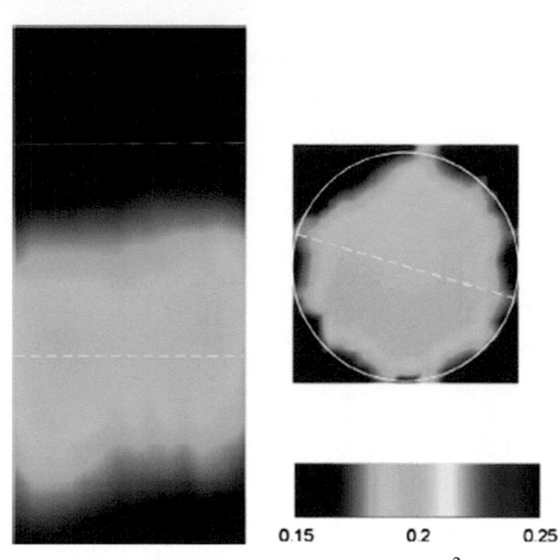

Figure 25.20 The image of a transient pulse (water: 21.7 kg/m^2 s; air: 0.61 kg/m^2 s). Figure reproduced from (Wang, Marashdeh, Motil, & Fan, 2014) with permission from.

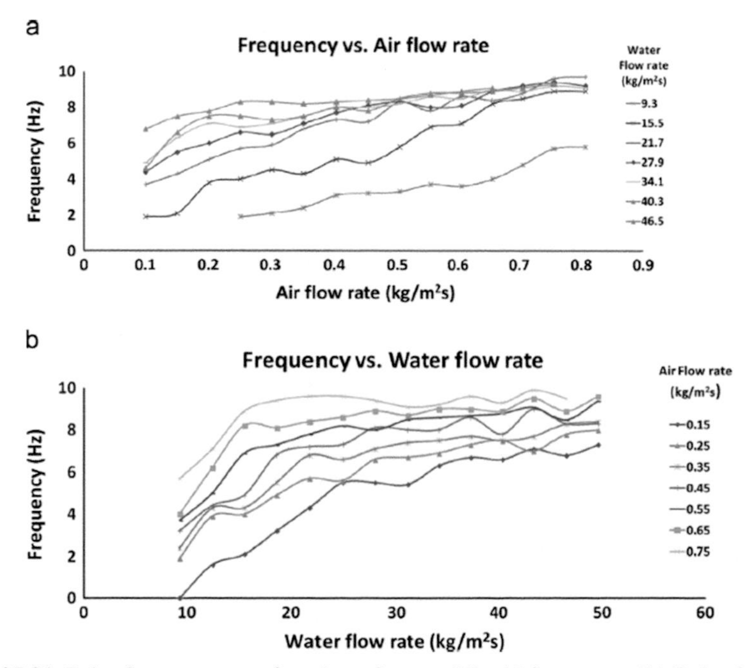

Figure 25.21 Pulse frequency as a function of gas and liquid flow rates. (A) Pulse frequency versus air flow rate and (B) pulse frequency versus water flow rate.
Figure reproduced from (Wang, Marashdeh, Motil, & Fan, 2014) with permission from.

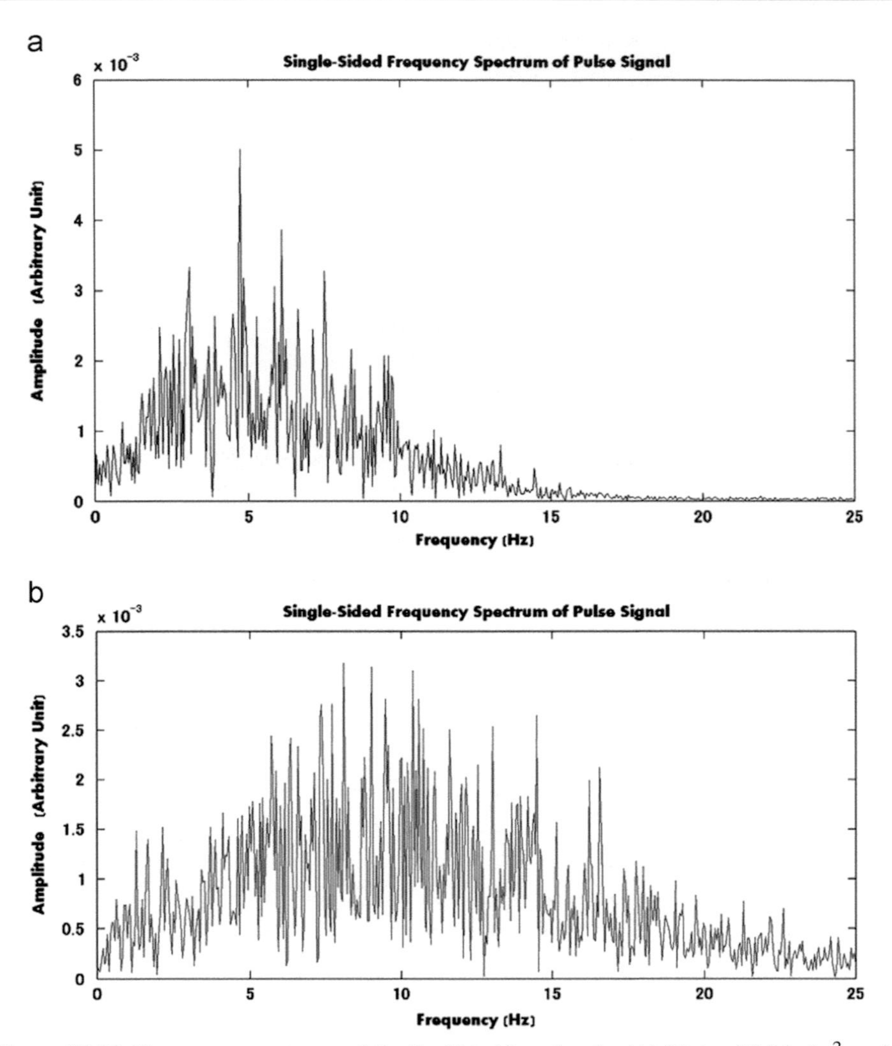

Figure 25.22 Frequency spectrums of the liquid holdup signals. (A) Water: 15.5 kg/m^2 s; air: 0.25 kg/m^2 s and (B) water: 46.5 kg/m^2 s; air: 0.76 kg/m^2 s.
Figure reproduced from (Wang A., Marashdeh, Motil, & Fan, 2014) with permission from.

phases. Additionally, resolution is increasing through adaptive style sensors and electronics. There is high demand in industry for real-time affordable sensors which can distinguish multiple phases in harsh conditions to automate and optimize large portions of transportation, conveying, processing, and production. Some industries looking to apply the advantages of ECT technology are geothermal power plants, oil and gas drilling, oil and gas production, and bulk solids transportation. ECT technology is advancing rapidly to address these needs.

25.6 Source of further information

For more information on surface polarization and electric field behavior in emulsions, see *Encyclopedia of Emulsion Technology, Volume 4*, edited by Daniel Schuster.

References

Chandrasekera, T. C., Wang, A., Holland, D. J., Marashdeh, Q., Pore, M., Wang, F., et al. (2012). A comparison of magnetic resonance imaging and electrical capacitance tomography: An air jet through a bed of particles. *Powder Technology*, 86–95.

Fan, L., & Zhu, C. (2005). *Principles of gasesolid flows*. Cambridge: Cambridge University Press.

Halow, J. S., Fasching, G. E., Nicoletti, P., & Spenik, J. L. (1993). Observations of a fluidized bed using capacitance imaging. *Chemical Engineering Science*, 643–659.

Halow, J., & Nicoletti, P. (1992). Observations of fluidized bed coalescence using capacitance imaging. *Powder Technology*, 255–277.

Huang, S. M., Plaskowski, A. B., Xie, C. G., & Beck, M. S. (1989). Tomographic imaging of two-component flow using capacitance sensors. *Journal of Physics E: Scientific Instruments, 173.*

Makkawi, Y. T., & Wright, P. C. (2004). Electrical capacitance tomography for conventional fluidized bed measurements—remarks on the measuring technique. *Powder Technology*, 142–157.

Pugsley, T., Tanfara, H., Malcus, S., Cui, H., Chaouki, J., & Winters, C. (2003). Verification of fluidized bed electrical capacitance tomography measurements with a fibre optic probe. *Chemical Engineering Science*, 3923–3934.

Rautenbach, C., Mudde, R. F., Yang, X., Melaaen, M. C., & Halvorsen, B. M. (2013). A comparative study between electrical capacitance tomography and time-resolved X-ray tomography. *Flow Measurement and Instrumentation*, 34–44.

Wang, A., Marashdeh, Q., & Fan, L.-S. (2014). ECVT imaging of 3D spiral bubble plume structures in gas-liquid bubble columns. *The Canadian Journal of Chemical Engineering*, 2078–2087.

Wang, F., Marashdeh, Q., Fan, L. S., & Warsito, W. (2010). Electrical capacitance volume tomography: Design and applications. *Sensors*, 1890–1917.

Wang, A., Marashdeh, Q., Motil, B. J., & Fan, L.-S. (2014). Electrical capacitance volume tomography for imaging of pulsating flows in a trickle bed. *Chemical Engineering Science*, 77–87.

Wang, F., Marashdeh, Q., Wang, A., & Fan, L. S. (2012). Electrical capacitance volume tomography imaging of three-dimensional flow structures and solids concentration distributions in a riser and a bend of a gas-solid circulating fluidized bed. *Industrial & Engineering Chemistry Research*, 10968–10976.

Wang, D., Xu, M., Marashdeh, Q., Straiton, B., Tong, A., & Fan, L.-S. (2018). Electrical capacitance volume tomography for characterization of gas–solid slugging fluidization with geldart group D particles under high temperatures. *Industrial and Engineering Chemistry Research*, 2687–2697.

Applications of AI and possibilities for process control

Saba Mylvaganam
Department of Electrical Engineering, IT and Cybernetics, Faculty of Technology, Natural Sciences and Maritime Sciences, University of South-Eastern Norway, Campus Porsgrunn, Norway

26.1 Introduction

Different sensor modalities, e.g., resistive, capacitive, inductive, radiometric, optical, etc., are used in various control scenarios involving thousands of sensors in many applications. In selected control loops involving slowly varying process parameters, these sensors deliver data, which are processed using physics-based modeling and used in control of the processes.

From sensors feeding data to physics-based models, typical process parameters can be extracted yielding visual information. When the sensors generate visual information, as is the case with ultrasonic, X-ray, MR, or IR imaging, pixels are generated, providing visual information, which can also be used to interpret the process parameters needed for process control.

We build upon the basic concepts presented in Chapter 13 in this book on "Machine Learning Process Information from Tomography Data," dealing with four steps, viz. S1 − Sensing energy excitation of space/time material distribution of interest, S2 − Detection of corresponding response, S3 − Inversion of response data to estimate reconstructed material distribution, often as an image, and S4 − Interpretation. We see these steps as a sequence of operations and extend them with some additional steps in a closed loop for using enhanced insight into the system behavior based on the artificial intelligence gained for control purposes.

Generally, modern process engineers are dealing with probes (sensors!), parameters, and pixels and phenomena, the omnipresent four P's, as illustrated in Figs. 26.1 and 26.2. Visual information is useful for understanding the consequences of a given situation in the processes. Process-related parameters can be used to interpret the situation and automate the decision process using dedicated control loops. Visual information delivering the information in the form of pixels can also lead to estimation of parameters needed for process control.

Data fusion for enhanced process information has been studied in the process tomography earlier with more focus on hardware solutions (Hoyle et al., 2001; West & Williams, 1999). Process imaging for process control has been addressed in (Scott & McCann, 2005).

Industrial Tomography. https://doi.org/10.1016/B978-0-12-823015-2.00014-5

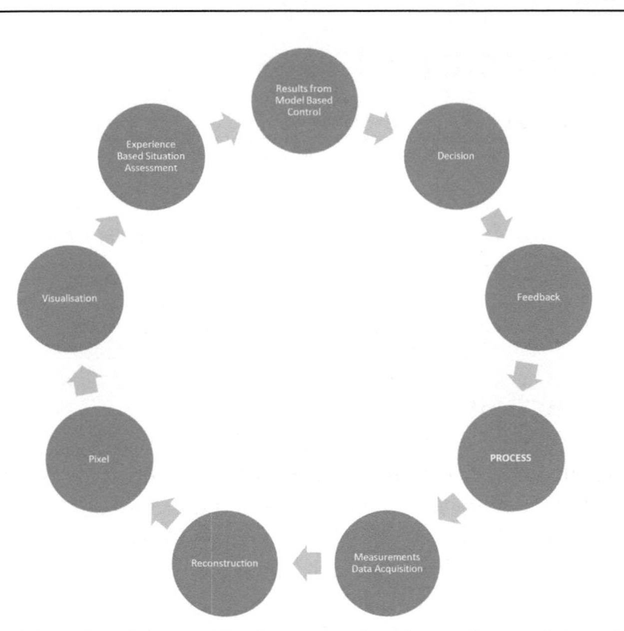

Figure 26.1 Pixel-based decision making in a control cycle, as discussed in the Introduction to this book.

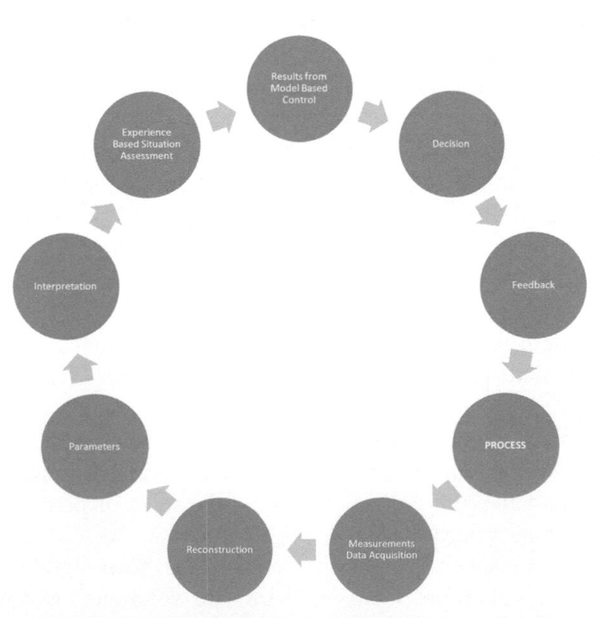

Figure 26.2 Parameters characteristic to the process from measurement data, as discussed in the Introduction to this book.

26.2 Artificial intelligence

With conventional sensors delivering continuously vast amount of process data and the arrays of sensors in process tomographic modules producing still more amount of data, there is a growing potential for the data hungry models using AI and Machine Learning (ML) algorithms.

AI techniques are based on emulating human decision-making processes, whereas ML uses algorithms to learn about the processes using vast amount of data. These are data hungry algorithms, deployed in process control scenario as schematically represented in Fig. 26.3. In the current developmental trend in the process industries, there is a big rush in implementing AI/ML in many stages running from the process floor to the top-management level.

First, principles and machine learning virtual flowmeters are discussed with many references addressing virtual flow metering (first principles VFM, data-driven VFM) and physical multiphase flow metering (MPFM) in (Bikmukhametov & Jäschke, 2020). Real-field applications of AI techniques involving neural networks of different types and architecture are also described in this paper. The approach presented in this chapter is a combination of MPFM and data-driven VFM.

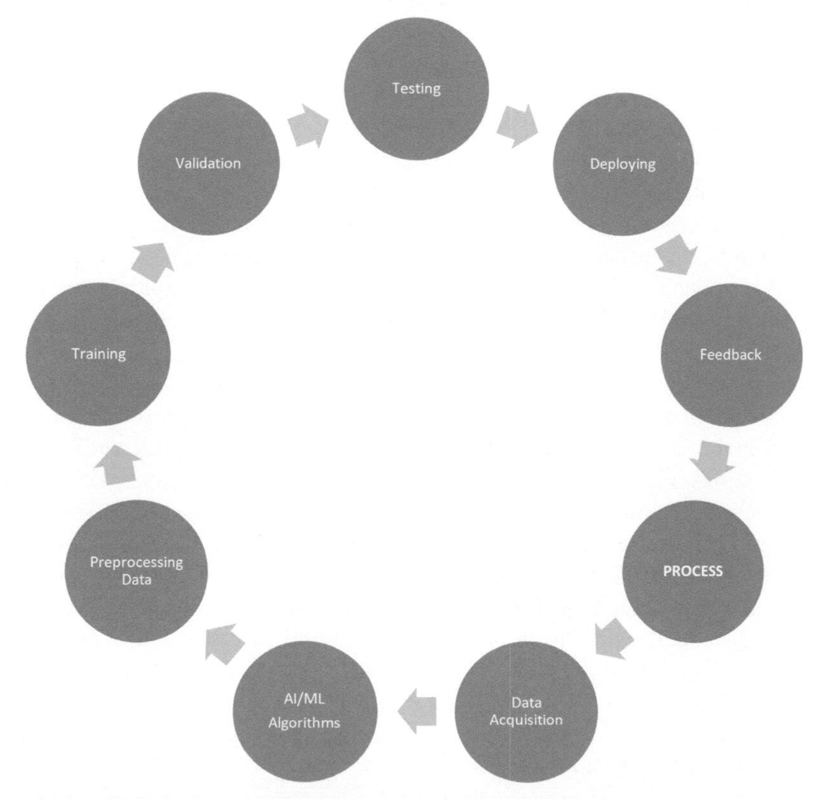

Figure 26.3 AI/ML in the control cycle involving data hungry algorithms with frequent "tweaking" based on the latest update of the data.

Process tomography/tomometry can assist the measurement and control engineers in the process industries with valuable data about the process using noninvasive and nonintrusive sensors and the fusion of data from these sensors. Electrical Impedance Tomography (EIT)/Tomometry (EITm), Electrical Capacitance Tomography (ECT)/Tomometry (ECTm), and Electrical Resistance Tomography (ERT)/Tomometry (ERTm) have been successfully applied in the identification of different flow regimes and material distributions in multiphase flow involving liquids and gases as well as air and particulates. Abnormal and dangerous flow conditions in multiphase flow processes involving unusual distribution of materials and their velocities in the process vessels or pipelines can in certain cases lead to hazardous conditions, which must be detected or preferably predicted with swift and suitable response to alleviate the incumbent disasters to personnel and property. Process tomography with the combination of sensor data fusion can lead to process operations with low maintenance costs and reduced security risks to personnel. Flow conditions in processes involving multiphase flows with liquids, gas, and particulates are flow regimes, volume fractions of the different phases, size of bubbles and the frequencies of their occurrences, size of slugs and the frequencies of their occurrences, etc. Estimation of these parameters on-line with nonintrusive sensing will help the process designers to develop sensor systems, data acquisition from these, and algorithms for alerting the process engineers, who can without delay adjust process parameters such as flow of the individual components of flow or even shut down the flow by deactivating the appropriate actuators such as pumps, valves, throttles, etc.

Simplified models use tell-tale values of few critical parameters, e.g., from a process tomographic module, as their inputs can help the process engineers to make decisions, to decide on suitable and swift sequence of actions. For making such decisions when the processes are running, measurement and/or soft sensor-based Model Free Adaptive Control (MFAC) algorithms are used to decide on suitable control actions (Chaminda, Yan, & Mylvaganam, 2012). These methods after undergoing multiple tests can be used to design MFAC algorithms which can be supplementary to conventional control algorithms.

Processes with multiphase flow involving mixture of liquids, gases, and particulates have very seldom a homogenous mixture of all these individual phases. High pressure variations involving slugs of oil/water and gas or air and particulates can lead to hazardous situations. Early warning of process parameters of flow regimes, sizes of slugs, etc., in a mix indicating impending danger, can help to avoid serious damages. The process tomographic systems and techniques described in this chapter show the possibilities of realizing MFAC for safer operation of processes with multiphase flows.

Combining conventional control algorithms with AI based on multimodal sensors and physical models can help to achieve reliable and real-time information of processes, which can help to prevent disasters like the Deepwater Horizon oil spill in the Mexican Gulf which led to loss of lives, properties, and long-term effects due to a combination of technical, management, and procedural problems still haunting the environment in a vast area around the platform, Fig. 26.4.

Figure 26.4 Mexican Gulf Deepwater Horizon Platform in flames – process failure leading to catastrophic accident - a major HSE incident with repercussions even today. Possibly a combination of technical, management and procedural malfunctioning and problems: Deepwater Horizon oil spill disaster with loss of lives, property, and environmental effects - avoidable with a combination of conventional control and AI based control, involving technical, management and procedural measures (Temming, 2020).
Photo Courtesy, Deepwater Horizon - Wikipedia, From Wikipedia: File: Deepwater Horizon offshore drilling unit on fire 2010.jpg - Wikimedia Commons. https://commons.wikimedia.org/wiki/File:Deepwater_Horizon_offshore_drilling_unit_on_fire_2010.jpg.

26.3 Multiphase flow processes for testing AI techniques

26.3.1 Multiphase rig for two-phase (air/water) flow

In many industrial processes, the materials flowing through pipes exist in various phases. Multiphase flow characterization is important in flow measurement studies and is a difficult task. The behaviors of multiphase flows are often categorized into different flow regimes. While some regimes are wanted, others can indicate flows that might be harmful, whether to process operation, infrastructure, or in extreme cases, safety.

It is hard to clearly define the transitions between the different flow regimes in multiphase flows. Also, there may be different ways to define their specific observed characteristics. However, we will use the following definitions:

- Stratified flow occurs when both liquid and gas flow rates are low. The two materials are totally separated and flow smoothly without noticeable ripples on the separation surface.
- As the gas flow rate increases, and big or small oscillations start forming on the surface, the flow is characterized as wavy.

- When the gas flow rate is increased exceedingly, moving away from wavy flow, a new flow regime called annular flow occurs.
- Then some mist is formed, and the liquid starts to coat/adhere to the walls. All these regimes can be categorized as continuous flow.
- When the liquid flow rate increases, uneven large waves are formed at various time intervals, filling the whole pipe cross-section, periodically changing, forming a flow regime categorized as intermittent flow.
- With intermittent flows with lower gas flow rates, the large waves in multiphase flows are called plugs.
- As the gas flow rate increases and the liquid bodies gather significant amount of gas bubbles, large waves called slugs are formed.

Inferential methods and soft sensing techniques using AI are useful in estimating process states, predicting process behavior, and measuring abstract conditions with relevant parameters not discernible with hard sensors. Recently, the use of such techniques has increasingly provided new avenues for measurement and control in the process industries, in which AI and machine learning have become part of the toolbox of the process engineers.

A simplified P&ID diagram of a multiphase flow loop for generating desired flow regimes with air, water, and oil as the phases in the flow is shown in Fig. 26.5, which shows the sensors and actuators relevant for monitoring and generating the relevant flow regimes.

Two Coriolis flow meters ("FT") with an uncertainty of ± 0.01 kg/min are located immediately before the ECT module, for measuring mass flow rates of air and water. Differential pressure transmitters ("PDT") are used for monitoring the pressure drop between the ends of the test section with the ECT module. The gamma radiation meter ("GD") estimates the phase fraction over the pipe cross-section based on the differing absorption coefficients of air, water, and oil. The control valves and pumps constitute the actuators in the system and adjust the amount of each phase to attain various

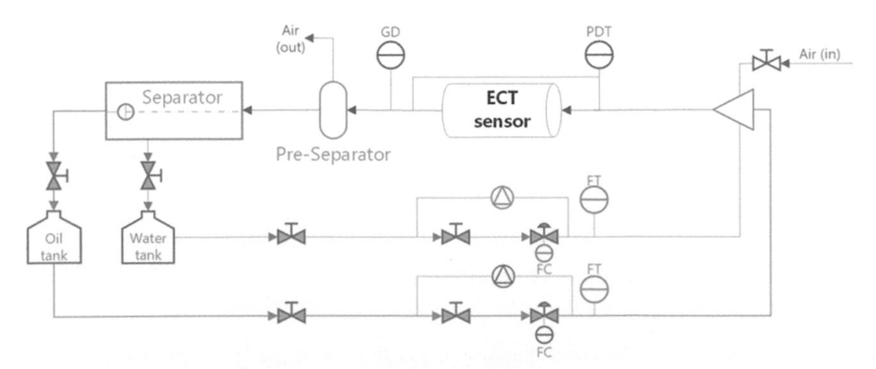

Figure 26.5 Piping and Instrumentation Diagram (P&ID) for the flow rig with ECT-module, sensors, and actuators in the vicinity of the ECT module. Coriolis flow meters ("FT"); Differential pressure transmitters ("PDT"); Gamma radiation meter ("GD").

compositions of the phases, i.e., fractions of oil, water, and air. From extensive tests run with different fractions of oil, water, and air, different desired flow regimes can be generated in the rig. For the two-phase flow studies with air and water, the test matrix with the details of air and water flow rates (WFRs) leading to various flow regimes is presented in Johansen, Grande Østby, Dupré, and Mylvaganam (2018).

A more detailed schematic of the sensor suite is given in Fig. 26.6 with some of the sensors delivering data to be used in the data fusion using AI techniques in this chapter.

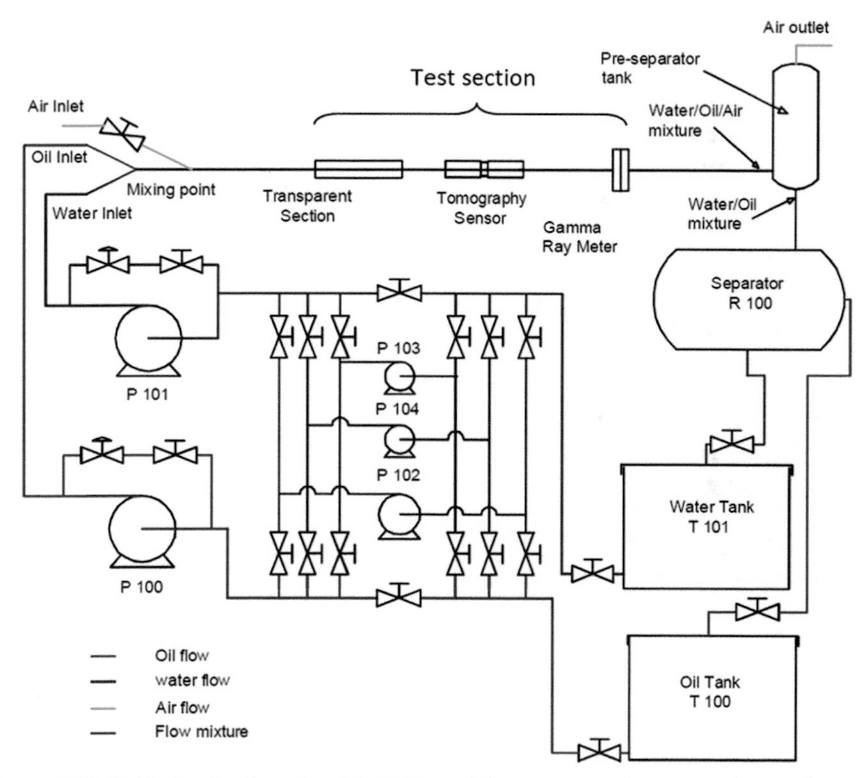

Figure 26.6 P&ID for the flow rig with ECT-module, sensors, and actuators for varying three phases in the flow. Transparent section used for high-speed camera and video recording and observing flow regimes.

26.3.2 Particulate flow

A fluidized bed column (FBC) used for observing fluidization of particles and their flows in a vertical pipe section with focus on the flow regimes is shown in Fig. 26.7. The focus in the usage of ECT module in these tests is to identify the flow regimes fixed bed, fluidization, and slugging. For these three flow regimes, starting point of fluidization and air inflow velocity for generating bubbling flow can be observed through the vertical transparent pipe section.

Based on these observations, control methods based on air inflow velocity can be developed for generating different flow regimes. The vertical FBC with the twin plane ECT module is shown in Fig. 26.7 (left) with a corresponding schematic diagram showing the major modules in Fig. 26.7 (right).

A review on the use of ECT in monitoring fluidized beds is found in Wang & Yang, 2020. ECT was used in pixel-based correlation to estimate the velocity components of two-phase solid/gas flow using ECT. The pixel by pixel correlation method for consecutive frames in a given sensor plane of the ECT module was used to trace the particle velocity profile in the transverse direction and the estimates were verified using LDA (Laser Doppler Anemometry) (Datta, Dyakowski, & Mylvaganam, 2007).

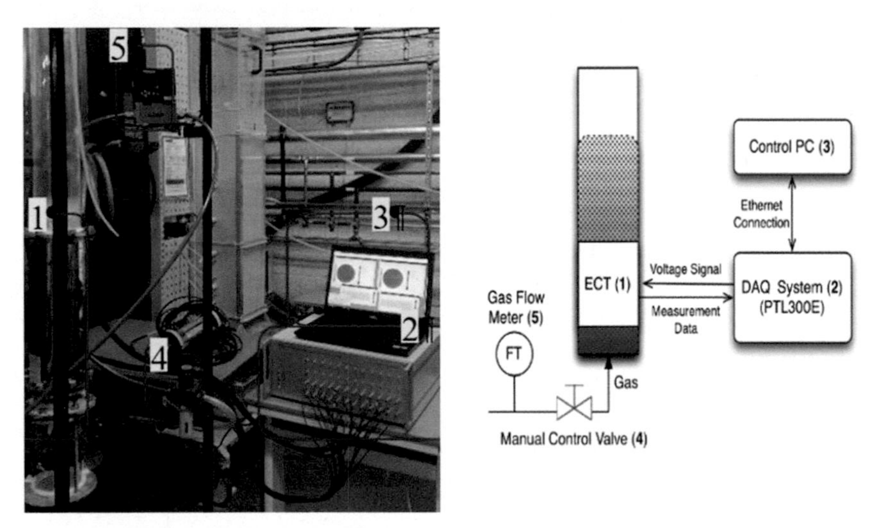

Figure 26.7 Fluidized bed column (FBC) and ECT-module (left). Schematic diagram of the FBC with installed twin plane ECT sensor (right). Numbers in photo and schematic represent the same module.

26.3.3 AI techniques in identifying flow regimes in multiphase flow

Flow regime studies have been done in many fields involving multiphase flows. In Fig. 26.8, flow regimes observed through the transparent section of the pipe section are shown in Fig. 26.6 are shown. Similar observations were used in designing experiments for creating different regimes to gather ample data for usage with AI techniques in determining flow regimes nonintrusively using ECT sensor modules.

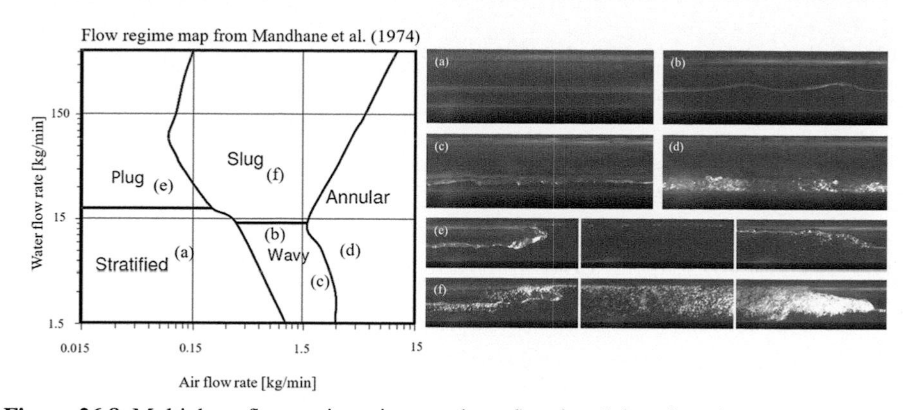

Figure 26.8 Multiphase flow regimes in two-phase flow involving air and water with high-speed images form flow studies using 15 m long test section of 56 mm inner diameter. High-speed camera images through transparent windows in the test section show (a) stratified, (b) low frequency wavy, (c) high frequency wavy, (d) annular, (e) end, middle and start of a plug, and (f) end, middle and start of a slug. Mass flow rates shown in classification chart, adapted from (Mandhane, Gregory, & Aziz, 1974) for flow regimes, are from Coriolis meters (FTs) shown in Fig. 26.6.

26.4 AI techniques relevant for process control

Two different approaches based on AI techniques will be used in determining flow regimes based on multiphase flow experiments. The techniques used are fuzzy logic, neural networks, and fuzzy neural network. The focus in this chapter is the application of these techniques and not on introducing these techniques.

In the context of fuzzy logic—based AI inference, the principle is based on fuzzifying the input variables and employing fuzzy logic rules based on linguistic variables which are combined to generate outputs in defuzzified form. The inference mechanism in the context of multiphase flow is shown in Fig. 26.9(a) followed by its implications for a fuzzy logic—based AI algorithm with the basic architecture schematically illustrated in Fig. 26.9(b).

Similarly, a neural network is illustrated for detecting flow regimes in Fig. 26.9(c) with some ECT measurements from 12 electrodes (66 measurements) attached to the multiphase flow loop as shown in Fig. 26.6 as inputs and the flow regimes as outputs.

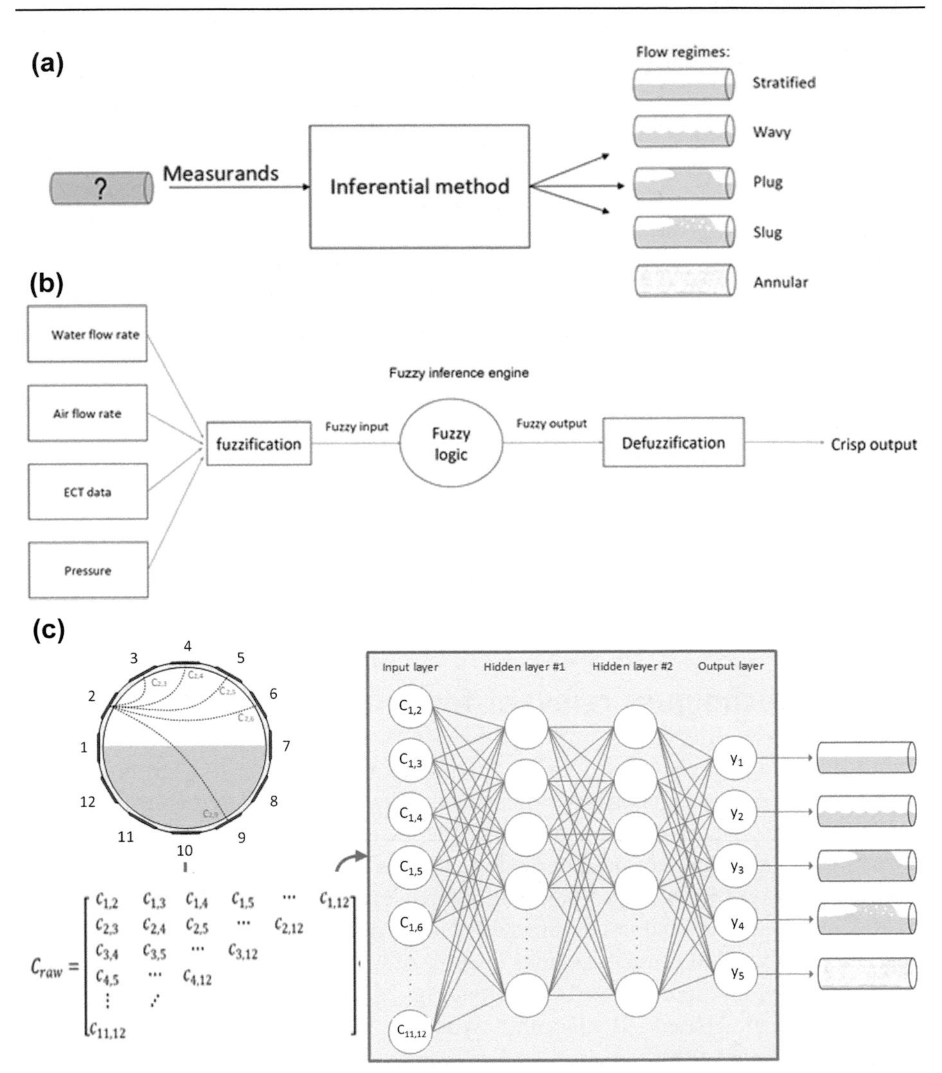

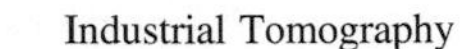

Figure 26.9 (a) Schematic of inferential method based on AI using different measurands; in this study electrical capacitance values using ECT. (b) Fuzzy logic—simplified fuzzification and defuzzification scheme using some of the measurands shown in Fig. 26.6 including ECT sensor module. (c) Schematic for an Artificial Neural Network (ANN) for flow regime identification in multiphase flow studies. C_{ij} $(i, j = 1,2,3, \ldots)$ are different capacitance values from the electrical capacitance tomograph (ECT) in use with the multiphase flow loop shown in Fig. 26.6.

26.4.1 Results from AI techniques using fuzzy logic and fuzzy neural network

A set of experiments run using the multiphase flow loop with the data from the sensors shown in Fig. 26.6 generated data from many sensors used in data fusion and the development of AI algorithms. The volume ratio was calculated using the algorithm in the ECT module. The variables used are air and WFRs and pressure, the values of which are shown in Fig. 26.10(a). An extended measurement campaign to acquire ample data was run using the test matrix shown in Fig. 26.10(b), which also shows the flow regime generated based on the speed of the two phases air and water which were controlled using inlet valve and the pump P101 shown in Fig. 26.6. Fig. 26.11(a) shows an overview of the measured flow rates of air and water and pressure along with the volume ratio on the y-axis and the measurement series as indicated in Fig. 26.10(a). These plots of all the variables on the same graph help to realize the fuzzification and the rules shown in Fig. 26.11(a). The fuzzification as indicated in Fig. 26.11(a) and the application of the rules shown in the same figure lead to the fuzzified output variables as the flow regimes and their membership functions selected by the user, shown in Fig. 26.11(b).

The output variables after the defuzzification are shown by the red (gray in print version) bars in the output variable (shown in blue [dark gray in print version] after the defuzzification) indicating the correct output based on this AI technique employing fuzzy logic with Fig. 26.12(a) indicating wavy, Fig. 26.12(b) indicating stratified, and Fig. 26.12(c) indicating annular flow regimes.

Fig. 26.12(a) indicates the output (red [gray in print version] vertical bar in the blue region indicating the crisp output value) as wavy confirmed by observation.

Fig. 26.12(b) indicates the output (red vertical bar in the blue region indicating the crisp output value) as stratified confirmed by observation.

Fig. 26.12(c) indicates the output (red vertical bar in the blue region indicating the crisp output value) as annular confirmed by observation.

Using the ANFIS in MATLAB, the measurement data obtained from running two-phase flow in the rig shown in Fig. 26.6, the flow regimes are indicated in Fig. 26.13(a). The numbers along the x-axis corresponds the experiment number given in Fig. 26.10(a).

AI techniques using the ANFIS algorithm give the output shown in Fig. 26.13(b) correctly predicting the flow regimes already reaching low error by epoch 80.

The method presented here uses the volume ratio estimated from the ECT modules and the air flow rate, WFR and pressure in the sensor fusion for the fuzzy logic−based AI algorithm, whereas an algorithm based on image fusion is described in Wang, Jia, & Wang, 2019. Using wire-mesh sensors in gas-liquid flow studies in horizontal pipelines, fuzzy logic algorithms have been used succesfully in identifying flow regimes in (Wiedemann, Döß, Schleicher, & Hampel, 2019), which has also list of references to publications dealing with fuzzy, neural, and fuzzy-neural systems for fusing sensor data in multiflow studies.

(a) Input/Output data determnation in/out of the Fuzzy system

Session	#	INPUTS: Water [kg/min]	Air [kg/min]	Pressure [mbar]	Volume ratio [%]	OUTPUT: Observed regime
SESSION 1	1	15,00	0.0	1.27	28.99	Stratified
	2	15,00	0.1	1.28	29.14	Stratified
	3	15,00	0.2	1.33	28.67	Stratified
	4	15,00	0.3	1.47	27.29	Stratified/wavy
	5	15,00	0.4	1.80	22.58	Wavy
	6	15,00	0.6	2.36	15.14	Wavy/slug
	7	15,00	0.8	3.29	5.31	Wavy/slug
	8	15,00	1.0	4.11	7.61	Wavy/slug
	9	15,00	1.5	6.98	3.53	Slug
	10	15,00	2.0	10.66	-0.02	Slug
	11	15,00	2.5	14.82	-3.15	Slug
	12	15,00	3.0	20.21	-6.69	Slug/annular
	13	15,00	3.5	27.22	-9.12	Annular
SESSION 2	14	7,80	0.0	1.07	13.44	Stratified
	15	7,00	0.1	1.04	12.51	Stratified
	16	7,00	0.2	1.07	12.04	Stratified
	17	7,00	0.3	1.12	11.81	Stratified
	18	7,00	0.4	1.19	11.71	Stratified
	19	7,00	0.6	1.37	10.31	Stratified
	20	7,00	0.8	1.92	9.14	Stratified
	21	7,00	1.0	2.36	5.13	Stratified/wavy
	22	7,00	1.5	4.42	1.02	Wavy
	23	7,00	2.0	7.52	-1.56	Wavy
	24	7,00	2.5	10.87	-4.05	Wavy/annular
	25	7,00	3.0	14.54	-5.97	Wavy/annular
	26	7,00	3.5	19.29	-8.42	Annular
	27	7,00	3.7	21.45	-9.45	Annular

(b)

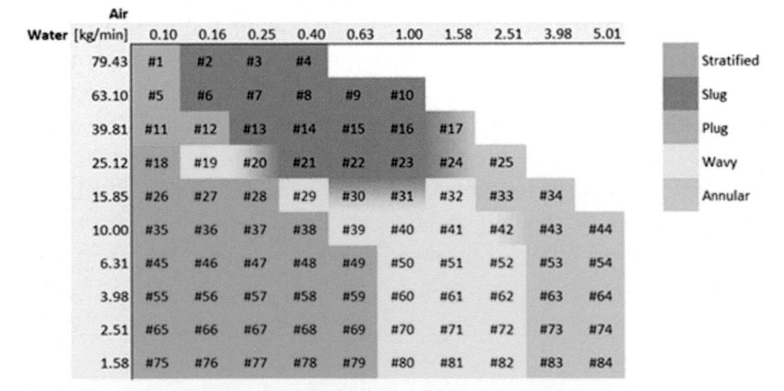

Figure 26.10 (a) Air and water flow rate with different flow regimes with selected parameters for fuzzy logic control. (b) Test matrix with details of 84 two-phase flow experiments for the different flow velocities and volume fractions of water and air. The numbers represent the different experimental runs with the corresponding flow regimes the color code in the inset legend.

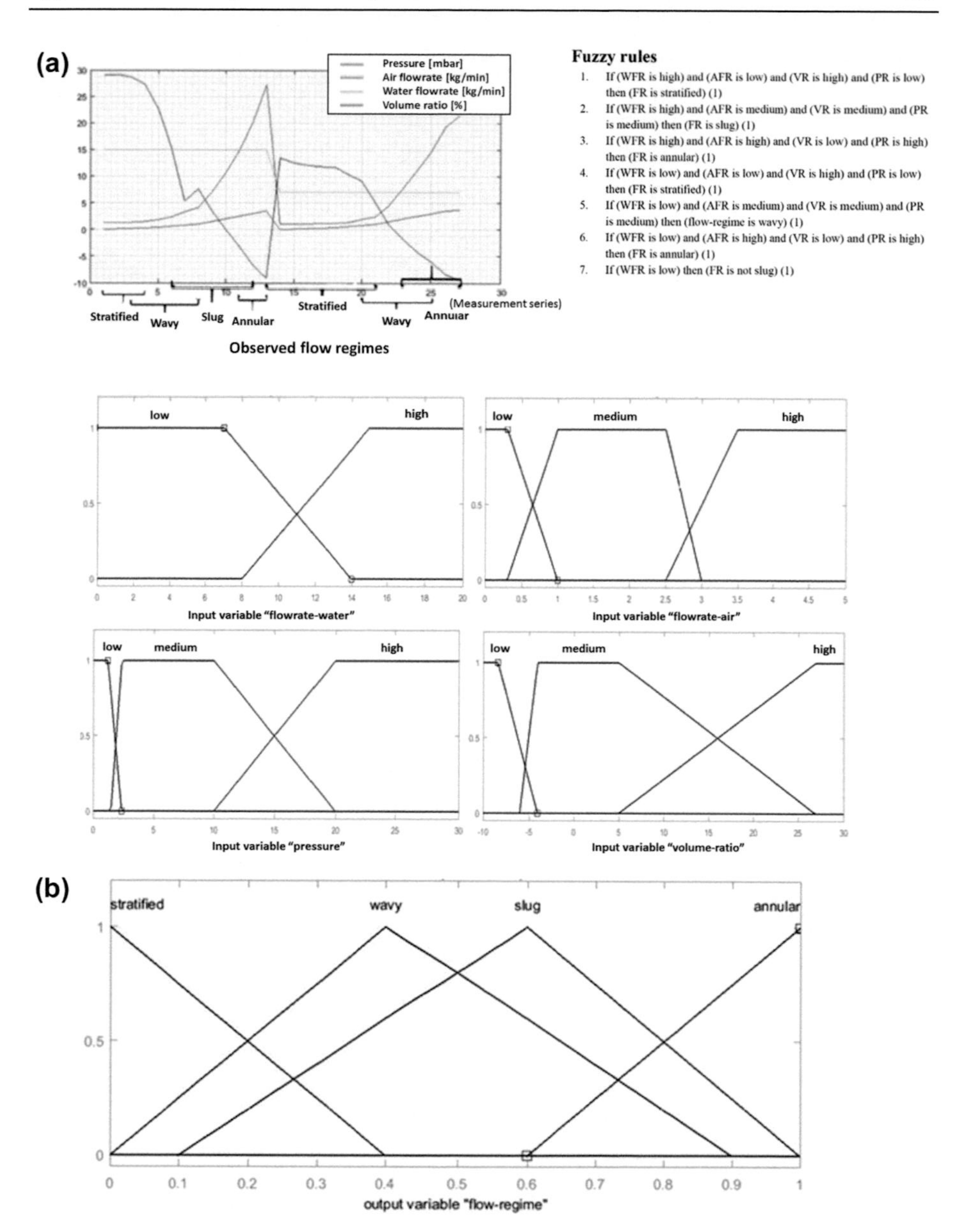

Figure 26.11 (a) Fuzzy rules generated based on flow regimes observed at different input conditions, as indicted in Fig. 26.10 with pressure and flow rates of water and air. (b) Fuzzy output based on flow regimes observed at different input conditions, as indicated in Fig. 26.10 with pressure and flow rates of water and air used as fuzzy inputs shown in Fig. 26.11(a).

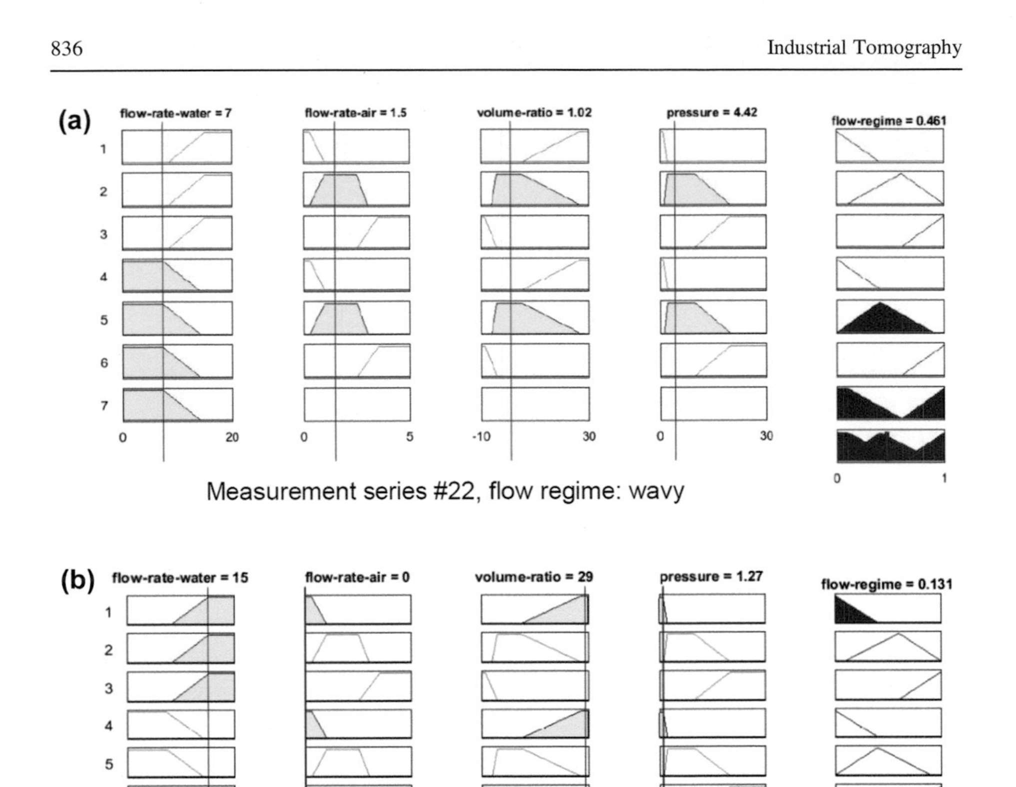

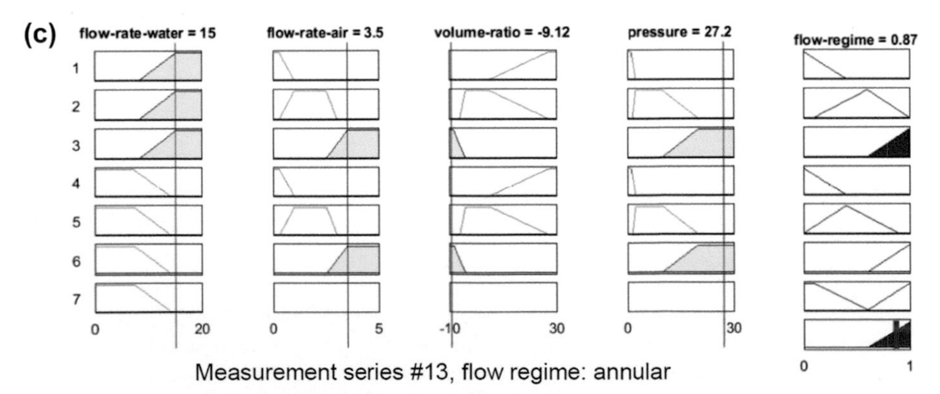

Figure 26.12 (a) Fuzzy logic output based on the fuzzy logic control; note the flow regime (wavy) indicated by the red bar in the defuzzification section with blue shaded area. (b) Fuzzy logic output based on the fuzzy logic control; note the flow regime (stratified) indicated by the red bar in the defuzzification section with blue shaded area. (c) Fuzzy logic output based on the fuzzy logic control; note the flow regime (annular) indicated by the red bar in the defuzzification section with blue shaded area.

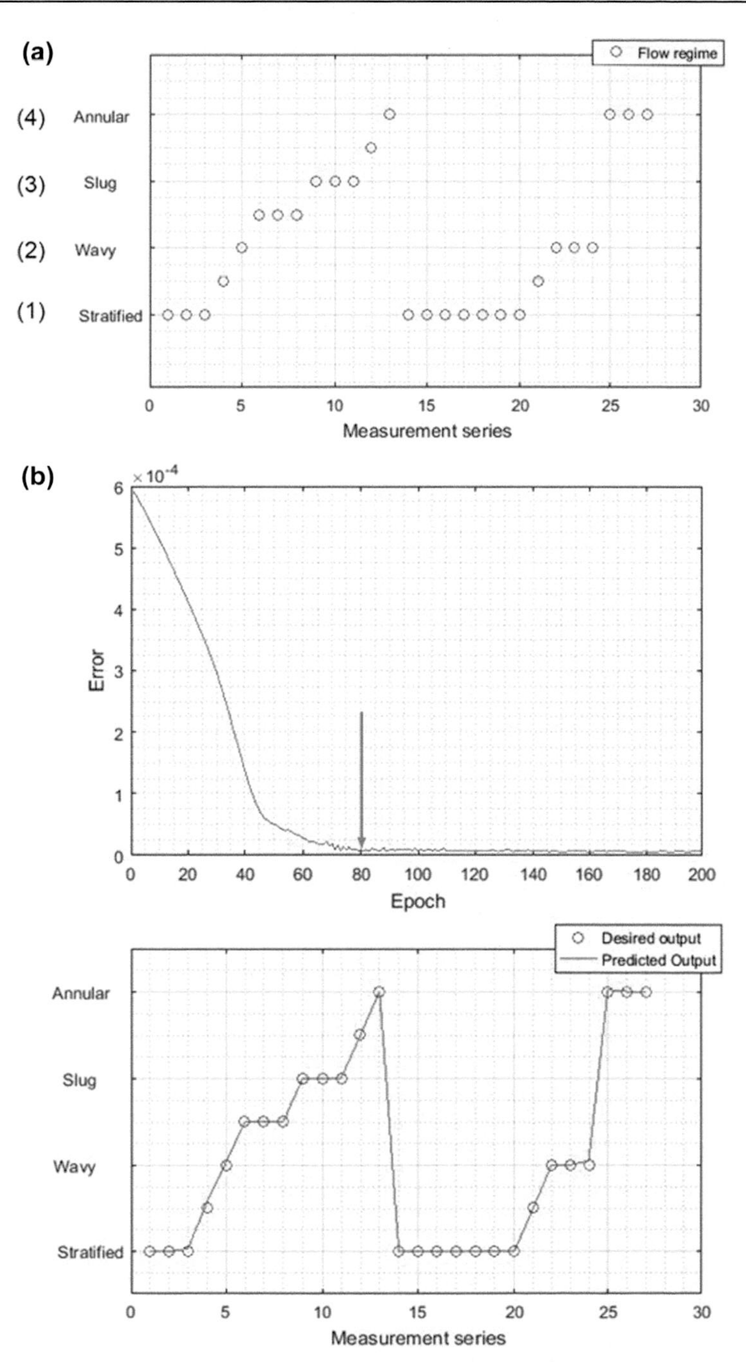

Figure 26.13 (a) Flow regimes observed based on Fig. 26.10 in different experimental runs, the numbers along x-axis corresponding to those given in Fig. 26.10. (b) The flow regimes detected by the ANFIS control unit with the error and the number of epochs.

26.4.2 Results from AI techniques using neural networks—LSTM

Neural networks have been used in process tomographic applications (Warsito & Fan, 2001; York, Ukpong, Mylvaganam, & Yan, 2012). In this section, the raw capacitance values are processed using AI techniques to investigate the multiphase flow and identify different flow regimes. The results presented are from two papers in the ninth World Congress in Process Tomography, WCIPT 2018 (Johansen et al., 2018; Yan & Mylvaganam, 2018).

As indicated in Fig. 26.9(c), the 66 measured capacitances from an ECT module with 12 electrodes can be used as inputs to a neural network. LSTM (Long Short-Term Memory) network was tested with the architecture shown in Fig. 26.14. The outputs are annular, plug, slug, stratified, and wavy flow regimes. The success of this AI technique can be seen in the performance of the networks using data from the test matrix used to run extensive data for the LSTM application as illustrated in Figs. 26.15 and 26.16.

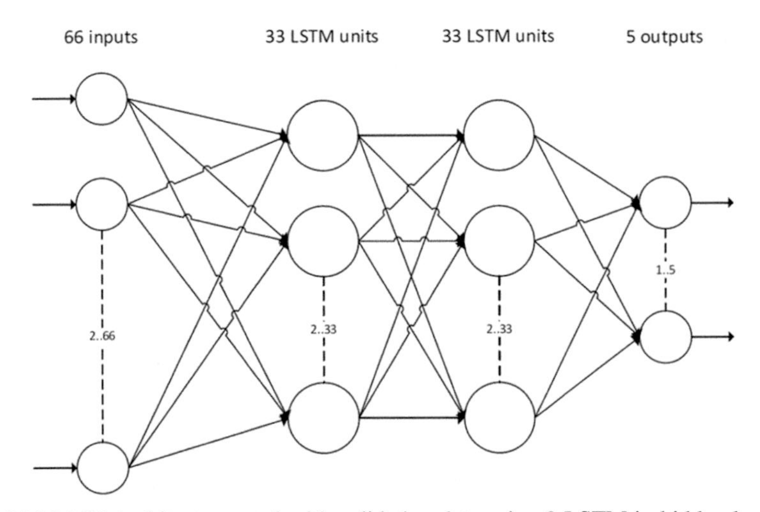

Figure 26.14 NN Architecture used with validation data using 2 LSTM in hidden layers; 66 capacitance values as inputs; 5 FRI (Flow Regime Identification) outputs representing annular, plug, slug, stratified and wavy flow regimes, (Johansen et al., 2018).

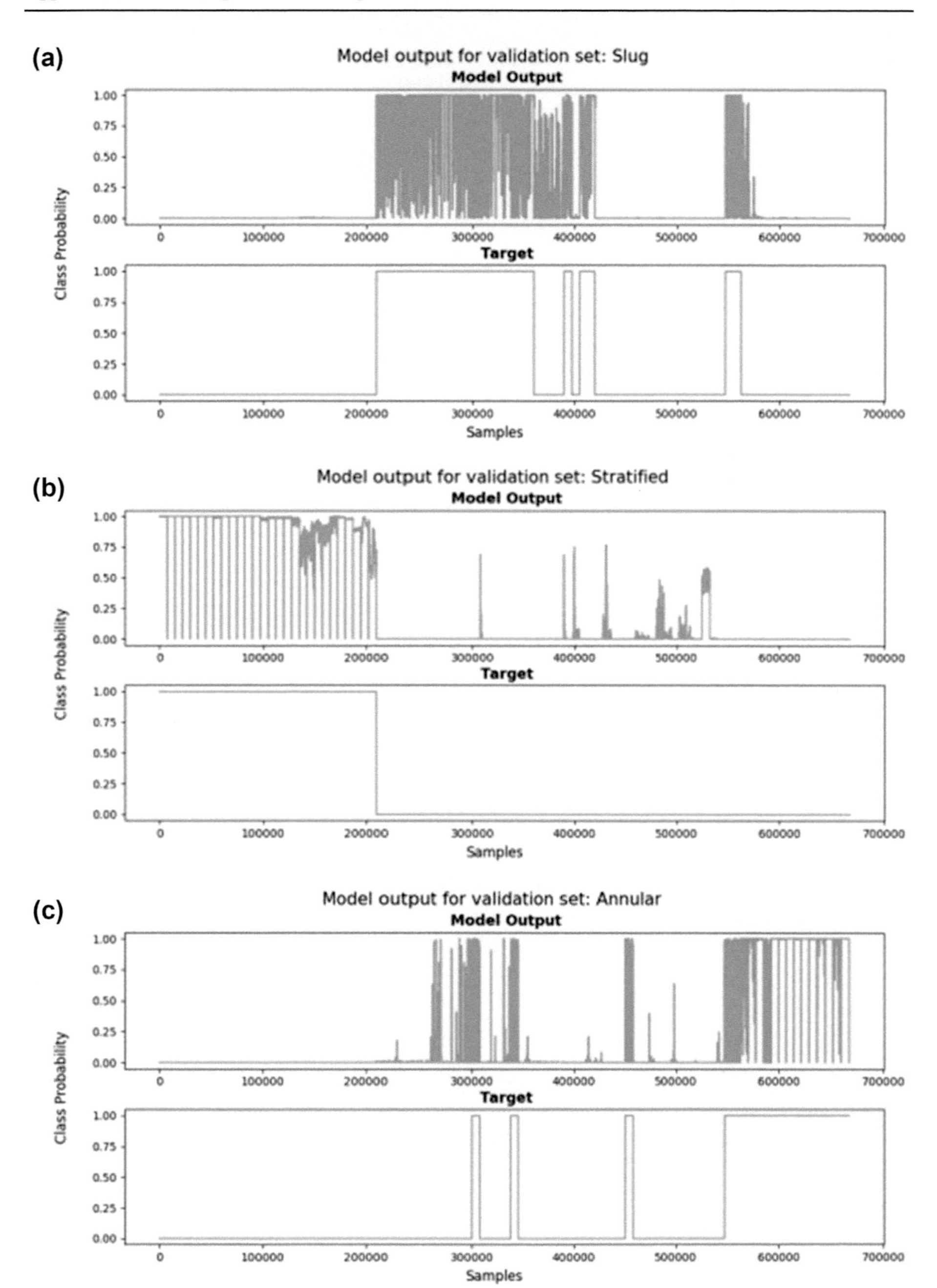

Figure 26.15 cont'd

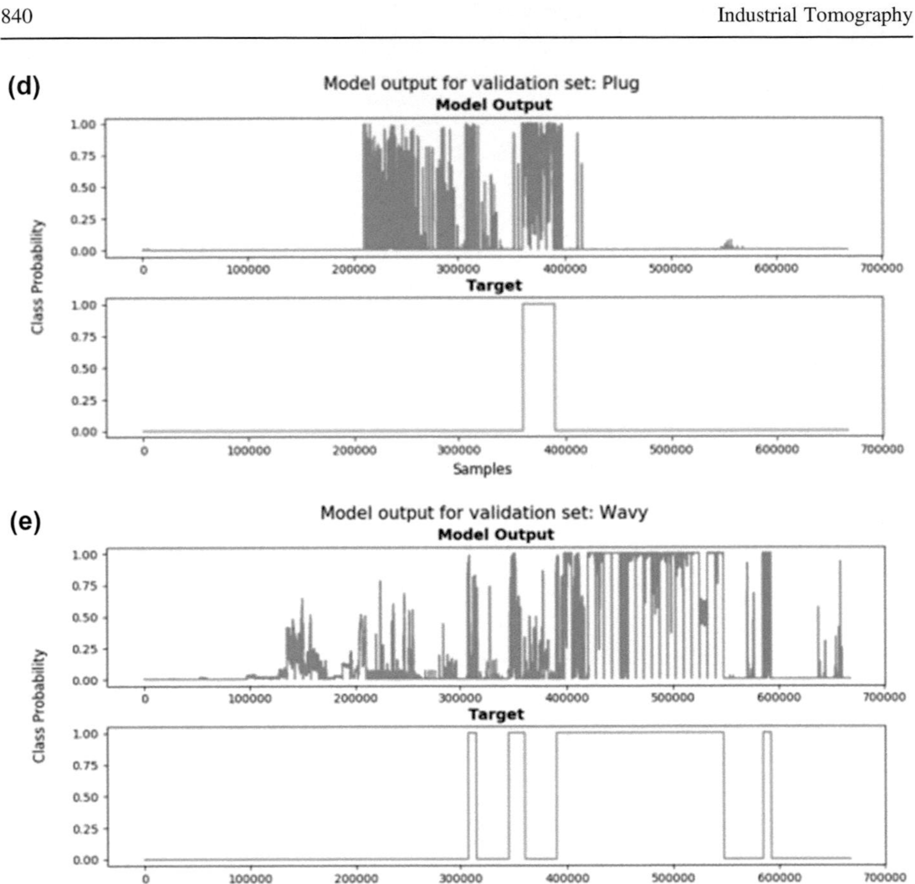

Figure 26.15 Output and target with respect to the time series for experiment #1 through #84 (excluding #19). Showing the overall model performance for (a) slug, (b) stratified, (c) annular, (d) plug, and (e) wavy flow regimes respectively, (Johansen et al., 2018).

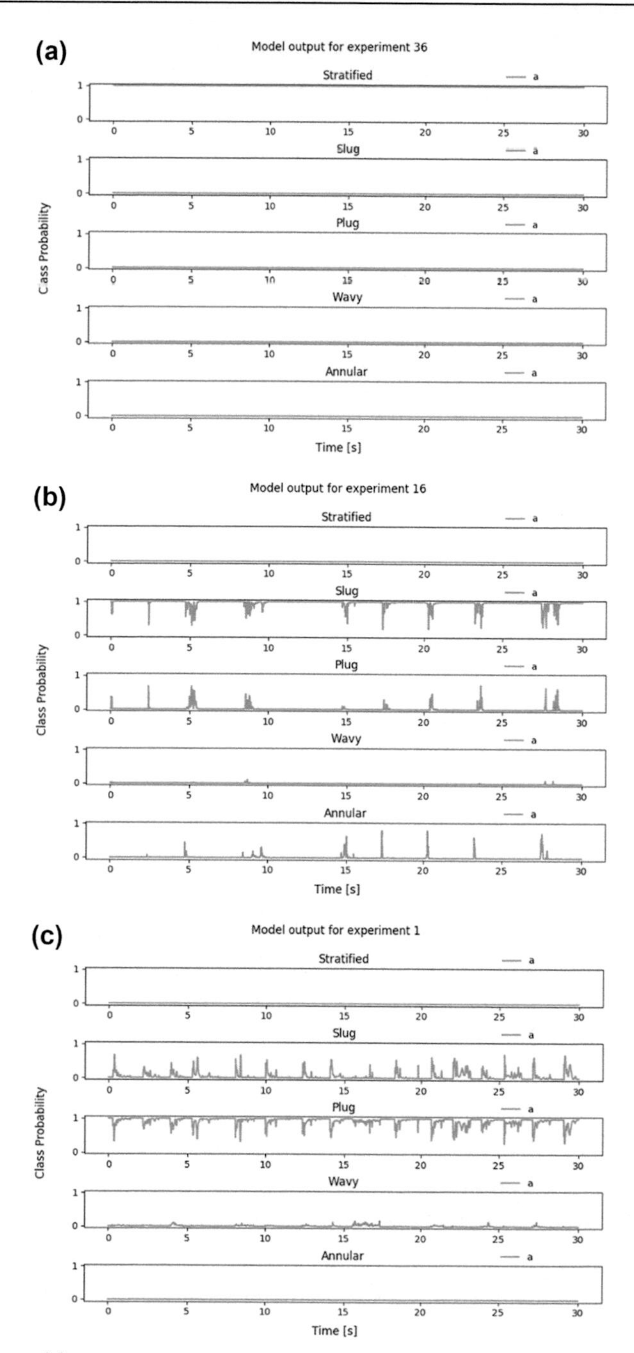

Figure 26.16 cont'd

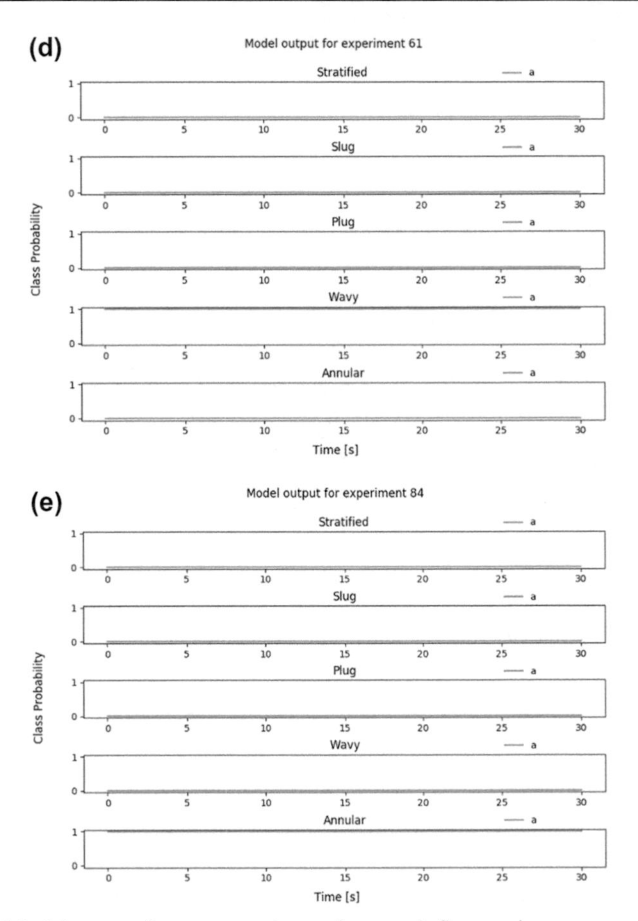

Figure 26.16 Model output for one experiment from each flow regime corresponding to Fig. 26.10(a). Anomalies/ambiguities possibly due to varying flow regimes in intermittent periods. (a) For experiment #36, correct flow regime as stratified. (b) For experiment #16, correctly identified most of the time as slug; at certain intervals identified as plugs. (c) For experiment #1, correctly identified most of the time plug; at certain intervals identified as slugs. (d) For experiment #61, correct flow regime most of the time as wavy. (e) For experiment #84, correct flow regime most of the time as annular, (Johansen et al., 2018).

26.4.3 Results from AI techniques using support vector machines—fluidized bed columns

Commonly encountered flow regimes in particulate flow in fluidized bed columns (FBC) are shown in Fig. 26.17(a), (Yan, 2016). Out of these flow regimes shown in Fig. 26.17(a), three types of regimes, bubbling, slugging, and turbulent to pneumatic conveying have been studied using twin plane ECT module. One SVM can be used to separate two clusters. A cascade of SVMs can be used to separate multiple clusters. Two linear SVMs are used to classify 'bubbling/not bubbling" and 'slugging/turbulent." 12 experiments were conducted to obtain the stacked images shown in Fig. 26.17(b).

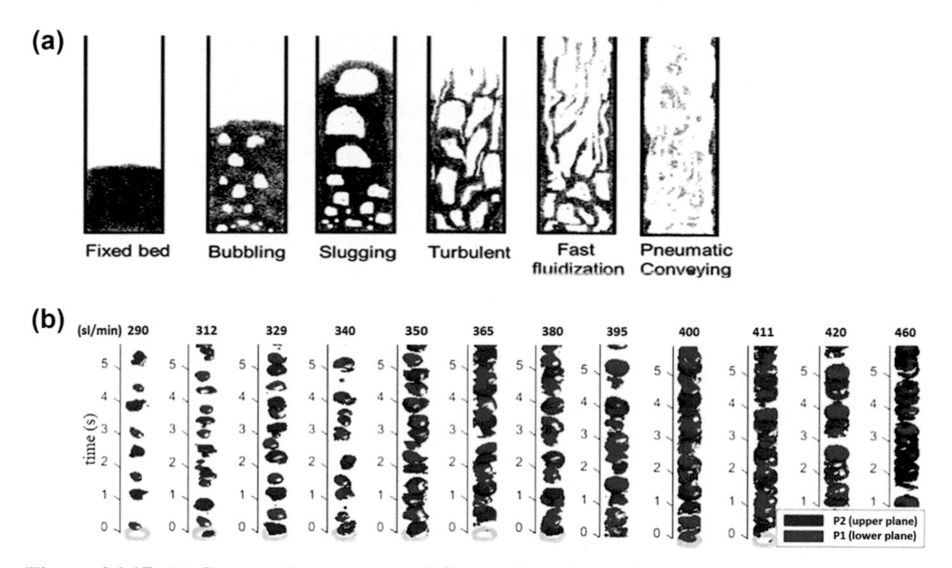

Figure 26.17 (a) Commonly encountered flow regimes in particulate flow in fluidized bed columns (FBC), Yan (2016). (b) Reconstructed bubble images with airflow varying from 290 to 340 SLM. 'Blue' indicates the bubble generated from the lower ECT sensor plane (p1, relatively close to the bottom of FBC and air inlet area, Fig. 26.7); 'Red' indicates the bubble generated from upper ECT sensor plane (p2).
Adapted from Yan et al., 2020.

Before applying the cascade of SVMs, each SVM must be trained separately. Equations from (26.1) to (26.11) are used in the training and validation of the data sets (Yan, Viumdal & Mylvaganam, 2020).

Each input to the SVM is organized as follows:

$$vr_i = \left[vr_{1,i}, vr_{2,i}\right] \in R^{m\times 2} \tag{26.1}$$

where $vr_{1,i}$ and $vr_{2,i}$ are the averaged and normalized volume ratio from one second (50 frames) measurements; the subscript, i indicates the experimental number; 1, 2 indicates the sensor plane, P1 or P2; m is the total number of averaged measurements used.

Grouping the data set of normalized volume ratios into two different sets for SVM training and validation purposes:

$$vr_{T,i} = vr_i(1:T,:) \in R^{T\times 2} \tag{26.2}$$

$$vr_{V,i} = vr_i(1:V,:) \in R^{V\times 2} \tag{26.3}$$

where the subscripts T and V are selected training and validation data size from each experiment, respectively.

Then the training data set for each flow regime is formed:

$$VR_{BT} = \left[vr_{T,1}; vr_{T,2}; vr_{T,3}\right] \in R^{3 \cdot T \times 2} \tag{26.4}$$

$$VR_{ST} = \left[vr_{T,4}; vr_{T,5}\right] \in R^{2 \cdot T \times 2} \tag{26.5}$$

$$VR_{ET} = \left[vr_{T,6}; vr_{T,7}; \ldots; vr_{T,12}\right] \in R^{7 \cdot T \times 2} \tag{26.6}$$

where the subscription T indicates the volume ratio data selected for training; B indicates the volume ratio data from bubbling measurements (corresponding to first to third experiments, represented in Fig. 26.17(b)); the subscript S indicates data from measurements involving slugging (corresponding to fourth and fifth experiments represented in Fig. 26.17(b)); and E indicates data from turbulent pneumatic conveying (corresponding to 6th to 12th experiments).

Then the training data sets and its corresponding targets for the two SVMs are selected:

$$\begin{cases} T_{\text{SVM1}} = [VR_{BT}; VR_{ST}; VR_{ET}] \\ I_d = [1; -1, -1] \end{cases} \tag{26.7}$$

$$\begin{cases} T_{\text{SVM2}} = [VR_{BT}; VR_{ST}; VR_{ET}] \\ I_d = [1; 1, -1] \end{cases} \tag{26.8}$$

where T_{SVM1} and T_{SVM2} are the training data sets for SVM1 and SVM2, respectively; I_d is the corresponding training targets, i.e., desired regions.

Similarly, the validation data for each trained SVM are formed:

$$VR_{BV} = \left[vr_{V,1}; vr_{V,2}; vr_{v,3}\right] \in R^{3 \cdot V \times 2} \tag{26.9}$$

$$VR_{SV} = \left[vr_{V,4}; vr_{V,5}\right] \in R^{2 \cdot V \times 2} \tag{26.10}$$

$$VR_{EV} = \left[vr_{V,6}; vr_{V,7}; \ldots; vr_{V,12}\right] \in R^{7 \cdot T \times 2} \tag{26.11}$$

with the subscript V indicating the validation data sets.

Fig. 26.18 shows all the testing samples (from experiments No. 1, 2, 3, 4, 5, 7, 8, 9, 12 with the stacked ECT images in Fig. 26.17) are classified correctly.

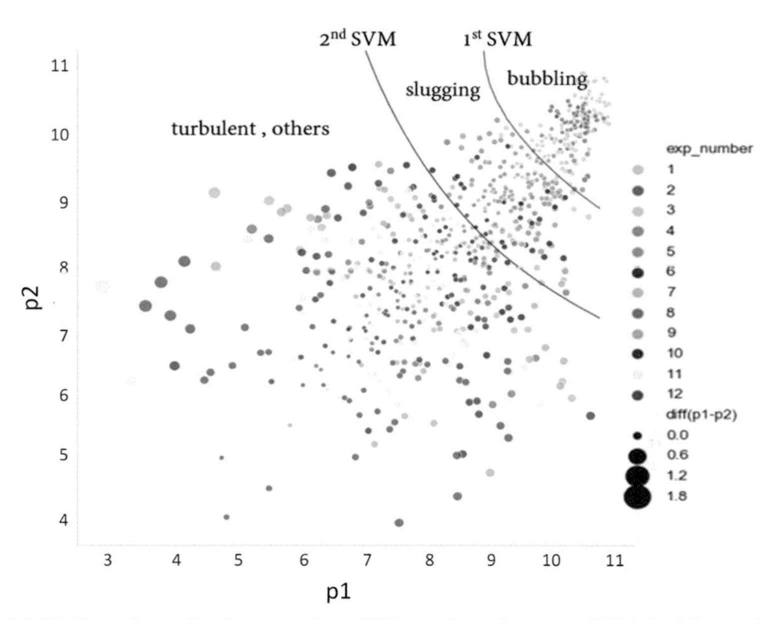

Figure 26.18 Overview of trained stacked SVM results; where 1st SVM (red hyperplane) is used to classify "bubbling" and others; 2nd SVM (blue hyperplane) is used to classify between "slugging" and "turbulent, others" (Yan et al., 2020).

26.5 Possibilities for AI-assisted control

A possible control mechanism based on data fusion with other sensor modalities is presented in Fig. 26.19. A combination of outputs from the ECT module and other sensors can be used to control the flow loop to avoid unwanted and possibly hazardous combinations of flow regimes involving high frequency slugs.

By using ECTm data as inputs to NN and pressure measurements, a model free adaptive controller (MFAC) can be implemented in processes involving multiphase flows. In a conventional model—based predictive control (MPC), the functioning of the MPC heavily relies on the accuracy of the model and control algorithms. In fact, in many applications, due to the substantial number of parameters involved in multiphase flow-based process, designing an MPC to tackle all flow conditions is a big challenge. In comparison, Fig. 26.19 presents the configuration of an MFAC based on NN, which does not require any dedicated model. By changing the strategy from MPC to MFAC control or to a workable blend of MPC and MFAC, the process engineer will have an increased advantage in preventing hazards, such as the one discussed in the introduction to this chapter and the effects of which are shown in Fig. 26.4.

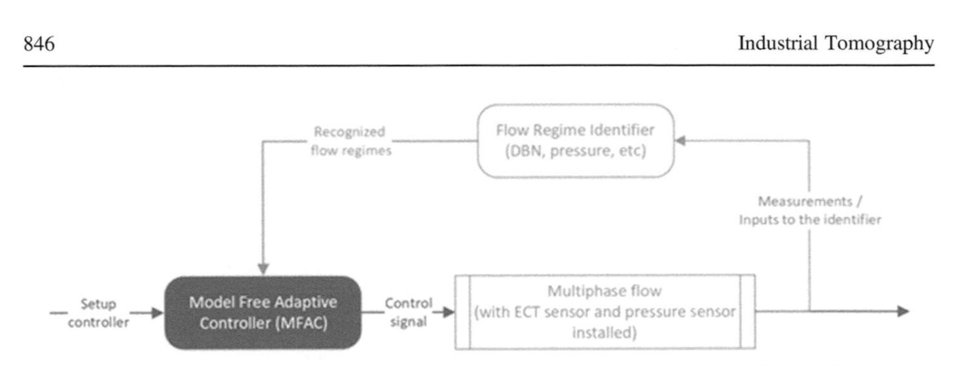

Figure 26.19 Flow chart of multiphase flow controller design scenario based on a signal plane ECT sensor using AI algorithms (Yan et al., 2020).

Similarly, an AI-assisted control mechanism can be realized using cascaded SVMs, which can assist the system in achieving the desired flow regimes in the fluidized bed, which are crucial in applications involving, e.g., mixing, chemical reactions, drying, etc.

Using variations in ECT image pixel data, a stack of 3D images of bubbles is reconstructed using time series from both sensor planes in a twin plane ECT module. From these reconstructed 3D images, the difference between various flow regimes can be observed and identified with good accuracy. This is a method based on image processing.

The bubble velocity and bubble frequency are estimated using the volume ratio data estimated from the tomograms obtained from the twin-pane ECT module shown in Fig. 26.7. The results obtained from this method show this method can be applied to identify bubble characteristics and bubble coalescence numerically. Further, using a stack SVM, the main fluidizing flow regimes can be identified through clustering the pattern from the averaged volume ratio from both sensor planes. The stack SVM classification/identification results can deliver a unique output, such as a single number (index), e.g., “1” for bubbling, “2” for slugging, and so on. Therefore, the stacked SVM method has the potential for applications in real-time control of processes involving FBC.

These results give useful information in identifying bubble velocity, frequency, location, coalescence and fluidization, and flow regime identification. Possible concept for an automatic in-line real-time control strategy of processes involving FBC based on AI is illustrated in Fig. 26.20.

Fig. 26.20 presents an AI-based control strategy by fusing ECT data with selected sensor data available in the process. The system architecture suggested in Fig. 26.20 has been realized in some processes. A cross-platform application involving LabVIEW and DELTA V has been realized in a robotic application. This technique can be used with AI-based process control involving process tomographic data too. MATRIKON OPC server can be replaced by AZURE cloud.

Fig. 26.21 flow chart of a MFAC with the main components—design scenario based on ECT sensor and other sensors, Courtesy, Aleksander Tokle Poverud, USN. A scenario relevant to the process industries. ECT-Electrical Capacitance Tomograph, PT-Pressure Transmitter, FT- Flow Transmitter etc. Python is increasingly used

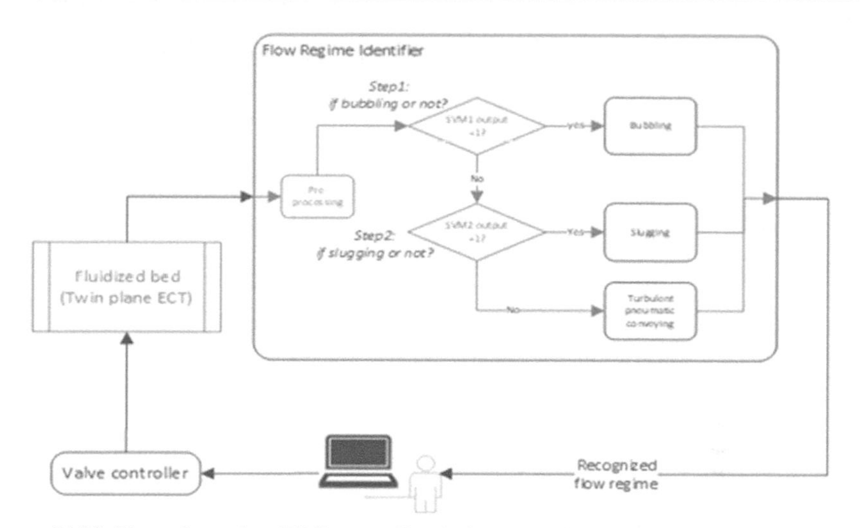

Figure 26.20 Flow chart of an FBC controller design scenario based on ECT sensor based on some machine-learning algorithms.

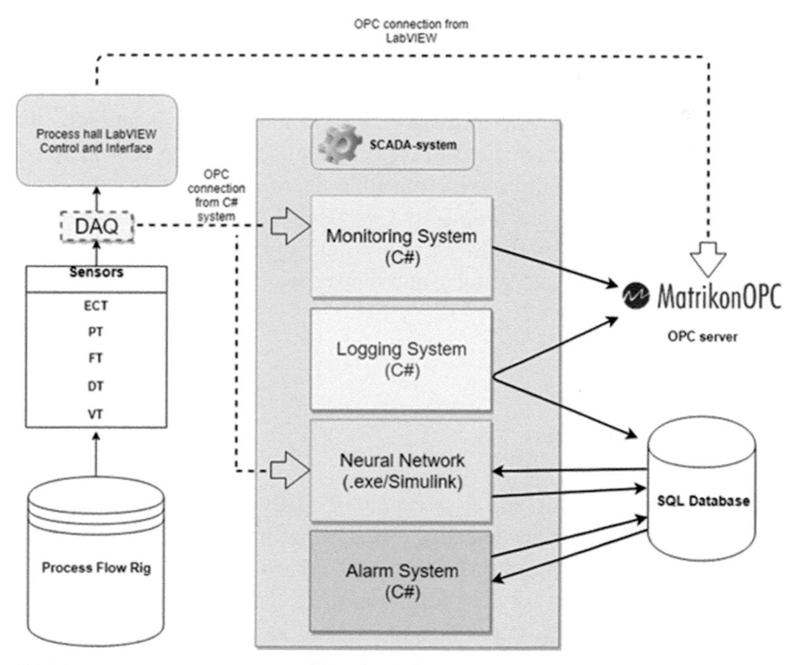

Figure 26.21 Flow chart of a MFAC with the main components - design scenario based on ECT sensor and other sensors, Courtesy, Aleksander Tokle Poverud, USN. A scenario relevant to the process industries. ECT-Electrical Capacitance Tomograph, PT-Pressure Transmitter, FT-Flow Transmitter etc. Python is increasingly used in the process industries for some of the tasks listed here.

Stages of Model building and usage	Data Collection & Preparation 1	Building ML Models (SANDBOX) 2	Running, Training, Evaluation 3	Deploying 4	Prediction 5
Platforms /Programs (examples)	Spark, Kafka	Jupyter, data Dataiku, Keras, H_2O.ai	Spark, TensorFlow, MXNet	Spark, TensorFlow, H_2O.ai	Spark, TensorFlow, H_2O.ai
On Premises / Public Cloud /Hybrid Cloud **Languages** - C++, Python, R, Julia/JuliaSim, JavaScript, Scala, Go, Perl (possible to use with MATLAB bindings) **Operating Systems: Windows, macOS, Linux**					

Figure 26.22 A scenario for implementing AI/ML in five stages (1 to 5) involving pre-processing data, building algorithms in sandbox, training, validating, and deploying in the field. Major actors are involved in providing Cloud services and in running ML algorithms, e.g., Amazon, Google, HP Enterprises etc. Python is increasingly used in the process industries for some of the tasks listed here (Yan et al., 2020).

in the process industries for some of the tasks listed here. Once the architecture is selected and the data flow is organized, a strategy based on edge to cloud computing using advanced and fast AI techniques can be executed as shown in Figs. 26.21 and 26.22. Figs. 26.21 and 26.22 helps to see the stages described in Figs. 26.1−26.3 and points out some of the currently available platforms for AI-based analysis which can be used in the manner portrayed in Fig. 26.21. The AI program platforms shown in Fig. 26.22 are some of the many available in the very dynamic field of AI and machine learning. Recently, a high-performance programming environment, JuliaSim, combining programming languages Julia and Modelica enabling integration of traditional modeling and simulation with machine learning was released Rackauckas et al., 2021. In Fig. 26.22, according to Google ML guidelines, between stages 1 and 2, checks are needed for identifying any data imbalances, between 3 and 4 for ensuring fair treatment of all groups and between 4 and 5 understanding model behavior on real data.

26.6 Future trends

Using ECT modules in multiphase flow rig and the FBCs for identifying flow regimes and using tomographic data in detecting operational hazards, this chapter presented some possibilities of incorporating AI in the of control of these processes. With the ever-increasing use of cloud services for computing and developments in near sensor signal processing techniques, commonly described as edge to cloud computing, AI methods can enhance MFAC using raw data from the sensors used in the process and data from process tomographic modules. As shown in Figs. 26.18 and 26.19, routines based on MFAC can be integrated in the already existing systems used for process measurements and control.

The approach presented in this chapter using MFAC falls under the class of "black-box models," used in complex nonlinear process modeling and can be enhanced by explainable machine learning approaches involving "gray box models," with inputs from well-proved models, at least partly describing the process. In an earlier work, ECT data have been used to enhance the CFD modeling of three phase flow (Pradeep et al., 2012). Recently, many researchers are working on Physics-Guided Neural Networks (PGNNs) and Physics-Informed Neural Networks (PINNs), referred to as hybrid models enhancing for explainability of the results obtained using ML/AI algorithms, Karniadakis et al. (2021). Data from process tomographic modules have also the often quoted attributes called "5 V's": Value ("Data is money"!, achieving new results, insight, clear improvements in the process control in the current context), Volume (stored and real-time data in the sizes up to few TB), Velocity (in batch, real time, streaming etc.), Variety (stuctured, unstructured, different formats, image/video/multimedia), and Veracity (refering to quality and credibility with associated uncertainties, latency and synchronus/asynchronus data particularly with multimodal tomograph). AI enabled data gives the process engineer tools for gaining hindsight, insight and foresight using descriptive (what happened?), diagnostic (why did it happen?), predictive (what will happen?) and prescriptive analytics (how can we make it happen?) as described in Gartner Analytic Ascendency Model, all of which are useful for control purposes with appropriate human input (human in the loop) followed by decision-making ending up in a suitable action.

26.7 Sources of further information

AI and machine learning are currently covered in different disciplines at varying depth, though the applications are many and implemented using the programs and platforms shown in Fig. 26.21. There are many AI and machine learning on-line courses. AI for process control has been addressed in a book, almost 3 decades back (Boullart, Krijgsman, & Vingerhoeds, 1992), covering fuzzy logic, neural network, and object-oriented approaches in AI-based process control. One traditionally well-known book on neural networks and SVM is Haykin (2009). Zadeh's original papers (Zadeh, 1965, 1968, 1994) on fuzzy logic are good sources for gaining insight. An overview of fuzzy logic is presented in a simple exposition in Kosko and Isaka (1993). Artificial intelligence and control are discussed with some of the latest developments in Russel (2019), Russel and Norvig (2009). The techniques with multimodal tomography described in Hoyle et al. (2001) can be useful in achieving sensor data fusion for AI techniques in process control.

For sensor near processing enabling edge to cloud computing, applications of FPGA arrays and processing can lead to VFM using MPFM as discussed in Bikmukhametov and Jäschke (2020), thus mirroring the scenario portrayed in Fig. 26.21. Hardware and software developments involving GPU-accelerated computing, e.g., CUDA GPUs NVIDIA Developer, and IPUs (Intelligent Processing Units) available from various vendors, e.g., IPU Products (graphcore.ai), in the market will also find their way into process tomographic/tomometric applications addressing the 5 V's mentioned in Section 26.6. These developments will help the process engineers in including process tomographic/tomometric data in process control applications and enabling AR (Augmented Reality) for visualisation and supervision of processes even with their handled devices. Further information on principle and application of process tomography can be found from Beck et al. (1993), Dickin and Wang (1996), Plaskowski et al. (1995), Scott and Williams (1995), and Williams and Beck (1995). Practical applications with many well-structured examples are found in Fuzzy Logic Toolbox and Statistics and Machine Learning Toolbox of MATLAB. Many applications using Python are found in the software platforms shown in Fig. 26.21.

Acknowledgments

The results used in the AI models used in this chapter are based on multiphase flow experiments with an array of multimodal sensors including the EIT, ECT, and GRM done in the process lab at the University of South-Eastern Norway (USN) with the help of Senior Engineer Fredrik Hansen and former and current bachelor, master students, and Ph.D. researchers. This chapter contains excerpts from our publications (WCIPT, 2018; Johansen et al., 2018; Yan and Mylvaganam, 2018) and (IFAC, 2020; Yan et al., 2020). We acknowledge the organizers of the conferences of WCIPT (2018) and IFAC (2020) for their copyright arrangements facilitating the reuse of excerpts from our papers presented in these conferences in this chapter.

References

Beck, M. S., Campogrande, E., Morris, M., Williams, R. A., & Waterfall, R. C. (1993). *Tomographic techniques for process design and operation*. Computational Mechanics Publications.

Bikmukhametov, T., & Jäschke, J. (2020). First principles and machine learning virtual flow metering: A literature review. *Journal of Petroleum Science and Engineering, 184*, 106487. https://doi.org/10.1016/j.petrol.2019.106487

Boullart, L., Krijgsman, A., & Vingerhoeds, R. A. (1992). *Application of artificial intelligence in process control*. Elsevier.

Chaminda, P. G. V., Yan, R., & Mylvaganam, S. (2012). Neural network-based interface level measurement in pipes using peripherally distributed set of electrodes sensed symmetrically and asymmetrically. *IEEE Transactions on Instrumentation and Measurement, 61*(9).

Datta, U., Dyakowski, T., & Mylvaganam, S. (2007). Estimation of particulate velocity components in pneumatic transport using pixel based correlation with dual plane ECT. *Chemical Engineering Journal, 130*(2).

Dickin, F., & Wang, M. (1996). Electrical resistance tomography for process applications. *Measurement Science and Technology, 7*, 247.

Haykin, S. (2009). *Neural networks and learning machines* (3rd ed.). Pearson.

Hoyle, B. S., Jia, X., Podd, F., Schlaberg, H. I., Tan, H. S., Wang, M., et al. (2001). *Design and application of a multi-modal process tomography system*. Measurement Science and Technology.

Johansen, R., Grande Østby, T., Dupré, A., & Mylvaganam, S. (2018). Long short-term memory neural networks for flow regime identification using ECT. In *9th World congress on industrial process tomography, Bath, UK.*

Karniadakis, G. E., Kevrekidis, I. G., Lu, L., et al. (2021). Physics-informed machine learning. *Nature Reviews Physics, 3*, 422–440. https://doi.org/10.1038/s42254-021-00314-5

Kosko, B., & Isaka, S. (1993). Fuzzy logic. *Scientific American, 269*(1).

Mandhane, J. M., Gregory, G. A., & Aziz, K. (1974). A flow pattern map for gas—liquid flow in horizontal pipes. *International Journal of Multiphase Flow, 1*(4).

Pradeep, C., Yan, R., Vestol, S., Melaaen, M., & Mylvaganam, S. (2012). Electrical capacitance tomography (ECT) and gamma radiation meter for comparison with and validation and tuning of CFD modeling of multiphase flow. *Measurement Science and Technology, 25*, 45–50. https://doi.org/10.1109/IST.2012.6295569. In press.

Plaskowski, A., Beck, M. S., Thorn, R., & Dyakowski, T. (1995). *Imaging of industrial flows: Applications of electrical process tomography*. IoP.

Rackauckas, C., et al. (2021). *Composing Modeling and Simulation with Machine Learning in Julia, TR2021-114, The 14th International Modelica Conference Linköping.*

Russell, S. (2019). *Human compatible: Artificial intelligence and the problem of control. United states*. https://www.matrikonopc.com/ (Accessed on 17 March 2021).

Russell, S., & Norvig, P. (2009). *Artificial intelligence: A modern approach*. Prentice Hall.

Scott, D. M., & Williams, R. A. (Eds.). (1995). *Frontiers in industrial process tomography*. AIChE-Eng. Foundation.

Scott, D. M., & McCann, H. (Eds.). (2005). *Process imaging for automatic control*. CRC Press-Taylor & Francis.

Temming, M. (2020). The Deepwater Horizon oil spill spread much farther than once thought, ScienceNews, the Deepwater Horizon oil spill spread farther than once thought. *Science News* (Accessed on 2 March 2021).

Tokle Poverud, A. (2019). *Flow-analytics using multiphase flow rig with multimodal sensor suite – with focus on void fraction, water-cut and flow regimes* (Master thesis). University of South-Eastern Norway.

Wang, Q., Jia, X., & Wang, M. (2019). Fuzzy logic based multi-dimensional image fusion for gas–oil-water flows with dual-modality electrical tomography. *IEEE Transactions on Instrumentation and Measurement, 69*(5).

Wang, H., & Yang, W. (2020). Application of electrical capacitance tomography in circulating fluidised beds – a review. *Applied Thermal Engineering, 176*.

Warsito, W., & Fan, L. S. (2001). Neural network based multi-criterion optimization image reconstruction technique for imaging two- and three-phase flow systems using electrical capacitance tomography. *Measurement Science and Technology, 12*(12).

West, R. M., & Williams, R. A. (1999). Opportunities for data fusion in multi-modality tomography, virtual centre for industrial process tomography. In *1st World congress on industrial process tomography* (pp. 195–200).

Wiedemann, P., Döß, A., Schleicher, E., & Hampel, U. (2019). Fuzzy flow pattern identification in horizontal air-water two-phase flow based on wire-mesh sensor data. *International Journal of Multiphase Flow, 117*, 153–162. https://doi.org/10.1016/j.ijmultiphaseflow.2019.05.004

Williams, R. A., & Beck, M. S. (Eds.). (1995). *Process tomography – principles, techniques and applications*. Butterworth-Heinemann, ISBN 0-7506-0744-0.

Yan, R. (2016). *Usage of process tomographic techniques in study of flow dynamics in fluid and particulate flow* (PhD thesis). University of South-Eastern Norway.

Yan, R., Viumdal, H., & Mylvaganam, S. (2020). *Process tomography for model free adaptive control (MFAC) via flow regime identification in multiphase flows*. IFAC.

Yan, R., & Mylvaganam, S. (2018). Flow regime identification with single plane ECT using deep learning. *9th World Congress on Industrial Process Tomography, Bath, UK*. In press.

York, T. A., Ukpong, A., Mylvaganam, S., & Yan, R. (2012). Parameter estimation from tomographic data using self-organising maps. In *Proceedings IST − IEEE International Conference on Imaging Systems and Techniques, Proceedings. IEEE* (pp. 112−116).

Zadeh, L. A. (1965). Fuzzy sets. *Information and Control, 8*(3).

Zadeh, L. A. (1968). Fuzzy algorithms. *Information and Control, 12*(2).

Zadeh, L. A. (1994). The role of fuzzy logic in modeling, identification and control. *Modeling and Identification and Control, 15*(3).

Diverse tomography applications 27

Jiabin Jia and Yong Bao
University of Edinburgh, Edinburgh, United Kingdom

27.1 Introduction

Exploring the unknown never stops in the process tomography research. Practical industrial application challenges are also pushing process tomography forward, motivating continuous innovation of process tomography, and maintaining the lifeline of process tomography. In turn, the development of process tomography enables a wider range of applications. In this chapter, some emerging applications of process tomography are reviewed to promote the diversity of process tomography. In the context of electrical tomography, distribution of dispersed phase in the packing column is monitored and quantified. 3D cell spheroid is imaged in a miniature bioassay. Electrical conductivity change is used smartly to map fabric pressure and recognize hand gesture. In the context of acoustic tomography, grain storage temperature, wood trunk decay, and concrete defect are detected according to a variation of sound speed in the different medium. In the context of optical tomography, the problem of restricted optical observation window is solved by using a single light field camera to deliver multiple line-of-sight measurements for temperature imaging. Diverse tomography applications were reported regularly (Aw, Rahim, Rahiman, Yunus, & Goh, 2014; Bera, 2017; Goh, Rahim, Rahiman, & Hafiz, 2016; Goh, Ruzairi, Hafiz, & Tee, 2017; Ma & Soleimani, 2017; Perrone, Lapenna, & Piscitelli, 2014; Wahab et al., 2015; Wang et al., 2018; Wang & Yang 2020; Yao & Takei 2017; Zhang, Wang, Yang, & Wang, 2014). We believe there will be many more eye-opening diverse tomography applications in the years to come.

27.2 Packed column monitoring with electrical tomography

Random or structured packing is commonly used in the chemical reactors to increase the contact area between the gas and liquid phases, and thus improve reaction efficiency. However, quantifying the gas void fraction in the packed bed reaction columns is a new challenge for electrical tomography, because packing materials add complexity to measurement. The presence of the packing materials significantly alters the preferred travel path of fluids. Packing introduces a stationary third solid phase, which must be considered for in the measurement. Using electrical tomography to measure gas void fraction and distribution of gas–liquid flow in a hollow pipe has

Industrial Tomography. https://doi.org/10.1016/B978-0-12-823015-2.00027-3

been applied in many multiphase flow studies. Electrical tomography should be a suitable imaging tool for packed column monitoring.

To find gas void fraction in the cocurrent packed column shown in Fig. 27.1, a procedure below is followed. First, the column was filled with liquid only and electrical impedance tomography (EIT) is utilized to detect mutual impedances of all electrode pairs as a reference. Second, the packing material is added into liquid and the mutual impedances are measured again. The changes in these mutual impedances with respect to the reference are used to obtain the volume fraction of the packing. After gas is introduced into the packed column, the changes in mutual impedances caused by the presence of gas are measured the third time. The changes in these mutual impedances are used to obtain the total fraction of the packing material and gas. At last, the gas void fraction is obtained by subtracting the void fraction of the packing from the total fraction. The accuracy of this procedure is validated in the simulation and experiment study (Wang, Jia, Yang, Buschle, & Lucquiaud, 2018).

The representative local gas void fractions with different packing materials are plotted in Fig. 27.2 to reveal the statistical property of the gas bubble distributions in the column. The gas flow rate was kept at 1.09 L/min in each case. The color bar on each figure represents the gas void fraction. Fig. 27.2a shows the 3D gas void fraction distribution in the ERT sensing plane cross-section. In the packing free column, gas bubbles predominantly pass through the center of the column. The bell-shaped gas void fraction distribution is centrally symmetric. Plastic pall rings do not appear to change the distribution of gas void fraction much as shown in Fig. 27.2b; however, the magnitude of the gas void fraction is larger compared to the packing free system. Plastic pall rings hinder the upward motion of gas bubbles; therefore, the residence time of gas bubbles is prolonged and the gas void fraction increases accordingly.

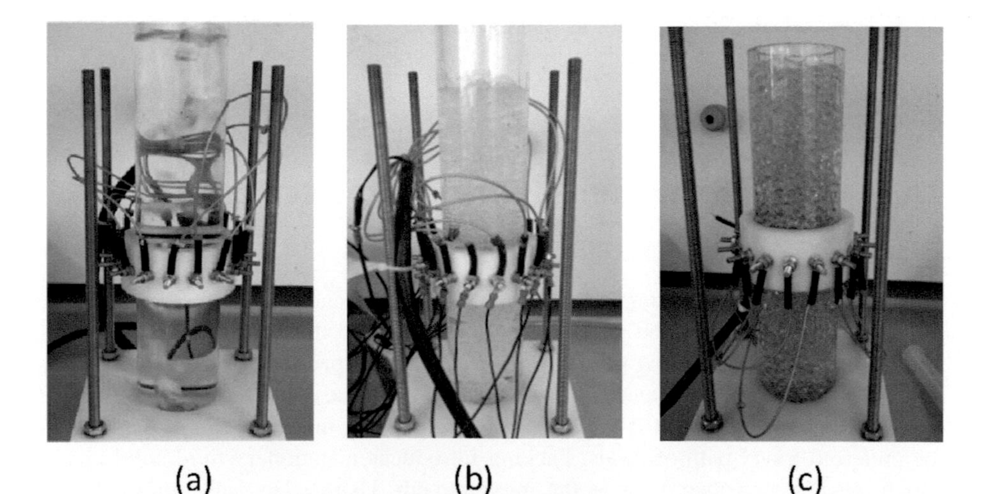

Figure 27.1 Cocurrent gas liquid column with EIT sensor. (a) packing free column, (b) pall rings packed bubble column, and (c) glass beads packed bubble column.
Reproduced with permission from the copyright holder, Elsevier.

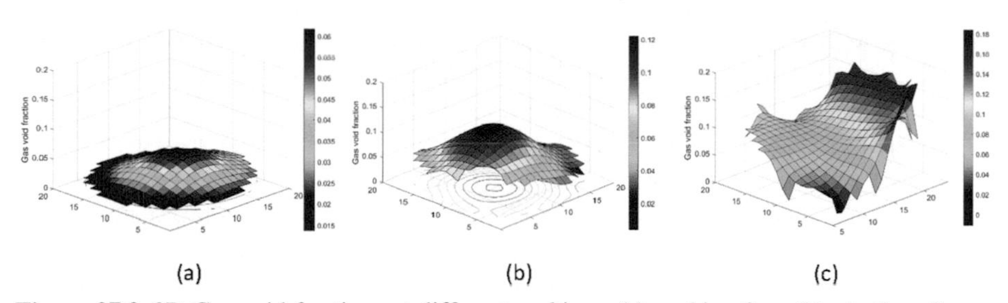

Figure 27.2 3D Gas void fractions at different packings. (a) packing free, (b) plastic pall rings packing, (c) glass beads packing.
Reproduced with permission from the copyright holder, Elsevier.

The contours on the x−y plane of Fig. 27.2b demonstrate the symmetry of gas bubble distribution. When considering the glass bead packing case, Fig. 27.2c indicates the packing significantly changes the distribution of gas bubbles in the column. The gas distribution is not centrosymmetric as previous two cases and has a large variation across the plane. As glass beads are randomly packed, it is possible that sparse and dense spots are formed in a random pattern, which could create arbitrary gas paths.

At each gas flow rate and packing condition, the EIT system collected 10,000 frames of images at an image rate of 125 frames/s, which was the slowest rate but has the highest systemic measurement accuracy. The data were split into 10 subsets with 1000 frames per subset. In each subset, the void fraction average was computed from the 1000 frames. The standard deviation of the 10 averages was calculated and represented using a box plot. The corresponding experimental results are shown in Fig. 27.3. It can be seen that the overall gas void fraction has a strong linear relationship with the gas volumetric flow rate in both the packing free column and packed columns, despite the different slopes with different packing condition. This phenomenon

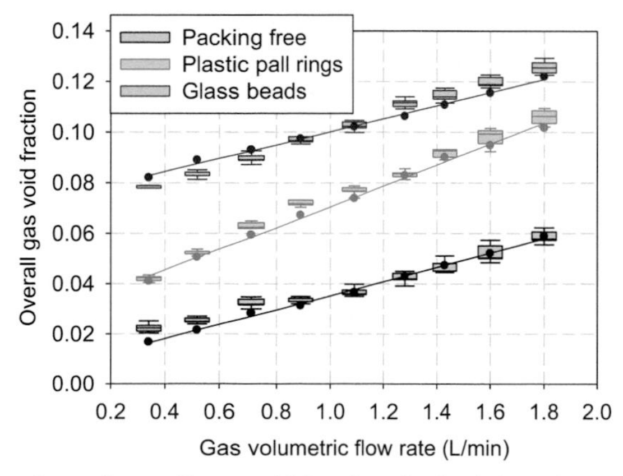

Figure 27.3 Comparison of overall gas void fraction obtained from the conventional bed expansion method and the EIT method.
Reproduced with permission from the copyright holder, Elsevier.

coincides with the findings in a number of previously published study (Moshtari, Babakhani, & Moghaddas, 2009). The overall gas void fraction in the glass beads packed column is always larger than that in the plastic pall rings packed column. The overall gas void fraction in the packing free column has the smallest value in each gas volumetric flow rate. The larger overall magnitude of the overall gas void fraction is attributed to the fact that the length of gas pathway is prolonged due to the packing.

These results are also found by Niranjan and Pangarkar (1984) who determined the gas phase void fraction in a packed bubble column by measuring the dispersion height for a given superficial gas velocity and the height of a clear liquid after the gas flow had been stopped. Because the liquid depth fluctuated dramatically, in particular, at the large gas flow rate, like 1.43 L/min, 10 liquid levels for each packing and gas flow rate condition were taken and averaged to calculate the overall gas void fraction. These values were shown in the box plots in Fig. 27.3 and compared with each average overall gas void fraction (solid dots) obtained from the EIT. In general, two sets of gas overall void fraction values had a good agreement, which proved the validity of the EIT method.

In concurrent flow, the conductive liquid is the continuous phase, which is an ideal scenario for EIT. In countercurrent packed column, liquid becomes dispersed phase, so Electrical Capacitance Tomography (ECT) is a right choice to analyze the distribution of a liquid phase and to quantify the liquid hold-up (Wu et al., 2018). Structured packings are tower internals that are used in separation processes such as absorption, distillation, and liquid—liquid extraction. The combination of large surface areas, low gas pressure drops, and high separation efficiencies makes structured packing ideal for large-scale atmospheric gas absorption processes, such as amine-based postcombustion CO_2 capture. In many industrial settings, it is not feasible to manage two points calibration. A single point calibration was developed in that work, which is beneficial for a wider industrial application.

2D reconstructed tomograms for a range of liquid loads are shown in Fig. 27.4. The tomograms show that liquid hold-up in the packing increases with liquid load, and a banding like liquid distribution forms at higher liquid loads (Fig. 27.4d—f). The reason for relatively poor liquid distribution is because the ECT sensor is located at the height of the first packing section and liquid predominantly travels along grooves to the column wall.

Inspection of the stained packing section after testing confirmed the banding flow pattern imaged by ECT. The diameter of a circular stain at the top of the packing (Fig. 27.5d) was similar with the diameter of the liquid distributor. A banding pattern along the length (Fig. 27.5a—c at the bottom of the packing (Fig. 27.5e) matched the orientation of the banding in Fig. 27.4. The experiment results demonstrate the potential of ECT for real-time liquid distribution imaging and local liquid hold-up measurement in the countercurrent packed column.

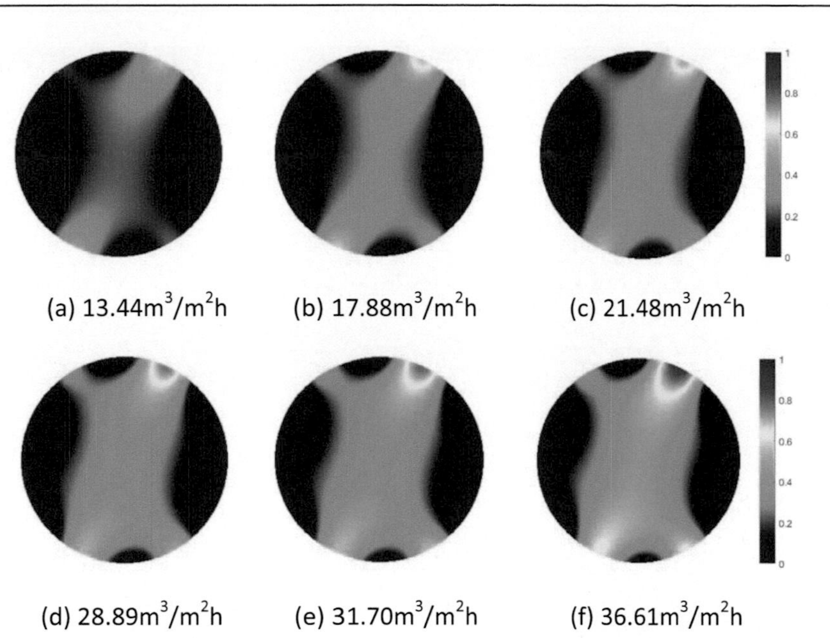

Figure 27.4 Reconstructed liquid distribution at different liquid load (0.1 mS/cm). (a) 13.44 m^3/m^2h. (b) 17.88 m^3/m^2h. (c) 21.48 m^3/m^2h. (d) 28.89 m^3/m^2h. (e) 31.70 m^3/m^2h. (f) 36.61 m^3/m^2h.
Reproduced from an open access resource.

27.3 3D Cell spheroid imaging by electrical impedance tomography

3D spheroid culture models are closer to an in vivo like morphology; it can better reflect the biological mechanisms of cell migration, differentiation, and viability. The 3D tumor cells are considered to have higher radio- and chemo resistance than cancer cells cultured at 2D because of the 3D cell—cell and cell—matrix interactions. For tissue engineering applications, it will be highly desirable to monitor in real time over long-term cell growth and differentiation, together with tissue formation in 3D scaffolds (Wu, Yang, Bagnaninchi, & Jia, 2018).

When cell culture is performed over days, the conductivity of the culture medium is likely to change continuously over time because of changes in dissolved gas content, and pH change associated with the formation of cell by-products. In such a case, the homogeneous reference is not available for time-difference EIT imaging. The frequency-difference EIT imaging can overcome this challenge, as it reconstructs images of the conductivity spectrum of tissues, which is only relevant to the frequency response of the cells. This method is more robust to the disturbance in the culture environment and is suitable to be applied in the long-term cellular assay.

In the frequency-difference EIT, the frequency range was set as 10—100 kHz and the voltage measurements at 10 kHz were used as a baseline reference. Figs. 27.6

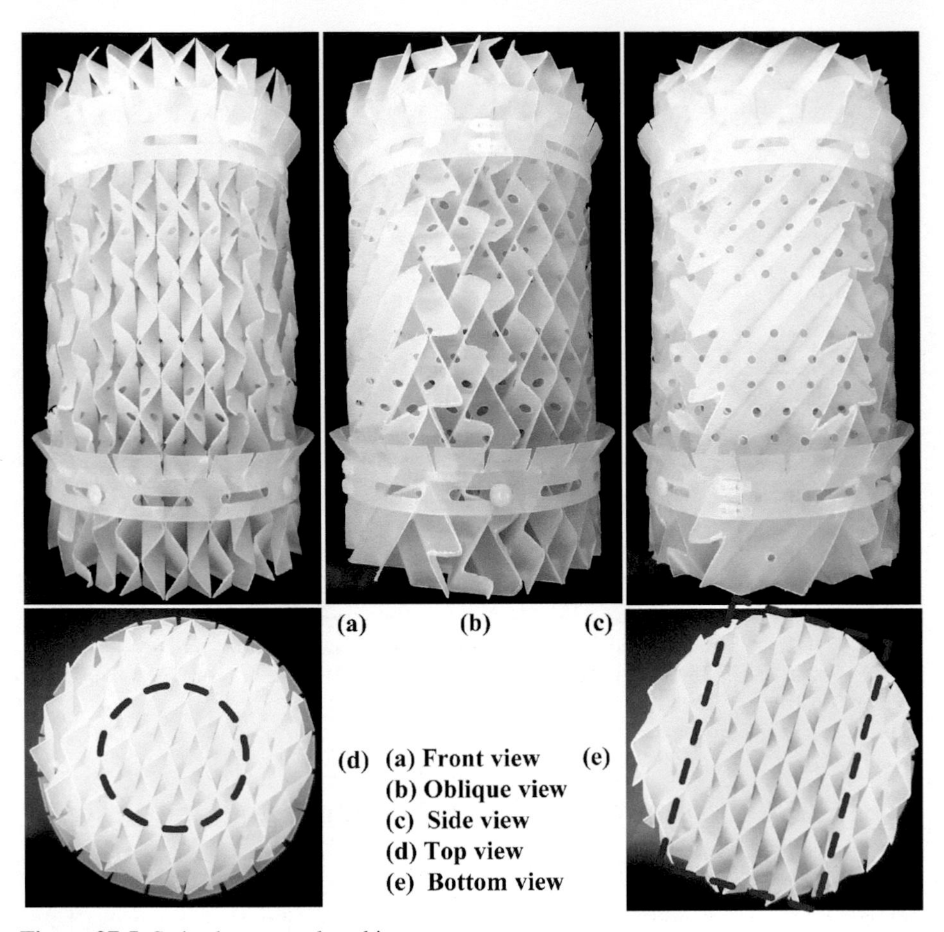

Figure 27.5 Stained structured packing.
Reproduced from an open access resource.

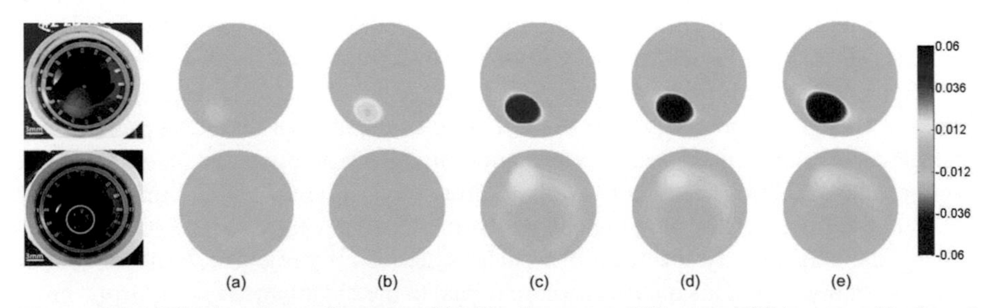

Figure 27.6 With baseline reference at 10 kHz, frequency-difference EIT images of hydrogel samples at (A) 20 kHz, (B) 40 kHz, (C) 60 kHz, (D) 80 kHz, and (E) 100 kHz. The first row is the cell-loaded hydrogel sample and the second row is the blank hydrogel.
Reproduced from an open access resource.

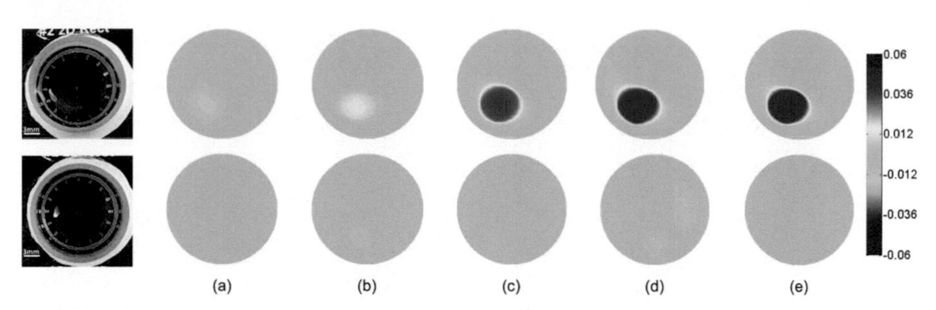

Figure 27.7 With baseline reference at 10 kHz, frequency-difference EIT images of AlgiMatrix scaffold samples at (a) 20 kHz, (b) 40 kHz, (c) 60 kHz, (d) 80 kHz, and (e) 100 kHz. The first row is the cell-loaded scaffold sample and the second row is the blank scaffold.
Reproduced from an open access resource.

and 27.7 show the frequency-difference images of the hydrogels and scaffolds. In Fig. 27.6, the conductivity of the cell-loaded hydrogel sample increases from 0.007 at 20 kHz to 0.063 at 100 kHz. Since the conductivity variation of the hydrogel is similar to that of the culture medium, hydrogel is not visible on the reconstructed images. The conductivity increase displayed is mainly attributed to the interfacial polarization across the cellular plasma membranes and their interactions with the extra and intracellular electrolytes. In Fig. 27.7, the conductivity of the cell-loaded scaffold sample increases from 0.006 at 20 kHz to 0.048 at 100 kHz, which is less than that of the hydrogels. It is because some cells are deposited on the bottom of the well through the pores of the scaffold before cell attachment, whereas all the cells in the suspension are encapsulated inside the hydrogel. Therefore, the actual cell concentration in the hydrogel sample is higher than that in the scaffold sample.

Cell mortality caused by a chemical insult can be reconstructed in real time through its correlated conductivity variations in the spheroid (Wu, Zhou, Yang, Jia, & Bagnaninchi, 2018). Experiment with MCF-7 spheroids was carried out to validate the feasibility of EIT for real-time imaging. The MCF-7 cells were cultured in the incubator for 6 days using optimized liquid overlay method to form spheroids in a radius of around 1.2 mm (Fig. 27.8). In order to monitor the dynamic course of cell death, the spheroids were introduced to the miniature sensor with 1.2 mL 2% Triton X-100 solution (experimental group). Triton X-100 can cause cell death in a short time. In the control group, the spheroids were introduced to the miniature sensor with 1.2 mL High Glucose (HG) culture medium. Reference was taken immediately after that. Both the experimental group and the control group have two samples. The cellular metabolic viability assay was operated on the spheroids to verify the EIT results.

Fig. 27.9a shows the reconstructed tomographic images for the response of the MCF-7 spheroids in the 2% Triton X-100 solution (experimental group) and the HG culture medium (control group). Images show the difference of conductivity between the reference and the conductivity at the selected time points. As expected, the conductivity of the spheroids in the Triton X-100 solution has a significant increase, while the conductivity of the spheroids in HG culture medium remains unchanged.

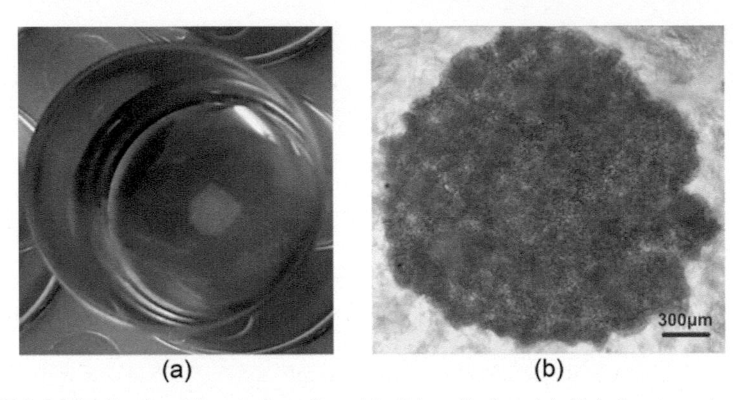

Figure 27.8 MCF-7 spheroid sample cultured in 24-well plate (a) digital camera image and (b) optical contrast microscopy image.
Reproduced from an open access resource.

In the experimental group, the conductivity of the spheroids starts to increase at about 2 min until it reaches a plateau between 0.04 and 0.05 at 22 min. The conductivity increase is caused by the destruction of the insulating cell membranes after cell death. This result is consistent with the previous MCF-7 chemical response data in 2D (Brischwein et al., 2006; Schwarzenberger et al., 2011), but a longer response time was observed, likely due to the 3D structure of the spheroids. The standard deviation around the mean could be explained by the individual differences between two spheroids in morphological characteristics, cell concentration, and initial cell viabilities. In Fig. 27.9B, the chemical insult over time is reconstructed in real time with a small fluctuation. Comparing the fluorescence value (RLU) in the control group and the experimental group shows that the viabilities of the MCF-7 spheroids almost drop to zero after a 30 min treatment in the 2% Triton X-100 solution (Fig. 27.9C).

The results show that the viability and integrity of the 3D cell spheroids in their cultural environments can be monitored in real time by the EIT. Although its spatial resolution is not as high as other imaging techniques, such as confocal and fluorescence microscopy, it does have several distinct advantages over the existing techniques including cost, nondestruction, portability, high temporal resolution, and potential for multiplexing and long-term high throughput screening. Overall, EIT maintains the advantages of other impedimetric measurement techniques while overcoming the lack of spatial resolution. EIT imaging provides a promising potential for 3D drug screening and tissue engineering.

27.4 Fabrics pressure mapping using electrical impedance tomography

The purpose of this study is to carry out a preliminary investigation of applying electrical impedance analysis to predict the behavior of an electro-conductive knitted

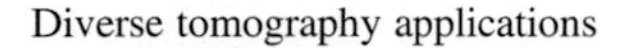

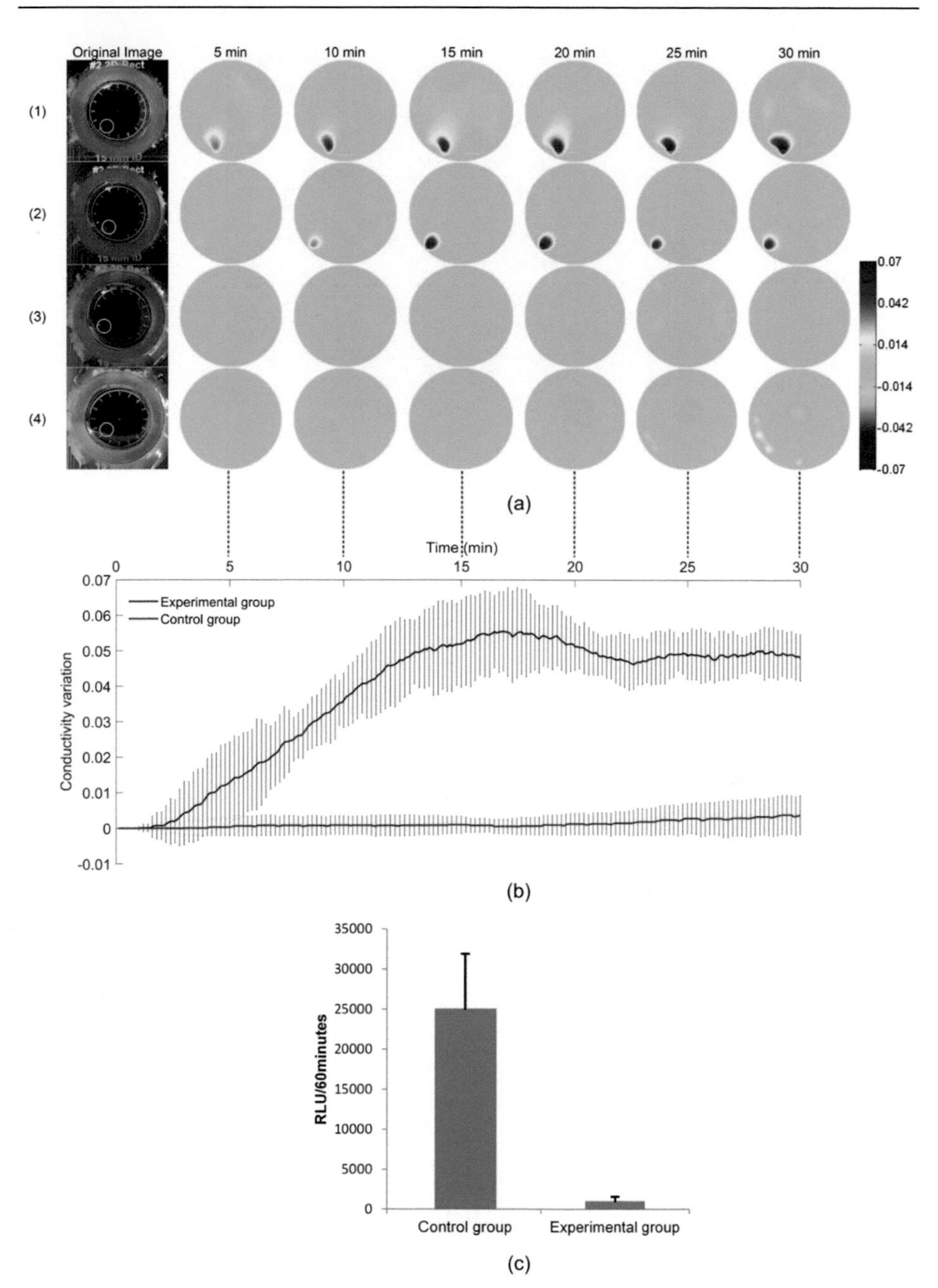

Figure 27.9 Reconstructed images for the spheroid samples in (a1−2) 2% Triton X-100 solution (experimental group) and (a3−4) HG culture medium (control group), (b) the conductivity variation of the spheroids ($n = 2$), and (c) the cellular metabolic viability assay for the spheroids.

Reproduced from an open access resource.

Figure 27.10 Fabric with 16 EIT electrodes.
Reproduced from an open access recourse.

structure. EIT was used as a mapping technique for deformation imaging in conductive knitted fabric (Duan, Taurand, & Soleimani, 2019; Yao, & Soleimani, 2012). The conductivity of the area changes when the fabric is stretched or compressed. The primary application of the pressure mapping is a touch sensor for robotics as well as biomechanical observations.

The electrical impedance of the sensor is measured at 16 peripheral points. The fabric diameter is 20 cm. 10 mA electric current with 2 kHz frequency is applied to the excitation electrodes. The fabric with a conductive material is developed by Less EMF, Inc. This material is a highly conductive (ρ approx. 1.0 V/sq unscratched) medical-grade silver-plated Nylon Dorlastan fabric with the ability to stretch in both directions. Fig. 27.10 shows the fabric and the electrode arrangement in the experiment. The conductivity image is reconstructed according to the measured impedances to represent the map of deformation within the conductive area.

Fig. 27.11 shows the deformation of the fabric as a result of multiple objects and the reconstruction of applied pressure. Fig. 27.11a shows the conductivity change after one finger applies pressure on the circular fabric sensing area. Fig. 27.11b demonstrates pressures applied through two fingers.

27.5 Hand gesture recognition using electrical impedance tomography

In biomedical application, EIT measures internal electrical impedance to infer the interior structure of the human body. A wearable, low-cost, and low-powered EIT system was built around an AD5933 impedance analyzer, which included a frequency generator with 1 Hz−100 kHz range and on-board 12-bit 1 MSPS ADC (Zhang & Harrison, 2015). A wristband with 32 stainless steel electrodes is wrapped around the forearm.

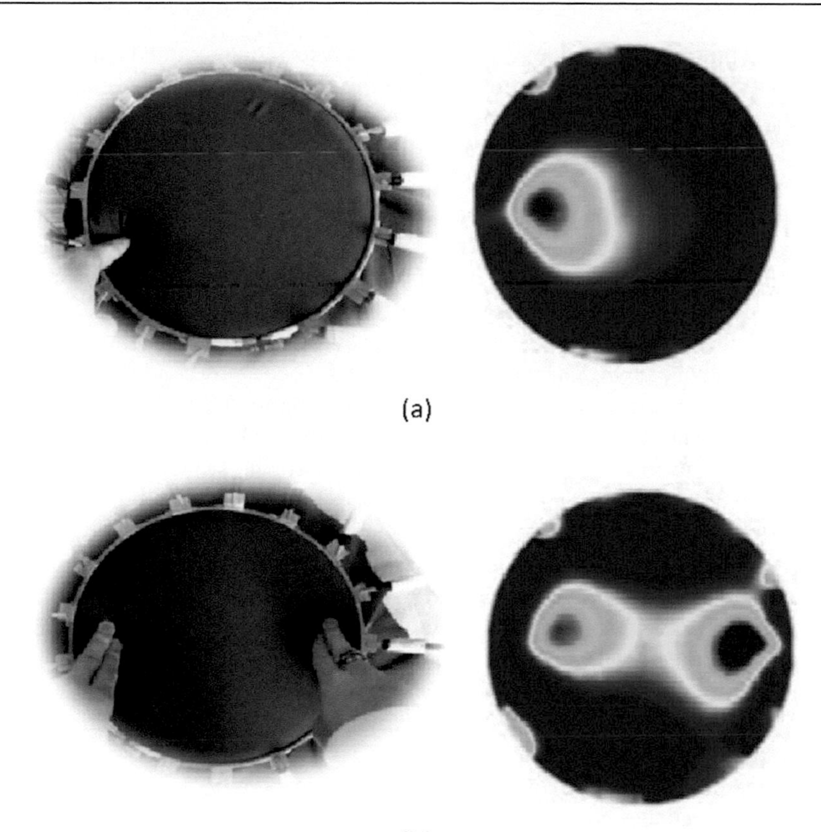

Figure 27.11 Pressure applied on fabrics (a) single pressure point, (b) double pressure points. Reproduced from an open access recourse.

Tomographic image is downsampled to 16 × 16 pixels as raw input features for a support vector machine. Four finger pinch gestures and seven hand gestures were classified. An average recognition accuracy of 94.3% across 10 participants was achieved (Zhang, Xiao, & Harrison, 2016). Hand gesture recognition based on the forearm EIT was extended to human−machine interface field, which establishes a link between humans and machines. In Wu's study (Wu, Jiang, Liu, Bayford, & Demosthenous, 2018), 40 bioimpedance values were measured and labeled for artificial neural networks training. 11 hand gestures and motions were recognized to control a prosthetic hand in real time.

27.6 Temperature monitoring in the stored grain using acoustic tomography

Food security has been and will continue to be a public concern worldwide. Increasing urbanization, climate change, and land use for nonfood crop production intensify these concerns of increasing food demands. While a lot of efforts have made to improve their agricultural production, postharvest loss does not receive the required attention. Every year one-third of the food produced is lost during the postharvest operations globally. Reducing the postharvest losses, especially in developing countries, could be the most cost-effective solution to increase the world's food availability and reduce food supply pressure.

Storage plays a vital role in the food supply chain, and several studies reported that maximum losses happen during this stage, especially the spoilage of bulk-stored grain. Temperature of the stored grain is an important factor for stored-grain deterioration detection, as the maximum temperature of the deterioration area is at least 10 K more than other areas. Grain has relatively low thermal diffusivity. Thus, the temperatures of grain bulk slowly change when heat transfer occurs by conduction. Therefore, the temperature measurement spatial resolution should be 0.5 m. For the local pointwise measurement, like thermocouples and temperature cables, a large number of sensors is required to meet the spatial resolution requirement. On the contrary, acoustic tomography has the ability to provide 2D temperature distribution images with only a few acoustic transducers placed around the sensing area. Owing to its fast response and noninvasive nature, acoustic tomography has great potential in grain deterioration detection. In 1997, Hickling studied the sound propagation characteristic in stored grain, which paved the way for quantitative grain temperature measurement using acoustic waves (Hickling, Wei, & Hagstrum, 1997). In 2012, Yan developed the acoustic tomography system for 2D grain temperature imaging (Yan, Chen, Zhou, & Liu, 2012), which gave the example for the potential of acoustic tomography method.

The measurement principle of acoustic tomography is to firstly reconstruct the temperature-dependent sound speed distribution from the time-of-flight (ToF) measurements. Then the corresponding temperature distribution can be calculated from the reconstructed sound speed. The major challenge in employing acoustic tomography for grain temperature field monitoring is the forward modeling. Accurate modeling of the relationship between grain temperature and sound travel time in stored grain is essential for tomographic reconstruction. Similar to other travel time tomography problem, the key to forward modeling is the sound propagation ray path. According to Hickling's study, stored grain can be considered a porous medium and sound is propagated principally through the gas in the passageways between the grain kernels. Different from the conventional acoustic tomography, the sound propagation ray path in the stored grain is not a straight line and cannot be easily determined based on sound speed map. To solve this problem, Yan used the speed conversion coefficient λ to convert measured sound velocity in stored grain c_G into the sound velocity of free space c_L with the same gaseous medium and temperature. Therefore, the measured sound velocity distribution in stored grain is firstly reconstructed using the conventional straight ray model, then temperature can be obtained with the use of speed conversion coefficient.

Generally, the forward problem of acoustic tomography is defined as follows:

$$\tau = Ls \tag{27.1}$$

where $\boldsymbol{s}$ is an N×1 vector that describes the slowness distribution. Element $s_j = 1/c_{G,j}(T)$ is the slowness in the j-th pixel. $\boldsymbol{L}$ is an M×N ray length matrix and its element $l_{i,j}$ is the segment length for each ray path across one pixel. $\boldsymbol{\tau}$ is an M×1 vector that represents the ToF measurement. N is the number of pixels and M is the number of ToFs.

Then temperature can be calculated as follows:

$$T_j = \frac{c_{L,j}}{R\gamma} = \frac{\lambda c_{G,j}}{R\gamma} \tag{27.2}$$

where $R = 287$ J kg^{-1} K^{-1} is the gas constant and specific heat ratio $\gamma = 1.4$.

The inverse problem of acoustic tomography is usually ill-posed, which means small error in the measurement will result in large perturbation in the reconstructed image. Acoustic tomography in stored grain is no difference. To solve the ill-posed inverse problem, commonly used regularization or subspace projection method can be useful. A lab-scale system was developed for experimental validation. As shown in Fig. 27.12, the transducer array consists of 16 transceivers to cover a 1.2 m by 1.2 m square sensing area. A broadband acoustic chirp (200−1500 Hz) is used as the excitation signal and semiparallel data collection scheme is applied for real-time monitoring.

According to the experiment result shown in Fig. 27.13, the temperature distributions with different hot spot locations in soybeans were well reconstructed using the acoustic tomography system. The maximum temperature of an artificial hot spot is 19 K more than the surrounding cool grain, and the acoustic temperature measurement method can accurately reconstruct the location and maximum temperature of hot spots. Therefore, it can be concluded that acoustic temperature measurement for stored grain is suitable for temperature distribution monitoring, particularly in the prediction of the temperature anomalies in stored grain.

Figure 27.12 A lab-scale acoustic tomography system for stored grain temperature monitoring.
Reproduced with permission from the copyright holder, Elsevier.

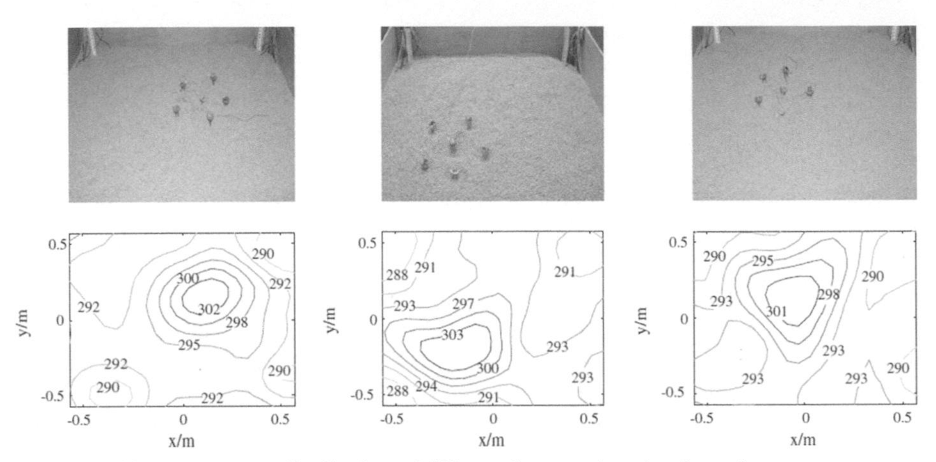

Figure 27.13 Temperature distribution of different hot spot location in soybeans. Reproduced with permission from the copyright holder, Elsevier.

27.7 Tree decay detection by acoustic tomography

Wood quality assessment is a crucial issue in the forestry industry. Significant efforts have been made toward robust Non-Destructive Evaluation (NDE) of wood properties of individual trees and assessing wood quality by stand and forest. Fast, accurate, and diagnostic NDE techniques can help to improve tree management, grow higher quality wood, and increase profits for the forest industry. Commercial NDE techniques use acoustic waves, electromagnetic waves, and mechanical microdrillings. Among all these measurement modalities, acoustic wave measurement is mostly used in practical applications due to its low equipment and implementation cost.

Recently, researchers have shown an increased interest in applying acoustic tomography in tree decay detection (Arciniegas, Prieto, Brancheriau, & Lasaygues, 2014). Different from the conventional single line of sight measurement, acoustic tomography can provide fruitful information about the interior structure. With the detailed tomographic images, researchers and wood manufacturers can better evaluate raw wood materials, like standing trees, stems, and logs.

Compared with other tomography applications, the sensing area for tree decay detection is usually irregular. This irregular sensing area requires a flexible transducer array with fast and easy implementation ability, and accurate sensor positioning calibration scheme for the tomography reconstruction. Wang in 2009 utilized the PiCUS Sonic Tomography system (Argus Electronic Gmbh, Rostock, Germany) for tree decay detection (Wang, Wiedenbeck, & Liang, 2009), as shown in Fig. 27.14.

The tomographic reconstruction of acoustic tomography for tree decay detection has no difference from other applications. Straight ray model and regularized linear inversion algorithms can be used for real-time tomographic imaging. As shown in Fig. 27.15, acoustic tomography is capable of detecting various internal structural defects in trees, including heartwood decay, internal cracks, and ring shake. However, due to the lack of high spatial resolution, acoustic tomography tends to overestimate the size of the defect, and cannot reliably detect small sapwood decay and insect holes.

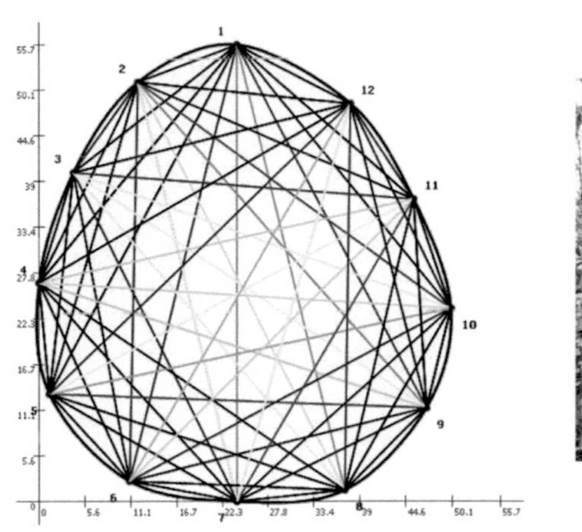

Figure 27.14 Sensor array setup of PiCUS Sonic Tomography system.
Reproduced from an open access resource.

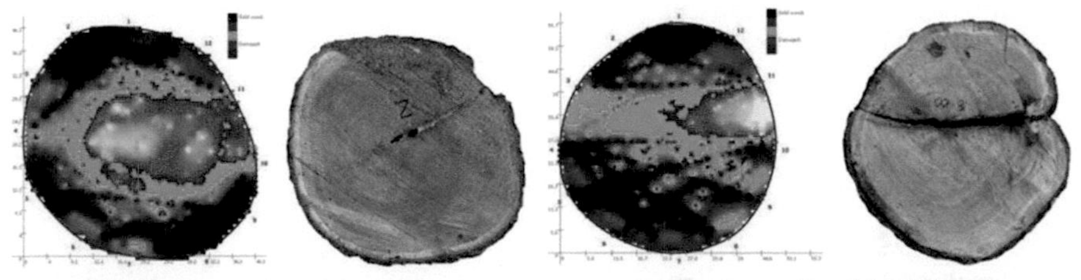

(a) Cross-section No. 2-50

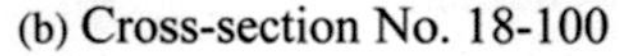

(b) Cross-section No. 18-100

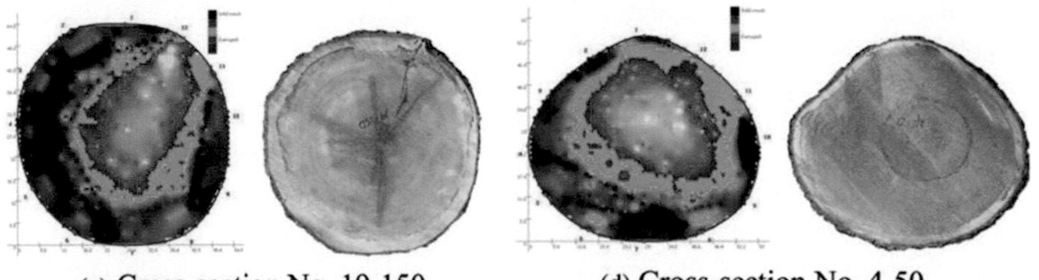

(c) Cross-section No. 19-150 (d) Cross-section No. 4-50

Figure 27.15 Reconstructed images for tree decay detection.
Reproduced from an open access resource.

Advanced tomographic inversion algorithm can be applied for better image quality. Liu (Liu & Li, 2018) proposed a hybrid wave propagation model which conducts curved ray inversion to improve the image quality, as shown in Fig. 27.16. Espinosa, Prieto, Brancheriau and Lasaygues (2020) developed a reconstruction algorithm for more detailed identification and quantification of the inner state of the anisotropic structure of the trunk. The proposed method takes into account the orthotropy property of wood material to perform ray tracing and use a polynomial approximation for dimensional reduction during image reconstruction. Du improved detection accuracy by utilizing the deep learning techniques and contour constraint under sparse sampling

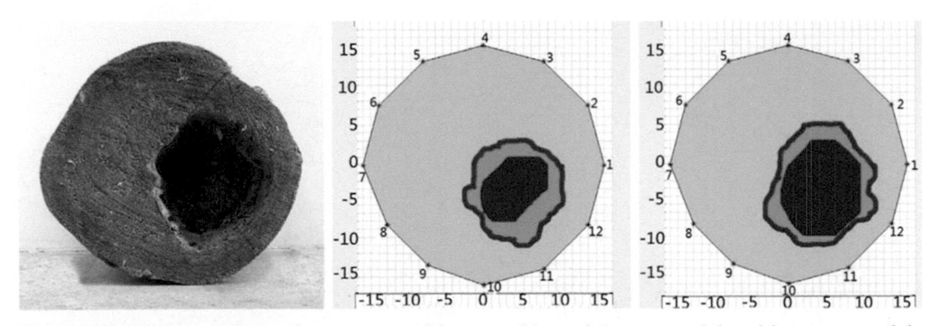

Figure 27.16 Comparison of reconstructed image with straight ray model and bent ray model. Reproduced with permission from the copyright holder, Elsevier.

(Du, Li, Feng, & Hu, 2019). As a result, the imaging accuracy maintained at more than 90%, even when the number of used sensors is decreased from 12 to 6.

Besides, according to Li, Wang, Wang and Allison (2012), the ToF measurements may lack the sensitivity to low-velocity features of decayed areas and has limited capability in detecting early stages of decay in trees. Therefore, additional measurement modalities may be necessary for the early stage decay detection.

27.8 Concrete defect detection by acoustic tomography

With the advantages of nonradiative and low equipment cost, acoustic tomography is a promising NDE technique for concrete interior structure imaging (Mita & Takiguchi, 2018; Rens, Transue, & Schuller, 2000). The measurement principle of acoustic tomography is to send acoustic signals through the sensing area and visualize the internal structure based on the corresponding interactions with the received signals. There are three major interactions amenable to acoustic measurements: the sound speed, the attenuation, and the scattering. Due to the high modeling complexity and computational cost of scattering tomography, acoustic tomography based on the sound speed and attenuation are more frequently used in concrete applications.

The forward model of the sound speed and attenuation imaging are defined as follows:

$$\begin{aligned} \tau &= Ls \\ \mathbf{r} &= L\alpha \end{aligned} \tag{27.3}$$

where $s \in \mathbb{R}^{N\times 1}$ and $\alpha \in \mathbb{R}^{N\times 1}$ represent the slowness (inverse of wave velocity) and attenuation coefficient distribution, respectively. $L \in \mathbb{R}^{M\times N}$ is the ray length matrix and its element $l_{i,j}$ the segment length for the i-th ray path across the j-th pixel. $\tau \in \mathbb{R}^{M\times 1}$ and $r \in \mathbb{R}^{M\times 1}$ represent the ToF and sound amplitude attenuation measurements. N is the number of pixels and M is the number of ray path.

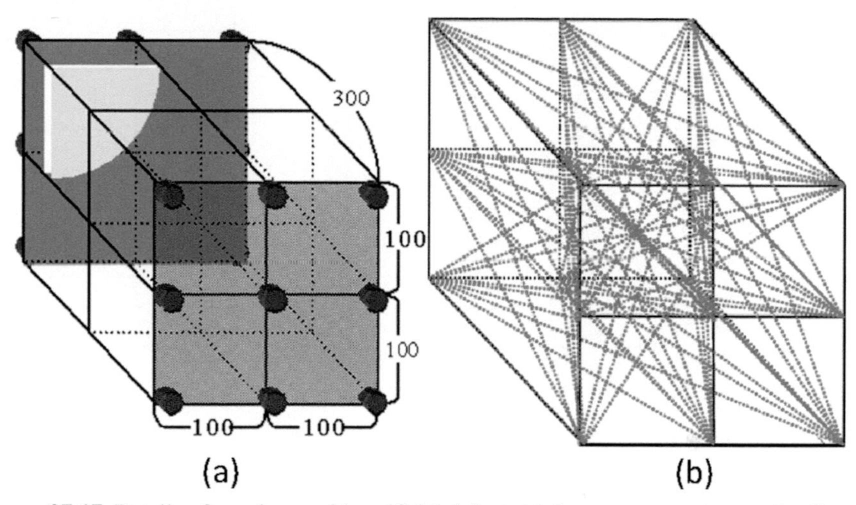

Figure 27.17 Details of specimen with artificial defect. (a) Sensor arrangement and cell discretization and (b) ray path coverage.
Reproduced with permission from the copyright holder, Elsevier.

A typical measurement setup is shown in Fig. 27.17.

The major challenge in acoustic tomography is to determine the actual ray path from transmitter to receiver. Knowledge of the ray path is necessary to accurately build the ray length matrix $\boldsymbol{L}$ for tomographic reconstruction. Due to the refraction effect of sound propagation in heterogeneous material, the ray paths may bend around the inclusion of low velocity. Therefore, acoustic tomography imaging becomes a nonlinear inverse problem, which solves both the ray tracing problem to determine the actual ray path and the tomographic reconstruction of sound speed distribution and attenuation. To prevent large computational cost for ray tracing, a variety of methods have been developed to solve this nonlinear inverse problem, including using isoparametric coordinate projections (Chai, Momoki, Kobayashi, Aggelis, & Shiotani, 2011), Dijkstra's algorithm (Zielińska, & Rucka, 2020), and network theory (Perlin & de Andrade Pinto, 2019).

Chai compared the reconstruction quality of acoustic travel time tomography and attenuation tomography (Chai et al., 2011). As shown in Fig. 27.18, the amplitude of ultrasound propagating in concrete undergoes a much greater change compared to the delay in travel time due to inhomogeneity, which indicates that the sound attenuation has higher sensitivity for evaluating the soundness of concrete.

According to the experimental results shown in Fig. 27.19, the attenuation tomography effectively visualizes the location of defect embedded in concrete. It is worth noticing that a unified system with only one array of ultrasonic transducers can collect both the attenuation and ToF measurement simultaneously from the received signal waveform. Combining the two modes of measurement can provide richer information to image the concrete structure in a comprehensive way.

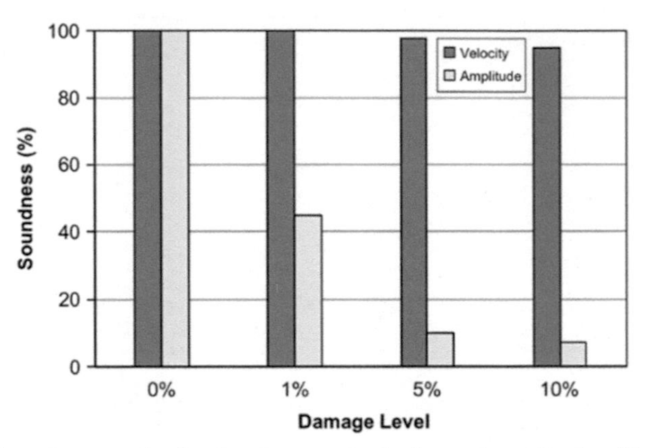

Figure 27.18 Soundness evaluation by change of velocity and amplitude with regards to damage level.
Reproduced with permission from the copyright holder, Elsevier.

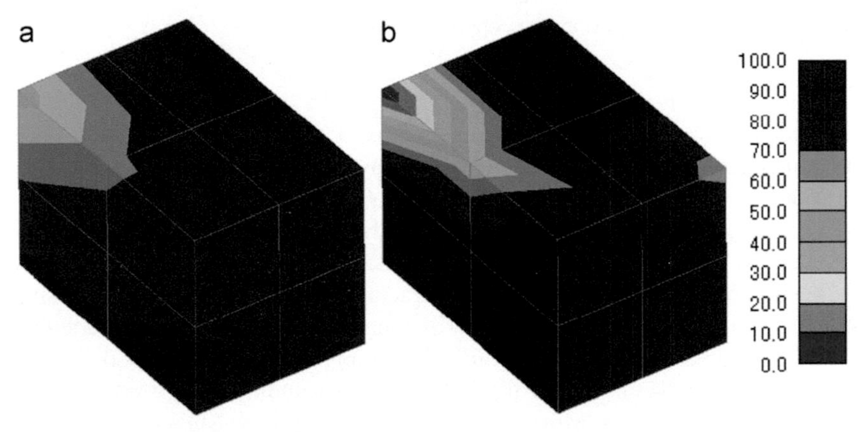

Figure 27.19 Tomography visualization results for the portion containing defect (unit in % of soundness) (a) velocity tomography, and (b) attenuation tomography.
Reproduced with permission from the copyright holder, Elsevier.

27.9 Temperature monitoring using single light field camera

High-fidelity temperature imaging is particularly demanded in gas turbine combustion diagnosis. The temperature distributions directly reflect the heat transfer and radiation, and more importantly, reveal combustion efficiency and the temperature-dependent emissions, such as NOx and CO. A vast majority of effort have been made to develop accurate and reliable imaging technique for temperature field, including radiative imaging techniques (Hossain, Lu, & Yan, 2012) and laser-based diagnostics (Liu, Cao, Lin, Xu, & McCann, 2018). Compared to the laser-based method, radiative imaging utilizes the CCD camera to provide much better spatial resolution in the temperature images.

Most of the imaging system requires multiple transducers placed on the boundary of the sensing area. However, for the high temperature and high pressure gas turbine combustion environment, the number and size of measurement windows/optical access is very restricted. To solve this problem, Sun et al. (2016), Sun, Moinul Hossain, Xu, & Zhang, (2018) developed a temperature monitoring system using the single light field camera. Different from the conventional camera, the light field camera can record the radiation intensity as well as direction information of different ray simultaneously. Therefore, this one camera system can provide multiple line-of-sight measurements for temperature imaging. An example is shown in Fig. 27.20.

Generally, the forward modeling of light is

$$\mathbf{I}_{ccd} = \mathbf{G}\mathbf{I}_B \tag{27.4}$$

where $\boldsymbol{I_{ccd}}$ denotes the intensity distribution at CCD camera, $\boldsymbol{I_B}$ denotes the monochromatic intensity of blackbody radiation, and $\boldsymbol{G}$ is the optical thickness matrix.

The optical thickness matrix is determined based on the ray path from the sensor pixel to the flame, which can be traced based on the pinhole camera model, as shown in Fig. 27.21.

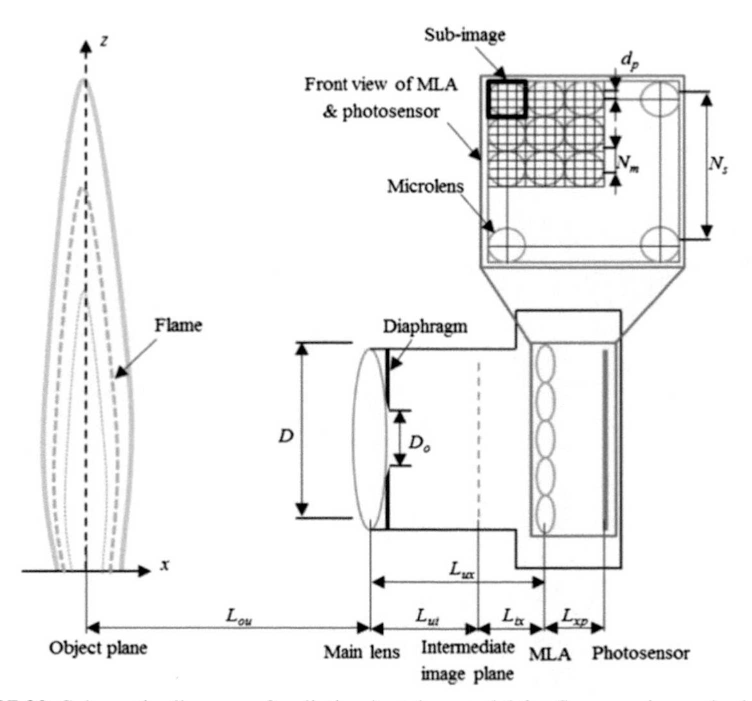

Figure 27.20 Schematic diagram of radiative imaging model for flames using a single light field camera.
Reproduced with permission from the copyright holder, Elsevier.

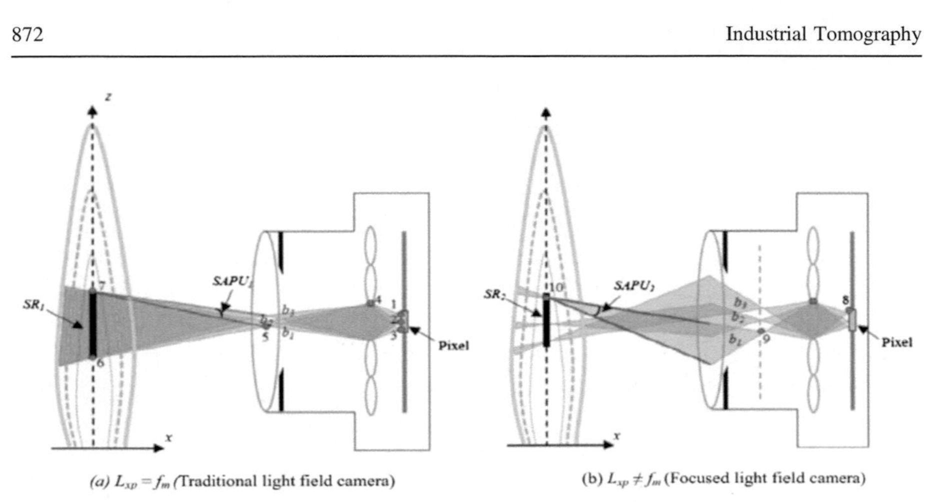

(a) $L_{xp} = f_m$ (Traditional light field camera) (b) $L_{xp} \neq f_m$ (Focused light field camera)

Figure 27.21 Schematic diagram of ray tracing in focused light field camera. Reproduced with permission from the copyright holder, Elsevier.

By solving Eq. (27.5), temperature can be calculated from the $\boldsymbol{I_B}$

$$T = c_2/\lambda \ln\left[c_1 / \left(\lambda^5 \pi \mathbf{I}_B + 1\right)\right] \tag{27.5}$$

Experiments result is shown in Fig. 27.22, which indicates that the proposed method is capable of reconstructing 3D flame temperature field.

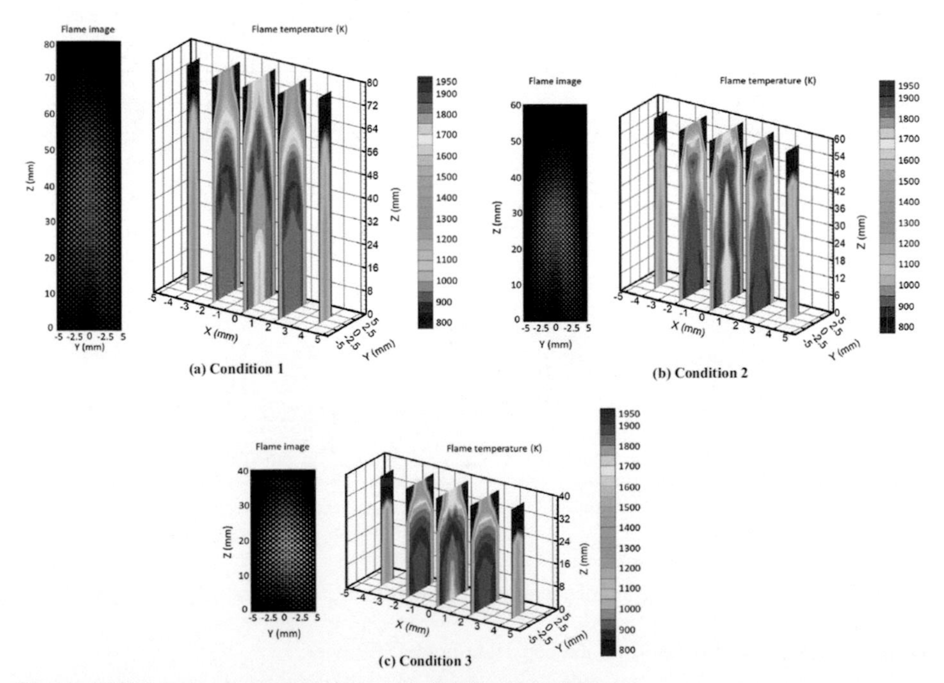

Figure 27.22 Reconstructed temperature distributions of flame. Reproduced with permission from the copyright holder, Elsevier.

27.10 Conclusion

Since process tomography has been invented, it never stops exploring diverse applications. Thanks to the emerging applications, the new development of process tomography is triggered in turn. It is for sure that more eye-opening studies are out there. More challenging and exciting demands in industrial and academic research sectors are waiting for process tomography to explore.

References

Arciniegas, A., Prieto, F., Brancheriau, L., & Lasaygues, P. (2014). Literature review of acoustic and ultrasonic tomography in standing trees. *Trees, 28*, 1559–1567.

Aw, S. R., Rahim, R. A., Rahiman, M. H. F., Yunus, F. R. M., & Goh, C. L. (2014). Electrical resistance tomography: A review of the application of conducting vessel walls. *Powder Technology, 254*, 256–264.

Bera, T. K. (2017). Applications of electrical impedance tomography (EIT): A short review, OP conference series: Materials science and engineering. In *3rd International conference on communication systems (ICCS-2017) 14–16 October 2017, Rajasthan, India* (Vol. 331).

Brischwein, M., Herrmann, S., Vonau, W., Berthold, F., Grothe, H., Motrescu, E. R., et al. (2006). Electric cell-substrate impedance sensing with screen printed electrode structures. *Lab on a Chip, 6*, 819–822.

Chai, H., Momoki, S., Kobayashi, Y., Aggelis, D., & Shiotani, T. (2011). Tomographic reconstruction for concrete using attenuation of ultrasound. *NDT & E International, 44*, 206–215.

Duan, X., Taurand, S., & Soleimani, M. (2019). Artificial skin through super-sensing method and electrical impedance data from conductive fabric with aid of deep learning. *Scientific Reports, 9*, 8831.

Du, X., Li, J., Feng, H., & Hu, H. (2019). Stress wave tomography of wood internal defects based on deep learning and contour constraint under sparse sampling. *International Conference on Intelligent Science and Big Data Engineering*, 335–346.

Espinosa, L., Prieto, F., Brancheriau, L., & Lasaygues, P. (2020). Quantitative parametric imaging by ultrasound computed tomography of trees under anisotropic conditions: Numerical case study. *Ultrasonics, 102*, 106060.

Goh, C. L., Rahim, A., Rahiman, R. F., & Hafiz, M. (2016). Process tomography of gas-liquid flow in a vessel: A review. *Sensor Review, 36*, 287–302.

Goh, C. L., Ruzairi, A. R., Hafiz, F. R., & Tee, Z. C. (2017). Ultrasonic tomography system for flow monitoring: A review. *IEEE Sensors Journal, 17*, 5382–5390.

Hickling, R., Wei, W., & Hagstrum, D. W. (1997). Studies of sound transmission in various types of stored grain for acoustic detection of insects. *Applied Acoustics, 50*, 263–278.

Hossain, M. M. M., Lu, G., & Yan, Y. (2012). Optical fiber imaging based tomographic reconstruction of burner flames. *IEEE Transactions on Instrumentation and Measurement, 61*, 1417–1425.

Liu, C., Cao, Z., Lin, Y., Xu, L., & McCann, H. (2018). Online cross-sectional monitoring of a swirling flame using TDLAS tomography. *IEEE Transactions on Instrumentation and Measurement, 67*, 1–11.

Liu, L., & Li, G. (2018). Acoustic tomography based on hybrid wave propagation model for tree decay detection. *Computers and Electronics in Agriculture, 151*, 276–285.

Li, L., Wang, X., Wang, L., & Allison, R. B. (2012). Acoustic tomography in relation to 2D ultrasonic velocity and hardness mappings. *Wood Science and Technology, 46*, 551–561.

Ma, L., & Soleimani, M. (2017). Magnetic induction tomography methods and applications: A review. *Measurement Science and Technology, 28*.

Mita, N., & Takiguchi, T. (2018). Principle of ultrasonic tomography for concrete structures and non-destructive inspection of concrete cover for reinforcement. *Pacific Journal of Mathematics for Industry, 10*.

Moshtari, B., Babakhani, E. G., & Moghaddas, J. S. (2009). Experimental study of gas hold-up and bubble behavior in gas-liquid bubble column. *Petroleum Coal, 51*, 27–32.

Niranjan, K., & Pangarkar, V. G. (1984). Gas holdup and mixing characteristics of packed bubble columns. *Chemical Engineering Journal, 29*, 101–111.

Perlin, L. P., & de Andrade Pinto, R. C. (2019). Use of network theory to improve the ultrasonic tomography in concrete. *Ultrasonics, 96*, 185–195.

Perrone, A., Lapenna, V., & Piscitelli, S. (2014). Electrical resistivity tomography technique for landslide investigation: A review. *Earth-Science Reviews, 135*, 65–82.

Rens, K. L., Transue, D. J., & Schuller, M. P. (2000). Acoustic tomographic imaging of concrete infrastructure. *Journal of Infrastructure Systems, 6*, 15–23.

Schwarzenberger, T., Wolf, P., Brischwein, M., Kleinhans, R., Demmel, F., Lechner, A., et al. (2011). Impedance sensor technology for cell-based assays in the framework of a high-content screening system. *Physiological Measurement, 32*, 977–993.

Sun, J., Xu, C., Zhang, B., Hossain, M. M., Wang, S., Qi, H., et al. (2016). Three-dimensional temperature field measurement of flame using a single light field camera. *Optics Express, 24*, 1118–1132.

Sun, J., Moinul Hossain, M., Xu, C., & Zhang, B. (2018). Investigation of flame radiation sampling and temperature measurement through light field camera. *International Journal of Heat and Mass Transfer, 121*, 1281–1296.

Wahab, Y. A., Rahim, R. A., Rahiman, M. H. F., Aw, S. R., Yunus, F. R. M., Goh, C. L., et al. (2015). Non-invasive process tomography in chemical mixtures-A review. *Sensors and Actuators B: Chemical, 210*, 602–617.

Wang, H. G., Che, H. Q., Ye, J. M., Tu, Q. Y., Wu, Z. P., Yang, W. Q., et al. (2018). Application of process tomography in gas-solid fluidised beds in different scales and structures. *Measurement Science and Technology, 29*.

Wang, H., Jia, J., Yang, Y., Buschle, B., & Lucquiaud, M. (2018). Quantification of gas distribution and void fraction in packed column using electrical resistance tomography. *IEEE Sensors, 18*, 8963–8970.

Wang, X., Wiedenbeck, J., & Liang, S. (2009). Acoustic tomography for decay detection in black cherry trees. *Wood and Fiber Science, 41*, 127–137.

Wang, H., & Yang, W. (2020). Application of electrical capacitance tomography in circulating fluidised beds-A review. *Applied Thermal Engineering, 176*, 115311.

Wu, H., Buschle, B., Yang, Y., Tan, C., Dong, F., Jia, J., et al. (2018). Liquid distribution and fraction measurement in counter current flow packed column by electrical capacitance tomography. *Chemical Engineering Journal, 353*, 519–532.

Wu, Y., Jiang, D., Liu, X., Bayford, R., & Demosthenous, A. (2018). A human-machine interface using electrical impedance tomography for hand prosthesis control. *IEEE Transaction on Biomedical Circuits and Systems, 12*, 1322–1333.

Wu, H., Yang, Y., Bagnaninchi, P., & Jia, J. (2018). Real-time monitoring of 3D cell spheroids by electrical impedance tomography. *Analyst, 143*, 4189–4198.

Wu, H., Zhou, W., Yang, Y., Jia, J., & Bagnaninchi, P. (2018). Exploring the potential of electrical impedance tomography for tissue engineering applications. *Materials, 11*, 1–11.
Yan, H., Chen, G., Zhou, Y., & Liu, L. (2012). Primary study of temperature distribution measurement in stored grain based on acoustic tomography. *Experimental Thermal and Fluid Science, 42*, 55–63.
Yao, A., & Soleimani, M. (2012). A pressure mapping imaging device based on electrical impedance tomography of conductive fabrics. *Sensor Review, 32*, 310–317.
Yao, J., & Takei, M. (2017). Application of process tomography to multiphase flow measurement in industrial and biomedical fields: A review. *IEEE Sensors Journal, 17*, 8196–8205.
Zhang, Y., & Harrison, C. (2015). Tomo: Wearable, low-cost, electrical impedance tomography for hand gesture recognition, UIST '15. In *Proceedings of the 28th annual ACM symposium on user interface software & technology, November* (pp. 167–173).
Zhang, W., Wang, C., Yang, W., & Wang, C. H. (2014). Application of electrical capacitance tomography in particulate process measurement-A review. *Advanced Powder Technology, 25*, 174–188.
Zhang, Y., Xiao, R., & Harrison, C. (2016). Advancing hand gesture recognition with high resolution electrical impedance tomography, UIST '16. In *Proceedings of the 29th annual symposium on user interface software and technology, October, 2016* (pp. 843–850).
Zielińska, M., & Rucka, M. (2020). Detection of debonding in reinforced concrete beams using ultrasonic transmission tomography and hybrid ray tracing technique. *Construction and Building Materials, 262*, 120104.

Index

Note: 'Page numbers followed by "f" indicate figures and "t" indicate tables.'

J

K

L

N

Printed in the United States
by Baker & Taylor Publisher Services